T0343494

Hydropedology

Hydropedology

Synergistic Integration of Soil Science and Hydrology

Edited by

Henry Lin

The Pennsylvania State University

AMSTERDAM • BOSTON • HEIDELBERG • LONDON
NEW YORK • OXFORD • PARIS • SAN DIEGO
SAN FRANCISCO • SINGAPORE • SYDNEY • TOKYO

Academic Press is an imprint of Elsevier

Academic Press is an imprint of Elsevier
225 Wyman Street, Waltham, MA 02451, USA
Radarweg 29, PO Box 211, 1000 AE Amsterdam, The Netherlands
The Boulevard, Langford Lane, Kidlington, Oxford, OX51GB, UK

First edition 2012

British Library Cataloguing in Publication Data
A catalogue record for this book is available from the British Library

Library of Congress Cataloging-in-Publication Data
A catalog record for this book is available from the Library of Congress

ISBN: 978-0-12-386941-8

Printed and bound by CPI Group (UK) Ltd, Croydon, CR0 4YY
Transferred to Digital Printing, 2013

Contents

Part III
Advances in Modeling, Mapping, and Coupling

Color version of figures in this book can be found at http://www.elsevierdirect.com/v2/companion.jsp?ISBN=9780123869418

Preface

In the era of sustainability movement, soil and water are two critical natural resources that are fundamental to ecosystem services and human well-being. Soils have been often referred to as the final frontier of science, and the vadose zone has also been called the last hydrologic frontier. Remarkably, the interaction between soil and water creates the fundamental interface between the abiotic and the biotic worlds, enables life of all kinds on Earth, and harbors the largest biodiversity underground. Yet, our understanding of the complexity of soil and water interactions in the landscape remains surprisingly limited, and our models for predicting water flow and chemical transport in real-world soils remain shaky. A related dilemma also appears in our scientific community: we now have more and more people studying soils, but we have fewer and fewer people knowing what soils really are. In the increasingly technology-dominated society and highly specialized disciplinary education, fewer and fewer people have ever touched a real soil profile in the field. This, unfortunately, hinders the in-depth understanding of the complexity and dynamics involved in a wide variety of soils and hydrologic processes in the landscape, and leads to unrealistic assumptions or negligence of sophisticated architecture of real-world soils and its fundamental controls on flow and transport processes in the field.

The goal of this book is to stimulate a new perspective in our basic thinking and approach to understanding and predicting soil and water interactions in the real-world across space and time and to explore enhanced means of promoting concerted and synergistic efforts from soil science and hydrology, along with other related disciplines. There is an established body of knowledge to be summarized and critically evaluated, knowledge gaps and integrated frameworks to be identified, and innovative approaches and new theories to be developed. This book demonstrates that hydropedology is a needed and promising interdisciplinary field that has significance in advancing both soil science and hydrology and can guide more effective field data acquisition, knowledge integration, and model-based prediction of complex landscape-soil-water-vegetation relationships across scales. Hydropedology emphasizes in situ soils in the landscape, where distinct pedogenic features (e.g. structure, macropore, and horizonation), environmental variables (e.g. climate, landforms, and organisms), and anthropogenic impacts (e.g. land use and management) interact and determine the storages, fluxes, and pathways of energy and mass transfer in the landscape. Synergies are expected by bridging pedology – a unique subdiscipline of soil science and geoscience – with soil physics, hydrology, and other related bio- and geosciences to advance the integrated studies of the Earth's Critical Zone.

This book is organized into three parts: I) overviews and fundamentals, II) case studies and applications, and III) advances in modeling, mapping, and coupling. The volume showcases some new ideas and approaches that do not necessarily fit conventional theories or practices. In Part I, after an overview chapter of what hydropedology entails, fundamental soil architecture across scales is discussed. This is followed by two chapters on preferential flow as related to pedological features and plant rooting systems, respectively. Two chapters on wet soils – hydric soils and subaqueous soils – are then presented, followed by a chapter on soil-mantled hillslope evolution and another chapter on thermodynamic limits of Critical Zone processes. In Part II, six case studies of hydropedology across different soils and landscapes from the United States to Australia are presented. This is followed by a chapter on the use of hydropedology as a powerful tool for environmental policy and regulations. In Part III, ways to advance soils and hydrologic modeling and mapping, as well as coupling hydropedology with biogeochemistry and ecohydrology, are highlighted. The last chapter provides a summary and some future outlooks on hydropedology. The total of 24 chapters in this volume covers a wide range of topics that are centered around addressing fundamentals and building bridges to better understand interactive pedologic and hydrologic processes across space and time. While selected figures are presented in color plates in the print version of this book, all figures – if they were made originally in color – are presented in color in the online version of this book. Each chapter of the book is individually indexed and is accessible in online database search.

In completing this book, I am indebted to a number of people. First and foremost, I thank the contributors for their interests and commitments in providing their expertise and insights that have made this book possible. I am grateful to all the peer reviewers who donated their valuable time and professional insights to ensure the quality of all chapters submitted. These individuals are acknowledged alphabetically in the following: Richard Arnold, Johan Bouma, Erin Brooks, Todd Caldwell, James Doolittle, Michael Gooseff, Li Guo, Marcus Hardie, Chi-Hua Huang, Nick Jarvis, Lixin Jin, James Katie, Raj Khosla, Peter Kinnell, Eric Kirby, Peter Kleinman, Del Levia, Allan Lilly, David Lindbo, Jintao Liu, Lifang Luo, Richard MacEwan, Tyson Ochsner, Toby O'Geen, Yakov Pachepsky, Gary Parkin, Jonathan Phillips, Marty Rabenhorst, Jan Seibert, Mark Thomas, James Thompson, Harold M. van Es, Larry Wilding, Michael Young, and A-Xing Zhu. Thanks also go to Christine Minihane, Acquisition Editor of Elsevier, for initiating this project, and to Jill Cetel, Editorial Project Manager of Elsevier, for her patience and editorial assistance in bringing this project to completion. Poulouse Joseph, Project Manager of Book Publishing Division of Elsevier, is also thanked for his professional helps in the production stage of this book.

Finally, my love and gratefulness go to my beloved wife, Jan, and wonderful kids, Alice and Jimmie, for all their patience and support throughout this project.

Henry Lin
March 24, 2012
State College, PA

Foreword

Ecosystem services are the flow of materials and energy that provide the vital constituents of all human well-being. These flows were classified into four categories by the Millennium Ecosystem Assessment as the supporting processes such as soil formation and nutrient cycling; the services of the provisioning of food, fiber, fuel, and fresh water; the regulating services associated with climate, water, and the atmosphere; plus the cultural services around esthetic appeal, recreational pursuits, and heritage preservation.

Ecosystem services flow from natural capital stocks of the materials and energies that form the ecological infrastructures of our terrestrial landscapes.

Two of our most valuable natural capital stocks are those of our soils and terrestrial waters.

Soils and waters are the *sine qua non* of our life on Earth. The benefits that we receive from the ecosystem services flowing from our soils and waters are invaluable and irreplaceable. The nexus of these teeming flows occur in the Critical Zone of the first few meters above and below the Earth's surface.

To sustain the present value of these soil- and water-derived ecosystem services and to enable adaptation to the exigencies that will be forced on us by climate change, we need to understand better the interactive biophysical and geochemical processes associated with water transport and storage in the Critical Zone of the globe's diverse soils.

How can we do this? By better understanding the hydropedology of our landscapes – That is how.

Hydropedology seeks to link better the hard-won knowledge from our soil scientists about our soil architecture, its functioning, and its distribution across our landscapes, with the understanding gained by hydrologists over the last few centuries of how the water cycle is linked to and modified by our diverse landscapes. There is a natural synergy between pedology and hydrology that this emerging discipline is exploiting.

This book shows us the way as to how this can be done. The book details the measurement and modeling dualism of hydropedology. The generation of new knowledge and understanding from the interactions between measurement and modeling of the hydropedology of our landscapes will provide us with better land-management practices to ensure the sustaining of the ecosystem services that we rely on, plus it will guide us in the development of policies and regulations that can protect and enhance the value of the ecological infrastructures comprising the stocks of our waters and soils.

Part I of the book examines fundamental hydrological aspects of the soil's architecture, biophysical functioning, and geochemical processes. These chapters have a strong emphasis on measurement-derived knowledge at the pore and pedon scale. The chapters of Part II upscale this measurement emphasis to the landscape through a series of case studies. A concluding chapter in Part II details how this knowledge can be incorporated into environmental policies and regulations. The chapters of the final Part III focus on the spatiotemporal modeling of the link between pedology and hydrology and show how these interactions may be mapped across the landscape so that we can better understand and manage the natural capital stocks of our soils and waters – for today and for our future generations.

Our soils and waters support human well-being. Hydropedology is an emerging interdisciplinary science that seeks to understand better the processes and services flowing from the nexus of soil and water interactions in the Critical Zone of the Earth's surface. This book provides great insight to the rapid advances that are being made.

Brent Clothier
Science Group Leader,
Systems Modelling, Plant & Food Research,
Palmerston North, New Zealand

Part I

Overviews and Fundamentals

Chapter 1

Hydropedology: Addressing Fundamentals and Building Bridges to Understand Complex Pedologic and Hydrologic Interactions

Henry Lin[1]

ABSTRACT

Hydropedology is an emerging interdisciplinary science that focuses on two fundamental questions: 1) How do soil architecture and the associated distribution of soils over the landscape exert a first-order control on hydrologic processes (and related biogeochemical dynamics and ecological functions)? 2) How do hydrologic processes (and the associated transport of energy and mass) influence soil genesis, evolution, variability, and function across space and time? The first question is related to the soil's role in water quantity and quality, while the second question is linked to the water's role in soil quantity and quality. Hydropedology is in a unique position to build bridges that are important to gain new insights into subsurface processes, including: 1) connecting fast and slow processes to soil function and soil formation, 2) bridging soil structural and landscape units to deterministic patterns and stochastic variability, and 3) linking mapping with monitoring and modeling to geographic and functional characterizations. Various examples are used in this overview chapter to illustrate how hydropedology can address the above-stated questions and how it can build the needed bridges. Selected frontiers in hydropedology applications and education are then highlighted. It is hoped that hydropedology can bring a new perspective in our basic thinking and approach toward understanding and predicting interactive pedologic and hydrologic processes across scales in the real world.

[1]Dept. of Ecosystem Science and Management, Pennsylvania State Univ., University Park, PA, USA. Email: henrylin@psu.edu

Hydropedology, Edited by H. Lin. DOI: 10.1016/B978-0-12-386941-8.00001-0

1. INTRODUCTION

To introduce the concept of hydropedology, let us start with a simple analogous and yet revealing question: "*What is the difference between a cow and the ground beef?*" To illustrate the fundamental difference between naturally formed *soil profiles* in the landscape and ground-sieved *soil materials* in the laboratory, Kubiena (1938) has insightfully pointed out that a crushed or pulverized sample of soil is as akin to a natural soil profile as a pile of debris is to a building. The ground beef is no longer a true living entity and has the original cow's structure destroyed, while a cow is a living organism that has intricate internal architecture for its proper functioning (Fig. 1a). Hence, the traditional way of studying soils using ground-sieved soil materials (or even beach sands or glass beads) and isolated soil cores or columns (especially repacked ones) can create significant deviations from the field reality, where distinct pedogenic features (such as structures, macropores, and horizons) and open boundary conditions (which change with climate, biota, and human activities) prevail.

In situ understanding of flow and transport processes in natural soils is critical to successful land use/management and effective scaling from point observations to landscape processes (Fig. 1). However, subsurface heterogeneity is ubiquitous in the real world, from croplands to pasturelands and to forestlands (Figs. 2 and 3). Heterogeneity here differs from randomness: the

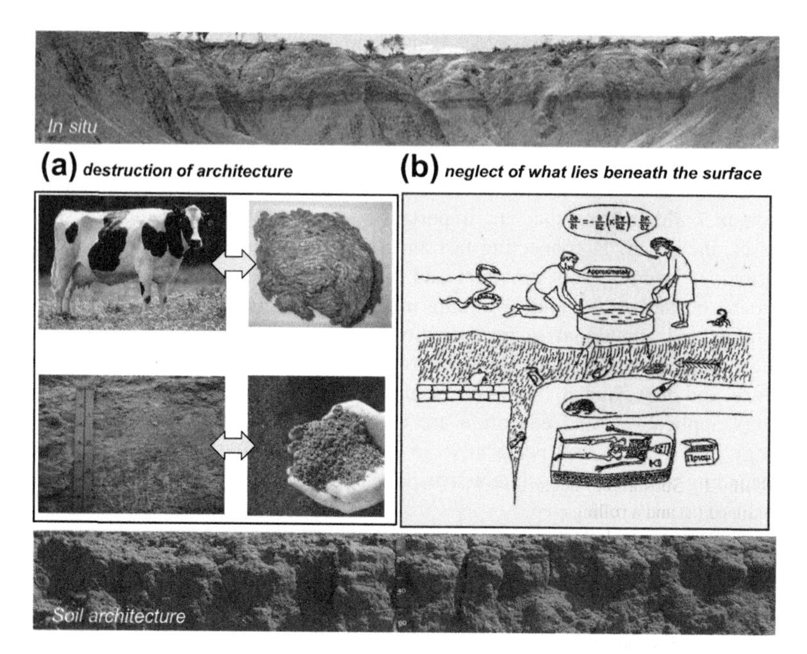

FIGURE 1 Two common problems in studying soils: (a) destruction of soil architecture and (b) neglect of what lies beneath the soil surface. *(The two photos at the top and bottom panels are courtesy of Alfred Hartemink; the diagram on the right is adopted from Hillel 1998). (Color version online and in color plate)*

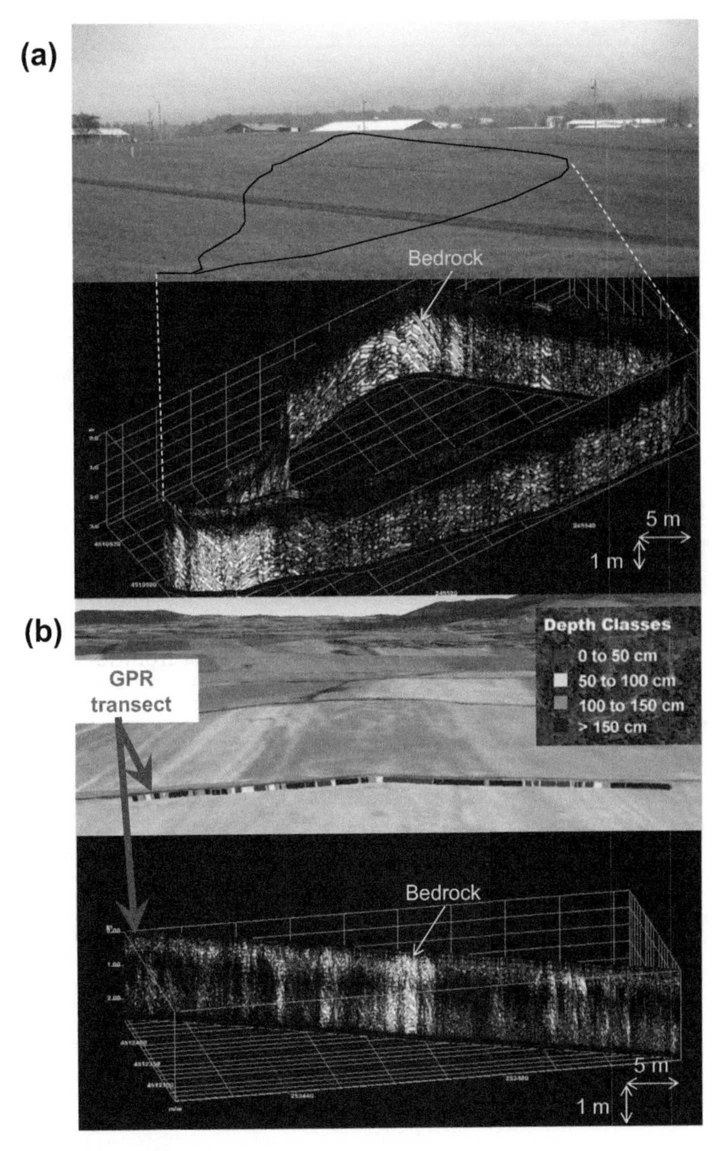

FIGURE 2 Subsurface heterogeneity and soil thickness variation across a gently-sloping pastureland (a) and a rolling cropland (b) in central Pennsylvania, as revealed by ground-penetrating radar (GPR) transects. In both radargrams, white areas indicate strong radar reflections, suggesting limestone bedrock (with bedrock orientation visible with bends). The color line in (b) shows four classes of depth to bedrock (soil thickness) based on the interpretation of the radargram. The soils in both locations are dominated by the complex associations of shallow Opequon series and deep Hagerstown series. Note that the vertical scale is exaggerated as compared to the horizontal scale for the purpose of display clarity. *(Color version online and in color plate)*

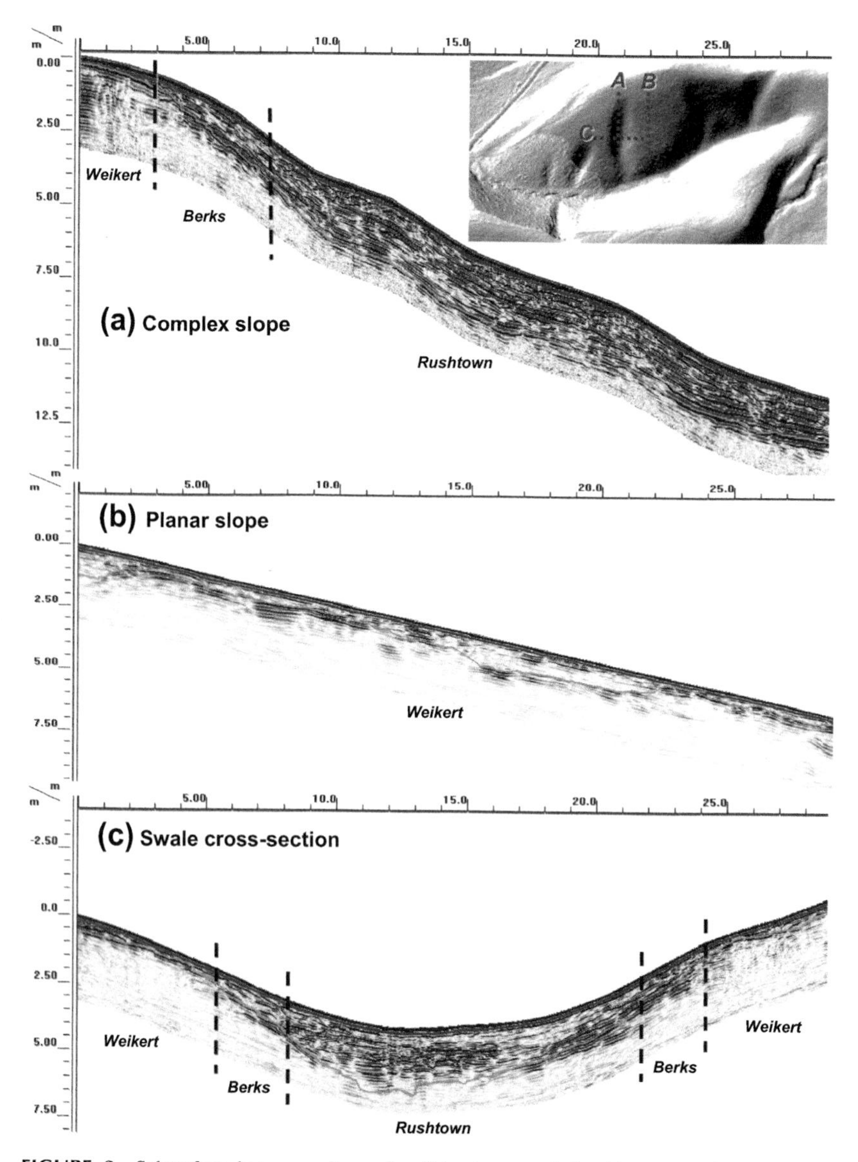

FIGURE 3 Subsurface heterogeneity and soil-landscape relationship across three types of steeply-sloping hillslopes in the forested Shale Hills catchment (shown in the inset) in central Pennsylvania, as revealed by ground-penetrating radar (GPR) transects: (a) a complex hillslope; (b) a planar hillslope; and (c) a cross-section of a swale. In all three radargrams, green curves indicate the soil-bedrock (fractured shale) interface based on the interpretation of the radargrams. The soil catenae in (a) and (c) are also shown, with the soil series differentiated mainly by depth to bedrock; however, soil drainage, wetness condition, and soil horizonation are also different among these soil series (see Fig. 11 and Lin et al., 2006, for more details of these soils). The vertical dash lines approximately separate the three soil series along the hillslopes, which show graduate transient from one soil type to another. Note that the vertical scale is exaggerated as compared to the horizontal scale for display clarity purpose. *(Color version online and in color plate)*

former is associated with order while the latter is linked to disorder. While enormous subsurface heterogeneity is daunting and challenging to fully comprehend, heterogeneity itself is an important prerequisite for a system to evolve and to stay "alive." Because of heterogeneity of all kinds, evolutionary processes have been made possible by the flow of energy, mass, entropy, and information that are driven by various gradients (e.g. Tiezzi, 2006; Lin, 2011a). Indeed, biodiversity is rooted in the enormous variability of life conditions in space and time (Jørgensen, 2006).

Hidden in the complex subsurface heterogeneity are various architectural controls that may be discerned from the pore scale to the landscape scale (Fig. 4). Such hierarchically organized heterogeneity governs the direction, distribution, and efficiency of energy, matter, entropy, and information transfers in the subsurface. For example, Fig. 4 illustrates that water flow and solute transport in real-world soils are bound to underlying architectural controls across a range of scales, which have been revealed by various studies and techniques.

Overall, two fundamental characteristics of soils in the landscape are essential to understanding and predicting real-world pedologic and hydrologic processes (Fig. 1):

1) *Architecture*: This refers to the entirety of how the soil is organized from the microscopic to megascopic levels (Fig. 5), including soil particles and pores, fabrics and aggregates, vertical layering and lateral distribution, and various interfaces within soil profiles and between soils and their external environments.
2) *In situ*: This refers to the original position of soils in the landscape, including the soil–landscape relationship, the environmental setting for soil functioning, and open soil system boundaries in the field, which are dynamic with various natural or anthropogenic perturbations.

Soils are three-dimensional bodies in the landscape, with different arrangements of vertical horizons and lateral variability (Figs. 1–4). Because soils are hidden underground, the human tendency of "*out of sight, out of mind*" has caused the neglect or downplay of soil complexity. For example, when measuring infiltration at the soil surface, soil physicists/hydrologists are not always aware of what lies hidden inside the soil profile (Fig. 1b), which often results in hard-to-interpret field data using idealized mathematical theories. Visualizing the opaque subsurface (especially nondestructively) remains a big challenge. Nevertheless, it is clear that the subsurface architecture has fundamental controls on hydrophysical and biochemical processes in real-world soils across scales. This has been clearly demonstrated in a wide variety of studies and using a wide range of techniques (e.g. dye tracing, thin section, sensor network, ground-penetrating radar, electromagnetic induction, X-ray computed tomography, and scanning electron microscopy, as illustrated in Fig. 6).

FIGURE 4 Illustrations of soil architecture across scales: (a) Landscape-scale soilscape: aerial photo of a Priarie Pothole region in North Dakota (left) showing closed depressions (dark areas) dispersed across the glacial landscape, where depressions are discharge areas and light-color soils are recharge areas (photo courtesy of Jimmie Richardson); glaciated mountain landscape in Colorado (right) showing outwash terrace, end moraine, and mountain slope, each with a contrasting soil type (photo courtesy of Doug Wysocki); (b) hillslope-scale cross-section: loess, colluvium over pre-Illinoian till in Lincoln, NE (left) showing pedologic features with stratigraphic significance (photo courtesy of Phil Schoeneberger); drumlin with interior exposed in Delta County, MI (right) showing the topographic highs of subglacial origin, the interdrumlins areas of glacial outwash, and the organics accumulated in wet areas following deglciation (photo courtesy of Doug Wysocki); (c) pedon-scale macromorphology: three contrasting soil orders with various soil architectures – Inceptisol (left), Vertisol (center), and Spodosol (right) (photo courtesy of USDA-NRCS); (d) thin-section micromorphology: surface horizon of an Ustert (left) that has been in >30 years pasture after cultivation for row-crops, showing well-developed subangular blocky peds (black), numerous pores (white) around peds, and high biologic activity consisting of worm casts and roots (root are circular, yellowish bodies associated with voids). Frame length is 4 mm taken in plane-polarized light mode (photo courtesy of Larry Wilding); a crusted (sealed) surface of an Ustalf (right), showing light-colored seal of clay and fine silt (upper half) with lower half illustrating porous nature of unsealed sandy matrix. Frame length is 1 mm taken with backscatter SEM mode *(Photo courtesy of Larry Wilding). (Color version online and in color plate)*

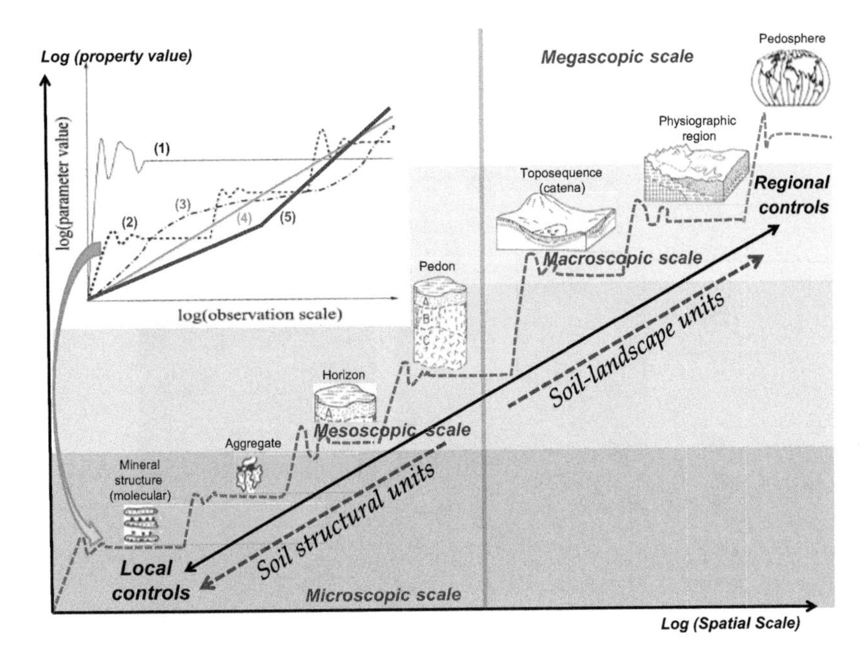

FIGURE 5 A hierarchical framework for bridging scales from the molecular level to the global pedosphere. Soil structural units refer to soil architecture within a pedon, while soil-landscape units refer to soil architecture in the landscape. Keys to bridging the multiple scales include changing dominant controls, structural interfaces, and two-way feedbacks (where upper-scale constraints lower-scale processes and lower-scale feedbacks to upper-scale dynamics). The inset shows five common models of soil spatial heterogeneity and related scaling schemes: (1) macroscopic homogeneity (thin line), (2) discrete hierarchy (dashed line), (3) continuous hierarchy (dashed dotted line), (4) classical fractal (thin straight line), and (5) multi-fractal (thick straight lines). (Various graphs of soil entities are taken from Wiliding, 2000; the inset is modified from Vogel and Roth, 2003). *(Color version online)*

In situ study of soils and their architectural complexity has traditionally been part of the pedology domain. As a unique subdiscipline of soil science and geoscience, pedology has been the integrated study of soils as natural or anthropogenically-modified landscape entities, encompassing soil physical, chemical, biological, and anthropogenic processes. Historically, pedology has focused on soil morphology, genesis, classification, mapping, and interpretation (Wilding, 2000; Buol et al., 2003; Bockheim et al., 2005), but in recent years has received elevated attention in monitoring and modeling (e.g. Reuter and Bell, 2003; Lin et al., 2006a; Minasny et al., 2008). The emphasis of pedology has now shifted from classification and inventory to understanding and quantifying spatially–temporally variable processes upon which the water cycle and ecosystems depend, and to marketing this information to the public domain (Lin et al., 2005, 2006b). Although the term pedology has sometimes been mixed with the general term of soil science, the field-based study of soils,

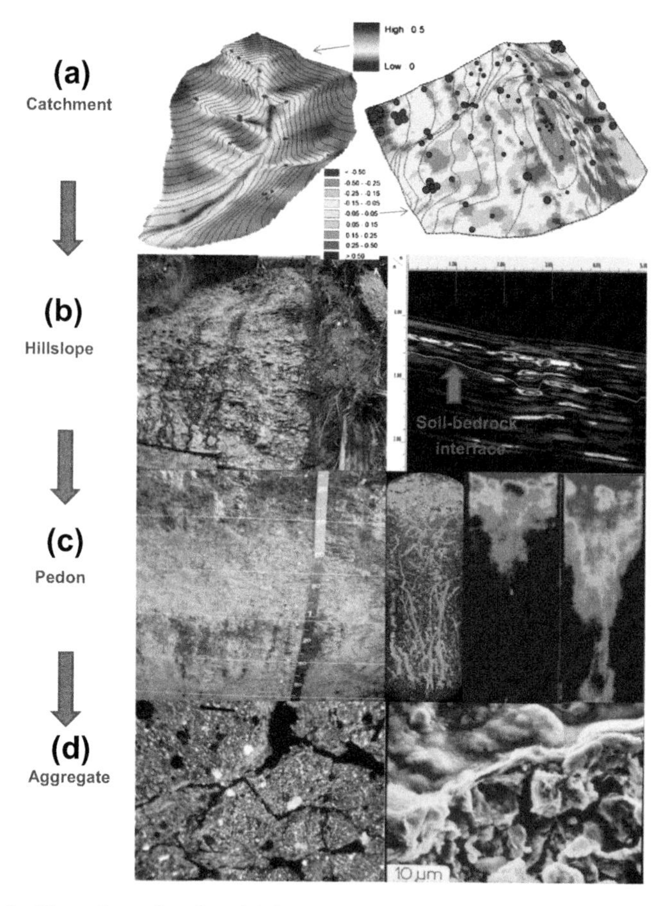

FIGURE 6 Illustrations of preferential flow and transport patterns in various field soils across scales and related techniques used: (a) Catchment scale: total soil moisture storage (m) in sola determined from >100 monitoring sites in the 7.9-ha forested Shale Hills catchment (left); relative difference (compared to the overall mean) in soil apparent electrical conductivity in the 19.5-ha Kepler Farm (right), obtained from electromagnetic induction survey across the entire landscape; (b) hillslope scale: blue-dyed water flow along a soil–bedrock interface (left) in an excavated hillslope in the Maimai catchment in New Zealand (photo courtesy of Chris Graham); subsurface lateral flow in a planar hillslope with shallow soil (right) in the Shale Hills catchment, revealed by repeated ground-penetrating radar scans (white and pink areas indicate water, with strong radar reflections); (c) pedon scale: fingering front from brilliant-blue dye-tracer experiment (left) showing the dye having moved through the blocky structured Bt horizon and accumulated below into numerous fingers (photo courtesy of Marios Sophocleous); macropore network (consisting of earthworm borrows, root channels, and others) in a soil column of 30-cm long (right) and preferential solute transport through this column, revealed by X-ray computed tomography; (d) aggregate scale: clay coatings (golden color) along ped surfaces (left) resulting from clay translocation from upper soil in an Alfisol, under cross-polarized light mode in thin section, with the frame length of 2.5 mm; scanning electron microscopy (SEM) of a clay coating (right) showing remarkable change in clay orientation and porosity between clay coating at ped surface and ped interior *(Photo courtesy of Larry Wilding.) (Color version online and in color plate)*

their architecture and formation, and their relation to the landscape have been the hallmark of pedologic study (Bockheim et al., 2005). Up to this date, pedology remains the only unique subdiscipline of soil science with its own theories and systematics, while the other subdisciplines of soil science have been derived (for the most part) from other scientific fields such as applied mathematics, physics, chemistry, biology, and others in the study of soils (or, more often than not, soil materials, i.e. the ground beef rather than the cow itself).

The emerging interest in the Earth's Critical Zone has sparked renewed interest in pedology (e.g. Wilding and Lin, 2006; Richter and Mobley, 2009). The Critical Zone has been defined by the National Research Council (NRC, 2001a) as the heterogeneous, near-surface terrestrial environment from the top of the vegetation down to the bottom of the aquifer, where complex interactions involving water, soil, rock, air, and organisms regulate the natural habitat and determine the availability of nearly every life-sustaining resource. How rocks weather into soils and how soil profiles evolve over time have historically been part of the pedologic study (e.g. Dokuchaev, 1893; Jenny, 1941; Joffe, 1949; Birkeland, 1974). By extending the top 1–2 m soil profile (largely the root zone) down to saprorite or the whole regolith (e.g. Cremeens et al., 1994) and quantifying fluxes at various spatial and temporal scales (e.g. Richardson and Vepraskas, 2001), pedologists have now teamed up with soil physicists, hydrologists, geologists, geomorphologists, geochemists, ecologists, and many other discipline scientists to better understand and quantify processes and patterns in the Earth's Critical Zone (e.g. Anderson et al., 2004; Lin et al., 2005, 2008; Gburek et al., 2006; Amundson et al., 2007; McDaniel et al., 2008; Bouma et al., 2011; Swarowsky et al., 2011; Zacharias et al., 2011; Banwart et al., 2011; and various chapters of this book).

At the same time, hydrogeosciences have also encountered a new intellectual paradigm that emphasizes the connection between the hydrosphere and the other components of the Earth system (NRC, 1991, 2012). Water has been recognized to be at the heart of both the causes and effects of climate change, and water has also been recommended as a unifying theme for understanding complex environmental systems (e.g. NRC, 1998, 1999). While hydrogeology, hydrometeorology, and ecohydrology have been well recognized, an important missing piece of the puzzle is the interface between the hydrosphere and the pedosphere. It is interesting to note that the soil has been referred to *the final frontier* of science (Sugden et al., 2004; Baveye et al., 2011) and the vadose zone has also been called *the last hydrologic frontier* (Nielsen, 1997). It is amazing that, over 500 years after Leonardo Da Vinci, the ground beneath our feet is still as alien as a distant planet. Yet, the interaction between soil and water beneath our feet has created the fundamental interface between the abiotic and the biotic worlds and has served as the crucible to terrestrial life (including the largest belowground biodiversity on Earth) (Hillel, 2008). In fact, water flux into and through soils in the landscape is the essence of soil functions that support diverse ecosystem services, which resembles in a way that blood circulates in a human body (Bouma, 2006). However, there are a number of limitations in classical flow theories for applications in real-world soils that have been increasingly recognized, including:

1) *Dynamic boundary and nonuniform flow*: Most classical flow theories have been based on the assumption of fixed boundary conditions (e.g. some kind of control exerted on a flow system from its boundary), while real flow system in natural soils is generally open with dynamic boundary (Beven, 2006; Bejan, 2007). Furthermore, the cross-sectional area of flow is often assumed to be a constant in classical flow theories, i.e. flow is uniform across the entire flow area and thus the flow can be approximated as 1D flow (Jury and Horton, 2004). In reality, heterogeneity in field soils leads to spatially and temporally variable cross-sectional area of flow (e.g. Gish and Shirmohammadi, 1991).
2) *Transient and nonsteady state*: Steady-state flow has been the focus of flow theories in soil physics/hydrology, whereas transient flow has been far less developed (Jury and Horton, 2004). In addition, soil hydraulic properties (e.g. van Genuchten parameters) are considered as stable, not time-dependent in nearly all existing hydrologic models (Šimůnek et al., 2003). In the real world, irregular and nonuniform water inputs from precipitation or irrigation, plus frequent biological and anthropogenic perturbations, can significantly alter flow regimes and soil hydraulic properties over time (Lin, 2010a).
3) *Flow pathways and patterns*: Where water flows and how that may change with time in the heterogeneous subsurface (i.e. flow configuration evolution) is a first-order question to be answered before flux of water can be effectively quantified. Traditionally, hydrologic processes have been conceptualized within the continuum domain (i.e. fluid in a porous medium is distributed throughout – and completely fills – the space it occupies, which is much the same way as atmospheric and oceanic circulation and the distribution of stresses in solids) (Bejan, 2007). However, heterogeneities, structures, fractures, roughness, and organisms in the subsurface make the field soil deviate significantly from such a continuum assumption (Lin, 2010a).

Both soil and hydrologic sciences are now at a critical threshold of advancing frontiers and exploring breakthroughs. Pedology and hydrology are scientific disciplines that are inherently associated with the landscape perspective and share many common interests (Lin et al., 2006b). However, these two disciplines have been largely disconnected in the past and "stereotype" visions have existed between pedologists and hydrologists (e.g. pedologists use "funny" names to describe soils and make many empirical statements about soil functions that are not necessarily supported by measurements, whereas hydrologists make unrealistic assumptions of homogeneity and isotropy about soils in their "sand-tank" models that clearly do not reflect real-world conditions). Even within soil science, pedology and soil physics have not been well connected (Lin et al., 2005; Braudeau and Mohtat, 2009), leading to a number of long-standing or emerging research areas in soil physics that are all field oriented (Jury et al., 2011) and the need to revitalize pedology through soil hydrology (Lin et al., 2006c). Integrating pedologic and hydrologic expertise is both necessary and synergistic toward the holistic understanding of the landscape–soil–water–vegetation relationship in the Critical Zone. However, the lack of convergence of pedology and hydrology/soil physics in the past has resulted in many knowledge gaps, such as:

- How to predict preferential flow pathways and their dynamics at different scales, their interfaces with the soil matrix, and their significance in different types of soils and landscapes?
- Where, when, and how water moves through different parts of the landscape and its impacts on soil physical, chemical, and biological processes and subsequently spatial–temporal patterns of soil distributions and soil functions?
- How to bridge multiple scales from microscopic (e.g. pores and aggregates) to mesoscopic (e.g. pedons and catenas), macroscopic (e.g. catchments and watersheds), and megascopic (e.g. regional and global) levels for coupled soil and hydrologic processes?
- How to close data gaps between soils databases (including soil surveys and maps) and hydrologic input parameters needed for models and how to quantify the associated uncertainty?
- How to stratify (or classify) a landscape into functional units that have similar pedologic and hydrologic processes and properties (e.g. similar flow patterns and transport mechanisms) and how to optimize monitoring or sampling locations so that effective upscaling can be achieved?

Consequently, hydropedology has emerged in recent years as a new interdisciplinary science that attempts to close such knowledge gaps (e.g. Lin, 2003; Lin et al., 2005, 2006b). Time is now ripe for synergistic integration of pedologists' expert knowledge of in situ soil architecture and the soil–landscape relationship with hydrologists' and soil physicists' quantitative skills and mathematical rigor of flow and transport theories. Such integration is facilitated by modern advances in monitoring, modeling, and mapping technologies as well as growing awareness of the need for interdisciplinary studies.

2. HYDROPEDOLOGY: FUNDAMENTAL QUESTIONS AND ILLUSTRATIVE EXAMPLES

Hydropedology emphasizes the linkage between the pedon and the landscape paradigms (Figs. 7 and 8). By integrating pedology, soil physics, hydrology,

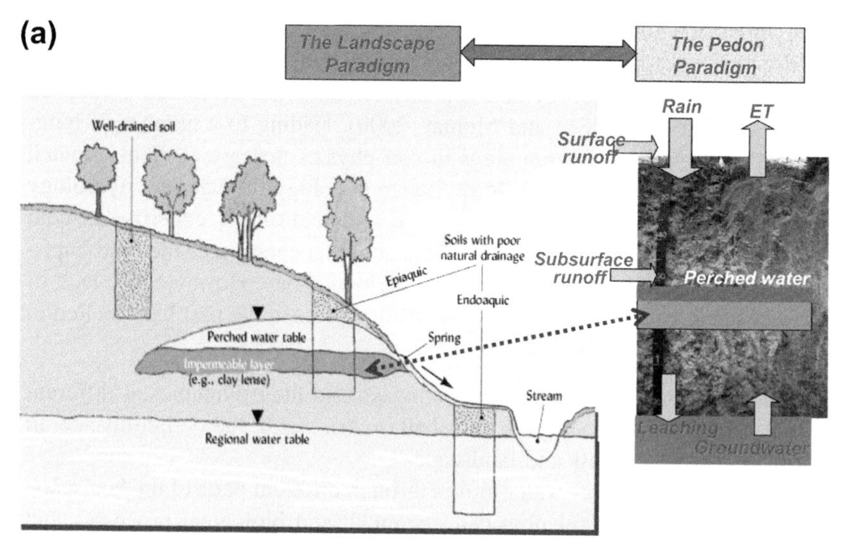

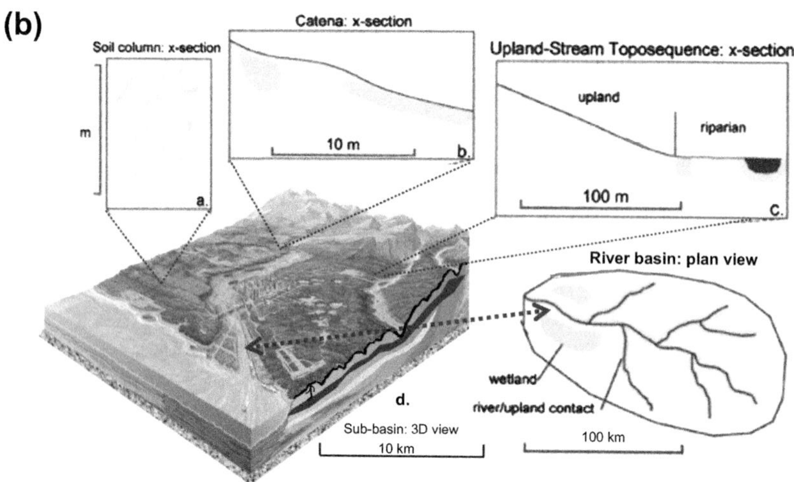

FIGURE 7 Connecting the pedon and the landscape paradigms through hydropedology: (a) cross-section of a typical hillslope, showing the regional and perched water tables in relation to three types of soils – one well-drained and two poorly drained (from Brady and Weil, 2004). The soil containing the perched water table is wet in the upper part, but may be unsaturated below the impermeable layer (this is called epiaquic), while the soil saturated by the regional water table is called endoaquic. Common hydrologic processes that may occur are illustrated in the zoom-in soil profile; (b) biogeochemical hot spots (e.g. disproportionally high denitrification as compared to the surrounding area) that may occur in the landscape at multiple scales: a. along root channels in the soil profile, b. in depressional wet spots along the catena, c. at the upland–stream interface in the toposequence, and d. in wetland or upland-river contact in the subbasin *(Modified from McClain et al., 2003) (Color version online)*

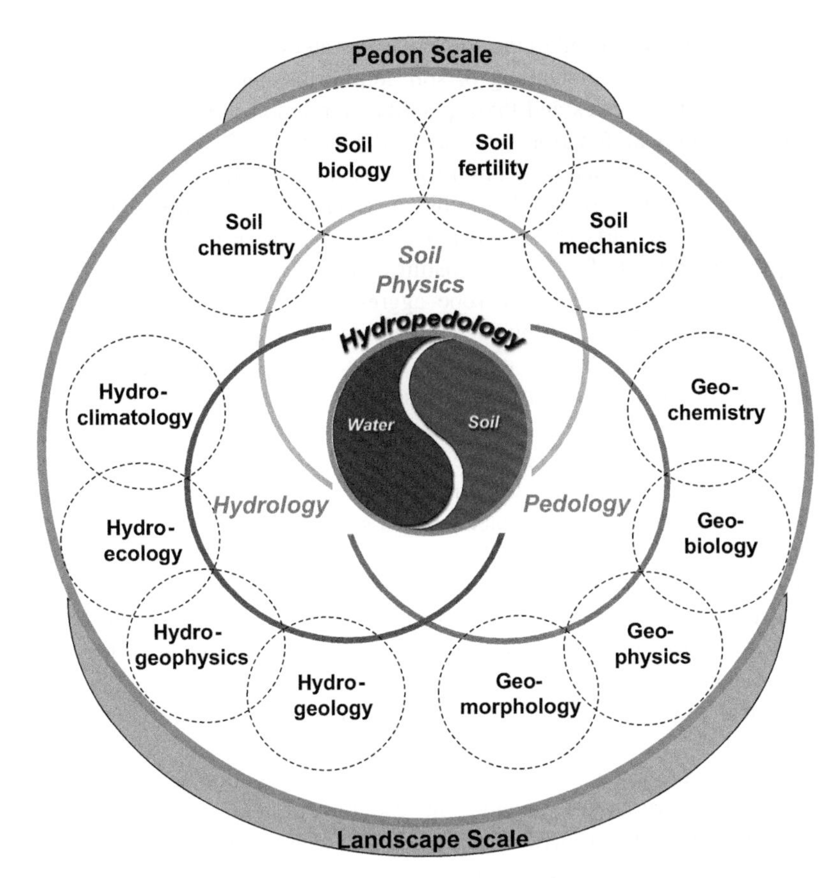

FIGURE 8 Interdisciplinary nature of hydropedology, with pedology, soil physics, and hydrology as its cornerstones. Pedology is an integrated Earth science, interfacing with other branches of geosciences; soil physics deals with the physical properties and processes of the soil, with an emphasis on the state and transport of matter (especially water) and energy in the soil, and is linked to other branches of soil science; hydrology deals with the hydrologic cycle across scales, and includes various subdisciplines. Both pedology and hydrology are inherently associated with the landscape perspective, while soil physics has traditionally focused on the soil profile scale. Green circles in the diagram symbolize life and ecosystems at large. *(Modified from Lin, 2003). (Color version online)*

along with other related bio- and geosciences, hydropedology addresses the following two fundamental questions (Lin et al., 2008; Lin, 2011b):

1) *How do soil architecture and the associated distribution of soils over the landscape exert a first-order control on hydrologic processes (and related biogeochemical dynamics and ecological functions)?*
2) *How do hydrologic processes (and the associated transport of energy and mass) influence soil genesis, evolution, variability, and function across space and time?*

The first question is related to the soil's role in water quantity and quality. This calls for adequate understanding of soil architecture and the soil–landscape relationship across scales and their quantitative relationships with hydrologic and biogeochemical functions. As detailed in Chapter 2 of this book, soil architecture represents a holistic, multicomponent, and multiscale organization of soil systems across scales. It exhibits hierarchical complexity and encompasses interlinked parts of solids, pores, and interfaces at each scale. For example, diverse interfaces (e.g. coatings on ped surfaces or pore edges, macropore–matrix, soil–root, microbe–aggregate, horizon–horizon, and soil–rock interfaces) are often triggers for preferential flow of water, chemicals, gases, and microbes, as well as active zones for biogeochemical reactions (Fig. 7). Numerous water-restricting subsoil horizons and features (e.g. fragipan, duripan, placic, petrocalcic, ortstein, paralithic and lithic contacts, and lamellae) have been identified in the U.S. Soil Taxonomy (Soil Survey Staff, 2010), which can act as aquitards or aquicludes to downward-moving water, resulting in seasonal perched water table in soil profiles or lateral subsurface throughflow in hillslopes (e.g. Gburek et al., 2006; McDaniel et al., 2008). Although considerable knowledge on soil structure has been amassed over the last decades, a comprehensive theory and an effective means of quantifying soil architecture across scales remain elusive (see Chapter 2 of this book for more details). Noninvasive investigations (e.g. X-ray computed tomography and geophysical tools) and continuous mapping (e.g. spectroscopy and remote sensing) have been increasingly employed to reveal soil architectural complexity and its impacts on flow and transport across scales (see examples in Figs. 2–4). However, how to enhance the resolution of field investigations, how to translate image data to hydrologic properties of interest, and how to quantitatively link soil architectural parameters to hydrologic models remain a challenge. Chapters 2–4 of this book further elaborate on this subject.

The second question above is linked to the water's role in soil quantity and quality. This includes the impact of hydrology on 1) soil formation, thickness, and distribution; 2) the nature, origin, and dynamics of soil pore space; and 3) soil function, quality, and related processes. How water moves through soils – along with associated shrink–swell, freeze–thaw, saturation–desaturation, oxidation–reduction, dissolution–precipitation, adsorption–desorption, and other processes – dictates the type of soils formed in a given environment, the dynamic evolution of soil heterogeneity, the stability or transformation of soil materials and structures, and the feedback mechanism of soil functions. For example, recharge hydrology removes materials from soil profiles, while discharge hydrology adds materials to soil profiles (Richardson and Vepraskas, 2001). Hydrology has been widely recognized as an important driving force of soil formation and evolution (e.g. Jenny, 1941; Rode, 1947; Daniels and Hammer, 1992; Buol et al., 2003; Schaetzl and Anderson, 2005). This is because all of the five natural soil-forming factors (climate, organisms, parent material, topography, and time) affect and are effected by hydrology, and all of the four general soil-forming processes (additions, deletions, transformations, and translocations) involve water in significant ways (Lin et al., 2005). Much like "*one cannot ignore the role of groundwater in performing geologic work*" (Domenico and Schwartz, 1998), water in the soil zone cannot be ignored in soil formation and soil function. Daniels and Hammer (1992) noted that "*we cannot study soil development and propose grand schemes of soil genesis without a concurrent study of the soil–landscape hydrologic system.*" It is important to note that hydrologic impacts on soils are largely accumulative effects (over a period of time), and hydropedology considers both current and past hydrologic impacts on soils. Chapters 5–8 of this book further elaborate on some of the related subjects.

Hydropedology is apparently linked to a number of disciplines (Fig. 8), with pedology, soil physics, and hydrology being its "cornerstones." These cornerstone disciplines have shared many common interests, particularly in areas related to the flow and transport of mass and energy through soils and landscapes. Although traditionally these three disciplines have had contrasting focuses and approaches in their investigations, the time is now ripe for their synergistic integration (for a review on this perspective, see Lin et al., 2005).

In the following, various aspects of hydropedology related to the above two fundamental questions are illustrated using examples from upland soils to subaqueous soils, and are linked to various chapters in this book.

2.1. Soil Architecture and Hillslope/Catchment Hydrology

Soil architecture exhibits strong impacts on hillslope/catchment hydrology from the soil surface down to water-restricting layer or the soil–bedrock interface (Figs. 6 and 7). In many cropped or pastured landscapes, soil surfaces may be covered with crusts or sealing (biological and/or physical origins), whereas in forested landscapes the soil surface is generally covered with litter layers, which may be hydrophilic or hydrophobic depending on the dryness of soil organic matter. Significant biological crusts could emerge at the soil surface even in the initial stage of ecosystem development, which, if not taken into account, can lead to the failure of classical hydrologic models in predicting hillslope/catchment discharge. For example, given soil texture and topography as inputs, ten conceptually different hydrologic models (Catflow, CMF, CoupModel, Hill-Vi, HYDRUS-2D, NetThales, SIMULAT, SWAT, Topmodel, and WaSiM-ETH) all failed to predict catchment discharge in an artificial sandy catchment in Germany, and none of these model simulations came even close to the observed water balance for the entire 3-year period (Holländer et al., 2009). Discharge was mainly predicted by these models as subsurface discharge with little direct runoff; in reality, surface runoff was a major flow component throughout the entire catchment despite the fairly coarse soil texture because of significant biological crusts (Gerwin et al., 2011). Chapter 16 of this book by Rinderer and Seibert has called for more dialogs between experimentalists and modelers, including the enhanced use of soils information (including hard and soft data) in hydrologic modeling.

Water-restricting subsoil horizons and the soil–bedrock interface also play an important role in hillslope/catchment hydrology, which may induce perched water table in soil profiles and promote lateral preferential flow (e.g. Freer et al., 2002; McGlynn et al., 2002; Gburek et al., 2006; McDaniel et al., 2008). For instance, dense and slowly permeable fragipans occupy ~16.5 million ha soils in the U.S. (mostly in the east and southeast; Lin et al., 2008). These are important controlling factors in variable source area hydrology. For example, Needelman et al. (2004) have shown that hillslopes with fragipan-containing soils produced considerably more runoff than did those adjacent soils without fragipans. Monitoring in the Palouse Region of northern Idaho and eastern Washington by McDaniel et al. (2008) also showed that the spatial distribution of saturation-excess runoff is largely controlled by the distribution of fragipan and its varying depth from the surface. On the other hand, for vast texture-contrasting soils in Australia, Hardie (2011) has showed that preferential flow could prevent the development of perched water table and subsurface lateral flow when soils were dry. Wilding et al. (Chapter 9 of this book) and Brooks et al. (Chapter 10 of this book) provide further evidence of subsoil restrictive layers in hillslope hydrology. Jarvis et al. (Chapter 3 of this book) have synthesized extensive experimental evidence accumulated over the last 20–30 years to emphasize the linkage between pedological features and preferential flow in various soil types of the world.

2.2. Soil Mapping and Landscape Hydrologic and Biogeochemical Functions

Soil distribution patterns in the landscape, including different kinds and thickness of soil horizons and how they are organized into various soil profiles (see examples in Figs. 1–4), have strong controls on surface and subsurface hydrologic processes. Precision mapping of soil–landscape relationships and spatially-explicit and quantitative soils data have been increasingly in demand for site-specific applications (such as landscape hydrology and precision agriculture). Figures 2 and 3 illustrate contrasting soil thickness (depth to bedrock) distributions in a typical pastureland, cropland, and forestland in central Pennsylvania, USA, where subsurface flow patterns have been linked to the distribution of soil types across the landscapes (e.g. Lin et al., 2006a; Zhu and Lin, 2009; Zhu et al., 2010a). Figure 6a shows the dominant control of soil distribution pattern on soil moisture storage throughout a forested catchment, where shallow and drier soils are mostly distributed in ridgetops and planar hillslopes, while thick and wet soils are located in swales and the valley floor (see Chapter 12 of this book for more details). Using time-lapse ground-penetrating radar with real-time soil moisture monitoring, Zhang et al. (submitted for publication) revealed that infiltration wetting front and preferential flow pattern differed considerably between contrasting soil types (see Chapter 13 of this book for more details). Using repeated electromagnetic induction surveys together with depth to bedrock and terrain, Zhu et al. (2010b) improved the precision soil mapping across an agricultural landscape, which captured the dynamics of soil moisture pattern and related subsurface flow. Various case studies presented in Chapters 9–14 of this book further demonstrate the close linkage between soil–landscape distributions and water flow patterns. Thompson et al. (Chapter 21 of this book) discuss the interactions and applications of digital soil mapping for hydropedology.

Since soil hydrology often triggers so-called hot spots and hot moments of biogeochemical dynamics (i.e. soil patches that show disproportionately high reaction rates relative to the surrounding matrix, or short time periods that exhibit disproportionately high reaction rates relative to longer intervening time periods) (McClain et al., 2003), soil distribution pattern and subsurface flowpaths are important in understanding the spatial pattern of biogeochemical dynamics and ecological functions across the landscape (Fig. 7). For instance, Pennock et al. (2010) demonstrated that soils in the Prairie Pothole Region wetlands were loci of high potential greenhouse gas emissions and that strong interactions between hydrology, soils, and water chemistry controlled the emissions of N_2O and CH_4. Using examples from carbon and nitrogen cycles, McClain et al. (2003) showed that hot spots often occur where hydrologic flowpaths converge with substrates or other flowpaths containing complementary or missing reactants, whereas hot moments occur when episodic hydrologic flowpaths reactivate and/or mobilize accumulated reactants. Vidon et al. (2010) emphasized transport-driven hot spots and hot moments by focusing on rainfall events and flow paths that lead to dissolved organic carbon flushing. Transport-driven hot spots are known to be important to transport labile solutes from uplands to streams (e.g. Vidon and Hill, 2004; Vidon et al., 2010). Therefore, intense precipitation events or snowmelt periods can lead to transport-driven hot moments, while transport-driven hot spots may be defined by preferential flowpaths that channelize water and solutes through the soil into groundwater or stream water. D'Amore et al. (Chapter 11 of this book) demonstrated that regional dissolved organic matter biogeochemistry was related to hydropedology, and Castellano et al. (Chapter 22 of this book) further discuss the necessity of coupling hydropedology and biogeochemistry in carbon and nitrogen cycling.

2.3. Wet Soils and Carbon Sequestration

With growing interests in wetlands and hydric soils, many studies have underlined the importance of soil morphology in interpreting soil hydrology (e.g. Vepraskas and Sprecher, 1997; Hurt et al., 1998; Rabenhorst et al., 1998; Richardson and Vepraskas, 2001). Soil hydromorphology deals with soil morphological features (especially redoximorphic features) caused by water saturation and their relation to hydrologic conditions. Redox features (formerly called mottles and low chroma colors) are formed by the processes of alternating reduction–oxidation due to saturation–desaturation and the subsequent translocation or precipitation of Fe and Mn compounds in soils (Hurt et al., 1998). Because soil morphology is sensitive to water regime in soils, and water table in hydric soils in particular, it could be used to estimate soil moisture regime and water table status, as demonstrated in many studies (e.g. Coventry and Williams, 1984; Evans and Franzmeier, 1986; Galusky et al., 1998; Severson et al., 2008). The depth to gleying, for instance, has long been used as an indicator of the mean position of wet-season water table (Franzmeier et al., 1983), while the depths to redox concretions and depletions have been linked to water-table fluctuations (Vepraskas, 1992). In the UK, in the absence of direct measurement, soils can be assigned to one of six soil wetness classes that describe the height and duration of water logging based on soil profile features including the depth to a slowly permeable layer and depth to gleying (Lilly and Matthews, 1994; Lilly et al., 2003). It should be noted, though, that relict redox features formed under past hydrologic condition or paleo-climate could be difficult to interpret and thus cautions should be exercised (Hurt et al., 1998; Rabenhorst et al., 1998). The rules and cautions in using soil morphology to infer soil hydrology are discussed in Chapter 5 of this book by Vepraskas and Lindbo.

The interconnectedness between soils and water quality in subaqueous environments, and the growing pressures on estuarine and freshwater bodies, has given rise to increased attention to soils under permanently inundated water called subaqueous soils (which have been traditionally treated as sediments). The observation and realization that subaqueous soil horizons are the result of pedogenic processes have led to a new definition of soils by the U.S. Department of Agriculture that accommodates subaqueous soils (see Chapter 6 of this book by Rabenhorst and Stolt). Estuaries are believed to have the highest primary productivity of all ecosystems, yet soils in estuarine environments have been largely overlooked in carbon (C) sequestration studies (Jespersen and Osher, 2007; Erich et al., 2010). Jespersen and Osher (2007) have suggested that systematically quantifying and dating C in estuarine soils can provide valuable data for use in regional and global C budgets and climate models, because missing C sinks in the global C budget may likely include those C sequestrated in subaqueous soils. As an illustration, Jespersen and Osher (2007) quantified organic C stored in the top 1 m of soils from the Taunton Bay estuary in Hancock County, Maine, and found that the organics were protected from decomposition by anaerobic conditions, leading to 136 Mg C ha^{-1} that is much greater than the C content in the top 1 m of Maine's upland soils. Over the geological time, clays formed in soils in the past that were washed from land into offshore waters are now believed to have altered the global C cycle, with huge impacts on climate during the Neoproterozoic (Kennedy et al., 2006). The importance of C in arctic soils is another globally significant C reservoir that may be threatened by global warming – if permafrost thaws then it will release trapped C and can further exacerbate global warming (Zimov et al., 2006). A comprehensive discussion on the pedogenesis, mapping, and applications of subaqueous soils is presented in Chapter 6 of this book, and Chapter 11 further discusses regional C biogeochemistry in relation to hydropedology in North American perhumid coastal temperate rainforest.

2.4. Hydrology as a Main Driving Force of Soil Change and a Factor of Soil Formation

Many factors affect soil development, but none is more important than the abundance, flux, pathways, and seasonal distribution of water (Daniels and Hammer, 1992). This has been demonstrated in numerous studies throughout the world (e.g. Rode, 1947; Coventry, 1982; Miller et al., 1985; Knuteson et al., 1989; Thompson et al., 1998; Richardson and Vepraskas, 2001; Jenkinson et al., 2002). With perhaps few exceptions, specific pedogenic processes all involve water in various ways, particularly in view of the cumulative effects of water fluxes through soil profiles over long periods of time. Note that it is not the total amount of water falling on the soil or total soil moisture content, but the actual amount of water that has passed through the soil profile that is the most important in fueling pedogenesis. Effective precipitation for pedogenesis thus includes the amount, frequency, intensity, duration, and chemistry of water infiltrated, stored, and percolated through soil profiles. Such an effect is accumulative over time in nature. Chadwick et al. (2003) noted that "*even though leaching is a function of rainfall, it is more directly a function of effective moisture in relation to the volume of water-holding pore space in the soil and element loss should be considered in relation to leaching intensity.*" Over pedogenic time, weathering consumes large pools of primary minerals and advanced weathering-stage soils are formed only if hydrologic removals of solutes outpace renewals that can come from weatherable minerals or atmospheric deposition (Fimmen et al., 2007). The exponential decline or humped models of soil production rate vs. increasing soil thickness also reflect the role of water in soil formation (Heimsath et al., 1997, 2009). The reduction of weathering rate with soil thickening is related to the exponential decrease of temperature amplitude with increasing depth and also the exponential decrease in average water penetration (for freely drained soils). In humped models, the fastest weathering rate occurs under an intermediate soil thickness because under thin soil or exposed bedrock water tends to runoff, thus reducing the chance of bedrock decomposition (Gilbert, 1877; Heimsath et al., 2009). The importance of hydrology and geomorphology in soil production is further elaborated in Chapter 7 of this book by Heimsath. Chapter 19 by Smouelian et al. also discusses the quantitative modeling of pedogenesis and the role of hydrology.

Groundwater also has significant impacts on soil formation and soil function (Fig. 7). Jenny (1941) recognized the importance of groundwater in soil formation, but included it under the topography factor. Rode (1947) has suggested adding groundwater, together with surface water and soil water, as a new soil-forming factor. Miller et al. (1985) and Henry et al. (1985) demonstrated the importance of groundwater flow and water-table depth on the genesis, characteristics, and distribution of soils within a hummocky landscape in Saskatchewan. Miller and Chanasyk (2010) also showed the strong impacts of groundwater on the distribution and features of Chernozems in southern Alberta, and suggested that past groundwater, climatic, and environmental conditions need to be considered to explain the genesis of some relict soils. Fitzpatrick et al. (1992) showed that yellow and gray duplex soils in Australia have transformed to saline sulfidic marsh soils in catchments where rising saline water tables have resulted from land clearing. Chapter 14 of this book by MacEwan et al. provides three case studies that demonstrate the intimate link between groundwater and hydropedology and its importance in guiding appropriate land use.

3. HYDROPEDOLOGY'S UNIQUE CONTRIBUTIONS: BRIDGING TIME, SPACE, AND SYSTEMS

Hydropedology spans spatial scale from the soil pore to the pedosphere and temporal scale from contemporary processes to deep geologic time; thus, hydropedology is in a unique position to help build bridges that are important to closing the knowledge gaps discussed above. Lin (2003) and Lin et al. (2005) have discussed the role that hydropedology plays in bridging disciplines, scales, data, and education. Lin (2010b) further described the opportunities in coupling hydropedology with related bio- and geosciences toward the holistic study of the Earth's Critical Zone (e.g. coupling hydropedology with ecohydrology to improve the linkage of belowground and aboveground processes, and coupling hydropedology with biogeochemistry to enhance the identification of hot spots and hot moments in element cycling).

Here, three important bridges that hydropedology is in a unique position to build are highlighted (Fig. 9), including 1) connecting fast and slow processes, 2) bridging soil structural and landscape units, and 3) linking mapping with monitoring and modeling.

3.1. Connecting Fast and Slow Processes: Soil Function and Soil Formation

Hydropedology encapsulates the co-evolution of fast and slow changes in multiphase soil systems (Fig. 10), where fast and cyclic *soil functioning processes* (SFPs) involve mostly liquids, gases, and biota (in which circulating water is a key), while slow and irreversible *specific pedogenic processes* (SPPs) involve predominantly solids (Targulian and Goryachkin, 2004; Lin, 2011a).

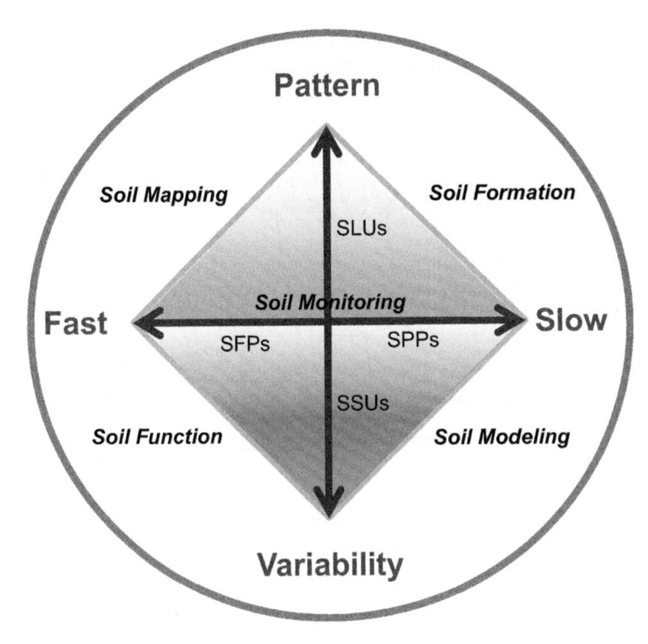

FIGURE 9 The dimensions of time and space in understanding complex soil systems: In the horizontal direction, fast soil functioning processes (SFPs) are linked to slow specific pedogenic processes (SPPs); in the vertical direction, smaller-scale variability is connected to larger-scale pattern via soil structural units (SSUs) and soil-landscape units (SLUs). Soil monitoring is essential for understanding soil function and soil formation, and is linked to soil variability and soil pattern through soil modeling and soil mapping. *(Color version online)*

Soil formation mostly refers to trend changes over long time periods, while soil functions concern mostly periodical and random changes within shorter time scales. However, the short- and long-term changes of soils are intertwined, with interactions and feedbacks between the two and accumulative and threshold effects in soil evolution (Fig. 10). Each specific pedogenic process is characterized by a set of solid-phase pedogenic features formed over hundreds to thousands or more years, while soil functioning processes are dominated by diurnal, seasonal, and annual changes (Targulian, 2005). As noted by Rode (1947), many soil functioning processes and related cycles are not completely closed, and many input–output fluxes are not necessarily balanced in open, dissipative soil systems. Such non-closed cycles and off-balanced fluxes of soil functioning processes generate residual solid-phase products in soil profiles over time. Each single cycle (e.g. seasonal perched water table) may generate only a micro-amount of transformed or newly formed solid products, which may hardly be detected; but being produced repeatedly over a long time these micro-amounts could gradually accumulate into macro-amounts that can be detected morphologically and/or analytically (Targulian and Goryachkin, 2004; Lin, 2011a). Such residual products generated through specific pedogenic

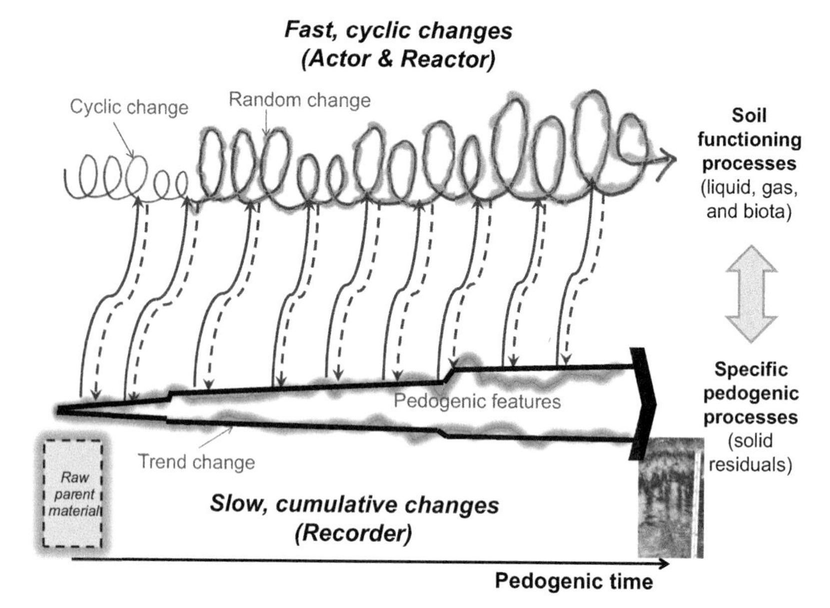

FIGURE 10 Schematic of the co-evolution of fast and slow changes in complex soil systems over time, with interactions (solid small arrows) and feedbacks (dashed small arrows). Fast and cyclic soil functioning processes (actors and reactors) involve mostly liquids, gases, and biota, while slow and irreversible specific pedogenic processes (recorders) involve predominantly solids. Three types of soil change – irregular (or random) variability, cyclic fluctuations, and trend changes – are illustrated. *(Modified from Lin, 2011a). (Color version online)*

processes can feedback to alter soil functioning, mostly in a gradual manner, but some could lead to threshold changes in the soil system (Lin, 2011a).

Both long-term soil formation and short-term soil function are strongly controlled by soil moisture and their accumulative effects over time. For example, infiltration, leaching, and drainage may be considered as short-term soil functioning processes, but repeated actions of such processes over time could lead to various specific pedogenic processes (which will depend on site conditions, geological materials, and other soil-forming factors), such as (Buol et al., 2003):

- Eluviation (albic horizon formation),
- Illuviation (argillic horizon formation),
- Decalcification (reactions that remove calcium carbonate),
- Salinization (accumulation of soluble salts),
- Pedolization (chemical migration of aluminum and iron and/or organic matter),
- Gleization (reduction of iron under poor drainage), and
- Rubification (release of iron and formation of iron oxides under good drainage).

Because of diverse types of soils formed around the globe and the heterogeneity in each of these soil types, there is a need to link the Dokovchiav–Jenny's theory of soil formation (Dokuchaev, 1893; Jenny, 1941) with the Darcy–Buckingham's law of water flow in variably-saturated soils (Darcy, 1856; Buckingham, 1907), whereby different types of soils may exhibit contrasting flow mechanisms, pathways, and patterns. This linkage, if successfully accomplished, can significantly improve flow and transport modeling and the quantification and prediction of soil change over short- and long-term time scales. Lin (2010a) has taken a first step in this direction by exploring the links between the principles of soil formation and flow regimes. The reconciliation of geological and biological processes (vastly different in time scales) is essential to understanding the complexity and evolution of the soil system (Lin, 2010b).

3.2. Bridging Soil Structural and Landscape Units: Deterministic Pattern and Stochastic Variability

Soil structural unit (SSU) here refers to soil architecture within a pedon (including soil particles, pores, fabrics, aggregates, horizons, and other organizational features in a soil profile), while *soil–landscape unit* (SLU) refers to soil architecture in the landscape (such as catenae, soilscapes, soil sequences, and soil zones) (Figs. 5 and 11) (see Chapter 2 of this book for more details on this topic). The linkage between soil structural and landscape units represents the soil–landscape relationship (Fig. 11) that has been traditionally studied by pedologists, including block diagrams, cross sections, and soil maps used to portray the general distribution patterns of soil types in relation to landforms and geology formations (e.g. Hole, 1976; Lin et al., 2005; Indorante, 2011). Lin et al. (2005) suggested to add hydrologic information (e.g. water-table dynamics, water flowpaths, hydric soils, and restrictive soil layers) to enhance the use of the classical soil–landscape relationship information, which could significantly improve the value of soil survey products and provide more useful framework of the landscape–soil–water relationship in different geographic regions for hydrologic modeling.

Through identifying myriads of soil patterns and understanding how heterogeneity occurs at different scales, hydropedology can help demystify the seemly mind-boggling variability of field soils. For example, two general controls of soil variability may be identified (Fig. 5): 1) regional controls, e.g. zonal soil patterns expressed over large areas that are the results of climatic and vegetative gradients, and 2) local controls, i.e. factors that vary over short distances, such as topography and parent material. The hierarchical organization of soils and landscapes can then be used to discern deterministic and stochastic components at each scale, with the deterministic portion being dictated by the underlying soil architectural control at that scale (such as soil profile layers or catenas) and the stochastic portion being randomized processes (such as rainfall inputs or biological activities). To illustrate this, water flow at

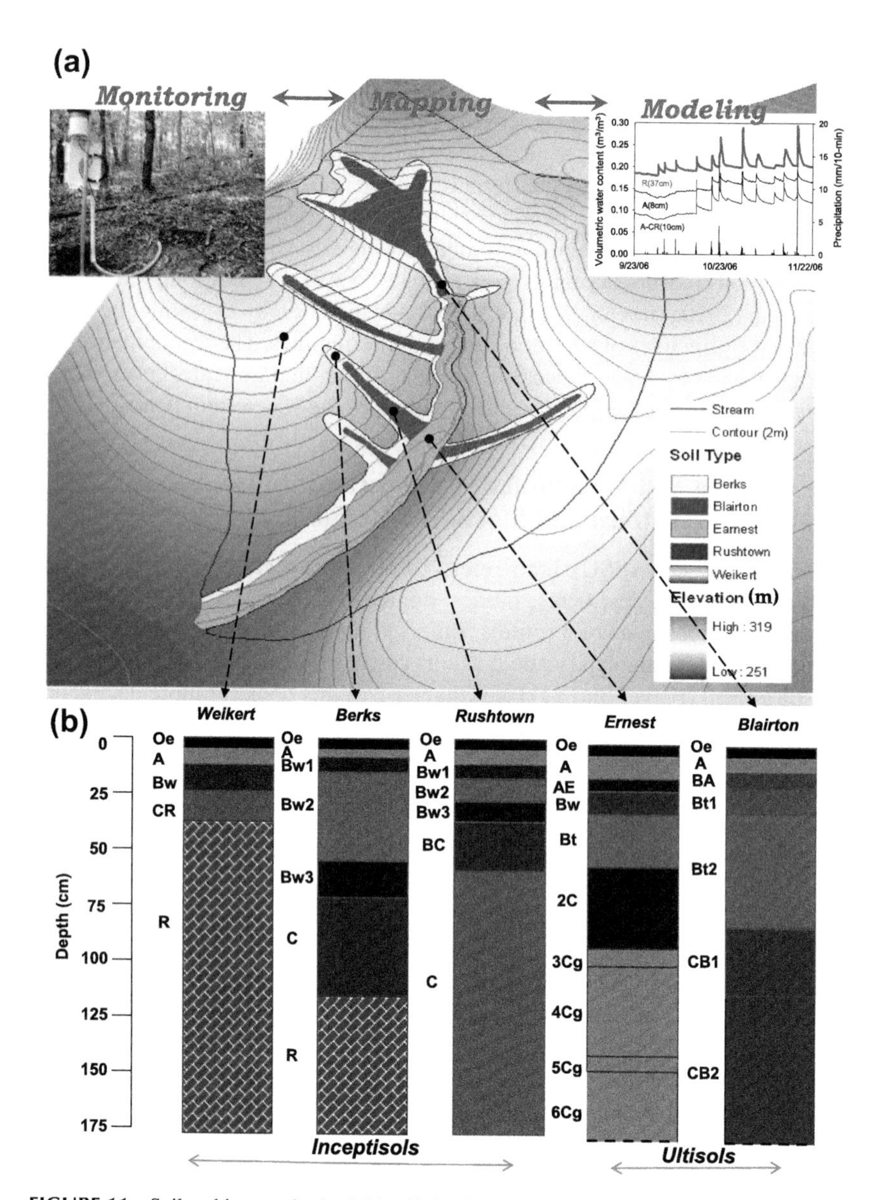

FIGURE 11 Soil architecture in the 7.9-ha Shale Hills Catchment, showing (a) a detailed, high-intensity soil map across the entire catchment (illustrating soil-landscape units) and (b) soil profiles of the five soil series identified (illustrating soil structural units). See Fig. 3 and Lin et al. (2006) for more details of these soils. This schematic also illustrates the bridge of mapping with monitoring and modeling for hydropedologic study in this catchment. *(Color version online and in color plate)*

the microscale is controlled by capillarity and laminar flow through individual soil pores and around peds; as scale increases, flow often becomes controlled by impeding layers in soil profiles, then accumulation of water downslope as dictated by soil catena and associated bedrock; then the flow is routed through a stream network, the formation of which is controlled by soil–landscape relationships (Lin et al., 2006b).

Some approaches can be used to advance the interpolation or extrapolation of point-based data to larger areas (Fig. 5): 1) Direct imaging and quantification of the architecture of soils and hydrologic systems through noninvasive techniques (such as geophysical tools) and continuous mapping (such as remote sensing) and 2) Utilization of soil architecture-forming processes that can be identified, quantified, and understood – the knowledge of which already exists and is continuously being updated in various branches of soil science, especially pedology. However, such knowledge has not yet been well utilized in hydrology and hydrologic modeling. Sommer et al. (2003) suggested an integrated method for soil–landscape analysis, where a hierarchical expert system for multidata fusion of inquiries, relief analysis, geophysical measurements, and remote-sensing data was developed, which was further combined with the soil-forming factorial model and the scaling framework of Vogel and Roth (2003) to address soil variability across scales. In their system, soil variability was separated at every scale into 1) a scale-typical and predictable part and 2) a random part that becomes structure at the next lower scale level. Such a combined characterization of underlying structural controls at different scales and the knowledge of soil architecture formation can improve the modeling and prediction of soil processes and the linkage of soil formation and soil function.

3.3. Linking Mapping with Monitoring and Modeling: Integrated Systems Approach

Hydropedology promotes the union of soil maps and soil functions. As Jenny (1941) noted, "*The goal of soil geographer is the assemblage of soil knowledge in the form of a map. In contrast, the goal of the 'functionalist' is the assemblage of soil knowledge in the form of a curve or an equation... Clearly, it is the union of the geographic and the functional method that provides the most effective means of pedological research.*" Consequently, soil mapping must go beyond the classical taxonomy-centered exercises and consider soil functional characterizations as well as soil variability quantifications. A soil map can no longer be a static product; rather, soil changes (especially under anthropogenic alternations) and functional units (such as hydrologic functions) should be developed.

The link between mapping and monitoring/modeling, especially in an iterative fashion with feedbacks, is important for systems integration across space and time (Fig. 11). Bridging mapping with monitoring and modeling can help the functional quantification of soil maps as well as the interpolation or extrapolation of point-based data. Monitoring and/or modeling can also permit dynamic mapping and ground truthing for soil maps. On the other hand, mapping addresses spatial heterogeneity, provides optimal site selection for monitoring/sampling, and facilitates spatially distributed modeling. Note that much effort by non-pedologists could be hampered when soil distribution and processes are not well understood. Site selection for monitoring or sampling and the design of modeling may not achieve optimal outcomes if the subsurface soil architecture is not well understood. Thus, understanding and mapping soil patterns and quantifying soil architecture across scales are some of key elements for advancing hydropedology.

The iterative loop of mapping, monitoring, and modeling (3M) has been suggested as a hydropedologic approach to landscape study (Fig. 11; Lin et al., 2005). This is because we normally monitor a pedon to collect point data and model a landscape to predict areal processes; the key connecting these two efforts is mapping. The iteration of this 3M provides an adaptive strategy and continuous refinement of models and monitoring networks as knowledge and database are accumulated over time. For example, Groffman et al. (2010) suggested an adaptive approach to managing whole agricultural landscape for environmental quality to deal with many uncertainties and constantly changing world, including 1) monitoring of landscape-scale initiatives, 2) comprehensive multidisciplinary modeling, and 3) mapping of the interface between source and sink areas of nonpoint pollution, including hydrologic connections (flow paths), hydric soils, and lower-order stream channels. Monitoring is a key component of this adaptive management; however, it is often difficult to design and implement monitoring programs because of cost constraints and other difficulties such as sampling and analytical methodologies (Lovett et al., 2007; Lindenmayer and Likens, 2010). Nevertheless, unless monitoring is conducted at the proper temporal and spatial scales with the aid of appropriate mapping and/or modeling, it is impossible to tell if new management approaches are working or not and why (Groffman et al., 2010). Effective adaptive systems thus require iterative looping among monitoring, modeling, and mapping, with feedbacks between them to achieve optimal outcomes over time.

4. FRONTIERS OF HYDROPEDOLOGY APPLICATIONS AND EDUCATION

While best known for its role in providing water and nutrients to support plant growth and ecosystems, soils also play critical roles in a number of other functions in the Earth's Critical Zone (Fig. 12). The soil is the natural integrator of diverse functions, providing a central link in the multiscale and interdisciplinary study of the Critical Zone (Lin, 2010b). Indeed, six grand environmental challenges identified by the U.S. National Research Council (2001b) are directly germane to soil and water resources: land-use dynamics, hydrologic forecasting, biogeochemical cycles, climate variability, ecosystem functions,

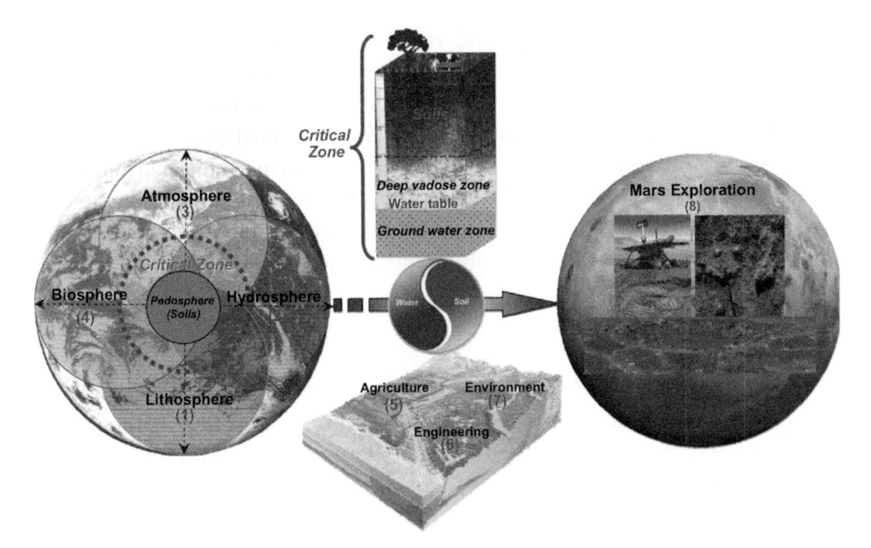

FIGURE 12 The "7 + 1" roles of hydropedology from the Earth's Critical Zone to Mars exploration (modified from Lin, 2005): (1) soil is a natural recorder of the Earth's history through its formation and evolution under the influence of climate, organism, parent material, relief, time, and human impact; (2) soil is a freshwater storage and transmitting mantle in the Earth's Critical Zone; (3) soil is a gas and energy-regulating geoderma in the land–atmosphere interface; (4) soil is the foundation of diverse ecosystems; (5) soil is a living porous substrate essential for plant growth, animal production, and food supply; (6) soil is a popular material for a variety of engineering and construction applications; (7) soil is a great natural remediation and buffering medium in the environment; and (8) soil is a possible habitat for extraterrestrial life – if any can be found – and serves as a frontier in extraterrestrial explorations for signs of liquid water and life. *(Color version online and in color plate)*

and environmentally influenced disease. In this section, selected frontiers of hydropedology applications and education are briefly highlighted, with links to relevant chapters in this book.

4.1. Hydropedology Applications

4.1.1. Coupling Hydropedology and Biogeochemistry

Biogeochemical cycles and ecological functions are tightly coupled with both hydrologic processes and soil reservoirs. Coupling physical/hydrologic and biogeochemical/ecological processes in the Critical Zone offers tremendous opportunities to advance our integrated understanding of energy, water, carbon, nutrient, and sediment dynamics. Example research opportunities include:

- Identification of hot spots and hot moments of biogeochemical processes triggered by soil hydrologic conditions, including carbon and nitrogen cycling and biogeochemical reaction rates limited by transport and pathway in the field (see Chapter 22 of this book);

- Quantification of pedogenesis and weathering rates across environmental gradients and the integration of chemical and isotopic data with hydrologic and pedologic models to understand and predict the creation and evolution of the Critical Zone (see Chapters 7 and 19 of this book);
- Investigation of soil microbial distribution pattern in the landscape and its in situ response to changing climate and land use, the degree of which is strongly influenced by the spatial–temporal variability of water and nutrients (such as those related to preferential flow pathways) (Bundt et al., 2001; Kay et al., 2007).

4.1.2. Coupling Hydropedology and Ecohydrology

Ecohydrology and hydropedology are two interdisciplinary fields that can be coupled to improve the linkage between aboveground and belowground processes. For example, Li et al. (2009) demonstrated important connection between aboveground stemflow and belowground preferential flow in desert shrubs, revealing important interrelationships between hydropedology and ecohydrology and how soil moisture and plant growth influence each other. Example frontier topics in this area include:

- Connecting multiple scales from plant stomata to individual plant, patch, and ecosystem, and identifying self-organizing principles between aboveground and underground processes at different scales (see Chapter 23 of this book);
- Co-evolution of soil and vegetation and related optimality of water, nutrient, and energy between the aboveground and the underground. This has an important connection to soil moisture that modulates the complex dynamics of climate–soil–water–vegetation relationships (Eagleson, 2002);
- Quantitative framework for determining the dynamics of soil water in humid ecosystems (especially wetlands), which presents contrasting features to water-limited ecosystems, where lateral flow, water-table fluctuation, and root water update are intertwined (Rodriguez-Iturbe et al., 2007).

4.1.3. Coupling Hydropedology and Hydrogeology/ Hydrogeophysics

Integration of hydropedology and hydrogeology can provide a more holistic framework of subsurface flow and transport from the land surface down to the aquifer. In addition, coupling hydropedology and hydrogeophysics is an area that has not yet been well explored. Example frontier topics include:

- Integrated modeling and prediction of flow and transport in structured soils and fractured rocks, which remain vital in toxic and nuclear-waste disposal and chemical spill sites. The bridging of pedon-scale observations and landscape-scale phenomena, together with adequate incorporation of soil architecture, rock fracture, and preferential flow, can enhance solutions to groundwater contamination and remediation (see Chapter 14 of this book);

- One area of uncertainty in groundwater contamination remediation and cleanup is the zone of local water-table fluctuation (which includes the capillary fringe) often referred to as the "smear zone." This uncertainty is a reflection of the complex interaction and variability of hydraulic, chemical, thermal, and microbial processes and conditions that affect the chemistry and mobility of contaminants (Nicholson and Fuhrmann, 2008);
- Many questions remain regarding how best to transfer geophysical signals to soil hydrologic properties of interest (Rubin and Hubbard, 2005). One way to get around this inversion problem is to insert pedologic insights (e.g. soil types, layers, and structural phases) to help constrain or zone the geophysical inversion (see Chapter 13 of this book).

4.1.4. Hydropedology and Land Use

How natural soils take in water upon precipitation under various climates offers clues as to "what" can best be done and "where" with the lowest risks and the greatest opportunities for land use and management (Bouma, 2006). In the meantime, communications with engineers and sociopolitical entities are critical for realizing optimal land-use goals (see Chapter 15 of this book). Example related issues include:

- Sustainable storm water management requires a combination of traditional "hard" engineering approaches with "soft," ecological engineering aimed at utilizing natural systems to provide needed services and balancing man-made structures with natural ecosystems;
- Scientifically sound and socio-economically feasible trading of water quality, water quantity, and carbon at different scales remain to be better developed, including how these trading impacts the physical reality of soil and water resources in a watershed;
- Use of soils for wastewater disposal, sewerage treatment, and biosolids application will become increasingly important. For example, in Australia, wastewater irrigation is driving much of sustainable irrigation practices, and wastewater irrigation is also the only irrigation activity regulated by environmental protection agencies (M. Hardie, personal communication).

4.1.5. Hydropedology and Extraterrestrial Explorations

Knowledge of the soil and its forming processes has a unique contribution to extraterrestrial explorations in search of water and life. For example, the findings of the NASA's Mars Exploration Rover mission and many other recent missions have created a brand-new window of opportunity for study of extraterrestrial soils (Targulian et al., 2010) and exohydrology (NRC, 2012). Some related research frontiers may include:

- How the weathering engine on Mars has transformed the protolith into various Martian soils and what may be the role of water and climate in the genesis and evolution of Martian soils that could yield insights into Martian history?
- Is there possible hydromorphism in Martian soils, and, if so, what that may imply? On Earth, the study of paleohydropedology (paleopedology + paleohydrology) has demonstrated valuable historical records of the past environment and ancient landscape–soil–water–vegetation–climate relationship (e.g. Ashley and Driese, 2000);
- How the soil–landscape relationship on Mars may be best investigated and interpreted to understand possible signature of water and even life in the past or current Martian environment?

4.2. Hydropedology Education

The interdisciplinary emphasis of education in the twenty-first century makes hydropedology a timely addition to the education of the next generation of soil scientists, hydrologists, environmentalists, and natural resource managers who can solve interdisciplinary problems. Hydropedology promotes integrated fieldwork, laboratory studies, and computer modeling to understand complex landscape–soil–hydrology relationships. In particular, fieldwork is the foundation of all, and the logical sequence of hydropedologic work should go from the field to the laboratory and modeling, and then back to the field. An adequate understanding of this sequence has a profound impact on how we educate future generations of hydropedologists (Lin et al., 2005).

An interesting and yet alarming paradox in our scientific community is worth noticing here: We now have more and more people studying soils, but we have fewer and fewer people knowing what soils really are. Many young and energetic researchers who have been trained in various disciplines are tackling complex issues related to soils, but fewer and fewer of them have ever touched a real soil profile in the field. This could hinder their in-depth understanding of the complexity and dynamics involved in soil processes, leading to often simplified and unrealistic assumptions about real-world soils or simple ignorance of soil heterogeneity. Many hydrologists, geochemists, engineers, and alike also appear to be insensitive to soil layering when examining soil profiles or installing sensors in the field. Some pedologists already voiced concerns two decades ago about losing much of what distinguishes pedologists as they became more laboratory oriented at the expense of fieldwork (e.g. Daniels, 1988; Jacob and Nordt, 1991). Butler (1980) has clearly pointed out that soil survey is "*one of the basic technologies of soil science*," and field-based pedology and soil survey training should be a basic skill for anyone who wishes to truly understand real-world soils. Observations have shown that those who are most effective educators in the classroom are those who have had soil survey experience, so they understand many of the challenges presented in this chapter.

In addition, the dearth of effective and compelling visualizations of the complex subsurface (see illustrations in Figs. 1–4) limits the general public's and students' interest and understanding of real-world soils. To advance hydropedology, we must challenge the conventional mistake of "*If you can't see it, it doesn't matter*" and to look underneath the surface for sustainability. Thus, improved means of revealing and visualizing subsurface heterogeneity and complex soil architecture can help advance hydropedology education and applications.

5. SUMMARY

This chapter provides an overview of hydropedology and its fundamental characteristics. In situ soil architecture and the soil–landscape relationship are two essential features for understanding flow and transport processes in real-world soils across scales. Building upon pedology as a unique and integrated study of soils as natural or anthropogenically-modified landscape entities, hydropedology seeks to understand the complex and dynamic behavior and distribution of soil–water interactions in contact with mineral and biological materials in terrestrial environments. It focuses on two fundamental process-oriented questions and three important bridges that can help close knowledge gaps in the integrated study of the landscape–soil–hydrology–vegetation relationship across space and time. By connecting fast and slow processes, bridging soil structural and landscape units, and linking mapping with monitoring and modeling, hydropedology can holistically integrate soil formation and soil function, connect deterministic pattern and stochastic variation, and link soil heterogeneity (space) and soil change (time).

Exciting opportunities abound to advance the frontiers of hydropedology research, education, and applications, as highlighted in this chapter and further discussed in the following chapters of this book. Important steps and approaches for future progress in hydropedology are discussed in the last chapter (Chapter 24), including the need to 1) overcome conceptual and technological bottlenecks; 2) develop holistic, multiscale, and quantitative modeling frameworks; 3) build a global alliance of integrated databases and observatories; 4) apply hydropedology in societally important issues; and 5) improve interdisciplinary education of the next generation of soil scientists and hydrologists.

ACKNOWLEDGMENTS

The author thanks Drs. Larry Wilding, Marcus Hardie, and Todd Caldwell for their review comments that have helped improve the quality of this paper.

REFERENCES

Amundson, R., Richter, D.D., Humphreys, G.S., Jobbágy, E.G., Gaillardet, J., 2007. Coupling between biota and Earth materials in the critical zone. Elements 3, 327–332.

Anderson, S., Blum, J., Brantley, S., Chadwick, O., Chorover, J., Derry, L., Drever, J., Hering, J., Kirchner, J., Kump, L., Richter, D., White, A., 2004. Proposed initiative would study Earth's weathering engine. EOS 85, 265–272.

Ashley, G.M., Driese, S.G., 2000. Paleopedology and paleohydrology of a volcaniclastic paleosol interval: implications for early pleistocene stratigraphy and paleoclimate record, Olduvai Gorge, Tanzania. J. Sediment Res. 70, 1065–1080.

Banwart, S., Stefano, M., Bernasconi, Bloem, J., Blum, W., Brandao, M., Brantley, S., Chabaux, F., Duffy, C., Kram, P., Lair, G., Lundin, L., Nikolaidis, N., Novak, M., Panagos, P., Ragnarsdottir, K.V., Reynolds, B., Rousseva, S., de Ruiter, P., van Gaans, P., van Riemsdijk, W., White, T., Zhang, B., 2011. Assessing soil processes and function across an International network of critical zone observatories: research hypotheses and experimental design. Vadose Zone J. 10, 974–987.

Baveye, P.C., Rangel, D., Jacobson, A.R., Laba, M., Darnault, C., Otten, W., Radulovich, R., Camargo, F.A.O., 2011. From dust bowl to dust bowl: soils are still very much a frontier of science. Soil Sci. Soc. Am. J. 75, 2037–2048.

Bejan, A., 2007. Constructal theory of pattern formation. Hydrol. Earth Syst. Sci. 11, 753–768.

Beven, K., 2006. Searching for the holy grail of scientific hydrology: $Q_t = H\underset{\leftarrow}{S}\,\underset{\leftarrow}{R}\,(\Delta t)A$ as closure. Hydrol. Earth Syst. Sci. 10, 609–618.

Birkeland, P.W., 1974. Pedology, Weathering, and Geomorphological Research. New Year, Oxford.

Bockheim, J.G., Gennadiyev, A.N., Hammer, R.D., Tandarich, J.P., 2005. Historical development of key concepts in pedology. Geoderma 124, 23–36.

Bouma, J., 2006. Hydropedology as a powerful tool for environmental policy research. Geoderma 131, 275–286.

Bouma, J., Droogers, P., Sonneveld, M.P.W., Ritsema, C.J., Hunink, J.E., Immerzeel, W.W., Kauffman, S., 2011. Hydropedological insights when considering catchment classification. Hydrol. Earth Syst. Sci. 15, 1909–1919.

Brady, N.C., Weil, R.R., 2004. Elements of the Nature and Properties of Soils, second ed. Pearson-Prentice Hall, Upper Saddle River, NJ.

Braudeau, E., Mohtat, R.H., 2009. Modeling the soil system: bridging the gap between pedology and soil-water physics. Glob. Planet. Change 67, 51–61.

Buckingham, E. 1907. Studies on the movement of soil moisture. U.S. Dept. of Agri. Bur. of Soils Bulletin 38.

Bundt, M., Jäggi, M., Blaser, P., Siegwolf, R., Hagedorn, F., 2001. Carbon and nitrogen dynamics in preferential flow paths and matrix of a forest soil. Soil Sci. Soc. Am. J. 65, 1529–1538.

Buol, S.W., Southard, R.J., Graham, R.C., McDaniel, P.A., 2003. Soil Genesis and Classification, fifth ed. Iowa State University Press, Ames. IA.

Butler, B.E., 1980. Soil Classification for Soil Survey. Clarendon Press, Oxford, UK.

Chadwick, O.A., Gavenda, R.T., Kelly, E.F., Ziegler, K., Olson, C.G., Elliott, W.C., Hendricks, D.M., 2003. The impact of climate on the biogeochemical functioning of volcanic soils. Chem. Geol. 202, 195–223.

Coventry, R.J., 1982. The Distribution of Red, Yellow and Grey Earths in the Torrens Creek Area, Central North Queensland. Australian Journal of Soil Res. 20, 1–14.

Coventry, R.J., Williams, J., 1984. Quantitative relationships between morphology and current soil hydrology in some Alfisols in semiarid tropical Australia. Geoderma 33, 191–218.

Cremeens, D.L., Brown, R.B., Huddleston, J.H., 1994. Whole Regolith Pedology. SSSA Special Publication #14. Soil Sci. Soc. Am. Inc., Madison, WI.

Daniels, R.B., 1988. Pedology, a field or laboratory science? Soil Sci. Soc. Am. J. 52, 1518–1519.

Daniels, R.B., Hammer, R.D., 1992. Soil Geomorphology. John Wiley, New York.

Darcy, H. 1856. Les Fontaines Publiques de la Ville de Dijon, Dalmont, Paris. (English translation by Patricia Bobeck).

Dokuchaev, V.V., 1893. The Russian Steppes and Study of Soil in Russia, Its Past and Present (Translated by J.M. Crawford). Ministry of Crown Domains, St. Petersburg, Russia.

Domenico, P.A., Schwartz, F.W., 1998. Physical and Chemical Hydrogeology, second ed. Wiley, New York.

Eagleson, P.S., 2002. Ecohydrology: Darwinian expression of vegetation form and function. Cambridge University Press, Cambridge, UK.

Erich, E., Drohan, P.J., Ellis, L.R., Collins, M.E., Payne, M., Surabian, D., 2010. Subaqueous soils: their genesis and importance in ecosystem management. Soil Use Manag. 26, 245–252.

Evans, C.V., Franzmeier, D.P., 1986. Saturation, aeration, and color patterns in a toposequence of soils in North-central Indiana. Soil Sci. Soc. Am. J 50, 975–980.

Fimmen, R.L., Richter, D.deB., Vasudevan, D., Williams, M.A., West, L.T., 2007. Rhizogenic Fe-C redox cycling: a hypothetical biogeochemical mechanism that drives crustal weathering in upland soils. Biogeochemistry 87, 127–141. DOI 10.1007/s10533-007-9172-5.

Fitzpatrick, R.W., Naidu, R., Self, P.G., 1992. Iron deposits and microorganisms occurring in saline sulfidic soils with altered soil water regime in the Mt. Lofty Ranges, South Australia. Catena Suppl. 21, 263–286.

Franzmeier, D.P., Yahner, J.E., Steinhardt, G.C., Sinclair Jr., H.R., 1983. Color patterns and water-table levels in some Indiana soils. Soil Sci. Soc. Am. J. 47, 1196–1202.

Freer, J., McDonnell, J.J., Beven, K.J., Peters, N.E., Burns, D.A., Hooper, R.P., Aulenbach, B., Kendall, C., 2002. The role of bedrock topography on subsurface storm flow. Water Resour. Res. 38 (12), 1269. doi:10.1029/2001WR000872.

Galusky, L.P., Rabenhorst, M.C., Hill, R.L., 1998. Toward the development of quantitative soil morphological indicators of water table behavior. In: Rabenhorst, M.C., Bell, J.C., McDaniel, P.A. (Eds.), Quantifying Soil Hydromorphology. SSSA Special Pub. #54. Soil Sci. Soc. Am. Inc, Madison, WI, pp. 77–93.

Gburek, W.J., Needelman, B.A., Srinivasan, M.S., 2006. Fragipan controls on runoff generation: hydropedological implications at landscape and watershed scales. Geoderma 131, 330–344.

Gerwin, W., Schaaf, W., Biemelt, D., Winter, S., Fischer, A., Veste, M., Hüttl, R.F., 2011. Overview and first results of ecological monitoring at the artificial watershed Chicken Creek (Germany). Phys. Chem. Earth 36, 61–73.

Gilbert, G.K., 1877. Report on the Geology of the Henry Mountains. US Geol. Surv., Washington DC.

Gish, T.J., Shirmohammadi, A. (Eds.), 1991. Preferential Flow. Proceedings of the National Symposium. Dec. 16–17, 1991, Chicago, IL. American Society of Agri. Engineers, St. Joseph, MI.

Groffman, P.M., Gold, A.J., Duriancik, L., Richard Lowrance, R., 2010. From connecting the dots to threading the needle: the challenges ahead in managing agricultural landscapes for environmental quality. In: Nowak, P., Schnepf, M. (Eds.), Managing Agricultural Landscapes for Environmental Quality II: Achieving More Effective Conservation. Soil and Water Conservation Society, Ankeny, IA.

Hardie, M., 2011. Effect of antecedent soil moisture on infiltration and preferential flow in texture contrast soils. Ph.D. Dissertation, Univ. of Tasmania, Australia.

Heimsath, A.M., Dietrich, W.E., Nishiizumi, K., Finkel, R.C., 1997. The soil production function and landscape equilibrium. Nature 388, 358–361.

Heimsath, A.M., Fink, D., Hancock, G.R., 2009. The 'humped' soil production function: eroding Arnhem Land, Australia. Earth Surf. Process. Landforms 34, 1674–1684.

Henry, J.L., Bullock, P.R., Hogg, T.J., Luba, L.D., 1985. Groundwater discharge from glacial and bedrock aquifers as a soil salinization factor in Saskatchewan. Can. J. Soil Sci. 65, 749–768.

Hillel, D., 1998. Environmental Soil Physics. Academic Press, San Diego, CA.

Hillel, D., 2008. Soil in the Environment – Crucible of Terrestrial Life. Academic Press, San Diego, CA.

Hole, F.D., 1976. Soils of Wisconsin. Wis. Geol. Nat. Hist. Surv. Bull. 87, Soils Ser. 62. Univ. of Wisconsin Press, Madison, WI.

Holländer, H.M., Blume, T., Bormann, H., Buytaert, W., Chirico, G.B., Exbrayat, J.-F., Gustafsson, D., Hölzel, H., Kraft, P., Stamm, C., Stoll, S., Blöschl, G., Flühler, H., 2009. Comparative predictions of discharge from an artificial catchment (Chicken Creek) using sparse data. Hydrol. Earth Syst. Sci. 13, 2069–2094.

Hurt, G.W., Whited, P.M., Pringle, R.F. (Eds.), 1998. Field Indicators of Hydric Soils in the United States. Ver. 4.0. USDA-NRCS, Ft. Worth, TX.

Indorante, S.J., 2011. The art and science of soil-landscape block diagrams: examples of one picture being worth more than 1000 words. Soil Surv. Horiz. 52, 89–94.

Jacob, J., Nordt, L., 1991. Soil and landscape evolution: a paradigm for pedology. Soil Sci. Soc. Am. J. 55, 1194.

Jenkinson, B.J., Franzmeier, D.P., Lynn, W.C., 2002. Soil hydrology on an end moraine and a dissected till plain in west-central Indiana. Soil Sci. Soc. Am. J. 66, 1367–1376.

Jenny, H., 1941. Factors of Soil Formation – A System of Quantitative Pedology. McGraw-Hill, NY.

Jespersen, J.L., Osher, L.J., 2007. Carbon storage in the soils of a Mesotidal Gulf of Maine estuary. Soil Sci. Soc. Am. J. 71, 372–379. doi:10.2136/sssaj2006.0225.

Joffe, J.S., 1949. Pedology, second ed. Pedology Publications, New Brunswick, NJ.

Jury, W.A., Horton, R., 2004. Soil Physics, sixth ed. Wiley.

Jury, W.A., Or, D., Pachepsky, Y., Vereecken, H., Hopmans, J.W., Ahuja, L.R., Clothier, B.E., Bristow, K.L., Kluitenberg, G.J., Moldrup, P., Simunek, J., van Genuchten, M.Th, Horton, R., 2011. Kirkham's Legacy and contemporary challenges in soil physics research. Soil Sci. Soc. Am. J. 75, 1589–1601.

Jørgensen, S., 2006. Preface. In: Tiezzi, E. (Ed.), Steps Towards an Evolutionary Physics. WIT Press, Boston.

Kay, D., Edwards, A.C., Ferrier, R.C., Francis, C., Kay, C., Rushby, L., Watkins, J., McDonald, A.T., Wyer, M., Crowther, J., Wilkinson, J., 2007. Catchment microbial dynamics: the emergence of a research agenda. Progr. Phys. Geogr. 31, 59–76.

Kennedy, M., Droser, M., Mayer, L.M., Pevear, D., Mrofka, D., 2006. Late Precambrian Oxygenation; Inception of the clay mineral Factory. Science 311, 1446–1449.

Knuteson, J.A., Richardson, I., Patterson, D.D., 1989. Pedogenic carbonates in a Calciaquoll associated with a recharge wetland. Soil Sci. Soc. Am. J. 53, 495–499.

Kubiena, W.L., 1938. Micropedology. Collegiate Press, Ames, IA.

Li, X.Y., Yang, Z.P., Zhang, X.Y., Lin, H.S., 2009. Connecting ecohydrology and hydropedology in desert shrubs: stemflow as a source of preferential water flow in soils. Hydrol. Earth Syst. Sci. 13, 1133–1144.

Lilly, A., Matthews, K.B., 1994. A soil wetness class map for Scotland: new assessments of soil and climate data for land evaluation. Geoforum 25, 371–379.

Lilly, A., Ball, B.C., McTaggart, I.P., Horne, P.L., 2003. Spatial and temporal scaling of nitrous oxide emissions from the field to the regional scale in Scotland. Nutr. Cycl. Agroe 66, 241–257.

Lin, H.S., 2003. Hydropedology: bridging disciplines, scales, and data. Vadose Zone J. 2, 1–11.

Lin, H.S., 2005. From the Earth's Critical Zone to Mars exploration: can soil science enter its golden age? Soil Sci. Soc. Am. J. 69, 1351–1353.

Lin, H.S., 2010a. Linking principles of soil formation and flow regimes. J. Hydrol. 393, 3–19.

Lin, H.S., 2010b. Earth's Critical Zone and hydropedology: concepts, characteristics, and advances. Hydrol. Earth Syst. Sci. 14, 25–45.

Lin, H.S., 2011a. Understanding soil change across time scales. Soil Sci. Soc. Am. J. 75, 2049–2070.

Lin, H.S., 2011b. Hydropedology: towards new insights into interactive pedologic and hydrologic processes in the landscape. J. Hydrol. 406, 141–145.

Lin, H.S., Bouma, J., Wilding, L., Richardson, J., Kutilek, M., Nielsen, D., 2005. Advances in hydropedology. Adv. Agron. 85, 1–89.

Lin, H.S., Kogelmann, W., Walker, C., Bruns, M.A., 2006a. Soil moisture patterns in a forested catchment: a hydropedological perspective. Geoderma 131, 345–368.

Lin, H.S., Bouma, J., Pachepsky, Y., Western, A., Thompson, J., van Genuchten, M. Th., Vogel, H., Lilly, A., 2006b. Hydropedology: synergistic integration of pedology and hydrology. Water Resour. Res. 42, W05301. doi:10.1029/2005WR004085.

Lin, H.S., Bouma, J., Pachepsky, Y., 2006c. Revitalizing pedology through hydrology and connecting hydrology to pedology. Geoderma. 131, 255–256.

Lin, H.S., Brook, E., McDaniel, P., Boll, J., 2008. Hydropedology and Surface/Subsurface runoff processes. In: Anderson, M.G. (Ed.), Encyclopedia of Hydrologic Sciences. John Wiley & Sons, Ltd DOI: 10.1002/0470848944.hsa306.

Lindenmayer, D.B., Likens, G.E., 2010. Improving ecological monitoring. Trends Ecol. Evol. 25, 200–201.

Lovett, G.M., Burns, D.A., Driscoll, C.T., Jenkins, J.C., Mitchell, M.J., Rustad, L., Shanley, J.B., Likens, G.E., Haeuber, R., 2007. Who needs environmental monitoring? Front. Ecol. Environ. 5, 253–260.

McClain, M.E., Boyer, E.W., Dent, C.L., Gergel, S.E., Grimm, N.B., Groffman, P.M., Hart, S.C., Harvey, J.W., Johnston, C.A., Mayorga, E., McDowell, W.H., Pinay, G., 2003. Biogeochemical hot spots and hot moments at the interface of terrestrial and aquatic ecosystems. Ecosystems 6, 301–312.

McDaniel, P.A., Regan, M.P., Brooks, E., Boll, J., Barndt, S., Falen, A., Young, S.K., Hammel, J.E., 2008. Linking Fragipans, Perched Water Tables, and Catchment-scale Hydrological Processes. Catena. doi:10.1016/j.catena.2007.05.011.

McGlynn, B.L., McDonnell, J.J., Brammer, D.D., 2002. A review of the evolving perceptual model of hillslope flowpaths at the Maimai catchments, New Zealand. J. Hydrol. 257, 1–26.

Miller, J.J., Chanasyk, D.S., 2010. Soil characteristics in relation to groundwater for selected Dark Brown Chernozems in southern Alberta. Can. J. Soil Sci. 90, 597–610.

Miller, J.J., Acton, D.F., Starnaud, R.J., 1985. The effect of groundwater on soil formation in a Morainal landscape in Saskatchewan. Can. J. Soil Sci. 65, 293–307.

Minasny, B., McBratney, A.B., Salvador-Blanes, S., 2008. Quantitative models for pedogenesis — a review. Geoderma 144, 140–157.

National Research Council (NRC), 1991. Opportunities in the Hydrologic Sciences. National Academy Press, Washington, D.C.

National Research Council (NRC), 1998. Decade-to-Century-Scale Climate Variability and Change: a Science Strategy. National Academy Press, Washington, D.C.

National Research Council (NRC), 1999. New Strategies for America's Watersheds. National Academy Press, Washington, D.C.

National Research Council (NRC), 2001a. Basic Research Opportunities in Earth Science. National Academy Press, Washington, D.C.

National Research Council (NRC), 2001b. Grand Challenges in Environmental Sciences. National Academy Press, Washington, D.C.

National Research Council (NRC), 2012. Challenges and Opportunities in the Hydrologic Sciences. National Academy Press, Washington, D.C.

Needelman, B.A., Gburek, W.J., Petersen, G.W., Sharpley, A.N., Kleinman, P. J.A., 2004. Surface runoff along two agricultural hillslopes with contrasting soils. Soil Sci. Soc. Am. J. 68, 914–923.

Nicholson, T.J., Fuhrmann, M., 2008. Mobility of radionuclides in the smear zone. In: Lin (Ed.), The First International Conference on Hydropedology Proram and Abstracts. Penn State Univ., University Park, PA, p. 72.

Nielsen, D.R., 1997. A challenging frontier in hydrology – the vadose zone. In: Buras, N. (Ed.), Reflections on Hydrology – Science and Practice. American Geophysical Union, Washington, DC, pp. 203–226.

Pennock, D., Yates, T., Bedard-Haughn, A., Phipps, K., Farrell, R., McDougal, R., 2010. Landscape controls on N_2O and CH_4 emissions from freshwater mineral soil wetlands of the Canadian Prairie Pothole region. Geoderma 155, 308–319.

Rabenhorst, M.C., Bell, J.C., McDaniel, P.A., 1998. Quantifying Soil Hydromorphology. SSSA Special Pub. #54. Soil Sci. Soc. Am. Inc, Madison, WI.

Reuter, R.J., Bell, J.C., 2003. Hillslope hydrology and soil morphology for a wetland basin in south-central Minnesota. Soil Sci. Soc. Am. J. 67, 365–372.

Richardson, J.L., Vepraskas, M.J., 2001. Wetland soils: genesis, hydrology, landscapes, and classification. LEWIS, CRC Press, Boca Raton, FL.

Richter, D.D., Mobley, M.L., 2009. Monitoring Earth's critical zone. Science 326, 1067–1068.

Rode, A.A., 1947. The Soil Forming Process and Soil Evolution. Israel Program for Scientific Translations, Jerusalem. (Translated by J.S. Joffe, 1961.)

Rodriguez-Iturbe, I., D'Odorico, P., Laio, F., Ridolfi, L., Tamea, S., 2007. Challenges in humid land ecohydrology: interactions of water table and unsaturated zone with climate, soil, and vegetation. Water Resour. Res. 43, W09301. doi:10.1029/2007WR006073.

Rubin, Y., Hubbard, S. (Eds.), 2005. Hydrogeophysics. Springer, Dordrecht, The Netherlands.

Schaetzl, R., Anderson, S., 2005. Soils – Genesis and Geomorphology. Cambridge, UK.

Severson, E.D., Lindbo, D.L., Vepraskas, M.J., 2008. Hydropedology of a coarse-loamy catena in the lower Coastal Plain, NC. Catana 73, 189–196.

Šimůnek, J., Jarvis, N.J., van Genuchten, M. Th., Gärdenäs, A., 2003. Review and comparison of models for describing non-equilibrium and preferential flow and transport in the vadose zone. J. Hydrol. 272, 14–35.

Soil Survey Staff, 2010. Keys to Soil Taxonomy, eleventh ed. United States Department of Agriculture – Natural Resources Conservation Service, Washington, DC.

Sommer, M., Wehrhan, M., Zipprich, M., Weller, U., Zu Castell, W., Ehrich, S., Tandler, B., Selige, T., 2003. Hierarchical data fusion for mapping soil units at field scale. Geoderma 112, 179–196.

Sugden, A., Stone, R., Ash, C., 2004. Ecology in the underworld. Science 304, 1613.

Swarowsky, A., Dahhgren, R.A., Tate, K., Hopmans, J.W., O'Geen, A.T., 2011. Catchment-scale soil water dynamics in a Mediterranean oak woodland. Vadose Zone J. 10, 800–815.

Targulian, V.O., 2005. Elementary pedogenic processes. Eurasian Soil Sci. 38, 1255–1264.

Targulian, V., Mergelov, N., Gilichinsky, D., Sergey, S., Nikita, D., Goryachkin, S., Ivanov, A., 2010. Dokuchaev's soil paradigm and extraterrestrial "soils". Paper presented at the 19th World Congress of Soil Science, Division Symposium 1.1, Astropedology extending soil science to other planets. Brisbane, Australia, 1–6 August 2010.

Targulian, V.O., Goryachkin, S.V., 2004. Soil memory: types of record, carriers, hierarchy and diversity. Rev. Mex. Cienc. Geol. 21, 1–8.

Thompson, J.A., Bell, J.C., Zanner, C.W., 1998. Hydrology and hydric soil extent within a Mollisol catena in Southeastern Minnesota. Soil Sci. Soc. Am. J 62, 1126–1133.

Tiezzi, E., 2006. Steps towards an Evolutionary Physics. WIT Press, Boston.

Vepraskas, M.J. 1992. Redoximorphic Features for Identifying Aquic Conditions, North Carolina Agricultural Research Service Technical Bulletin No. 301, NC, USA.

Vepraskas, M.J., Sprecher, S.W. (Eds.), 1997. Aquic Conditions and Hydric Soils: The Problem Soils. SSSA Special Pub. #50. Soil Sci. Soc. Am. Inc, Madison, WI.

Vidon, P., Hill, A.R., 2004. Denitrification and patterns of electron donors and acceptors in eight riparian zones with contrasting hydrogeology. Biogeochemistry 71, 259–283.

Vidon, P., Allan, C., Burns, D., Duval, T.P., Gurwick, N., Inamdar, S., Lowrance, R., Okay, J., Scott, D., Sebestyen, S., 2010. Hot spots and hot moments in riparian zones: potential for improved water quality management. J. Am. Water Resour. Assoc. 46, 278–298.

Vogel, H.J., Roth, K., 2003. Moving through scales of flow and transport in soil. J. Hydrol. 272, 95–106.

Wilding, L.P., 2000. Pedology. In: Sumner, M.E. (Ed.), Handbook of Soil Science. CRC Press, Boca Raton, FL, pp. E-1–E-4.

Wilding, L.P., Lin, H.S., 2006. Advancing the frontiers of soil science towards a geoscience. Geoderma 131, 257–274.

Zacharias, S., Bogena, H., Samaniego, L., Mauder, M., Fuß, R., Pütz, T., Frenzel, M., Schwank, M., Baessler, C., Butterbach-Bahl, K., Bens, O., Borg, E., Brauer, A., Dietrich, P., Hajnsek, I., Helle, G., Kiese, R., Kunstmann, H., Klotz, S., Munch, J.C., Papen, H., Priesack, E., Schmid, H.P., Steinbrecher, R., Rosenbaum, U., Teutsch, G., Vereecken, H., 2011. A network of terrestrial environmental observatories in Germany. Vadose Zone J. 10, 955–973.

Zhang, J., Lin, H.S., Doolittle, J. Soil layering and preferential flow impacts on seasonal changes of GPR signals in two contrasting soils. Geoderma, submitted for publication.

Zhu, Q., Lin, H.S., 2009. Simulation and validation of concentrated subsurface lateral flow paths in an agricultural landscape. Hydrol. Earth Syst. Sci. 13, 1503–1518.

Zhu, Q., Lin, H.S., Doolittle, J., 2010a. Repeated electromagnetic induction surveys for determining subsurface hydrologic dynamics in an agricultural landscape. Soil Sci. Soc. Am. J. 74, 1750–1762.

Zhu, Q., Lin, H.S., Doolittle, J., 2010b. Repeated electromagnetic induction surveys for improving soil mapping in an agricultural landscape. Soil Sci. Soc. Am. J. 74, 1763–1774.

Zimov, S.A., Edward Schuur, A.G., Stuart Chapin III, F., 2006. Permafrost and the global carbon budget. Science 312, 1612–1613.

Chapter 2

Understanding Soil Architecture and Its Functional Manifestation across Scales

Henry Lin[1]

ABSTRACT

A system's architecture is the backbone of its functions, which governs the pathways and patterns of its mass and energy storages and fluxes. Soils have complex architectural organizations as a result of natural soil-forming processes and anthropogenic impacts. To better understand, measure, model, and use soils information for hydrologic, biogeochemical, and ecologic applications, a holistic concept of soil architecture is needed. Soil architecture here refers to the entirety of how the soil is organized from the microscopic to the megascopic scales. It exhibits hierarchical organization and encompasses interlinked components of solids, pores, and interfaces at each scale. Two general levels of organization are commonly recognized, with the soil–landscape relationship being the key to link the two: 1) soil architecture within a soil profile (termed soil structural units), such as particles, pores, aggregates, horizons, and pedons; and 2) soil architecture in the landscape (termed soil-landscape units), such as catenae, soilscapes, soil sequences, and soil zones. Hierarchical frameworks of soil mapping, modeling, and their connections, as well as the formation and pattern of soil architecture at different scales, are reviewed and discussed in this chapter. A new era of soils research is called for to improve the understanding and quantification of soil architecture across scales and its quantitative links to soil functions. This will require new concepts and technological breakthroughs to permit soils to be studied without undue disturbance of their original architecture. Toward this goal, five needs are suggested: conceptual advance, scaling framework, technological breakthrough, quantitative formulation, and co-evolution of soil–water in the anthropocene.

1. CONCEPT OF SOIL ARCHITECTURE AND ITS SIGNIFICANCE

While soil aggregates have been well studied for decades (especially in pedology, soil physics, and soil biology), a comprehensive theory and an

[1]Dept. of Ecosystem Science and Management, Pennsylvania State Univ., University Park, PA, USA. Email: henrylin@psu.edu

Hydropedology, Edited by H. Lin. DOI: 10.1016/B978-0-12-386941-8.00002-2

effective means of quantifying the *totality* of a soil's internal organization is missing. This has hindered the advancement of the systems view of soils and the coupling of soil structure into flow and reactive transport models (e.g. Bouma et al., 1979. Also see various chapters of this book). It has also constrained the effectiveness of scaling various soil processes and properties. A related fundamental issue is that heterogeneous and structured soils may not be effectively treated as a continuum for fluid flow; rather, soil architecture at different scales imposes specific pathways and complex patterns for flow and reaction in natural soils (e.g. Bouma, 1992; Flühler et al., 1996; Lin, 2010).

Soil architecture, as used here, refers to the *entirety* of how the soil is organized across scales from the microscopic to the megascopic levels (Fig. 1).

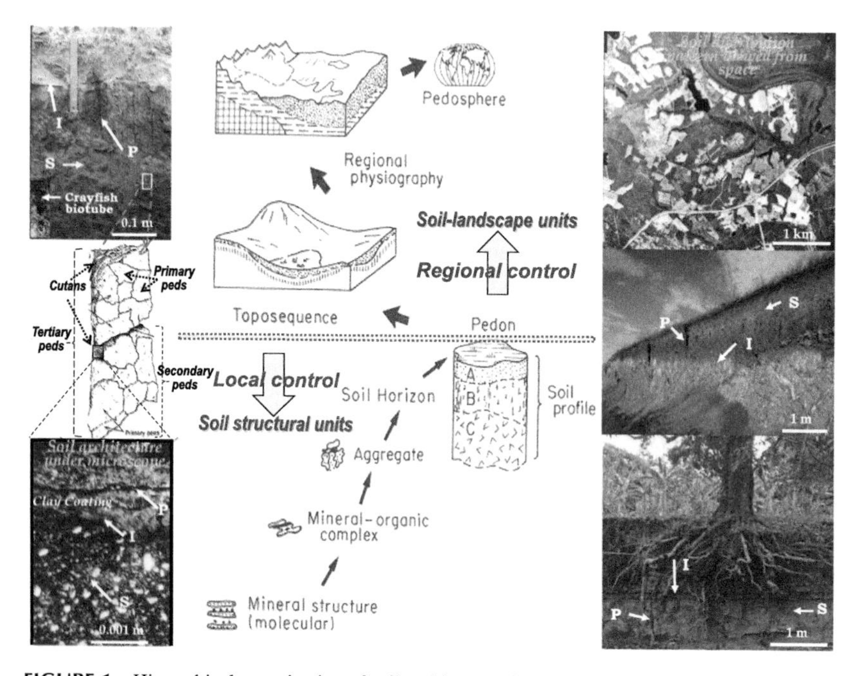

FIGURE 1 Hierarchical organization of soil architecture from the microscopic to the megascopic levels. This hierarchy may be grouped into two general levels: 1) soil architecture within a soil profile (termed soil structural units), such as particles, pores, aggregates, horizons, and pedons, which are largely influenced by local controls (such as parent material and land use), and 2) soil architecture in the landscape (termed soil–landscape units), such as catenae, soilscapes, soil sequences, and soil zones, which are generally linked to regional factors (such as climate and vegetation). At each scale illustrated with photos, three interlinked components of soil architecture are highlighted using arrows (S – solids, P – pores, I – interfaces). Sources: the central sketch of soil architectural hierarchy is from Wilding (2000); the left sketch of soil peds is from Brewer (1964); the thin section at the lower left corner is courtesy of Larry Wilding; the lower right corner soil profile with tree roots is from http://www.flickr.com/photos/aaronescobar/2569091622/in/set-72157602046046491; the right middle soil profile cross-section is courtesy of Phil Schoeneberger; the upper left is an Alfisol profile from Texas Coastal Plain, and the upper right is an aerial photo showing the soil-landscape in South Carolina Coastal Plain, both are courtesy of Henry Lin. *(Color version online and in color plate)*

FIGURE 2 Tight coupling between soil architecture and biological factors (S – solids, P – pores, I – interfaces): The left photo is a hydric soil from the Coastal Plain in North Carolina that shows many fine roots distributed in the pore space, the soil matrix, and soil aggregates (photo courtesy of David Lindbo); the right photo is a forest soil from east Texas that shows a native earthworm (red coiled ball) and medium and small tree roots, which influence soil architecture by macropore and aggregation formation (photo courtesy George Damoff and Kenneth Farrish). In both photos, the reddish vs. grayed colors show the oxidized vs. reduced soil matrix caused by heterogeneous and dynamic soil moisture. *(Color version online and in color plate)*

It encompasses three interlinked components at each scale of its hierarchical organization (see illustrations in Figs. 1 and 2):

1) *Solid components* ("building blocks"), such as matrix (soil particles and fabrics), aggregates (the type, strength, and size of peds, and the size distribution and stability of aggregates), horizons (different kinds and thicknesses of layers), pedons (three-dimensional soil bodies with a specific arrangement of soil horizons), and catenae (chains of related soils with primary difference in drainage due to topography).
2) *Pore space* ("openings" between building blocks), including the size distribution, connectivity, tortuosity, density, morphology, and edge of pores and their networks, as well as fluids and biota contained within the "living" pore space.
3) *Interfaces* (within solids or pores and in between them), including structural surface features (such as clay films, organic coatings, sand or silt coats, stress surfaces, and slickensides) and interfaces among various parts of the multiphase soil system (such as the interfaces of macropore–matrix, soil–root, microbe–aggregate, horizon–horizon, unsaturated-saturated, and soil–rock).

The above concept of soil architecture represents a holistic, multicomponent, and multiscale view of a soil system's entire internal organization. This is much broader than the classical concept of soil structure, which has exclusively focused on soil aggregates or peds (i.e. naturally formed soil aggregates separated by planes of weakness).

According to Hasegawa and Warkentin (2006), the term "soil architecture" was used first by E.J. Russell (1912) to describe different pore sizes, interconnected with solids that are available for attachment of water, chemicals, and microorganisms. Warkentin (2008) equated soil architecture to soil structure as the arrangement of voids and solids in soils, but preferred the term soil architecture because of its emphasis on spaces for soil functions.

Pore space is an essential part of soil architecture (see illustrations in Fig. 3). Indeed, heterogeneous soils are "living" space for diverse organisms, resulting in the largest underground biodiversity on Earth. The tight connection between soil architecture and biological factors is clearly illustrated in various photos in Figs. 1–3. However, pores (and related soil functions associated with pore

FIGURE 3 Illustrations of pore space and its complex network as an integral part of soil architecture. Left: Two intact soil columns (10-cm diameter and 30-cm long) visualized under X-ray demonstrate contrasting origins, sizes, and geometries of pores. The first column shows many earthworm borrows, root channels, and other types of macropores in an agricultural soil developed from limestone. The pink areas are entrapped air bubbles in this otherwise completely saturated soil column. The second column shows highly irregular pore space in between a large amount of rock fragments in a forest soil derived from shale. Right: Two photos illustrate the horizontal and vertical view of soil architecture and related pore space in an Alfisol (Mollic Natrustalf) from the University of Tasmania Farm in Australia (scale division is 10 cm) (photos courtesy of Marcus Hardie). The upper photo shows large pore space exposed after sand infillings in between coarse columnar peds in the B horizon have been removed by a garden blower. The lower photo shows vertical shrinkage cracks (indicated by arrows) in the soil profile after a dye infiltration experiment. *(Color version online and in color plate)*

space) have been largely neglected in U.S. soil surveys, leading to the missing of pores in the traditional concept of soil structure in the U.S. But pore features have been an integral part of soil structure in Australia (e.g. Brewer, 1964), Canada (e.g. Coen and Wang, 1989), Europe (e.g. Hodgson, 1997), and some other countries.

Although soil pore space and soil aggregation are closely interrelated, numerous soils have inter-, intra-, and/or trans-pedal pores that are not represented well by aggregates or pedality. These pores, formed by biological activities (e.g. root channels and animal borrows), physical processes (e.g. desiccation and freeze–thaw), or chemical reactions (e.g. dissolution or binding), are essential to hydrologic and biogeochemical processes in soils. In addition, the classical soil structure concepts have all been constrained to a specified soil horizon, and thus do not capture important interfaces between different soil layers or continuous pore features across horizons (such as earthworm borrows or root channels). Hence, the use of the term soil architecture here is meant to broaden the perspective across scales and to enhance the representation of the soil's entire organization.

Traditionally, pedology has focused on the solid part of the soil system as solids contain the products of weathering and the records of pedogenesis, while soil physics/hydrology has focused on the pore space in which mass and energy transfers occur. Soil biology and biogeochemistry, on the other hand, have focused on soil aggregates and interfaces as the living space for microorganisms and the reaction front for biogeochemical dynamics. It is obvious that the integrated view of both solids and pores, together with interfaces within and between them, will provide a more complete understanding of the whole system.

The importance of soil architecture has long been recognized in soil science and hydrology; however, its adequate quantification and incorporation into models have notoriously lagged behind (e.g. Letey, 1991; Bouma, 1992; Young et al., 2001; Lin et al., 2005a; Baveye, 2006; Jury et al., 2011. Also see various chapters of this book). For example, bypass flow (free water moving along macropores through unsaturated soils) is an important hydrologic process in structured soils, but has not yet been considered in flow theory (e.g. Bouma et al., 1979). This phenomenon is related to a number of factors, including:

- Inconsistent and fragmented concepts of soil structure used by various researchers;
- Overemphasis on ground and sieved soil materials and soil texture in the past, which has downplayed the role of soil architecture;
- Inadequate consideration of interfaces where many physical, chemical, or biological processes occur;
- Lack of adequate tools/methods to directly quantify soil architecture and its link to soil functions, especially noninvasively and in situ; and
- Lack of a comprehensive theory of the formation, evolution, quantification, and modeling of soil architecture across scales.

Because a system's architecture is the backbone of its functions, and because ground beef is no longer representative of a cow's functions (see discussion on this in Chapter 1 of this book), it has become imperative that a new era of soils research must be based on soils architecture and its quantitative relationships with various soil functions. For example, seven soil functions have been defined in the Soil Protection Strategy of the European Union (CEC, 2006; Bouma, 2010). As is well recognized from past research, soil architecture and its spatial–temporal variability fundamentally influence nearly all soil processes (e.g. infiltration, erosion, leaching, aeration, chemical reactions, solute transport, root penetration, microbial respiration, trace gas emission, and mechanical strength). In the meantime, rising environmental awareness and demands for precision natural resources management require that we know not only the quantity and quality of soil and water, but also *where* and *when* a problem may occur and how to prevent or remediate it effectively with limited resources. Consequently, finer spatial and temporal resolutions in our understanding of complex soil systems and greater ability of pinpointing actual flow pathways and reaction fronts are needed.

In the past, components of soil architecture have been evaluated largely by indirect methods that correlate soil architectural components to various soil properties or processes of interest (e.g. using soil water retention curve to infer soil pore-size distribution). Thin sections and various microscopes have been used to examine undisturbed soil architecture using small samples. In recent decades, advanced technologies have been increasingly utilized to provide more direct and nondestructive observations and quantifications of soil solids, pore space, and their interfaces at different scales, including environmental scanning electron microscopy (ESEM), X-ray or gamma-ray computed tomography (CT), nuclear magnetic resonance (NMR), ground-penetrating radar (GPR), electromagnetic induction (EMI), electrical resistivity tomography (ERT), proximal or remote sensing, and others (Fig. 4; see illustrations in other chapters of this book). These technologies vary widely in their capacities to reveal soil architectural complexity, including detectable soil features, sample size, penetrating depth, allowable environmental setting, spatial resolution, temporal frequency, and cost. Continued advances in these and new technologies are highly desirable to ease the technological bottleneck in studying soils without undue disturbance of their original architecture. At the same time, linking the use of advanced technologies to soil functionality is also critical so that these technologies would not become abstract glorification or provide only pretty images without quantitative functions.

Improved knowledge of soil architecture formation and evolution, together with continued advances in measurement and modeling techniques, can fundamentally improve our perception and prediction of soils. In fact, sufficient knowledge has already accumulated over past decades that allows us to begin

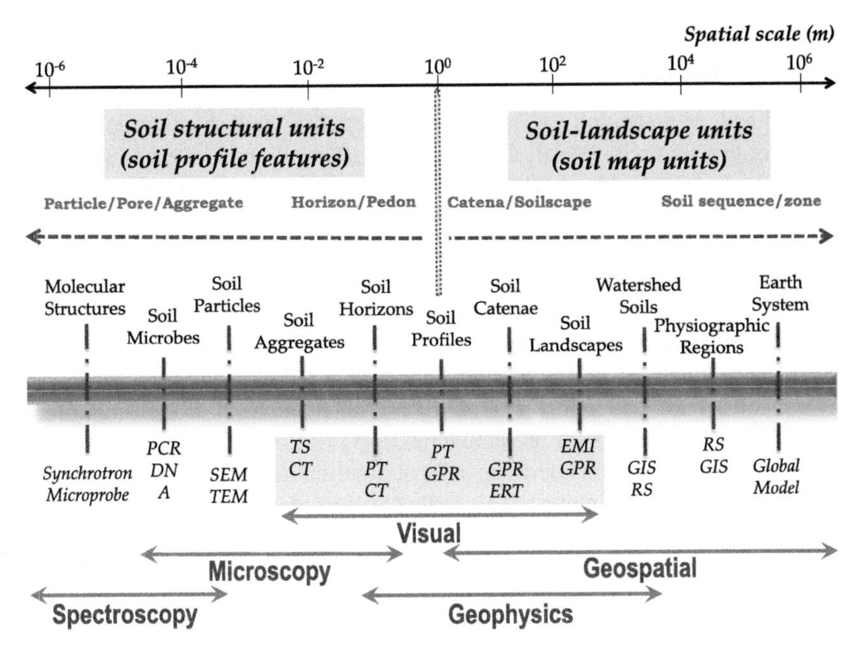

FIGURE 4 Hierarchical organization of soil architecture from the molecular level to the global level, and common tools/techniques used to reveal soil architecture at each level. The light blue shaded tools represent intermediate scales that connect soil structural units within a profile and soil–landscape units in the landscape. PCR: polymerase chain reaction; DNA: deoxyribonucleic acid; SEM: scanning electron microscopy; TEM: transmission electron microscopy; TS: thin section; CT: computed tomography; PT: photography; GPR: ground-penetrating radar; EMI: electromagnetic induction; ERT: electrical resistivity tomography; GIS: geographical information system; RS: remote sensing. *(Color version online)*

synthesizing various aspects of soil architecture. The following sections of this chapter are a first step in this direction. Meanwhile, noninvasive and high-resolution imaging technologies will be improved continuously and can make the quantification of soil architecture more feasible and affordable in the future. History has shown that many leap-forwards in medical, military, oil/gas/mineral explorations, and space exploration technologies have brought a suite of imaging, scanning, geophysical, and geospatial techniques to civilian use, including the study of soils (Fig. 4). Much like sensor networks – which were a dream for monitoring watershed hydrology some 20 years ago – have now become widely available. It is conceivable that, in the not too distant future, more powerful imaging and measurement techniques and devices could allow us to significantly advance the understanding of the opaque soil. This is driven in part by human's inclination to visualize things, as a picture is worth more than a thousand words. Combined with other advances in science, this can lead to much improved quantification and prediction of spatially and temporally variable processes in all soils.

2. HIERARCHICAL MULTISCALE FRAMEWORKS OF SOIL ARCHITECTURE

Hierarchical organization of soil architecture has long been recognized, including the hierarchy of soil aggregates (e.g. Tisdall and Oades, 1982), soil heterogeneity (e.g. Wilding, 2000), and soil image (e.g. Vogel and Roth, 2003). Earlier studies have also suggested a discrete hierarchy of representative elementary volume (REV) in soils, where the REV is a local property related to a given level of soil structural units (e.g. Cushman, 1990; Bouma, 1992; Vogel et al., 2002). For instance, Anderson and Bouma (1973) and Lauren et al. (1988) showed that the REV for measuring soil hydraulic properties should be a function of soil structural units, and thus standardized fixed sample volume for diverse soils could lead to incomparable data. Vogel and Roth (2003) suggested a "scaleway" as a means for predictive modeling of flow and transport in soils based on the explicit consideration of spatial structure that is assumed to be present at any scale of interest. This scaleway considers three features: 1) the structure of the medium, which must be known, 2) corresponding effective material properties, and 3) a process model at the scale of interest. Despite tremendous efforts in the past, the development of scaling relations, defining effective hydraulic properties, and quantifying soil structure–function relations remain the top three challenges in contemporary soil physics/hydrology research (Jury et al., 2011).

Two hierarchical frameworks could be used to help organize our understanding of soil modeling and mapping as well as their connections (Figs. 5–6). Soil mapping hierarchy here refers to the different levels of depicting soil spatial distribution patterns over landscapes of varying sizes, while soil modeling hierarchy portrays soil physical, chemical, and/or biological processes at different scales (Lin and Rathbun, 2003). Explicit links between these two hierarchies are important to connect soil architectures and functions, which have not yet been established. This is in part due to the mismatch of scales between spatial mapping and process-based modeling (Fig. 5). In soil mapping, polygon aggregation or disaggregation of map units are involved, which are determined irrespective of a process-based model. In contrast, in modeling soil processes, upscaling or downscaling of model inputs or outputs are considered, which are defined generally in the context of a specific model. In addition, the term *scale* carries different meanings between mapping and modeling. In cartography, map scale refers to the ratio of map to reality, and the scale becomes smaller as spatial information is aggregated into larger areas; whereas in the modeling arena, scale is often used in a colloquial sense, so large scale loosely refers to a large area extent. To improve the linkage between soil mapping and soil modeling, connecting digital soil mapping with hydropedology is needed (see Chapter 21 of this book for further discussion on this topic). It is encouraging that such a need is being recognized in recent published work (e.g. Schmidt et al., 2010; Bouma et al., 2011; Terribile et al., 2011).

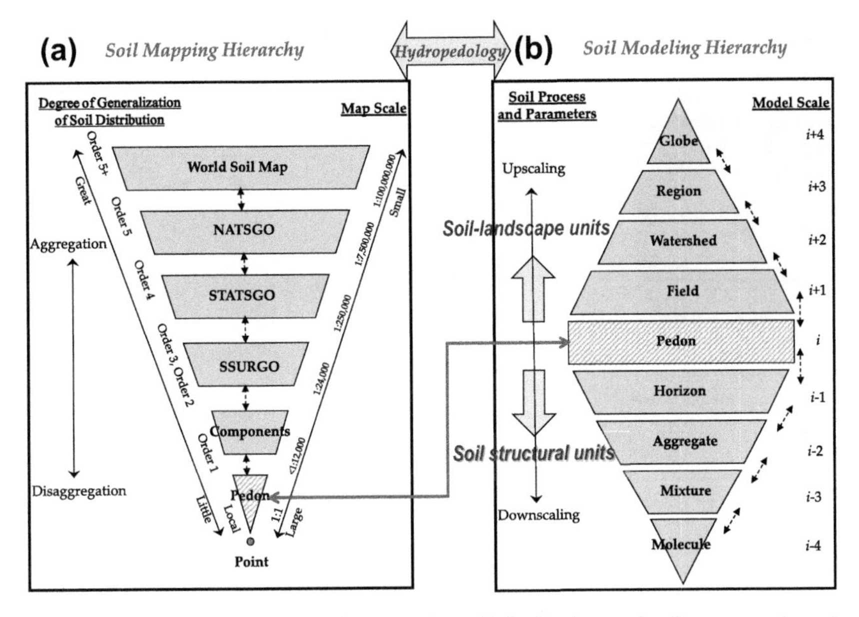

FIGURE 5 Hierarchical frameworks for bridging soil distributions and soil processes through hydropedology (modified from Lin and Rathbun, 2003): (a) soil mapping hierarchy for depicting soil spatial distributions at different map scales and (b) soil modeling hierarchy for simulating dynamic soil processes at different model scales. The SSURGO, STATSGO, and NATSGO are county, state, and national level digital soil geographic databases, respectively, which represent different intensities and scales of mapping (see text for more details). The pedon is the basic unit for pedogenic studies. The soil–landscape relationship is the key to soil mapping, and soil structural units are key to understanding soil processes within a pedon. *(Color version online)*

Another constraint in connecting soil architecture and soil function across scales is the gap between discrete scales used in mapping or modeling. Although there are approximations to connect the scales of attribute-based mapping (such as spatial interpolation from point observations to areal coverage using kriging) or process-based modeling (such as the extrapolation of smaller-area model parameters to larger areas through regionalization), significant gaps remain for *direct* translation from one scale to another. At present, no single satisfactory theory exists that is suitable for spatial aggregation (or upscaling), disaggregation (or downscaling), and temporal inference (or prediction) of soils and hydrologic information across scales (Lin et al., 2006; Jury et al., 2011). The major complementary approaches include scaling via discrete or continuous hierarchies, continuous spatial variation models of fractal and geostatistics, and Bayesian multiscale modeling (e.g. Vogel and Roth, 2003; Pachepsky et al., 2003; Lin and Rathbun, 2003).

Because different factors or processes may dominate at different scales and emerging phenomena may occur as the scale changes, a hierarchical set of

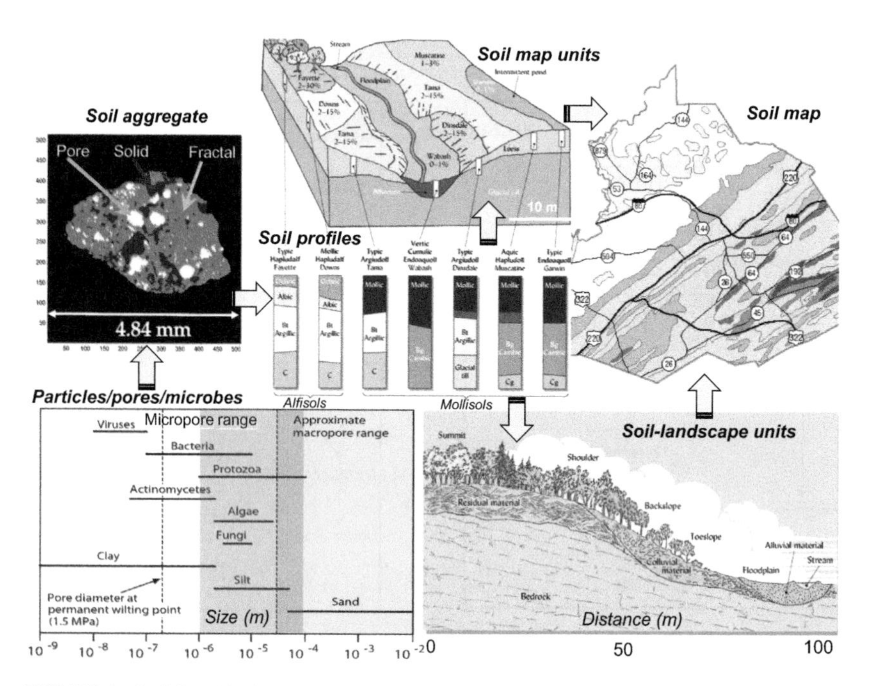

FIGURE 6 Building blocks of soil architecture from particles to landscapes as indicated by arrows. The smallest components inside a soil aggregate are compared in the lower left diagram *(modified from Buchan and Flury, 2007)*, where approximate macropore range (gray area) has been studied using X-ray CT. The enhanced CT resolution (highlighted in red) can now reach 1 μm, enabling the direct observations of finer microscopic features in soils. The aggregate shown in the upper left is an image obtained from X-ray CT, showing the pore-solid-fractal model (see text for more details). The fractal phase (orange color) is the porous matrix consisting of complex surfaces, likely dominated by organics. The catena and soil profile diagram in the upper middle and the soil–hillslope transect in the lower right are from Brady and Weil (2004). The general soil map (1:316,800, 5th-order) of Centre County, Pennsylvania (only a portion is shown) in the upper right is courtesy of the USDA-NRCS. *(Color version online and in color plate)*

models and maps may be a way forward; however, these models and maps need to be linked meaningfully, including the consideration of 1) structural interfaces and boundary shift as the scale changes and 2) constraint-feedback mechanisms between upper and lower scales in the hierarchy. Another important factor to recognize is that both the *mechanism* and the *magnitude* of soil variability should be considered simultaneously when dealing with scaling (Wilding, 2000; Lin et al., 2005a; Sommer, 2006). As Nielsen and Wendroth (2003) pointed out, while a versatile and powerful set of statistical/geostatistical tools exist for diagnosing spatially and temporally variable field observations, we have to explore the cause of variation and to improve and expand pedologic concepts. The efforts made by pedologists, soil physicists, hydrologists, and others on soil variability and scaling have not converged well in the past. Thus, it is hoped that hydropedology can facilitate such synergism.

While the scale of soil architecture ranges from the microscopic to the megascopic levels, two general categories of scales could be used to facilitate the linkage of soil architectural hierarchy from individual soils to the landscape (Figs. 1, 4 and 5): 1) soil architecture within a soil profile (termed soil structural units), and 2) soil architecture in the landscape (termed soil-landscape units). These concepts are further explained in the following sections.

2.1. Soil Architecture Within a Soil Profile (Soil Structural Units)

This category may be further grouped into: 1) the scale of particles, pores, and aggregates, which are organized around peds of different types and sizes (including associated pore space and interfaces), and 2) the scale of horizons and profiles, which are centered around pedons composed of different types and thicknesses of soil horizons (Figs. 5 and 6). The first level is fundamental to flow and transport at the microscopic level (such as advection, diffusion, and dispersion) as well as biogeochemical processes (such as mineral weathering, microbial activities, nutrient cycling, and carbon sequestration). The second level is the basic unit for pedologic studies, including pedogenesis, classification, morphology, mapping, and interpretations.

An important outcome of pedogenesis is the formation of soil aggregates of various kinds (Fig. 7). "*Ped*" is a unique term in soil science and "*pedology*" captures that uniqueness well. Generally speaking, the strength and expression of pedality increase with pedogenic time, and the degree of pedality is usually expressed most strongly at the surface and decreases with depth, resulting in a general trend of increasing ped size and decreasing ped grade with increasing soil depth. However, human activities and other disturbance could significantly alter soil aggregates in both positive and negative ways.

Soil horizonation is another important outcome of pedogenesis. The various kinds and thicknesses of soil horizons and how they are arranged in pedons reflect past and current soil-forming environment, which serve as the basis for classifying soils and understanding soil functions (Fig. 6). Soil horizons are recognized in the field according to soil morphological features, such as color, texture, consistence, structure, ped coatings, nodules or concretions, root and pore distribution, rock fragment, cementation, pH, carbonate content, salinity, and boundary characteristics (Soil Survey Staff, 1993). Quantitative diagnostic horizons are used in the U.S. *Soil Taxonomy* to classify soils, and their key features are summarized in Table 1, including a number of water-restrictive horizons (such as fragipan, duripan, and petrocalcic horizon). Note that each diagnostic soil horizon (and each individual soil taxonomic class) encompasses a specific *range* of soil properties.

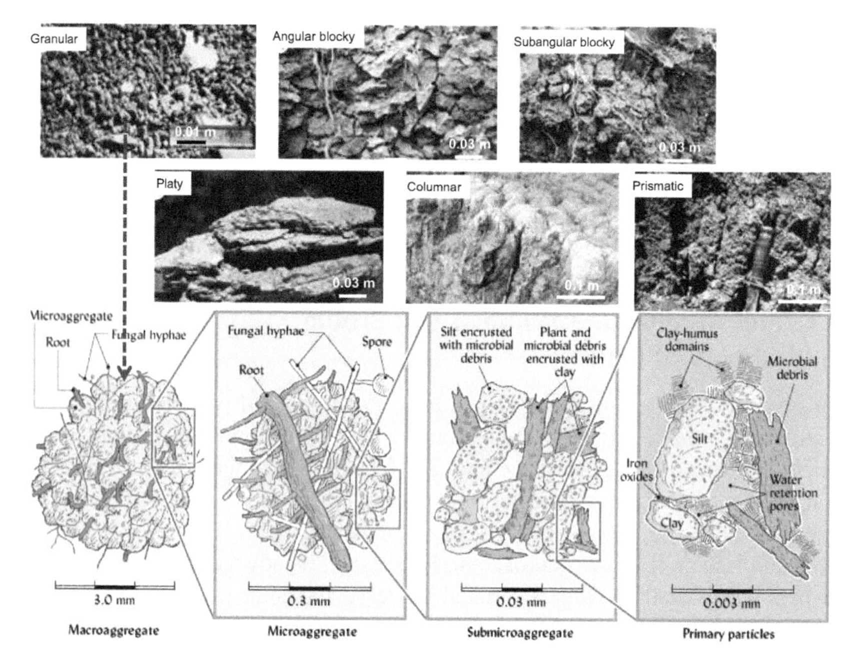

FIGURE 7 Common types of peds (upper panel) and the hierarchy of soil aggregates (lower panel). Larger aggregates are often composed of an agglomeration of smaller aggregates. Note the kind of sticky network formed from fungal hyphae, fine roots, plant and microbial debris, clay, humus, and Fe or Al oxides in the aggregates of various sizes. *(Modified from Brady and Weil, 2004) (Color version online)*

2.2. Soil Architecture in the Landscape (Soil–landscape Units)

This category may also be grouped into two levels: 1) soil catenae and other soilscapes at the local scale, and 2) soil zones and other soil patterns (such as soil sequences) at the regional scale. Soil distribution patterns at the local and regional scales are generally depicted on soil maps of various scales or soil–landscape block diagrams or cross-sections (see examples in Figs. 6, and 8–11). The related terms are defined below:

- *Catena* is a chain of related soils along a hillslope in which the distribution of the soils is a function of elevation and slope (Milne, 1936a,b). Bushnell (1943) conceived of the catena as a topohydrosequence of soils developed from the same parent material with primary difference in drainage due to relief (Fig. 10).
- *Soilscape* is a generic term that refers to the arrangement of soil components of a landscape (Figs. 8–9) (Hole, 1976). The term *soil cover pattern* (or soil cover structure) instead of soilscape has also been used (mainly in Russia) (Fridland, 1976), emphasizing the heterogeneous spatial distribution of soils that are due to factors that vary over relatively short distances (such as relief and parent material).

TABLE 1 Diagnostic Soil Horizons Identified in the U.S. Soil Taxonomy and Their Key Features. Specific Criteria Used Can be Found in Soil Survey Staff (2010)

Diagnostic Horizons	Generic Horizon	Key Features
Diagnostic Surface Horizons		
Anthropic	A	Human-modified mollic-like horizon, with high available P
Folistic	O or A	Thick organic soil material or high organic carbon in Ap horizon saturated for <30 days in normal years
Histic	O	Very thick organic soil material or high organic carbon in Ap horizon saturated for >30 days in normal years
Melanic	A	Thick, black, >6% organic carbon with andic soil properties
Mollic	A	Thick, dark-colored, high base saturation, strong structure
Ochric	A	Too light-colored, low organic content or thin to be Mollic
Plaggen	A	Human-made sodlike horizon created by years of manuring
Umbric	A	Same as Mollic except low base saturation (<50%, acidic)
Diagnostic Subsurface Horizons		
Agric	A or B	Silt, clay, and humus accumulation under cultivation just below plow layer
Argillic	Bt	Silicate clay accumulation
Cambic	Bw, Bg	Weakly developed B, with change or alternation by physical movement or by chemical reactions, generally nonilluvial
Duripan	Bqm	Hardpan, strongly cemented by silica
Fragipan	Bx	Dense pan, seemingly cemented when dry, brittle when moist, and slake when submerged in water. Typically has redox features and bleached prism ped faces
Glacic layer	Bf or Cf	Massive ice or ground ice in the form of ice lenses or wedges

(Continued)

TABLE 1 **Diagnostic Soil Horizons Identified in the U.S. Soil Taxonomy and Their Key Features. Specific Criteria Used Can be Found in Soil Survey Staff (2010)—Cont'd**

Diagnostic Horizons	Generic Horizon	Key Features
Glossic	E	Tongue resulting from the degradation of an argillic, kandic, or natic horizon
Kandic	Bt	Argillic, kaolinite clays (low activity)
Natric	Btn	Argillic, high in sodium, columnar or prismatic structure
Ortstein	Bhm	Cemented by humus and aluminum
Oxic	Bo	Highly weathered, primarily mixture of Fe, Al oxides and non-sticky-type silicate clays
Permafrost		A perennially frozen horizon remaining below 0 °C for ≥2 years in succession
Petrocalcic	Ckm	Carbonate cemented horizon
Petrogypsic	Cym	Gypsum cemented horizon
Placic	Csm	Thin pan cemented by iron (or iron and manganese) and organic matter
Spodic	Bh, Bs	Organic matter, Fe and Al oxide accumulation

- *Soil zones* refer to regional soil distribution patterns over large geographic areas that are associated with particular climatic and vegetative zones, which are thought to be related to regional or higher classification categories of soils (Fig. 11) (Baldwin et al., 1938; Fridland, 1976).
- *Soil sequences* are related soils that are expressed by gradual changes thought to be the result from a gradient in a dominant soil-forming factor, including climo-, bio-, topo-, litho-, and chrono-sequences. Generally, climo-, bio-, and chrono-sequences are best observed over relative large areas, while topo- and litho-sequences are more often considered in local soilscapes (Schaetzl and Anderson, 2005).

The wide applications of the catena concept have been complicated by parent material variations and climatic differences, because strictly speaking the catena concept is limited to uniform parent material and similar climate (SSSA, 2001). Thus, soilscape is used as a more generic term that includes catenae and other local soil distribution patterns over short distances. Note that the term

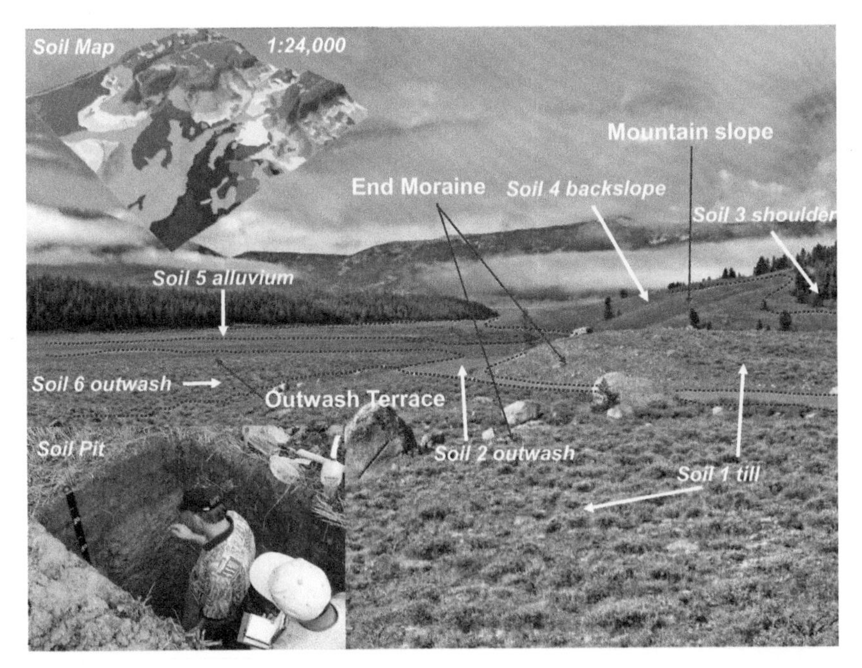

FIGURE 8 From soil pits to soil maps: Delineation of soil map units based on the paradigm of soil–landscape relationships. Illustrated here are six soil types in a glaciated mountain landscape in Colorado identified in the 2nd order soil mapping. *(Landscape Photo Courtesy of Doug Wysocki) (Color version online and in color plate)*

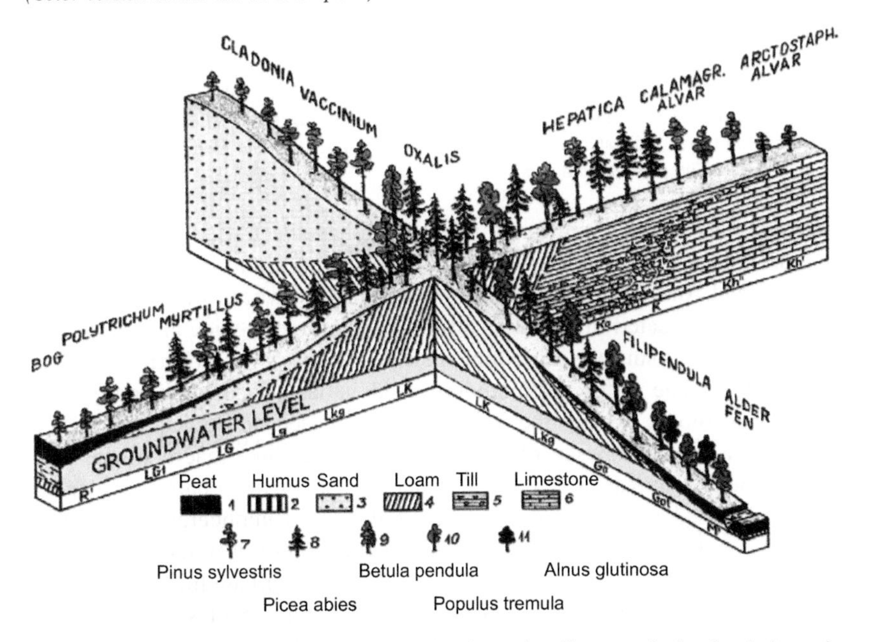

FIGURE 9 Two crossing soilscapes in Estonia showing main soil types and related main forest site types. Soil type symbols indicated at the bottom of each soilscape (from L to M and from Kh to R) follow the classification of World Reference Base (2001). *(From Uuemaa et al., 2008) (Color version online)*

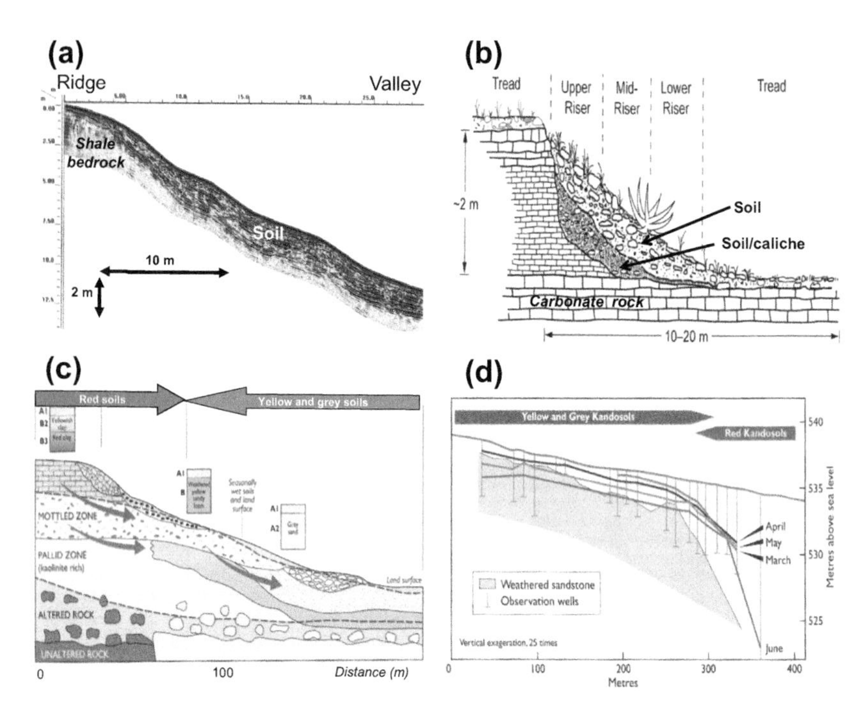

FIGURE 10 Two pairs of contrasting catenae from different geographic regions: (a) soil thickness increases downslope in the Shale Hills Catchment in central Pennsylvania (green curve indicates the soil–bedrock interface); (b) soil thickness decreases downslope in central Texas with stair-step topography developed from limestone (from Wilding, 2007); (c) a common soil catena along an eroding hillslope in Australia, showing iron mobilization and water flow pathways downslope; and (d) a toposequence controlled by the weathered substrate near Torrens Creek, Queensland, Australia, showing an opposite trend compared to (c) in terms of the distribution of soil colors, soil types, and water table *(Modified from Coventry, 1982 and McKenzie et al., 2004). (Color version online and in color plate)*

landscape may mean two different views (both have been used in soil science literature): 1) a section or portion of the land, which could be large or small (Schaetzl and Anderson, 2005), and 2) a land mosaic where a cluster of local ecosystems is repeated in a similar form over a kilometer-wide area or a land surface that a human eye can comprehend in a single view (Forman, 1995; SSSA, 2001). The latter is commonly used in landscape architecture and landscape ecology, where the term *region* is used to indicate a broad geographical area composed of landscapes with the same macroclimate and tied together by human activities (Forman, 1995). In this chapter, the term landscape is used generically (as is common in soil science literature), and the term soil–landscape is used to imply combined local and regional levels.

Soil survey and mapping (see an illustration in Fig. 8) are the most widely used means of depicting soil–landscape relationships. Soil surveyors map soils with a conceptual model of soil–landscape relationships using air photo

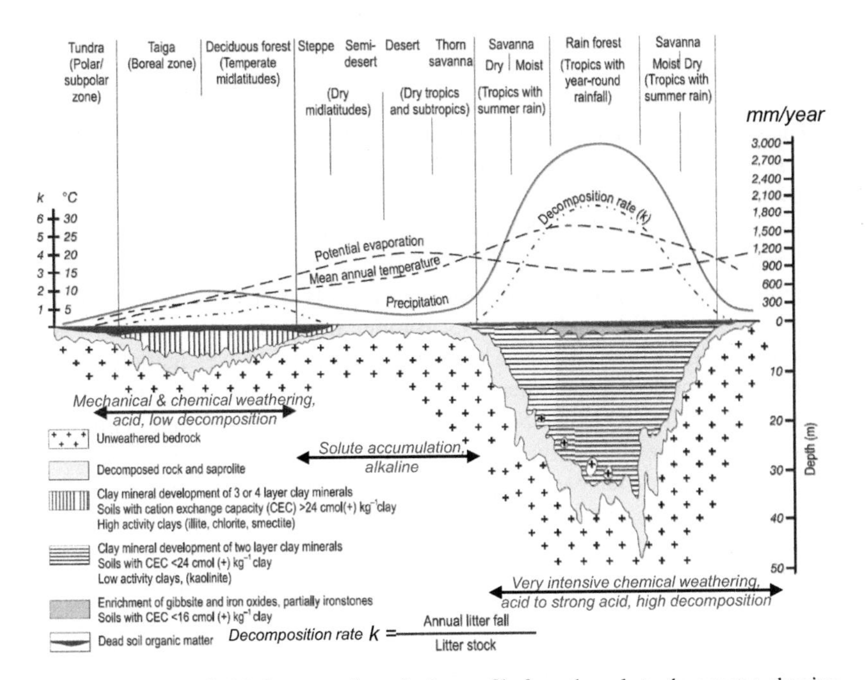

FIGURE 11 general global pattern of weathering profile from the pole to the equator, showing a latitudinal transect of the major climatic zones and associated soil thickness and layering. *(Modified from Schultz, 2005) (Color version online)*

interpretation and collated information on the soil and its relations with landform, geology, vegetation, and land use (Hudson, 1992). To confirm interpretation, field observations are made at a selected number of locations chosen by soil surveyors using formal knowledge (mainly soil-forming theory) and intuitive judgment (expert experience) (Soil Survey Staff, 1993). Soil map units (generally composed of soil series or other taxonomic components above the series level, plus phase criteria and slope class) are then used to represent local or regional soil–landscape delineations (rather than *individual soil kinds*), depending on survey scale (Soil Survey Staff, 1993; Wysocki et al., 2000; Buol et al., 2011). This is because soils and landscapes are tightly coupled and virtually every delineation in traditional soil surveys includes other soil components or miscellaneous areas that are not identified in the name of a map unit because of limitations imposed by the intensity or scale of mapping (Soil Survey Staff, 1993). Many of these included soils are either too small to be delineated separately at a given soil survey order or deliberately included in another map unit to avoid excessive detail (Soil Survey Staff, 1993). Therefore, spatial variation inherently exists within all soil map units, but the degree of variation depends on soil survey order, map scale, kind of map unit, or purpose of survey (Fig. 5a). In the past, variation within soil map units has been described qualitatively in vague terms (Arnold and Wilding, 1991). Thus, quantification of soil map unit purity is an important improvement needed in modern soil surveys and mapping (e.g. Arnold and Wilding, 1991; Lin et al., 2005a,b).

There are five orders of soil surveys in the U.S. (Fig. 5a), ranging from the 1st order for detailed mapping (minimum delineation size $\leq$1 ha, 1:15,840 or larger cartographic scale, mapping units mostly consociations of phases of soil series) to the 5th order for very general mapping (minimum delineation size 252–4000 ha, 1:250,000 or smaller cartographic scale, mapping units largely associations consisting of two or more dissimilar soil components) (Soil Survey Staff, 1993; Buol et al., 2011). Although most soil maps available in the U.S. have been made with field investigations, some have been compiled from other sources. Such maps include (Fig. 5a): 1) generalized STATSGO (State Soil Geographic Database) soil maps compiled by aggregating the delineations of existing soil surveys or SSURGO (Soil Survey Geographic Database) maps, or NATSGO (National Soil Geographic Database) soil maps aggregated from STATSGO maps, and 2) schematic soil maps synthesized from information on soil-forming factors and/or any other available information. The latter merges with the 5th order (exploratory) soil surveys without a sharp distinction (Soil Survey Staff, 1993).

It is important to recognize that traditional soil surveys in the U.S. have been designed for general land-use planning purposes at the county or regional levels. In recent years, with growing demands for site-specific applications and high-resolution soils data, 1st order or high-intensity soil surveys have become more and more needed. Thus, other means of mapping soils, such as the application of geostatistics (e.g. Heuvelink and Webster, 2001), the use of proximal or remote sensing (e.g. McKenzie and Ryan, 1999), and spatial modeling through terrain and other geospatial analyses (e.g. Burrough and McDonnell, 1998), have been increasingly used, leading to the current active interest in digital soil mapping (Zhu et al., 2001; McBratney et al., 2003; Lagacherie et al., 2007). It is, however, important to keep in mind that field experience (including soft data and tacit knowledge), ground-truthed or hard data, and process-based understanding have been the hallmark of pedology (e.g. Daniels, 1988; Arnold and Wilding, 1991; Jacob and Nordt, 1991; Schaetzl and Anderson, 2005; Lin et al., 2005a). Goryachkin (2005) noted that the loss of valuable field experience would be a serious threat to pedology as a fundamental science. Bouma (2010) has made a plea to find better ways to close the "knowledge chain" from the K2 (qualitative and empirical soil survey knowledge) to the K5 (quantitative and mechanistic process knowledge).

3. FORMATION AND PATTERN OF SOIL ARCHITECTURE

Soil architecture is both a consequence of and a prerequisite for many critical processes regulating soil functions. Understanding soil architecture-forming processes can help identify deterministic patterns and stochastic variations of soil behaviors under various environmental conditions. In the following sections, we first present a generic interpretation of soil architecture formation from a thermodynamic point of view. We then discuss specific pedogenic processes leading to the formation and change of soil architecture from the pore/aggregate scale to the horizon/pedon scale and the soilscape/regional scale.

3.1. A Generic Interpretation of Soil Architecture Formation

Fundamental thermodynamics can provide a general explanation of soil architecture formation and evolution (Lin, 2010, 2011). This is because pedogenesis is an energy-consuming process, and the simultaneous occurrence of dissipation and organization has long been recognized in pedogenesis (e.g. Hole, 1961; Volobuyev, 1963; Runge, 1973; Smeck et al., 1983; Johnson and Watson-Stegner, 1987; Johnson et al., 1990; Addiscott, 1995; Rasmussen et al., 2005; Lin, 2010).

Associated with energy dissipation, a dual partitioning occurs during pedogenesis (Lin, 2010, 2011):

1) Organizing processes, such as aggregation, humus accumulation, horizonation, flocculation, illuviation–eluviation, and secondary mineral formation; and
2) Dissipating processes, such as aggregate degradation, humus decomposition, homogenization, dispersion, mixing, and primary mineral weathering.

The first set of processes facilitates the formation of soil structural units, while the second set of processes tends to generate soil matrix; the two combined form soil architectures of various kinds. This dual partitioning of pedogenesis is consistent with the theory of "dissipative structure" proposed by Prigogine (1967), which suggests that a system forms its structure away from thermodynamic equilibrium by energy dissipation.

Soil systems with aggregates, horizons, and profiles formed over time represent more and more ordered states than their precursors (until a soil starts to degrade or when a regressive process occurs). Thus, soil formation and evolution can be conceptualized as consuming energy and exporting entropy (i.e. degraded energy) in order to "grow" and preserve its internal order in order to function (Lin, 2011). As energy dissipates and entropy exports, complex structures are formed through the soil system's self-organization in order to increase the efficiency in processing available free energy (see Chapter 8 of this book). Self-organization signifies a process of attraction and repulsion in which the internal organization of the soil system increases in complexity without being guided by an outside source. Soils utilize highly ordered solar energy (through either direct short-wave radiation or indirect photosynthetic organic matter) that fuels pedogenesis, and then return to the environment degraded energy to maintain or further develop its internal order to function.

It is interesting that, for the most part, solar radiation tends to concentrate soil constituents at the soil surface (through processes such as evaporation, transpiration, photosynthesis, biocycling, and leaf falls), whereas gravity tends to move soluble and suspended soil components further down into or out of the soil profile (Smeck et al., 1983). As a result of these two opposite tendencies, some soil components have a bimodal distribution in soil profiles (i.e. one peak in the soil surface and another peak near the solum bottom). Furthermore, the soil surface interfaces with the atmosphere and interacts most significantly with terrestrial biosphere; thus, soil surface horizons tend to accumulate energy and matter (e.g. organic carbon) (unless disturbed or eroded). On the other hand, the soil solum bottom interfaces with the lithosphere, which generally restricts further vertical downward movement of water and materials and, in sloping landscapes, permits lateral flow of energy and mass from the surrounding areas; thus, the bottom portion of the solum or the upper portion of the substratum or saprolite tends to accumulate energy and mass. This leads the middle portion of many soil profiles to be possibly the least active zone in soil processes (e.g. Runge and Riecken, 1966; Smeck et al., 1983; Flühler et al., 1996; Lin et al., 2008). This has important implications for understanding soil architecture and modeling soil functions.

3.2. Soil Architecture at the Aggregate and Pore Scales

There are many factors influencing the production or degradation of soil aggregates, including 1) *materials* (parent materials, mineral composition, organic matter, clay, water, and inorganic cementing agents), 2) *biology* (plant roots, microorganisms, earthworms, insects, and all the by-products of any digestive, reproductive macro- and microorganisms), 3) *climate* (wetting–drying, freezing–thawing, rain drop impacts, and temperature), and 4) *human activities* (tillage, compaction, irrigation, fertilization, contamination, and many others). In particular, biofactors such as earthworms and roots are vital for the formation of stable soil aggregates (Figs. 1, 2 and 7). Glomalin (named after the Glomales order of fungi; Wright et al., 1996) is a glycoprotein produced abundantly on hyphae and spores of arbuscular mycorrhizal fungi in soils and roots. It permeates organic matter, binding it to silt, sand, and clay particles, thus forming clumps of soil aggregates and keeping stored soil carbon from escaping (Fig. 7). Another increasingly dominant factor in influencing the production or degradation of soil aggregates is human management activities, which has been recently reviewed comprehensively by Bronick and Lal (2005).

Flocculation by di- and trivalent cations and aggregation by various cementing materials (organic and inorganic) are accelerated under the cyclic influences of wetting–drying, freezing–thawing, clay translocation and coating on aggregate surfaces, and various biological activities (Brady and Weil, 2004). Water has significant impacts on soil aggregation formation and stabilization. Water is needed in forming soil aggregates by facilitating flocculation and aggregation; however, water may also cause aggregation deterioration in two ways (Hillel, 1998): 1) the hydration effect of water, which causes the disruption of aggregates through the processes of swelling and the exploding of entrapped air, if soil moisture changes drastically or stays saturated for a long period of time, and 2) the impact of failing rain drops on exposed soil surface, which exerts a significant dispersive action on aggregates. The dispersed particles can then be carried into soil pores, causing increased compaction and decreased porosity. This is particularly evident when intense rain destroys the structure of the topsoil and forms a dense, nearly impervious surface known as a crust or sealing (Hillel, 1998; Brady and Weil, 2004).

Traditionally, soil aggregates have been evaluated by pedologists in the field using morphological descriptions or thin sections, while soil physicists and soil biologists have employed wet and dry sieving, elutriation, and sedimentation to conduct aggregate-size distribution and aggregate stability, aiming to measure the percentage of water-stable aggregates in the soil and the extent to which the finer mechanical separates are aggregated into coarser fractions (Jury and Horton, 2004). In the absence of direct quantification, soil aggregation has been frequently evaluated by methods that correlate it to soil properties such as saturated hydraulic conductivity, water retention parameters, infiltration rate, gas diffusion rate, and others. The belief that soil structure can be expressed as the aggregate-size distribution and the degree of aggregate stability has served as the dominant paradigm in soil physics for much of the past research on soil structure (Hillel, 1998; Jury and Horton, 2004). However, as discussed in Section 1 of this chapter, aggregation alone is insufficient to capture the totality of soil architecture (see demonstrations in Figs. 1–3).

Soil architecture at the pore and aggregate scale has been suggested as having fractal characteristics, meaning self-similarity over a certain range of scales (e.g. Anderson et al., 1998; Bartoli et al., 1998). Thus, fractal mathematics (geometrical fractals or probabilistic fractals; Baveye and Boast, 1998) has been applied to soil aggregate-size, pore-size, as well as particle-size and other soil structural unit related properties (such as soil water retention, hydraulic conductivity, and preferential flow) (Baveye et al., 1999; Pachepsky et al., 2000). The fundamental equation applying to all fractals is the number–size relationship (Mandelbrot, 1982):

$$N(r) = k r^{-D}, \tag{1}$$

where $N(r)$ is the number of elements of size equal to r (unit length or "yardstick"), k is the number of initiators of unit length (called "prefactor"), and D is the fractal dimension.

However, even though the fractal dimensions of two objects are the same, their structures could be quite different (Pendleton et al., 2005; Luo and Lin, 2009). Thus, another parameter, termed lacunarity, is necessary. The joint distribution of fractal dimension and lacunarity can better discriminate soil architecture (Zeng et al., 1996; Luo and Lin, 2009). Lacunarity measures the deviation of a geometric object from the translational invariance or homogeneity and can be considered as a scale-dependent index of heterogeneity

(Mandelbrot, 1982; Plotnick et al., 1993). The prefactor k in Eqn (1) is associated with the lacunarity, which reflects the fraction of space occupied by the feature of interest, the degree of dispersion and clustering, the presence of self-similarity or randomness, and the existence of hierarchical structure (Plotnick et al., 1993). Lacunarity has been applied to differentiate the structures and spatial patterns of both fractals and nonfractals, such as structural patterns associated with different landscapes and land-uses (Plotnick et al., 1993) as well as pore geometry in heterogeneous porous media (Kim et al., 2007).

Perrier et al. (1999) proposed a generalized pore-solid-fractal (PSF) model (Fig. 6), which is shown to exhibit either a fractal or a nonfractal pore surface depending on the model parameters. In the PSF model, the fractal dimension, D, is expressed as

$$D = d + \log(1 - P - S)/\log n, \tag{2}$$

where d is a given Euclidean dimension, P is the proportion of pore phase, S is the proportion of solid phase, and n is the inverse of the similarity ratio. A third phase, labeled as fractal (F) phase (where $P + S + F = 1$), is the proportion for the next stage of partitioning that exhibits a self-similar manner (Gibson et al., 2006). Note that such a fractal phase is different from the fractal dimension indicated in Eqn (1). With the exception of two special cases corresponding to a solid mass fractal and a pore mass fractal, the PSF model displays symmetric power law or fractal pore-size and solid-size distributions (Perrier et al., 1999).

In spite of an impressive body of literature on fractal applications in soil science, this field of research remains to be further developed (Baveye and Boast, 1998). As Vogel and Roth (2003) pointed out, it is not obvious that soils should exhibit fractal properties, because the self-similarity of soil architecture would suggest some self-similarity in the formation of soil architecture. However, quite different processes are generally involved in soil architecture formation and each process may introduce a scale of its own, as discussed in this section.

Reconstruction, visualization, and quantification of 3-D pore networks are essential to correlating soil pore space to physical, chemical, and biological functions and to modeling their dynamics under different environmental conditions and land uses. Advanced imaging techniques, particularly X-ray CT, have become increasingly attractive for achieving this goal (e.g. Anderson et al., 1990; Perret et al., 1999, 2000; Lindquist, 2002; Luo et al., 2008). Mathematical morphology (Serra, 1982) has been used to quantify the characteristics of 3-D earthworm burrows, including pore-size distribution, length, connectivity, and branching intensity (Capowiez et al., 1998; Pierret et al., 2002; Bastardie et al., 2003). Perret et al. (1999) reconstructed the 3-D images of soil macropores in intact soil columns and calculated the number of macropore networks, length, tortuosity, hydraulic radius, numerical density of networks, and their connectivity. Luo et al. (2010a,b) introduced an improved protocol to quantify soil pore networks based on X-ray CT images and have applied that to the study of land-use impacts on soil pore networks. Continued efforts to quantify the patterns of pore networks and their links to soil functions in different soils and land uses will help improve the predictive modeling of flow and transport.

3.3. Soil Architecture at the Horizon and Pedon Scales

Additions, losses, transformations, and translocations of materials are four general soil-forming processes leading to the formation of various soil horizons. Generally, a series of pedogenic processes are bundled together to form a particular soil profile under a given set of environmental conditions. These specific pedogenic processes include a wide variety, some of which are exemplified below (for a more complete list and the explanation of the terms listed below, see Schaetzl and Anderson, 2005 or Buol et al., 2011).

- *Additions*: Enrichment, littering, cumulization, and melanization.
- *Losses*: Leaching, runoff, and erosion.
- *Transformations*: Synthesis, decomposition, humification, mineralization, ripening, gleization, podzolization, desilication, resilication, paludization, braunification, rubification, ferrugation, loosening, and hardening.
- *Translocations*: Eluviation, illuviation, gleization, calcification, decalcification, salinization, desalinization, alkalization, dealkalization, lessivage, pedoturbation, podzolization, desilication, melanization, leucinization, braunification, rubification, and ferrugation.

Schaetzl and Anderson (2005) noted that the processes of organizing or reorganizing soil fine particles without significant alteration of their composition are weakly reflected in the recognized specific pedogenic processes. Sometimes, only loosening and hardening of soil materials (Buol et al., 2011) are mentioned. There has been virtually no attempt to distinguish between qualitatively different processes involved in soil structure formation (Targulian and Goryachkin, 2004), especially in subsoils (Buol et al., 2011). This is a knowledge gap that has significance for understanding and predicting soil architecture formation and function.

Many factors influence the formation or degradation of soil horizons, including all soil-forming factors and human activities. Critical among the environmental factors are soil moisture and temperature regimes. For instance, a well-drained condition will permit soil development to proceed to a more highly-weathered stage with more differentiated horizons, whereas a poorly-drained condition restricts this process. Hence, local topography and geological condition can significantly alter soil horizonation within a given climatic zone. Once formed, there are many possible pedogenic pathways for a soil to evolve over time, including progressive, regressive, or static processes (Johnson and Watson-Stegner, 1987; Phillips, 1993; Schaetzl and Anderson, 2005). Regressive cycles of soil formation often occur as a result of a sudden change in soil-forming factors (e.g. erosion and sedimentation), which can also produce major markers in soil profiles (especially in regions with a marked history of climate change). The self-organization capacity of a soil system may also have a limit. The considerable retardation or end of this self-organization is referred to as the climax, quasi-climax, or steady state (dynamic equilibrium) of the functioning soil system (Targulian, 2005). Yaalon (1971) called the excessive clay film development in argillic horizon "self-terminating," which results in reduced capacity to transmit matter and energy vertically.

How soil properties change with depth is controlled fundamentally by soil architecture at the horizon and pedon scale, which is a function of soil type and specific soil property. While the depth function of soil properties could be complicated, the following are some commonly observed trends:

- *Exponential decrease*: for example, soil carbon and nitrogen contents, which are generally associated with biological activities that are most active in surface soils.
- *Exponential decrease and then increase*: this pattern is generally related to a restrictive layer where water and nutrients may accumulate and thus become enriched.
- *Exponential increase and then decrease*: for example, clay and chemical accumulations in B horizons where leaching and translocation of materials from the upper part accumulate in the middle part of the profile.
- *Graduate increase*: this is related to weathering gradient and is often due to mobile elements not yet completely leached out of the soil, such as cations in semi-arid soils.
- *Nearly constant*: for example, texture or pH in weakly-developed soils or highly-weathered soils.
- *Multiple peaks:* two or more enriched zones in the profile, such as organic carbon in Spodosols, or in soils with buried horizons.

Changes in soil saturated hydraulic conductivity (K_{sat}) and drainable porosity with soil depth can illustrate the importance of pedologic knowledge in hydrologic modeling: These two important soil hydraulic properties are commonly assumed to exponentially decline with increasing soil depth in hydrologic models (e.g. Beven and Kirkby, 1979; McDonnell, 2003; Weiler et al., 2005). This holds true in many sola from the organic-rich top layers to the more compacted or restrictive subsoil layers (Kendall et al., 1999; Brooks et al., 2004, 2007). However, some soils have higher K_{sat} in lower horizons, in soil horizons beneath water-restrictive layer, or in highly weathered or fractured C horizons (saprolite), leading to the trend of exponential decline and then increase, or other more complicated pattern of K_{sat} in the soil profile (e.g. West et al., 2008; Lin et al., 2008). Therefore, following specific horizonation is beneficial as it is characteristic for every soil.

3.4. Soil Architecture at the Soilscape and Regional Scales

The spatial distribution of soils is so closely coupled with the landscape that soilscapes or soil–landscape units are generally used to characterize soil maps instead of individual soil kinds (Soil Survey Staff, 1993; Wysocki et al., 2000; Park et al., 2001; Hupy et al., 2004). Figure 9 illustrates two crossing soilscapes that display clear association between soil types and topography, as well as parent materials, hydrology, and vegetation. Soil–landscape models thus have been widely used to predict a number of soil attributes using terrain, geology, hydrology, vegetation, and other environmental data, such as the mapping of soil series, drainage class, A horizon depth, solum thickness, presence or absence of a specific soil horizon (such as fragipan or E horizon), soil texture, soil moisture, hydric soils, water-holding capacity, organic carbon content, and nutrient contents (e.g. Bell et al., 1992; Gessler et al., 1995, 2000; Thompson et al., 1997; McKenzie and Ryan, 1999; Gobin et al., 2001; Ziadat, 2005. Also see Chapter 21 of this book). Figures 8–10 illustrate the close association between soil types and topography, as well as parent materials, hydrology, and vegetations.

Catenary soil distribution often occurs in response to the way water runs down a hillslope and is a manifestation of the interrelationship between soils and geomorphic processes (Hall and Olson, 1991; Moore et al., 1993; Thompson et al., 1997; Lin et al., 2005a,b). Catenae are also called toposequences or hydrosequences, especially in depositional landscapes, where hydrology exerts a more dominant control on soil distribution pattern. The real significance of catenae lies in the recognition of linked soil and geomorphic processes, especially those driven by water movement downslope. Fluxes of water, solutes, and sediments, and water-table depth (drainage condition) typically differ in soils along a catena (Sommer and Schlichting, 1997; Schaetzl and Anderson, 2005).

Catenae have been recognized in a variety of areas under different climatic conditions and have played an important role in soil, landform, and hydrology studies (e.g. Gerrard, 1981; Daniels and Hammer, 1992; Jenkinson et al., 2002; Reuter and Bell, 2003). Sommer and Schlichting (1997) distinguished three archetypes of catenae based on mass balance and hydrologic regime: 1) *transformation catenae*: only transformation processes occur, showing no gains or losses of soil component; 2) *leaching catenae*: with losses in at least part of the catena and no accompanying elemental gains in other parts; and 3) *accumulation catenae*: showing gains in at least part of the catena but no losses elsewhere in the catena. However, catenae in different environments may exhibit markedly different relationships between soil, topographic position, and hydrologic property. Below are three pairs of catenae that illustrate this point (Fig. 10):

- Opposite trends of soil thickness and soil wetness distribution along catenae are illustrated in Fig. 10a and b. In the humid forested catchment developed from shale in the Ridge and Valley physiographic region of the USA, soil thickness and wetness generally increase downslope from the hilltop to the valley floor in convergent hillslopes (Fig. 10a). In contrast, in the rugged Hill Country of central Texas with stair-step topography developed from limestone, soil thickness and wetness decrease from the hilltop (upper riser

with steeper slopes) to the hill bottom (tread with flatter slope) (Fig. 10b). Chapters 9 and 13 of this book provide further details of these two landscapes.
- In many low-relief landscapes of humid regions, proximity of a water table to the soil surface increases with distance away from stream or drainage way. An example is the broad, flat, low-relief Atlantic coastal plain in the southeastern USA. Poorly-drained soils are found toward the centers of the broad interstream divides, while well-drained soils are restricted to the edges of the flats and slopes closest to the streams and estuaries (Daniels et al., 1971, 1984). The opposite trend is seen in the higher-relief landscapes associated with the nearby Piedmont region. Here, the water-table's proximity to the soil surface decreases with increasing distance away from the streams or drainage ways. As a result, well-drained soils occupy the uplands and poorly-drained soils occupy the lower slope positions near the streams (Daniels et al., 1984).
- Contrasting soil distribution patterns can occur in catenae depending on local hydrology being controlled by surface or subsurface topography (Fig. 10c and d). A common soil catena along an eroding hillslope in Australia is shown in Fig. 10c. This catena can be explained by the classical catenary model where surface topography controls the hydrologic regime (from well-drained to poorly-drained soils along downslope gradient). In contrast, the topography of an underlying weathered rock substrate with low-permeability (shown in Fig. 10d) controls the subsurface hydrology and soil distribution in another Australian landscape, resulting in poorly-drained to well-drained soil distribution along downslope gradient (Coventry, 1982).

Regional soil patterns are controlled by gradients or zones of dominant soil-forming factors. In early soil classification systems (e.g. Baldwin et al., 1938), zonal, azonal, and intrazonal soils were used to classify soils based on climatic zones. Zonal soils, including pedocals (soils of dry regions with Ca accumulation) and pedalfers (soils of humid regions with Al and Fe enriched in subsoils), are associated with distinct climatic and vegetative zones and thus have broad geographic distributions, such as Tundra soils, Desert soils, Brown soils, Chernozem soils, Prairie soils, Podzol soils, Laterite soils, and others (Baldwin et al., 1938). In contrast, intrazonal soils are those dominated by local or particular factors of soil formation such as salts (e.g. Solonetz soils) or high water tables (e.g. Bog soils), whereas azonal soils refer to those with weakly developed or incomplete profiles (i.e. young soils), such as Alluvial soils and Dry sands. All of these zonal, azonal, and intrazonal soils have very different architectures.

Strakhov (1967) and Schultz (2005) presented a general global pattern of soil weathering profile from the pole to the equator (Fig. 11), showing a transect across major climatic zones (aboveground energy, water, and biomass inputs) and associated soil thickness and layering features (belowground mineral weathering and carbon accumulation). These broad zonal patterns also correspond to terrestrial ecozones (Schultz, 2005). While generally correct, this simplified trend of weathering and soil architecture across the globe may create misconceptions, which ignore various local variability, climate change, and denudation rate adjustment (Schaetzl and Anderson, 2005).

The classical soil-forming factorial model (Jenny, 1941) has suggested possible soil sequences resulting from the gradual changes in a dominant soil-forming factor over areas of different sizes, giving rise to climo-, bio-, litho-, topo-, and chrono-sequences of soils. These soil sequences have been widely studied and have contributed notably to the understanding of soil distribution patterns (Amundson et al., 1994; Buol et al., 2011). With the exception of toposequences and lithosequences at the local level, grouping individual soils into sequences over large areas have also produced questions or even doubts (for a comprehensive discussion on this, see Schaetzl and Anderson, 2005). The main challenge here is the ecological adage that "everything affects everything else" in the natural environment, which rules out singular factors and requires a holistic study of soil systems. Chronosequence (assumes comparable pedogenic changes through time) is perhaps the most common of the sequences derived from the soil-forming factorial model. However, a number of issues require carefully-chosen soil sequence for such comparison, including the determination of time zero, possible polygenetic pathways, spatial heterogeneity of other soil-forming factors, sensitivity to initial condition, and certain level of constancy of other factors (Schaetzl and Anderson, 2005; Lin, 2011). Both climosequence and biosequence are difficult to isolate (Schaetzl and Anderson, 2005); thus, climo-biosequence is much more common and easier to find, which is linked to soil zones discussed above in association with particular climatic and vegetative zones.

4. CONCLUDING REMARKS

A new era of soils research is needed to focus on soil architecture across scales and its quantitative links to soil functions. Toward this end, several needs are suggested below based on the review and synthesis provided in this chapter:

1) *Conceptual advance*: It is important to recognize that soils distributed across the land are unlike the atmosphere or the ocean (which may be mixed readily and a continuum approach is suited to model its fluid dynamics). Real-world soils exhibit clear architectural organizations that control the pathways and patterns of mass and energy flow and transport at different scales. Thus, a combination of deterministic and stochastic approaches would be a more appropriate way to couple structural and random components of the soil system. Another conceptual advance needed is to improve the holistic understanding of soils as a complex system through coupled processes and interactions rather than isolated individual factors.
2) *Scaling framework*: The soil–landscape relationship is a key to link soil architecture within a profile and across a landscape, which impacts the scaling of soil processes and properties. The complexity of interwoven

regional soil-forming environment and local soil variability needs to be better conveyed and quantified in soil maps and databases. The co-evolution of fast and slow processes in the multiphase soil system also needs to be better incorporated into models and scaling.

3) *Technological breakthrough*: Innovative quantification of soil architecture and its impacts on soil functions is a key to advance hydropedology. A bottleneck exists for in situ high resolution (e.g. submeter to cm) and noninvasive mapping/imaging of soil architecture. Geophysical and remote-sensing methods are promising, but their spatial–temporal resolutions for subsurface investigations and related inversion algorithms need to be improved. It is hopeful that, in not too distant future, more powerful field tools and techniques will permit a breakthrough understanding of the opaque soil system.
4) *Quantitative formulation*: A significant gap remains between soil architectural parameters and field-measured soil functions at different scales. Pore space is a new arena for pedologic research, as it emphasizes the nature, origin, and significance of pore space in various soil functions. However, how to link pedologic features, pore characteristics, and soil functions at different scales is a challenging and yet critical question to be answered. The link between soil maps and soil functions is also needed, especially quantitatively, but it is largely missing.
5) *Co-evolution of soil–water in the anthropocene*: Water plays a significant role in forming peds, horizons, pedons, catenae, and soil zones, which in turn impact hydrologic processes. The co-evolution of soil and water and the related life factors is essential to hydropedology. Humans have significant roles in impacting this co-evolution and soil architecture, both constructively and destructively, but their damage is a severe form of soil degradation. Good soil aggregates, horizons, and profiles take a long time to form, but can be destroyed or eliminated quickly by compaction, tillage, erosion, pavement, or other forms of anthropogenic soil destruction. Thus, protecting desirable soil architecture is a key to sustainable soil resources in the anthropocene.

The emerging global network of Critical Zone Observatories (Lin et al., 2011) would be good locations where some of the above suggestions could be tried out. Critical Zone Observatories represent an opportunity for significant advances across soil science, hydrology, geomorphology, geophysics, biogeochemistry, and other disciplines active in terrestrial ecosystem research. Soil architecture and soil functions are essential to these interdisciplinary and multiscale studies.

ACKNOWLEDGMENTS

The author thanks Drs. Richard Arnold, Johan Bouma, and James Doolittle for their helpful review comments that have improved the quality of this paper.

REFERENCES

Addiscott, T.M., 1995. Entropy and sustainability. Eur. J. Soil Sci. 46, 161–168.

Amundson, R., Harden, J., Singer, M., 1994. Factors of Soil Formation: A Fiftieth Anniversary Retrospective. SSSA Special Publication #33. Soil Sci. Soc. Am., Inc., Madison, WI.

Anderson, A., McBratney, A.B., Crawford, J.W., 1998. Applications of fractals to soil studies. Adv. Agron. 63, 1–76.

Anderson, J.L., Bouma, J., 1973. Relationships between hydraulic conductivity and morphometric data of an argillic horizon. Soil Sci. Soc. Am. Proc. 37, 408–413.

Anderson, S.H., Peyton, R.L., Gantzer, C.J., 1990. Evaluation of constructed and natural soil macropores using X-Ray computed-tomography. Geoderma 46, 13–29.

Arnold, R.W., Wilding, L.P., 1991. The need to quantify spatial variability. In: Mausbach, M.J., Wilding, L.P. (Eds.), Spatial Variabilities of Soils and Landforms. SSSA Special Pub. #28. Soil Sci. Soc. Am., Inc., Madison, WI, pp. 1–8.

Baldwin, M., Kellogg, C.E., Thorp, J., 1938. Soil classification. In: Yearbook of Agriculture: Soils and Men. U.S. Dept. Agri., U.S. Fovt. Printing Office, Washington, DC, pp. 979–1001.

Bartoli, F., Dutartre, Ph., Gomendy, V., Niquet, S., Dubuit, M., Vivier, H., 1998. Fractals and soils structure. In: Baveye, P., Parlange, J.-Y., Stewart, B.A. (Eds.), Fractals in Soil Science. Advances in Soil Science. CRC Press, Boca Raton, FL, pp. 203–232.

Bastardie, F., Capowiez, Y., Dreuzy, J.-R., Cluzeau, D., 2003. X-ray tomographic and hydraulic characterization of burrowing by three earthworm species in repacked soil cores. Appl. Soil Ecol. 24, 3–16.

Baveye, P., 2006. Comment on "Soil structure and management: a review" by C.J. Bronick and R. Lal. Geoderma 134, 231–232.

Baveye, P., Boast, C.W., 1998. Concepts of "fractals" in soil science: demixing apples and oranges. Soil Sci. Soc. Am. J. 62, 1469.

Baveye, P., Parlange, J.-Y., Stewart, B.A., 1999. Fractals in Soil Science. Advances in Soil Science. CRC Press, Boca Raton, FL.

Bell, J.C., Cunningham, R.L., Havens, M.W., 1992. Calibration and validation of a soil landscape model for predicting soil drainage class. Soil Sci. Soc. Am. J. 56, 1860–1866.

Beven, K.J., Kirkby, M.J., 1979. A physically based variable contributing area model of basin hydrology. Hydrol. Sci. Bull. 24, 43–69.

Bouma, J., 1992. Effects of soil structure, tillage, and aggregation upon soil hydraulic properties. In: Wagenet, R.J., Bavege, P., Stewart, B.A. (Eds.), Interacting Processes in Soil Science. Lewis Publishers, Boca Raton, FL, pp. 1–36.

Bouma, J., 2010. Implications of the knowledge paradox for soil science. Adv. Agron. 106, 143–171.

Bouma, J., Jongerius, A., Schoonderbeek, D., 1979. Calculation of saturated hydraulic conductivity of some pedal clay soils using micromorphometric data. Soil Sci. Soc. Am. J. 43, 261–264.

Bouma, J., Droogers, P., Sonneveld, M.P.W., Ritsema, C.J., Hunink, J.E., Immerzeel, W.W., Kauffman, S., 2011. Hydropedological insights when considering catchment classification. Hydrol. Earth Syst. Sci. 15, 1909–1919.

Brady, N.C., Weil, R.R., 2004. Elements of the nature and properties of soils, second ed. Pearson-Prentice Hall, Upper Saddle River, NJ.

Brewer, R., 1964. Fabric and Mineral Analysis of Soils. Robert E. Krieger Publishing Company, Huntington, NY.

Bronick, C.J., Lal, R., 2005. Soil structure and management: a review. Geoderma 124, 3–22.

Brooks, E.S., Boll, J., McDaniel, P.A., 2004. A hillslope-scale experiment to measure lateral saturated hydraulic conductivity. Water Resour. Res. 40, W04208. doi:10.1029/2003WR002858.

Brooks, E.S., Boll, J., McDaniel, P.A., 2007. Distributed and integrated response of a GIS-based hydrologic model in the eastern Palouse region, Idaho. Hydrol. Process. 21, 110–122.

Buchan, G.D., Flury, M., 2007. Pathogens: transport by water. In: Trimble, S.W. (Ed.), Encyclopedia of Water Science, second ed. Taylor and Francis Group LLC, New York.

Buol, S.W., Southard, R.J., Graham, R.C., McDaniel, P.A., 2011. Soil Genesis and Classification, sixth ed. Iowa State University Press, Ames. IA.

Burrough, P., McDonnell, R., 1998. Principles of Geographic Information Systems. Oxford.

Bushnell, T.M., 1943. Some aspects of the soil catena concept. Soil Sci. Soc. Amer. Proc. 7, 466–476.

Capowiez, Y., Pierret, A., Daniel, O., Monestiez, P., Kretzschmar, A., 1998. 3D skeleton reconstructions of natural earthworm burrow systems using CAT scan images of soil cores. Biol. Fertil. Soils 27, 51–59.

Commission of the European Community (CEC), 2006. Directive of the European Parliament and of the Council, establishing a framework for the protection of soil and amending Directive 2004/35/EC. Brussels.

Coen, G.M., Wang, C., 1989. Estimating vertical saturated hydraulic conductivity from soil morphology in Alberta. Can. J. Soil Sci. 69, 1–16.

Coventry, R.J., 1982. The distribution of red, yellow and grey Earths in the Torrens Creek area, central North Queensland. Aust. J. Soil Res. 20, 1–14.

Cushman, J.H., 1990. An introduction to hierarchical porous media. In: Cushman, J.H. (Ed.), Dynamics of Fluids in Hierarchical Porous Media. Academic Press, London, pp. 1–6.

Daniels, R.B., 1988. Pedology, a field or laboratory science? Soil Sci. Soc. Am. J. 52, 1518–1519.

Daniels, R.B., Hammer, R.D., 1992. Soil Geomorphology. John Wiley, New York.

Daniels, R.B., Gamble, E.E., Nelson, L.A., 1971. Relations between soil morphology and water-table levels on a dissected North Carolina coastal plain surface. Soil Sci. Soc. Am. Proc. 35, 781–784.

Daniels, R.B., Kleiss, H.J., Buol, S.W., Byrd, H.J., Phillips, J.A., 1984. Soil Systems in North Carolina. Bulletin 467. North Carolina Agricultural Research Service, Raleigh, North Carolina.

Flühler, H., Durner, W., Flury, M., 1996. Lateral solute mixing processes – a key for understanding field-scale transport of water and solutes. Geoderma 70, 165–183.

Forman, R.T.T., 1995. Land Mosaics. The Ecology of Landscapes and Regions. Cambridge University Press, Cambridge, UK.

Fridland, V.M., 1976. Pattern of the Soil Cover. Israel Program for Scientific Translation. Keter Publishing House, Jerusalem Ltd.

Gerrard, A.J., 1981. Soils and Landforms – An Integration of Geomorphology and Pedology. George Allen & Unwin, London.

Gessler, P.E., Chadwick, O.A., Chamran, F., Althouse, L., Holmes, K., 2000. Modeling soil–landscape and ecosystem properties using terrain attributes. Soil Sci. Soc. Am. J. 64, 2046–2056.

Gessler, P.E., Moore, I.D., McKenzie, N.J., Ryan, P.J., 1995. Soil landscape modeling and spatial prediction of soil attributes. Int. J. Geogr. Inf. Syst. 9, 421–432.

Gibson, J.R., Lin, H., Bruns, M.A., 2006. A comparison of fractal analytical methods on 2-and 3-dimensional computed tomographic scans of soil aggregates. Geoderma 134, 335–348.

Gobin, A., Campling, P., Feyen, J., 2001. Soil-landscape modeling to quantify spatial variability of soil texture. Phys. Chem. Earth B 2, 41–45.

Goryachkin, S.V., 2005. Studies of the soil cover patterns in modern soil science: approaches and tendencies. Eurasian Soil Sci. 38, 1301–1308.

Hall, G.F., Olson, C.G., 1991. Predicting variability of soils from landscape models. In: Mausbach, M.J., Wilding, L.P. (Eds.), Spatial Variabilities of Soils and Landforms. Soil Science Society of America, Madison, WI, pp. 9–24. Special Publication 28.

Hasegawa, S., Warkentin, B.P., 2006. The changing understanding of physical properties of soils: water flow, soil architecture. In: Warkentin, B.P. (Ed.), Footprints in the Soil. People and Ideas in Soil History. Elsevier, Amsterdam, The Netherlands, pp. 339–366.

Heuvelink, G.B.M., Webster, R., 2001. Modelling soil variation: past, present, and future. Geoderma 100, 269–301.

Hillel, D., 1998. Environmental Soil Physics. Academic Press, NY.

Hodgson, J.M., 1997. Soil Survey Field Handbook: Describing and Sampling Soil Profiles, third ed. Soil Survey Tech. Monograph #5, Soil Survey and Land Research Centre, Silsoe, England.

Hole, F.D., 1961. A classification of pedoturbations and some other processes and factors of soil formation in relation to isotropism and anisotropism. Soil Sci. 91, 375–377.

Hole, F.D., 1976. Soils of Wisconsin. The University of Wisconsin Press, Madison, WI.

Hudson, B.D., 1992. The soil survey as paradigm-based science. Soil Sci. Soc. Am. J. 56, 836–841.

Hupy, C.M., Schaetzl, R.J., Messina, J.P., Hupy, J.P., Delamater, P., Enander, H., Hughey, B.D., Boehm, R., Mitroka, M.J., Fashoway, M.T., 2004. Modeling the complexity of different, recently deglaciated soil landscapes as a function of map scale. Geoderma 123, 115–130.

Jacob, J., Nordt, L., 1991. Soil and landscape evolution: a paradigm for pedology. Soil Sci. Soc. Am. J. 55, 1194.

Jenkinson, B.J., Franzmeier, D.P., Lynn, W.C., 2002. Soil hydrology on an end moraine and a dissected till plain in west-central Indiana. Soil Sci. Soc. Am. J. 66, 1367–1376.

Jenny, H., 1941. Factors of Soil Formation—A System of Quantitative Pedology. McGraw-Hill, NY.

Johnson, D.L., Keller, E.A., Rockwell, T.K., Dembroff, G.R., 1990. Dynamic pedogenesis – new views on some key soil concepts, and a model for interpreting quaternary soils. Quatern. Res. 33, 306–319.

Johnson, D.L., Watson-Stegner, D., 1987. Evolution model of pedogenesis. Soil Sci. 143, 349–366.

Jury, W.A., Or, D., Pachepsky, Y., Vereecken, H., Hopmans, J.W., Ahuja, L.R., Clothier, B.E., Bristow, K.L., Kluitenberg, G.J., Moldrup, P., Simunek, J., van Genuchten, M. Th., Horton, R., 2011. Kirkham's Legacy and contemporary challenges in soil physics research. Soil Sci. Soc. Am. J. 75, 1589–1601.

Jury, W.A., Horton, R., 2004. Soil Physics, sixth ed. Wiley.

Kendall, K.A., Shanley, J.B., McDonnell, J.J., 1999. A hydrometric and geochemical approach to test the transmissivity feedback hypothesis during snowmelt. J. Hydrol. 219, 188–205.

Kim, J.W., Perfect, E., Choi, H., 2007. Anomalous diffusion in two-dimensional Euclidean and prefractal geometrical models of heterogeneous porous media. Water Resour. Res. 43, W01405. doi:101029/2006WR004951.

Lagacherie, P., McBratney, A.B., Voltz, M. (Eds.), 2007, Digital Soil Mapping. An Introductory Perspective. Developments in Soil Science, vol. 31. Elsevier, Amsterdam, The Netherlands.

Lauren, J.G., Wagenet, R.J., Bouma, J., Wosten, J.H.M., 1988. Variability of saturated hydraulic conductivity in a Glossaquic Hapludalf with macropores. Soil Sci. 145, 20–28.

Letey, J., 1991. The study of soil structure—Science or art. Aust. J. Soil Res. 29, 699–707.

Lin, H.S., 2010. Linking principles of soil formation and flow regimes. J. Hydrol. 393, 3–19.

Lin, H.S., 2011. Understanding soil change across time scales. Soil Sci. Soc. Am. J. 75, 2049–2070.

Lin, H.S., Rathbun, S., 2003. Hierarchical frameworks for multiscale bridging in hydropedology. In: Pachepsky, Y., Radcliffe, D., Selim (Eds.), Scaling Methods in Soil Physics. CRC Press, Boca Raton, FL, pp. 353–371.

Lin, H.S., Bouma, J., Wilding, L., Richardson, J., Kutilek, M., Nielsen, D., 2005a. Advances in hydropedology. Adv. Agron. 85, 1–89.

Lin, H.S., Wheeler, D., Bell, J., Wilding, L., 2005b. Assessment of soil spatial variability at multiple scales. Ecol. Model. 182, 271–290.

Lin, H.S., Bouma, J., Pachepsky, Y., Western, A., Thompson, J., van Genuchten, M. Th., Vogel, H., Lilly, A., 2006. Hydropedology: synergistic integration of pedology and hydrology. Water Resour. Res. 42, W05301. doi:10.1029/2005WR004085.

Lin, H.S., Brook, E., McDaniel, P., Boll, J., 2008. Hydropedology and Surface/Subsurface runoff processes. In: Anderson, M.G. (Ed.), Encyclopedia of Hydrologic Sciences. John Wiley & Sons, Ltd DOI: 10.1002/0470848944.hsa306.

Lin, H.S., Hopmans, J., Richter D., 2011. Interdisciplinary sciences in a global network of Critical Zone Observatories. Vadose Zone J. 10, 781–785.

Luo, L.F., Lin, H.S., 2009. Lacunarity and fractal analyses of soil macropores and preferential transport using Micro X-ray computed tomography. Vadose Zone J. 8, 233–241.

Luo, L.F., Lin., H.S., Hellack, P., 2008. Quantifying soil structure and preferential flow in intact soil using X-ray computed tomography. Soil Sci. Soc. Am. J. 72, 1058–1069.

Luo, L.F., Lin, H.S., Li, S.C., 2010a. Quantification of 3-D soil macropore networks in different soil types and land uses. J. Hydrol. 393, 53–64.

Luo, L.F., Lin, H.S., Schmidt, J., 2010b. Quantitative relationships between soil macropore characteristics and preferential flow and transport. Soil Sci. Soc. Am. J. 74, 1929–1937.

Mandelbrot, B.B., 1982. The Fractal Geometry of Nature. W. H. Freeman and Company, New York.

McBratney, A.B., Santos, M.L.M., Minasny, B., 2003. On digital soil mapping. Geoderma 117, 3–52.

McDonnell, J.J., 2003. Where does water go when it rains? Moving beyond the variable source area concept of rainfall-runoff response. Hydrol. Process. 17, 1869–1875.

McKenzie, N., Jacquier, D., Isbell, R., Brown, K., 2004. Australian Soils and Landscapes: An Illustrated Compendium. CSIRO Publishing, Collingwood.

McKenzie, N.J., Ryan, P.J., 1999. Spatial prediction of soil properties busing environmental correlation. Geoderma 89, 67–94.

Milne, G., 1936a. Normal erosion as a factor in soil profile development. Nature 138, 548–549.

Milne, G. 1936b. A provisional soil map of East Africa. East Africa Agric. Res. Stn., Amani Memoirs, No. 34.

Moore, I.D., Gessler, P.E., Nielsen, G.A., Peterson, G.A., 1993. Soil attribute prediction using terrain analysis. Soil Sci. Soc. Am. J. 57, 443–452.

Nielsen, D.R., Wendroth, O., 2003. Spatial and Temporal Statistics: Sampling Field Soils and Their Vegetation. Catena Verlag, Cremlingen-Destedt.

Pachepsky, Y.A., Radcliffe, D.E., Selim, H.M. (Eds.), 2003. Scaling Methods in Soil Physics. CRC Press, Boca Raton, FL.

Pachepsky, Y.A., Crawford, J.C., Rawls, W.J. (Eds.), 2000. Fractals in Soil Science. Developments in Soil Sci, vol. 27. Elsevier, Amsterdam, The Netherlands.

Park, S.J., McSweeney, K., Lowery, B., 2001. Identification of the spatial distribution of soils using a process-based terrain characterization. Geoderma 103, 249–272.

Pendleton, D.E., Dathe, A., Baveye, P., 2005. Influence of image resolution and evaluation algorithm on estimates of the lacunarity of porous media. Phys. Rev. E 72 (4) Art. No.041306.

Perret, J., Prasher, S.O., Kantzas, A., Langford, C., 1999. Three-dimensional quantification of macropore networks in undisturbed soil cores. Soil Sci. Soc. Am. J. 63, 1530–1543.

Perret, J., Prasher, S.O., Kantzas, A., Langford, C., 2000. A two-domain approach using CAT scanning to model solute transport in soil. J. Environ. Qual. 29, 995–1010.

Perrier, E., Bird, N., Rieu, M., 1999. Generalizing the fractal model of soil structure: the pore-solid fractal approach. Geoderma 88, 137–164.

Phillips, J.D., 1993. Stability implications of the state factor model of soils as a nonlinear dynamical system. Geoderma 58, 1–15.

Pierret, A., Capowiez, Y., Belzunces, L., 2002. 3D reconstruction and quantification of macropores using X-ray computed tomography and image analysis. Geoderma 106, 247–271.

Plotnick, R.E., Gardner, R.H., Oneill, R.V., 1993. Lacunarity indexes as measures of landscape texture. Landscape Ecol. 8, 201–211.

Prigogine, I., 1967. Introduction to Thermodynamics of Irreversible Processes, third ed. John Wiley, New York, NY.

Rasmussen, C., Southard, R.J., Horwath, W.J., 2005. Modeling energy inputs to predict pedogenic environments using regional environmental databases. Soil Sci. Soc. Am. J. 69, 1266–1274.

Reuter, R.J., Bell, J.C., 2003. Hillslope hydrology and soil morphology for a wetland basin in south-central Minnesota. Soil Sci. Soc. Am. J. 67, 365–372.

Runge, E.C.A., 1973. Soil development sequences and energy models. Soil Sci. 115, 183–193.

Runge, E.C.A., Riecken, F.F., 1966. Influence of natural drainage on distribution and forms of phosphorus in some Iowa prairie soils. Soil Sci. Soc. Am. Proc. 30, 624.

Russell, E.J., 1912. Soil Conditions and Plant Growth. Longmans Green, London.

Schaetzl, R., Anderson, S. 2005. Soils – Genesis and Geomorphology. Cambridge, UK.

Schultz, J., 2005. The Ecozones of the World. The Ecological Divisions of the Geosphere, second ed. Springer, Berlin.

Serra, J., 1982. Image Analysis and Mathematical Morphology. Academic Press, London.

Smeck, N.E., Runge, E.C.A., Mackintosh, E.E., 1983. Dynamics and genetic modeling of soil systems. In: Wilding, L.P., et al. (Eds.), Pedogenesis and Soil Taxonomy. Elsevier, New York, pp. 51–81.

Schmidt, K., Behrens, T., Friedrich, K., Scholten, T., 2010. A method to generate soilscapes from soil maps. J. Plant Nutr. Soil Sci. 173, 163–172.

Soil Science Society of America (SSSA), 2001. Glossary of Soil Science Terms. Soil Sci. Soc. Am., Madison, WI.

Soil Survey Staff, 1993. Soil Survey Manual. U.S. Dept. Agri. Handbook No. 18, U.S. Government Printing Office, Washington, DC.

Soil Survey Staff, 2010. Keys to Soil Taxonomy, eleventh ed. United States Department of Agriculture – Natural Resources Conservation Service, Washington, DC.

Sommer, M., 2006. Influence of soil pattern on matter transport in and from terrestrial biogeosystems—A new concept for landscape pedology. Geoderma 133, 107–123.

Sommer, M., Schlichting, E., 1997. Archetypes of catenae in respect to matter a concept for structuring and grouping catenae. Geoderma 76, 1–33.

Strakhov, N.M., 1967. Principles of Lithogenesis. Oliver and Boyd, Edinburgh, UK.

Targulian, V.O., 2005. Elementary pedogenic processes. Eurasian Soil Sci. 38, 1255–1264.

Targulian, V.O., Sokolova, T.A., 1996. Soil as a bio-abiotic natural system; a reactor, memory and regulator of biospheric interactions. Eurasian Soil Sci. 29, 34–47.

Terribile, F., Coppola, A., Langella, G., Martina, M., Basile, A., 2011. Potential and limitations of using soil mapping information to understand landscape hydrology. Hydrol. Earth Syst. Sci. 15, 3895–3933.

Thompson, J.A., Bell, J.C., Butler, C.A., 1997. Quantitative soil-landscape modeling for estimating the areal extent of hydromorphic soils. Soil Sci. Soc. Am. J. 61, 971–980.

Tisdall, J.M., Oades, J.M., 1982. Organic matter and water-stable aggregates in soils. J. Soil Sci. 33, 141–163.

Uuemaa, E., Roosaare, J., Kanal, A., Mander, U., 2008. Spatial correlograms of soil cover as an indicator of landscape heterogeneity. Ecol. Indicat. 8, 783–794.

Vogel, H.J., Roth, K., 2003. Moving through scales of flow and transport in soil. J. Hydrol. 272, 95–106.

Vogel, H.J., Cousin, I., Roth, K., 2002. Quantification of pore structure and gas diffusion as a function of scale. Eur. J. Soil Sci. 53, 465–473.

Volobuyev, V.R., 1963. Ecology of Soils. Academy of Sciences of the Azerbaidzan SSR. Institute of Soil Science and Agrochemistry. Israel Program for Scientific Translations, Jerusalem (Translated into English by A. Gourevich, 1964.).

Warkentin, B.P., 2008. Soil structure: a history from tilth to habitat. Adv. Agron. 97, 239–272.

Weiler, M., McDonnell, J.J., Tromp van Meerveld, I., Uchida, T., 2005. Subsurface Stormflow runoff Generation processes. In: Anderson, M.G. (Ed.), Encyclopedia of Hydrological Sciences. Wiley, pp. 1719–1732.

West, L.T., Abreu, M.A., Bishop, J.P., 2008. Saturated hydraulic conductivity of soils in the Southern Piedmont of Georgia, USA: field evaluation and relation to horizon and landscape properties. Catena 73, 174–179.

Wilding, L.P., 2000. Pedology. In: Sumner, M.E. (Ed.), Handbook of Soil Science. CRC Press, Boca Raton, FL, pp. E-1–E-4.

Wilding, L.P., 2007. Hydrological attributes and treatment capabilities of "caliche and related soils" pertinent to on-site sewage facilities. Texas On-site Wastewater Treatment Research Council Final Report, 1–77. Contract number 582-3-55760.

World Reference Base (WRB), 2001. World Soil Resources Reports 94. Lectures Notes on the Major Soils of the World. In: Driessen, P., Deckers, J., Spaargaren, O., Nachtergaele, F. (Eds.). Food and Agriculture Organization of the United Nations, Rome, p. 334.

Wright, S.F., FrankeSnyder, M., Morton, J.B., Upadhyaya, A., 1996. Time-course study and partial characterization of a protein on hyphae of arbuscular mycorrhizal fungi during active colonization of roots. Plant Soil 181, 193–203.

Wysocki, D., Schoeneberger, P., LaGarry, H., 2000. Geomorphology of soil landscapes. In: Sumner, M.E. (Ed.), Handbook of Soil Science. CRC Press, Boca Raton, FL, pp. E-5–E-39.

Yaalon, D.H., 1971. Soil-forming processes in time and space. In: Yaalon, D.H. (Ed.), Paleopedology: Origin, Nature and Dating of Paleosols. Halsted Press, A division of John Wiley & Sons, Inc, New York, NY, pp. 29–39.

Young, I.M., Crawford, J.W., Rappoldt, C., 2001. New methods and models for characterising structural heterogeneity of soil. Soil Till. Res. 61, 33–45.

Zeng, Y., Gantzer, C.J., Payton, R.L., Anderson, S.H., 1996. Fractal dimension and lacunarity of bulk density determined with X-ray computed tomography. Soil Sci. Soc. Am. J. 60, 18–1724.

Zhu, A.X., Hudson, B., Burt, J., Lubich, K., Simonson, D., 2001. Soil mapping using GIS, expert knowledge, and fuzzy logic. Soil Sci. Soc. Am. J. 65, 1463–1472.

Ziadat, F., 2005. Analyzing digital terrain attributes to predict soil attributes for a relatively large area. Soil Sci. Soc. Am. J. 69, 1590–1599.

Chapter 3

Preferential Flow in a Pedological Perspective

Nicholas J. Jarvis,[1,*] Julien Moeys,[1] John Koestel[1] and John M. Hollis[2]

ABSTRACT

Preferential flow is often thought of as a spatially and temporally highly variable and random process and, as such, essentially unpredictable. This chapter discusses the experimental evidence accumulated during the last 20–30 years that suggests that such a view is too pessimistic. The chapter first presents a brief overview of methods to quantify preferential flow in soil and the characteristics of pore systems that control it. The influence of basic soil properties and site attributes on soil structure and preferential flow is then discussed. This is followed by an overview of the dominant flow and transport regimes found in the major soil types of the world, which is based on a comprehensive review and synthesis of the extensive literature. In the context of predictive applications of preferential flow models, the use of both continuous and class pedotransfer functions is then discussed. This is followed by a brief presentation of case studies dealing with the development and application of pesticide risk assessment and management tools that employ hydropedological concepts and approaches to account for macropore flow impacts at catchment and regional scales. The chapter concludes with a discussion of uncertainty and how to deal with it, and also makes some recommendations for future research directions that can help fill some of the remaining knowledge gaps.

1. INTRODUCTION

Preferential flow and transport occurs in the soil unsaturated zone when vertical velocities in a small fraction of the total pore volume are much faster than rates of lateral equilibration of water pressures and/or solute concentrations with the surrounding water-filled pores (Beven and Germann, 1982; Flühler et al., 1996). This physical nonequilibrium occurs at two distinct scales: (i) heterogeneous, funnel, or finger flow in soil matrix pores at the Darcy or pedon scale

[1]Department of Soil & Environment, SLU, Box 7014, 750 07 Uppsala, Sweden
[2]58, St. Annes Rd. London Colney, St. Albans, Herts. AL2 1LJ, UK
[*]Corresponding author: Email: nicholas.jarvis@slu.se

Hydropedology, Edited by H. Lin. DOI: 10.1016/B978-0-12-386941-8.00003-4

due to macroscopic differences in hydraulic characteristics resulting, for example, from spatial variation in texture (e.g. depositional lenses, layers, and bedding planes, e.g. Kung, 1990), bulk density, the presence of large stones and rocks (Bogner et al., 2008a) or water repellency (e.g. Ritsema et al., 1993) and (ii) flow in large, continuous, structural pores (macropores) induced by an extreme discontinuity in permeability at the pore scale (Jarvis, 2007a). Three distinct types of macropores are recognized: biopores formed by plant roots and macrofauna, fissures created by swell/shrink or freeze/thaw, and irregular inter-aggregate voids created by tillage implements.

Qualitatively accurate descriptions of rapid nonequilibrium water flow in structured soil were reported as long ago as the late-19th century (Lawes et al., 1882). However, it was first in the 1960s and 1970s (Thomas & Phillips, 1979; Beven and Germann, 1982) that experimental observations of preferential flow and the resulting effects on hydrologic response and patterns of solute displacement began to challenge the established paradigm that water flow in soil is uniform. The often dramatic effects of preferential flow on contaminant leaching coupled with the growing realization that it is a widespread phenomenon have stimulated the development of many models that can account for nonequilibrium flow and transport. Initially, these models were developed as research tools with a view to further understand the processes, so little attention was paid to how their parameters could be estimated. Model applications have mostly been restricted to column and plot experiments at well-investigated sites, where input parameters can be derived by a combination of direct measurements and calibration (Köhne et al., 2009a,b). However, many public authorities and stakeholders need models and decision-support tools that can account for the effects of preferential flow at the larger scales (e.g. farms, catchments, or even regions) that are relevant for management (Vanclooster et al., 2004). Preferential flow models must then be used to make predictions, for example, of contaminant fluxes associated with groundwater recharge or discharge to surface waters, without direct measurements of input parameters or site data for calibration. The question is: can this be done with acceptable uncertainty?

This is certainly an inherently difficult task. Macropore flow is highly sensitive to initial and boundary conditions (Kluitenberg and Horton, 1990; Zehe and Blöschl, 2004; McGrath et al. 2009; Larsbo, 2011) and the geometry and properties of conducting macropores are also 'a priori' unknown and are known to be highly variable, both spatially and temporally (Jury and Flühler, 1992). Indeed, many researchers have questioned whether macropore flow is an essentially random and unpredictable process (e.g. Beven, 1991). However, we do not need to know the characteristics and behavior of individual macropores, only their spatially integrated effects on flow and transport at larger scales. In this respect, experimental evidence accumulated during the last 30 years suggests that macropore flow effects on solute transport at the soil horizon and pedon scale can often be explained in terms of measurable soil properties or observable soil morphology (e.g. Bouma and Wösten, 1979; Flury et al., 1994;

Vervoort et al., 1999; Vanderborght et al., 2001; Ersahin et al., 2002; Roulier and Jarvis, 2003; Akhtar et al., 2003; Mooney and Morris, 2008). One qualitative example is shown in Fig. 1, which shows the dye staining pattern obtained on a silty clay soil in central Sweden (Haplic Cambisol, IUSS, 2006), which has been under unmanaged grass for at least 50 years (Jarvis et al., 2010). Uniform flow was observed in the *Ah* horizon, which is rich in organic matter and has a fine crumb structure induced by earthworm bioturbation. Preferential flow, mostly along aggregate faces, but also in a few earthworm channels, was generated at the interface with the *Bw* horizon, which has a strong blocky structure.

The second scale gap (pedon to landscape) can only be bridged if the distribution of pedons is sufficiently predictable across the landscape and internal variation in their functional response is small compared to the differences between them (Webster, 2000; Heuvelink and Webster, 2001). In the context of solute transport, Addiscott and Mirza (1998) suggested that...'*larger areas of land should behave in a more determinate and therefore predictable way than small volumes of soil*'. To illustrate this contention, they drew an analogy with denitrification in soil, which together with macropore flow is probably the most often cited example of a highly variable and essentially unpredictable soil process. At the small scale, denitrification is difficult to predict because the microsites ('hot spots') responsible for most of the nitrous oxide emission are highly variable spatially, their locations are unknown, and their activity is very sensitive to small variations in soil wetness. However, predictive relationships for denitrification are much easier to establish at the landscape scale, since soil wetness, which is the major control on denitrification, can be assessed from topography and soil type (Groffman and Tiedje, 1989; Lilly et al., 2003). Using a range of examples from catchment hydrology, Seyfried and Wilcox (1995) also made the point that the significance of

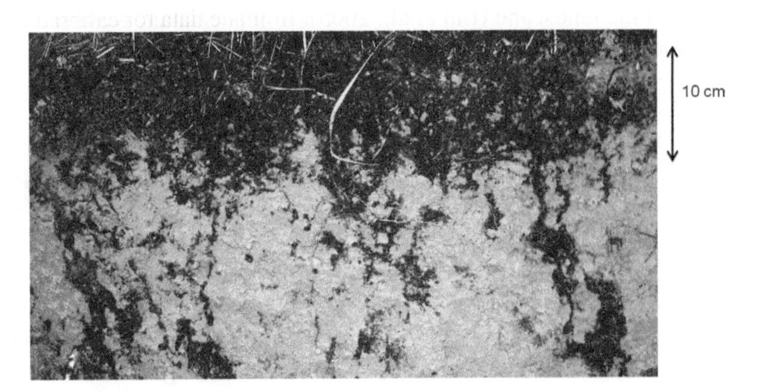

FIGURE 1 Photograph of a vertical soil profile at Skogsvallen, near Uppsala, after application of Brilliant Blue dye (see Jarvis et al., 2010 for details of the site and the experiments, picture courtesy of Dr. Mats Larsbo, SLU, Soil & Environment, Uppsala, Sweden).

deterministic variation has been underestimated in the hydrological sciences, even though it dominates hydrologic variability at larger scales. The deterministic nature of preferential flow at the landscape scale is illustrated by Cammeraat (1992) and Cammeraat and Kooijman (2009) who investigated the complex interrelationships of biotic and abiotic factors that control hydrologic response in a deciduous woodland catchment in Luxembourg. The soils, which are developed in weathered marl, have porous loamy *A* and *E* horizons overlying finer-textured and less permeable *B* horizons. Two distinct ecosystems were identified, with contrasting hydrologic response: forest dominated by beech (*Fagus sylvatica*) is found on drier, slightly coarser-textured soils, which have a lower pH and little macrofaunal activity. Matrix flow was found to dominate water discharging from these areas. Oak–hornbeam (*Quercus robur–Carpinus betulus*) woodland is found in wetter depression areas in the catchment, with slightly more clayey soil (E. Cammeraat, pers. comm.) that cracks readily in the topsoil and also supports much larger earthworm populations. The discharge response to rainfall in these areas is dominated by preferential flow in interconnected macropore and pipe systems, generated by perched water tables developing at the interface between *A*/*E* and *B* horizons.

This example illustrates the fact that determinism at the landscape scale arises because recognizable soil types or pedons (or more generally, 'structural elements'; Vereecken et al., 2007) display characteristic flow and transport patterns, which evolve as a function of biotic (vegetation, land use and management, and macrofaunal activity) and abiotic (soil moisture status and temperature) factors under the influence of local parent material, topography, and climate. Indeed, it is striking that the same factors that have traditionally been considered by pedologists to determine soil formation (Jenny, 1941) also influence preferential flow through their effects on structure-forming and degrading processes (Fig. 2). This reasoning suggests that hydropedological approaches may offer a good way to bridge the scale gaps between the soil pore, the pedon, and the landscape (Lin et al., 2005). In this chapter, we first briefly discuss the most widely used models, experimental techniques, and quantitative

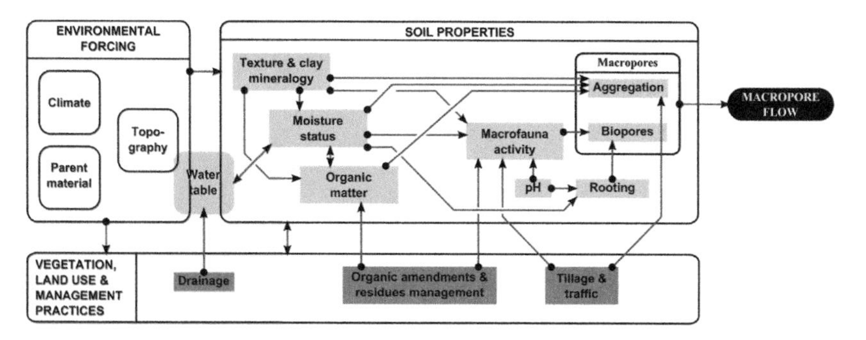

FIGURE 2 Controls on soil structure and macropore flow: a complex web of interactions between soil-forming factors (driving forces), soil properties, and land management.

indicators that can be used to characterize preferential flow and transport. We then consider the mechanisms of preferential flow generation and the characteristics of soil pore systems that sustain it. This is followed by a discussion of how site factors, soil properties, and pedogenic processes influence soil structure and susceptibility to preferential flow. Finally, we describe several spatial modeling tools that deal with preferential flow and the methodologies that were developed to parametrize them.

2. QUANTIFYING PREFERENTIAL FLOW

It is beyond the scope of this chapter to describe in-depth the various experimental and modeling approaches that have been adopted to quantify preferential flow and transport. Only a brief description of some commonly used methods is given here, and the reader is referred to recent reviews of available experimental techniques (Allaire et al., 2009) and models of preferential flow (Feyen et al., 1998; Šimůnek et al., 2003; Gerke, 2006; Köhne et al., 2009a,b). A two-region concept that relies on effective ('lumped') parameters as surrogates for the unknown geometry and properties of the macropore system has been widely adopted (Köhne et al., 2009a,b). The mobile–immobile advection–dispersion equation (MIM; Van Genuchten and Wierenga, 1976) assumes steady-state flow in a mobile fraction of the pore space and a first-order mass exchange with the remaining immobile fraction. This model has been frequently used to interpret the results of solute transport experiments carried out under steady flow conditions, but is less well suited to transient conditions in the field. Dual-permeability models (e.g. Larsbo et al., 2005; Šimůnek and van Genuchten, 2008) are better suited for this purpose, since they apply continuum model concepts (e.g. Richards' equation and the advection–dispersion equation) to two mobile, interacting, pore regions.

The determination of breakthrough curves on columns of undisturbed soil is probably the most widely used experimental technique to characterize solute transport. The higher time moments of breakthrough curves, which reflect spread and asymmetry, should in principle be good indicators of the degree of physical nonequilibrium (Valocchi, 1985; Jacobsen et al., 1992). The skewness of the curve in particular is often suggested as an indicator of the degree of preferential transport (e.g. Stagnitti et al., 2000). However, the higher temporal moments are very sensitive to even slight truncation (i.e. termination of the experiment before all the solute has leached) of the breakthrough curve (Young and Ball, 2000; Jawitz, 2004), something which is difficult to avoid in practice. Other quantitative measures derived from breakthrough curves have also been proposed as indicators of the degree of preferential flow, which may be more robust, including the column Peclet or Brenner number (Hatano et al. 1992; Vervoort et al., 1999), as well as nonparametric measures such as the pore volumes drained at the maximum concentration (Ren et al., 1996; Comegna, et al., 1999; McLeod et al., 2008; Jarvis et al., 2009), early quantiles of solute arrival times (Knudby and Carrera, 2005; Le Goc et al., 2010; Koestel et al., 2011a), or integral measures such as the holdback factor, defined as the amount of original fluid remaining in a column when one pore volume of displacing fluid has entered (Danckwerts, 1953; Rose, 1973).

Dye staining experiments have also been widely used to visualize and characterize preferential flow patterns (e.g. Kung, 1990; Flury et al., 1994). These methods give a snapshot of flow heterogeneity frozen at one moment in time, rather than a spatially integrated measure of flux as a function of time. Measurement of tracer breakthrough curves can also be combined with dye staining to identify flow pathways and thus maximize the information obtained from column experiments (e.g. Hatano et al., 1992; Morris and Mooney, 2004; Jarvis et al., 2008). Soil monoliths equipped with multiple outflow collectors at the base give information on both temporal and spatial variation in flow and transport (Stagnitti et al., 1999; Bejat et al., 2000; de Rooij and Stagnitti, 2002), but there are fewer reported examples of such experiments. It should also be noted that smaller soil columns (c. < 20 cm diameter) may be too small to capture properly the spatial scale of finger or funnel flow, and perhaps also macropore flow in coarse-structured subsoil. Instead, experiments at the scale of large monoliths or soil profiles and trenches in the field (m^2 scale) may be required (e.g. Kung, 1990; Vanderborght et al., 2001).

The degree of preferential flow can be quantified from photographic images of dye staining patterns at horizon and pedon scales by calculating simple indices that characterize flow pattern heterogeneity, for example fractal dimensions, measures of information entropy or complexity or other empirical statistics (Hatano et al., 1992; Olsson et al., 2002; Öhrström et al., 2002; Bogner et al., 2008a,b; Van Schaik, 2009; Wang et al., 2009). Photographic techniques have a resolution which is usually not sufficient to identify the geometry of individual pores and pore networks. In contrast, more detailed high-resolution 2D images of pore-scale structures can be obtained with thin-sectioning methods (e.g. Bouma et al., 1977; Walker and Trudgill, 1983; Ringrose-Voase, 1996; Lipiec et al., 1998; Prado et al., 2009), but only for relatively small samples. The main uncertainty with both approaches is the fact that each image is only a two-dimensional representation of what is really a three-dimensional pattern, which can give rise to misleading impressions of the pore system and the prevailing flow processes. Dyes also adsorb more or less strongly to soil constituents (Ketelsen and Meyer-Windel, 1999; Schwartz et al., 1999; Kasteel et al., 2002) and are therefore not perfect tracers. In a similar way, it is difficult to interpret unambiguously anion breakthrough curves in column experiments, due to exclusion in negatively charged soils or sorption in variably charged soils. Another feature which is common to many tracer and dye staining experiments is that they are often carried out under surface ponding or at high water application rates corresponding to rain events with long return periods (Jarvis, 2007a). Although these conditions would be relevant for irrigated agriculture, it is difficult to extrapolate from the results of such experiments to the range of natural weather conditions experienced in the field, since preferential flow is extremely sensitive to boundary conditions (e.g. Kluitenberg and Horton, 1990).

X-ray computed tomography (CT) overcomes many of the disadvantages of photography and thin sectioning, enabling 3D imaging of pore structure on samples large enough to be representative of the typical spatial patterns of soil macropores and at resolutions comparable with conventional thin sectioning (Taina et al., 2008). Quantitative descriptors of pore networks and soil structure can be related to transport behavior if solute breakthrough experiments are carried out in the same columns used for imaging (e.g. Kasteel et al., 2000; Perret et al., 2000; Luo et al., 2008; Luo et al., 2010; Köhne et al., 2011). Electrical resistance tomography (Olsen et al., 1999) can be used to obtain direct images of tracer transport (providing the water content and temperature are constant) at larger scales and thus is well suited to studies of Darcy-scale heterogeneous flow (Koestel et al., 2009). However, the technique cannot resolve preferential transport at the pore scale.

3. FACTORS AFFECTING PREFERENTIAL FLOW

This section attempts to answer two questions, which are critically important as a basis for predictive modeling: (i) Considering the initial and boundary conditions that 'trigger' the process, what attributes of soil pore systems control susceptibility to preferential flow? (ii) How do site factors and fundamental soil properties affect these key pore system attributes?

3.1. Soil Pore Systems, Hydraulic Properties, and Preferential Flow

Many experiments have demonstrated that the presence of continuous networks of larger macropores (> c. 300–500 μm in diameter) allows pore-scale nonequilibrium flow and transport (Jarvis, 2007a). A lack of well-developed interconnected networks of smaller macropores and mesopores is also critically important in promoting macropore flow. This is because the presence of such networks reduces the likelihood that soil water potentials increase sufficiently to 'trigger' flow in larger macropores under transient boundary conditions in the field and they also help to dissipate vertical fluxes in macropores by increasing rates of lateral convective and diffusive mass exchange (e.g. Luo et al., 2008). Thus, broadly generalizing, soils characterized by a distinct bimodal pore-size distribution, with large vertically continuous macropores embedded in a relatively slowly permeable soil matrix, are susceptible to nonequilibrium flow and transport. These soils are characterized by abrupt increases in hydraulic conductivity, K, across a narrow range of pressure potentials, ψ, close to saturation (e.g. Jarvis, 2008; Fig. 3).

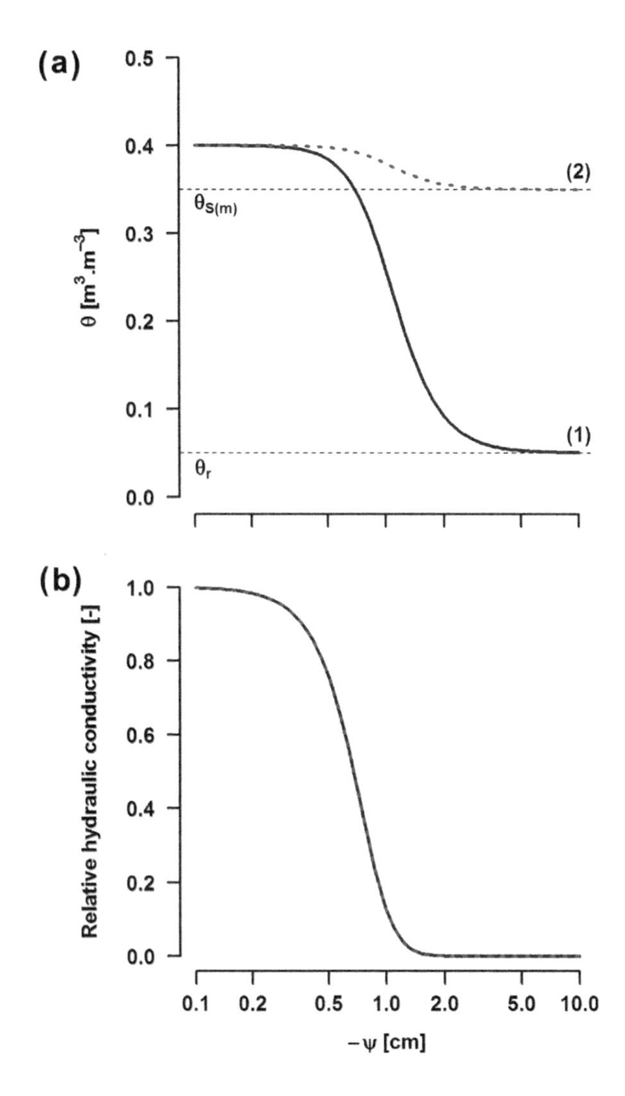

FIGURE 3 Schematic diagram of characteristic near-saturated hydraulic functions for water-repellent sand (1, solid line) and macroporous clay soil (2, dashed line): (a) water retention and (b) hydraulic conductivity. $\theta_{s(m)}$ indicates the saturated matrix water content in the clay soil, while θ_r is the residual water content in the sand. The relative hydraulic conductivity as a function of pressure potential is identical in both soils.

The concept of the 'structure hierarchy' (Fig. 4) is widely employed by both soil physicists and biologists as a useful model of soil structure (Tisdall and Oades, 1982; Hadas, 1987; Dexter, 1988; Oades and Waters, 1991; Pachepsky and Rawls, 2003). Aggregates of a given size consist of smaller subunits separated by planes of weakness. In turn, these subunits are comprised of even smaller aggregates, and so on. The lower the order in the structure hierarchy,

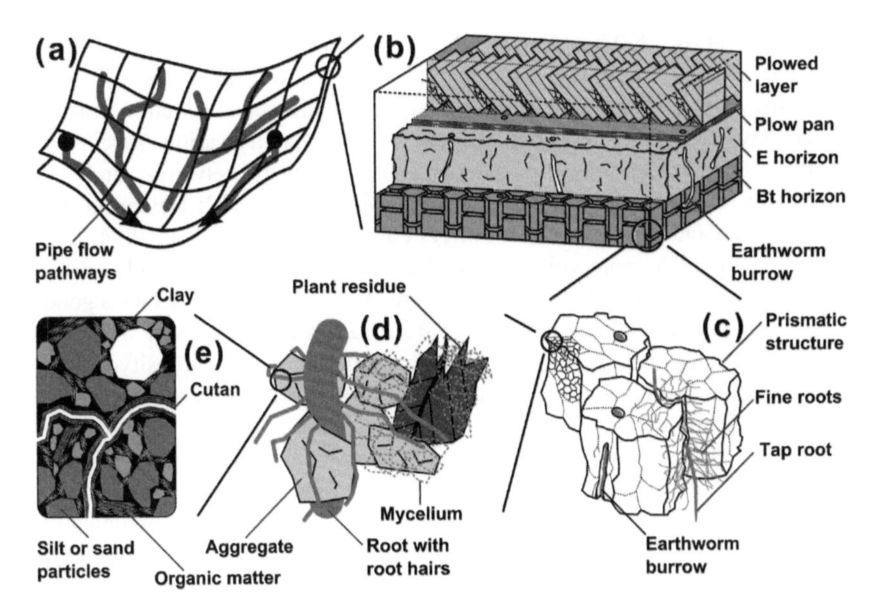

FIGURE 4 The soil structure hierarchy moving down through characteristic scales: (a) the hillslope, (b) the pedon, (c) peds/fissures and large biopores, (d) aggregation, and (e) fabric.

the denser and stronger are the aggregates, since they exclude the pores between the aggregates of all higher orders (Hadas, 1987; Horn et al., 1994). Pedologists also make use of the concept of the soil structure hierarchy, as it is embodied in the FAO terminology for soil structure description. In addition to classifying the size of the main structural elements (very fine through to very coarse), the FAO system recognizes the grade of structure, which reflects the degree of structural development at smaller scales and also the presence of aggregate surfaces (cutans) with characteristics distinct from the bulk soil. These clay and organic-rich coatings and linings on aggregate and macropore surfaces are often less permeable and/or more hydrophobic than the bulk soil (Hallett and Young, 1999; Gerke and Köhne, 2002; Köhne et al. 2002) and therefore also promote nonequilibrium flow. Thus, 'weak' structure is characterized by the fact that…'*when gently disturbed, the soil material breaks into a mixture of few entire aggregates, many broken aggregates, and much material without aggregate faces*', whereas a strong structure is reflected in a…'*prominent arrangement of natural surfaces of weakness. When disturbed, the soil material separates mainly into entire aggregates. Aggregate surfaces generally differ markedly from aggregate interiors*' (FAO, 2006). The pore-space hierarchy complements the aggregate hierarchy (Elliott and Coleman, 1988) such that larger aggregates are associated with larger, more widely spaced, and more continuous inter-aggregate fissures. Thus, strong macropore flow should be expected in strongly structured soils with vertically continuous macropores embedded in a soil with a poorly developed or degraded structure hierarchy (Jarvis, 2007a). Conversely, weak nonequilibrium flow can be expected in weakly structured soils characterized by a well-developed structure hierarchy. As fractals are 'self-similar' objects of hierarchical structure, fractal theory may be a suitable conceptual framework to quantify various aspects of the pore system hierarchy, the geometry of macropore networks, and the propensity for preferential flow (e.g. Brakensiek et al., 1992; Hatano et al., 1992; Hatano and Booltink, 1992; Perret et al., 2003; Peyton et al., 1994; Luo and Lin, 2009). The pore-volume fractal dimension, which is strongly correlated with macroporosity (Lipiec et al., 1998; Perret et al., 2003), seems a good predictor of preferential flow (Hatano et al., 1992). Pore connectivity will limit preferential transport in 'massive' soil horizons of small macroporosity. Under these limiting conditions, the spectral fractal dimension (Anderson et al., 1996) might be a useful complementary predictive indicator.

The lack of a structural hierarchy can explain the occurrence of finger and funnel flow in single-grain sandy soils: they can be conceptualized as comprising one or more material 'domains' each of which is macroscopically very homogeneous, with a narrow pore-size distribution. Different flow regimes (e.g. homogeneous flow, finger flow, and funnel flow) may develop depending on the spatial arrangement, inclination, and continuity of these domains (Wang et al., 1998). Water repellency in sandy soils effectively steepens the wetting soil water characteristic and thus also the $K(\psi)$ function (Bachmann et al., 2007), which generates preferential finger flow (Deurer and Bachmann, 2007). Thus, soils susceptible to either macropore flow or finger flow are both characterized by abrupt jumps in conductivity across a narrow range of pressure potentials (see Fig. 3); only the scale is different: in the case of macropore flow, this $K(\psi)$ 'jump' occurs at the microscopic scale for a small change of volumetric water content, since macropores occupy only a few percent of the total porosity. In contrast, fingering takes place in a homogeneous soil of narrow pore-size distribution, such that the $K(\psi)$ jump relates to the macroscopic (Darcy) scale, with the total pore volume partitioned into one nearly saturated part (the finger) and one part that remains dry.

The structural hierarchy can also be extended to the larger hillslope scale (Fig. 4). In humid climates, lateral preferential flow in interconnected pipe systems can dominate stream response in catchments with steep slopes and permeable soils overlying hard rock or otherwise slowly permeable substrates (Weiler and McDonnell, 2007; Anderson et al., 2009; Cammeraat and Kooijman, 2009). They develop by a self-organizing process from internal erosion of macropores such as old root channels and burrows created by macrofauna (e.g. earthworms, moles, and mice) during periods of lateral downslope flow at the saturated interface between the soil and the less permeable substrate (McDonnell, 1990; Sidle et al., 2001; Nieber et al., 2006).

Clothier et al. (2008) argued that we need to better understand the balance or 'tradeoff' between the economic costs of preferential flow (e.g. pollution of water bodies) and the benefits ('ecosystem services') generated by soil structure (improved nutrient cycling, better aeration etc.). In this respect, if we define 'good' structure as a well-developed structure hierarchy (Dexter, 1988) and accept the arguments presented above that the impact of nonequilibrium preferential flow is greatest in soils where the structure hierarchy is poorly developed or degraded, then there should be no conflict between maintaining good soil structure and thus sustainable crop production on the one hand and minimizing environmental pollution arising from preferential flow on the other (see also Elliott and Coleman, 1988; de Jonge et al., 2009). Simple 'back of the envelope' calculations suggest that the global value of 'good' soil structure is c. US$300 billion (Clothier et al., 2008). This makes it imperative that we improve our understanding of the site attributes and factors that control soil structure and preferential flow.

3.1.1. What Factors Control the Structure Hierarchy?

A complex web of interacting biological, chemical, and physical processes, acting through time, determines the expression of soil structure and the potential for preferential flow at any given site. Figure 5, taken from Jarvis

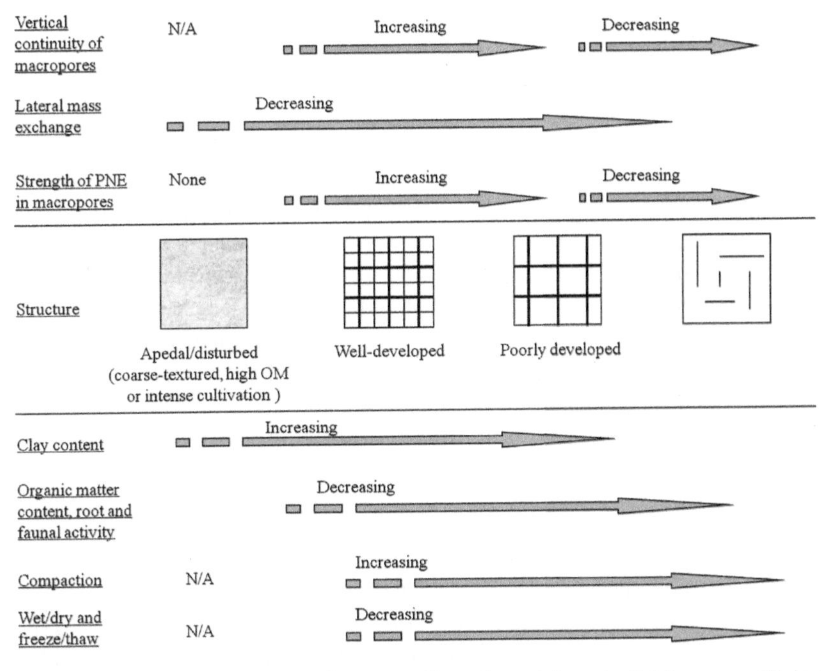

FIGURE 5 Factors affecting the soil structure hierarchy and the potential for nonequilibrium flow and transport in macropores (Jarvis, 2007a). The center of the diagram shows for illustrative purposes four classes of structural development. The bottom third shows how the various factors of structure formation and degradation influence structural development, while the top third illustrates the probable consequences for nonequilibrium water flow and solute transport in macropores (PNE = physical nonequilibrium, OM = organic matter).

(2007a), attempts to summarize how the structure hierarchy is affected by important soil and site factors and the likely consequences for preferential flow. These factors of soil structure formation and degradation have been covered in several detailed reviews (Dexter, 1988; Elliott and Coleman, 1988; Kay, 1990; Oades, 1993; Horn et al. 1994; Angers and Caron 1998; Six et al., 2004) and are therefore only briefly discussed below.

Soils with clay contents greater than about 15% that are subject to drying periods usually exhibit moderate to strong aggregate structure (Horn et al., 1994). Aggregation develops from the failure planes that develop in soil when stresses due to drying and clay shrinkage locally overcome soil cohesion. The desiccation of artificial clays devoid of life will only produce a dense matrix interspersed with a few wide cracks. Thus, physical processes alone are insufficient to explain the structure hierarchy: biological activity and soil organic materials play major roles in stabilizing these failure planes and enhancing the aggregation hierarchy. Thus, topsoils are generally characterized by a better developed structure hierarchy than subsoils because biological activity is more intense and wet/dry (and freeze/thaw) cycles are more frequent and severe (Southard and Buol, 1988).

Micro-aggregates composed of stable organic matter–clay mineral complexes are generally considered to be the fundamental building blocks of structure (i.e. the lowest level in the hierarchy, Fig. 4) for most soils (Tisdall and Oades, 1982), although organic matter seems to play a less important role for aggregation in old, highly weathered, soils dominated by oxides (Oades and Waters, 1991). Dexter et al. (2008) showed that some physical properties of the soil matrix were correlated with the amount of stable complexed organic matter, the maximum amount of which is set by the clay and perhaps also the fine silt content (Hassink, 1997). Conversely, more labile and non-complexed organic carbon (i.e. microbial and plant exudates, as well as living biological tissues such as fungal hyphae and plant roots) maintains structural stability at higher levels in the hierarchy, binding together micro-aggregates into larger macro-aggregates (Fig. 4; Tisdall and Oades, 1982; Six et al., 2004). Thus, the addition of fresh organic material and its decomposition by microorganisms has been shown to promote aggregation and aggregate stability (Gerzabek et al., 1995; Watts et al., 2001; Loveland and Webb, 2003). A decrease in soil organic matter content, and especially the younger more labile fraction, is associated with a loss of macro-aggregation and a degradation of the structural hierarchy, as reflected in a reduction of soil 'friability' (Watts and Dexter, 1998). Thus, soil structure is dependent on the carbon balance, which is strongly affected by land management. For example, Schlüter et al. (2011) applied high-resolution X-ray tomography to core samples taken from long-term field trials to demonstrate that the connectivity of networks of smaller macropores (< 0.3 mm in diameter) was poorer in nonfertilized plots that were depleted of soil carbon. Compared to natural vegetation or managed grassland, soil organic matter contents, especially the labile fraction (e.g. Hassink, 1997), decrease under continuous arable farming due to the export of carbon in harvests and stimulation of mineralization by tillage disturbance. A loss of organic matter means that arable soils are also more easily compacted by traffic (Soane, 1990; Watts and Dexter, 1997). Repeated traffic compaction and fragmentation by annual cultivation in arable agriculture combine to degrade the structural hierarchy and coarsen the soil structure in the tilled layer (Hadas, 1987; Lipiec et al. 1998; Roger-Estrade et al., 2004). For these reasons, stronger preferential flow is often found in arable soils (Walker and Trudgill, 1983; Jarvis et al., 1991; Jarvis et al., 2007; Luo et al., 2010), although heavy machinery traffic or poaching by cattle can also severely degrade soil structure and enhance preferential flow in intensively managed grasslands (Dreccer and Lavado, 1993).

Land use also exerts a first-order control on the abundance of different types of macropores. For example, root channels are abundant in undisturbed forest soils, where they often dominate preferential flow (e.g. Sollins and Radulovich, 1988; Bachmair et al., 2009; Lange et al., 2009). Trees and bushes may also concentrate a significant fraction of precipitation into stemflow, which is then directed preferentially along large roots deep into the soil (Johnson and Lehmann, 2006). Root biopores are much less important in cultivated arable topsoils as a result of tillage disruption. Here, the main preferential flow pathways are the void spaces between dense aggregates and clods created by tillage implements, which may also be partially infilled with looser soil or straw (Coquet et al., 2005; Kasteel et al., 2007; Jarvis et al., 2008). However, some crop plants (e.g. alfalfa) produce large tap roots that can significantly enhance water flow and solute transport deep into the subsoil (e.g. Mitchell et al., 1995; Perillo et al., 1999). As roots are the main agents of soil drying, they also strongly influence the development of macro-aggregate structure, especially in fine-textured soils (Kay, 1990; Angers and Caron, 1998) where they tend to grow preferentially along aggregate faces or in other large macropores, such as earthworm channels (Hatano et al., 1988; van Noordwijk et al., 1993; Stewart et al., 1999).

The channels created by deep-burrowing 'anecic' earthworm species represent potentially major conduits for preferential flow deep into the subsoil (e.g. Ehlers, 1975; Germann et al., 1984; Edwards et al., 1992). The extent to which this potential is realized will depend sensitively on local weather and climate conditions. For example, strong preferential flow has been demonstrated in grassland soil with abundant macrofaunal channels under high-intensity rain or ponding, whereas these large earthworm biopores may not conduct water at the smaller flow rates characteristic of shorter return period rainfall events (e.g. Shipitalo et al., 1993; Vogeler et al., 1997; Bachmair et al., 2009). 'Endogeic' earthworm species live predominantly in the topsoil where they create temporary burrow systems by feeding on organic matter ingested with soil. Under favorable conditions, endogeic earthworm populations can turn over up to 10–15% of the topsoil mass per year (Curry and Schmidt, 2007).

This intense 'bioturbation' creates a fine crumb-like structure and thus a well-developed structure hierarchy (Six et al., 2004) which, in contrast to the activity of anecic species, may reduce susceptibility to macropore flow (see Fig. 1; Jarvis et al., 2010). Earthworm populations are generally favored by land-use systems that minimize disturbance and improve litter supply such as forest, grassland, orchards, or no-till arable (Edwards and Lofty, 1982; Whalen, 2004; Smith et al., 2008; Lindahl et al., 2009). In addition to the effects of land use, the abundance of earthworms is also affected by basic soil properties such as pH and texture (Joschko et al., 2006; Lindahl et al., 2009) and site attributes such as soil wetness (Cannavacciuolo et al., 1998). Earthworms have been widely studied because they are ubiquitous in almost all environments except deserts. However, other macrofaunal 'ecosystem engineers' (Jones et al., 1994) also affect soil structure and preferential flow, including ants (Wang et al., 1994), termites (Léonard and Rajot, 2001), and even cranefly larvae (Holden and Gell, 2009).

Different land-use systems and management practices affect not only types and abundance of macropores but also the potential for preferential finger flow. For example, the long-term use of organic amendments in arable farming produces organic-rich topsoils ('Plaggen') that can become water repellent when dry and therefore prone to fingering (Vanderborght et al., 2001). In general, undisturbed soils (e.g. grassland, orchards, forests, and no-till arable) are more prone to develop water repellency when dry (Hallett et al., 2001; Doerr et al., 2006; de Jonge et al., 2007; Jarvis et al., 2008; Blanco-Canqui and Lal, 2009) due to the lack of disturbance and the larger soil organic carbon contents. In natural or less intensively managed ecosystems, substantial differences in soil water repellency have even been noted under different plant species, due to the contrasting properties of their litter and root exudates (Verheijen and Cammeraat, 2007).

3.2. Pedogenesis and Preferential Flow

Figure 2 illustrates schematically how the factors of soil formation (Jenny, 1941) interact to influence soil structure and the potential for preferential flow over spatial scales ranging from fields (soil management and cropping) to hillslopes (soil moisture redistribution) and regions (e.g. climatic factors) and over time scales ranging from seasons (e.g. soil structure evolution after tillage) to thousands of years (e.g. landscape and soil-forming factors). Generalizing, macro-aggregation and macropore systems are strongly influenced by factors operating at shorter time scales, from seasons to decades (e.g. tillage, cropping, and variations in macrofaunal populations), especially in the topsoil, while soil texture, micro-aggregation, and mesopore networks are more stable, being more affected by factors operating over historical time scales from decades to thousands of years (e.g. weathering of parent material, clay translocation, carbon turnover and humus accumulation, and bioturbation). Some of these aspects are now considered in a pedogenic perspective, emphasizing the medium to long-term factors which control micro-aggregation and mesopore networks, and thus the ease with which macropore flow is both initiated and dissipated. The analysis makes use of the WRB system for soil classification (IUSS, 2006) as a conceptual framework. It focuses on the scale of the pedon (see Fig. 4) as this forms the critical link between processes operating at the lower orders of the structure hierarchy and those operating at the hillslope and landscape level. Table 1 presents a tentative summary of this discussion, by listing the dominant flow regimes expected for the major reference soil groups of WRB making use of the simple classification system proposed by Weiler and Flühler (2004).

In young soils, parent material characteristics are often so dominant that they alone can determine the propensity for preferential flow. A good example is the case of Andosols developing in parent material of volcanic origin. Allophanic clay minerals produced from the weathering of volcanic tephra have a high affinity for organic carbon, which is effectively sequestered in stable complexes (Parfitt, 2009). Andosols are therefore highly porous with large organic matter contents and a fine crumb structure. Despite often large clay contents, they show little propensity for preferential flow (Aislabie et al., 2001; McLeod et al., 2001; Lohse and Dietrich, 2005), or at most only weak nonequilibrium flow related to the immobile water trapped in micro-aggregates (Prado et al., 2009). In contrast, soils developing in fine-textured fluvial deposits (Fluvisols) are often strongly structured, especially after drainage for agriculture, and are therefore highly prone to macropore flow (Bouma and Dekker, 1978; Brown et al., 2000; Aislabie et al., 2001; Scorza et al., 2004). Macropore flow is highly unlikely in coarse-textured young depositional soils (e.g. Fluvisols and Arenosols), but they can be prone to funnel or finger flow depending on soil water repellency and the spatial arrangement and continuity of the depositional layers (e.g. Ritsema et al., 1993; McLeod et al., 2001; Wessolek et al., 2008). For example, Kung (1990) and Mayes et al. (2003) showed that funnel flow can be generated by discontinuities in inclined sedimentary layers, while Vanderborght et al. (2001) showed that effective lateral mixing in two laminated Arenosols characterized by alternating thin horizontal layers of finer and coarser material (lamellae) led to a convective–dispersive transport regime.

Regosols are a diverse group of soils, but they are young and aggregation is little developed, so the propensity for preferential flow will depend on the presence or absence of biopores, which depends primarily on land use (e.g. Vanderborght et al., 2001; Zehe and Flühler, 2001; Jacques et al., 2002). Shallow soils over hard rock (Leptosols) are also effectively young soils, either because they really are young, or more often because the underlying rock is so hard that weathering is slow, or because the dissolution of pure calcareous rock by rainfall takes a very long time to build up a significant thickness of mineral residue, or because erosion on steep slopes keeps pace with weathering. Leptosols are also weakly aggregated coarse or

TABLE 1 Likely Dominant Flow Regimes in the Reference Soil Groups of the World (IUSS, 2006)

Soil-forming factors (dominant identifier)	Reference soil group	Approximate equivalent in soil taxonomy[a]	Dominant flow regime[b]	Comments[c]
Organic	Histosols	Histosols	I (II, IIIa)	II = fingering due to water repellency in drained soils during drought; IIIa crack flow in drained soils during drought
Human influence	Anthrosols, Technosols	Anthrepts, Arents		Very diverse group of soils
Shallow	Cryosols	Gelisols		No information, but funnel flow likely related to seasonal ice wedge relicts
	Leptosols	Shallower (<25 cm) or very stony lithic subgroups of Entisols	I, II, IIIa	As for Regosols: limited soil development, so flow regime will depend on parent material, texture, and land cover
Influenced by water	Vertisols	Vertisols	IIIc	Weak interaction with matrix due to high clay content. The 2:1 clay mineralogy enhances swell/shrink
	Fluvisols	Fluvents, Fluvaquents	I, IIIc	I = coarse-textured/laminated sediments IIIc = drained, fine-textured sediments

TABLE 1 Likely Dominant Flow Regimes in the Reference Soil Groups of the World (IUSS, 2006)—cont'd

Soil-forming factors (dominant identifier)	Reference soil group	Approximate equivalent in soil taxonomy[a]	Dominant flow regime[b]	Comments[c]
	Solonetz	Natric great groups	IIIb,c	IIIb,c = strong columnar structure in subsoil. Topsoil class will depend on soil properties and land use
	Solonchaks	Salids, Halaquepts	I, IIIa,b,c	Texture dependent: IIIc = strong macropore flow in clayey soils (e.g. Moreno et al., 1995; Armstrong et al., 1996)
	Gleysols	Endoaquic great groups	I, IIIa,b	I = undrained, III = drained for agriculture; subclass depends on texture: a = medium textured; b = fine textured
Fe/Al chemistry	Andosols	Andisols	I (II)	Fine crumb structure
	Podzols	Spodosols	I, II	II = tongues (e.g. due to tree wind-throw), fingering (water repellency)
	Plinthosols	Plinthic great groups & subgroups	I, II, IIIa	I = upper horizons, II = heterogeneous flow through discontinuous plinthic horizons, IIIa = plinthic horizon (slow macropore flow)

(Continued)

TABLE 1 Likely Dominant Flow Regimes in the Reference Soil Groups of the World (IUSS, 2006)—cont'd

Soil-forming factors (dominant identifier)	Reference soil group	Approximate equivalent in soil taxonomy[a]	Dominant flow regime[b]	Comments[c]
	Nitisols	Some soils in: Kandic great groups of Alfisols & Ultisols; Oxic subgroups of Inceptisols; Oxisols	IIIc	Strong fine blocky structure
	Ferralsols	Most Oxisols	I, II	Stable micro-aggregation
Stagnant water	Planosols	Albaquults	I, IIIc	I = *A* and eluviated horizons, IIIc = clayey subsoil
	Stagnosols	Epiaquic great groups	IIIa,b,c	Depending on texture/bulk density; a = medium textured; b = fine textured; c = very fine textured
Organic matter accumulation	Chernozems Kastanozems Phaozems	Mollisols	I, IIIa,b,c	I = mollic topsoil horizon, IIIa,b = cambic, argic subsoil horizon (class is texture dependent), IIIc = vertic properties in subsoil
Accumulation of less soluble salts	Gypsisols Durisols	Gypsic suborders & great groups of Aridisols and Gypsic subgroups of Inceptisols. Duric suborders & great groups of		No information was found, but flow regimes are likely similar to Calcisols

TABLE 1 Likely Dominant Flow Regimes in the Reference Soil Groups of the World (IUSS, 2006)—cont'd

Soil-forming factors (dominant identifier)	Reference soil group	Approximate equivalent in soil taxonomy[a]	Dominant flow regime[b]	Comments[c]
		Aridisols & durinodic subgroups of Entisols & Inceptisols		
	Calcisols	Calcic suborders & great groups of Aridisols and calcic subgroups of Inceptisols	I, II, IIIa,b	II = heterogeneous flow in stony or water repellent soils. IIIa,b = macropore flow may occur above the calcic horizon, depending on the presence of biopores
Illuviation	Albeluvisols	Glossic great groups & subgroups of Alfisols	II, IIIb	II = tonguing; IIIb = argic horizon (high-activity clay)
	Alisols, Luvisols	Non-Kandic Ultisols Non-Kandic Alfisols	I, IIIb	I = A/E horizons, IIIb = argic horizon (high-activity clay)
	Acrisols, Lixisols	Kandic Ultisols Kandic Alfisols	I, IIIa	I = *A/E* horizons, IIIa = argic horizon (low-activity clay and stable micro-aggregation)
Young soils	Umbrisols	Humic subgroups of Inceptisols except Eutric great groups; Humaqueptic subgroups of Aquents	I, IIIa	I = organic-rich topsoil; IIIa = stagnic or fine-textured Cambic subsoil

(*Continued*)

TABLE 1 Likely Dominant Flow Regimes in the Reference Soil Groups of the World (IUSS, 2006)—cont'd

Soil-forming factors (dominant identifier)	Reference soil group	Approximate equivalent in soil taxonomy[a]	Dominant flow regime[b]	Comments[c]
	Arenosols	Psamments	I, II	I = loose or laminated material; II = finger flow in water-repellent soils, funnel flow in soils with discontinuous or inclined layering
	Cambisols	Inceptisols except gypsic, durinodic & calcic subgroups	IIIa,b,c	Depending on texture
	Regosols	Entisols except durinodic subgroups	I, II, IIIa	Diverse group of soils. Limited soil development, so flow regime will depend on parent material, texture and land cover

[a] *Refers to the U.S. Soil Taxonomy. It is not possible to make exact correlations between soil classes in the two system as the definitions of differentiating criteria are usually slightly different. The correlations are thus only approximate. The identified flow regimes and comments are intended to be valid for the WRB system.*
[b] *Dominant flow regimes in the soil unsaturated zone, with secondary regimes indicated in brackets. According to Weiler and Flühler's (2004) classification system: I = homogeneous flow, II = heterogeneous flow, III = macropore flow (a = strong, b = medium, c = weak interaction with the matrix).*
[c] *See text for supporting references.*

medium-textured soils with limited profile development, so the strength of preferential flow should depend primarily on the presence of biopores, soil stoniness, and, in dry climates, the development of water repellency. In humid climates, Leptosols will also tend to have high organic matter contents due to the shallow rooting depth, which should further reduce susceptibility to preferential flow. Thus, the dominant flow regimes are likely to be homogeneous flow, heterogeneous flow, and, at high rainfall rates, macropore flow in any biopores that may be present (Calvo-Cases et al., 2003; Bogner et al., 2008a; Van Schaik, 2009).

As soils develop, physico-chemical weathering of parent materials over time influences the propensity for preferential flow. The shrink–swell process associated with the presence of 2:1 lattice type (high activity) clay minerals, either inherited from soil parent materials or weathered from them, promotes the development of distinct (cambic) subsoil horizons with some aggregate structure. Broadly, the propensity for preferential flow in such Cambisols varies with soil texture, with weak preferential flow found in sandy loams through to strong macropore flow in clays (e.g. Jarvis et al., 1991; Akhtar et al., 2003; Kim et al., 2004; Dörfler et al., 2006). Since Cambisols are still relatively young soils, soil structure and the propensity for preferential flow still closely reflect soil texture and the nature of the parent material (Kim et al., 2004). Depending on the climate, very clayey soils can show such a strong ability to swell and shrink with changes in moisture content that the process dominates their development. These Vertisols (or soils with vertic properties) are highly susceptible to macropore flow (e.g. Kissel et al., 1973; Nobles et al., 2004), although a lack of pore continuity can restrict flow in fully swollen soil, especially if the structure is damaged by poor management practices (e.g. Crescimanno et al., 2007). Under climate regimes where there is a period of precipitation excess (i.e. zero soil moisture deficit) of more than two to three months, medium- to fine-textured soils or those with dense parent material (usually of glacial origin) often exhibit only a weakly developed coarse structure in the subsoil. This inhibits the downward percolation of excess moisture resulting in seasonal saturation within and above these layers. These Stagnosols usually show moderate to strong preferential flow (e.g. Brown et al., 1995; Jensen et al., 1998; Lange et al., 2009), both along root and faunal channels and also in fissures, as highlighted by their gray, iron-depleted surfaces. Soils in which such layers are overlain by much coarser-textured horizons (Planosols) can show quite extreme contrasts in flow regimes between the upper sandy horizons characterized by uniform flow or fingering and the underlying fine-textured layers which are highly susceptible to macropore flow (Zumr et al., 2006; Hardie et al., 2011).

In climates with excess moisture, at least during some part of the year, the process of clay illuviation generates, with time, soils with texturally contrasting layers (e.g. Luvisols, Alisols, Lixisols, and Acrisols) in which the upper, clay-depleted horizons are only weakly structured and less susceptible to preferential flow than the underlying clay-enriched and more strongly structured argic (Bt) horizon (e.g. Vervoort et al., 1999; Schwartz et al., 1999; Shaw et al., 2000; Vanderborght et al., 2001; Franklin et al., 2003; Nobles et al., 2004). In an extreme form (Albeluvisols), the lower part of the upper clay-depleted layers also becomes pale and depleted of iron, forming tongues or narrow wedges penetrating deep into the underlying clay-enriched horizon, which can constitute preferential flow pathways through the soil (Vanderborght et al., 2001). As physico-chemical weathering intensifies, particularly in warm and moist tropical and subtropical climates, 2:1 lattice clays are converted to 1:1 lattice (low activity) clays and amorphous sesquioxidic compounds are liberated. Depending on the amount and type of clays present, this process may produce soils with well-developed aggregate structure (Nitisols) and a strong tendency to preferential flow (Heng et al., 1999; McLeod et al., 2004) or soils with more weakly structured argic horizons (Lixisols and Acrisols) in which preferential flow is less dominant but still occurs (Shaw et al., 2000, Franklin et al., 2003; McLeod et al., 2004). On the oldest, most stable land surfaces, soils with 1:1 lattice clays and relatively large amounts of sesquioxidic compounds develop. Under conditions of fluctuating groundwater, soils with massive reddish and pale mottled subsoil horizons develop (Plinthosols) in which water movement is slow but can occur along preferential pathways if the layer is not uniformly compact (Carlan et al., 1985; Shaw et al., 1997). In free draining locations, Ferralsols are found that are characterized by reddish/yellowish horizons with very stable micro-aggregation. These soils are, in principle, not especially susceptible to nonequilibrium flow and transport since lateral mass exchange is very efficient and the matrix permeability may be larger than the clayey texture would suggest (e.g. Anamosa et al., 1990; Alavi et al., 2007; Tang et al., 2009). However, despite these features, preferential flow has been observed in Ferralsols under the near-saturated and saturated conditions that commonly occur in the wet tropical rain forest climate zone, both through inter-aggregate zones of high permeability and in root biopores (e.g. Rao et al., 1974; Reichenberger et al., 2002; Laabs et al., 2002; Renck and Lehmann, 2004; Wickel et al., 2008).

Climate/vegetation interactions often dominate pedogenesis. For example, podzols develop due to the migration of Fe, Al, and organic compounds under the acid conditions found under coniferous forest or heath vegetation, especially those growing in temperate or boreal climates on sandy parent materials. Little macropore flow can be expected in these soils due to their sandy texture and a lack of faunal biopores due to the acid conditions. However, heterogeneous flow and fingering due to soil water repellency seems common (Vanderborght et al., 2001, Deurer et al., 2001). Soils that develop under the natural grassland found in continental ‘steppe’ climates (e.g. Phaeozems, Chernozems, and Kastanozems) represent another well-known example of the importance of vegetation as a soil-forming factor. These soils have accumulated relatively large amounts of organic matter in their *A* horizons, which are often rather deep. This gives them a friable crumb structure and limits preferential flow in these upper layers, at least under natural conditions. However, conversion to continuous arable farming can increase the risk of preferential flow by degrading and coarsening the topsoil structure (Khitrov et al., 2009). Moderate to strong preferential flow can also occur in the subsoil, through earthworm biopores and rodent burrows infilled with loose soil material (Khitrov et al., 2009) or along aggregate surfaces in horizons with argic (e.g. Luvic Phaozems, Ersahin et al., 2002; Bedmar et al., 2008) or vertic (Tallon et al., 2007) characteristics. Umbrisols (and soils with Umbric horizons) represent a final example of soil types whose genesis is governed by climate/vegetation interactions. These soils accumulate significant amounts of acid organic matter because of slow biological turnover under acid conditions, low temperature, surface wetness or a combination of these factors. They develop in cool, wet, mainly acid conditions, typically on upland massifs with montane forests and grasslands of low nutritional value. As with the grassland Steppe soils, the large organic matter content gives them a friable crumb structure that limits preferential flow (Rawlins et al. 1997), although it may be significant in the subsoil of the most developed types with stagnic or fine-textured cambic horizons.

Predominantly organic soils (Histosols) and soil horizons that develop under permanently wet conditions for reasons either of climate or topography usually show little propensity for preferential flow (Brown et al., 2000; Roulier and Jarvis 2003). However, if effectively drained, organic-rich soils can crack or become hydrophobic during dry seasons and this may generate preferential flow (Kätterer et al., 2001; Schwärzel et al., 2002; Stutter et al., 2005). Soils in low-lying areas or topographic depressions in the landscape that are permanently affected by groundwater (Gleysols) have subsoil horizons that usually show little structural development and only weak preferential flow (Holden, 2009). However, again, if effectively drained for agriculture, they are increasingly subject to the process, depending on their clay and organic matter contents (Dörfler et al., 2006). Solonetz soils develop in fine loamy or clayey sediments usually with impeded drainage in low-lying landscapes in dry continental 'steppe' climates. These soils have a very strong columnar aggregate structure in the subsoil, which is highly susceptible to macropore flow (Dreccer and Lavado, 1993).

At the other extreme of the soil moisture gradient, plant primary production and thus the input of organic carbon to soil are extremely limited in arid environments, so that physical processes dominate the development of aggregate structure. Since weathering is slow, Regosols in arid landscapes are often relatively coarse textured, with a limited potential for preferential flow, at least in the initial stages of soil development. However, Meadows et al. (2008) showed that pedogenesis of the Av (vesicular or Yermic) horizon in a desert pavement soil resulted in a transition from matrix-dominated flow to macropore flow due to the incorporation of fine eolian dust, which produces a finer texture, thereby enhancing swell–shrink behavior and structural development. Also, root biopores have been shown to contribute to preferential flow under plants growing in sparsely vegetated desert ecosystems (e.g. Devitt and Smith, 2002), albeit under extreme conditions (e.g. surface flooding).

Finally, it should be noted that causal relationships between climate and vegetation/land use factors on the one hand, and soil structure and macropore flow on the other, may not always be obvious due to polygenetic soil evolution (Johnson and Watson-Stegner, 1987), which gives rise to relict soil morphological features and properties that are unrelated to current climate or land use, reflecting instead historical conditions. For example, Shaw et al. (1997) and Franklin et al. (2003) found that solute transport patterns in Acrisols and Plinthosols in southern USA were still influenced by the natural forest that existed prior to clearance for agriculture. Relatively uniform solute transport was found through a loamy organic-rich *BA* horizon with an open, porous, and finely aggregated fabric resulting from macrofaunal bioturbation, while preferential flow occurred through the highly weathered *Bt* and *Btv* (plinthic) horizons, where pathways were confined mostly to channels created by old tree roots and deep burrowing soil animals.

4. USING HYDROPEDOLOGICAL PRINCIPLES TO SUPPORT PREFERENTIAL FLOW MODELING AT THE LANDSCAPE SCALE

A considerable worldwide experimental effort during recent decades has generated an extensive amount of data on preferential flow and transport characteristics in a wide range of contrasting soils at horizon, pedon, and plot scales. As we have seen, the propensity for preferential flow is determined by the interaction between soil parent material, which determines the texture and mineral content of soil horizons, the duration and intensity of physico-chemical weathering processes that control clay mineralogy, translocation processes that redistribute materials (e.g. clays) within the soil profile, climate, and vegetation cover (in both a short-term and historical perspective) through its effects on soil organic matter and macrofaunal activity and man's activities encompassing aspects of land use and soil disturbance such as traffic and tillage. One important challenge now is to synthesize all this knowledge into robust and useable parametrization methodologies that can support model predictions of flow and transport at the landscape scale.

In practice, for spatial applications, model parameters cannot be measured or derived by calibration. Instead, they must be estimated from widely available survey data (e.g. soil type, soil texture, organic matter content, and pH) using statistical methods or pedotransfer functions (Bouma et al., 1996; Wösten et al., 2001; McBratney et al., 2002). There are two fundamental types of pedotransfer function: model parameters can either be estimated as continuous functions of one or more independent variables, for example using multiple linear regression (continuous pedotransfer functions), or they can be fixed at single values for the different classes (e.g. USDA texture class) of a classification system (class pedotransfer functions). Wösten et al. (1995) compared model predictions of four functional aspects of soil behavior (soil workability, soil aeration, and nonreactive and reactive solute transport) using continuous and class pedotransfer functions for soil hydraulic functions. They found some significant systematic differences between the two approaches, but these were mostly very small and considered to be irrelevant in practice.

4.1. Continuous Pedotransfer Functions for Estimating Preferential Flow Parameters

Although the use of pedotransfer functions has a long history, most work has focused on estimation of hydraulic properties, especially water retention, from soil texture, while with some notable exceptions (e.g. Lin et al., 1999; Pachepsky et al., 2006; Lilly et al., 2008), the effects of soil structure have often been ignored. Similarly, only a few studies have considered pedotransfer approaches for solute transport parameters. Dispersivity in the single-region advection–dispersion equation (ADE) has been found to increase with clay content for a limited data set (Perfect et al., 2002), while in a larger data-mining exercise, Vanderborght and Vereecken (2007) attributed an increase of dispersivity with flow rate in fine-textured soils to the occurrence of preferential flow. Continuous pedotransfer functions have also been developed for the two additional parameters required by the dual-porosity MIM. Thus, both Vervoort et al. (1999) and Shaw et al. (2000) found that soil horizons with larger clay contents had weaker mass exchange and smaller mobile water contents giving stronger preferential transport. Similarly, Jarvis et al. (2007) showed that the mass transfer parameter in the dual-permeability model MACRO (Larsbo et al., 2005) was significantly correlated with clay and organic matter content in arable topsoils. Mass exchange was weaker and preferential transport stronger in soils of large clay content and small organic matter content. Only a few studies have attempted to develop pedotransfer functions for the additional parameters needed to describe the hydraulic properties of the macropore region in dual-permeability models (Šimůnek, et al. 2003). A dependence of the saturated macropore hydraulic conductivity on macroporosity has been demonstrated many times (e.g. Germann and Beven, 1981; Rawls et al., 1993). More recent research suggests that clay content is also a good predictor of saturated macropore conductivity, with larger values found in fine-textured and therefore generally more strongly structured soils (Børgesen et al., 2006; Jarvis, 2008). Conversely, the saturated hydraulic conductivity of the soil matrix is smaller in clay soils (Smettem and Bristow, 1999; Jarvis et al., 2002). Thus, although the number of studies that have derived pedotransfer functions for model parameters is limited, their results are consistent with one another and seem broadly in agreement with the known effects of soil properties on soil structure development.

4.2. Class Pedotransfer Functions, Classification Systems, and Vulnerability Mapping

Conceptual models that attempt to synthesize and generalize hydropedological knowledge and process understanding in the form of simple classification systems may offer a flexible and robust alternative to direct model-dependent approaches based on continuous pedotransfer functions. One advantage of classification systems is that 'class' pedotransfer functions can be used to link them to any model considered appropriate for the purpose at hand. They also make it easier to exploit existing soil maps and survey databases and categorical, qualitative, 'soft' data such as land-use categories and soil types as predictor variables (Bouma, 1991; Bouma et al., 1996; Lin et al., 2005). Diagnostic horizons and subordinate characteristics may be especially useful categorical variables in this respect (Bouma, 1991), since they reflect the integrated effects of pedogenic processes operating at any site through time and therefore carry significant information regarding the extent and nature of preferential water flow and solute transport.

Quisenberry et al. (1993) were the first to propose a system to classify the potential of mapped soil types for preferential flow. However, the classification system was rather limited in scope, covering the dominant soils of South Carolina. Jarvis et al. (2009) described a classification scheme designed to support spatial model predictions of pesticide losses to surface water and groundwater, which is more ambitious and generic in scope. The scheme, which takes the form of a simple decision tree, was derived from a combination of simple 'rules of thumb' derived from a review of the literature (Jarvis, 2007a) and statistical analyses of factors influencing soil aggregation and the abundance of anecic earthworms (Lindahl, 2009; Lindahl et al., 2009). The scheme groups soil horizons into one of four classes with respect to susceptibility to macropore flow on the basis of widely available survey data (e.g. climate zone, land-use category, FAO horizon designation, USDA textural class, organic carbon content, bulk density, and pH). Predictions of the scheme were compared with the results of solute breakthrough experiments collated from the literature. The mean values of the selected indicator of preferential flow (pore volumes drained at the peak concentration) differed significantly between the classes, but the classification explained only 30% of the overall variation. However, much of the unexplained variation could be attributed to within-class variation in clay content (Jarvis et al., 2009).

The results of Jarvis et al. (2009) provide some systematic and objective evidence that the effects of macropore flow can be predicted from soil survey and land-use data. However, the data set used to test the model was quite small and the approach very much biased toward European soils and agricultural settings, since it was developed to support pesticide risk assessment in the EU. Recently,

FIGURE 6 Decision tree to support predictions of the effects of preferential flow, extended and modified from Jarvis et al. (2009). For definition of Andic/Vitric material see IUSS (2006), for master horizon types and horizon suffixes see FAO (2006). Coarse texture = sand or loamy sand; fine texture = clay, silty clay, clay loam, or silty clay loam according to the USDA system; and foc is the organic carbon content. Primary tillage refers to plowing, while secondary tillage refers to more intensive operations using implements that pulverize the soil (harrows and rotovators). Classes I, II, III, and IV imply 'negligible', 'low' 'moderate,' and 'high' impact of preferential flow in terms of accelerating flow/transport velocities through the horizon. Referring to the classification by Weiler and Flühler (2004) that is used in Table 1, class I can be equated with homogeneous flow, class II with heterogeneous flow or weak macropore flow, and classes III and IV with moderate and strong macropore flow, respectively.

Koestel et al. (2011b) greatly increased the number of tracer breakthrough curves included in the database, which has enabled further testing of the scheme proposed by Jarvis et al. (2009). Figure 6 shows an updated version of the classification, which has been refined and further generalized by including some

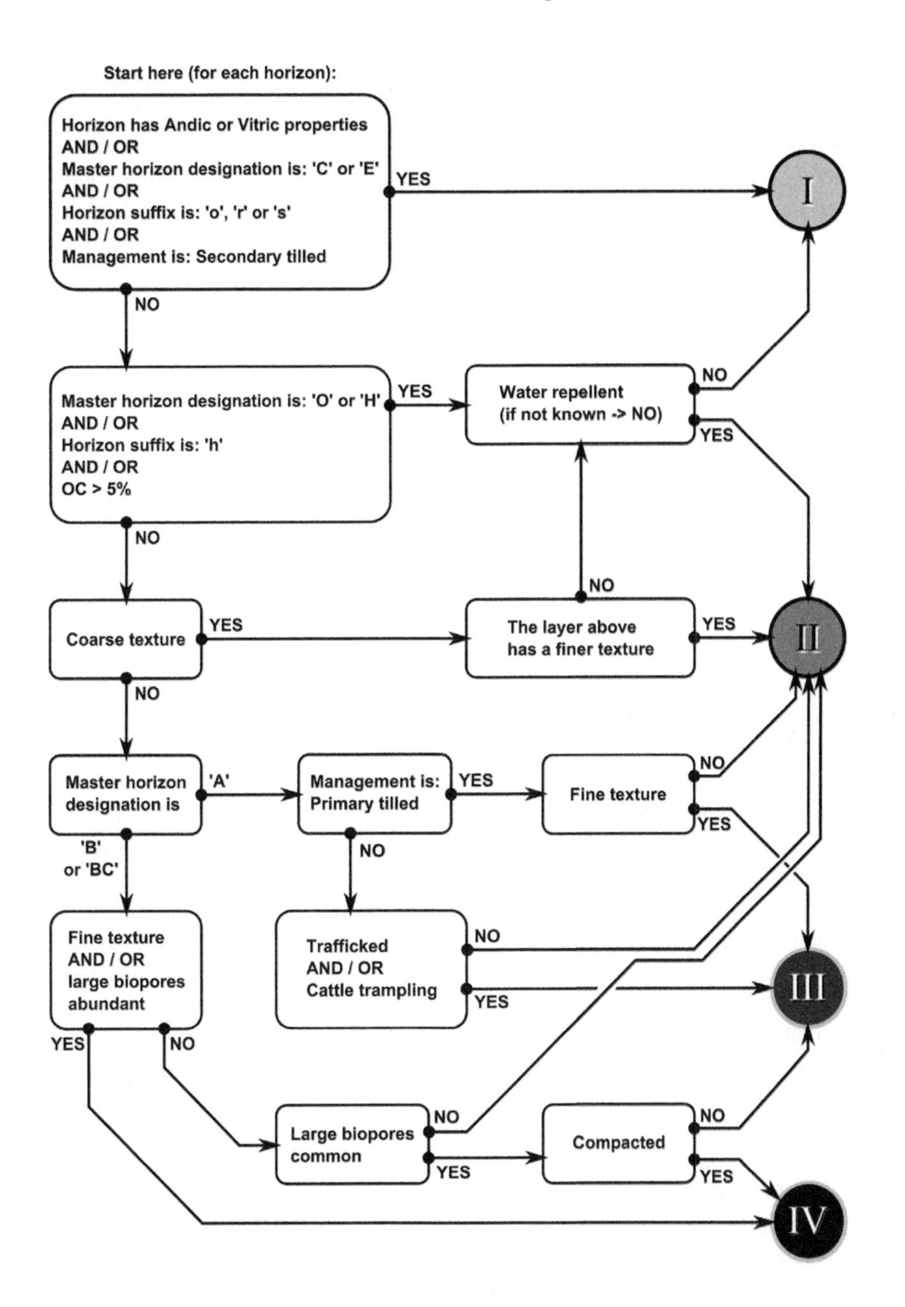

additional subordinate horizon characteristics and material properties (FAO, 2006) as predictor variables. This new version should better reflect the limited potential for preferential flow in soils where iron and aluminum chemistry affects soil structural development through stable micro-aggregation (e.g. Andosols and Ferralsols). In addition, '*E*' horizons are now explicitly considered to lack the potential for preferential flow, since they are generally unstructured (see Fig. 6). Furthermore, the scope has been widened to include preferential finger flow generated by two important mechanisms (water repellency or textural discontinuity at a horizon boundary). In this respect, the modified scheme implicitly assumes that the effects of finger flow can be equated with weak macropore flow (see Weiler and Flühler, 2004) in terms of accelerating flow and transport velocities. Obviously, water repellency varies temporally as a function of soil water content (as does structural porosity in swelling/shrinking clay soils), but such details are beyond the scope of this simple scheme. However, as a first approximation for sandy soils, the answer to the question as to whether the horizon is water repellent could be made dependent on land-use and management practices (e.g. affirmative for undisturbed sites such as no-till arable, forest, or grassland). The scheme requires a submodel to predict the occurrence of large biopores (abundant, common, or rare) in subsoil (see Fig. 6). Jarvis et al. (2009) proposed such a classification for anecic earthworm biopores based on the work of Lindahl et al. (2009). For completeness, this could easily be complemented by considering a class of plants/crops with large tap roots (e.g. alfalfa, trees, etc.) that also produce large biopores.

Figure 7 shows the results of an evaluation of the performance of the scheme making use of the large database on tracer breakthrough experiments collated by Koestel et al. (2011b). A subset of the data ($n = 203$) representing experiments carried out on undisturbed soil under similar conditions (column lengths between 5 and 40 cm, flow rates >1 cm h^{-1} and/or pressure heads within the column > -6 cm) was used. It should be noted that not all aspects of the scheme could be tested with this data set. Specifically, we assumed that the small column sizes and saturated and near-saturated experimental conditions precluded the generation of finger flow and we also ignored the potential effects of harrowing, since most publications do not report soil management practices in such detail. The pore volumes drained when 5% of the applied tracer had leached ($p_{0.05}$) was used a robust indicator of preferential flow (Koestel et al., 2011a). Analysis of variance suggests that the overall model performance is acceptable (p-value <0.0001), although there are no significant differences (at p-value $= 0.05$) between the mean $p_{0.05}$ values of class IV and classes III and II. This may be partly due to the fact that very few ($n = 11$) of the samples in the database are placed in class IV. Further research is clearly needed, but it is possible that only a three-class scheme can be justified by the data and classes III and IV can be combined. Figure 7 shows that there is also considerable unexplained within-class variation. Nevertheless, ca. 44% of the overall variance is explained by the model, which is better than the performance of the

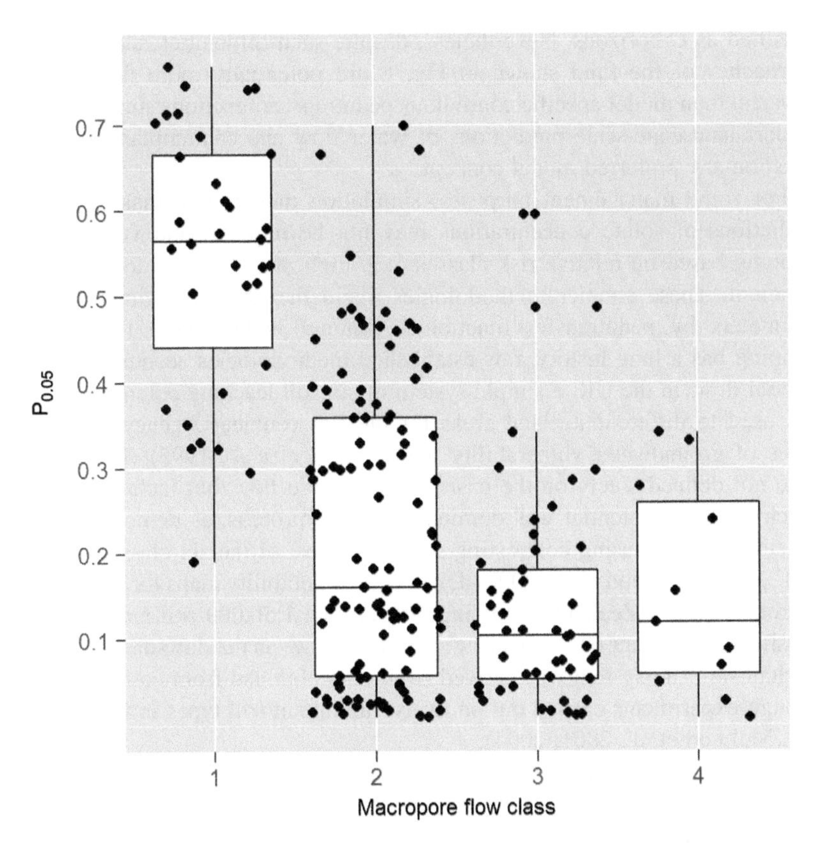

FIGURE 7 Predicted macropore flow class compared with the degree of preferential flow measured in short column breakthrough curve experiments; as indicated by $p_{0.05}$, the pore volumes drained when 5% of the applied solute amount has leached. The center line is the median, the box denotes the 25th and 75th percentiles and the 'whiskers' represent 1.5*interquartile range.

original scheme developed on the limited data set (Jarvis et al., 2009). Obviously, a scheme such as that shown in Fig. 6 involves gross simplifications and limitations, not least the coarse nature of the classification (Jarvis et al., 2009) and the fact that flow and transport regimes in any given horizon are always to some extent influenced by the properties and flow regimes in overlying horizons (e.g. Flury et al., 1994; Vanderborght et al., 2001; Capuliak et al., 2010). The subjective nature of pedological descriptions can also lead to error. One illustrative example that we noted is the ambiguous definition and inconsistent interpretation of what constitutes a *C* horizon (Tandarich et al., 2002). The model assumes that *C* horizons are not susceptible to preferential flow (i.e. class I), since according to the FAO definition they have not been subjected to pedogenic processes. However, some geological materials such as fractured saprolite or glacial till that are known to be prone to preferential flow (e.g. Jørgensen et al., 1998; Mayes et al., 2003; Helmke et al., 2005) are sometimes classified as *C* horizons. Nevertheless, despite such difficulties, classification approaches of the kind shown in Fig. 6 are potentially more flexible and powerful than model-specific continuous pedotransfer functions since they can support landscape-scale predictions of water flow and contaminant transport based on any preferred model concept.

For some management purposes, simulation models that make absolute predictions of solute concentrations may not be required and vulnerability mapping based on relative risk classes (e.g. high, medium, and low) may be sufficient. These simpler methodologies avoid the need to estimate model parameters by pedotransfer functions. Although vulnerability (or hazard) mapping has a long history, few established methodologies account for preferential flow. In the UK, a simple system of six soil leaching potential classes was used to differentiate land areas for the Environment Agency's national series of groundwater vulnerability maps (Palmer et al., 1995). The classes were not defined solely on the basis of macropore flow, but included two in which leaching potential was dominated by the process, as demonstrated in a lysimeter study using contrasting soils from five of the six classes (Brown et al., 2000). McLeod et al. (2008) developed vulnerability maps for leaching of microbes in New Zealand by linking the national 1:50,000 soil map to a risk classification system that considers macropore flow as the dominant transport mechanism. The system was derived from data obtained from column breakthrough experiments carried out on twelve dominant soil types in the country (e.g. McLeod et al., 2001; 2004).

Qualitative vulnerability assessments involve many uncertainties and difficulties. For example, it is not easy to deal with strongly layered soils, where the flow regime may change significantly within the profile (Table 1). Also, the spatial pattern of leaching risks will be different for different solutes of interest (for example, nitrate compared with pesticides, or even for pesticide compounds of contrasting properties; Stenemo et al., 2007). Thus, in principle, there is no unique groundwater vulnerability map which will be appropriate for all solutes. Instead, quantitative spatial modeling is required to predict actual risks for each solute of interest.

4.3. Landscape-Scale Modeling Tools: Case Studies in Pesticide Risk Assessment

In recent years, regulatory demands in Europe (e.g. the EU Water Framework Directive and the 'Thematic Strategy for Sustainable Use of Pesticides') have prompted the development of spatial modeling tools to support pesticide risk assessment and management that can account for all significant loss pathways, including preferential flow. For example, Balderacchi et al. (2008) used the network of pedotransfer functions (pedotransfer system) developed by Stenemo and Jarvis (2007) for the dual-permeability model MACRO to support a catchment and regional-scale risk assessment tool for pesticide leaching to groundwater in northern Italy. Spatially distributed simulations with dual-permeability models run in real time in this way demand considerable computer resources. To circumvent this practical problem, Stenemo et al. (2007) used artificial neural networks to encapsulate the pedotransfer system developed by Stenemo and Jarvis (2007) in a simulation meta-model of MACRO. Lindahl et al. (2008) used this meta-model to investigate the feasibility of using site-specific applications to reduce pesticide leaching to groundwater in the presence of uncertainty in spatial patterns of degradation. At the national scale, Holman et al. (2004) developed a simulation meta-model of MACRO linked to a simpler attenuation factor model for the vadose zone to create a spatially distributed modeling system for predicting pesticide losses to groundwater in England and Wales. The model parameters were estimated using a combination of measured data for 'benchmark soils', expert judgment and some of the pedotransfer functions and rules developed for MACRO by Jarvis et al. (1997). The meta-model, in the form of 'look-up' tables, was based on the results of over 4000 MACRO model simulations for a range of soil types, climate regimes, pesticide properties, and application patterns. Linking the meta-model to existing spatial databases of soil, climate, and compound-specific doses allowed prediction of the concentration of pesticide leaching at 1 m depth. Testing the system against measured data from the lysimeter studies of Brown et al. (2000) showed that it successfully predicted qualitative aspects of leaching (relative ranking), but that absolute values of pesticide leaching losses were relatively poorly predicted. At the national scale, groundwater monitoring data collected by the UK Environment Agency was used to show that the system predicted reasonably well both the relative ranking and realistic regional patterns of leaching of atrazine, isoproturon, chlorotoluron, and lindane (Holman et al., 2004).

The modeling tools described above consider only one flow pathway in the landscape (leaching to groundwater) and generally make predictions of pesticide concentrations for a given depth in the soil (the lower boundary of the model). Modeling tools that explicitly consider different flow pathways in the landscape (leaching, surface runoff, and flow to subsurface drainage systems) are needed to properly distinguish which water bodies are at risk (i.e. surface water or groundwater) and to correctly identify appropriate mitigation strategies. In part to address this issue, a comprehensive hydropedological modeling system for assessing pesticide impacts on water quality at catchment and regional scales was recently developed in the EU project FOOTPRINT. The system is conceived as a simulation meta-model (as 'look-up' tables for the models MACRO and PRZM) based on a large number of pre-defined generic scenarios that characterize the complete spectrum of European agro-environmental conditions (Centofanti et al., 2008). The scenarios were created by intersecting spatial data layers in a GIS representing climatic zones, land cover (arable, pasture, permanent crops, and nonagricultural areas), crop types (soft wheat, winter barley, sugar beet, grain maize, etc.), and soil types (identified from their hydrological, textural, and organic matter profiles). The data set created in this way comprises approximately 1.7 million polygons representing individual areas with the same type of land use, cropping, climatic zone, and soil map unit. Each unique combination of land use, cropping, climatic zone, and soil map unit represents a single agro-environmental scenario in which the local soil is defined from a range of soil types with a defined percentage probability of occurrence and, for those scenarios that have a partly or wholly arable land use, a defined range of annual crops with an estimated percentage probability of occurrence. The scenarios represent the spatial variation and heterogeneity of the European agricultural landscape and, using their associated data on weather, soil physical, soil hydrological, and crop growth characteristics enable a European-scale parametrization of the MACRO and PRZM models using both continuous (e.g. matrix hydraulic properties in MACRO) and class pedotransfer functions (e.g. macropore flow parameters using the classification scheme described by Jarvis et al., 2009).

A key component of the scenarios, with respect to model parametrization, is the grouping of European soil types into a limited number of FOOTPRINT Soil Types (FSTs) based on their hydrological, textural, and organic matter (i.e. sorption potential) characteristics. Each of the 373 defined FSTs is characterized by a set of land-use specific soil properties and, based on a simplification and integration of the HOST (Boorman et al., 1995; Schneider et al., 2007) and French CORPEN (Groupe "diagnostic" du CORPEN) conceptual hydrological models, placed into one of 14 hydrologic classes that differentiate soils according to the dominant pathways of water movement in the landscape. This hydrological model component is critical because (i) it enables the surface runoff component of fast response to be separated from the throughflow and drainage components, thus facilitating an improved parametrization of the PRZM model for estimating surface run-off losses (Hollis, 2007) and (ii) it is used to define the lower boundary condition in MACRO, which determines the partitioning of excess water between recharge to groundwater and discharge to surface water via throughflow or drainflow. Thus, conceptual hydrologic classification systems like that in FOOTPRINT provide the important link between soil types and predictions of flow processes and pathways in the landscape (Schoeneberger and Wysocki, 2005; Lin et al., 2006; Schmocker-Fackel et al., 2007).

5. CONCLUDING REMARKS

From this review, we conclude that even though macropore types and numbers may be variable across short distances (and also at short time scales), spatial patterns of preferential flow at the landscape scale are far from being completely random, but instead show a clear deterministic component, because recognizable diagnostic soil horizons, soil materials, and pedons tend to display characteristic flow and transport patterns. One important reason for this is that the smaller pores in the 'structure' hierarchy, which regulate the initiation of macropore flow and its dissipation within the soil, are relatively stable and predictable. This is because they depend on soil properties (e.g. clay content and mineralogy, and organic matter content), pedogenic processes (e.g. weathering of parent material, clay translocation, and carbon turnover), and site factors (climate, topography, parent material, and land use) that are generally expressed uniformly across larger areas of land and at long temporal scales, ranging from decades and centuries to many thousands of years.

Thus, the synthesis of the literature presented in this chapter suggests that preferential flow regimes should be predictable for many practical purposes, despite the complexity of the web of interacting factors and processes that govern soil structure and the potential for preferential flow in soil (Fig. 2). However, very few studies have attempted to test such a claim by quantifying the predictability of preferential flow. For limited data sets, the parameters of dual-porosity and dual-permeability models have been correlated with basic properties such as clay and organic matter contents (Vervoort et al., 1999; Shaw et al., 2000). A simple decision tree presented by Jarvis et al. (2009) to classify susceptibility to macropore flow has also been successfully tested against a limited data set of solute breakthrough experiments taken from the literature. A new version of this scheme is presented here (Fig. 6), which should represent an improvement. Quantitative meta-analyses of literature data may also help us to more clearly identify what we know and what we do not know about how soil properties and site attributes control preferential flow. This kind of research should be greatly facilitated by the further development of the comprehensive database on solute transport experiments collated from the literature by Koestel et al. (2011b). Hopefully, this database can be made freely available to researchers through the internet, as has been done, for example, for soil hydraulic properties (e.g. the UNSODA database) and can be continuously updated. One difficulty to overcome, which we also encountered here in trying to present a coherent picture of flow and transport regimes in the major soils of the world, is that too many solute transport studies fail to report basic soil and site data. We recommend that reviewers and editors of papers that report experiments on water flow and solute transport insist that authors provide full information on the soil classification (according to the WRB, in addition to any national classification), basic soil properties such as texture, organic matter, pH, and bulk density for each horizon in the profile, key morphological properties (e.g. gleying) and details of past and present land use, tillage, and other relevant management practices. This failure to report basic soil and site data is probably because many soil physicists and hydrologists have focused their attention and interest on the flow and transport processes *per se*, with little collaboration with pedologists who study the properties and genesis of the natural body (the soil) in which these processes are taking place. As suggested by Lin et al. (2005), hydropedology has an important role to play in bridging the gaps between these disciplines.

Prompted by stakeholder and regulatory demands, several spatial modeling tools have been developed that deal with preferential flow as a major transport pathway, despite the large predictive uncertainty that undeniably exists. Apart from model error itself, parameter estimates are also subject to great uncertainty. Pedotransfer functions are often only derived from relatively small data sets, and the dangers of statistical extrapolation to situations outside the 'training' data are well known. Furthermore, deterministic modeling for a given simulation scenario ignores spatial variation in soil properties. How we deal with this uncertainty depends strongly on the purpose and use of the modeling tool (Jarvis, 2007b). For example, the use of worst-case parameter estimates and Monte Carlo uncertainty analyses are appropriate methods when estimating environmental exposure in the context of pesticide regulation (e.g. Stenemo and Jarvis, 2007). However, for risk management, it is not sufficient to treat uncertainty as random stochastic variation, since it is important to know more precisely where in the landscape soils susceptible to preferential flow and contaminant losses are located. This is a downscaling problem: the soil properties and site factors that determine the risk of preferential leaching are reasonably well understood, but the data to support risk mapping and modeling may only be available at a coarse spatial resolution compared with the limited spatial extent of the areas of land at risk (Leu et al., 2004). The possibilities of combining traditional sources of data with novel remote and proximal sensing technologies (e.g. electrical conductivity) and fuzzy classification techniques (McBratney et al., 2000) to improve the spatial resolution and robustness of risk mapping should be investigated.

The development of risk assessment tools is driven by end-user demands. Scientists working for public authorities need predictive decision-support tools to help them effectively manage and mitigate risks to the environment to meet regulatory requirements. In discussions with these practitioners, soil scientists have consistently emphasized the significant impacts of preferential flow on surface water and groundwater quality. Not surprisingly, practitioners have taken these pleas to heart, and are now asking researchers to supply them with modeling tools that can account for this process. We cannot then reply that we know too little and the uncertainties are too large, and that they should wait several more decades for the outcome of more research. We have to apply the knowledge we already have, not least because the alternatives are unacceptable: decision makers tend to take overly 'precautionary' decisions in the face of unquantified uncertainty (Jarvis, 2007b), which would place unnecessary restrictions and costs on industry and land managers, or what is worse, they might use inappropriate modeling tools that do not account for preferential flow and therefore underestimate risk. In this respect, hydropedological approaches offer considerable promise in supporting the parametrization of preferential flow and transport models. Of course, such model systems should be continuously tested against a wide range of data for their ability to match both measured parameter values and overall system response. On the basis of such tests, they can be iteratively refined and improved.

REFERENCES

Addiscott, T.M., Mirza, N.A., 1998. New paradigms for modelling mass transfers in soils. Soil Tillage Res. 47, 105–109.

Aislabie, J., Smith, J.J., Fraser, R., McLeod, M., 2001. Leaching of bacterial indicators of faecal contamination through four New Zealand soils. Aust. J. Soil Res. 39, 1397–1406.

Akhtar, M.S., Richards, B.K., Medrano, P.A., deGroot, M., Steenhuis, T.S., 2003. Dissolved phosphorus from undisturbed soil cores: related to adsorption strength, flow rate or soil structure? Soil Sci. Soc. Am. J. 67, 458–470.

Alavi, G., Dusek, J., Vogel, T., Green, R.E., Ray, C., 2007. Evaluation of dual-permeability models for chemical leaching assessments to assist pesticide regulation in Hawaii. Vadose Zone J. 6, 735–745.

Allaire, S.E., Roulier, S., Cessna, A.J., 2009. Quantifying preferential flow in soils: a review of different techniques. J. Hydrol. 378, 179–204.

Anamosa, P.R., Nkedi-Kizza, P., Blue, W.G., Sartain, J.B., 1990. Water movement through an aggregated, gravelly oxisol from Cameroon. Geoderma 46, 263–281.

Anderson, A.E., Weiler, M., Alila, Y., Hudson, R.O., 2009. Dye staining and excavation of a lateral preferential flow network. Hydrol. Earth Syst. Sci. 13, 935–944.

Anderson, A.N., McBratney, A.B., FitzPatrick, E.A., 1996. Soil mass, surface, and spectral fractal dimensions estimated from thin section photographs. Soil Sci. Soc. Am. J. 60, 962–969.

Angers, D.A., Caron, J., 1998. Plant-induced changes in soil structure: processes and feedbacks. Biogeochem. 42, 55–72.

Armstrong, A.S.B., Rycroft, D.W., Tanton, T.W., 1996. Seasonal movement of salts in naturally structured saline-sodic clay soils. Agric. Water Manag. 32, 15–27.

Bachmair, S., Weiler, M., Nützmann, G., 2009. Controls of land use and soil structure on water movement: lessons for pollutant transfer through the unsaturated zone. J. Hydrol. 369, 241–252.

Bachmann, J., Deurer, M., Arye, G., 2007. Modeling water movement in heterogeneous water-repellent soil: 1. Development of a contact angle dependent water-retention model. Vadose Zone J. 6, 436–445.

Balderacchi, M., di Guardo, A., Vischetti, C., Trevisan, M., 2008. The effect of crop rotation on pesticide leaching in a regional pesticide risk assessment. Environ. Sci. Technol. 42, 8000–8006.

Bedmar, F., Costa, J.-L., Giménez, D., 2008. Column tracer studies in surface and subsurface horizons of two typic argiudolls. Soil Sci. 173, 237–247.

Bejat, L., Perfect, E., Quisenberry, V.L., Coyne, M.S., Haszler, G.R., 2000. Solute transport as related to soil structure in unsaturated intact soil blocks. Soil Sci. Soc. Am. J. 64, 818–826.

Beven, K., Germann, P., 1982. Macropores and water flow in soils. Water Resour. Res. 18, 1311--1325.

Beven, K., 1991. Modeling preferential flow: an uncertain future?. Proc. National Symp. In: Gish, T.J., Shirmohammadi, A. (Eds.), Preferential Flow. ASAE, St. Joseph, Michigan, USA, pp. 1–11.

Blanco-Canqui, H., Lal, R., 2009. Extent of soil water repellency under long-term no-till soils. Geoderma 149, 171–180.

Bogner, C., Engelhardt, S., Zeilinger, J., Huwe, B., 2008a. Visualization and analysis of flow patterns and water flow simulations in disturbed and undisturbed tropical soils. In: Beck, E., Bendix, J., Kottke, I., Makeschin, F., Mosandl, R. (Eds.), Gradients in a Tropical Mountain Ecosystem of Ecuador. Ecol. Studies, vol. 198. Springer-Verlag, Berlin Heidelberg, pp. 387–396.

Bogner, C., Wolf, B., Schlather, M., Huwe, B., 2008b. Analyzing flow patterns from dye tracer experiments in a forest soil using extreme value statistics. Eur. J. Soil Sci. 59, 103–113.

Boorman, D.B., Hollis, J.M., Lilly, A., 1995. Hydrology of Soil Types: a Hydrologically-based Classification of the Soils of the United Kingdom. Inst. Hydrol. Rep. No. 126, Wallingford, UK, pp. 137.

Børgesen, C.D., Jacobsen, O.H., Hansen, S., Schaap, M.G., 2006. Soil hydraulic properties near saturation, an improved conductivity model. J. Hydrol. 324, 40–50.

Bouma, J., Jongerius, A., Boersma, O., Jager, A., Schoonderbeek, D., 1977. The function of different types of macropores during saturated flow through four swelling soil horizons. Soil Sci. Soc. Am. J. 41, 945–950.

Bouma, J., Dekker, L.W., 1978. A case study on infiltration into dry clay soil I. Morphological observations. Geoderma 20, 27–40.

Bouma, J., Wösten, J.H.L., 1979. Flow patterns during extended saturated flow in two undisturbed swelling clay soils with different macro-structures. Soil Sci. Soc. Am. J. 43, 16–22.

Bouma, J., 1991. Influence of soil macroporosity on environmental quality. Adv. Agron. 46, 1–37.

Bouma, J., Booltink, H.W.G., Finke, P.A., 1996. Use of soil survey data for modeling solute transport in the vadose zone. J. Environ. Qual. 25, 519–526.

Brakensiek, D.L., Rawls, W.J., Logsdon, S.D., Edwards, W.M., 1992. Fractal description of macroporosity. Soil Sci. Soc. Am. J. 56, 1721–1723.

Brown, C.D., Hodgkinson, R.A., Rose, D.A., Syers, J.K., Wilcockson, S.J., 1995. Movement of pesticides to surface waters from a heavy clay soil. Pestic. Sci. 43, 131–140.

Brown, C.D., Hollis, J.M., Bettinson, R.J., Walker, A., 2000. Leaching of pesticides and bromide through lysimeters from five contrasting soils. Pest Manag. Sci. 56, 83–93.

Calvo-Cases, A., Boix-Fayos, C., Imeson, A.C., 2003. Runoff generation, sediment movement and soil water behaviour on calcareous (limestone) slopes of some Mediterranean environments in southeast Spain. Geomorphology 50, 269–291.

Cammeraat, E.L.H., Kooijman, A.M., 2009. Biological control of pedological and hydro-geomorphological processes in a deciduous forest ecosystem. Biologia 64, 428–432.

Cammeraat, L.H., 1992. Hydro-geomorphological Processes in a Small Forested Catchment: Preferred Flow Paths of Water. PhD thesis. Univ. Amsterdam, Netherlands, pp. 146.

Cannavacciuolo, M., Bellido, A., Cluzeau, D., Gascuel, C., Trehen, P., 1998. A geostatistical approach to the study of earthworm distribution in grassland. Appl. Soil Ecol. 9, 345–349.

Capuliak, J., Pichler, V., Flühler, H., Pichlerová, M., Homolák, M., 2010. Beech forest density control on the dominant water flow types in andic soils. Vadose Zone J. 9, 747–756.

Carlan, W.L., Perkins, H.F., Leonard, R.A., 1985. Movement of water in a plinthic paleudult using a bromide tracer. Soil Sci. 139, 62–66.

Centofanti, T., Hollis, J.M., Blenkinsop, S., Fowler, H.J., Truckell, I., Dubus, I.G., Reichenberger, S., 2008. Development of agro-environmental scenarios to support pesticide risk assessment in Europe. Sci. Total Environ. 407, 574–588.

Clothier, B., Green, S.R., Deurer, M., 2008. Preferential flow and transport in soil: progress and prognosis. Eur. J. Soil Sci. 59, 2–13.

Comegna, V., Coppola, A., Somella, A., 1999. Nonreactive solute transport in variously structured soil materials as determined by laboratory-based time domain reflectometry (TDR). Geoderma 92, 167–184.

Coquet, Y., Coutadeur, C., Labat, C., Vachier, P., van Genuchten, M.T., Roger-Estrade, J., Šimůnek, J., 2005. Water and solute transport in a cultivated silt loam soil: 1. Field observations. Vadose Zone J. 4, 573–586.

Crescimanno, G., de Santis, A., Provenzano, G., 2007. Soil structure and bypass flow processes in a vertisol under sprinkler and drip irrigation. Geoderma 138, 110–118.

Curry, J.P., Schmidt, O., 2007. The feeding ecology of earthworms – a review. Pedobiologia 50, 463–477.

Danckwerts, P.V., 1953. Continuous flow systems – distribution of residence times. Chem. Eng. Sci. 2, 1–13.

Dexter, A.D., 1988. Advances in characterization of soil structure. Soil Till. Res. 11, 199–238.

Dexter, A.D., Richard, G., Arrouays, D., Czyż, E.A., Jolivet, C., Duval, O., 2008. Complexed organic matter controls soil physical properties. Geoderma 144, 620–627.

de Jonge, L.W., Moldrup, P., Jacobsen, O.H., 2007. Soil-water content dependency of water repellency in soils: effect of crop type, soil management, and physical-chemical parameters. Soil Sci. 172, 577–588.

de Jonge, L.W., Moldrup, P., Schjønning, P., 2009. Soil infrastructure, interfaces and translocation processes in inner space ("Soil-it-is"): towards a road map for the constraints and crossroads of soil architecture and biophysical processes. Hydrol. Earth Syst. Sci. 13, 1485–1502.

de Rooij, G.H., Stagnitti, F., 2002. Spatial and temporal distribution of solute leaching in heterogeneous soils: analysis and application to multisampler lysimeter data. J. Contam. Hydrol. 54, 329–346.

Deurer, M., Duijnisveld, W.H.M., Böttcher, J., Klump, G., 2001. Heterogeneous solute flow in a sandy soil under a pine forest: evaluation of a modeling concept. J. Plant Nutr. Soil Sci. 164, 601–610.

Deurer, M., Bachmann, J., 2007. Modeling water movement in heterogeneous water-repellent soil: 2. A conceptual numerical simulation. Vadose Zone J. 6, 446–457.

Devitt, D.A., Smith, S.D., 2002. Root channel macropores enhance downward movement of water in a Mojave desert ecosystem. J. Arid Environ. 50, 99–108.

Doerr, S.H., Shakesby, R.A., Dekker, L.W., Ritsema, C.J., 2006. Occurrence, prediction and hydrological effects of water repellency amongst major soil and land-use types in a humid temperate climate. Eur. J. Soil Sci. 57, 741–754.

Dörfler, U., Cao, G., Grundmann, S., Schroll, R., 2006. Influence of a heavy rainfall event on the leaching of [^{14}C]isoproturon and its degradation products in outdoor lysimeters. Environ. Poll. 144, 695–702.

Dreccer, M.F., Lavado, R.S., 1993. Influence of cattle trampling on preferential flow paths in alkaline soils. Soil Use Manag. 9, 143–148.

Edwards, C.A., Lofty, J.R., 1982. The effect of direct drilling and minimal cultivation on earthworm populations. J. Appl. Ecol. 19, 723–734.

Edwards, W.M., Shipitalo, M.J., Traina, S.J., Edwards, C.A., Owens, L.B., 1992. Role of *Lumbricus terrestris* (L.) burrows on quality of infiltrating water. Soil Biol. & Biochem 24, 1555–1561.

Ehlers, W., 1975. Observations on earthworm channels and infiltration on tilled and untilled loess soil. Soil Sci. 119, 242–249.

Elliott, E.T., Coleman, D.C., 1988. Let the soil work for us. Ecol. Bulletins 39, 23–32.

Ersahin, S., Papendick, R.I., Smith, J.L., Keller, C.K., Manoranjan, V.S., 2002. Macropore transport of bromide as influenced by soil structure differences. Geoderma 108, 207–223.

FAO, 2006. Guidelines for Soil Description, fourth ed. Food & Agricultural Organization of the United Nations (FAO), Rome, Italy, p. 97.

Feyen, J., Jacques, D., Timmerman, A., Vanderborght, J., 1998. Modelling water flow and solute transport in heterogeneous soils: a review of recent approaches. J. Agric. Eng. Res. 70, 231–256.

Flühler, H., Durner, W., Flury, M., 1996. Lateral solute mixing processes – a key for understanding field-scale transport of water and solutes. Geoderma 70, 165–183.

Flury, M., Flühler, H., Jury, W.A., Leuenberger, J., 1994. Susceptibility of soils to preferential flow of water: a field study. Water Resour. Res. 30, 1945–1954.

Franklin, D.H., West, L.T., Ratcliffe, D.E., Hendrix, P.F., 2003. Characteristics and genesis of preferential flow paths in a piedmont ultisol. Soil Sci. Soc. Am. J. 71, 752–758.

Gerke, H.H., Köhne, J.M., 2002. Estimating hydraulic properties of soil aggregate skins from sorptivity and water retention. Soil Sci. Soc. Am. J. 66, 26–36.

Gerke, H.H., 2006. Preferential flow descriptions for structured soils. J. Plant Nutrit. Soil Sci. 169, 382–400.

Germann, P., Beven, K., 1981. Water flow in soil macropores. III A statistical approach. J. Soil Sci. 32, 31–39.

Germann, P.F., Edwards, W.M., Owens, L.B., 1984. Profiles of bromide and increased soil moisture after infiltration into soils with macropores. Soil Sci. Soc. Am. J. 48, 237–244.

Gerzabek, M.H., Kirchmann, H., Pichlmayer, F., 1995. Response of soil aggregate stability to manure amendments in the Ultuna long-term soil organic matter experiments. Zeitschr. Pflanzer. Boden 158, 257–260.

Groffman, P.M., Tiedje, J.N., 1989. Denitrification in north temperate forest soils: spatial and temporal patterns at the seasonal and landscape scales. Soil Biol. Biochem. 21, 613–620.

Hadas, A., 1987. Long-term tillage practice effects on soil aggregation modes and strength. Soil Sci. Soc. Am. J. 51, 191–197.

Hallett, P.D., Young, I.M., 1999. Changes to water repellence of soil aggregates caused by substrate-induced microbial activity. Eur. J. Soil Sci. 50, 35–40.

Hallett, P.D., Baumgartl, T., Young, I.M., 2001. Subcritical water repellency of aggregates from a range of soil management practices. Soil Sci. Soc. Am. J. 65, 184–190.

Hardie, M.A., Cotching, W.E., Doyle, R.B., Holz, G., Lisson, S., Mattern, K., 2011. Effect of antecedent soil moisture on preferential flow in a texture-contrast soil. J. Hydrol. 398, 191–201.

Hassink, J., 1997. The capacity of soils to preserve organic C and N by their association with clay and silt particles. Plant and Soil 191, 77–87.

Hatano, R., Iwanaga, K., Okajima, H., Sakuma, T., 1988. Relationship between the distribution of soil macropores and root elongation. Soil Sci. Plant Nutr. 34, 535–546.

Hatano, R., Booltink, H.W.G., 1992. Using fractal dimensions of stained flow patterns in a clay soil to predict bypass flow. J. Hydrol. 135, 121–131.

Hatano, R., Kawamura, N., Ikeda, J., Sakuma, T., 1992. Evaluation of the effect of morphological features of flow paths on solute transport by using fractal dimensions of methylene blue staining pattern. Geoderma 53, 31–44.

Helmke, M.F., Simpkins, W.W., Horton, R., 2005. Fracture-controlled nitrate and atrazine transport in four Iowa till units. J. Environ. Qual. 34, 227–236.

Heng, L.K., Tillman, R.W., White, R.E., 1999. Anion and cation leaching through large undisturbed soil cores under different flow regimes. I. Experimental results. Aust. J. Soil Res. 37, 711–726.

Heuvelink, G.B.M., Webster, R., 2001. Modelling soil variation: past, present and future. Geoderma 100, 269–301.

Holden, J., 2009. Topographic controls upon soil macropore flow. Earth Surface Proc. Landforms 34, 345–351.

Holden, J., Gell, K.F., 2009. Morphological characterization of solute flow in a brown earth grassland soil with cranefly larvae burrows (leatherjackets). Geoderma 152, 181–186.

Hollis, J.M., 2007. Modelling runoff inputs to surface waters: present state and future focus. In: Del Re, A.A.M., Capri, E., Fragoulis, G., Trevisian, M. (Eds.), Environmental Fate and Ecological Effects of Pesticides. Proceedings XIII Symposium on Pesticide Chemistry. Universita Cattolica del Sacro Cuore, Piacenza, Italy, pp. 416–425.

Holman, I., Dubus, I.G., Hollis, J.M., Brown, C.D., 2004. Using a linked soil model emulator and unsaturated zone leaching model to account for preferential flow when assessing the spatially distributed risk of pesticide leaching to groundwater in England and Wales. Sci. Tot. Environ. 318, 73–88.

Horn, R., Taubner, H., Wuttke, M., Baumgartl, T., 1994. Soil physical properties related to soil structure. Soil Till. Res. 30, 187–216.

IUSS Working group WRB, 2006. World Reference Base for Soil Resources 2006, second ed. World Soil Resources Reports no. 103, FAO, Rome, p. 128.

Jacobsen, O.H., Leij, F.J., van Genuchten, M.Th., 1992. Parameter determination for chloride and tritium transport in undisturbed lysimeters during steady flow. Nordic Hydrol 23, 89–104.

Jacques, D., Šimůnek, J., Timmerman, A., Feyen, J., 2002. Calibration of Richards' and convection-dispersion equations to field-scale water flow and solute transport under rainfall conditions. J. Hydrol. 259, 15–31.

Jarvis, N.J., Bergström, L., Dik, P.E., 1991. Modelling water and solute movement in macroporous soil. II. Chloride leaching under non-steady flow. J. Soil Sci. 42, 71–81.

Jarvis, N.J., Hollis, J.M., Nicholls, P.H., Mayr, T., Evans, S.P., 1997. MACRO_DB: a decision-support tool to assess the fate and mobility of pesticides in soils. Environ. Modell. Softw. 12, 251–265.

Jarvis, N.J., Zavattaro, L., Rajkai, K., Reynolds, W.D., Olsen, P.-A., McGechan, M., Mecke, M., Mohanty, B., Leeds-Harrison, P.B., Jacques, D., 2002. Indirect estimation of near-saturated hydraulic conductivity from readily available soil information. Geoderma 108, 1–17.

Jarvis, N.J., 2007a. A review of non-equilibrium water flow and solute transport in soil macropores: principles, controlling factors and consequences for water quality. Eur. J. Soil Sci. 58, 523–546.

Jarvis, N., 2007b. Dealing with uncertainty in decision-support tools for exposure assessments of pesticide leaching to groundwater. In: Del Re, A.A.M., Capri, E., Fragoulis, G., Trevisan, M. (Eds.), Environmental Fate and Ecological Effects of Pesticides. Univ. Cattolica del Sacro Cuore, Piacenza, Italy, pp. 10–21.

Jarvis, N., Larsbo, M., Roulier, S., Lindahl, A., Persson, L., 2007. The role of soil properties in regulating non-equilibrium macropore flow and solute transport in agricultural topsoils. Eur. J. Soil Sci. 58, 282–292.

Jarvis, N.J., Etana, A., Stagnitti, F., 2008. Water repellency, near-saturated infiltration and preferential solute transport in a macroporous clay soil. Geoderma 143, 223–230.

Jarvis, N.J., 2008. Near-saturated hydraulic properties of macroporous soils. Vadose Zone J. 7, 1302–1310.

Jarvis, N.J., Moeys, J., Hollis, J.M., Reichenberger, S., Lindahl, A.M.L., Dubus, I.G., 2009. A conceptual model of soil susceptibility to macropore flow. Vadose Zone J. 8, 902–910.

Jarvis, N.J., Taylor, A., Larsbo, M., Etana, A., Rosén, K, 2010. Modelling the effects of bioturbation on the re-distribution of ^{137}Cs in an undisturbed grassland soil. Eur. J. Soil Sci., 61, 24–34.

Jawitz, J.W., 2004. Moments of truncated continuous univariate distributions. Adv. Water Resour. 27, 269–281.

Jenny, H., 1941. Factors of Soil Formation – a System of Quantitative Pedology. McGraw-Hill, New York.

Jensen, M.B., Hansen, H.-C., Hansen, S., Jørgensen, P., Magid, J., Nielsen, N.-E., 1998. Phosphate and tritium transport through undisturbed subsoil as affected by ionic strength. J. Environ. Qual. 27, 139–145.

Johnson, D.L., Watson-Stegner, D., 1987. Evolution model of pedogenesis. Soil Sci. 143, 349–366.

Johnson, M.S., Lehmann, J., 2006. Double-funneling of trees: stemflow and root-induced preferential flow. Ecosci. 13, 324–333.

Jones, C.G., Lawton, J.H., Shachak, M., 1994. Organisms as ecosystem engineers. Oikos 69, 373–386.

Jørgensen, P.R., McKay, L.D., Spliid, N.Z.H., 1998. Evaluation of chloride and pesticide transport in a fractured clayey till using large undisturbed columns and numerical modelling. Water Resour. Res. 34, 539–553.

Joschko, M., Fox, C.A., Lentzsch, P., Kiesel, J., Hierold, W., Krück, S., Timmer, J., 2006. Spatial analysis of earthworm biodiversity at the regional scale. Agric. Ecosyst. Environ. 112, 367–380.

Jury, W.A., Flühler, H., 1992. Transport of chemicals through soil: mechanisms, models, and field applications. Adv. Agron. 47, 141–201.

Kasteel, R., Vogel, H.-J., Roth, K., 2000. From local hydraulic properties to effective transport in soil. Eur. J. Soil Sci. 51, 81–91.

Kasteel, R., Vogel, H.-J., Roth, K., 2002. Effect of non-linear adsorption on the transport behaviour of brilliant blue in a field soil. Eur. J. Soil Sci. 53, 231–240.

Kasteel, R., Garnier, P., Vachier, P., Coquet, Y., 2007. Dye tracer infiltration in the plough layer after straw incorporation. Geoderma 137, 360–369.

Kätterer, T., Schmied, B., Abbaspour, K.C., Schulin, R., 2001. Single- and dual-porosity modelling of multiple tracer transport through soil columns: effects of initial moisture and mode of application. Eur. J. Soil Sci. 52, 25–36.

Kay, B.D., 1990. Rates of change in soil structure under different cropping systems. Adv. Soil Sci. 12, 1–52.

Ketelsen, H., Meyer-Windel, S., 1999. Adsorption of brilliant blue FCF by soils. Geoderma 90, 131–145.

Khitrov, N.B., Zeiliger, A.M., Goryutkina, N.V., Omel'chenko, N.P., Nikitina, N.S., Utkaeva, V.F., 2009. Preferential water flows in an ordinary chernozem of the Azov plain. Eurasian Soil Sci. 42, 757–768.

Kim, J.-G., C-Chon, M., J- Lee, S., 2004. Effect of structure and texture on infiltration flow pattern during flood irrigation. Environ. Geol. 46, 962–969.

Kissel, D.E., Ritchie, J.T., Burnett, E., 1973. Chloride movement in undisturbed swelling clay soil. Soil Sci. Soc. Am. J. 37, 21–24.

Kluitenberg, G.J., Horton, R., 1990. Effect of solute application method on preferential transport of solutes in soil. Geoderma 46, 283–297.

Knudby, C., Carrera, J., 2005. On the relationship between indicators of geostatistical, flow and transport connectivity. Adv. Water Resour. 28, 405–421.

Koestel, J., Kasteel, R., Kemna, A., Esser, O., Javaux, M., Binley, A., Vereecken, H., 2009. Imaging brilliant blue stained soil by means of electrical resistivity tomography. Vadose Zone J. 8, 963–975.

Koestel, J., Moeys, J., Jarvis, N.J., 2011a. Evaluation of nonparametric shape measures for solute breakthrough curves. Vadose Zone J. 10. doi:10.2136/vzj2011.0010.

Koestel, J., Moeys, J., Jarvis, N.J, 2011b Meta-analysis of the effects of soil properties, site factors and experimental conditions on preferential solute transport. Hydrol. Earth Syst. Sci. Discuss., 8, 10007–10052.

Köhne, J.M., Gerke, H.H., Köhne, S., 2002. Effective diffusion coefficients of soil aggregates with surface skins. Soil Sci. Soc. Am. J. 66, 1430–1438.

Köhne, J.M., Köhne, S., Šimůnek, J., 2009a. A review of model applications for structured soils: (a) water flow and tracer transport. J. Contam. Hydrol. 104, 4–35.

Köhne, J.M., Köhne, S., Šimůnek, J., 2009b. A review of model applications for structured soils: (b) pesticide transport. J. Contam. Hydrol. 104, 36–60.

Köhne, J.M., Schlüter, S., Vogel, H.-J., 2011. Predicting solute transport in structured soil using pore network models. Vadose Zone J. 10, 1082–1096.

Kung, K.-J.S., 1990. Preferential flow in a sandy vadose zone. I. Field observation. Geoderma 46, 51–58.

Laabs, V., Amelung, W., Pinto, A., Zech, W., 2002. Fate of pesticides in tropical soils of Brazil under field conditions. J. Environ. Qual. 31, 256–268.

Lange, B., Lüescher, P., Germann, P.F., 2009. Significance of tree roots for preferential infiltration in stagnic soils. Hydrol. Earth Syst. Sci. 13, 1809–1821.

Larsbo, M., 2011. An episodic transit time model for quantification of preferential solute transport. Vadose Zone J. 10, 378–385.

Larsbo, M., Roulier, S., Stenemo, F., Kasteel, R., Jarvis, N.J., 2005. An improved dual-permeability model of water flow and solute transport in the vadose zone. Vadose Zone J. 4, 398–406.

Lawes, J.B., Gilbert, J.H., Warington, R., 1882. On the Amount and Composition of the Rain and Drainage Waters Collected at Rothamsted. William Clowes & Sons Ltd., London.

Le Goc, R., de Dreuzy, J.R., Davy, P., 2010. Statistical characteristics of flow as indicators of channeling in heterogeneous porous and fractured media. Adv. Water Resour. 33, 257–269.

Léonard, J., Rajot, J.L., 2001. Influence of termites on runoff and infiltration: quantification and analysis. Geoderma 104, 17–40.

Leu, C., Singer, H., Stamm, C., Müller, S., Schwarzenbach, R., 2004. Variability of herbicide losses from 13 fields to surface water within a small catchment after a controlled herbicide application. Env. Sci. Technol. 38, 3835–3841.

Lilly, A., Ball, B.C., McTaggart, I.P., Horne, P.L., 2003. Spatial and temporal scaling of nitrous oxide emissions from the field to the regional scale in Scotland. Nutr. Cycling Agroecosyst. 66, 241–257.

Lilly, A., Nemes, A., Rawls, W.J., Pachepsky, Y.A., 2008. Probabilistic approach to the identification of input variables to estimate hydraulic conductivity. Soil Sci. Soc. Am. J. 72, 16–24.

Lin, H.S., McInnes, K.J., Wilding, L.P., Hallmark, C.T., 1999. Effects of soil morphology on hydraulic properties. II. Hydraulic pedotransfer functions. Soil Sci. Soc. Am. J. 63, 955–961.

Lin, H.S., Bouma, J., Wilding, L.P., Richardson, J.L., Kutílek, M., Nielsen, D.R., 2005. Advances in hydropedology. Adv. Agron. 85, 1–89.

Lin, H.S., Kogelmann, W., Walker, C., Bruns, M.A., 2006. Soil moisture patterns in a forested catchment: a hydropedological perspective. Geoderma 131, 345–368.

Lindahl, A.M.L., Söderström, M., Jarvis, N.J., 2008. Influence of input uncertainty on prediction of within-field pesticide leaching risks. J. Contam. Hydrol. 98, 106–114.

Lindahl, A.M.L., 2009. Sources of Pesticide Losses to Surface Waters and Groundwater at Field and Landscape Scales. PhD thesis no. 2009:50. Acta Univ. Agric. Sueciae, SLU, p. 72.

Lindahl, A.M.L., Dubus, I.G., Jarvis, N.J., 2009. A site classification scheme to predict the abundance of the deep-burrowing earthworm *Lumbricus terrestris* L. Vadose Zone J. 8, 911–915.

Lipiec, J., Hatano, R., Slowińska-Jurkiewicz, A., 1998. The fractal dimension of pore distribution patterns in variously-compacted soil. Soil Till. Res. 47, 61–66.

Lohse, K.A., Dietrich, W.E., 2005. Contrasting effects of soil development on hydrological properties and flow paths. Water Resour. Res. 41, W12419. doi:10.1029/2004WR003403.

Loveland, P., Webb, J., 2003. Is there a critical level of organic matter in the agricultural soils of temperate regions: a review. Soil Till. Res. 70, 1–18.

Luo, L., Lin, H., Halleck, P., 2008. Quantifying soil structure and preferential flow in intact soil using X-ray computed tomography. Soil Sci. Soc. Am. J. 72, 1058–1069.

Luo, L., Lin, H., 2009. Lacunarity and fractal analyses of soil macropores and preferential transport using micro-X-ray computed tomography. Vadose Zone J. 8, 233–241.

Luo, L.F., Lin, H., Schmidt, J., 2010. Quantitative relationships between soil macropore characteristics and preferential flow and transport. Soil Sci. Soc. Am. J. 74, 1929–1937.

Mayes, M.A., Jardine, P.M., Mehlhorn, T.L., Bjornstad, B.N., Ladd, J.L., Zachara, J.M., 2003. Transport of multiple tracers in variably saturated humid region structured soils and semi-arid region laminated sediments. J. Hydrol. 275, 141–161.

McBratney, A.B., Odeh, I.O.A., Bishop, T.F.A., Dunbar, M.S., Shatar, T.M., 2000. An overview of pedometric techniques for use in soil survey. Geoderma 97, 293–327.

McBratney, A.B., Minasny, B., Cattle, S.R., Vervoort, R.W., 2002. From pedotransfer functions to soil inference systems. Geoderma 109, 41–73.

McDonnell, J.J., 1990. A rationale for old water discharge through macropores in a steep, humid catchment. Water Resour. Res. 26, 2821–2832.

McGrath, G.S., Hinz, C., Sivapalan, M., 2009. A preferential flow leaching index. Water Resour. Res. 45, W11405.

McLeod, M., Aislabie, J., Smith, J., Fraser, R., Roberts, A., Taylor, M., 2001. Viral and chemical tracer movement through contrasting soils. J. Environ. Qual. 30, 2134–2140.

McLeod, M., Aislabie, J., Ryburn, J., McGill, A., 2004. Microbial and chemical tracer movement through granular, ultic and recent soils. New Zeal. J. Agric. Res. 47, 557–563.

McLeod, M., Aislabie, J., Ryburn, J., McGill, A., 2008. Regionalizing potential for microbial bypass flow through New Zealand soils. J. Environ. Qual. 37, 1959–1967.

Meadows, D.G., Young, M.H., McDonald, E.V., 2008. Influence of relative surface age on hydraulic properties and infiltration on soils associated with desert pavements. Catena 72, 169–178.

Mitchell, A.R., Ellsworth, T.R., Meek, B.D., 1995. Effect of root systems on preferential flow in swelling soil. Comm. Soil Sci. Plant Anal. 26, 2655–2666.

Mooney, S.J., Morris, C., 2008. A morphological approach to understanding preferential flow using image analysis with dye tracers and X-ray computed tomography. Catena 73, 204–211.

Moreno, F., Cabrera, F., Andreu, L., Vaz, R., Martin-Aranda, J., Vachaud, G., 1995. Water movement and salt leaching in drained and irrigated marsh soils of south-west Spain. Agric. Water Manag. 27, 25–44.

Morris, C., Mooney, S.J., 2004. A high-resolution system for the quantification of preferential flow in undisturbed soil using observations of tracers. Geoderma 118, 133–143.

Nieber, J.L., Steenhuis, T.S., Walter, T., Bakker, M., 2006. Enhancement of seepage and lateral preferential flow by biopores on hillslopes. Biologia 61, S225–S228.

Nobles, M.M., Wilding, L.P., McInnes, K.J., 2004. Pathways of dye tracer movement through structured soils on a macroscopic scale. Soil Sci. 169, 229–242.

Oades, J.M., Waters, A.G., 1991. Aggregate hierarchy in soils. Aust. J. Soil Res. 29, 815–828.

Oades, J.M., 1993. The role of biology in the formation, stabilization and degradation of soil structure. Geoderma 56, 377–400.

Öhrström, P., Persson, M., Albergel, J., Zante, P., Nasri, S., Berndtsson, R., Olsson, J., 2002. Field-scale variation of preferential flow as indicated from dye coverage. J. Hydrol. 257, 164–173.

Olsen, P.A., Binley, A., . Henry-Poulter, S., Tych, W., 1999. Characterizing solute transport in undisturbed soil cores using electrical and X-ray tomographic methods. Hydrol. Proc. 13, 211–221.

Olsson, J., Persson, M., Albergel, J., Berndtsson, R., Zante, P., Öhrström, P., Nasri, S., 2002. Multiscaling analysis and random cascade modeling of dye infiltration. Water Resour. Res. 38. doi:10.1029/2001WR000880.

Pachepsky, Y.A., Rawls, W.J., 2003. Soil structure and pedotransfer functions. Eur. J. Soil Sci. 54, 443–451.

Pachepsky, Y.A., Rawls, W.J., Lin, H., 2006. Hydropedology and pedotransfer functions. Geoderma 131, 308–316.

Palmer, R.C., Holman, I.P., Lewis, M.A., 1995. Guide to Groundwater Vulnerability Mapping in England and Wales. HMSO, London, 1995, p. 46.

Parfitt, R.L., 2009. Allophane and imogolite: role in soil biogeochemical processes. Clay Miner 44, 135–155.

Perfect, E., Sukop, M.C., Haszler, G.R., 2002. Prediction of dispersivity for undisturbed soil columns from water retention parameters. Soil Sci. Soc. Am. J. 66, 696–701.

Perillo, C.A., Gupta, S.C., Nater, E.A., Moncrief, J.F., 1999. Prevalence and initiation of preferential flow paths in a sandy loam with argillic horizon. Geoderma 89, 307–331.

Perret, J.S., Prasher, S.O., Kantzas, A., Langford, C., 2000. A two-domain approach using CAT-scanning to model solute transport in soil. J. Environ. Qual. 29, 995–1010.

Perret, J.S., Prasher, S.O., Kacimov, A.R., 2003. Mass fractal dimension of soil macropores using computed tomography: from the box-counting to the cube-counting algorithm. Eur. J. Soil Sci. 54, 569–579.

Peyton, R.L., Gantzer, C.J., Anderson, S.H., Haeffner, B.A., Pfeifer, P., 1994. Fractal dimension to describe soil macropore structure using X ray computed tomography. Water Resour. Res. 30, 691–700.

Prado, B., Duwig, C., Márquez, J., Delmas, P., Morales, P., James, J., Etchevers, J., 2009. Image processing-based study of soil porosity and its effect on water movement through andosol intact columns. Agric. Water Manag. 96, 1377–1386.

Quisenberry, V.L., Smith, B.R., Phillips, R.E., Scott, H.D., Nortcliff, S., 1993. A soil classification system for describing water and chemical transport. Soil Sci. 156, 306–315.

Rao, P.S.C., Green, R.E., Balasubramanian, V., Kanehiro, Y., 1974. Field study of solute movement in a highly aggregated oxisol with intermittent flooding. II. Picloram. J. Environ. Qual. 3, 197–202.

Rawlins, B.G., Baird, A.J., Trudgill, S.T., Hornung, M., 1997. Absence of preferential flow in the percolating waters of a coniferous forest soil. Hydrol. Proc. 11, 575–585.

Rawls, W.J., Brakensiek, D.L., Logsdon, S.D., 1993. Predicting saturated hydraulic conductivity utilizing fractal principles. Soil Sci. Soc. Am. J. 57, 1193–1197.

Reichenberger, S., Amelung, W., Laabs, V., Pinto, A., Totsche, K.U., Zech, W., 2002. Pesticide displacement along preferential flow pathways in a Brazilian oxisol. Geoderma 110, 63–86.

Ren, G.-L., Izadi, B., King, B., Dowding, E., 1996. Preferential transport of bromide in undisturbed cores under different irrigation methods. Soil Sci. 161, 214–225.

Renck, A., Lehmann, J., 2004. Rapid water flow and transport of inorganic and organic nitrogen in a highly aggregated tropical soil. Soil Sci. 169, 330–341.

Ringrose-Voase, A.J., 1996. Measurement of soil macropore geometry by image analysis of sections through impregnated soil. Plant and Soil 183, 27–47.

Ritsema, C.J., Dekker, L.W., Hendrickx, J.M.H., Hamminga, W., 1993. Preferential flow mechanism in a water repellent sandy soil. Water Resour. Res. 29, 2183–2193.

Roger-Estrade, J., Richard, G., Caneill, J., Boizard, H., Coquet, Y., Defossez, P., Manichon, H., 2004. Morphological characterisation of soil structure in tilled fields: from a diagnosis method to the modelling of structural changes over time. Soil Till. Res. 79, 33–49.

Rose, D.A., 1973. Some aspects of hydrodynamic dispersion of solutes in porous materials. J. Soil Sci. 24, 285–295.

Roulier, S., Jarvis, N.J., 2003. Modeling macropore flow effects on pesticide leaching: inverse parameter estimation using microlysimeters. J. Environ. Qual. 32, 2341–2353.

Schlüter, S., Weller, U., Vogel, H.-J., 2011. Soil-structure development including seasonal dynamics in a long-term fertilization experiment. J. Plant Nutr. Soil Sci. 174, 395–403.

Schmocker-Fackel, P., Naef, F., Scherrer, S., 2007. Identifying runoff processes on the plot and catchment scale. Hydrol. Earth Syst. Sci. 11, 891–906.

Schneider, M.K., Brunner, F., Hollis, J.M., Stamm, C., 2007. Towards a hydrological classification of European soils: preliminary test of its predictive power for the base flow index using river discharge data. Hydrol. Earth Syst. Sci. 11, 1–13.

Schoeneberger, P.J., Wysocki, D.A., 2005. Hydrology of soils and deep regolith: a nexus between soil geography, ecosystems and land management. Geoderma 126, 117–128.

Schwartz, R.C., McInnes, K.J., Juo, A.S.R., Cervantes, C.E., 1999. The vertical distribution of a dye tracer in a layered soil. Soil Sci. 164, 561–573.

Schwärzel, K., Renger, M., Sauerbrey, R., Wessolek, J., 2002. Soil physical characteristics of peat soils. J. Plant Nutr. Soil Sci. 165, 479–486.

Scorza, R.P., Smelt, J.H., Boesten, J.J.T.I., Hendriks, R.F.A., van der Zee, S.E.A.T.M., 2004. Preferential flow of bromide, bentazon and imidacloprid in a Dutch clay soil. J. Environ. Qual. 33, 1473–1486.

Seyfried, M.S., Wilcox, B.P., 1995. Scale and the nature of spatial variability: field examples having implications for hydrologic modeling. Water Resour. Res. 31, 173–184.

Shaw, J.N., West, L.T., Truman, C.C., Radcliffe, D.E., 1997. Morphologic and hydraulic properties of soils with water restrictive horizons in the Georgia coastal plain. Soil Sci. 162, 875–885.

Shaw, J.N., West, L.T., Radcliffe, D.E., Bosch, D.D., 2000. Preferential flow and pedotransfer functions for transport properties in sandy Kandiudults. Soil Sci. Soc. Am. J. 64, 670–678.

Shipitalo, M.J., Edwards, W.M., Redmond, C.E., 1993. Comparison of water movement and quality in earthworm burrows and pan lysimeters. J. Environ. Qual. 23, 1345–1351.

Sidle, R.C., Noguchi, S., Tsuboyama, Y., Laursen, K., 2001. A conceptual model of preferential flow systems in forested hillslopes: evidence of self-organization. Hydrol. Proc. 15, 1675–1692.

Šimůnek, J., Jarvis, N.J., van Genuchten, M.T., Gärdenäs, A., 2003. Review and comparison of models for describing nonequilibrium and preferential flow and transport in the vadose zone. J. Hydrol. 272, 14–35.

Šimůnek, J., van Genuchten, M.T., 2008. Modeling nonequilibrium flow and transport processes using HYDRUS. Vadose Zone J. 7, 782–797.

Six, J., Bossuyt, H., Degryze, S., Denef, K., 2004. A history of research on the link between (micro)aggregates, soil biota, and soil organic matter dynamics. Soil Till. Res. 79, 7–31.

Smettem, K.R.J., Bristow, K.L., 1999. Obtaining soil hydraulic properties for water balance and leaching models from survey data: 2. Hydraulic conductivity. Aust. J. Agric. Res. 50, 1259–1262.

Smith, R.G., McSwiney, C.P., Grandy, A.S., Suwanwaree, P., Snider, R.M., Robertson, G.P., 2008. Diversity and abundance of earthworms across an agricultural land-use intensity gradient. Soil Till. Res. 100, 83–88.

Soane, B.D., 1990. The role of organic matter in soil compactibility: a review of some practical aspects. Soil Till. Res. 16, 179–201.

Sollins, P., Radulovich, R., 1988. Effects of soil physical structure on solute transport in a weathered tropical soil. Soil Sci. Soc. Am. J. 52, 1168–1173.

Southard, R.J., Buol, S.W., 1988. Subsoil blocky structure formation in some North Carolina paleudults and paleaqults. Soil Sci. Am. J. 52, 1069–1076.

Stagnitti, F., Li, L., Allinson, G., Phillips, I., Lockington, D., Zeiliguer, A., Allinson, M., Lloyd-Smith, J., Xie, M., 1999. A mathematical model for estimating the extent of solute- and water-flux heterogeneity in multiple sample percolation experiments. J. Hydrol. 215, 59–69.

Stagnitti, F., Allinson, G., Morita, M., Nishikawa, M., Ii, H., Hirata, T., 2000. Temporal moments analysis of preferential solute transport in soils. Environ. Modell. Assess. 5, 229–236.

Stenemo, F., Lindahl, A., Gärdenäs, A., Jarvis, N.J., 2007. Meta-modelling of the pesticide fate model MACRO for groundwater exposure assessments using artificial neural networks. J. Contam. Hydrol. 93, 270–283.

Stenemo, F., Jarvis, N.J., 2007. Accounting for uncertainty in pedotransfer functions in exposure assessments of pesticide leaching to groundwater. Pest Manag. Sci. 63, 867–875.

Stewart, J.B., Moran, C.J., Wood, J.T., 1999. Macropore sheath: quantification of plant root and soil macropore association. Plant Soil 211, 59–67.

Stutter, M.I., Deeks, L.K., Billett, M.F., 2005. Transport of conservative and reactive tracers through a naturally structured upland podzol field lysimeter. J. Hydrol. 300, 1–19.

Taina, I.A., Heck, R.J., Elliot, T.R., 2008. Application of X-ray computed tomography to soil science: a literature review. Can. J. Soil. Sci. 88, 1–20.

Tallon, L.K., Si, B.C., Korber, D., Guo, X., 2007. Soil wetting state and preferential transport of *Escherichia coli* in clay soils. Can. J. Soil Sci. 87, 61–72.

Tandarich, J.P., Darmody, R.G., Follmer, L.R., Johnson, D.L., 2002. Historical development of soil and weathering profile concepts from Europe to the United States of America. Soil Sci. Am. J. 66, 335–346.

Tang, G., Mayes, M.A., Parker, J.C., Yin, X.L., Watson, D.B., Jardine, P.M., 2009. Improving parameter estimation from column experiments by multi-model evaluation and comparison. J. Hydrol. 376, 567–578.

Thomas, G.W., Phillips, R.E., 1979. Consequences of water movement in macropores. J. Environ. Qual. 8, 149–156.

Tisdall, J.M., Oades, J.M., 1982. Organic matter and water-stable aggregates in soils. J. Soil Sci. 33, 141–163.

Valocchi, A., 1985. Validity of the local equilibrium assumption for modeling sorbing solute transport through homogeneous soils. Water Resour. Res. 21, 808–820.

Van Genuchten, M.T., Wierenga, P.J., 1976. Mass transfer studies in sorbing porous media. Part I. Analytical solutions. Soil Sci. Soc. Am. J. 40, 473–480.

van Noordwijk, M., de Ruiter, P.C., Zwart, K.B., Bloem, J., Moore, J.C., van Faassen, H.G., Burgers, S.L.G.E., 1993. Synlocation of biological activity, roots, cracks and recent organic inputs in a sugar beet field. Geoderma 56, 277–286.

Van Schaik, N.L.M.B., 2009. Spatial variability of infiltration patterns related to site characteristics in a semi-arid watershed. Catena 78, 36–47.

Vanclooster, M., Boesten, J., Tiktak, A., Jarvis, N., Kroes, J., Muñoz-Carpena, R., Clothier, B., Green, S., 2004. On the use of unsaturated flow and transport models in nutrient and pesticide management. In: Feddes, R.A., de Rooij, G.H., van Dam, J.C. (Eds.), Unsaturated-zone Modeling: Progress, Challenges and Applications. Kluwer Academic Publishers, pp. 331–361.

Vanderborght, J., Vanclooster, M., Timmerman, A., Seuntjens, P., Mallants, D., Kim, D.-J., Jacques, D., Hubrechts, L., Gonzalez, C., Feyen, J., Diles, J., Deckers, J., 2001. Overview of inert tracer experiments in key Belgian soil types: relation between transport and soil morphological and hydraulic properties. Water Resour. Res. 37, 2873–2888.

Vanderborght, J., Vereecken, H., 2007. Review of dispersivities for transport modeling in soils. Vadose Zone J. 6, 29–52.

Vereecken, H., Kasteel, R., Vanderborght, J., Harter, T., 2007. Upscaling hydraulic properties and soil water flow processes in heterogeneous soils: a review. Vadose Zone J. 6, 1–28.

Verheijen, F.G.A., Cammeraat, L.H., 2007. The association between three dominant shrub species and water repellent soils along a range of soil moisture contents in semi-arid Spain. Hydrol. Proc. 21, 2310–2316.

Vervoort, R.W., Radcliffe, D.E., West, L.T., 1999. Soil structural development and preferential solute flow. Water Resour. Res. 35, 913–928.

Vogeler, I., Scotter, D.R., Green, S.R., Clothier, B.E., 1997. Solute movement through undisturbed soil columns under pasture during unsaturated flow. Aust. J. Soil Res. 35, 1153–1163.

Walker, P.J.C., Trudgill, S.T., 1983. Quantimet image analysis of soil pore geometry: comparison with tracer breakthrough curves. Earth Surf. Proc. Land. 8, 465–472.

Wang, D., Norman, J.M., Lowery, B., McSweeney, K., 1994. Nondestructive determination of hydrogeometrical characteristics of soil macropores. Soil Sci. Soc. Am. J. 58, 294–303.

Wang, K., Zhang, R., Hiroshi, Y., 2009. Characterizing heterogeneous soil water flow and solute transport using information measures. J. Hydrol. 370, 109–121.

Wang, Z., Feyen, J., Ritsema, C.J., 1998. Susceptibility and predictability of conditions for preferential flow. Water Resour. Res. 34, 2169–2182.

Watts, C.W., Dexter, A.R., 1997. The influence of organic matter in reducing the destabilization of soil by simulated tillage. Soil Till. Res. 42, 253–275.

Watts, C.W., Dexter, A.R., 1998. Soil friability: theory, measurement and the effects of management and organic carbon content. Eur. J. Soil Sci. 49, 73–84.

Watts, C.W., Whalley, W.R., Longstaff, D.J., White, R.P., Brooke, P.C., Whitmore, A.P., 2001. Aggregation of a soil with different cropping histories following the addition of organic materials. Soil Use Manag. 17, 263–268.

Webster, R., 2000. Is soil variation random? Geoderma 97, 149–163.

Weiler, M., Flühler, H., 2004. Inferring flow types from dye patterns in macroporous soils. Geoderma 120, 137–153.

Weiler, M., McDonnell, J.J., 2007. Conceptualizing lateral preferential flow and flow networks and simulating the effects on gauged and ungauged hillslopes. Water Resour. Res. 43, W03403. doi: 10.1029/2006WR004867.

Wessolek, G., Schwärzel, K., Greiffenhagen, A., Stoffregen, H., 2008. Percolation characteristics of a water repellent sandy soil. Eur. J. Soil Sci. 59, 14–23.

Whalen, J.K., 2004. Spatial and temporal distribution of earthworm patches in corn field, hayfield and forest systems of southwestern Quebec, Canada. Appl. Soil Ecol. 27, 143–151.

Wickel, A.J., van de Giesen, N.C., Sá., T.D.d.A., 2008. Stormflow generation in two headwater catchments in eastern Amazonia. Brazil, Hydrol. Proc. 22, 3285–3293.

Wösten, J.H.M., Finke, P.A., Jansen, M.J.W., 1995. Comparison of class and continuous pedotransfer functions to generate soil hydraulic characteristics. Geoderma 66, 227–237.

Wösten, J.H.M., Pachepsky, Ya.A., Rawls, W.J., 2001. Pedotransfer functions: bridging the gap between available basic soil data and missing soil hydraulic characteristics. J. Hydrol. 251, 123–150.

Young, D.F., Ball, W.P., 2000. Column experimental design requirements for estimating model parameters from temporal moments under nonequilibrium conditions. Adv. Water Resour. 23, 449–460.

Zehe, E., Flühler, H., 2001. Slope scale variation of flow patterns in soil profiles. J. Hydrol. 247, 116–132.

Zehe, E., Blöschl, G., 2004. Predictability of hydrologic response at the plot and catchment scales: role of initial conditions. Water Resour. Res. 40, W10202. doi: 10.1029/2003WR002869.

Zumr, D., Dohnal, M., Hrnčíř, M., Císlerová, M., Vogel, T., Doležal, F., 2006. Simulation of soil water dynamics in structured heavy soils with respect to root water uptake. Biologia 61, S320–S323.

Chapter 4

Preferential Flow Dynamics and Plant Rooting Systems

Peter F. Germann,[1,*] Benjamin Lange[2] and Peter Lüscher[3]

ABSTRACT

Roots of some plant species are well known in agronomy and forestry for their power to ameliorate poorly structured soils by forming long-lasting macropores that improve infiltration, drainage, and aeration. However, the Richards (1931) equation – the most frequently applied approach to soil hydrology – is not particularly well suited to deal with macropore flow because capillary potential and the content of sessile water are pivotal. Alternative approaches are presented that deal more directly with flow and the associated porosity. One of them dates back to the 1920s. More recent developments stress viscosity and mobile water resulting in a two-parameter Stokes approach to flow in permeable media. The necessary relationships are briefly introduced. Further, the parameters of Stokes flow were correlated with root densities that were determined in situ in the same volumes of soils with varying degrees of stagnic properties. Infiltration improved with increasing root density up to a maximum of about 7500 m m^{-3}. Higher root densities did not increase macropore flow presumably because other soil structures became more important. The example illustrates the capacity of the Stokes flow approach to relate soil properties with hydropedological parameters that are derived from basic principles.

1. INTRODUCTION

The study of soil architecture affecting hydrologic processes across spatio-temporal scales is still challenging. There is a continuing unbridled want for functionally relating the structure with the hydrology of soils as, for instance, Gerke et al. (2010) have documented. Here, a more pragmatic approach is

[1]Prof. Em. Soil Science Section, Department of Geography, University of Bern, Hallerstrasse 12, 3012 Bern, Switzerland

[2]Research Associate, Swiss Federal Institute for Forest, Snow and Landscape Research, Zürcherstrasse 111, 8903 Birmensdorf, Switzerland

[3]Research Leader, Swiss Federal Institute for Forest, Snow and Landscape Research, Zürcherstrasse 111, 8903 Birmensdorf, Switzerland

*Corresponding author: Email: pf.germann@bluewin.ch

Hydropedology, Edited by H. Lin. DOI: 10.1016/B978-0-12-386941-8.00004-6

pursued in that soil architecture is inferred from the interpretation with Stokes flow theory of rapid water-content variations that are due to controlled infiltration. The focus of the chapter is on roots that enhance infiltration and drainage in soils with stagnic properties.

The beneficial effects of plant roots on soil functions are well known in agricultural engineering. Roots of some species form macropores even in stagnic soil horizons that enhance infiltration and drainage, and ultimately aeration. Roots of reeds, for instance, are well adapted to stagnic soils and might be instrumental to soil development at large. As another example, Leuenberger (1935) wrote about farmers in the valley of the river Gürbe (a tributary of the Aare river near Bern, Switzerland), who were encouraged during the drainage works from 1917 to 1923 to plant cabbage for the amelioration of the former swamp soils. As a long-lasting consequence, producing cabbage and its processing to sauerkraut are still the region's trademark. More recently, Sustainable Agricultural Network (2007) offers numerous practical examples of cover crops that are suited to reduce detrimental effects of naturally or man-made compacted soils. However, scientific investigations demonstrating directly the impact of roots on infiltration and drainage are far less abundant than the numerous practical advices. The reports by Angers and Caron (1998) about the co-evolution of root systems with soil structures, and by Mitchell et al. (1995) about alfalfa roots that produced stable macropores in swelling clay soils indicate the importance of roots on soil functions. However, there is still little information available about the structures thus formed and their quantitative impact on hydropedology.

The Forest Hydrological Paradigm (FHP) claims that forests reduce peak flows in steep first-order catchments. The paradigm was used to justify countless reforestation projects in the Swiss Alpine region that were completed between 1870 and 1950. Compared with other vegetation canopies, reforestations were thought to increase depth, density, and sustainability of root systems. The presumed increase of root density, mainly in soils with stagnic horizons, was thought to enhance rapid infiltration through macropores, eventually mitigating flash floods. Research in support of the FHP was initiated on the scales of small watersheds as well as soil profiles. The paired watershed approach was initiated in 1903 when runoff monitoring began in the completely forested Sperbelgraben and in the nearby partially forested Rappengraben. Stähli et al. (2011), while presenting the data records, also summarized the century-long experience. Processes at the small catchment scale are not further pursued here, and preference is given to the early research at the profile scale with Burger's (1922) comparative measurements of infiltrability and air capacity.

The two examples from agronomy and forestry point to the formation of macropores through roots as, for instance, Beven and Germann (1982) have postulated. Macropores are thought of as conduits supporting rapid preferential flow. Although the quest was clear at least since Burger's (1922) investigations not many quantitative approaches have evolved that look more closely at the relationship between root density and infiltration-drainage behavior. This is because there are hardly any hydropedological procedures available that parameterize macropore flow, thus hinting at an epistemic no-man's land.

At least four requirements for promising research schemes are distilled from previous attempts to demonstrate the effects of root density on soil's infiltrability, drainability, and water and air capacities:

1) Experimental protocols should apply to in situ investigations at the soil profile scale;
2) Experiments should be nondestructive with respect to soil hydrology, allowing for repeated infiltration under various initial and boundary conditions;
3) Significant hydrological parameters should emerge for testing their quantitative relationships with root densities;
4) Root densities and soil hydrological parameters should be determined in the same soil volumes.

This chapter reviews briefly approaches to rapid infiltration and drainage at the soil profile scale. Then it introduces Burger's (1922) soil hydrological method of assessing infiltrability and air capacity. Third, a water-content wave approach is presented including its application to the comparison of infiltration patterns in a stagnic soil under crop rotation and under an eight-year old plantation of *Miscanthus sinesis*. Fourth, the Stokes flow theory is summarized to the point of deriving its two flow parameters from infiltration experiments. Finally, root densities in a stagnic forest soil are correlated with the Stokes flow parameters that were determined in situ at the profile scale.

2. APPROACHES TO RAPID INFILTRATION

Infiltration in today's soil hydrology is typically dealt with various hews of the Richards (1931) equation that stresses capillary potentials ψ m and sessile water contents θ m^3 m^{-3}. The potential ψ is considered to be strongly related to $\theta(\psi)$ (i.e. retention curve) and the hydraulic conductivity $k(\psi \; or \; \theta)$ (m s^{-1}). Richards (1931) delimited his approach to capillary flow: "*When the conditions for equilibrium under gravity … are fulfilled, the velocity and acceleration of the capillary liquid are everywhere zero…. which means that the force arising from the pressure gradient just balances gravity. … If this condition does not obtain there will be a resultant water moving force and in general there will be capillary flow.*" (p.322). The steep drop of hydraulic conductivity from saturation, k_{sat}, of two orders of magnitude vis-à-vis the feeble decrease of ψ from 0 to -0.1 m indicates gravity-dominated flow in unsaturated porous media as, for instance, Germann and Beven (1981a) have shown. These conditions do not necessarily comply with Richards (1931) capillary flow.

Early approaches to infiltrabitilty and storage catered to soil–water management. Veihmeyer (1927), for example, was concerned with optimizing irrigation of Californian orchards while introducing the concept of field capacity. Burger (1922) in his quest to support the FHP compared infiltrability and air capacity in soils under mature forests and under agriculture. However, the capillarity-dominated Richards (1931) equation also dominated the concepts of soil water relationships. Alternative ideas evolved, like Horton's infiltration capacity in the early 1930's according to Beven (2004). However, the alternatives could not withstand the popularity of the Richards equation that increased again since the late 1950s with the advent of ever-faster computers and increasingly sophisticated numerical codes, as HYDRUS (Radcliffe and Simunek, 2010) may demonstrate. However, Schmalz et al. (2003) modeled with the HYDRUS-2D code infiltration into a 2-m-deep sand tank with but modest success, and Germann and Hensel (2006), using the same code, were not able to simulate the high wetting front velocities observed in undisturbed soils. Dual-porosity approaches to macropore flow evolved from the Richards equation to account for rapid infiltration as Gerke and van Genuchten (1993) have introduced. Arora et al. (2011) inserted artificial macropores with various areal densities into columns of undisturbed soil while investigating water flow and bromide transport. They concluded from inverse modeling that more complex models are required than the dual-porosity model and the mobile–immobile model. In addition, the model MACRO of Jarvis et al. (1991) assesses infiltration into the matrix with the Richards equation while routing the excesses of matrix flow along effective macropores with a kinematic wave approach (Jarvis, 1994). Flows along macropores that are imbedded in a nonsaturated matrix may lead to the so-called nonequilibrium flow in the context of the Richards equation as Jarvis (2007) has reviewed.

On the other hand, relaxing the soil hydrologic ψ relationships and concentrating on mobile water may lead away from the epistemic flaw and towards novel approaches to macropore flows. Stokes flow, for instance, provides such an alternative. The approach considers that gravity is exclusively equilibrated by viscosity during infiltration as, for instance, Gemann and al Hagrey (2008) have demonstrated.

3. INFILTRATION AND AIR CAPACITY ACCORDING TO BURGER

Burger (1922) compared infiltrability and air capacity of forest soils with agricultural soils. For that, he drove a beveled 0.1-m-high steel cylinder with a 1-L volume flush into the soil. A mesh was laid on the soil surface inside the cylinder and a second cylinder with the same dimensions was tightly pushed on the rim of the first one. The upper one was completely filled with water, and the infiltration time IT (s) was recorded for the water level to drop to the soil surface. Thus, Burger's infiltrability is

$$IB = [0.1(\mathrm{m})/IT]. \tag{1}$$

Infiltration runs on the same sample immediately following the first one increased infiltration times.

Burger (1922) also determined the air capacity of undisturbed soil cores that were sampled with, and stored within the same cylinders. The samples were water saturated in a trough by gradually raising the water level. The cylinders were sealed under water and weighed, placed on a standardized gravel bed for draining to constant weight, and weighed again. The weight differences were converted to water volumes that were then expressed as part of the soil volumes and dubbed the soils' air capacity, *AC*. Benguerell's (1998) analysis of all recovered data from Burger's (1922, 1927, 1937) investigations are displayed in Fig. 1 showing that high infiltration rates occur in soils with high air capacities.

Burger's (1922) approach with a variable head at the soil surface, and a moving and geometrically ill-defined wetting front within the soil, is difficult to cast into a theoretical framework that leads to a general experimental framework. Thus, Burger's approach never got beyond the status of a test, and modern concepts ousted it in the 1950s. However, there is late scientific merit to his efforts although Burger did not further interpret the data. Germann and Beven (1981b) found the relationship of

$$IB = 3.27 \cdot AC^{2.4 \pm 0.33}, \tag{2}$$

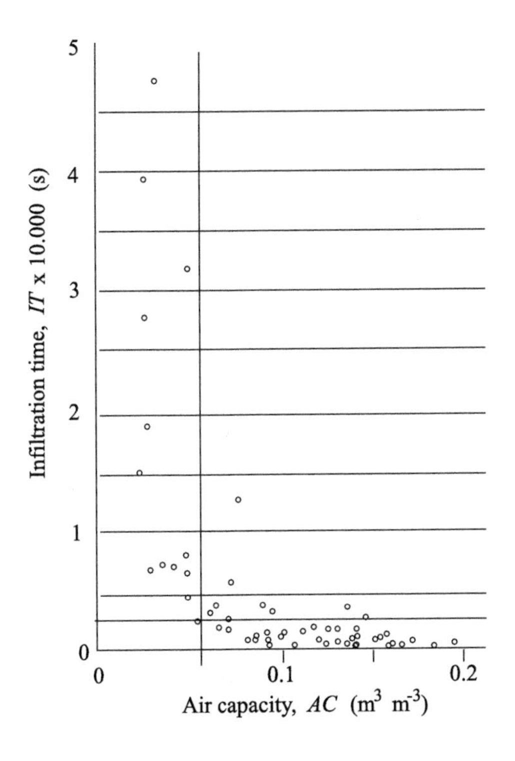

FIGURE 1 Burger's infiltration times IT (s) to infiltrate 0.1 m of water column vs. air capacity AC ($m^3\ m^{-3}$). An approximate threshold is indicated at $AC = 0.06$ below which IT increases drastically. *(From Benguerel, 1998)*

with a coefficient of determination of $r^2 = 0.77$. The result is close to the expected exponent of 2.0 when applying Hagen–Poiseuille flow (Germann and Hensel, 2006). Moreover, Fig. 1 demonstrates a drastic increase of *IT* when *AC* drops below approximately 0.06 $m^3\ m^{-3}$. Flühler (1973) found that air-filled porosity has to be greater than about 0.05 $m^3\ m^{-3}$ for sufficient oxygen supply. The juxtaposition indicates that Burger's approach applies to the assessment of continuous and efficient transport systems of both, gas and water flows when the macropores are either filled with soil air or water. Burger's (1922) concept is possibly worth revisiting in view of today's methods.

Burger was able to demonstrate with this simple infiltration test that soils under mature forests always absorbed higher water volumes at faster rates when compared with arable and pasture soils nearby. The results were probably biased due to careful site selection prior to the measurements. With the first part of the FHP thus proven and with the expectation to demonstrate the soil hydrologic benefits of reforestations, he continued the measurements in the newly established forests. However, frustrated Burger (1927) noted that *IB* increased only in very few areas of forty-year-old reforestations. He considered that the four decades since the beginning of the large-scale reforestation projects was probably too short a period for the general development of hydraulically significant soil structures.

However, follow-up politically oriented reports ignored Burger's scientific disenchantment. There is the case of a prominent forest administrator, who praised the marvelous hydrologic effects of fifty-year-old reforestations in the Canton of Fribourg (Switzerland). He also mentioned the minor drawback that previously poorly drained stagnic soils with neutral chemical reactions under pasture converted into very permeable and, unfortunately, acidic Spodosols (USDA, 1999) under the new forests (Jungo, 1941). The administrator, either knowingly or naïvely (or probably both), thus drew political profit from the mosaïc-like distributions of Spodosols adjacent to Aquolls and Aquepts (USDA, 1999) within the major reforestation project in his Canton of Fribourg.

Frequently, forests are still tagged with the unchallenged FHP although today's investigations propose a more differentiated view. Lüscher and Zürcher (2003), for instance, cautiously lead away from the general applicability of the FHP and recommend an independent look at pedo-hydrological aspects when assessing a forest's flood-mitigating capacity.

4. HYDROPEDOLOGICAL ROOT EFFECTS ASSESSED WITH PATTERNS OF WATER-CONTENT WAVES

A water-content wave, WCW, is the response of the soil water content at some depth in the profile to a pulse of sprinkler irrigation. The pulse is applied during a finite period of time and with a constant rate that is usually close to the infiltration potential. A WCW may be recorded, for instance, with TDR equipment. Figure 2a depicts a time series of water content, $\theta(z,t)$ $m^3\ m^{-3}$,

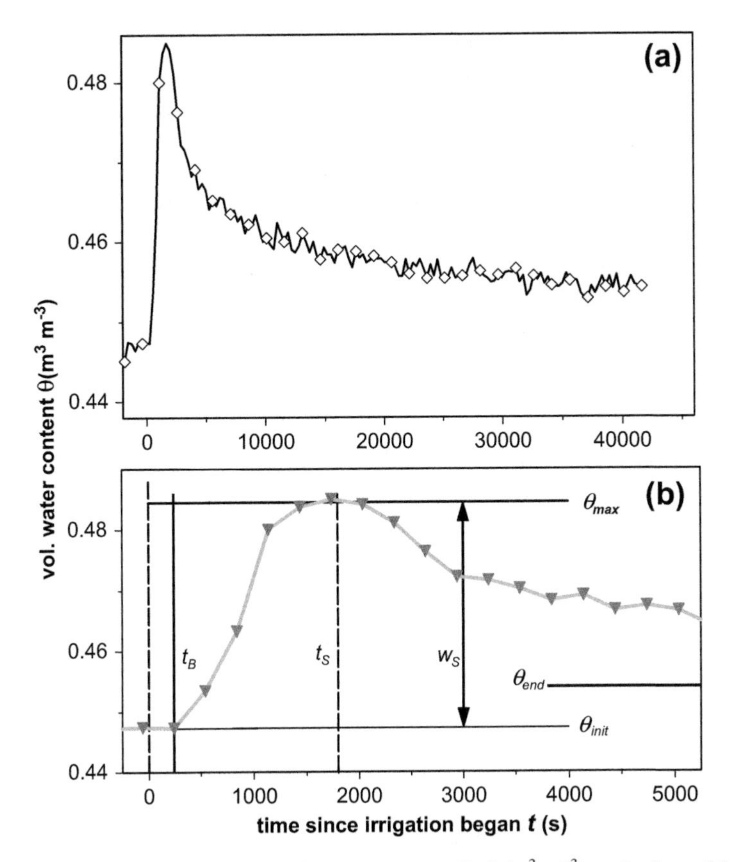

FIGURE 2 (a) Time series of volumetric water contents, $\theta(z,t)$ ($m^3\ m^{-3}$), at depth $z = 0.3$ m in a clay loam under forest. Duration and rate of sprinkler irrigation were 0.5 h and 80 mm h^{-1}, i.e. 2.22×10^{-5} ms^{-1}. (b) Blow-up of first 5000 s of Fig. 2a, showing the typical features of a water-content wave, WCW: t_B and t_S are the arrival time of the wetting front at $z = 0.3$ m and end of sprinkling; θ_{init} and θ_{max} are the volumetric water contents prior to the arrival of the wetting front and at the steady-state maximum, while θ_{end} is the water content at the end of the recorded WCW at $t = 42{,}500$ s, Fig. 2a. The WCW's amplitude is $w_S = \theta_{max} - \theta_{init}$. Note that θ_{max} was always inferior to saturation. *(Data from Bretscher, 2001)*

monitored over a period of more than 11 h after the onset of sprinkling that was applied during ½ h with a rate of 80 mm h^{-1} (equals 2.22×10^{-5} m s^{-1}).

A WCW, as shown more closely in Fig. 2b, that is due to unhampered rapid infiltration features typically a steep increase from the initial water content, θ_{init}, to a plateau of constant water content, θ_{max} (both $m^3\ m^{-3}$), that is followed by a concave tail. The constant water content at the plateau indicates steady state while the concave tail is the relaxation of the input pulse. The times t_B and t_S present the first increase of the water content and the end of sprinkling, while θ_{end} $m^3\ m^{-3}$ is the water content at the end of the recorded

WCW at $t = 42{,}500$ s (see Fig. 2a). The first significant increase of water content is interpreted as the arrival time of the WCW at the depth of measurement. The amplitude of the WCW is $w_S = \theta_{max} - \theta_{init}$ that is considered the mobile water moving through macropores rather independently of the surrounding matrix. In particular, the WCW is considered not being much influenced by the antecedent soil water status.

Germann (1999) distinguished patterns that deviated from the WCW (Fig. 2b). For instance, a delayed and gradual increase of the WCW indicates flow through a poorly structured horizon; a convex tail is related with water perching a short distance beneath the depth of measurement, while an S-shaped increase to a constant water content without a dropping tail a short period after the cessation of irrigation is indicative of capillarity-dominated infiltration. Mdaghri Alaoui et al. (1997) and Germann et al. (2002) provide further examples of interpreted WCW patterns, while the following section demonstrates how to detect enhanced infiltrability that is most likely due to reed roots.

Jäggi (2001) irrigated plots of 1 m × 1 m in a field in crop rotation that was sown with clover at the time of experimentation, and in the adjacent field that carried an eight-year old plantation of the reed *Miscanthus sinensis*. The soil at both sites was an Aquult (USDA, 1999) with a stagnic horizon below the depth of 0.4 m. The parent material was a fluvial deposit. Sprinkling at both sites was during 1 h with rates of 75 mm h^{-1}. TDR probes were horizontally installed at depths 0.1, 0.2, 0.3, 0.4, and 0.5 m and volumetric water contents θ were recorded with TDR equipment at intervals of 300 s. For details, see Germann et al. (2002).

Figure 3 shows the four passings of the two WCWs, one under *Miscanthus* and the other under clover at depths 0.4 and 0.5 m. The time series were standardized by subtracting θ_{init} from the respective $\theta(z,t)$. Irrigation lasted longer than the time period depicted; thus the graphs cannot show the tails. The WCW under *Miscanthus* shows a modest increase of about 0.01 to 0.01.5 $m^3\ m^{-3}$ at both depths. The WCW arrived at 0.4 m between 120 and 420 s, and at 0.5 m between 420 and 720 s after the beginning of sprinkling. The WCW under clover shows a steep increase of about 0.04 $m^3\ m^{-3}$ at 0.4 m, where it arrived between 600 and 900 s after the onset of sprinkling. However, the WCW increased only gradually to about 0.01 $m^3\ m^{-3}$ at the depth of 0.5 m, and arrived there between 1800 and 2100 s.

Pattern interpretation is straight forward. The WCW under *Miscanthus* moved unhampered through the compacted horizon while under clover the WCW got perched between the depths of 0.4 and 0.5 m, as the higher amplitude at the 0.4-m depth indicates when compared with the *Miscanthus* WCW. In addition, the WCW under clover cannot really form at the 0.5-m depth, which is a sign of delayed infiltration into the compacted horizon. Because of the two contrasting canopies, it is concluded that roots of *Miscanthus sinensis* were able to penetrate the compacted horizon, thus forming hydrological significant

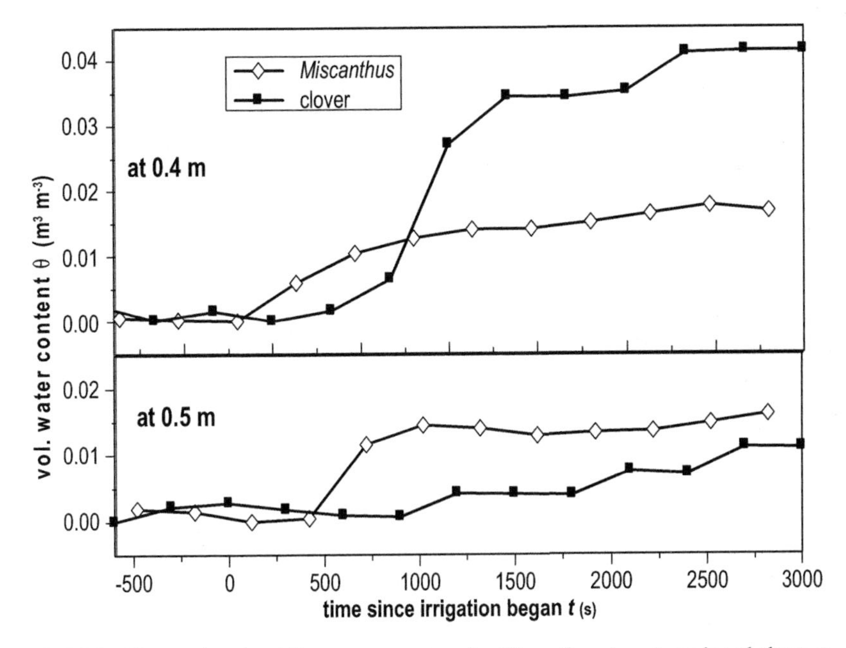

FIGURE 3 Time series of mobile water contents under *Miscanthus sinensis* reeds and clover at depths of 0.4 and 0.5 m. *(From Jäggi, 2001)*

macropores. Moreover, average wetting front velocities evolve from dividing the depths of measurement by the WCWs' corresponding arrival times. The averages of the estimated arrival time ranges yield wetting front velocities under *Miscanthus* of 1.5 and 1 mm s^{-1} at the depths of 0.4 and 0.5 m. Under clover at the 0.4-m depth the wetting front velocity amounts to 0.5 mm s^{-1}, while it is not definable at the 0.5-m depth. Lower wetting front velocities indicate more pronounced flow restrictions that are due to thinner water films, as it will be demonstrated in the following section. All three wetting front velocities score in the upper half of the 215 wetting velocities of Germann and Hensel (2006).

5. ELEMENTS OF A STOKES APPROACH TO MACROPORE FLOW

This section introduces the basic elements of Stokes flow in permeable media as a tool to quantify WCWs similar to the one in Fig. 2. Only the key relationships are given here, while the detailed derivations were previously published. The approach applies directly to the routing of single input pulses as they are due, for instance, to controlled sprinkling. The approach is amenable to the theory of kinematic waves, thus permitting to model any arbitrary sequence of input pulses through superposition.

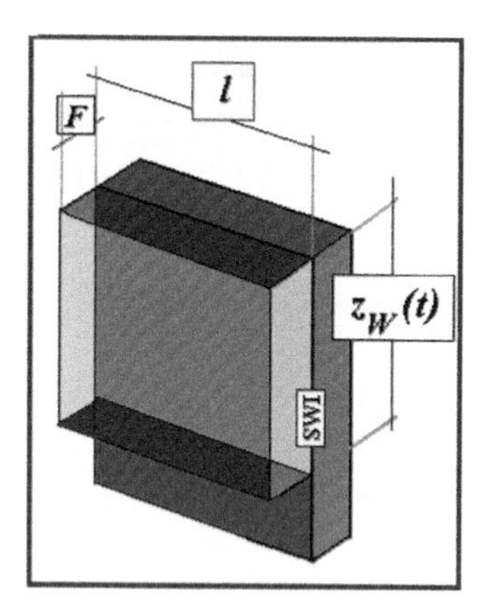

FIGURE 4 Schematic representation of Stokes flow, where F m and l m are the film's thickness and contact length projected onto the horizontal cross-sectional area, $z_W(t)$ is the depth to the wetting front as a function of time, and SWI is the solid–water interface.

Stable flow in water films along the walls of macropores is now considered. Figure 4 illustrates a water film of thickness F m flowing along a solid–water interface, SWI, where no slip occurs i.e. the innermost water layer – *lamina* in Latin – sticks to the wall. All the other *laminae* of the film glide one over the other with their velocities increasing towards the air–water interface, AWI, hence laminar flow. Gravity is the only driving force and the viscous force is the only restricting flow. These conditions qualify for Stokes flow. The flow velocity v (m s^{-1}) is constant because the two forces balance each other. Moreover, the macropores are assumed to be wide enough for the film thickness F to react without constraints on the rate q (m s^{-1}) of the water input to the soil surface. Thus, the volume flux density along macropores may vary within the range of $0 < q \leq k_{sat}$ (m s^{-1}), where the upper limit is hydraulic conductivity at saturation. The wetting front position is $z_W(t)$ m, while l m is the contact length of the water film with the resting parts of the solid–water–air system i.e. the projection of the SWI onto the horizontal cross-sectional area A (m^2).

The velocity of the wetting front is:

$$v = \frac{z_W(t)}{t}, \tag{3}$$

where t (s) is the time elapsed since the beginning of water input to the soil surface. The specific contact length is:

$$L = \frac{\sum_A l}{A}, \tag{4}$$

while the mobile volumetric water content becomes

$$w = \frac{\sum_A l \cdot F \cdot z_W(t)}{A \cdot z_W(t)} = L \cdot F \cdot \tag{5}$$

Moreover, Germann et al. (2007), for instance, presented the following relationships:

$$v = F^2 \cdot \frac{g}{3 \cdot \mu}, \tag{6}$$

where g (=9.81 m s^{-2}) and μ (=10^{-6} m^2 s^{-1}) are acceleration due to gravity and kinematic viscosity of the water. The film thickness F in Eqn (6) depends only on the wetting front velocity v that is viewed as the minimum width of continuous 'macropores'. Hincapié and Germann (2009a) present examples of film thicknesses. From volume balance requirements, it follows that the volume flux density is:

$$q = w \cdot v = F^3 \cdot L \cdot \frac{g}{3 \cdot \mu}. \tag{7}$$

The celerity is the velocity of any change that rides on the water film that is,

$$c = \frac{dq}{dw} = F^2 \cdot \frac{g}{\mu} = 3 \cdot v \cdot \tag{8}$$

Thus, L and F are the two parameters that determine one flow constellation in a macroporous medium. Any two of the three variables q, v, and w suffice to estimate the parameters from measurements at the system (i.e. Darcy) scale. From Eqn (7), it follows that:

$$q = v^{3/2} \cdot L \cdot \left(\frac{3 \cdot \eta}{g}\right)^{1/2}. \tag{9}$$

The macropore flow hypothesis assumes pores that are wide enough to permit any film thickness F to form within $0 < q < q_{max}$ ($\leq k_{sat}$), thus leading to the condition of constant L i.e. $dL/dq = 0$. Hincapié and Germann (2009b) provide an example of $[q \propto v^{3/2}]$ from sprinkling on a column of an undisturbed forest soil with rates of 5, 10, 20, and 40 mm h^{-1}. The coefficient of determination was $r^2 = 0.95$. However, the macropore flow hypothesis restricts Stokes flow in permeable media, and it demands testing whenever an analysis requires $dL/dq = 0$.

A rectangular input pulse is given by its volume flux density and its duration, q and t_S, (s), for instance, the time lapsed from the beginning to the end of sprinkler irrigation. The abrupt end at t_S releases a draining front that moves with the celerity c. The wetting front intercepts the draining front at time and depth:

$$t_I = \frac{3}{2} \cdot t_S, \tag{10}$$

$$z_I = F^2 \cdot t_S \cdot \frac{g}{2 \cdot \eta}. \tag{11}$$

Once the draining front has arrived at a depth in the range of $0 < z_W(t) \leq z_I$, the input pulse starts to relax according to:

$$w(z_W, t) = L \cdot F \cdot \left(\frac{t_D(z_W) - t_S}{t - t} \right)^{1/2}, \tag{12}$$

while the arrival time of the draining front at z_W within $0 < z_W(t) \leq z_I$ is:

$$t_D(z_W) = t_S + \frac{c}{z_W}. \tag{13}$$

Eqns (12) and (13) represent the trailing wave of the Stokes approach. Germann (1985) and Germann et al. (2007) provide details for deriving Eqns (3)–(13). They also present the relationships for routing the input pulse beyond z_I.

The parameters of Eqns (3)–(13) are now estimated from the WCW of Fig. 2 that was measured with TDR equipment at $z_W = 0.3$ m. One possible procedure takes the following steps:

1) Standardize the WCW according to $w(z_W, t) = \theta(z_W, t) - \theta_{\text{end}}$ with the notion that $(\theta_{\text{init}} - \theta_{\text{end}})$ are water losses due to capillarity.
2) Set $(L\ F)$ equal to $(\theta_{\max} - \theta_{end})$ from the data.
3) Match Eqn (12) to the data with $t_D(z_W)$ as the only free matching factor.
4) Compute c and v from Eqns (13) and (8).
5) Compute F from Eqn (6).
6) Compute q from either Eqn (7) or Eqn (9).

Eqns (3)–(13) are analytical expressions, thus permitting to start the analysis from any one of them, providing the opportunity to design the accompanying experiments to particular situations.

Figure 5a shows the result of the procedures (1) to (6). The modeled trailing wave matches the data well. However, $t_W(z)$ had to be increased markedly compared with t_B in Fig. 2 in order to fit $t_D(z)$ and the time scale in general. The fact that $t_W >> t_B$ is most likely due to few channels allowing faster wetting fronts.

Stokes flow is amenable to the theory of kinematic waves according to Lighthill and Witham (1955). In particular, superposition of single kinematic waves allows their spatio-temporal addition to a new composite kinematic wave under the consideration that individual waves lose their identity after joining. Germann et al. (2007) applied Stokes flow to a multitude of paths by arbitrarily dividing the increasing limb of a WCW into rivulets, and routing each of them according to the Eqns (3)–(13) before superposition. Figure 5b illustrates the procedure applied to six rivulets that were derived from the WCW of Fig. 2. The trailing wave does not deviate considerably from the one presented in Fig. 5a, hinting that matching a single wave to data may suffice in many applications. The rivulet approach thus may primarily serve as a research

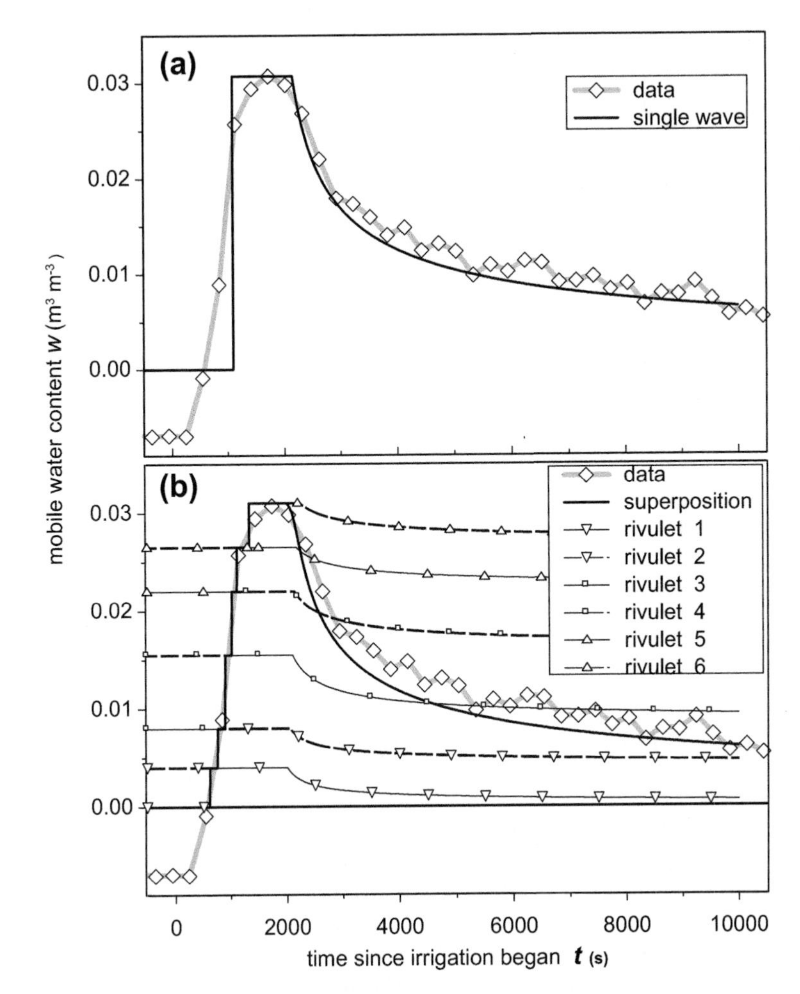

FIGURE 5 (a) Single-wave approach to the data and (b) matching data with six rivulets and superposition. Data are the same as in Fig. 2.

tool that may produce adequate relations between the pore and soil morphology and the Stokes flow parameters.

6. EFFECTS OF TREE ROOT DENSITY ON STOKES FLOW

6.1. Introduction

A decision tool for forest managers about species selection and forest treatment in view of flood mitigation in torrential catchments is the ultimate goal of investigating roots of forest trees affecting the hydrology of soils. The intermediate results of Lange et al. (2009) offered here deal with the relationship between tree root densities and infiltrability in soils with various degrees of aneroxia (i.e. various degrees of oxygen depletion due to slow drainage vis-à-vis high amounts of rainfall).

Data collection was concerned with:

1) In situ morphological interpretation of the soil profiles at 16 sites within a typical spruce–fir–beech forest on Aquolls and Dystrochrepts, delineating the soil horizons and separating them into five groups, according to their well-visible susceptibility of low redox potentials.
2) In situ experimental determination of the Stokes flow parameters in each soil horizon.
3) Laboratory determination of the root densities of samples taken from each horizon.

Data interpretation followed the three steps of:

1) Establishing relationships between root density as independent variables and the flow parameters as the dependent variables.
2) Inserting the relationships of the flow parameters vs. root density into the flow equation, yielding volume flux density q as a function of root density R.
3) Assessing the effects of $q(R)$ on the five groups of morphologically discriminated soil horizons.

6.2. Site Selection, Soils, and Horizons

The field site was located in the Flysch region of the canton Bern (Switzerland) at about 1000 m a.s.l., where mean annual precipitation is about 1600 mm. The forest stand is composed of Norway spruce (*Picea abies*), the most abundant tree species, mixed with silver fir (*Abies alba*), and a few solitary trees of European beech (*Fagus sylvatica*) (Lange et al., 2009). The plant ecologists Ellenberg and Klötzli (1972) described the site as *Bazzanio-Abietetum* that includes the three tree species and shrubs like blueberries (*Vaccinium myrtillus*).

Sixteen sites within a forest stand were identified for detailed investigations. The soil profiles reaching to depths of approximately 1 m are classified as Aquolls, Aqualfs, Udalfs, and Dystrochrepts, according to USDA (1999). Small ridges and depressions characterize the microtopography, and infiltrability at some sites is reduced due to clay-rich horizons. The morphological survey of the 16 profiles yielded a total of 67 horizons that were categorized in the following five horizon groups, HG:

HG1: Well-aerated top soil
HG2: Well-aerated subsoil
HG3: Subsoil showing few rusty mottles
HG4: Hydromorphic subsoil containing numerous rusty and some greenish, bluish, and grayish mottles
HG5: Anoxic subsoil with dominating greenish, bluish, and grayish mottles.

6.3. In situ Sprinkler Infiltration

An area of 1 m × 1 m at each of the 16 sites was rained on three times with 70 mm h^{-1}, i.e. 2×10^{-5} m s^{-1} that corresponds to the region's annual hourly maxima with a return period of about 160 years. Each site was sprinkler irrigated three times with 23-h intervals between the applications. One TDR probe was horizontally installed in the center of each horizon.

The first runs encountered the actual distribution of initial soil moisture within the profile, the second runs started close to satiation, while initial soil moisture of the third runs usually were marginally higher than those of the second runs. Neither ponding nor surface runoff was observed despite the rather high application rates. Only results from the third runs are presented here for better comparison among the 16 plots. They yielded 67 time series of $\theta(z,t)$, and the 67 WCWs were interpreted according to Fig. 5 and Eqns (3)–(13).

6.4. Tree Root Density

After the irrigation–drainage experiments, soil cores with diameters of 0.1 m and lengths of 0.2 m were extracted with a mechanical corer from underneath the sprinkled area at the depths of the TDR probes. Root extraction was by washing the core samples in a 1-mm sieve. Root morphology was analyzed with winRHIZO (V4.1c; Regent Instruments Inc., Quebec, Canada). The root parameters were root length R per unit volume of soil (m m^{-3}), surface area, diameter, and volume, but only R is here pursued. Figure 6 shows the cumulative distribution of R deduced from the 80 samples. About 10% of the samples belong to $R > 10{,}000$ m^{-2}.

6.5. Relationships between Infiltrability and Root Density

Figures 7 and 8 show the relationships of L and F vs. R. Thin F and long L means dispersed flow in a great number of diffusive voids as in a well

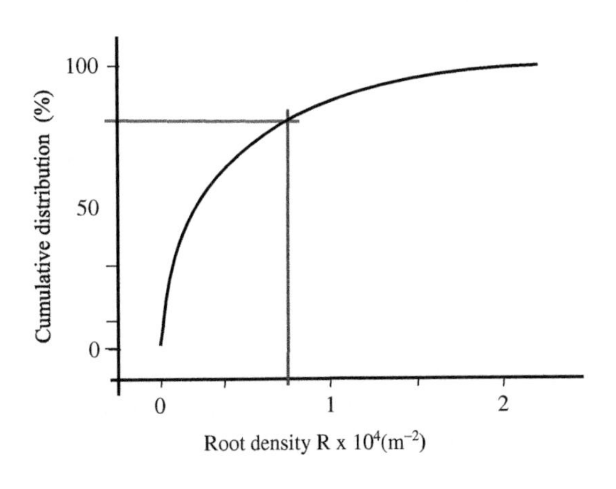

FIGURE 6 Cumulative relative distribution of root lengths R (m^{-2}) per unit soil volume derived from the 67 samples taken from 16 sites. (For R_{opt} see Figs. 7–9).

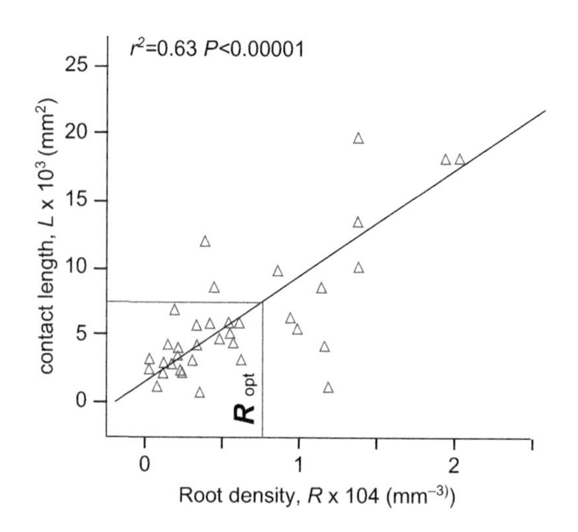

FIGURE 7 Contact length, L vs. root density, R. (For R_{opt} see Figs. 6–9).

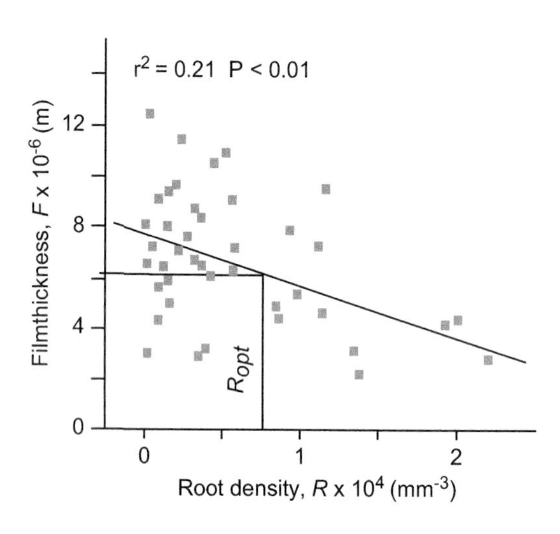

FIGURE 8 Film thickness, F vs. root density, R. Also indicated is R_{opt} (see Figs. 6–9).

developed A horizon. In contrast, thick F and short L indicate relatively few flow paths along distinct channels. Linear regressions produced the relationships of:

$$L(R) = 0.75(\text{m})\cdot R + 2000(\text{m}^{-1}), \tag{14}$$

$$F(R) = 2.1 \times 10^{-10}(\text{m}^3)\cdot R + 8 \times 10^{-6}(\text{m}), \tag{15}$$

with 65 degrees of freedom and coefficients of determination of $r^2(L,R) = 0.63$ and $r^2(F,R) = 0.21$. One wishes coefficients that were closer to unity; however, both are significant outmost at the 1%-error level.

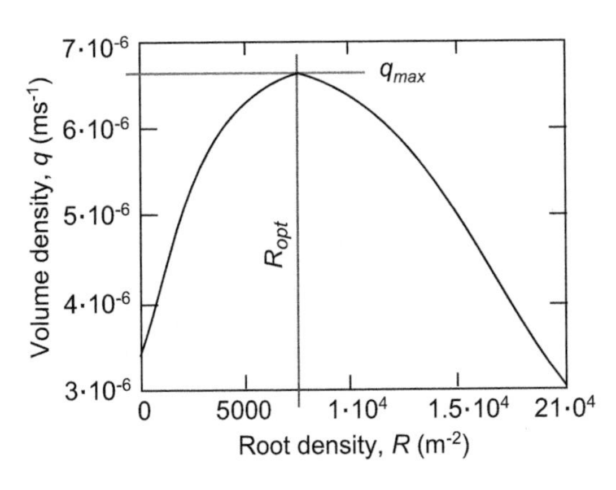

FIGURE 9 Volume flux density, q (m s^{-1}) as function of root density, R (m^{-2}), Eqn (16). The maximum volume flux density at $q_{\max} = 6.6 \times 10^{-6}$ m s^{-1} determines the optimal root density of $R_{\text{opt}} = 7500$ m^{-2}, Figs. 6–10.

Introducing Eqns (14) and (15) into Eqn (7) yields q as function of R:

$$q(R) = \frac{g}{3 \cdot \eta} \cdot F(R)^3 \cdot L(R). \tag{16}$$

Figure 9 illustrates the forth-power relationship of $q(R)$ with a maximum at $q_{\max} = 6.6 \times 10^{-6}$ m s^{-1} (=24 mm h^{-1} or approximately 1 inch h^{-1}) at $R_{\text{opt}} = 7500$ m^{-2}. In stagnic soil horizons, Eqn (16) suggests a high sensitivity of flow rates with respect to low-root densities. Their impact decreases and vanishes beyond R_{opt}, where other soil properties dominate flow. Figure 6 shows that in about 80% of our cases $R < R_{\text{opt}}$. Eqn (16) and Fig. 9 were derived solely with the parameters R, and F and L from the Stokes approach, Eqns (3)–(13). See Lange et al. (2009) for details.

6.6. Morphological Soil Horizons and Tree Root Densities

Figure 10 displays R across the five horizon groups HG1 to HG5. Their field morphological categorization seems reasonable despite some statistical overlapping root densities. The figure also depicts R_{opt} according to Fig. 9. The soil

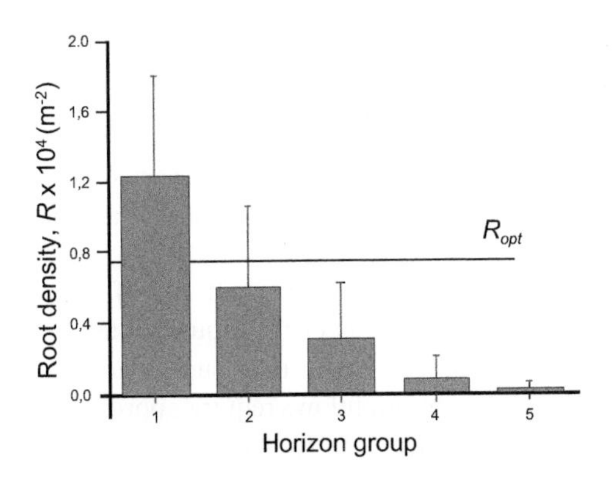

FIGURE 10 Root density R (m m^{-3}) vs. horizon groups, HG, and depicting R_{opt} =7500 m^{-2} from Fig. 9.

hydrological capacities in all but HG1 can be improved through increasing root density, for instance, through favoring fir trees at this kind of sites. However, a limit is expected towards HG5 because even the growth of fir roots is limited in the hostile anoxic environment.

The statistically stronger expression $L(R)$, Eqn (14), and the weaker relationship of $F(R)$, Eqn (15), led nonetheless to a reasonable range of the core statement $q(R)$, Eqn (16) and Fig. 9. It provides a base for the improvement of infiltrability and water storage in stagnic soils that still may support tree growth. Figure 10 illustrates well the co-emergence of root densities and soil structures in the sense of Angers and Caron (1998), when considering a gradual development from HG5 towards HG2.

7. CONCLUDING REMARKS

The quest to scientifically support the Forest Hydrological Paradigm motivated Burger's (1922, 1927, 1937) investigations into infiltrability and air capacity. They were based on the concept of relating the flow rate from in situ infiltration with a measure for the volume fraction of large pores. He graphically compared the two parameters from some samples but without further considerations. In order to produce comparable data sets, he had to stick strictly to the protocol he had established early on categorizing the approach as a reasonable test. Thus, he was not in a position to substantially contribute to the methodological progress in emerging hydropedology. However, re-interpretation of Burger's (1922) accomplishments in view of Stokes flow at large, and the condition for sufficient aeration in stagnic soils in particular, moves his test procedure closer to wider applicability.

However, progress was along other paths. There are solid arguments for the advent of the capillarity-dominated Richards (1931) equation that fascinates us in several ways. It deals with the seepage of soil moisture in all directions; the focus on potentials relates elegantly with water aspects in plant physiology, micro-meteorology, and the vadose zone down to shallow water tables. Research evolved interdependently in these water-related subject areas. Moreover, the soil hydrological functions of retention curves and hydraulic conductivity were determined in numerous laboratories and became available in great numbers. Their combinations with in situ time-dependent recordings of capillary potentials allowed spatio-temporal estimations of soil moisture fluxes and storage as long as capillary potentials remain above –70 kPa. Agricultural engineering picked up the concept for designing irrigation and drainage schemes.

However, hydrologists dealing with groundwater flow and runoff from catchments were not always enthused by the Richards (1931) equation as, for instance, Beven and Germann (1982) compiled. Here, it is suggested that particular soil hydrological processes like preferential flows require approaches that are decoupled from capillarity and sessile water in Richards' sense.

Stokes approach to flow in permeable media, as briefly introduced here, is a candidate for dealing with aspects of preferential flow. The flux-controlled boundary relates directly with infiltration as the transgression of water to below ground from above ground, where it is exposed to pressures that are at least atmospheric. The viable concept of Stokes flow may lead to progress in various directions. The inclusion of superposition and its relation to root densities are two examples of its versatility, and Lange et al. (2011), as another example, assessed with it the initiation of perching of shallow water tables.

ACKNOWLEDGMENTS

We thank two anonymous reviewers who were instrumental in shaping the presentation. Henry Lin's continuing effort in bringing our endeavor to fruition is deeply appreciated.

REFERENCES

Angers, D.A., Caron, J., 1998. Plant-induced changes in soil structure: processes and feedbacks. Biogeochemistry 45, 55–72.

Arora, B., Mohanty, B.P., McGuire, J.T., 2011. Inverse estimation of parameters for multidomain flow models in soil columns with different macropore density. Water Resour. Res. 47, W04512. doi:10.1029/2010WR009451.

Benguerel, P., 1998. Burgers Daten. BSc.-Thesis, Dept. of Geography, Univ. Bern, unpubl.

Beven, K.J., Robert, E., 2004. Horton's perceptual model of infiltration. Hydrol. Process. 18, 3447–3460.

Beven, K., Germann, P., 1982. Macropores and water flow in soils. Water Resour. Res. 18, 1311--1325.

Bretscher, I. 2001. Präferenzielles Fliessen in einem Pseudogley. MSc.-Thesis, Faculty of Science, Univ. Bern, unpubl.

Burger, H., 1922. Physikalische eigenschaften von wald- und freilandböden. Mitt. Schweiz. Centralanst. Forstl. Vers'wes XIII. Band (1. Heft), 3–221.

Burger, H., 1927. Physikalische eigenschaften von wald- und freilandböden. II. Mitteilung: einfluss der durchforstung auf die physikalischen eigenschaften der waldböden. Mitt. Schweiz. Centralanst. Forstl. Vers'wes XIV. Band (2. Heft), 201–250.

Burger, H., 1937. Physikalische eigenschaften von Wald- und freilandböden. V. Mitteilung: entwässerungen und aufforstungen. Mitt. Schweiz. Anst. Forstl. Vers'wes. XX. B (1. Heft), 5–100.

Ellenberg, H., Klötzli, F., 1972. Waldgesellschaften und waldstandorte der schweiz. Mitt. Eidg. Anst. Forstl. Vers'wes 48, 587–930.

Flühler, H., 1973. Sauerstoffdiffusion im boden. Mitt. Eidg. Anst. Forstl. Vers'wes 49 (2), 125–250.

Gerke, H.H., van Genuchten, R., 1993. A dual-porosity model for simulating the preferential movement of water and solutes in structured porous media. Water Resour. Res. 29, 305–319.

Gerke, H.H., Germann, P., Nieber, J., 2010. Preferential and unstable flow: from the pore to the catchment scale. Vadose Zone J. 9, 207–212. doi:10.2136/vzj2010.0059.

Germann, P., 1985. Kinematic wave approach to infiltration and drainage into and from soil macropores. Trans. Am. Soc. Agric. Eng. 28 (3), 745–749.

Germann, P., 1999. Makroporen und präferenzielle Sickerung (Chapter 2.7.1). In: Blume, H.-P., Felix-Henningsen, P., Fischer, W.R., Frede, H.-G., Horn, R., Stahr, K. (Eds.), Handbuch der Bodenkunde 6.Erg.Lfg.7/99, pp. 1–14.

Germann, P., Hensel, D., 2006. Poiseuille flow geometry inferred from velocities of wetting fronts in soils. Vadose Zone J. 5, 867–876.

Germann, P., Helbling, A., Vadilonga, T., 2007. Rivulet approach to rates of preferential infiltration. Vadose Zone J. 6, 207–220. doi:10.2136/vzj2006.0115.

Germann, P., Beven, K., 1981a. Water flow in soil macropores. I. An experimental approach. J. Soil Sci. 32, 1–13.

Germann, P., Beven, K., 1981b. Water flow in soil macropores. III. A statistical approach. J. Soil Sci. 32, 31–39.

Germann, P., al Hagrey, S.A., 2008. Gravity-driven and viscosity-dominated infiltration into a full-scale sand model. Vadose Zone J. 7 (4), 1160–1169.

Germann, P.F., Jäggi, E., Niggli, T., 2002. Rate, kinetic energy and momentum of preferential flow estimated from in-situ water content measurements. Europ J. Soil Sci. 53 (4), 607–618.

Hincapié, I., Germann, P., 2009a. Abstraction from infiltrating water content waves during weak viscous flows. Vadose Zone J. 8, 996–1003. doi:10.2136/vzj2009.0012.

Hincapié, I., Germann, P., 2009b. Impact of initial and boundary conditions on preferential flow. J. Contam. Hydrol. 104, 67–73.

Jäggi, E., 2001. Strukturverbesserung eines verdichteten Bodens mit Chinaschilf *Miscanthus sinensis*. MSc.-thesis, Faculty of Science, Univ. Bern, unpubl.

Jarvis, N.J., 2007. A review of non-equilibrium water flow and solute transport in soil macropores: principles, controlling factors and consequences for water quality. Eur. J. Soil Sci. 58, 523–546. doi:10.111/j.1365-2389.2007.00915.x, 2007.

Jarvis, N.J., 1994. The MACRO Model Version 3.1-Technical Description and Sample Simulations. Reports and Dissertations No. 19. Department of Soil Science, Swedish University of Agriculutral Sciences, Uppsala, Sweden.

Jarvis, N.J., Jansson, P.E., Dick, P.E., Messing, I., 1991. Modelling water and solute transport in macroporous soils. I. Model description and sensitivity analysis. J. Soils Sci. 42, 59–70.

Jungo, J., 1941. Fünfzig jahre aufforstungen in den tälern der aergera, des höllbachs und der sense. Bull. Soc. Frib. Sci. Nat. 35, 114–128.

Lange, B., Lüscher, P., Germann, P., 2009. Significance of tree roots for preferential infiltration in stagnic soils. Hydrol Earth Syst. Sci. 13, 1809–1821.

Lange, B., Germann, P.F., Lüscher, P., 2011. Runoff-generating processes in hydromorphic soils on a plot scale: free gravity-driven versus pressure-controlled flow. Hydol. Process 25, 873–885.

Leuenberger, W., 1935. Das Gürbetal. Vogt-Schild, A.G. Solothurn (Switzerland), p. 156.

Lighthill, G.J., Witham, G.B., 1955. On kinematic waves. I. Flood movement in long rivers. Proc. Roy. Soc. A 229, 281–316 (British).

Lüscher, P., Zürcher, K., 2003. Waldwirkung und Hochwasserschutz: Eine Differenzierte Betrachtungsweise ist Angebracht (Forest Effects and Flood Protection: Pledge for a More Detailed View). In: Landesanstalt für Wald- und Forstwirtschaft, LWF-Bericht, vol. 40. Bayer, Freising. 30–33.

Mdaghri Alaoui, A., Germann, P.F., Lichner, L., Novak, W., 1997. Preferential transport of water and ^{131}Iodide in a clay loam assessed with TDR- and kinematic wave techniques and boundary-layer flow theory. Hydrol. Earth Syst. Sci. 4, 813–822.

Mitchell, A.R., Ellsworth, T.R., Meek, B.D., 1995. Effect of root systems on preferential flow in swelling soil. Commun. Soi Sci. Plant Anal. 26, 2655–2666.

Radcliffe, D., Simunek, J., 2010. Soil Physics with HYDRUS. CRC-Press.

Richards, L.A., 1931. Capillary conduction of liquids through porous mediums. Physics 1, 318–333.

Schmalz, B., Lenartz, B., van Genuchten, M.T., 2003. Analysis of unsaturated water flow in a large sand tank. Soils Sci. 168 (1), 3–14.

Stähli, M., Badoux, A., Ludwig, A., Steiner, K., Zappa, M., Hegg, C., 2011. One Century of Hydrological Monitoring in Two Small Catchments with Different Forest Coverage. Environmental Monitoring and Assessment. doi:10.1007/s10661-010-1757-0.

Sustainable Agricultural Network (SAN), 2007. Managing Cover Crops Profitably. Handbook Series No. 3, third ed. National Agricultural Library, Beltsville, MD, p. 244.

USDA, 1999. Soil Taxonomy. US Dept. Agric. Soil Conserv. Serv., Washington DC.

Veihmeyer, F.J., 1927. Some factors affecting the irrigation requirements of deciduous orchards. Hilgardia 2, 125–291.

Chapter 5

Redoximorphic Features as Related to Soil Hydrology and Hydric Soils

M.J. Vepraskas[1,*] and D.L. Lindbo[1]

ABSTRACT

Hydromorphic features are used to identify where saturation and anaerobic conditions have occurred in soils. These features include redoximorphic features (Fe/Mn based), carbon accumulations (often as darkened A horizons, coated grains, and/or muck), and a sulfur-based feature formed by H_2S. Hydromorphic features form when four specific conditions occur in the soil: 1) organic matter must be present, 2) microorganisms must be actively respiring and oxidizing organic matter, 3) the soil must be saturated, and 4) dissolved oxygen must be removed from the soil water. There are seven rules for formation and interpretation of these features presented. These rules can be applied within a given profile or toposequence to relate the morphologies to hydrology. However, to identify the specific hydrologic conditions of a site appropriate instrumentation must be in place in order to calibrate features to duration and frequency of saturation and reduction. Relict hydromorphic features may remain when hydrologic conditions are changed. However, identifying such features in soils is difficult and must be done cautiously.

1. INTRODUCTION

Hydromorphic features form in soils that are saturated and anaerobic. They include the redoximorphic features, layers of organic soil material, and the odor produced by hydrogen sulfide gas. The hydromorphic features are commonly used to identify soils, and specifically the soil layers, which have been periodically saturated and anaerobic. They are useful because direct measurement of long-term hydrology is time consuming and expensive, and a one-day observation of water-table depth is generally useless because it cannot account

[1]Department of Soil Science, North Carolina State University, Box 7619, Raleigh, NC 27695-7619, USA

[*]Corresponding author: Email: sscmjv@ncsu.edu

Hydropedology, Edited by H. Lin. DOI: 10.1016/B978-0-12-386941-8.00005-8

for seasonal or annual variation. Hydromorphic features are relatively permanent, and can be used as indicators of soil saturation over the long term (years to decades). They are useful for identifying wetland soils, as well as for evaluating soils for on-site waste treatment and dispersal, and in elucidating paleoenvironmental changes (Kemp, 1999). Although hydromorphic features show whether a soil has been saturated and anaerobic, they do not directly indicate the duration of saturation and reduction without calibration to the water-table levels and oxidation–reduction potentials on a site-by-site basis. They may also be used to compare relative duration of saturation between soils within a given region (Lindbo et al., 2010).

The major objectives of this chapter are 1) to define major types of features and review how they are formed, 2) to illustrate how the features can be interpreted using seven rules for interpretation, 3) to relate the hydric soil indicators to hydrology in several landscapes, and 4) to discuss how to identify and interpret relict hydromorphic features.

2. HYDROMORPHIC FEATURES

2.1. Definitions and Types

The features highlighted here are the major ones used to identify aquic conditions, depth to seasonal saturation, and hydric soils (Soil Survey Staff, 2010). We will focus on wetland soils (hydric soils) here, because the hydromorphic features that have been defined for wetland identification are the most comprehensive set of hydromorphic features available (USDA-NRCS, 2010). In the U.S., state and federal laws protect wetlands from being filled in or drained (National Research Council, 1995). These jurisdictional wetlands are identified on the basis of three parameters: wetland hydrology, hydrophytic plants, and hydric soils (Environmental Laboratory, 1987). Hydric soils are defined as soils that formed under conditions of saturation, flooding, or ponding long enough during the growing season to develop anaerobic conditions in the upper part (USDA-NRCS, 2010). The upper part of the soil is considered generally to be the upper 30 cm of loamy and clayey soils, and the upper 15 cm of sandy soils.

Hydric soils are identified by looking for "field indicators" which are generally layers within 30 cm of the surface that have formed under saturated and reduced conditions. Hydric soil field indicators encompass the three major groups of hydromorphic features which are labeled for the elements that have formed them: Fe/Mn-based features, carbon-based features, and sulfur-based features (USDA-NRCS, 2010).

2.1.1. Fe/Mn-based Features (Redoximorphic Features)

Redoximorphic features form through the reduction, movement, and reoxidation of Fe and Mn oxides and hydroxides. There are three categories of redoximorphic features: redox depletions, redox concentrations, and the reduced matrix. The formation of these features was discussed by Vepraskas (2000, 2004) and will not be covered here. Because they form by either the accumulation or removal of Fe/Mn compounds, redoximorphic features have characteristic colors: redox depletions are gray; redox concentrations tend to be red, brown, or yellow; and the reduced matrix appears gray in situ but will turn red, brown, or yellow when exposed to air. As a result of these characteristics, redoximorphic features are normally identified on the basis of their Munsell color as described by the Munsell color system (Cleland, 1921).

Redox depletions form when Fe(III) has been reduced to Fe(II), dissolved, and moved out of a volume of soil. These features usually have Munsell values of 4 or more and chromas of 2 or less (although features of chroma 3 and sometimes 4 may also be formed by the same processes), and have a chroma lower than the matrix.

Redox concentrations form when reduced Fe has accumulated and is then oxidized. They generally have a chroma higher than that of the matrix because of their higher Fe contents. There are three types of redox concentrations: nodules and concretions, masses, and pore linings. The nodules and concretions are hard, stone-like accumulations of Fe and Mn oxides. Iron masses are soft accumulations that lie within the matrix, while pore linings are soft accumulations that lie along or around a root channel or structural crack.

The reduced matrix appears to be a redox depletion when viewed in the soil, but it changes to a redox concentration when exposed to air. This redoximorphic feature contains sufficient amounts of reduced Fe that changes color when exposed to air as the reduced Fe is oxidized. This color change is the primary manner by which a reduced matrix is identified. For the Fe to be in a reduced form the soil must be saturated and anaerobic at the time the reduced matrix is described. Therefore, unlike redox depletions and concentrations, the reduced matrix is a temporary feature.

Redoximorphic features are common and are probably the most widely used features to identify soils that are saturated and anaerobic. However, there must be sufficient Fe in the soil for the features to form.

2.1.2. Carbon-based Features

Carbon-based features take a variety of forms, but all develop when organic C accumulates under anaerobic conditions. A comprehensive description of the features available for hydric soil identification was presented by USDA-NRCS (2010). There are three groups of features which vary depending on the amount of organic C present, the thickness of the soil layers, and Munsell color: organic soil materials, mucky mineral layers, and a dark-colored (organic rich) mineral soil material. Organic soil materials contain more than 12% organic C (Fig. 1). Carbon-based features in this category include Histosols, Histic epipedons, and layers of sapric material. Mucky mineral layers are present when organic

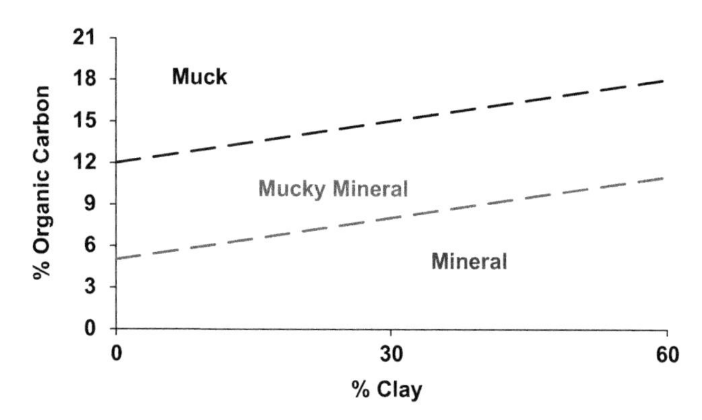

FIGURE 1 Relationship of percent organic carbon content to percent clay content for muck, mucky mineral, and mineral soil materials (*From USDA-NRCS, 2010*).

C levels are between 5 and 12% depending on clay content. These horizons are mineral but are described as "mucky mineral" or "mucky modified textured" layers. The last group, those with a dark surface, occurs when organic C accumulates in amounts less than 5%. Generally, these C-based features are identified by color as shown by Munsell values of 3 or less and chromas of 1 or less (USDA-NRCS, 2010).

Hydromorphic features composed of organic C are most commonly used for identification of hydric soils, because they are commonly found in wetland areas rather than uplands. The C-based features require several characteristics to identify them as indicators of hydric soils because while organic C accumulates under saturated and anaerobic conditions it can also accumulate when temperatures are below biological zero (5 °C) (Everett, 1983), and where large amounts of leaf litter are deposited on the soil surface (i.e. Folists). There may not be a direct connection between the occurrence of a C-based feature and saturation and anaerobic conditions, and so the thickness, color, state of decomposition, and soil temperature regime are used to relate the organic materials to the periods of saturation and anaerobic conditions needed for a soil to be considered a hydric soil (Table 1). In most cases, when the organic soil material is in a layer with a thickness of 20 cm or more then it will usually be found in a soil that has been saturated and anaerobic for a period long enough to meet the requirements of a hydric soil (USDA-NRCS, 2010). However, to ensure accuracy, the soil scientist should have additional evidence that saturation does occur. When organic soil layers are less than 20 cm thick, then soil temperature, color, state of decomposition, and, in some cases, the color of the underlying soil material are used to determine if the material is formed in hydric soil conditions. As shown in Table 1, sapric materials with value of 3 or less and chroma of 1 or less are usually related to hydric conditions when found in thickness <20 cm. Layers

TABLE 1 Examples of Organic C Layers that are Used to Identify Hydric Soils in the U.S. A Map Showing all Land Resource Regions (LRRs) in the U.S. is Presented in USDA-NRCS (2010)

Name	Layer thickness	State of decomp.	Color	Aquic conditions	Land resource regions used	States in the U.S. in LRR
Histosol	41 cm	Any	Any	Assumed	Any	All
Histic epipedon	20–40 cm	Any	Any	Known to occur	Any	All
Black histic	20–40 cm	Any	Hue of 10YR or yellower, value 3 or less, chroma 1 or less	Assumed	Any	All
2 cm muck	2 cm or more	Sapric	Value 3 or less, chroma 1 or less	Assumed	M and N	Southern MN to northern AL
1 cm muck	1 cm or more	Sapric	Value 3 or less, chroma 1 or less	Assumed	D, F, G, H, P, and T	NC coastal plain
Muck presence	Any	Sapric	Value 3 or less, chroma 1 or less	Assumed	U, V, and Z	Southern FL

as thin as 1 cm can be used in warm areas such as southern Florida, while in southern Minnesota the organic layer must be 2 cm or thicker.

While Histosols can be soils that were saturated and anaerobic regardless of soil temperature, the use of the remaining C-based features to identify saturated and anaerobic conditions is restricted to certain temperature conditions. For example, a layer of sapric material (muck) that is $\geq$2 cm thick is interpreted as indicating saturation and anaerobic conditions in Iowa (mesic temperature regime), where winter soil temperatures fall below 5 °C, only when the Munsell value is 3 or less and the chroma is 1 or less. However, in Florida (thermic temperature regime) where temperatures are above 5 °C year-round, then a sapric layer of any thickness (with same color requirements) indicates saturation and anaerobic conditions, because the sapric material can only be preserved under anaerobic conditions.

2.1.3. S-based Features

The reduction of sulfate creates hydrogen sulfide (H_2S) gas which has a distinct odor of rotten eggs. Detection of this odor in a soil identifies this feature. Hydrogen sulfide gas can only be formed and thus detected when the soil is saturated and anaerobic. In contrast to the C-based and Fe-based features, production of H_2S gas in quantities that can be detected (smelled) is relatively rare and most commonly occurs in coastal soils in contact with saltwater. As a result, this feature is not used as frequently as the others to identify saturation and anaerobic conditions in soil.

2.2. Formation

2.2.1. Four Factors Needed for Formation

The three groups of features noted above all form in different ways. However, they all require that the soil be saturated long enough to develop reduced or anaerobic conditions. The development of anaerobic conditions occurs when four basic conditions are met: 1) Organic matter must be present, 2) microorganisms must be actively respiring and oxidizing organic tissues, 3) the soil must be saturated, and 4) dissolved oxygen must be removed from the soil pore water. These four factors must all be present for a soil to develop anaerobic conditions. If even one is missing then the soil will remain aerobic, and hydromorphic features indicative of saturation and reduction will not form.

Microorganisms create the anaerobic conditions through the oxidation–reduction (redox) reactions they use to get their energy through respiration. Microorganisms will oxidize organic compounds and reduce various electron acceptors. The organic matter sources appear to be primarily root tissues, because redox depletions frequently develop around root channels (Vepraskas, 2004). Other organic residues may be used (e.g. dead leaves) if they are buried. Dissolved organic C may also be an important source of oxidizable organic C. While the biochemistry of the process is complex, the oxidation of organic materials by microorganisms is important because the organic matter supplies the electrons which are used to reduce the oxygen which creates the anaerobic conditions in saturated soils, and later reduces the Fe/Mn oxides and hydroxides as well as sulfates that will form the Fe-based and S-based indicators.

It can be assumed that if decomposable organic tissues are present in the soil, then microbes should be active as long as the soil temperature is suitable. It is generally assumed that suitable temperatures occur when the soil temperature is above biological zero or 5 °C. Saturation is required to stop atmospheric oxygen from entering the soil. If oxygen is present, and is resupplied to the soil from the atmosphere, then it will be the dominant electron acceptor used by the microorganisms. Saturating the soil by filling the pores with water effectively blocks atmospheric O_2 from entering the soil.

Although a soil is saturated, the water can contain dissolved oxygen and this must be removed for the soil to become anaerobic. This will happen in time if the previous three factors are met. It occurs fastest when the water is stagnant. Moving water prolongs the amount of time needed to achieve anaerobic conditions (Dusek et al., 2008).

2.2.2. *Importance of Microsites or "Hotspots"*

When the four aforementioned conditions are met, then anaerobic conditions should develop in a soil. As might be expected, the soils first become anaerobic in the small volumes of soil which contain the organic tissues that the microbes are actively oxidizing such as dead roots or buried leaves (Fig. 2). These microsites or "hotspots" of biological activity are where the four conditions needed for development of anaerobic conditions are met (Parkin, 1987). Figure 2a shows the live roots prior to saturation. Once the water table is raised to the surface, some roots die and are oxidized by bacteria (Fig. 2b). The black color around the roots defines the hotspots and in this case is formed by FeS, which is a combination of reduced Fe and S (Fig. 2b). The reduction occurs around the roots and extends into the soil matrix for about 1 cm. Over time (Fig. 2c), the black FeS color is seen to extend into the matrix, possibly due to dissolved organic compounds produced in the hotspot diffusing into the matrix and allowing bacteria there to reduce Fe and S. As the water table is lowered, oxygen enters the soil and oxidizes both the Fe and S and the black color disappears. Although they cannot be seen in the photographs shown, Fe concentrations had formed around the root channel surfaces, with redox depletions occurring behind the Fe concentrations but along the root channel.

Figure 3 illustrates a simple process of how a redox depletion and concentration could form in a soil. This process may occur around a dead root which forms the locus for a hotspot, as shown in Fig. 2.

3. INTERPRETATION

3.1. Rules for Interpreting Redoximorphic Features

Redoximorphic features are the most common form of hydromorphic feature that occurs in soils that are saturated and reduced. They are also the most widely used features for evaluating saturation and anaerobic conditions because they form at any depth in the soil where some organic tissues are found, as opposed to the C-based and S-based features which are used primarily near the surface to identify hydric soils. Understanding how redoximorphic features relate to saturation and reduction can be confusing, and so we propose a set of rules to help with their interpretation in the field. The interpretation of these features can be relatively straightforward if the soil scientist understands their formation. The following rules for interpretation are simple guidelines. There

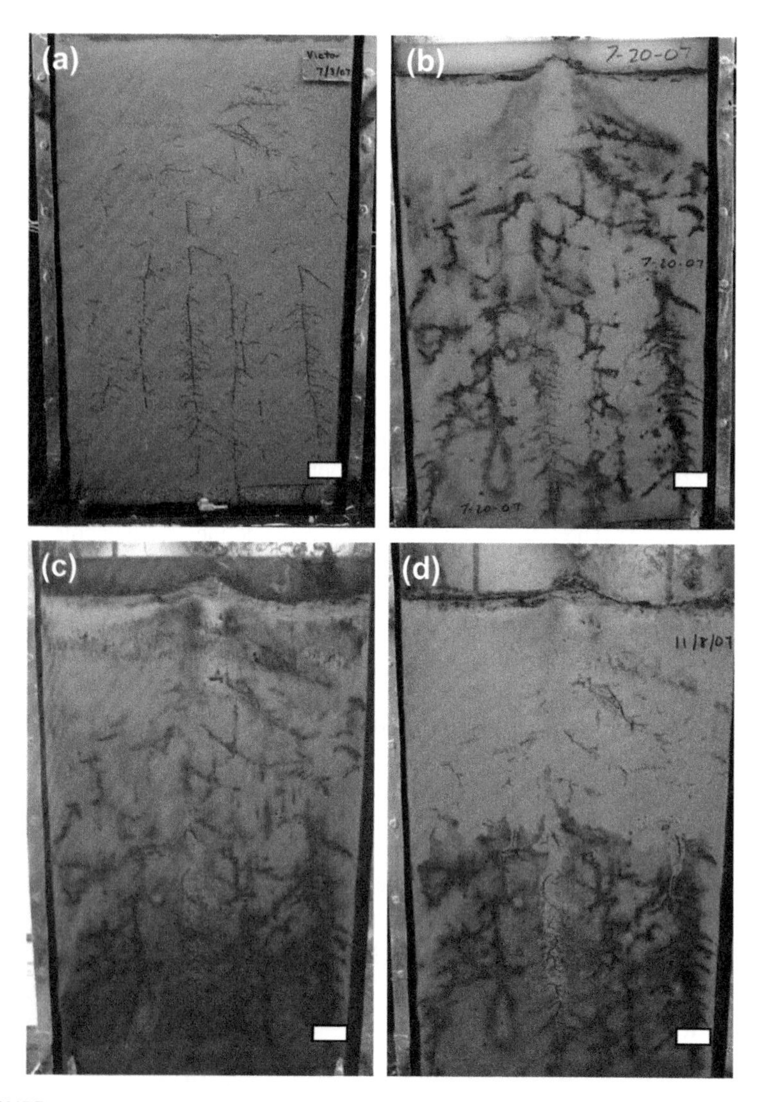

FIGURE 2 Illustration of microsite (hotspot) formation around decomposing roots. Pond pine roots (*Pinus serotina* L) are shown in boxes (30 cm wide × 60 cm tall × 5 cm deep) filled with sand. In (a), pond pine roots were grown at field capacity moisture until the roots reached the bottom of the box. The soil in the box was saturated and ponded water was kept on the surface. After 17 days of saturation (b), some roots died and reducing conditions developed in the rhizospheres surround the dead roots shown by the black color which is produced by FeS. The black areas are the microsites which develop in the rhizospheres around the decomposing organic tissue. After 100 days (c), reducing conditions have extended from the microsites into the soil matrix. The soil was drained (d) after 100 days, and as oxygen penetrates downward into the soil, the black color is lost as the FeS is oxidized. The oxidized Fe formed pore linings along root channels, while redox depletions formed in the original microsites (not shown). White bars are 4 cm long. *(Color version online and in color plate)*

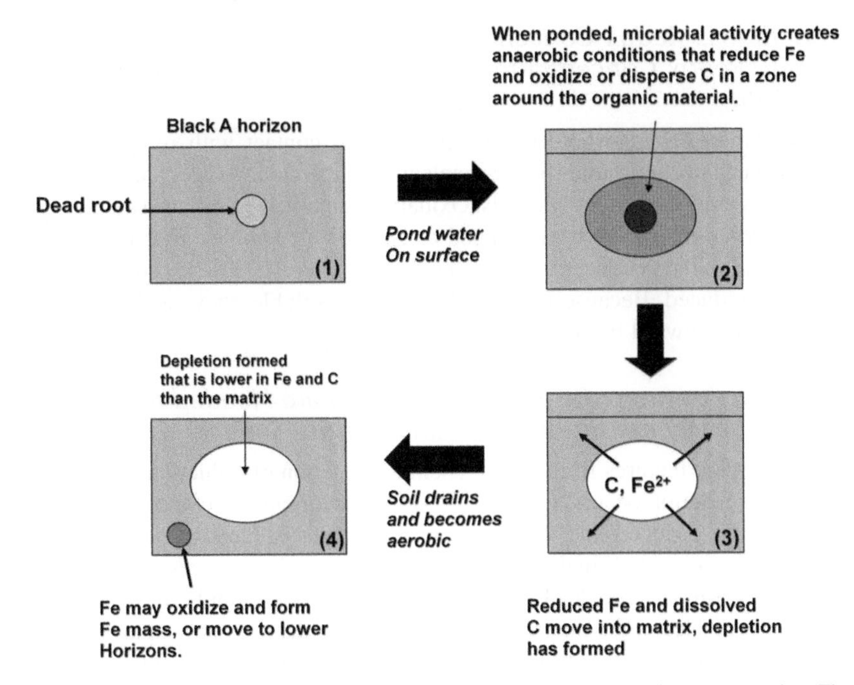

FIGURE 3 Illustration showing formation of a redox depletion and redox concentration. The formation begins around a dead root (seen in cross section) that will form the locus of a microsite. Bacteria begin to oxidize the root tissue. When the soil saturates, microbial respiration reduces dissolved oxygen in the soil water to make the soil anaerobic. Eventually, Fe is reduced in the microsite (2) that developed around the root tissue where the microbes are active. The reduced Fe diffuses into the matrix (3). Organic C that is bound to the soil particles and gives the soil a black color, can also be dispersed when Fe is reduced and it too moves out of the microsite. The loss of Fe and C leaves the microsite a gray color. When the soil drains, the reduced Fe will oxidize to form an Fe mass, or possibly a pore lining. The gray microsite is an Fe depletion. The process shown is illustrated for an A horizon, but can occur in any soil horizon where roots or organic tissues occur. *(Modified from Vepraskas et al., 2006)*

are always exceptions, but, in general, these represent the most common conditions.

3.1.1. Rule 1: Redox Depletions Usually Form where Roots Grow—Around Root Channels or Cracks—when the Four Conditions are Met. If Depletions Occur where Roots could not Grow then Features may not be Redoximorphic Features

This rule pertains to subsoil layers where most of the organic C needed by the microbes for creating reducing conditions is supplied by the roots. Isolated gray spots in the soil matrix will usually not be redox depletions if roots could not enter there. While there are exceptions to any rule, it is the authors' experience that roots will supply most of the C needed to create redox depletions.

3.1.2. Rule 2: Redox Concentrations form where Oxygen was Present. Redox Concentrations Form by the Oxidation of Fe(II) and Mn(II)

The oxidation occurs where these ions came in contact with oxygen. This oxidation can occur around root channels when molecular oxygen enters these large pores after the water table falls. Oxidation can also occur in the soil matrix where oxygen may have been trapped as isolated air bubbles after the water table rose. The redox concentrations do not show where the Fe they contain was originally reduced. Because Fe(II) and Mn(II) are soluble, they can be carried upward or downward in the soil.

3.1.3. Rule 3: Longer Periods of Saturation are Needed to Form Depletions Deeper in the Soil than Near the Surface

Redoximorphic features form by the chemical reaction of reduction. As noted earlier, this reaction occurs where bacteria oxidize organic C compounds. The amount of Fe reduced and the speed at which it is reduced depend on the amounts of oxidizable C in the soil. Assume, for example, that redox depletions will only form around dead roots, because the roots supply most of the oxidizable C. Near the soil surface roots are abundant, closely spaced, and distributed throughout the A horizon. The microbial population should be large and well distributed throughout the horizon, particularly around the roots. When the A horizon is saturated, the large microbial population quickly depletes the soil of oxygen and reducing conditions may develop within 5 days of the onset of saturation (He et al., 2003). Redox depletions near the surface should be plentiful because they formed around the abundant root mass (He et al., 2003).

On the other hand, deeper (i.e. below the A horizon to the C horizon) in the soil the organic C levels are low because there are fewer roots and spacing between roots increases. The microbial population is small and focused only around the roots. Thus, when deeper horizons are saturated, the reduction occurs first around the roots, and the soil at a distance from the roots remains in an oxidized condition. In such cases, redox depletions will only form around the roots, and the remaining soil matrix will retain a high chroma color (Morgan and Stolt, 2006; Severson et al., 2008).

3.1.4. Rule 4: Water Tables Rise above Depth of Redox Depletions

As noted earlier, redox depletions form when Fe is reduced and removed from a portion of the soil, as shown in Fig. 3. These processes begin once the horizon is saturated and atmospheric oxygen is prevented from entering the horizon. However, the time required to reduce Fe ranges from <5 days to nearly 50 days following saturation, and depends on the amount of organic C and temperature (He et al., 2003). Once the microbes reduce Fe it must diffuse away from zone where the depletion will form. The time required for this has not been well documented. However, under ideal conditions depletion formation can occur in about 7 consecutive days in constructed wetlands (Vepraskas et al., 2006). Brief periods of saturation lasting <7 days are too short to allow formation of redox depletions. Depletions do not occur in every soil horizon that has been saturated; they occur where the horizon is saturated long enough for the depletion-forming processes to occur.

3.1.5. Rule 5: Durations of Saturation cannot be Predicted from the Abundance of Redoximorphic Features Unless they have been Calibrated with On-Site Measurements of Saturation

In general, the longer a soil horizon is saturated then the more redox depletions it should contain. However, it should be clear following the discussion for Rules 3 and 4 that saturation by itself is not the only factor causing redox depletions to form. As a result, there is no simple relationship between duration of saturation and abundance of redox depletions. In other words, one cannot estimate how long a soil horizon is saturated simply by looking at the abundance of redox depletions. It is possible to calibrate the duration of saturation to depletion abundance, but such relationships are site dependent and vary with soil depth (He et al., 2003).

3.1.6. Rule 6: Stratified Soils (Strongly Contrasting Textures) can Form Redoximorphic Features when "Unsaturated"

Saturated soils have water pressures that are zero or positive. This water can flow out of the soil and into wells or large holes in the soil. In unsaturated soils, water pressures are negative or less than atmospheric. This water is held in the soil under a tension or suction force. It does not flow out of the soil into large holes. Iron reduction will occur under anaerobic conditions which frequently develop following saturation of a soil horizon. Saturation is required because the large soil pores must be filled with water to prevent oxygen from moving along them into the soil. In fine-textured soil materials that are not penetrated by large (e.g. >1 mm dia.) root channels, water contents may become high enough to fill all pores in the soil matrix, and prevent the entrance of oxygen, even though the water pressure is negative. In these cases, Fe reducing conditions have developed and redox depletions formed. Water content can remain high, particularly when the fine-textured material overlies sands. Vepraskas et al. (1974) found that anaerobic conditions can develop in silty clay loam materials over sands even when the silty clay loam material is unsaturated. In such cases, root channels and interpedal cracks will remain air-filled such that redox concentrations develop along them. The fine-textured matrix, however, can be anaerobic due to a water content that is close to saturation and effectively excludes atmospheric oxygen from penetrating into the matrix.

Redox depletions will be found in the matrix of such soils. This condition is believed to occur only in stratified soil materials with fine-textured material overlying coarser-textured material such as a coarse sand. The high water content buildup in the fine-textured material occurs because the water is held in place under a suction that prevents downward movement until the soil becomes fully saturated.

3.1.7. Rule 7: Longer Periods of Saturation are Needed to Form Depletions in Coarse Textured Soils than in Finer Textured Soils

As previously discussed, redoximorphic features form when all four specific conditions needed for iron reduction are met in the soil. Rule 3 states that as soil depth increases the amount of organic carbon decreases; thus, the speed of redoximorphic feature formation decreases. Sandier soils are often lower in organic carbon as compared to finer-textured soils in the same landscape; therefore, redox depletions are likely to form more slowly in sandy soils. Furthermore, sandy soils drain faster and have a greater permeability than finer-textured soils. This causes a greater oxygen flux that can further inhibit or slow the rate of Fe reduction. The combination of lower carbon and greater amounts oxygen means that depletions in sandy soils form more slowly than finer-textured soils. Another way to put it is that given the same percentage of depletions, those in a sandy soil may indicate a longer duration of saturation as compared to those in a finer-textured soil (Lindbo et al., 2006; Morgan and Stolt, 2006; Severson et al., 2008). For example, a sandy soil with 2% redox depletions might be saturated for 80% (cumulative) of the year, whereas a sandy clay loam with 2% redox depletions might saturate for 27% (cumulative) of the year. It is still important to remember that the specific frequency and duration of saturation cannot be determined just by the percentage of the features and that the features need to be calibrated from site to site.

3.2. Hydric Soil Field Indicators

While it is difficult to relate individual hydromorphic features to exact durations of saturation and reduction, it has been possible to define "sets" of features that indicate when minimum periods of saturation and reduction have been exceeded. The most comprehensive group of such sets of hydromorphic features is termed the "hydric soil field indicators" (USDA-NRCS, 2010). Hydric soil field indicators are layers of soil material containing diagnostic morphological features that have formed under saturated and anaerobic conditions. They are used to identify hydric soils that occur in wetlands that are protected by state and federal laws. Examples of hydric soil field indicators are listed in Table 2. The field indicators are rigidly defined on the basis of a soil layer's: organic matter content, matrix color, colors of redoximorphic features, depth, and thickness.

TABLE 2 A List of Selected Hydric Soil Field Indicators Used to Identify Wetlands that are Protected by Federal and State Laws in the U.S. The Field Indicators Shown have been Developed by the USDA-NRCS (2010), and Full Color Photographs are Shown in that Publication

Symbol	Name	Required soil material	Description
A1	Histosol	Organic	Classifies as a Histosol (except Folists) or as a Histel (except Folistel). Soil must have 40 cm or more or organic soil material in the upper 80 cm. Material may be fibric, hemic or sapric.
A2	Histic epipedon	Organic	Histic epipedons are surface horizons between 20 and 40 cm thick composed of organic soil material. A histic epipedon that is underlain by mineral soil material with chroma 2 or less. Aquic conditions or artificial drainage are required.
S1	Sandy mucky mineral	Sands and loamy sands	A mucky modified sandy mineral layer 5 cm or thicker starting within 15 cm of the soil surface. Soil organic C concentrations must be between 5 and 12%.
S7	Dark surface	Sands and loamy sands	A layer 10 cm or thicker starting within the upper 15 cm of the soil surface and with a matrix value of 3 or less and chroma of 1 or less. At least 70% of the visible soil particles must be covered, coated, or similarly masked with organic material. The matrix color of the layer directly below the dark layer must have chroma of 2 or less.
F3	Depleted matrix	Loams or clays	A layer 15 cm or thicker, beginning within 25 cm of the soil surface, with 60% of the matrix having a chroma of 2 or less and value of 4 or more. Redox concentrations are

(Continued)

TABLE 2 **A List of Selected Hydric Soil Field Indicators Used to Identify Wetlands that are Protected by Federal and State Laws in the U.S. The Field Indicators Shown have been Developed by the USDA-NRCS (2010), and Full Color Photographs are Shown in that Publication—cont'd**

Symbol	Name	Required soil material	Description
			required when the layer is in A or E horizons. Contrast of the concentrations must not be faint, and the abundance of the concentrations must be 2% or more.
F8	Redox depressions	Loams and clays	In closed depressions subject to ponding, 5% or more distinct or prominent redox concentrations occurring as soft masses or pore linings in a layer that is 5 cm or thicker and is entirely within the upper 15 cm of the soil.
F13	Umbric surface	Loams and clays	A layer 25 cm or thicker starting within 15 cm of the soil surface that has value 3 or less and chroma 1 or less in which the lower 10 cm has the same colors as those described above, or any other color that has chroma of 2 or less.

Vepraskas and Caldwell (2008) estimated the saturation requirements needed to form selected field indicators for landscapes in North Carolina. Field studies were conducted by monitoring water-table levels at locations having field indicators. The water-table data were then used to calibrate a hydrologic model for each soil location to estimate daily water-table levels over a 40-year period. Results are shown in Table 3. Wetland hydrology conditions occurred when the water table was within 30 cm of the soil surface, or above the surface, for periods of 14 days or more in a year. As shown in Table 3, the field indicators composed of organic C met wetland hydrology conditions in a greater proportion of years than did field indicators composed of redoximorphic features. In addition, the organic C layer thickness correlated to the site's wetness as indicated by the increasing proportion of years meeting wetland hydrology.

TABLE 3 Summary of the Relationship between Hydric Soil Field Indicators and Selected Hydrologic Variables. The Relative Wetland-Hydrology Class is a Suggested Ranking of the Field Indicators on the Basis of their Relationship to Wetland Hydrology. Such a Table would Allow Wetland Delineators to Estimate Various Hydrologic Components using Field Indicators

Type of feature	Hydric soil field indicator	Proportion of years wetland hydrology met (% of years)	Average duration of saturation during growing season (days yr^{-1})	Average duration of ponding during growing season (days yr^{-1})
Organic C	Histosol (A1)	100	228	139
	Histic epipedon (A2)	"	178	67
	Sandy mucky mineral (S1), Dark surface (S7)		115	3
	Umbric surface (F13)	95	40	ND[‡]
Redoxmorphic	Depleted matrix (F3)			
	Redox depressions (F8)	87	29	"
None[†]	No indicator present	3	5	"

[†]*Soils saturated within 30 cm of surface for less than 50% of years do not meet the requirements for wetland hydrology.*
[‡]*ND means not determined.*
(Vepraskas and Caldwell, 2008).

4. LANDSCAPE RELATIONS

The redoximorphic features and pedon interpretation rules discussed previously form the basis for the following discussion of landscape and hydrology interpretations. Although the redoximorphic features are the same regardless of the landscape in which they form, their overall pattern of distribution within and between pedons on a landscape can be used to make broader hydrologic interpretations.

A landscape can be divided into two broad hydrologic zones: 1) recharge zone and 2) discharge zone (Fig. 4) (Heath, 1980). The recharge zone is characterized by a downward movement of water from soil surface (unsaturated) to groundwater (saturated). In the discharge zone, water moves from groundwater toward the surface. A flow through zone may also occur on some landscapes where groundwater reaches the surface but the flow lines are essentially parallel to the ground surface. As water moves it carries with it dissolved constituents such as nutrients, dissolved organic matter, and reduced Fe and Mn. The latter of these constituents is critical to redoximorphic feature formation as discussed previously. In a general sense, Fe and Mn are leached from the soils in the recharge zone and should result in soils where redox depletions are more common than redox concentrations. In the discharge zones reduced Fe and Mn move to the surface where they are oxidized (Lindbo and Richardson, 2000; Rhoton et al., 2002) and should form more concentrations than in recharge areas (Hayes and Vepraskas, 2000).

Complicating this simple model is the reality that water tables are seldom static and there is considerable water movement within the capillary fringe (Abit et al., 2008). The result is a redistribution of Fe within a profile as well as on a landscape scale. This is best demonstrated by the distribution of redox features in a moderately well-drained profile. Starting at the surface, redox concentrations occur first, followed by depletions and concentrations, and leading eventually to a depth where depletions are in sufficient amount to

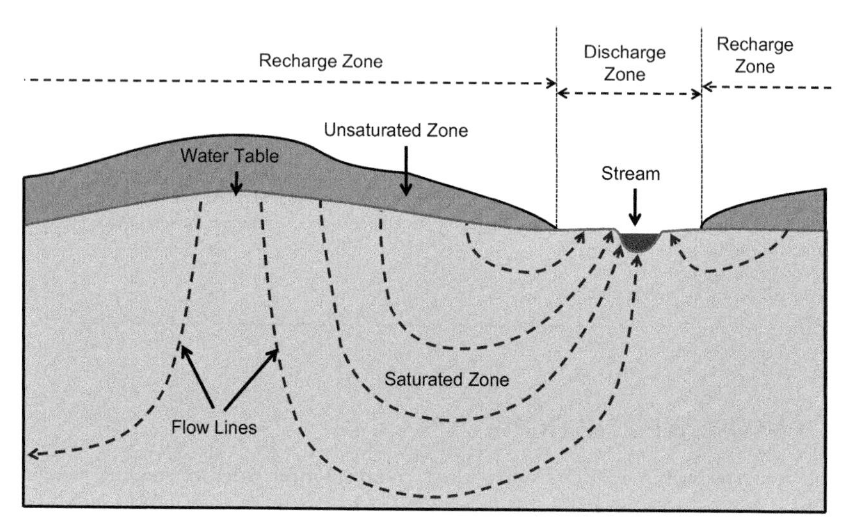

FIGURE 4 Idealized schematic of recharge and discharge zones on a landscape typical in sloping landscape in humid regions *(From Heath, 1980). (Color version online)*

constitute the matrix (e.g. depleted matrix and hydric soil indicators) (Morgan and Stolt, 2006; Severson et al., 2008; West et al., 2008). The concentrations in the upper portion of the profile represent the zone of greatest water-table fluctuations where any reduced Fe moves up and is oxidized in the more aerobic zone. These features are akin to the concentrations observed in the landscape-scale discharge zones.

4.1. Sloping Landscapes – Endosaturation

Perhaps the most straightforward and the most investigated redox–hydropedologic relation is that of an endosaturated toposequence on slopes greater than 2% (Fig. 5). Endosaturation is defined as continuous saturation to a depth of 2 m from the top of the water table. This differs from episaturation (landscape relations discussed later in this chapter) where there is a zone of saturation within the upper 2 m of the soil that is underlain by an unsaturated zone. Veneman et al. (1998) summarized the general morphologic relations to drainage class. Since drainage classes are regionally defined, the discussion that follows will be based on internal free water occurrence classes (ifw) as defined in the *Soil Survey Manual* (Soil Survey Division Staff, 1993). Soils with very deep ifw contain redox depletions (<2 chroma) at depths generally greater than 150 cm. They may contain redox depletions >2 chroma and redox concentrations above 150 cm. Deep ifw soils contain <2 chroma depletions and are saturated between 100 and 150 cm. They often have matrix chromas that are slightly higher, chroma 6 or 8 in upper horizons as compared to matrix chromas of 4 or 6 in very deep ifw in their upper horizons. They may also have a darker A horizon, suggesting more organic matter accumulation. The moderately deep ifw soils have 2 chroma depletions shallower in the profile, occurring between

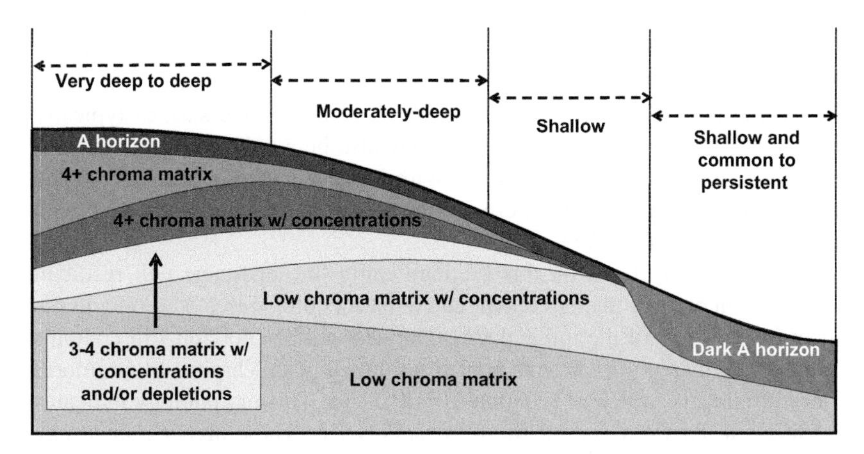

FIGURE 5 Typical toposequence on a sloping site (>2%) in a humid environment such as eastern North America (not to scale). *(Color version online)*

50 and 100 cm. Matrix colors trend toward lower chromas with increasing depth below the first occurrence of redox depletions. Eventually the entire horizon may become a depleted matrix. Moving lower on the landscape the shallow ifw soils occur next. In these low chroma depletions occur closer to the surface, between 25 and 50 cm. Also, at this depth, redox concentrations are common and may extend up to the surface. The A horizon may also be darker when compared to deeper ifw soils. A depleted matrix occurs closer to the surface in shallow ifw soils as well. The very shallow ifw soils have low chroma depletions and or matrices at depths from 0 to 25 cm and occur at the lowest point on the landscape. The subsoils directly below the A horizon or within 25 cm of the surface have depleted matrices dominated by redox depletions with chromas of 2 or less. These soils have the darkest A horizons on the landscape. As the cumulative saturation increases the very shallow ifw soils have an A horizon >25 cm thick.

The above relations focus on the depth at which saturation occurs but do not directly suggest duration or frequency of saturation (Veneman and Pickering, 1984; Soil Survey Division Staff, 1993; Veneman et al., 1998; Ticehurst et al., 2007). Nevertheless, understanding the morphological relations is useful in local land-use planning (Ticehurst et al., 2007). As discussed previously in the rules, duration and frequency need to be determined at a regional scale if not at the local scale (Morgan and Stolt, 2006; Severson et al., 2008).

4.2. Flat Landscape – Endosaturation

Unlike sloping landscapes where the very deep and deep ifw soils are on the high points of the landscape, the "high points" on a flat landscape are not pronounced, and ifw classes and soil color patterns change with distance to a river or stream. This landscape is common in the lower coastal plain region of the southeastern U.S. where the uplands consist of broad coastal plain flats (interstream divides) that are dissected by rivers. Histosols or soils with histic epipedons are found in the middle of the flats which are farthest from the rivers (Daniels et al., 1999). The mineral soils below the organic surface typically have a ≤2 chroma matrix. This matrix may also be a reduced matrix if sufficient Fe remains in the soil solution. With increasing depth gley hues of the matrix may occur, suggesting more prolonged saturation and possibly stagnant water (Daniels et al., 1978).

Small rises of 30 cm on this predominantly flat landscape will result in a decrease in organic matter content and a matrix color with a 3 chroma in the upper 30 cm or so of the mineral soil. The near-surface horizon still contains common to many ≤2 chroma depletions and higher chroma (4 or more) concentrations if sufficient Fe remains in the soil. This morphology indicates a change in ifw from very shallow to shallow ifw. These higher points may result from underlying geomorphic features (large scale) or to tree throw and resulting cradle-knoll topography (small scale).

4.2.1. Red Edge on Flat Landscapes

Broad flat landscapes do have natural drainage ways, streams, or rivers associated with them. These drainage ways are spaced widely apart, often >1 km. They have a distinct toposequence that has been referred to as the red edge or dry edge (Daniels and Gamble, 1967; Daniels et al., 1971) (Fig. 6). In this toposequence, the highest point on the landscape is also a poorly drained profile with depleted, reduced, and or gleyed matrices near the surface. The surface may also be organic or a low chroma, low value (black) mineral material. Moving toward the drainage way and on the upper shoulder the upper 30 cm of the mineral material becomes ≤3 chroma with common to many ≤2 chroma depletions. Few redox concentrations are present at this point, suggesting that the profile is in the recharge zone with the dominant direction of water flow being down and toward the drainage way. At the shoulder position the ifw class moves toward moderately deep to deep depending on the relief of the associated drainage way. If the relief is greater than 150 cm the profile will take on a similar morphology to the very deep ifw soil discussed in the sloping landscape; however, the deeper horizons remain gleyed. The deeper gleyed horizons in this landscape suggest long term if not continuous saturation (Daniels et al., 1978). Below the shoulder and into the foot and toeslope position the matrix becomes lower chroma with both common depletions and concentrations. In this zone, the concentrations are due to groundwater discharge, assuming that the groundwater

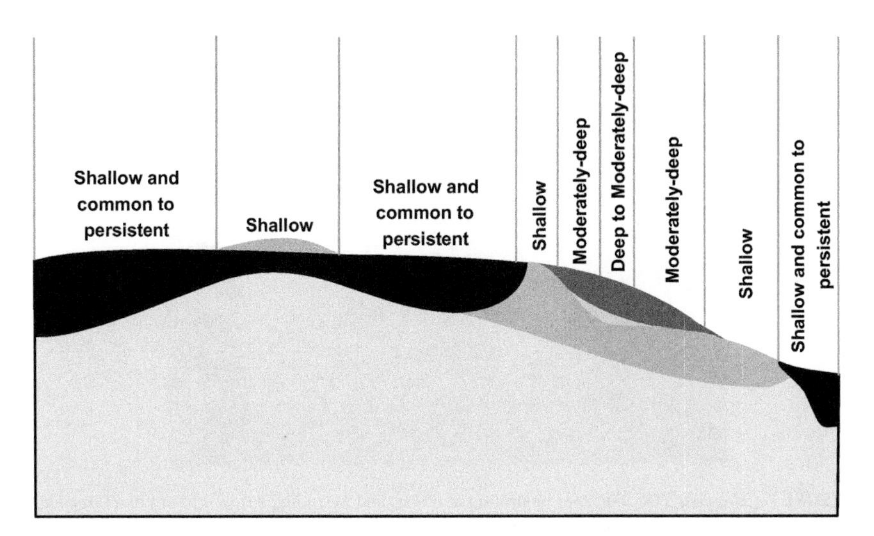

FIGURE 6 Typical distribution on a flat landscape, broad interstream divide (<2% slope) with a "dry-edge" or "red-edge" in a humid environment such as the Southeastern Coastal Plain of North America (From Daniels and Gamble, 1967 and Daniels et al., 1971). The soils at the break in slope near the right side of the diagram have deeper saturation due to the natural drainage. Slopes in this part of the landscape exceed 2%; thus, the morphologic pattern is similar to that seen in Fig. 5 but occur over a short horizontal distance (not to scale). *(Color version online)*

contains sufficient reduced Fe that oxidizes when it comes back to the surface. The source of the reduced Fe is the recharge zones on the landscape where the soils have been saturated and reduced (Hayes and Vepraskas, 2000).

4.3. Episaturation (Perched Water Table)

Episaturation or perched water tables occur when a zone of saturation occurs above an unsaturated zone or horizon (Fig. 7). This is commonly found where a slowly permeable or restrictive horizon (e.g. fragipans – Bx and Btx; ortstein – Bsm and Bhsm; cemented – Bm; and argilic with expansive clay – Btss) underlies a more permeable one. Saturation then occurs in the overlying more permeable horizon. The duration of saturation, organic matter content, and frequency of saturation will all affect the specific pattern of redoximorphic

FIGURE 7 Fragipan (Btx horizon) beginning at 80 cm and extending below 150 cm in a Grenada Series (fine-silty, mixed, active, thermic oxyaquic Fraglossudalfs), Crocket County, TN. Commonly, episaturation occurs in the upper Btx and extends into the E′ and lower B horizon (approximately 45–80 cm). This zone is characterized by common redox concentrations (soft masses) and depletions associated with channels. The matrix may be low chroma (depleted) as well. Within and below the fragipan, depletions associated with channels and Fe masses are common but the matrix no long has a low chroma. *(Photo by F. E. Rhoton, Retired, USDA-ARS.) (Color version online and in color plate)*

features present as previously discussed. One key difference is the change in the features with depth below the zone of saturation. In endosaturated conditions, redox depletions increase with depth until they dominate the matrix. In episaturation this is not the case. Below the zone of saturation redox depletions decrease in abundance and are associated with large water-conducting pores (Lindbo et al., 1995, 2000; McDaniel and Falen, 1994; Lindbo and Veneman, 1993; Payton, 1993). As depth increases depletions will give way to concentrations and eventually to horizons with few to no redoximorphic features if no saturation occurs deeper in the profile. Often, soils with episaturation have a deeper zone of saturation below the episaturated zone. In this case, the pattern of redox features discussed in the endosaturation section is likely to occur.

Another case where episaturation can occur is where a fine-textured horizon overlies a coarse-textured horizon (see Rule 6). In this situation, the pattern of the redoximorphic features is slightly different from those in the previous section. In the fine-textured material the structural peds become saturated and reduced more commonly (quickly) yet the macropores and channels remain unsaturated. The matrix appears depleted with redox concentrations occurring along pores, channels, and ped faces. These redox concentrations form as reduced Fe from the ped interior diffuses to the pore edges and is subsequently oxidized in proximity to the air-filled void (Vepraskas et al., 1974; Veneman et al., 1976; Clothier et al., 1978; Morgan and Stolt, 2006). Overall, the pattern in the fine-textured horizon appears similar to poorly drained soils or soils with a long duration of saturation. If this is a case of episaturation, the underlying coarser material is devoid of redoximorphic features (Clothier et al., 1978). If, on the other hand, the underlying horizon is similarly depleted and showed redoximorphic features (Vepraskas, 2004), the profile is assumed to be endosaturated.

4.3.1. Anthrosaturation

Paddy soils are intentionally flooded and ponded, most often for cultivation of rice, and are referred to as having anthrosaturation. Although they can be viewed as having episaturation, they do have a distinctive pattern of redoximorphic features due to a long duration of saturation and reduction and their unique set of agronomic practices (Table 4) (Hseu and Chen, 1996,1999; Zhang and Gong, 2003; Kyuma, 2004).

The plants grown in paddy soils typically have Fe-coated pore linings (e.g. oxidized rhizospheres (USDA NRCS, 2010) associated with living roots (Chen et al., 1980; Xu and Zhu, 1983)). They form as reduced Fe present in the pore water encounters dissolved oxygen introduced by the actively respiring plant roots. In the presence of the dissolved oxygen, the reduced Fe oxidizes forming pore linings around the root channels. If reduced Mn is present, it will oxidize in a less reducing zone, often closer to the living root or on the proximal edge of the pore lining. Reduced Mn is translocated deeper into the soil as it will remain in solution longer than reduced Fe. The result is Mn concentrations as masses,

TABLE 4 Description and Interpretation of Redoximorphic Features in Anthrosaturated Soil

Depth	Duration of saturation and reducing conditions	Redoximorphic features	Matrix
Surface	Long	Fe pore linings (oxidized rhizospheres); Fe pore linings	Depleted
Saturated interface	Long	Fe pore linings (oxidized rhizospheres); Fe pore linings	Depleted
Unsaturated interface	Intermittent	Fe pore linings (oxidized rhizospheres); Fe and Mn pore linings; Fe masses Fe depletions adjacent to channels	
Below unsaturated interface	Short to none	Mn pore linings; Fe, Fe/Mn, and Mn masses	
Deep in profile	none	Fe masses	

(After Hseu and Chen, 1999; Kyuma, 2004).

nodules, and pore linings at the interface between the overlying saturated zone and underlying unsaturated zone of the paddy soil (Kyuma, 2004). Nodules and masses are more common as depth increases to the saturated–unsaturated soil interface and where saturation levels fluctuate the most (Hseu and Chen, 2001; Kyuma, 2004). The matrix of paddy soils is similar in appearance to a depleted matrix and may also be a reduced matrix as the Fe may not be translocated or leached from the saturated zone (Hsue et al., 1999; Zhang and Gong, 2003; Kyuma, 2004).

4.4. Flow-through Landscapes – Depressions and Vernal Pools

Landscapes containing numerous closed depressions exhibit a complex distribution of features due to the flow path of water in and out of the depression (Fig. 8) (Evans and Freeland, 2000). Soils in the depressions in the recharge zone are dominated by leaching and redox depletions. In the flow-through areas, accumulation of carbon as a thick A horizon is common and both depletions and concentrations are present. In the discharge zone, the soils show more concentrations as Fe(II) is moved back to the surface and is oxidized. In subhumid climates in addition to Fe-based features, carbonate and saline features are observed. Recharge zone soils are leached with regard to carbonates in the center of the depression while soils at the edge of the depression are enriched with carbonates and have a relatively high chroma

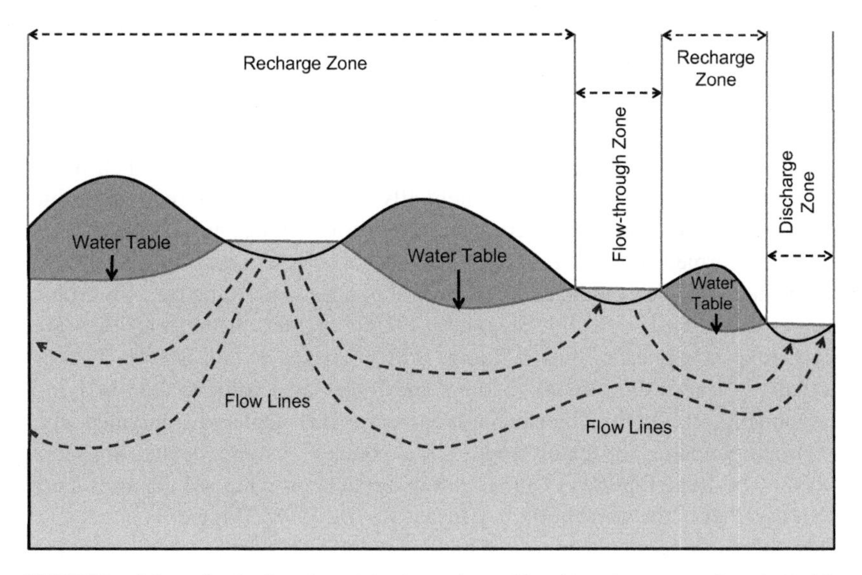

FIGURE 8 Schematic of a flow-through landscape in a subhumid environment such as the Prairie pothole region in the northern Great Plains, North America. *(From Evans and Freeland, 2000 and Richardson et al., 1994) (Color version online)*

matrix in the B horizon similar to that observed in the red edge soils. The flow-through zone shows carbon accumulation as a thick A horizon as well as having an overall calcareous profile. Soils in the discharge zone are dominated by accumulations of carbonates, gypsum, and saline features as these dissolved materials move with groundwater back to the surface (Arndt and Richardson, 1989; Richardson et al., 1994).

5. CONTEMPORARY VERSUS RELICT FEATURES

Deciding whether a feature has formed contemporaneously due to present hydrology or is a relict of a previous moisture regime can be difficult. Identification begins with an understanding of the landscape and hydrology. One can begin to suspect that a given feature is relict by first looking for evidence that the hydrology has changed. For example, if a site has been artificially drained with extensive ditching or there is evidence of stream incision, or the features occur on upper stream terraces, then it is possible that the observed features could be relict. If no evidence of hydrologic alteration exists, it is less likely that features are relict. In the process of calibrating percent gray color (redox depletions) to a saturation event index (SEI), He et al. (2003) observed that at depth (45 cm) 10% gray color indicates an SEI of 0.1. This corresponds to a single saturation event of 21 days or longer, occurring once every 10 years. This does not mean the feature is relict; it means the feature may only represent periodic saturation.

In addition to observing changes in hydrology, identification or interpretation of relict features is accomplished by using a detailed morphologic evaluation of patterns of features, by use of hand lens, thin sections, etc. (Tucker et al., 1994; Greenberg and Wilding, 1998; D'Amore et al., 2004; Lindbo et al., 2010). Several relationships need to be established when interpreting redoximorphic features as potentially relict. First, if the redoximorphic feature is related to current structural features (pores, voids, planes, and others) it is likely formed in situ, (Tucker et al., 1994; Greenberg and Wilding, 1998; Lindbo et al., 2000) as indicated in Rule 1 above. Second, if the redoximorphic features are coated with illuvial material (clay, organic matter, etc) they are likely relict (Vepraskas and Wilding, 1983; Tucker et al., 1994). This is particularly true if the illuvial coatings are of the same color as the overlying horizon (Fig. 9). On the other hand, if there are clay depletions, nodules with irregular boundaries, or features with diffuse boundaries, the features are more likely to be contemporary. Contemporary, relict, or inherited features may occur together in the same horizon (Busacca et al., 1998; Muggler et al., 2001). When this is observed, the presence of contemporary features should be used to suggest current hydrologic conditions.

The morphology of the boundaries of redoximorphic features, masses in particular, is often used to differentiate between relict and contemporary

FIGURE 9 Photo of a redoximorphic feature that is interpreted as being "relict". Note the brown clay coating in the channel. If this was still an actively forming redoximorphic feature it is assumed that the Fe coating and coloring the clay would be reduced and the color would be lower chroma. *(Color version online and in color plate)*

features. The presence of a diffuse boundary of a mass is often considered to be indicative of a contemporary feature (Tucker et al., 1994; Suddhiprakaru and Kheoruenromne, 1994; Costantini et al., 2007). In another study, the combination of pore linings (hypocoatings) and masses in the same horizon indicates that the features are contemporary (Greenberg and Wilding, 1998).

Diffuse boundaries suggest a contemporary feature, whereas an abrupt boundary suggests the feature is no longer forming (Tucker et al., 1994). This does not always mean the nodule has been transported (Stolt et al., 1994; Lindbo et al., 2000; D'Amore et al., 2004). Stolt et al. (1994) found that certain nodules having abrupt boundaries may form in place if there are isolated areas of contrasting particle size within a horizon. Nodules that have a distinctly different internal matrix or internal pedofeatures as compared to their surrounding matrix are more likely to be inherited (from parent material) or transported from another source (King et al., 1990; Mücher and Coventry, 1994; Tucker et al., 1994; Muggler and Buurman, 2000), moved by biological mixing (Bouma et al., 1990) or moved by argilloturbation (McCarthy et al., 1998), or they may be degradation remnants (Lindbo et al., 2000). Also, if the nodule is coated with clay it is likely to be a relict feature (Vepraskas et al., 1994; Tucker et al., 1994).

The roundness of the nodules may also be used to assess whether it is contemporary and forming, or relict and degrading. Some researchers concluded rounded nodules are related to current hydrology, especially if drainage is restricted (Schwertmann and Fanning, 1976; Richardson and Hole, 1979; Elless and Rabenhorst, 1994; Elless et al., 1996), whereas others conclude rounded nodules are transported and not formed in situ (McCarthy et al., 1998). Degrading nodules often have sand or silt grains protruding from their boundaries and will have irregular shapes (Tucker et al., 1994). They may also have depletions around their outer edges (Mücher and Coventry, 1994). Tucker et al. (1994) indicated that the more pronounced the degradation of the nodules, the greater the degree of current saturation.

In terms of interpretation, determining if a site has relict or contemporary hydromorphic features is difficult and no single parameter works in all cases. In order to be completely confident regarding current hydrology, observation wells (piezometers) are required. As indicated in Rule 5, the exact relationship between redoximorphic features and hydrology can only be established through on-site measurements.

6. SUMMARY AND CONCLUSIONS

Hydromorphic features form in saturated, anaerobic soils and are used to identify where such conditions occur in soils. The most widespread use of these features is for identification of jurisdictional wetlands in the U.S. and in site evaluation for determining depth to seasonal saturation. Three categories of hydromorphic features have been identified: redoximorphic features (Fe/Mn based); carbon accumulation often as darkened A horizons, coated grains, and muck; and a sulfur-based feature – the presence of H_2S. Hydromorphic features form when four specific conditions are met in the soil: 1) organic matter must be present, 2) microorganisms must be actively respiring and oxidizing organic material, 3) the soil must be saturated, and 4) dissolved oxygen must be removed from the groundwater. There are seven rules for formation and interpretation of these features. These rules can be applied within a given profile or toposequence to relate the morphologies to hydrology in a general sense. However, if the specific hydrologic conditions of a site need to known, the site must be appropriately instrumented in order to calibrate features to duration and frequency of saturation and reduction. Distribution of hydromorphic features on landscapes varies dependent on landscape position as well as general slope. In the case of epi- and anthrosaturation, the distribution of hydromorphic features can identify the extent of the saturated and unsaturated zone in the profile and on the landscape if the above rules are followed. Relict hydromorphic features are those that are present in soil but which cannot form under current conditions due to a change in hydrology that prevents anaerobic conditions from developing. While a number of characteristics have been studied to identify relict redoximorphic features, it is difficult to determine with certainty that such features are relict unless the interpretation is justified by studying the soil's current hydrology.

REFERENCES

Abit, S.M., Amoozegar, A., Vepraskas, M.J., Niewoehner, C.P., 2008. Solute transport in the capillary fringe and shallow groundwater: field evaluation. Vadose Zone J. 7, 890–898.

Arndt, J.L., Richardson, J.L., 1989. Geochemical development of hydric soil salinity in a North Dakota prairie pothole wetland system. Soil Sci. Soc. Am. J. 53, 848–855.

Bouma, J., Fox, C.A., Miedema, R., 1990. Micromorphology of hydromorphic soils: applications for soil genesis and land evaluation. In: Douglas, L.A. (Ed.), Soil Micromorphology: a Basic and Applied Science. Developments in Soil Science 19. Elsevier, Amsterdam.

Busacca, A., Cremaschi, M., 1998. The role of time versus climate in the formation of deep soils of the Apennine fringe of the Po Valley, Italy. Quat. Int. 51/51, 95–107.

Chen, C.C., Dixon, J.B., Turner, F.T., 1980. Iron coatings on rice roots: morphology and models of development. Soil Sci. Soc. Am. J. 44, 1113–1119.

Cleland, T.M., 1921. The Munsell Color System: a Practical Description with Suggestions for its Use. Munsell Color Co., Boston, MA. www.applepainter.com/ (Available online at: verified 8 January 2012).

Clothier, B.E., Polluk, J.A., Scotter, D.R., 1978. Mottling in soil profiles containing a coarse-textured horizon. Soil Sci. Am. J. 42, 761–763.

Costantini, E.A.C., Priori, S., 2007. Pedogenesis of plinthite during early Pliocene in the Mediterranean environment: case study of a buried paleosol at Podere Renieri, central Italy. Catena 71, 425–443.

D'Amore, D.V., Stewart, S.R., Huddleston, J.H., 2004. Saturation, reduction, and the formation of iron-manganese concretions in the Jackson-Frazier Wetland. Oregon. Soil Sci. Soc. Am. J. 68, 1012–1022.

Daniels, R.B., Gamble, E.E., 1967. The edge effect in some ultisols in the North Carolina coastal plain. Geoderma 1, 117–124.

Daniels, R.B., Gamble, E.E., Nelson, L.A., 1971. Relations between soil morphology and water table levels on a dissected North Carolina Coastal plain surface. Soil Sci. Soc. Am. Proc. 35, 781–784.

Daniels, R.B., Buol, S.W., Kleiss, H.J., Ditzler, C.A., 1999. Soil Systems in North Carolina. Tech Bull. 314. North Carolina State University, p. 118.

Daniels, R.B., Gamble, E.E., Wheeler, W.H., Gilliam, J.W., Wiser, E.H., Welby, C.W., 1978. Water Movement in Surficial Coastal Plain Sediments, Inferred from Sediment Morphology. Tech. Bull. No. 243. NC Agricultural Exp. Station, Raleigh.

Dusek, J., Picek, T., Cizkova, H., 2008. Redox potential dynamics in a horizontal subsurface flow constructed wetland for wastewater treatment: diel, seasonal and spatial fluctuations. Ecol. Eng. 34 (3), 223–232.

Elles, M.P., Rabenhorst, M.C., 1994. Micromorphological interpretations of redox processes in soils derived from Triassic red bed parent materials. In: Ringrose-Voase, A.J., Humphreys, G.S. (Eds.), Soil Micromorphology: Studies in Management and Genesis. Developments in Soil Science, vol. 22. Elsevier, Amsterdam, pp. 171–178.

Elles, M.P., Rabenhorst, M.C., James, B.R., 1996. Redoximorphic features in soils of the Triassic culpepper basin. Soil Sci. 161, 58–69.

Environmental Laboratory, 1987. Corps of Engineers Wetlands Delineation Manual. Technical Report Y-87–1. U.S. Army, Corps of Engineers, Washington, DC.

Evans, C.V., Freeland, J.A., 2000. Wetland soils of basins and depressions of glacial terrains. In: Richardson, J.L., Vepraskas, M.J. (Eds.), Wetland Soils: Their Genesis, Morphology, Hydrology, Landscapes, and Classification. CRC Press, Boca Raton, FL, pp. 251–266.

Everett, K.R., 1983. Histosols. In: Wilding, L.P., Smeck, N.E., Hall, G.F. (Eds.), Pedogenesis and Soil Taxonomy, II. The Soil Orders. Elsevier, Amsterdam.

Greenberg, W.A., Wilding, L.P., 1998. Evidence of contemporary and relict redoximorphic features of an Alfisol in East-Central Texas. In: Rabenhorst, M.C., Bell, J.C., McDaniel, P.A. (Eds.), Quantifying Soil Hydromorphology. SSSA Special Publication 54, pp. 227–246. Soil Sci. Soc. Am. Madison, WI. USA.

Hayes Jr., W.A., Vepraskas, M.J., 2000. Morphological changes in soils produced when hydrology is altered by ditching. Soil Sci. Soc. Am. J. 64, 1893–1904.

He, X, Vepraskas, M.J., Lindbo, D.L., Skaggs, R.W., 2003. A method to predict soil saturation frequency and duration from soil color. Soil Sci. Soc. Am. J. 57, 961–969.

Heath, R.C. 1980. Basic elements of groundwater hydrology with reference to conditions in North Carolina. U.S. Geological Survey, Water Resources Investigations, Open-File Report no. 80–44, pp. 87.

Hseu, Z.Y., Chen, Z.S., 2001. Quantifying soil hydromorphology of a rice-growing ultisol toposequence in Tawain. Soil Sci. Soc. Am. J. 65, 270–278.

Hseu, Z.Y., Chen, Z.S., 1999. Micromorphology of redoximorphic features of subtropical anthraquic ultisols. Food Sci. Ag. Chem. 1, 194–202.

Hseu, Z.Y., Chen, Z.S., Wu, Z.D., 1999. Characteristics of placic horizons in two subalpine forest inceptisols. Soil Sci. Soc. Am. J. 63, 941–947.

Hseu, Z.Y., Chen, Z.S., 1996. Saturation, reduction, and redox morphology of seasonally flooded Alfisols in Taiwan. Soil Sci. Soc. Am. J. 60, 941–949.

Kemp, R.A., 1999. Micromorphology of loess-paleosol sequences: a record of palaeoenvironmental change. Catena 35, 179–196.

King, H.B., Torrance, J.K., Bowen, L.H., Wang, C., 1990. Iron concentrations in a typic dystrochrept in Taiwan. Soil Sci. Soc. Am. J. 54, 462–468.

Kyuma, K., 2004. Paddy Soil Science. Kyoto University Press and Trans Pacific Press, Melbourne, Australia, p. 280.

Lindbo, D.L., Veneman, P.L.M., 1993. Morphology of selected Massachusetts fragipan soils. Soil Sci. Soc. Am. J. 57, 437–442.

Lindbo, D.L., Richardson, J.L., 2000. Hydric soils and wetlands in riverine systems. In: Richardson, J.L., Vepraskas, M.J. (Eds.), Wetland Soils: Their Genesis, Morphology, Hydrology, Landscape, and Classification. CRC Press, Boca Raton, FL Ch. 12.

Lindbo, D.L., Severson, E.D., Lanier, G.K., Vepraskas, M.J., 2006. Correlation of redoximorphic features to hydrology. Abstract of World Congress of Soil Science, Philadelphia PA.

Lindbo, D.L., Stolt, M.H., Vepraskas, M.J., 2010. Formation and interpretation of redoximorphic features. In: Stoops, G. (Ed.), Micromorphological Features of Soil and Regolith. Elsevier Publ. Amsterdam.

Lindbo, D.L., Rhoton, F.E., Bigham, J.M., Jones, F.S., Smeck, N.E., Hudnall, W.H., Tyler, D.D., 1995. Loess toposequences in the lower Mississippi River Valley I: fragipan morphology and identification. Soil Sci. Soc. Am. J. 59, 487–500.

Lindbo, D.L., Rhoton, F.E., Bigham, J.M., Smeck, N.E., Hudnall, W.H., Tyler, D.D., 2000. Fragipan degradation and nodule formation in glossic Fragiudalfs of the lower Mississippi river valley. Soil Sci. Soc. Am. J.

McCarthy, P.J., Martini, I.P., Leckie, D.A., 1998. Use of micromorphology for palaeoenvironmental interpretation of complex alluvial paleosols: an example from the Mill creek formation (Albian), southwestern Alberta, Canada. Palaeogeogr. Palaeoclimatol. Palaeoecol. 143, 87–110.

McDaniel, P.A., Falen, A.L., 1994. Temporal and spatial patterns of episaturation in a Fragixeralf landscape. Soil Sci. Soc. Am. J. 58, 1451–1457.

Morgan, C.P., Stolt, M.H., 2006. Soil morphology-water table cumulative duration relationships in southern New England. Soil Sci. Soc. Am. J. 70, 816–824.

Muggler, C.C., Buurman, P., 2000. Erosion, sedimentation and pedogenesis in a polygenetic Oxisol sequence in Minas Gerais, Brazil. Catena 41, 3–17.

Muggler, C.C., van Loef, J.J., Buurman, P., van Doesburg, J.D.J., 2001. Mineralogical and (sub) microscopic aspects of iron oxides in polygenetic Oxisols from Minas Gerais, Brazil. Geoderma 100, 147–171.

Mücher, H.J., Coventry, R.J., 1994. Soil and landscape processes evident in hydromorphic grey earth (Plintusalf) in semiarid tropical Australia. In: Ringrose-Voase, A.J., Humphreys, G.S. (Eds.), Soil Micromorphology: Studies in Management and Genesis. Developments in Soil Science 22. Elsevier, Amsterdam. pp. 221–232.

National Research Council, 1995. Wetlands: Characteristics and Boundaries. National Academy Press, Washington, DC.

Parkin, T.B., 1987. Soil microsites as a source of denitrification variability. Soil Sci. Soc. Am. J. 51, 1194–1199.

Payton, R.W., 1993. Fragipan formation in argillic brown earths (Fragiudalfs) of the Milfield Plain, north-east England. III. Micromorphology, SEM and EDXRA studies of Fragipan degradation and the development of Glossic features. J. Soil Sci. 44, 725–739.

Pickering, E.W., Veneman, P.L.M., 1984. Moisture regimes and morphological characteristics in a hydrosequence in central Massachusetts. Soil Sci. Soc. Am. J. 48, 113–118.

Rhoton, F.E., Bigham, J.M., Lindbo, D.L., 2002. Properties of iron oxides in streams draining the loess uplands of Mississippi. Appl. Geochem. 17, 409–419.

Richardson, J.L., Hole, F.D., 1979. Mottling and iron distribution in a Glossoboralf-Haplaquoll hydrosequence on a glacial moraine in northwestern Wisconsin. Soil Sci. Soc. Am. J. 552–558.

Richardson, J.L., Arndt, J.L., Freeland, J., 1994. Wetland soils of the prairie potholes. In: Sparks, D.L. (Ed.), Advances in Agronomy, vol. 52. Academic Press, San Diego, CA, pp. 121–171.

Schwertmann, U., Fanning, D.S., 1976. Iron-manganese concretions in a hydrosequence of soil in loess in Bavaria. Soil Science Society of America Journal 40, 731–738.

Severson, E.D., Lindbo, D.L., Vepraskas, M.J., 2008. Hydropedology of a coarse-loamy catena in the lower Coastal Plain. Catena 73, 189–196.

Soil Survey Division Staff, 1993. Soil Survey Manual. USDA Handbook No. 18. US Gov. Print. Off, Washington DC, p. 437.

Soil Survey Staff, 2010. Keys to Taxonomy, eleventh ed. USDA-NRCS, p. 338.

Stolt, M.H., Ogg, C.M., Baker, J.C., 1994. Strongly contrasting redoximorphic patterns in Virginia valley and ridge paleosols. Soil Sci. Soc. Am. J. 58, 477–484.

Suddhiprakarn, A., Kheoruenromme, I., 1994. Fabric features in laterite and plinthite layers in ultisols in northeast Thailand. In: Ringrose-Voase, A.J., Humphreys, G.S. (Eds.), Soil Micromorphology: Studies in Management and Genesis. Developments in Soil Science 22. Elsevier, Amsterdam, pp. 51–64.

Ticehurst, J.L., Cresswell, H.P., McKenzie, N.J., Glover, M.R., 2007. Interpreting soil and topographic properties to conceptualize hillslope hydrology. Geoderma 137, 279–292.

Tucker, R.J., Drees, L.R., Wilding, L.P., 1994. Signposts old and new: active and inactive redoximorphic features; and seasonal wetness in two Alfisols of the gulf coast region of Texas, U.S.A. In: Ringrose-Voase, A., Humphreys, G.S. (Eds.), Soil Micromorphology: Studies in Management and Genesis. Developments in Soil Science 22. Elsevier, Amsterdam. pp. 99–106.

USDA-NRCS, 2010. Field indicators of hydric soils in the United States, Version 7.0. In: Vasilas, L.M., Hurt, G.W., Noble, C.V. (Eds.). USDA, NRCS in Cooperation with the National Technical Committee for Hydric Soils, Fort Worth, TX.

Veneman, P.L.M., Vepraskas, M.J., Bouma, J., 1976. The physical significance of soil mottling in a Wisconsin toposequence. Geoderma 15, 103–118.

Veneman, P.M., Lindbo, D.L., Spokas, L.A., 1998. Soil moisture and redoximorphic features, a historical perspective. In: Rabenhorst, M.C., et al. (Eds.), Quantifying Soil Hydromorphology. SSSA Spec. Pub. 54, pp. 1–24.

Vepraskas, M.J., 2000. Morphological features of seasonally reduced soils. In: Richardson, J.L., Vepraskas, M.J. (Eds.), Wetland Soils: Genesis, Hydrology, Landscapes, and Classification. Lewis Publ. Boca Raton, FL, pp. 163–182.

Vepraskas, M.J., 2004. Redoximorphic Features for Identifying Aquic Conditions. Tech. Bull. 301. NC Agric. Res. Serv. Raleigh, NC.

Vepraskas, M.J., Wilding, L.P., 1983. Albic neoskeletans in argillic horizons as indicators of seasonal saturation. Soil Sci. Soc. Am. J. 47, 1202–1208.

Vepraskas, M.J., Caldwell, P.V., 2008. Interpreting morphological features in wetland soils with a hydrologic model. Catena 73, 153–165.

Vepraskas, M.J., Baker, F.G., Bouma, J., 1974. Soil mottling and drainage in a Mollic Hapludalf as related to suitability for septic tank construction. Soil Sci. Am. Proc. 38, 497–501.

Vepraskas, M.J., Wilding, L.P., Drees, L.R., 1994. Aquic conditions for soil taxonomy: concepts, soil morphology, and micromorphology. In: Ringrose-Voase, A.J., Humphreys, G.S. (Eds.), Soil Micromorphology: Studies in Management and Genesis. Developemnts in Soil Science 22. Elsevier, Amsterdam. pp. 117–131.

Vepraskas, M.J., Richardson, J.L., Tandarich, J.P., 2006. Dynamics of redoximorphic feature formation under controlled ponding in a created riverine wetland. Wetlands 26 (2), 486–496.

West, L.T., Abreu, M.A., Bishop, J.P., 2008. Saturated hydraulic conductivity of soils in the Southern Piedmont of Georgia, USA: Field evaluation and relation to horizon and landscape properties. Catena 73, 174–179.

Xu, Q., Zhu, H., 1983. The characteristics of spotted horizons in paddy soils. Acta Pedologica Sinica 20, 53–59.

Zhang, G.L., Gong, Z.T., 2003. Pedogenic evolution of paddy soils in different soil landscapes. Geoderma 115, 15–29.

Chapter 6

Subaqueous Soils: Pedogenesis, Mapping, and Applications

M.C. Rabenhorst[1,*] and M.H. Stolt[2]

ABSTRACT

Although occasional references have been made to subaqueous soils during the last one and half centuries, only during the past two decades has a full recognition been granted to soils permanently covered by water. Early subaqueous soil research focused on overcoming the limitations of observing, describing, sampling, and mapping soils under water. Technological advances enabled the acquisition of high-quality bathymetry, and the development of subaqueous digital elevation models (DEMs) has led to the recognition and naming of subaqueous landforms. Observation of soil–landscape associations and confirmation that the soil–landscape paradigm could be applied in subaqueous environments opened the door for the inventory and mapping of permanently submersed soils. The observation and realization that subaqueous soil horizons were the result of pedogenic processes led to a new definition of soils by the US Department of Agriculture that accommodated subaqueous soils. Following the recognition and mapping of subaqueous soils, a number of new soil survey applications began to be developed mostly for estuarine systems. These soil interpretations addressed needs such as restoration of submerged aquatic vegetation, potential hazards from dredging, issues related to the mooring of vessels, and siting productive areas for shellfish aquaculture. More recent work has begun to address concerns emerging in freshwater subaqueous soils. The interconnectedness between water quality and soils in subaqueous environments as well as the growing pressures on estuarine and freshwater bodies suggest continued needs for solutions to issues associated with subaqueous soils.

1. INTRODUCTION

Subaqueous soils (SASs) are generally understood to be those soils that are permanently inundated by water and, usually, where the overlying water

[1]Dept. of Environmental Sci. & Technology, Room 1109, H.J. Patterson Hall, University of Maryland, College Park, MD 20742-5821

[2]Department of Natural Resources Science, 112 Kingston Coastal Institute, 1 Greenhouse Road, University of Rhode Island, Kingston, RI 02881

[*]Corresponding author: Email: mrabenho@umd.edu

Hydropedology, Edited by H. Lin. DOI: 10.1016/B978-0-12-386941-8.00006-X column is no more than a few meters deep. It is obvious, therefore, that subaqueous soils are intimately tied to water resources. They have been recognized mostly in coastal and estuarine settings, where the pioneering research on SAS was initiated (Demas, 1993). More recently, however, some pedologists have begun to evaluate SAS in inland settings where the water bodies overlying the SAS are mainly fresh (Bakken and Stolt, 2010; Erich et al., 2009; Erich et al., 2010b).

Many important economic and recreational resources are tied to shallow-water coastal environments. The land margins surrounding coastal areas have experienced unprecedented development pressures, as vacationers annually flock to resorts and getaways within view or easy access to coastal waters. At the same time, estuaries and nearshore coastal waters have become recognized as critical nursery zones for fin and shellfish. Over the last several decades, both environmental pressures and the heightened awareness of their importance have invigorated efforts to properly manage estuarine resources. One tangible example has been renewed efforts to reestablish submersed aquatic vegetation (Churchill et al., 1978; Short, 1988; Short, 1993; Reid et al., 1993; Kopp and Short, 2000; Merkel, 2004). Most efforts to improve estuaries have focused on water-quality parameters, which is understandable, given serious deleterious effects caused by eutrophication of water bodies. Nevertheless, sediments or SASs have traditionally received little attention. As regulations governing water quality and best management practices have been implemented, water conditions have improved. At the same time, there has been a significant research interest in SASs. Together, these have contributed to a growing awareness of the importance of SASs in managing estuaries as seen in partnerships such as the Mapcoast project in Rhode Island (Mapping Partnership for Coastal Soils and Sediments) (Payne and Turenne, 2009).

What we refer to as subaqueous soils (SASs) have been traditionally recognized as sediments and framed within the purview of coastal geology. As such, variations in properties with depth have been primarily considered to be the result of sedimentary processes with little attention to pedogenic alterations. Nevertheless, the geological community generated a significant body of information regarding (primarily surficial) properties of SASs. Whether or not sediments beneath shallow water should be considered soil has been the topic of debate for many years. There are some who have argued that the upper limit of soil must be the atmosphere, implying that materials permanently submerged by water are excluded. Others have advocated that the upper boundary should include shallow water (Demas, 1993).

When the first edition of *Soil Taxonomy* was published (Soil Survey Staff, 1975), the articulated definition of soils required a link between soil and plants, stating that soils must support or be capable of supporting plant growth. However, this document also stated explicitly that the upper limit of soil can be shallow water, in which case at its margins it grades to deep water.

There was, however, no mention of what constituted "shallow" or "deep" water. Most individuals understood this to mean that marshes supporting emergent vegetation were soils, but when water became too deep for emergent vegetation, the soil boundary had been reached. Some, such as pioneer George Demas, were unsatisfied with this interpretation, and advocated for the recognition of soils supporting submerged aquatic vegetation (SAV) (Demas, 1993; Demas et al., 1996). The new discussion (and proposals) initiated by Demas came to fruition when the 2nd edition of *Soil Taxonomy* was published (Soil Survey Staff, 1999).

The new definition of soil included several key concepts. First, links to plant growth were weakened while pedogenetic concepts were strengthened. If a soil contained recognizable horizons formed through pedogenetic processes, then it was no longer required to have the ability to support rooted plants in a natural environment. Second, and of particular importance here, soils were permitted to be permanently covered by as much as 2.5 m of water. However, this 2.5 m depth was represented as the likely maximum depth at which rooted plants could continue to grow. This new definition and explanation of soil opened the door for the recognition of subaqueous soils by USDA, and, shortly thereafter, several official series descriptions (OSDs) of subaqueous soils were approved (USDA-NRCS, 2011a). These were used by Demas in a pilot scale project in Sinepuxent Bay, MD (Demas, 1998).

2. PEDOGENESIS OF SUBAQUEOUS SOILS

2.1. Traditional Models

The generalized model of Simonson (1959) summarizing all pedogenic processes as additions, losses, transfers, or transformations can be applied to subaqueous systems and can be used to explain the development of soil horizons in subaqueous soils. For example, both the additions of organic materials through the growth of SAV and detrital additions from algae, when combined with the translocating activity of benthic organisms and the microbial transformations of the organic matter, can largely explain the formation and occurrence of A horizons in the surface of SASs. Similarly, we observe the additions of Fe oxides that are sorbed to mineral (primarily clay) particles of detrital sediments that are transformed via the microbial reduction of sulfate to sulfide, into the authigenic sulfide minerals (mainly pyrite) within the upper horizons of SASs (Demas and Rabenhorst, 1999).

Arguably the most commonly recognized and cited model of pedogenesis is the state factor model of Jenny (1941), where he described soils forming as a function of five primary factors that include climate, organisms, parent material, relief, and time. One might argue that the impact of climate is limited because rainfall has essentially no impact on permanently saturated soils; however, temperature differences may significantly affect SAS genesis.

Because SASs tend to be quite young in age (mostly late Holocene) the influence of the nature of the geological parent sediment is little altered and strongly apparent in the soils. The activity of organisms through production and addition of organic matter, turbation, or mixing in the upper zone and through aeration of the upper horizon are all important in SAS horizon differentiation. Of particular significance in SASs is Jenny's recognition of the importance of relief or topography. While some of the topography-related factors primarily described by Jenny (such as erosion and proximity of the water table) are relatively unimportant in SAS systems, variations in soils as a function of landform and water depth (both tied to topography) are quite important.

2.2. Early Investigations and Concepts

According to Hansen (1959), the first use of the terms describing subaqueous soils were introduced by VonPost (1862). He described "gyttja" and "dy" as two contrasting soil types, both of which formed under water. Although his concept of soil differed from our current one, he was, nevertheless, the first to use this terminology.

Almost a century following von Post, Mortimer (1950) made reference to the sediments of freshwater lakes as "underwater soils". Although he described a number of depth functions showing changes in biogeochemical constituents with proximity to or distance from the "soil" or sediment surface, he described no concepts that were distinctly pedological other than to use the term soil.

In 1953, the renowned soil scientist Kubiena proposed a comprehensive soil classification system for Europe that included a place for what he called "the neglected sub-aqueous soils" (Kubiena, 1953). In this system, the SASs were placed into two main categories: (1) the true subaqueous soils that were always covered with water and did not form peat and (2) the peat-forming soils, which were mostly Histosols in emergent wetlands, bogs, or forests, but that would not be considered subaqueous by current understanding. He did carry forward the terms gyttja and dy from van Post, and also introduced the terms "protopedon" and "sapropel" for two additional classes of SASs. These terms, however, are not currently used in either *Soil Taxonomy* (Soil Survey Staff, 2010) or the World Reference Base (IUSS Working Group WRB, 2006) and thus lack clear definition. Kubiena also introduced the use of particular horizonation for SAS profiles, including (A)C – soils that do not have a distinct humus layer (an A horizon); AC – those that do have a distinct humus layer; and AG – those with a humus layer underlain by a gleyed horizon. Although Kubiena was the first to develop a classification system for subaqueous soils, there is no evidence that this classification system is currently in use anywhere. Drawing largely upon Kubiena's (1953) work, Muckenhausen (1965) included SASs within the classification system developed for the soils of Germany. Nevertheless, there is no evidence that such soils were recognized to any significant extent or included within soil mapping efforts.

Within the framework of discussing the chemistry of wet soils, Ponnamperuma (1972) argued the case that "submerged" soils (those that are permanently or substantially under water) should be considered to be soils for the following six reasons: (1) they are formed from soil materials, (2) pedogenic processes are occurring within them, (3) they contain OM and organisms, (4) subaqueous soil bacteria are similar to those found in subaerial soils, (5) soil horizons are observed and present, and (6) there are variations in texture, mineralogy, and OM content (as commonly observed in subaerial soils). Items 2 and 5 are of particularly interest to pedologists because they form the basis for the current definition of soil (Soil Survey Staff, 1999).

2.3. Contributions of George Demas and Others

Despite over a century of (at least occasional) references to work describing SASs, most American soil scientists failed to recognize systems that were covered by water too deep to support emergent wetland vegetation as soils (Soil Survey Staff, 1975). This changed in the 1990s as a result of the pioneering work conducted and articulated by the late George Demas (Demas, 1993; Demas, 1998). While working as a soil survey project leader and mapping soils in tidal marshes surrounding coastal lagoons behind mid-Atlantic barrier islands, he observed SAV and other organisms inhabiting the soils permanently submerged beneath the tidal waters. He therefore concluded that soil mapping should continue into these environments and this became the basis for his work on the SAS of Sinepuxent Bay, MD (Demas, 1998). For these efforts, he was recognized, receiving the Emil Truog award from the Soil Science Society of America and also the Secretary's Honor Award for Scientific Research from the USDA.

A number of milestones were achieved through Demas' work that launched additional contributions toward the development of SASs. One of the daunting problems facing scientists working in the subaqueous environment is the inability to visually observe the subaqueous land surface, and the general lack of access to high-quality topography (bathymetry) that is so heavily utilized in subaerial studies. While good-quality color infrared photography was available, even under ideal conditions, visibility was limited to about 1 m below the water surface. Using technology just coming online at the time, Demas was able to join a research-grade fathometer with a real-time GPS while at the same time recording tide levels to permit correction for the tide (Demas and Rabenhorst, 1998). This enabled development of a high-quality bathymetric DEM for Sinepuxent Bay. Today, relatively inexpensive bottom profilers (fish finders) with built-in GPS are used to construct bathymetric maps (August and Costa-Pierce, 2007; Payne, 2007).

Most soil mapping efforts are based upon the idea that within a given region (climate, parent material, soil age, etc.), soils occurring on a particular landform are expected to be similar to soils occurring in other areas but on similar landforms. Using the bathymetric data, Demas (1998) constructed contour maps of the subaqueous landscape. These maps were joined with aerial photography, so that subaqueous landforms could be identified and named in a manner similar to subaerial landforms. This development led to the recognition that the soil–landscape paradigm could be applied in subaqueous systems. For example, within the mid-Atlantic coastal lagoons, it was demonstrated that those soils formed on the storm surge washover fan flats bear a great deal of similarity to each other, while those soils forming on lagoon bottom landforms also share a great many similarities, but that the soils of these two landforms are highly contrasting.

About the time that Demas was completing his dissertation, Mike Bradley was beginning his work on SASs in Ninigret Pond, a coastal lagoon in Rhode Island (Bradley, 2001). Bradley's work added substantially to the development of soil–landscape models in coastal lagoons (Bradley and Stolt, 2003). He also was able to demonstrate that subaqueous landforms change relatively little except in the most dynamic portions of the estuary (Bradley and Stolt, 2002), thus alleviating the concerns of some, that SAS maps would need to be redone after every major storm or hurricane. One of the issues that emerged from Bradley's work was the need to use common and recognizable terms to describe subaqueous landforms and features. This led to the development of a glossary for subaqueous geomorphic terms (National Cooperative Soil Survey, 2005; Stolt et al., 2004). Bradley also was the first to begin work on important ecological interpretations and utilization of SASs with a particular focus on eelgrass (*Zostera marina*) restoration (Bradley and Stolt, 2006).

In addition to tackling subaqueous soil mapping, Demas documented the pedogenic processes leading to the formation of soil horizons. Approximately 100 pedons were described in the pilot project in Sinepuxent Bay where mostly A, C, O, and transitional horizons were recognized in the soil profiles. As subaqueous soil sampling and descriptions expanded outside of Sinepuxent Bay, buried (submerged) soils with B horizons (formed in a different pedogenic setting) were encountered including spodic (Ellis, 2006), argillic, and cambic horizons (Erich et al., 2010b). Demas (1998) placed each of the pedons he characterized within the existing soil classification system at that time; essentially, all fit within various classes of Aquents in Soil Taxonomy (Soil Survey Staff, 1975). Recognizing that these permanently submersed soils differed markedly from subaerial soils in the Aquents class, discussions began on modifying Soil Taxonomy to better accommodate SASs. These discussions were led primarily by the Northeast National Cooperative Soil Survey partners. In 2010, new taxa in the Entisol (Wassents) and Hisotol (Wassists) orders were added to Soil Taxonomy to accommodate subaqueous soils (Soil Survey Staff, 2010). Demas (1998) proposed six soil series to accommodate mapping of SASs in Sinepuxent Bay: Trappe, Whittington, Tizzard, Southpoint, Sinepuxent, and Fenwick (which, as a tribute to his efforts, was later renamed

Demas following Dr. Demas' untimely death in 1999). Web soil survey maps of coastal lagoons and embayments in Rhode Island and Connecticut (Fig. 6) use another 9 soil series (USDA-NRCS, 2011a).

As an integration and summary of his work, Demas proposed a modification of Jenny's state factor equation that was adapted to include additional factors important to subaqueous pedogenesis. In addition to drawing heavily on Jenny (Jenny, 1941), this approach incorporated factors proposed by Folger (1972) related to sediment characteristics. However, furthermore, factors addressing the character of the overlying water column and also the possible contribution by catastrophic events were also incorporated (Demas and Rabenhorst, 2001).

Following the work of Demas and Bradley, studies began in Delaware and in Maine. In Delaware, Cary Coppock began mapping Rehoboth Bay which represented a significantly larger area (3,300 ha) than previously had been examined in other subaqueous investigations (Coppock and Rabenhorst, 2003). In addition, Coppock attempted to refine soil landscape models for mid-Atlantic coastal lagoons. In Taunton Bay, Maine, Chris Flannagan worked toward developing soil–landscape models in a dramatically different setting than the coastal lagoons in southern New England and the mid-Atlantic (Osher and Flannagan, 2007). Jen Jesperson followed Flannagan's work by using the Taunton Bay SAS map to assess carbon storage and sequestration in these subaqueous soils (Jesperson and Osher, 2007).

In July 2004, several researchers working in SASs hosted a workshop in Delaware to compare notes and to acquaint interested colleagues with the state of the science related to SASs and participants came from along the entire Atlantic coast (Rabenhorst et al., 2004). As a result of these training efforts, many new SAS projects began emerging and, at present, there have probably been more than 20 SAS studies completed or underway. In August 2010, a second national workshop on SASs was held in Narragansett, RI, with attendees from across the country. Interest in SASs initially focused upon coastal and estuarine areas, but there is an increasing awareness and interest in freshwater systems (Bakken and Stolt, 2010; Erich et al., 2010b).

3. MAPPING OF SUBAQUEOUS SOILS

In the last decade, the soil has become recognized as an essential component of the Earth's "critical zone," and ecosystem managers are requiring soil resource information in order to best address important issues. Thus, while water-quality issues have dominated estuarine management concerns, there is growing awareness that SASs also are important. Acquiring and maintaining a soil data layer that includes spatial information as well as characterization data are more important than ever. The particular value of a pedological soil map (in contrast to the more common sediment mapping data) is that it carries three-dimensional information rather than just two-dimensional data on the upper zone or surface horizon.

3.1. Unique Challenges and Solutions to Working in SASs

Soil mapping activities have been underway for a hundred years; so one would think that protocols would all be in place and well tested. Regarding SASs, however, there are several unique challenges that stand to impede rapid progress. These challenges include the following: (1) the landscape surface is obscured from view, either by visual observation or by satellite or aerial imagery, (2) sampling of soils must be accomplished with the soil surface being covered with water, and (3) soil samples which have formed or long persisted in a subaqueous environment have the potential to experience dramatic changes in chemical or mineralogical characteristics following exposure to air.

One of the most useful tools available for dealing with the submerged landscape is aerial photography. Provided imagery is collected during times when water is clear, some important subaqueous landscape features can be observed in either B&W or color IR photos. However, even under good water conditions, the limit of observation is usually about a meter or perhaps a little more. As a result, an alternate approach is needed to acquire information where the water depth is greater. Good-quality bathymetry, especially when the data can be manipulated with various computer programs, can permit good assessment of the subaqueous land-surface topography and landforms. Although NOAA archives and manages bathymetric information for most US estuaries (these sources should be explored), high-quality, high-resolution data sets are rarely available. Therefore, researchers often must first collect a bathymetric data set in order to reconstruct and analyze the subaqueous land surface. Advancing technology has made this a far simpler task than it was even a decade ago. There are commercially available fathometers at a reasonable cost that either contain or will allow interface with a GPS so that X, Y, Z data (Lat, Long, and depth) can be collected, downloaded, and analyzed. Surveying to connect with a local elevation benchmark will permit one to assign elevation data relative to an appropriate datum (such as NAVD88). The primary complication in coastal systems is that the collection of bathymetric data usually extends over the course of one or more (sometimes many) tidal cycles, which impose a cyclical (though often irregular) deviation to the raw data. The solution usually is to install one or more tide gauges in the water body during the period when bathymetric data are collected (Payne, 2007). Most GPS or fathometers will place a date–time stamp on the data that can be tied back to the tide gauge data, so that appropriate corrections can be made.

3.2. Sample Collection

The traditional tool of the soil scientist is the bucket auger, and, surprisingly, this tool has been effectively used to observe, describe, and map SASs. It does have, however, a number of limitations. In most cases, it requires that the operator be in the water and standing on the subaqueous soil surface. Usually, the operator plants themself in a fixed position during the entirety of the boring, which is necessary for repeatedly finding the hole and proceeding with the boring. This also will normally limit the depth of water in which borings can be made to about 1–1.5 m (depending on the height of the operator). In cases where the water is particularly clear, a bucket auger can sometimes be used from a boat, but this usually limits the depth of the soil that can be observed.

As samples are collected (a bucket at a time), they can be placed either on a boat or on a floating raft made for the purpose. As is commonly the case when using a bucket auger, there is some mixing of samples across horizons. For routine description and mapping, this may not be problematic. To keep the sample intact, a closed-sided bucket should be used. There are some samples that may be difficult to extract using a bucket auger, either because they are too sandy or because they are fine textured but too soft (high n-value). In either case, samples may tend to flow out of the bottom of the bucket before they can be retrieved. The modified bucket auger designed by Darmody for work in tidal marshes may be useful in such circumstances (Darmody et al., 1976).

Vibracoring represents technology already available from the coastal geology community (Fig. 1). A vibracorer consists of a high-frequency concrete vibrating head (designed to vibrate air out of poured concrete before it hardens) that is attached with a clamp to a length of aluminum irrigation pipe. When the high-frequency vibrations are passed through the irrigation pipe to the saturated soil, the vibration liquefies a thin (mm-thick) layer around the tube that allows fairly easy insertion of the pipe downward into the soil when some weight (such as the operator's body weight) is applied. Once the tube is inserted in the soil, water is added to fill the rest of the pipe (if not already filled), and the pipe is tightly sealed/plugged at the top (to prevent the material from being pulled back out of the tube during extraction by forces of suction). The pipe is removed from the soil with a winch or chainfall attached to a tripod. This procedure necessitates that the vibracorer be mounted on a boat or other working platform and that the vessel be made secure by using multiple (usually 3 or 4) anchors. Once the core is extracted, it can either be opened at the time or it can be sealed and stored and opened when back on shore. Opening of aluminum cores is facilitated by using electric sheet metal shears. Typically, two cuts are made down opposite sides of the core and a strong thin wire is pulled down the length of the core cutting the core in half before the two halves are split apart with the two halves of the aluminum pipe, as shown in Fig. 2.

One important benefit of vibracoring is that an intact core up to several meters in length can be obtained so that there is little or no mixing between soil horizons and with the integrity of the soil material itself largely preserved. This allows for a more accurate description of the soil, including the horizon boundaries. A second benefit is that the cores are sealed and protected from air and from oxidation until they are opened. Therefore, a large number of cores can be collected and taken back to shore for description at a later time. However, there are times when one may want to know immediately the nature

FIGURE 1 Sampling a subaqueous soil with a vibracore sampler. Jim Turenne (pictured here in Wickford Harbor, RI) is using a chainfall and tripod on pontoon boat to remove a 10 cm diameter aluminum core (7.5 cm cores are also used). The vibrating head is clamped to the top of the core and runs to a gas-powered vibrator (not shown). The high-speed vibrations allow the core to be pushed easily into the subaqueous soil. Prior to removal, the remainder of the core (that not filled with soil) is filled with water and sealed with a cap (red). The core is either opened on the boat and described and sampled immediately or sealed at the bottom and stored for transport back to a refrigerator in the lab for description and sampling at a later time. *(Photo by Margot Payne) (Color version online)*

of the soil sampled, in which case the opening and description of the core are done onboard the research vessel. Another important benefit of the vibracorer is that it is useful in a wide range of soil materials, including both those that are sandy and those that are fine textured.

One disadvantage of vibracoring is that substantial time is required to secure the boat or platform before sampling can begin. However, once this preliminary setup is complete, extraction of a core can usually be accomplished within 10–30 min. Another problem that is often observed is compaction, collapse, or what is sometimes referred to as "rot" of the core. This can be detected when the length of the extracted core is less than the depth of the soil

FIGURE 2 Three soil profiles collected by vibracoring in Rehoboth Bay, Delaware, USA, that demonstrate dramatic differences among soils associated with various subaqueous landforms. Core A is a Sulfic Psammowassent (Demas series) sampled on a storm surge washover fan flat; core B is a fine-silty Typic Sulfiwassent (Truitt series proposed) sampled on a lagoon bottom; core C is a sandy over loamy Haplic Sulfowassent (Tizzard series) sampled on a submerged mainland flat. *(Color version online and in color plate)*

from which it was sampled. Careful measurements must be made to determine how much compaction has occurred during the vibracoring process. Depending on the nature of the soil and the equipment being used, this can range from as little as 5% to as much as 30%. When the amount of compaction is small, it is usually assumed that the compaction was uniform across horizons. However, when compaction is greater, this may not be a good assumption.

A third tool that is useful in sampling soft (high *n*-value) materials is the Macaulay sampler (Jowsey, 1966). This auger is designed to collect undisturbed half cores in 50 cm increments by inserting the auger into the soil to the desired depth and then rotating the auger a half turn. A flat vane provides radial stability so that the half core can be collected during rotation. After extraction, the core is rotated back exposing the half core (Fig. 3). Multiple extensions can be added to this instrument to allow for deep sampling in deeper water, over the side of a boat. Because the materials sampled are soft,

FIGURE 3 A fifty-cm half core of a subaqueous soil collected using a Macaulay sampler while researchers remained within the vessel. Note that variations between soil horizons are preserved and readily viewed in cores collected using this technique. *(Photo courtesy of Jonathan Bakken) (Color version online and in color plate)*

there is no need for the sampler to reenter the soil at the same location. The sampling is rapid and easy, although there is some delay when extensions are added. The design precludes any vertical compaction and that, combined with the half core design, makes it particularly useful for collecting samples for bulk density determinations. The Macaulay auger allows immediate observation for morphological description and sampling. The primary limitation of the Macaulay auger is that it can only be used in relatively soft materials, as one must be able to push the auger into the soil. Therefore, it is not usable in sandy soils or finer textured materials that are not soft (i.e. higher density and firmer consistence).

3.3. Sample Storage Problems with SASs

The permanent submergence of SASs by water affects the properties of the soils themselves. Saturation largely excludes oxygen, except from the uppermost zone (mm to cm) where it can enter via diffusion (and sometimes aided by bioturbation). As a result, most SASs are strongly anaerobic as evidenced by the presence of sulfide minerals and also by low measured Eh values, especially below the upper few cm of the soil (Fig. 4). Many have formed in brackish or saline environments and those will have sulfate as a significant anion that affects the products of redox processes.

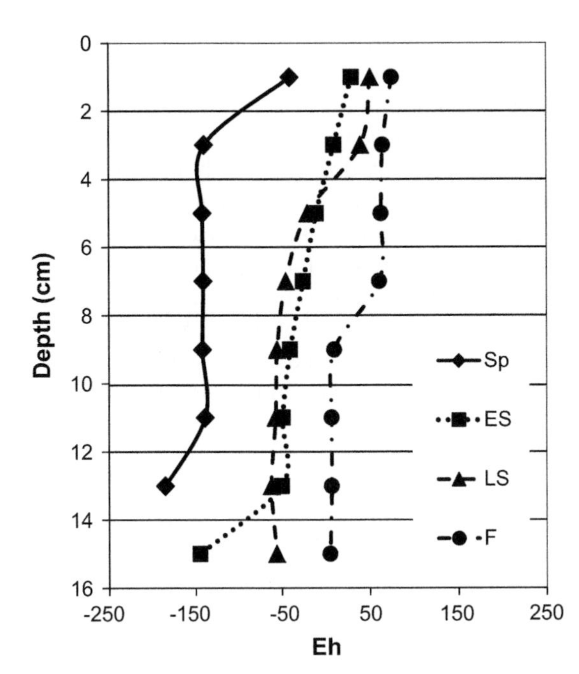

FIGURE 4 Redox potential of a Shoal soil from Wickford Harbor, Rhode Island. Monitoring periods were: Sp = Spring; ES = Early Summer; LS = Late Summer; F = Fall. *(From Payne, 2007)*

Some properties of SASs change very little or not at all, while other properties change dramatically. In addition, some of the properties that change dramatically for SASs from estuarine systems may be relatively stable for SASs in fresh systems. In general, physical properties of SASs are stable. Properties such as particle size or bulk density would not be expected to change as a result of exposure to air or to oxidation. Nevertheless, some additional steps may be required to conduct particle size analyses due to the presence of (usually chloride) salts in the samples from the seawater, or from sulfate salts that can form from the oxidation of sulfide during the storage of samples under oxidizing conditions. This can usually be addressed by adding a step at the beginning of the analysis using dialysis to remove the salts. In addition to physical properties that are stable, there are some chemical properties that are also fairly stable. Total organic C would be an example of a chemical property not expected to change much upon exposure to air.

In contrast to the stability of physical properties, many of the chemical properties of SASs (especially those from estuarine settings) can change dramatically if exposed to (and especially if stored in) an oxidizing environment. Those properties that are usually stable for most subaerial soils[1], but

1. Exceptions to this would be soils of tidal marshes (we consider these subaerial soils rather than subaqueous soils), which also often have similar potential problems as do SAS.

which may change most dramatically with SASs, include pH, sulfide content, and EC (soluble salts). Most of these changes are driven by the presence and oxidation of sulfide minerals that commonly form in SASs of estuarine environments. The upper horizons of SASs may contain iron monosulfides (FeS), which are especially labile and can rapidly oxidize chemically in a matter of minutes upon exposure to oxygen. Fortunately, when present, monosulfides usually represent a small portion of the total sulfide minerals in a sample. Pyrite (FeS_2) is usually the dominant sulfide mineral found in SASs and can occur in quantities of up to several percent as S. Pyrite can be oxidized microbially under moist oxidizing conditions over a period of days to weeks. During the oxidation of pyrite, both sulfuric acid is generated and sulfate salts are formed (Rabenhorst and Fanning, 2002; Rabenhorst et al., 2002). Under prolonged storage, pH can drop to less than 3 (Fig. 5).

3.4. Addressing Sample Storage Problems

Most of the problems associated with changes in SAS soil properties upon storage are related to chemical or microbiologically driven changes tied to exposure to oxygen. Therefore, the general principles governing protection against these kinds of changes in soil properties are to slow down the chemical reaction, protect against exposure to oxygen, or both. Slowing down the

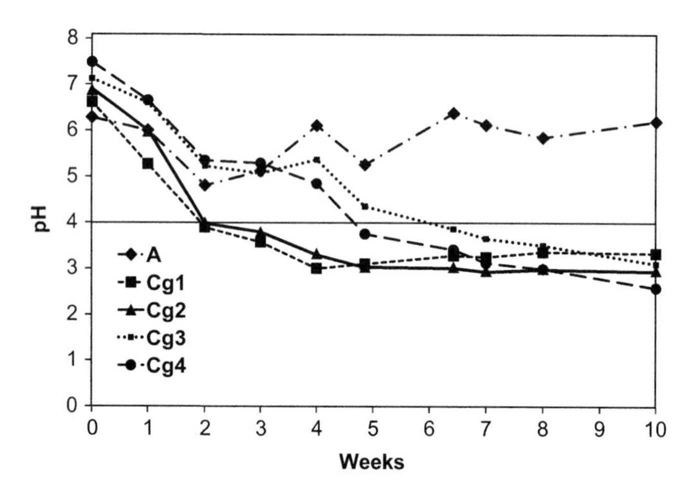

FIGURE 5 Change in pH over time of samples from subaqueous soil horizons incubated under moist oxidizing conditions. Samples that contain significant Fe sulfide minerals, such as pyrite, will become more acid with time. If the pH drops a minimum of 0.5 units to a pH below 4.0, the materials are considered to be sulfidic (Soil Survey Staff, 2010). Depending on the quantity of sulfides present and the buffering capacity of the soils, some samples drop below pH 4 within two weeks (Cg2 and Cg3 horizons), while others may take longer (Cg1 and Cg4). The A horizon may not have significant accumulations of sulfide minerals and was still at a pH of approximately 6 after 10 weeks. *(Data from Balduff, 2007)*

chemical and microbial reactions causing changes can be accomplished by cooling the sample. If there will be only a short period before analytical work is undertaken, samples can be placed on ice in a cooler. If it will be longer before analyses can be run, then samples can be frozen using either dry ice or liquid nitrogen. An intermediate strategy is to chill the samples on ice while samples are being collected, and then freeze the samples upon returning to the lab.

Some strongly reduced and labile phases such as Fe(II) or sulfides can oxidize fairly quickly when exposed to oxygen. Placing samples into a zip-lock plastic bag and squeezing to exclude air before sealing can be somewhat effective in protecting against exposure to oxygen. This may work for short periods of time, and may be most effective in protecting less labile phases such as disulfides such as pyrite (which are mainly oxidized by microbial activity). It should be noted that over time, small quantities of oxygen can diffuse through plastic bags even when fully sealed. The method described above may not be adequate for more highly labile phases such as monosulfides. If samples are thought to contain monosulfides, then it may be necessary to place them into heavy-duty zip-lock bags and to sparge repeatedly with N_2 gas before sealing. The sparging removes oxygen from storage bags prior to sealing, and thus is more effective. Probably the most effective means of protection is to combine sparging with N_2 gas, followed by freezing in the field.

4. APPLICATIONS OF SUBAQUEOUS SOIL INFORMATION

Many of the subaqueous soil efforts over the last decade have focused on addressing the issues raised in the previous section. The result has been progress in the development of subaqueous soil maps. The first area where subaqueous soil mapping was incorporated into the USDA-NRCS soil information system was in Rhode Island (Fig. 6) where a seamless soil survey provides information for both subaerial and subaqueous portions of the landscape. The definition of soil was changed to include subaqueous soils (Soil Survey Staff, 1999) as resource scientists began to recognize the importance of these shallow-water soil resources for their habitats, structure, and associated ecosystem functions. With most subaqueous soil efforts having concentrated on developing field and laboratory methodology (Bradley, 2001; Coppock and Rabenhorst, 2003; Demas, 1998; Ellis, 2006; Flannagan, 2005; Jesperson, 2006; Jesperson and Osher, 2007; Payne, 2007), the breadth of information relating subaqueous soil type with resource use and management is quite limited. Because soils are classified using properties related to use and management (Soil Survey Staff, 1999), soils having the same classification are expected to have similar use and management interpretations, and these interpretations should be applicable on a regional scale. Considering that nearly two-thirds of the world's population lives in coastal areas (Trenhaile, 1997), and that shallow-water habitats are highly valued and heavily used resources, the need for subaqueous soil use and management interpretations is paramount. To date, most of the interpretive studies have focused on submerged aquatic

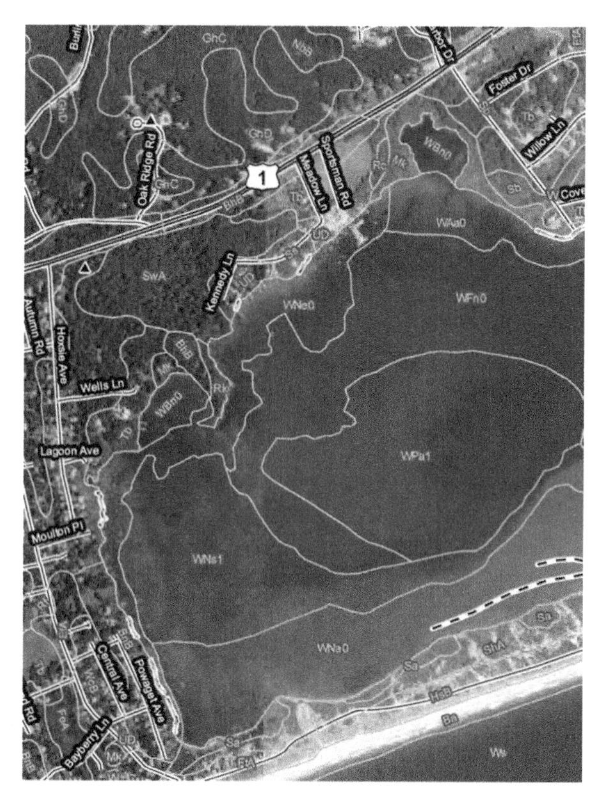

FIGURE 6 A soil survey map from Washington County, RI, that includes the west end of Ninigret Pond showing delineations of both subaerial and subaqueous soil map units *(From USDA-NRCS, 2011b)*. *(Color version online and in color plate)*

vegetation (SAV) restoration and carbon accounting (Table 1) in estuarine environments. Other studies have investigated relationships between soil properties and shellfish aquaculture, heavy metal accumulation, water quality, dredge placement, and mooring types. As the science progresses, a wide range of use and management interpretations are expected to be developed (August and Costa-Pierce, 2007; Payne and Turenne, 2009). These interpretations will aid in ecosystem management, sustainability, and conservation efforts.

4.1. Carbon Accounting

With the concern over increasing carbon dioxide emissions leading to global warming, significant interest has emerged in the role of soil systems as a carbon sink. These interests have led to numerous studies that focus on quantifying soil carbon sinks in subaerial systems. Although carbon in forested and emergent wetland soils has been well studied, investigations focused on estuarine soils as

TABLE 1 Summary of Use and Management Subaqueous Soils Publications

Use and management interpretation	Location	Publications
Submerged aquatic vegetation restoration	Ninigret Pond, RI; Indian River Inlet, FL; Chincoteague Bay, MD; Potters Pond, RI; Quonochontaug Pond, RI	(Bradley, 2001; Bradley & Stolt, 2006; Fischler, 2006; Balduff, 2007; Pruett, 2010)
Carbon accounting	Taunton Bay, ME; Greenwich Bay, RI; Wickford Harbor, RI; Little Narragansett Bay, RI; Chincoteague Bay, MD; Ninigret Pond RI	(Jesperson, 2006; Balduff, 2007; Jesperson and Osher, 2007; Payne, 2007; Pruett, 2010)
Water-quality assessment	Greenwich Bay, RI	(Payne, 2007)
Dredging	Rhode Island lagoons and embayments	(Salisbury, 2010)
Mooring Placement	Little Narragansett Bay, CT & RI	(Surabian, 2007)
Shellfish aquaculture	Ninigret and Quonochontaug Ponds, RI	(Salisbury, 2010)
Heavy metal accumulation	Rhode Island lagoons and embayments	(Pruett, 2010)

carbon sinks have been largely overlooked (Chmura et al., 2003; Jesperson and Osher, 2007; Thom et al., 2003). Considering that the subaqueous component may occupy as much as 90% of an estuary, these areas likely represent a significant and unaccounted for carbon sink.

Several recent pedologic studies have investigated the distribution of carbon in estuarine subaqueous soils (Table 1). Many subaqueous soils show the typical SOC with depth patterns that are observed in subaerial soils. Since most of the carbon is added at the soil surface as detritus from plants and animals in the subaqueous soil system, surface horizons typically have the highest SOC contents and values decrease with depth (Balduff, 2007; Jesperson, 2006; Pruett, 2010). Because subaqueous landscape units are quite dynamic, buried horizons are often encountered and carbon content may vary dramatically with depth (Jesperson and Osher, 2007) much like subaerial floodplain systems (Fig. 7).

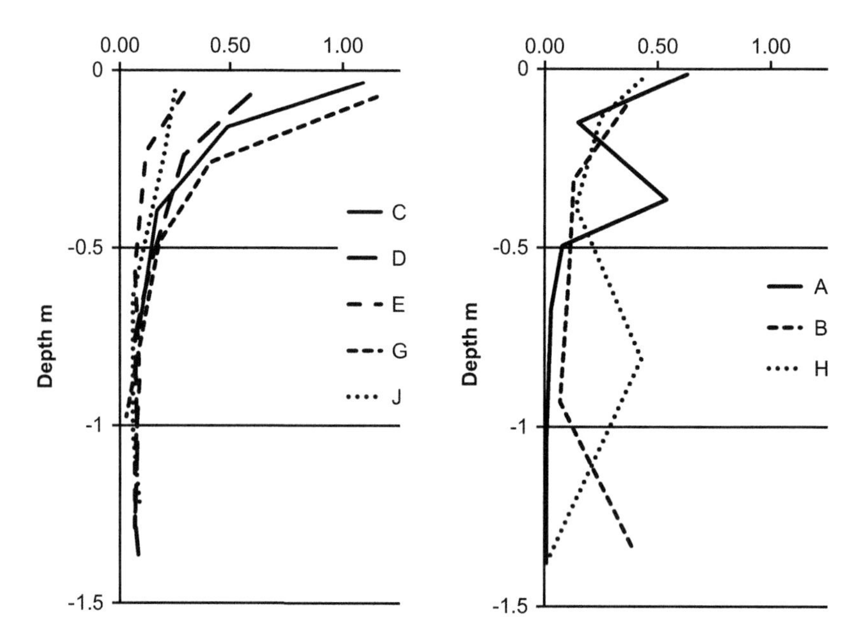

FIGURE 7 Soil OC depth functions in subaqueous soils in Sinepuxent Bay, MD. Representative pedons C, D, E, G, and J show typical depth patterns that are observed in many subaerial soils where SOC decreases regularly with depth. In contrast, the fluvial character of representative pedons A, B, and H shows an irregular decrease in SOC with depth, much like soils in subaerial floodplain systems. *(Modified from Demas, 1998)*

The quantity of carbon stored in subaqueous soil systems is dependent upon a number of factors. Jesperson and Osher (2007) and Payne (2007) investigated carbon pools in embayments, while Balduff (2007) and Pruett (2010) measured pools in coastal lagoons. Pools were calculated to a meter depth in order to compare pools to adjacent subaerial soils. Similar relationships between soil types and carbon pools were found between the embayments and lagoons. Estuarine subaqueous soils were found to have SOC pools equivalent in magnitude to those of subaerial systems and the quantity of SOC stored varied as a function of soil and landform (Fig. 8). The highest SOC pools were observed in areas of the estuary where sea-level rise had buried former tidal marshes, and observed SOC values averaged over 300 Mg ha^{-1} (Balduff, 2007; Pruett, 2010), which are as much as fresh and tidal marsh soil systems. Finer-textured subaqueous soils occurring in coves and lagoon and fluvial–marine bottoms have SOC pools (135–200 Mg ha^{-1}) equivalent to poorly drained mineral soils (Jesperson and Osher, 2007; Payne, 2007; Pruett, 2010). Carbon pools in higher-energy, sandier soil–landscape units, such as shoals and shorefaces in embayments or flood tidal delta or washover fans, had the lowest carbon pools (22–75 Mg ha^{-1}) (Pruett, 2010). The magnitude of these pools was comparable to average subaerial Udipsamments (see Kern, 1994).

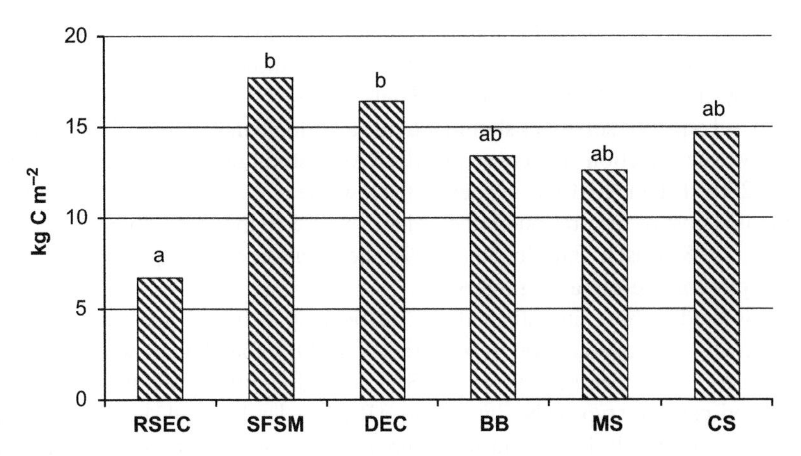

FIGURE 8 Organic C stored in various subaqueous soil map units in Taunton Bay, ME. Bars with the same letter indicate no significant difference at $\alpha = 0.05$. RSEC = Recently submerged edges and coves; SFSM = submerged fluvial streams and marshes; DEC = deep edge and cove; BB = bay bottom; MS = mussel shoal; CS = channel shoulder. *(Data from Jesperson and Osher, 2007)*

Therefore, those involved in global carbon accounting should consider including soil carbon stocks from subaqueous soils as well as subaerial soils.

4.2. Submerged Aquatic Vegetation Restoration

Submerged aquatic vegetation constitutes important rooted plants in coastal ecosystems (Thayer et al., 1984). The ecosystems that these submerged grasses help to create are highly productive, and provide habitat for a variety of finfish, shellfish, shrimp, and benthic infauna (Bertness et al., 2001), while also functioning as traps for sediment and pollutants from the water column (Ginsburg and Lowenstam, 1958; Harlin et al., 1982). Habitats for SAV have been declining over the last several decades as a result of coastal eutrophication, water pollution, increased water temperatures, increased sedimentation rates, wasting disease, algae blooms, and other stressors (Hemminga and Duarte, 2000; Short et al., 1996). In response to such declines, many SAV restoration projects have been implemented (Churchill et al., 1978; Kopp and Short, 2000; Merkel, 2004; Reid et al., 1993; Short, 1988; Short, 1993). Success rates for these projects, however, are quite variable (Short et al., 2002), with average success rates around 30% (Fonseca et al., 1998).

Poor site selection is often cited as the major factor limiting the success of SAV restoration projects (Calumpong and Fonseca, 2001). Site-selection models are often used for determining locations for restoration. These models are based on parameters such as wave action, light attenuation, and substrate type. The most commonly used model was developed by Short et al. (2002) to quantitatively assess potential restoration sites. Only three broad classes of estuarine substrate types are used in this model, and all are based solely on sediment grain size. Since SAV health is affected by both physical and chemical properties such as sulfide levels, organic matter content, and sediment-reducing conditions (Bradley and Stolt, 2006; Holmer and Nielsenj, 1997; Koch, 2001; Wicks et al., 2009), subaqueous soil maps should provide additional information that can help improve site selection and, consequently, SAV restoration success (Bradley and Stolt, 2006; Pruett, 2010). Considering the low rate of restoration success, and an estimated cost of restoration on the order of $40–$80 thousand dollars per acre (Short, 1988; Busch and Golden, 2009), the creation of subaqueous soil map could pay for itself in a single project.

Few studies have investigated the relationship between SAV distribution, health, and subaqueous soil type. Bradley (2001) found that percent cover of eelgrass (*Zostera marina*, a common SAV) varied significantly with subaqueous soil–landscape unit type in Ninigret Pond, a Rhode Island coastal lagoon. Eelgrass beds were primarily located in low-energy depositional basins such as lagoon bottom, shallow lagoon bottom, and barrier cove units, and that eelgrass beds from these units propagated soils on nearby flood tidal delta slope and washover fan slope soil–landscape units. A decade later, Pruett (2010) found similar relationships between eelgrass cover and soil–landscape units in Ninigret Pond.

The distributions of eelgrass relative to soil type observed by Bradley (2001) and Pruett (2010) suggested that certain soils are most likely to provide optimal soil conditions for eelgrass establishment. To test this, Pruett (Pruett, 2010) studied eelgrass health and transplant success as a function of soil type in Ninigret and two other coastal lagoons in Rhode Island. She observed that soil–landscape units supporting the most eelgrass, and with the highest eelgrass growth rates, did not have the highest health index or success rates for eelgrass restoration, thus decoupling the relationship between the presence of eelgrass and the potential for restoration. Ratios of aboveground to belowground biomass were lowest for the soils with greatest restoration success, suggesting that soil properties may be a limiting factor in restoration success. The literature (Koch, 2001; Wicks et al., 2009) suggests that soils with SOC levels greater than 3–4% typically have little restoration success. Pruett (2010) suggested that elevated SOC levels may be indicative of a strongly reducing environment (eelgrass stressor) with elevated sulfides (eelgrass stressor and/or phytotoxin) that leads to poor plant health, inhibiting SAV restoration. Based on these studies, Pruett (2010) suggested that the ideal soils for transplant were soils where SOC levels are below 3% (SOM <5%), soil sulfide levels are relatively low, and water depths are great enough to support SAV. These soil types are found in the transitional landscape units (i.e. flood tidal delta slope and washover fan slope) that separate deep low-energy basins from sandy flats. Eelgrass that is restored on these units may propagate to the other surrounding units. An example where SAS data were included in models predicting locations for successful SAV growth is shown in Fig. 9.

Balduff (2007) compared the subaqueous soil system map she made of Chincoteague Bay, Maryland, with the published aerial distributions of

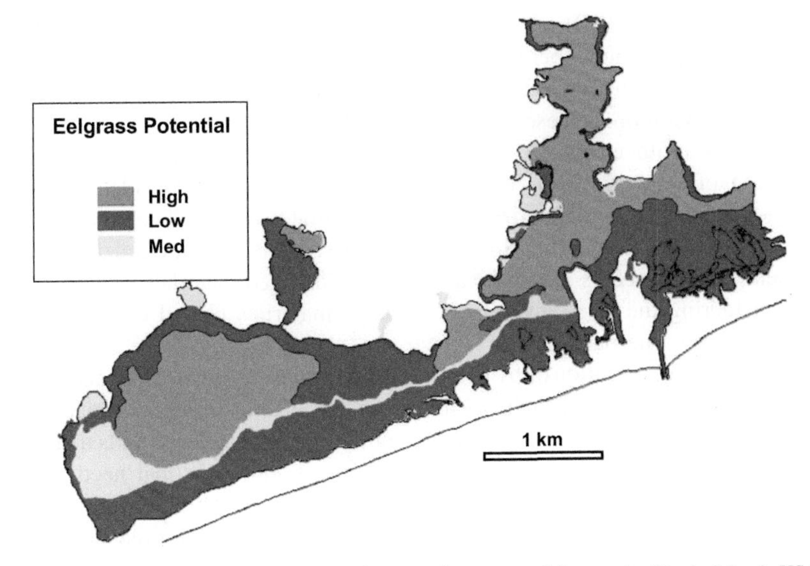

FIGURE 9 Eelgrass potential interpretative map for a coastal lagoon in Rhode Island, USA (Ninigret Pond). Potential ratings were based on soil properties and bathymetric data. Boundaries are based on a subaqueous soil survey of Ninigret Pond. *(From MapCoast, 2011) (Color version online and in color plate)*

eelgrass. Using information from the literature, Balduff (2007) created eelgrass suitability classes based on the soil texture, SOC content, and sulfide levels. Using these criteria, each subaqueous soil map unit was assigned a suitability class (slight, moderate, or severe) and the interpretive map was compared with eelgrass maps produced by the Virginia Institute of Marine Studies (Orth et al., 2005). Map units with the slight ratings (low sulfides, low SOC, and sandy textures) had the highest eelgrass cover, suggesting a relationship between soil type and eelgrass distribution. These results support the conclusions drawn by Pruett (2010) that the best soils for restoration are those with relatively low sulfide levels and higher redox potentials.

4.3. Dredging and Dredge Placement

Dredging is a common practice in estuarine environments where channels and inlets need maintenance for navigation. For example, more than 300 million cubic meters of material are dredged by the US Army Corp of Engineers each year (Winfield and Lee, 1999). Dredged materials are potentially a valuable resource, provided that good-quality materials are identified and dredged separately. Because the dredging or disposal of some materials can lead to environmental problems, knowledge of the physical and chemical properties of the dredge material is required before the materials can be placed on land (Brandon and Price, 2007; Cheng, 1986; Winfield and Lee, 1999; Yozzo et al., 2004).

Winfield and Lee (1999) proposed that dredge material may be analyzed for grain size, water content, permeability, pH, salinity, and contaminant content to identify potential environmental impacts if the materials are placed on land. However, in addition to these parameters, the potential for the creation of acid-sulfate soils (with extremely low pH values; <4.0, metal toxicity, and the formation of salts) from the oxidation of sulfides must be considered when dredge materials are characterized. Figure 5 illustrates how acid conditions can develop when sulfide-bearing SAS materials are placed into an oxidizing environment, such as occurs during land disposal of dredged materials.

Considering the large quantities of dredge materials placed on land each year, and the onerous task of effectively characterizing these materials, understanding the relationships between dredge material properties and their source or location would help to minimize the efforts needed to meet the challenge. The best approach may be to identify those subaqueous soils that can be safely dredged and placed on the land surface and those that need to be managed to minimize environmental impact. Although a number of studies have examined the environmental impacts of applying marine dredged materials on land from a soils' perspective (Cheng, 1986; Demas et al., 2004; Fanning et al., 2004; McMullen, 1984), studies that have examined relationships between subaqueous soil type and the application of dredge materials on the land surface are quite limited.

Salisbury (2010) simulated upland placement of dredge materials using a mesocosm experiment. Dredged materials were obtained from eight different soil mapping units within embayments and coastal lagoons. Mesocosms were placed out of doors and leachate was analyzed for pH, conductivity, and sulfate content over time. Salts washed out of the dredged material fairly quickly such that the EC of the leachate dropped below 5 dS m^{-1} within 10 months. Two trends in leachate pH and sulfate content were observed as a function of soil type. Leachate from finer-textured dredged materials from low-energy landscape units that classify within Sulfiwassent great groups showed a large drop in pH ($pH \leq 4.0$) associated with sulfide oxidation and the creation of acid-sulfate conditions. These conditions persisted for the duration of the two-year experiment. Leachate from coarser-textured materials of high-energy landscape units that classified as Psammo, Fluvi, and Haplo Wassent great groups typically had pH values that stayed between 6 and 8, suggesting that acidity was of little concern in these dredge materials. In both materials, the leachate sulfate content decreased over time, but sulfate content of leachate from the finer-textured materials was higher due to high levels of sulfides found in these soils. These mesocosm experiments suggest that subaqueous soil maps can be used to identify dredge materials that could potentially cause environmental problems if placed on land and these materials should be managed to accommodate these problems.

Metals such as Cd, As, Pb, Zn, Cu, and Cr are brought into coastal ecosystems through rivers and streams, outfall pipes, and atmospheric deposition. These metals may become hazardous if they enter the food chain via bottom-dwelling organisms and submerged vegetation that are consumed by fish, birds, and mammals, eventually reaching the human population (Ridgeway and Shimmield, 2002). Many of these metals are incorporated into coastal soils such that these soils often function as important sinks for these metals (Ridgeway and Shimmield, 2002; Ruiz-Fernandez et al., 2009; Santschi et al., 1984). In reducing environments such as subaqueous soils, sulfides may serve as a sink through precipitation with the contaminant metals (Griffin et al., 1989). When these precipitates are placed on land as dredge materials and allowed to oxidize, soluble metal sulfates may be produced (Calvert and Ford, 1973; McMullen, 1984; Tack et al., 1995). The solubility of these metals is increased as the pH is lowered during sulfuricization (Fanning, 2002). The low pH results in elevated levels of metals in solution, creating toxic conditions to plants and leading to the leaching of iron, which can clog pipes and stain rivers red from iron precipitation (Brady and Weil, 2007; Calvert and Ford, 1973; McMullen, 1984). At low soil pH (<4.0) heavy metal solubility increases, making these metals available for plant uptake where the metals may enter the food chain (Brady and Weil, 2007; Tack et al., 1995). Subaqueous soils may contain large amounts of sulfides and heavy metals (Cheng, 1986; Fanning et al., 2004; McMullen, 1984; Payne, 2007). Thus, understanding the distribution of metals relative to subaqueous soil type may be an important use and management interpretation if the soils are to be dredged and placed on the land surface.

There are numerous studies demonstrating elevated levels of metals in estuarine subaqueous systems. For example, in Rhode Island alone studies have been conducted by Goldberg et al. (1977), Santschi (1980), Santschi et al. (1984), Nixon et al. (1986), Corbin (1991), Nixon (1991), Bricker (1993), and Ford (2003). Few studies have taken a systematic approach to examine metal distribution by examining metal concentration in subaqueous soils. Pruett (2010) investigated metal (Zn, Pb, As, Cu, and Cr) concentrations in samples collected from surface horizons of subaqueous soils in three coastal lagoons and three embayments in Rhode Island. Ninety percent of 91 samples analyzed were above regional soil background levels for at least one of the metals studied. Pruett (2010) found that Pb and Zn concentrations were typically above regional soil background levels in subaqueous soils in all estuaries and that embayments had higher concentrations of Cr and Cu than coastal lagoons. Although proximity to contaminant sources (subaerial runoff) was found to be an important determinant in the spatial distribution of high metal concentrations, the distribution of metal concentrations was found to vary significantly by subaqueous soil type. Silt plus clay, SOC content, and sulfidic materials correlated positively with high metal concentrations, suggesting that organic carbon and sulfides serve as a sink for the metals. Concentrations of Pb and Zn were significantly greater in Hydrowassents and Sulfiwassents great groups than Haplowassents and Psammowassents great groups. Metals were found in high-enough concentrations to negatively affect benthic ecosystems primarily in nonsandy soil types. Pruett (2010) concluded that because subaqueous soil types are defined by their inherent chemical and physical properties, and metal concentrations vary significantly with soil type, a subaqueous soil survey could be used to identify potential areas of metal contamination.

4.4. Moorings and Docks

Marinas and mooring fields are used throughout the range of water bodies to secure and launch boats. Subaqueous soils serve as the foundation for the marina docks or moorings (permanent anchor to which boats are secured in a harbor). How well the mooring or dock functions is dependent upon the bearing capacity or n-value of the subaqueous soil. The n-value is a parameter used to generally describe bearing strength of soil materials (Soil Survey Division Staff, 1993; Soil Survey Staff, 1999). Although originally conceived as a quantitative measure relating soil moisture, particle size data, and organic matter (Pons and Zonneveld, 1965), it is often used in a more qualitative sense. Soils considered slightly fluid are assigned a n-value of 0.7–1, whereas those that are moderately or highly fluid (with very low bearing strength) are assigned n-values >1 (Soil Survey Division Staff, 1993), and would be considered to have high n-values.

Typic Sulfiwassents and Hydrowassents typically have high n-value materials in the upper 50 cm of the soil surface. Surabian (2007) suggested that mushroom anchors would work best in these high n-value soils because these moorings sink into the low-bearing-capacity soils and are kept in place by surface area and suction forces. For most other subaqueous soils, deadweight anchors are best suited for securing moorings (Surabian, 2007).

4.5. Shellfish Aquaculture

Shellfish are an important commercial and recreational resource of estuaries. Bivalves are one of the oldest and most productive segments of US aquaculture. A combination of factors, including overfishing, mortalities due to disease, lack of successful sets, as well as intense coastal development and accompanying pollution, have severely impacted bivalves in US coastal waters (Harvell et al., 1999; Tarnovsky, 2003). These declines have heightened activities toward bivalve aquaculture and restoration. Worldwide, the aquaculture industry continues to develop and expand. As a commodity the value of aquaculture products per acre typically far exceed those of traditional agriculture. For example, the average value of Rhode Island shellfish aquaculture products (oysters and clams) was estimated at nearly 13 thousand dollars per acre (Alves, 2007). Although shellfish aquaculture has been quite successful, few data are available to resource managers for determining the most productive areas for aquaculture farms. Considering the cash value of these aquaculture products, and the increase in growth of the aquaculture industry, developing an understanding of these relationships is critical to cost-effective management of these resources. A subaqueous soil survey would provide an excellent systematic approach to identify the most cost-effective areas for shellfish aquaculture.

Several studies have focused on shellfish aquaculture and substrate-type relationships (Grizzle and Lutz, 1989; Grizzle and Morin, 1989; Pratt, 1953; Pratt and Campbell, 1956; Wells, 1957). Although soils in these studies were crudely classified (i.e. sand or mud), the investigations found a relationship between growth and particle size. The earliest investigations found that as particle size decreased, shellfish (clams) growth also decreased (Pratt, 1953; Pratt and Campbell, 1956). This relationship was attributed to the inhibition of feeding mechanisms in the finer-textured material (Pratt and Campbell, 1956). More recent studies (Grizzle and Lutz, 1989; Rice, 1992; Rice and Pechenik, 1992) found a similar relationship with higher growth rates in clams associated with sandier substrates. Their explanation was that sandier soils are associated with higher current velocities and thus greater food availability (seston). These studies suggest that subaqueous soil type may serve as a proxy for determining areas of favorable seston fluxes, and could thereby be used to predict areas of the estuary with the highest potential for shellfish growth.

To test the effectiveness of a subaqueous soil survey for identifying areas with the highest potential for shellfish growth, Salisbury (2010) investigated shellfish growth on five different subaqueous soil–landscape units in two coastal lagoons in Rhode Island. At the subgroup level, the studied soils were classified as Typic Sulfiwassents, Haplic Sulfiwassents, Sulfic Psammowassents, Fluventic Psammowassents, and Aeric Haplowassents. Oysters (*Crassostrea virginica*) were grown in trays resting on the bottom and hard shell clams (*Mercenaria mercenaria*) were grown directly in the soil. Oyster growth rates varied between years and estuaries, suggesting that environmental factors influenced growth rates. Multiple regression analysis suggested that 53% of the variance in oyster growth among soils could be explained by grain size, organic carbon, and calcium carbonate content. Growth rates of both species were significantly correlated with increases in sand content. Analysis of variance showed that there was a significant difference in growth rates among soil types. Less than 25% of the oysters grown on Typic Sulfiwassent soils located on lagoon bottom landscape units reached harvest size (7.5 cm) by the end of the two growing seasons. In contrast, 44–73% of the oysters grown on the other soil types reached harvest size after two growing seasons. Although these studies were limited to two lagoons, the results clearly suggest that subaqueous soil survey data can provide an effective approach for identifying the best areas for shellfish aquaculture.

4.6. Freshwater Applications

The growing interest in subaqueous soils has been driven by the need for tools to manage shallow aquatic systems and resources at an ecosystem scale. As described above, most of the mapping and interpretation work has been focused on coastal subaqueous soils. Similar interpretations and approaches are needed for freshwater systems (Bakken and Stolt, 2010; Erich et al., 2010a). The reasons for this are many. Local, state, and federal agencies are trying to manage the effects of elevated levels of nutrients such as phosphorus and nitrogen that have entered surface waters. At the same time in these freshwater aquatic systems, there are issues related to accumulated sediment, contaminants such a metals, herbicides, and pesticides, and the proliferation of invasive species such as milfoil. At the present time, there are essentially no maps of the littoral zone for simply understanding these systems as habitats or as a resource. Studies in coastal systems have found that estuarine subaqueous soils contain significant pools of soil organic carbon (in most cases, more than adjacent subaerial soils). It is unknown what are the carbon pools and sequestration rates in freshwater subaqueous soils. Future work is needed to address such issues and questions.

5. SUMMARY AND CONCLUSIONS

Over the last two decades, a collection of investigations, primarily along the eastern seaboard of the United States, has resulted in a better understanding of the properties, distribution, and extent of soils that formed under permanent shallow water. Most of the early subaqueous soil research focused on the methods and approaches to sample, characterize, and map the permanently submersed landscapes and soils. Of particular importance was the development of bathymetric maps to identify underwater landscape units and the establishment of soil–landscape relationships. These studies led to the recognition that permanently submersed substrates develop in a systematic response to their environment and the associated physical, chemical, and biological processes (pedogenesis). As such, the definition of soils was changed to include subaqueous soils. In addition, new taxa were added to Entisols and Histosols in Soil Taxonomy to accommodate the special properties of subaqueous soils. With characterization and mapping approaches in place, subaqueous soil investigations turned to establish how soil surveys could be used as a resource management tool guiding use and management decisions for coastal, estuarine, and littoral ecosystems. Of particular interest to coastal managers and scientists were issues of carbon accounting, dredging effects, dredge materials placement, mooring type and siting, SAV restoration, and shellfish aquaculture and conservation. Although these applied soil studies demonstrate important advancements, the results represent applications within a limited window of geographic and estuarine settings. Therefore, additional studies, in different climates and settings, should be undertaken to test these relationships and further develop the link between soils and particular interpretations. Most of the SAS studies have focused on estuarine systems. Thus, additional work is especially needed within the range of freshwater systems. Although two-thirds of the US population lives in coastal areas and uses these resources for everything from transportation to sustenance to recreation, the expansion of soil survey efforts into subaqueous systems has been limited. Examining the value of subaqueous soil systems relative to the cost of mapping and inventorying these resources should lead to broad-scale soil mapping of subaqueous environment and the development of interpretive tools for managing this new resource inventory.

ACKNOWLEDGMENTS

The assistance of Mike Bradley, Maggie Payne, and Jim Turenne is gratefully acknowledged. This work was supported by the Maryland and Rhode Island Agricultural Experiment Stations.

REFERENCES

Alves, D., 2007. Aquaculture in Rhode Island 2007 Yearly Status Report. Coastal Resources Management Council, Wakefield, RI.

August, P., Costa-Pierce, B., 2007. Mapping submerged habitats: A new frontier. 41 Degrees North 4:3.

Bakken, J., Stolt, M.H., 2010. Freshwater subaqueous soil survey investigations and applications annual meeting of the soil science society of America. Soil Science Society of America, Long Beach, CA.

Balduff, D.M., 2007. Pedogenesis, Inventory and Utilization of Subaqueous Soils in Chincoteague Bay. Maryland, University of Maryland, College Park.

Bertness, M.D., Gaines, S.D., Hay, M.E., 2001. Marine Community Ecology. Sinauer Associates, Inc., Sunderland, MA.

Bradley, M.P., 2001. Subaqueous Soils and Subtidal Wetlands in Rhode Island. M.S. Thesis. Univ. of Rhode Island, Kingston.

Bradley, M.P., Stolt, M.H., 2002. Evaluating methods to create a base map for a subaqueous soil inventory. Soil Sci. 167, 222–228.

Bradley, M.P., Stolt, M.H., 2003. Subaqueous soil-landscape relationships in a Rhode Island estuary. Soil Sci. Soc. Am. J. 67, 1487–1495.

Bradley, M.P., Stolt, M.H., 2006. Landscape-level seagrass-sediment relations in a coastal lagoon. Aquat. Bot. 84, 121–128.

Brady, N.C., Weil, R.R., 2007. The Nature and Properties of Soils, fourteenth ed. Prentice Hall.

Brandon, D.L., Price, R.A., 2007. Summary of Available Guidance and Best Practices for Determining Suitability of Dredged Material for Beneficial Uses. ERDC EL TR-07–27. U.S. Army Engineer Research and Development Center, Vicksburg, MS.

Bricker, S.B., 1993. The history of Cu, Pb, and Zn inputs to Narragansett Bay, Rhode Island as recorded by salt-marsh sediments. Estuaries and Coasts 16, 589–607.

Busch, K., Golden, R.R., 2009. Large-Scale Restoration of Eelgrass (Zostera marina) in the Patuxent and Potomac Rivers, Maryland. Final Report to National Oceanic and Atmospheric Administration Grant Award Number: NA03NMF4570470. Maryland Department of Natural Resources Annapolis, MD.

Calumpong, H.P., Fonseca, M.S., 2001. Chapter 22: Seagrass transplantation and other restoration methods. In: Short, F.T., Coles, R.G. (Eds.), Global Seagrass Research Methods. Elsevier Science B.V., Amsterdam, The Netherlands.

Calvert, D.V., Ford, H.W., 1973. Chemical properties of acid-sulfate soils recently reclaimed from Florida marshland. Soil Sci. Soc. Am. J. 37, 367–371.

Cheng, D.G., 1986. Acid Sulfate Soils in Baltimore Harbor Dredged Materials. I. Survey of Soils and Associated Natural Resources. II. Liming Studies. M.S. Thesis, Univ. of Maryland, College Park, MD.

Chmura, G.L., Anisfeld, S.C., Cahoon, D.R., Lynch, J.C., 2003. Global carbon sequestration in tidal saline wetland soils. Global Geochemical Cycles 17, 1111.

Churchill, A.C., Cok, A.E., Riner, M.I., 1978. Stabilization of Subtidal Sediments by the Transplantation of the Seagrass *Zostera marina* L. Rept. # NYSSGP-RS-78–15. New York Sea Grant, Stony Brook, NY.

Coppock, C., Rabenhorst, M.C., 2003. Subaqueous Soils of Rehoboth Bay, DE: Soil Mapping in a Coastal Lagoon Annual Meeting Soil Science Society of America, Denver, CO.

Corbin, J., 1991. Metal Pollution in Narragansett Bay. M.S. Thesis, Univ. of Rhode Island, Kingston, RI.

Darmody, R.G., Rabenhorst, M.C., Foss, J.E., 1976. Bucket auger modification for tidal marsh sampling. Soil Sci. Soc. Am. J. 40, 321–322.

Demas, G.P., 1993. Submerged soils: a new frontier in soil survey. Soil Survey Horizons. Soil Survey Horizons 34, 44–46.

Demas, G.P., 1998. Subaqueous soil of Sinepuxent Bay, Maryland. Ph.D. Dissertation, University of Maryland, College Park. (Ed.), Proceedings of the 16th World Congress on Soil Science, Montpellier, France. August 20–26, 1998.

Demas, G.P., and Rabenhorst, M.C., 1998. Subaqueous soils: a resource inventory protocol. Proceedings of the 16th World Congress on Soil Science, Montpellier, France. August 20–26, 1998. Sym #17, on CD. Published by the International Soil Science Society (ISSS).

Demas, G.P., Rabenhorst, M.C., 1999. Subaqueous soils: Pedogenesis in a submersed environment. Soil Sci. Soc. Am. J. 63, 1250–1257.

Demas, G.P., Rabenhorst, M.C., 2001. Factors of subaqueous soil formation: a system of quantitative pedology for submersed environments. Geoderma 102, 189–204.

Demas, G.P., Rabenhorst, M.C., Stevenson, J.C., 1996. Subaqueous soils: a pedological approach to the study of shallow-water habitats. Estuaries 19, 229–237.

Demas, S.Y., Hall, A.M., Fanning, D.S., Rabenhorst, M.C., Dzantor, E.K., 2004. Acid sulfate soils in dredged materials from tidal Pocomoke Sound in Somerset County, MD, USA. Aust. J. Soil Res. 42, 537–545.

Ellis, L.R., 2006. Subaqueous Pedology: Expanding Soil Science to Near-Shore Subtropical Marine Habitats. PhD. Dissertation. Univ. of Florida, Gainesville, FL.

Erich, E., Drohan, P., Lupton, M., Boyer, E., Brooks, R., 2010a. A Hydropedological Perspective of Mercury Distribution in Soils of the Black Moshannon Lake Drainage Basin Annual Meeting of the Soil Science Society of America. Soil Science Society of America, Long Beach, CA.

Erich, E., Drohan, P., Lupton, M., Lindeburg, K., Boyer, E., Bishop, J., 2009. The Extent and Characterization of Freshwater Subaqueous Soils of Black Moshannon Lake, Pennsylvania Soil Science Society of America Annual Meeting November 2009, vol. 3. Pittsburg, PA.

Erich, E., Payne, M., Surabian, D., Collins, M.E., Drohan, P.J., Ellis, L.R., 2010b. Subaqueous soils: their genesis and importance in ecosystem management [electronic resource]. Soil Use Manage. 26, 245–252.

Fanning, D.S., 2002. Acid sulfate soils. pp. 1–22, In: Lal, R. (Ed.), Encyclopedia of Soil Science. Marcel Dekker. http://www.dekker.com/servlet/product/productid/E-ESS.

Fanning, D.S., Coppock, C., Orndorff, Z.W., Daniels, W.L., Rabenhorst, M.C., 2004. Upland active acid sulfate soils from construction of new Stafford County, Virginia, USA, Airport. Aust. J. Soil Res. 42, 527–536.

Flannagan, C.T., 2005. Subaqueous Soil Survey of Taunton Bay, Maine. M.S. Thesis. Univ. of Maine, Orono, ME.

Fischler, K.C. 2006. Observations and characterization of subaqueous soils and seagrasses in a recently constructed habitat in the Indian River Lagoon, Florida. M.S. Thesis, University of Florida, Gainesville, FL.

Folger, D.W., 1972. Characteristics of Estuarine Sediments of the United States USGS Professional Paper, vol. 742. US Dept. of Interior, Washington, DC.

Fonseca, M.S., Kenworthy, W.J., Thayer, G.W., 1998. Guidelines for the Conservation and Restoration of Seagrasses in the United States and Adjacent Waters National Oceanic and Atmospheric Administration (NOAA). Coastal Ocean Office, Silver Spring, MD.

Ford, K.H., 2003. Assessment of the Rhode Island Coastal Lagoon Ecosystem. PhD. Dissertation. Univ. of Rhode Island, Kingston, RI.

Ginsburg, R.N., Lowenstam, H.A., 1958. The influence of marine bottom communities on the depositional environment of sediments. J. Geol. 66, 310–318.

Goldberg, E.D., Gamble, E., Griffin, J.J., Koide, M., 1977. Pollution history of Narragansett Bay as recorded in its sediments. Estuarine Coastal Mar. Sci. 5, 549–561.

Griffin, T.M., Rabenhorst, M.C., Fanning, D.S., 1989. Iron and trace metals in some tidal marsh soils of the Chesapeake Bay. Soil Sci. Soc. Am. J. 53, 1010–1019.

Grizzle, R.E., Lutz, R.A., 1989. A statistical model relating horizontal seston fluxes and bottom sediment characteristics to growth of *Mercenaria mercenaria*. Mar. Biol. 102, 95–105.

Grizzle, R.E., Morin, P.J., 1989. Effect of tidal currents, seston, and bottom sediments on growth of *Mercenaria mercenaria*: results of a field experiment. Mar. Biol. 102, 85–93.

Hansen, K., 1959. The terms gyttja and dy. Hydrobiologia 13, 309–315.

Harlin, M.M., Thorne-Miller, B., Boothroyd, J.C., 1982. Seagrass-sediment dynamics of a flood-tidal delta in Rhode Island (U.S.A.). Aquat. Bot. 14, 127–138.

Harvell, C.D., Kim, K., Burkholder, J.M., Colwell, R.R., Epstein, P.R., Grimes, D.J., Hofmann, E.E., Lipp, E.K., Osterhaus, A.D., Overstreet, R.M., Porter, J.W., Smith, G.W., Vasta, G.R., 1999. Emerging marine diseases–climate links and anthropogenic factors. Science 285, 1505–1510.

Hemminga, M.A., Duarte, C.M., 2000. Seagrass Ecology. Cambridge University Press, New York, NY.

Holmer, M., Nielsenj, S.L., 1997. Sediment sulfur dynamics related to biomass-density patterns in *Zostera marina* (eelgrass) beds. Mar. Ecol. Prog. Ser. 146, 163–171.

IUSS Working Group WRB, 2006. World Reference Base for Soil Resources 2006, second ed. FAO, Rome.

Jenny, H., 1941. Factors of Soil Formation: A System of Quantitative Pedology. McGraw-Hill, New York.

Jesperson, J.L., 2006. Organic Carbon in the Subaqueous Soils of a Mesotidal Maine Estuary: An Investigation of Quantity and Source. M.S. Thesis. Univ. of Maine, Orono, ME.

Jesperson, J.L., L.J. Osher., 2007. Carbon storage in the soils of a mesotidal gulf of maine estuary Soil Sci. Soc. Am. J. 71, 372–379.

Jowsey, P.C., 1966. An improved peat sampler. New Phytol. 65, 245–248.

Kern, J.S., 1994. Spatial patterns of soil organic carbon in the contiguous United States. Soil Sci. Soc. Am. J. 58, 439–455.

Koch, E.W., 2001. Beyond light: physical, geological, and geochemical parameters as possible submersed aquatic vegetation habitat requirements. Estuaries 24, 1–17.

Kopp, B.S., Short, C.A., 2000. New Bedford Harbor Eelgrass Restoration Project. Progress Report to the New Bedford Harbor Trustees Council. Jackson Estuarine Laboratory, Durham, NH.

Kubiena, W.L., 1953. The Soils of Europe. T. Murby, London.

MapCoast, 2011. Eelgrass Potential Interpretative Map for Ninigret Pond. Static maps. Mapping Partnership for Coastal Soils and Sediments. http://www.ci.uri.edu/projects/mapcoast/data.html#Static.

McMullen, M.C., 1984. Adaptability of Selected Conservation Plant Species in Relation to pH and Electrical Conductivity of Active Acid Sulfate Soils in Baltimore Harbor Dredged Materials. M.S. Thesis. Univ. of Maryland, College Park, MD.

Merkel, K.W., 2004. Experimental Eelgrass Transplant Program. Investigations for On-site Eelgrass Mitigation. Final Report to California Department of Transportation. Merkel & Associates Inc., Parametrix. Inc., San Francisco, CA.

Mortimer, C.H., 1950. Underwater 'soils': a review of lake sediments. J. Soil Sci. 1, 63–73.

Muckenhausen, E., 1965. The soil classification system of the Federal Republic of Germany. Pedologie 3, 57–74.

National Cooperative Soil Survey, 2005. Glossary of Terms for Subaqueous Soils, Landscapes, Landforms, and Parent Materials of Estuaries and Lagoons [Online]. Available by http://nesoil.com/sas/Glossary-Subaqueous%20Soils.pdf.

Nixon, S., 1991. Recent metal inputs to Narragansett Bay. Final Report, Narragansett Bay Project. Rhode Island Department of Environmental Management, Providence, RI.

Nixon, S.W., Hunt, C.D., Nowicki, B.N., 1986. The retention of nutrients (C, N, P), heavy metals (Mn, Cd, Pb, Cu), and petroleum hydrocarbons in Narragansett Bay. In: Lasserre, P., Martin, J.M. (Eds.), Biogeochemical Processes at the Land-Sea Boundary. Elsevier Oceanography Series, forty third ed., vol. 43. Elsevier, New York, NY.

Orth, R.J., Wilcox, D.J., Nagey, L.S., Owens, A.L., Whiting, J.R., Kenne, A.K., 2005. 2004 Distribution of Submerged Aquatic Vegetation in Chesapeake Bay and Coastal Bays. Spec. Sci. Rep. No. 146. Virginia Inst. of Mar. Sci.

Osher, L.J., Flannagan, C.T., 2007. Soil/landscape relationships in a mesotidal Maine estuary. Soil Sci. Soc. Am. J. 71, 1323–1334.

Payne, M.K., 2007. Landscape-Level Assessment of Subaqueous Soil and Water Quality in Shallow Embayments in Southern New England. M. S. Thesis. Univ. of Rhode Island, Kingston, RI.

Payne, M.K., Turenne, J., 2009. Mapping the "new frontier" of soil survey: Rhode Island's Mapcoast Partnership. Soil Survey Horizons 50, 86–89.

Ponnamperuma, F.N., 1972. The chemistry of submerged soils. Adv. Agron. 24, 29–96.

Pons, L.J., Zonneveld, I.S., 1965. Soil Ripening and Soil Classification; Initial Soil Formation of Alluvial Deposits with a Classification of the Resulting Soils, vol. 13. Int. Institute for Land Reclamation and Improvement, Pub., Wageningen.

Pratt, D.M., 1953. Abundance and growth of *Venus mercenaria* and *Callocardia morhuana* in relation to the character of bottom sediments. J. Mar. Res. 12, 60–74.

Pratt, D.M., Campbell, D.A., 1956. Environmental factors affecting growth in *Venus mercenaria*. Limnol. Oceanogr. 1, 2–17.

Pruett, C.M., 2010. Interpretations of Estuarine Subaqueous Soils: Eelgrass Restoration, Carbon Accounting, and Heavy Metal Accumulation. M.S. Thesis. Univ. of Rhode Island, Kingston, RI.

Rabenhorst, M.C., Fanning, D.S., 2002. Acid Sulfate Soils – Problems. pp. 23–26; Encyclopedia of Soil Science. Marcel Dekker. http://www.dekker.com/servlet/product/productid/E-ESS.

Rabenhorst, M.C., Fanning, D.S., Burch, S.N., 2002. Acid Sulfate Soils - Formation. pp. 14–18; Encyclopedia of Soil Science. Marcel Dekker. http://www.dekker.com/servlet/product/productid/E-ESS.

Rabenhorst, M.C., King, P., Stolt, M.H., Osher, L.J., Coppock, C., Flannagan, C.T., 2004. National Workshop on Subaqueous Soils Annual Meeting of the Soil Science Society of America, Seattle, WA.

Reid, R.N., MacKenzie, C.L., Vitaliano, J.J., 1993. A Failed Attempt to Re-establish Eelgrass in Raritan Bay (New York/New Jersey), pp. 93–97, Vol. Northeast Fisheries Science Center Reference Document. NOAA/National Marine Fisheries Service, Highland, NJ.

Rice, M.A., 1992. The Northern Quahog: the Biology of *Mercenaria mercenaria*. Rhode Island Sea Grant, Narragansett, RI.

Rice, M.A., Pechenik, J.A., 1992. A review of the factors influencing the growth of the northern quahog, *Mercenaria mercenaria* (Linnaeus, 1758). J. Shellfish Res. 11, 279–287.

Ridgeway, J., Shimmield, G., 2002. Estuaries as repositories of historical contamination and their impact on shelf seas. Estuarine Coastal Shelf Sci. 55, 903–928.

Ruiz-Fernandez, A.C., Frignani, M., Hillaire-Marcel, C., Ghaleb, B., Arvizu, M.D., Raygoza-Viera, J.R., Paez-Osuna, F., 2009. Trace metals (Cd, Cu, Hg, and Pb) accumulation recorded in the intertidal mudflat sediments of three coastal lagoons in the Gulf of California, Mexico. Estuaries Coasts 32, 551–564.

Salisbury, A.R., 2010. Developing subaqueous soil interpretations for Rhode Island estuaries. M.S. Thesis. Univ. of Rhode Island, Kingston, RI.

Santschi, P.H., 1980. A revised estimate for trace metal fluxes to Narragansett Bay: a comment. Estuar. Coast. Mar. Sci. 11, 115–118.

Santschi, P.H., Nixon, S., Pilson, M., Hunt, C., 1984. Accumulations of sediments, trace metals (Pb, Cu) and hydrocarbons in Narragansett Bay, Rhode Island. Estuarine Coastal Shelf Sci. 19, 427–449.

Short, F.T., 1988. Eelgrass-scallop research in the Niantic River, Westford-East Lyme, Connecticut Shellfish Commission, Final Report. Jackson Estuarine Laboratory, Durham, NH.

Short, F.T., 1993. The Port of New Hampshire Interim Mitigation Success Assessment Report. Report to the New Hampshire Department of Transportation. Jackson Estuarine Laboratory, Durham, NH.

Short, F.T., Burdick, D.M., Granger, S., Nixon, S.W., 1996. Long-term decline in eelgrass, *Zostera marina* L. linked to increased housing development. In: Kuo, J., et al. (Eds.), Seagrass Biology: Proceedings of an International Workshop, Rottnest Island, Western Australia. Sciences UWA, Nedlands, Western Australia, pp. 25–29.

Short, F.T., Davis, R.C.R.C., Kopp, B.S., Short, C.A., Burdick, D.M., 2002. Site-selection model for optimal transplantation of eelgrass *Zostera marina* in the northeastern US. Mar. Ecol. Prog. Ser. 227, 253–267.

Simonson, R.W., 1959. Outline of a generalized theory of soil genesis. Soil Sci. Soc. Am. Proc. 23, 152–156.

Soil Survey Division Staff, 1993. Soil Survey Manual. USDA-SCS. Agriculture Handbook 18. US Government Printing Office, Washington, DC.

Soil Survey Staff, 1975. Soil Taxonomy: A Basic System of Soil Classification for Making and Interpreting Soil Surveys. Agriculture Handbook 436. US Government Printing Office, Washington, DC.

Soil Survey Staff, 1999. Soil Taxonomy. U.S. Dept. Agric. Handbook No. 436, second ed. US Govt. Printing Office, Washington, DC.

Soil Survey Staff, 2010. Keys to Soils Taxonomy, eleventh ed. USDA-NRCS, Washington, DC.

Stolt, M.H., Bradley, M.P., Rabenhorst, M.C., King, P., Osher, L.J., 2004. Developing Terminology for Subaqueous Soils: Landforms and Landscape Units Annual Meeting of the Soil Science Society of America, Seattle, WA.

Surabian, D.A., 2007. Moorings: an interpretation from the coastal zone soil survey of Little Narragansett Bay, Connecticut and Rhode Island. Soil Survey Horizons 48, 90–92.

Tack, F.M., Callewaert, O.W.J.J., Verloo, M.G., 1995. Metal solubility as a function of pH in a contaminated, dredged sediment affected by oxidation. Environ. Pollut. 91, 199–208.

Tarnovsky, M., 2003. Maryland Oyster Population Status Report. 2002 Fall Survey. Maryland Department of Natural Resources, Annapolis, MD.

Thayer, G.W., Kenworthy, W.J., Foseca, M.S., 1984. The ecology of eelgrass meadows of the Atlantic coast: a community profile, pp. FWS/OBS-84/02, In: U. S. F. W. Service, (Ed.). US Govt. Printing Office, Washington, D.C.

Thom, R.M., Borde, A.B., Williams, G.D., Woodruff, D.L., Southard, J.A., 2003. Climate change and seagrasses: climate-linked dynamics, carbon limitation and carbon sequestration. Gulf Mex. Sci. 21, 134.

Trenhaile, A.S., 1997. Coastal Dynamics and Landforms. Clarendon Press, Oxford, UK.

USDA-NRCS. 2011a. Official Soil Series Descriptions (OSD) with series extent mapping capabilities. http://soils.usda.gov/technical/classification/osd/index.html.

USDA-NRCS. 2011b. Web Soil Survey. http://websoilsurvey.nrcs.usda.gov/app/HomePage.htm.

VonPost, H., 1862. Studier ofver nutidans kopregena jordbildningar, gytta, torf, mylla. Klg. Sv. Vetensk. Akad. Handl. 4.

Wells, H.W., 1957. Abundance of the hard clam *Mercenaira mercenaria* in relation to environmental factors. Ecology 38, 123–128.

Wicks, E.C., Koch, E.W., O'Neil, J.M., Elliston, K., 2009. Effects of sediment organic content and hydrodynamic conditions on the growth and distribution of *Zostera marina*. Mar. Ecol. Prog. Ser. 378, 71–80.

Winfield, L.E., Lee, C.R., 1999. Dredged Material Characterization Tests for Beneficial Use Suitability. DOER Technical Notes Collection (TN DOER-C2). U.S. Army Engineer Research and Development Center, Vicksburg, MS.

Yozzo, D.J., Wilber, P., Will, R.J., 2004. Beneficial use of dredged material for habitat creation, enhancement, and restoration in New York-New Jersey Harbor. J. Environ. Manage. 73, 39–52.

Chapter 7

Quantifying Processes Governing Soil-Mantled Hillslope Evolution

Arjun M. Heimsath[1]

ABSTRACT

This chapter presents an overview of how field-based methods quantify the processes shaping upland, soil-mantled landscapes. These methods have been applied across diverse field areas, ranging from the tropical sandstones of northern Australia to the alpine granites of the Sierra Nevada in California. In all cases, the landscapes examined through such work are relatively gently sloping with a generally continuous soil mantle. Soil on such upland landscapes is distinctly defined to be the physically mobile layer derived primarily from the underlying parent material with organic content from native flora and fauna. These upland soils are distinguished from agriculture or lowland soils by the convex-up, hilly topographies that are the focus of this study. Parent material is generally saprolite, i.e. weathered bedrock that retains relict rock structure and is physically immobile. Processes shaping such landscapes include the physical and chemical processes that weather the parent material, and the processes moving the soil downslope. These processes are quantified using several different field-based methods. In situ produced cosmogenic nuclides (^{10}Be and ^{26}Al) are measured in the parent material directly beneath the mobile soil mantle and define the relationship between soil production rates and the overlying soil thickness. The same cosmogenic nuclides are measured in detrital sediments sampled from local channels to quantify basin-averaged erosion rates. Mobile and trace elements are measured in both the parent material and the soils to define chemical weathering rates and processes. Short-lived, fallout-derived isotopes (^{210}Pb, ^{137}Cs, and ^{7}Be) are measured in soil profiles to quantify sediment-transport processes and short-term erosion rates. Parent material strength (competence), or resistance to shear, is measured using a hand-held shear vane as well as a cone penetrometer. The methodology for some results and the connections drawn from this diverse toolbox are summarized in this chapter. An important conclusion connecting parent material strength to the physical processes transporting soil and the chemical processes weathering the parent material emerges with the observation that parent material strength increases with overlying soil thickness and, therefore, the weathered extent of the saprolite. This observation highlights the importance of quantifying

[1]School of Earth and Space Exploration, Arizona State University, AZ, USA.
Email: Arjun.Heimsath@ASU.edu

Hydropedology, Edited by H. Lin. DOI: 10.1016/B978-0-12-386941-8.00007-1

hillslope hydrologic processes at the same locations where the measurements described herein are made. Specifically, by understanding the hydrologic pathways that help drive the weathering processes of upland soils and saprolites, we will gain considerably greater insight into how the processes described in this chapter drive landscape evolution.

1. INTRODUCTION

Quantifying the rates of Earth surface processes across soil-mantled landscapes is critical for many disciplines (Anderson, 1994; Bierman, 2004; Dietrich et al., 2003; Dietrich and Perron, 2006). Balances between soil production and transport determine whether soil exists on any given landscape, as well as how thick it might be. Soil presence and thickness, in turn, help support much of the life that we humans are familiar with, play important roles in the hydrologic cycle, and are coupled with processes that impact the atmosphere. Similarly, sustainability of the Earth's soil resource under growing human population pressures also depends on the balance between soil production and erosion: agriculture depends on soil. Fully quantifying the conditions that will lead to a depletion of the soil resource is as important, in my opinion, as quantifying the degradation of our drinking water. As such, the near-surface environment that includes soil and its parent material is known as the Critical Zone (CZ) and is the focus of exciting new research initiatives, including the NSF-supported Critical Zone Observatory program (Amundson et al., 2007; Anderson et al., 2004; Anderson et al., 2007; Brantley et al., 2007). Over the last ten years, significant progress in the field (e.g. Anderson and Dietrich, 2001; Burkins et al., 1999; White et al., 1996; Yoo et al., 2007), lab (e.g. White and Brantley, 2003), and with modeling (e.g. Minasny and McBratney, 2001; Mudd and Furbish, 2006) has enabled a new level in understanding how forces of sediment production and transport shape landscapes and help determine the viability of the Earth's soil resource (Montgomery, 2007). Despite this progress, there are significant gaps in our understanding of this interface between humans, the atmosphere, biosphere, and lithosphere. One of the most significant gaps is quantifying what controls soil thickness (Anderson et al., 2007; Brantley et al., 2007), while another is the continued quantification of the sediment-transport relationships (*cf* Geomorphic Transport Laws, Dietrich et al., 2003; Heimsath et al., 2005). The focus of this chapter is on soil-mantled, upland landscapes to provide an overview of how we quantify the rates and processes active in shaping such landscapes. This chapter is not a comprehensive review of the field, but does serve well to review some of the key methods and results applied on these landscapes.

Hilly and mountainous landscapes around the world are mantled with soil. In regions where external sources of sediment (e.g. eolian and glacial deposition) are absent or negligible, the soil mantle is typically produced from the underlying bedrock. Gilbert (1877) first suggested that the rate of soil production from the underlying bedrock is a function of the depth of the soil mantle. We term this rate law the soil production function (Heimsath et al., 1997) – defined as the relationship between the rate of bedrock conversion to soil and the overlying soil thickness. This soil depth that sets the rate of soil production is a result of the balance between the soil production and erosion. If local soil depth is constant over time, the soil production rate equals the erosion rate, which equals the lowering rate of the land surface. Quantifying the soil production function therefore furthers our understanding of the evolution rates of soil-mantled landscapes (Anderson and Humphrey, 1989; Rosenbloom and Anderson, 1994; Dietrich et al., 1995; Heimsath et al., 1997). Heimsath et al. (1997, 1999, 2000) reported spatial variation of erosion rates, suggesting that the landscapes were out of the state of dynamic equilibrium, as first conceptualized by Gilbert (1877, 1909) and then Hack (1960). Such a condition, where the landscape morphology is time-independent, was an important assumption for landscape evolution models. Showing that upland landscapes are actually evolving with time helped spur a new level of understanding for how landscapes change under climatic and tectonic forcing.

On actively eroding hilly landscapes, characterized by ridge and valley topography (Fig. 1), the colluvial soil mantle is typically thin and is produced and transported by mechanical processes. Tree throw, animal burrowing, and similar processes, such as freeze–thaw and shrink–swell cycles, convert in-place bedrock to a mobile, often rocky, soil layer that is then transported downslope by the same actions (Lutz and Griswold, 1939; Lutz, 1940; Lutz, 1960; Hole, 1981; Mitchell, 1988; Matsuoka, 1990; Schaetzl and Follmer, 1990; Norman et al., 1995; Paton et al., 1995). On steep slopes, shallow landsliding also transports

FIGURE 1 Photograph of a typical hilly, soil-mantled landscape that is the focus of work described in this chapter. Point Reyes, California. *(Photo by the author) (Color version online)*

material downslope, and may also play a role in producing soil. While such processes are aided directly, and even accelerated, by chemical weathering of the bedrock, they are able to produce soil from bedrock irrespective of its weathered state. Early quantification of the soil production function focused on low-gradient topography developed on relatively homogeneous bedrock, where simple rate laws can characterize the geomorphic processes (Heimsath et al., 1997, 1999, 2000, 2009). Methods applied in those studies were also applied to steep, landslide-dominated landscapes to extend our understanding of the connections between soil production, erosion, and landscape evolution (Heimsath et al., 2001a; Heimsath et al., 2012).

In this chapter, I review the conceptual framework used to quantify soil production and hillslope weathering rates, and explore the field measurements that we make to connect processes governing hillslope hydrology with soil production and transport. Specifically, I will review the mass-balance approach used to model landscape evolution and identify the key field-based parameters that we quantify with direct measurements. These measurements include (1) cosmogenic nuclides (^{10}Be and ^{26}Al) to quantify soil production and catchment-averaged erosion rates, (2) major and trace element analyses to quantify the weathered state and weathering rate of the parent material and soil, (3) short-lived isotopes (^{137}Cs and ^{210}Pb) and optically stimulated luminescence (OSL) to quantify soil transport processes, and (4) cone penetrometer and shear vane measurements to quantify the competence of the parent material underlying the actively eroding soil mantle (i.e. its resistance to physical disruption). This chapter is deliberately focused on expanding upon the methods that we use to examine hilly and mountainous landscapes that are continuously mantled with soil, which is actively produced from the underlying parent material. Instead of focusing on any specific field site, which our various papers do, I will present representative results in a more general way to enable the discussion that will conclude this chapter. To help provide context for these results drawn from different landscapes and studies, I provide a brief overview of how they are all related following the summary of the conceptual framework and methods.

2. CONCEPTUAL FRAMEWORK AND METHODS

2.1. Soil Production and Chemical Weathering

Consider a depth profile at a hillcrest, extending from the soil surface to unweathered bedrock (Fig. 2). The layer of physically and/or chemically altered material (saprolite and soil) atop crystalline bedrock can be millimeters to tens of meters thick depending on the landscape in question and the external driving forces eroding the landscape. A number of definitions are found throughout the literature for these terms; here, 'soil' is the physically mobile material that is produced by the mechanical disruption of the underlying

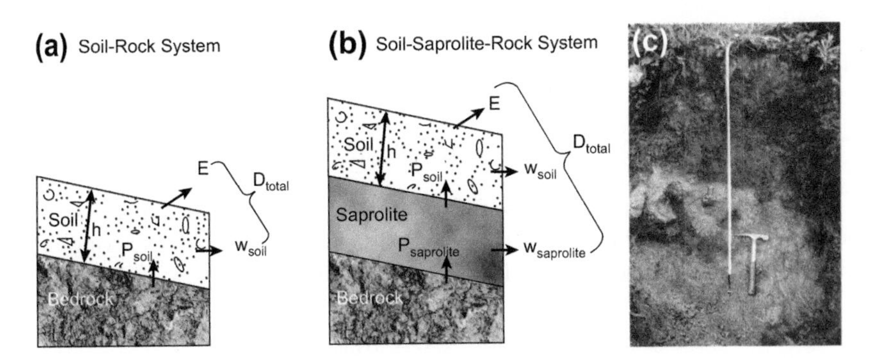

FIGURE 2 Conceptual cross section of soil–saprolite–bedrock showing fluxes and symbols. Soils may be mechanically produced from bedrock (a) or weathered saprolite (b). Based on conservation of mass Eqns (1) and (2) in text, if soils have a steady-state thickness, then soil production (P_{soil}) is balanced by weathering (W_{soil}) and erosion (E). If saprolite thickness attains a similar steady state, then saprolite production (P_{sap}) is balanced by mass loss from the saprolite profile by weathering (W_{sap}) and the mechanical conversion to soil (P_{soil}). Note that total denudation rate (D_{total}) is calculated from the sum of losses, and reflects only erosion and weathering in soil in Fig. 1a, but accounts for the additional losses due to saprolite weathering in Fig. 1b. (c) Photograph of a typical soil-weathered bedrock boundary with gopher burrow shown penetrating the light-colored saprolite. *(From Dixon et al., 2009a)*

bedrock (Fig. 2b), or saprolite (Fig. 2c). Saprolite is the nonmobile, weathered mantle produced by chemical alteration of the bedrock that lies beneath. At the upper boundary, saprolite is incorporated into the mobile soil column through physical disruption by soil production mechanisms such as tree throw, macro- and microfauna burrowing, and frost cracking. At the lower boundary, saprolite is produced from parent bedrock by chemical dissolution and mineral transformation. While mineralogical changes and alteration occur throughout the saprolite column, much of the chemical mass loss can occur as a discrete weathering front near the bedrock boundary (Frazier and Graham, 2000; Buss et al., 2004; Fletcher et al., 2006; Lebedeva et al., 2007; Buss et al., 2008). We focus specifically on the mass loss due to chemical processes (here termed chemical weathering) and physical processes (here termed erosion) on soil-mantled hillslopes. By these definitions, weathering within saprolite is purely chemical, while soils evolve by both chemical transformations and physical disruption and erosion.

Using basic mass conservation principals, any change in soil mass, expressed as the product of soil density (ρ_{soil}) and change in soil thickness (h), reflects the mass rates (determined with the appropriate bulk density measurements times the rate) of soil production (P_{soil}), erosion (E), and weathering (W_{soil}), such that

$$\rho_{soil}\frac{\partial h}{\partial t} = P_{soil} - E - W_{soil}. \tag{1}$$

If soil thickness (h) is constant over time $\left(\frac{\partial h}{\partial t} = 0\right)$, then the rate of soil mass loss equals the rate of soil production:

$$P_{\text{soil}} = E + W_{\text{soil}}. \tag{2}$$

The sum of all rates of mass loss can be termed the total denudation rate (D_T), and these losses occur by physical erosion of soil and chemical weathering in both the soil (W_{soil}) and the saprolite (W_{sap}), such that (Fig. 2c)

$$D_T = E + W_{\text{total}} = E + W_{\text{soil}} + W_{\text{sap}}. \tag{3}$$

Recalling mass conservation for steady-state soil thickness Eqn (2), and Eqn (3) can be written as

$$D_T = P_{\text{soil}} + W_{\text{sap}}. \tag{4}$$

It is important to note from Eqn (4) that the soil production rate (P_{soil}) is smaller than the total denudation rate (D_T) in a landscape that experiences chemical weathering in the saprolite. Furthermore, P_{soil}, which can represent a rate of mass flux, can be converted to a landscape lowering rate in units of length per time by dividing by saprolite density if the analyses are done in mass loss per time. D_T cannot as easily be converted into a lowering rate because of its associated saprolite weathering term in Eqn (4); saprolite weathering is observed to be isovolumetric for the studies examined through this conceptual framework, and mass is lost from the saprolite without a corresponding change in volume.

Rates of chemical weathering in catchments have commonly been quantified using stream solute data. These measurements provide a valuable quantification of instantaneous weathering, but provide limited insight into landscape evolution due to their short measurement timescales (1–10 y) and because they integrate mass losses from all points within the catchment, thus treating a watershed as a black box. Riebe et al. (2003a) developed a method to calculate chemical weathering rates in actively eroding terrains by coupling a mass-balance approach using immobile elements (Brimhall et al., 1992; Brimhall and Dietrich, 1987) to rates of landscape lowering derived from cosmogenic radionuclides (CRNs). CRNs such as ^{10}Be and ^{26}Al provided a tool to measure surface rates of denudation over longer timescales (10^3–10^5 y) than solute measurements (1–10 y). These longer timescales, though still only a fraction of the evolutionary period of some landscapes, are more relevant to studies examining the influence of external forcing (climate and tectonics) on landscape change.

CRN concentrations in a sample of rock, saprolite, or soil record the rate of surface denudation processes that removed overlying mass, and therefore the rate at which the sample approaches the land surface (Lal, 1991):

$$D_{\mathrm{CRN}} = \frac{P_0 \Lambda}{N}. \tag{5}$$

Here, the CRN-derived surface denudation rate (D_{CRN}; g cm^{-2} y^{-1}) is a function of the measured CRN concentration in quartz (atoms g^{-1}), the CRN production rate at the surface (P_0: atoms g^{-1} y^{-1}), which decreases exponentially with depth, and the CRN attenuation length (Λ; g cm^{-2}). A complete version of Eqn (5), used to calculate rates in this chapter, is presented and discussed thoroughly by Balco et al. (2008); however, the simplified version shown by Eqn (5) is sufficient for the purposes of our discussion. For ^{10}Be, the most widely used CRN for determining landscape denudation rates, the penetration depths of cosmic rays, assuming a mean attenuation of 160 cm^2 kg^{-1}, are ~140 cm through soil and ~60 cm through rock (Balco et al., 2008). Nuclide concentrations, therefore, record near-surface mass removal within the top few meters of the Earth's surface, and therefore will reflect both chemical and physical losses within soil (typically less than 2 m thick); however, mass losses due to chemical weathering at the bedrock–saprolite boundary are likely to occur at deeper depths than recorded by CRNs. These losses remain unknown and highlight an outstanding research area for future work. A sample of soil, saprolite, or rock will, therefore, have a ^{10}Be concentration that reflects the rate at which overlying soil was removed. Assuming local steady-state soil thickness (Eqn (2)), this rate is equivalent to the soil production rate (Heimsath et al., 1997, 1999; Heimsath, 2006):

$$D_{\mathrm{CRN}} = P_{\mathrm{soil}}. \tag{6}$$

Here, we caution that the generic term denudation (D_T) should not be used to specifically denote CRN-derived rates of landscape lowering. CRN-derived rates do not capture the result of deep saprolite weathering, and thus only record a portion of total denudation on deeply weathered landscapes. As a consequence, we distinguish between soil production rates (P_{soil}) that equal CRN-derived surface denudation rates (D_{CRN}) and total denudation rates (D_T) that reflect combined rates of mass loss in soil and saprolite due to physical and chemical processes. To fully quantify the potential mass loss from deep saprolite chemical weathering, we expand upon previously developed methods for determining soil production, erosion, and chemical weathering rates.

The enrichment of an immobile element in a weathered product relative to the parent material can be used to calculate the fraction of mass that was lost to chemical weathering (Brimhall et al., 1992; Brimhall and

Dietrich, 1987). Riebe et al. (2001b) termed this relationship the chemical depletion fraction (CDF):

$$\mathrm{CDF} = \left(1 - \frac{[I]_p}{[I]_w}\right), \tag{7}$$

where the subscripts 'p' and 'w' refer to concentrations of the immobile element (I) in the parent material and weathered material, respectively. For accurate calculation of the CDF, several important conditions must be met: homogeneous parent material, chemical immobility of the reference element, and minimal chemical weathering during lateral soil transport, which is also an area of active research. Additionally, we must assume no inputs of the immobile element by volcanic ash or continental dust, although this highlights the need for further studies to fully constrain such potential inputs.

Equation (7) can be used to represent the chemical depletion fraction due to soil weathering, saprolite weathering, or total weathering processes. We term these respective depletion fractions the $\mathrm{CDF_{soil}}$, $\mathrm{CDF_{sap}}$, and $\mathrm{CDF_{total}}$, and each is written as Eqn (7), replacing the generic immobile element 'I', with the element zirconium, which is immobile in most weathering environments (e.g. Green et al., 2006).

Following Riebe et al. (2003a), the total chemical weathering rate is the product of the total denudation rate and the total CDF:

$$W_{\mathrm{total}} = D_T \times \left(1 - \frac{[\mathrm{Zr}]_{\mathrm{rock}}}{[\mathrm{Zr}]_{\mathrm{soil}}}\right) = D_T \times \mathrm{CDF_{total}}. \tag{8}$$

Here, the $\mathrm{CDF_{total}}$ represents the fraction of total denudation (D_T) that occurs by all chemical losses (in both the saprolite and the soil). Assuming steady-state soil and saprolite thickness, the saprolite weathering rate can be similarly written (Dixon et al., 2009a). The soil weathering rate is then the calculated difference between W_{total} and W_{sap}.

Calculations of chemical weathering and erosion rates require the measurement of CRN-derived soil production rates. Taking this into account, we rearrange the equations for erosion and weathering in terms of CRN-derived P_{soil} (Eqn (5)). The weathering rate of soil (W_{soil}) is calculated as the product of P_{soil} and the $\mathrm{CDF_{soil}}$:

$$W_{\mathrm{soil}} = D_T \times \left(1 - \frac{[\mathrm{Zr}]_{\mathrm{saprolite}}}{[\mathrm{Zr}]_{\mathrm{soil}}}\right) = P_{\mathrm{soil}} \times \mathrm{CDF_{soil}}. \tag{9}$$

Here, the $\mathrm{CDF_{soil}}$ represents the fraction of original saprolite mass lost due to chemical weathering. The erosion rate (E) can then be calculated as the difference between P_{soil} and W_{soil} (following Eqn (2)).

Lastly, the rate of saprolite weathering (W_{sap}) is calculated by returning to basic principles regarding conservation of mass for immobile elements. For a chemically immobile element such as Zr, the conservation of mass equation can be written as

$$P_{sap} \times [Zr]_{rock} = P_{soil} \times [Zr]_{saprolite} = E \times [Zr]_{soil}. \quad (10)$$

Here, P_{sap} (Fig. 2b) represents the rate conversion of rock to saprolite, and is mathematically equivalent to the total denudation rate (D_T) assuming a steady-state regolith thickness. Solving for total denudation yields

$$D_T = P_{sap} = P_{soil}\left(\frac{[Zr]_{saprolite}}{[Zr]_{rock}}\right). \quad (11)$$

Substituting Eqn (11) into a form of Eqn (8) for saprolite then gives us the equation for the saprolite weathering rate (W_{sap}):

$$W_{sap} = P_{soil} \times \left(\frac{[Zr]_{saprolite}}{[Zr]_{rock}} - 1\right). \quad (12)$$

Summing calculated weathering and erosion rates from these equations allows one to calculate the total denudation rates (Eqn (3)) (Dixon et al., 2009a). These equations differ from those of Riebe et al. (2003a) by the definition that CRN-derived rates reflect only soil production rates, and not total denudation rates in regions mantled by saprolite.

2.2. Soil Transport

The conceptual framework used for our studies is based on the equation of mass conservation for physically mobile soil overlying its parent material (Carson and Kirkby, 1972; Dietrich et al., 1995). Typically, the boundary between soil and the underlying weathered (or fresh) bedrock is abrupt and can be defined within a few centimeters. Soil is produced and transported by mechanical processes, and soil production rates decline exponentially with depth (Heimsath et al., 1997, 2000). The transition from soil-mantled to bedrock-dominated landscapes occurs when transport rates are greater than production rates (Anderson and Humphrey, 1989) and two transport functions are typically used to model landscape evolution (Braun et al., 2001; Dietrich et al., 2003). The slope-dependent transport law has its basis in the characteristic form of convex, soil-mantled landscapes assumed to be in equilibrium, and has some field support (e.g. McKean et al., 1993; Roering et al., 2002). A nonlinear, slope-dependent transport law also has its roots in morphometric observations, and has recent support with the veracity of assuming landscape equilibrium (Roering et al., 1999), experimental constraints (Roering et al., 2001), or for landscapes where postfire ravel processes are thought to dominate (Gabet et al., 2003; Roering and Gerber, 2005; Lamb et al., 2011).

Mechanistically, soil transport should depend on soil thickness as well as slope, as suggested by quantification of freeze–thaw (e.g. Anderson, 2002; Matsuoka and Moriwaki, 1992), shrink–swell (e.g. Fleming and Johnson, 1975), viscous or plastic flow (e.g. Ahnert, 1976), and bioturbation processes (Gabet, 2000). In each case, disturbance processes set the mobile soil thickness and modulate soil transport by setting the magnitude of slope-normal displacement, while slope sets the downslope component of the gravitational driving force. Depth-dependent flux is also suggested by the velocity profiles from segmented rod studies and modeling (Young, 1960; Young, 1963). For flux to be proportional to the depth–slope product is an idea that has never been tested despite being suggested almost 40 years ago (Ahnert, 1967).

Recently, Furbish (2003) suggested that for transport due to dilational effects of biotic activity, the vertically averaged volumetric flux density (i.e. vertically averaged velocity), $\overline{q}_s$ (L t^{-1}), is proportional to land surface gradient, S. The depth-integrated flux per unit contour distance is then the product of soil thickness, H, and this flux density

$$H\overline{q}_s = -K_h HS, \tag{13}$$

where K_h (L t^{-1}) is a transport coefficient, assumed to be constant. We tested this transport relation using estimates of local depth-integrated flux, $H\overline{q}_s$, obtained by integrating the soil production rate between flow lines at three distinct field sites (Heimsath et al., 2005), and found support for the depth-dependent transport relationship (Furbish et al., 2009). Importantly, our work quantifying the soil production function using both CRNs and the morphometric test of plotting topographic curvature against soil thickness also supports the slope-dependent transport relationship (Heimsath et al., 1997, 1999, 2000, 2006). Clearly, more direct quantification of the soil transport processes is needed and some of our studies are pursuing this actively.

Long-standing efforts to determine average erosion rates across agricultural landscapes involve the construction of small-scale runoff plots (Roels, 1985), monitoring of suspended sediment concentrations in streams draining the landscapes (Steegen et al., 2001), and, with recent analytical breakthroughs, through the use of sediment tracers (Matisoff et al., 2001) or short-lived isotopes (Wilson et al., 2003; Matisoff et al., 2002a,b). Collection and analyses of soil from runoff plots or suspended sediment analyses remain problematic primarily because of the difficulties with ensuring a closed system, while injection of tracers onto agricultural lands immediately imparts a short-term measurement bias. Attempts to quantify sediment-transport processes include segmented rod studies (Young, 1960), tephra deposit mapping (Roering et al., 2002), detailed measurement of bioturbation (Black and Montgomery, 1991; Gabet, 2000), optically stimulated luminescence dating of individual quartz grains (Heimsath et al., 2002), and short-lived isotope measurements (Wallbrink and Murray, 1996; Walling and He, 1999; Kaste et al., 2007; Dixon et al., 2009a). Despite the extensive effort across disciplines, quantifying sediment-transport processes and rates remains elusive (Dietrich et al., 2003).

During the middle of the twentieth century, the detonation of nuclear weapons in the atmosphere injected a host of artificial radionuclides into the environment with half-lives ranging from days to decades. While atmospheric weapons were tested in various countries from the 1940s until the late 1970s, the vast majority of fallout in the central United States was deposited between 1955 and 1967 (Cambray et al., 1989; Simon et al., 2004). These fallout radionuclides offer a unique tool for determining time-integrated erosion rates on a landscape with only one or a few visits to the site (Brown et al., 1981a; Brown et al., 1981b; Zhang et al., 1994; Quine et al., 1997; Walling et al., 1999; Walling et al., 2002; He and Walling, 2003; Porto et al., 2003; Fornes et al., 2005). Fallout radionuclides are used extensively in erosion and sediment-transport studies on both agricultural and forested landscapes. Atmospherically delivered ^{210}Pb ($T_{1/2} = 22.3$ y, $^{210}Pb_{ex}$, "in excess" of that supported by direct decay of ^{222}Rn in soil), cosmogenic ^{7}Be ($T_{1/2} = 53$ d), and weapons-derived ^{137}Cs ($T_{1/2} = 30.1$ y) and ^{241}Am ($T_{1/2} = 432$ y) can be used alone or simultaneously to quantify and trace erosional processes (Wallbrink and Murray, 1996; Walling et al., 1999; Whiting et al., 2001), date and source sediments (Appleby and Oldfield, 1992), and determine sediment transit times (Bonniwell et al., 1999). Fallout radionuclides are useful geomorphic tools because of their unique atmospheric source term, but the technique relies on the assumption that the radionuclides are geochemically immobile and thus effective particle tracers.

Relative rates of soil mixing can be evaluated through measurements of fallout nuclide activities in soil profiles (Dörr, 1995; Tyler et al., 2001). Because these atoms are deposited at the soil surface, mixing processes can increase the dispersion of nuclides with depth and the overall downward transport rate. For example, on agricultural landscapes, where tilling homogenizes the soil to the depth of the plow layer, a unique, well-mixed ^{137}Cs profile captures the process (Walling et al., 1999). Field evidence suggests that some soils are homogenized naturally by burrowing organisms (Heimsath et al. 2002; Black and Montgomery, 1991), which would mix fallout isotopes down to a depth governed by the flora and fauna at the site. However, in some undisturbed forest soils where bioturbation is less obvious, the transport mechanisms can be more elusive and difficult to quantify. By measuring the vertical distribution of fallout radionuclides in soils and calculating diffusion coefficients, we can quantify sediment-transport mechanisms and mixing rates that are operating on the short but important timescale of 10–100 years.

The techniques of relating fallout radionuclide profiles and inventories to absolute and relative erosion rates rely on finding a "noneroding" reference location to compare to eroding or aggrading sites (Lowrance et al., 1988). Nearly all erosion studies by other research groups compare soils sampled at cultivated sites with those sampled from an undisturbed field or forest. We examined soil mixing and transport processes across our soil-mantled hillslopes using fallout radionuclides and by observations of biological activity in the field. We measured $^{210}Pb_{excess}$ (here, shortened to ^{210}Pb) and ^{137}Cs nuclide activities with depth and calculated total inventories, to gain insight into mixing (e.g. Kaste et al., 2007; Dixon et al., 2009a) and soil erosion mechanisms (e.g. Kaste et al., 2006; O'Farrell et al., 2007). The nuclide profile depth was defined as the soil depth at which greater than 95% of the cumulative nuclide activity, or inventory, was obtained. We used nuclide profiles to determine the degree of physical mixing, and calculated a mixing coefficient by the best-fit exponential curve to an advection–diffusion equation (Kaste et al., 2007).

Steady-state profiles of ^{210}Pb provide insight into mixing and soil transport over short timescales (10^2–10^3 years). The depth distribution of ^{210}Pb in soils can be described by the steady-state solution to the advection–diffusion equation (e.g. He and Walling, 1996; Kaste et al., 2007):

$$A(z) = A_0 \exp\left[\frac{v - \sqrt{v^2 + 4\lambda D}}{2D}(z)\right], \tag{14}$$

where '$A(z)$' is the nuclide activity at a specific depth (in Bq cm^{-3}), 'A_0' is the activity at the surface, 'v' is the downward advection velocity due to leaching (cm y^{-1}), 'λ' is radioactive decay (y^{-1}), and 'D' is a diffusion-like mixing coefficient (cm^2 y^{-1}). Advection rates have previously been measured using the depth of concentration of weapons-derived ^{137}Cs, which was delivered to soils as a thermonuclear bomb product between 1950 and 1970, peaking in 1964 (e.g. Kaste et al., 2007). We were sometimes unable to determine clear subsurface peaks in ^{137}Cs activity profiles that correspond to this delivery. Instead, in such cases, we calculated diffusion-like mixing coefficients by assuming that advection plays a minimal role in subsurface nuclide redistribution. We then modeled a best-fit diffusion equation to each profile by minimizing the sum of residuals. Importantly, the timescale of such quantification is only appropriate for recently (~10 to $<$100 years) active processes.

2.3. Parent Material Strength

The depth dependency of soil production rates followed Gilbert's early reasoning that rock disintegration was fastest under thin soils and slowed where bedrock emerged at the surface or was deeply mantled with soil. Gilbert believed that on a steady-state landscape, where erosion is equaled by soil production, highest erosion rates would be under thin soils and lowest rates would be at the extremes because (i) exposed bedrock would limit transport to the slow rate of exposed and coherent rock and (ii) thick soil cover would develop for low transport rates. While this conceptual framework persisted (Carson and Kirkby, 1972) and was explored extensively (Ahnert, 1967; Anderson and Humphrey, 1989; Cox, 1980; Dietrich et al., 1995; Heimsath et al., 1997; Minasny and McBratney, 1999), the underlying mechanisms governing it remain fundamentally untested. The sharp boundary between saprolite and soil as well as the biogenic disruption by gopher burrowing, ants, earthworms, and burrowing wombats (Australia) are widely observed, highlighting the role of parent material weatherability in setting the upper bound for soil production, as well as biogenic role in downslope transport. Additionally, quantifying parent material strength at locations where soil production and transport rates as well as chemical weathering rates are known helps make the connection between hillslope hydrology and the geomorphic processes driving landscape evolution.

Specifically, the parent material resistance to the physical processes that convert it to transportable sediment is likely to reflect the intersection of several aspects of forces driving landscape evolution. First, tectonic forces set the base level that upland landscape evolution is thought to be responding to. Second, climatic forces set both the biota (i.e. flora and fauna) providing the physical forces of hillslope weathering, and the chemical weathering parameters that break down minerals and remove mass in solution. Third, the combination of physical and chemical weathering processes is likely to determine just how weathered (and therefore resistant to shear) the parent material becomes. For example, active bioturbation that mixes a soil thoroughly also helps convey water and organic acids to the soil-parent material boundary. Conversely, a landscape lacking bioturbation and experiencing active overland flow processes is likely to have a smaller degree of contact between meteoric waters and the parent material. Understanding hillslope hydrologic pathways are, therefore, critically important for coupling the physical and chemical processes that we have focused on thus far. Quantifying all of the different processes and parameters that factor into how hillslopes are evolving thus enables us to better predict how landscapes are likely to change with the changing driving forces of climate, tectonics, and anthropogenic.

We apply independent, field-based measurements to quantify the strength (i.e. resistance to shear) of the parent material underlying the mobile soil layer. Quantifying strength can be done in a number of ways. Selby (1980, 1985) suggests a semiquantitative classification of rock mass strength for crystalline bedrock. Although this method is comprehensive and appropriate for rocky slopes, it is difficult to compare with soil-mantled slopes where we are interested in the saprolite resistance to physical disruption. Several studies have explored the physical characteristics of granitic and metamorphic rocks and saprolites, but have mostly focused on groundwater saturation effects and slope stabilities (e.g. Jiao et al., 2005; Johnson-Maynard et al., 1994; Jones and Graham, 1993; Kew and Gilkes, 2006; Schoeneberger et al., 1995). It is important to note that the complexities inherent in quantifying the strength of competent rocky slopes are lessened by the fact that we are interested in quantifying the resistance of a relatively continuous layer of weathered material that is overlain by a relatively smooth surface mantle of soil.

We use two instruments to assess saprolite strength: a dynamic-cone penetrometer and a shear vane tester (Herrick and Jones (2002) and GEONOR). We also experimented with Schmidt hammers (see recent review article and results in Viles et al. (2011)), but their dependence on grain boundaries and limited applicability to competent bedrock meant that we could not use them for saprolitic parent material (i.e. anything experiencing even the slightest degree of weathering). The penetrometer measures the compressibility of a material in units of strikes per depth, which can be converted to kPa when calibrated with the shear vane tester. The shear vane tester measures the shear strength in kilopascals (kPa). We used a Geonor H-60 handheld vane tester (GEONOR) with an instrument range from 0 to 260 kPa. The standard procedure is to place the end of the vane normal to the surface and insert it about 5 cm below the surface to minimize any edge effects, then turn the handle clockwise at a constant rate until the lower and upper portions of the instrument move in unison, indicating that the maximum shear strength has been obtained. Other studies have explored the correlation between these two instruments at different sites (Bachmann et al., 2006; Zimbone et al., 1996). The shear strength of a soil is directly related to the normal stress applied (McKyes, 1989); therefore, an instrument that applies a force normal to the surface can be used to estimate a relative shear strength. Several field seasons of experimenting with these instruments led us to a procedure that appears to yield reproducible and reliable results (Byersdorfer, 2006; Johnson, 2008).

3. FIELD SITE SUMMARY

To tie together the methods outlined above, I draw upon studies across a few selected field areas examined by my research group. It is beyond the scope of this chapter to go into all the details, or expand upon all of the specific assumptions and caveats for each of these study areas, but there are key aspects of the sites that I summarize here to help navigate the connections between the conceptual framework and the representative results. Recall that a key assumption used to develop the conceptual framework is the assumption that soil thickness does not change with time locally. This does not mean that soil thickness is uniform across the landscape, but it does mean that our initial site selection was motivated by convincing ourselves that local soil thickness remains constant (excluding the stochastic perturbations associated with bioturbation that are relatively quickly "healed"). A similarly important criterion for site selection was that the processes we observe today reflect the processes driving landscape evolution over the timescales captured by our methodologies (tens to tens of thousands of years). Three field areas were, therefore, the initial focus of our studies: Tennessee Valley (TV), California, Point Reyes (PR), California, and Nunnock River (NR), southeastern Australia. Each of these field sites has smoothly convex-up soil-mantled ridges separated by unchanneled, "zero-order" swales and also had independent studies, suggesting that the impact of the Pleistocene–Holocene transition on local biota was not severe. Two of these field sites (PR and NR) have granitic parent material, while TV has metasedimentary rock. The relative ease of working on granitic parent material drove further studies to focus on granitic landscapes, where possible.

Following the honing of methodology across these field sites, we expanded our studies to include an examination of the role of climatic forcing. For that work we used a climate transect in the southern Sierra Nevada Mountains (Dixon et al., 2009a,b) and I summarize some of that work here by focusing on the climate end members of the alpine region of Whitebark (WB) and the oak-grassland region at Blasingame (BG). We worked in parallel to examine the role of tectonic forcing by focusing on a series of sites across the San Gabriel Mountains of California (SG) (DiBiase et al., 2010; Heimsath et al., 2012); however, I draw only upon the coupling between parent material strength and soil production for that site here. I do also draw briefly upon work done in a New England Forest (NE) to highlight some differences between processes active on the upland landscapes focused on here and a temperate forest that experienced Pleistocene glaciation. Table 1 summarizes the field sites used to draw the representative results shown here, and does not include other sites that I make only brief note of (e.g. Oregon Coast Range and northern Australia). It also does not list the methods used at or key results from these sites that are not raised in this paper.

4. REPRESENTATIVE RESULTS AND DISCUSSION

4.1. Soil Production

The first quantification of the soil production function was from Tennessee Valley (TV), California, and utilized both ^{10}Be and ^{26}Al measured in the saprolite beneath a range of soil depths (Heimsath et al., 1997, 1999). Further work on both granite and metasedimentary parent material supported the newly developed methodology (Heimsath et al., 2000, 2001a,b) and showed a wide range of application for this transport relationship. More recent work helps illustrate the results well (Heimsath et al., 2005, 2010). Soil production rates from 13 samples from the bedrock–soil interface at Point Reyes (PR) define a clear exponential decline of soil production with soil depth such that the variance-weighted best-fit regression (Fig. 3a) is

$$\varepsilon(H) = (88 \pm 6)\cdot e^{-(0.017\pm0.001)H}, \quad (15)$$

where soil production, $\varepsilon(H)$, is in meters per million years and soil depth, H, is in centimeters. Average erosion rates, inferred from three samples of

TABLE 1 Summary of Field Sites Reported in this Chapter

Field site (abbreviation)	Methods used	Key results	References
Tenn. Valley (TV)	CRNs, short-lived isotopes	Soil production, depth-dependent transport	Heimsath et al. (1997, 1999, 2005); Kaste et al. (2007)
Point Reyes (PR)	CRNs, chem., strength	Soil production, chem. weathering, parent material strength, depth-dependent transport	Heimsath et al. (2005)
Nunnock River (NR)	CRNs, chem., short-lived isotopes, strength	Soil production, chem. weathering, parent material strength, depth-dependent transport	Heimsath et al. (2000, 2005, 2006, 2010); Kaste et al. (2007)
Blasingame (BG)	CRNs, chem., short-lived isotopes	Soil production, chem. weathering, transport processes	Dixon et al. (2009a,b)
Whitebark (WB)	CRNs, chem., short-lived isotopes	Soil production, chem. weathering, transport processes	Dixon et al. (2009a,b)
New England (NE)	Short-lived isotopes	Lack of soil mixing, short vs. long term *E*.	Kaste et al. (2007)
San Gabriels (SG)	CRNs, strength	Soil production, parent material strength	Heimsath et al. (2012)

stream sand from the creek draining the study area, determine a basin-averaged rate of 62 m/m.y., similar to Tennessee Valley (Fig. 1b). Lower erosion rates from two tors at Point Reyes are consistent with granitic corestone emergence shown at Nunnock River (NR) (Heimsath et al., 2000). Comparison of the soil production functions reveals remarkable similarity in form with relatively little scatter around the Point Reyes regression line (Fig. 3b). Similar to the Nunnock River and Tennessee Valley sites, these data, combined with the spatial variation of depth data from Fig. 4a, discussed below, show spatially variable rates of soil production, indicating a landscape out of long-term dynamic equilibrium. This is not surprising given the proximity of the San Andreas fault to the site, and the southern California origin of the Point Reyes Peninsula (Heimsath et al., 2005).

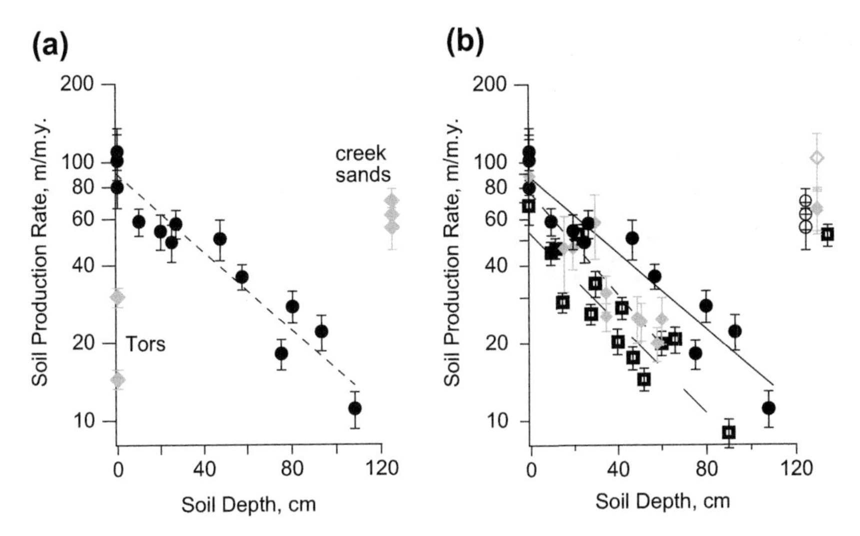

FIGURE 3 Soil production functions. (a) Point Reyes (PR). Solid black circles are averages of rates from both ^{10}Be and ^{26}Al, and error bars represent all errors propagated through nuclide calculations, i.e. uncertainty in atomic absorption, accelerator mass spectrometry, bulk density and soil depth measurements, and attenuation length of cosmic rays. Gray diamonds are erosion rates from outcropping tors on ridge crests. (b) Point Reyes data shown with results from Nunnock River (open black squares, long-dashed regression line, Heimsath et al., 2000) and Tennessee Valley (gray diamonds, short-dashed regression line, Heimsath et al., 1997); open symbols labeled "creek sands" are average rates from detrital cosmogenic nuclide concentrations from each site. *(From Heimsath et al., 2005)*

The Point Reyes results are evidence for the applicability of the soil production function as a transport relationship (cf. Dietrich et al., 2003) characterizing hilly, soil-mantled landscapes. Comparison of functions from these different sites offers the potential for untangling the connections between erosion rates, climate, and tectonics. Specifically, the relationship slope quantifies the depth dependency of the soil-producing mechanisms. Similarity in slope (Fig. 3b) reveals that soil production rates are halved beneath 35 cm of soil and down to a tenth of their respective maxima beneath 115 cm, roughly the maximum soil depth across divergent ridges. Biogenic processes are dominant across each site, are controlled by climate, and potential variations in their effects across the landscapes may be responsible for this order-of-magnitude difference in lowering rates. The absolute magnitude of such disequilibrium is quantified by the slope as well as the intercept of the soil production function, the latter potentially set by the underlying strength, or resistance to erosion, of the parent material. Such disequilibrium would suggest a long-term flattening of noses, a trend that is likely to be offset by the periodic evacuation of the convergent hollows such that local base level to the hillslopes is reset, and ridge–valley topography is maintained (Dietrich et al., 1995) (Pelletier and Rasmussen, 2009).

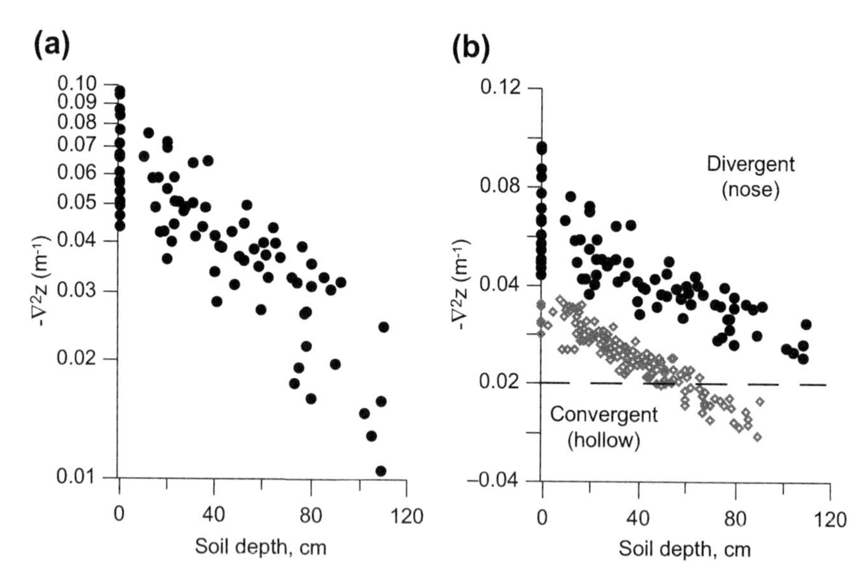

FIGURE 4 Negative curvature versus vertical soil depth. Curvature calculated as in Heimsath et al. (1999) and is proxy for soil production if local soil depth is constant with time. (a) Black circles from individual soil pits at Point Reyes with exponential negative curvature axis. (b) Black circles same as Figure 4a with open gray diamonds from Nunnock River (Heimsath et al., 2000). Tennessee Valley data from Heimsath et al. (1997, 1999) overlay Nunnock River data values and range and are not included here for plot clarity. Black dashed line separates divergent from convergent topography. *(From Heimsath et al., 2005)*

4.2. Chemical Weathering

To connect the soil production function to the chemical weathering processes active across hilly landscapes, we focus on the southern Sierra Nevada Mountains, California (Dixon et al., 2009a,b). We applied our diverse methodology across several field sites ranging from the low elevation oak-grassland landscape at Blasingame (BG) to the alpine region near the crest of the western Sierra at Whitebark (WB). Hillslope soil production rates average 82 ± 10 ton/km^2/a (37 ± 4 m/Ma; mean $\pm$ std error) at the low elevation BG and 52 ± 5 ton/km^2/a (24 ± 2 m/Ma) at high elevation WB site. At BG, soil production rates decrease with increasing soil thickness (Fig. 5a) and distance from the slope crest (Fig. 5b), as observed in other temperate landscapes (e.g. Figs 3a and b). At WB, such a relationship is not observed.

Chemical weathering results in an average net loss of $24 \pm 4\%$ of the soil mass at both sites, calculated as the average CDF_{soil}, and CDF values are not significantly different at the two sites. Dahlgren et al. (1997) observed that clay content of the low elevation soils exceeds that of WB soils by a factor of two. This suggests a discrepancy in how soil weathering intensity is recorded by CDF and clay abundance. The CDF quantifies net elemental losses; however,

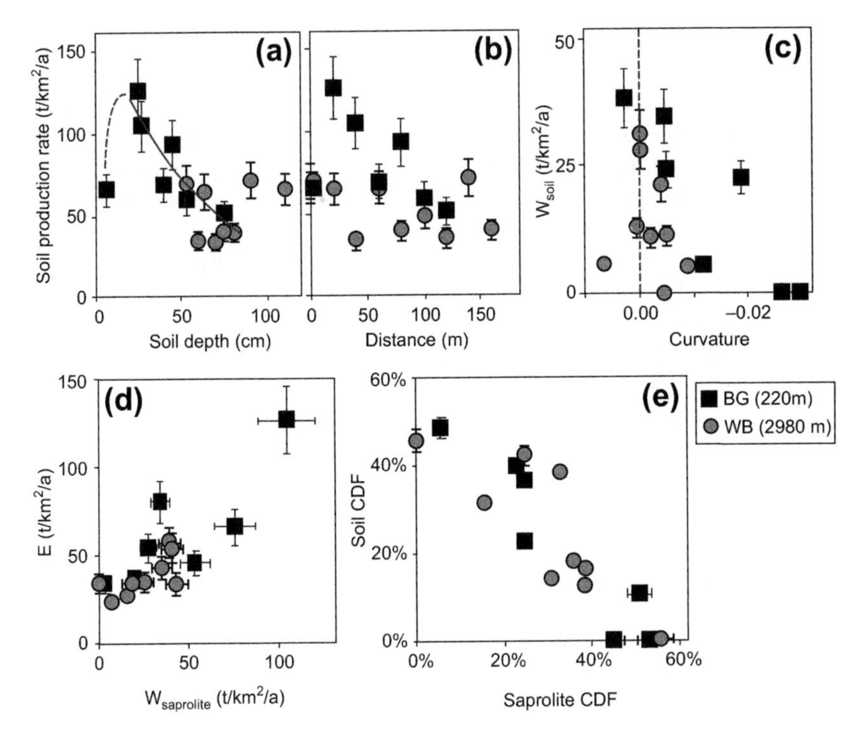

FIGURE 5 Differences in erosion and weathering between high- (Whitebark, WB) and low-elevation (Blasingame, BG) field sites from Dixon et al. (2009b) study. Average ^{10}Be-derived, soil production rates are higher at BG than WB. At BG, these rates decrease with (a) soil thickness and (b) distance from the ridge crest. (c) Soil chemical weathering rates (W_{soil}) decrease with increasing convexity (negative curvature) at BG, and insignificantly at WB. (d) Physical erosion rates (e) increase with the chemical weathering rate of saprolite ($W_{saprolite}$) at both sites. Average rates of erosion are faster at the warmer and drier BG, compared to the colder and wetter WB, although saprolite weathering rates are not significantly different. (E) Soil and saprolite weathering extents, shown by chemical depletion fractions (CDF) are negatively correlated.

secondary mineral formation is the balance between chemical dissolution of primary minerals and the leaching of weathering products. Potential mass loss may exceed net mass loss at low elevation, due to secondary mineral development and retention, in agreement with previous observations of the low leaching potential of clay minerals in these soils compared to high elevation soils (Dahlgren et al., 1997). Thus, the total chemical alteration at BG site is greater despite similar net losses to WB.

At both sites, CDF_{sap} data indicate that saprolite weathering is a large portion of the total weathering losses, averaging $31 \pm 4\%$ and reaching values as high as 56% of the original rock mass. Saprolite weathering rates average 46 ± 13 ton/km^2/a at BG and 25 ± 5 ton/km^2/a at WB. Physical erosion rates at low and high elevations average 64 ± 12 ton/km^2/a and 38 ± 4 ton/km^2/a, respectively. At BG, soil chemical weathering rates decline with increasing convexity (Fig. 5c). Furthermore, physical erosion and saprolite weathering rates at both sites are positively correlated (Fig. 5d), and a strong negative relationship exists between the chemical weathering extents of soils and saprolites (Fig. 5e). In the Discussion section, we explore implications of these data following the quantification of transport processes and the parent material strength.

Chemical weathering facilitates physical erosion by the dissolution of primary minerals, reducing the competency of rock and increasing erodibility. Our data are among the first to quantify links between saprolite weathering and physical erosion. Physical erosion rates increase with saprolite weathering rates (Fig. 5d). These data suggest that physical erosion is dependent on the chemical weathering extent and rate of bedrock, since weathered saprolite is more easily detachable and mobilized into the overlying soil column. Furthermore, soil chemical weathering rates at low elevation decline with increasing convexity (Fig. 2c), and the intensity of chemical weathering in soils and saprolites is inversely related (Fig. 2e). Soil weathering may be low where saprolite weathering is high due to faster erosion that reduces soil residence times (e.g. Anderson, 2002). As water and sediment are shed off divergent areas of the landscape, decreased water–soil interaction could further result in decreased chemical weathering of soils. Conversely, similar CDF_{total} across the Sierras may support the idea that weathering of parent bedrock is limited by the supply of fresh minerals, rather than reaction rates (e.g. Riebe et al., 2000; Riebe et al., 2001a; West et al., 2005). Saprolite weathering in the Sierras, indicated by saprolite CDFs and rates, is controlled by processes not clearly identified from our study, but is possibly linked to climate, moisture availability, and hillslope morphology. Our data suggest that soil weathering is limited by the availability of fresh minerals, and is therefore low when saprolite has previously depleted this supply.

Second, and perhaps most importantly, these data show an example of the strong feedbacks between physical erosion and chemical weathering at both Sierran study sites, despite broad differences between the climates and the soil production and transport mechanisms. Other authors have reported positive correlations between soil chemical weathering and erosion and have suggested that physical erosion sets the pace for chemical weathering (e.g. Riebe et al., 2003b; Riebe et al., 2004) in soil-mantled terrain. Our data indicate saprolite weathering and erosion are positively linked (Fig. 5d), and soil weathering is reduced where both saprolite weathering (Fig. 5e) and landscape convexity (Fig. 5c) are high. In summary, our data suggest that saprolite weathering controls erosion and weathering of the overlying soil by depleting primary minerals, decreasing rock competence, and increasing mobility of weathered material. Since chemical weathering of the saprolite accounts for such significant mass loss from these landscapes, we suggest that not accounting for it leads to missing a critical aspect of erosion–weathering feedbacks.

4.3. Soil Transport

Soil depths vary spatially across the divergent areas of the Point Reyes (PR) landscape such that topographic curvature declines exponentially with increasing soil thickness (Fig. 4a), although a linear decline similar to the observations from the Nunnock River (NR) and Tennessee Valley (TV) field sites cannot be ruled out (Fig. 4b). Using curvature as a proxy for soil production (Heimsath et al., 1999), these data support the soil production function defined by Eqn (15), assuming an independently documented (Reneau and Dietrich, 1990, 1991) "linear diffusivity" of 30 cm^2 yr^{-1}. The clear linear (versus exponential) decline of curvature with increasing soil depths at NR suggested, however, that a linear slope-dependent transport model did not adequately capture the transport mechanisms, prompting modeling (Braun et al., 2001) and optically stimulated luminescence (Heimsath et al., 2002) studies highlighting the importance of soil thickness in controlling transport.

We plot depth-integrated flux, determined by integrating soil production rates downslope, against the depth–slope product across all field areas to test the nonlinear transport law of Eqn (13) (Fig. 6a). We observe linear

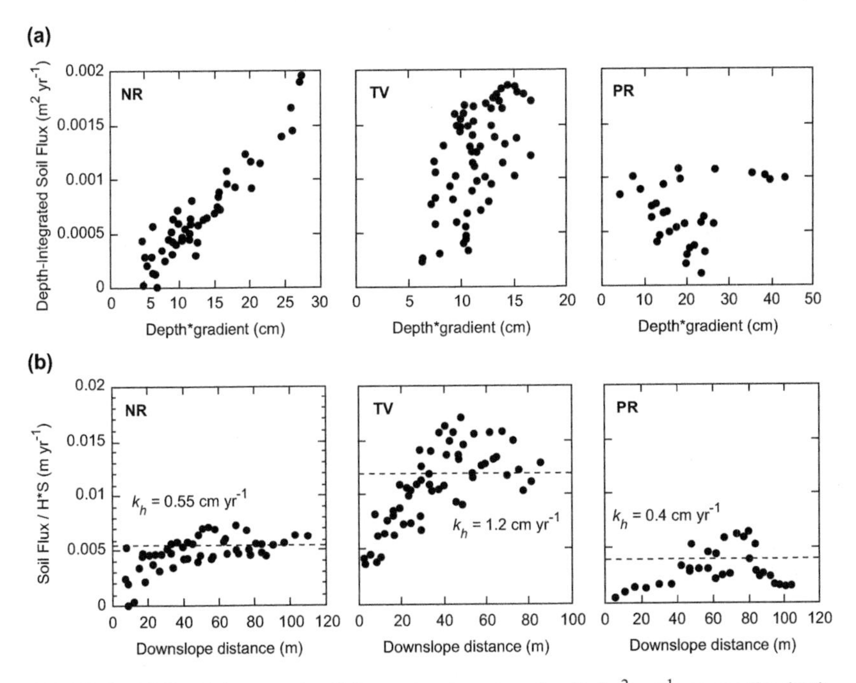

FIGURE 6 (a) Depth-integrated soil flux per unit contour length (m^2 yr^{-1}) versus the depth–slope product (cm) for all field sites as described in Heimsath et al. (2005). (b) Depth-integrated flux divided by depth–slope product versus downslope distance. K_h value, determined by fitting data shown in A, is dashed line. NR, Nunnock River; TV, Tennessee Valley; PR, Point Reyes.

increases of soil flux with increasing depth–slope product for both the NR and TV field sites, offering strong support for the depth-dependent transport relationship. Data from NR support Eqn (13) with a transport coefficient, K_h, equal to 0.55 cm yr^{-1}. Roughly equating this coefficient to the linear "diffusivity" with an average soil thickness for Nunnock River of 65 cm yields a coefficient of 36 cm^2 yr^{-1}, which is remarkably similar to the 40 cm^2 yr^{-1} reported by Heimsath et al. (2000). Reversing the process for the TV and PR data, which have independently determined linear "diffusivities" of 50 and 30 cm^2 yr^{-1} and average soil thicknesses of 30 and 50 cm, yields depth-dependent transport coefficients of 1.7 and 0.6 cm yr^{-1}, respectively. Using the data from TV and PR (Fig. 6A), K_h values of 1.2 and 0.4 cm yr^{-1} are determined.

This comparison of transport coefficients places the depth-dependent transport flux within the context of the more familiar slope-dependent transport framework and supports the applicability of a linear transport law for low-gradient, convex landscapes. Plotting flux against gradient shows, however, that a linear relationship does not reflect all the data (Data Repository; see footnote 1). Our test for depth-dependent transport thus involves two sets of complementary plots: $H\overline{q}_s$ versus HS (Fig. 6a) and $H\overline{q}_s(HS)^{-1}$ versus downslope distance, X (Fig. 6b). If Eqn (1) is correct, the first plots (Fig. 6a) should show a linear increase, with slope equal to K_h and zero origin – in the absence of covariance of H or S with distance X. However, because $H\overline{q}_s$ must, by definition, increase with X, these plots might exhibit spurious covariance with the depth–slope product. Thus, the importance of the inset plots (Fig. 6b) is to remove any such covariance such that the data should be homoscedastic about a flat line equal to K_h (dashed line) to support Eqn (1). We observe, instead, an increase with distance close to the ridge crests, followed by a tendency to flatten roughly around the K_h values. Potential explanations for the increase include the unknown role of chemical weathering or a nonconstant, covariant transport coefficient that we have not yet quantified.

Fallout radionuclide activity–depth profiles and field observations reveal distinct differences in sediment-transport processes at the climate end members of our Sierra Nevada field site. Vegetative density is lowest at the high elevation WB, with an average of 83% bare soil versus 4% at BG. Low vegetative cover and high precipitation at WB result in low soil resistance to surface water flow and rain drop splash (e.g. Prosser and Dietrich, 1995). Rills began ~40 m downslope from the crest at WB. These have an upslope spacing of 23 m decreasing to an average spacing of 9 m at ~60 m from the slope crest. No rilling was evident at BG. At both sites, bioturbation is evident in soils, and gopher burrows were observed parallel to the ground surface and as deep as the soil–saprolite interface. Mapping gopher burrow density indicates that the burrowing activity at WB is 53% that of BG (Dixon et al., 2009a).

Penetration depths of ^{210}Pb and ^{137}Cs increase linearly with burrowing activity at both sites (Fig. 7a), suggesting that bioturbation – through physical transport and altered hydrology – redistributes nuclides to depth in the soil. Assuming nuclide profiles form primarily by diffusion-like processes, the average diffusive mixing coefficient of soils is 0.28 ± 0.05 cm^2/a at BG, greater than the average 0.15 ± 0.02 cm^2/a at WB (Fig. 7b). ^{210}Pb inventories do not vary consistently with topography at BG; however, at WB, inventories are lowest where slopes are steepest and have the greatest upslope contributing area (Fig. 7c,d). Upslope contributing area has the strongest negative correlation with nuclide inventory at WB, suggesting soil loss scales with

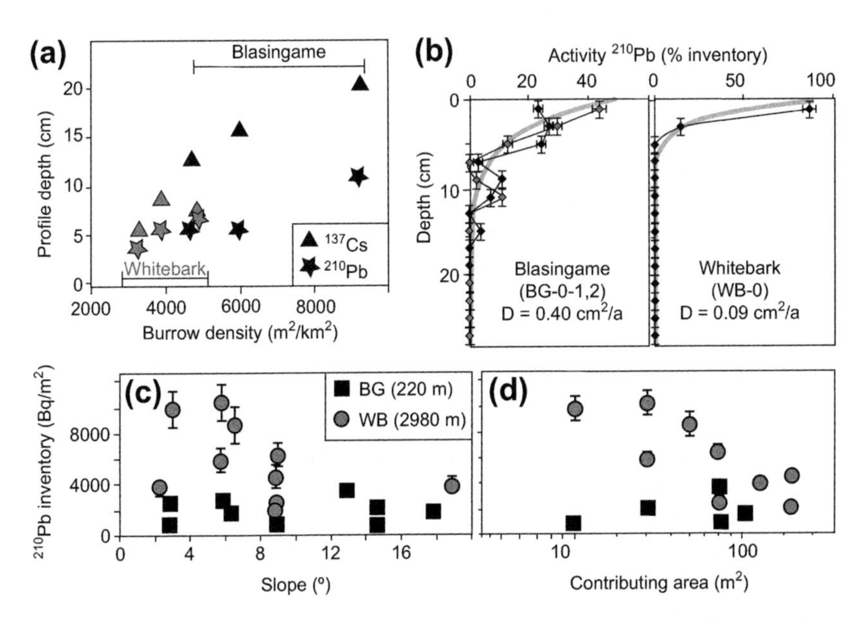

FIGURE 7 (a) Surface-burrowing activities from three transects at BG and WB from Dixon et al. (2009b) increase with associated profile depths for fallout nuclide $^{210}Pb_{excess}$ and ^{137}Cs. "Profile depth" is defined as the soil depth at 95% cumulative nuclide inventory. (b) Fallout profiles show nuclide activity versus depth for hillcrests at BG (two profiles shown are 2 m apart) and WB (one profile) and are deeper at BG. We calculated diffusion-like mixing coefficients for each profile (shown by broad gray line) by the best fit to the diffusion equation, Eqn (14), where '$A(z)$' and 'A_0' are nuclide activity at depth ('z') and surface, respectively, 'λ' is nuclide decay. Here, we assume advection velocity ('V') is zero. Diffusive mixing coefficient of hillcrests are shown in the figure, and average hillslope values at each site are 0.28 ± 0.05 cm^2/yr at BG and 0.15 ± 0.02 cm^2/yr at WB. Inventories (c, d) of $^{210}Pb_{excess}$ and ^{137}Cs for downslope soils at low-elevation BG (gray squares) and high-elevation WB (black circles). Inventory data points reflect those calculated from individual soil profiles and activities of additional bulk soil samples gathered downslope. (c) Nuclide inventories at high elevation are lower at high slopes; however, no statistically significant correlation exists. (d) At BG, inventories do not change markedly, while at WB, nuclide inventories decrease with distances downslope and increasing contributing area.

discharge (e.g. Kaste et al., 2006). This correlation, in agreement with the observation of rills, shows that overland flow plays an important role in soil transport at WB.

While overland flow may play a dominant role in sediment transport at WB, it likely has little impact on soil production. Spatial patterns of soil production are distinctly different at the two sites (Fig. 5a), and an apparent soil production function at low elevation is consistent with production mechanisms such as rooting and bioturbation, which are expected to be depth dependent (e.g. Heimsath et al., 1997, 2000, 2005). It is possible that the absence of a trend between P_{soil} and depth at the higher elevation, WB, site is due to soil depths temporarily out of local steady state; and, indeed, the two deepest samples are anomalies in the sampled transect. More likely, the absence of depth-dependent soil production at WB (Fig. 5a) and the differences in hillslope patterns of erosion and weathering (Fig. 5b,c) suggest that a different mechanism is dominant at the high elevation site. With average annual temperatures of 3.9 °C, freeze–thaw may occur at WB; however, this process is also likely depth dependent (e.g. Anderson, 2002). Furthermore, freeze–thaw is likely not a dominant soil production or transport process given that rills are prominent on the land surface and that soil thicknesses typically exceed one meter. Biotite hydration and oxidation may occur at depth in saturated soils during spring snowmelt; however, our data do not speak directly to this mechanism and further research is needed to explain what processes ultimately create these thick high elevation soils.

The steady-state distribution of $^{210}Pb_{ex}$ in soils can be described by solving the traditional advection–diffusion model (e.g. DeMaster and Cochran, 1982), where the amount of $^{210}Pb_{ex}$ in a particular volume of soil (A, in Bq cm^{-3}) is controlled by the initial amount (A_0), an advection term, which is defined here as downward leaching of nuclides sorbed to colloidal materials, as expressed above in Eqn (14). Radioactive equilibrium is a reasonable assumption at our sites: they have remained free from anthropogenic disturbance for at least 3 half-lives of ^{210}Pb, precipitation (which can control deposition) has remained relatively constant during the 20th century, and measured total inventories are consistent with those predicted by deposition models. By modeling the distribution of a single isotope in the soil column, it is not possible to solve for unique values of v and D, and it is unrealistic to ignore either leaching or diffusion-like processes in soils. While our goal is to quantify D to measure sediment mixing, we need to constrain advection that might occur. For this, we use the distribution of weapons-derived nuclides (^{137}Cs and ^{241}Am) in the soil profile, which had a pulse-like input to soils during the 1950s and 1960s, with a strong global depositional maxima in 1963–1964 (US ERDA, 1977). Using the precise position of the subsurface concentration maxima of weapons-fallout in our soils, we determine v (Fig. 8).

While advection–diffusion models are the most common method for describing the depth distribution of fallout nuclides (Walling et al., 1999 and references therein), an alternative model for describing radionuclide

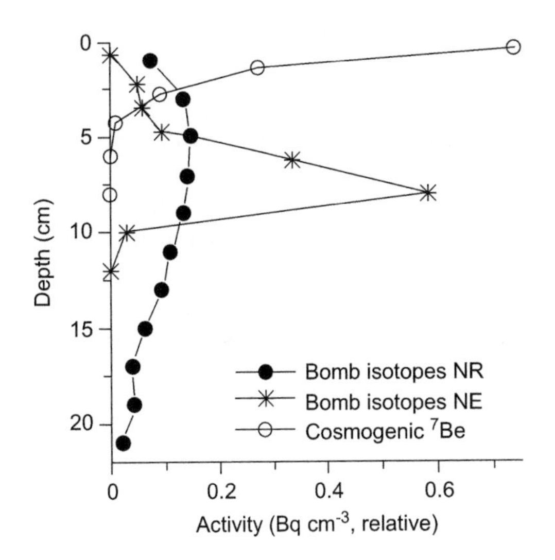

FIGURE 8 Depth profiles of cosmogenic ^{7}Be and weapons-fallout isotopes at New England (NE) and Nunnock River, Bega Valley (BV) field sites as described in Kaste et al. (2007). ^{7}Be is largely (>80%) retained in the upper 2 cm of soil. Weapons-fallout isotopes display a sharp subsurface peak at NE because of limited mixing; at BV, similar isotopes are more dispersed presumably from bioturbation. We calculate v by using the position of the subsurface maxima and assume that it tracks the 1963 deposition spike: v ranged from 0.1 to 0.2 at NE, and from 0.08 to 0.12 at BV and MC.

depth profiles may be required for sites that do not fall into the reasonably well-constrained conceptual framework outlined above. For example, in New England (NE) >70% of the total $^{210}Pb_{ex}$ inventory resides in organic matter and the soils we examined showed no evidence of physical mixing. These, therefore, provided an important comparison with the well-mixed upland soils that are more typical for the landscapes selected for our work. Organic matter decomposition must govern the concentration depth profiles of $^{210}Pb_{ex}$ in O horizons to some extent, since the half-life of ^{210}Pb is nearly an order of magnitude higher than the half-life or organic matter. As organic matter decomposes, $^{210}Pb_{ex}$ concentrations will increase from relative carbon loss. A modified constant initial concentration (CIC) model (developed elsewhere for sedimentary environments; Appleby and Oldfield, 1992) can be applied to our measured $^{210}Pb_{ex}$ profiles, assuming that litterfall and ^{210}Pb flux are relatively constant over time, and that the contribution of root mass in the O horizon is relatively low. We posit that the only difference between the CIC model and our data is an exponential-based organic matter decomposition function. The decomposition functions that best explained our observed $^{210}Pb_{ex}$ concentration-depth profiles from NE can be evaluated using additional data on C cycling developed by others for similar forests (Currie and Aber, 1997; Berg, 2000).

We maximized the fit on the upper 15–20 cm of soil, the portion of the soil with the majority of the nuclide inventory and the portion most likely to be influenced by organisms. At NR and TV, the advection–dispersion model described the $^{210}Pb_{ex}$ profiles well, as total misplaced inventory was typically <18%. Dispersion coefficients can then be used to calculate mixing time constants τ for a soil of thickness L ($\tau = L^2D^{-1}$). A soil 35 cm thick at NR is "turned over", or mixed every 1200 y, and a similar thickness of soil at TV is mixed every 660 y. If we use the mixing timescale for a soil of 35-cm thickness at NR and assume that a particle travels 0.5 L during τ, we calculate grain-scale vertical displacement velocities of 1–2 cm per century. Surprisingly, these values are relatively consistent with the 1–4 cm per century grain-scale velocities calculated at the same site by luminescence dating (Heimsath et al., 2002).

In contrast with NR and TV, a simple advection–dispersion model did not fit the data well in NE; our best model-data fits had misplaced inventory >30% (Fig. 9). In addition, the D values calculated here were on the order of 1 cm^2 y^{-1}, a rate inconsistent with the preservation of the weapons spike (Fig. 8). Because the upper 5–10 cm of soil is the O horizon at our NE sites, which is usually >70% organic matter and very porous (~75%), the initial infiltration of rainwater during intense storms could result in subsurface $^{210}Pb_{ex}$ peaks (Fig. 9c). Profiles of cosmogenic ^{7}Be would capture this process well because of its short half-life and its association with large rain events (Olsen et al., 1985). However, episodic vertical transport that might accompany intense precipitation events appears to be low at all three sites: ^{7}Be activity declines rapidly with depth and the upper 2 cm of soil typically retains >80% of the ^{7}Be inventory (Fig. 8). ^{7}Be was completely retained in the upper 1–2 cm of soil at NR and TV.

Physical soil mixing rates determined with short-lived isotopes correlate well with physical denudation rates measured by others (Fig. 10). Specifically, short-term soil mixing rates are highest at TV, where Heimsath et al. (1997) reported landscape denudation rates approaching 0.1 mm y^{-1} from cosmogenic nuclide analysis. On the other hand, physical denudation rates measured by

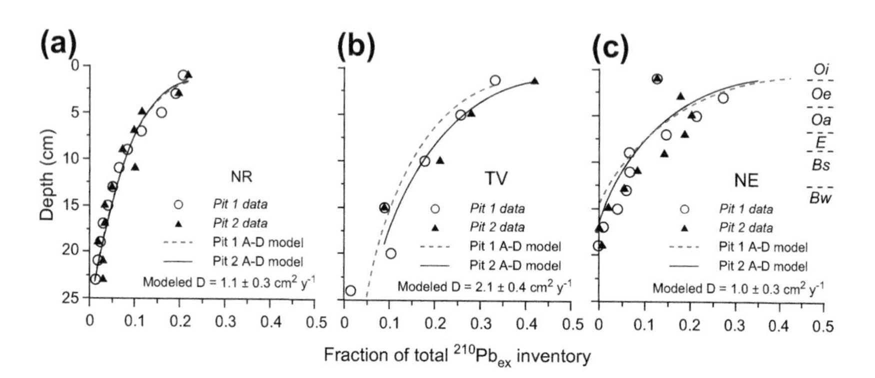

FIGURE 9 $^{210}Pb_{ex}$ data and advection–diffusion (a–d) model given by Eqn (14); v was calculated using weapons-derived isotopes (Fig. 8); we then solved for a D that best matched our data: 80–90% agreement at BV and Point Reyes, Marin County (MC), sites of Heimsath et al. (2005), in NE fits were typically <70%. Appendix 1 of Kaste et al. (2007) has more details on NE fits. Soil horizons observed at New England shown on far right; no equivalent horizon boundaries exist at Bega Valley and Marin County. Oi, fresh litter; Oe, moderately decomposed litter; Oa, humus; E, strongly leached zone; Bs, zone with strong illuviation of iron oxides; Bw, chemically altered zone with weak illuviation.

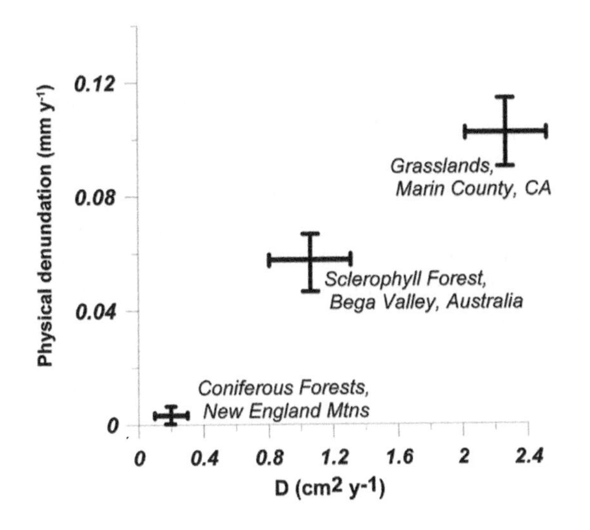

FIGURE 10 Short-term physical soil mixing rates calculated in Kaste et al. (2007) vs. landscape denudation measured by sediment traps and cosmogenic isotopes in Heimsath et al. (1997; 2000). A maximum D of 0.2 cm^2 yr^{-1} for NE soils was determined by using the shape of the ^{241}Am peak (plots similar to Fig. 7) and an instantaneous pulse-like input source assumption.

sediment traps in central NE are orders of magnitude lower (Likens and Bormann, 1995), and we conclude from our nuclide profiles that short timescale diffusion-like processes play a negligible role in mixing the soil layer here. This suggests that relatively short-term soil mixing processes may limit erosion rates, and that there is a rough steady state between the continuous processes of soil mixing and long-term landscape evolution across contrasting field settings. These results are interesting because tree-throw and landslides, which can occur episodically on timescales >100 y, are thought to be a significant sediment-transport mechanism (Dietrich et al., 2003). Because the $^{210}Pb_{ex}$ depth-concentration profiles are not described well with an A–D model, our work demonstrates quantitatively that freeze–thaw does not drive diffusion-like soil mixing in New England. The persistence of an insulating snowpack cover during the winter months probably minimizes the frequency of soil freezing cycles here (Likens and Bormann, 1995). The presence of a fibrous, porous organic horizon in northern NE that undergoes short-term mixing only via decomposition (and not random stirring) may protect mineral soil, and probably limits physical denudation here. Using short-lived isotopes in this way demonstrates quantitatively how process rates can vary in different geological settings on a timescale that is traditionally difficult to capture. Furthermore, we show that short-term diffusion-like processes are significant, and can play an important role in landscape evolution and the fate of any element delivered to the land surface.

4.4. Parent Material Strength

I now attempt to pull together all of the above results and discussion into a final field-based measurement that is intended to provide a key mechanistic parameter for hillslope evolution. At several of our field sites, we used both the cone penetrometer and the shear vane described above to quantify parent material resistance to shear. Other studies explored the correlation between these two instruments at different sites (Bachmann et al., 2006; Zimbone et al., 1996). Bachmann et al. specifically addressed the appropriateness of using a dynamic-cone penetrometer in soils to measure a horizontal shear stress that is a relative measure for the in situ strength state. The shear strength of a soil is directly related to the normal stress applied (McKyes, 1989); therefore, an instrument that applies a force normal to the surface enables the estimate a relative shear strength. In fact, Bachmann et al. (2006) found that the vertically measured compression strength (strikes/cm) obtained at the final soil depth linearly related to the horizontal shear stress measured by the shear vane tester. We confirmed this with our own measurements and determined a linear fit between dynamic-cone penetrometer and shear vane tester measurements that we used to convert strikes/cm from the cone penetrometer to kilopascals (Fig. 11).

Using both the shear vane and the cone penetrometer in soil pits dug across three different field sites first reported elsewhere (Heimsath et al., 2000, 2005, 2006, 2012), we quantify depth dependence for parent material strength (Fig. 12a). For each of the three field sites used for this aspect of our work, we occupied the soil pits used for our cosmogenic nuclide and chemical weathering sampling studies and measured the competence of the parent material immediately beneath the mobile soil layer. The shallowest soils were observed in the rugged landscape of the San Gabriel Mountains, CA (SG) (Heimsath et al., 2012), where there is a relatively steep increase in parent material strength with increasing soil depth (black diamonds in Fig. 12a). Samples from

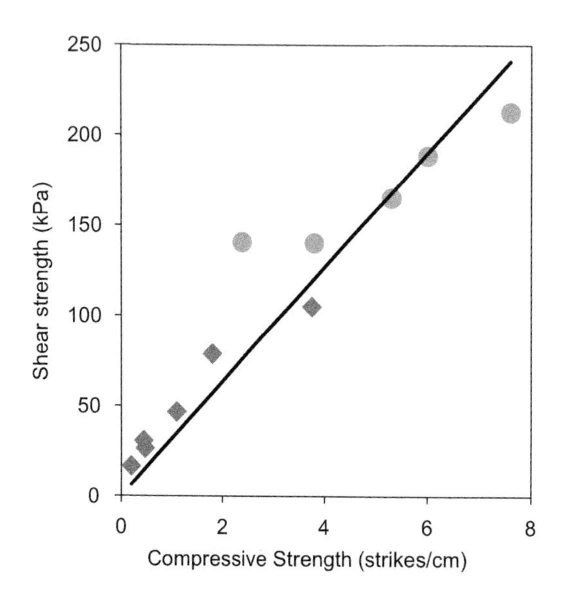

FIGURE 11 Linear relationship between shear vane tester and dynamic-cone penetrometer. Good linear fit confirms principal stress theory such that the vertically measured compression strength (strikes/cm) obtained at the final soil depth is identical in size with the horizontal shear stress measured by the shear vane tester. Green filled circles from Byersdorfer (2006) and blue filled diamonds from Johnson (2008). Black line is fit to both data sets, shear strength (kPa) = 32 × compressive strength (strikes/cm); $R^2 = 0.86$. *(Color version online)*

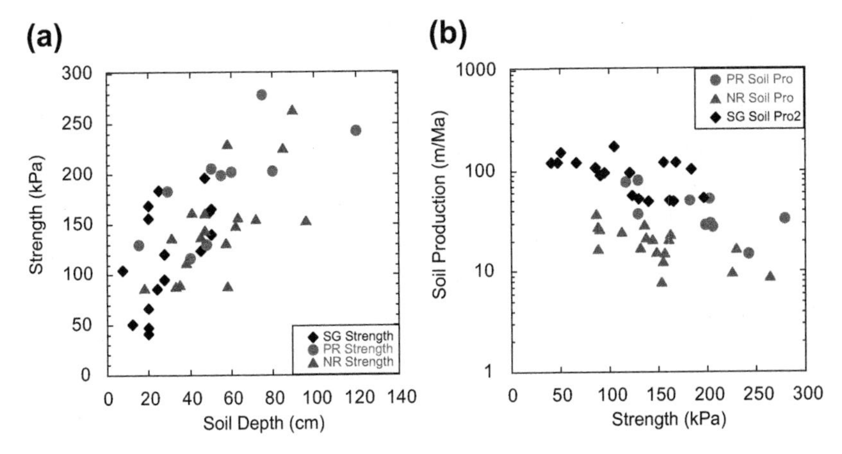

FIGURE 12 (a) Shear strength of the parent material beneath the active soil layer for three of the field sites discussed here: San Gabriel Mountains (SG), California, described in DiBiase et al. (2010) and Heimsath et al. (2012); Point Reyes (PR), California, just north of photograph in Figure 1 and discussed in Heimsath et al. (2005); Nunnock River (NR), Bega Valley, Australia, described initially in Heimsath et al. (2000) and several papers since then. Strength of the saprolite is measured with both a shear vane and a cone penetrometer across the full range of soil depths – strength vs. depth and increases across all sites with increasing soil thickness; (b) soil production rates calculated with the soil production functions reported in Heimsath et al. (in review) for SG, Heimsath et al. (2005 for PR) and Heimsath et al. (2000) for NR vs. the shear strength measured beneath the same soil thicknesses as shown in (a). *(Color version online)*

the Nunnock River, Australia, site (Heimsath et al., 2000, 2006) (NR; red triangles) overlay the SG measurements and the Point Reyes, California, samples (PR; blue circles) (Heimsath et al., 2005). Each of the three field sites shows a distinct increase of parent material strength with increasing overlying soil thickness. Additionally, when taken together, these data suggest a broader relationship between parent material resistance to shear and the soil depth that mantles the sample locations.

In related work that I did not expand upon in this chapter, we showed that the chemical weathering rates in saprolite decreased beneath increasing soil thicknesses (Burke et al., 2007, 2009). Those findings were counterintuitive to our observations of more highly weathered saprolite corresponding to thicker overlying soil depths, but made sense when examined from the perspective that the higher soil production rates found beneath shallower soils should correspond to higher chemical weathering rates (Dixon et al., 2009a,b). The impact on physical weathering processes of the slower weathering rate beneath thicker soils is quantified with our measurements of parent material strength (Fig. 12a), and the observation of higher clay contents being found beneath thicker soils (Burke et al., 2009). We suggest that these higher clay contents and the more deeply weathered saprolites (i.e. with higher chemical indices of alteration as discussed in Burke et al. (2007)) are actually more resistant to physical shear, rather than being more easily mobilized by physical weathering processes. While this may seem to be the opposite of what we might predict, the observations from a wide range of weathered granite saprolites, across very different field areas, seem to indicate a very clear relationship between parent material competence and overlying soil depth.

Exploring the implications of this relationship between strength and overlying soil thickness is the final connection made with the different measurements summarized in this chapter. Each of the field sites discussed, as well as several that were not, has shown clear soil production functions quantified using cosmogenic nuclides (^{10}Be and, in some cases, ^{26}Al). Eqn (15) reports one such equation for the Point Reyes (PR), California, field area. Similar equations were reported for the NR (Heimsath et al., 2000, 2006) and SG (Heimsath et al., 2012) field sites and are not repeated here. Each of these soil production functions defines a strikingly clear decrease of soil production rates with increasing soil thickness. Fig. 12a shows a clear increase in strength of the same parent material used to quantify the soil production functions with increasing soil thickness. If we use the relationships quantified with our independent field measurements we can plot soil production as a function of parent material strength (Fig. 12b) to suggest an inferred dependence between soil production and competence. Data from each of the three field sites reveal separate, but overlapping, relationships, and the data taken together suggest that there may be a broadly applicable dependence between parent material strength and soil production processes. Although the relationship suggested by this figure makes intuitive sense, it has never been documented before.

5. CONCLUSION

The relationships suggested by Fig. 12 provide an exciting way to tie together the very different aspects of quantifying hillslope processes presented in this chapter. These aspects appear to be broadly applicable to landscapes that do not experience glacial processes and have not had significant eolian deposition and we deliberately focus on the physically mobile soils that form a thin mantle across upland landscapes. These studies, and the summary provided by this chapter, also provide some clear paths that we are attempting to follow to help more fully understand how hilly and mountainous landscapes evolve under different conditions. Our conceptual framework for examining such landscapes connects denudation, driven by external forcing, to both physical and chemical processes acting on the soils and the underlying parent material. Soil production processes are primarily physical (burrowing fauna, disrupting flora, freeze–thaw, and wetting–drying), and also depend inherently on the chemical weathering experienced by the parent material. We have found that both soil production and chemical weathering rates decrease with increasing soil thickness. We also show that the extent, or degree, of weathering increases beneath thicker soils and suggest that this increase in clay content contributes to the increase in parent material competence observed beneath thicker soils. This increase in competence corresponds to a decrease in soil production rates at the same sites and we speculate that quantifying clay content more completely will lead to a more definitive connection.

Processes transporting the soil play an important role in connecting the incoming precipitation with the underlying parent material; namely, we have used short-lived fallout-derived isotopes to quantify the physical mixing processes that are active across these landscapes. At all the sites where we have made these measurements, we have documented a transition from dominant physical mixing in the soil column to a dominance of overland flow with increasing distance downslope. This transition from mixing to overland flow corresponds to the transition from thin to thick soils because all of our field sites have increasing soil depths downslope. The increase in overland flow processes downslope may mean that the parent material beneath thicker soils is in contact with soil water that has traveled farther through soil and is, therefore, not as able to weather as the water reacting with the parent material beneath the thinner, upslope soils. While this suggestion is speculative at best, it presents a very clear link between hillslope hydrological processes acting across these landscapes and the physical and chemical processes that I have summarized in this chapter. If we can make all the observations presented in this chapter on landscapes where the hillslope hydrology is well known, or vice versa, there will be excellent opportunities to test this speculation and lead to a truly comprehensive understanding of how upland landscapes evolve.

ACKNOWLEDGMENTS

This chapter reflects the enormous amount of work done by my graduate students over the last decade. Jim Kaste helped hone and develop the short-lived isotope methodology applied here to quantifying soil transport processes and rates. Ben Burke helped initiate the studies connecting soil production rates and processes with saprolite weathering. Jean Dixon took all these methods further and integrated them with extensive work on quantifying the chemical weathering signals on upland landscapes. Elizabeth Johnson and Joel Beyersdorfer lifted the heavy cone penetrometer across many hillslopes and contributed to the strength studies. Recent application of all of these methods in the San Gabriel Mountains benefited greatly from the work of Roman DiBiase, Matt Jungers, and Josh Landis. My sincere thanks to all of them.

REFERENCES

Ahnert, F., 1967. The role of the equilibrium concept in the interpretation of landforms of fluvial erosion and deposition. In: Macar, P. (Ed.), L'evolution Des Versants. University of Liege, Liege, pp. 23–41.

Ahnert, F., 1976. Brief description of a comprehensive three-dimensional process-response model of landform development. Zeitschrift fur Geomorphologie, Supp. Band 25, 29–49.

Amundson, R., Richter, D.D., Humphreys, G.S., Jobbagy, E.G., Gaillardet, J., 2007. Coupling between biota and earth materials in the critical zone. Elements 3 (5), 327–332.

Anderson, R.S., Humphrey, N.F., 1989. Interaction of weathering and transport processes in the evolution of arid landscapes. In: Cross, T.A. (Ed.), Quantitative Dynamic Stratigraphy. Prentice-Hall, Englewood Cliffs, N.J, pp. 349–361.

Anderson, R.S., 1994. Evolution of the santa cruz mountains, California, through tectonic growth and geomorphic decay. J. Geophys. Res. Solid Earth 99 (B10), 20161–20179.

Anderson, R.S., 2002. Modeling the tor-dotted crests, bedrock edges, and parabolic profiles of high alpine surfaces of the Wind River Range, Wyoming. Geomorphology 46, 35–58.

Anderson, S.P., Dietrich, W.E., 2001. Chemical weathering and runoff chemistry in a steep headwater catchment. Hydrolog. Process. 15 (10), 1791–1815.

Anderson, S.P., Blum, J.D., Brantley, S.L., Chadwick, O., Chorover, J., Derry, L.A., Drever, J.I., Hering, J.G., Kirchner, J.W., Kump, L.R., Richter, D.D., White, A.F., 2004. Proposed initiative would study earth's weathering engine. EOS Trans. Am. Geophys. Union 85 (28), 265–269.

Anderson, S.P., von Blanckenburg, F., White, A.F., 2007. Physical and chemical controls on the critical zone. Elements 3 (5), 315–319.

Appleby, P.G., Oldfield, F., 1992. Application of lead-210 to sedimentation studies. In: Ivanovich, M., Harmon, R.S. (Eds.), Uranium-series Disequilibrium. Oxford University Press.

Bachmann, J., Contreras, K., Hartge, K.H., MacDonald, R., 2006. Comparison of soil strength data obtained in situ with penetrometer and with vane shear test. Soil Tillage Res. 87, 112–118.

Balco, G., Stone, J.O., Lifton, N.A., Dunai, T.J., 2008. A complete and easily accessible means of calculating surface exposure ages or erosion rates from ^{10}Be and ^{26}Al measurements. Quaternary Geochronology 3, 174–194.

Berg, B., 2000. Litter decomposition and organic matter turnover in northern forest soils. Forest Ecol. Manag. 133, 13–22.

Bierman, P.R., 2004. Rock to sediment – slope to sea with Be-10 – rates of landscape change. Annu. Rev. Earth Planet. Sci. 32, 215–255.

Black, T.A., Montgomery, D.R., 1991. Sediment transport by burrowing animals, Marin county, California. Earth Surf. Process. Landforms 16, 163–172.

Bonniwell, E.C., Matisoff, G., Whiting, P.J., 1999. Determining the times and distances of particle transit in a mountain stream using fallout radionuclides. Geomorphology 27, 75–92.

Brantley, S.L., Goldhaber, M.B., Ragnarsdottir, K.V., 2007. Crossing disciplines and scales to understand the critical zone. Elements 3 (5), 307–314.

Braun, J., Heimsath, A.M., Chappell, J., 2001. Sediment transport mechanisms on soil-mantled hillslopes. Geology 29, 683–686.

Brimhall, G.H., Dietrich, W.E., 1987. Constitutive mass balance relations between chemical composition, volume, density, porosity, and strain in metasomatic hydrochemical systems: results on weathering and pedogenesis. Geochim. Cosmochim. Acta 51, 567–587.

Brimhall, G.H., Chadwick, O.A., Lewis, C.J., Compston, W., Williams, I.S., Danti, K.J., Dietrich, W.E., Power, M.E., Hendricks, D., Bratt, J., 1992. Deformational mass transport and invasive processes in soil evolution. Science 255, 695–702.

Brown, R.B., Cutshall, N.H., Kling, G.F., 1981a. Agricultural erosion indicated by Cs-137 redistribution 1. Levels and distribution of Cs-137 activity in soils. Soil Sci. Soc. Am. J. 45 (6), 1184–1190.

Brown, R.B., Kling, G.F., Cutshall, N.H., 1981b. Agricultural erosion indicated by Cs-137 redistribution. 2. Estimates of erosion rates. Soil Sci. Soc. Am. J. 45 (6), 1191–1197.

Burke, B.C., Heimsath, A.M., White, A.F., 2007. Coupling chemical weathering with soil production across soil-mantled landscapes. Earth Surf. Process. Landforms 32, 853–873.

Burke, B.C., Heimsath, A.M., Dixon, J.L., Chappell, J., Yoo, K., 2009. Weathering the escarpment: chemical and physical rates and processes, South-Eastern Australia. Earth Surf. Process. Landforms 34, 768–785.

Burkins, D.L., Blum, J.D., Brown, K., Reynolds, R.C., Erel, Y., 1999. Chemistry and mineralogy of a granitic, glacial soil chronosequence, Sierra Nevada Mountains, California. Chem. Geol. 162 (1), 1–14.

Buss, H.L., Brantley, S.L., Sak, P.B., White, A.F., 2004. Mineral dissolution at the grantie-saprolite interface. In: Wanty, R.B., Seal, R.R.I. (Eds.), 11th International Symposium on Water-Rock Interaction 11. Taylor and Francis, Saratoga Springs, NH, pp. 819–823.

Buss, H.L., Sak, P.B., Webb, S.M., Brantley, S.L., 2008. Weathering of the rio blanco quartz diorite, Luquillo Mountains, Puerto Rico: Coupling oxidation, dissolution, and fracturing. Geochim. Cosmochim. Acta 72, 4488–4507.

Byersdorfer, J.P., 2006. Correlating saprolite strength, soil production rate and chemical weathering in granitic terrain: Point Reyes, CA and Nunnock River, Australia. MS Thesis, Dartmouth College, Department of Earth Sciences.

Cambray, R.S., Playford, K., Lewis, F.N.J., Carpenter, R.C., 1989. Radioactive Fallout in Air and Rain: Results to the End of 1988. UK Atomic Energy Authority, Report no. AERE-R 10155, Environ. Med. Sci. Div. Harwell Lab, Oxfordshire.

Carson, M.A., Kirkby, M.J., 1972. Hillslope Form and Process. Cambridge University Press, New York, p. 475.

Cox, N.J., 1980. On the relationship between bedrock lowering and regolith thickness. Earth Surf. Process. Landforms 5, 271–274.

Currie, W.S., Aber, J.D., 1997. Modeling leaching as a decomposition process in humid montane forests. Ecology 78 (6), 1844–1860.

Dahlgren, R.A., Boettinger, J.L., Huntington, G.L., Amundson, R.G., 1997. Soil development along an elevational transect in the western Sierra Nevada, California. Geoderma 78, 207–236.

DeMaster, D.J., Cochran, J.K., 1982. Particle mixing rates in deep-sea sediments determined from excess ^{210}Pb and ^{32}Si profiles. Earth Planet. Sci. Lett. 61, 257–271.

DiBiase, R.A., Whipple, K.X., Heimsath, A.M., Ouimet, W.B., 2010. Landscape form and millennial erosion rates in the San Gabriel Mountains, CA. Earth Planet. Sci. Lett. 289, 134–144.

Dietrich, W.E., Perron, J.T., 2006. The search for a topographic signature of life. Nature 439, 411–418.

Dietrich, W.E., et al., 2003. Geomorphic transport laws for predicting landscape form and dynamics. In: Wilcock, P., Iverson, R. (Eds.), Prediction in Geomorphology. American Geophysical Union, Washington, D.C, pp. 103–132.

Dietrich, W.E., Reiss, R., Hsu, M.-L., Montgomery, D.R., 1995. A process-based model for colluvial soil depth and shallow landsliding using digital elevation data. Hydrolog. Process. 9, 383–400.

Dixon, J.L., Heimsath, A.M., Amundson, R., 2009a. The Critical Role of Climate and Saprolite Weathering in Landscape Evolution. Earth Surf. Process. Landforms 34, 1507–1521.

Dixon, J.L., Heimsath, A.M., Kaste, J., Amundson, R., 2009b. Climate-driven processes of hillslope weathering. Geology 37, 975–978.

Dörr, H., 1995. Application of Pb-210 in Soils. J. Paleolimnol. 13 (2), 157–168.

Fleming, R.W., Johnson, A.M., 1975. Rates of seasonal creep of silty clay soil. Q. J. Eng. Geol. 8, 1–29.

Fletcher, R.C., Buss, H.L., Brantley, S.L., 2006. A spheroidal weathering model coupling porewater chemistry to soil thicknesses during steady-state denudation. Earth Planet. Sci. Lett. 244, 444–457.

Fornes, W.L., Whiting, P.J., Wilson, C.G., Matisoff, G., 2005. ^{137}Cs-derived erosion rates in an agricultural setting: the effects of model assumptions and management practices. Earth Surf. Process. Landforms 30 (9), 1181–1189.

Frazier, C.S., Graham, R.C., 2000. Pedogenic transformation of fractured granitic bedrock, southern California. Soil Sci. Soc. Am. J. 64, 2057–2069.

Furbish, D.J., 2003. Using the dynamically coupled behavior of land-surface geometry and soil thickness in developing and testing hillslope evolution models. In: Wilcock, P., Iverson, R. (Eds.), Prediction in Geomorphology, vol. 135. American Geophysical Union, Washington, D.C, Geophysical Monograph, pp. 169–182.

Furbish, D.J., Haff, P.K., Dietrich, W.E., Heimsath, A.M., 2009. Statistical description of slope-dependent soil transport and the diffusion-like coefficient. J. Geophys. Res. 114, F00A05, doi: 10.1029/2009JF001267.

Gabet, E.J., 2000. Gopher bioturbation: field evidence for nonlinear hillslope diffusion. Earth Surf. Process. Landforms 25, 1419–1428.

Gabet, E.J., Reichman, O.J., Seabloom, E.W., 2003. The effects of bioturbation on soil processes and sediment transport. Ann. Rev. Earth Planet. Sci. 31, 249–273.

GEONOR, Instructions for Use, Inspection Vane Tester, H-60. GEONOR, INC, Milford, PA. Available at http://www.geonor.com/field_vane_testing.html

Gilbert, G.K., 1877. Report on the Geology of the Henry Mountains (Utah). United States Geological Survey, Washington, D.C.

Gilbert, G.K., 1909. The Convexity of Hilltops. Journal of Geology 17 (4), 344–350.

Green, E.G., Dietrich, W.E., Banfield, J.F., 2006. Quantification of chemical weathering rates across an actively eroding hillslope. Earth Planet. Sci. Lett. 242, 155–169.

Hack, J.T., 1960. The interpretation of erosional topography in humid temperate regions. Am. J. Sci. 258A, 80–97.

He, Q., Walling, D.E., 2003. Testing distributed soil erosion and sediment delivery models using Cs-137 measurements. Hydrolog. Process. 17 (5), 901–916.

He, Q., Walling, D.E., 1996. Interpreting particle size effects in the adsorption of Cs-137 and unsupported Pb-210 by mineral soils and sediments. J. Environ. Radioact. 30 (2), 117–137.

Heimsath, A.M., 2006. Eroding the land: steady-state and stochastic rates and processes through a cosmogenic lens. Geol. Soc. Am. Spec. Pap. 415, 111–129.

Heimsath, A.M., Chappell, J., Fifield, K., 2010. Eroding Australia: Rates and Processes from Bega Valley to Arnhem Land. Geological Society, London. Special Publications, 346: 225–241.

Heimsath, A.M., Chappell, J., Dietrich, W.E., Nishiizumi, K., Finkel, R.C., 2000. Soil production on a retreating escarpment in southeastern Australia. Geology 28, 787–790.

Heimsath, A.M., Chappell, J., Dietrich, W.E., Nishiizumi, K., Finkel, R.C., 2001a. Late Quaternary erosion in southeastern Australia: a field example using cosmogenic nuclides. Quaternary Int. 83–5, 169–185.

Heimsath, A.M., Chappell, J., Finkel, R.C., Fifield, L.K., Alimanovic, A., 2006. Escarpment erosion and landscape evolution in southeastern Australia. v. Penrose Conference Series. Geol. Soc. Am. Spec. Pap. 398, 173–190.

Heimsath, A.M., Chappell, J., Spooner, N.A., Questiaux, D., 2002. Creeping soil. Geology 30, 111–114.

Heimsath, A.M., Dietrich, W.E., Nishiizumi, K., Finkel, R.C., 1997. The soil production function and landscape equilibrium. Nature 388, 358–361.

Heimsath, A.M., Dietrich, W.E., Nishiizumi, K., Finkel, R.C., 1999. Cosmogenic nuclides, topography, and the spatial variation of soil depth. Geomorphology 27, 151–172.

Heimsath, A.M., Dietrich, W.E., Nishiizumi, K., Finkel, R.C., 2001b. Stochastic processes of soil production and transport: erosion rates, topographic variation, and cosmogenic nuclides in the Oregon coast range. Earth Surf. Process. Landforms 26, 531–552.

Heimsath, A.M., Furbish, D.J., Dietrich, W.E., 2005. The illusion of diffusion: field evidence for depth-dependent sediment transport. Geology 33, 949–952.

Heimsath, A.M., Hancock, G.R., Fink, D., 2009. The 'humped' soil production function: eroding Arnhem Land, Australia. Earth Surf. Process. Landforms 34, 1674–1684.

Heimsath, A.M., DiBiase, R.A., Whipple, K.X., 2012. Soil production limits and the transition to bedrock dominated landscapes. Nature Geoscience 5 (5), 210–214.

Herrick, J.E., Jones, T.L., 2002. A dynamic cone penetrometer for measuring soil penetration resistance. Soil Sci. Soc. Am. J. 66, 1320–1324.

Hole, F.D., 1981. Effects of animals on soil. Geoderma 25, 75–112.

Jiao, J.J., Wang, X.-S., Nandy, S., 2005. Confined groundwater zone and slope instability in weathered igneous rocks in Hong Kong. Eng. Geol. 80, 71–92.

Johnson-Maynard, J., Anderson, M.A., Green, S., Graham, R.C., 1994. Physical and hydraulic-properties of weathered granitic rock in Southern California. Soil Sci. 158 (5), 375–380.

Johnson, E.R., 2008. Rocks Matter: Exploring the Intersection of Lithology and Tectonics. MS Thesis, Dartmouth College Department of Earth Sciences.

Jones, D.P., Graham, R.C., 1993. Water-holding characteristics of weathered granitic rock in chaparral and forest ecosystems. Soil Sci. Soc. Am. J. 57 (1), 256–261.

Kaste, J.M., Heimsath, A.M., Bostick, B.C., 2007. Short-term soil mixing quantified with fallout radionuclides. Geology 35, 243–246.

Kaste, J.M., Heimsath, A.M., Hohmann, M., 2006. Quantifying sediment transport across an undisturbed prairie landscape using cesium-137 and high resolution topography. Geomorphology 76, 430–440.

Kew, G., Gilkes, R., 2006. Classification, strength and water retention characteristics of lateritic regolith. Geoderma 136, 184–198.

Lal, D., 1991. Cosmic ray labeling of erosion surfaces: in situ nuclide production rates and erosion models. Earth Planet. Sci. Lett. 104, 424–439.

Lamb, M.P., Scheingross, J.S., Amidon, W.H., Swanson, E., Limaye, A., 2011. A Model for Fire-Induced Sediment Yield by Dry Ravel in Steep Landscapes. J. Geophys. Res. 116, F03006, doi:10.1029/2010JF001878.

Lebedeva, M.I., Fletcher, R.C., Balashov, V.N., Brantley, S.L., 2007. A reactive diffusion model describing transformation of bedrock to saprolite. Chem. Geol. 244, 624–645.

Likens, G.E., Bormann, F.H., 1995. Biogeochemistry of a Forested Ecosystem. Springer-Verlag, New York, p. 160.

Lowrance, R., McIntyre, S., Lance, C., 1988. Erosion and deposition in a field forest system estimated using cesium-137 activity. J. Soil Water Conservat. 43 (2), 195–199.

Lutz, H.J., Griswold, F.S., 1939. The influence of tree roots on soil morphology. Am. J. Sci. 258, 389–400.

Lutz, H.J., 1940. Disturbance of forest soil resulting from the uprooting of trees. Bull. Yale Univ. Sch. Forest 45, 1–37.

Lutz, H.J., 1960. Movement of rocks by uprooting of forest trees. Am. J. Sci. 258, 752–756.

Matisoff, G., Bonniwell, E.C., Whiting, P.J., 2002a. Radionuclides as indicators of sediment transport in agricultural watersheds that drain to Lake Erie. J. Environ. Qual. 31 (1), 62–72.

Matisoff, G., Bonniwell, E.C., Whiting, P.J., 2002b. Soil erosion and sediment sources in an Ohio watershed using beryllium-7, cesium-137, and lead-210. J. Environ. Qual. 31 (1), 54–61.

Matisoff, G., Ketterer, M.E., Wilson, C.G., Layman, R., Whiting, P.J., 2001. Transport of rare earth element-tagged soil particles in response to thunderstorm runoff. Environ. Sci. Technol. 35 (16), 3356–3362.

Matsuoka, N., 1990. The rate of bedrock weathering by frost action: field measurements and a predictive model. Earth Surf. Process. Landforms 15, 73–90.

Matsuoka, N., Moriwaki, K., 1992. Frost heave and creep in the sor rondane mountains. Arctic Antarct. Alpine Res. 24, 271–280.

McKean, J.A., Dietrich, W.E., Finkel, R.C., Southon, J.R., Caffee, M.W., 1993. Quantification of soil production and downslope creep rates from cosmogenic ^{10}Be accumulations on a hillslope profile. Geology 21 (4), 343–346.

McKyes, E., 1989. Agricultural Engineering Soil Mechanics. In: Developments in Agricultural Engineering, vol. 10. Elsevier, Amsterdam, p. 292.

Minasny, B., McBratney, A.B., 1999. A rudimentary mechanistic model for soil production and landscape development. Geoderma 90 (1–2), 3–21.

Minasny, B., McBratney, A.B., 2001. A rudimentary mechanistic model for soil formation and landscape development II. A two-dimensional model incorporating chemical weathering. Geoderma 103, 161–179.

Mitchell, P., 1988. The influences of vegetation, animals and micro-organisms on soil processes. In: Viles, H.A. (Ed.), Biogeomorphology. Basil Blackwell, New York, pp. 43–82.

Montgomery, D.R., 2007. Soil erosion and agricultural sustainability. Proc. Natl. Acad. Sci. U S A 104 (33), 13268–13272.

Mudd, S.M., Furbish, D.J., 2006. Using chemical tracers in hillslope soils to estimate the importance of chemical denudation under conditions of downslope sediment transport. J. Geophys. Res. Earth Surf. 111 (F2).

Norman, S.A., Schaetzl, R.J., Small, T.W., 1995. Effects of slope angle on mass movements by tree uprooting. Geomorphology 14, 19–27.

O'Farrell, C.R., Heimsath, A.M., Kaste, J.M., 2007. Quantifying hillslope erosion rates and processes for a coastal California landscape over varying timescales. Earth Surf. Process. Landforms 32, 544–560.

Olsen, C.R., et al., 1985. Atmospheric fluxes and marsh-soil inventories of Be-7 and Pb-210. J. Geophys. Res. Atmos. 90 (ND6), 487–495.

Paton, T.R., Humphries, G.S., Mitchell, P.B., 1995. Soils: A New Global View. UCL Press Limited, London, p. 213.

Pelletier, J.D., Rasmussen, C., 2009. Quantifying the climatic and tectonic controls on hillslope steepness and erosion rate. Lithosphere 1, 73–80.

Porto, P., Walling, D.E., Tamburino, V., Callegari, G., 2003. Relating caesium-137 and soil loss from cultivated land. Catena 53 (4), 303–326.

Prosser, I.P., Dietrich, W.E., 1995. Field experiments on erosion by overland flow and their implication for a digital terrain model of channel initiation. Water Resour. Res. 31, 2867–2876.

Quine, T.A., Govers, G., Walling, D.E., Zhang, X.B., Desmet, P.J.J., Zhang, Y.S., Vandaele, K., 1997. Erosion processes and landform evolution on agricultural land – new perspectives from caesium-137 measurements and topographic-based erosion modelling. Earth Surf. Process. Landforms 22 (9), 799–816.

Reneau, S.L., Dietrich, W.E., 1990. Depositional history of hollows on steep hillslopes, coastal Oregon and Washington. Natl. Geogr. Res. 6 (2), 220–230.

Reneau, S.L., Dietrich, W.E., 1991. Erosion rates in the southern oregon coast range: evidence for an equilibrium between hillslope erosion and sediment yield. Earth Surf. Process. Landforms 16 (4), 307–322.

Riebe, C.S., Kirchner, J.W., Finkel, R.C., 2003a. Long-term rates of chemical weathering and physical erosion from cosmogenic nuclides and geochemical mass balance. Geochim. Cosmochim. Acta 67, 4411–4427.

Riebe, C.S., Kirchner, J.W., Finkel, R.C., 2003b. Sharp decrease in long-term chemical weathering rates along an altitudinal transect. Earth Planet. Sci. Lett. 6393, 1–14.

Riebe, C.S., Kirchner, J.W., Finkel, R.C., 2004. Erosional and climatic effects on long-term chemical weathering rates in granitic landscapes spanning diverse climate regimes. Earth Planet. Sci. Lett. 224, 547–562.

Riebe, C.S., Kirchner, J.W., Granger, D.E., Finkel, R.C., 2000. Erosional equilibrium and disequilibrium in the Sierra Nevada, inferred from cosmogenic Al and Be in alluvial sediment. Geology 28, 803–806.

Riebe, C.S., Kirchner, J.W., Granger, D.E., Finkel, R.C., 2001a. Minimal climatic control on erosion rates in the Sierra Nevada, California. Geology 29, 447–450.

Riebe, C.S., Kirchner, J.W., Granger, D.E., Finkel, R.C., 2001b. Strong tectonic and weak climatic control of long-term chemical weathering rates. Geology 29, 511–514.

Roels, J.M., 1985. Estimation of soil loss at a regional scale based on plot measurements – some critical considerations. Earth Surf. Process. 10, 587–595.

Roering, J.J., Almond, P., Tonkin, P., McKean, J., 2002. Soil transport driven by biological processes over millenial time scales. Geology 30 (12), 1115–1118.

Roering, J.J., Gerber, M., 2005. Fire and the evolution of steep, soil-mantled landscapes. Geology 33, 349–352.

Roering, J.J., Kirchner, J.W., Dietrich, W.E., 1999. Evidence for non-linear, diffusive sediment transport on hillslopes and implications for landscape morphology. Water Resour. Res. 35 (3), 853–870.

Roering, J.J., Kirchner, J.W., Sklar, L.S., Dietrich, W.E., 2001. Hillslope evolution by nonlinear creep and landsliding: an experimental study. Geology 29, 143–146.

Rosenbloom, N.A., Anderson, R.S., 1994. Hillslope and channel evolution in a marine terraced landscape, Santa Cruz, California. J. Geophys. Res. 99 (B7), 14,013–14,029.

Schaetzl, R.J., Follmer, L.R., 1990. Longevity of treethrow microtopography: implications for mass wasting. Geomorphology 3, 113–123.

Schoeneberger, P.J., Amoozegar, A., Buol, S.W., 1995. Physical property variation of a soil and saprolite continuum at 3 geomorphic positions. Soil Sci. Soc. Am. J. 59 (5), 1389–1397.

Selby, M.J., 1980. A rock mass strength classification for geomorphic purposes: with tests from Antarctica and New Zealand. Zeitschrift fur Geomorphologie N. F. 24 (1), 31–51.

Selby, M.J., 1985. Hillslope Materials and Processes. Oxford University Press, Oxford, England.

Simon, S.L., Bouville, A., Beck, H.L., 2004. The geographic distribution of radionuclide deposition across the continental US from atmospheric nuclear testing. J. Environ. Radioact. 74 (1–3), 91–105.

Steegen, A., Govers, G., Takken, I., Nachtergaele, J., Poesen, J., Merckx, R., 2001. Factors controlling sediment and phosphorus export from two Belgian agricultural catchments. J. Environ. Qual. 30 (4), 1249–1258.

Tyler, A.N., Carter, S., Davidson, D.A., Long, D.J., Tipping, R., 2001. The extent and significance of bioturbation on Cs-137 distributions in upland soils. Catena 43 (2), 81–99.

U.S. ERDA (United States Energy Research and Development Administration), 1977, Final Tabulation of Monthly 90Sr Fallout Data: 1954–1976: New York, Environmental Measurements Laboratory, Report HASL-329.

Viles, H., A., Goudie, A., Grab, S., Lalley, J., 2011. The use of the schmidt hammer and equotip for rock hardness assessment in geomorphology and heritage science: a comparative analysis. Earth Surf. Process. Landforms 36, 320–333.

Wallbrink, P.J., Murray, A.S., 1996. Distribution and variability of Be-7 in soils under different surface cover conditions and its potential for describing soil redistribution processes. Water Resour. Res. 32 (2), 467–476.

Walling, D.E., He, Q., 1999. Using fallout lead-210 measurements to estimate soil erosion on cultivated land. Soil Sci. Soc. Am. J. 63 (5), 1404–1412.

Walling, D.E., He, Q., Blake, W., 1999. Use of Be-7 and Cs-137 measurements to document short- and medium-term rates of water-induced soil erosion on agricultural land. Water Resour. Res. 35 (12), 3865–3874.

Walling, D.E., Russell, M.A., Hodgkinson, R.A., Zhang, Y., 2002. Establishing sediment budgets for two small lowland agricultural catchments in the UK. Catena 47 (4), 323–353.

West, A., Galy, A., Bickle, M., 2005. Tectonic and climatic controls on silicate weathering. Earth and Planetary Science Letters 235 (1–2), 211–228.

White, A.F., Brantley, S.L., 2003. The effect of time on the weathering of silicate minerals: why do weathering rates differ in the laboratory and field? Chem. Geol. 202 (3–4), 479–506.

White, A.F., Blum, A.E., Schulz, M.S., Bullen, T.D., Harden, J.W., Peterson, M.L., 1996. Chemical weathering rates of a soil chronosequence on granitic alluvium. 1. Quantification of mineralogical and surface area changes and calculation of primary silicate reaction rates. Geochim. Cosmochim. Acta 60 (14), 2533–2550.

Whiting, P.J., Bonniwell, E.C., Matisoff, G., 2001. Depth and areal extent of sheet and rill erosion based on radionuclides in soils and suspended sediment. Geology 29, 1131–1134.

Wilson, C.G., Matisoff, G., Whiting, P.J., 2003. Short-term erosion rates from a Be-7 inventory balance. Earth Surf. Process. Landforms 28 (9), 967–977.

Yoo, K., Amundson, R., Heimsath, A.M., Dietrich, W.E., Brimhall, G.H., 2007. Integration of geochemical mass balance with sediment transport to calculate rates of soil chemical weathering and transport on hillslopes. J. Geophys. Res. Earth Surf. 112 (F2).

Young, A., 1960. Soil movement by denudational processes on slopes. Nature 188, 120–122.

Young, A., 1963. Some field observations of slope form and regolith, and their relation to slope development. Trans. Inst. Br. Geogr. 32, 1–29.

Zhang, X.B., Quine, T.A., Walling, D.E., Li, Z., 1994. Application of the cesium-137 technique in a study of soil-erosion on gully slopes in a yuan area of the loess plateau near Xifeng, Gansu Province, China. Geogr. Ann. Phys. Geogr. 76 (1–2), 103–120.

Zimbone, S.M., Vickers, A., Morgan, R.P.C., Vella, P., 1996. Field investigations of different techniques for measuring surface soil shear strength. Soil Technol. 9 (1–2), 101–111.

Chapter 8

Thermodynamic Limits of the Critical Zone and their Relevance to Hydropedology

Axel Kleidon,[1,*] Erwin Zehe[2] and Henry Lin[3]

ABSTRACT

The physical, chemical, and biological transformations in the Earth's Critical Zone – from the atmosphere–vegetation interface down to the bottom of the aquifer – shape soils, vegetation, landscapes, and biogeochemical cycling. To sustain these transformations, free energy – the part of energy that can be used to perform physical, chemical, or biological work – needs to be generated and supplied continuously. This chapter provides an overview of the basis to characterize the Critical Zone as a thermodynamic system and illustrates the relevance of upper thermodynamic limits to Critical Zone processes. We first describe how the relevant forms of energy are formulated in thermodynamic terms and how the laws of thermodynamics restrict the direction and magnitude of these transformations. These principles are then illustrated by three examples of how thermodynamic limits restrict (1) turbulent exchange between the land surface and the atmosphere, (2) mechanical weathering from periodic heating and cooling of the ground, and (3) the transport of sediments by river flow. We then outline the relevance of these considerations to hydropedology in the context of the drivers for pedogenesis and the generation and maintenance of preferential flow in soils and landscapes. We conclude with a summary and a brief prospectus for future work relating to thermodynamic limits and optimality related to these limits within the Critical Zone and its applications in hydropedology.

1. INTRODUCTION

The Earth's Critical Zone (CZ) comprises the part of the continents from the bedrock of the upper lithosphere to the lower boundary of the atmosphere, and

[1]Max-Planck-Institute for Biogeochemistry, Hans-Knöll-Str. 10, 07745 Jena, Germany
[2]Karlsruhe Institute of Technology, Karlsruhe, Germany
[3]Dept. of Ecosystem Science and Management, Penn State University, University Park, PA, USA
[*]Corresponding author: Email: akleidon@bgc-jena.mpg.de

Hydropedology, Edited by H. Lin. DOI: 10.1016/B978-0-12-386941-8.00008-3

includes soils, vegetation, landscapes, and aquifers (NRC, 2001). It is the place where a myriad of processes add, transform, and remove energy and materials. The processes within the CZ can be categorized broadly into physical, chemical, and biological processes. Physical processes include the physical breakdown of rocks into smaller pieces and the mechanical removal of materials due to erosion and sediment transport. The chemical transformations include the weathering of primary rock minerals into dissolved ions in the soil water and the formation of secondary minerals. Biotic processes include the productivity of the terrestrial vegetation, which adds organic material to soils, microbial activity within the soil, which transforms material within the CZ, and fauna activities that further transform organic and inorganic materials within the CZ.

What these transformations have in common is that they require free energy to make them happen. Free energy is energy that can be used to perform work (in engineering, this is sometimes referred to as "exergy," but here we stay with the terminology used in physics). Work is needed to lift sediments and accelerate them within streams, to break the molecular bonds of rock minerals to create smaller grains, to chemically dissolve rock minerals, and so on. The rate at which work is performed is described by the physical term "power." It requires the generation of free energy, or power, to maintain these transformations, and hence the activity within the CZ. This is where thermodynamics becomes relevant, as it provides the framework to quantify generation and transfer rates of free energy and the fundamental limits to these.

The purpose of this chapter is to provide an overview of thermodynamics and how it can be applied to characterize and quantify the functioning of the CZ. Thermodynamics has long been used for particular processes in the CZ, but its foundation is mostly hidden in the well-established concepts of soil hydrology (e.g. matric potential) and geochemistry (e.g. affinity). The basis of thermodynamics is, however, general and can be applied to practically all types of processes (e.g. Kondepudi and Prigogine, 1998). This thermodynamic basis allows us to express CZ processes in terms of how they transform energy of different forms and it sets the rules and limits for conversions of one form of energy to another. Hence, thermodynamics provides a unifying framework to consistently describe not only the details at the process level, but also how these processes interact within the CZ as well as with the environment at large at the systems level.

When we want to describe the CZ as a thermodynamic system, an important difference to classical thermodynamics is that we deal with a system that is driven and maintained in a state far from thermodynamic equilibrium (Lin, 2010a,b, 2011; Rasmussen et al., 2011). This disequilibrium state is reflected by the sustained presence of gradients in thermodynamic variables, such as temperature, gravitational potential, or affinity, which allow for the generation of free energy to fuel CZ activities. These gradients can only be maintained through external forcing. If we were to "switch off" the external forcing, then processes would deplete gradients, and the system would eventually evolve toward a uniform state of thermodynamic equilibrium in which transformations would no longer exist. The exchange fluxes of the CZ with the Earth system at large thus play a vital role to understand the activity and limits on CZ processes and reflect that the CZ is a thermodynamic system that is maintained far from equilibrium.

This chapter is organized around a series of questions. The first question deals with somewhat technical aspects of how the CZ is described as a thermodynamic system (i.e. a clearly delineated space of the natural world for which the interior dynamics as well as the exchange to the surroundings is described by the physical theory of thermodynamics). The next question concerns how the laws of thermodynamics are relevant for providing direction and limits to processes in the CZ. This is followed by three specific examples that illustrate the relevance of these thermodynamic limits. Then, we discuss how thermodynamics may help us gain a better understanding of some current topics in hydropedology, specifically how the intriguing concept of entropy may relate to order and structure within the CZ and how thermodynamics may improve the characterization of pedogenesis and preferential flow. We conclude with a brief summary and outlook on future applications.

2. HOW IS THE CRITICAL ZONE DESCRIBED AS A THERMODYNAMIC SYSTEM?

To describe the CZ as a thermodynamic system, we need to delineate the boundary of the system CZ before we describe different types of energy involved in CZ processes. The exchange fluxes across the system boundary are then formulated in terms of how these affect different forms of energy within the CZ.

2.1. Delineating the Boundaries of the Critical Zone

The delineation of the system's boundary is in principle arbitrary, but important, as it defines the place at which the exchange fluxes take place. Figure 1 shows a schematic diagram of the upper lithosphere, the land surface, the ocean, and the lower atmosphere, and the delineation of the boundary of the CZ that we use here. The upper boundary is drawn just above the surface of the soil and the vegetative cover at the interface to the lower atmosphere above the canopy. The lower boundary is at the lower end of the regolith at the interface with groundwater or weathering front. This system boundary corresponds approximately to the definition of the CZ (NRC, 2001).

Note that these boundaries are not necessarily easily drawn in the field because of their highly heterogeneous nature. In addition, these boundaries are not static, as continental uplift and erosion affect the elevation, and changes in vegetative cover can affect the height and location of the boundary to the atmosphere. The lower boundary is drawn at bedrock, so that groundwater is

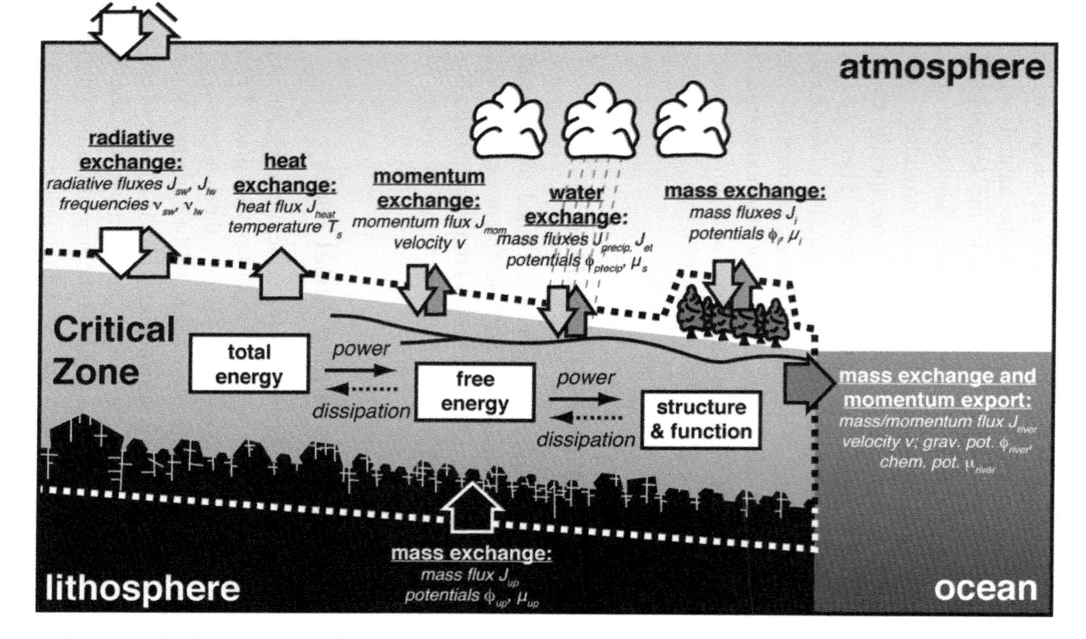

FIGURE 1 Schematic diagram of the Critical Zone (CZ) as a thermodynamic system: the system boundary (dotted lines), the characterization of the total energy within the CZ, and the exchange fluxes across the boundary that alter the total energy of the CZ. By expressing the exchange in terms of conjugate variables (see Table 1), these are consistently formulated as changes of different forms of energy. Radiative exchange is described by radiative fluxes (J_{sw}, J_{lw}) in terms of photon fluxes (dN/dt) and their respective frequencies υ. Heat fluxes J_{heat} directly describe the change in thermal energy, and the respective temperature describing the associated entropy exchange. The heat flux from the interior is excluded here due to its small value compared to the surface heat fluxes. Momentum exchange is expressed by a momentum flux $J_{mom} = dp/dt$ (which can be interpreted as a force) and the respective velocity v. Mass exchanges are expressed in terms of mass fluxes $J_{mass} = dm/dt$ (e.g. precipitation J_{precip}, evapotranspiration J_{et}, river discharge J_{river}, and continental uplift J_{up}) and either their gravitational potential φ or their chemical potentials μ. Gradients in the individual contributions to the total energy drive the generation of free energy (work performed in time, or power) and the dynamics of the CZ, which in turn relate to formation and maintenance of structure and function within the CZ.

considered as part of the system. The lateral boundaries are nontrivial due to spatial heterogeneity at different scales (Blöschl and Sivapalan, 1995), but could be drawn, e.g. at the boundaries of catchments, basins, or continents, depending on the scope of evaluation.

2.2. Forms of Energy of the Critical Zone

In thermodynamics, the sum of all forms of energy of a system is referred to as the total energy of the system, U. The total energy includes thermal energy (or heat, such as the heat stored in the soil profile), geopotential energy associated with topographic height, kinetic energy (such as water flow), binding energies (such as those found in rock minerals or soil water bound in the soil matrix), and chemical energies (such as biomass, organic matter, and different minerals). Note that this definition of the total energy is somewhat different from the common definition of the internal energy of a thermodynamic system, which typically excludes the macroscopic contributions of kinetic energy to the total energy of the system. As these macroscopic contributions play important roles for CZ processes, we need to account for all forms of energy, but this leads to some differences in the formulations compared to classical thermodynamics.

In thermodynamics, these different forms of energy are expressed as pairs of conjugate variables (Table 1). A pair of conjugate variables consists of one intensive variable that is independent of the size of the system (i.e. independent of its volume and/or mass) and an extensive variable that depends on the size of the system. Heat is described by temperature T (the intensive variable) and entropy S (the extensive variable). The thermodynamic entropy S is a material property that describes how much energy is dispersed among different microstates, which can be formalized in more precise terms in the context of statistical mechanics. For instance, the water molecule in vapor can move, vibrate, or rotate more freely and has therefore many more modes, or microstates, to store energy than a water molecule that is bound in a liquid or solid. Hence, the specific entropy (i.e. entropy per unit mass) of water vapor is higher than that of liquid or solid water. Potential energy is described as the product of the gravitational potential φ ($= g \cdot z$, with g being the gravitational acceleration and z being topographic height) and mass m. Kinetic energy is described by the velocity v and the respective momentum p ($= m \cdot v$). Binding energies are formulated in terms of the chemical potential μ and mass m, where the chemical potential is defined as the change of binding energy content (U_{BE}) with respect to the change in mass, i.e. $\mu = \partial U_{BE}/\partial m$. Chemical potentials can be used to describe a variety of energy forms, including the energy in molecular bonds in solids, capillary and adhesive forces (although these can also be formulated more specifically in terms of surface tension and surface area), and partial pressures of compounds. Unsaturated or supersaturated air can, for instance, be expressed in terms of a chemical potential, with the relationship

TABLE 1 **Different Types of Energy Relevant to the Critical Zone, their Characterization in Terms of Pairs of Conjugate Variables, and Illustrative Examples. The Product of the Intensive Variable (Independent of System Size) and the Corresponding Extensive Variable (Dependent of System Size) Yields the Expression for the Associate form of Energy**

Form of energy	Intensive variable	Extensive variable	Example
Radiative energy	Photon energy $h\,\upsilon$	Number of photons N	Solar and terrestrial radiation
Heat (thermal energy)	Temperature T	Entropy S	Ground heat storage
Kinetic energy	Velocity v	Momentum p	Wind, eolian transport, river flow
Potential energy	Gravitational potential φ	Mass m	Topography
Binding energy	Chemical potential μ	Mass m	Soil water, minerals, osmotic potential, vapor pressure deficit
Chemical energy	Affinity A	Extent of reaction ξ	Biomass, soil organic matter
Electrical energy	Charge Q	Voltage V	Charged soil particles

$\mu = R_v\, T \ln RH$, with R_v being the gas constant for water vapor and RH being the relative humidity of the air. Note that binding energies are typically negative, as it requires energy to "unbind" mass. The chemical energy associated with chemical compounds is formulated in terms of the affinity A and the extent of the chemical reaction ξ. These quantities relate directly to the chemical potentials and concentrations of the related compounds except for being constrained by the stoichiometry of the reaction. Other forms of energy, e.g. electric energy or magnetic energy, can be formulated equivalently, are relevant, e.g. for estimating the role of charged stationary soil particles, but are not considered here.

Note that the common pair of pressure–volume ($p\ V$) is not listed in Table 1 because it refers mostly to a form of work, rather than to a particular form of energy. This is illustrated by the following example: When the air near the surface is heated, e.g. by absorption of radiation, air expands and performs $p\ V$ work. This work is typically associated with an increase in the gravitational potential of the air, that is, some of the heating $d(T\ S) > 0$ results in an accompanying change $d(\varphi\, m) > 0$ through $p\, V$ work. Hence, in terms of actual changes in energy, the $p\, V$ work term is already accounted for in the change in potential energy. For other processes, e.g. deformation processes in the soil such as shrinking–swelling, $p\, V$ work is relevant as well although it needs to be carefully considered whether the associated change in energy is already accounted for by another form of energy.

A change in the total energy of the CZ is then expressed in terms of changes associated with each form of energy, i.e.

$$dU = d(T\,S) + \sum_i d(\varphi_i m_i) + \sum_i d(p_i v_i) + \sum_i d(\mu_i m_i) + \sum_i d(A_i \zeta_i) + \ldots, \tag{1}$$

where i stands for different compounds (e.g. air, vapor, water, sediment, different minerals, and organics) and ... represents additional forms of energy. Changes in different contributions can occur due to exchange of energy or mass with the surroundings or due to internal dynamics. Examples for a change in energy associated with exchange fluxes is the absorption of radiation, which adds heat ($d(T\,S) > 0$), so that $dU > 0$, and evaporation, which removes water vapor from the system ($d(\mu_i\ m_i) < 0$), thereby decreasing the overall total energy ($dU < 0$). Examples of energy changes associated with internal dynamics is the generation of kinetic energy associated with surface runoff, which increases the kinetic energy ($d(v_i\, p_i) > 0$) at the expense of the potential energy ($d(\varphi_i\, m_i) < 0$), and frictional dissipation of water flow, which releases heat ($d(T\,S) > 0$) at the expense of kinetic energy ($d(v_i\, p_i) < 0$). Both examples of internal dynamics leave the total energy of the system unaffected (i.e. $dU = 0$).

What is not included in this sum is radiant energy, that is, energy associated with radiative exchange before it is converted to heat by absorption or from heat by emission. For abiotic processes, the radiative flux mostly acts as a heat source; that is, it becomes relevant once radiation is absorbed and converted into heat, or when heat is removed by the emission of radiation. Absorbed radiation appears as a term $d(T\,S) > 0$, while emission of radiation enters the sum as a term $d(T\,S) < 0$. Photosynthesis – and photochemistry in general – does not operate on heat, but uses radiative energy directly to generate chemical energy. By this, photosynthesis is able to generate chemical free energy differently from solar radiation compared to heat engines (the way by which most abiotic processes generate forms of free energy). In principle, this can be formulated using thermodynamics as well, but a specific treatment goes beyond the scope of this chapter as it would involve the detailed description of photochemical processes. Most processes in the following deal with conversions of heat and forms of energy that were originally derived from heat, except when we deal with biotic activity, which we include in the above formulation merely as a process that generates chemical free energy in the form of $d(A\,\xi) > 0$.

2.3. Exchange Fluxes of the Critical Zone

The exchange fluxes and the respective conjugate variables at which the exchange takes place are shown in Fig. 1. A complete formulation of the exchange between the CZ and other Earth systems in terms of conjugate variables is necessary in order to quantify how these exchange processes provide or remove energy for the activity within the CZ. Estimates of the associated magnitudes are provided in Section 5.

With the atmosphere, the CZ exchanges radiation of different frequencies, momentum, heat, water in different phases, solids (dust and aerosols), and trace gases. The exchange of radiation at the top of the CZ consists of incoming solar radiation with a relatively high frequency, and terrestrial radiation of relatively low frequencies, which is exchanged in both directions. Once absorbed or emitted, it adds (or removes) heat ($d(T\,S) > 0$ and $d(T\,S) < 0$, respectively). The momentum exchange ($d(v_{Air}\,p_{Air}) > 0$) with the atmosphere relates to the velocity gradient between the boundary layer of the atmosphere and the land surface, which causes turbulent exchange of heat ($d(T\,S)$, with the sign depending on the temperature gradient) and mass. Hydrological fluxes exchange potential and chemical energy between the atmosphere and the CZ. Precipitation adds water at a certain height to the CZ, thereby adding potential energy to the CZ ($d(\varphi_{H_2O}\,m_{H_2O}) > 0$). As precipitated water was desalinated when evaporated from the ocean, that adds chemical free energy to the precipitated water ($d(\mu_{H_2O,Liquid}\,m_{H_2O}) > 0$) that can be used to dissolve minerals. Evapotranspiration removes water from the CZ ($d(\mu_{Vapor}\,m_{Vapor}) < 0$). The momentum flux at the boundary to the atmosphere results in a force acting on the surface yielding surface stress, which is utilized for the work needed to lift and export solids, such as aerosol and dust, to the atmosphere ($d(\varphi_{Solid}\,m_{Solid}) < 0$). Trace gas exchange concerns particularly the exchange of CO_2 and O_2 that is associated with photosynthesis and drives most of the biotic activity within the CZ. The exchange is described in the form of chemical energy, i.e. $d(\mu_{CO_2} m_{CO_2})$ and $d(\mu_{O_2} m_{O_2})$, and relates mostly to the partial pressures of the compounds within the atmosphere. The exchange can proceed in either direction, depending on the particular gradients in partial pressure at the boundary.

With the ocean, the CZ exchanges mostly through continental discharge of river flow and the dissolved and suspended material in the discharge. Continental discharge exports potential and kinetic energy of the flowing water ($d(\varphi_{H_2O}\,m_{H_2O}) < 0$ and $d(v_{H_2O}\,p_{H_2O}) < 0$), and also the potential, kinetic, and binding energies of the suspended and dissolved material.

The exchange of the CZ with the lithosphere is mostly through the uplift of continental mass, which adds potential energy and, indirectly, mass to the CZ ($d(\varphi_{Rock}\,m_{Rock}) > 0$). Meanwhile, the leaching of soil water and recharge of groundwater reduce potential, chemical, and thermal energies from the CZ to the interface with the lithosphere.

Given the annual mean magnitude of solar radiative heating in the order of 240 W m^{-2}, the heat exchange with the lithosphere in the order of <0.1 W m^{-2} and the ocean can usually be neglected.

3. HOW DOES THERMODYNAMICS IMPOSE DIRECTION AND LIMITS ON THE DYNAMICS OF THE CRITICAL ZONE?

With the boundary, boundary fluxes, and the total energy content of the CZ now being described in thermodynamic terms, the critical question is how thermodynamics can help us to better understand the dynamics within the CZ. In the following, we first review the first and second laws of thermodynamics, how these apply to systems maintained in disequilibrium, and how these act as constraints to the direction and the extent of energy conversions.

3.1. The First and Second Laws of Thermodynamics in Closed Systems

The first law of thermodynamics states that the change in total energy $\mathrm{d}U$ in a closed system (a system that only exchanges energy, but no mass) is balanced by the net addition of heat $\mathrm{d}Q$ and the work done by the system $\mathrm{d}W$:

$$\mathrm{d}U = \mathrm{d}Q - \mathrm{d}W, \tag{2}$$

with the convention that work is positive when work is performed by the system, and negative when work is performed on the system.

The second law makes a statement about the direction into which processes evolve. In classical thermodynamics, a change in entropy $\mathrm{d}S$ is defined by the change that occurs when a certain amount of heat $\mathrm{d}Q$ is added or removed from the equilibrium system at a certain temperature T:

$$\mathrm{d}S = \mathrm{d}Q/T. \tag{3}$$

The second law then states that the total entropy of an isolated system, one that does not exchange energy or mass, can only increase, i.e.

$$\mathrm{d}S \geq 0. \tag{4}$$

The implication of the second law is illustrated by the following thought experiment. Consider two large reservoirs of heat within a system that are at different temperatures, T_h and T_c, with $T_h > T_c$. The two reservoirs as a whole form an isolated system that does not exchange heat or mass with the surroundings. If a certain amount of heat $\mathrm{d}Q$ is removed from the first, warmer reservoir, then the entropy of this reservoir is decreased by $\mathrm{d}S_h = -\mathrm{d}Q/T_h$. When this heat is added to the second, colder reservoir and is fully mixed, then the entropy of the second system is increased by $dS_c = dQ/T_c$. The total entropy of whole system then changes by

$$dS_{total} = dS_h + dS_c = dQ(1/T_c - 1/T_h) = dQ(T_h - T_c)/T_h T_c. \quad (5)$$

Since $T_h > T_c$, $dS_{total} > 0$, the entropy of the whole system is increased by the exchange of heat dQ. In other words, the flow of heat from the warmer reservoir to the colder reservoir is a manifestation of the second law. If the heat were to flow in the opposite direction, it would result in a negative, total change in entropy and thereby violate the second law. This is why entropy in the second law is also referred to as the arrow of time (Eddington 1928), as it dictates the temporal direction into which processes proceed. The magnitude of entropy production is also a direct measure for the intensity by which a thermodynamic gradient is destroyed, i.e. for the irreversibility of the process. In the following, we refer to the entropy produced by a process as the entropy production, σ, e.g. by heat transport.

3.2. The First and Second Laws of Thermodynamics for Open Systems

When we consider the first law for a system in which several forms of energy contribute to the total energy U of the system, then the first law implies that a change in one form of energy needs to be compensated by a change in another form of energy or by the exchange of energy and mass with the surroundings. When, for instance, we consider the changes in heat content ($T\,S$) in time, then the statement of the first law reads as

$$d(TS)/dt = dQ/dt - (P - D), \quad (6)$$

where the change in heat content $d(TS)$ is balanced by the net heating $J_{heat,net} = dQ/dt$ (an exchange flux with the surroundings) minus the net amount of work performed by heat through time ($P-D$). The net amount of work is represented by two contributions: the work performed by some form of heat engine, i.e. a process that draws work out of a heating gradient to generate free energy with a power $P = dW/dt$, and the heating resulting from the dissipation D of free energy, e.g. due to heat loss generated by friction.

To illustrate this, let us consider the addition of water by precipitation at some height within a catchment. The depletion of the potential energy of the surface water along a slope $d(\varphi m) < 0$ provides the power to generate kinetic energy associated with the runoff $d(pv) > 0$. In other words, a gradient in gravitational potential is depleted to generate a gradient in momentum. Once set into motion, friction dissipates this kinetic energy, i.e. $d(pv) < 0$, by generating dissipative heating $d(TS) > 0$. Here, the depletion of the momentum gradient generates a gradient in heating. Since essentially all processes within the CZ can be formulated in the form of conjugate variables, the dynamics are all about conversions between different forms of energy. The first law then simply provides an accounting rule to ensure that no energy is lost.

The second law imposes constraints and a direction on the energy conversions. Before it can be formulated, we need to first consider the entropy balance of the system to account for the entropy that is produced by irreversible processes within the system as well as the entropy exchange with the surroundings associated with the exchange fluxes of heat, mass, and other forms. A change in the total entropy of the system dS_{total}/dt is then balanced by the total entropy produced within the system and the net exchange of entropy $J_{entropy,net}$ associated with energy and mass exchanges with the surroundings:

$$dS_{total}/dt = \sum_i \sigma_i + J_{entropy,net}, \quad (7)$$

where σ_i is the entropy produced by different, irreversible processes within the system. In this expression, the second law translates into the requirement that $\sigma_i \geq 0$. Hence, the second law provides a general direction into which processes within a system proceed.

3.3. The Second Law and Directions for Processes within the Critical Zone

Because many of the processes within the CZ are not necessarily driven by temperature gradients (e.g. water fluxes within the soil are driven by gradients in soil water potential), it is not directly apparent that these processes are ultimately driven by heating gradients and that they end up in heat, and thus governed by the first and second law of thermodynamics. At the very large, planetary scale, practically all processes are driven by solar radiation and by the depletion of the interior heat content that is produced by radioactive decay and crystallization of the core in the Earth's interior. Solar radiation is the radiation emitted at the very high surface temperature of the Sun of about 5800 K and is then partially absorbed by the Earth system. The resulting heat is transformed through the myriad of Earth system processes and emitted back to space, but at a much lower temperature of about 255 K on average. This direction of downgrading radiation from the low entropy associated with the high emission temperature of the Sun with higher frequencies to the much higher entropy associated with lower radiative temperature of the Earth with much lower frequencies reflects the direction imposed by the second law. Likewise, the high temperatures of the Earth's interior compared to the moderate temperature at the Earth's surface provide the temperature gradients to fuel geophysical, geochemical, and geobiological processes of the interior Earth and the lithosphere. Ultimately, these are the thermodynamic drivers of the Earth system dynamics, and it is the depletion of these gradients that reflects the direction governed by the second law. Even photosynthesis, which is not directly driven by temperature gradients but rather by the electronic absorption of visible radiation that is nevertheless associated with the high emission temperature of the Sun, follows this direction prescribed by the second law.

Processes within the CZ also reflect this direction prescribed by the second law. For some processes that are directly driven by heating differences, this direction is easy to see. For instance, the ground heat flux is driven by temperature gradients within the soil profile. As shown by the example in Section 3.1, such a heat flux produces entropy, that is, it increases the entropy of the system by leveraging out the difference in temperature. Likewise, the turbulent transfer of sensible heat follows and depletes temperature gradients between the soil surface and the lower atmosphere.

With other processes in the CZ, the immediate direction imposed by the second law is not quite so apparent. For example, precipitation does not seem to be directly related to a heating gradient. To see the reflection of the second law for these processes, we need to follow through the chain of energy conversions that start from a heating gradient, and is eventually transformed into the form of free energy of interest (e.g. the potential energy associated with surface water at a certain elevation fed by precipitation) until it ultimately ends up in a form of dissipative heating. This is not apparent when we focus on the CZ without the context of how the Earth system at large generates free energy and transfers parts of it into the CZ. This is illustrated in Fig. 2. The potential free energy imported into the CZ by precipitation originates from hydrologic cycling. The hydrologic cycle is largely driven by atmospheric motion, which is driven by differences in radiative heating. So the influx of free energy associated with precipitation originates from radiative heating gradients. Likewise, the topographic gradients within the CZ originate from the potential energy generated by continental uplift. This is fueled by the replenishment of the continental crust and lifting through plate tectonics, which is driven by heating gradient between the Earth's interior and surface. The second law is also reflected in the direction of the chemical transformations from primary minerals of parent rocks to soil minerals during soil formation. In their seminal work, Volubuyev and Ponomarev (1977) and Volubuyev (1983) identified consistent trends toward lower Gibbs free energy and higher specific entropies during pedogenesis (Fig. 3).

FIGURE 2 Physical means by which the planetary forcings of solar radiation and Earth interior heating result in free energy input to the Critical Zone (CZ) by a series of energy conversions. The upper part of the diagram shows how radiative heating gradients are converted by the atmospheric heat engine into atmospheric motion. Large-scale motion in turn acts to dehumidify the atmosphere, which provides the free energy associated with hydrologic cycling in the form of potential and chemical free energy that is imported into the CZ in form of precipitation. The lower part of the diagram shows how heating gradients in the Earth interior – caused by secular cooling and radioactive decay – maintain motion in the form of mantle convection and plate tectonics. This motion is needed to sustain the uplift of continental crust, which imports free energy into the CZ associated with the topography and geological materials. The details of the CZ dynamics are shown in Fig. 1. *(From Kleidon (2010), and Dyke et al., 2011).*

Two examples are given here to show how the free energy imported into the CZ feeds irreversible processes that convert free energy into heat, thereby unavoidably producing entropy. The first example considers the conversion of motion into heat. Once motion is generated, e.g. as river flow, its associated

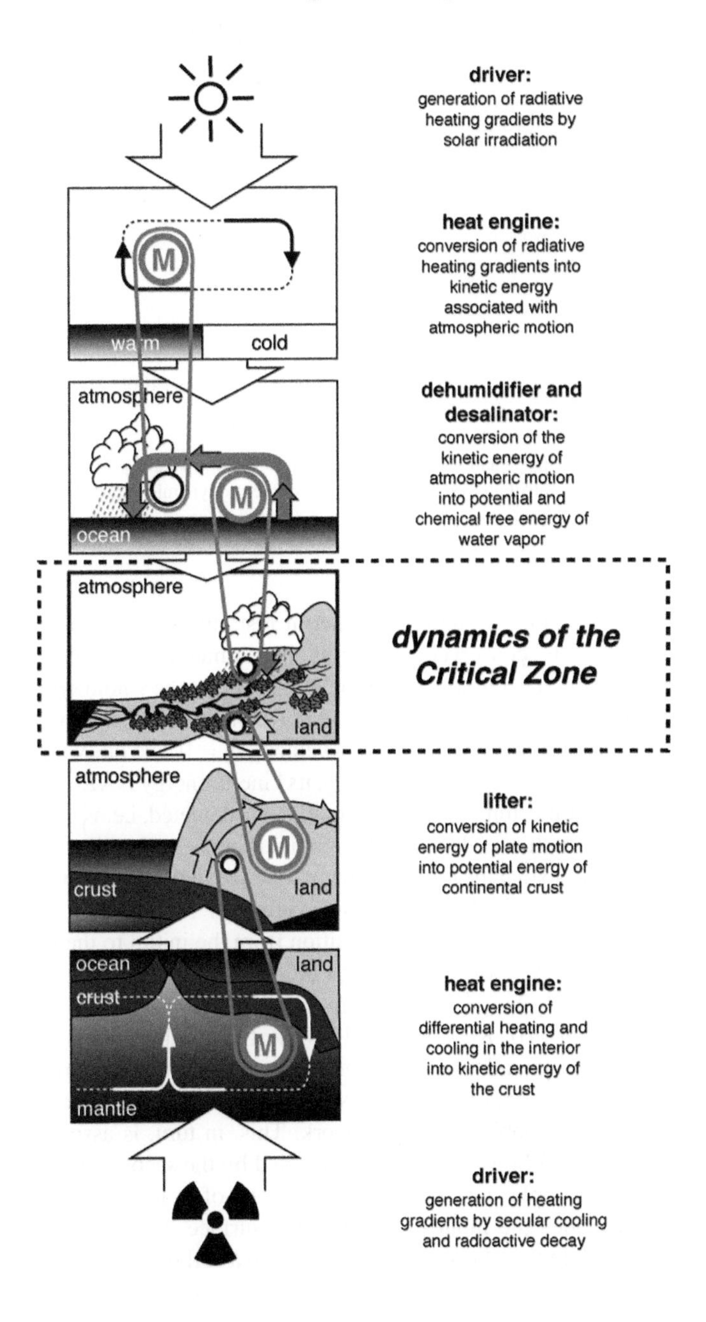

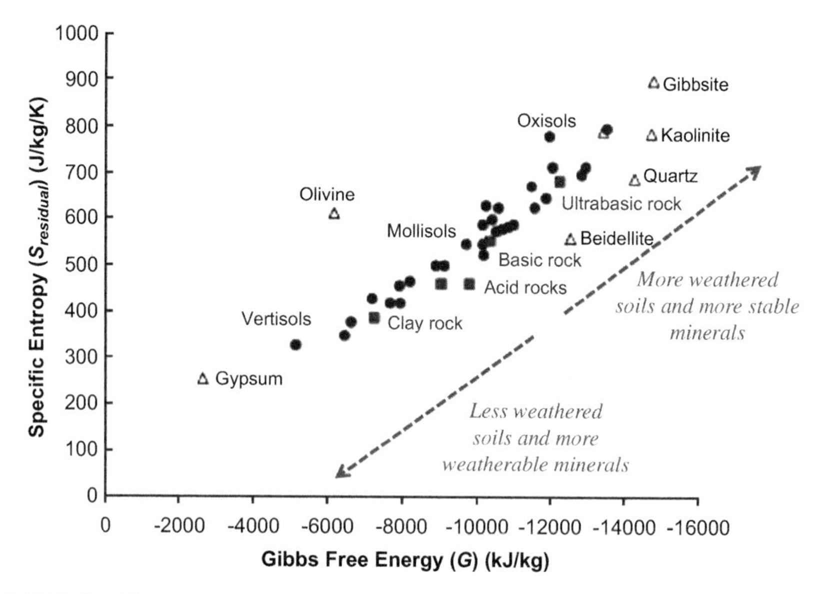

FIGURE 3 Gibbs free energy and specific entropy of weathering residuals using mineralogical composition and total chemical analysis based on standard thermodynamic data for different soils, minerals, and rocks *(Modified from Minasny et al., 2008, based on the data of Volobuyev and Ponomarev, 1977, Volobuyev et al., 1980, and Yatsu, 1988). (Color version online)*

momentum will spread, transfers it to water at rest, and takes sediment along. While momentum is conserved, gradients in momentum are depleted by momentum diffusion (i.e. the spread of momentum across the whole system), and so is the kinetic energy. This can easily be seen when we consider two reservoirs of water, one with mass m_1 and at rest with a velocity $v_1 = 0$, and one with mass m_2 and in motion with a velocity $v_2 > 0$. Initially, its kinetic energy is $KE = 1/2\, m_2\, v_2^2$. In a final state of equilibrium, the velocity gradient is depleted, i.e. $v_1' = v_2'$ (with the prime referring to this final state). Momentum is conserved while the velocity gradient is depleted, so that the initial momentum, $m_2\, v_2$, needs to balance the final momentum, $m_1 v_1' + m_2 v_2'$. This yields an expression for the final velocities, $v_1' = v_2' = m_2/(m_1 + m_2)v_2$. In the transition from the initial to the final state, the kinetic energy of the system decreased from KE to $KE' = 1/2\, m_1 v_1'^2 + 1/2\, m_2 v_2'^2 = m_2/(m_1 + m_2)KE$, with the difference $KE{-}KE' = m_1/(m_1 + m_2)\ KE$ dissipated as heat. In other words, as fluid dynamical processes spread momentum within the fluid and to its surroundings, the generated free energy is inevitably dissipated into heat, that is, in a form of energy that is less available to perform work. This, in turn, is associated with a higher entropy and follows the direction dictated by the second law.

The second example considers the movement of water within the soil. Precipitation provides the input of water at the surface. With this input, free energy is provided, in that the water is not bound to a soil matrix (i.e. it does not have negative binding energy), it has potential energy associated with the elevation of the surface above the sea level, and it is desalinated. This disequilibrium with the mineral composition of the crust provides chemical free energy that is exploited in the chemical weathering of soils. As water infiltrates the soil, its potential energy is depleted. This depletion proceeds faster when it gains some kinetic energy associated with flow (which will eventually be dissipated into heat by friction). When water starts binding to the soil matrix, it gains (negative) binding energy, which is balanced by the release of heat of wetting. This binding energy is described by the well-known matric potential of soil water. Diffusive redistribution of water along gradients of chemical (matric) and gravitational potential within the soil is then directed toward lowering its total energy toward a minimum. As the sum of the binding and potential energy is reduced, the surplus energy is released as heat. Depending on the relative wetness of the soil, the reduction of the total energy may be achieved either by a tighter binding to the soil matrix – this would reduce the total energy by further increasing the (negative) binding energy – or by draining into deeper soil depths – this would reduce the total energy by further lowering its potential energy. A state of local equilibrium is reached when the sum of energies is minimized for a given mass of soil water, which corresponds to the absence of gradients in the total soil water potential. Hence, the downward flux of water into the soil involves the depletion of the free energy associated with the potential energy of unbound rainwater to a state of lower potential and stronger (negative) binding energies and the release of heat.

The upward transfer of soil water associated with evaporation (or transpiration) also follows thermodynamic gradients and the direction imposed by the second law. The process of evaporation requires heat that is needed for the phase transition of water from liquid to gaseous form. This heat is required to break up the binding energy between water molecules (mostly in the form of hydrogen bridges) to convert liquid water into water vapor and, to a much lesser extent, to overcome the binding energy of water to the soil matrix. The driver for this upward transport is, however, not the heat that is involved in the phase transition, but rather found in the unsaturated state of the atmosphere. To understand how free energy is associated with unsaturated air, we first note that the atmospheric vapor content is in a state of thermodynamic equilibrium with an open, free water surface (i.e. the shape of water that involves the least amount of binding energy associated with surface tension) when it is saturated. It is in this state in which the combined free energy of heat and the intermolecular binding energy among the two phases of water is minimized. No further gradient exists that could drive the conversion of one phase into another and thereby deplete gradients between heat and binding energy (in thermodynamics, this is referred to as the Gibbs–Duhem relationship). When we now consider the evaporation of a certain amount of water $\mathrm{d}m$, it involves the increase in the binding energy of bound water to free water by $\mu_s \, \mathrm{d}m$, the reduction of the heat content of the surface by $L \, \mathrm{d}m$ (with L being the heat of vaporization), and the increase of energy associated with the addition of water to the atmosphere by $d(\mu_a\, m_a) = d\mu_a\, m_a + \mu_a\, dm_a$. The overall changes in entropy involve the decrease of entropy of the surface by the removal of heat $L\, dm/T_s$, the decrease of material entropy of the soil $S_l\, dm$ (with S_l being the molar entropy of liquid water), and the increase of entropy of the atmosphere by $S_{v,a}\, dm$ (with $S_{v,a}$ being the molar entropy of water vapor at the relative humidity *RH* of the air). Through the evaporation of d*m*, the overall entropy is increased by $dS = S_{v,a}dm - S_l dm - Ldm/T_s$. If we view the evaporation process in two steps (with $dS = dS_1 + dS_2$), with the first step consisting of a reversible phase transition from liquid to vapor at the surface (with $dS_1 = 0$), and a second step involving the irreversible mixing of the released vapor with the unsaturated air of the atmosphere ($dS_2 > 0$), we then gain the following expressions for the two contributions: For the reversible phase change, we obtain $dS_1 = S_{v,s}dm - S_l dm - Ldm/T_s = 0$, which yields the "trivial" insight that the latent heat of vaporization is given by the differences in molar entropies times the temperature at which the phase transition occurs (this is how the heat of vaporization is defined). The second term yields the entropy associated with the mixing of the near saturated air at the surface with the unsaturated air above: $dS_2 = (S_{v,s} - S_{v,a})dm = -R_v \ln(RH)dm$. Note that $dS_2 > 0$ since $\ln(RH) < 0$. Another way to look at this increase in entropy due to water vapor mixing is to interpret this increase in entropy to reflect the free expansion of water vapor from saturation to the vapor pressure of the unsaturated air. Hence, the upward transport of water from the soil to the atmosphere follows the thermodynamic gradients that are mostly associated with binding energies, and overall results in the production of entropy as required by the second law.

Both these examples highlight how transformations of free energy by hydrologic fluxes into the CZ depend on the strength of the hydrologic cycle. It is the joint action of the hydrologic cycle to lift moisture to a certain elevation above sea level that adds potential energy to the CZ, and the dehumidification of the air that provides a thermodynamic gradient and hence free energy to drive upward hydrologic fluxes within the CZ. The strength of both of these aspects is set by the strength of the atmospheric heat engine. Hence, we need to consider the CZ in the context of the functioning of the Earth system at large to fully understand and appreciate the direction that the second law provides for processes within the CZ.

3.4. The Second Law and Maximum Power Limits

In addition to setting a direction, the second law also imposes the upper bound on how much work can be performed, and thereby on the generation rate of free energy. In classical thermodynamics, this limit is known as the Carnot limit of a perfect heat engine. The starting point for the derivation of the Carnot limit is to ask what the maximum value of dW is in Eqn (2) that is compatible with the second law. To derive this limit, imagine a heat engine that is driven by a heat flux J_{in} that is added at a temperature T_{in}, it rejects a heat flux J_{out} at a temperature T_{out}, and draws a certain amount of power $P = dW/dt$ from it. To derive an upper limit that serves as a "ceiling" for evolutionary dynamics, we assume a steady state in which the heat content of the system does not change in time, i.e. $dT/dt = 0$ and $dS/dt = 0$. The first law then tells us that

$$J_{in} = J_{out} + P. \tag{8}$$

In steady state in which the entropy no longer changes in time, i.e. $dS/dt = 0$, the entropy balance of this engine consists of the import of entropy by the heating of J_{in}/T_{in}, the export of entropy by the cooling of J_{out}/T_{out}, and the entropy produced within the engine σ:

$$J_{in}/T_{in} - J_{out}/T_{out} + \sigma = 0. \tag{9}$$

The power drawn from the engine is not associated with an entropy flux as it constitutes the generation of free energy (when this free energy is dissipated in steady state, i.e. $P = D$, it produces entropy at a rate of P/T). The power (i.e. the generation of free energy) is constrained by Eqn (9), because the second law dictates that $\sigma \geq 0$. The best case is obtained when $\sigma = 0$, i.e. when there are no dissipative processes taking place within the heat engine. Then, we can use the entropy balance Eqn (9) to express J_{out} in terms of J_{in} and the temperatures (with $J_{out} = J_{in}\ T_{out}/T_{in}$), and use the first law to get the upper limit on $P\ (= J_{in} - J_{out})$ as

$$P = J_{in} - J_{out} \leq J_{in}\ (1 - T_{out}/T_{in}) = J_{in}(T_{in} - T_{out})/T_{in}. \tag{10}$$

Since the Carnot efficiency $\eta = (T_{in} - T_{out})/T_{in}$ at the surface is in the order of 30 K/300 K $\approx$ 10% or less, the typical generation rates of free energy of about 10 W m^{-2} or less in the climatological mean (see Table 2) are much smaller than the terms within the climatological surface energy balance (e.g. in comparison to the mean solar radiative surface heating of 160 W m^{-2}; Kiehl and Trenberth, 1997) even though power is decisive for driving the dynamics of the system.

When we consider processes within the CZ, we often cannot make the assumption that there is no entropy production within the engine and we cannot assume that the driving gradient remains fixed (Kleidon, 2010). For instance, the generation of motion in the atmospheric boundary layer, driven by the gradient in heating between the land surface and the atmosphere aloft, competes with the depletion of this gradient by radiative exchange. We can derive an equivalent expression as the Carnot limit if we assume that the entropy production within the engine is a process that is driven by the temperature gradient $(T_{in} - T_{out})$ in a linear way. We slightly rewrite the first law as a heat balance:

$$J_{in} = J_d + J_{ex}, \tag{11}$$

where $J_d = k_d\ (T_{in} - T_{out})$ is an irreversible process within the engine with a diffusion coefficient k_d, and J_{ex} is the heat flux utilized by the engine to generate work. When we apply the Carnot limit merely to the heat flux J_{ex}, we get the extractable power P_{ex}:

$$P_{ex} = J_{ex}\ (T_{in} - T_{out})/T_{in} = J_{ex}(J_{in} - J_{ex})/(k_d T_{in}), \quad (12)$$

when we use the heat balance and the expression for J_d to express the temperature difference in terms of J_{in} and J_{ex}. This expression has a maximum value for an optimum heat flux $J_{ex} = J_d = 1/2\ J_{in}$ of

$$P_{ex,max} = J_{in}^2/(4k_d T_{in}). \quad (13)$$

To get an idea for what this expression means, let us consider the Carnot limit applied to the heat flux J_{in}. Then, we would get the total power of the system to be

$$P_{tot} = J_{in}(T_{in} - T_{out})/T_{in} = J_{in}^2/(k_d T_{in}), \quad (14)$$

which is four times greater than the limit of $P_{ex,max} = 1/4\ P_{tot}$ established above and expressed by Eqn (13). Hence, the actual upper limit on free energy generation in the presence of competing processes is actually much lower than the Carnot limit. Some examples on the application of this to processes within the CZ are given below.

Note that this maximum power limit is essentially equivalent to the well-established maximum power principle in electrical engineering. This limit establishes the maximum electrical power that can be drawn from a generator. The correspondence is established by noting that the involved conjugate variables are charge Q (with the electric current $I = dQ/dt$ corresponding to heat fluxes J) and voltage difference U (corresponding to the temperature difference $T_{in} - T_{out}$).

It is also relevant to note that the generation of free energy and the associated dynamics within the system are able to deplete the driving gradient faster than in their absence. This can be seen when the driving temperature gradient, $T_{in} - T_{out}$, is formulated by using the expression for J_d and the first law, which yields

$$T_{in} - T_{out} = (J_{in} - J_{ex})/k_d. \quad (15)$$

This gradient decreases with greater J_{ex}. Hence, the generation of free energy and the associated dynamics deplete the temperature gradient faster, thereby producing entropy at a higher rate. Given that the maximum power state sets the limit to the strongest dynamics as it involves the strongest generation rate (and dissipation rate) of free energy in steady state, it implies that the gradient is depleted as much as thermodynamically possible.

Hence, the maximum power limit extends the implications of the second law in that dynamics do not just follow the direction by the second law, but that they may proceed in this direction at the fastest possible pace. Interactions and feedbacks play a central role in this maximization, and are reflected in the faster depletion of the driving gradient and the resulting low, but maximized, efficiency of free energy generation. The theoretical basis for the maximum power limit may be found as a special case of the proposed general principle of maximum entropy production (e.g. Dewar 2005, Kleidon et al. 2010, Dewar 2010).

4. HOW ARE THERMODYNAMIC LIMITS RELEVANT TO PROCESSES OF THE CRITICAL ZONE?

To illustrate the relevance of thermodynamic limits to the activity of the CZ, we use three simple models that illustrate these limits for turbulent exchange between the land surface and the atmosphere, for mechanical weathering derived from periodic heating and cooling of the land, and for work derived from river flow to drive sediment transport. The setup of the models as well as the results of these are shown in Fig. 4 and described in the following. Each example closes with a simple estimate of the maximum power that can be derived from the particular forcing for a typical setting. The relevance of these estimates is then discussed further in a broader context in Section 5.

4.1. Maximum Power Limit for Turbulent Exchange with the Atmosphere

The first example deals with the turbulent exchange of mass through the upper boundary of the CZ into the atmosphere (Fig. 4a). This turbulent exchange is associated with the generation of kinetic energy that allows for the transport of sensible and latent heat, as well as trace gas exchange and eolian transport from the surface into the atmosphere. The primary driver for this turbulent exchange is the gradient in heat that is generated by the heating of the land surface by the absorption of solar radiation and the radiative cooling of the atmosphere aloft. The gradient in heat $d(T\ S)$ between the surface and the atmosphere is utilized to generate mechanical work that is then reflected in a gradient in kinetic energy $d(p\ v)$.

It is easy to show that a maximum exists for the power that can be derived from this heating gradient. We consider the local surface energy balance in simplified form in steady state (i.e. $dT/dt = 0$):

$$0 = J_{sw} - J_{lw} - J_{ex}, \tag{16}$$

where J_{sw} is the heating of the surface by the absorption of solar radiation, $J_{lw} = k_r (T_s - T_a)$ is the linearized net emission of terrestrial radiation with a radiative coefficient k_r and depends on the temperatures of the surface and the atmosphere, T_s and T_a, respectively. The heat flux J_{ex} is the flux utilized

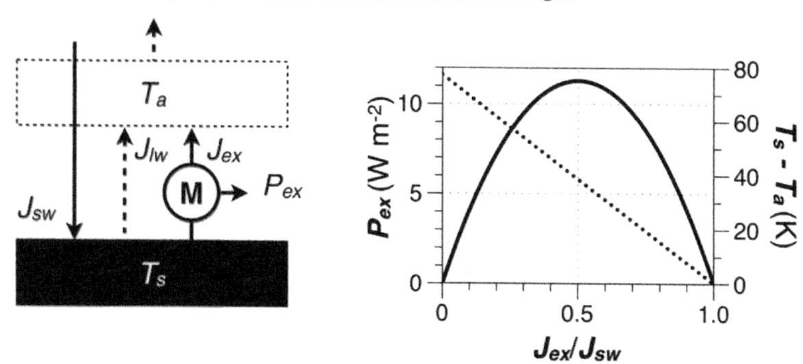
(a) maximum power for turbulent exchange
Ta
Jlw
Jex
M
Pex
Jsw
Ts
Pex (W m-2)
Ts - Ta (K)
Jex/Jsw
0
5
10
0
0.5
1.0
0
20
40
60
80

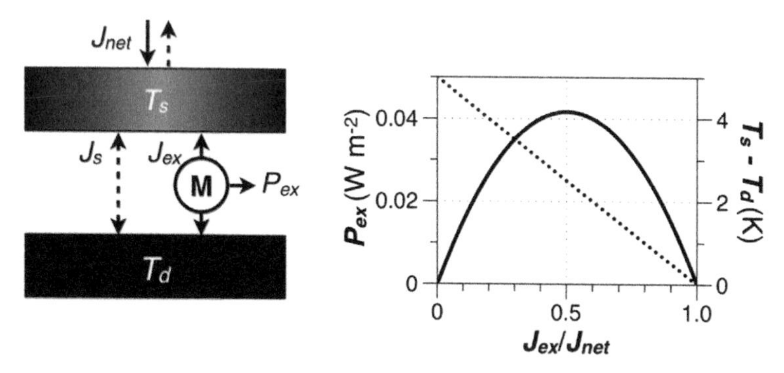
(b) maximum power for mechanical weathering
Jnet
Ts
Js
Jex
M
Pex
Td
Pex (W m-2)
Ts - Td (K)
Jex/Jnet
0
0.02
0.04
0
0.5
1.0
0
2
4

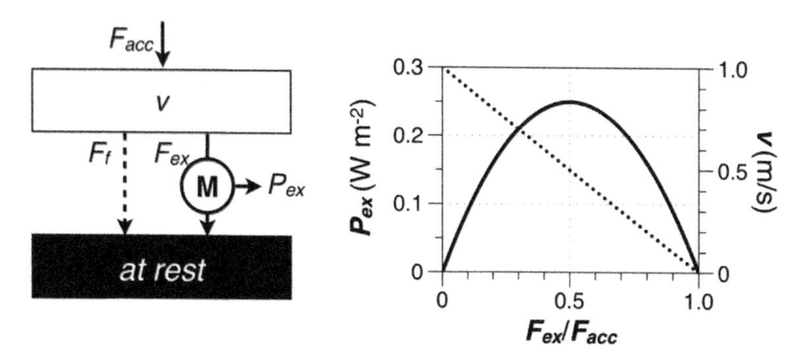
(c) maximum power and sediment transport
Facc
v
Ff
Fex
M
Pex
at rest
Pex (W m-2)
v (m/s)
Fex/Facc
0
0.1
0.2
0.3
0
0.5
1.0
0
0.5
1.0

by an imagined heat engine that generates the work for motion. The maximum power P_{ex} that can be generated by this engine is given by the Carnot limit:

$$P_{ex} = J_{ex}(T_s - T_a)/T_s. \qquad (17)$$

When we use the surface energy balance to express the temperature gradient $(T_s - T_a)$ in terms of the heat fluxes (J_{sw} and J_{ex}), we obtain

$$P_{ex} = J_{ex}(J_{sw} - J_{ex})/(k_r T_s). \qquad (18)$$

This expression is quadratic in J_{ex} and has a maximum value (when we neglect the slight dependence of T_s on the fluxes in the denominator). By $dP_{ex}/dJ_{ex} = 0$ we obtain the optimum heat flux as $J_{ex,opt} = J_{lw} = 1/2\, J_{sw}$, which yields a maximum power of

$$P_{ex,max} = J_{sw}^2/(4k_r\, T_s) = 1/4\, P_{tot}(J_{ex} = 0), \qquad (19)$$

where $P_{tot}(J_{ex} = 0)$ is the potential power in the case of radiative equilibrium.

Given a mean surface solar forcing of $J_{sw} = 168\ \mathrm{W\,m^{-2}}$, the optimum turbulent heat flux yielding maximum power is $J_{ex,opt} = 85\ \mathrm{W\,m^{-2}}$, which compares fairly well to the estimated global mean flux of 99 $\mathrm{W\,m^{-2}}$ (Kiehl and Trenberth 1997) when also considering that the contribution of forced convection by large-scale wind fields and variations in the greenhouse effect

FIGURE 4 Maximum power limits for (a) turbulent exchange between the land surface and the atmosphere, (b) mechanical weathering derived from periodic heating and cooling of the land, and (c) mechanical work derived from water flow to drive sediment transport. In the diagrams, the "M" marks an "engine" that generates free energy out of a gradient. In the graphs, the solid lines show power and the dotted lines show the respective driving gradient as a function of the extraction flux in relation to the input. The first example (a) shows how mechanical power is derived from the differential heating between the surface (with a temperature T_s) and the atmosphere (with temperature T_a) by solar radiation J_{sw}. The resulting temperature gradient is depleted by longwave radiative transfer J_{lw} and by the extraction of heat J_{ex} for the heat engine. With increasing heat extraction J_{ex}, the temperature gradient $(T_s - T_a)$ is depleted, thereby restricting the maximum possible power to intermediate values and resulting in a low maximum efficiency of free energy generation. The second example (b) deals with the maximum power that can be extracted within the ground from periodic net heating J_{net} at the surface. Here, the temperature gradient forms between the surface (with T_s) and the deep soil (with T_d) and is depleted by heat conduction J_s and heat extraction J_{ex}. The third example (c) deals with the maximum power that can be extracted from a momentum gradient between a fluid moving at velocity v and a surface at rest. The momentum gradient is generated by an accelerating force F_{acc} (or a momentum influx from the surroundings) and depleted by friction F_f and momentum extraction. Despite the difference in processes and gradients, all examples here show similar maximum power limits. The similarity arises because the extraction of power needs to be directed to enhance the depletion of the driving gradient, as dictated by the second law. This results in a fundamental trade-off, with a greater flux resulting in a smaller driving gradient, hence resulting in a maximum power state.

that would affect the value of k_r are not included in this maximum estimate. With an empirical value of $k_r = 2\,\mathrm{W\,m^{-2}\,K^{-1}}$ and a typical temperature of $T_s = 300\,\mathrm{K}$, the maximum power to drive convective motion is in the order of $10\,\mathrm{W\,m^{-2}}$.

4.2. Maximum Power Limit for Mechanical Weathering

The second example relates to the maximum mechanical work that can be derived from periodic heating and cooling of the land and that could be used, for instance, to physically weather rocks (Fig. 4b). We consider a periodic heating flux with a period τ of the form $J_{\mathrm{net}}(t) = \Delta J_{\mathrm{net}} \sin 2\pi t/\tau$, where ΔJ_{net} is the amplitude of the forcing and where we exclude the mean forcing. We neglect the changes in heat content at the surface, and balance this heat flux with the ground heat flux J_s and the heat flux J_{ex} extracted to derive work:

$$0 = J_{\mathrm{net}}(t) - J_s - J_{\mathrm{ex}} = \Delta J_{\mathrm{net}} \sin 2\pi t/\tau - k_s(T_s - T_d) - J_{\mathrm{ex}}. \quad (20)$$

The energy balance can be used to express the temperature gradient ΔT within the soil as a function of the net heating, the soil thermal conductivity k_s, and the heat flux J_{ex}:

$$\Delta T(t) = T_s - T_d = (\Delta J_{\mathrm{net}} \sin 2\pi t/\tau - J_{\mathrm{ex}})/k_s. \quad (21)$$

This temperature gradient within the soil is used to derive the work, so that the extracted power is given by

$$P_{\mathrm{ex}}(t) = J_{\mathrm{ex}} \Delta T(t)/T_s = J_{\mathrm{ex}}(\Delta J_{\mathrm{net}} \sin 2\pi t/\tau - J_{\mathrm{ex}})/(k_s T_s). \quad (22)$$

This expression is essentially equivalent to the above example, and has a maximum value of

$$P_{\mathrm{ex,max}}(t) = J_{\mathrm{net}}(t)^2/(4k_s T_s) \quad (23)$$

for an optimum heat flux $J_{\mathrm{ex,opt}}(t) = 1/2\,\Delta J_{\mathrm{net}} \sin 2\pi t/\tau$.

When these expressions are averaged over the time period τ, they yield a mean flux of $J_{\mathrm{ex,opt,avg}} = 0$ and a mean power of $P_{\mathrm{ex,max,avg}} = \Delta J_{\mathrm{net}}^2/(8k_s T_s)$. With a thermal conductivity in the order of $1\,\mathrm{W\,m^{-1}\,K^{-1}}$, a vertical distance in the order of 1 m, and $T_s \approx 300\,\mathrm{K}$, this expression informs us that only a fraction of the heating amplitude can be extracted for work, and the greater the amplitude, the disproportionally greater the maximum work that can be extracted. What this means is that for an amplitude of radiative heating in the order of $\Delta J_{\mathrm{net}} = 10\,\mathrm{W\,m^{-2}}$, about $0.04\,\mathrm{W\,m^{-2}}$ of work can be performed. This work depends strongly on this amplitude, and this, in turn, is consistent with the common, empirical findings that mechanical weathering rates are the highest in environments with greater variations in temperature (Peltier 1950).

4.3. Maximum Power Limit for Sediment Transport

The third example we use here concerns the maximum power that can be drawn out of river flow to drive sediment transport (Fig. 4c). Here, the potential energy of runoff $d(m\ \varphi)$ is in part converted into kinetic energy of water flow $d(p\ v)$ by an accelerating force F_{acc}, and we ask how much of this kinetic energy can be utilized to perform work on the sediment. We start with the momentum balance of water flow in steady state (i.e. $dp/dt = 0$):

$$0 = F_{acc} - F_f - F_{ex}, \tag{24}$$

where $F_f = k\ v$ is a linearized friction force (for simplicity we consider the linear form, an equivalent maximum power state can also be derived for other forms, e.g. quadratic) and F_{ex} is the force that extracts momentum from the flow to work on the sediment.

The power that is drawn by the force F_{ex} is given by

$$P_{ex} = F_{ex} v. \tag{25}$$

Similar to the above, we can now use the momentum balance to replace v in terms of the forces F_{acc} and F_{ex}:

$$P_{ex} = F_{ex}(F_{acc} - F_{ex})/k. \tag{26}$$

Again, we get an expression that is quadratic in F_{ex} and has a maximum value. This value is obtained by $dP_{ex}/dF_{ex} = 0$ to yield an optimum drag force of $F_{ex} = F_{fric} = 1/2\ F_{acc}$ and a maximum power expression of

$$P_{ex,max} = F_{acc}^2/(4k) = 1/4 P_{tot}(F_{ex} = 0). \tag{27}$$

What this expression tells us is that – as before – only a fraction of the stream power contained in the water flow can maximally be converted to perform work on sediment transport. It also tells us that work on the sediment results in a reduced flow velocity. This is directly evident from Eqn (24), which yields $v = (F_{acc} - F_{ex})/k$, i.e. v decreases with increasing values of F_{ex}. This upper limit naturally does not tell us how this maximum is achieved by the flow, and leaves out many of the specifics, such as threshold processes or the mode of sediment transport (suspension, saltation, or creep). A key to understand how this maximum power limit can be achieved is to recognize that since the driving force for runoff is proportional to slope, the power that can be extracted from runoff depends on the square of the slope. This nonlinear dependence of the extracted power on the slope is a central ingredient to understand why preferential flow structures are energetically favored, as we will elaborate in Section 5.4.

5. HOW DOES THERMODYNAMICS RELATE TO STRUCTURE AND FUNCTION OF THE CRITICAL ZONE?

In this section, we first provide a first-order estimate of the different sources of free energy that is made available to the processes within the CZ. We then make the connection between the performed work and the presence of structures and give three examples to illustrate how thermodynamics could help better understand the structure, function, and evolution of the CZ.

5.1. Simple Estimates for Free Energy Sources for Critical Zone Processes

To get an impression of the relative magnitude by which different processes deliver and generate free energy for the activity of CZ processes, a simple and global estimate is provided in Table 2. These estimates are all given as climatological mean values in units of W, that is, rates of free energy generation per unit time. These estimates are given here merely for providing an order of magnitude, whereas the actual numbers are quite uncertain. Note that all generation rates of free energy listed in Table 2 are in the order of $<1–2\ W\ m^{-2}$, which is much smaller than the typical order of magnitude associated with *heat* fluxes, such as radiative heating, and fluxes of sensible and latent heat, which are typically in the order of $100\ W\ m^{-2}$.

The first line in Table 2 describes the generation of chemical free energy from solar radiation by photosynthesis. This estimate is based on the gross primary productivity of the terrestrial biosphere of about 120 GtC/yr, which reflects the conversion of the free energy of solar radiation into the chemical free energy associated with sugars and the release of oxygen (Dyke et al., 2011; Kleidon, 2010). Note that this considerable contribution of chemical free energy generation of 150 TW within the CZ does not result from a large energy flux, but rather from the utilization of a fraction of the large amount of free energy associated with solar radiation that is lost when solar radiation is converted into heat. While this free energy is primarily used to build and maintain biomass, some fraction of it yields chemical free energy within the CZ, e.g. in the form of exudates or organic matter.

The next two sources listed in Table 2 are "heat engine" type processes that convert differential heating into work. The first "heat engine" type process is atmospheric convection, which is driven by the vertical gradient in radiative heating that is maintained between the surface and the atmosphere aloft. As shown in the example 1 of maximum power limits above, this gradient can generate up to ~1400 TW over land. This source of free energy is used mostly to expand air, generate potential energy, and a fraction of which yields kinetic energy associated with convective motion within the atmosphere. These processes take place in the atmosphere; hence, it is only a fraction of the kinetic energy that is in contact with the surface and generates shear stress available for

TABLE 2 Global Estimate of the Magnitudes by which Different Forms of Free Energy are Imported and Generated for CZ Processes in the Climatological Mean. The First Column States the Process Involved in Generating Free Energy, the Second and Third Columns Describe the Forms of Energy that the Specified Process Involves, and the Last Column Provides the Order of Magnitude for the Rate at which Free Energy is Generated (in Units of Terawatts, $1\ TW = 10^{12}\ W$; 1 TW Corresponds to about $7\ mW\ m^{-2}$ of Land Surface)

Process	Original form of energy	Resulting form of energy	Estimated magnitude globally
Photosynthesis	Radiative	Chemical	152 TW
Convection	Heat	Kinetic, potential	<1400* TW
Atmospheric boundary layer dissipation	Kinetic	Potential, kinetic	<50* TW
Periodic heating	Heat	Binding, chemical	<50 TW
Precipitation	Potential, chemical	Potential + kinetic, chemical	<13 TW <0.15 TW
Evapotranspiration	Potential, chemical	Kinetic, potential, chemical	<170* TW
Continental uplift	Potential, chemical	Potential, mechanical, chemical	<1* TW

The numbers marked with (*) describe free energy generation outside the CZ, of which a fraction is available to drive exchange fluxes in the CZ (After Kleidon, 2010 and Dyke et al., 2011).

CZ processes. The associated transfer of momentum acts as a drag force that can lift particles, such as aerosols or dust, and can set surface elements into motion (e.g. the swaying of canopies). Free energy is also potentially imported into the CZ by the shear stress associated with the kinetic energy of large-scale atmospheric motion. The number of <50 TW is based on climate model simulations that quantify the upper limit on extractable wind power (Miller et al., 2011). While the study of Miller et al. (2011) focused on limits to wind power, the approach taken in the study serves as a general estimate on how much of the kinetic energy of large-scale motion within the atmospheric boundary layer over land can possibly be converted into other forms.

The second "heat engine" type process is the periodic heating and cooling of the CZ at the surface (example 2 in Section 4), which can generate less than 50 TW of free energy. This number is based on an estimate with a land surface model that simulates soil heat conduction at the global scale (F. Gans pers. comm). As discussed above, this free energy can be used either to lift soil or to mechanically break up rocks and thereby reduce their (negative) binding energy. This estimate is also an upper estimate, in that it considers the upper limit of work generation at the surface, where the ground heat flux and temperature variations are the greatest. As the depth to bedrock increases during soil development, the ground heat flux and the temperature variations at depth decrease, and so does the upper limit on the mechanical work that could be derived for mechanical weathering.

Precipitation at a given elevation imports free energy into the CZ in the form of its associated potential energy and chemical free energy (since rainwater is desalinated freshwater). This potential energy flux is given by the mass flux of precipitation times the gravitational potential associated with the topographic height, and was estimated by Kleidon (2010) to be about 13 TW. As shown in example 3 above, only a fraction of these 13 TW can be utilized further to drive lateral transport processes on land. The chemical free energy of precipitation is estimated by the work needed to desalinate seawater upon evaporation from the ocean (Kleidon, 2010), which is <0.15 TW. This sets the upper limit on how much chemical work can be performed by dissolving minerals as the water gets into contact with the continental crust.

The unsaturated air of the atmospheric boundary layer provides free energy to drive the evapotranspiration flux from the CZ to the atmosphere. The associated work is estimated using climate model simulations of Kleidon (2008), in which the associated entropy production was estimated to be about 4.2 mW m^{-2} K^{-1} on land. Using a mean surface temperature of 288 K and scaling this number to the total continental area yields an estimate of 170 TW. Some of this power is consumed by mixing within the atmospheric boundary layer, and some fraction can be used to perform work in the CZ, particularly in the form of lifting water against gravity.

Continental uplift adds free energy to the CZ in terms of generating potential energy of the bulk mass of the CZ. Using a simple model, Dyke et al. (2011) estimated the associated power to be less than 1 TW. This low rate of free energy generation results from the very small interior heat flux, which constrains all free energy generation within the interior (including kinetic energy associated with mantle convection and plate motion) to be less than 40 TW (Dyke et al. 2011). Hence, the rate of chemical free energy generation, which is not included in the <1 TW estimate, must also be very small.

The global estimates of the free energy sources shown in Table 2 provide a few important insights (despite their rough and preliminary nature):

1) The order of magnitude at which work is being performed in the CZ is generally much smaller than the rates of heating. This is to be expected from thermodynamics, as free energy generation rates have to be significantly smaller than heating rates to conform to the second law of thermodynamics. While these rates are not directly relevant for heat fluxes, these are nevertheless critical because they reflect the dynamics of the CZ that, in the end, alter the heat and mass exchange fluxes of the CZ with the environment.

2) Biotic activity generates free energy at a substantial rate, particularly when noting that this free energy is available as chemical free energy. Abiotic ways to generate free energy, for instance, through hydrologic cycling, are much less efficient as these involve several conversion steps from a heat engine to motion to chemical free energy with low conversion efficiencies. Hence, the estimates given in Table 2 substantiate the dominant role of biotic activity in shaping the form and function of the CZ in a quantitative and fundamental way.
3) Hydrological fluxes provide the most important source of free energy to drive downgradient transport of mass within the CZ. The other potential sources for lateral transport are convective and large-scale motion. While playing a role in arid regions for the transport of dust, this source diminishes once the surface is covered by vegetation. Hence, in densely vegetated regions, the free energy associated with precipitation provides essentially the only means for effective lateral transport of continental mass.

What is not shown in the global estimates in Table 2 is the regional variation in their magnitude. Photosynthetic rates are much higher in the tropical regions than in arid or polar regions. Generation rates associated with precipitation are highest in regions of high rainfall and topographic gradients. Periodic heating and cooling are strongest in the continental polar regions and weakest in the tropics. We would therefore expect that these estimates vary geographically and in terms of their relative importance. Geographic differences in the form and function of the CZ may then be attributable to the geographic differences of the forms of free energy that sustain its activity. We explore this in the following, first in general terms regarding the relationship of generation rates of free energy with structure, and then in more detail regarding pedogenesis.

5.2. Nonuniformity and Structures as Reflections of Past Work

The association of work performed and structure is fairly straightforward. If no free energy was imported or generated within the CZ, the second law would imply that the processes within the CZ would smooth out all gradients over time, resulting in a uniform state in which no structure and activity exist. Nonuniformity, the presence of gradients, variability in space and time, heterogeneity, and structures are hence the consequence of work done on the thermodynamic system. In fact, these aspects of nonuniformity relate directly to the integrated net amount of work done, that is, by how much a gradient was generated, to what extent it was depleted by dissipation and used by another process to convert it to some other form of free energy. We can formalize this description by expressing the free energy content A associated with a thermodynamic variable at a time t as the integration of the generation rate (or power) P, the dissipation rate D, and the extraction rate X to other forms of energy over the past from an initial time $t' = 0$ to the time $t' = t$:

$$A(t) = \int_0^t (P(t') - D(t') - X(t'))\mathrm{d}t'. \tag{28}$$

Let us briefly illustrate this expression. A topographic gradient is formed by the lifting of continental mass – resulting from the work performed by interior processes. This rate of lifting would be reflected in the integrated power $\int P(t')\,\mathrm{d}t'$. The depletion of this process by erosion and sediment transport would constitute $\int D(t')\,\mathrm{d}t'$, while this form of free energy is not transformed to other processes (i.e. $X = 0$ for this example). The resulting topographic gradient that we can observe is described by the present-day value of A, and reflects the integrated past work of lifting and dissipation of the topographic gradient by erosion and sediment transport.

When we think of the formation of soils, the net amount of work performed is reflected in the unbinding and chemical alterations of rocks into smaller grains and dissolved forms as well as their removal. Likewise, preferential flow structures within the soil – such as macropores – as well as river networks are structures associated with nondiffusive flow that need to be generated and maintained by performing work. This leans toward the formulation of a working hypothesis: not just gradients, but also structures in general, in space and time, are intimately linked to the past net amount of work – the integrated difference of generation and depletion of free energy over the past – that processes have performed. This is similar to the concept of embodied energy in ecology (e.g. Odum, 1988).

Because of their direct relevance to topics in hydropedology, we explore the latter two examples in more detail in the following.

5.3. Thermodynamics and Structure – Pedogenesis as an Example

The soil pedon reflects the physical, chemical, and biological alterations that took place in the past that transformed parent material into soil particles, secondary minerals, aggregates, and horizons. These transformations were fueled by the addition of heat, mass, and free energy of different forms, and the soil profile reflects how much of these different types were added, have been transformed along the direction dictated by the second law, and how much of these have been exported in solid, dissolved, or gaseous forms. Hence, the soil profile reflects the combined action of processes that perform work and create structure, and those that deplete free energy and reduce structure. This broad categorization of organizing and dissipating processes during pedogenesis has long been recognized (e.g. Hole, 1961; Smeck et al., 1983; Johnson and Watson-Stegner, 1987; Johnson et al., 1990; Addiscott, 1995; Rasmussen et al., 2005; Lin, 2010a, b).

The processes that are involved in pedogenesis can be further distinguished between processes of physical and mechanical nature that affect physical soil structure – in terms of particle distribution, preferential flow structure, and alike – from chemical ones that alter the chemical composition of the soil material. Physical and mechanical transformations convert material into successively smaller particles or relate to the physical redistribution of material within the pedon. The rate at which these physical processes take place depends on the rate at which work is performed, e.g. to overcome the binding forces among minerals (thereby increasing the value of U), or to relocate materials within the pedon. This work is derived, for instance, from periodic heating (Section 4.2), freeze–thaw cycles (which amplify the maximum work of periodic heating), and shrinking–swelling cycles, and also involves physical work done by root systems or burrowing animals in cracking and lifting soils (which must come at an energetic cost and needs to be accompanied by the consumption of carbohydrates).

Chemical weathering results from exothermic chemical reactions that convert chemical free energy into heat. In other words, a decrease in chemical free energy, $\mathrm{d}(A\ \xi) < 0$, is compensated by an increase in heat, $\mathrm{d}(T\ S) > 0$. The increase in heat ($\mathrm{d}(T\ S) = \mathrm{d}T\ S + T\ \mathrm{d}S$) is reflected in both, an increase in temperature $\mathrm{d}T$ and an increase in the specific entropy $\mathrm{d}S$ of the products. Hence, as weathering progresses, the soil accumulates weathered material with a higher specific entropy in time. Volobuyev and Ponomarev (1977) showed some interesting patterns among different groups of soils (Fig. 3): one group is accompanied by a decrease in Gibbs free energy and an increase in specific entropy of weathering residuals, leading to more weathered soils with large quantities of minerals and oxides that reflect higher intensity of leaching and more resistance to further weathering (such as Oxisols, Ultisols, and Spodosols), and the other group is just the opposite showing more "reactive" soils that are less weathered (such as Vertisols, Histosols, and Aridisols). Continued physical and chemical weathering through time results in the accumulation of more acidic and stable minerals within the soil profile such as quartz (sand particles), secondary clay minerals (such as kaolinite, smectite, and illite), and Fe and Al oxides (such as hematite, goethite, and gibbsite), which have higher specific entropy.

Both of these two basic trends in pedogenesis, toward smaller particle sizes and toward weathered material with higher specific entropies, are altered by energy and material exchanges of the CZ with the environment. Here, energy plays a central role. The addition of mass, mostly by precipitation, organic material, and continental uplift, is associated with the delivery of potential and chemical free energy, while the generation of kinetic free energy associated with leaching or runoff provides the means to accelerate the translocation or removal of soil material in suspended and dissolved forms. Hence, the characteristics and function of the pedon should relate strongly to the magnitudes of the different processes at work (as estimated globally in Table 2) and how these vary geographically. The description of these thermodynamic drivers of pedogenesis is incomplete if only mass and energy balances are considered. The related conjugate variables are needed to express each of the mass fluxes in energetic terms. This formulation in energetic terms is the foundation to describe how much free energy is imported into the system with the exchange of mass, and it is this free energy that drives the dynamics and transformations within the system.

5.4. Thermodynamics and Function – Flow Networks as Examples

The functional implications of the presence of structure in soils can also be consistently formulated using thermodynamics. The presence of structure within a system – represented by nonuniformity in contrast to a uniform state without structure – results in different dynamics of CZ processes. In fact, it is likely to affect the thermodynamic behavior of the processes in such a way that the dynamics are accelerated, with greater rates of free energy generation, transfer, and dissipation, and, ultimately, faster depletion of the driving gradient. In other words, it is through the presence of structure that the overall dissipative activity of the system is enhanced, thereby accelerating the progression into the direction dictated by the second law. This interpretation of structure and dissipative behavior of the system is similar to the common notion in nonequilibrium thermodynamics that dissipative structures relate to a system's state far from thermodynamic equilibrium (Prigogine, 1955, 1967; Kondepudi and Prigogine, 1998).

To illustrate the effect of structure on the thermodynamics of a system in concrete terms, we consider the example of surface runoff, as illustrated in Fig. 5. In this example, we compare the conversion rates of free energy of water to sediment flow of a uniform slope (Fig. 5a) to the case of preferential flow in a simple channel (Fig. 5b). Both slopes shown in Fig. 5 have the same overall mean slope. Even though these slopes may have the same steady-state water mass balance, they must differ in their rate of power generation, as reflected by the momentum at which water is exported and the rate of sediment export. This difference reflects solely the deviations from the mean slope, which directly relate to the presence of the structure in the system.

The runoff generation on the slope provides the free energy to drive the dynamics of overland flow and sediment transport. The generated power in this process relates directly to the example of maximum power for sediment transport discussed in Section 4.3. There, it was shown that the maximum possible power to drive sediment transport depends on the driving force as F_{acc}^2

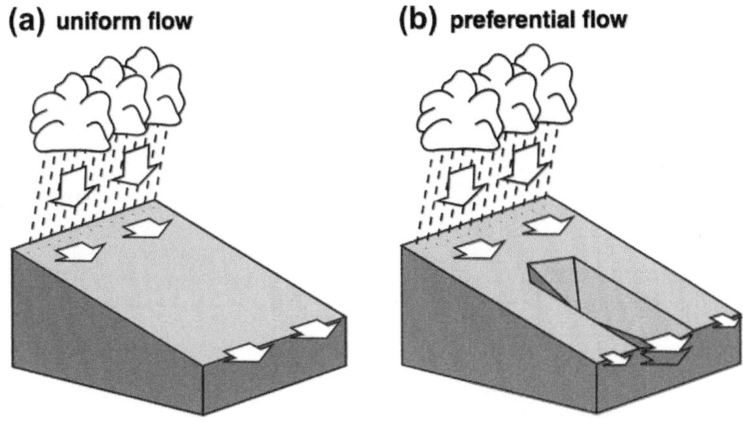

attribute	uniform flow	preferential flow
gradient	uniform slope	non-uniform slope
power	uniform gradient yields uniform acceleration and power	locally steepened gradient enhances acceleration and disproportionally enhanced power, thereby overall generating more power for sediment transport
frictional dissipation	uniform flow results in uniform, high frictional dissipation	preferential flow lowers frictional dissipation within structure, enhancing sediment export
sediment export	low, uniform sediment export across slope	high sediment export by preferential flow
gradient dissipation	weak	stronger

FIGURE 5 Comparison of (a) a uniform slope with a uniform, sheet flow and (b) a slope with a channel structure and preferential flow, in terms of power generation, transfer, and dissipation. The differences of the two slopes as characterized in thermodynamic terms are summarized in the accompanying table *(From Kleidon et al., in prep.).*

Eqn (27). As the driving force F_{acc} relates directly to the slope ∇z, a locally steepened slope yields disproportionally more power for sediment transport than a uniform slope. This can be illustrated by comparing the power generated by a uniform slope (with the variables used to refer to this case with an index "u") to the case of differing gradients but the same mean slope (with index "s"). For simplicity, let us consider the slope being represented by two terms, ∇z_1 and ∇z_2, with the mean slope given by $\nabla z_{avg} = 1/2(\nabla z_1 + \nabla z_2)$. In the uniform case, the generated power is $P_u \propto \nabla z_{1,u}^{\;2} + \nabla z_{2,u}^{\;2} = 2\nabla z_{avg}^{\;2}$ because $(\nabla z_{avg} = \nabla z_{1,u} = \nabla z_{2,u})$. In the case with structure, let us write $\nabla z_{1,s} = \nabla z_{avg} + \Delta z$ and $\nabla z_{2,s} = \nabla z_{avg} - \Delta z$. The generated power now is $P_s \propto \nabla z_{1,s}^{\;2} + \nabla z_{2,s}^{\;2} = 2\nabla z_{avg}^{\;2} + 2\Delta z^2$; that is, the generated power on the nonuniform slope is greater than that for the uniform slope, $P_s > P_u$, by a term that is proportional to Δz^2. From this simple example we may generalize that the nonuniformity associated with structure generates more power that can be used to expand the structure further, thus resulting in a positive feedback on structure formation (Fig. 6; see more details in Section 5.5).

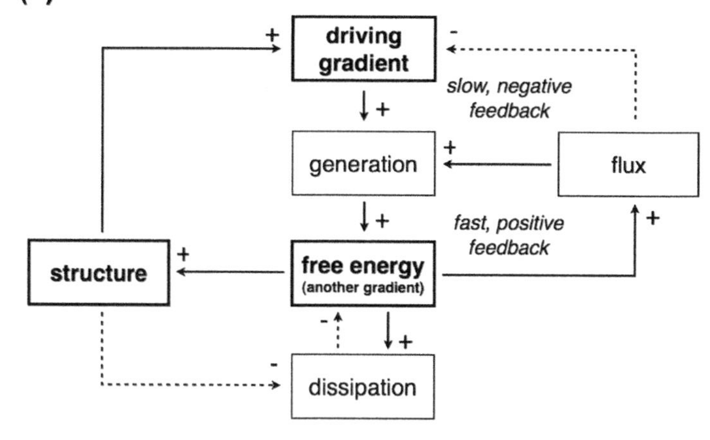

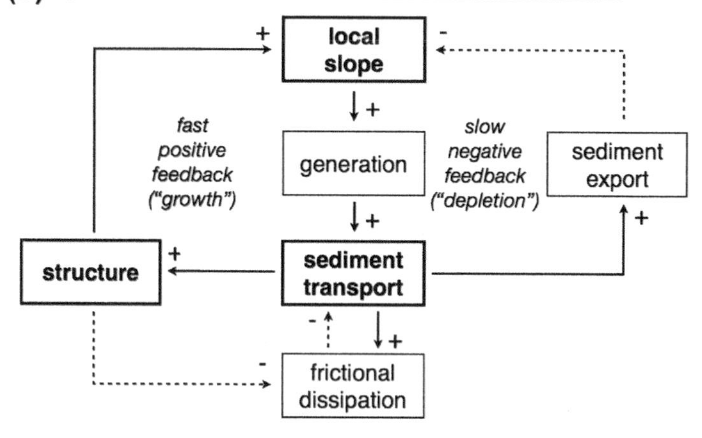

FIGURE 6 Feedback diagrams to illustrate the relationship and feedbacks loops associated with gradient dissipation, free energy generation, and structure (a) in general and (b) specifically applied to river network structures (as in Fig. 5). Solid lines with "+" indicate positive influences, e.g. a larger driving gradient results in a greater generation rate. Dashed lines with "−" represent negative influences, e.g. an enhanced rate of gradient dissipation results in a smaller driving gradient.

Furthermore, the locally steepened slope at the boundaries of the structure comes at the expense of a less steepened slope further down within the structure. The frictional loss and work are reduced for the flow within the resulting channel because the contact area of the fluid to the surface is reduced when compared to sheet flow (i.e. the reduction in the ratio of contact area to flow volume reduces the value of k in Eqn (24)). With this, water (and sediment) can be exported faster by the structure (note, however, that momentum export is not included in Eqn (24)). As a consequence of this reduction in friction and work on the sediment, this structure becomes more persistent in time.

To summarize this example, while the mean properties and forcings of the two configurations shown in Fig. 5 look the same, they differ in their efficiency to deplete the gradient in gravitational potential. This is achieved by the nonuniformity associated with the structure present in Fig. 5b. Locally steepened slopes generate more power, which can extend the structure, and more preferential flow can lower the value of k, thereby further reducing the "waste" of frictional dissipation. Hence, the association of minimizing energy dissipation of fractal flow networks, as shown by flow networks in living organisms (West et al., 1997, 1999) as well as minimizing energy expenditure in river networks (Rodriguez-Iturbe and Rinaldo, 2001), can be reinterpreted to be a reflection of maximizing the power associated with building the structures. This maximization, in turn, dissipates the driving gradient faster than that without preferential flow.

The main characteristics of the structure, steepened gradients at the boundary and minimized dissipation within the structure, are a common phenomenon in nonequilibrium systems. It is, for instance, also observed in convection cells. There, the temperature gradients are steepened at the cell boundary, while within the convection cell, the gradient almost vanishes (Schneider and Kay, 1994). Just as the thermal convection cell likely acts to maximize heat transfer between two heat reservoirs and thereby dissipate the heating gradient at the fastest possible rate, so does a river network structure likely act to maximize sediment transfer, thereby depleting topographic gradients as fast as possible.

Hence, this insight regarding structures and their effects on power generation should extend to a whole range of structures relevant to hydropedology. Similar to river networks, flow networks are abundant in soils, such as root branching networks, mycorrhizal mycelial networks, animal borrowing networks, networks of cracks and fissures, and others. Free energy inputs provide the means to form flow networks in soils, and networks provide a means of minimizing frictional losses, thereby maximizing structure formation. Like the free energy associated with water flow on land that creates dendritic stream networks, water flowing through soils creates network-like flow paths in the subsurface. As water moves through soils, changes in soil texture, structure, organic content, mineral species, biological activities, and other features will modify the resistance to the flow, causing change in flow path to allow water to follow the path of least resistance, thus resulting in a preferential flow network that has the least global flow resistance.

If this line of reasoning is substantiated, the preferential flow characteristics of a soil profile could be an important indicator of the profile's stage of development. Some evidence has suggested that a similarity may exist between river dendritic structures and subsurface preferential flow networks. For example, Deurer et al. (2003) found that the drainage network in a sandy soil under a coniferous forest in north Germany closely resembled the drainage network of mountainous streams. They also found that the fractional area of the entire profile occupied by the network decreased exponentially with depth. By transporting organic substances and soil particles, networks in soils will progressively change hydraulic properties along these transport paths. Repeated action and reaction in a variety of planes of weakness (such as the space in between peds) lead to the formation of soil structure and preferential flow pathways, which are evidenced by numerous observable soil morphological features (such as thickened clay films or other coatings on ped surfaces in many soils).

Deurer et al. (2003) suggested that networks seem to be a scale-invariant functional structure for water and solute transport from the pore scale up to the catchment scale. Depending on the governing physical processes, these networks may exhibit different topologies. Overall, flow and transport networks in soils are formed by forcing the soil formation, mainly climate and organisms. Cycles of wetting and drying, freezing and thawing, shrinking and swelling, coupled with organic matter accumulation and decomposition, biological activities, and chemical reactions lead to the formation of diverse soil aggregates and pore networks in the subsurface. In particular, plant roots, burrowing animals, and mycorrhizal fungi are active in creating networks in soils. The origin, dynamics, and recurrent patterns of self-organization of preferential flow networks in the subsurface have become the subjects of recent research and model development (e.g. Lehmann et al., 2007; James and Roulet, 2007; Weiler and McDonnell, 2007; Michaelides and Chappell, 2009; Zehe et al., 2010).

5.5. Thermodynamics and Evolution of the CZ

Thermodynamics should also inform us about the evolutionary dynamics of the CZ. Specifically, there should be some generalities involved in how processes evolve to states of maximum power, as such limits should have two feedbacks in common (Fig. 6a). A driving gradient (e.g. heating gradient or potential gradient) provides the source for the generation of free energy, which is associated with the creation of a gradient in one of the conjugate variables of another form of energy (e.g. velocity gradient is associated with kinetic energy). This generation of free energy is associated with a flux (e.g. heat flux or momentum flux). As free energy is generated, the associated flux is increased, which enhances the generation rate. This can be seen in Eqn (12), where P_{ex} would initially increase with J_{ex} as long as the temperature gradient is not affected by much. This constitutes a fast, positive feedback on the generation rate of free energy (marked "fast positive feedback" in Fig. 6a). On the other hand, the flux increases the depletion of the driving gradient, which results in a decrease in generation rate. The depletion of the driving gradient is accomplished by the depletion of a reservoir of the associated form of energy, so that the dynamics of depleting the gradient should proceed at a slower rate than the generation of free energy (otherwise, the dynamics would not get started in the first place). This forms a slow, negative feedback on the generation rate of free energy (marked "slow negative feedback" in Fig. 6a). The evolutionary dynamics toward a maximum power state could be understood as a result from these two competing feedbacks that act on different timescales.

The resulting free energy (box "free energy" in Fig. 6a) represents the dynamics that shapes the accelerated depletion of the driving gradient. The positive feedback mentioned above is realized by the formation of structures (box "structure" in Fig. 6a), which is actually associated with two positive feedbacks: First, the reduction of frictional losses of the flow within the structure should enhance the maintenance of free energy and the structure. Second, the structure is associated with a local steepening of the driving gradient at the structure boundary, which should enhance the local free energy generation rate within the structure and enhance its growth. The overall dynamics, however, all relate to the accelerated depletion of the driving gradient. This depletion shapes the ultimate and inevitable negative feedback that sets the limits to the long-term dynamics and structure formation of the system.

These feedbacks between free energy generation, structure, and gradient depletion should be quite general and are now illustrated in more detail as these apply to river network structures (Fig. 6b). As discussed in Section 5.4, runoff generation on a slope provides the means to generate free energy to drive sediment transport and export from the slope. The presence of preferential flow, in the form of channel networks, enhances these dynamics. Once a perturbation in the form of a local steepening of the slope occurs on an initially uniform slope, this provides the means to disproportionally enhance power generation locally. Thereby, sediment transport is locally enhanced, resulting in the growth of the perturbation and the initiation of structure formation. The enhanced sediment export from the perturbation acts to steepen the slope further, resulting in the upslope growth of the structure. As the structure grows further into the slope, the sides framing the channel steepen, become more susceptible to perturbations, and thereby favor the spread of the structure across the slope. Thus, the loop of "local slope steepening"–"enhanced sediment export"–"structure formation"–"further steepening" constitutes the positive feedback on the growth of the structure ("fast positive feedback" in Fig. 6b). The channeling of the flow within the structure results in a reduction of the ratio of contact area to flow volume, which reduces frictional loss within the structure, thereby enhancing sediment transport and export. This constitutes a positive feedback on the maintenance and persistence of the structure. These dynamics overall result in the enhanced export of sediment and thereby the depletion of the slope that drives the dynamics. Hence, this "depletion" negative feedback limits the dynamics of free energy generation and the formation and maintenance of structures. The specific limits to structure formation, e.g. in terms of the steepness of the slope, will depend on the material properties that are involved in sediment transport (e.g. the thresholds and work needed for sediment detachment). These aspects would need to be further explored in more detail and related more closely to previous related concepts, such as minimum energy dissipation (West et al., 1999), minimum energy expenditure (Rodriguez-Iturbe and Rinaldo, 2001), and the emergent property of hydraulic selection (Philipps, 2010).

6. SUMMARY AND OUTLOOK

This chapter provides an overview of how thermodynamics can be used to describe the Critical Zone in terms of its structure, functioning, dynamics, and its interaction with the environment. At the center of this formulation is the description of CZ processes and fluxes in terms of different forms of energy (heat, potential, kinetic, binding, and chemical energies). To do so, it requires the full description of CZ processes in terms of sets of conjugate variables, which complements the description of energy, mass, and momentum balances. It is through the pairs of conjugate variables that the respective exchange fluxes are expressed as fluxes of energy. The first and second laws of thermodynamics then set the direction and limits for the activity of CZ processes. For this activity, it is not the absolute amount of energy that is important, but rather gradients that can be converted into free energy. Free energy, the fraction of the total energy that can be converted into physical or chemical work, is the "fuel" for driving the CZ dynamics, such as generating flow, lifting mass, or transporting sediments, thereby creating new gradients in thermodynamic variables at the expense of other gradients. The second law imposes the constraint that these conversions typically proceed with an efficiency of much less than one, so that each energy conversion is associated with an inevitable loss of free energy in the form of heat (because of friction and other irreversible processes). Irrespective of the involved complexity of the processes and interactions within the CZ, these dynamics ultimately deplete thermodynamic driving gradients, and thereby follow the direction imposed by the second law of thermodynamics.

The direction imposed by the second law can be determined in a straightforward way, as is the case of heat conduction within the soil that follows and depletes temperature gradients directly. This direction can also be accomplished in a more complex way in which the generation of free energy fuels dynamics that accelerate the depletion of the thermodynamic driving gradient. This was qualitatively illustrated here in the context of preferential flow structures, specifically channel flow and river networks that accelerate the depletion of topographic gradients by maximizing water flow and sediment export, and in terms of the feedbacks involved in the evolutionary dynamics. In addition to the overall direction, the second law poses additional constraint on the generation rates of free energy. In classical thermodynamics, this constraint yields the Carnot limit of a heat engine, but in settings more characteristic for CZ processes, this constraint corresponds to the maximum power limit that is typically much lower than the Carnot limit.

The overview given in this chapter is a first step toward a full thermodynamic description of the CZ. It shows how the processes within the CZ are intimately coupled to each other by the flow and transformation of energy and how thermodynamics provides fundamental insights into directions, limits, evolutionary dynamics, and the fundamental importance of structures. To develop this perspective further, we need to think more in terms of free energy when measuring and modeling CZ processes. We need to think of the typical variables of CZ processes in the context of the types of energy they represent, processes in terms of which types of energy they convert, exchange processes in terms of which forms of energy they provide or export, and structures in terms of which types of energy conversions they result from and which types of energy conversions they affect. By doing so, we should gain a more profound understanding of the CZ and how it interacts with the Earth system that will enable us to build better predictive models of the CZ.

ACKNOWLEDGMENTS

AK acknowledges financial support through the Helmholtz Alliance "Planetary Evolution and Life." The authors thank Jonathan Phillips and Tyson Ochsner for constructive reviews.

REFERENCES

Addiscott, T.M., 1995. Entropy and sustainability. Eur. J. Soil Sci. 46, 161–168.

Bloschl, G., Sivapalan, M., 1995. Scale issues in hydrological modeling – a review. Hydrol. Process. 9, 251–290.

Deurer, M., Green, S.R., Clothier, B.E., Böttcher, J., Duijnisveld, W.H.M., 2003. Drainage networks in soils: a concept to describe bypass-flow pathways. J. Hydro. 272, 148–162.

Dewar, R.C., 2005. Maximum entropy production and the fluctuation theorem. J. Phys. A 38, L371–L381.

Dewar, R.C., 2010. Maximum entropy production as an inference algorithm that translates physical assumptions into macroscopic predictions: don't shoot the messenger. Entropy 11, 931–944.

Dyke, J.G., Gans, F., Kleidon, A., 2011. Towards understanding how surface life can affect interior geological processes: a non-equilibrium thermodynamics approach. Earth Syst. Dynam. 2, 139–160.

Eddington, A.S., 1928. The Nature of the Physical World. Macmillan, London.

Hole, F.D., 1961. A classification of pedoturbations and some other processes and factors of soil formation in relation to isotropism and anisotropism. Soil Sci. 91, 375–377.

James, A.L., Roulet, N.T., 2007. Investigating hydrologic connectivity and its association with threshold change in runoff response in a temperate forested watershed. Hydrol. Process. 21, 3391–3408.

Johnson, D.L., Watson-Stegner, D., 1987. Evolution model of pedogenesis. Soil Sci. 143, 349–366.

Johnson, D.L., Keller, E.A., Rockwell, T.K., Dembroff, G.R., 1990. Dynamic pedogenesis – new views on some key soil concepts, and a model for interpreting quaternary soils. Quatern. Res. 33, 306–319.

Kiehl, J.T., Trenberth, K.E., 1997. Earth's annual global mean energy budget. Bull. Am. Met. Soc. 78, 197–208.

Kleidon, A., 2008. Entropy production by evapotranspiration and its geographic variation. Soil and Water Res. 3 (S1), S89–S94.

Kleidon, A., 2010. Life, hierarchy, and the thermodynamic machinery of planet Earth. Phys. Life Rev. 7, 424–460.

Kleidon, A., Malhi, Y., Cox, P.M., 2010. Maximum entropy production in environmental and ecological systems. Phil. Trans. Roy. Soc. B 365, 1297–1302.

Kleidon, A., Zehe, E., Ehret, U., Scherer, U. Thermodynamics, maximum power, and the dynamics of preferential river flow structures on continents, in prep. for submission to Hydrol. Earth Syst. Sci.

Kondepudi, D., Prigogine, I., 1998. Modern Thermodynamics – from Heat Engines to Dissipative Structures. Wiley, Chichester.

Lehmann, P., Hinz, C., McGrath, G., Tromp-van Meerveld, H.J., McDonnell, J.J., 2007. Rainfall threshold for hillslope outflow: an emergent property of flow pathway connectivity. Hydrol. Earth System Sci. 11, 1047–1063.

Lin, H.S., 2010a. Linking principles of soil formation and flow regimes. J. Hydrol. doi:10.1016/j.jhydrol.2010.02.013.

Lin, H.S., 2010b. Earth's critical zone and hydropedology: concepts, characteristics, and advances. Hydrol. Earth System Sci. 14, 25–45.

Lin, H.S., 2011. Three principles of soil change and pedogenesis in time and space. Soil Sci. Soc. Am. J. 75, 1–22.

Michaelides, K., Chappell, A., 2009. Connectivity as a concept for characterizing hydrological behavior. Hydrol. Process. 23, 517–522.

Miller, L.M., Gans, F., Kleidon, A., 2011. Estimating maximum global land surface wind power extractability and associated climatic impacts. Earth Syst. Dynam. 2, 1–12.

Minasny, B., McBratney, A.B., Salvador-Blanes, S., 2008. Quantitative models for pedogenesis—a review. Geoderma 144, 140–157.

National Research Council (NRC), 2001. Basic Research Opportunities in Earth Science. National Academy Press, Washington, D.C.

Odum, H.T., 1988. Self-organization, transformity, and information. Science 242, 1132–1139.

Peltier, L.C., 1950. The geographic cycle in periglacial regions as it is related to climatic geomorphology. Ann. Am. Assoc. Geogr. 40, 214–236.

Philipps, J.D., 2010. The job of the river. Earth Surf. Process. Landforms 35, 305–313.

Prigogine, I., 1955. Introduction to Thermodynamics of Irreversible Processes. John Wiley, New York, NY.

Prigogine, I., 1967. Introduction to Thermodynamics of Irreversible Processes, third ed. John Wiley, New York, NY.

Rasmussen, C., Southard, R.J., Horwath, W.J., 2005. Modeling energy inputs to predict pedogenic environments using regional environmental databases. Soil Sci. Soc. Am. J. 69, 1266–1274.

Rasmussen, C., Troch, P.A., Chorover, J., Brooks, P., Pelletier, J., Huxman, T., 2011. An open system framework for integrating critical zone structure and function. Biogeochemistry. DOI: 10.1007/s10533-010-9476-8.

Rodriguez-Iturbe, I., Rinaldo, A., 2001. Fractal River Basins: Chance and Self-organization. Cambridge University Press, Cambridge, UK.

Schneider, E.D., Kay, J.J., 1994. Life as a manifestation of the second law of thermodynamics. Math. Comput. Model 19, 25–48.

Smeck, N.E., Runge, E.C.A., Mackintosh, E.E., 1983. Dynamics and genetic modeling of soil systems. In: Wilding, L.P., et al. (Eds.), Pedogenesis and Soil Taxonomy. Elsevier, New York, pp. 51–81.

Volobuyev, V.R., 1983. Thermodynamic basis of soil classification. Sov. Soil Sci. 15, 71–83.

Volobuyev, V.R., Ponomarev, D.G., 1977. Some thermodynamic characteristics of mineral associations of soils. Sov. Soil Sci. 9, 1–11.

Volobuyev, V.R., Ponomarev, D.G., Mikailov, F.D., 1980. Relation between the thermodynamic functions of soils, their mineral-composition and infiltration capacity. Sov. Soil Sci. 12, 210–212.

Weiler, M., McDonnell, J.J., 2007. Conceptualizing lateral preferential flow and flow networks and simulating the effects on gauged and ungauged hillslopes. Water Resour. Res. 43, W03403. doi:10.1029/2006WR004867.

West, G.B., Brown, J.H., Enquist, B.J., 1997. A general model for the origin of allometric scaling laws in biology. Science 276, 122–126.

West, G.B., Brown, J.H., Enquist, B.J., 1999. The fourth dimension of life: fractal geometry and allometric scaling of organisms. Science 284, 1677–1679.

Yatsu, E., 1988. The Nature of Weathering. An Introduction. Sozosha, Tokyo.

Zehe, E., Blume, T., Günter, B., 2010. The principle of 'maximum energy dissipation': a novel thermodynamic perspective on rapid water flow in connected soil structures. Philos. Trans. Roy. Soc. B Biol. Sci. 365, 1377–1386.

Part II

Case Studies and Applications

Chapter 9

Hydropedology in Caliche Soils Weathered from Glen Rose Limestone of Lower Cretaceous Age in Texas

Larry P. Wilding,[1,*] Maria M. Nobles,[2] Bradford P. Wilcox,[1] Charles M. Woodruff, Jr.[3] and Henry Lin[4]

ABSTRACT

Weathering of limestone bedrock of differing physical and chemical stability results in caliche (carbonate-rich) soils that commonly have stepped riser and tread microterrains formed on the Glen Rose Limestone (Lower Cretaceous) of Central Texas, USA. Locally, these stair-stepped landforms result in striking differences in soil/regolith properties, vegetative patterns, and hydrologic functions. Changes in landscape attributes from risers to treads include decreases in vegetative cover, slope gradient, soil thickness, infiltration rate, water-storage capacity, biological activity, and soil organic carbon content. This is accompanied by corresponding increases in surface runoff and erosion from riser to tread. Carbonate contents increase from surface horizons to subsoils reflecting processes of dissolution, leaching, and translocation by transient water within the vadose zone. Mollisols on upper and middle risers serve as local recharge sites, while discharge and runoff occur on lower risers and subjacent treads dominated by Inceptisols. However, repeating steps from upper to lower landform positions present multiple barriers to runoff and subsurface flow. Incident water enters a series of disjunct perched water tables that fill and drain according to prevailing weather conditions and vegetative consumptive use. Thus, riser and tread microforms are essentially independent hydropedologic units on stepped hillslopes. They have limited interconnectivity among multiple, ephemeral water tables that form above aquitards during seasonal wet periods. During these periods, water logging and associated anoxic conditions above aquitards provide a secondary buffer to water

[1]Dept. of Soil and Crop Sciences, Texas A&M University, College State, TX, 77843, USA

[2]Dept. of Natural Resources and Environmental Sciences, Alabama A&M University, 4900 Meridian Street, Normal, AL 35672, USA

[3]Woodruff Geologic Consulting, Inc., 711 West 14th Street, Austin, TX 78701, USA

[4]Dept. of Ecosystem Science and Management, The Pennsylvania State University, University Park, PA, USA

[*]Corresponding author: Email: wilding@tamu.edu

Hydropedology, Edited by H. Lin. DOI: 10.1016/B978-0-12-386941-8.00009-5

quality by denitrification. Important hydrologic implications are long mean residence times of water and solutes (assuming only localized bedrock jointing and thus minimal fissure/ fracture flow), multiple cycles of water storage and bioremediation, on-land retention of eroded sediments, and high consumptive use by evapotranspiration. Dye-staining tracers using Br indicate that most of the soil matrix and coarse fragments are involved in water and solute transport. Preferential flow is limited to along roots, root channels, consolidated calcic materials, and open joint planes in bedrock. This chapter demonstrates the importance of integrated hydropedologic approach (integrating geology, soils, landform, and hydrologic processes) to address the effectiveness of caliche soils for the treatment and disposal of on-site wastewaters in environmentally sensitive limestone landscapes.

1. INTRODUCTION

Land use in limestone regions of Central Texas with high carbonate (caliche) soils is increasingly shifting from traditional ranchland to residential and industrial development. Caliche soils are extensive in many regions of Texas and west of the 97th meridian where as much as 10–50% of the total land area is composed of carbonate-rich soils (Wilding et al., 2001). With increased land-use pressures, water management, conservation, and environmental protection become critical issues to landowners, developers, regulatory authorities, and the lay public (Woodruff et al., 1992; Wilcox et al., 2007; Woodruff and Wilding, 2008). Special attention is being given to aquifer recharge and contamination. Until recently, little was known regarding the hydrologic properties or bioremediation potential of caliche soils (Wilding et al., 2001). Specifically, limited knowledge was available about soil moisture regimes, water retention capacity, pathways of water and solute movement, hydraulic conductivity, caliche bioremediation, and recharge into subjacent limestone aquifers (Woodruff et al., 1992; Wilcox et al., 2007; Woodruff and Wilding, 2008; Nobles et al., 2010). For example, early concepts of the Brackett soil series (one of the major caliche soils mapped in the Hill County in Central Texas) were soils characterized as thin, stony, strongly calcareous, low infiltration, low organic carbon, low water retention, high surface runoff, minimal bioremediation, and slight pedogenic development (Werchan et al., 1974). In fact, many lay public and even some geoscience professionals questioned whether "soils" existed in these landscapes, or whether the land surface simply represented a geologic regolith without the classical biological activity attributed to soils (Woodruff and Wilding, 2008). This misrepresentation was propagated in the Travis County Soil Survey Report where the Brackett series was described as "*...shallow, well drained soils... underlain by interbedded limestone and marl*" (Werchan et al., 1974).

It is reported that caliche landscapes in the Hill Country of Central Texas contribute approximately 85% of the water that recharges the Edwards aquifer along the Balcones Fault Zone (Slade et al., 1984; Woodruff, 1984; Woodruff and

Abbott, 1979). This aquifer serves as the major groundwater supply for residential, industrial, and agricultural developments along the Balcones Escarpment in Central Texas (Woodruff and Wilding, 2008). Further, it directly or indirectly governs the quality and quantity of many surface and subsurface waters that flow toward the Gulf Coast region well beyond the escarpment margin.

To assist the local needs of treating wastewater on-site and protecting groundwater quality, we conducted initial fieldwork and laboratory analyses from 1991 to 1995 under the auspices of Barton Creek Properties (Stratus Properties) west of Austin, TX. From these initial studies, soil–landform relationships and hydrologic concepts of stepped hillsides were conceptualized (Woodruff et al., 1992). During initial and later studies, multiple sites in the Hill Country of Central Texas (Fig. 1) were instrumented with piezometers and tensiometers to monitor soil moisture status (Woodruff et al., 1992; Wilding, 2007; Wilcox et al., 2007), establish hydrologic flow patterns with dye tracers (Nobles et al., 2010), measure in situ infiltration rates (Woodruff et al., 1992; Wilcox et al., 2007), determine in situ runoff/erosion rates (Wilcox et al., 2007; Wilding, 2007; Woodruff and Wilding, 2008), and verify soil morphological patterns associated with riser–tread couplets in over 100 backhoe trenches excavated normal to stepped and nonstepped hillslope terrain.

This work, coupled with earlier studies on caliche soils in the Grande Prairie Land Resource Region of Texas (West et al., 1988a,b,c), demonstrated the mechanisms and processes of flow and transport in caliche soils and subjacent substrates, soil–landform relationships, differentiation of soil (pedogenic) and bedrock (lithogenic) carbonates, translocations and transformations of carbonates relative to hydrology, and morphological, physical, chemical, biological, and hydrologic properties of caliche and related soils. From morphological properties, hydrologic pathways may be inferred. For example, translocation of carbonates provides a good indicator of water flow directions and pathways. Where the carbonates accumulate as soft masses in voids, along peds (aggregate surfaces) as filaments, threads and coatings, on lower surfaces of coarse fragments, as pendants, and/or as cements along limestone joint planes, vertical preferential flow is prominent. However, when carbonates occur in laminar layers on top of limestone bedrock surfaces, horizontal water movement is prominent (Wilding et al., 1997), and the bedrock serves as an aquitard (restricts vertical water movement). Dissolution of disseminated fine-grained carbonates from the whole soil matrix of caliche surface horizons and their transport to lower Bk horizons provide good evidence for matrix flow with little or no preferential transport. Finally, recrystallization and micritization of indigenous carbonates within a given horizon or biotic precipitation of carbonates imply primarily short-range transport processes (West et al., 1988a; Rabenhorst et al., 1984; Monger et al., 1991).

The purpose of this chapter was to summarize biophysiochemical properties of caliche soils and elucidate hydrologic processes across caliche soil landscapes

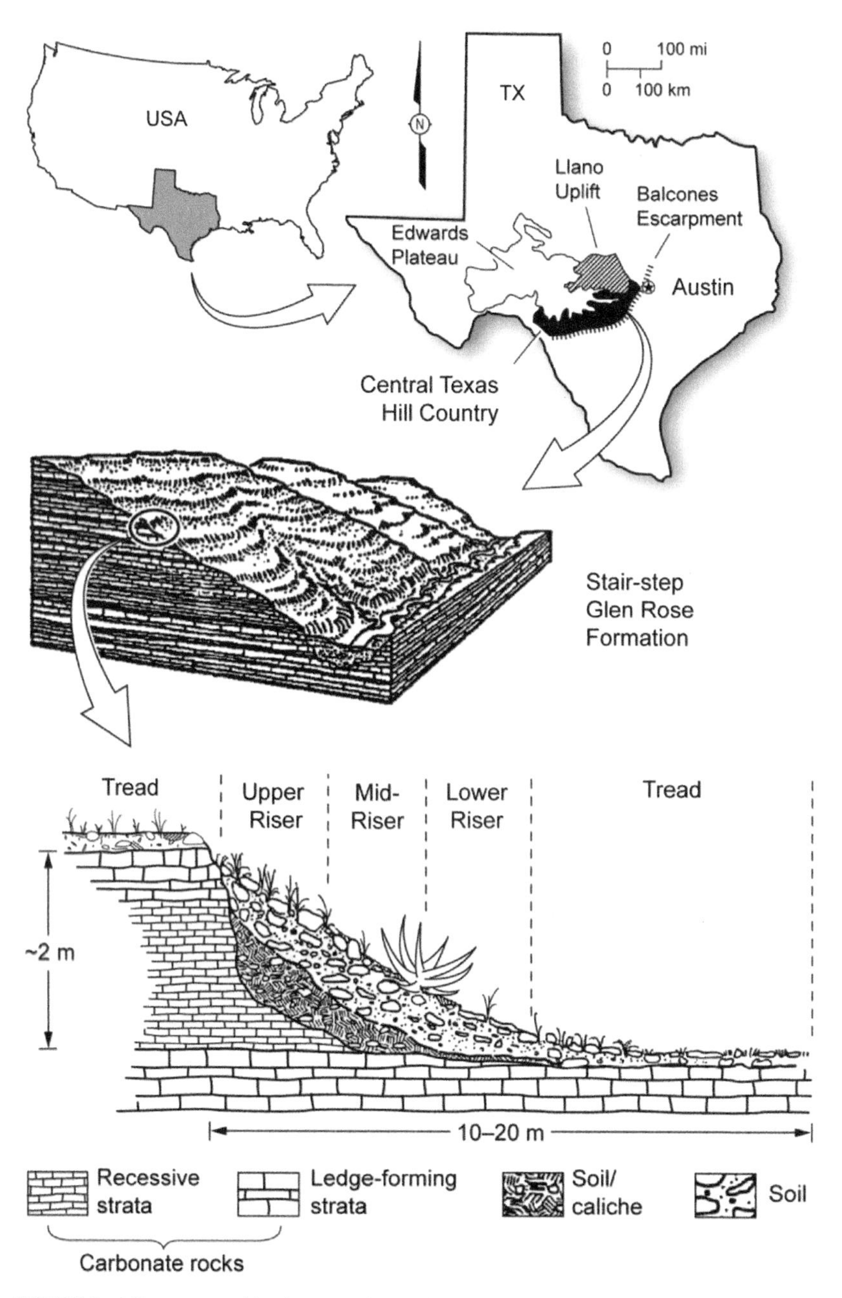

FIGURE 1 Microtopographic elements of central Texas hill country, in context with schematic of stepped terrain, and regional physiographic and geological features. *(Modified from Woodruff et al., 1992, and Wilcox et al., 2007)*

in the Hill Country of Central Texas. This chapter demonstrates the importance of integrated hydropedologic approach (i.e. combining geology, soils, landform, and hydrology) to address the effectiveness of caliche soils for treatment and disposal of on-site wastewater in environmentally sensitive landscapes. This has the benefit of helping to develop more effective guidelines to evaluate the probable performance of caliche and related soils/regoliths for on-site wastewater treatment systems and to provide science-based environmental regulatory decisions. The chapter also illustrates some general relationships among landscapes, soils, and hydrology that are counterintuitive, which may apply to other places where similar soil–landscape state factors exist. In the following, we first briefly describe geological and landscape setting of the study area. After the description of the materials and methods employed in our studies, we detail the soil–landscape attributes and hydraulic properties along the stepped riser–tread microterrains, which are followed by the observation and quantification of flow and transport pathways through dye tracing. We then highlight a number of misperceptions and environmental implications from the findings of this study. The chapter ends with a summary of important lessons learned.

2. GEOLOGICAL AND LANDSCAPE SETTING

The geologic setting of the study area includes a single rock unit, the Glen Rose Limestone. This unit crops out west of the main line of displacement of the Balcones Fault Zone in Central Texas and is the predominant bedrock unit forming the dissected edges of the Edwards Plateau region (Central Texas Hill Country, Fig. 1).

The Glen Rose Limestone is the dominant member of the Trinity Division that composes the basal part of the Comanche Series (Lower Cretaceous) of Texas (Young, 1972). This rock unit consists of interbeds of well-indurated limestone and dolomite. Varying susceptibility to weathering and erosion of these calcareous interbeds has produced the stepped terrain that is characteristic of the Central Texas Hill Country. In western Travis County, which includes the western parts of Austin, the base of the Glen Rose Limestone is not exposed, but this rock unit includes a vertical section of more than 200 m in this area. Elsewhere in Central Texas, total thickness of this unit may approach 300 m.

Glen Rose Limestone strata that weather into stepped landscape compose a mosaic of lithofacies representing various depositional environments along a transgressive shoreline during the early Cretaceous time (Stricklin et al., 1971). Throughout its thick section, the Glen Rose Limestone was deposited within a few meters of sea level. Thus, the interbeds composing this unit were formed in shoreface/lagoon/strand-plain/tidal flat environments similar to the modern Gulf shorelands along the coast of South Texas (Baffin Bay and Laguna Madre). The depositional environment was probably arid to semi-arid, and the terrain was likely of low relief, as evidenced by the paucity of terrigenous sands and muds within the overall Glen Rose section. Nonetheless, local occurrences of calcareous clays play an important role in determining rock fabric, which in turn, determines the permeability and weathering characteristics of the various interbeds. The intervals that formed in lagoonal environments consist of multiple, discontinuous clay partings that impart a nodular or "marly" aspect to those interbeds as they weather. These heterogeneous strata form the "risers" of the stepped terrain. In contrast, limestone and dolomites that contain few or no clay partings (thus, a more homogeneous fabric) are expressed as ledges that cap the risers and form the bases of the treads of the stair-stepped hills.

The Glen Rose Limestone is a locally important aquifer in Central Texas. It does not typically comprise an integrated system of porosity and permeability, but rather is compartmentalized with limited vertical connectivity. Water-holding and groundwater-transmitting properties vary markedly with the more clayey units forming aquitards, and the more homogeneous limestones and dolomites forming local zones of higher permeability. Ironically, the interbeds with clay partings that retard underground flow are the same intervals that, upon weathering, form regoliths noted for storage of incident waters along steeply sloping risers, whereas the more resistant units that underlie the treads are not receptive to recharge from the surface.

The occurrence of the Glen Rose Limestone along the dissected margins of the Edwards Plateau makes up a vital component in the hydrology, ecology, and land use of Central Texas. It constitutes the chief substrate unit underlying the contributing watersheds that supply runoff to the down-faulted Edwards aquifer along the Balcones Escarpment. Moreover, the Glen Rose Limestone underlies suburban areas west of cities and towns along the Balcones Escarpment, and this Hill Country landscape with its "stair-step hills" and dissected valleys is highly prized as home sites. Despite its importance, few published studies, except those of the authors cited herein, have focused on the Glen Rose Limestone in terms of its land-resource attributes, soil conditions, and hydrologic functions. Studies that integrate bedrock, soils, landform, and hydrologic processes would be worthwhile in areas underlain by carbonate rocks in subhumid to semi-arid environments. In the U.S., potentially similar substrates and soils may occur in the stepped terrain of the Flint Hills of Kansas.

Stepped and nonstepped terrains of the Brackett (loamy, carbonatic, thermic, shallow Typic Haplustepts) and Volente (fine, mixed, active, thermic Pachic Haplustolls) soil series mapping units served as the major foci for this study. Soils within the Bracket series mapping units are well-drained upland soils, while soils in the Volente series mapping units are also well drained but found on terraces and alluvial floodplains near the valley floor. These caliche and closely allied soils encompass broad ranges in geographic, geologic, climatic, topographic, soil, and vegetation habitats. Over the region where these series are mapped, mean annual precipitation ranges from 610 to 840 mm (*ustic* mean annual soil moisture regime) and mean annual soil temperature from 18

FIGURE 2 (a) Aerial and (b) oblique photos of a stepped landform in the central Texas Hill Country. *(Modified from Woodruff and Wilding, 2008)*

to 21 °C (*thermic* mean annual soil temperature regime) (Soil Survey Staff, 1999).

The tread and riser soil complex on stepped hillslopes creates distinctive vegetative and phototonal patterns that mimic the hillslope contours (Fig. 2). Darker-colored contour bands are associated with risers, while lighter bands illustrate ledge-former bedrock underlying treads (Fig. 2a). Woody plants and the tall grasses are concentrated on the deeper soils on the risers. The shallow soils on the treads are mainly capable of supporting the smaller grasses and forbs that do not require access to deeper soil water. The woodland is predominately Ashe juniper (*Juniperus ashei*) and liveoak (*Quecus virginiana*). In the better-managed sites, common grass species are little bluestem (*Schizachyrium scoparium*), Indian grass (*Sorghastrum nutans*), and sideoats grama (*Bouteloua curtipendula*). In heavier grazed sites, Wright's three awn (*Aristida wrightii*), hairy grama (*Bouteloua hirsute Lag.*), red grama (*Bouteloua trifida Thurb.*), seep muhly (*Muhlenbergia lindheimeri*), cedar sedge (*Carex planostachys*), and other short grasses become more common.

3. MATERIALS AND METHODS

3.1. Field Methods

From 1991 to 1995, case studies in stepped caliche landforms were conducted west of Austin, TX (Fig. 1), to define soil/regolith conditions, landform relationships, and hydrologic processes in 26 trench exposures (Woodruff et al., 1992; Woodruff and Wilding, 2008). During this period stepped riser–tread soil–landform patterns were discovered, soil moisture regimes were quantified using in situ tensiometers, infiltration studies were conducted with a mobile infiltrometer, runoff and erosion magnitudes were monitored in microcatchments under natural rainfall events, and the synthesis of these initial data sets were compiled (Woodruff et al., 1992). Later, Wilding (2007), Wilcox et al. (2007), and Woodruff and Wilding (2008) amplified on these initial case studies using expanded data sets from subsequent consultancy investigations.

Overall, more than 100 trenches were excavated from 1991 to 2006 by backhoes where Brackett and Volente soil series were mapped (Wilding, 2007). Each trench, 10–30 m long, was excavated perpendicular to riser and tread elements to bedrock refusal that ranged from nil to over 3 m. Similar excavations were also made perpendicular to slope gradients on nonstepped caliche landforms of the Brackett mapping units and on soils of terraces and alluvial floodplains of the Volente mapping units. From these trench exposures, more than 1000 pedons were described for soil morphological properties, and 100 pedons sampled and analyzed for selected physical and chemical properties.

At selected trench locations, soil infiltration rates, surface runoff, soil erosion, soil matric potential, and dye-staining tests were determined. Infiltration rates and dye-staining tests were run at the same site or within 1–2 m proximity of each other while other field measurements were not all common to a given location. The methodologies we employed are summarized in the following.

3.1.1. Soil Stratigraphy and Sampling

The rationale for trench excavations in chosen locations was to provide direct observations representative of the range of soil/geomorphology conditions in stepped terrains from upland summits to valley floors. In identifying trench locations, considerations were made of vegetative cover, slope gradient, step expression, stoniness, degree of soil degradation, and backhoe access. At each trench, a series of soil profile logs with elevation control were made laterally along the trench cross section from riser to tread at intervals sufficiently close to capture the horizon stratigraphy and its spatial variability. Properties observed included thickness and type of horizonation (A, Bk, Bkm, Crk, and R), soil color, coarse fragment content (hard and soft rock percentages), soil texture, soil structure, pedogenic carbonate forms and patterns, root distribution patterns, soil consistency, and horizon boundary. These attributes help differentiate pedogenic from lithogenic factors, including carbonate redistribution, hydrologic dynamics, and degree of soil development and weathering. Standards of the National Cooperative Soil Survey Program were followed for sampling and soil morphological descriptions (Soil Survey Staff, 1993).

Selective sampling strategy was designed to permit collection of a minimum data set that could be scientifically interpreted and transferred with validity to other soil areas not analyzed. Selective sampling was conducted by genetic horizonation (i.e. mollic, calcic, petrocalcic, and restrictive layers), as horizons with the same genetic horizonation would yield similar physical, chemical, and biological attributes, and function similarly within the limited geographical area of caliche soils studied. For example, genetic horizonation served as pedotransfer function to extend the database of biophysiochemical attributes beyond the sampled populous (Bouma, 1989; Soil Survey Staff, 1993, 2010). This approach minimized the labor and financial expenditures required to gain a reliable understanding of the whole landscape's soil–regolith attributes across the caliche stepped and nonstepped terrains.

3.1.2. Terminal Infiltration Rates

Infiltration rates were determined in the field using a mobile needle infiltrometer to simulate natural rainfall as described in detail by Blackburn et al. (1974), Wilcox et al. (2007), and Wilding (2007). Simulated rainfall was applied to plots of about 0.33 m^2 at rates of 10–35 cm/h for 30–60 min until steady-state runoff occurred. Steady-rate runoff was determined from the volume of the runoff leaving the infiltration plot every 5 min during the measurements. The difference between application rate and steady-state runoff rate was considered the terminal infiltration rate as this term is used in this paper. The high rates of water application were to simulate short duration storm events.

A total of 110 infiltration tests were conducted on mapping units of the Brackett and Volente series under various antecedent soil moisture conditions. This resulted in infiltration tests on 5 summits, 63 stepped sites (28 treads and 35 risers), 33 nonstepped upland sites, 5 terrace sites (includes alluvial uplands with similar hydrologic function), and 4 floodplain sites. All infiltration tests were run without prewetting, but 17 sites were run at antecedent moisture conditions followed by runs 24 h later. Mean differences among terminal infiltration rates of sites run under antecedent soil moisture and then 24 h later were less than 1.5 cm/h (Wilding, 2007). These relatively small differences are likely due to the very low shrink/swell potentials of caliche soils with stable pores under variable soil moisture conditions. All comparisons of infiltration rates reported here were for sites run under antecedent soil moisture contents.

3.1.3. Soil Matric Potential

To verify soil moisture status on stepped landforms, a total of 15 sites (7 risers, 2 riser/treads, and 6 treads) were instrumented at various depths with ceramic cup tensiometers to measure soil matric potential. Details of installation and value adjustments are presented by Wilding (2007) and Wilcox et al. (2007). Tensiometers were placed in Bk horizons and above aquitard bedrock layers in a sequence of decreasing elevation. Data were collected weekly (until soil matric potential reached below –900 cm water tension) and also within a few hours or immediately after all rainfall events for about three years (1992–1995). According to the 150-year meteorological record of annual rainfall for Austin, 1992 was one of the wettest years on record (1168 mm), 1993 one of the driest years (673 mm), and 1994 (1006 mm) and 1995 (864 mm) were above the long-term mean of 841 mm. Time series plots of soil matric potentials in reference to rainfall events permitted the calculation of the percentage of time when soils were in states of saturation, moist and very moist, and slightly moist and dry at specific soil depths (Wilcox et al., 2007; Wilding, 2007).

3.1.4. Runoff and Erosion

A total of 15 sites (7 risers, 2 riser/treads, and 6 treads) were also instrumented with microcatchments on multiple steps within the study area to monitor naturally occurring runoff and erosion using methods reported by Wilding (2007), Wilcox et al. (2007), and Woodruff and Wilding (2008). The microcatchments were about 0.3 m^2. They were connected to a downslope storage device to collect runoff from microcatchments under natural rainfall events. Rainfall was monitored at each microcatchment with either a manual or tipping bucket rain gauge. Aliquots of the runoff (100 ml sample) were measured for suspended sediments as described by Wilding (2007) and Woodruff and Wilding (2008). At one site, seven microcatchments were installed in a sequence to measure runoff and erosion from a single transect corresponding to riser, riser/tread, and tread microtopographic elements (Wilcox et al., 2007). Between 1992 and 1995, rainfall and runoff were monitored after each rainfall event, for a total duration of 804 days.

3.1.5. Dye-Staining Tests

A total of 18 sites (7 risers, 4 treads, 3 alluvial floodplains, and 3 upland summits) were selected to determine solute movement in soils of varying properties and landscape positions. Aqueous solution of $CaBr_2$ (30 g L^{-1}) was added into metal frames according to the procedure described in Nobles et al. (2010). Two to three vertical faces were then excavated after each dye application in parallel to the slope gradient for a total of 47 sections. Indicator solution, prepared according to the method suggested by Lu and Wu (2003), was subsequently sprayed on each soil face until the dark blue patterns of a Prussian blue complex (indicating the presence of Br^-) developed on the exposed soil faces. These faces were then digitally photographed and analyzed.

Resulting Br^- tracer patterns were converted to binary format with Adobe Photoshop (Adobe Systems Incorporated, San Jose, CA). Horizon areas stained by Br^- tracer were analyzed and compared based on the ratios of the total horizon stained area, as well as horizontal and vertical tracer distribution distances according to the methods reported in Nobles et al. (2010).

3.2. Laboratory Analyses

Field samples collected were subsequently analyzed for particle-size distribution, bulk density, water retention, shrink/swell potential (COLE), carbonate content, pH, organic carbon content, cation exchange capacity, and calcium carbonate equivalent according to procedures outlined in Soil Survey Staff (2004), Wilding (2007), and in methods used at Texas A&M Soil Characterization Laboratory (http://soildata.tamu.edu/methods.pdf). In selected cases, rock fragments from A, Bk, Bkm, Cr, and R horizons, and oriented clods were sampled for microfabric features, bulk density, water retention, and mineralogy.

Particle-size distribution was obtained using the pipette method with dispersion in sodium hexametaphosphate; cation-exchange capacity with a mechanical variable-rate extractor using 1 N NaOAc (pH 8.2); soil reaction (pH) with an electronic pH meter on 1:1 soil-to-water mixtures; calcium carbonate equivalent was calculated from calcite and dolomite percentages using a gasometrical procedure; total carbon was determined by dry combustion and organic carbon as the difference of total carbon and inorganic carbon (calcium carbonate equivalent); and water retention and bulk density on Saran-coated clods under soil matric potentials of –0.33 and –15 bars to represent upper and lower limits of plant-available water. Water retention difference (WRD) is defined as the volumetric fraction of water retained in the whole soil in terms of cm^3 water/ cm^3 soil, or cm water/cm soil after correction for coarse fragment volumes (Soil Survey Staff, 2004, 2011). Reductions in WRD values were made proportional to the fractional percentage that coarse fragments contribute to total soil volumes. For our work this correction was only made for hard coarse limestone fragments that were too cohesive to be broken by hand and exhibited limited water retention attributes. Cumulative water retention difference (CWRD) represents the total plant-available water (cm) held in the soil per given horizon or profile thickness. It represents the product of WRD (cm water/ cm soil) multiplied by soil thickness (cm) for each horizon in the profile. These CWRD values are then summed to obtain CWRD for the whole soil (cm water/ total soil thickness).

For both field and laboratory analyses, there was considerable spatial variability from horizon to horizon and from site to site, even on comparable landform elements. For this reason, statistics for central tendencies and the dispersion of the data are given in the form of box plots. These plots give the mean, median, and 5, 10, 25, 75, 90, and 95 percentiles for data distributions. In some instances where data sets are less robust (where the number of observations are ≤ 8), only the mean, median, 25 and 75 percentiles are given in box plots. Tables 1 and 2 present central tendency statistics including the mean, standard error of mean, standard deviation, coefficient of variation, and number of observations for each of the box plots presented in this paper. Key statistics are presented to compute significant differences among parameter means using a Student's *t*-test should that information be of interest. For example, if the difference between two means being compared exceeds twice the sum of standard errors of means being compared, then the means can be expected to be statistically different at $p < 0.05$ level (Wilding and Drees, 1983).

4. SOIL-LANDSCAPE FEATURES AND HYDRAULIC PROPERTIES

The soil and landform attributes collected in the field and confirmed by laboratory analyses serve to comprehensively define the biophysiochemical properties of caliche soils, understand their hydraulic functions, and determine

TABLE 1 Spatial Variability of Selected Soil Physical and Chemical Attributes among Different Landforms (X = Mean, SE = Standard Error of Mean, SD = Standard Deviation, CV = Coefficient of Variation, and *n* = Number of Observations)

Attribute	Landforms	X	SE	SD	CV (%)	(*n*)	
Soil Thickness (cm)	Summits	45	7.8	19	42	6	
	Risers	92	8.1	48	52	35	
	Treads	33	2.5	15	45	34	
	Nonstepped	104	4.9	41	40	71	
	Terraces	111	23.2	66	59	8	
	Floodplains	98	27.0	66	67	6	
	All	**84**	**3.9**	**50**	**59**	**160**	
Cumulative Water Retention	Summits	4.1	0.7	1.8	43	6	
Difference (cm)		Risers	8.8	1.0	5.7	65	35
	Treads	4.3	0.4	2.3	53	34	
	Nonstepped	16.3	0.9	7.8	48	71	
	Terraces	13	2.8	8	59	8	
	Floodplains	14.7	5.2	12.8	87	6	
	All	**11.4**	**0.7**	**8.3**	**72**	**160**	
Infiltration Rates (cm/hr)	Summits	10.7	3.7	8.3	5	5	
	Risers	9.8	0.9	5.2	53	35	
	Treads	5.0	0.5	2.4	47	28	
	Nonstepped	12.4	0.6	3.7	30	33	
	Terraces	11.3	1.1	2.4	21	5	
	Floodplains	6.8	1.1	2.2	33	4	
	All	**9.5**	**0.5**	**5.1**	**53**	**110**	
Organic Carbon (kg/m^2/soil profile thickness)	Summits	5.5	1.2	3.0	55	6	
	Risers	11.0	1.1	6.5	59	35	
	Treads	6.3	0.7	3.9	62	34	
	Nonstepped	21.2	1.0	8.6	40	71	
	Terraces	17.0	11.0	3.8	64	8	
	Floodplains	14.2	2.1	5.3	37	6	
	All	**14.7**	**0.8**	**9.6**	**65**	**160**	

their capacity to mitigate negative environmental consequences such as groundwater contamination and surface runoff.

With few exceptions (i.e. infiltration rates and nonstepped terrain data), field and laboratory data summarized in box plots represent only those pedons that were sampled for laboratory assays. This means that many of the pedons that were described in trenches for morphological properties were not included in these box plots. For nonstepped landforms, CEC was calculated from an algorithm developed for caliche soils in this study relating CEC, OC, and clay (described in Section 4.1.8). Likewise, volumetric water retention (WRD, and

TABLE 2 Spatial Variability of Selected Soil Physical and Chemical Attributes among Different Horizons (X = Mean, SE = Standard Error of Mean, SD = Standard Deviation, CV = Coefficient of Variation, n = Number of Observations). Data for Water Retention and Bulk Density of Soft Fragments are from this Study and that of Grand Prairie Land Resource Region of Texas (West, 1986)

Attribute	Horizons	X	SE	SD	CV (%)	(n)
Organic Carbon (%)	A, A1	3.1	0.1	1.5	47	103
	A2, A3	2.2	0.2	1.0	44	34
	2Bk	0.9	0.1	0.5	54	89
	3Cr, Crk	0.9	0.2	1.2	125	30
Calcium Carbonate Equiv. (%)	A, A1	61.6	2.2	22.3	26	103
	A2, A3	70.6	2.4	14.1	20	35
	2Bk	81.4	1.1	10.0	12	89
	3Cr, Crk	81.0	2.2	11.9	15	30
Coarse Fragments (%)	A, A1	29	2.4	24	85	103
	A2, A3	25	4.2	25	100	35
	2Bk	18	2.1	20	115	89
	3Cr, Crk	47	3.8	30	64	63
Clay (%)	A, A1	21.1	1.0	10.4	49	103
	A2, A3	20.2	1.6	9.4	47	35
	2Bk	19.6	0.9	8.9	45	89
	3Cr, Crk	17.7	1.5	8.4	48	30
Cation-Exchange Capacity (meq/100g soil)	A, A1	21.5	1.4	11.1	52	66
	A2, A3	18.5	1.7	9.0	49	29
	2Bk	8.8	0.7	5.2	59	59
	3Cr, Crk	8.0	1.1	5.3	66	21
Water Retention (% vol.)	A, A1	23.4	1.9	10.7	46	30
	A2, A3	17.3	2.1	7.6	44	13
	2Bk	14.9	1.5	7.9	53	29
	3Cr, Crk	12.2	5.1	12.4	102	6
	R	8.4	2.5	5.2	62	3
	Hard CF's	9.7	3.2	5.7	57	3
	Soft CF's	19.7	2.6	9.4	47	13
Bulk Density (g/cc) at −0.33 bar tension	A, A1	1.27	0.1	0.27	20	30
	A2, A3	1.42	0.1	0.26	18	13
	2Bk	1.66	0.1	0.31	18	29
	3Cr, Crk	1.80	0.1	0.21	11	6
	R	2.09	0.1	0.21	9	3
	Hard CF's	2.13	0.1	0.21	10	4
	Soft CF's	1.68	0.1	0.3	47	18

CWRD) were reported as expected values for nonstepped terrains from field soil textural classes according to the USDA-NRCS protocol.

Spatial variation of the attributes presented in Tables 1 and 2 for the Brackett and Volente mapping units can generally be described as highly variable with coefficients of variation (CVs) ranging from 45% to 105% (Wilding and Drees, 1983). With some exceptions, most of the CV values center around 50%, which means broad dispersion of data sets (illustrated in box plots). Bulk density and calcium carbonate equivalent (CCE) are much less variable than the other attributes measured, with their CVs ranging from about 10 to 25% (Table 2). One of the most variable properties observed is coarse fragment content (CF). Here, the CVs ranged from 85 to 115%. No systematic relationship was found in CVs among different landscape units, which were of similar magnitudes as among different soil horizons. This suggests that stratifying soils by landform units or by horizonation did not significantly decrease the spatial variability of the attributes investigated.

4.1. Stepped Landform–Soil Relationships

Sedimentary strata having alternating rock properties commonly exhibit repetitive ranges of chemical and physical stability. Weathering of these strata results in a stepped hillslope terrain, and such riser and tread microtopographic elements are typical of the Glen Rose Limestone (Woodruff et al., 1992; Wilding, 2007; Wilcox et al., 2007; Woodruff and Wilding, 2008). These microterrain units yield striking differences in soil depth, soil/geomorphic processes, and their hydrologic function (Fig. 1). Soil distribution patterns, soil development, biogeochemical properties, and hydrologic functions are controlled locally by repetitious sequences of riser–tread microtopographic units (risers typically are 1–2 m high and treads 10–20 m long). The variable geometry of the steps is dependent upon the thickness, composition, and resistance to weathering of the interbedded limestone/dolomitic strata; the greater the proportion of strata having a more heterogeneous fabric, the greater the step relief and the thicker soils on risers. In contrast, strata having a more homogeneous fabric form resistant ledges above and below risers and form the resistant substrate underlying treads as seen in aerial and oblique scenes of this stepped landscape (Fig. 2).

Relatively thick regolith/soil sequences are set as disjunct wedges into the stepped terrain. Typical changes in soil attributes observed along a traverse from the riser to tread couplets include 1) decreases in slope gradient, soil thickness, permeability, water-storage capacity, and biological activity (total organic carbon sequestered) and 2) increases in soil runoff and erosion on treads (Fig. 3).

Slope gradients are always greater on risers (25–40%) than treads (5–10% or less). In spite of the high slope gradient on upper and middle risers, infiltration rates averaged 11 cm/h, mean plant-available water was 9 cm for entire soil profile, surface runoff averaged 5% of mean annual precipitation, and sediment erosion rate was around 55 kg/ha/yr (Woodruff and Wilding, 2008;

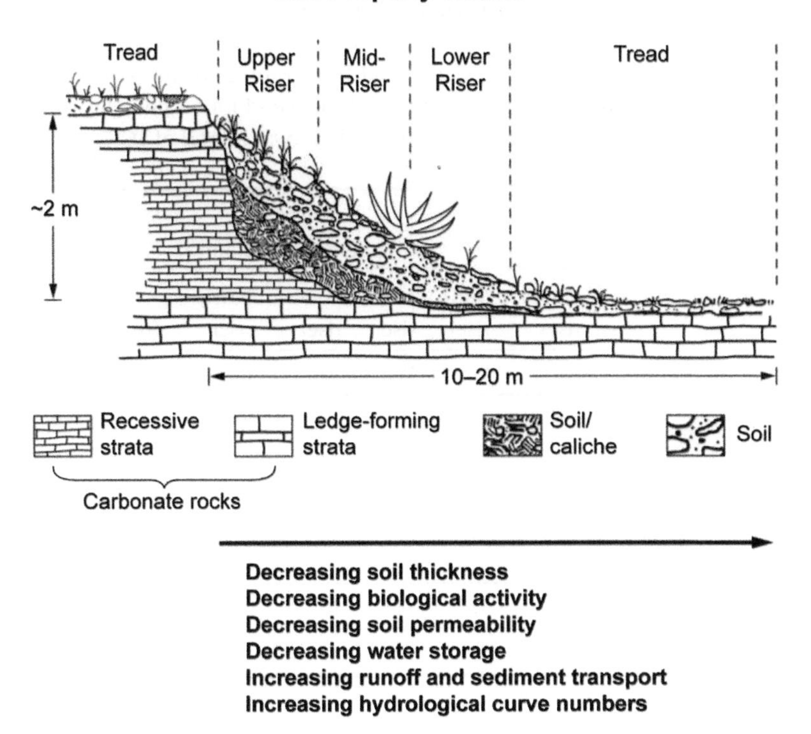

FIGURE 3 General soil property trends from riser to tread in a schematic of stepped landform. *(Modified from Wilding, 2007)*

Wilding, 2007). This is in strong contrast to lower risers and low gradient treads where infiltration rates averaged about 5 cm/h, mean plant-available water was only 3 cm for entire soil profile, surface runoff reached 31% of mean annual precipitation, and erosion rate jumped to about 315 kg/ha/yr (Woodruff and Wilding, 2008). Similar data for a series of measurements made on a representative single step are given in Table 3 (Wilcox et al., 2007). These data suggest that upper and middle risers are the preferred sites for application of treated wastewaters with minimal environmental hazards for surface runoff and off-site transport.

4.1.1. Horizon Stratigraphy

Horizon stratigraphy is spatially diverse in the study area but illustrates repeating patterns of soil-microtopographic elements (Fig. 4a). For risers, horizon stratigraphy is commonly A, 2Bk, 3Cr, and 4R. Sometimes a mixed transitional horizon occurs between A and Bk horizons (identified as A/Bk or Bk/A) and Cr or R horizons (identified as Bk/Cr, Cr/Bk, Bk/R, or R/Bk). These are horizons that contain physical mixtures of two different genetic horizons.

TABLE 3 Soil and Hydrologic Characteristics for a Sequence of Runoff Plots Positioned along a Transect of Riser and Tread Stepped Landform. Replicate Measurements were Made of all Topographic Positions Except the Upper Riser. Runoff and Erosion were Monitored for an 804-Day Period from 1992 to 1995. During this Time there was a Total of 1625 mm of Rainfall

Soil and hydrologic attribute	Topographic position						
	Upper riser	Mid-riser	Mid-riser	Lower riser	Lower riser	Tread	Tread
		Rep 1	Rep 2	Rep 1	Rep 2	Rep 1	Rep 2
Runoff (% of precipitation)	2	11	28	32	24	33	12
Erosion rate (kg/ha/yr)	10	35	285	1795	1510	480	275
Terminal Infiltration rate (cm/hr)	14.8	11.0	7.7	2.9	4.7	0.0	7.7
Slope (%)	26	38	36	16	26	4	6
Soil Depth (cm)	66	97	91	41	66	10	15

(Modified from Wilcox et al., 2007)

The A horizon is a zone of organic matter enrichment and formed in colluvium of hard ledge-former rock fragments and transported pedisediments or weathering products from upslope treads. The Bk subsoils are loci of pedogenic carbonate enrichment into weathered caliche regolith. Here, accumulation of pedogenic carbonates reflects carbonate dissolution, translocation, and immobilization from upper to lower horizons. They are also loci where in situ micritization of indigenous fine-grained disseminated carbonates is prominent (West et al., 1988a). At the base of the soil are root and water-restrictive Cr and/or R layers. These layers also commonly accumulate pedogenic carbonates (Crk and Rk) that plug joint planes and fissures in the limestone bedrock. Petrocalcic horizons (Bkm), when present, are superposed above less permeable Crk and Rk horizons. Riser stratigraphy denotes a sequence of pedogenically developed horizons that have highest organic matter accumulations, infiltration rates, biological activity, root proliferation, and most intense soil weathering conditions that decrease with depth. Vertical root proliferation and fluid transport essentially cease at contacts with petrocalcic and bedrock restrictive layers. Ephemeral aquitards thus develop above these restrictive layers during seasonal wet periods.

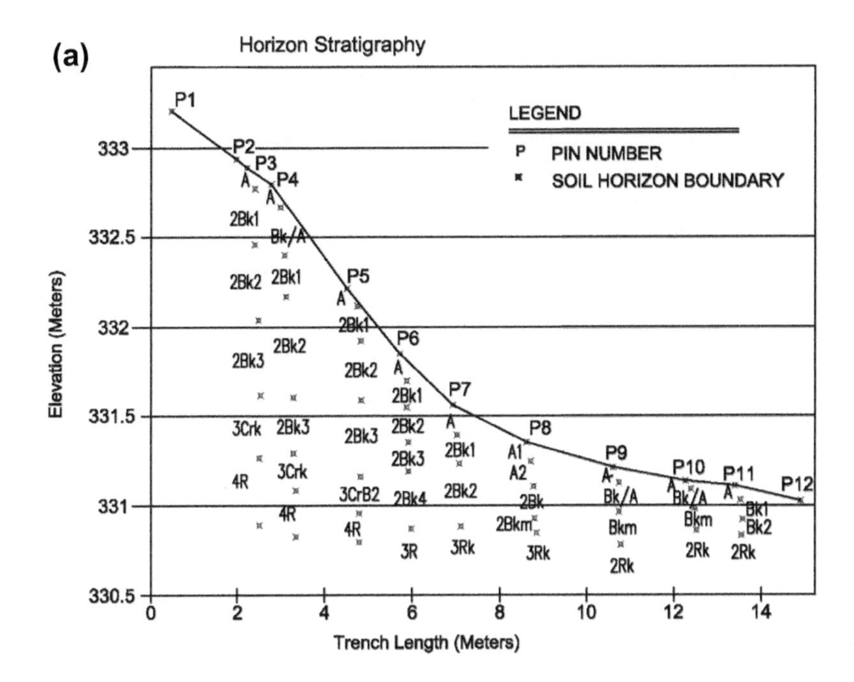

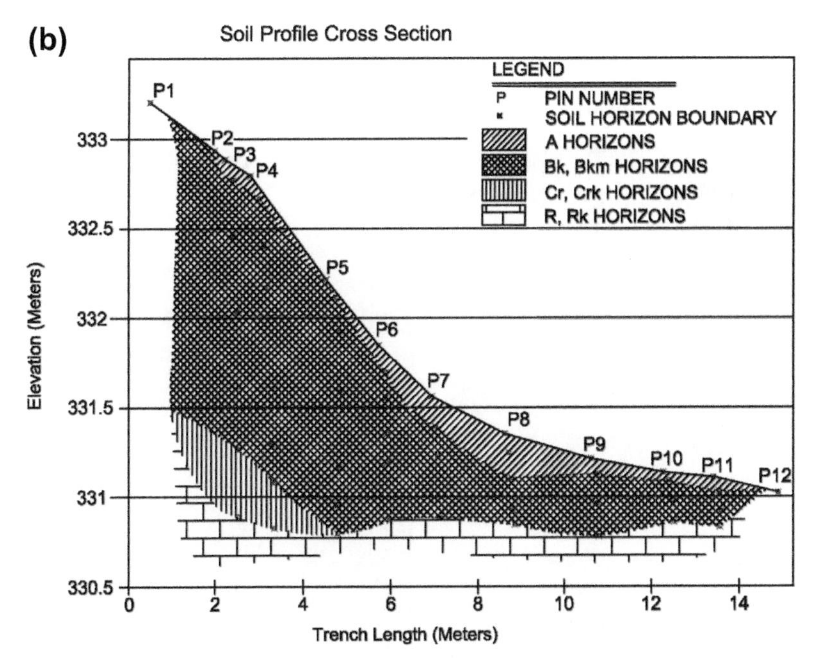

FIGURE 4 (a) Horizon stratigraphy and (b) soil profile cross section at trench 4 of the rocky creek watershed utility, L.P., Travis County, TX. *(Courtesy of Dennis Lozano, P.E., Murfee Engineering Company, Inc., Austin, Texas)*

Horizon stratigraphy on treads (A, Bk, 2Ck, and/or 2Rk) is often similar to risers but much thinner, and the A and Bk horizons are usually formed from limestone residuum and transported pedisediments rather than colluvium. The same kind of processes involving dissolution and translocation of carbonates, accumulation of OC, formation of soil structure, and development of soil color occur in treads as in risers, but soils are less intensively weathered, translocation of carbonates less strongly expressed, and horizon thickness is significantly compressed. This is related to lateral flow being greater in treads than vertical transport, with surface runoff and erosion maximized on sparsely vegetated, crusted, gently sloping tread soil surfaces.

Soil stratigraphy logs taken at close lateral intervals were helpful in reconstructing soil profile cross sections within trenches (Fig. 4a). By linking all A horizons, Bk and Bkm horizons, Cr and Crk layers, and R and Rk layers along the riser–tread transect, zones of organic carbon enrichment in A horizons, inorganic pedogenic carbonate enrichment in Bk and Bkm horizons, and restrictive bedrock aquitard layers in Cr, Crk, R, and Rk horizons can be clearly depicted (Fig. 4b).

4.1.2. Soil Taxonomy Classification

Soils of upper and middle risers are generally classified as Udic Calciustolls (Fig. 5a), but Petrocalcic Calciustolls are sometimes present on lower riser positions. Petrocalcic placement reflects sufficient accumulation of pedogenic

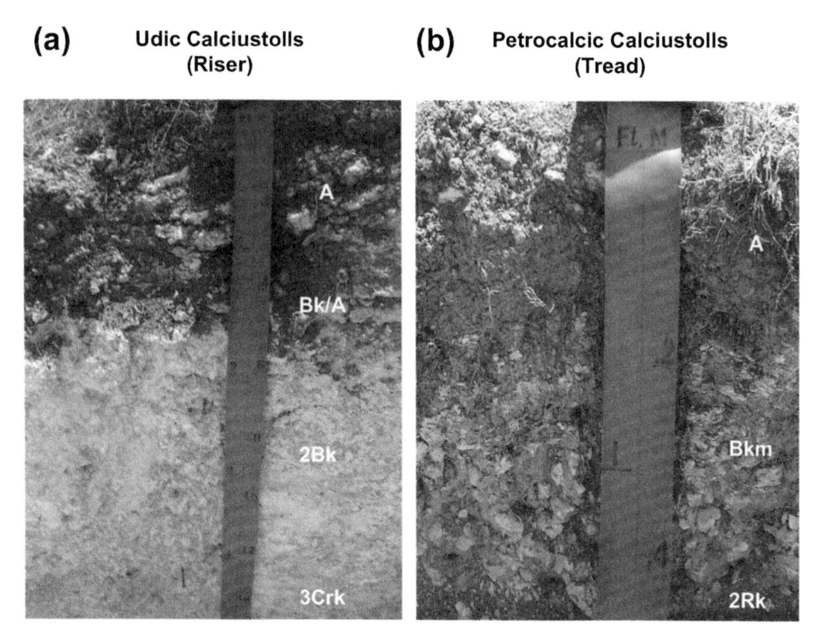

FIGURE 5 (a) Udic Calciustolls on riser at pin 4 and (b) Petrocalcic Calciustolls on tread at pin 9 in trench illustrated in Fig. 4. *(Color version online and in color plate)*

carbonates within 100 cm of the surface to form indurated Bkm horizons that are laterally extensive. Thinner soils on treads are classified as Lithic Calciustolls, Petrocalcic Calciustolls (Fig. 5b), or Lithic Haplustepts and Petrocalcic Haplustepts. Inceptisols (Haplustepts) contain insufficient OC to be classed as Mollisols (Calciustolls). The Lithic class means bedrock is $\leq$50 cm of the soil surface. Soil Taxonomy (Soil Survey Staff, 1999, 2010) family textures are quite variable depending on coarse fragment (CF) contents and fine-earth textures. They include skeletal ($\geq$35% CFs) and nonskeletal ($\leq$35% CFs) classes of loamy, fine loamy, coarse loamy, silty, and clayey family textures. More of the soils on risers are skeletal while soils on treads are generally nonskeletal with loamy or silty textures. The family mineralogy placement is carbonatic ($\geq$40% CCE) and mean annual soil temperature regime is thermic ($\geq$15 °C to <22 °C).

Soils of summits are classified as Lithic Calciustolls, Petrocalcic Calciustolls, Lithic Haplustepts, and Petrocalcic Haplustepts depending on depth to bedrock, amounts of sequestered organic carbon, and degree of expression of calcic and petrocalcic horizons. The classification of these soils, family placements, and function in terms of surface runoff are similar to that of treads in stepped hillslopes.

Soils on terraces and floodplains within the Volente mapping units are classified as Petrocalcic Calciustolls, Vertic Calciustolls, Pachic Calciusolls, and Cumulic or Pachic Haplustepts. Most of the terrace soils have a petrocalcic horizon within 100 cm of the surface, are fine textured, and have dark-colored surface horizons. Terrace soils serve as buffers to water and solute transport from upland overland flow. Soils in valley fill and alluvial deposits of floodplains commonly have thick dark-colored surface horizons with or without subsoil calcic horizons. Soils of valley floors are fine loamy and clayey in upper sola and have skeletal textures in lower sola depending on transport and sediment accretion processes.

4.1.3. Soil Thickness

Soil thickness is highly variable across caliche landforms, but clear trends are evident from thinner soils on upland summits and treads and thicker soils on stepped risers, non-stepped terrain, terraces, and valley floors (Fig. 6). The thickness of upland soils is bedrock controlled and dependent on the relative resistance to weathering of indurated recessive and ledge-former bedrock strata. On risers, soils have formed in strata having heterogeneous fabrics (nodular carbonates, moderate amounts of noncarbonate mineral occlusions, clay partings, bioturbation, burrows, macrofossils, etc.) that usually develop moderately thick to thick Bk horizons in caliche regoliths. Conversely, on more gently sloping summit and tread landforms, more homogeneous ledge-former strata crop out with only a thin, discontinuous soil cover. Sensitivity to weathering is related to the diversity of bedrock composition (carbonate mineralogy and noncarbonate constituents), textures, fabrics, bioturbation, porosity, fissuring, and fracturing (Woodruff and Wilding, 2008). Once exposed, heterogeneous

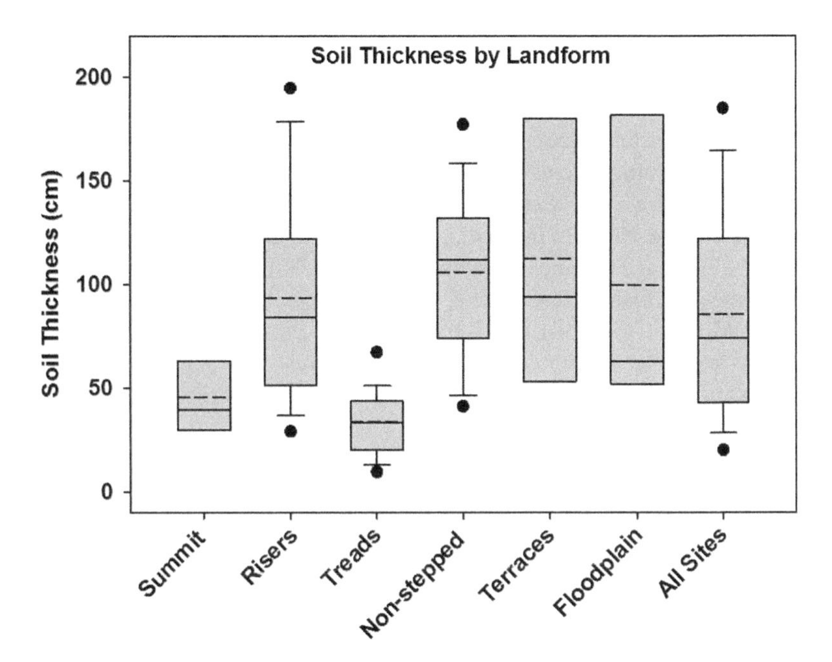

FIGURE 6 Box plots for soil thickness in different landforms. With sufficient data a box and whisker diagram present the median (solid line) and mean (dashed line); and the 5% (lower dot), 10% (lower whisker), 25% (lower edge of box), 75% (upper edge of box), 90% (upper whisker), and 95% (upper dot) distribution of data. Number of observations for each box plot is given in Table 2.

recessive substrata weather more rapidly via exfoliation, hydration, dissolution, oxidation, shrink/well, and other biological/physical/chemical processes that promote bedrock disaggregation and disintegration. The weathered products resemble marls, but unweathered heterogeneous carbonate strata are well-indurated consolidated bedrocks.

Soil thickness and corresponding soil morphological properties indicate and govern hydrologic processes from summits to valley floors. This is discussed in detail later in this chapter. Briefly, variations in soil and bedrock composition on stepped hillslope terrains favor a series of local recharge/discharge sites that present multiple barriers to surface and subsurface water flow.

4.1.4. Soil Organic Carbon (OC)

Caliche soils are favorable for OC sequestration. This fact, along with other evidences of faunal and floral activities, strongly suggests a pedogenic, rather than lithogenic, origin of caliche (Wilding, 1993; Woodruff and Wilding, 2008). Values of OC are high despite light soil colors, especially in the A and upper Bk horizons, and remain near 1% at the base of the soil in and above restrictive bedrock layers (Fig. 7a). The stability of OC in caliche soils is attributed to calcium carbonate-rich substrates that form stable calcium

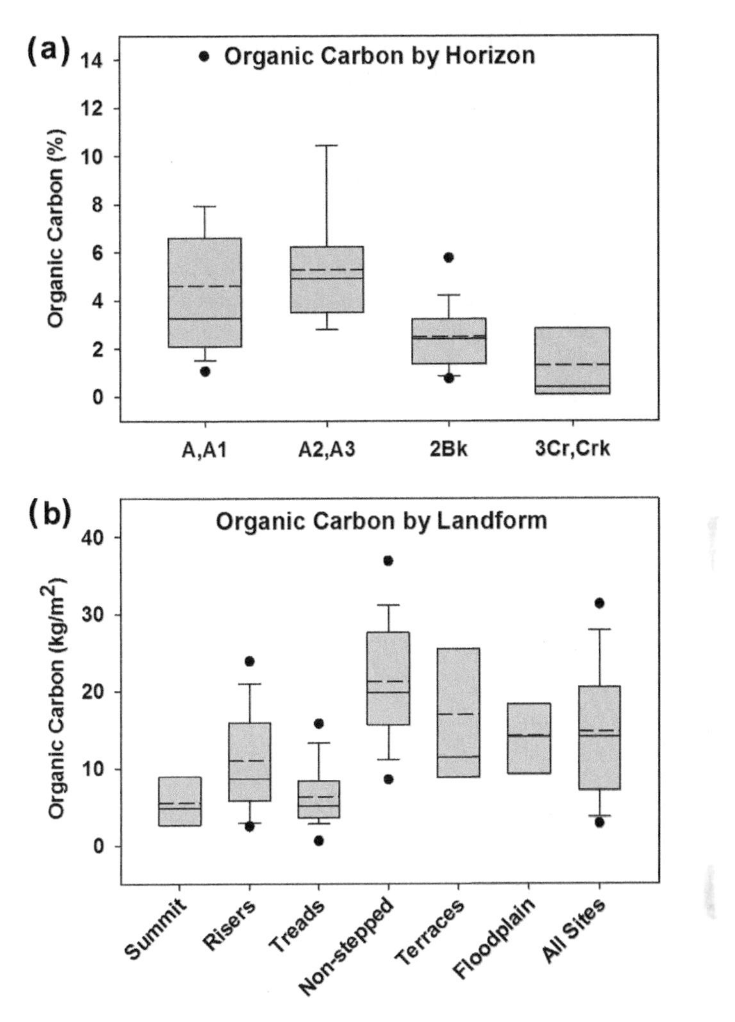

FIGURE 7 Box plots of (a) organic carbon content (weight percentage) in different soil horizons, and (b) total organic carbon content (kg/m^2) sequestered in soils in different landforms. Number of observations for each box plot is given in Tables 2 and 3.

humates. Additionally, OC is entrapped within pedogenic carbonates, micropores, and reactive clay mineral interlayers that restrict access by microorganisms. Total OC accumulations in the entire soil profile (Fig. 7b) are comparable to those in Alfisols, Vertisols, and Mollisols in more humid regions of Texas and elsewhere (Drees et al., 2001; Nordt and Wilding, 2009). However, total OC values reported by Bohn (1976) for North American soils of comparable orders are generally larger than those reported for caliche soils herein. In our study, highest values are on risers, terraces, and nonstepped landforms, while summit and tread soils contain the least OC. Lower OC accumulations on summit and tread soils reflect lower biomass production, higher oxidation rates, greater drought stress, and greater runoff and erosion on these thinner soils, as identified in this study.

Organic carbon enhances nutrient retention and chemicals that may otherwise be lost with leaching from these fragile and infertile soils. Even with high OC contents, caliche soils are commonly deficient in plant-available nitrogen, phosphorus, and micronutrients (Wilding, 2007). However, high OC contents favor microbial growth and other biotic activity that promote stable soil structure and a porosity through which infiltration rates are sustained, soil water stored, and bioremediation achieved. Higher OC contents in the A horizons partially explain their lower mean bulk densities (~1.3 to 1.4 g/cm^3, Table 2) and more favorable biotic environment.

4.1.5. Calcium Carbonate Equivalent (CCE)

The CCE in caliche soils is high and reflects development from calcite and dolomite-rich bedrocks. The CCEs increase from about 50 to 70% in surface horizons to about 80–90% or more in parent materials (Fig. 8a). Higher CCE values in substrates reflect in part translocation of carbonates from the A to Bk

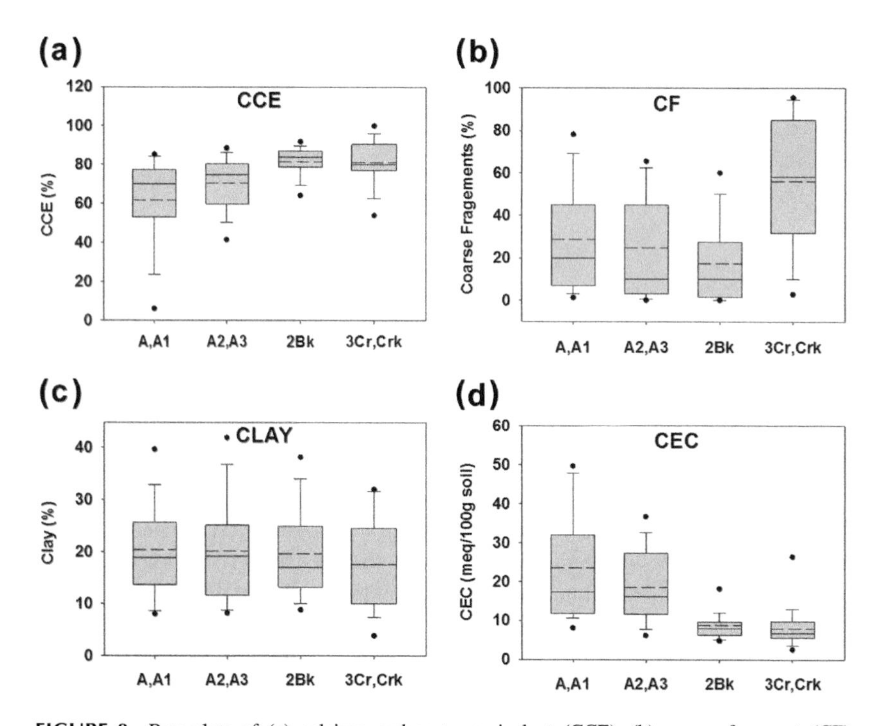

FIGURE 8 Box plots of (a) calcium carbonate equivalent (CCE), (b) coarse fragment (CF) content, (c) clay content, and (d) cation-exchange capacity (CEC). Number of observations for each box plot is given in Table 3.

or Bkm horizons in addition to differences in parent materials among these horizons. Segregated forms of carbonates that support such interhorizon translocation have been well documented (Wilding et al., 1997; Wilding, 2007) but many disseminated fine-grained carbonates are also of pedogenic origin and formed by inorganic and/or biological processes (West et al., 1988a; Rabenhorst et al., 1984; Monger et al., 1991; Wilding et al., 1997). Translocated carbonate cements decrease soil porosity and permeability in the lower Bk and Bkm horizons; they also plug joint planes and fissures in the Crk and Rk layers (Wilding et al., 1997). These aquitards limit transport of water and chemicals to subjacent aquifers and become physical and chemical buffers to groundwater contamination. Evidence of chemical reduction at the aquitard interface is redoximorphic iron oxides zoned along bedding planes in the lower Bk, Bkm, and Crk restrictive layers. These redoximorphic markers form in response to oxidation/reduction conditions during alternating periods of drainage and formation of ephemeral water tables above aquitards (Vepraskas, 1992). Nitrate movement is also buffered by denitrification processes in saturated zones above aquitards. Phosphorus, microelements, and toxic heavy metals are immobilized and become less plant available in soils with high pH and CCE values (Adriano, 1986; Mortvedt et al., 1999; Wilding, 2007). Further, the fact that many disseminated carbonates are pedogenic (up to 80%), fine-grained, and are expected to have higher specific surface areas than indigenous disseminated carbonates, make them strong candidates to be more chemically reactive (West et al., 1988a; Rabenhorst et al., 1984).

High CCEs promote low shrink/swell potentials in caliche soils with the values of coefficient of linear extensibility (COLE) ranging from nil to <0.04 (Wilding, 2007). One reason for such low COLE values in caliche soils is because 10–50% (mean of 25%) of the total clay fraction is carbonate clay (West, 1986). This reduces the shrink/swell potential of noncarbonate clays. Calcium and Mg are the dominant exchangeable cations that help maintain soil structure in a flocculated state with low sodium adsorption ratios. Soil reaction in all horizons is buffered in the pH range of 7.0–8.5 by high CCEs. Landscapes are physically stable and exhibit little danger of mass movement or slump, even on steep risers of up to 45% slope gradient. Moderately high infiltration rates are maintained in caliche soils independent of soil texture because of flocculated soil systems that have negligible shrink/swell capacity.

4.1.6. Coarse Fragments (CF)

Coarse fragments of limestone/dolostone are extensive and highly variable in caliche soils (Fig. 8b). For example, CF content in the A, Bk, and Cr horizons ranged from 0 to 90%, with mean values of 25–50%. The high content of CFs, especially those that can be broken by hand, contributes not only to higher infiltration rates but also to higher plant-available water. Studies by Wilding (2007) and West (1986) have shown that moderately cemented soft CFs (sampled mostly from Bk and Cr horizons) hold on average 20% volumetric plant-available water (Table 19-2). Hard CFs (sampled from A and Bk horizons) contain on average about 8% plant-available water content. Hard CFs are derived mostly from the disintegration of moderately- to well-indurated Bkm, Cr, and R horizons (Wilding, 2007). Variability in porosity and water retention in bedrock strata and CFs reflect initial rock porosity and degree of weathering. Similar data for Cr and R layers confirm estimates of volumetric plant-available water in CFs, except the bulk density and water retention for soft CFs more closely approximate Bk horizons (Table 2).

Based on –0.33 bar bulk density values shown in Table 2 and using 2.71 g/cm^3 for calcite particle density, the mean total porosity of soft CFs is about 38% and that of hard CFs about 21%. The mean volumetric water retention at –0.33 bar tension are 38.3 and 19.6% for soft and hard CFs, respectively, while that at –15 bar tension are 17.3 and 11.3% for soft and hard CFs, respectively. Calculations using the capillary tension equation [r (radius of pore in mm) = 1.5/h (tension in cm of water)] suggest that essentially all the pores in these CFs have radii of <4.4 μm. Those with larger pore radii contribute <2% to the total porosity. Most of the macropores in CFs are associated with horizontal planar voids in the bedrock (Wilding et al., 1997). This means that on average about 21% of the pores in soft CFs have radii between 4.4 and 0.1 μm with 17% <0.1 μm. Similar data for hard CFs are about 8% between 4.4 and 0.1 μm with 11% <0.1 μm. Such porosity attributes of CFs enhance water-storage capacities on a total soil basis.

About half or slightly more of the total CFs observed in the field were soft CFs. They are considered to be reactive soil materials, and function like the fine-earth fraction. Hence, in descriptions of caliche soils volume estimates were made of both soft and hard CFs, and only the hard CFs that could not be broken by hand were used in determining skeletal content and corrections in biophysiochemical parameters to a CF-free basis. Coarse fragments have added value to soil bioremediation because they would increase mean residence time for solutions entering fragments, without inducing high infiltration rates.

4.1.7. Particle-Size Distribution

Soil textures are commonly gravelly or very gravelly (sometimes extremely gravelly) loam, clay loam or silt loam on summits, risers, and treads. Finer textures in downslope terraces and alluvial floors reflect sediment transport and concentration of insoluble clay-rich carbonate weathering products. For upland soils, essentially all of the sand, silt, and a significant proportion (25–50%) of the clay separates are dolomitic or calcitic in mineralogy. When the bedrock lithology is dolomitic, the fine-earth fraction is commonly silt or silt loam reflecting the preponderance of dolomite silt-size rhombs released from the bedrock upon partial weathering.

Mean clay percentages are about 20% for all caliche soil horizons (Fig. 8c), while that of silt are 45% and total sands 35%. No depth functions are apparent for any particle separates in these soils. This likely reflects relatively weak weathering intensities accompanied by high spatial variability vertically and laterally. These caliche soils contain a high proportion of sand and silt which provides optimal porosity conditions for water infiltration, redistribution, storage, and microbial activity, soil oxidation, and bioremediation.

4.1.8. Cation Exchange Capacity (CEC)

The CEC of soils is controlled by OC content, noncarbonate clay content, carbonate content (CCE), and hydrated iron/aluminum oxyhydroxides. Figure 8d illustrates that the highest CECs are in surface horizons (20–25 meq/100 g soil) and decrease with depth to <10 meq/100 g soil in the Bk and Crk horizons. In surface horizons, OC contributes the most to CEC while in subsoils the noncarbonate clay fraction contributes most to CEC. The chemical sorption capacity as a function of organic carbon and clay contents yielded the following relationship for caliche soils in this study: $CEC = -4.039 + 0.639$ (clay %) $+ 4.326$ (OC %), with an R^2 of 0.81 ($n = 72$) (Wilding, 1993). Charge contributions to CEC predicted from this equation for our soils would be about 430 meq/100g OC and 65meq/100g clay. These values are close to those obtained by West (1986) for similar soils in the Grand Prairie region of Texas (440 meq/100 g OC and 67 meq/100 g clay).

Clay contents coupled with high OC provide very favorable CECs for retention of nutrients and compounds for plant uptake and subsequent bioremediation. Most of the charge in these soils is negative, and hence the soil will adsorb and release cations to the soil solution in proportion to the chemical equilibrium of these constituents in solution. Nutrients in soils and wastewater, except for nitrates, will be either adsorbed to the soil clay and organic matter constituents or precipitated as slowly soluble products on carbonate and sesquioxide mineral grain surfaces.

4.2. Soil Hydrologic Functions

4.2.1. Soil Water Retention

Caliche soils in this study have low to moderate plant-available water retention held at matric potentials between –0.33 and –15 bars. Figure 9a shows that such plant-available water for the fine-earth fraction of different horizons averaged between 10 and 25%.

The mean cumulative water retention difference (CWRD) for soils on different landforms is shown in Fig. 9b. For summits and treads, the CWRD values are low, with means of about 4 cm for total soil profile. For risers, the CWRD is low to moderate, but about twice the values as compared to summits and treads. For soils on terraces and floodplains, the CWRD values are moderate to high, with mean values of about 15 cm for total soil profile.

Upland caliche soils with limited CWRDs (summits and treads) are especially subject to overland flow during periods of high-intensity rainfall because their low storage capacities are easily exceeded by infiltrating water. When this happens, runoff from these sites recharges downslope riser, terrace, and

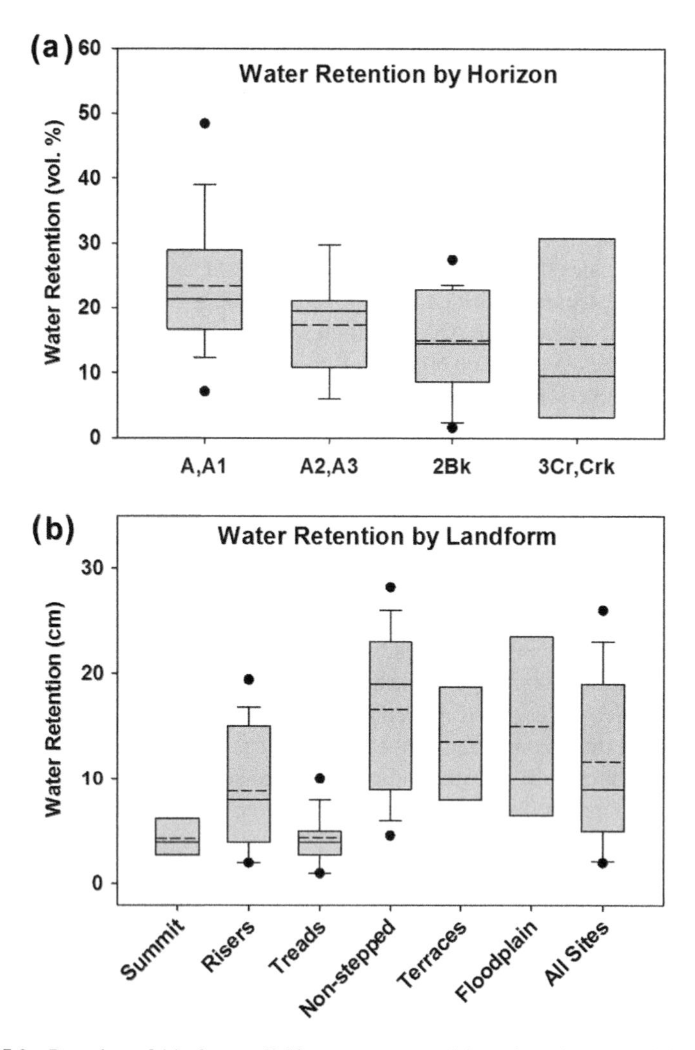

FIGURE 9 Box plots of (a) plant available water content (% by volume between –0.33 and –15 bars) in the fine-earth fraction (<2 mm) in different horizons of caliche soils, and (b) cumulative water retention difference (cm water/soil profile thickness) for soils in different landforms. Number of observations for each box plot is given in Tables 2 and 3.

floodplain soils that have higher water-storage capacities, higher plant cover, and greater consumptive water use.

4.2.2. Terminal Infiltration Rate

Mean terminal infiltration rates for soils on different landscape elements are presented in Fig. 10. Infiltration rates are variable but higher for soils with well-established vegetative cover, organic mulch, vigorous rooting, and with well-aggregated surface horizons than for bare soil conditions that are eroded,

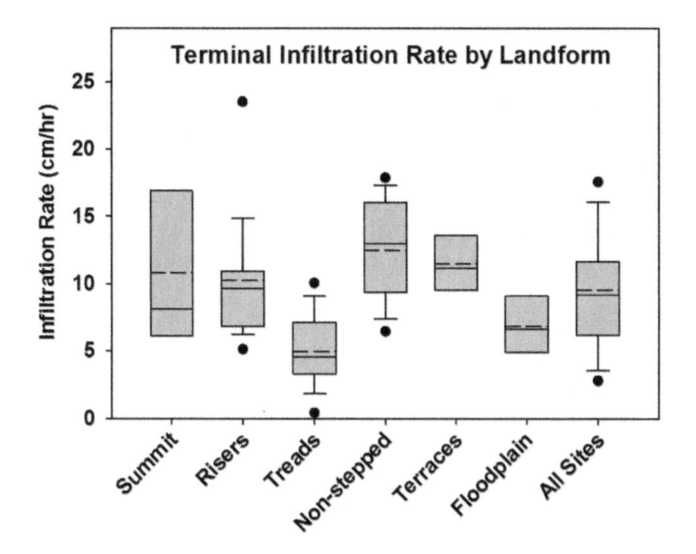

FIGURE 10 Box plots for terminal infiltration rates of soils in different landforms. Number of observations for each box plot is given in Table 2.

compacted, or sealed with inorganic and/or organic crusts. Mean infiltration rates ranged from a low of about 5 cm/h on treads to a high of about 12 cm/h on terraces and nonstepped uplands. Rates are commonly about twice as great for risers as compared to treads, though overlap occurs if crusted and sealed surface conditions occur on given sites. For all infiltration tests run across all sites in this study, 90% were between 2.5 and 14 cm/h (Fig. 10).

However, as discussed previously, high CFs in caliche soils are not as negative an attribute as if they were siliceous CFs; this is because both hard and soft CFs absorb significant volumes of water, contribute to water storage, and enhance residency time for fluids absorbed in limestone materials. The increased rate of fluid transport with increased gravel content does not generate a risk of insufficient contact time to promote effective bioremediation of infiltrating water, as CF contents of up to 60% do not result in excessively high infiltration rates in these soils. Furthermore, there are multiple buffers at the base of the soil to restrict the water movement, including partially cemented calcic and petrocalcic horizons, widespread areas of weakly jointed and slowly permeable limestone bedrock, infillings of joint and fracture planes of the soil–bedrock interface with carbonate cements, and fewer root volumes available for preferential flow. Chemical reduction of nitrates is also probable at the base of the soil with aquitards and wet soil conditions.

4.2.3. Runoff and Erosion

In our study of selected riser–tread sequences, runoff monitored from natural rainfall events is consistent with patterns observed from the infiltration experiments (Table 3). Runoff as a percentage of precipitation increased downslope from upper riser to lower riser, and runoff from the lower risers was comparable to that of treads. Runoff from the two tread plots, however, was markedly different (Table 3) because of different surface covers. The high-runoff-producing tread (replicate 1) was mostly bare soil and rock, while the other tread (replicate 2) had good vegetation cover. In marked contrast to that from the upper and mid-risers, erosion was much higher at the lower risers, which combine relatively low infiltration rates with that of steeper slopes than treads.

These observations of runoff, along with the infiltration measurements, indicate that most of the surface runoff generated in this landscape comes from the lower risers and treads. During the observation period, around 1/3 of rainfall ran off the lower riser and tread plots. The lower risers also appear to be the main source areas for sediment transport. However, relatively little sediment comes from the even steeper upper and mid-risers.

Despite gentle slopes, residual soil cover and pedisediments on treads are subject to rapid erosion, as most incident water is shunted off these surfaces. Runoff dominates because of the slab-like, weakly jointed, resistant substrate, and thin soils with low waterstorage capacities, as well as degraded, algal encrusted surfaces that impede infiltration (Wilding et al., 1997). Moreover, during extensive rainy periods, the riser–tread interfaces may discharge perched water stored within the regolith, thereby causing erosion near those interfaces. Evidence for erosion on treads is indicated by sparse vegetation and the local presence of small isolated pinnacles of soils protected by tufts of grass that stand several cm above the intervening, denuded ground. Juniper stumps with exposed roots commonly occur and provide further evidence for rapid erosion on treads (Marsh and Marsh, 1993).

The riser–tread hillslopes are composed of an alternating sequence of sources and sinks that facilitate both surface and subsurface flow. During periods of high rainfall, overland flow is captured by downslope risers, stored, and ultimately used by vegetation or slowly released by "seeps" at the base of the riser. We hypothesize that while traversing from summits to down-gradient terraces and floodplains, water flow occurs along a cascading pathway from step to step. This would result in multiple buffers for water-quality remediation and numerous sediment traps for transported particulates. Soils in the lower topographic positions serve as important buffers restricting overland flow from direct entry into freshwater streams and river systems. While this conceptual model seems plausible, it has not yet been completely verified with large-scale experimental testing.

Nevertheless, it is clear that hydrologic processes and water balance across this landscape are heavily governed by soil thickness and its water-holding capacity. Because caliche soils commonly have low to moderate water-holding capacities, buffers in the soil system that mitigate surface runoff can be swamped and lead to flash flooding during high-intensity short-duration storm events, especially under antecedent wet soil conditions. This limitation deserves vigilant attention by local authorities responsible for watershed management and natural hazard mitigation in these landscapes.

4.2.4. Soil Moisture Dynamics

Time series plots of soil matric potential and associated water-table depths are presented for two adjacent risers separated topographically by an elevation of 8.5 m (Fig. 11a, b). Similar time series and water-table data are available for 12 other sites in the study area but for space limitations are not included herein. From these data, computations were made of the percentage of time during different seasons when ephemeral perched water tables were present in caliche soils (Fig. 12), and the probability of encountering a perched water table above an aquitard at various soil depths (Fig. 13).

Ephemeral perched water tables have high spatial and seasonal variability from riser to riser. Over the monitoring period, the percentage of time that saturated conditions existed in these soils varied from ~2% up to ~30% (Fig. 12). The highest values were during seasonal wet periods of late fall, winter, and early spring (Figs 11 and 12). Perched water tables occurred

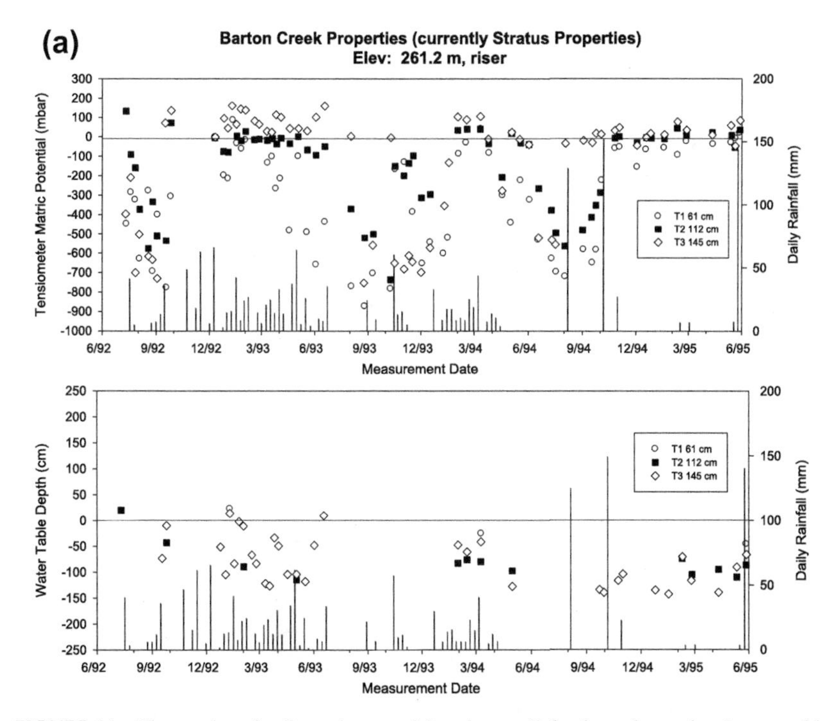

FIGURE 11 Time series of soil matric potential at three soil depths and associated water table and daily precipitation for (a) a riser at site 5 (Barton Creek Properties) and (b) an adjacent downslope riser at site 6 (shown in next page).

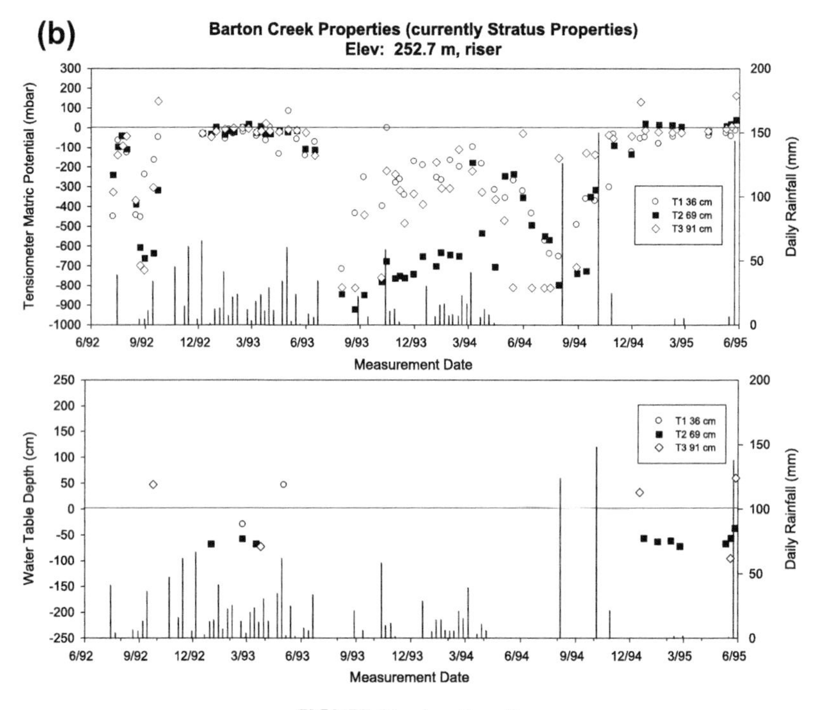

FIGURE 11 (*continued*).

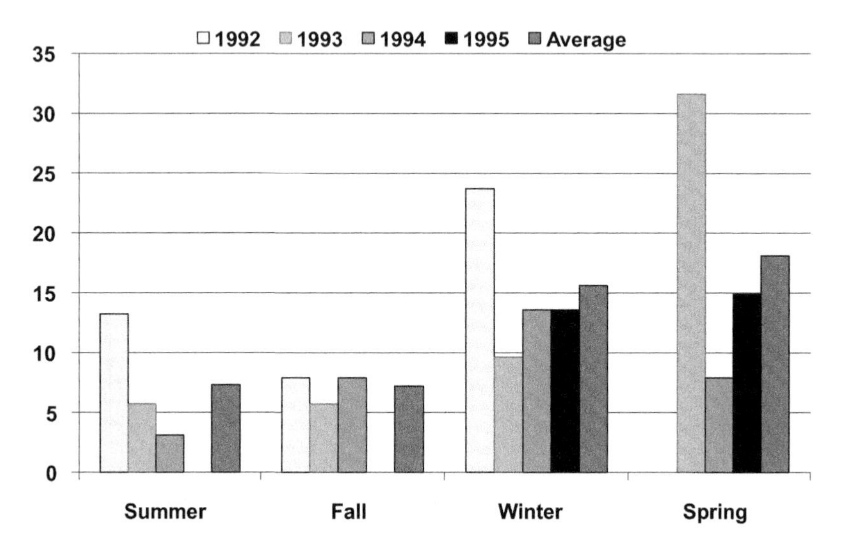

FIGURE 12 Percentage of seasonal saturation computed from matric potential data for seven sites at Barton creek properties monitored during 1992–1995 (total 804 days).

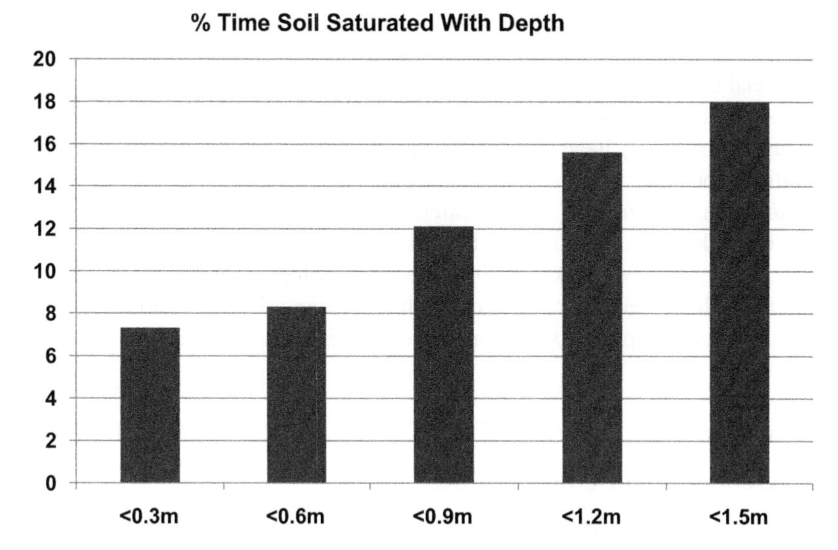

FIGURE 13 Percentage of time when soil was saturated (ephemeral perched water table) below soil surface based on matric potential monitored from 1992 to 1995 (total 804 days).

more frequently immediately overlying aquitards (less permeable and partially cemented calcic and petrocalcic horizons and weakly fractured–jointed and plugged limestone bedrock). The percentage of time water tables occurred in risers increases with soil depth from ~8% at <30 cm to ~18% at <150 cm (Fig. 13). It is less probable to have a perched water table near the soil surface where evapotranspiration and vegetative consumption are the greatest. Rapid movement of soil water through permeable mollic and upper calcic horizons also favors the formation of a perched water table where drainage loss through restrictive horizons is slower than recharge from overlying horizons.

Within a given riser there is strong evidence for quick increases in soil water matric potential following rainfall events (Fig. 11). This indicates rapid movement and recharge of infiltrating surface water into step risers via the soil matrix and macropores (root channels, animal biopores, rock fragments, and aggregate ped surfaces). Water travels into the subsoil to depths of less permeable aquitard zones. However, no systematic increase in soil wetness was found from upslope to downslope risers in this stair-stepped landform (Wilcox et al., 2007). At some sites (e.g. Fig. 11a), the uppermost riser had the highest percentage of times with perched water tables and very moist soil conditions. This supports the notion that perched water tables are not closely interlinked. Connectivity among perched water tables is limited from upslope to downslope risers. Likewise, the time series plots of downslope tensiometers and water tables do not reflect proportional increasing hydrostatic heads with decreasing elevations, which should be the case if the water tables were closely interlinked (Fig. 11b).

Perched multiple water tables at different depths in the same site may indicate confined water tables. Hydrostatic heads associated with some perched water tables lend credence to this interpretation and further support the fact that steps are essentially independent hydropedologic units. It is possible that noncontinuous data sets may have prevented us from capturing the most transient flow conditions, but this would not invalidate major inferences drawn from these measurements. From soil color patterns, it can be inferred that seasonal oxidation/reduction conditions occur in the lower vadose zones where perched water tables form (Vepraskas, 1992). Redoximorphic concentrations and iron depletions along ped surfaces and bedding planes are direct morphological evidence of contemporaneous chemical reduction above aquitard layers.

5. PATHWAYS OF WATER FLOW AND SOLUTE TRANSPORT

5.1. General Field Observations

Numerous field observations showed that the dye stained the majority of A and upper Bk horizons including rock fragments in an all-matrix pattern. There was relatively little preferential flow along ped surfaces, root or worm channels, or along rock surfaces. Coarse limestone fragments that could be broken by hand were wetted with Br^- tracer throughout the fragment interior, indicating that water flowed into these soft rocks. In contrast, many of the hard limestone rock fragments were wetted along the rock surface and the dye tracer had penetrated into the rock matrix only about 2–3 cm (Nobles et al., 2010; Fig. 16d). This is an important finding because it supports the observation that coarse fragments in caliche soils, along with the fine-earth fraction of these soils, can serve as an effective water-storage media and increase plant-available moisture storage of these droughty soils. The fact that preferential flow was limited as compared to all matrix flow in most of the soils tested in this study indicates that the caliche soils and a significant portion of rock fabrics contribute to biophysiochemical activities. At greater depths in the Bk, Bkm, and into Cr restrictive layers, there was more evidence of preferential flow around more strongly consolidated Bk and Bkm caliche materials, and short-distance penetrations (a few cm) into restrictive layers, especially along occasional root pathways and open joint fissures and fractures.

Another pertinent point from these observations was that for many of the riser positions, the Br^- tracer did not reach the substratum limestone rock interface. The tracer moved not only below the infiltration cell but also beyond the limits of the cell at lateral distances of 0.3–1 m. This suggests that the soils had decreased permeability with depth and both the lower Bk or Bkm horizons and the restrictive Cr and R horizons acted as aquitards to vertical water movement. In other cases, particularly in some mid-riser and tread landform positions, the dye tracers reached the limestone interface, but when the bedrock was excavated several cm deeper, it was apparent that the dye was restricted to the interface and rarely was observed below the interface. Exceptions to this generality are where roots penetrated along fissures and fractures in the bedrock.

5.2. Bromide Tracer Distributions in Soils

In a comparison of riser and tread sites, areas stained by Br^- tracer in the A horizons were similar for both microtopographic positions (Fig. 14a,b). However, in the 2Bk and 3C/Cr/R horizons, areas stained by Br^- tracer in tread sites were significantly higher compared with risers (Fig. 14a,b). This is not surprising because the volume of dye tracer applied to soils in both microtopographic positions was the same and soils on treads were characterized by shallower depths to subjacent restrictive layers. This is supported by the analysis of vertical transport of Br^- tracer that was significantly greater in riser than tread sites (about 1.0 m in risers and 0.5 m in treads; Fig. 15a). Hence, depth to restrictive horizons (averaged 1.2 m in risers and 0.4 m in treads) largely controlled the shallow depth of penetration of the dye in treads. On the average, Br^- penetrated about 1 cm into restrictive layers along treads before transport was terminated (Fig. 16c). Dye tracer solution apparently perched on

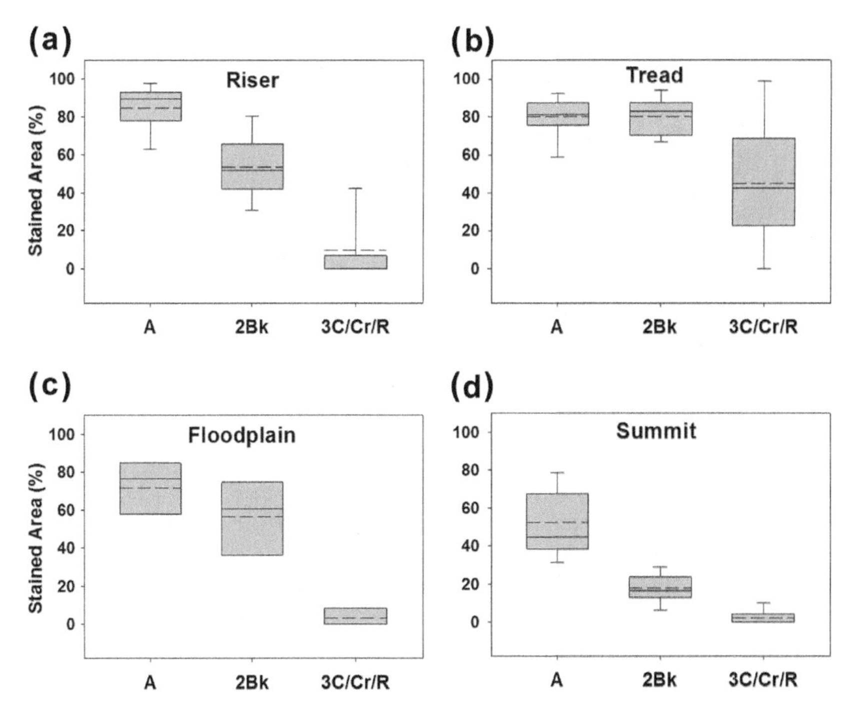

FIGURE 14 Areas of soil horizons stained by Br^- tracer, as percentage of total horizon area, in: (a) riser sites ($n = 18$); (b) tread sites ($n = 9$); (c) alluvial floodplain sites ($n = 7$); and (d) upland summit sites ($n = 9$).

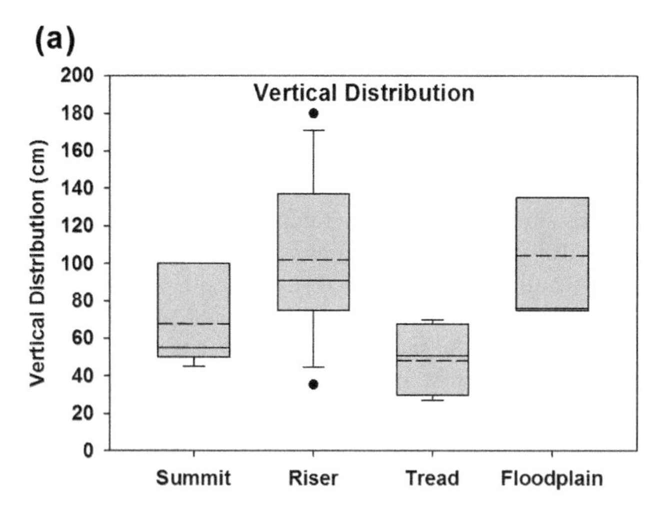

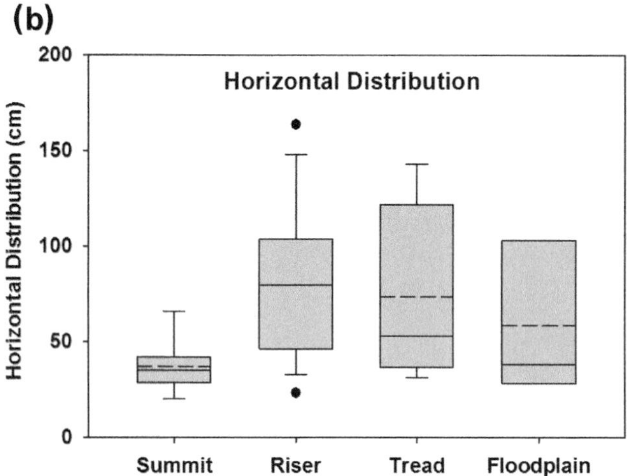

FIGURE 15 (a) Vertical distribution depth of Br^- tracer into the soils studied ($n = 18$); (b) Horizontal distribution distance of Br^- tracer beyond the zone of the infiltration cell ($n = 18$).

restrictive layers fostering greater Bk staining and associated downslope lateral movement along the 2Bk and/or 3C/Cr/R horizon interfaces. This is supported by earlier observations of horizontal transport along restrictive horizons in caliche soils by Hennessy et al. (1983). A strong horizontal component for Br^- movement was also observed in riser sites (Figs. 15b and 16a). While no statistically significant differences in horizontal Br^- distribution was found between riser and treads in this study, in risers Br^- often moved in a downslope plume along steep slope gradients.

Areas stained by Br^- tracer in A horizons were higher in floodplain sites than in upland summit sites, although the differences were not statistically

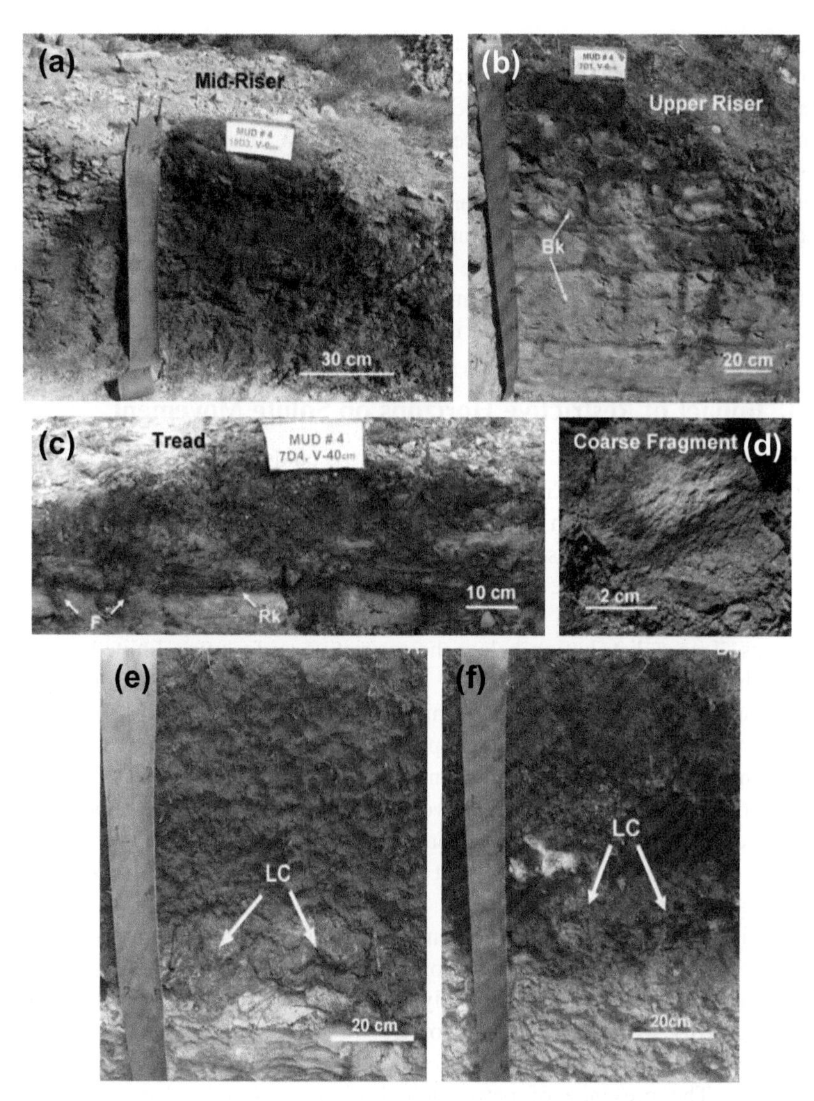

FIGURE 16 Illustration of Br^- tracer movement in risers (a and b), tread (c), limestone coarse fragment (d), and in Petrocalcic Calciustolls of low-level (e) and high-level (f) terraces. Photos illustrate (a) all matrix flow of tracer in middle riser site above a restrictive Cr limestone layer; (b) preferential flow of tracer along joint planes in weakly weathered limestone soil/regolith of upper riser site; (c) tracer transport along and into limestone bedrock (Rk) and through vertical discontinuities in fractured bedrock (F) on a tread site; (d) penetration of dye tracer for 2–4 cm into a hard limestone coarse fragment: (e) tracer transport in soil with petrocalcic horizon and laminar cap (LC) of low-level terrace of the Volente mapping unit in Travis county municipal utility district No. 4; and (f) tracer transport in soil with petrocalcic horizon and laminar cap (LC) of high-level terrace of the Volente mapping unit in rocky creek watershed utility, L.P. *(Figures 16c and 16e are modified from Nobles et al., 2010) (Color version online and in color plate)*

significant (Fig. 14c,d). This is most likely related to the crusts and compaction commonly present in surface horizons of summit sites. In 2Bk horizons, areas stained by Br^- tracer were significantly higher in floodplain as compared to summit sites. No significant differences in Br^- distribution were observed in restrictive 3C/Crk/R horizons (Fig. 14c,d). In both cases, very little of the dye solution reached the restrictive horizons at the base of these soils. Bromide moved further vertically in floodplain sites (average depth 0.9 m), as compared to shallower summit sites (average depth 0.6 m) (Fig. 14a), but no significant differences were found in horizontal Br^- distributions because of the high spatial variability from site to site in this parameter.

5.3. Impact of Restrictive Horizons on Solute Movement

Vertical distribution of Br^- in all sites across the landscape was correlated with depth to restrictive 3C/Crk/R horizons ($r = 0.7$, $n = 47$). Bromide reached limestone rock interface in some mid-riser, floodplain, summit, and many tread landform positions. In these sites, vertical Br^- tracer movement was restricted by the soil–rock interface, while the presence of restrictive 3C/Cr/R horizons contributed to increased horizontal Br^- movement along the restrictive surfaces. Tracer distribution below the restrictive surface was limited to root channels, unplugged joint planes, and discontinuities with infillings of carbonate cements, with dye flow terminating at the dead-end voids (Fig. 16c). In riser sites at the interface with upslope ledge-former limestone outcrops, preferential flow was observed along joint planes in the soil–rock fabric (Fig. 16b). Here, limestone beds were weakly weathered into moderately cohesive 2Bk horizons. When more prominent horizontal bedding was reached at the base of the soil, tracer transport decreased rapidly and terminated before reaching hard limestone interface.

In sites with pedogenically cemented petrocalcic horizons with a laminar cap overlying hard recessive limestone (low-level terrace site, Fig. 16e), or a laminar cap developed in gravelly outwash deposits (high-level terrace site, Fig. 16f), vertical movement of Br^- was limited to less than 15 cm into the petrocalcic horizon that acted as an effective aquitard to solute transport. The presence of petrocalcic horizon materials also resulted in significant horizontal Br^- movement along the interface with the restrictive horizons.

Limited vertical tracer transport and strong horizontal flow along restrictive horizons show evidence that these restrictive horizons provide strong aquitards to rapid solute movement and transport to groundwater; they also serve as drivers in the development of ephemeral seasonal perched water tables. These findings are supported by Wilding (2007) and Wilcox et al. (2007) who found saturated or very moist conditions to persist at the base of the riser soil above petrocalcic horizons and bedrock interfaces for a period of up to 5.5 months. Vertical groundwater recharge in these upland components of the landscape would be restricted to flow through unfilled fractures in bedrock materials (Wilcox et al., 2007; Woodruff and Wilding, 2008).

5.4. Hydrologic Implications of Dye Tracer Observations

During large seasonal rainfall events, deep and moderately deep upper and middle riser soils behave as hydrologic recharge zones. Due to their greater soil depths, high limestone gravel content, and potential for plant-available water storage, riser soils are capable of longer mean solute residence times, reduced solute transport to groundwater, and enhanced bioremediation potential above limestone bedrock or strongly cemented caliche aquitards. In lower riser and tread soils, seasonal matrix saturation may lead to discharge along seepage zones above restrictive horizons (Fig. 16c). When this happens, considerable horizontal flow is generated along aquitard surfaces with increased lateral through flow and surface runoff (Wilding, 2007; Wilcox et al., 2007).

6. MISCONCEPTIONS AND ENVIRONMENTAL IMPLICATIONS

This hydropedologic case study of the caliche hillslope terrains challenges conventional wisdom about caliche soils in the region. Misconceptions propagated by some professional geoscientists, engineers, environmental regulators, and the lay public include the following (Woodruff et al., 1992; Wilding, 2007; Woodruff and Wilding, 2008):

- Soils are either thin or absent across the stepped caliche landscapes;
- Soils are thickest on gently sloping treads and thinnest on steeply sloping risers (presumably because of geomorphic stability inferences invoked);
- Soils are low in organic carbon because they are light colored;
- Soils lack sufficient biotic activity for bioremediation of wastes;
- Adverse environmental impacts are expected to decrease with decrease in slope gradient;
- Most rainfall runs off the surface rapidly as overland flow and results in severe flash-flooding hazards downslope;
- High rates of overland flow imply immediate adverse water-quality effects in downstream areas;
- High gravel contents (>30%) negatively affect water-storage capacity and result in rapid flow to groundwater;
- Clay textures are not suitable for on-site wastewater treatment media;
- Amended soil materials must be added to the soil surface for suitable landscaping purposes; and
- Surface slope gradients must be <15% to prevent runoff of treated wastewater effluents.

These misconceptions do not take into consideration the current knowledge of hydropedology of caliche soils as discussed in this chapter and landform–soil–water relationships. They disregard internal buffers such as aquitards that limit aquifer recharge and enhance mean residence time for water residing in the vadose zone for plant consumptive evapotranspiration. They fail to differentiate the unique differences in physical and chemical behaviors between noncarbonate and carbonate-rich soil systems. They do not consider the landscape scale at which hydrologic processes function across these caliche soils. The management scale in this stepped landscape is on the order of meters from risers to treads. Lastly, they fail to appreciate utilizing native soil surfaces rather than amended "soil" materials for landscaping in residential communities, thus enhancing surface runoff, erosion, and overland flow with amended materials (Brandt and Wilding, 2009). Misconceptions have promoted mismanagement of land and water resources across a range of scales from individual home sites to entire watersheds. Because of this, adverse impacts from mismanagement commonly extend downstream to the recharge zone of the Edwards aquifer (Woodruff and Wilding, 2008).

Contrary to conventional wisdom, moderately deep to deep caliche soils occupy the steepest slope gradients on stepped risers of these landforms. Steep slope gradients of upper and middle risers are the most stable geomorphic sites and least subject to runoff and erosion. In contrast, most surface runoff (and much erosion) originates from lower risers and gently sloping treads. When overland flow occurs, most is from upslope treads and continues to downslope risers, thereby promoting further infiltration and retention and diminution of overland flow. Retained water is cycled through biophysiochemical soil transformers that mitigate adverse water-quality effects by consumptive evapotranspiration, bioremediation, and chemical sorption. Entrained materials – either suspended sediment or traction debris – are deposited on or beneath riser surfaces where water percolates into the porous caliche regolith. In this way, multiple, disjunct, near-surface, porous, and permeable zones occur as a cascading system of hydrologic buffers across a dissected, generally steeply sloping terrain (Woodruff et al., 1992; Wilcox et al., 2007; Wilding, 2007; Woodruff and Wilding, 2008). High volumes of soft and hard limestone CFs enhance water retention of these soils without resulting in undesirably high leaching rates (Nobles et al., 2010). In short, water retained in these vadose zone buffers maintains a locally dense and diverse cover of upland plants, and still more water is retained owing to interception by the various layers of vegetation – from tree canopy to ground cover. These various hydrologic buffers combine to lessen the peaks of flood hydrographs (Woodruff and Wilding, 2008).

The biophysiochemical behavior of caliche soils presents additional buffers to chemical contamination of surface and subsurface water supplies. Moderate to high OC contents coupled with highly reactive clay minerals provide caliche soils with moderate to high chemical adsorption capacities. The high carbonate status of caliche buffers soil reaction in the alkaline range (pH 7.5–8.3). This decreases plant availability and uptake of most heavy metals and micronutrients. Carbonates also serve as templates for phosphate precipitation and adsorption that decreases its activity and uptake. Phosphorus transport, along with other sorbed chemicals on sediments, is primarily by soil erosion.

However, the latter is mitigated by flocculation of clays under high calcium soil systems, and compartmental sediment traps in multiple downslope steps. Nitrate transfer into subsurface aquifers is buffered by soil and bedrock aquitards, and by biochemical reduction to nitrogen gas during periods of soil saturation and chemical reduction above the aquitards.

Even with multiple-buffered stepped terrains, flash floods are common in the Hill Country of Central Texas. This is due to high-intensity short-duration storm events that deliver sufficient volumes of rainwater to override and compromise the buffering effects of these caliche soils. Low to moderate water-storage capacity and the "overprint" of residential development that generates increasingly large areas of nonporous or slowly porous anthropogenic soil cover are two major contributors. It is, thus, pertinent to consider this matter in designing best management strategies for on-site disposal of treated wastewater effluents such that dosing frequencies and rates do not exceed the ability of the soil and vegetative cover to accommodate additional hydrologic inputs. Furthermore, as noted by Marsh and Marsh (1995), flash flooding is enhanced under urban conditions because much of the rainfall is shunted into drainage ditches, storm sewers, and culverts with little or no soil retention possible. Also, Leopold (1968) points out that under urban conditions greater volumes of water are conveyed more rapidly to surface streams, thus increasing peaks of flood hydrographs while decreasing lag times for these peaks.

Backhoe trench excavations were the only suitable means of investigating rocky and stony limestone caliche soils with sufficient exposure to accurately differentiate pedogenic from lithogenic soil–landscape attributes. Rather serendipitously, we discovered early in our study a micropattern of soil–regolith conditions that repeat themselves across the riser–tread motif of stepped terrains. Without trench exposures it is not possible to accurately record soil–geomorphic–hydrologic attributes of these landscapes or to develop the appropriate hydropedologic understanding for transfer to other stepped limestone terrains. The parent material composition of riser units was found to be transported colluvial materials superposed uncomformably over weathered limestone residuum that rests upon ledge-former limestone strata. In contrast, tread units are composed of pedisediments transported from upslope risers, admixed with limestone residuum, and superposed uncomformably over ledge-former limestone strata. This reality led to vastly different interpretations of the soil–regolith attributes influencing landforms, soil thickness patterns, water flow and solute transport along slope gradients, spatial variability in soil horizonation, and pedogenesis.

7. SUMMARY

The soil is the first line of defense against water resources pollution. We have discussed unique attributes of caliche soils in stepped hillslopes of Central Texas in this regard. Many of the soil attributes that favor biophysiochemical filtering of water resources are poorly understood. Lacking an accurate soil resource database, well-intended environmental public ordinances are developed that incorrectly address environmental concerns. This chapter highlights the intimate connection among landform, soils, and hydrology across caliche soil–landscapes in the Hill Country of Central Texas. It is apparent that this connection is essential to address the effectiveness of water resources management, and, in our case study, the effectiveness of caliche soils for on-site wastewater treatment and disposal in environmentally sensitive landscapes.

Extensive field investigations carried out in this study enabled the depiction of the general patterns of soil properties and hydrologic functions along the stepped riser–tread catena, which were contrary to conventional perceptions. The integrated landscape-based hydropedologic approach offers important insights into the real-world heterogeneous subsurface, and the nature of interactive pedologic, geologic, biologic, and hydrologic processes across the landscape. Such information is critical to successful environmental management and regulations. In spite of all the environmental buffers in caliche soils, hydrologic limitations in these systems deserve vigilant attention by authorities responsible for watershed management and natural hazard mitigation.

ACKNOWLEDGMENTS

We wish to acknowledge financial support for this work through the Texas On-site Wastewater Treatment Research Council (Contract No. 582-3-55760), the USDA-Natural Resources Conservation Service (Agreement No. 69-7442-5-680), and consultancies through several Municipal Utility Districts and Barton Creek Properties (Stratus Properties) in Travis County, Texas. Further, we wish to acknowledge the gratis help of Dennis Lozano, P.E., Murfee Engineering Company, Inc., Austin, Texas, for his drafting of the soil stratigraphy cross sections (Fig. 4). We are grateful to Robert Bazan, Dennis Lozano, Mariano Moreno, Philip Taucer, and Kristin Wilcox for their assistance in the fieldwork. The synthesis of physical and chemical databases by Donna Prochaska, Soil and Crop Sciences Department, Texas A&M University is also gratefully acknowledged.

REFERENCES

Adriano, D.C., 1986. Trace Elements in the Terrestrial Environment. Springer-Verlag, New York.

Blackburn, W.H., Meeuwig, R.O., Skau, C.M., 1974. Mobile infiltrometer for use on rangeland. J. Range Manage. 27, 322–323.

Bohn, H.L., 1976. Estimate of organic carbon in world soils. Soil Sci. Soc. Am. J. 40, 468–470.

Bouma, J., 1989. Using soil survey data for quantitative land evaluation. Adv. Soil Sci. 9, 177–213.

Brandt, J.E., Wilding, L.P., 2009. Imported soils and hydrological consequences in the Austin area: how these soils may be functioning as impervious cover. In: WoodruffJr., C.M., Slade, R.M. (Eds.), Trip Coordinators. Urban Hydrology of Austin, TX, Some Halloween Tricks and Treats. Field Trip Guidebook 31. Austin Geological Society, pp. 67–80.

Drees, L.R., Wilding, L.P., Nordt, L.C., 2001. Reconstruction of inorganic and organic carbon sequestration across broad geoclimatic regions. In: Lal, R., McSweeney, K. (Eds.), Soil Carbon Sequestration and the Greenhouse Effect. SSSA Special Publication. No. 57. Soil Science Society of America, Madison, WI, pp. 155–172.

Hennessy, J.T., Gibbens, R.P., Tromble, J.M., Cardenas, M., 1983. Water properties of caliche. J. Range Manage. 36 (6), 723–726.

Leopold, L.B., 1968. Hydrology for Urban Land Planning—A Guidebook on the Hydrologic Effects of Urban Land Use. U.S. Geological Survey. Circular 554.

Lu, J., Wu, L., 2003. Visualizing bromide and iodide water racer in soil profiles by spray method. J. Environ. Qual. 32, 363–367.

Marsh, W.M., Marsh, N.L., 1993. Juniper trees, soil loss, and local runoff process. In: Woodruff Jr., C.M., Marsh, W.M., Wilding, L.P. (Eds.), Soils, Landforms, Hydrologic Processes, and Land-Use Issues—Glen Rose Limestone Terrains, Barton Creek Watershed. Society of Independent Professional Earth Scientists, Central Texas chapter, Travis County, Texas (Field report and guidebook). pp. 4-1–4-14.

Marsh, W.M., Marsh, N.L., 1995. Hydromorphic considerations in development planning and stormwater management, Central Texas Hill Country, USA. Environ. Manage. 19, 693–702.

Monger, H.C., Daugherty, L.A., Lindemann, W.C., Liddell, C.M., 1991. Microbial precipitation of pedogenic calcite. Geology 19, 997–1000.

Mortvedt, J.J., Murphy, L.S., Follet, R.H., 1999. Fertilizer Technology and Application. Meister Publishing, Willoughby, Ohio.

Nobles, M.M., Wilding, L.P., Lin, H.L., 2010. Flow pathways of bromide and brilliant Blue FCF tracers in caliche soils. J. Hydrol. 393, 114–122.

Nordt, L.C., Wilding, L.P., 2009. Organic carbon stocks and sequestration potential of vertisols in the coast prairie major land resource area of Texas. In: Lal, R., McSweeney, K. (Eds.), Soil Carbon Sequestration and the Greenhouse Effect, second ed. SSSA Special Publication 57. Soil Science Society of America, Madison, WI, pp. 159–168.

Rabenhorst, M.C., Wilding, L.P., West, L.T., 1984. Identification of pedogenic and lithogenic carbonates using stable carbon isotope and microfabric analyses. Soil Sci. Soc. Am. J. 48, 125–132.

Slade Jr., R.M., Dorsey, M.E., Stewart, S.L., 1984. Hydrology and Water Quality of the Edwards Aquifer Associated with Barton Springs in the Austin Area. US Geological Survey Water-Resources Investigation, Texas. Report 86-4036.

Soil Survey Staff, 1993. Soil Survey Manual. Handbook No. 18. U.S. Govt. Printing Office, Washington, D.C, pp. 292–293.

Soil Survey Staff, 1999. Soil Taxonomy: A Basic System of Soil Classification for Making and Interpreting Soil Surveys, second ed. U.S. Department of Agriculture, Natural resources conservation service, Agricultural handbook No. 436. pp. 1–869.

Soil Survey Staff, 2004. In: Burt, R. (Ed.), Soil Survey Laboratory Methods Manual. Soil Survey Investigations. Report No. 42. Version 4.0. U.S. Department of Agriculture, Natural Resources Conservation Service. http://soils.usda.gov/technical/lmm/.

Soil Survey Staff, 2010. Keys to Soil Taxonomy, eleventh ed. U.S. Dept. of Agriculture, Natural Resources Conservation Service, pp. 1–338.

Soil Survey Staff, 2011. In: Burt, R. (Ed.), Soil Survey Laboratory Information Manual. Soil Survey Investigations Report No. 45, Version 2.0. U.S. Department of Agriculture, Natural Resources Conservation Service. ftp://ftp-fc.sc.egov.usda.gov/NSSC/Lab_Info_Manual/SSIR_45.pdf

Stricklin Jr., F.L., Smith, C.I., Lozo, F.E., 1971. Stratigraphy of Lower Cretaceous Trinity Deposits of Central Texas. The University of Texas at Austin. Bureau of Economic Geology Report of Investigations. No. 71.

Vepraskas, M.J., 1992. Redoximorphic Features for Identifying Aquic Conditions. Tech. Bull. 301. NC Agric. Res. Serv., Raleigh. (Revised and reprinted 1994, 1996, 2000).

Werchan, L.E., Lowther, A.C., Ramsey, R.N., 1974. Soil Survey of Travis Country, Texas. United States Department of Agriculture, Soil Conservation Service in cooperation with Texas Agricultural Experiment Station, p. 123.

West, L.T., 1986. Genesis of soils and carbonate enriched horizons associated with soft limestones in Central Texas. Ph.D. diss., Texas A & M Univ., College Station (Diss. Abstr. 86–15009).

West, L.T., Wilding, L.P., Hallmark, C.T., 1988c. Calciustolls in Central Texas. II. Genesis of calcic and petrocalcic horizons. Soil Sci. Soc. Am. J. 52, 1731–1740.

West, L.T., Wilding, L.P., Stahnke, C.R., Hallmark, C.T., 1988b. Calciustolls in Central Texas. I. Parent material uniformity and hillslope effects on carbonate-enriched horizons. Soil Sci. Soc. Am. J. 52, 1722–1731.

West, L.T., Drees, L.R., Wilding, L.P., Rabenhorst, M.C., 1988a. Differentiation of pedogenic and lithogenic carbonate forms in Texas. Geoderma. 43, 271–287.

Wilcox, Bradford P., Larry Wilding, P., Woodruff Jr., C.M., 2007. Soil and topographic controls on runoff generation from stepped landforms in the Edwards Plateau of Central Texas. Geophys. Res. Lett. 34, 2424. doi:10.1029/2007GL030860.

Wilding, L.P., 2007. Hydrological Attributes and Treatment Capabilities of "Caliche and Related Soils" Pertinent to On-site Sewage Facilities, Texas On-site Wastewater Treatment Research Council Final Report. Contract number 582-3-55760. pp. 1–77.

Wilding, L.P., 1993. Soils of Tributary Sub-Basins in the Barton Creek Watershed-Implications for a Reappraisal of Hill Country Soils, *in* Soils, Landforms, Hydrologic Processes, and Land-Use Issues-Glen Rose Limestone Terrains, Barton Creek Watershed, Travis County, Texas: Society of Independent Professional Earth Scientists, Central Texas Chapter, Austin, Texas, Field Report and Guidebook (revised from field trip on 5 December 1992), pp. 3-1–3-12.

Wilding, L.P., Drees, L.R., 1983. Spatial variability and pedology. In: Wilding, L.P., Smeck, N.E., Hall, G.F. (Eds.), Pedogenesis and Soil Taxomomy: 1. Concepts and Interactions. Elsevier, New York, pp. 83–116.

Wilding, L.P., Woodruff Jr., C.M., Owens, P.R., 2001. Caliche Soils as a Filter Medium for Treatment and Disposal of Wastewater. Texas Natural Resource Conservation Commission, Final Report to Texas On-Site Wastewater Treatment Research Council (MC-178), Contract No. 582-1-83209. pp. 1–77.

Wilding, L.P., Drees, L.R., Woodruff Jr., C.M., 1997. Mineralogy and microfabrics of limestone soils on stepped landscapes in Central Texas. In: Shoba, S., Gerasimova, M., Miedema, R. (Eds.), Soil Micromorphology: Studies on Soil Diversity, Diagnostics, and Dynamics. Proceedings of X Int. Working Meeting on Soil Micromorphology, Moscow, Russia. July, 1996, pp. 204–218.

Woodruff Jr., C.M., Wilding, L.P., 2008. Bedrock, soils, and hillslope hydrology in the Central Texas hill country, USA: implications on environmental management in a carbonate-rock terrain. Environ. Geol. 55, 605–618. DOI:10.1007/s00254-007-1011-4; (Published on-line, 2007.

Woodruff Jr., C.M., Marsh, W.M., Wilding, L.P., 1992. Soils Landforms, Hydrologic Processes, and Land Use Issues–Glen Rose Limestone Terrains, Barton Creek Watershed, Travis County, Texas. Field Report and Guidebook (Revised 1993). Society of Independent Professional Earth Scientists, Austin, Texas (Central Texas Chapter).

Woodruff Jr., C.M., 1984. Water budget analysis for the area contributing recharge to the Edwards Aquifer, Barton Springs Segment. In: Woodruff Jr., C.M., Slade Jr., R.M. (Eds.), Hydrogeology of the Edwards Aquifer—Barton Springs Segment, Travis and Hays Counties. Austin Geological Society Guidebook 6, Texas, pp. 36–42.

Woodruff Jr., C.M., Abbott, P.L., 1979. Drainage basin evolution and aquifer development in a karstic limestone terrain, south-central Texas. USA. Earth Surf. Process 4, 319–334.

Young, K., 1972. Mesozoic history, Llano region. In: Barnes, V.E., Bell, W.C., Clabaugh, S.E., CloudJr., P.E., McGehee, R.V., Rodda, P.U., Young, K. (Eds.), Geology of the Llano Region and Austin Area—Field Excursion. The University of Texas at Austin, pp. 41–46. Bureau of Economic Geology. Guidebook 13.

Chapter 10

Hydropedology in Seasonally Dry Landscapes: The Palouse Region of the Pacific Northwest USA

Erin S. Brooks,[1,*] Jan Boll[2] and Paul A. McDaniel[3]

ABSTRACT

The Palouse region of the Pacific Northwest USA provides a unique opportunity to study relationships between soil morphology and hydrologic processes. Regionally east–west gradients in mean annual precipitation and temperature generally account for spatial variability in major soil types. However, at the hillslope scale, the steep topography of the Palouse induces unique microclimates leading to extreme variability in effective precipitation. Argillic and fragipan horizons in the north-facing, wetter, microclimates lead to seasonal perched water and rapid subsurface lateral flow accelerating the eluviation of clays in albic Esoil horizons and driving variable source area hydrology. Soil slips (i.e. rapid mass wasting events) below deep snow drifts and long-term loss of top soil as a result of excessive erosion through runoff and tillage processes both are involved in positive feedback mechanisms that drive the hydrology of the landscape. Development of this conceptual understanding of Palouse region hydropedology has had profound implications for generating effective solutions to regional erosion, water quality, and groundwater recharge problems.

1. INTRODUCTION

The Palouse region comprises ~9000 km^2 of eastern Washington and northern Idaho and is part of a larger area of thick eolian soils in the northwestern USA

[1]Dept. Bio. and Ag. Eng., University of Idaho, Moscow, ID 83844-2060, USA
[2]Dept. Bio. and Ag. Eng., University of Idaho, Moscow, ID 83844-3006, USA
[3]Soil and Land Resources Division, Dept. of Plant, Soil, and Entomological Sciences, University of Idaho, Moscow, ID 83844-2330, USA
[*]Corresponding author: Email: ebrooks@uidaho.edu

Hydropedology, Edited by H. Lin. DOI: 10.1016/B978-0-12-386941-8.00010-1

(Busacca, 1989). The Palouse is one of the most unique, easily recognizable landscapes around the world. Palouse soils provide some of the highest dryland white wheat yields throughout the world (WWC, 2009). The Palouse landscape has been described as a photographer's dream (http://www.washingtonstatetours.com/palouse.asp) and has been the subject of a National Geographic Magazine cover story entitled "A Paradise Called the Palouse" (Austin, 1982). The close relationship between soil morphology and hydrologic processes in the Palouse region is a fascinating example of hydropedology. In this chapter, we describe the complex processes that have led to both regional and hillslope-scale variability in soil morphology, particularly soils having hydrologically restrictive layers (e.g. argillic and fragipan horizons). We will also demonstrate how soil scientists, hydrologists, and engineers have integrated knowledge of soil formation and morphology (i.e. pedology) with the understanding of hydrologic processes at the plot/field scale to the hillslope/catchment scale to better manage water resources. The interactions between hydrologic processes and soils in the Palouse region are an example of the hydropedology paradigm outlined in Lin (2003) and Lin et al. (2008), and serve to illustrate the utility of this approach in gaining a comprehensive understanding of the region's water/environmental quality issues.

2. ENVIRONMENTAL PROBLEMS

Much of the original soil and hydrologic research in the Palouse region was motivated by efforts to reduce soil erosion associated with dryland farming practices. In 1978, the average annual erosion rate on cultivated cropland in the Palouse River basin was 31 tonnes/ha (14 tons/acre), with some slopes having experienced erosion rates of 200–450 tonnes/ha (90–200 tons/ac) in a single winter season (USDA, 1978). Since the region was first cultivated about 100 years ago, all the original top soil has been lost from 10% of the cropland, and one-fourth to three-fourths of the original top soil has been lost from 60% of the cultivated area in the basin (USDA, 1978). The 1978 report estimated that the loss of top soil has reduced overall wheat yield from the region by 20%. Through widespread adoption of conservation tillage practices to increase infiltration and decrease soil erodibility at the soil surface, erosion in the region has been greatly reduced and stream sediment loads are significantly declining (Ebbert and Roe, 1998; McCool and Roe, 2005; Kok et al., 2009; Brooks et al., 2010).

A recent growing concern over steadily declining water supplies in the primary drinking water aquifer in the Palouse has motivated scientists and engineers to study subsurface hydrologic flow paths. Since the development of widespread farming in the region where initial settlers found artesian aquifers producing over 14,000,000 L of water per day in 1885, water levels in these aquifers have been steadily dropping at 0.3–0.5 m per year (Robischon, 2006, 2008). In 1992 the major communities and universities in the region agreed to limit pumping increases to 1% per year in an attempt to stabilize groundwater levels. Despite nearly stable pumping rates over the last 15 years, water levels in the deep Grand Ronde aquifer have continued to decline. Future planning for long-term growth of regional communities requires a sound understanding of landscape hydrology and aquifer recharge. The spatial variability of soil properties and the inter-relationships between soil morphology and hydrologic function are essential aspects of the landscape that must be understood.

As in most river basins in the world, the Palouse region faces challenges of improving surface water quality. The presence of hydrologically restrictive layers in some Palouse soils (e.g. argillic and fragipan horizons) enhances the downslope lateral movement of water leading to a greater potential for the delivery of fertilizers and other contaminants to streams through by-pass flow and saturation-excess mechanisms (McDaniel et al., 2008). Scientists are investigating the impact of alternative management scenarios such as reduced fall application of fertilizers on crop yields and variable rate application of fertilizers through precision farming as a means to reduce losses of fertilizers and improve stream health. These alternative management practices also benefit from a thorough understanding of the inter-relationships between soil properties and hillslope hydrologic flow paths.

3. PALOUSE SOIL-FORMING FACTORS AND SOIL MORPHOLOGY

Through the pioneering work of Dokuchaiev, pedologists affirm that soil formation and development are related to five primary factors: climate, organisms, relief, parent material, and time (CLORPT) (Jenny, 1941). These soil-forming factors can be used to describe general spatial variability in soil properties across a landscape. We first describe the importance of these factors at the regional scale and then in greater detail how the hydrology of the region creates a feedback mechanism, which enhances each of these factors leading to greater hillslope-scale variability in soil morphology.

3.1. Regional-Scale Variability in Soil Morphology

The Palouse (Busacca, 1989) is located within the Northwestern Wheat and Range Region and is a subset of the Palouse and Nez Perce Prairie region as described by the Major Land Resource Areas delineation (USDA-NRCS, 2006) (Fig. 1). The elevation of the region ranges from 500 m in the west to 900 m in the east. This east–west gradient is reflected in many of the dominant soil and climatic features.

Unlike landscapes of most humid regions, the Palouse region is characterized by a xeric moisture regime (Soil Survey Staff, 2006), a seasonally dry climate characterized by cool, moist winters and warm, dry summers. In Moscow, ID, located in the eastern part of the Palouse region, ~70% of the annual precipitation is received during the months of November through May

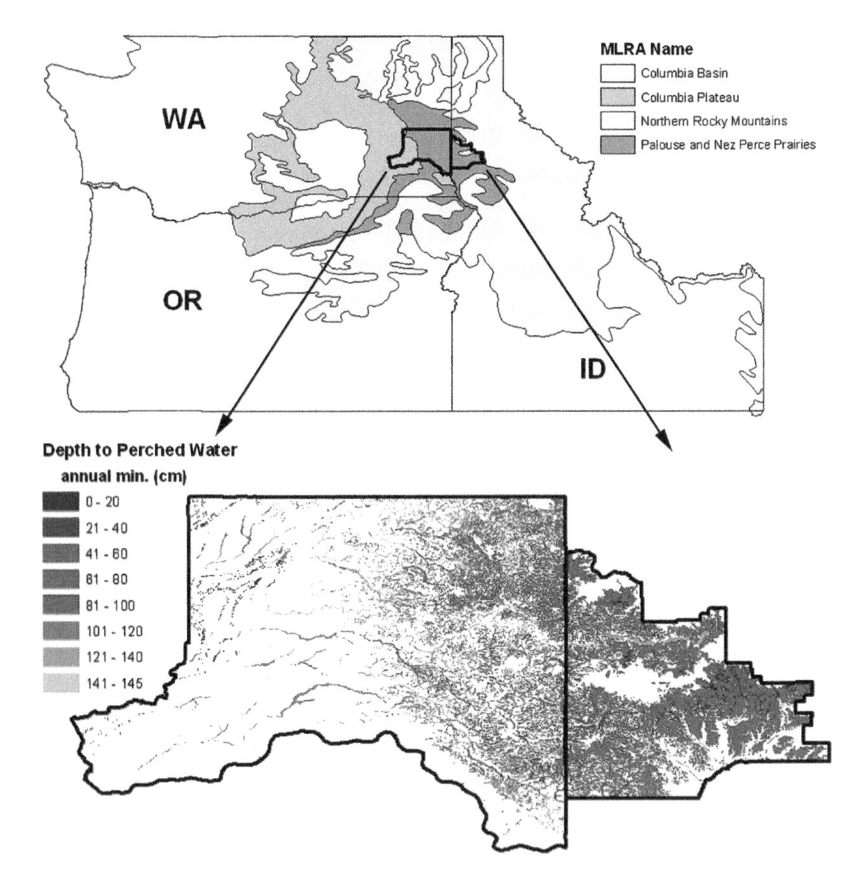

FIGURE 1 The Palouse region as defined by the major land resource area (USDA-NRCS, 2006) in blue. The inset picture shows depth to annual minimum perched water table depth defined as *"The shallowest depth to a wet soil layer (water table) at any time during the year expressed as centimeters from the soil surface, for components whose composition in the map unit is equal to or exceeds 15%"* by Whitman county, WA (Donaldson, 1980) and Latah county, ID (Barker, 1981) soil surveys. *(Color version online and in color plate)*

when evapotranspiration is low. As a result the effective precipitation, defined by the difference between monthly precipitation and evapotranspiration, is highest during winter months.

Despite the relative uniformity of the parent material, soils formed in Palouse loess are not homogenous silt loam profiles (McDaniel and Hipple, 2010). The Palouse loess was blown in from south-westerly winds and soil survey maps reveal a clear gradation in particle size with distance from the sediment source with relatively coarse particles settling out in western regions and deposition of fine particles in the eastern regions. Palouse soils were formed through a series of deposition events often resulting in multiple buried paleosols (Busacca, 1989).

One of the most dominant soil-forming factors used to describe the variability in soil type across the Palouse is mean annual effective precipitation or often simply the mean annual precipitation. Soils located in regions with greater effective precipitation experience greater leaching of carbonates and have well-developed prevalent evidence of clay eluviation and subsequent illuviation in argillic or fragipan horizons.

In the same manner as elevation, mean annual precipitation and mean annual temperature exhibit a clear east–west gradient where the western regions experience the least precipitation and the greatest mean annual temperatures (Fig. 2). Generally, Haploxerolls are found in the drier, western parts of the Palouse receiving ~450 mm or less of annual precipitation with mean annual air temperatures greater than 8 °C (Donaldson, 1980; Busacca, 1989). With low annual effective precipitation and minimal leaching, there is little evidence of clay illuviation in the subsoil and calcium carbonate accumulation is more prevalent. Argixerolls are found in the ~450–700-mm precipitation zone, and have subsurface horizons characterized by hydro-consolidation and clay

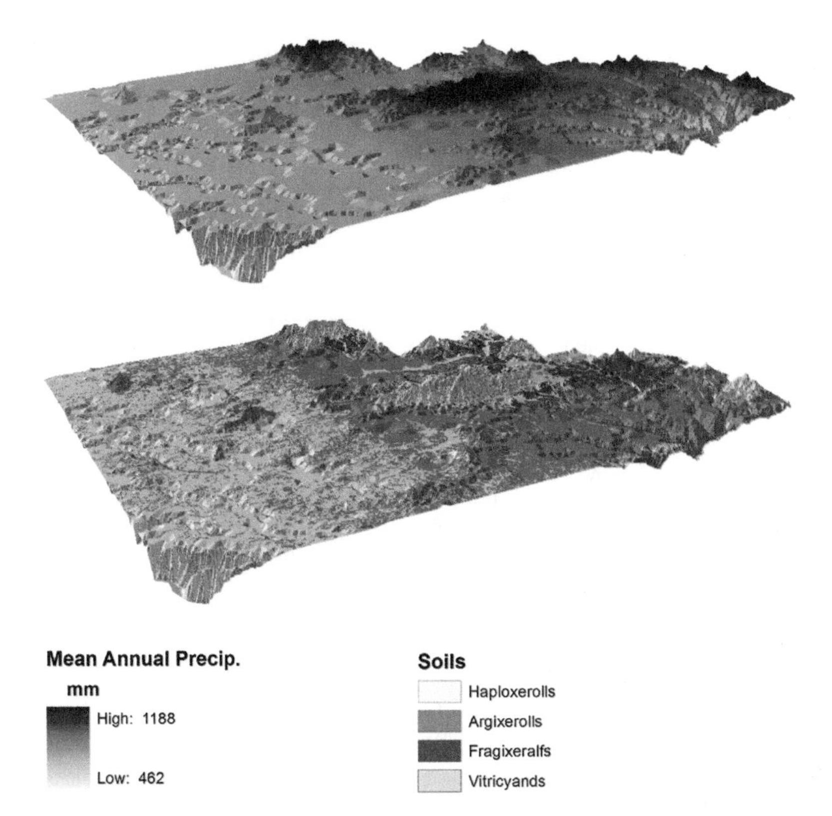

FIGURE 2 Mean annual precipitation in mm (Daly et al., 2008) and major soil types in Whitman county, WA and Latah county, ID regions of the Palouse. *(Color version online and in color plate)*

translocation (argillic horizons); well-developed Fragixeralfs with dense, brittle fragipans are common in the >700-mm precipitation zone (Donaldson, 1980; Barker, 1981; Busacca, 1989; Soil Survey Staff, 2006) (Fig. 3). The more strongly developed argillic and fragipan horizons in soils of the intermediate and higher-precipitation zones are thought to be part of a late Pleistocene paleosol that can be traced across the region (Kemp et al., 1998) and is estimated to be between ~20,000 and 40,000 years before present (Sweeney et al., 2007). These older argillic and fragipan horizons are overlain by horizons that have developed in younger loess over the past ~15,000 years (McDaniel and Hipple, 2010).

The argillic and fragipan horizons greatly restrict water flow and root development. They have bulk density values of ~1.65 Mg m^{-3} and have

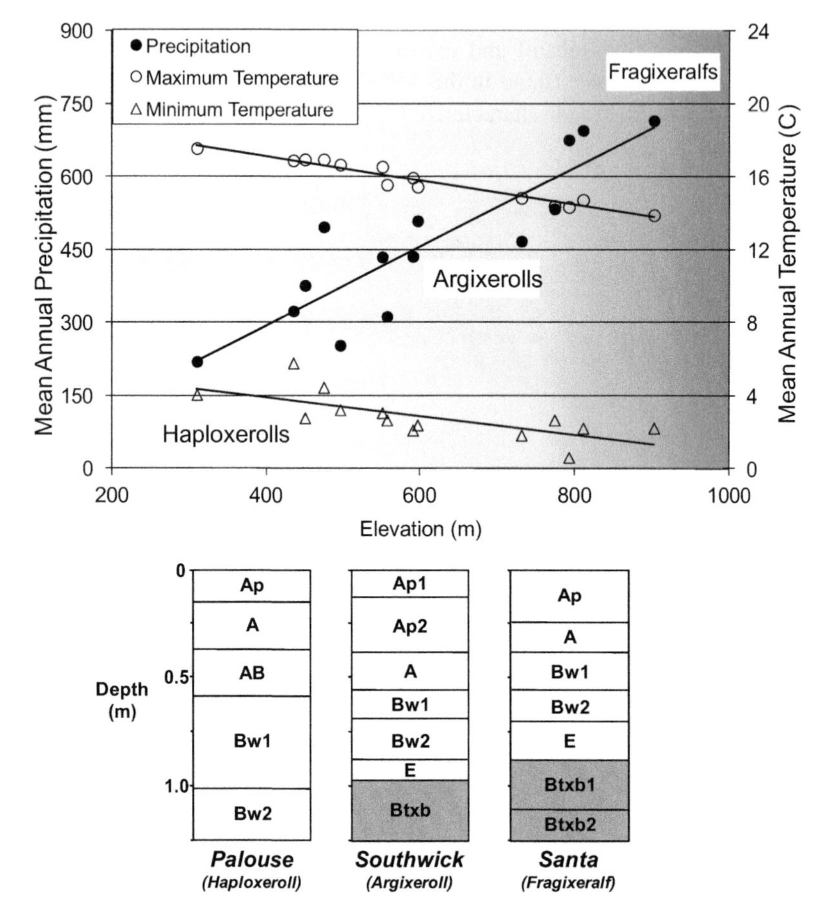

FIGURE 3 Relationships between mean annual precipitation, maximum temperature, minimum temperature, and elevation. Haploxerolls, Argixerolls, and Fragixeralfs are often found in regions with characteristics associated with yellow-, green-, and red-colored areas, respectively. *(Color version online)*

poorly developed macropore networks (McDaniel et al., 2001). As a result, saturated hydraulic conductivity (K_{sat}) values are extremely low, often several orders of magnitude less than those of surface horizons. Laboratory measurements of intact soil cores indicate that the Ap horizons of Argixerolls and Fragixeralfs have mean K_{sat} values of 0.74 and 1.29 m day^{-1}, while mean K_{sat} values for argillic and fragipan horizons of these soils are 0.1 and 0.06 m day^{-1} (McDaniel et al., 2001). Hillslope-scale measurements indicate that K_{sat} of Ap horizons can be as high as 15.1 m day^{-1} (Brooks et al., 2004).

One of the major hydrologic consequences of argillic and fragipan horizons when coupled with relatively high winter precipitation is an extensive network of seasonal perched water tables (PWTs). These PWTs can be present for most of the winter and early spring months especially in the eastern parts of the region, which have relatively high mean annual precipitation (>700 mm) (Fig. 1).

3.2. Hillslope-Scale Variability in Soil Morphology

Although regional-scale patterns in soil type generally can be described well by spatial variability in effective precipitation, a high degree of complexity exists at the hillslope scale. At this local scale, topography plays a major role in soil formation.

Most ridges in the Palouse are located in southeasterly–northwesterly direction with long, southwest-facing slopes ranging in steepness up to 35%; slope gradients on shorter, steeper northeast-facing slopes are 45–55% in places (Rockie, 1951). Total local relief ranges from 30 m to 60 m (100–200 ft). Winter wheat is harvested from these extreme slopes using combines with specially designed levelers, which prevent them from overturning.

Early investigators quickly realized the influence of topography on local variability in effective precipitation and soil morphology (Lotspeich and Smith, 1953; Weaver, 1914). Since most of the precipitation in the region falls during winter months much of this precipitation falls as snow, often resulting in large snow drifts on northeast-facing slopes. In addition to snow drifting on northeast-facing slopes, the incident shortwave solar radiation also is much less on these north-facing slopes than on south-facing slopes often delaying melting and evaporation until later in the season. Lotspeich and Smith (1953) in their description of the Palouse Catena identified the following three local microclimates: ridge tops, north-facing slopes, and south-facing slopes. The north-facing slopes with their low solar radiation and redistribution of snow represent the wettest, coolest, microclimate. The ridge tops which are often blown clean of snow represent a dry, windy, microclimate. The south-facing slopes which have the greatest incident solar radiation represent a microclimate with relatively high evapotranspiration rates. Rockie (1951) estimated that north-facing slopes can receive as much as 6–10 times the effective precipitation as ridge tops (Fig. 4).

This topographically induced variability in effective precipitation leads to hillslope-scale variability in soil development (e.g. eluviation/illuviation and carbonate leaching). In the intermediate precipitation zone, it is not unusual for south-facing slopes to have soils with calcium carbonate accumulations similar to those on the drier, western edge of the Palouse. North-facing slopes can have soils with argillic horizons similar to those found on the moister, eastern edge of the Palouse (Lotspeich and Smith, 1953; Donaldson, 1980) (Fig. 4).

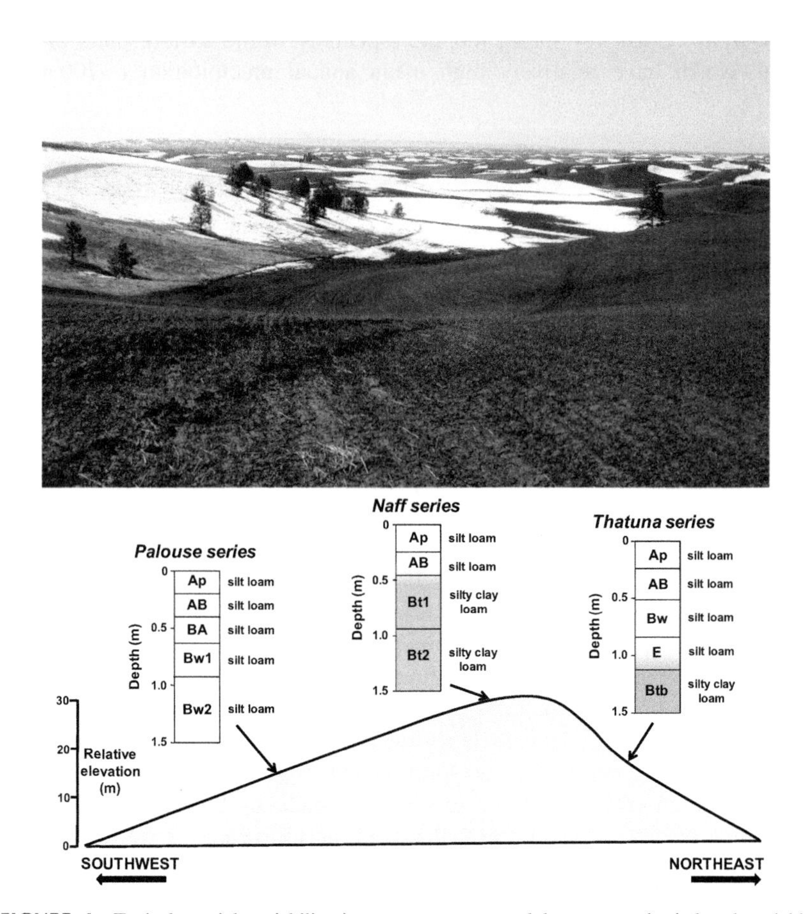

FIGURE 4 Typical spatial variability in snow cover caused by topography-induced variable snowmelt rates and snow drifting in the Palouse. This topographically induced variability in effective precipitation often leads to hillslope-scale variability in soil development with Argixerolls (e.g. Thatuna series soils) found on north-facing slopes and Haploxerolls (e.g. Palouse series soils) found on south-facing slopes. *(Color version online)*

Topography has a major influence on wind patterns. Wind affects snow drifting, and hence variable hydrologic conditions, as well as soil movement and erosion. Tillage can accentuate the movement of soil downslope because of the manner of cultivation across slopes using moldboard plows or one-way disks. Topography has a major influence on wind patterns in the region and results in large noticeable snow drifts each winter and spring. These same wind patterns undoubtedly affected the manner in which soil was redistributed during and immediately following rapid loess deposition events. Years of wind erosion and tillage erosion have removed the top soil from many ridge tops exposing more resistant, less permeable, clay-enriched subsoils or argillic horizons. Tillage erosion occurs when the furrow from a moldboard plow or one-way disk is thrown downhill. Kaiser (1961) dramatically demonstrated tillage erosion losses from the edge of a Palouse field of 0.75 m (2.5 ft) from 1911 to 1942 and another of 0.6 m (2 ft) of soil by 1959 at the same field site.

4. POSITIVE FEEDBACK MECHANISMS BETWEEN SOIL MORPHOLOGY AND HYDROLOGIC PROCESSES

Soil formation and hydrologic processes are clearly interrelated in the Palouse region. Soil structural features largely control subsurface water movement through a hillslope and long-term patterns in soil water movement largely control soil formation. An important focus of hydropedology is to develop a clear understanding of these processes in order to improve management of soil and water resources. In this section, we describe some of the feedbacks between soil morphology and the hydrology of the Palouse.

Much of the characteristic topography of the Palouse has been shaped and formed through the frequent occurrence of shallow soil slips on north-facing slopes (Rockie, 1934; Donaldson, 1980). This is a positive feedback following the extensive accumulation of snow (i.e. snow drifts) on steep north-facing slopes, which provides a relatively large amount of effective precipitation at a concentrated point in the landscape (Fig. 4). These north-facing slopes tend to have argillic horizons which greatly restrict vertical water movement. Hence, the concentration of effective precipitation at a point initiates PWTs immediately above this argillic horizon. As the pore pressure increases, a shear face is established resulting in sudden failure of the hillslope. The probability of soil failure increases with the steepness of the slope.

Another hydrologically induced positive feedback mechanism is the formation of albic horizons above argillic and fragipan horizons. Albic E-horizons are light-colored soil layers which, due to water movement, have become depleted in clay (i.e. eluviation) and result in enrichment of clay in underlying argillic (Bt) and fragipan (Btx) horizons. Clearly, water movement drives this soil-forming process. In the Palouse region, however, horizontal rather than vertical translocation of clays occurs within E-horizons as a result of the steep topography. It has been suggested that the development of albic E-horizons in the Palouse region is controlled more by saturated subsurface lateral flow, which occurs above these argillic/fragipan horizons (Lotspeich and Smith, 1953; Rieger and Smith, 1955; Busacca, 1989).

Recent studies have confirmed that subsurface lateral flow can be a dominant hydrologic process in these steep landscapes with a shallow depth to a restricting layer. Using an isolated 18 m by 35 m hillslope plot located on a 20% slope in a 0.7-m thick fragipan soil, Brooks et al. (2000) measured 88% of the total spring precipitation leaving the plot as saturated subsurface lateral flow. Tracer studies also confirmed substantial subsurface lateral flow above these argillic and fragipan horizons (Reuter et al., 1998; McDaniel et al., 2008).

The positive feedback between subsurface lateral flow and the development of E-horizons leads to the hypothesis that characteristics of the E-horizon can be an indicator for subsurface hydrology. Locations of a hillslope having well-developed, thick, albic E-horizons should indicate significant subsurface lateral flow whereas locations with little lateral flow would have thin, faint, albic E layers. Although no comprehensive study has been conducted to adequately confirm this hypothesis, Lotspeich and Smith (1953) generally reported non-existent E-horizons on ridge tops, which form and quickly thicken as the slope increases down to an inflection point on north-facing slopes where the E-horizon thins as the slope gradually flattens (see Fig. 2 in Lotspeich and Smith, 1953). In a study on the eastern Palouse where soil profiles were characterized using soil cores taken from a 10 m × 15 m grid (Brooks et al., 2007a; McDaniel et al., 2008), the close correlation can be seen between topography, effective precipitation, and E-horizon thickness (Fig. 5). Both this study and the Lotspeich and Smith (1953) study suggests that there is a close correlation between magnitude of subsurface lateral flow and albic layer thickness.

An added complexity in the development of albic layers in the Palouse region is that most of these argillic and fragipan horizons may once have been surface soil horizons which were subsequently buried by younger loess. Stratigraphic investigations have revealed the presence of as many as 19 buried or relic soils in a 26-m deep loess section, creating a 'layer cake' effect within the Palouse hills (Busacca, 1989). Therefore, the formation of these argillic and fragipan horizons may be uniquely associated with the hydrology present during inter-depositional periods. Since the topography of these buried surface soil horizons may have been different from current topography, inferring the presence and absence of an albic horizon and/or argillic/fragipan horizons by topography alone may be misleading. In some cases, the spatial distribution of these restrictive layers and the thickness of these layers appear to have no correlation with the present-day topography. It is possible that these fragipan and argillic horizons were formed

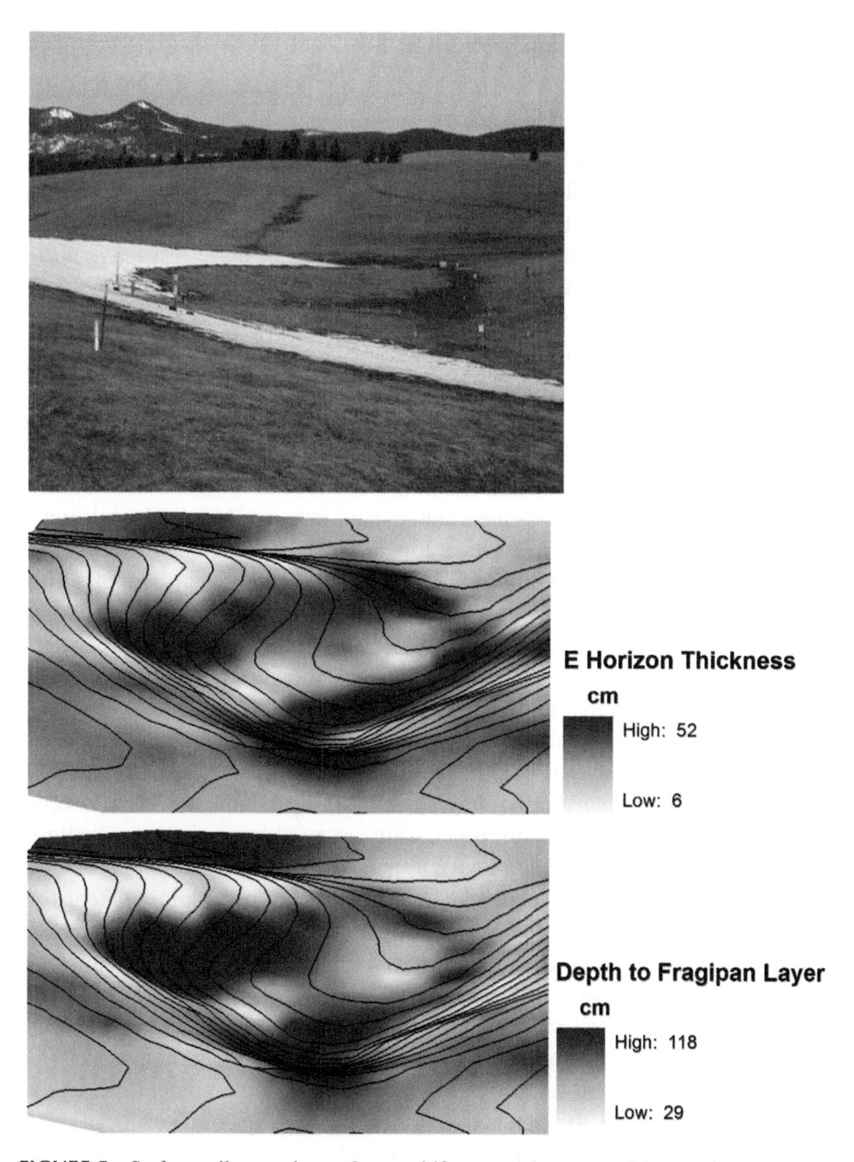

FIGURE 5 Surface soil saturation and snow drift patterns in a perennial grass field located in a Fragixeralf from the eastern Palouse region (Brooks et al., 2007a). The E-horizon thickness and depth to the fragipan soil horizon are also shown as a 3D plot from the same perspective as the actual image. *(Color version online and in color plate)*

under different annual precipitation regimes than today. However, stratigraphic evidence indicates that the relative distributions of precipitation and temperature have been consistent throughout the development of these soils (Busacca, 1989; O'Geen et al., 2005).

4.1. Effects of Argillic and Fragipan Soils on Hydrologic Processes

Since there is feedback between subsurface lateral flow and soil development in soils having an argillic or fragipan horizon, it is important to understand the dominant features controlling the initiation, persistence, and magnitude of subsurface lateral flow within a hillslope. It is also particularly important for soil and water management to understand the impact of these restrictive horizons not only on subsurface lateral flow but also on generation of surface runoff and erosion. In the next section, we describe the important roles of drainable porosity, depth to a restrictive layer, and saturated lateral hydraulic conductivity on subsurface lateral flow.

4.1.1. Drainable Porosity

One of the important characteristics of Palouse loess is a high water holding capacity, which makes the soils ideal for dryland farming. However, soils with argillic or fragipan horizons have a low drainable porosity, and this has significant consequences for subsurface lateral flow and surface runoff. The drainable porosity of a soil is a measurement of the volume of water that drains from soil pores for a given drop in water-table depth. Drainable porosity in deep, freely draining soils is often defined as the difference between the saturated moisture content and field capacity moisture content (equivalent to 3.33 m of tension). In shallow soils with a restrictive layer, however, free drainage does not occur as it does in deep soils. During the wet season a PWT forms, which on sloping ground drains primarily by lateral flow or has very little drainage on relatively flat ground. Thus, at equilibrium in these shallow soils, soil water tension above the PWT may only reach 1 m as opposed to 3.33 m, typically defined as equivalent to field-capacity moisture content for a deep soil. The drainable porosity in shallow soils, therefore, is smaller than is typically assumed for free-draining deep soils. In shallow soils, the moisture content at any given point in a profile is closely related to the depth to the perched water table. Soil moisture measurements made at a catchment near Troy, ID, in the eastern Palouse region by Brooks et al. (2007a) at both 0.10 m and 0.20 m below the soil surface indicate that the moisture content is linearly related to the depth to the perched water table, see Fig. 6. The linear relationships in Fig. 6 indicates that soil moisture only drops 0.12 $m^3\ m^{-3}$ from full saturation at 0.50 $m^3\ m^{-3}$ to 0.38 $m^3\ m^{-3}$ for a 0.55-m drop in perched water table depth. In contrast, this 0.12-$m^3\ m^{-3}$ drainable porosity for a 0.55-m deep soil is nearly half the theoretical drainable porosity based on tension of 3.33 m at field capacity for the same soil in the absence of a perched water table, 0.20 $m^3\ m^{-3}$.

An important consequence of soils with low drainable porosity is that subsurface lateral flow is initiated frequently, even from very small storms. Perched water table fluctuations in soils having a small drainable porosity can

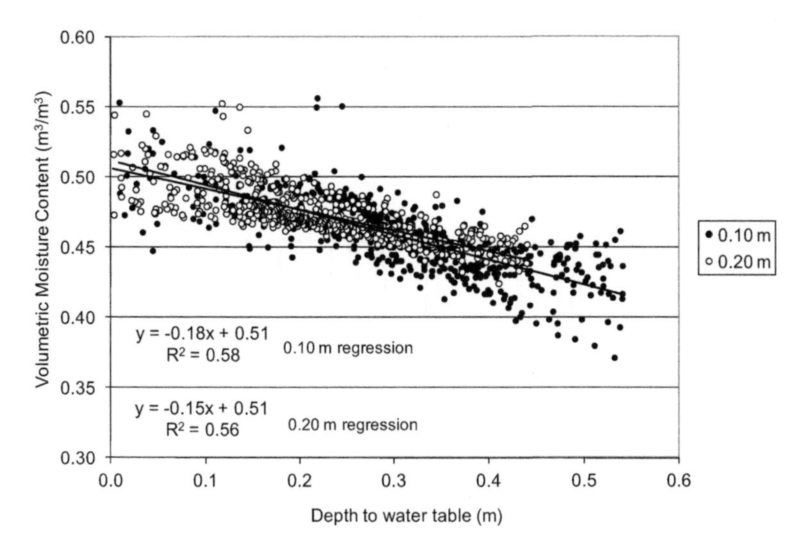

FIGURE 6 Drainable porosity relationship defined by simultaneous measurements of volumetric moisture content and depth to a perched water table in a Fragixeralf from the eastern Palouse region (Brooks et al., 2007a). Numbers in legend represent the depth where the soil moisture content was measured. *(Reproduced by Permission from John Wiley & Sons, Ltd.)*

be very dynamic and surface runoff will be more 'flashy' relative to soils having a greater drainable porosity (Dunne et al., 1975) (Fig. 7). McDaniel et al. (2001) measured a 0.6-m increase in PWT height within one day at the Troy, ID catchment. Brooks et al. (2007a) showed that a 0.20 m-change in water table resulted from only a 0.01-m change in total water storage for a fragipan soil. Once generated, these fluctuating PWTs are present for as long as six months during the winter and early spring months (Fig. 7) and can generate surface runoff and erosion multiple times during the season (McDaniel et al., 2001).

4.1.2. Depth to the Restricting Layer

Like drainable porosity, depth to the restricting layer is an important characteristic that strongly controls the initiation of PWTs and subsequently subsurface lateral flow and saturation-excess runoff. Shallow soils, such as those found in eroded ridge-top soils, have a low water storage capacity and therefore generate runoff more frequently than deeper soils. Figure 8 presents the sensitivity of simulated saturation-excess runoff to depth to the restricting layer using the water erosion prediction project (WEPP) model (version 2006.5) (Dun et al., 2009). In this exercise, surface runoff was predicted for an identically shaped hillslope having a 2% slope, covered in perennial grass cover, using 30-year daily climate data generated using the CLIGEN model for each specific location. As seen in Fig. 8, surface runoff increases as soil depth decreases. Interestingly, regions of the country having xeric moisture regimes (i.e. wet winters

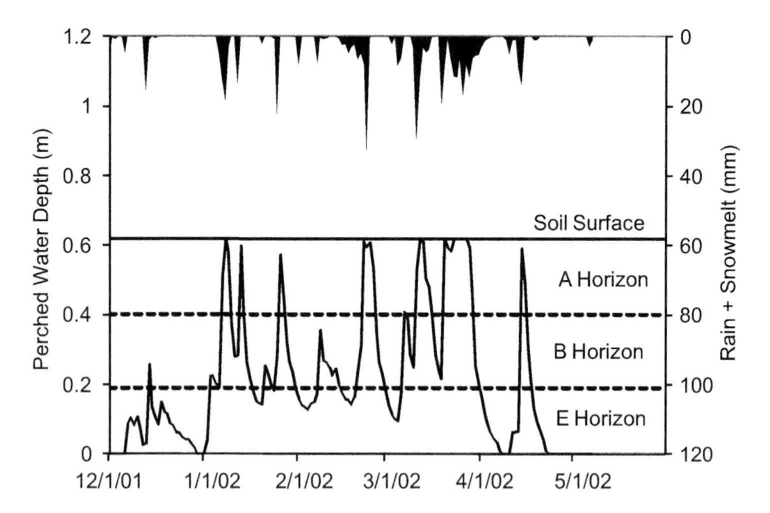

FIGURE 7 Perched water table dynamics in a Fragixeralf from the eastern Palouse region. Data points represent PWT height above the hydraulically restrictive fragipan during the 2001–02 water year. Daily total input to the soil profile (i.e. daily measured rain plus simulated daily snowmelt using the SMR model (Brooks et al., 2007a) is also provided as a black area chart from the upper axis.

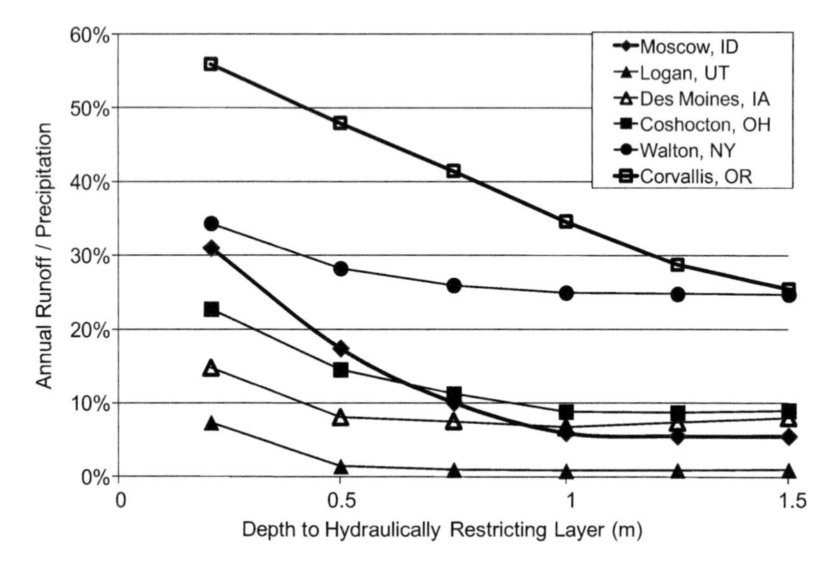

FIGURE 8 Simulated sensitivity of mean annual percent runoff to the depth to a hydraulically restrictive layer for various climate regions using the water erosion prediction project (WEPP) model. Runoff was simulated for a 100-m hillslope seeded to perennial grass, having a slope steepness of 2%, with a silt loam soil having an impermeable restrictive layer at various depths.

and dry summers) tend to be much more sensitive to soil depth (i.e. Moscow, ID and Corvallis, OR) than regions with dry winters and wet summers (i.e. Des Moines, IA). This simple analysis shows that the amount of runoff in the eastern edge of the Palouse region (i.e. Moscow, ID) increases nearly six fold for a soil depth of 0.2 m compared to a soil depth greater than 1 m. Brooks et al. (2007b) measuring PWT fluctuations on a shallow ridge-top soil confirmed these findings and observed that saturation on ridge-top soils can be a source of water, which initiates downslope rill erosion. Another positive feedback mechanism in the Palouse region is the cycle of long-term soil erosion which reduces the depth to argillic or fragipan horizons, and this, in turn, leads to increased runoff, erosion, and further reduction in soil depth.

4.1.3. Lateral Saturated Hydraulic Conductivity

As PWTs rise into near surface soil horizons with greater macroporosity, the magnitude of subsurface lateral flow increases greatly. The hillslope-scale experiment conducted in the Palouse by Brooks et al. (2004) provided the opportunity to compare the effects of measurement scale on the magnitude of K_{sat} with depth. Using outflow measurements and PWT profiles, a decay function representing a decrease in K_{sat} with depth was identified. Comparisons with small-scale K_{sat} measurements using a Guelph Permeameter, laboratory experiments on soil cores, and estimates from the published soil survey indicated that the hillslope-scale K_{sat} was 15, 5 and 2 times greater than the small-scale measurements in the A, B, and E horizons (Fig. 9), respectively. Thus,

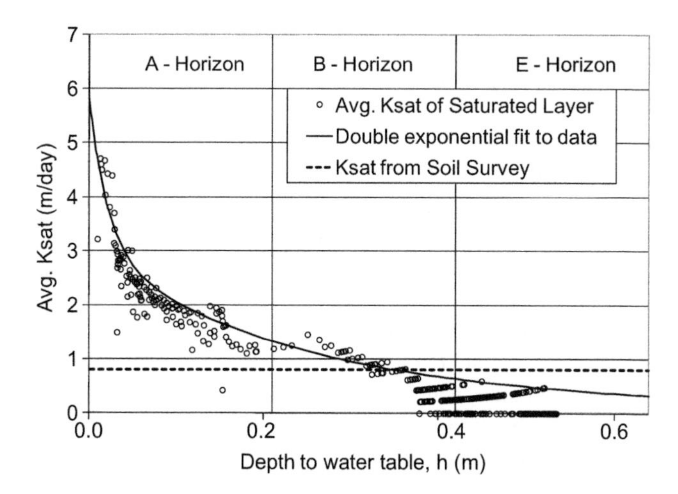

FIGURE 9 Measured lateral saturated hydraulic conductivity versus depth from the soil surface to the perched water table from an 18-m long hillslope plot in a Fragixeralf from the eastern Palouse region (Brooks et al., 2004). The fragipan soil horizon was located at a depth of 0.64 m. Saturated hydraulic conductivity from the Latah county, ID soil survey (Barker, 1981) is also indicated on the figure. *(Reproduced by Permission of the American Geophysical Union.)*

field scale lateral transport via PWTs may be an order-of-magnitude greater than predicted using smaller-scale measurements.

Hillslope-scale tracer studies in the Palouse region have shown that relatively high transport rates occur via PWTs. Maximum transport of applied Br^- was shown to decrease with annual precipitation across the region, with rates of 86, 50, and 35 cm d^{-1} measured at sites receiving 830, 700, and 610 mm of mean annual precipitation, respectively (Reuter et al., 1998). In another study, Br^- tracer was also found to move ~9 m d^{-1} via a PWT (McDaniel et al., 2008). This rapid movement is attributed to preferential flow through macropores, possibly of biological origin. In all cases, the most rapid transport of tracers occurs when PWTs are present in the more permeable surface horizons.

4.1.4. Variable Source Area Hydrology

In the moister regions of the Palouse, substantial subsurface lateral flow occurs on steep slopes and converges at flatter, toe-slope positions. This leads to prolonged periods of saturation-excess runoff (Brooks et al., 2007a) following variable source area hydrology concepts (Hewlett and Hibbert, 1967; Dunne and Black, 1970; Dunne et al., 1975). Saturation-excess runoff occurs when the total water added to the soil profile exceeds the soil's water-storage capacity, a situation that occurs when the soil profile is fully saturated (Hewlett and Hibbert, 1967; Dunne and Black, 1970; Dunne et al., 1975). Shallow ridge-top soils generate frequent runoff because the soil profile has very little capacity to store the water. However, for deeper soils (i.e. soils having a restrictive layer below 0.5 m depth), subsurface lateral flow drives the spatial distribution of surface runoff (Fig. 5). With the relatively low rainfall intensities in the Palouse region, infiltration-excess runoff (Horton, 1933, 1940) occurs much less frequently than does saturation-excess runoff. This is especially true with fields under conservation tillage. The spatial distribution of argillic and/or fragipan horizons, their depth below the soil surface, and their topographic position control saturation-excess runoff generation at hillslope and catchment scales.

5. IMPLICATIONS FOR SOIL AND WATER MANAGEMENT IN THE PALOUSE

Improved understanding of the close feedback between soil morphology and hydrologic processes in the Palouse region has significant implications for addressing regional erosion, water quality, and groundwater recharge problems. The management implications for each of these environmental problems are discussed in the following sections.

Excessive soil erosion has historically been the major environmental problem in the region. Much progress has been made to reduce soil erosion through conversion from intensive farming practices that pulverize the soil leaving an erodible soil surface to conservation tillage systems. These systems increase soil organic matter and surface residue, resulting in increased infiltration and reduced runoff and erosion (Kok et al., 2009; Brooks et al., 2010). However, increased infiltration adds more water to the soil profile which increases the likelihood of development of a PWT in soils with argillic or fragipan horizons. Once this PWT develops, subsurface lateral flow is initiated and increases the likelihood and occurrence of saturation-excess runoff in converging zones of the landscape. The consequence is greater potential for gullies to form in these saturated regions. Over the last 10 years the region has begun addressing gully formation through installation of gully plugs (Idaho NRCS Conservation Practice Standard 638). These plugs are designed to intercept and remove surface runoff before gully formation is initiated.

Improving the water quality of lakes and streams is also a pressing need in the Palouse region. Application of hydropedology can lead to improved strategies for use of agrichemicals and a reduction in transport of pollutants to streams. The primary management strategy to reduce the transport and delivery of agrichemicals is to schedule the timing and amount of agrichemicals applied such that the risk of delivery to streams is minimized. These management decisions require a fundamental understanding of the dominant hydrologic flow paths within a landscape and the characteristics of the particular chemical constituent (Walter et al., 2000). Management practices which aim to reduce particulate-bound chemicals should focus on areas in the landscape having the greatest erosion whereas soluble chemicals should focus on runoff-producing areas (Walter et al., 1979).

Conservation tillage in the Palouse has greatly reduced soil erosion and therefore has reduced the delivery of particulate bound chemicals by minimizing infiltration-excess runoff and increasing surface cover. However, this practice has only minimally affected the generation of saturation-excess runoff and, as discussed above, may actually have led to increased runoff and increased delivery of soluble chemicals in converging zones of a landscape (e.g. toe-slopes). Although it is commonly understood that soil erosion increases with slope steepness, since lateral flow increases with slope steepness, saturation-excess runoff is least likely to occur on the steepest slopes. Therefore, in landscapes where the risk of infiltration-excess runoff is minimal, it may be less risky to apply a chemical on a steep slope rather than on a flat slope in order to minimize the risk that this chemical will be transported by surface runoff. However, it should also be recognized that despite limited surface runoff on steep sections of the slope, subsurface lateral transport can be significant (e.g. as high as 9 m/day; McDaniel et al., 2008). Management decisions therefore must be made which consider the various flow-delivery mechanisms and balance the relative risk of delivery throughout the landscape. In general, minimizing the transport of soluble chemicals is best achieved in the Palouse by avoiding excessive application of agrichemicals and nutrients in the fall season in landscape positions that are likely to produce saturation-excess runoff. In the future, as more land owners in the Palouse acquire variable rate fertilizer application technology there will be greater opportunities to develop management plans which take into account the spatial and temporal variability of these dominant hydrologic delivery mechanisms at the field scale.

One of the most pressing needs in the Palouse is to replenish declining drinking water supplies and identify potential aquifer-recharge flow paths. Through research over the last ten years, it has become apparent that the probability of groundwater recharge flow paths through soils having argillic or fragipan soils is very low. Research using natural tracers, hydrometric measurements, and stratigraphic observations indicate that multiple restrictive soil layers significantly affect groundwater recharge. Where buried soil layers consist of argillic and/or fragipan horizons, long pore-water residence times suggest that these uplands do not serve as active recharge zones (O'Geen et al., 2002; Dijksma et al., 2011). In valleys, the layers of buried soils are thinner and not as strongly developed. Consequently, valley soilscapes likely play a more active role than uplands in local recharge processes (O'Brien et al., 1996; O'Geen et al., 2003).

6. FUTURE DIRECTION OF HYDROPEDOLOGY RESEARCH IN THE PALOUSE

Although decades of research in the Palouse region have demonstrated the importance of understanding the linkages between hydrology and pedology, there are many more opportunities to better understand feedback mechanisms and apply this knowledge to manage environmental problems of the region. We provide here a few of the pressing research questions for future hydropedology research:

- How will continued long-term subsurface lateral flow affect the structure of albic, argillic, and fragipan horizons and their hydraulic properties? Will these layers continue to develop and result in even greater reductions in hydraulic conductivity? Will the transition to no-till practices lead to further stripping of clays and increase the importance of managing subsurface lateral flow?
- How much of the variability in soil morphology within a hillslope can be attributed to hydrologic processes from the current climate, or does the soil structure suggest that the soil developed under a different climate regime? What were the hydrologic processes that first generated these argillic soils? How did the topography of these earlier buried soil horizons affect the development of these soils?
- Are there other identifiable soil features other than the albic layers (i.e. redoximorphic features) that would suggest preferential subsurface lateral flow paths within a hillslope (Zhu and Lin, 2009)? How can we make better use of remote sensing to improve spatial soil mapping?
- With the current understanding of the dominant soil-forming factors and the hydrologic feedback mechanisms, how accurately can digital soil models

predict the location of argillic and fragipan soils and the depth to these hydrologically restricting layers? At what spatial scale can soil models improve upon current soil survey models?

- How can the feedback mechanisms described in this chapter be best incorporated and described in detailed soil surveys? The hydrology of a landscape reflects the underlying spatial soil structure and soil physical characteristics. The hydrologic response of a watershed as identified by mean residence time, peak runoff rates, surface saturation patterns, hydrologic connectivity, and other measureable features therefore should be useful for characterizing the soil structure of the upslope catchment. Better describing the hydrology of a landscape should improve our understanding of the underling soil structural control and improve our ability to predict the effects of various management practices on the hydrologic response of the catchment.
- Can digital soil-mapping techniques be used to improve parameterization and the predictive accuracy of distributed hydrologic models? Improved classification of soil morphology and spatial soil structure at a catchment scale should reduce the uncertainty in identifying soil physical parameters and improve the predictive ability of hydrologic models.

These and other research questions, if answered, will provide watershed managers with improved tools and information, allowing them to take a more targeted approach to the management of environmental problems of the Palouse region.

7. SUMMARY

The Palouse region provides an interesting case study of the coupled processes of soil morphology and hydrology. Topography in this region affects local variability in effective precipitation and soil morphology, causing strong interrelation of soil formation and hydrologic processes. On a regional scale, the distribution of soils is relatively well described by variability in climate, particularly mean annual precipitation. Palouse soils have formed in a sequence of loess deposits, and have developed slowly permeable argillic and fragipan horizons in those parts of the region receiving greater annual precipitation. At the hillslope-scale, soil structural features largely control subsurface water movement and long-term patterns in soil water movement in turn largely control and accelerate soil formation. The steep hillslope topography of the Palouse creates unique microclimates dictated by variability in incident solar radiation and redistribution of precipitation through snow drifting that results in extreme variability in effective precipitation. Variability in effective precipitation means that each microclimate is associated with unique hydrologic processes. The presence of argillic and fragipan horizons in areas of greater effective precipitation leads to seasonal PWTs that support rapid downslope subsurface lateral flow on the steep slopes. The high lateral flow provides a positive feedback mechanism that results in the continued formation of albic E-horizons. Spatial variability in soil depth and convergence of this lateral flow predominantly drives the generation of saturation-excess runoff.

Through the conceptual understanding of the Palouse region aided by hydropedology, soil and water resource management has become more effective and spatially accurate. Rather than uniform, widespread management, reduction in overall sediment load is achieved through practices targeted to land types most susceptible to erosion. Based on the runoff-generation mechanisms in the landscape, farmers can avoid fertilizer and pesticide application in critical management zones during periods of the year having the greatest risk of transport. A conceptual understanding of the spatial distribution of hydrologically restrictive argillic and fragipan horizons has assisted managers in identifying potential groundwater recharge zones. Hydropedologic research in the Palouse region remains active and will likely result in more spatially explicit soil maps further improving the effectiveness of watershed management in the region.

REFERENCES

Austin, B., 1982. A paradise called the palouse. Natl. Geogr. 161 (6), 798–819.

Barker, R.J., 1981. Soil Survey of Latah County Area, Idaho. USDA—Soil Conservation Service. U.S. Govt. Print. Office, Washington, DC.

Brooks, E.S., Boll, J., McDaniel, P.A., 2004. A hillslope-scale experiment to measure lateral saturated hydraulic conductivity. Water Resour. Res. 40, W04208. doi:10.1029/2003WR002858.

Brooks, E.S., Boll, J., McDaniel, P.A., 2007a. Distributed and integrated response of a GIS-based hydrologic model in the eastern Palouse region, Idaho. Hydrol. Process 21, 110–122.

Brooks, E.S., Yao, C., Boll, J., Chen, S., McCool, D., 2007b. Improving Tillage Systems for Minimizing Erosion, STEEP Final Project Report. Available at (accessed 02.02.10.). Solutions to Environmental and Economic Problems. Univ. of Idaho, Moscow, Washington State Univ., Pullman; Oregon State Univ., Corvallis. http://pnwsteep.wsu.edu/annualreports/2007/pdf/Boll_Improving_tillage.pdf

Brooks, E.S., Boll, J., Snyder, A.J., Ostrowski, K.M., Kane, S.L., Wulfhorst, J.D., Van Tassel, L.W., Mahler, R., 2010. Long term sediment loading trends in the Paradise creek watershed. J. Soil Water Conserv., 65:6-331-341, doi:10.2489/jswc.65.6.331

Brooks, E.S., McDaniel, P.A., Boll, J., 2000. Hydrologic Modeling in Watersheds of the Eastern Palouse: Estimation of Subsurface Flow Contributions. Presented at the 2000 Pacific Northwest Region Meeting in Richland, WA, Paper No. PNW2000-10. ASAE, 2950 Niles Rd. St. Joseph, MI 49085-9659.

Busacca, A.J., 1989. Long quaternary record in eastern Washington, USA, interpreted from multiple buried paleosols in loess. Geoderma 45, 105–122.

Daly, C., Halbleib, M., Smith, J.I., Gibson, W.P., Doggett, M.K., Taylor, G.H., Curtis, J., Pasteris, P.A., 2008. Physiographically-sensitive mapping of temperature and precipitation across the conterminous United States. Int. J. Climatol. DOI: 10.1002/joc.1688.

Dijksma, R., Brooks, E.S., Boll, J., 2011. Groundwater recharge in Pleistocene sediments overlying basalt aquifers in the Palouse Basin, USA: modeling of distributed recharge potential and identifying water pathways. Hydrogeol. J. 19, 489–500 (DOI 10.1007/s10040-010-0695-9).

Donaldson, N.C., 1980. Soil Survey of Whitman County, Washington. USDA—Soil Conservation Service. U.S. Govt. Print. Office, Washington, DC.

Dun, S., Wu, J.Q., Elliot, W.J., Robichaud, P.R., Flanagan, D.C., Frankenberger, J.R., Brown, R.E., Xu, A.C., 2009. Adapting the water erosion prediction project (WEPP) model for forest applications. J. Hydrol. 466 (1–4), 46–54.

Dunne, T., Black, R.D., 1970. Partial area contributions to storm runoff in a small New England watershed. Water Resour. Res. 6, 1296–1311.

Dunne, T., Moore, T.R., Taylor, C.H., 1975. Recognition and prediction of runoff-producing zones in humid regions. Hydrological Sci. Bull. 20, 305–327.

Ebbert, J.C., Roe, R.D., 1998. Soil Erosion in the Palouse River Basin: Indications of Improvement. U.S. Geological Survey Fact Sheet FS-069–98, on line at URL. http://wa.water.usgs.gov/ccpt/pubs/fs.069-98.html, (accessed 05.05.06.).

Hewlett, J.D., Hibbert, A.R., 1967. Factors affecting the response of small watersheds to precipitation in humid regions. In: Sopper, W.E., Lull, H.W. (Eds.), Forest Hydrology. Pergamon Press, Oxford, pp. 275–290.

Horton, R.E., 1933. The role of infiltration in the hydrological cycle. Trans. Am. Geophys. Union 14, 446–460.

Horton, R.E., 1940. An approach to the physical interpretation of infiltration capacity. Soil Sci. Soc. America Proc. 4, 399–417.

Jenny, H., 1941. Factors of Soil Formation – A System of Quantitative Pedology. McGraw Hill, New York.

Kaiser, V.G., 1961. Historical land use and erosion in the Palouse—a reappraisal. Northwest Sci. 35 (4), 139–153.

Kemp, R.A., McDaniel, P.A., Busacca, A.J., 1998. Genesis and relationship of macromorphology and micromorphology to contemporary hydrological conditions of a welded Argixeroll from the Palouse. Geoderma 83, 309–329.

Kok, H., Papendick, R., Saxton, K., 2009. STEEP: impact of long-term conservation farming research and education in Pacific Northwest Wheatlands. J. Soil Water Cons. 64 (4), 253–264.

Lin, H., 2003. Hydropedology: bridging disciplines, scales, and data. Vadose Zone J. 2, 1–11.

Lin, H.S., Brooks, E., McDaniel, P., Boll, J., 2008. Hydropedology and surface/subsurface runoff processes. In: Anderson, M.G. (Ed.), Encyclopedia of Hydrologic Sciences. John Wiley & Sons, Ltd. DOI: 10.1002/0470848944.hsa306

Lotspeich, F.B., Smith, H.W., 1953. Soils of the Palouse loess: I. The Palouse catena. Soil Sci. 76, 467–480.

McCool, D.K., Roe, R.D., 2005. Long-term erosion trends on cropland in the Pacific Northwest. Presented at the 2005 Pacific Northwest Section Meeting of the ASAE, Lethbridge, Alberta, Canada, Sept. 22–24, 2005. ASAE Section Meeting Paper No. PNW05-1002. St. Joseph, Mich., p. 17.

McDaniel, P.A., Hipple, K.W., 2010. Mineralogy of loess and volcanic ash eolian mantles in Pacific Northwest (USA) landscapes. Geoderma 154, 438–446.

McDaniel, P.A., Gabehart, R.W., Falen, A.L., Hammel, J.E., Reuter, R.J., 2001. Perched water tables on Argixeroll and Fragixeralf hillslopes. Soil Sci. Soc. Am. J. 65, 805–810.

McDaniel, P.A., Regan, M.P., Brooks, E., Boll, J., Barndt, S., Falen, A., Young, S.K., Hammel, J.E., 2008. Linking Fragipans, Perched Water Tables, and Catchment-Scale Hydrological Processes. Catena. doi:10.1016/j.catena.2007.05.011.

O'Brien, R., Keller, C.K., Smith, J.L., 1996. Multiple tracers of shallow ground-water flow and recharge in hilly loess. Ground Water 34, 675–682.

O'Geen, A.T., McDaniel, P.A., Boll, J., 2002. Chloride distributions as indicators of vadose zone stratigraphy in Palouse loess deposits. Vadose Zone J. 1, 150–157.

O'Geen, A.T., McDaniel, P.A., Boll, J., Brooks, E., 2003. Hydrologic processes in valley soilscapes of the eastern Palouse Basin in northern Idaho. Soil Sci. 168, 846–855.

O'Geen, A.T., McDaniel, P.A., Boll, J., Keller, C.K., 2005. Paleosols as deep regolith: implications for groundwater recharge across a loessial climosequence. Geoderma 126, 85–99.

Reuter, R.J., McDaniel, P.A., Hammel, J.E., Falen, A.L., 1998. Solute transport in seasonal perched water tables in loess-derived soilscapes. Soil Sci. Soc. Am. J. 62, 977–983.

Rieger, S., Smith, H.W., 1955. Soils of the Palouse loess: II. Development of the A2 horizon. Soil Sci. 79, 301–319.

Robischon, S., 2006. 2002–2005 Palouse Groundwater Basin Use Report. Palouse Basin Aquifer Committee, Moscow, Idaho. www.webs.uidaho.edu/pbac/Annual_Report/Final_PBAC_Annual_Report_2006_hi_res.pdf.

Robischon, S., 2008. 2007 Palouse Ground Water Basin; Water Use Report. Palouse Basin Aquifer Committee, Moscow, Idaho. www.webs.uidaho.edu/pbac/Annual_Report/Final_PBAC_Annual_Report_2007_low_res.pdf.

Rockie, W.A., 1934. Snowdrifts and the Palouse topography. Geogr. Rev. 24, 380–385.

Rockie, W.A., 1951. Snowdrift erosion in the Palouse. Geogr. Rev. 41, 34–48.

Soil Survey Staff, 2006. Keys to Soil Taxonomy, tenth ed. United States Department of Agriculture – Natural Resources Conservation Service, Washington, DC.

Sweeney, M.R., Gaylord, D.R., Busacca, A.J., 2007. Evolution of eureka flat: a dust-producing engine of the Palouse loess, USA. Quatern. Int. 162–163, 76–96.

USDA, 1978. Palouse Cooperative River Basin Study: Soil Conservation Service, Forest Service, and Economics and Cooperative Service. U.S. Gov. Print. Office, Washington, DC, p. 182.

USDA-NRCS, 2006. Land Resource Regions and Major Land Resource Areas of the United State, the Caribbean, and the Pacific Basin, US Dept. Agric. Handbook 296. US Department of Agriculture –Natural Resources Conservation Service, Washington, DC, p. 682.

Walter, M.T., Walter, M.F., Brooks, E.S., Steenhuis, T.S., Boll, J., Weiler, K.R., 2000. Hydrologically sensitive areas: variable source area hydrology implications for water quality risk assessment. J. Soil Water Conserv. 3, 277–284.

Walter, M.F., Steenhuis, T.S., Haith, D.A., 1979. Nonpoint source pollution control by soil and water conservation practices. Trans. Am. Soc. Agric. Eng. 22 (5), 834–840.

Weaver, J.E., 1914. Evaporation and plant succession in southeastern Washington and adjacent Idaho. Plant World 17, 273–294.

WWC, 2009. Washington Wheat Facts 2008–2009. on line at URL. Washington Wheat Commission. http://admin.aghost.net/images/E0177801/2008WF4WebSmHomepage.pdf (accessed Dec. 2011).

Zhu, Q., Lin, H.S., 2009. Simulation and validation of concentrated subsurface lateral flow paths in an agricultural landscape. Hydrol. Earth Syst. Sci. 13, 1503–1518.

Chapter 11

Hydropedology of the North American Coastal Temperate Rainforest

David V. D'Amore,[1,*] Jason B. Fellman,[2] Richard T. Edwards,[1] Eran Hood[2] and Chien-Lu Ping[3]

ABSTRACT

The North American perhumid coastal temperate rainforest (NCTR), which extends along the coastal margin of British Columbia to southeast Alaska, is characterized by intense orographic precipitation caused by Pacific storm systems striking coastal mountains. This precipitation regime influences the development of soil and vegetation communities, which in turn influence the transfer of terrestrial carbon and nitrogen (primarily in dissolved forms) and other nutrients to aquatic ecosystems. These terrestrial subsidies of dissolved organic matter (DOM) are increasingly being recognized as an ecological function important to freshwater and marine ecosystems. The concentration and quality of DOM exported from the NCTR soils vary according to soil types (e.g. wetland vs. upland soils) that have formed along a topographic gradient from flat organic soils to steep mineral soils. Hydropedology provides a template to evaluate the functions associated with water movement through soils along this gradient. Five key issues of hydropedology – structure, function, scale, integration, and disturbance – were used as a framework to present the information gathered through multiyear research at the Juneau hydrologic observatory in the NCTR. We demonstrated how hydropedology can be used to elucidate mechanisms that influence DOM production and export in the NCTR. This information provides a foundation for modeling terrestrial DOM production and export under varying environmental conditions.

[1]Forest Service, U.S. Department of Agriculture, Pacific Northwest Research Station, Juneau, AK 99801, USA

[2]Environmental Science Program, University of Alaska Southeast, Juneau, AK 99801, USA

[3]School of Natural Resources and Agricultural Sciences, University of Alaska, Fairbanks, AK 99775, USA

[*]Corresponding author: Email: ddamore@fs.fed.us

Hydropedology, Edited by H. Lin. DOI: 10.1016/B978-0-12-386941-8.00011-3

1. THE USE OF HYDROPEDOLOGY IN THE NORTH AMERICAN TEMPERATE BIOME

1.1. The North American Coastal Temperate Rainforest Biome

The North American coastal temperate rainforest (CTR) biome extends along the coastal margin of the Pacific Northwest of the U.S.A, British Columbia, and Alaska (Fig. 1). The perhumid northern portion of the CTR, the NCTR, is composed of the southeastern Alaskan panhandle and the north coast of British Columbia and represents the largest unpopulated and unmanaged portion of the CTR. The NCTR is characterized by abundant precipitation from North Pacific storm systems that confront coastal mountains leading to intense orographic precipitation that influences terrestrial and aquatic ecosystems. Diverse terrestrial habitats and large fluxes of material to the coastal margin through plentiful surface water channels are defining attributes of the region. The NCTR has great ecological value for its abundant wildlife and diverse ecosystems that provide habitat for many endemic species of plants and animals (Cook et al., 2006). There is also an emerging appreciation for the transfer of dissolved organic matter (DOM) from terrestrial to freshwater and marine ecosystems along coastal margins such as the NCTR (Muller-Karger et al., 2005).

1.2. Applying Hydropedologic Techniques to Understand Terrestrial and Aquatic Processes in the NCTR

The flow of water is the most fundamental component in the development and maintenance of ecosystem functions in the NCTR, yet the details of water storage, transport, and delivery remain poorly understood at several spatial scales. This is despite the fact that the flux of freshwater from terrestrial systems to the Gulf of Alaska is approximately twice the annual discharge of the Mississippi River (Neal et al., 2010). The NCTR is a challenging environment for developing a spatially distributed model of hydrologic and biogeochemical function due to the close association of diverse ecosystem types (Fig. 2). The NCTR contains over 3000 watersheds $\geq$121.4 ha that drain directly into saltwater and can be delineated from the USFS level 6 hydrologic unit code (HUC) watershed boundary layer. These watersheds have formed from a variety of different combinations of geology, weather, and vegetation, creating the need for adequate replicate systems for quantifying water and nutrient fluxes and testing resulting catchment nutrient flux models.

Establishing models of how terrestrial material influences downstream aquatic ecosystems is a priority for management of terrestrial and stream resources. Hydropedology is the multiscale investigation of the source, storage, flow path, residence time, availability, and spatio-temporal distribution of water in soils that occur in the Earth's Critical Zone (Lin, 2003). Hydropedology is

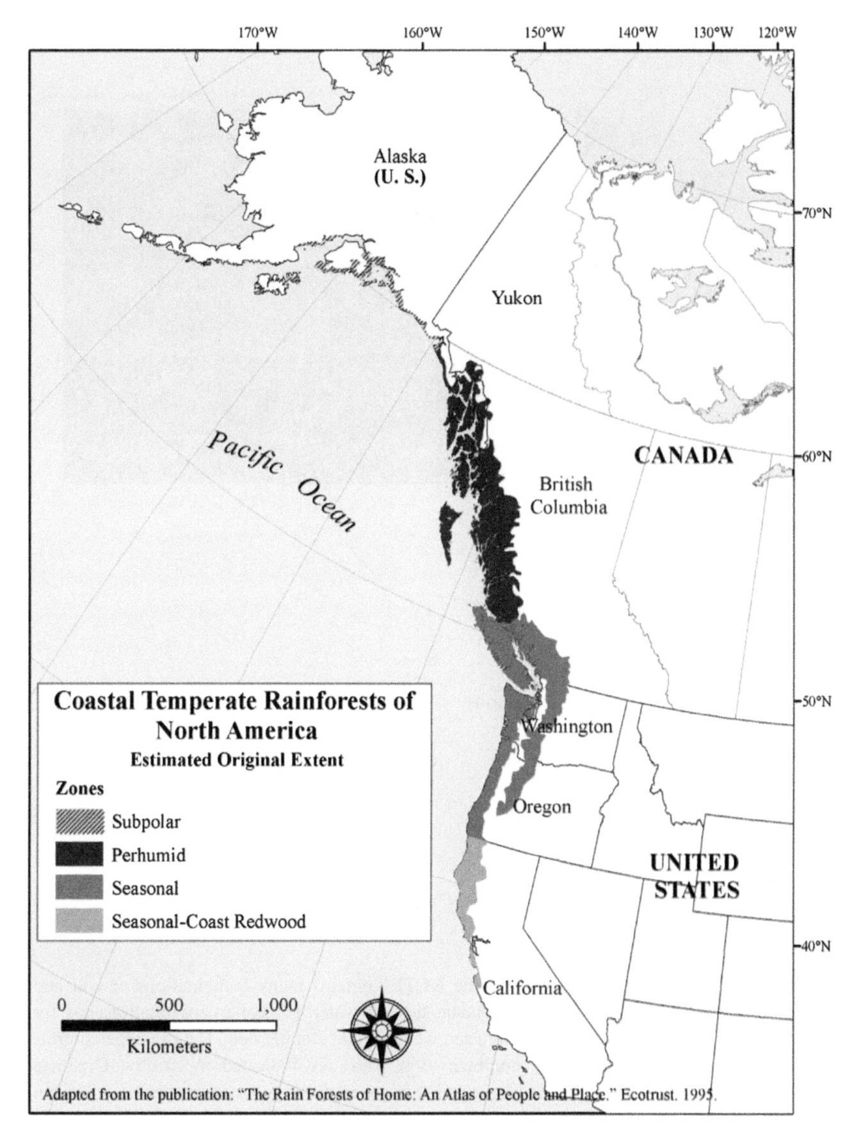

FIGURE 1 The extent and delineation of zones of the coastal temperate rainforest along the North American Pacific coast.

also the study of how the transport of materials and energy by water can be quantified and understood (Lin et al., 2006a,b). The application of hydropedology in the NCTR can increase our understanding of how the flow of material through terrestrial subsystems (i.e. different soil types) is related to biogeochemical fluxes at the watershed scale. Hydropedology can be used to

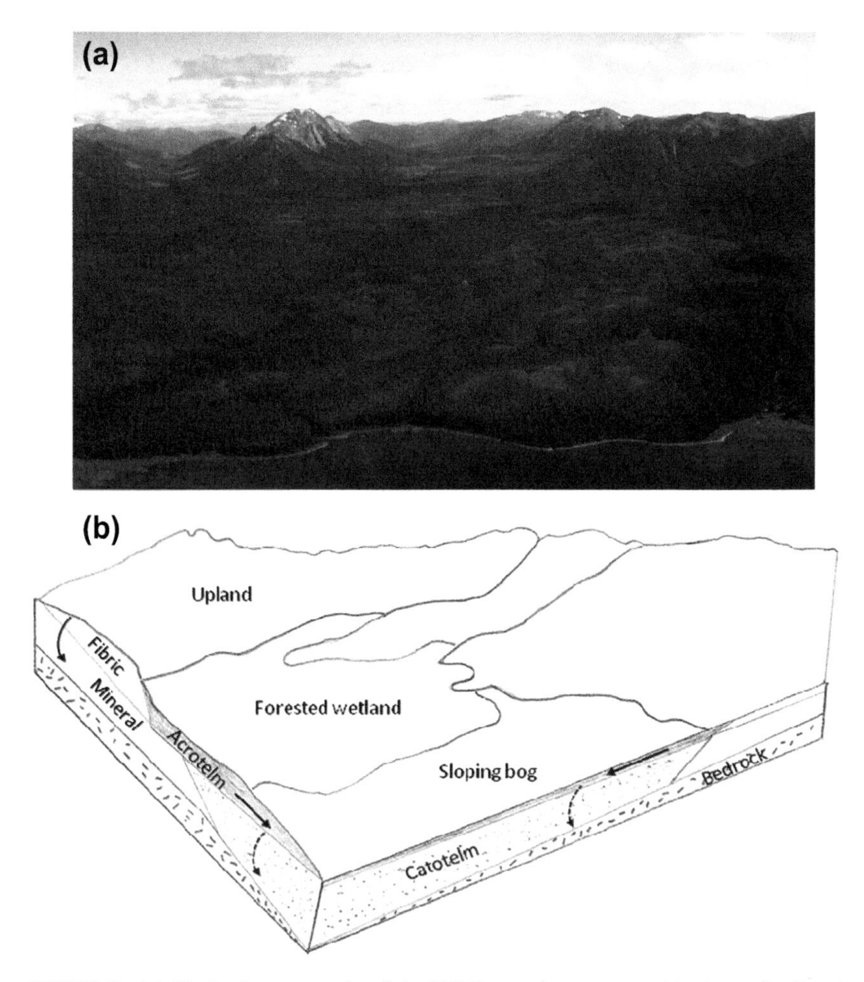

FIGURE 2 (a) The landscape mosaic of the NCTR contains many combinations of soil and vegetation woven together through the abundant flow of water. (b) Landscape relationships for three hydropedologic units are: Upland, forested wetland, and sloping bog. The drainage controls for each unit are illustrated and the dominant flow paths are indicated by arrows. Drainage controls: Upland = bedrock; Forested wetland and sloping bog = Lower organic horizon (catotelm).

create a conceptual framework to link soils with watershed biogeochemical function. There are two essential components of the hydropedologic approach that can be applied to the NCTR to address the challenge imposed by the diverse landscape. The first is landscape stratification using soil pedological mapping techniques. The second is the determination of hydrologic functions associated with these pedological units. With proper delineation, the variability in underlying biogeochemical processes can be minimized to capture the overall landscape response among units for development of robust watershed biogeochemical models. Testing models that describe terrestrial-aquatic biogeochemical linkages is very important given the increasing need to identify and predict the effects of changes in climate and land use on watershed nutrient cycles.

In this chapter, we present a framework for understanding terrestrial and aquatic interactions based on the hydropedologic research vision (Lin et al., 2006a). The framework is derived from data gathered at a soil hydrologic observatory in the NCTR that relies on the extensive soil ecosystem classification available through the Tongass National Forest (Tongass) soil resource inventory (USDA Forest Service, 1996). Our design for the hydrologic observatory uses mapped geomorphic and vegetative patterns present on the landscape as a first approximation of functional units. We define similar soil geomorphic and vegetative functional units as hydropedologic units, which are similar, but not necessarily the same as soil map units. Hydropedologic units are a key to adequately attributing the influence of terrestrial hydrologic and biogeochemical functions on watershed outputs from specific areas of the landscape. We use these hydropedologic units to test the relationship between the distinct functional responses in hydrodynamics and associated biogeochemical transport and transformation. We then examine whether these landscape units provide adequate resolution of cumulative biogeochemical and hydrologic changes to extrapolate and aggregate these functions throughout the entire watershed. Finally, we present some tools for refining landscape stratification using topographic models for soil moisture.

2. SETTING AND DETAILS OF THE NCTR HYDROLOGIC OBSERVATORY

2.1. Geology and Climate

The accretion of terranes through the subduction of the Pacific plate under the North American continent combined with geologic faulting, deposition, and intrusion has formed a diverse assemblage of physiographic divisions along the extended island archipelago of the NCTR (Gehrels and Berg, 1994). Climate shifts and glaciation through several glacial epochs (Mann and Hamilton, 1995) left a pattern of mountain summits, fjordlands, wide glacial valleys, widespread deposits of glacial drift, and varying stream dissection along hillslopes. These postglacial physiographic landscape features established the structural foundation for development of stream and vegetation patterns. The slope and parent material imposed by the geologic and geomorphic history in a watershed influence the trajectory of soil hydrologic patterns and the associated development of soils and plant communities.

The rainforest climate of the NCTR is cool and wet across a wide range of latitude. Rainforests have at least 1400 mm of annual precipitation and cool annual temperatures (<5 °C) (Alaback, 1996). The CTR contains four rainforest subzones: warm temperate, seasonal, perhumid, and subpolar (Alaback, 1991; Fig. 1). The warm temperate and seasonal subzones have fire as a source of stand replacement while fire is very infrequent in the perhumid and subpolar zones. Annual precipitation increases and potential evapotranspiration decreases from the south to north along the coast. As a result, the interaction between moisture and temperature shifts from limiting growth in the south to limiting decomposition in the north. Lower decomposition has resulted in enhanced organic matter accumulation at these northern latitudes (Alaback and McClellan, 1992).

2.2. Soils and Vegetation

Landform and climate are the principal components that determine soil geomorphic associations and plant distribution in the NCTR. Groundbreaking studies of landscape evolution and vegetation community development were undertaken in postglacial landscapes of the NCTR (Chandler, 1942; Crocker and Major, 1955; Ugolini, 1968; Chapin et al., 1994). These early models of ecosystem development used the glacial chronosequence as a template for studying the dynamics of vegetation feedbacks to soils and the initiation of peatland as a climax state due to the development of impermeable spodic horizons in the absence of disturbance (Ugolini and Mann, 1979). Implicit in these models is the accumulation of water leading to waterlogging of soils and subsequent development of deep organic accumulations. However, these models did not address seasonal cycles of water flow and accumulation in the soils and consequent changes in soils and vegetation that impacted the entire watershed.

Young, recently deglaciated areas developed soil after initial plant colonization and the accumulation of large amounts of organic matter. Organic acid leaching can develop incipient spodic horizons in soils after 200 years, but mature Spodosols take 500–1000 years to form well-developed spodic horizons (Chandler, 1942). The Spodosols are common on moderately- to well-drained soils and dominate upland landforms (Fig. 2). Histosols are common in areas of low slope that form the NCTR's abundant wetlands. The original glacial chronosequence studies established idealized principles for landscape evolution, but there is much parent material variability across the extended island archipelago. Soils of the region formed in the postglacial Holocene deposits composed of glacial drift, colluvium, alluvium, and bedrock. In addition, soils were periodically disturbed by windthrow, which provides a new mix of parent material for subsequent soil formation (Bormann et al., 1995). Often modal concepts of the mapped soil series are not expressed throughout the mapping unit and a broad range of characteristics exists. Therefore, rather than relying on ideal expressions of soil type, we rely on general soil geomorphic relationships at higher orders in taxonomic groups to identify soil types across broad landscapes.

Forbs and alders initially colonized postglacial soils and were followed by conifers during the first half of the Holocene. Organic accumulations enabled the spread of Sitka spruce (*Picea sitchensis* (Bong.) Carr) and western hemlock (*Tsuga heterophylla* (Raf.) Sarg.) across the landscape from various glacial refugia (Carrara et al., 2007). A climate shift to colder, wetter conditions approximately 4000 years ago promoted the development of extensive peatlands and the increase in redcedar (*Thuja plicata* (Donn ex D.Don)) and yellow-cedar (*Callitropsis nootkatensis* (d. Spach)) (Heusser, 1960). The current distribution of vegetation communities among different soil types is a response to varying seasonal soil saturation that led to well-drained or waterlogged soils (Neiland, 1971).

Plant distribution patterns and ecology were strong influences in establishing landscape classification guides in the NCTR (Viereck et al., 1992; Meidinger and MacKinnon, 1989). The latest landscape stratification established subsection boundaries for the entire NCTR and neighboring parts of Canada using the National Hierarchical Framework of Ecological Units (ECOMAP; Cleland et al., 1997; Nowacki et al., 2001). The ECOMAP delineation was guided by physiography, lithology, and surficial geology which are all factors related to the land's ability to process water. ECOMAP is much more detailed than the major land resource area (MLRA) map that has two delineations in the U.S. NCTR, the Alexander Archipelago-Gulf of Alaska Coast and the Southern Alaska Coastal Mountains (USDA, 2004).

2.3. Establishing a Hydropedologic Observatory in the NCTR

We established a hydrologic observatory to implement a working model consistent with the research vision for hydropedology (Lin et al., 2006a). We chose watersheds in three different ecological subsections as delineated by ECOMAP as our core sites to address the higher-order hydrogeomorphic control (Fig. 3). Peterson watershed, which is drained by Peterson Creek, is in the Stephens Passage glaciomarine terrace subsection and is composed primarily of slowly permeable glaciomarine sediments (Miller, 1973) along with bedrock outcrops that occur on moderate to low slopes (Fig. 3). Peterson represents watershed types dominated by wetlands (53% of watershed area). In contrast, McGinnis watershed, which is drained by McGinnis Creek, is primarily composed of recently deglaciated areas within the Boundary Ranges Icefield subsection and has low wetland coverage (<5% of watershed area). Eaglecrest watershed, which is drained by Fish Creek, is composed of intrusive volcanic and sedimentary rock in the Stephens Passage volcanic subsection and represents a mix of physiographic features from alpine to lowland wetlands.

The highly resolved ECOMAP landtype-phase classification (Cleland et al., 1997), which provides information on the distribution of fine-scale hydrologic classes, is not available in the NCTR. The best-available information on soil hydrology is the drainage class assignment of Tongass soil map units based on soil morphology (Fig. 4). The Tongass soil resource inventory provides

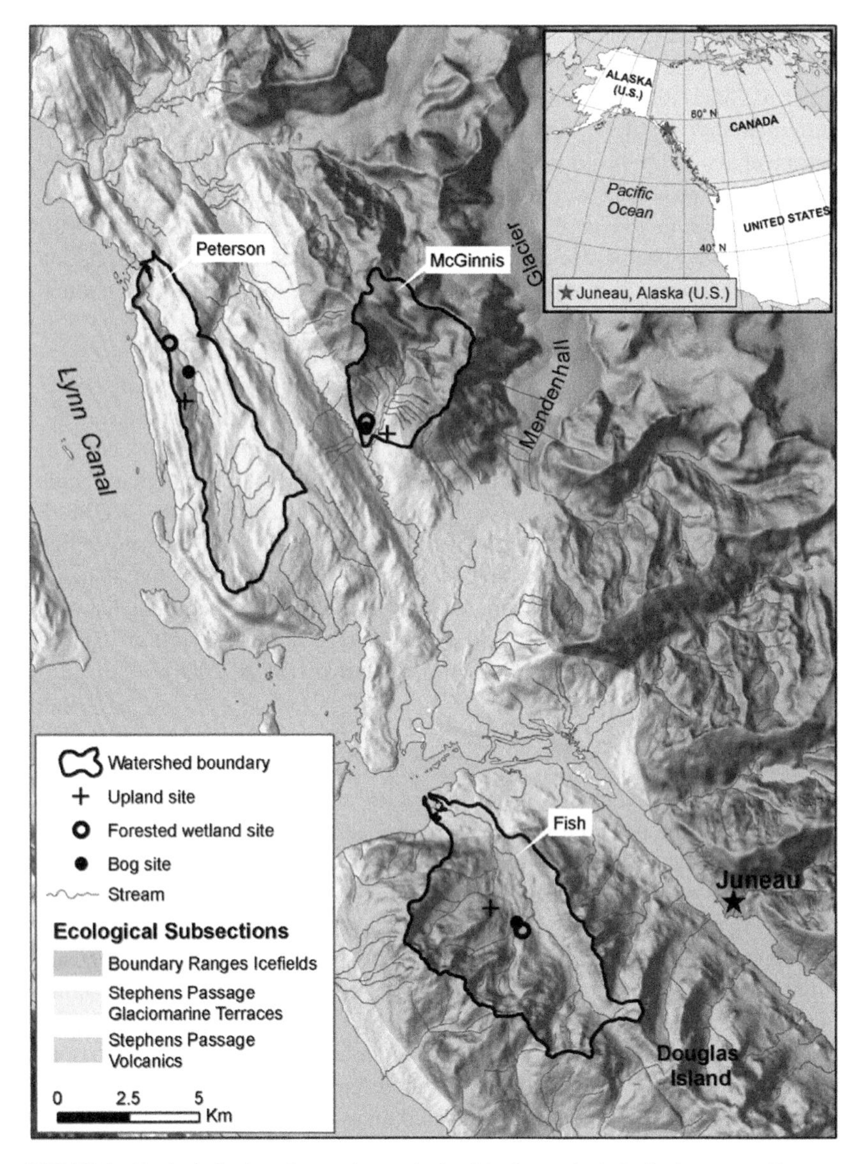

FIGURE 3 Ecological subsections and watersheds of the Juneau hydrologic observatory. Peterson Creek and McGinnis Creek watersheds include the uplifted marine terraces and boundary ranges Icefields, which are located on the mainland. The Eaglecrest (Fish Creek) watershed is located on the Stephens passage volcanics located on Douglas Island. *(Color version online and in color plate)*

a discrete approach (see Park and van de Geisen, 2004) to subdivide the coarse subsection structure available in the ECOMAP (Nowacki et al., 2001; Fig. 3) into units that capture small-scale spatial variability within homogeneous geological formations. The discrete mapping approach recognizes the presence

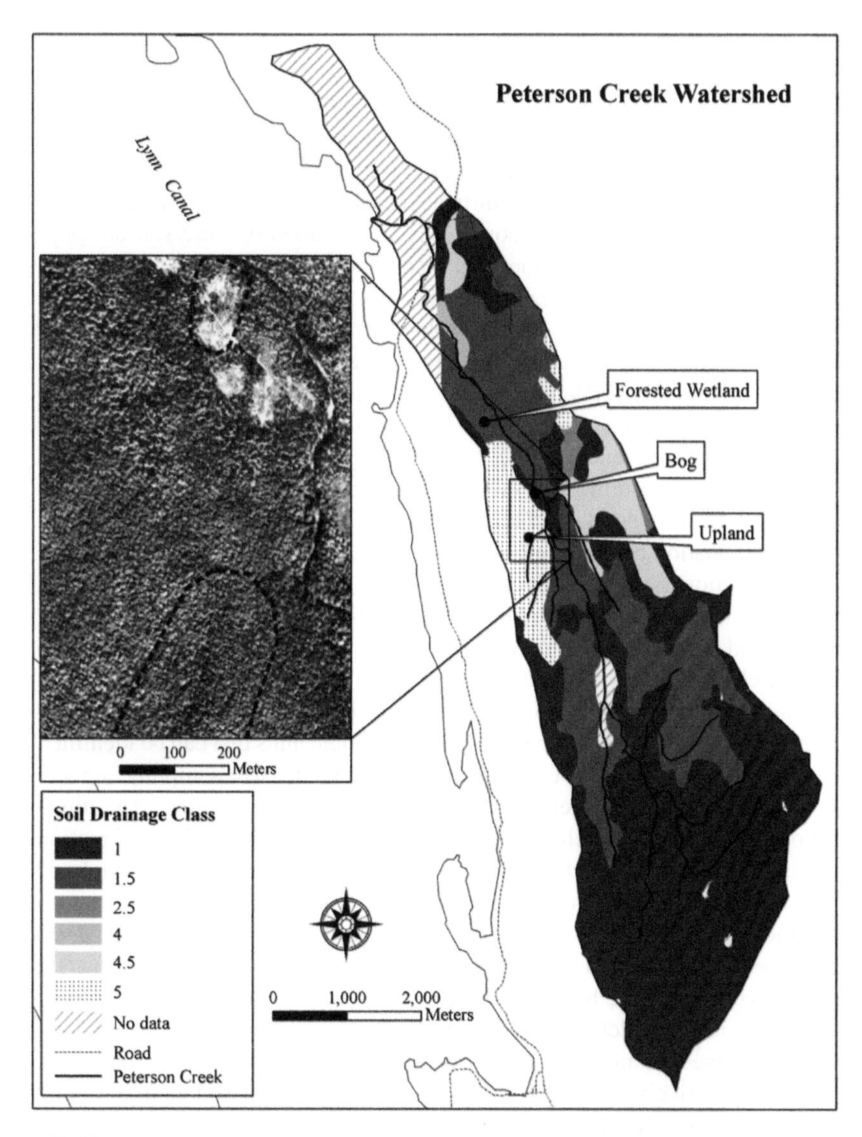

FIGURE 4 Soil drainage classes delineated in Peterson Creek watershed by average soil map unit composition. The aerial photograph shows outlines of the upland and sloping bog hydropedologic units. Soil drainage classes: 1 = very poorly drained; 2 = poorly drained; 3 = somewhat poorly drained; 4 = moderately well drained; 5 = well drained. Note that all classes are not represented due to averaging by soil map unit composition of soil drainage classes.

of a distinct spatial arrangement associated with pedologic and geomorphic structure that influences hydrologic gradients present on the landscape.

Tongass soil map units are closely associated with ecosystem types defined on landforms and were based primarily on soil–vegetation relationships during soil and ecosystem mapping (USDA Forest Service, 1996). Therefore, the Tongass soil resource inventory provides the building blocks for a preliminary hydropedologic framework across the region with spatially explicit maps of soil–vegetation assemblages associated with order 3 and 4 soil surveys. However, there are many variants to established soil series due to the heterogeneity caused by small-scale disturbance, such as windthrow and landslides. Soil map unit associations and complexes are common because soil development can vary substantially due to the small-scale variation of soil-forming factors caused by these localized disturbances. To address this variability, we integrate wetlands into the hydropedologic framework as they are a clear landscape feature and an obvious choice for a soil/vegetation class as they cover approximately 21% of the NCTR. The readily available landscape classification of wetlands across the U.S. portion of the NCTR through the National Wetland Inventory (USFWS, 2009) makes applying this ecosystem classification quite useful. The NWI mapping does not capture many of the wet forests that do not meet the jurisdictional requirements of wetland delineation, so the combination of soil maps and wetland delineations provides a means to capture functional units on the landscape.

We use the term hydropedologic unit to define the combination of the soil and wetland map unit with biogeochemical function. The hydropedologic units had soil characteristics that were hypothesized to control soil hydrology and influence the overall hydrologic and biogeochemical behavior of the subcatchment (Table 1). We chose three hydropedologic units that can be identified through soil surveys and wetland inventories for intensive sampling: upland mineral soils, forested wetland soils, and peatland bog soils. The three hydropedologic units are closely associated with the soil map unit designation and associated drainage class (Table 1). Therefore, similar hydropedologic units were located in each of the subsections (Fig. 3). Three individual subcatchments of each type were delineated within the watershed blocks by soil and vegetation maps and verified in the field to assure that they adequately represented the selected hydropedologic units (Table 1; Fig. 3). The hydrologic control section was defined by the presence and depth of the acrotelm and catotelm in the wetland soils (D'Amore et al., 2010). The upland soils were located on relatively steep hillslopes defined by a lack of soil saturation and bedrock hydrologic control (Fig. 2). The subcatchments within each watershed were delineated to locate flow-gauging weirs to measure chemical export from each site to test the hypothesis that hydrologic fluctuations were related to biogeochemistry of the catchment.

Wetlands are roughly mapped by the National Wetland Inventory (NWI), but low resolution and inaccuracies dictate that field observations of soil patterns be used to more accurately delineate the boundaries of those classes in the field. We use the NWI nomenclature for classes of forested wetland and upland. We have established a "sloping bog" class as a modification to the NWI nomenclature for the palustrine emergent class that combines the features

TABLE 1 **Watershed Location, Hydropedologic Units, and Soil Attributes for Experimental Catchments in the Juneau, AK Area. Drainage Class is based on Modal Soil Pedon Morphological Characteristics from the Tongass National Forest Soil Resource Inventory. Drainage Control is based on Either the Presence of a Slowly Permeable Subsurface Aquitard (Catotelm and Bedrock) or a Zone of Saturated Lateral Flow near the Soil Surface (Acrotelm)**

Watershed	Hydropedologic unit	Soil classification	Drainage class	NWI category*	Drainage control
Peterson	Sloping bog	Typic Cryohemist	1.0	PEM	Catotelm
	Forested wetland	Histic Cryaquept	1.5	PF	Acrotelm
	Upland	Lithic Haplocryod	5.0	UP	Bedrock
McGinnis	Sloping bog	Typic Cryohemist	1.0	PEM	Catotelm
	Forested wetland	Terric Cryohemist	2.5	PF	Acrotelm
	Upland	Typic Humicryod	4.5	UP	Bedrock
Eaglecrest	Sloping bog	Typic Cryohemist	1.0	PEM	Catotelm
	Forested wetland	Terric Cryohemist	2.5	PF	Acrotelm
	Upland	Typic Haplocryod	5.0	UP	Bedrock

** National Wetland Inventory (NWI) categories: PEM = palustrine emergent; PF = palustrine forested; UP = upland (Cowardin et al., 1979).*

associated with fens and bogs. Sloping bogs have water flow derived from groundwater, similar to fens, which flows through vegetation communities more commonly associated with bogs. The slope bog is recognized in the Canadian wetland classification (NWWG, 1998), but is not identified in the NWI. Therefore, we have established a class of hydropedologic unit that can be used across the entire NCTR.

3. A FRAMEWORK FOR INTEGRATED HYDROPEDOLOGIC STUDIES IN THE NCTR

The hydropedologic framework can be used to test the strength of association of hydrologic fluctuations and biogeochemical transformations within the provisional hydropedologic units through plot-based studies that quantify terrestrial–aquatic biogeochemical fluxes. The hydropedologic units can be used to estimate biogeochemical fluxes from entire watersheds by extrapolation of fluxes across similar hydropedologic units delineated by the soil resource inventory. Our integrated research fills a need to understand soil hydrology and watershed biogeochemical fluxes in the NCTR. Five key issues provided guidelines for describing our hydropedologic research: structure, function, scale, integration, and disturbance (Lin et al., 2006a). We use these issues as a framework to present the information gathered through our research at the Juneau hydrologic observatory and evaluate the potential of future hydropedologic research in the NCTR.

3.1. What are the Hierarchical Structures of Soil and Water in the NCTR?

The NCTR provides an excellent natural laboratory for the hydropedologic model. It is the largest intact contiguous expanse of coastal temperate rainforest in the world and contains thousands of watersheds largely devoid of major human disturbances. These watersheds contain a wide range of relative proportions of our defined units (e.g. 5–95% wetland) allowing us to verify watershed biogeochemical models across a broad geographic area. The application detailed here provides a first approximation for defining hydropedologic units and using these units to describe biogeochemical cycles at the watershed scale. Thus, our examples provide a hierarchal structure to help explain biogeochemical cycling in individual hydropedologic units with the ultimate goal of identifying biogeochemical signatures of the entire watershed.

The storage, flux, pathway, and residence time of water are distinctly different within our three classes, confirming that they span multiple dimensions of variability across the landscape. The NCTR has a moisture excess during most of the year, but the balance between precipitation and evapotranspiration creates conditions that influence soil profile saturation patterns (Fig. 5). The water tables among the three units were stratified as expected from the initial landscape partitioning where deep water tables were measured in upland mineral soils and shallow water tables were measured in forested wetland and sloping bog soils (Fig. 6). Although this relationship was expected, the consistency among the replicate units (e.g. three sloping bog sites) supported the premise that these units provide adequate first approximations of general water-table fluctuations across the drainage gradient.

The water-table depth in the mineral soils is limited by impermeable bedrock and slightly weathered regolith. Water moves through the soils on the 'upland' landscape positions, but there are fluctuations within the profile

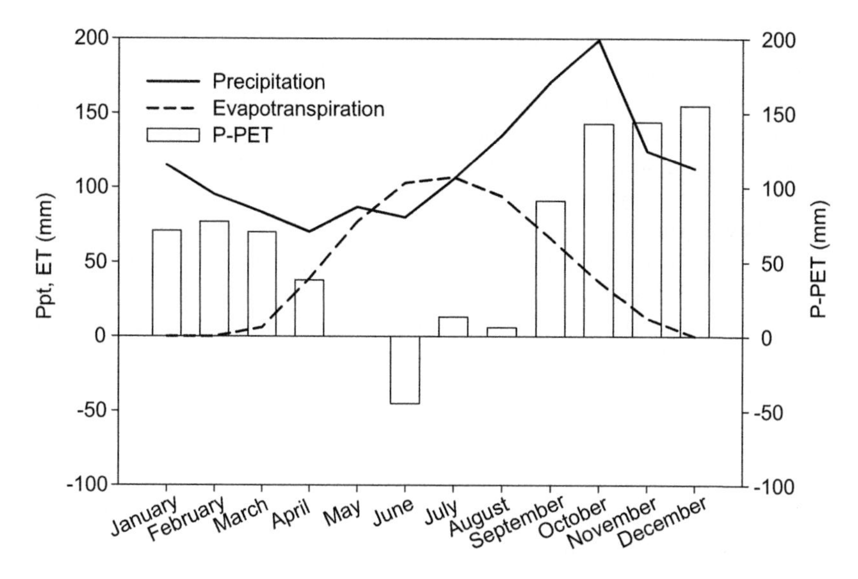

FIGURE 5 Patterns of annual precipitation and evapotranspiration in the Juneau, Alaska area. Precipitation, evapotranspiration, and the calculated moisture surplus or deficit (Precipitation [P]–Potential evapotranspiration [PET]) are shown over an annual cycle. Data were derived from the potential evapotranspiration estimates of Patric and Black (1968).

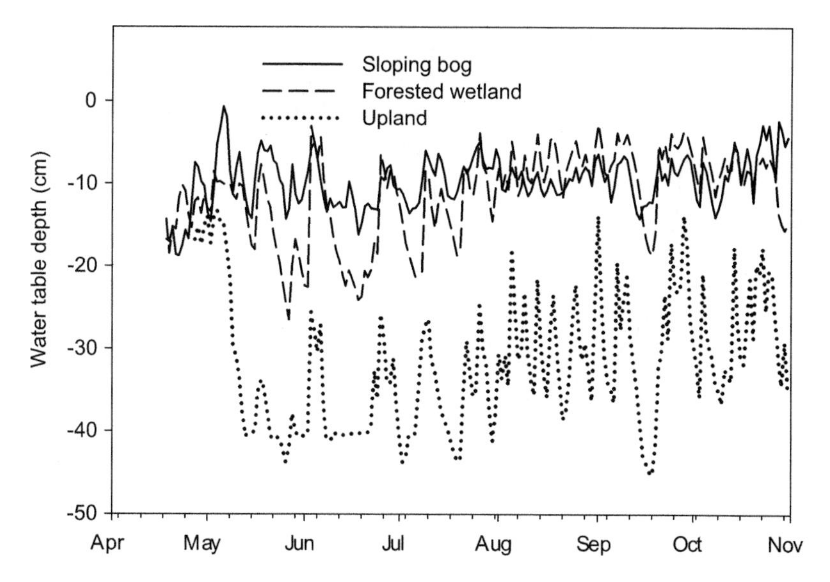

FIGURE 6 Water-table fluctuations in uplands, forested wetlands, and sloping bogs. Data from three replicates of each hydropedologic unit, which were measured during 2006–07 (data adapted from D'Amore, 2011). The period of moisture deficit relative to the surface can be seen most clearly in the water-table depression of the forested wetland site from late May to late July.

consistent with a groundwater flux across the bedrock interface. Recharge is the dominant water-flow pathway downward through the fibric surface horizon and spodic mineral horizon toward the impermeable bedrock. After it rains, water infiltrates down through the soil profile as recharge with no overland or near-surface flow and is then redirected laterally across the bedrock interface (Fig. 2).

The flow regime of the forested wetland and sloping bog is regulated by the depth of the permeable, surface acrotelm horizon (D'Amore et al., 2010). There are two distinct cycles within these peatland soils: a distinct water-table drawdown in June–August and near-surface saturation in the fall during higher rainfall (Fig. 7). The near-surface saturation (<10 cm) is quite similar within the two wetland soils with nearly complete saturation of the soil profile, but forest wetland water tables drop faster and to deeper depths during drawdown periods. The water-table drawdown indicates the presence of an unknown water export pathway, either a much deeper zone of flow within the forested wetland soils or evapotranspiration by trees. Rainfall events create subsurface flow and pressure heads within the soil matrix that lead to ephemeral increases in water-table height and incursions into the unsaturated soil zone (i.e. acrotelm) from the saturated zone below. These incursions are much more

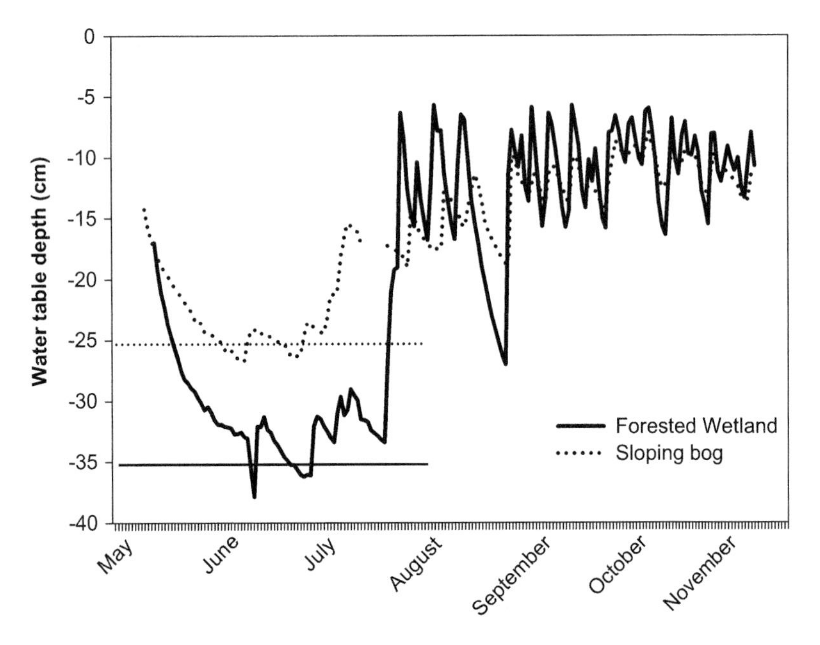

FIGURE 7 The depth and duration of the water table measured in the acrotelm (surface organic horizon) of the forested wetland and sloping bog at the McGinnis Creek watershed from May to November of 2004 (adapted from D'Amore et al., 2010). Horizontal lines represent depth of the acrotelm (aerobic surface horizon) in each soil during the annual period of moisture deficit.

dramatic in the forested wetland due to the deeper acrotelm. The depth and duration of aerobic conditions in the acrotelm are similar to other regions with extensive peatlands (Holden and Burt, 2003; Worrall et al., 2002, 2003), but the frequency and magnitude of water-table fluctuations are more dynamic in NCTR peatlands.

Understanding how soil saturation interacts with biogeochemical transformations within each hydropedologic unit provides insight into material transfers and biogeochemical transformations within a watershed. The duration and fluctuation of water tables in peatlands have been clearly linked to biogeochemical shifts (Strack et al., 2008). Although aerobic surface horizons persist in wetland soils even during periods of heavy rainfall (D'Amore et al., 2010), redox potential varied with soil saturation at the water-table interface (Fig. 8). The responsiveness of redox potential to changes in soil saturation suggests highly dynamic shifts in microbial metabolism at the interface between the acrotelm and catotelm. The anaerobic–aerobic boundary is a highly reactive zone for microbial activity and, hence, biogeochemical transformations (Gutknecht et al., 2006). Redox potential measurements are a surrogate measure for potential biogeochemical transformations within soils and riparian sediments (Miller et al., 2006). For example, the soils are a rich source of reduced DOM, which is an important substrate for microbial communities that oxidize and alter the nature of dissolved organic carbon (DOC) in soil and soil solution. The

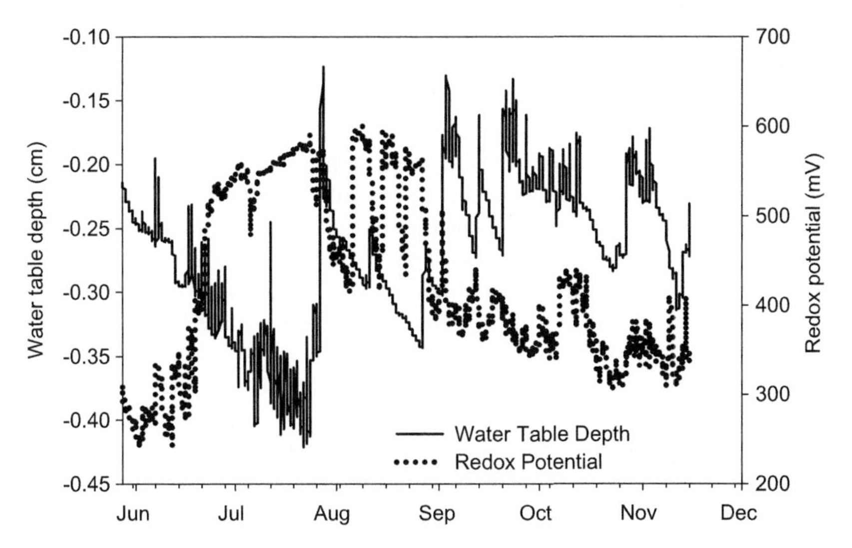

FIGURE 8 Fluctuations in water-table depth and redox potential at in the McGinnis forested wetland site from June to November 2004. Redox potential was measured with permanently installed platinum electrodes in the lower part of the surface organic horizon (acrotelm). See D'Amore et al., 2010 for detailed methods.

sequential processing of DOM by soil microbes contributes to electron transfers and is closely tied to the degradation of organic acids and other chemical transformations such as denitrification (Miller et al., 2009). The combination of water-table fluctuations and redox potential shifts provides a means to identify discrete zones of fine-scale biogeochemical activity in soils. The hydrodynamics of the hydropedologic units provide a means to distribute water-table fluctuations to broader landscapes through extrapolation of function to similar mappable units.

3.2. How can Function be Integrated into the Structure of the Soil Hydrologic System of the NCTR?

Flow-path integration describes how the flow path of water, and the material entrained in the flow path, integrates stream and landscape heterogeneity (Fisher and Welter, 2005). Flow-path integration highlights how small-scale biogeochemical transformations that occur in three-dimensional space are integrated into net cumulative fluxes from the watershed (Fisher et al., 2004). The key concept in this approach is how nutrient retention and cycles associated with different ecosystem types control the composition of water flowing through the sequence of structural forms encountered on the landscape (Giblin et al., 1991; Fisher et al., 2007). Hydropedology and flow-path integration emerge from two different disciplines, but both strive to understand the linkage between the production, transformation, and export of DOM from terrestrial to aquatic ecosystems. Interdisciplinary concepts such as these are not often applied by a broad and diverse group of scientists, but remain within the discipline from which they have emerged. However, these concepts and the ability to integrate terrestrial and aquatic biogeochemical research are highly desirable in the NCTR. The concepts of terrestrial nutrient cycles and exports to aquatic ecosystems can provide a means to address the effects of alterations to functions in both systems and related maintenance of coastal marine food webs. Ultimately, the system must be viewed as a unified flow path to achieve this goal (Fisher et al., 2004). The hydropedologic approach is a compromise that uses discrete units to quantify the continuous flow of material to aquatic ecosystems.

The hydropedologic units are tools for organizing drainage units by function. The observed water-table fluctuations and redox potential shifts within hydropedologic units can be linked with biogeochemical cycling and potentially to landscape unit function. The hydropedologic units are useful for describing DOM export from terrestrial to aquatic environments in the NCTR. In the same way, the hydropedologic unit approach can also provide insight into the quality of the DOM produced and exported from the landscape, which has important implications for productivity in downstream aquatic ecosystems.

The distinction between upland and wetland is a good predictive distinction for determining the DOM flux from a watershed (Mulholland and Kuenzler, 1979; Mulholland, 2003). Soil attributes, such as soil C:N, have also been used effectively to predict watershed DOM export across a broad array of systems (Aitkenhead and McDowell, 2000). Our watershed stratification provides a good estimate for DOM concentrations (measured as DOC) using the wetland distinction. There is a strong relationship between wetland extent and DOC concentrations in NCTR streams (D'Amore et al., submitted for publication; Fig. 9), which is also apparent in the subcatchment outlet streams for the hydropedologic units (Fig. 10). The hydrodynamics and intensity of the biogeochemical exchange in the acrotelm differ between the sloping bog and forested wetland provide an explanation for the local strength of the association between depth to water table and soil DOC concentrations (D'Amore et al., 2010).

Integrating physical and biogeochemical cycles is essential in creating a useful tool to understand the effect of soil hydrologic cycles on DOM export. Surface organic horizons in both mineral and organic soils provide highly conductive flow paths that facilitate the transfer of water and DOM through the soil and into surface water channels. Deeper, organic horizons have much lower conductivities and transfer water through the soil more slowly allowing for longer soil residence times for DOM and associated nutrients. The internal drainage of mineral soils shifts the chemical transformations from anaerobic to

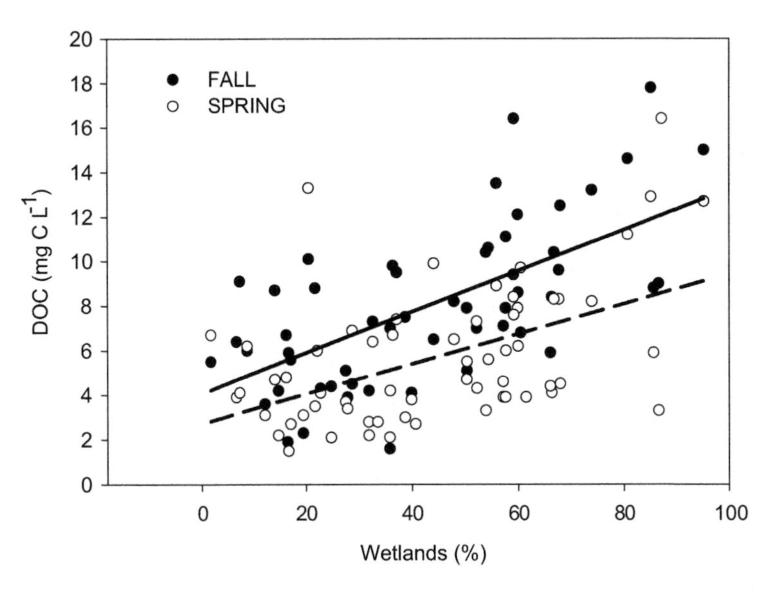

FIGURE 9 Relationship between the extent of wetlands and the concentration of dissolved organic carbon (DOC) in streams in the Alaskan portion of the northern coastal temperate rainforest.

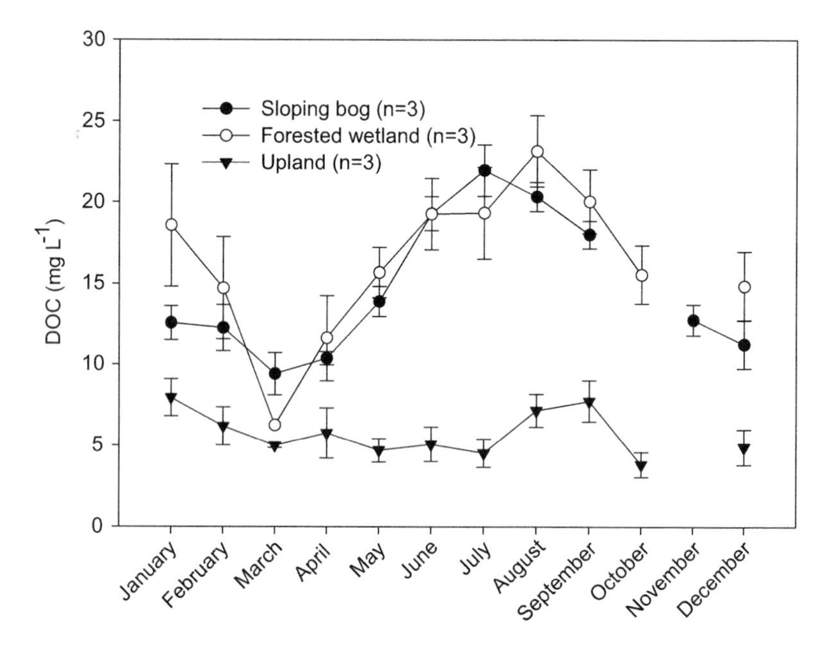

FIGURE 10 Monthly concentrations of dissolved organic carbon (DOC) measured in outlet streams draining the sloping bogs, forested wetlands, and uplands in the Juneau, AK area in 2006–07 (adapted from D'Amore, 2011). Means (±1 standard error) for measurements taken every 1–4 weeks are averaged by month for each hydropedologic unit.

aerobic conditions, which is in direct contrast to peatland soils. These conditions vary seasonally, as warmer periods and similar or reduced rainfall alter the late spring and early summer soil saturation dynamics.

Dissolved organic matter characterization provides a means for distinguishing biogeochemical signatures and potential ecological function among the hydropedologic units (Fellman et al., 2008). For example, ordination analysis clearly distinguishes the hydropedologic units into distinct classes based on the relationship between protein-like and humic-like fluorescence components (Fig. 11). These differences in the chemical properties of DOM are reflected in DOM bioavailability, as significant seasonal differences are observed both within and among hydropedologic units (Fig. 12). The movement of water through the forested wetland is controlled by the depth of the permeable surface peat (acrotelm) and the intensity and duration of rainfall (D'Amore et al., 2010). The sloping bog has a much smaller acrotelm and the water flow is dominated by the near-surface, slowly permeable catotelm. The upland soils are dominated by water flow downward in the profile and transport across the slowly permeable bedrock or paralithic contact. The distinct quality of DOM and the differences in dominant hydrologic flow paths among the hydropedologic units suggest that

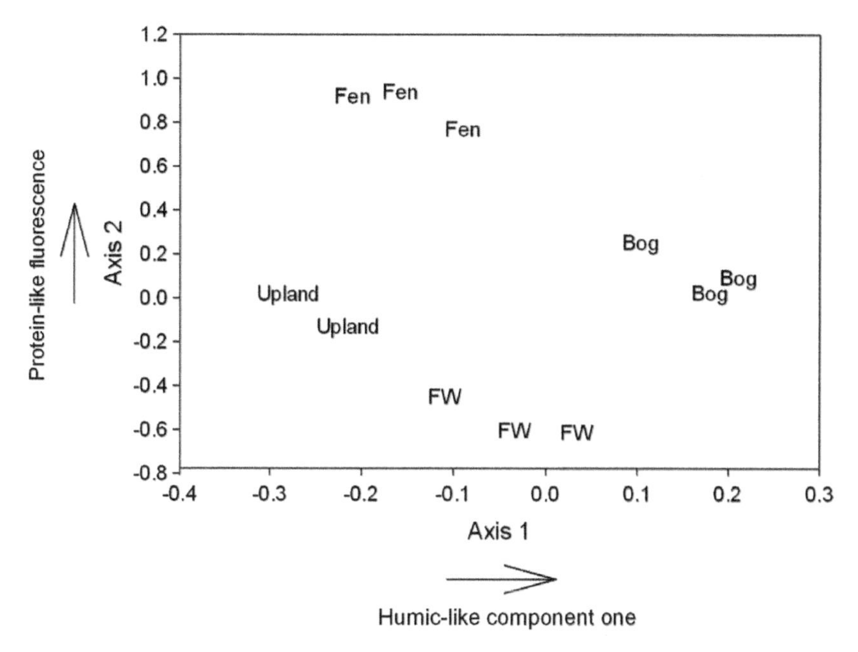

FIGURE 11 Ordination of hydropedologic units according to component composition modeled with parallel factor analysis (PARAFAC). Axis 1 is derived from humic components and axis 2 is derived from protein components identified through PARAFAC of the dissolved organic matter in soil solution. FW = forested wetland.

the aggregated collection of hydropedologic units has the potential to alter watershed-scale biogeochemical processes by influencing patterns in labile DOM delivery to streams.

Differences in chemical export within hydropedologic units and seasonal variability among these units have implications for the biogeochemical function of the entire watershed (Fellman et al., 2009b). For instance, low biotic demand and short soil residence time from predominantly surface soil flow paths lead to a pulse of labile DOM during spring snowmelt for wetland sites, but less so for upland sites (Fig. 12). In the wetland sites especially, DOC concentrations can be quite low during the spring snowmelt because of simple dilution in relation to the mass of water. Competition between DOM transport and transformation is ultimately mediated by soil hydrologic flow rate, or contact time (Randerson et al., 2002). Thus, reduced DOM concentrations may also result from a stable water table that restricts organic matter mineralization and diminishes the pool of leachable DOM.

During the summer growing season, water-table drawdown occurs in the bog and forested wetland and water moves predominantly through deep and less conductive flow paths (Fig. 7). High biotic demand for labile DOM by soil microbes and long soil residence time results in the delivery of mainly

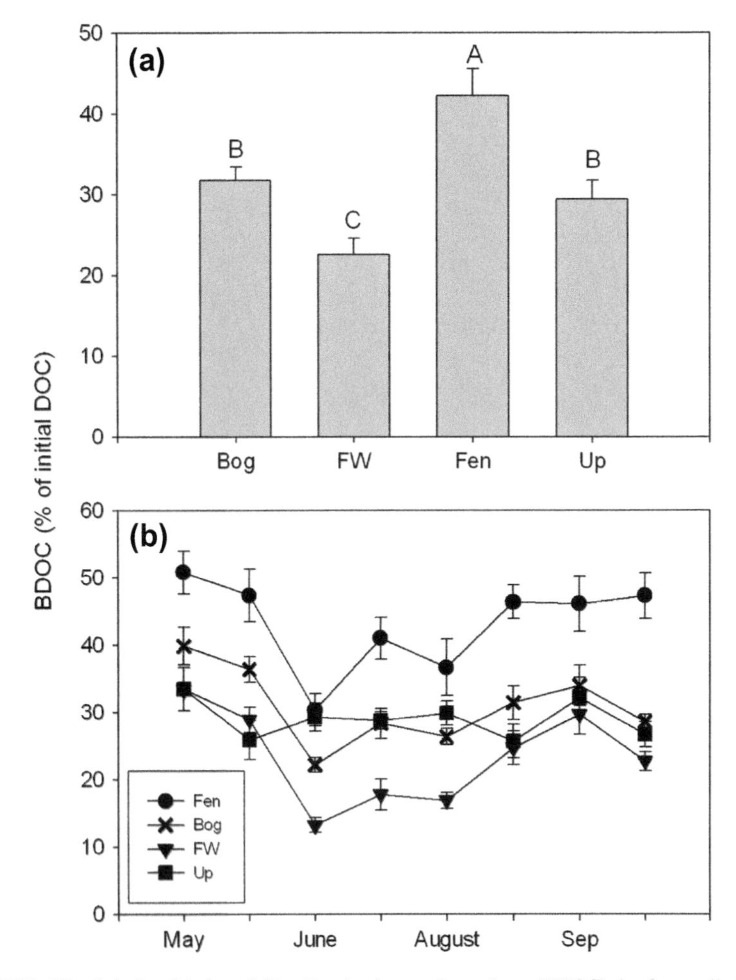

FIGURE 12 Solution biodegradable dissolved organic carbon (BDOC) in four soil types. Patterns of (a) average BDOC in soil and (b) time series for the four soil types collected across the range of sample dates. Significant differences among soil types are indicated by different capital letters above the columns; error bars indicate ±1 SE and $N = 3$ for all soil types. Abbreviations: FW, forested wetland and Up, upland forest. *(Adapted from Fellman et al., 2008)*

recalcitrant DOM to streams from wetland sites. Consequently, the quality and biodegradability of the streamwater DOM may vary dramatically with source material and season among the distinct units (Fellman et al., 2009a). These observed differences in DOM biogeochemistry clearly demonstrate that these hydropedologic units may have different functional roles in terms of biogeochemistry within the watershed.

Our work along the hydropedologic gradient has also provided new insights into the role of discrete landscape units for providing labile organic matter to aquatic ecosystems. In particular, research in the NCTR has demonstrated the importance of wetlands for the export of labile DOM that can support stream heterotrophic productivity (Fellman et al., 2008, 2009a). This highlights a fundamental change in our view of wetland ecosystems, which have typically been seen as sources of highly degraded dissolved material. The recognition of the importance of terrestrial nutrient subsidy to aquatic systems has been an important shift in our understanding of material flows in the NCTR. The transfer of material from the terrestrial to aquatic ecosystems in the NCTR provides important nutrient subsidies for both freshwater and marine ecosystems (Hood et al., 2009; Fellman et al., 2010). This finding also illustrates the importance of understanding how hydropedologic factors influence terrestrial biogeochemical cycles which can then be linked to soil-water export pathways of dissolved reactive materials.

3.3. What is the Potential for Scaling Functions across the Complex Landscape of the NCTR?

We must derive soil–geomorphic relationships from existing data such as soil map unit classifications to establish initial hypotheses related to functional responses in the absence of high-resolution topographic maps. Finely delineated terrestrial landscape attributes can be determined using high-resolution morphometric analyses such as light detection and ranging (LiDAR) imagery. However, LiDAR mapping is not available on much of the NCTR. In any case, the delineation of these highly resolved landscape attributes (e.g. 1 m^2) generally exceeds our ability to accurately measure the functional attributes of such small features. To constructively stratify the landscape, a broad grouping of 'subject' landscape types was chosen to represent hydropedologic units. The resolution of current digital elevation models (DEMs) and wetland and vegetation maps do not support a finer parsing of the landscape than the three fundamental types we used. Finer distinctions could be made of these types, but because our intent was to upscale these fundamental units to the watershed scale, we used classes that could readily be predicted using available GIS data layers. These landscape units occur at different locations along the hydrologic gradient, and so represent rational ways to partition the landscape to study the functional linkages between soils and hydrology.

The highly dissected nature of the landscape of the NCTR is evident from coalescing flow on hillslopes during storms, and the integrated response of this water delivery system is quite profound. For instance, during intense storms, streamwater DOC concentrations often dramatically increase as terrestrial source pools and hydrologic flow paths are different between baseflow and stormflow (Fellman et al., 2009b). As soils become saturated, water moves through shallow soil layers or at the acrotelm/catotelm interface (D'Amore et al., 2010) and differing flow paths can entrain DOM with different chemical properties. Thus, temporally transient water fluxes influence shifts in biogeochemical patterns. It is also possible that DOM export during storms accounts for a large fraction of the annual watershed flux as found in studies from a similar humid region in the United Kingdom (Worrall et al., 2002, 2003), which highlights the importance of individual storms for watershed biogeochemical budgets. Partitioning the watershed into hydropedologic units and understanding their biogeochemical response to high flow events could be used to improve annual estimates of watershed-scale and regional DOM fluxes.

The hydropedologic stratification provides a framework for downscaling models of climate drivers to plots for modeling soil hydrology in connection with regional weather patterns. The parameter-elevation regressions on independent slopes model (PRISM Climate Group, 2010) provides estimates of precipitation and temperature at a resolution of 0.8 km. Precipitation and temperature data analyzed by PRISM models can be used to estimate evapotranspiration, which strongly influences the accumulation and flow of water in soils during the growing season. This information can provide support for developing watershed or regional water budgets by constraining estimates of fluctuations in soil water storage. The potential influence of deviations in soil saturation can also be applied to modeling plant responses across watersheds to guide vegetation-monitoring programs.

The distribution of frequently saturated soils can predict the presence of plant communities, which is especially important in the face of shifting populations of plants due to climate change (Hennon et al., 2011). Hydric soils have been successfully used as a predictor variable to describe differences in timber volume classes across the Tongass (Caouette and DeGayner, 2005). However, a hydric soil class merely assigns a wet or dry state, but does not provide any details of how the underlying soil nutrient turnover may influence the specific vegetation assemblages. The hydropedologic approach provides a framework to design experiments to examine small-scale processes in soils and the broader control on vegetation structure. This can then be linked to the biogeochemical properties such as DOM bioavailability to complete an assessment of the feedback mechanisms and interactions among all these key elements in vegetation response to watershed biogeochemical characteristics. Yellow-cedar is a particularly good example of a potential application of this approach because its distribution is closely linked to soil moisture patterns. The yellow-cedar tree is not obligated to saturated soils (D'Amore and Hennon, 2006) but follows a niche strategy to persist in marginal conditions with low nutrient turnover and nearly saturated soil conditions (D'Amore et al., 2009). The prediction of future yellow-cedar decline will rely on accurate spatial models of soil saturation as well as changes in those patterns over time (Hennon et al., 2012).

3.4. How can Information from Diverse Research Studies be Integrated into a Hydropedologic Model in the NCTR?

We provide a framework for predicting biogeochemical cycling in catchments that is similar to the geochemical catena (Johnson et al., 2000; Palmer et al., 2004). The influence of soil hydrology on biogeochemical cycles has not been clearly articulated in the coastal temperate rainforest research. Therefore, a working model using this approach and implementation of an integrated hydrologic observatory is useful for addressing goals associated with biogeochemical research in the NCTR. Our study outlines a method to demonstrate how linking terrestrial and aquatic research through the use of well-defined hydropedologic units can provide benefits to both fields. Traditionally, these disciplines have shared similar biogeochemical concepts such as the study of limiting nutrients, but measurements and perspectives regarding applications of these measurements vary (Grimm et al., 2003). For example, aquatic ecologists are now exploring the potential for wetlands to impact stream productivity beyond lakes and ponds. Another key example is the ability to derive information regarding wetland function from structural aspects of wetland ecology (Bridgham et al., 1996).

The hydropedologic unit delineation can provide a link between disciplines as water flow is the key nexus between terrestrial and aquatic environments. Soil functional attributes can be applied to hydropedologic units for estimates of specific aspects of biogeochemical functions among individual hydropedologic units and these functions can be extrapolated to entire watersheds. For example, quantifying seasonal soil saturation across different hydropedologic units can be applied to key nutrient turnover measures. Denitrification rates are influenced by soil factors such as texture and moisture content which control soil saturation and redox potential (Pinay et al., 2003). A hydropedologic model provides a means to estimate zones of high denitrification potential, such as riparian zones and seeps, based on the stratification of the landscape according to soil factors. Using this approach, replicated studies can be designed to quantify denitrification rates that will enable more accurate estimates of nitrogen fluxes from specific areas within a watershed. The spatial resolution of these hot spots for denitrification can be combined with other N flux pathways that can then be appropriately scaled to estimate the overall watershed nitrogen flux.

3.5. How does Anthropogenic Disturbance Interact with Hydropedologic Functions?

The ecosystems of the NCTR are primarily driven by small-scale natural and anthropogenic disturbance over short time scales. Because fire is uncommon, wind is the primary driver of change in forest structure through disturbance (Kramer, 2001). The major anthropogenic disturbance in the NCTR has been timber harvest and to a lesser extent the construction of roads, especially through wetlands. In the southeast Alaskan portion of the NCTR, approximately 220,000 ha of forest have been logged (USDA Forest Service, 2008). It is unclear how harvesting and road building have affected watershed hydrology and the delivery of water and materials to streams. Timber harvest covers approximately 3% of the total area of the Tongass and appears not to exert a large influence on the overall landscape. However, individual landscape features and soil types such as karst formations have been heavily used for timber harvest. The loss of large stature trees can alter evapotranspiration in stands and the subsequent timing and delivery of water from soils to streams. Although this type of change affects surface and subsurface water flow, its effects on soil attributes including soil development and biogeochemical cycling are unknown. The change in trajectory of soil development following disturbance for individual landscape units should be considered in overall hydropedology models.

Road networks are closely associated with timber harvest in the NCTR. The impact on groundwater quality and flow along with the alteration of material delivered to streams due to the presence of roads are not well understood. Therefore, road construction and associated impacts to watershed hydrologic cycles remain a concern for forest managers. The few existing hydrologic studies on roads have not identified any major impacts on peatland (Kahklen and Moll, 1999) or mineral soil (McGee, 2000) hydrology. However, linear road features have potentially disrupted flow paths and created biogeochemical interactions between soil water and crushed rock that may reduce the residence time of water within soils before delivery to the stream channel. Further, the increased rock–water interaction may alter the biogeochemical signature through oxidation of DOM and altered pH.

4. FUTURE APPLICATIONS AND CONCLUSIONS

The sheer volume of water draining through the NCTR and the magnitude of dissolved material export underscores the need for a process-based understanding of what regulates water and material fluxes. The large number of watersheds, their heterogeneity, and the isolated nature of the region make prediction and management a challenge in the NCTR. A systematic modeling approach that is both scalable and representative of the diverse range of watersheds will allow land managers to make predictions at both local and regional scales. We suggest that hydropedology offers an opportunity to address these needs in the NCTR.

We posed the hypothesis that hydropedologic units encompass soil hydrology and associated biogeochemical behavior at a resolution that balances fine-scale variability with larger-scale reproducible measureable outputs. Hydropedologic units are a logical link between regional-scale climate models and reach-scale predictions of habitat response because they are fundamental functional units that can be readily mapped within a watershed. Our review synthesizes the close association between DOM with the different units as an example of a key biogeochemical process. We provide evidence that the interaction of soil hydrologic and biogeochemical transformations controls the unique attributes of DOM in catchments. Our work quantified the magnitude of water-table fluctuations and identified the role that hydrodynamics played as a key characteristic that controls DOM and at scales detectable in a watershed. Because these units represent a broad class of various soil map units that are available across the NCTR, we have provided a tool for extrapolating biogeochemical functions across landscapes such as that illustrated in Fig. 2.

Although there is a growing understanding of the interaction among topography, subsurface structure, and flow control (Lin et al., 2006b; Lin, 2010), we are just beginning to conduct experiments and understand flow regimes within the soils of the NCTR. The complexity of forested watershed hydrology is related to the landscape geomorphology. Resolving the opaque nature of the subsurface with integrated models of water and material flows is a high priority. Landscape units that are operationally defined by quantitative geomorphometry are much more useful in delineating areas of water flow and coalescence. Soil geomorphic models rely on accurate and preferably high-resolution digital elevation measurements along with good information on soil moisture patterns for calibration. The current challenge is to create a more refined model linking soil geomorphic associations and biogeochemical function to flow paths delineated with topographic models. A model of subsurface flow paths we created using the existing 60-m DEM of the NCTR provides a satisfactory resolution for the Peterson Creek drainage, with a good correspondence between predicted groundwater levels and existing hydropedologic units (Fig. 13). The application of high-resolution imagery can greatly improve the ability to delineate hydropedologic units and design hypothesis tests of functions. Once more accurate topographic imagery is available for a greater portion of the landscape of the NCTR, we can link the two approaches and create higher-resolution functional maps.

Hydropedology is a valuable tool that will continue to play an important role in terrestrial and aquatic research in the NCTR. The main conclusions we can draw from the current state of the research in the NCTR are:

1) Hydropedology is a flexible and robust approach that infuses soil and wetland delineations with function by providing a link between soil hydrology and DOM biogeochemistry;
2) Hydropedology provides an approach to scale up biogeochemical fluxes from individual hydropedologic units to the watershed scale;
3) Hydropedologic units provide a response variable for predicting functional alterations in watersheds from changes in land use and climate in the NCTR.

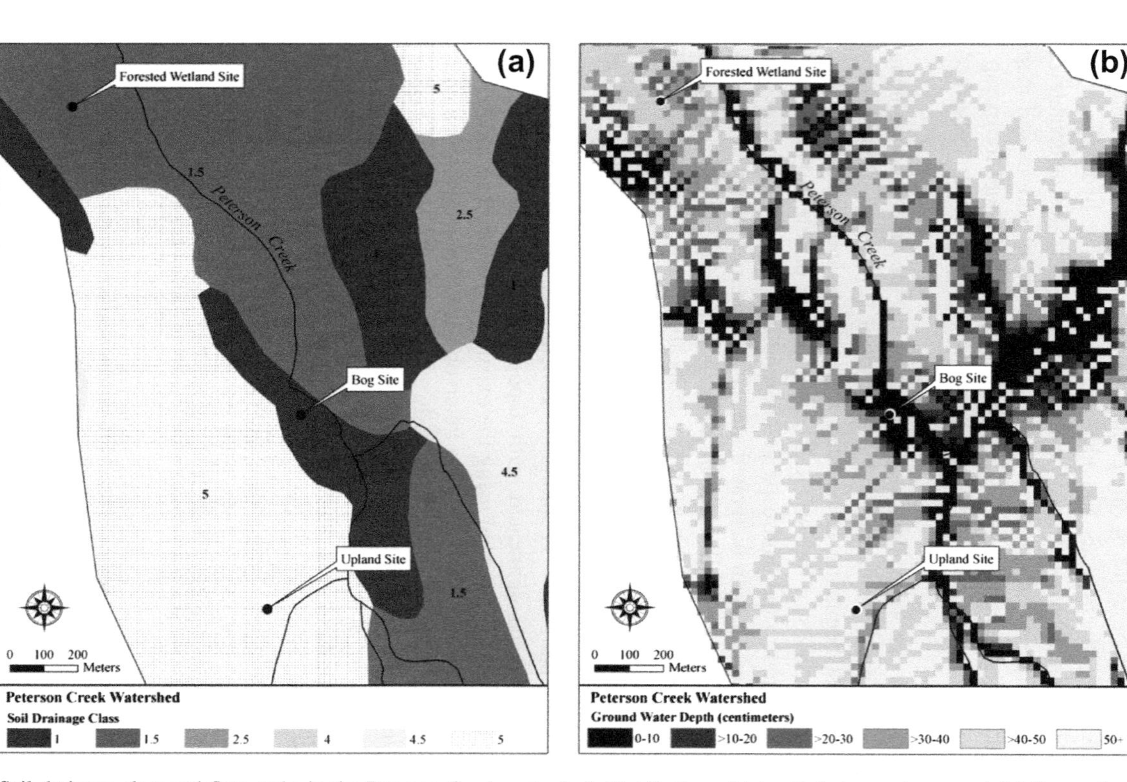

FIGURE 13 Soil drainage class and flow paths in the Peterson Creek watershed. Distribution of (a) soil drainage classes and (b) flow paths predicted from a compound topographic index model. Soil drainage classes are: 1 = very poorly drained; 2 = poorly drained; 3 = somewhat poorly drained; 4 = moderately well drained; 5 = well drained. Note that not all classes are represented due to averaging by soil map unit composition of soil drainage classes. Groundwater depth is estimated in six 10-cm depth classes arrayed in 20-m pixels through the use of a downscaled 60-m digital elevation model.

ACKNOWLEDGMENTS

We would like to acknowledge the contribution of Frances Biles to map production and Pat Dryer for assistance with map production and hydrologic mapping.

REFERENCES

Aitkenhead, J., McDowell, W., 2000. Soil C:N ratio as a predictor of annual DOC flux at local and global scales. Glob. Biogeochem. Cycles 14, 127–138.

Alaback, P.B., 1991. Comparative ecology of temperate rainforests of the Americas along analogous climatic gradients. Rev. Chil. Hist. Nat. 64, 399–412.

Alaback, P.B., 1996. Biodiversity patterns in relation to climate: the coastal temperate rainforests of North America. In: Lawford, R.G., et al. (Eds.), High-latitiude Rainforests and Associated Ecosystems of the West Coast of the Americas. Ecological Studies, vol. 116. Springer, NY, pp. 105–133.

Alaback, P.B., McCellan, M.H., 1992. Effects of global warming on managed coastal ecosystems of western North America. In: Mooney, H.A., et al. (Eds.), Earth System Responses to Global Change: Contrasts between North and South America. Academic Press, New York, NY, pp. 299–327.

Bormann, B.T., Spaltenstein, H., McClellan, M.H., Ugolini, F.C., Cromack, J.K., Nay, S.M., 1995. Rapid soil development after windthrow disturbance in pristine forests. J. Ecol. 83, 747–756.

Bridgham, S.D., Pastor, J., Janssens, J.A., Chapin, C.T., Malterer, J., 1996. Multiple limiting gradients in peatlands: a call for a new paradigm. Wetlands 16, 45–65.

Caouette, J.P., DeGayner, E.J., 2005. Predictive mapping for tree sizes and densities in southeast Alaska. Landsc. Urban Plan. 72, 49–63.

Carrara, P.E., Ager, T.A., Baichtal, J.F., 2007. Possible refugia in the Alexander Archipelago of southeastern Alaska during the late Wisconsin glaciation. Can. J. Earth Sci. 44, 229–244.

Chandler, R.F., 1942. The time required for podzol formation as evidenced by the Mendenhall glacial deposits near Juneau, Alaska. Soil Sci. Soc. Am. Proc. 7, 454–459.

Chapin, F.S., Walker, L.R., Fastie, C.L., Sharman, L.C., 1994. Mechanisms of primary succession following deglaciation at Glacier Bay, Alaska. Ecol. Monogr. 64, 149–175.

Cleland, D.T., Avers, P.E., McNab, W.H., Jensen, M.E., Bailey, R.G., King, T., Russell, W.E., 1997. National hierarchical framework of ecological units. In: Boyce, M.S., Haney, A. (Eds.), Ecosystem Management Applications for Sustainable Forest and Wildlife Resources. Yale Univ. Press, New Haven, CT, pp. 181–200.

Cook, J.A., Dawson, N., MacDonald, S.O., 2006. Management of highly fragmented systems: the north temperate Alexander Archipelago. Biol. Conserv. 133, 1–15.

Cowardin, L.M., Carter, V., Golet, F.C., LaRoe, E.T., 1979. Classification of Wetlands and Deepwater Habitats of the United States. U.S. Dept. Int., Fish and Wildl. Serv., Washington, DC.

Crocker, R.L., Major, J., 1955. Soil development in relation to vegetation and surface age at Glacier Bay, Alaska. J. Ecol. 45, 169–185.

D'Amore, D.V. 2011. Hydrologic Controls on Carbon Cycling in Alaskan Coastal Temperate Rainforest Soils. Ph.D. Diss. Fairbanks: University of Alaska Fairbanks.

D'Amore, D.V., Hennon, P.E., 2006. Evaluation of soil saturation, soil chemistry, and early spring soil and air temperatures as risk factors in yellow-cedar decline. Glob. Change Biol. 12, 524–545.

D'Amore, D.V., Fellman, J.B., Edwards, R.T., Hood, E., 2010. Controls on dissolved organic matter concentrations in soils and streams from a forested wetland and sloping bog in southeast Alaska. Ecohydrology 3, 249–261.

D'Amore, D.V., Hennon, P.E., Schaberg, P.G., Hawley, G., 2009. The adaptation to exploit nitrate in surface soils predisposes yellow-cedar to climate change-induced decline and enhances the survival of redcedar. For. Ecol. Manag. 258, 2261–2268.

Fellman, J.B., D'Amore, D.V., Hood, E., Boone, R.D., 2008. Fluorescence characteristics and biodegradability of dissolved organic matter in forest and wetland soils from coastal temperate watersheds in southeast Alaska. Biogeochemistry 88, 169–184.

Fellman, J.B., Hood, E., D'Amore, D.V., Edwards, R.T., 2009a. Seasonal changes in the chemical quality and biodegradability of dissolved organic matter exported from soils to streams in coastal temperate rainforest watersheds. Biogeochemistry 95, 277–293.

Fellman, J.B., Hood, E., Edwards, R.T., D'Amore, D.V., 2009b. Changes in the concentration, biodegradability and fluorescent properties of dissolved organic matter during stormflows in coastal temperate watersheds. J. Geophys. Res. Biogeosci. 114, G01021.

Fellman, J.B., Spencer, R.G.M., Hernes, P.J., Edwards, R.T., D'Amore, D.V., Hood, E., 2010. The impact of glacier runoff on the biodegradability and biochemical composition of terrigenous dissolved organic matter in near-shore marine ecosystems. Mar. Chem. 121, 112–122.

Fisher, S.G., Welter, J.R., 2005. Flowpaths as integrators of heterogeneity in streams and landscape. In: Lovett, G.M., et al. (Eds.), Ecosystem Function in Heterogeneous Landscapes. Springer-Verlag, New York, NY, pp. 311–328.

Fisher, S.G., Heffernan, J.B., Sponseller, R.A., Welter, J.R., 2007. Functional ecomorphology: feedbacks between form and function in fluvial landscape ecosystems. Geomorphology 89, 84–96.

Fisher, S.G., Sponseller, R.A., Heffernan, J.B., 2004. Horizons in stream biogeochemistry: flowpaths to progress. Ecology 85, 2369–2379.

Gehrels, G.E., Berg, H.C., 1994. Geology of southeastern Alaska. In: Plafker, G., et al. (Eds.), The Geology of North America, The Geology of Alaska. Geological Society of America, pp. 451–467.

Giblin, A.E., Nadelhoffer, K.J., Shaver, G.R., Laundre, J.A., McKerrow, A.J., 1991. Biogeochemical diversity along a riverside toposequence in arctic Alaska. Ecol. Monogr. 61, 415–435.

Grimm, N.B., Gergel, S.E., McDowell, W.H., Boyer, E.W., Dent, C.L., Groffman, P., Hart, S.C., Harvey, J., Johnson, C., Mayorga, E., McClain, M.E., Pinay, G., 2003. Merging aquatic and terrestrial perspectives of nutrient biogeochemistry. Oecologia 442, 485–501.

Gutknecht, J.L.M., Goodman, R.M., Balser, T.C., 2006. Linking soil process and microbial ecology in freshwater wetland ecosystems. Plant Soil 289, 17–34.

Hennon, P.E., D'Amore, D.V., Schaberg, P.G., Wittwer, D.T., Shanley, C.S., 2012. Shifting climate, altered niche, and a dynamic conservation strategy for yellow-cedar in the North Pacific Coastal Rainforest. Bioscience 62, 147–158.

Heusser, C.J., 1960. Late-Pleistocene Environments of North Pacific North America. Amer. Geogr. Soc. Spec. Publ. 35, New York, NY.

Holden, J., Burt, T.P., 2003. Hydrological studies on blanket peat: the significance of the acrotelm-catotelm model. J. Ecol. 91, 86–102.

Hood, E., Fellman, J., Spencer, R.G.M., Hernes, P.J., Edwards, R., D'Amore, D., Scott, D., 2009. Glaciers as a source of ancient and labile organic matter to the marine environment. Nature 24, 1044–1048.

Johnson, C.E., Ruiz-Mendez, J.J., Lawrence, G.B., 2000. Forest soil chemistry and terrain attributes in a catskills watershed. Soil Sci. Soc. Amer. J. 64, 1804–1814.

Kahklen, K., Moll, J., 1999. Measuring Effects of Roads on Groundwater: Five Case Studies. U.S.D.A. Forest Service, Technology and Development Program. SDTDC 9977-1801, San Dimas, CA.

Kramer, M.G. 2001. Maritime Windstorm Influence on Soil Process in a Temperate Rainforest. Ph.D. diss. Oregon State University, Corvallis.

Lin, H., 2003. Hydropedology: bridging disciplines, scales, and data. Vadose Zone J. 2, 1–11.

Lin, H.S., 2010. Linking principles of soil formation and flow regimes. J. Hydrol. 393, 3–19.

Lin, H., Bouma, J., Pachepsky, Y., Western, A., Thompson, J., van Genuchten, R., Vogel, H.J., Lilly, A., 2006a. Hydropedology: synergistic integration of pedology and hydrology. Water Resour. Res. 42, W05301. doi:10.1029/2005WR004085.

Lin, H.S., Kogelmann, W., Walker, C., Bruns, M.A., 2006b. Soil moisture patterns in a forested catchment: a hydropedological perspective. Geoderma 131, 345–368.

Mann, D.H., Hamilton, T.D., 1995. Late Pleistocene and Holocene paleoenvironments of the North Pacific Coast. Quaternary Sci. Rev. 14, 449–471.

McGee, K.L., 2000. Effects of Forest Roads on Surface and Subsurface Flow in Southeast Alaska. M.S. Thesis. Oregon State University, Corvallis.

Meidinger, D., MacKinnon, A., 1989. Biogeoclimatic classification-the system and its application. In: Ferguson, D.E., et al. (Eds.), Proceedings: Land Classifications based on Vegetation: Applications for Resource Management. U.S.D.A. Forest Service Gen. Tech. Rep. INT-257, Moscow, ID, pp. 215–222.

Miller, R.D., 1973. Gastineau Channel Formation: a Composite Glaciomarine Deposit Near Juneau, Alaska. U.S. Geologic. Surv. Bull. 1394, Washington, DC.

Miller, M.P., McKnight, D.M., Cory, R.M., Williams, M.W., Runkel, R.L., 2006. Hyporheic exchange and fulvic acid redox reactions in an alpine stream/wetland ecosystem, Colorado Front Range. Environ. Sci. Technol. 40, 5943–5949.

Miller, M.P., McKnight, D.M., Chapra, S.C., Williams, M.W., 2009. A model of degradation and production of three pools of dissolved organic matter in an alpine lake. Limnol. Oceanogr. 54, 2213–2227.

Mulholland, P., 2003. Large-scale patterns in dissolved organic carbon concentration, flux, and sources. In: Findlay, S., Sinsabaugh, R. (Eds.), Aquatic Ecosystems: Interactivity of Dissolved Organic Matter. Elsevier Science, pp. 139–159.

Mulholland, P., Kuenzler, E., 1979. Organic carbon export from upland and forested wetland watersheds. Limnol. Oceanogr. 24, 960–966.

Muller-Karger, F.E., Varela, R., Thunell, R., Luerssen, R., Hu, C., Walsh, J.J., 2005. The importance of continental margins in the global carbon cycle. Geophys. Res. Lett. 32, L01602. doi:10.1029/2004GL021346.

National Wetlands Working Group (N.W.W.G.), 1988. Wetlands of Canada. Environment Canada, Sustainable development branch, Ottawa, Ontario, Canada. Ecological Land Classification Series 24.

Neal, E.G., Hood, E., Smikrud, K., 2010. Contribution of glacier runoff to freshwater discharge into the Gulf of Alaska. Geophys. Res. Lett. 37, L0604. doi:10.1029/2010GL042385.

Neiland, B.J., 1971. The forest-bog complex in Southeast Alaska. Vegetation 22, 1–63.

Nowacki, G., Krosse, P., Fisher, G., Brew, D., Brock, T., Shephard, M., Pawuk, W., Baichtal, J., Kissinger, E., 2001. Ecological Subsections of Southeast Alaska and Neighboring Areas of Canada. U.S.D.A., Forest Service, Alaska Region, Tech. Publ. R10-TP-75.

Palmer, S.M., Driscoll, C.T., Johnson, C.E., 2004. Long-term trends in soil solution and stream water chemistry at the Hubbard Brook experimental forest: relationship with landscape position. Biogeochemistry 68, 51–70.

Park, S.J., van de Giesen, N., 2004. Soil-landscape delineation to define spatial sampling domains for hillslope hydrology. J. Hydrol. 295, 28–46.

Patric, J.H., Black, P.E., 1968. Potential Evapotranspiration and Climate in Alaska by Thorntwaite's Classification. U.S.D.A. Forest Service, Pacific Northwest Forest and Range Experiment Station Research Paper PNW-71. Juneau, AK.

Pinay, G., O'keefe, T., Edwards, R.T., Naiman, R.J., 2003. Potential denitrification activity in the landscape of a Western Alaska drainage basin. Ecosystems 6, 336–343.

PRISM Climate Group. 2010. http://www.prism.oregonstate.edu/, Corvallis: Oregon State University.

Randerson, J.T., Chapin, F.S., Harden, J., Neff, J.C., Harmon, M., 2002. Net ecosystem production: a comprehensive measure of net carbon accumulation by ecosystems. Ecol. Appl. 12, 937–947.

Strack, M., Waddington, J.M., Bourbonniere, R.A., Buckton, E.L., Shaw, K., Whittington, P., Price, J.S., 2008. Effect of water table drawdown on peatland dissolved organic carbon export and dynamics. Hydrol. Proc. 22, 3373–3385.

USDA Forest Service, 1996. Landforms of the Alaska region classification guide. Alaska Region.

USDA Forest Service, 2008. Tongass Young-Growth Management Strategy Tongass National Forest, Region 10.

USDA Natural Resources Conservation Service, 2004. Land Resource Regions and Major Land Resource Areas of Alaska. http://www.ak.nrcs.usda.gov/technical/lrr.html.

USFWS, 2009. Classification of Wetlands and Deepwater Habitats of the United States. National Wetlands Inventory Website. U.S. Department of the Interior; Fish and Wildlife Service, Washington, D.C. http://www.fws.gov/wetlands.

Ugolini, F.C., 1968. Soil development and alder invasion in a recently deglaciated area of Glacier Bay, Alaska. In: Trappe, J.M., et al. (Eds.), Proceedings, Biology of Alder. U.S.D.A., Forest Service, Pacific Northwest Research Station, Portland, OR, Pullman, WA, pp. 115–139.

Ugolini, F.C., Mann, D.H., 1979. Biopedological origin of peatlands in southeast Alaska. Nature 281, 366–368.

Viereck, L.A., Dyrness, C.T., Batten, A.R., Wenzlick, K.J., 1992. The Alaska Vegetation Classification. U.S.D.A. Forest Service, Pacific Northwest Research Station, Gen. Tech. Rep. PNW-GTR-286.

Worrall, F., Burt, T., Adamson, J., 2003. Controls on the chemistry of runoff from an upland peat catchment. Hydrol. Proc. 17, 2063–2083.

Worrall, F., Burt, T.P., Jaeban, R.Y., Warburton, J., Shedden, R., 2002. Release of dissolved organic carbon from upland peat. Hydrol. Proc. 16, 3487–3504.

Chapter 12

Hydropedology in the Ridge and Valley: Soil Moisture Patterns and Preferential Flow Dynamics in Two Contrasting Landscapes

Ying Zhao,[1] Jialiang Tang,[2] Chris Graham,[3] Qing Zhu,[4] Ken Takagi[5] and Henry Lin[1,*]

ABSTRACT

This chapter investigates soil moisture spatial–temporal patterns and preferential flow dynamics in two contrasting landscapes in the Ridge and Valley Physiographic Region in eastern United States: 1) the Shale Hills is a steep-sloped forestland, 7.9 ha, with a ratio of elevation change over total area (E/A) of 6.8, and 2) the Kepler Farm is a more gently rolling cropland, 19.5 ha, with the E/A ratio of 1.2. Similarities and differences between the two landscapes were examined based on multiple years' soil and hydrologic monitoring. Through a series of statistical analyses, we found that: 1) Soil moisture spatial correlation length was an order of magnitude shorter in the Shale Hills (20–40 m) as compared to the Kepler Farm (100–153 m), suggesting a greater spatial variability within a shorter distance in the Shale Hills that was aligned with topography. Regression kriging with terrain attributes was found optimal for interpolating soil hydrologic properties in the Shale Hills, while ordinary kriging was optimal at the Kepler Farm; 2) Soil moisture spatial variability across the Shale Hills increased exponentially with catchment-wide wetness, whereas that increased linearly at the Kepler Farm. This was in part due to different landscape configurations and the E/A ratios; 3) The mean relative difference (MRD) of soil moisture was smaller at the Kepler Farm (−55 to 56%) than that in the Shale Hills (−80 to 81%) over the same time period analyzed in time stability analysis, reflecting the overall larger spatial

[1]Dept. of Ecosystem Science and Management, Pennsylvania State Univ., University Park, PA
[2]Institute of Mountain Hazards and Environment, Chinese Academy of Sciences, Chengdu, China
[3]Boise State University, Boise, Idaho
[4]Nanjing Institute of Geography and Limnology, Chinese Academy of Sciences, Nanjing, China
[5]Dept. of Earth Sciences, Boston University, Boston, MA
[*]Corresponding author: Email: henrylin@psu.edu

Hydropedology, Edited by H. Lin. DOI: 10.1016/B978-0-12-386941-8.00012-5

variation of soil moisture in the Shale Hills. Inversely, standard deviation of MRD was slightly larger in the Kepler Farm (4–46%) than that in the Shale Hills (3–28%), implying slightly stronger temporal dynamics in the Kepler Farm due to cropping impacts; 4) The primary Empirical Orthogonal Functions explained 76–90% variation of soil moisture in the Shale Hills, but only 31–67% variation at the Kepler Farm, reflecting the dominant control on soil moisture by terrain in the Shale Hills, while soil properties showed stronger influence on soil moisture at the Kepler Farm; 5) The Shale Hills had overall averaged 37% frequency of preferential flow occurrence during 175 rainfall events in 2007–2009, while the Kepler Farm had 19% during 139 rainfall events occurred in 2008–2010. This study illustrated the spatial complexity and temporal dynamics in soil moisture patterns and preferential flow dynamics that were controlled by the interactions of terrain, soil, and vegetation, and that such interactions changed with landscape characteristics, seasonal wetness, and soil depth.

1. INTRODUCTION

Soil moisture is a key state variable in earth system dynamics (Famiglietti et al., 1998) and is critical in hydropedologic studies (Lin et al., 2006a,b). "Where, when, and how" water moves through various soils in different landscapes and how water flow impacts soil processes and subsequent soil moisture patterns need to be better understood in real-world landscapes (Lin et al., 2006b). Soil moisture has often been reported to show spatial dependence as well as time stability (e.g. Shouse et al., 1995; Grayson et al., 2002; Zhao et al., 2010). Spatial dependency is commonly characterized by geostatistical analysis, while time stability analysis can be used to reveal persistent wet or dry sites within a landscape (Vachaud et al., 1985; Lin, 2006). One way to account for both spatial patterns and time persistence is Empirical Orthogonal Function (EOF) analysis, through which underlying stable patterns of soil moisture may be derived from large multidimensional datasets (Jawson and Niemann, 2007; Korres et al., 2010). For example, Jawson and Niemann (2007) showed that the spatial and temporal patterns of large-scale soil moisture can be described by few main EOFs that were related to soil texture, topography, and land use.

Because of the temporal dynamics of soil moisture and related hydrologic processes such as preferential flow, numerous studies have shown that static, terrain-derived indices alone rarely explained more than 50% of soil moisture variability (e.g. Famiglietti et al., 1998; Western et al., 1999; Takagi and Lin, 2012). Yoo and Kim (2004) concluded that no simple and unique mechanism can explain the evolution of soil moisture field in different landscapes. Although seasonality and local vs. nonlocal hydrologic fluxes exert strong influences on surface and near-surface soil moisture organization (Grayson et al., 1997; Western et al., 1999), less is known about the controls on subsurface soil moisture spatial pattern and how they may change with seasons and landscape settings. For instance, despite the importance of subsurface preferential flow in hydrologic processes at various spatial and temporal scales

(Lin et al., 2006b), controls on the initiation and location of subsurface preferential flow and its impacts on soil moisture pattern remain poorly understood. The frequency of occurrence and the control of preferential flow initiation in natural landscapes remain largely unclear.

To showcase hydropedologic study in different landscapes, this chapter is included to illustrate 1) similarities and differences in soil moisture spatial–temporal patterns and their underlying controls between two contrasting landscapes – a forestland and a cropland – in the Ridge and Valley Physiographic Region in eastern United States based on multiple years' monitoring datasets and 2) the frequency and controls of preferential flow occurrence and its links to soil types, landscape positions, and precipitation characteristics in these two landscapes. This case study demonstrates that topography and soil properties have intertwined effects on soil moisture variability and preferential flow dynamics, and that hydropedology provides a useful framework to enhance the understanding of complex landscape–soil–water–vegetation relationships.

2. MATERIALS AND METHODS

2.1. Study Areas and Soils

Two contrasting landscapes in the Ridge and Valley Physiographic Region in eastern United State were selected for this study: 1) the steeply-sloped Shale Hills, with a ratio of elevation change over total area (E/A) of 6.8, and 2) the gently-rolling Kepler Farm, with the E/A ratio of 1.2. The Shale Hills is a 7.9-ha V-shaped forested catchment, while the Kepler Farm is a 19.5-ha roughly inverse V-shaped cropped land approximately 6 km north of the Shale Hills (Fig. 1). Basic characteristics of the two landscapes are illustrated in Figs. 2 and 3, respectively, including slope, topographic wetness index, soil clay content, and depth to bedrock. Both landscapes have been described in detail by Lin et al. (2006b) and Zhu and Lin (2011), respectively. Thus, only a brief summary is provided below.

The Shale Hills elevation ranges from 256 m above sea level at the outlet of the catchment to 310 m at the highest ridge. Several species of maple (*Acer* spp.), oak (*Quercus* spp.), and hickory (*Carya* spp.) are typical deciduous trees found on the sloping areas and on ridges, while the valley floor is covered by eastern hemlock [*Tsuga canadensis* (L.) Carrière] coniferous trees (Takagi and Lin, 2011). The catchment is underlain by about 300-m-thick, steeply bedded, highly fractured Rose Hill Shale. Depth to bedrock ranges from <0.25 m on the ridgetop and upper side slopes to >2–3 m on the valley floor and swales. The soils were formed from shale colluvium or residuum, with many channery shale fragments throughout most of the soil profiles. The soils are generally silt loams and silty clay loams in texture, with some clay loams and sandy clay loams. Soil thickness, landscape position, and depth to redoximorphic features were the main criteria used to differentiate five soil series identified in the catchment

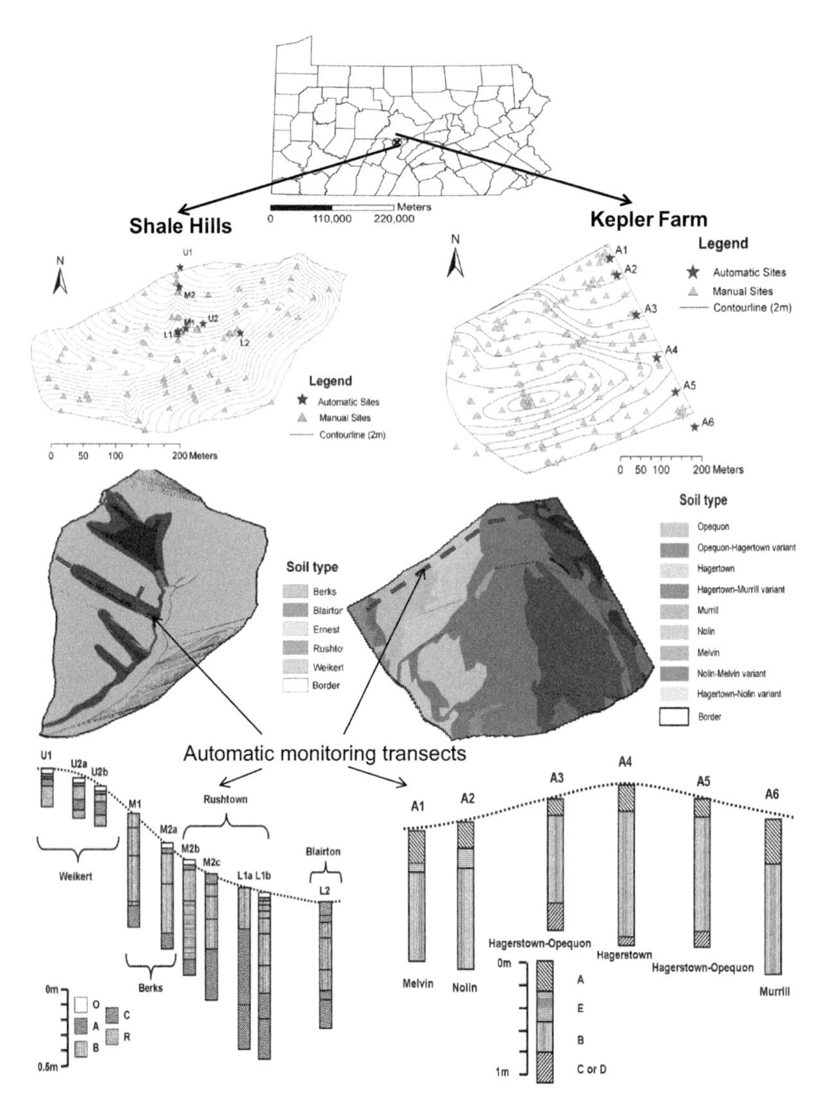

FIGURE 1 Locations of soil moisture monitoring sites in the Shale Hills (left) and the Kepler Farm (right), with the spatial distribution of sampling points, soil map, and the transect of real-time soil moisture monitoring in each landscape indicated. The bottom panel shows the general layout of automatic soil moisture monitoring transects, with sensor installation depths and relative topographic positions indicated. Horizontal lines in that panel indicate approximate soil moisture sensor positions within each soil profile, while box fill patterns indicate different soil horizons within the same soil profile (see Table 1 for more details). *(Color version online and in color plate)*

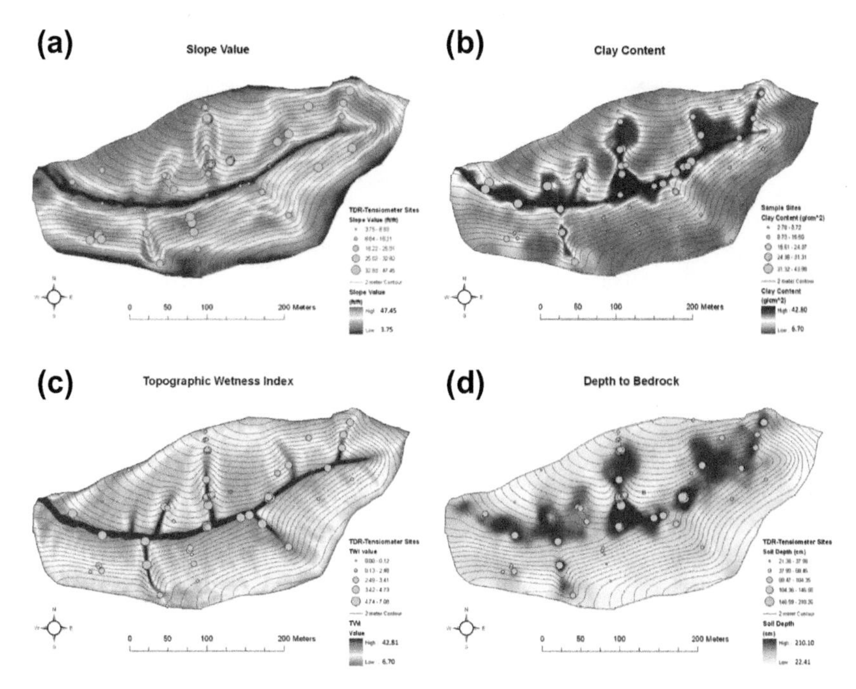

FIGURE 2 Maps of soil-terrain attributes in the Shale Hills generated using regression kriging: (a) slope, (b) clay content within solum, (c) topographic wetness index, and (d) depth to bedrock. *(Color version online and in color plate)*

through precision soil mapping (Lin et al., 2006b); their main features are summarized below:

1) The Weikert series (*loamy skeletal, mixed, active, mesic Lithic Dystrudepts*) is a shallow, well-drained soil on steep planar hillslopes and ridgetops, with depth to fractured shale bedrock less than 0.5 m;
2) The Berks series (*loamy skeletal, mixed, active, mesic Typic Dystrudepts*) is a moderately deep, well-drained soil formed on the toeslope positions and the side slopes of concave hillslopes, with 0.5–1 m depth to bedrock;
3) The Rushtown series (*loamy skeletal over fragmental, mixed, active, mesic Typic Dystrudepts*) is a very deep, excessively drained soil at the center of concave hillslopes, with >1 m depth to bedrock;
4) The Ernest series (*fine-loamy, mixed, superactive, mesic Aquic Fragiudults*) is a very deep, poorly to moderately well-drained soil on the valley floor and around a first-order stream; and
5) The Blairton series (*fine-loamy, mixed, active, mesic Aquic Hapludults*) is a very deep, moderately well-drained soil found at the east end of the catchment valley floor.

The Kepler Farm elevation ranges from 373 m at the footslope at the northern corner to 396 m at the ridgetop in the middle of the landscape. Typical crops

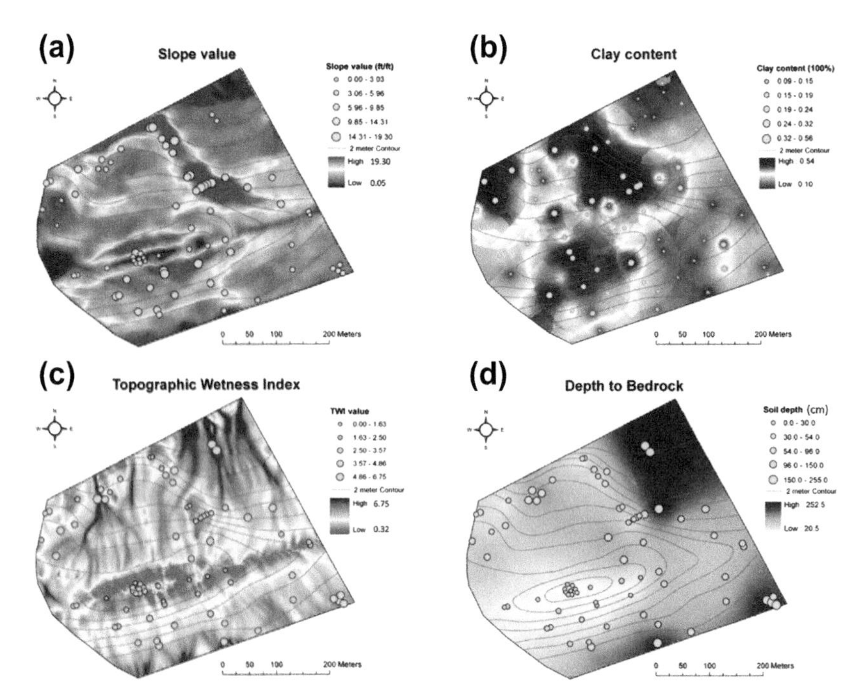

FIGURE 3 Maps of soil-terrain attributes in the Kepler Farm generated using ordinary kriging: (a) slope, (b) clay content within solum, (c) topographic wetness index, and (d) depth to bedrock. *(Color version online and in color plate)*

grown on this farm are corn (*Zea mays* L.), soybean (*Glycine max*.L.), and winter wheat (*Triticum aestivum* L. emend. Thell.), which vary from year to year and from one field to another. Depth to bedrock ranges from <0.25 m in the summit to >3 m in the footslope. Five soil series have been identified based on country-level general soil survey (Zhu and Lin, 2011):

1) The Hagerstown series (*fine-loamy, mixed, semiactive, mesic Typic Hapludalfs*) is a well-drained soil formed from limestone residuum, with the solum >1.0 m thick;
2) The Opequon series (*clayey, mixed, active, mesic Lithic Hapludalfs*) is a well-drained soil formed from limestone residuum, with the solum <0.5 m thick;
3) The Murrill series (*fine-loamy, mixed, semiactive, mesic Typic Hapludults*) consists of deep (>1 m), well-drained soils formed from sandstone colluvium with underlying residuum weathered from limestone;
4) The Nolin series (*fine-silty, mixed, active, mesic Dystric Fluventic Eutrudepts*) is a thick (>2 m), well-drained soil formed from alluvium washed from surrounding uplands, underlain by limestone; and
5) The Melvin series (*fine-silty, mixed, active, nonacid, mesic Fluvaquentic Endoaquepts*) is a thick (>2 m), poorly-drained soil formed from alluvium washed from surrounding uplands, underlain by limestone.

TABLE 1 Characteristics of Real-time Soil Moisture Monitoring Sites and Sensor Installation Depths in the Shale Hills and the Kepler Farm. The Actual Layouts of the Sites and Sensor Locations within Each Soil Profile are Shown in Fig. 1

Site	Landform	Soil series	Depth to bedrock (m)	Upslope contributing area (m^2)	Local slope (%)	Depth of soil moisture probes (cm) and their horizon (parenthesis)
Shale Hills						
U1	Ridgetop	Weikert	0.2	14	23.1	5 (Oe), 8 (A), 10 (A), 17 (C), 37 (R)
U2a	Upper hillslope	Weikert	0.4	66	28.3	5 (Oe), 8 (A), 21 (Bw), 31 (CR), 39 (R)
U2b	Upper hillslope	Weikert	0.4	66	28.3	5 (Oe), 8 (A), 15 (Bw), 28 (CR), 38 (R)
M1	Midslope in swale	Berks	>1.5	38	31.3	14 (Bw1), 41 (Bw2), 86 (Bw3), 90 (Bw3),111 (C)
M2a	Midslope in swale	Berks	1.1	676	35.5	5 (Oe), 10 (A), 40 (Bw2), 88 (Bw3), 103 (C)
M2b	Midslope in swale	Rushtown	1.5	676	35.5	5 (O), 10 (A), 40 (Bw2), 97 (BC), 112 (C)
M2c	Midslope in swale	Rushtown	1.5	676	35.5	10 (A), 22 (Bw1), 44 (Bw2), 73 (Bw3), 123 (3)
L1a	Lower in swale	Rushtown	>3.0	1122	10.7	8 (A), 18 (Bw), 39 (Bw3), 115 (C1), 156 (C2)
L1b	Lower in swale	Rushtown	>3.0	1122	10.7	5 (O), 8 (A), 12 (Bw1), 15 (Bw1), 22 (Bw2), 40 (Bw3), 68 (BC), 92 (BC), 122 (C1), 162 (C2)
L2	Valley	Blairton	>2.5	19054	4.2	13 (A), 20 (BA), 35 (Bt1), 66 (Bt2), 86 (Bt2), 95 (CB1)

(Continued)

TABLE 1 Characteristics of Real-time Soil Moisture Monitoring Sites and Sensor Installation Depths in the Shale Hills and the Kepler Farm. The Actual Layouts of the Sites and Sensor Locations within Each Soil Profile are Shown in Fig. 1—Cont'd

Site	Landform	Soil series	Depth to bedrock (m)	Upslope contributing area (m^2)	Local slope (%)	Depth of soil moisture probes (cm) and their horizon (parenthesis)
Kepler Farm						
A1	Valley	Melvin	>1.4	2204	1.36	5 (Ap1), 38 (E), 51 (BE), 65 (Bt1), 89 (Bt2)
A2	Near valley	Nolin	>1.6	864	4.26	6 (Ap1), 41 (E), 60 (BE), 94 (Bt1), 100 (Bt1)
A3	Upper slope	Hagerstown–Opequon	1.15	430	9.08	5 (Ap1), 33 (Bt1), 80 (Bt2), 115 (soil–bedrock), 128 (CR)
A4	Ridgetop	Opequon	1.6	7	8.43	6 (Ap1), 42 (BE), 84 (Bt1), 138 (Bt2), 160 (soil–bedrock)
A5	Upper slope	Hagerstown–Opequon	1.4	19	4.74	5 (Ap1), 30 (Bt1), 71 (Bt2), 121 (Bt3), 140 (soil–bedrock)
A6	Midslope	Murrill	>1.6	446	2.51	10 (Ap1), 40 (Ap2), 75 (Bt1), 102 (Bt2), 140 (Bt3)

Because of the general nature of the soil survey for the Kepler Farm (without a precision soil mapping), several variants of Hagerstown–Opequon, Hagerstown–Murrill, Hagerstown–Nolin, and Nolin–Melvin were identified in our earlier study to represent transition zones among the identified typical soil series (Zhu and Lin, 2011).

The Shale Hills and the Kepler Farm have a typical humid continental climate, with the area's mean monthly temperature of −3.7 °C (minimum) in January and 21.8 °C (maximum) in July, and an annual average precipitation of 980 mm (National Weather Service, State College, PA). The growing season of the area is from late May to middle October, which accounts for about 38% of the averaged annual precipitation.

2.2. Data Collections

Extensive soil moisture data were collected from both landscapes using manual and automatic methods from 2005 to 2010 (Tables 1 and 2). The manual datasets covered the entire landscapes (spatially extensive) but were temporally limited (approximately weekly to biweekly data), while the automatic datasets captured temporal dynamics (10 min time steps) but were spatially limited (only a few representative sites). Thus, we rely mostly on manual datasets for soil moisture spatial pattern analysis and on automatic datasets for preferential flow dynamics analysis. Detailed descriptions of both manual and automatic soil moisture datasets collected have been provided by Lin (2006) and Graham and Lin (2011) for the Shale Hills and Zhu and Lin (2011) for the Kepler Farm, respectively. Thus, only a brief summary is provided here.

At the Shale Hills, a total of 106 monitoring sites were installed throughout the catchment, while at the Kepler Farm a total of 145 sites were established throughout the landscape (Fig. 1). The site selections were based on the combined consideration of a number of factors, including soil types, landform units, terrain attributes, spatial scales, and others (see Lin, 2006; Zhu and Lin, 2011). At each site, a 0.051-m diameter PVC tube was installed into the soil profile to a maximum depth of 1.1 m (maximum length of the soil sampler used) or to the sampler refusal, whichever came first. Prior to the installation of the PVC tube, a slide hammer and hollow metal soil sampler were used to collect a 0.038-m diameter soil core that was saved for later description and analysis. The PVC tube was capped at the bottom with a watertight seal and a removable cap was placed on top to prevent water from entering the tube. During each measurement, volumetric soil water content (θ) was determined at each site at depths of 0.1, 0.2, 0.4, 0.6, 0.8, and 1.0 m (or to weathered bedrock at locations with shallower depth to bedrock) using TRIME-T3 time domain reflectometry (TDR), which has a tube access probe calibrated to the 0.051-m diameter PVC tube) (IMKO, Ettlingen, Germany). The TDR probe is 0.2 m long (with an effective measuring length of 0.18 m) and the midpoint of the probe was used to represent the measurement depth from the soil surface.

TABLE 2 Time Periods of Soil Moisture Data Collections from 2005 to 2010 Used in Various Analyses in This Study (shaded areas). SH: Shale Hills; KF: Kepler Farm; GS: Geostatistical Analysis; EOF: Empirical Orthogonal Function analysis; TS: Time Stability Analysis; PF: Preferential Flow Analysis

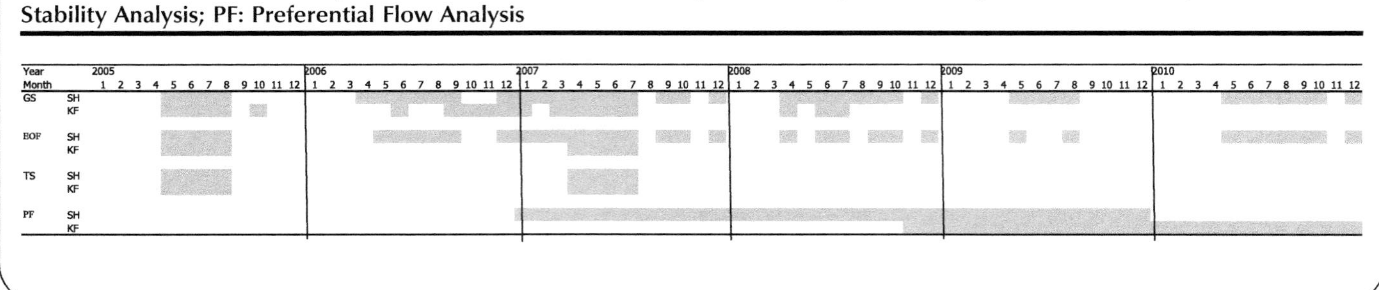

At each measurement depth, two or three readings were taken in orthogonal directions (probe rotated 90° between measurements), and the mean value was used. Measurement accuracy of the TRIME-T3 probe is ±2.0% in the range of 0–40% and ±3.0% in the range of 40–60% for most soils (IMKO, 2000). Prior to each round of measurement, the probe was calibrated in the laboratory with glass beads with known moisture content. Because of the diversity of soils in the two landscapes and the large number of sites monitored, we did not perform soil-horizon specific calibration of the probe. The actual number of measured locations varied from time to time due to personnel availability and weather conditions. The subsequent analysis included 36 sampling dates where at least 65 soil moisture monitoring locations were measured for the Shale Hills, and 8 sampling dates where at least 100 soil moisture monitoring locations were measured for the entire Kepler Farm (Table 2). We considered this as a reasonable trade-off between having a sufficient number of sites to adequately represent spatial coverage within each landscape and sufficient sampling days to have an appropriate temporal coverage over various seasons.

Automatic soil moisture data were collected in real time at 10-min intervals along a representative transect in each of the two landscapes (Fig. 1), with 10 sites in the Shale Hills and 6 sites in the Kepler Farm (Table 1). In each monitoring site, 4–10 depths were installed with a soil moisture sensor (the number depended on soil thickness and horizonation). At each site, a pit was excavated and capacitance-type probes (Decagon Devices, Pullman, WA) were installed in various soil horizons from the soil surface down to the soil–bedrock interface or as deep as excavated safely. The shallowest probe at each site ranged from 5 to 15 cm, while the deepest ranged from 37 to 162 cm, depending on the soil type (Table 1 and Fig. 1). Three representative sites in the upper, middle, and lower portions of the hillslope (sites U1, M1, and L2) in the Shale Hills, and all the sites at the Kepler Farm had one vertical profile instrumented. Other three sites in the Shale Hills had 2–3 vertical profiles instrumented in different slope segments: site U2 had probes installed at an upslope and a nearby downslope directions (labeled as U2a and U2b, respectively); site M2 was instrumented on the left, right, and center of the excavated pit (labeled as M2a, M2b, and M2c, respectively); and site L1 had probes installed on the left and right directions of the excavated pit (labeled as L1a and L1b, respectively). Further details on the soil moisture probe installation in the Shale Hills can be found in Lin and Zhou (2008).

In both study areas, soil water content was examined in relation to soil-terrain attributes that were anticipated to influence water flow and storage. Precipitation was measured in each landscape in real time. To analyze basic soil properties, soil cores (58 in the Shale Hills and 145 in the Kepler Farm) were first described using standard soil survey procedures, including horizon thickness, color, texture, structure, roots, and amount of redoximorphic features. Then bulk soil samples were used for physiochemical analysis by soil horizons. Prior to performing soil textural analysis, rock fragments (>2-mm sieve) were separated from the bulk sample and weighed. The <2-mm soil fraction was analyzed for soil particle size distribution and organic matter content. Depth to bedrock was determined in situ when collecting soil cores. A digital elevation model (DEM) of each landscape was used to derive primary topographic attributes (elevation, slope, curvature, and upslope contributing area) and one composite topographic attribute (topographic wetness index). These terrain attributes were extracted at all soil moisture monitoring sites using the coordinates obtained from the total station survey. More details of our field data collections and processing can be found in Takagi and Lin (2011) for the Shale Hills and in Zhu and Lin (2011) for the Kepler Farm.

2.3. Data Analyses

2.3.1. Statistical and Geostatistical Analyses

To study the evolution of soil moisture spatial structure, we conducted a geostatistical analysis that deals with how variance and covariance change with distance between observations (Zhao et al., 2011). A spatial structure is commonly modeled by fitting a variogram function to an empirical semivariogram $\gamma(h)$:

$$\gamma(h) = \frac{1}{2N(h)} \left\{ \sum_{\alpha=1}^{N(h)} [z(x_\alpha + h) - z(x_\alpha)]^2 \right\}, \tag{1}$$

where $N(h)$ is the number of pairs separated by the lag distance h, $z(x_\alpha)$ and $z(x_\alpha + h)$ are observation values at spatial locations x_α and $x_\alpha + h$, respectively. The $\gamma(h)$ is a plot of half the squared difference between pair observations (the semivariance) against their distance in space, averaged for a series of distance classes. A variogram function is defined by its model and three parameters: the sill (maximum semivariance), the range (the maximum distance at which pairs of observations influence each other), and the nugget effect (the variance within sampling units). The model most commonly used (e.g. spherical, exponential, or Gaussian) assumes that there is no spatial dependence for distances larger than the range.

In this study, we investigated the spatial variability of soil moisture storage in the entire solum (i.e. A and B horizons combined), which was calculated as

$$S = \sum_{i}^{n} \theta_i \times d_i, \tag{2}$$

where S is soil moisture storage for a particular site (m), n is the number of depths measured in a soil profile, θ_i is the volumetric moisture content at the ith depth, and d_i is the representative thickness of the ith depth interval.

2.3.2. Time Stability Analysis

Vachaud et al. (1985) suggested the concept of time stability to indicate the relative difference between an individual measurement of a soil property and its mean over all sampling points across the space collected at the same time. Time stability assessment uses $\theta_{i,j,t}$, the soil water content at location i in field j and time t, to calculate the field mean as

$$\bar{\theta}_{j,t} = \frac{1}{n_{j,t}} \sum_{i=1}^{n_{j,t}} \theta_{i,j,t}, \tag{3}$$

where $t = 1, 2, \ldots, n_t$ (the number of sampling dates), $j = 1, 2, \ldots, n_j$ (the number of fields), and $i = 1, 2, \ldots, n_{j,t}$ (the number of sampling points within field j at time t). The mean relative difference (MRD) over a monitoring period and its standard deviation (SD) for each sampling point were calculated by

$$\mathrm{MRD} = \frac{1}{n_t} \sum_{t=1}^{n_t} \frac{\theta_{i,j,t} - \bar{\theta}_{j,t}}{\bar{\theta}_{j,t}}, \tag{4}$$

$$\mathrm{SD} = \frac{1}{n_t - 1} \sum_{t=1}^{n_t} \left(\frac{\theta_{i,j,t} - \bar{\theta}_{j,t}}{\bar{\theta}_{j,t}} - \mathrm{MRD} \right)^2. \tag{5}$$

The MRD at a sampling point quantifies whether that location is wetter or drier than the field overall average. The SD of the MRD characterizes the variability of MRD at that location within the time period considered. In this study, to ensure a comparable time window for temporal stability analysis between the two contrasting landscapes, we selected five times between May and October in 2005 (i.e. growing period) and three times between March and May in 2007 (nongrowing period) during which both the landscapes had comparable datasets (Table 2).

2.3.3. Empirical Orthogonal Function Analysis

Empirical Orthogonal Function (EOF) analysis decomposes the observed variability of a dataset into a set of orthogonal spatial patterns (EOFs) or a set of time series called expansion coefficients (ECs) (Korres et al., 2010). This is done by transforming the original dataset into a new set of uncorrelated variables so that the first few of the new variables explain most of the variation in the original dataset. The resulting correlation matrix is real and symmetric, and therefore possesses a set of orthogonal eigenvectors with positive eigenvalues. If there are geophysical data maps that are time series with any $m \times n$ matrix, A, there exist uniquely two orthogonal matrices, U and V and a diagonal matrix L such that

$$A = U \times L \times V^{\mathrm{T}}, \tag{6}$$

where V^T is the transpose of a matrix V. Note that L is padded with zeros to make the square diagonal matrix into a $m \times n$ matrix. This also implies that L has at most $M = \min(m, n)$ nonzero elements. The columns of U are called the EOFs of A and the corresponding diagonal elements of L are called the eigenvalues. Each row of V serves as a series of time coefficients that describes the time evolution of the particular EOF. The map associated with an EOF represents a pattern, which is statistically independent and spatially orthogonal to the others. The eigenvalue indicates the amount of variance accounted for by the pattern. For EOF analysis here, we used the spatial anomalies of soil moisture dataset to exclude the temporal variation from consideration (Jawson and Niemann, 2007). The spatial anomalies are calculated by subtracting the mean soil moisture for a given sampling day from all the soil moisture observations collected on that day.

While single soil moisture pattern might be affected by random processes, significant EOFs represent stable patterns of a dataset. A small number of EOF patterns together can explain a large portion of the total variability in soil moisture for the subsequent correlation analysis; that is, the EOF pattern can correlate with the geophysical characteristics of a study area to determine possible dominant controls. The Pearson's correlation was used in the correlation between soil moisture and soil-terrain attributes. The EOF analyses were conducted using SPSS 13.0.1 (SPSS Inc., Chicago).

2.3.4. Preferential Flow Analysis

In this study, we defined preferential flow as a flow scenario that leads to a subsoil horizon responding to a rainfall input earlier than a soil horizon above it (Lin and Zhou, 2008; Graham and Lin, 2011). Intuitively, rainwater should infiltrate into soil and pass the upper soil first before it can reach a lower horizon. Therefore, when a soil moisture sensor buried in a subsoil responded to a storm event earlier than other sensors above it, we assume that a preferential flow event has occurred. In this case, the water has either bypassed the overlying soil via vertical preferential flow (e.g. macropore flow or finger flow) or transported to the deeper soil from upslope or sideslope areas via subsurface lateral flow, or both. We used the technique developed by Graham and Lin (2011) to analyze real-time soil moisture response to all rainfall events observed during our monitoring time periods, and classified each sensor location into the following three flow categories for each rainfall event:

1) *Sequential flow*, where sensors in all soil depths within the same site responded in sequence, with the surface sensor responding first, then the next deeper sensor, and so on through all of the sensors installed at a site;
2) *Preferential flow*, where at least one deeper sensor responded to precipitation input before a shallower sensor, or where shallow and deep sensors responded sequentially but a sensor in between did not respond at all;
3) *Nondetectable flow*, where sensors showed no response to precipitation input.

The frequency of preferential flow occurrence in the two landscapes was then compared. A total of 175 events that occurred in three years from 2007 to 2009 in the Shale Hills and 139 events that occurred in nearly two-and-a-half years from 2008 to 2010 in the Kepler Farm were summarized. In the meantime, possible underlying controls of preferential flow occurrence were explored using the following approach.

Precipitation indices and initial soil moisture indices were examined in relation to preferential flow occurrence. A total of 33 precipitation indices were derived from the time series data of rainfall, and 8–14 initial soil moisture indices (depending on the number of sensors installed at each site) were determined using initial soil moisture contents recorded right before a precipitation event. Part of these indices (significant ones) is listed in Table 3, but a full list can be found in Graham and Lin (2011). The precipitation indices were grouped into (with some overlaps) (i) antecedent precipitation, (ii) precipitation event characteristics, and (iii) precipitation timing. All the indices were sorted according to the three flow categories and their means were statistically compared using *t*-tests. When the mean of an index was significantly different for the set of precipitation events that resulted in preferential flow vs. those that resulted in sequential flow, it was considered as a significant control on preferential flow occurrence. Further details of this analysis can be found in Graham and Lin (2011).

3. RESULTS AND DISCUSSION

3.1. Soil Moisture Spatial–Temporal Patterns and their Controls

3.1.1. Spatial–Temporal Patterns of Soil Moisture

Representative soil moisture storage maps under relatively wet, moist, and dry conditions, along with their associated histograms and semivariagrams, are illustrated in Fig. 4 for the Shale Hills and the Kepler Farm, respectively. The general patterns of the landscape-wide wet-up and dry-down were roughly correlated with the overall distribution patterns of the soil types and terrain features in both landscapes (Figs. 2 and 3). Particularly evident were the following: 1) lower water content and low water storage portions in the landscapes were always associated with the shallowest soils distributed in the ridgetops (i.e. the Weikert soils in the Shale Hills and the Opequon soils at the Kepler Farm) and 2) higher water content and high water storage portions were always associated with the deepest soils located in the valley floors (i.e. the Ernest soils in the Shale Hills and the Melvin soils at the Kepler Farm).

The spatial structure of soil moisture in both landscapes, however, varied seasonally: during wet periods, higher sills and shorter correlation lengths were observed, while during dry periods smaller sills and longer correlation lengths appeared (Fig. 4). Between the two landscapes, soil moisture spatial correlation

TABLE 3 Comparisons of Precipitation and Initial Soil Moisture Indices for Events that Produced Preferential Flow vs. those that Produced Sequential Flow During the Entire Real-time Monitoring Period. Presented are p Values from t-tests. Shaded Cells Indicate the Indices with Significant Difference between Preferential Flow and Sequential Flow Events (95% Confidence Interval). The Boxes Indicate the Significant Controls in Regulating Preferential Flow Occurrence in the Two Contrasting Landscapes; See the Text for Explanation

Landscapes	Shale Hills										Kepler Farm					
Sites	U1	U2a	U2b	M1	M2a	M2b	M2c	L1a	L1b	L2	A1	A2	A3	A4	A5	A6
Index†																
Total ppt	0.81	0.19	0.31	0.77	0.00	0.16	0.91	0.02	0.06	0.01	0.63	0.00	0.00	0.00	0.66	0.00
ppt duration	0.29	0.03	0.92	0.21	0.01	0.01	0.21	0.26	0.37	0.65	0.85	0.00	0.01	0.04	0.90	0.00
Max ppt Intensity	0.09	0.17	0.52	0.22	0.96	0.04	0.01	0.70	0.44	0.74	0.91	0.00	0.54	0.01	0.29	0.00
ppt variance	0.17	0.53	0.42	0.16	0.25	0.46	0.07	0.62	0.53	0.08	0.79	0.00	0.85	0.00	0.54	0.00
ppt skew	0.02	0.00	0.78	0.38	0.00	0.00	0.01	0.35	0.82	0.02	0.94	0.78	0.38	0.86	0.03	0.22
ppt kurtosity	0.06	0.01	0.55	0.38	0.01	0.00	0.01	0.38	0.96	0.02	0.93	0.96	0.30	0.78	0.04	0.17
Time Since Last Event	0.37	0.02	0.36	0.42	0.56	0.52	0.99	0.18	0.17	0.25	0.75	0.01	0.02	0.65	0.85	0.00
Air Temperature	0.00	0.00	0.98	0.09	0.00	0.00	0.00	0.83	0.36	0.00	0.07	0.24	0.21	0.11	0.45	0.55
API1	0.02	0.61	0.38	0.01	0.91	0.43	0.00	0.22	0.78	0.11	0.79	0.01	0.99	0.16	0.36	0.00
API4	0.11	0.48	0.64	0.03	0.39	0.15	0.06	0.50	0.14	0.17	0.37	0.11	0.01	0.03	0.98	0.43
API7	0.16	0.83	0.35	0.15	0.33	0.22	0.37	0.59	0.03	0.25	0.38	0.91	0.10	0.03	0.85	0.45
API14	0.76	0.10	0.95	0.17	0.66	0.08	0.30	0.27	0.19	0.80	0.73	0.95	0.26	0.68	0.38	0.63
Initial average θ	0.00	0.00	0.01	0.67	0.29	0.10	0.81	0.00	0.02	0.41	0.08	0.02	0.38	0.27	0.49	0.78
Initial depth-weighted θ average θ	0.00	0.00	0.00	0.73	0.28	0.03	0.82	0.01	0.02	0.48	0.07	0.02	0.67	0.04	0.17	0.80
Horizon 1 initial θ	0.01	0.00	0.00	0.30	0.92	0.61	0.09	0.00	0.00	0.05	0.00	0.99	0.10	0.00	0.19	0.54
Horizon 2 initial θ	0.00	0.00	0.00	0.78	0.92	0.03	0.65	0.03	0.04	0.24	0.29	0.02	0.41	0.00	0.53	0.61
Horizon 3 initial θ	0.00	0.00	0.01	0.49	0.05	0.00	0.58	0.01	0.13	0.01		0.01	0.95	0.00	0.00	0.73
Horizon 4 initial θ	0.00	0.00	0.01	0.41	0.01	0.03	0.88	0.01	0.13	0.07	0.66	0.18	0.27	0.10	0.77	0.55
Horizon 5 initial θ	0.85	0.00	0.00	0.70	0.01	0.09	0.52		0.05	0.15	0.52	0.01	0.04	0.47	0.20	0.98

† ppt, precipitation; API1 to 14, antecedent precipitation index (API) for 1-, and 14-d; , volumetric soil moisture content; horizons 1 to 5 are numbered sequentially from the soil surface downward to the deepest soil depth where a soil moisture sensor was installed.

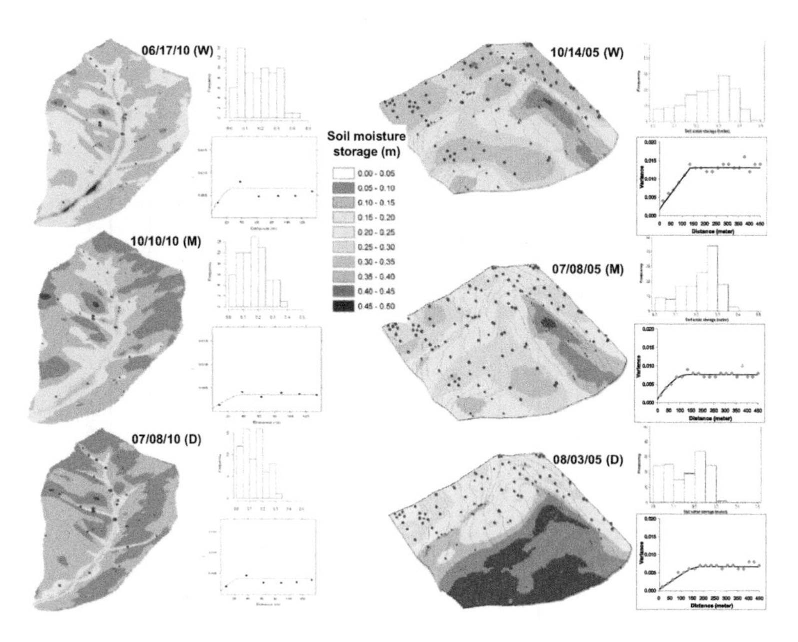

FIGURE 4 Representative maps of soil moisture storage with the solum (A and B horizons) and the associated histograms and semivariagrams under relatively wet (upper), moist (middle), and dry (lower) conditions in the Shale Hills (left) and the Kepler Farm (right), respectively. Maps were interpolated using regression kriging for the Shale Hills and ordinary kriging for the Kepler Farm (see the text for explanation). *(Color version online and in color plate)*

length was an order of magnitude shorter in the Shale Hills (20–40 m) than that at the Kepler Farm (100–153 m) throughout all soil depths monitored (Table 4), suggesting a stronger spatial variability within a shorter distance in the Shale Hills that was aligned with steeper slopes there. In the V-shaped Shale Hills with the ratio of elevation change (m) over the total area (ha) (E/A) being 6.8, soil moisture patterns exhibited clear alignment with topography, whereas that in the roughly inverse V-shaped Kepler Farm with the E/A ratio of 1.2, soil moisture alignment with topography was less obvious (Fig. 4). While the soil moisture spatial correlation length was about the same at different depths in the Shale Hills, it increased with depth at the Kepler Farm (Table 4). Interestingly, the differences in correlation length of several potential controlling factors of soil moisture were comparable between the two landscapes (Figs. 2 and 3). For example, the correlation range of slope, topographic wetness index, clay content, and depth to bedrock were all shorter (19–46 m) in the Shale Hills as compared to that at the Kepler Farm (96–252 m).

A related interesting outcome from the comparison of the two contrasting landscapes was that soil moisture spatial variability increased exponentially with catchment-wide wetness across all measurement depths in the Shale Hills, while that at the Kepler Farm was linear (Fig. 5). In the Shale Hills, steep slopes and highly permeable soils can easily channel water down the hillslope toward

TABLE 4 Summary of the Similarities and Differences in Soil Moisture Patterns and their Controls between the Shale Hills and the Kepler Farm

Landscape	Depth (cm)	Elevation range (m)	Area (ha)	Sample spacing (m)	Mean relative difference (MRD) (%)	Standard deviation of MRD (%)	Spatial correlation range (m)	Variability explained by first EOF (%)	Optimal kriging method	Relationship of soil moisture variation vs. mean wetness	Dominant soil moisture controls	Preferential flow frequency
Shale Hills	10				−45.7–70.4	5.1–23.6	21–37	75.5				
	20				−55.8–80.8	2.8–26.6	23–35	85.2				
	40				−79.9–80.6	2.5–26.8	26–40	89.7				
	60				−60.4–79.9	2.9–28.0	22–39	83.5				
	80				−68.5–52.7	2.5–27.1	20–32	87.2				
	100				−76.5–79.9	2.5–19.3	21–30	88.9				
	Range	*256–310*	*7.9*	*20*	*−80–81*	*3–28*	*20–40*	*76–90*	*Regression*	*Exponential*	*Topography > Soil*	*18–58%*
Kepler Farm	10				−26.4–15.4	7.2–34.6	100–132	31.1				
	20				−31.2–42.0	5.0–31.0	110–151	65.2				
	40				−48.0–56.4	4.6–45.8	108–145	65.5				
	60				−49.4–28.2	6.7–46.1	112–148	43.9				
	80				−54.5–36.6	4.0–34.3	123–150	64.7				
	100				−40.8–33.8	7.7–39.7	125–153	66.6				
	Range	*373–96*	*19.5*	*40*	*−55–56*	*4–46*	*100–153*	*31–66*	*Ordinary*	*Linear*	*Soil> Topography*	*6–32%*

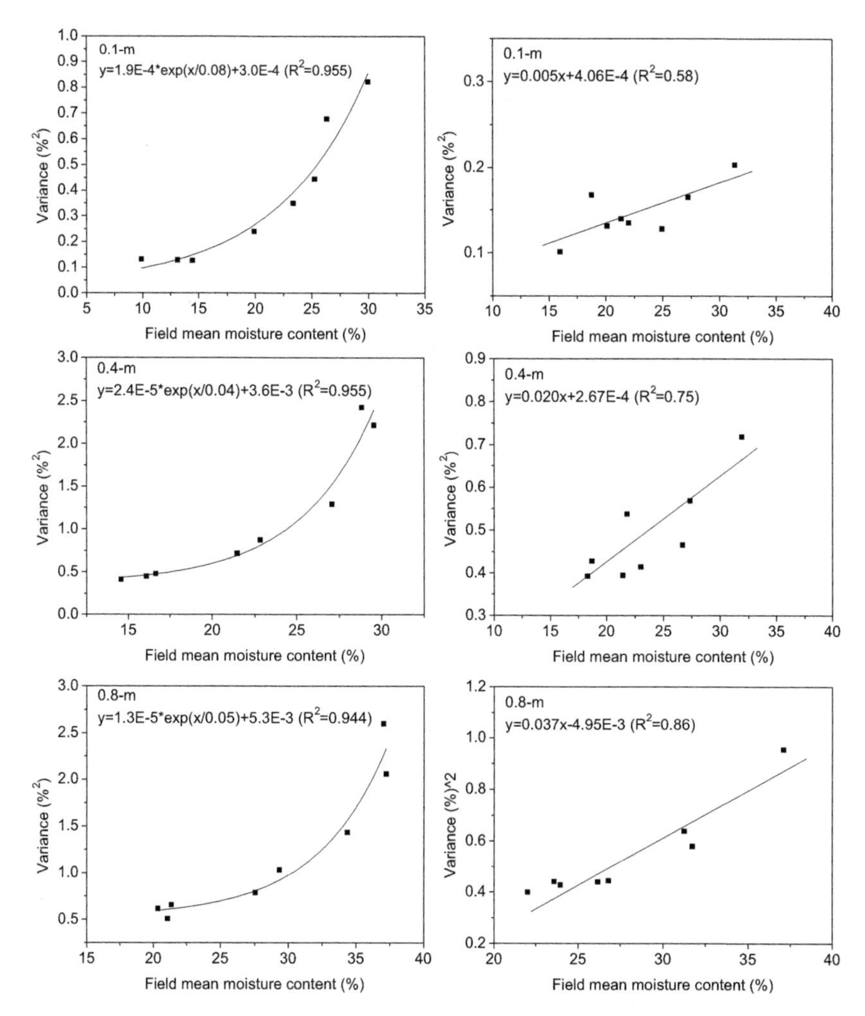

FIGURE 5 Soil moisture spatial variability (variance) in relation to landscape-wide mean moisture content at multiple depths (0.1 m, 0.4 m, and 0.8 m) in the Shale Hills (left) and the Kepler Farm (right), based on the manually-collected soil moisture data for the same time period from 2005 to 2007.

the swales and the valley floor during wet periods, while at the Kepler Farm gently-rolling landforms and less permeable soils result in more evenly distributed soil moisture across the whole landscape. As discussed by Takagi and Lin (2012), the presence of a shallow water table in the valley floor of the Shale Hills during wet periods was one driver of the exponential increase of soil moisture spatial variability.

Zhu and Lin (2010) examined optimal interpolation methods for both landscapes, and found that soil moisture was best interpolated using ordinary kriging at the Kepler Farm but in regression kriging was optimal in the Shale Hills. They concluded that landscape characteristics, including E/A ratio and parent material variation, were important in determining optimal interpolation method for soil moisture and various soil properties. Because of a smaller E/A ratio (1.2) and multiple parent materials at the Kepler Farm, ordinary kriging was shown to be optimal. In contrast, a larger E/A ratio of 6.8 and a single parent material resulted in regression kriging being generally preferable in the Shale Hills.

To determine temporal changes of soil moisture spatial patterns, we used the MRD and its SD from the time stability analysis to locate 1) temporally stable area that is representative of the overall landscape mean moisture condition and 2) temporally unstable area that may be linked to preferential flow occurrence. For the same time window used for this analysis between the two landscapes, there was a smaller span of MRD in soil moisture at the Kepler Farm (−55 to 56%) than that in the Shale Hills (−80 to 81%) (Fig. 6 and Table 4), but slightly higher SD at the Kepler Farm (4–46%) as compared to that in the Shale Hills (3–28%) (Fig. 7 and Table 4). For example, the MRD in the topsoil ranged from –45.7 to 70.4% in the Shale Hills, while MRD at the topsoil varied from –26.4 to 15.4% at the Kepler Farm, indicating larger spatial variation in the Shale Hills. At the same time, the SD of MRD ranged from 5.1 to 23.6% in the Shale Hills in the topsoil, but from 7.2 to 34.6% at the Kepler Farm topsoil, indicating slightly stronger temporal dynamics of soil moisture at the Kepler Farm. This was probably linked to crop growth at the Kepler Farm, as growing vs. nongrowing seasons and different crops grown there could have caused more changes in soil moisture over time as compared to the Shale Hills with a relatively stable vegetation cover. The spans of either MRD or SD of MRD were the highest in the intermediate soil depths (0.4–0.6 m) for both the landscapes, possibly implying the more likely occurrence of subsurface lateral flow at those soil depths. This is consistent with our earlier study results (Lin et al., 2006b; Zhu and Lin, 2009; Takagi and Lin, 2011). In contrast, the near-surface (<0.4 m) moisture fluctuated more seasonally, and the soil moisture patterns at deeper depths (>0.8 m) became more temporally persistent (i.e. smaller SD values).

3.1.2. Controls of Soil Moisture Spatial–Temporal Patterns

The EOF analysis showed that the soil moisture variability in both landscapes could be significantly explained by a few main EOFs (Table 5). In the Shale Hills, the first four EOFs together explained 87% of the total variability for 10-cm soil moisture, while the first EOF (EOF1) alone explained 76% of total variance. At the 20-cm depth, the EOF1 explained 85% of the total variability. With increasing soil depth, total variation explained by EOFs increased up to nearly 90% for EOF1 and up to 96% for the significant EOFs (Table 5). The surface soil variances explained by EOFs in the Shale Hills are higher than other reported studies (e.g. Jawson and Niemann, 2007; Korres et al., 2010;

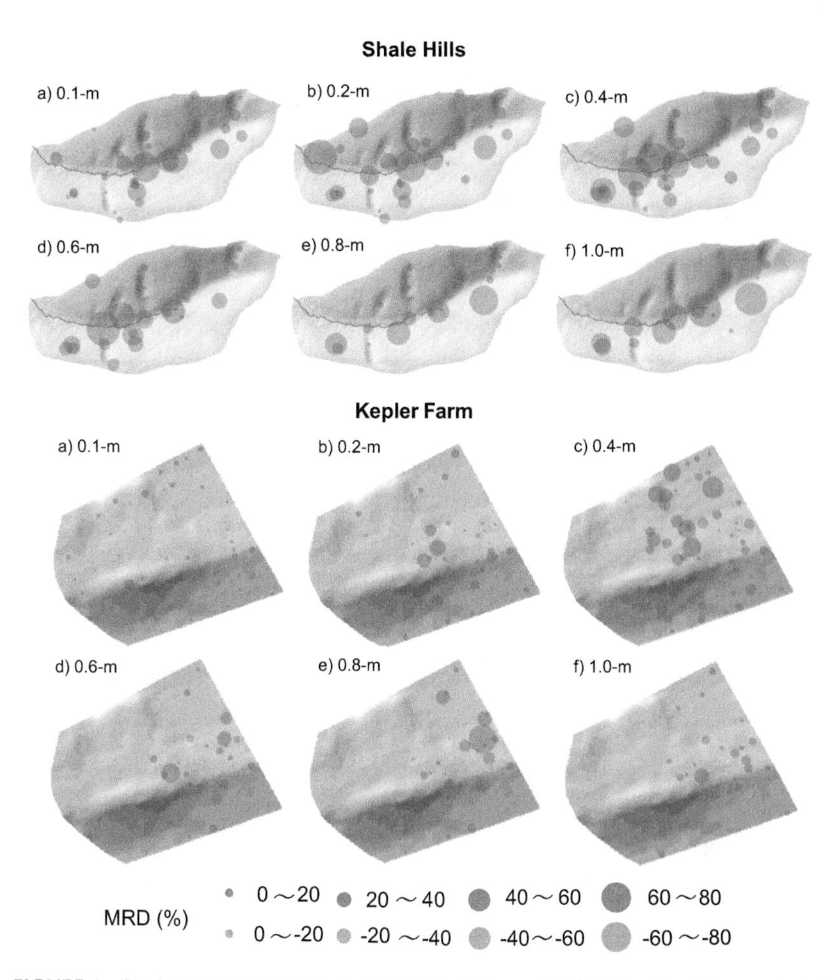

FIGURE 6 Spatial distribution of the mean relative difference (MRD) of soil moisture content at each depth (from 0.1 to 1 m) in the Shale Hills (upper) and the Kepler Farm (lower). Size of circle indicates the magnitude of MRD, with smaller circle indicating closer to spatially-averaged moisture content. Red circles indicate negative MRD (less than the overall landscape mean at a given depth), while blue circles indicate positive MRD (wetter than the overall landscape mean at a given depth). *(Color version online and in color plate)*

which found about 75% of surface soil moisture variability explained by the derived significant EOFs).

Compared to the Shale Hills, the soil moisture variability explained by EOFs at the Kepler Farm was much smaller (Table 5). The two significant EOFs explained 51% of the total variance in the topsoil moisture, while the EOF1 alone explained 31% of the total variability. With increasing soil depth, total variation explained by EOF1 increased to about 65% for most depths (except 60 cm with 44%) and up to 80% for the significant EOFs (Table 5).

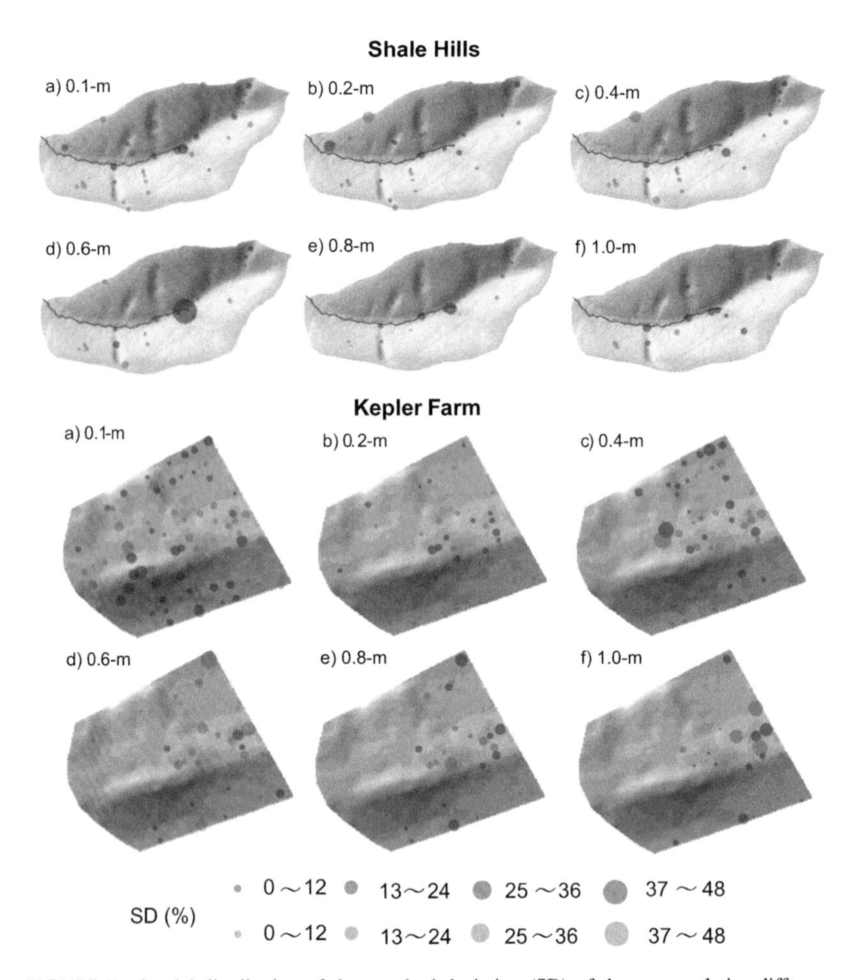

FIGURE 7 Spatial distribution of the standard deviation (SD) of the mean relative difference (MRD) of soil moisture content at each depth (from 0.1 to 1 m) in the Shale Hills (upper) and the Kepler Farm (lower). Size of circle indicates the magnitude of the standard deviation of MRD, with smaller circle indicating temporally more stable sites. Red circles indicate negative MRD (less than the overall landscape mean at a given depth), while blue circles indicate positive MRD (wetter than the overall landscape mean at a given depth). *(Color version online and in color plate)*

The spatial patterns determined from the EOF analysis were correlated with soil-terrain attributes for both landscapes (Table 5). In the Shale Hills, there were strong correlations between the EOFs of soil moisture and soil–topography properties at each measurement depth, but such correlation was much weaker overall at the Kepler Farm. Between the terrain and soil attributes in correlating with soil moisture, terrain features clearly outperformed soil properties in explaining the variability of soil moisture in the Shale Hills, while soil properties slightly outperformed terrain features at the Kepler Farm (Tables 5 and 6).

TABLE 5 Correlation Coefficients between Empirical Orthogonal Functions (EOFs) of Soil Moisture at Various Depths and Soil-terrain Attributes in the Shale Hills and the Kepler Farm. The Numbers in Parentheses After Each EOF Indicate the Cumulative Total Variability Explained. E: Elevation; SP: Slope Percent; UCA: Upslope Contribution Area; PC: Profile Curvature; DB: Depth to Bedrock; TWI: Topographic Wetness Index; RF: Rock Fragment; and SOC: Organic Matter Content

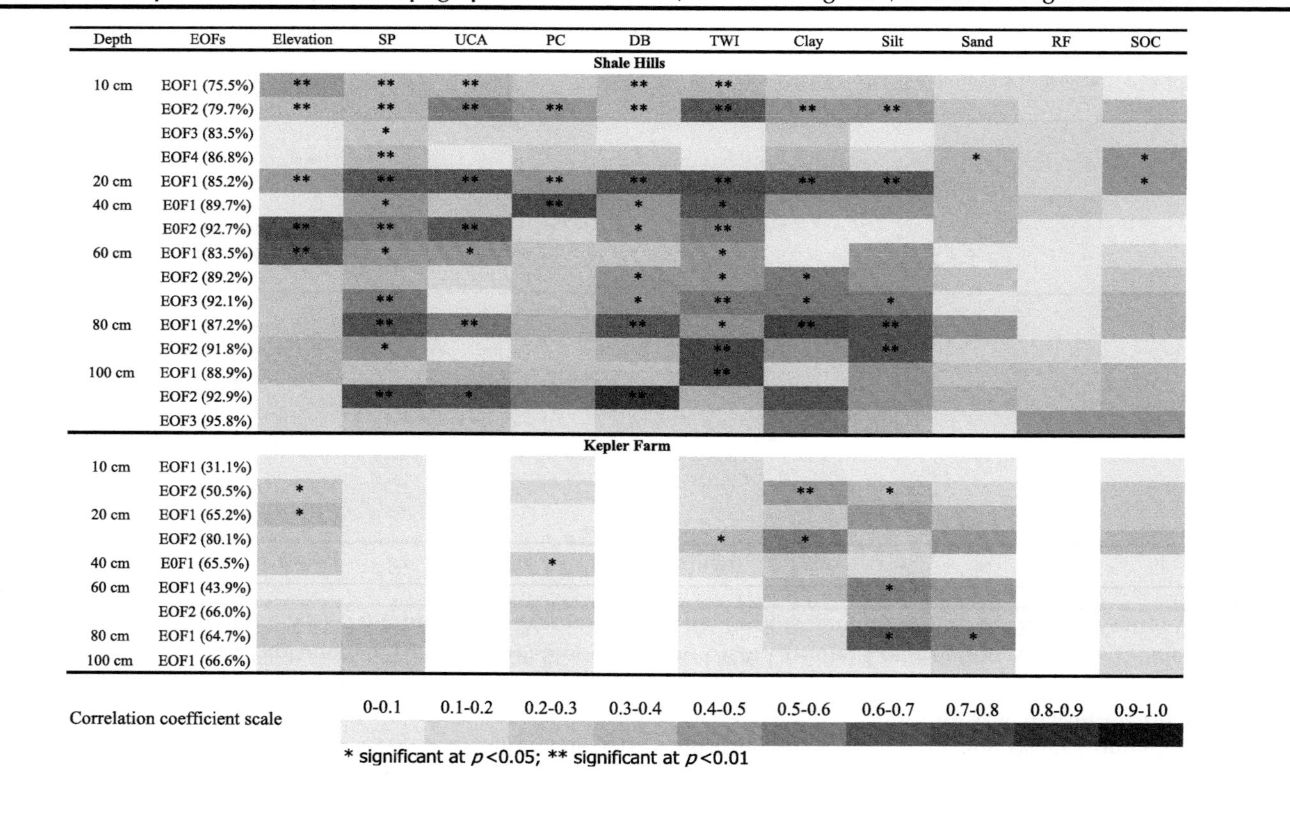

Depth	EOFs	Elevation	SP	UCA	PC	DB	TWI	Clay	Silt	Sand	RF	SOC
						Shale Hills						
10 cm	EOF1 (75.5%)	**	**	**		**	**					
	EOF2 (79.7%)	**	**	**	**	**	**	**	**			
	EOF3 (83.5%)		*									
	EOF4 (86.8%)		**							*		*
20 cm	EOF1 (85.2%)	**	**	**	**	**	**	**	**			*
40 cm	E0F1 (89.7%)		*		**	*	*					
	E0F2 (92.7%)	**	**	**		*	**					
60 cm	EOF1 (83.5%)	**	*	*			*					
	EOF2 (89.2%)					*	*	*				
	EOF3 (92.1%)		**			*	**	*	*			
80 cm	EOF1 (87.2%)		**	**		**	*	**	**			
	EOF2 (91.8%)		*				**		**			
100 cm	EOF1 (88.9%)						**					
	EOF2 (92.9%)		**	*		**						
	EOF3 (95.8%)											
						Kepler Farm						
10 cm	EOF1 (31.1%)											
	EOF2 (50.5%)	*						**	*			
20 cm	EOF1 (65.2%)	*										
	EOF2 (80.1%)						*	*				
40 cm	E0F1 (65.5%)				*							
60 cm	EOF1 (43.9%)								*			
	EOF2 (66.0%)											
80 cm	EOF1 (64.7%)								*	*		
100 cm	EOF1 (66.6%)											

Correlation coefficient scale: 0-0.1, 0.1-0.2, 0.2-0.3, 0.3-0.4, 0.4-0.5, 0.5-0.6, 0.6-0.7, 0.7-0.8, 0.8-0.9, 0.9-1.0

* significant at $p<0.05$; ** significant at $p<0.01$

TABLE 6 Correlation Coefficients between Empirical Orthogonal Functions (EOFs) of Soil Moisture at Various Depths and Soil-terrain Attributes, Grouped by Soil Wetness Conditions (Dry: Landscape Mean Volumetric Soil Water Content $\theta < 15\%$; Moist: $\theta = 15–22\%$; Wet: $\theta > 22\%$) in the Shale Hills Catchment and by Growing (G) (from Late May to Mid-October) and Nongrowing (UG) (the Rest of Year) Seasons at the Kepler Farm. The Numbers in Parentheses After Each EOF Indicate the Cumulative Total Variability Explained. E: Elevation; SP: Slope Percent; UCA: Upslope Contribution Area; PC: Profile Curvature; DB: Depth to Bedrock; TWI: Topographic Wetness Index; RF: Rock Fragment; and SOC: Organic Matter Content

Depth	Wetness /Season	EOFs	Elevation	SP	UCA	PC	DB	TWI	Clay	Silt	Sand	RF	SOC
Shale Hills													
10 cm	Dry	EOF1 (73.8%)		**			**	**					
	Moist	EOF1 (76.4%)	**	**	**	**	**	**	**	**			
	Wet	EOF1 (82.9%)	**	**	**	**	**	**	**	**			**
40 cm	Dry	EOF1 (92.5%)	**	**	**	**	**	**	**	**			
	Moist	EOF1 (86.7%)	**	**	**	*	**	**	**	**			
	Wet	EOF1 (90.8%)	*	**	**	**	**	**	*	*			
80 cm	Dry	EOF1 (90.9%)	**	**				**	**	**			
	Moist	EOF1 (88.6%)	**	**			*	**	**	**			
	Wet	EOF1 (90.1%)	*	**	*	*	*	**	*	*			
Kepler Farm													
10 cm	Growing	EOF1 (43.1%)											*
	Non-growing	EOF1 (46.2%)	**						**	*			
40 cm	Growing	EOF1 (70.7%)	*			*							
	Non-growing	EOF1 (82.3%)											
80 cm	Growing	EOF1 (75.2%)	*							**	*		
	Non-growing	EOF1 (90.1%)		**						*	*		

Correlation coefficient scale: 0-0.1, 0.1-0.2, 0.2-0.3, 0.3-0.4, 0.4-0.5, 0.5-0.6, 0.6-0.7, 0.7-0.8, 0.8-0.9, 0.9-1.0

* significant at $p<0.05$; ** significant at $p<0.01$

Among the soil properties examined in this study, depth to bedrock (which is corrected with topography in the Shale Hills) and texture had significant explanatory power of soil moisture variability in the Shale Hills, while texture was the main soil property that had significant explanatory power at the Kepler Farm. Elevation, slope, and curvature were negatively related to soil moisture content, while upslope contributing area, depth to bedrock, topographic wetness index, and percent silt and clay contents were positively related to soil moisture content. This is likely due to the fact that most soils with deep profiles in both landscapes were generally distributed in lower elevation and concave areas (i.e. valley floor and swales) where soil moisture was generally the wettest.

In the Shale Hills, the correlation coefficient values generally increased with soil depth for slope and percent silt and clay contents, whereas the highest values were observed at intermediate depths for curvature, depth to bedrock, and topographic wetness index (Table 5). This supports the observations from our earlier studies (Takagi and Lin, 2011) that subsurface lateral flow is important in controlling soil moisture dynamics at these intermediate depths. At the Kepler Farm, the influence of soil and terrain attributes on soil moisture did not show an obvious trend with soil depth in the EOF analysis (Tables 5 and 6).

To further examine the wetness/season-dependent relationships between soil moisture and soil-terrain attributes, the soil moisture data were grouped into relatively wet, moist, and dry conditions in the Shale Hills and growing and nongrowing seasons at the Kepler Farm (Table 6). In the Shale Hills, the primary EOFs at all depths under the wet conditions explained slightly higher amount of variance when compared to those from the complete dataset. In contrast, the primary EOFs under the nongrowing seasons in the Kepler Farm explained a much higher amount of variance than those from the complete dataset. This confirms that cropping has added another layer of complexity in influencing soil moisture spatial–temporal pattern at the Kepler Farm, as also suggested in our earlier study (Zhu and Lin, 2011).

3.2. Preferential Flow Dynamics

3.2.1. Frequency of Preferential Flow Occurrence

Preferential flow was common in both landscapes (Table 7), especially in late summer to early fall (data not shown). However, its occurrence varied from event to event and from site to site. Overall, preferential flow occurrence was more frequent in the Shale Hills (average 37% of total precipitation events) than that at the Kepler Farm (average 19% of total precipitation events) (Fig. 8 and Table 7). In the Shale Hills, preferential flow occurred at least in one or more monitoring sites for 90% of the total 175 events analyzed during the 3-yr period, with 50%, 38%, and 22% of all the events causing preferential flow to occur at 4, 5, and $\geq$6 sites (out of total 10 monitoring sites), respectively. At the Kepler Farm, preferential flow occurred at one or more sites for 59% of the total 139 events analyzed over the 2.5-yr period, with 25%, 14%, and 2% of all the

TABLE 7 Number of Precipitation Events and the Percentage of the Total Events Analyzed (in Parenthesis) Leading to Each of the Three Flow Regimes (Preferential Flow, Sequential Flow, and Nondetectable Flow) at each Real-time Monitoring Site in the Two Landscapes. Total Events Analyzed for the Shale Hills was 175 Over a Three-year Period from 2007 to 2009, while that for the Kepler Farm was 139 Over a Two-and-a-half-year Period from 2008 to 2010. The Monitoring Sites are Arranged in Topographic Sequence (see Fig. 1 for detail layouts)

	Upper slope			Middle slope				Lower slope		
Shale Hills	**U1**	**U2a**	**U2b**	**M1**	**M2a**	**M2b**	**M2c**	**L1a**	**L1b**	**L2**
Preferential Flow	93 (58.1%)	53 (30.3%)	74 (42.3%)	31 (17.7%)	91 (52.3%)	63 (36.2%)	36 (20.7%)	41 (24.1%)	95 (55.9%)	60 (34.3%)
Sequential Flow	38 (23.8%)	107 (61.1%)	71 (40.6%)	57 (32.6%)	35 (20.1%)	61 (35.1%)	93 (53.4%)	81 (47.6%)	32 (18.8%)	75 (42.9%)
Nondetectable flow	29 (18.1%)	15 (8.6%)	30 (17.1%)	87 (49.7%)	48 (27.6%)	50 (28.7%)	45 (25.9%)	48 (28.2%)	43 (25.3%)	40 (22.9%)
Kepler Farm	**A3**	**A4**	**A5**	**A6**				**A1**	**A2**	
Preferential flow	36 (25.9%)	45 (32.4%)	24 (17.3%)	19 (13.7%)				8 (5.8%)	29 (20.9%)	
Sequential flow	68 (48.9%)	32 (23.0%)	38 (27.3%)	39 (28.1%)				61 (43.9%)	36 (25.9%)	
Nondetectable flow	35 (25.2%)	62 (44.6%)	77 (55.4%)	81 (58.3%)				70 (50.4%)	74 (53.2%)	

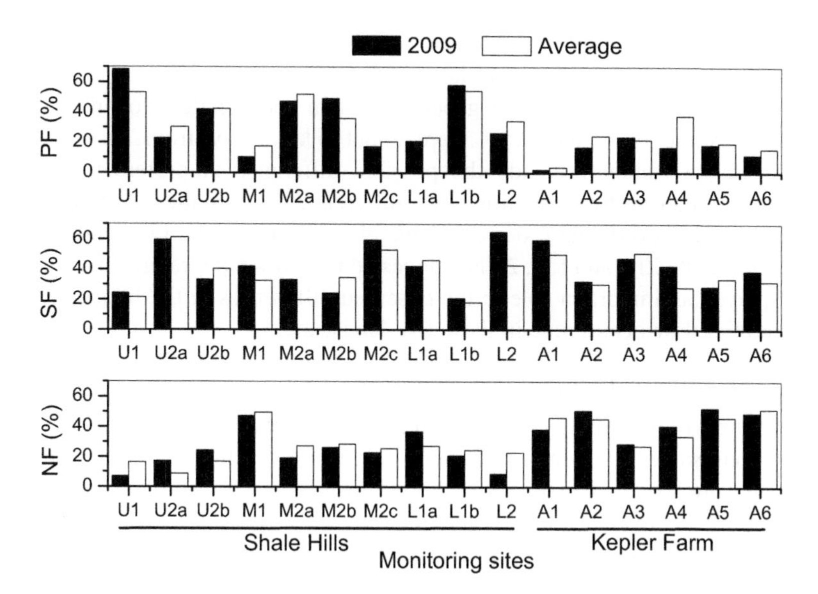

FIGURE 8 Comparisons of occurrence frequency (as percentage of total precipitation events) of three flow regimes – preferential flow (PF), sequential flow (SF), and nondetectable flow (NF) – in 2009 between the Shale Hills (left) and the Kepler Farm (right) based on the real-time soil moisture monitoring data. The overall averaged occurrence frequencies over the entire monitoring periods (2007–2009 for the Shale Hills and 2008–2010 for the Kepler Farm) are also shown.

events causing preferential flow to appear at 2, 3, and ≥4 sites (out of total 6 sites), respectively.

The site-specific frequency of preferential flow occurrence in the Shale Hills ranged from 18% (at a mid-slope site; M1) to 58% (at an upper sideslope site; U1), while three sites distributed along the hillslope (at an upper, a middle, and a lower portion of the hillslope; U1, M2a, and L1b) had the highest frequency of preferential flow occurrence (52–58% of the 175 events) (Table 7). In comparison, the frequency of preferential flow occurrence at the Kepler Farm ranged from 6% (at the valley site near stream; A1) to 32% (at the ridgetop site; A4), while the highest frequency was 21–32% (at the lower and upper positions of the hillslope; A2, A3, and A4) (Table 7).

For the entire real-time monitoring periods, the averaged frequency for the three flow regimes (preferential flow, sequential flow, and nondetectable flow) across all the monitoring sites were 37.0%, 37.7%, and 25.3%, respectively, in the Shale Hills, and were 19.3%, 32.9%, and 47.8%, respectively, at the Kepler Farm. In 2009 when both landscapes had comparable datasets, the averaged frequency for preferential flow, sequential flow, and nondetectable flow across all the monitoring sites were 36.3%, 40.5%,and 23.2%, respectively, in the Shale Hills, and were 15.0%, 41.6%, and 43.4%, respectively, at the Kepler Farm. There was no obvious trend in the frequency of preferential flow occurrence from year to year in both landscapes; however, year to year variability did exist, especially in the ridgetop sites (site U1 in the Shale Hills and site A4 at the Kepler Farm) (data not shown).

Note that overall rainfall was generally similar between the two landscapes studied, as the two landscapes were about 6 km away from each other. However, we cannot preclude the possible difference in preferential flow occurrence induced by vegetation cover difference between the forestland and the cropland. Nevertheless, our extensive observations in the two landscapes (e.g. Lin, 2006; Zhu and Lin, 2009) suggested that soil–topography differences have played an important role in the observed differences in the frequency of preferential flow occurrence in these two landscapes.

3.2.2. Dominant Controls of Preferential Flow Occurrence

The controls of preferential flow occurrence varied between the two landscapes and among sites within each landscape (Table 3). Nevertheless, some general patterns were observed. In the Shale Hills, skewness and kurtosis of precipitation were significant in controlling preferential flow occurrence at five or more sites (with greater skewness and kurtosis values favoring preferential flow occurrence, especially in the ridgetop and hillslope sites). Precipitation amount was significant in three sites (one middle and two lower portions of the hillslope), while precipitation duration and maximum intensity were significant in two to three sites located largely in the midslope of the monitored hillslope. Of the antecedent precipitation indices, only antecedent precipitation index for one day (API1) was a significant control at three sites located in the upper to middle portions of the hillslope. The air temperature was significant at six sites located throughout the hillslope, with preferential flow occurring at lower and midslope sites (L2, M2a, M2b, and M2c) when the temperature was lower (which corresponded to colder, wet periods in the area) and at upper hillslope sites (U2a and U1) when the temperature was higher (corresponding to warmer, dry periods in the area). The impact of air temperature was also linked to the influence of initial soil moisture on preferential flow occurrence, which showed significant controls throughout the hillslope and at various soil depths (Table 3). Preferential flow occurred more favorably under higher initial soil moisture in the lower part of the hillslope (sites L2, L1a, and L1b) but under lower initial soil moisture in the upper and middle parts of the hillslope (sites U2a, U2b, U1, M2a, and M2b). The ridgetop and upper planar hillslope sites in particular were sensitive to dry initial soil moisture for inducing preferential flow occurrence because of hydrophobicity, cracking, and shallowness of the soils there. Our earlier studies (Lin and Zhou, 2008; Graham and Lin, 2011) and other studies (e.g. Taumer et al., 2006) have also found correlation between dry soils and preferential flow occurrence.

In comparison, precipitation amount, duration, maximum intensity, and variance showed more significant impacts on the occurrence of preferential flow at the Kepler Farm, but not precipitation skewness and kurtosis (Table 3). Similar to the Shale Hills, antecedent precipitation indices were generally not significant in influencing preferential flow occurrence in this cropland.

However, unlike the Shale Hills, the air temperature did not show any impact on the occurrence of preferential flow at the Kepler Farm. The impact of initial soil moisture on the occurrence of preferential flow was also less significant at the Kepler Farm, with only two or three sites showing significant impacts at various soil depths (Table 3). Wetter initial soil moisture increased preferential flow occurrence in site A2 (near valley) but drier initial soil moisture promoted more preferential flow in site A4 (ridgetop).

In the two landscapes, we did not find evidence of direct topographic control on the absolute or relative frequency of preferential flow occurrence, suggesting that the effects of topography on preferential flow have been masked by other factors such as initial soil moisture, soil types, and soil properties. In both landscapes, the dominant controls of preferential flow occurrence at the upper part of the hillslopes tended to be dry initial soil moisture, hot air temperature, and the time of the year (late summer to early fall), factors that were correlated with soil hydrophobicity and cracking; while at the lower part of the hillslopes, wet initial soil moisture and total precipitation tended to induce more preferential flow. At the midslope part of the hillslopes, preferential flow occurrence was generally associated with longer precipitation events where the precipitation intensity distribution was more uniform as well as when initial soil was wet. In addition, precipitation events that resulted in widespread preferential flow across the landscapes generally had greater duration, amount, and average intensity from the start of the event to the time of maximum intensity. This was discussed more extensively in Graham and Lin (2011).

In this study, while the majority of the indices that were identified as significant controls of preferential flow occurrence can be easily derived from the precipitation record, the initial soil moisture indices cannot. As the first part of this chapter shows, soil moisture dynamics in both landscapes are controlled by a complex interplay among soil features, topographic attributes, precipitation dynamics, and vegetation water use; the site-specific soil moisture dynamics and the occurrence of preferential flow cannot be adequately predicted or modeled by precipitation and topography alone, as traditional hydrologic models have attempted to do. Instead, appropriate incorporation of soil types and soil properties, plus real-time monitoring of soil moisture, can greatly enhance such efforts.

4. SUMMARY AND CONCLUSIONS

Based on spatially extensive sampling and real-time monitoring of soil moisture from 2005 to 2010, this study has demonstrated the coupled impacts of topography, soil properties, and vegetation on soil moisture spatial–temporal patterns and preferential flow occurrence in two contrasting landscapes. The major differences and similarities between the forested Shale Hills and the cropped Kepler Farm are summarized in Table 4. Overall, topographic variation was more influential than soil properties in the Shale Hills in controlling soil moisture variability and preferential flow occurrence, while soil properties were more evident than the topography at the Kepler Farm. Consequently, terrain features clearly outperformed soil properties in explaining soil moisture variability in the Shale Hills, while soil properties slightly outperformed terrain attributes at the Kepler Farm in explaining soil moisture variability.

Subsurface preferential flow was common in both landscapes, but occurred about twice as frequently in the steep-sloped forestland as in the more gently rolling cropland. In both the landscapes, initial soil moisture played significant but different roles in regulating preferential flow occurrence in different hillslope positions, with higher initial moisture favoring preferential flow occurrence in the lower portion of the hillslopes while lower initial moisture favoring preferential flow occurrence in the upper part of the hillslopes. The effects of topography on preferential flow occurrence thus have been complicated by other factors including seasonal wetness condition, soil type, and soil properties.

Comparisons between the two contrasting landscapes have illustrated complex landscape–soil–water–vegetation relationships, and that hydropedology is important in enhancing the understanding and prediction of soil moisture patterns and preferential flow dynamics in different landscapes. Real-time monitoring of soil moisture and appropriate incorporation of soil types and soil properties into hydrologic modeling would certainly improve the reliability of prediction.

ACKNOWLEDGMENTS

This research has been supported by the USDA National Research Initiative (grant #2002-35102-12547), the USDA Higher Education Challenge Competitive Grants Program (grant #2006-38411-17202), and the U.S. National Science Foundation (grant #EAR-0725019). Assistance in field data collections from the Penn State Hydropedology Group is gratefully acknowledged.

REFERENCES

Famiglietti, J.S., Rudnicki, J.W., Rodell, M., 1998. Variability in surface moisture content along a hillslope transect: Rattlesnake Hill, Texas. J. Hydrol. 210, 259–281.

Graham, C., Lin, H.S., 2011. Controls and frequency of preferential flow occurrence at the Shale Hills Critical Zone Observatory: a 175 event analysis of soil moisture response to precipitation. Vadose Zone J. 10, 816–831.

Grayson, R.B., Western, A.W., Chiew, F.H.S., Blöschl, G., 1997. Preferred states in spatial soil moisture patterns: local and non local controls. Water Resour. Res. 33, 2897–2908.

Grayson, R.B., Blöschl, G., Western, A.W., McMahon, T.A., 2002. Advances in the use of observed spatial patterns of catchment hydrological response. Adv. Water Resour. 25, 1313–1334.

IMKO, 2000. TRIME TDR Probes. IMKO, Ettlingen, Germany.

Jawson, S.D., Niemann, J.D., 2007. Spatial patterns from EOF analysis of soil moisture at a large scale and their dependence on soil, landuse, and topographic properties. Adv. Water Resour. 30 (3), 366–381.

Korres, W.C., Koyama, N., Fiener, P., Schneider, K., 2010. Analysis of surface soil moisture patterns in agricultural landscapes using Empirical Orthogonal Functions. Hydrol. Earth Syst. Sci. 14, 751–764.

Lin, H.S., 2006. Temporal stability of soil moisture spatial pattern and subsurface preferential flow pathways in the Shale Hills Catchment. Vadose Zone J. 5, 317–340.

Lin, H., Zhou, X., 2008. Evidence of subsurface preferential flow using soil hydrologic monitoring in the Shale Hills catchment. Eur. J. Soil Sci. 59, 34–49.

Lin, H.S., Bouma, J., Pachepsky, Y., Western, A.W., Thompson, J.A., van Genuchten, M.T., Vogel, H., Lilly, A., 2006a. Hydropedology: synergistic integration of pedology and hydrology. Water Resour. Res. 42, W05301. doi: 10.1029/2005WR004085.

Lin, H.S., Kogelmann, W., Walker, C., Bruns, M.A., 2006b. Soil moisture patterns in a forested catchment: a hydropedological perspective. Geoderma 131, 345–368.

Shouse, P.J., Russell, W.B., Burden, D.S., Swlim, H.M., Sisson, J.B., van Genuchten, M.Th., 1995. Spatial variability of soil water retention functions in a silt loam soil. Soil Sci. 159, 1–12.

Takagi, K., Lin, H.S., 2011. Temporal evolution of soil moisture spatial variability in the Shale Hills catchment. Vadose Zone J. 10, 832–842.

Takagi, K., Lin, H.S., 2012. Changing controls of soil moisture spatial organization in the Shale Hills Catchment. Geoderma 173–174, 289–302.

Taumer, K., Stoffregen, H., Wessolek, G., 2006. Seasonal dynamics of preferential flow in a water Repellent soil. Vadose Zone J. 5, 405–411.

Vachaud, G., Passerat de Silans, A., Balabanis, P., Vauclin, M., 1985. Temporal stability of spatially measured soil water probability density function. Soil Sci. Soc. Am. J. 49, 822–828.

Western, A.W., Grayson, R.B., Blöschl, G., Willgoose, G.R., McMahon, T.A., 1999. Observed spatial organization of soil moisture and its relation to terrain indices. Water Resour. Res. 35, 797–810.

Yoo, C., Kim, S., 2004. EOF analysis of surface soil moisture field variability. Adv. Water Resour. 27, 831–842. doi:10.1016/j. advwatres.2004.04.003.

Zhao, Y., Peth, S., Horn, R., Hallett, P., Wang, X.Y., Giese, M., Gao, Y.Z., 2011. Factors controlling the spatial patterns of soil moisture investigated by multivariate and geostatistical analysis. Ecohydrology 4, 36–48.

Zhao, Y., Peth, S., Wang, X.Y., Lin, H.S., Horn, R., 2010. Temporal stability of surface soil moisture spatial patterns and their controlling factors in a semi-arid steppe. Hydrol. Proc. 24, 2507–2519.

Zhu, Q., Lin, H.S., 2009. Simulation and validation of concentrated subsurface lateral flow paths in an agricultural landscape. Hydrol. Earth Syst. Sci. 13, 1503–1518.

Zhu, Q., Lin, H.S., 2010. Interpolation of soil properties based on combined information of spatial structure, sample size and auxiliary variables. Pedosphere 20, 594–606.

Zhu, Q., Lin, H., 2011. Influence of soil, terrain, and crop growth on soil moisture variation from transect to farm scales. Geoderma 163, 45–54.

Chapter 13

Geophysical Investigations of Soil–Landscape Architecture and Its Impacts on Subsurface Flow

James Doolittle,[1] Qing Zhu,[2] Jun Zhang,[3] Li Guo[3] and Henry Lin[3,*]

ABSTRACT

Landscape–soil–hydrology relationships vary across landscapes and are often exceedingly complex. Characterizing these complex relationships at different spatial–temporal scales and their impacts on subsurface flow and transport is a major challenge in hydropedology. Soil–landscape relationships have been traditionally inferred from point-based pedologic observations. These observations are then often extrapolated to the landscape, but seldom measured or directly confirmed across the landscape. Data gaps thus exist at the intermediate scales of hillslopes and catchments in terms of actual subsurface soils distribution and their features (termed soil–landscape architecture here), which will affect the upscaling of point-based monitoring data or the downscaling of remote-sensing data. This chapter illustrates two commonly-used geophysical tools – ground-penetrating radar (GPR) and electromagnetic induction (EMI) – in revealing complex soil–landscape architecture and its impacts on subsurface flow. We investigated two contrasting landscapes – one cropland and one forestland – that are typical of the Ridge and Valley Physiographic Region in eastern United States. Repeated or time-lapsed GPR and EMI were emphasized to provide a better understanding of subsurface architecture and its effects on subsurface flow through soil moisture change over the time scales of seasons or precipitation/infiltration events. Our results showed that 1) relative difference in soil apparent electrical conductivity (EC_a) across the landscapes remained relatively stable over time, which corresponded to stable soil-landform units; 2) changes in EC_a over seasons within the same landscape indicated (to some degree) active zones of subsurface flow, which corresponded with simulated water flow paths and observed soil morphology; 3) south- vs. north-facing hillslopes and linear vs. concave slopes showed differences in EMI and GPR responses, which reflected the underlying differences in soil architecture;

[1]National Soil Survey Center, USDA-NRCS, Newtown Square, PA, USA
[2]Nanjing Institute of Geography and Limnology, Chinese Academy of Sciences, Nanjing, China
[3]Dept. of Ecosystem Science and Management, Pennsylvania State Univ., PA, USA
[*]Corresponding author: Email: henrylin@psu.edu

Hydropedology, Edited by H. Lin. DOI: 10.1016/B978-0-12-386941-8.00013-7

4) depth to bedrock was highly variable across the two landscapes, but predictable patterns in the forestland were revealed through extensive GPR surveys; and 5) subsurface preferential flow pathways and patterns were identified through time-lapsed GPR investigations, which showed significant differences between shallow and deep soils. This study demonstrates the potential of geophysical tools in easing the technological bottleneck of subsurface investigation and closing data gaps at the intermediate scales.

1. INTRODUCTION

At the intermediate scales of hillslopes and small catchments, the resolution of most remote-sensing techniques is too coarse to decipher soil architectures and hydrologic processes. Remote-sensing techniques also do not penetrate deeply into the soil (typically limited to top 5–6 cm). Point-sampling methods (such as soil pits, monitoring wells, and soil moisture probes) can provide detailed, but highly site-specific soil and hydrologic data. As the collection of point data is time consuming, labor intensive, costly, and generally destructive, it is therefore confined to a limited number of sampling points. Because of these limitations, soil and hydrologic properties and processes for larger landscapes covering widely spaced sampling points must be inferred. Such inferences are often based on deductions and simplified assumptions drawn from the parameters measured at the individual sampling points. The inability of remote-sensing and point-sampling methods to adequately characterize the complex subsurface at the intermediate scales fosters significant ambiguity in data interpolations and model predictions (Lin et al., 2006a; Lin, 2010).

Bridging data gaps that exist at the intermediate scales requires the adaptation and use of new technologies. Geophysical methods offer considerable advantages in this regard because of their economy, speed and ease of use, noninvasiveness, and their capacity for collecting spatially dense data relevant to soil and hydrologic properties and processes in a relatively short time. The use of geophysical methods to bridge measurement gaps at the intermediate scale is not new (Shield and Sopper, 1969). Over the last three decades, the use of geophysical methods has significantly increased in the studies of soils (Doolittle and Butnor, 2008), hydrology (Robinson et al., 2008; Slater and Comas, 2008), and landscapes (Van Dam, 2010). The most commonly used geophysical methods include time domain reflectometry (TDR), electromagnetic induction (EMI), ground-penetrating radar (GPR), magnetrometry, resistivity, and seismic. Each geophysical method measures a distinct physical property (e.g. electrical conductivity, magnetic susceptibility, or dielectric permittivity) of the subsurface without direct access to the sampled volume (Daniels et al., 2003; Allred et al., 2008). These measured physical properties can then be associated with specific soil and hydrologic properties. Advantages of geophysical methods include increased sampling density across landscapes, extension of information gathered from point-based measurements, and recognition of variability within remote-sensing footprints.

Selection of the most suitable geophysical method requires an understanding of the properties that influence the method's response, and whether and to what extent subsurface properties will affect the measured geophysical signals (Allred et al., 2008). Whereas surface geophysical methods allow more continuous spatial coverage than point-sampling approaches, they are limited in their capacity to resolve and characterize many subsurface features. Recent improvements in equipment functionality, data acquisition, processing, display, and interpretation programs, as well as integration with complementary technologies have encouraged the expanding use of geophysical methods in landscape, soil, and hydrologic investigations.

Ground-penetrating radar and electromagnetic induction are two noninvasive geophysical methods that have been commonly used in soil, hydrology, and landscape studies. Electromagnetic induction measures the bulk or apparent electrical conductivity (EC_a) of earthen materials. Apparent electrical conductivity is a depth-weighted, average measure of the soil's ability to conduct electrical currents over a specific depth interval (Greenhouse and Slaine, 1983). The electrical conductivity of soils is influenced by the type and concentration of ions in solution, the amount and type of clays in the soil matrix, the volumetric water content, and the temperature and phase of the soil water (McNeill, 1980). The EC_a of soils will increase with increases in soluble salts, water, and/or clay contents (Kachanoski et al., 1988; Rhoades et al., 1976). Variations in EC_a have been associated with changes in soil types, particle-size distribution, clay mineralogy, bulk density, cation exchange capacity, salinity, nutrients, organic matter content, and moisture content (Kachanoski et al., 1988; Waine et al., 2000; Anderson-Cook et al., 2002; Brevik and Fenton, 2003; King et al., 2005; Frogbrook and Oliver, 2007; Farahani and Flynn, 2007; Kravchenko, 2008). Electromagnetic induction thus has been used to measure and characterize soil water content, subsurface flow, depth to water table, and soil drainage (Kachanoski et al., 1990; Sheets and Hendrickx, 1995; Khakural et al., 1998; Scanlon et al., 1999; Kravchenko et al., 2002; Schumann and Zaman, 2003; Robinson et al., 2008; Zhu et al., 2010a,b).

Ground-penetrating radar is an impulse radar system designed for shallow (0–30 m) subsurface investigations. Ground-penetrating radars transmit short pulses of very high and ultra-high (from about 30 MHz to 1.2 GHz) frequency electromagnetic energy into the ground from an antenna. Compared with most other geophysical techniques, GPR provides a higher resolution of subsurface features. However, the effectiveness of GPR is highly site- and application-specific. Soils having high electrical conductivity rapidly attenuate radar energy, restrict penetration depth, and severely limit the effectiveness of GPR (Olhoeft, 1998; Doolittle and Butnor, 2008). In soils, the most significant conduction-based energy losses are due to ionic charge transport in the soil solution and electrochemical processes associated with cations on clay minerals (Neal, 2004). These losses can critically impact the effectiveness of GPR (Campbell, 1990; Olhoeft, 1998).

For over thirty years, GPR has been used to characterize soil and hydrology-related properties (Collins, 2008). In areas of coarse-textured soil materials, GPR has been used to chart water-table depths (Bohling et al., 1989; Smith et al., 1992; Iivari and Doolittle, 1994; Doolittle et al., 2006), estimate parameters for hydrologic models (Kowalsky et al., 2004; Doolittle et al., 2006), define recharge and discharge areas (Bohling et al., 1989), detect groundwater flow patterns (Iivari and Doolittle, 1994; van Overmeeren, 2004; Doolittle et al., 2006), and characterize near-surface hydropedologic features (van Overmeeren, 1998; Doolittle et al., 2006). In addition, GPR has been used to study preferential flow (Steenhuis et al., 1998; Gish et al., 2002; Freeland, 2008) and the movement of wetting fronts in sandy soils (Saintenoy et al., 2008). At the intermediate scale, GPR has been used to map spatial–temporal variations in soil water content of surface layers (Huisman et al., 2002, 2003; Lunt et al., 2005; Weihermüller et al., 2007; Redman, 2008). Hubbard et al. (2005) and Rubin and Hubbard (2006) have summarized the use of GPR to estimate water content, hydraulic conductivity, geochemistry, and lithofacies zonation.

The depth to bedrock and the topography of the bedrock surface affect the subsurface flow of water. Bedrock restricts, redirects, and concentrates the infiltration of water. Ground-penetrating radar has been used extensively to chart bedrock depths (Collins et al., 1989; Davis and Annan, 1989), changes in rock type (Davis and Annan, 1989), fractures, joints, bedding, and cleavage planes (Nascimento da Silva et al., 2004; Pipan et al., 2000; Porsani et al., 2005; Theune et al., 2006), and faults (Demanet et al., 2001). Ground-penetrating radar has also been used to study fractures, unloading or exfoliation joints, bedding and stress planes, and cavities of the underlying bedrock (Theune et al., 2006; Porsani et al., 2005; Al-fares et al., 2002).

With the advent of digital outputs and advanced data processing software, an emerging GPR method is the analysis of subsurface structures, distributions, and geometries from a three-dimensional (3D) perspective. Three-dimensional GPR allows the visualization of data volumes from different perspectives and cross-sections (Beres et al., 1999). This can assist the identification and characterization of subsurface architectures. In areas of electrically resistive materials, Grasmueck and Green (1996) noted that, compared with traditional two-dimensional (2D) GPR, 3D GPR can provide unrivaled resolution and detail of subsurface features. Beres et al. (1999) observed that 3D GPR often improves the definition of subsurface structural trends and results in more complete and less ambiguous interpretations than the traditional 2D GPR.

Unavoidably, the acquisition of data for 3D GPR images requires greater expenditures of time and other resources than the collection of 2D radargrams. To construct data for 3D GPR images, a relatively small area (generally $<50\ m^2$) is intensively surveyed with closely spaced (typically 0.1–0.5 m), parallel GPR traverse lines. The relatively dense network of traverse lines is necessary to resolve the geometries and sizes of different subsurface features and to prevent spatial aliasing of the data (Grasmueck and Green, 1996).

The additional resources needed to collect and process GPR data for 3D imaging are often compensated for by more comprehensive spatial coverage and higher resolution of subsurface features (Grasmueck and Green, 1996).

In 3D GPR, data from closely spaced, parallel lines are processed into a 3D image using processing software packages such as RADAN (GSSI, Salem, NH) or GPR-Slice (Geophysical Archaeometry Laboratory, Woodland Hills, CA)[1]. Once processed, arbitrary cross-sections, insets, and time slices can be extracted from the 3D dataset and viewed from nearly any angle. Some software packages allow the observer to rapidly travel through the entire data volume with animated imagery (Grasmueck, 1996). Interactive software packages permit the rapid display of any subsection or block within the surveyed grid. The flexibility of 3D visualizations can greatly facilitate the interpretation of many spatial relationships and the analysis of structural features.

When possible, the use of more than one geophysical method is recommended as each can provide complementary and expanded information, which leads to improved and more accurate interpretations. The methods should complement one another in respect to their ease of use, site coverage, sensitivity to subsurface properties, and resolution of subsurface features. At the intermediate scale, EMI and GPR have been used in tandem to provide subsurface information that complements point-sampling measurements (Stroh et al., 2001; Inman et al., 2002). For example, André et al. (2010) observed a strong agreement between EC_a pattern and spatial soil structural units imaged with GPR. While a one-time use of either EMI or GPR may be useful to map the subsurface heterogeneity, these methods are especially revealing for hydropedologic investigations when used repeatedly over the same area under different moisture conditions or in a time-lapsed approach over a period associated with infiltration or chemical inputs across the soil–landscape (e.g. Toy et al., 2010; Zhu et al., 2010a,b).

This chapter demonstrates the synergistic use of EMI and GPR in hydropedologic investigations in two contrasting landscapes in central Pennsylvania. Electromagnetic induction was used to provide lower resolution but broader spatial coverage of the entire landscapes based on EC_a patterns. Ground-penetrating radar was used to provide a higher resolution of the subsurface architecture along selected traverse lines and small grids. Data were collected using these geophysical methods in a repeated manner to provide information about dynamic processes involving soil moisture change and water flow in the subsurface. Repeated EMI or GPR surveys were conducted over different seasons or following precipitation or artificial infiltration events to identify possible subsurface flow pathways and patterns. Using ancillary information and tools such as global positioning system (GPS), real-time soil moisture monitoring, and in situ observations, data collected with these geophysical

1. Manufacturer's names are provided for specific information; use does not constitute endorsement.

methods were used to enhance the understanding of hydropedologic properties and processes. Specific objectives of this study included: 1) investigations of subsurface soil architecture at the intermediate scale of hillslopes and small catchments and the quantification of the subsurface variability and 2) identification of subsurface preferential flow pathways and patterns and their changes under different moisture conditions. The approaches illustrated and the information gained through this study can help ease the technological bottleneck for subsurface investigations and thus advance hydropedology.

2. MATERIALS AND METHODS

2.1. Study Areas

2.1.1. Kepler Farm

The 19.5-ha Kepler Farm is located in central Pennsylvania in a valley of the Ridge and Valley Physiographic Region of eastern U.S. (Fig. 1). This region is characterized by parallel ridges and valleys that are the erosional remnants of anticlines, synclines, and thrust blocks (USDA-NRCS, 2006). The site consists of several cultivated fields that extend across a prominent limestone ridge, which is situated roughly in the middle of the field. Within the study site, elevation ranges from 373 to 396 m. The depth to limestone bedrock ranges

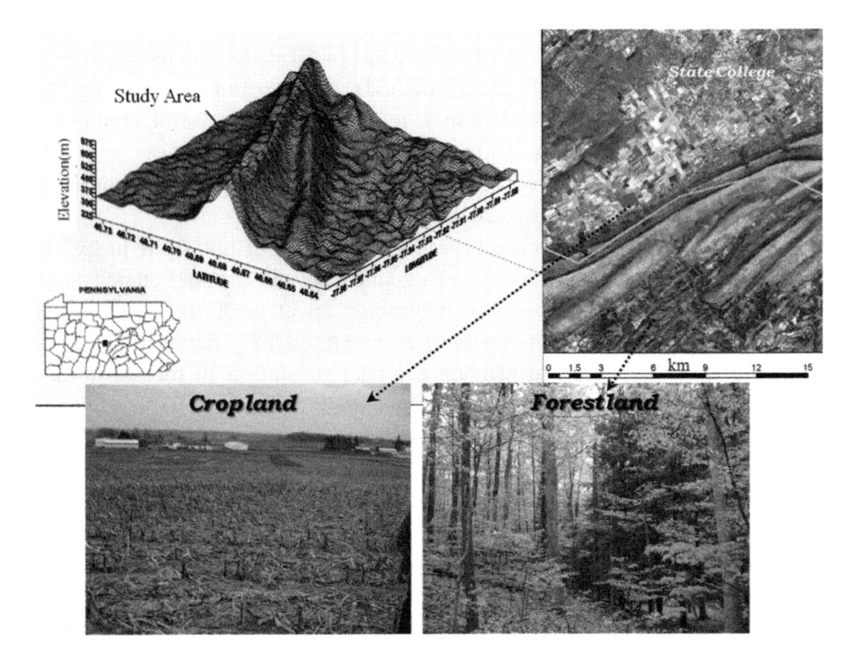

FIGURE 1 Location of the study areas in the Ridge and Valley Physiographic Region of central Pennsylvania. Two contrasting landscapes – a cropland and a forestland – are located on the opposite side of the Tussey Mountain ridge. *(Color version online)*

from <0.25 m on ridge tops to >3 m on foot slopes. The soil/limestone interface is pitted and has an irregular topography caused by both minor and major dissolution features. Joints, fractures, and dissolution features within the limestone bedrock create preferential flow pathways.

The soils on the Kepler Farm have formed in limestone residuum, sandstone colluvium, or alluvium materials. The well-drained, fine-textured Hagerstown (*fine, mixed, semiactive, mesic Typic Hapludalfs*) and Opequon (*clayey, mixed, active, mesic Lithic Hapludalfs*) soils formed in limestone residuum and dominate this landscape. These soils have clayey subsoils that restrict the flow of water and influence subsurface flow (Zhu and Lin, 2009). Depth to bedrock is shallow (<0.5 m) or deep (>1 m) for the Opequon and Hagerstown soils, respectively. Eighty percent of the study area is mapped as consociations and complexes of these two soils on the second-order soil map. The very deep (>1.5 m), well-drained, medium-textured Murrill (*fine-loamy, mixed, semiactive, mesic Typic Hapludults*) soils dominated the lower-lying sideslopes on the south-facing side, which are composed of colluvial materials weathered from sandstone and the underlying limestone residuum. These soils are mapped as consociations and make up about 10% of the study area. The very deep, medium-textured Nolin (*fine-silty, mixed, active, mesic Dystric Fluventic Eutrudepts*) and the very deep, poorly drained, medium-textured Melvin (*fine-silty, mixed, active, nonacid, mesic Fluvaquentic Endoaquepts*) soils have been formed in silty alluvium in a lowland located in the extreme northeastern portion of the study site.

2.1.2. Shale Hills

The 7.9-ha, forested Shale Hills Catchment is located about 5 km from the Kepler Farm on the opposite side of the Tussey Mountain (Fig. 1). The catchment is defined by a narrow, well-defined ridge line and is characterized by moderately steep slope gradients (up to 48%). Elevations range from 256 to 310 m. The catchment is incised into a shale ridge composed of the Rose Hill formation, a member of the Clinton Group of middle Silurian age (Berg et al., 1980). The Rose Hill formation consists principally of thinly bedded, highly fractured, folded, and faulted acid clay shale (Folk, 1960). The catchment is elongated in an east–west direction. Seven well-defined, linear swales of varying dimensions extend downslope and onto the valley floor; five along the south-facing slopes, and two along the north-facing slopes.

The second-order soil survey has the catchment largely mapped as the Berks–Weikert association, on steep (25–70%) slopes, and the Ernest silt loam, on 3–8% slopes. The catchment includes small areas of the Berks–Weikert shaly silt loam, on 15–25% slopes, and the Berks shaly silt loam, on 8–15% slopes. These soils contain large amounts of shale fragments (15–90% by volume) and are underlain by thinly bedded and highly fractured shale (Merkel, 1978). The low electrical conductivity of these soils makes them suited to GPR investigations. However, the steep, forested slopes make this terrain relatively inhospitable to geophysical surveys.

The well-drained, moderately deep (0.5–1.0 m) Berks (*loamy skeletal, mixed, active, mesic Typic Dystrudepts*) and shallow Weikert (*loamy skeletal, mixed, active, mesic Lithic Dystrudepts*) soils have formed in residual material weathered from acid shale on higher-lying slope components. The very deep, moderately well-drained to somewhat poorly drained Ernest (*fine-loamy, mixed, superactive, mesic Aquic Fragiudults*) soils formed in colluvium derived from acid shale along lower foot slopes that adjoin a stream channel. Although not recognized in the general soil survey, areas of Rushtown (*loamy skeletal over fragmental, mixed, active, mesic Typic Dystrudepts*) soils dominate the narrow, incised swales that descend the sideslopes to the valley floor. The very deep, excessively drained Rushtown soils formed in colluvial materials.

Lin et al. (2006b) showed that the soil delineations on the second-order soil map of this catchment are too coarse for hydropedologic studies and that some soil boundary lines are misplaced. They therefore developed a more detailed first-order soil map for the catchment. A high-intensity soil survey was completed for the catchment using standard grid-transect methods (Soil Survey Staff, 1993). Depth to bedrock and landscape position were the principal criteria used to differentiate various soils (Lin et al., 2006b). Within the catchment, the shallow Weikert soils dominate the higher-lying and more sloping linear and convex summit and sideslope components. The moderately deep Berks and the very deep Rushtown soils are in swales and a large portion of east end of the catchment. The very deep Ernest and moderately deep, somewhat poorly drained to moderately well-drained Blairton (*fine-loamy, mixed, active Aquic Hapludults*) soils were mapped on lower-lying foot slopes and along the stream channel.

Deciduous trees (mostly maple [*Acer* spp.], oak [*Quercus* spp.], and hickory [*Carya* spp.]) cover the higher-lying, more sloping areas of the catchment. Conifers (mostly eastern hemlock [*Tsuga Canadensis*]) cover the lower reach of the order-one stream channel that drains the catchment. The whole catchment floor is covered by a mantle of fallen leaves, tree limbs, and undergrowth, which is sporadically thick and impassable.

2.2. Geophysical Surveys

2.2.1. EMI

Geo-referenced EMI surveys were conducted across the study areas at various times using different meters, dipole orientations, and geometries. Meters used included the EM38, EM38-MK2, and EM31 manufactured by Geonics Limited (Mississauga, Ontario), and the Dualem-2 manufactured by Dualem (Milton, Ontario). A Trimble AG114 L-band differential GPS antenna (Trimble, Sunnyvale, CA) was used to georeference EC_a data collected with the EMI meters.[1] An Allegro CX field computer (Juniper Systems, North Logan, UT) was used to record and store both GPS and EC_a data. To help summarize the results of each EMI survey, SURFER for Windows (version 9.0) software

(Golden Software, Inc., Golden, CO) was used to construct images of the EC_a data. Grids of EC_a data were produced using kriging methods with an octant search.

At the Kepler Farm, the first survey was conducted in January 1997 and subsequent surveys were made between 2006 and 2009. In January 1997, two meters were used: the EM38 operated in vertical dipole orientation (EM38V) and the EM31 operated in vertical (EM31V) and horizontal (EM31H) dipole orientations. These pedestrian surveys, which were completed in the station-to-station mode using a 30-m grid interval, resulted in only 243 measurements in each survey. All subsequent surveys were completed on a mobile platform (meters mounted on a sled and towed behind an all-terrain vehicle) with meters operated in the continuous mode and its long axis parallel to the direction of travel. Measurements were collected at a rate of 1/sec. The second survey was completed in March 2006 with a Dualem-2 meter operated in both perpendicular geometry (EMPRP) and horizontal co-planar geometry (EMHCP). The measurement density of this survey was approximately 3×8 m, resulting in >5400 readings. The remaining six surveys were completed from January 2008 to April 2009. These repeated surveys were designed to capture the temporal variation of soil EC_a at different times of a year. The EM38 meter operated in vertical dipole orientation (EM38V) was used in all these six surveys. These surveys were conducted in the same manner, including the traverse paths, with the same measurement density as the survey done in 2006.

A calibration site was established on the ridgetop over an area with exposed bedrock at the Kepler Farm and along a ridge trail within the Shale Hills Catchment. These sites were selected because of their strong electrically resistive nature and very low EC_a value. Before each EMI survey, either the EM38 or the EM31 meter was properly calibrated (instrument zeroed) at the calibration site. The calibrations of EM38 and EM31 meters followed the procedures described by Geonics Limited (1998) and McNeill (1980). We also returned to the calibration site to check for instrument drift after each survey, but no significant drift was observed in our surveys. According to Taylor (2000), calibration of Dualem-2 is not required.

Changes in soil temperature influence EC_a reading. McNeill (1980) noted that EC_a will increase about 2 percent per °C increase in soil temperature. Therefore, it is advisable to "temperature correct" all EC_a measurements, especially when sites are revisited at different times of a year. All EC_a measurements in this study were corrected to a standard temperature of 25 °C using the formula derived by Sheets and Hendrickx (1995). At the Kepler Farm, air and soil temperatures at different depths at the time of each survey were measured in a nearby weather station. At the Shale Hills Catchment, a temperature probe was inserted to a depth of 50 cm at the calibration site. Sufficient warm-up time for the instrument was provided to reduce EMI instrument temperature drift (Robinson et al., 2003). Some of the EMI surveys were conducted during low air temperatures. However, at the Kepler Farm, actual soil temperatures at different depths monitored during the same time period showed that even as air temperature dropped below 0 °C, the soil temperature below 4 cm remained above freezing (Zhu et al., 2010a).

The sensitivity of EMI instruments to soil depth is dependent on soil electrical conductivity and its variation both spatially and vertically along a soil profile. According to Callegary et al. (2007), higher soil electrical conductivity will reduce the exploration depth of EMI instruments. Therefore, the effective exploration depth of a meter may vary at different times and locations within the same landscape. However, under conditions of low-induction numbers (LIN), the depth of exploration is considered independent of soil electrical conductivity (McNeill, 1980). In our study areas, soils are electrically resistive with EC_a values largely less than 25 mS/m (Zhu et al., 2010a). As a consequence, the depth sensitivity of EMI meters should approximate those predicted by the LIN approximation and manufacturer's specifications.

In the Shale Hills Catchment, only the EM38 and the EM38-MK2 meters were used. The size and lightweight of these instruments make them suited for pedestrian surveys on steeply sloping, forested terrains where limited underbrush and ground barriers exist. The meters were operated in the continuous mode with measurements recorded at a rate of 1/s. The long axes of the meter was orientated parallel to the direction of traverse.

2.2.2. GPR

The GPR unit used in this study was the Subsurface Interface Radar (SIR) System-3000 (SIR-3000), manufactured by Geophysical Survey Systems, Inc. (GSSI; Salem, NH).[1] The SIR-3000 consists of a digital control unit (DC-3000) with keypad, SVGA video screen, and connector panel. A 10.8-V lithium-ion rechargeable battery powers the system. The SIR-3000 weighs about 9 lbs (4.1 kg) and is backpack portable. Antennas with center frequencies of 200, 400, and 900 MHz were used at various times in this study. With an antenna, the SIR-3000 requires two people to operate. Daniels (2004) and Jol (2009) discussed the use and operation of GPR.

The RADAN for Windows (version 6.6) software program (GSSI)[3] was used to process the radargrams. Processing included header editing, setting the initial pulse to time zero, color table and transformation selection, distance normalization, range gain adjustments, signal stacking, and migration (see Daniels (2004) and Jol (2009) for discussions of these techniques). Some radargrams were corrected for changes in surface elevation. Using a process known as *surface normalization*, at each distance mark appearing on a radargram, an elevation was assigned to marker file and the vertical scale at that point was accordingly adjusted to changes in topography. Surface normalization helps to improve the interpretative quality of radargrams and the association of subsurface reflectors with landscape components.

Using the Interactive Interpretation Module of RADAN, depth to a subsurface reflector, which was inferred to be soil–bedrock contact in many cases of this study, was semi-automatically picked and numeric information on its position, depth, and signal amplitude was recorded in a layer file. Layer files were exported into Excel worksheets for documentation and analysis.

A relatively new development in GPR is its integration with GPS. When used with a Trimble AgGPS114 L-band differential GPS antenna (Trimble, Sunnyvale, CA), each radar scan can be geo-referenced. The synergistic use of these technologies permits the collection of large, tabular, geo-referenced GPR datasets that can be stored, manipulated, analyzed, and displayed in Google Earth and Geographic Information System. This greatly improves the utility and interpretation of GPR.

Ground-penetrating radar is a time-scaled system. The system measures the time that it takes for electromagnetic energy to travel from an antenna to an interface (e.g. bedrock, soil horizon, and stratigraphic layer) and back. To convert the travel time into a depth scale, either the velocity of pulse propagation or the depth to a reflector must be known. The relationships among depth (D), two-way pulse travel time (T), and velocity of propagation (v) are described by Daniels (2004):

$$v = 2D/T. \tag{1}$$

The velocity of propagation is principally affected by the relative dielectric permittivity (Er) of the profiled material(s) according to Daniels (2004):

$$\mathrm{Er} = (C/v)^2, \tag{2}$$

where C is the velocity of propagation in a vacuum (0.30 m/ns). Typically, velocity is expressed in meters per nanosecond (ns). In soils, the amount and physical state (temperature dependent) of water have the greatest effect on the Er and v.

Based on the two-way pulse travel time to a known subsurface reflector (metal plate) that was buried within the study areas, the velocity of propagation and the relative dielectric permittivity through the upper part of soil profiles were estimated using Eqns (1) and (2).

2.3. Time-Lapsed Geophysical Surveys

Within the Shale Hills Catchment, repetitive GPR surveys were conducted in small grids to observe temporal changes in subsurface reflections associated with soil moisture change over seasons or with water flow through soil profiles by artificially introduced water using either a line or a point source. To complete a GPR grid survey, multiple, closely spaced parallel traverses were completed across the grid with a 400 MHz antenna. Traverse lines were orientated essentially orthogonal to the slope. The interval between successive traverse lines was 10 cm.

Besides GPR signal changes over dry and wet seasons, two experiments are reported here: one in the shallow Weikert soil using a line source of water infiltration (a 1-m long, 10-cm deep trench) and the other in the deep Rushtown soil using a point source (a 15-cm deep, 10-cm diameter hole located within the grid). Repeated GPR scans were conducted along the same traverse lines prior to infiltration and at various times after the infiltration (e.g. 15, 45, and 75 min). All radargrams used to prepare the 3D images were subjected to the same processing procedures, which included header editing, setting the initial pulse to time zero, color table and transformation selection, and distance normalization. After initial processing, each set of radargrams were combined into a 3D radar file, which was migrated and stacked. Migration was used to reduce the noise associated with hyperbolic diffraction patterns and more properly align inclined interfaces.

3. RESULTS AND DISCUSSION

The effective use of GPR and EMI depends on a number of factors, including site conditions, application purposes, and applied methodologies. Different measured physical properties, spatial coverages, and sensing resolutions are involved in GPR and EMI, but both can be used for quantifying soil spatial variability and detecting possible subsurface flow paths – if used and interpreted appropriately. In this study, because of the different soil types and landscape features involved, EMI was better utilized in the 19.5-ha cropland (the Kepler Farm) to identify larger-area soil spatial variability patterns and general flow pathways in the subsurface over seasonal changes, while GPR was better utilized in the 7.9-ha forested catchment (the Shale Hills) to provide a finer resolution of the depth to bedrock and more specific flow paths through interfaces in soil profiles over either seasonal or event-based soil moisture changes. This is because most soils in the Kepler Farm have high clay contents that limited GPR application, while low and relatively invariable EC_a due to the underlying electrically resistive materials, plus challenging terrain with steep slopes and much undergrowth and trees limit pedestrian EMI surveys in the Shale Hills. In both landscapes, repeated or time-lapsed use of EMI or GPR was emphasized to provide more informative outcomes in understanding the soil–landscape architecture and its impacts on subsurface water flow.

In the following, we organize our results from several years of geophysical investigations in these two landscapes. Under each of the tools used, we organize the results around the two objectives of this study: 1) imaging subsurface soil architecture and quantifying its variability and 2) identifying subsurface flow pathways and their seasonal or event-based changes.

3.1. Electromagnetic Induction

3.1.1. Kepler Farm

The magnitude of soil EC_a across the 19.5-ha cropland changed over time, and no two EMI maps surveyed from 1997 to 2009 were exactly the same (Fig. 2a).

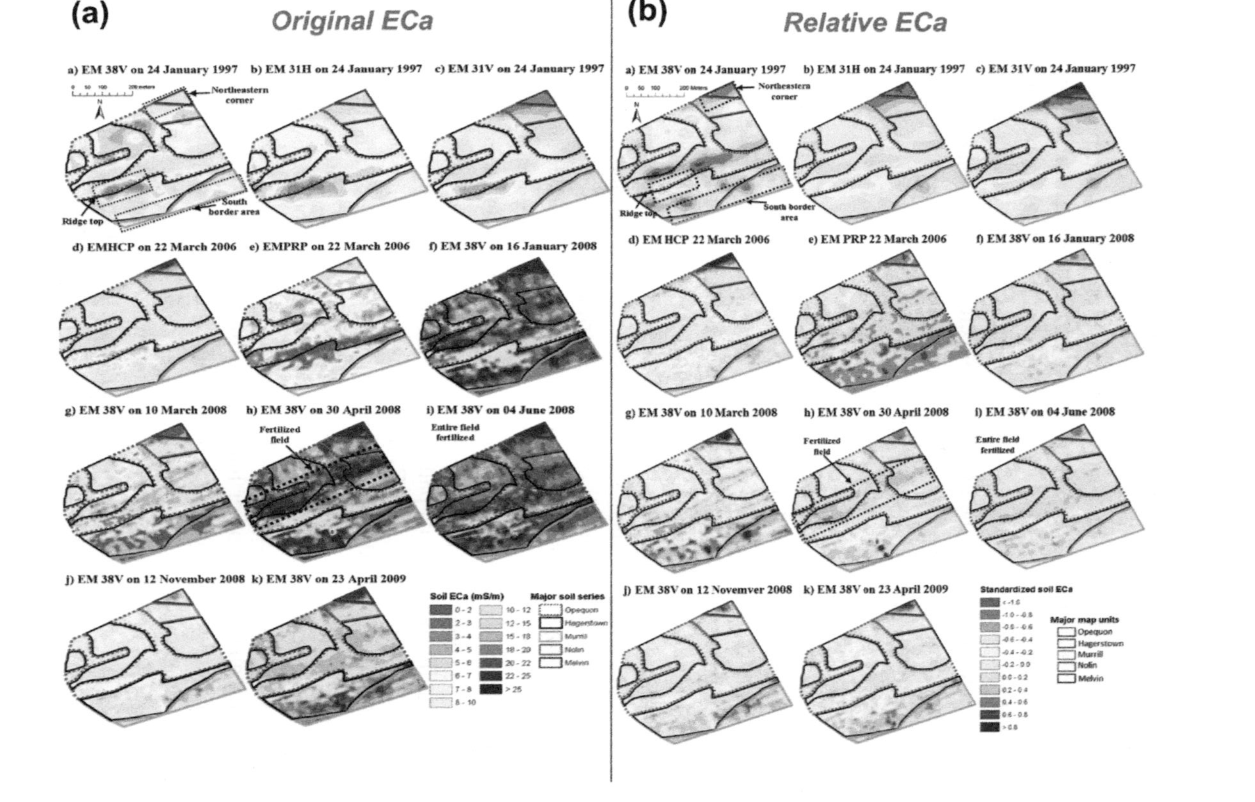

FIGURE 2 (a) Interpolated maps of soil apparent electric conductivity (EC_a) obtained from repeated EMI surveys conducted from 1997 to 2009 at the Kepler Farm. The date and meter used in each EMI survey are indicated at the top of each map; (b) standardized soil EC_a maps for each EMI survey. For each survey, the mean EC_a for the entire landscape was set as 0, and the difference between the measured EC_a and the mean is displayed on each map. Positive and negative values indicate higher and lower values than the overall averaged EC_a, respectively. On each map, the boundaries of soil delineations from the second-order soil map are superimposed. *(Modified from Zhu et al., 2010a.) (Color version online and in color plate)*

However, the relative differences among various areas within the same landscape, after EC_a values were standardized relative to their overall landscape means in each survey (Fig. 2b), appear to be relatively similar and temporally stable among the repeated EMI surveys (regardless of the instrument, instrument geometry, and survey time). This reflects the underlying control of soil-landscape units that are relatively stable over time. The only exception was the partial masking effects caused by the localized application of fertilizer (potash) on 30 April 2008 (map h in Fig. 2a). However, a whole field fertilizer application, as that done on 4 June, 2008 (map i in Fig. 2a), did not noticeably change the overall pattern of the relative EC_a differences among various areas within this landscape. In terms of the EC_a maps collected in 1997 (maps a–c in Fig. 2), their overall patterns were dictated by the coarser resolution of the data collected (i.e. 243 points interpolated in 30-m grid), thus accounting for some of the differences from the other relative EC_a maps (with >5400 points interpolated in 3-m grid).

The stable patterns on the standardized EC_a maps (Fig. 2b) for this landscape could be utilized to refine the second-order soil map, which did not capture well such stable soil-landscape units. For example, consistently low EC_a was measured along higher-lying ridgetop and shoulder slope areas that are associated with the well-drained, shallow Opequon soils (Figs. 2 and 3a). In contrast, high EC_a was consistently measured in the northeastern corner of the study area where the very deep, poorly drained Melvin soils occur (Figs. 2 and 3a) and where the water table was observed at or near the surface during wet seasons. The south-facing hillslope also has higher EC_a in nearly all the EMI surveys (Figs. 2 and 3). This area has lower elevations compared to the north-facing hillslope and consists of the very deep, medium-textured Murrill and deep, fine-textured Hagerstown soils.

Similar, temporally stable spatial patterns can also be observed in maps that summarize the relative difference of EC_a through a temporal stability analysis of the repeated EMI surveys, as shown by Zhu et al. (2010a) and illustrated here in Fig. 3b. In these maps, positive relative differences in soil EC_a are mostly observed in the northeastern corner and on south-facing hillslope areas, indicating that soil EC_a values in these areas were higher than the overall average for the entire landscape. On the other hand, the most negative relative difference occurs on higher-lying ridgetop areas, indicating lower soil EC_a than the landscape's overall average.

The EC_a data collected under relatively dry and wet conditions are further illustrated in Fig. 4, where a map showing the seasonal change between the two sampling periods shows seasonal wet spots and likely subsurface water flow paths (blue- and purple-colored areas). Higher EC_a values were recorded under wetter (average EC_a: 15.8 mS/m) than drier (average EC_a: 13.4 mS/m) soil moisture conditions. Zhu and Lin (2009) and Zhu et al. (2010a) attributed the greater spatial variability of EC_a during wetter periods in this landscape to the lateral redistribution of soil water caused by topographic gradients.

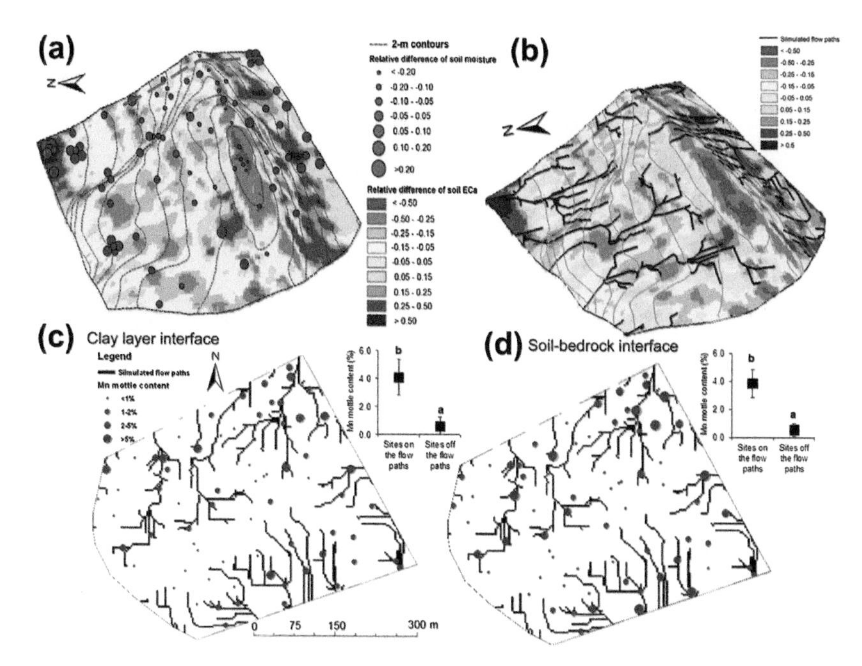

FIGURE 3 (a) A three-dimensional presentation of the relative difference of soil apparent electrical conductivity (EC_a) (color ramps) collected with the EM38V over a three-month period (March, April, and June 2008) in comparison with the total soil moisture storage in the top 1.1-m solum (dots). Negative values indicate lower (drier) than the overall average of the entire landscape, while positive values indicate higher (wetter) than the overall average (from Zhu et al., 2010a); (b) relative difference of soil EC_a collected with the EM38V over a relatively wet season (between 16 January and 10 March, 2008) in comparison with simulated subsurface lateral flow paths. The red-colored dashed line in (a) and (b) approximates the location of the GPR transect shown in Fig. 7. (c) and (d) depict the observed Mn content in the soil profiles at the clay layer interface (c) and the soil–bedrock interface (d) in relation to the simulated concentrated flow paths (from Zhu and Lin, 2009). In the insets, Mn content on and off the simulated flow paths is compared. Bars with different letters indicate significant statistical difference at the $p < 0.05$ level. *(Color version online and in color plate)*

The elevated EC_a in the spring was mainly attributed to wetter soil conditions. When the whole Kepler Farm was divided into south- and north-facing slopes using the ridgetop as a divide (Fig. 4), EC_a was higher and more variable on the south-facing slope in both sampling periods, but especially in the wet period. This is consistent with the soil distribution pattern discussed above.

Zhu and Lin (2009) simulated subsurface lateral flow paths for the Kepler Farm, and found that soil EC_a was higher and temporally more unstable in areas close to the simulated flow paths (Fig. 3). Based on over 145 cores extracted and described for this landscape, Zhu and Lin (2009) observed that subsurface lateral flow paths were associated with larger amounts of Mn in soil samples at the interface with the clayey subsoil and the underlying bedrock in this

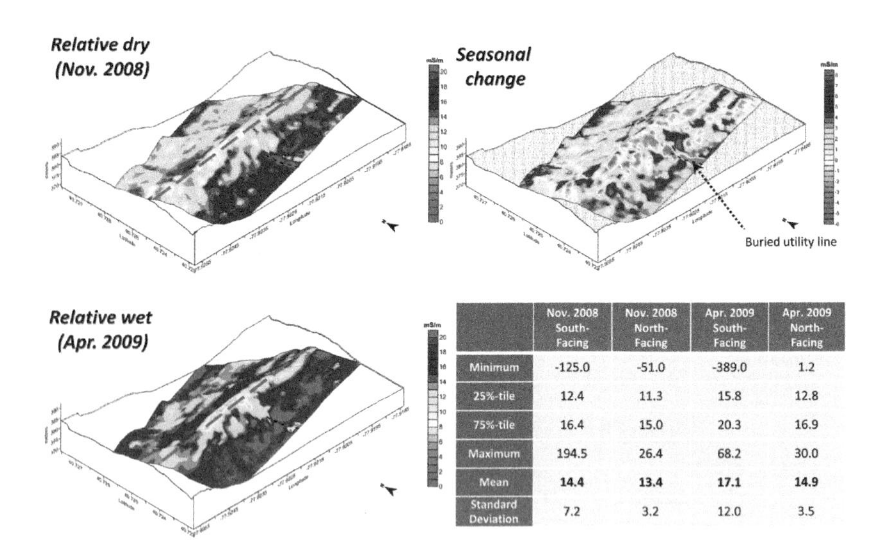

	Nov. 2008 South-Facing	Nov. 2008 North-Facing	Apr. 2009 South-Facing	Apr. 2009 North-Facing
Minimum	-125.0	-51.0	-389.0	1.2
25%-tile	12.4	11.3	15.8	12.8
75%-tile	16.4	15.0	20.3	16.9
Maximum	194.5	26.4	68.2	30.0
Mean	**14.4**	**13.4**	**17.1**	**14.9**
Standard Deviation	7.2	3.2	12.0	3.5

FIGURE 4 Spatial distribution of EC_a measured with the EM38 meter operated in the deeper-sensing VDO during relatively dry (Nov. 2008) and wet (Apr. 2009) periods (left) and their difference (right). The arrow on the right indicates an small area with a buried utility line that caused anomalous EC_a readings. The inset table shows the statistical summary of the EC_a values for each survey and dominant slope aspects (divided by the dashed line indicated on the left). *(Color version online and in color plate)*

landscape (Fig. 3c and d). Significant differences ($p < 0.05$) were observed in the relative change in soil EC_a for areas 0–5, 5–10, 10–15, and >15 m away from the simulated flow paths during the wetter period, but not during the drier period (Zhu et al., 2010a). Although this statistical analysis suggests that locations closer to simulated flow paths have increasingly higher soil EC_a, this pattern is not completely clear on the soil EC_a maps. Subsurface flow paths could not be pinpointed on the soil EC_a maps (Fig. 3) probably because of their limited spatial resolution and that of the EMI meters. However, in some areas with high relative differences in soil EC_a, denser network of simulated flow paths were observed (Fig. 3). Besides the spatial resolution issue, complex interactions among multiple, spatially varying soil properties and landscape features that influence soil EC_a readings, such as terrain attributes, depth to bedrock, soil texture, and management practices, could have also masked subsurface flow paths on the interpolated soil EC_a maps (Zhu et al., 2010a).

3.1.2. Shale Hills

The Shale Hills Catchment is characterized by exceedingly low and relatively invariable EC_a (Figs. 5 and 6). Within this catchment, the very low EC_a reflects the electrically resistive nature of the soil and bedrock, and the low ionic concentration of the soil solution. Reconnaissance EMI surveys conducted in October 2005 (dry soil conditions) and March 2006 (wet soil conditions)

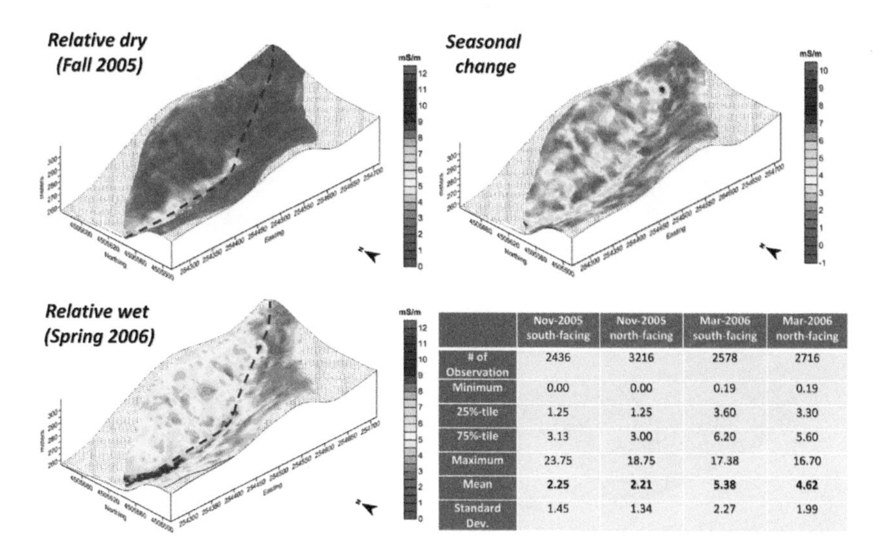

	Nov-2005 south-facing	Nov-2005 north-facing	Mar-2006 south-facing	Mar-2006 north-facing
# of Observation	2436	3216	2578	2716
Minimum	0.00	0.00	0.19	0.19
25%-tile	1.25	1.25	3.60	3.30
75%-tile	3.13	3.00	6.20	5.60
Maximum	23.75	18.75	17.38	16.70
Mean	**2.25**	**2.21**	**5.38**	**4.62**
Standard Dev.	1.45	1.34	2.27	1.99

FIGURE 5 Results of two EMI surveys conducted in the Shale Hills Catchment in October 2005 (dry condition) and March 2006 (wet condition) using EM38 meter with a density of about 750 measurements per ha. However, sampling was not uniform throughout the catchment and was not necessarily consistent between the two surveys because of steep terrain and trees making it challenging to traverse the landscape. On some lower-lying areas within the catchment, GPS data were either lost or degraded by multipath distortion and masking problems caused by the steep terrain and dense vegetation canopy. At the time of the October 2005 survey, soils were dry and stream flow was restricted to the lower reaches of the channel. Soils were moist at the time of the March 2006 survey, and stream flow was observed throughout the entire reach of the channel. The dashed line approximately separates the catchment into south- and north-facing hillslopes. The inset table shows the statistical summary of the EC_a values from each survey. *(Color version online and in color plate)*

showed that EC_a ranged from about 0 to 24 mS/m. However, over most of the catchment, EC_a did not vary by more than 4 mS/m. In spite of the low and relatively invariable EC_a, temporal differences in EC_a were noticeable in this landscape: higher and more variable EC_a data were collected in the wet period than that in the dry period (Fig. 5). In the EC_a map obtained in the wet period, several weakly expressed linear spatial patterns of relatively higher EC_a extend upslope from the stream channel and identify the general locations of swales.

Collectively, the spatial EC_a patterns shown in Fig. 5 suggest two major, temporally stable units in the catchment: the valley floor and higher-lying slope components. Lower EC_a values were recorded on the sideslopes and summit areas where the well-drained Weikert, Berks, and Rushtown soils are dominant, while higher EC_a values were measured along the valley floor where the somewhat poorly drained Ernest soils are dominant. While individual areas of Weikert, Berks, and Rushtown soils were indistinguishable with EMI at this landscape scale, they collectively form a distinct soil-landform unit having EC_a that ranges from 0 to 8 mS/m. Along the stream channel, persistently higher

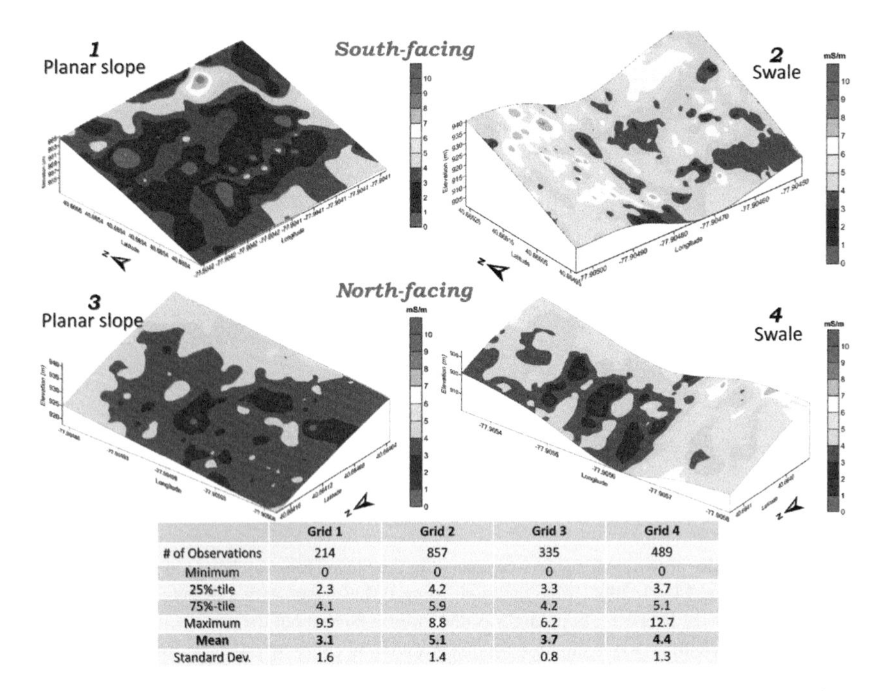

	Grid 1	Grid 2	Grid 3	Grid 4
# of Observations	214	857	335	489
Minimum	0	0	0	0
25%-tile	2.3	4.2	3.3	3.7
75%-tile	4.1	5.9	4.2	5.1
Maximum	9.5	8.8	6.2	12.7
Mean	3.1	5.1	3.7	4.4
Standard Dev.	1.6	1.4	0.8	1.3

FIGURE 6 Distribution of EC_a values across selected soil–landscape components (linear sideslope and concave swale) on south- and north-facing slopes in the Shale Hill Catchment. These grid surveys were completed in November 2010 under relatively dry soil moisture conditions. Grids 1 and 3 were located on plain sideslopes dominated by the shallow Weikert soils. Grids 2 and 4 were located in swales dominated by the deep Berks–Rushtown catena. Grids 1 and 2 and Grids 3 and 4 were located on south- and north-facing slopes, respectively. The inset table shows the comparison of EC_a values collected on each grid. *(Color version online and in color plate)*

patterns of EC_a, with a range from 5 to 24 mS/m, are apparent. The higher clay content of the Ernest soils lowers the permeability, retards deep drainage, and is responsible for the formation of perched high water table during wet periods. The higher clay content and the wetter soil conditions are assumed responsible for the higher EC_a along the stream channel.

More in-depth EMI surveys were conducted in distinct soil-landscape units within the catchment to capture short-range variability in EC_a caused by variations in soil and hydrologic properties. Four small grids were established on two selected soil-landscape units on both south- and north-facing hillslopes: 1) linear sideslopes dominated by the Weikert soils and 2) concave swales composed of the Berks–Rushtown catena (Fig. 6). The grids were variable in size and ranged from about 0.24 to 0.32 ha for the Weikert-dominated sites (grids 1 and 3), and from 1.05 to 1.78 ha for the Berks–Rushtown-dominated sites (grids 2 and 4). Each grid was surveyed uniformly with the EM38 meter along closely spaced traverse lines under relatively dry soil moisture conditions. As expected, the seasonally low soil moisture contents resulted in low and relatively invariable EC_a (Fig. 6). The two grids located on the Weikert-dominated linear sideslopes had a slightly lower average EC_a than the two grids located on the Berks–Rushtown catena in swales. Note that the data collected at the grid 1 contained a cluster of anomalous EC_a measurements caused by the EM38 meter passing too close to installed instruments as noted during the field data collection. When these values were excluded, the grids located on linear sideslopes showed a lower range in EC_a values than the other two grids located in the swales, which was attributed to shallow soils and low clay and moisture contents. For the surveys of the Berks–Rushtown-dominated swales (grids 2 and 4), higher EC_a measurements were recorded on the convex shoulder slopes into the swale. This pattern was especially evident on the western shoulder slopes into the swales and may reflect structural controls on soil moisture distribution.

3.2. Ground-Penetrating Radar

Knowledge of the depth, arrangement, and geometry of subsurface soil horizon interfaces and the soil–bedrock interface is important to understanding subsurface water flow. In the absence of outcrops or exposures, GPR is the tool of choice for noninvasively imaging subsurface architectures at the intermediate scales (Cagnoli and Russell, 2000; Dagallier et al., 2000).

3.2.1. Kepler Farm

Most soils in the Kepler Farm have high clay contents (Zhu et al., 2010a), and thus are considered to have low to moderate potential for GPR use (http://soils.usda.gov/survey/geography/maps/GPR/index.html). Because of this limitation, GPR was not used extensively in this crop field. Nevertheless, limited GPR transects conducted in the Kepler Farm have revealed a highly variable subsurface architecture. In a nearly 200-m long transect shown in Fig. 7, the shallow Opequon soils and deep Hagerstown soils are irregularly intermixed, which is due to the irregular distribution and weathering of the underlying limestone bedrock. Based on a total of 12,850 measurements of depth to bedrock interpreted from the radargram, the depth to bedrock along this transect ranged from nearly 0 to over 240 cm, with an overall average of 128 cm. Overall, based on standard soil depth classes used in soil surveys, this GPR traverse line is 25% shallow, 10% moderately deep, 16% deep, and 50% very deep to bedrock. The color-coded pattern of this GPR traverse line in Fig. 7a and the irregular radar reflection pattern in Fig. 7b clearly demonstrate the complexity of the subsurface architecture in this landscape, which has a significant impact on water flow. This complexity is also reflected in the spatial EC_a patterns shown in Fig. 3a and b along this traverse line. Zhu and Lin (2009) have also noted the importance of bedrock topography and clay layer interface

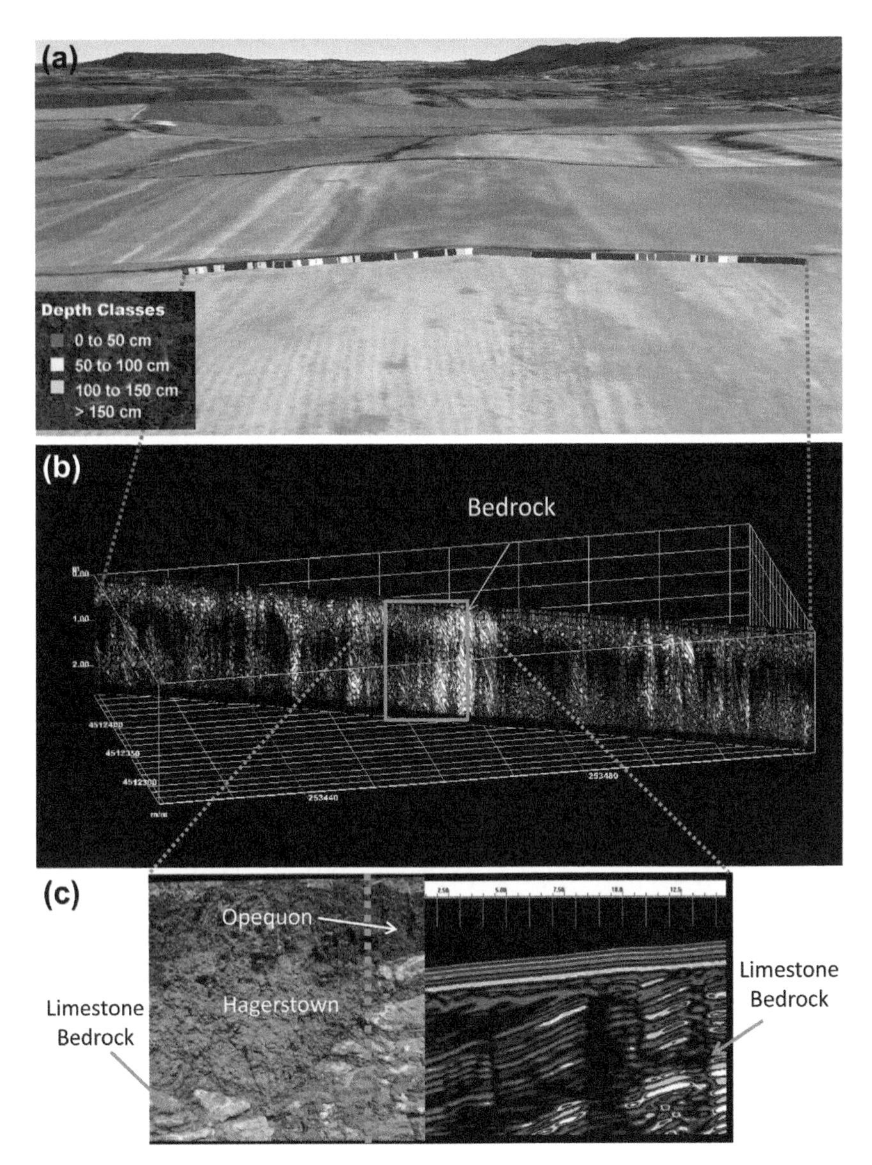

FIGURE 7 Google Earth image (a) of a nearly 200-m long GPR transect on the Kepler Farm (see its relative location on the farm in Fig. 3). Different colors are used to identify different soil depth classes. The depth to limestone bedrock was interpreted from the radargram (b). This area is mapped on the second-order soil map as Opequon–Hagerstown complex (3–8% slopes) and Hagerstown silt loam (3–8% slopes), where the shallow Opequon soils and the deep Hagerstown soils are irregularly intermixed – see an example soil profile photo shown in (c). Both of these soils were formed in residuum weathered from limestone. *(Color version online and in color plate)*

on subsurface flow pathways and patterns in this landscape, which are linked to the highly heterogeneous depth to bedrock revealed by the GPS transect in Fig. 7.

3.2.2. Shale Hills

The synergism of GPR and GPS allows the expedient mapping of the depth to bedrock at the intermediate scales. Fig. 8b shows a Google Earth image with several GPR transects collected with GPS in various parts of the Shale Hills Catchment. Color codes have been used to identify the depth to bedrock based on standard soil depth class criteria. The frequency distribution of the depth to bedrock for these transects is summarized in Table 1. A comparison of GPR interpreted vs. in situ observed (via augering or trenching) depth to bedrock is shown in Fig. 8d. As these data show, the depth to bedrock is quite variable across this catchment. However, after extensive GPR surveys across the catchment, combined with in situ measurements of the depth to bedrock, the following patterns have been recognized for this catchment (Fig. 8b and c):

1) Convex summit areas and linear upper sideslopes have principally shallow (<50 cm, colored white) and moderately deep (50–100 cm, colored red) soils, with the south-facing sideslopes (transects 1 and 2) having slightly deeper soil depths than the north-facing counterpart (transect 5);
2) The swales have a greater amount of very deep (>200 cm, colored blue) and deep (150–200 cm, colored green) soils (transects 3, 4, and 8);
3) Higher-lying sideslopes have mostly shallow to moderately deep soils (transect 7), while lower-lying sideslopes tend to have more moderately deep and deep soils (transect 6); and
4) The lower-lying and less sloping valley floor has chiefly deep and very deep soils (Fig. 8a and c).

A non-geo-referenced radargram from a 30-m transect that traversed the valley floor is shown in Fig. 8a. This radargram has been surface normalized to show the topography of the valley floor. Along this traverse line, the averaged depth to bedrock is 127 cm, with a range of 83–233 cm. Several prominent, discontinuous, linear reflectors are evident in the radargram between the 12 and 27 m distance marks. These high-amplitude (colored white and gray) reflectors are believed to represent layers of contrasting alluvium and/or colluvium. These layers differ in grain-size, density, and amount of coarse fragments. Such contrasting layers will influence the flow of water.

Additional GPR transects from the Shale Hills Catchment are illustrated in Fig. 9. These radargrams have been surface normalized to show the topography of the sideslope and swale. Fig. 9a is a radargram from a transect that traversed a linear, upper sideslope. In this radargram, the soil–bedrock

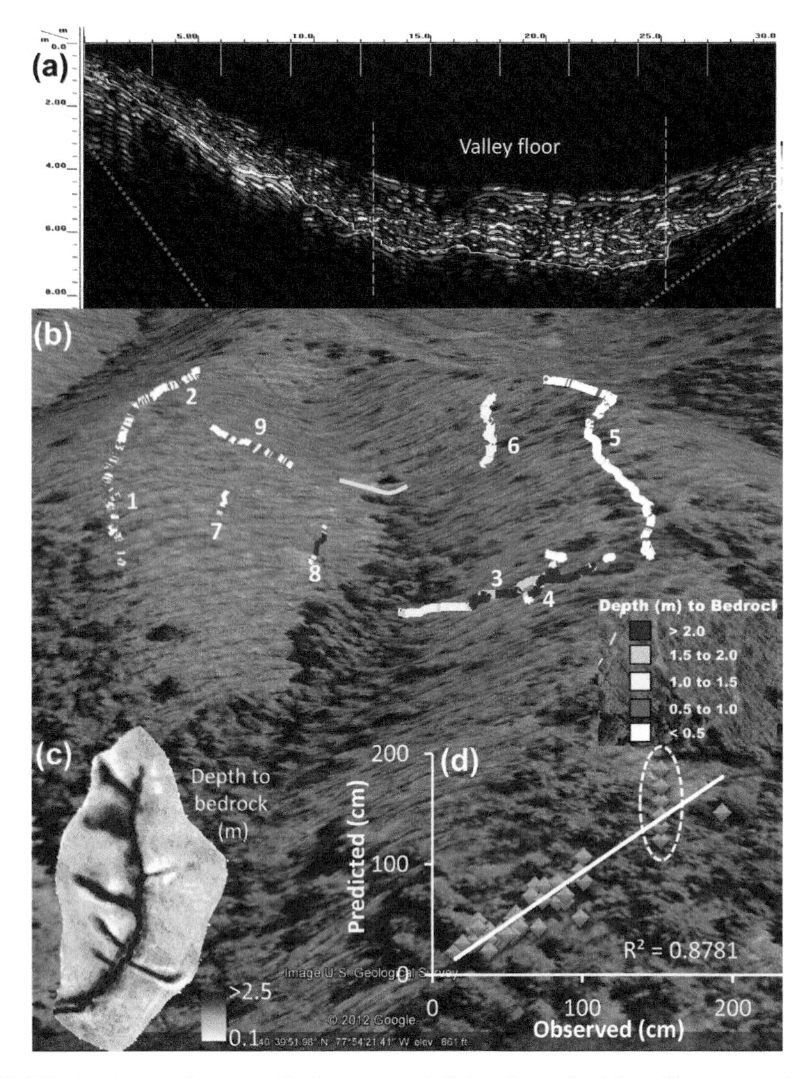

FIGURE 8 (a) A surface normalized radargram (relief of about 4.2 m) for a 30-m traverse line located near the beginning of the stream channel in the Shale Hills Catchment. The narrow, white-colored line on the radargram indicates the interpreted soil–bedrock interface. The two vertical, dashed lines approximate the edges of the valley floor; (b) nine GPR transects are color-coded to indicate the depth to bedrock as interpreted from the radargrams. Transects 1, 2, and 5 are along the perimeter trail of the catchment on convex summit areas; transect 3 is along the long axis and transects 4 and 8 are orthogonal to the long axis of swales (note the swales in the catchment are not evident on this Google Earth image); transect 6 is along a lower-lying sideslope; transect 7 is along the upper end of a swale; transect 9 is along a linear hillslope. (c) The depth to bedrock map for the entire Shale Hills Catchment compiled from over 300 point observations and over 50 GPR transects; and (d) a comparison of GPR interpreted (predicted) and in situ observed (ground truthed) depth to bedrock measurements, showing a reasonable linear fit with $R^2 = 0.88$. Note the data points within the circled area were all limited to 1.5 m – the maximum depth of the auger used in ground truthing. *(Color version online and in color plate)*

TABLE 1 Frequency Distribution (Fraction of Total Number of Observations) of the Depth to Bedrock for Eight GPR Transects Collected from the Shale Hills Catchment (See Fig. 8 for the Locations of These Transects). The Classes for Depth to Bedrock are Based on the Standard of U.S. Soil Survey

Transect #	1	2	3	4	5	6	7	8
Total # of Observations	14359	10458	12523	7120	16026	11082	1547	584
Depth to Bedrock (m)								
<0.5	0.47	0.49	0.06	0.28	0.85	0.02	0.69	0.30
0.5–1.0	0.51	0.48	0.12	0.06	0.15	0.19	0.17	0.19
1.0–1.5	0.02	0.02	0.17	0.07	0.00	0.78	0.08	0.09
1.5–2.0	0.00	0.00	0.28	0.22	0.00	0.01	0.06	0.07
>2	0.00	0.00	0.36	0.38	0.00	0.00	0.00	0.35

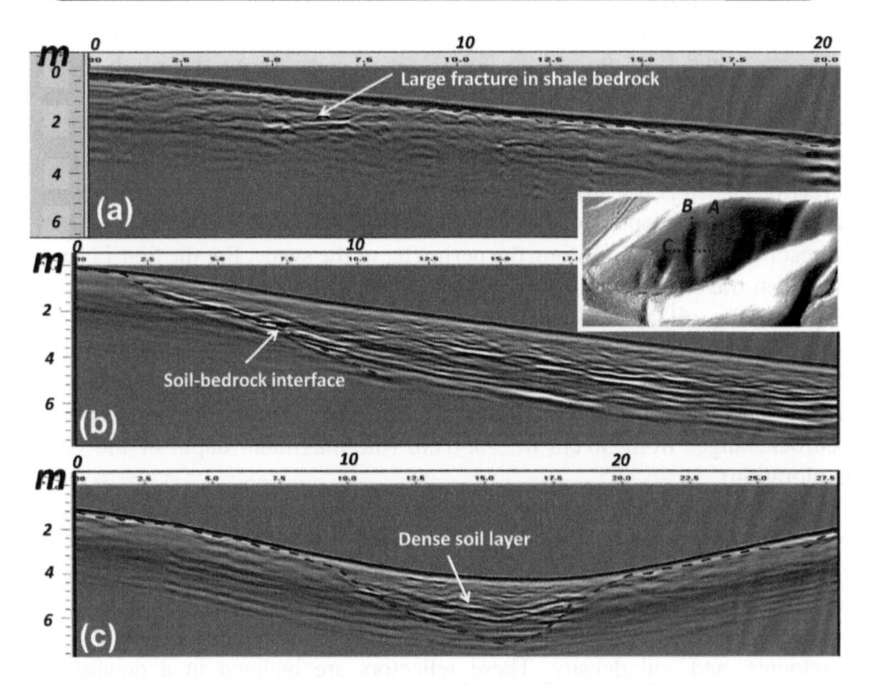

FIGURE 9 Three surface-normalized radargrams from the Shale Hills Catchment. Radargrams were collected along a linear sideslope (a), a concave swale (b), and a cross-section of a swale (c) that is orthogonal to the transect (b). The red-colored, segmented lines indicate the interpreted depth to the soil–bedrock interface. The inset image shows the catchment outline and the locations of the three transects. *(Color version online and in color plate)*

interface (approximated with a red-colored, segmented line) appears as moderate amplitude, subsurface reflection that directly underlies the surface reflection. Along this traverse line, the averaged interpreted depth to bedrock is 41 cm, with a range of 29–56 cm. While relatively shallow, the topography of the soil–bedrock interface appears wavy to irregular with noticeable flexures in the underlying bedrock and pockets of slightly deeper soil materials. This subsurface microtopography influences the distribution and flow of water.

In the radargrams shown in Fig. 9, below the soil–bedrock interface, there are few reflectors in the shale bedrock. Observed fractures and cleavages in the Rose Hill formation are very thin (0.1–1.3 cm) and closely spaced. In general, these fractures and cleavages are not adequately resolved with antenna wavelengths used in these transects (200 MHz antenna). These features produce scattering losses, which dissipate the radiate energy. Larger fractures and cleavages that are filled with contrasting materials are more detectable (see the fracture indicated by arrow in Fig. 9A). The detection of fracture and bedding plane often depends on the width and the moisture content of the material filling the discontinuity (Grasmueck, 1996; Olhoeft, 1998; Buursink and Lane, 1999; Lane et al., 2000).

Figure 9 also contains radargrams that were conducted parallel (Fig. 9b) and orthogonal (Fig. 9c) to the long axis of a swale on the south-facing hillslope. In both radargrams, the soil–bedrock interface has been approximated with a red-colored segmented line. In this soil catena, the depth to bedrock rapidly deepens toward the centerline of the swale and in downslope directions as soils transition from Weikert to Berks to Rushtown (which are differentiated by depth to bedrock). In general, the depth to bedrock is shallowest near the head and along the perimeter of the swale. For the first 11 m of the 27-m traverse line that was conducted along the centerline of the swale (Fig. 9b), the soil–bedrock interface can be followed in a downslope direction as it plunges to greater depths beneath a thickening column of colluvium. For the remaining 17 m of this traverse line, the colluvium–bedrock interface is below the depth of radar exploration (3.5 m). Along this traverse, the depth to bedrock ranges from 36 cm to >350 cm (the maximum depth of the GPR exploration).

In Fig. 9b, multiple, inclined, segmented, weak to high-amplitude, linear reflections are evident within the colluvium. These subsurface reflections represent interfaces separating contrasting layers of colluvium. The contrast among these layers is caused by differences in dielectric properties associated with differences in water content, particle-size distribution, amount of shale fragments, and soil density. These reflectors are inclined in a downslope direction and appear continuous over short distance. The presence, geometry, and continuity of these layers will influence water flow.

In Fig. 9c, the depth and the topography of the soil–bedrock interface along the Rushtown–Berks–Weikert catena are revealed and can be used to characterize water-restricting layers and flow paths (see the next section for more details). On this radargram, higher-lying portions are dominated by the shallow Weikert soils. As the swale is entered, the soil–bedrock interface plunges toward the centerline of the swale and the soil transition from the shallow Weikert to the very deep Rushtown over short distances. In this cross-section, the swale is filled with slightly wavy, segmented, linear reflectors that were later determined as soil layer of contrasting density. These reflectors represent contrasting layers of colluvium. These layers appear continuous over short distances (1–5 m). The presence, continuity, and geometry of these layers will affect the flow of water. Along this traverse line, the depth to bedrock averages 86 cm with a range from 15 to 241 cm. Based on the radar data, the depth to bedrock along this traverse line is 40% shallow (Weikert), 26% moderately deep (Berks), 15% deep, and 20% very deep (Rushtown) soils.

3.2.3. Time-Lapsed GPR for Detecting Subsurface Flow

Repeated or time-lapsed GPR can be used to characterize changes in soil water content over time, infer subsurface hydrologic processes, and study the impacts of soil architecture and flow dynamics on GPR signals. Subsurface flow is influenced by soil, stratigraphic, and lithologic layering, and complex macropore systems such as those generated by shrink–swell, animal borrows, and plant roots. While GPR cannot differentiate various types of macropores and its detection resolution is coarser than most macropores in soils, time-lapsed GPR can be used to capture the impacts of subsurface architecture on water flow over seasons and individual rainfall/infiltration events.

Zhang et al. (2012a,b) demonstrated the usefulness of GPR in combination with real-time soil water monitoring to reveal seasonal or event-based GPR signal changes in relation to soil profile architecture in the Shale Hills Catchment. They showed that GPR reflections from the soil–bedrock interface and the weathered–unweathered rock interface become more intermittent as the Weikert soil became wetter (Fig. 10). This scattering of GPR reflection under wet conditions is associated with the nonuniform distribution of water that had infiltrated the fractured shale bedrock (and possibly along tree root channels as well). This nonuniform distribution of water created hyperbolas and crisscrossing patterns along the soil/bedrock interfaces, which breaks the continuity of radar reflections. The 2D and 3D radargrams in Fig. 10C highlight the more broken appearance under wet than under dry conditions of the weathered–unweathered bedrock interfaces in the Weikert soil. In contrast, GPR reflections along the BC-C horizon interface became clearer as the very deep Rushtown soil became wetter (Fig. 10). This is due to a more uniform distribution of water along this water-restricting interface during wet conditions. The accumulating water led to a greater contrast in dielectric permittivity and thus stronger GPR reflections from the soil horizon interface.

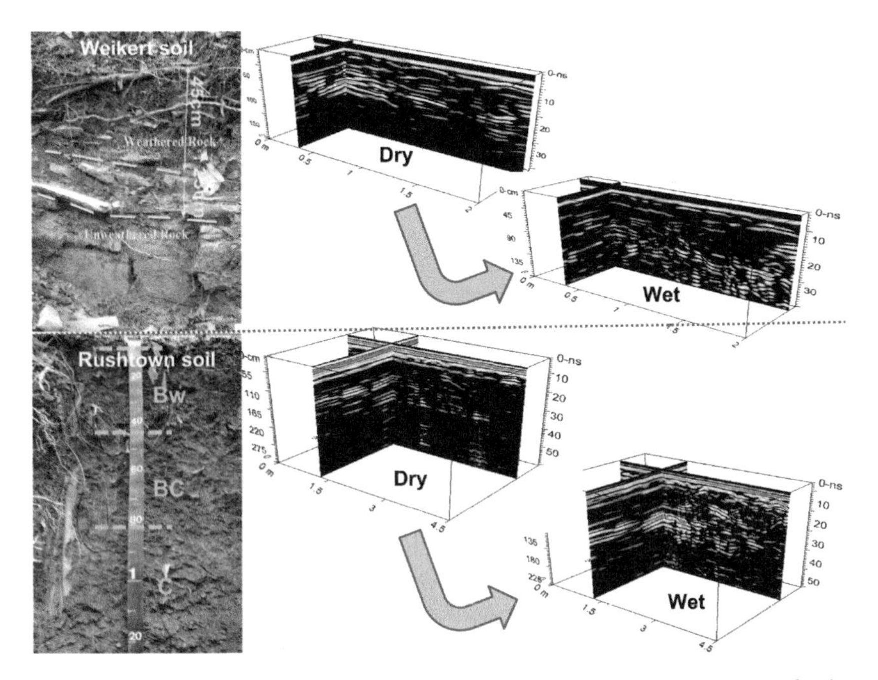

FIGURE 10 Photographs and the changes in 3D GPR images from dry to wet seasons for the shallow Weikert (upper panel) and the deep Rushtown (lower panel) soils and their horizonation. *(Color version online and in color plate)*

The GPR images of the Rushtown soil in Fig. 10C highlight the more continuous and stronger reflections from the BC-C interface under wet conditions.

Artificial infiltration experiments were used to demonstrate the use of time-lapsed GPR in capturing event-based flow dynamics in relation to subsurface architecture. Using a line source of infiltration into an area of Weikert soils, clear lateral flow through the soil and the fractured shale was captured by comparing radargrams collected before and after the introduction of water, as shown in Fig. 11.

Figure 12 contains two sets of four 3D GPR images illustrating the effects of water infiltration and redistribution from a hole in an initially dry Rushtown soil. These 3D images were prepared from multiple radargrams collected over a small grid site. Separate 3D images show reflection patterns: 1) prior to wetting, 2) 15 min after constant head, 3) 45 min after constant head, and 4) 75 min after constant head in the infiltration hole. In each of these radargrams, reflectors vary in amplitude and expression, and are assumed to represent different strata within the colluvium. Strata vary in grain-size distribution, rock fragments, moisture, and/or density. The higher the amplitude of the reflected signals (high-amplitude reflections are colored white, blue, green, and yellow in

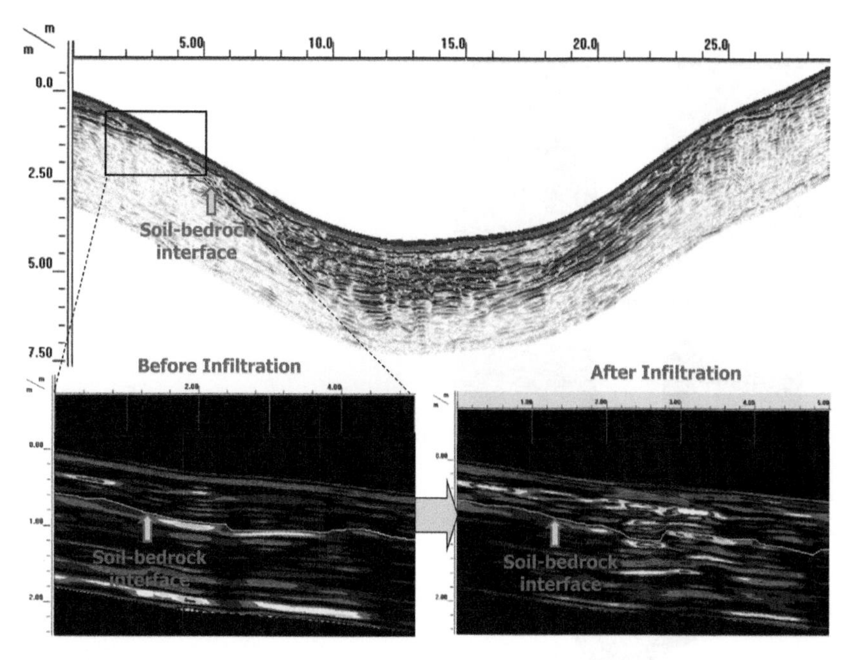

FIGURE 11 A surface-normalized radargram collected using a 200 MHz antenna across a swale (orthogonal to its long axis). The green-colored line approximates the interpreted soil–bedrock interface. The zoom-in images show a section of the hillslope before (left) and 15-min after (right) ten gallons of water infiltrated into the soil between the 2 and 3 m distance marks on the radargram. *(Color version online and in color plate)*

Fig. 12), the more abrupt and contrasting the difference in dielectric properties across the interfaces.

In Fig. 12, approximately 15 min after the attainment of a constant head in the infiltration hole, a high-amplitude reflection has appeared near the area labeled as "A." This reflection is believed to be caused by infiltrating water. In the same image, a relatively high-amplitude subsurface spatial pattern (labeled as "B") appears to have enlarged and moved downslope along the base of the cutout cube. Approximately 45 min after the infiltration, the previously recognized high-amplitude reflection ("B" in the 15-min image) appears to have weakened in amplitude (labeled as "D" in the 45-min image) and extended downward. At this moment in time, two subsurface interfaces (labeled as "E" and "F" in the 45-min image), which represent denser layers of colluvium, become highly expressed along the lower, front sidewall of the cube, suggesting that water has moved along these formerly drier surfaces. Approximately 75 min after the attainment of a constant head in the infiltration hole, a high-amplitude reflection reappears beneath the infiltration hole (labeled as "G" in the 75-min image). The two subsurface interfaces ("E" and "F") identified earlier are now less well expressed. This can be explained by the wetting front

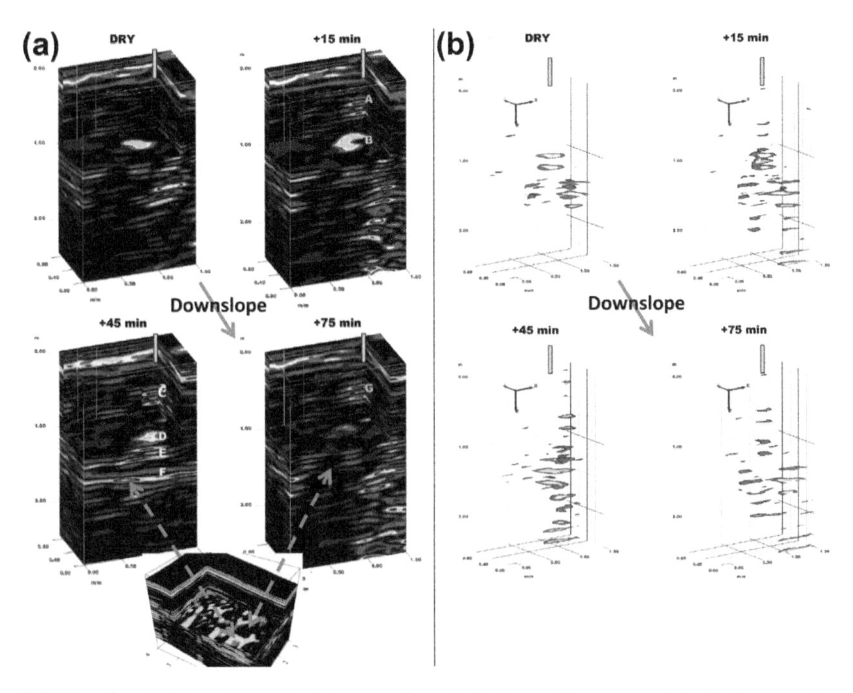

FIGURE 12 (a) Three-dimensional images of a grid site located in an area of the Rushtown soils. Water was introduced through a 5-cm diameter and 15-cm deep cylindrical hole (approximated by a green-colored column near the upper right-hand corner of each image) under the initial dry condition. The repeated GPR grid surveys were conducted before, and 15, 45, and 75 min after achieving constant head in the infiltration hole. The additional cutout of 3D GPR cube at the bottom shows a water-restricting dense soil layer that is heterogeneous and sloping. (b) Hilbert transformed 3D images that highlight the areas with strong radar reflections detected, which indicate water. *(Color version online and in color plate)*

moving laterally downslope above the denser subsurface layer and/or vertically downwards through the denser layer, which was observed to be heterogeneous with pockets and channels of less dense materials (so the overall denser layer is like a Swiss cheese).

The same composite radargrams are shown in Fig. 12b after Hilbert transformation. Hilbert transformation uses the magnitude of the return signal to decompose multiple hyperbolic reflections into more compact and representative forms (Daniels, 2004). Apparent on these transformed images is the vertical extension of high-magnitude signals after the infiltration experiment has begun, and later the subsequent partial decay of the signal magnitudes as water permeated the subsurface. The above flow patterns captured by the time-lapsed GPR suggest that water first has moved vertically through the soil until reaching a denser layer where it then moved laterally in a downslope direction, but also moved further down vertically through the less dense areas in the otherwise dense layer. Such interpretations are consistent with field observations and other GPR work done in this catchment (Zhang et al., 2012a,b).

4. SUMMARY AND CONCLUSIONS

Ground-penetrating radar can provide high-resolution subsurface information, which may be used to estimate depths to soil horizons and bedrock surfaces that restrict, redirect, and/or concentrate water flow. Compared with GPR, EMI provides lower resolution, but can be used to study soil and hydrologic properties across a broader spectrum of soil types and spatial extents, especially if used in a repeated manner in the same area. The synergistic use of these tools is demonstrated in this study designed to better understand 1) subsurface architecture and its spatial variability and 2) subsurface flow pathways and patterns in contrasting soils and landscapes. Within the two contrasting landscapes investigated, EMI characterized major spatial–temporal EC_a patterns that correlated with variations in soil-landform units and soil moisture dynamics. Ground-penetrating radar then provided higher-resolution spatial records in selected transects or grids, which illustrated the subsurface architecture of soils and bedrocks and their impacts on subsurface water flow.

Apparent electrical conductivity is often used as a surrogate measure for spatially-varying soil physiochemical properties that are difficult, time consuming, or expensive to measure. Because of the large number of measurements that can be obtained within relatively short time, maps derived from EC_a data provide a higher level of resolution than existing second-order soil maps. At the intermediate scales, spatial EC_a patterns provide a means for subdividing soil-landscapes into distinct units that reflect spatially-varying soil and hydrologic properties and processes. Such delineations can help improve the understanding of complex landscape–soil–hydrology relationships, and aid the selection of optimal location for monitoring or sampling. Ground-penetrating radar provides detailed information of the spatial variability of subsurface features, which, when combined with other techniques (e.g. soil moisture monitoring) and used in a time-lapsed manner, can be used to detect subsurface flow pathways and patterns.

The effective use of GPR and EMI, however, is highly site- and application-specific. In this study, because of different soil types and landscape features, EMI was better utilized at the Kepler Farm to identify soil spatial variability over a large area of clay-enriched soils, while GPR was better utilized at the Shale Hills to determine depth to bedrock and subsurface flow over smaller areas of medium-textured soils.

Within the Shale Hills, EC_a is low and relatively invariable because of the dominant control of the underlying electrically-resistive Rose Hill shale. In contrast, EC_a is more suitable for capturing soil variability at the more gently sloping Kepler Farm with its higher clay content soils and multiple parent materials. Stable patterns of relative EC_a in both landscapes are linked to different soil types and landform units. The highest EC_a occurs in soils formed in fine-textured deposits and in flat lower-lying, less-well drained areas, while the lowest EC_a occurs on sideslopes and summit areas composed of well-drained shallow soils. Ground-penetrating radar, on the other hand, worked well in the Shale Hills, producing relatively high-resolution images of the underlying soil and bedrock architectures and their spatial variability. Ground-penetrating radar displayed strong signal attenuation and restricted exploration depths in a large portion of the Kepler Farm. In the Shale Hills, GPR provided imagery showing the presence, depth, and continuity of soil, stratigraphic, and lithologic layers and inhomogeneities, implying possible water flow direction and pattern. As the depth to bedrock and the topography of the bedrock surface affect water flow, GPR records of water-restricting soil horizons and soil–bedrock interfaces are invaluable to hydropedologic studies.

The integration of pedology, soil physics, and hydrology into hydropedology has been facilitated by the increasing availability and improvements in geophysical tools. The integration of GPR and EMI with GPS and other technologies and the use of interactive interpretation software provide more effective approaches for characterizing soils at the hillslope and small catchment scales, as demonstrated in this study. These will help ease the technological bottleneck for subsurface investigations. The synergistic use of these technologies permits the collection of large geo-referenced geophysical datasets that can be displayed and analyzed in geospatial software, thus improving the utility of geophysical tools in hydropedologic investigations, especially at the catena and catchment scales.

ACKNOWLEDGMENTS

This research has been supported by the USDA National Research Initiative (grant #2002-35102-12547), the USDA Higher Education Challenge Competitive Grants Program (grant #2006-38411-17202), and the U.S. National Science Foundation (grant #EAR-0725019).

REFERENCES

Al-fares, W., Bakalowicz, M., Guerin, R., Dukhan, M., 2002. Analysis of the karst aquifer structure of the Lamalou area (Herault, France) with ground penetrating radar. Journal of Applied Geophysics 51, 97–106.

Allred, B.J., Ehsani, M.R., Saraswat, D., 2005. The impact of temperature and shallow hydrologic conditions on the magnitude and spatial pattern consistency of electromagnetic induction measured soil electrical conductivity. Transaction of the American Society of Agricultural Engineers 48 (6), 2123–2135.

Allred, B.J., Ehsani, M.R., Daniels, J.J., 2008. General considerations for geophysical methods applied to agriculture. In: Allred, B.J., Daniels, J.J., Ehsani, M.R. (Eds.), Handbook of Agricultural Geophysics. CRC Press, Taylor and Francis Group, Boca Raton, Florida, pp. 3–16.

Anderson-Cook, C.M., Alley, M.M., Roygard, J.K.F., Khosla, R., Noble, R.B., Doolittle, J.A., 2002. Differentiating soil types using electromagnetic conductivity and crop yield maps. Soil Science Society of America Journal 66 (5), 1562–1570.

André, F., Saussez, S., Van Duraient, R., 2010. High-resolution imaging of a vineyard in south France using ground penetrating radar and electromagnetic induction. In: Proceedings of the 13th International Conference on Ground Penetrating Radar, June, 21–25 2010. Institute of Electrical and Electronics Engineers (IEEE), Lecce, Italy, pp. 623–625. ISBN: 978-1-4244-4605-6.

Beres, M., Huggenberger, P., Green, A.G., Horstmeyer, H., 1999. Using two- and three-dimensional georadar methods to characterize glaciofluvial architecture. Sedimentary Geology 129, 1–24.

Berg, T.M., Geyer, A.R., Edmunds, W.E., 1980. Geologic Map of Pennsylvania. 4th Ser., Map 1. Pennsylvania Geologic Survey, Harrisburg, PA.

Bohling, G.C., Anderson, M.P., Bentley, C.R., 1989. Use of Ground Penetrating Radar to Define Recharge Areas in the Central Sand Plain. Technical Completion Report G1458-03. Geology and Geophysics Department, University of Wisconsin-Madison, Madison, Wisconsin.

Brevik, E.C., Fenton, T.E., 2003. Use of the Geonics EM-38 to delineate soil in a loess over till landscape, southwestern Iowa. Soil Survey Horizons 44, 16–24.

Buursink, M.L., Lane, J.W, 1999. Characterizing fractures in a bedrock outcrop using ground-penetrating radar at Mirror Lake, Grafton County, New Hampshire. In: Morganwalp, D.W., Buxton, H.T. (Eds.), U.S. Geological Survey Toxic Substances Hydrology Program–Proceedings of the Technical Meeting, Charleston, South Carolina, pp. 769–776, March 8–12, 1999–Volume 3 of 3–Subsurface Contamination from Point Sources: U.S. Geological Survey Water-Resources Investigations Report 99-4018C.

Cagnoli, B., Russell, J.K., 2000. Imaging the subsurface stratigraphy in the Ubehebe hydrovolcanic field (Death Valley, California) using ground-penetrating radar. Journal of Volcanology and Geothermal Research 96, 45–56.

Callegary, J.B., Ferre, T.P.A., Groom, R.W., 2007. Vertical spatial sensitivity and exploration depth of low-induction-number electromagnetic-induction instruments. Vadose Zone Journal 6, 158–167.

Campbell, J.E., 1990. Dielectric properties and influence of conductivity in soils at one to fifty megahertz. Soil Science Society of America Journal 54, 332–341.

Collins, M.E., 2008. History of ground-penetrating radar applications in agriculture. In: Allred, B.J., Daniels, J.J., Ehsani, M.R. (Eds.), Handbook of Agricultural Geophysics. CRC Press, Taylor and Francis Group, Boca Raton, Florida, pp. 45–55.

Collins, M.E., Doolittle, J.A., Rourke, R.V., 1989. Mapping depth to bedrock on a glaciated landscape with ground-penetrating radar. Soil Science Society of America Journal 53 (3), 1806–1812.

Dagallier, G., Laitinen, A.I., Malartre, F., Van Campenhout, I.P., Veeken, P.C.H., 2000. Ground penetrating radar application in a shallow marine Oxfordian limestone sequence located on the eastern flank of the Paris Basin, NE France. Sedimentary Geology 130, 149–165.

Daniels, D.J., 2004. Ground Penetrating Radar, second ed. The Institute of Electrical Engineers, London, United Kingdom.

Daniels, J.J., Allred, B., Collins, M., Doolittle, J., 2003. Geophysics in soil science. In: Lal, R. (Ed.), Encyclopedia of Soil Science. Marcel Dekker, Inc., New York, New York, pp. 1–5.

Davis, J.L., Annan, A.P., 1989. Ground-penetrating radar for high-resolution mapping of soil and rock stratigraphy. Geophysical Prospecting 37, 531–551.

Demanet, D., Evers, L.G., Teerlynck, H., Dost, B., Jongmans, D., 2001. Geophysical investigations across the Peel boundary fault (The Netherlands) for a paleoseismological study. Netherlands Journal of Geosciences 80 (3–4), 119–127.

Doolittle, J.A., Butnor, J.R., 2008. Chapter 6, soils, peatlands, and biomonitoring. In: Jol, H.M. (Ed.), Ground Penetrating Radar: Theory and Applications. Elsevier, Amsterdam, The Netherlands, pp. 179–202.

Doolittle, J.A., Jenkinson, B., Hopkins, D., Ulmer, M., Tuttle, W., 2006. Hydropedological investigations with ground-penetrating radar (GPR): estimating water-table depths and local ground water flow pattern in areas of coarse-textured soils. Geoderma 131 (3/4), 317–329.

Farahani, H.J., Flynn, R.L., 2007. Map quality and zone delineation as affected by width of parallel swaths with mobile agricultural sensors. Biosystems Engineering 96 (2), 151–159.

Folk, R.L., 1960. Petrography and origin of the Tuscarora, Rose Hill, and Keefer formations, lower and middle Silurian of eastern west Virginia. Journal of Sedimentary Research 30 (1), 1–58.

Freeland, R.S., 2008. Ground-penetrating radar mapping of near surface preferential flow. In: Allred, B.J., Daniels, J.J., Ehsani, M.R. (Eds.), Handbook of Agricultural Geophysics. CRC Press, Taylor and Francis Group, Boca Raton, Florida, pp. 337–344.

Frogbrook, Z.L., Oliver, M.A., 2007. Identifying management zones in agricultural fields using spatially constrained classification of soil and ancillary data. Soil Use and Management 23, 40–51.

Geonics Limited, 1998. EM38 Ground Conductivity Meter Operating Manual. Geonics Ltd., Mississauga, Ontario.

Gish, T.J., Dulaney, W.P., Kung, K-J.S., Daughtry, C.S.T., Doolittle, J.A., Miller, P.T., 2002. Evaluating use of ground-penetrating radar for identifying subsurface flow paths. Soil Science Society of America Journal 66, 1620–1629.

Grasmueck, M., 1996. 3-D ground-penetrating radar applied to fracture imaging in gneiss. Geophysics 61 (4), 1050–1064.

Grasmueck, M., Green, A.G., 1996. 3-D georadar mapping: looking into the subsurface. Environmental and Engineering Geoscience II (2), 195–220.

Greenhouse, J.P., Slaine, D.D., 1983. The use of reconnaissance electromagnetic methods to map contaminant migration. Ground Water Monitoring Review 3 (2), 47–59.

Hubbard, S.J., Chen, C., Williams, K., Peterson, J., Rubin, Y., 2005. Environmental and agricultural applications of GPR. In: Proceedings of the 3rd International Workshop on Advanced Ground Penetrating Radar (IWAGPR 2005). Elsevier Engineering Information, Inc, pp. 45–48.

Huisman, J.A., Snepvangers, J.C., Bouten, W.W., Heuvelink, W., 2002. Mapping spatial variation in surface water content: comparison of ground-penetrating radar and time domain reflectometry. Journal of Hydrology 269, 194–207.

Huisman, J.A., Hubbard, S.S., Redman, J.D., Annan, A.P., 2003. Measuring soil water content with ground penetrating radar: a review. Vadose Zone Journal 2, 476–491.

Iivari, T.A., Doolittle, J.A., 1994. Computer simulations of depths to water table using ground-penetrating radar in topographically diverse terrains. In: Kovar, K., Soveri, J. (Eds.), Ground Water Quality Management (Proceedings of GQM 93, International Association of Hydrological Sciences), Tallinn, Estonia, pp. 11–20, September 1993.

Inman, D.J., Freeland, R.S., Ammons, J.T., Yoder, R.E., 2002. Soil investigations using electromagnetic induction and ground-penetrating radar in southwest Tennessee. Soil Science Society of America Journal 66, 206–211.

Jol, H., 2009. Ground Penetrating Radar: Theory and Applications. Elsevier Science, Amsterdam, The Netherlands.

Kachanoski, R.G., Gregorich, E.G., van Wesenbeeck, I.J., 1988. Estimating spatial variations of soil water content using noncontacting electromagnetic inductive methods. Canadian Journal of Soil Science 68, 715–722.

Kachanoski, R.G., Van Wesenbeeck, I.J., Von Bertoldi, P., Ward, A., Hamlen, C., Kachanoski, R.G., 1990. Measurement of soil water content during three-dimensional axial-symmetric water flow. Soil Science Society of America Journal 54 (3), 645–649.

Khakural, R.G., Robert, P.C., Hugins, D.R., 1998. Use of non-contacting electromagnetic inductive methods for estimating soil moisture across landscapes. Communications in Soil Science and Plant Analysis 29 (11/14), 20055–22065.

King, J.A., Dampney, P.M.R., Lark, R.M., Wheeler, H.C., Bradley, R.I., Mayr, T.R., 2005. Mapping potential crop management zones within fields: use of yield-map series and patterns of soil physical properties identified by electromagnetic induction sensing. Precision Agriculture 6, 167–181.

Kowalsky, M.B., Finsterle, S., Rubin, Y., 2004. Estimating flow parameter distribution using ground-penetrating radar and hydrological measurements during transient flow in vadose zone. Advances in Water Resources 27, 583–599.

Kravchenko, A.N., 2008. Mapping soil drainage classes using topographic and soil electrical conductivity. In: Allred, B.J., Daniels, J.J., Ehsani, M.R. (Eds.), Handbook of Agricultural Geophysics. CRC Press, Taylor and Francis Group, Boca Raton, Florida, pp. 255–261.

Kravchenko, A.N., Bollero, G.A., Omonode, R.A., Bullock, D.G., 2002. Quantitative mapping of soil drainage classes using topographical data and soil electrical conductivity. Soil Science Society of America Journal 66, 235–243.

Lane, J.W., Buursink, M.L., Haeni, F.P., Versteeg, R.J., 2000. Evaluation of ground-penetrating radar to detect free-phase hydrocarbons in fractured rocks – results of numerical modeling and physical experiments. Ground Water 38 (6), 929–938.

Lin, H., 2010. Earth's Critical Zone and hydropedology: concepts, characteristics, and advances. Hydrology and Earth System Sciences 14, 25–45.

Lin, H.S., Bouma, J., Pachepsky, Y., Western, A., Thompson, J., van Genuchten, M. Th., Vogel, H., Lilly, A., 2006a. Hydropedology: synergistic integration of pedology and hydrology. Water Resource Research 42, W05301. doi: 10.1029/2005WR004085.

Lin, H.S., Kogelmann, W., Walker, C., Burns, M.A., 2006b. Soil moisture patterns in a forested catchment: a hydropedological perspective. Geoderma 131, 345–368.

Lunt, I.A., Hubbard, S.S., Rubin, Y., 2005. Soil moisture content estimation using ground-penetrating radar reflection data. Journal of Hydrology 307, 254–269.

McNeill, J.D., 1980. Electrical Conductivity of Soils and Rock; Technical Note TN-5. Geonics Limited, Mississauga, Ontario, Canada.

Merkel, E., 1978. Soil Survey of Huntingdon County, Pennsylvania. USDA-Soil Conservation Service in Cooperation with Pennsylvania State University College of Agriculture and the Pennsylvania Department of Environmental Resources and State Conservation Commission. US Government Printing Office, Washington DC.

Nascimento da Silva, C.C., de Medeiros, W.E., Jarmin de Sa, E.F., Neto, P.X., 2004. Resistivity and ground-penetrating radar images of fractures in a crystalline aquifer: a case study in Caicara farm – NE Brazil. Journal of Applied Geophysics 56, 295–307.

Neal, A., 2004. Ground-penetrating radar and its use in sedimentology: principles, problems and progress. Earth-Science Reviews 66, 261–330.

Olhoeft, G.R., 1998. Electrical, magnetic, and geometric properties that determine ground-penetrating radar performance. In: Proceedings, Seventh International Conference on Ground-Penetrating Radar. University of Kansas, Lawrence, Kansas, pp. 177–182. May 27–30, 1998.

Pipan, M., Baradello, L., Forte, E., Prizzon, A., 2000. GPR study of bedding planes, fractures and cavities in limestone. In: Noon, D. (Ed.), Proceedings Eight International Conference on Ground-Penetrating Radar. May 23–26, 2000. The University of Queensland, Goldcoast, Queensland, Australia, pp. 682–687.

Porsani, J.L., Elis, V.R., Hiodo, F.Y., 2005. Geophysical investigations for the characterization of fractured rock aquifer in Itu, SE Brazil. Journal of Applied Geophysics 57, 119–128.

Redman, J.D., 2008. Soil water content measurements using ground-penetrating radar surface reflectivity method. In: Allred, B.J., Daniels, J.J., Ehsani, M.R. (Eds.), Handbook of Agricultural Geophysics. CRC Press, Taylor and Francis Group, Boca Raton, Florida, pp. 317–322.

Rhoades, J.D., Raats, P.A., Prather, R.J., 1976. Effects of liquid-phase electrical conductivity, water content and surface conductivity on bulk soil electrical conductivity. Soil Science Society of America Journal 40, 651–655.

Robinson, D.A., Binley, A., Crook, N., Day-Lewis, F.D., Ferré, T.P.A., Grauch, V.J.S., Knight, R., Knoll, M., Lakshmi, V., Miller, R., Nyquist, J., Pellerin, L., Singha, K., Slater, L., 2008. Advancing process-based watershed hydrological research using near-surface geophysics: a vision for, and review of, electrical and magnetic geophysical methods. Hydrological Processes 22, 3604–3635.

Robinson, D.A., Lebron, I., Lesch, S.M., Shouse, P., 2003. Minimizing drift in electrical conductivity measurements in high temperature environments using the EM38. Soil Science Society of America Journal 68, 339–345.

Rubin, Y., Hubbard, S.J., 2006. Hydrogeophysics. Springer, Dordrecht, The Netherlands.

Saintenoy, A., Schneider, S., Tucholka, P., 2008. Evaluating ground penetrating radar use in water infiltration monitoring. Vadose Zone Journal 7, 207–214.

Scanlon, B.R., Paine, J.G., Goldsmith, R.S., 1999. Evaluation of electromagnetic induction as a reconnaissance technique to characterize unsaturated flow in an arid setting. Ground Water 37 (2), 296–304.

Schumann, A.W., Zaman, Q.U., 2003. Mapping water table depth by electromagnetic induction. Applied Engineering in Agriculture 19 (6), 675–688.

Sheets, K.R., Hendrickx, J.M.H., 1995. Noninvasive soil water content measurements using electromagnetic induction. Water Resources Research 31, 2401–2409.

Shield, R.R., Sopper, W.E., 1969. An application of surface geophysical techniques to the study of watershed hydrology. Journal of the American Water Resources Association 5, 37–49.

Slater, L., Comas, X., 2008. The contribution of ground penetrating radar to water resource research. In: Jol, H.M. (Ed.), Ground Penetrating Radar: Theory and Applications. Elsevier, Amsterdam, The Netherlands, pp. 203–236.

Smith, M.C., Vellidis, G., Thomas, D.L., Breve, M.A., 1992. Measurement of water table fluctuations in a sandy soil using ground penetrating radar. Transactions of the American Society of Agricultural Engineers 35, 1161–1166.

Soil Survey Staff, 1993. Soil Survey Manual. Soil Conservation Service, USDA Handbook 18. US Government Printing Office, Washington DC.

Steenhuis, T., Vandenheuvel, K., Weiler, K.W., Boll, J., Daliparthy, J., Herbert, S., Kung, K-J.S., 1998. Mapping and interpreting soil textural layers to assess agri-chemical movement at several scales along the eastern seaboard (USA). Nutrient Cycling in Agroecosystems (Netherlands) 50, 91–97.

Stroh, J.C., Archer, S., Doolittle, J.A., Wilding, L., 2001. Detection of edaphic discontinuities with ground-penetrating radar and electromagnetic induction. Landscape Ecology 16, 377–390.

Taylor, R.S., 2000. Development and Applications of Geometric-sounding Electromagnetic Systems. Dualem Inc, Milton Ontario.

Theune, U., Rokosh, D., Sacchi, M.D., Schmitt, D.R., 2006. Mapping fractures with GPR: a case study from Turtle Mountain. Geophysics 71 (5), B139–B150.

Toy, C.W., Steelman, C.M., Endres, A.L., 2010. Comparing electromagnetic induction and ground penetrating radar techniques for estimating soil moisture content. In: Proceedings of the 13th International Conference on Ground Penetrating Radar, June, 21–25 2010. Institute of Electrical and Electronics Engineers (IEEE), Lecce, Italy, pp. 606–611. ISBN: 978-1-4244-4605-6.

USDA-NRCS, 2006. Land Resource Regions and Major Land Resource Areas of the United States, the Caribbean, and the Pacific Basin. USDA Handbook 296.

Van Dam, R.L., 2010. Landform characterization using geophysics-recent advances, applications, and emerging tools. Geomorphology doi:10.1016/j.geomorph.2010.09.005.

van Overmeeren, R.A., 1998. GPR and wetlands of the Netherlands. In: Plumb, R.G. (Ed.), Proceedings of the Seventh International Conference on Ground-Penetrating Radar. May 27–30, 1998. Radar Systems and Remote Sensing Laboratory, University of Kansas, Lawrence, Kansas, pp. 251–258.

van Overmeeren, R.A., 2004. Georadar for hydrogeology. Fast Break 12 (8), 401–408.

Waine, T.W., Blackmore, B.S., Godwin, R.J., 2000. Mapping available water content and estimating soil textural class using electromagnetic induction. EurAgEng Paper No. 00-SW-044, Warwick, UK.

Weihermüller, L., Huisman, J.A., Lambot, S., Herbst, M., Vereecken, H., 2007. Mapping the spatial variation of soil water content at the field scale with different ground penetrating radar techniques. Journal of Hydrology 340, 205–216.

Zhang, J., Lin, H.S., Doolittle, J., 2012a. Subsurface lateral flow as revealed by combined ground penetrating radar and real-time soil moisture monitoring. Hydrological Processes (in revision).

Zhang, J., Lin, H.S., Doolittle, J., 2012b. Soil layering and preferential flow impacts on seasonal changes of GPR signals in two contrasting soils. Submitted to Geoderma (submitted for publication).

Zhu, Q., Lin, H.S., 2009. Simulation and verification of subsurface lateral flow paths in an agricultural landscape. Hydrology and Earth System Sciences 13, 1503–1518.

Zhu, Q., Lin, H.S., Doolittle, J., 2010a. Repeated electromagnetic induction surveys for determining subsurface hydrologic dynamics in an agricultural landscape. Soil Science Society of America, Journal 74, 1750–1762.

Zhu, Q., Lin, H.S., Doolittle, J., 2010b. Repeated electromagnetic induction surveys for improving soil mapping in an agricultural landscape. Soil Science Society of America, Journal 74, 1763–1774.

Chapter 14

Hydropedology, Geomorphology, and Groundwater Processes in Land Degradation: Case Studies in South West Victoria, Australia

Richard MacEwan,[1,*] Peter Dahlhaus[2] and Jonathon Fawcett[3]

ABSTRACT

Land cleared for agriculture in southeast Australia from the late 19th to mid-20th century has suffered various types and degrees of hydrologically driven degradation. Water erosion, landslides, waterlogging, and land salinization have occurred across a range of soils and landscapes. The principal mechanisms for these processes are well understood but the circumstances in which they manifest are not uniform; in each landscape situation the geological context, geomorphology, and soils are different and provide unique structural controls for the surface and subsurface hydrology. Prescriptions for management of soil and land degradation are only likely to be successful if appropriate conceptual models of these structural controls have been developed for each landscape under consideration. This chapter describes three case studies in Victoria, Australia, in which hydropedology, hydrogeology, and chemistry have been applied to the diagnosis of degradation processes and the prescription of land-management solutions.

1. INTRODUCTION

Many soil- and land-degradation processes, for example, erosion, salinity, waterlogging, soil structure decline, nutrient loss, and acidification, are hydrologically driven or depend on hydrologic conditions. Understanding

[1]Future Farming Systems Research Division, Department of Primary Industries, Bendigo 3554, Australia
[2]School of Science and Engineering, University of Ballarat, Ballarat 3357, Australia
[3]Sinclair Knight Merz, Bendigo 3552, Australia
[*]Corresponding author: Email: Richard.MacEwan@dpi.vic.gov.au

Hydropedology, Edited by H. Lin. DOI: 10.1016/B978-0-12-386941-8.00014-9

how water moves on, within and through the soil mantle, is therefore essential for diagnoses of why and where these processes occur. Knowledge of the underlying geology and hydrogeology adds a contextual dimension for catchment-scale water balances in which interaction between surface water and groundwater may be a factor. For example, the origin, quantity, and quality of water persisting at the soil surface can result from runoff, interflow, and return flow from the soil, high water table, high groundwater pressure, or a combination of these. Each origin and flow pathway has different implications for water quality with respect to solute and sediment load, and the manifestation of erosion, waterlogging, or soil salinity. Diagnosis of the causes of these degradation features should take account of the unique characteristics of soil, geomorphology, and geology in order to construct a conceptual model for the landscape under study, from which appropriate management can be devised to alleviate or prevent degradation. Misdiagnosis can readily lead to prescription of costly and ineffective solutions. Economic analysis of national investment into salinity research and management has highlighted the critical need for policies to be informed by science and effective management strategies (Commonwealth of Australia, 2004; Pannell, 2001).

Hydropedology has an essential role since soil is the component in the landscape along with land-use management and vegetation that determine all the partitioning of water-balance components affecting surface hydrology, subsurface water storage, drainage, and recharge to groundwater. However, unless the hydrogeological context is also brought into the picture and applied to the diagnosis of hydrologically influenced land-degradation processes, mistakes will be made. The converse is also true. We provide three case studies for landscapes that illustrate the importance of developing an appropriate conceptual model of hydrologic processes in the Critical Zone (NRC, 2001; Lin, 2010), in particular for implementation of policy and practice in land management.

2. BACKGROUND

The research reported in these case studies was largely carried out in the 1990s at a time when there was a national focus on salinity in Australia amid concerns that land-management practices were contributing to an increase in land degraded by salinity. This problem was well known in the Murray–Darling Basin irrigation districts, underlain by shallow, often saline, groundwater tables, but, outside these districts, in areas of nonirrigated, rainfed agriculture, 'dryland' salinity was recognized as an even greater and more extensive issue. Compilation of data on affected land and the potential for dryland salinity nationally suggested that 5.7 million hectares were at risk in 2000 and that, by 2050, this could extend to 17 million hectares (Commonwealth of Australia, 2001). A generalized model of a cleared landscape, changed eco-hydrology, increased groundwater recharge, mobilization of stored salts, rising groundwater, and discharge of saline groundwater prevails (Eberbach, 2003). In Australia, explanations for the occurrence and management of dryland salinity, at regional and catchment scales, are guided by delineation of groundwater flow systems linked to topographical characteristics that can be classified in relation to groundwater recharge and discharge (Coram et al., 2001; Walker et al., 2003). The prescribed land-management solution has been to establish perennial and high water-use vegetation in areas that can be identified as significantly contributing to groundwater recharge and to use eco-mimicry in agricultural systems to restore catchment water balances to pre-clearing conditions (Hatton and Nulsen, 1999). Subsequently, since the early days of the salinity program, this simple explanation and prescription has been challenged to accommodate additional data and information for specific landscapes (van Bueren and Price, 2004).

The landscape examples described in this chapter are all in southeastern Australia, to the south of the Great Dividing Range and outside the Murray–Darling Basin (Fig. 1). Although dryland salinity was seen as the major land-management issue at a political level (Nicholson et al., 1992), this was

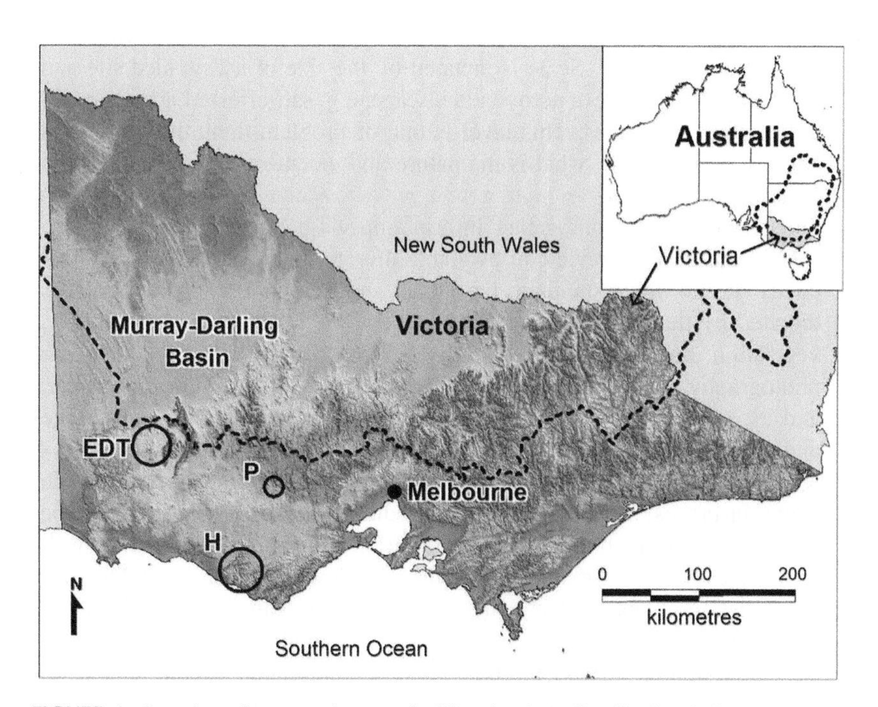

FIGURE 1 Location of case study areas in Victoria, Australia. Shading indicates elevation above sea level (dark = high; light = low). EDT = Eastern Dundas Tablelands; P = Pittong; and H = Heytesbury Settlement. The Murray–Darling Basin is bounded by the broken line.

invariably compounded by waterlogging and drainage problems which were more extensive than salinity and a more serious impediment to agriculture. In addition to salinity and waterlogging, one study area is highly prone to landslides and another has the potential to develop acid sulfate soils.

3. METHODS

The studies draw principally on field observations supplemented by installation of piezometers, excavation of soil pits, and supporting laboratory analyses for soil and water chemistry, including some dating of groundwater samples. The resulting conceptual models evolved with each successive investigation and form working hypotheses for processes in these landscapes. Details of these studies are contained in unpublished honors and master's theses cited in this chapter and two review papers (Dahlhaus and MacEwan, 1997; Dahlhaus et al., 2000). Each landscape model is open to modification as new data arise to augment the existing information.

3.1. Observations and Scale

Our landscape studies have three levels of detail, and multiple scales:

1) Primary observations of the conspicuous degradation effects or symptoms are made in the field. Scale is dictated by the size of a degraded site and the frequency or pattern across a landscape, e.g. salt-affected areas, eroded gullies, and landslides. This level is one of problem definition; it should answer the question, 'what is the nature and extent of degradation in this landscape?'
2) Secondary observations gather information via a broader reconnaissance survey and desktop review of available literature, maps, and imagery that contribute to understanding the local and regional context. Scale is dictated by that of existing maps e.g. geology, land systems, hydrology, vegetation, and land use. Additional spatial context is provided by air photography and satellite imagery (Landsat and MODIS) for land cover, and by airborne geophysics (gamma, magnetic, and gravity) for surface and deep geology. Point data may be available from previous soil surveys and from groundwater bores (piezometers and observation wells). In some instances, historical maps and diaries have helped to describe past conditions including early recognition of land degradation as well as changes in land use and land cover. This level of study should answer the question, 'what is the structure of the landscape within which the degradation is occurring?' At this stage, a hypothetical model needs to be formulated to explain the link between landscape structure and landscape hydrologic process. The model guides field investigations and appropriate site selection for hydrologic bores, or piezometers, and for soil-profile studies.

3) Tertiary observations are specific site investigations: soil pits and auger survey; groundwater bores and piezometry; water analysis and geochemistry; land-surface analysis; and interpretation in the form of hydrographs and toposequences. Deployment of ground-based geophysical surveys is a useful supplement to pits, augers, and bores: electromagnetic induction using EM38 and EM31 (Geonics Ltd., Ontario, Canada) for detection of subsurface differences between 0.5 and 5 m depth, gamma radio spectroscopy for 0–0.3 m depth, and shallow explosives (0.7 m depth) sensed by geophones to detect deeper regolith structure (faults).

Anecdotal information from landholders regarding the impact on farming enterprises and any actions they may have taken to remedy, or adapt to, the degradation problem can supplement all three stages and, in most cases, this information is the prime driver for the initial investigation.

3.2. Understanding the Processes – Principles

The surface and subsurface hydrology in each landscape is subject to structural controls in soil horizons, regolith, and hard-rock geology at different scales (Lin, 2003; Narasimhan, 2005).

Hydropedology investigates these controls in the soil profile, pedon, and soil mantle or uppermost part of the regolith. Classical soil-survey techniques (Soil Survey Division Staff, 1993; McDonald et al., 1990) are used to distinguish and map differences between soils. Soil-profile hydrologic controls can be interpreted with respect to texture profile, soil structure, and color (Bigham et al., 1993; Lindbo et al., 2010; Northcote, 1979, 1983; Pachepsky and Rawls, 2004; Vepraskas, 2004). A 3D conceptual model for soil–landscape hydrology links the soil point observations with land-surface features: 1D observations are made from auger samples and in situ measurements (e.g., of hydraulic conductivity); 2D descriptions are made of soil pit faces (depth plus lateral variability); soil maps, soil profiles, and terrain models of land surface can then be combined to create 2D and 3D toposequences (Conacher and Dalrymple, 1977).

Conacher and Dalrymple (1977) described a generic landscape model of nine land-surface units, each having unique soil/water/gravity relationships that may be applied to analysis and mapping of any land-surface catena (Conacher and Dalrymple, 1977 p.101–109). The nine-unit land-surface model, summarized in Table 1, provides a useful framework for interpreting the hydropedology of hillslopes and small catchments.

Not all recent Australian studies of soil and hillslope hydrology have used Conacher and Dalrymple's (1977) model, but similar principles have been applied to interpret soil and land-surface features: for example, Fritsch and Fitzpatrick (1994) proposed a morphologically based hydropedologic method to construct soil–water–landscape models, demonstrating this in a South Australian catchment to explain salinity and acidification processes; Park and van Giesen's (2004) adaptation of the nine-unit land-surface model, to

TABLE 1 The Nine Land-Surface Units and their Principal Distinguishing Features Summarized from Conacher and Dalrymple (1977, p 101–109)

	Land-surface unit	Dominant soil/water/gravity process and soil features
1	Interfluve	Interfluve; vertical soil–water movements; no lateral drainage. Shallow soils; laterally uniform; may be well drained to water logged depending on rainfall, spatial extent (no lateral drainage), and permeability of regolith.
2	Seepage slope	Mechanical and chemical eluviation by lateral subsurface soil-water movements. Soils variable laterally; presence of discrete gleyed horizons; bleached A2 or E horizons; piping and tunneling. Distinguished from unit 5 by having a relatively smooth surface.
3	Convex creep slope	Convex unit where soil creep dominates. Relatively uniform soils laterally; soils shallower than those in unit 2. Distinguished from units 5 by having a smooth convex surface.
4	Fall face	Steep; subject to rock fall and slide; no soil formation; spatial extent too small to map except at very large scales. Usually indicated by linear feature on maps.
5	Transportational midslope	Mass transport of soil downslope via flows, slumps, slides, raindrop impact, and surface wash. Distinguished from units 2 and 3 by irregular ground surface or by bare areas, rills, crusts, and evidence of sheet erosion.
6	Colluvial footslope	Active colluvial redeposition from upslope; low-angle fans; depositional lobes from mass movement as slides. Seepage zones; surface and subsurface gleying; heterogeneous soils thickening downslope. Organic matter, colloid, oxide, and salt enrichment relative to upslope units.
7	Alluvial toeslope	Redeposition of alluvial materials from up valley by overbank flow and non-channelized flash floods; buried soils (paleosols); layered additions of texturally well-sorted material. Soil drainage may be limited by soil water tables or by groundwater.
8	Channel wall	Lateral corrasion by stream action. Limited spatial extent.
9	Stream bed	Transportation of material down valley by stream action.

delineate soil–landscape units, improved estimates of average soil moisture content in hillslope hydrologic modeling; and Ticehurst et al. (2007) measured hydrologic components on a hillslope in New South Wales to validate a conceptual model constructed on the basis of soil-profile morphology and redoximorphic features.

It is important to note that the depth limit of the land-surface model is the 'base of the soil' (Conacher and Dalrymple, 1977, p.11) and is concerned with understanding spatial differences in dominant soil–water–transport processes influencing soil formation. Hillslope hydrology has influences from below the base of the soil; each of the recent Australian examples (cited above) incorporates the deeper hydrology to some degree. While a conceptual hydropedologic model for a study area is readily encompassed within hillslope and subcatchment boundaries, the hydrogeological effects on local catchment processes are often the result of geological structures with considerably greater spatial extent than the surface-water catchment. Hydrogeology encompasses the full regolith and hard-rock geology with respect to water storage, water movement, and water chemistry (Domenico and Schwartz, 1990; Fetter, 1980; Freeze and Cherry, 1979).

Toth's (1999) perspective on groundwater as a general geologic agent and his description of the hydrologic environment (Tóth, 1970) provide a framework that complements hydropedologic studies. For example, the spatial distribution and character of groundwater-related geomorphic features, such as springs, seeps, soil salinization, and groundwater-dependent ecosystems, can be explained more completely in their hydrogeological context.

4. CASE STUDIES

The Eastern Dundas Tablelands (EDT), Pittong, and the Heytesbury case study areas are in southwest Victoria, Australia (Fig. 1). The climate of the region is described as 'Mediterranean', having a winter-dominant rainfall, and hot dry summers. Mean annual rainfall is 500–700 mm in the EDT, 600–750 mm in the Pittong area, and 800–1100 mm in the Heytesbury area. The region is regarded as 'high rainfall' for farming systems. Both the EDT and Pittong areas are dominantly used for sheep grazing and fine wool production; the Heytesbury area is used for dairy farming.

Each case study is summarized with respect to the symptoms of degradation, the broad hydrogeological context, and a simple conceptual model in the form of a 2D schematic for slope processes combining hydrogeology and hydropedology. The unique distinguishing features and processes for each case, and conclusions concerning the required treatment or management of the soil and land degradation, are explained.

Soils in the study areas have generally been described and classified according to two Australian systems: the Factual Key of Northcote (1979) and the revised Australian Soil Classification of Isbell (2002). The Eastern Dundas Tablelands and the Pittong area are dominated by soils in the Duplex divisions of Northcote (1979) or texture-contrast soils: Chromosols, Kurosols, and Sodosols of Isbell (2002). USDA Soil Taxonomy equivalents of Duplex or texture-contrast soils are found in the Alfisol and Ultisol soil orders (Soil Survey Staff, 2010). The A horizons are generally sandy loams to loams; B horizon textures are dominated by clay (sandy clay to heavy clay). Generally, the difference in texture between A and B is extremely obvious and boundaries are abrupt (textural change occurring within 5–20 mm) or sharp (boundary <5 mm) (McDonald et al., 1990). These soil types frequently have prominent, often bleached, A2 horizons and develop perched water tables in winter.

Texture-contrast soils are also found in the Heytesbury area but are less developed and less dominant; Podosols (Isbell, 2002) or Spodosols (Soil Survey Staff, 2010) are extensively found on the plains; Dermosols and Sodosols (Isbell, 2002) or Alfisols (Soils Survey Staff, 2010) are found on the valley sides; Dermosols, Podosols, Hydrosols, and Vertosols (Isbell, 2002) or Aquepts, Aquods, Aqualfs, and Vertisols (Soil Survey Staff, 2010) are found on the valley floor.

4.1. Case Study: Eastern Dundas Tablelands

The Dundas Tablelands is an isolated dissected paleoplain located at the western end, and to the south, of the Great Dividing Range that defines the southern-most catchment boundary of the Murray–Darling Basin (Fig. 1). The predominant land use since European settlement in the 19th century is grazing for fine wool production. The tablelands are distinguished into western and eastern portions by the degree of dissection and persistence of the paleoplain, the division being marked by a major geological fault zone. The Eastern Dundas Tableland (EDT) is a slightly domed tableland, with two acid-volcanic lava domes and generally radial drainage patterns. The volcanic deposits are at least 150 m thick in places and have been dated as late Silurian to early Devonian. The regolith has an extensive duricrust or strongly ferruginized surface overlying variable depth of saprolite. The saprolite is kaolinized and has been considered as equivalent to the pallid zone of a lateritic profile and is several meters thick (Hills, 1939 cited in Joyce et al., 2003). Differential weathering of the duricrust takes the form of parallel red, iron-rich bands alternating with pale, clay-rich bands, and as subsurface horizons with irregular, rounded, coarse ferruginous gravel or pisoliths. The banded features have been referred to as 'tiger mottles' (MacEwan, 2005; Paine and Phang, 2005). Dissection of the plateau by rivers and streams has created shallow U-shaped valleys rarely deeper than 20 m. Munro (1998) reported that salinity was affecting 6000 ha, or 3.3 percent, of the EDT.

4.1.1. Degradation Features on the Eastern Dundas Tablelands

Soil degradation consists of irregular shaped scalds, devoid of vegetation, along drainage flats, at the breaks-of-slope, and along valley sides. Below the barren scalds, vegetation is dominated by species tolerant to salt and waterlogging such as Buck's Horn Plantain (*Plantago coronopus*), Water Buttons (*Cotula coronopifolia*), Annual Beard Grass (*Polypogon monspeliensis*), and Sea Barley Grass (*Critesion marinum*). Distinct areas of Fog Grass (*Holcus lanatus*) are found directly upslope of degraded discharge zones. Groundwater discharge occurs as free-flowing springs and as diffuse seeps. Iron staining, salt efflorescence, and sealing of the soil surface are common features of the diffuse seepage areas, and landowners have reported that these areas appeared to be expanding upslope as well as downslope.

4.1.2. Eastern Dundas Tablelands Hydrogeology

Research in the EDT included a major groundwater survey (Cossens, 1999). A series of projects (Jerinic, 1993; Woof, 1994; Cossens, 1999; Fawcett, 2004), that documented existing groundwater bores, installed new piezometers, and analyzed groundwater chemistry, established that there is a single, regional, groundwater system operating within the tablelands (Fig. 2). The groundwater flow system of the EDT is an example of a gravity-driven groundwater flow system within a regionally unconfined aquifer (Fawcett, 2004; Tóth, 1962, 1999). Within recharge areas, hydraulic heads decrease with depth and water flow is downward; within discharge areas, hydraulic heads increase upwards resulting in the convergence of flow paths, particularly along geological structures. Fawcett (2004) described the system as

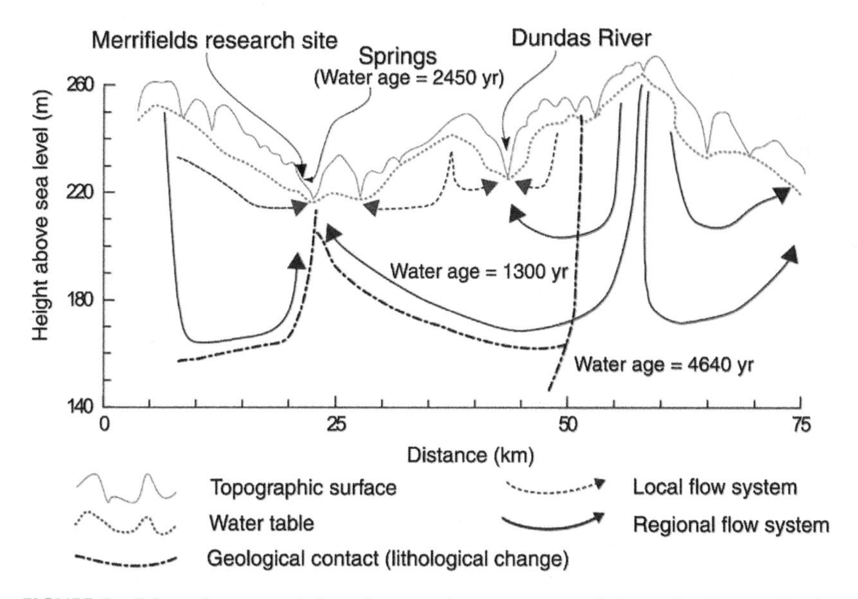

FIGURE 2 Schematic representation of a groundwater system underlying the Eastern Dundas Tablelands, showing flow paths, water age, salinity, and relationship to groundwater discharge (*From Fawcett, 2004*).

a fractured rock aquifer consisting of a series of nested flow systems increasing in flow length and water age with depth (Fig. 2). In the EDT, the spatial distribution of discharge zones (springs) is correlated with regional and local geological structures. Of 128 permanently flowing springs mapped, 73% are located within 500 m of a mapped fault line or a contact between two separate geological units. Point flowing springs across the EDT are recognized as primary landscape features that existed prior to European land clearance (Nathan, 1998; Fawcett et al., 2008).

4.1.3. Historical Analysis of the Eastern Dundas Tablelands

Research by Nathan (1998, 1999, 2000) established that salinity was present prior to European settlement and that the tree cover had, for a period, been denser post-settlement, than pre-settlement. Nathan used archived maps from the early exploration of the region, diaries, administrative records of settlement, and records of timber mills, the latter revealing the succession of tree species that had followed initial settlement. Prior to European settlement, the EDT was a grassy, open forest of *Eucalypt* spp., *Banksia* spp., and *Allocasuarina* spp., with *Banksia marginata* the dominant tree species (Nathan, 1998, 1999). By the 1900s, land clearing, sheep grazing, and a variable climate changed the dominant tree species to River Red Gum (*Eucalyptus camaldulensis*) and the dominant native grasses were Wallaby Grass (*Austrodanthonia* spp.), Kangaroo Grass (*Themeda triandra*) and Spear Grass (*Austrostipa* spp.). Today, less than three percent of the pre-European vegetation remains on the EDT.

Dahlhaus et al. (2000) reported that trend analysis of streamflow records for the Glenelg River from 1890 to 1950 indicates a decline in baseflow to the river for the 60-year period that tree clearing took place. On the basis of this hydrologic analysis and Nathan's tree-cover data, Dahlhaus et al. (2000) propose that groundwater recharge and discharge have not increased on the EDT as a result of settlement and tree clearing. However, landholder experiences and investigations by government since the 1980s have documented that degradation, primarily erosion around the saline discharge areas, has increased (GHCMA, 2005).

4.1.4. Hydropedology in the Eastern Dundas Tablelands

The soils and landscapes of the Dundas Tablelands have been described in the context of land-systems surveys by Gibbons and Downes (1964). Hydropedologic studies in the EDT have been reported by Brouwer and van de Graaf (1989), Fawcett (1996), Dahlhaus and MacEwan (1997), and Brouwer and Fitzpatrick (2000, 2002a, 2002b).

Soils of the EDT are dominated by texture-contrast soils (Isbell, 2002), generally having loamy A horizons, thick bleached A2 horizons, and sharp boundaries to well-structured mottled clay Bt horizons. Redoximorphic features are more prominent with depth, the redox concentrations being brighter, redder, and confined to ped interiors; ped faces are covered by thick cutans. The redox pattern and cutans are indicators of vertical and lateral throughflow of water and periodic saturation of interpedal pores. Clay mineralogy is dominantly kaolinitic. Strong prismatic and columnar structure occurs in subsoils of lower slope and valley floor soils. Many subsoils here are sodic (exchangeable sodium percentage >6%). Coarse ferruginous gravels are common in the A2 and in deeper B horizons of soils on the plateau and upper slopes. These gravels are a weathering product of the laterized or strongly ferruginized weathered volcanic regolith (Paine, 2003; Paine and Phang, 2005).

Fawcett (1996) conducted laboratory studies of water movement using large diameter (0.3 m) intact cores, extracted from 0 to 1.0 and 1.0 to 2.0 m depth. He concluded that recharge via preferential pathways such as macropores, particularly root channels and interpedal pores, was constrained by conditions deeper in the profile (below 1 m), which would restrict the rate of transmission of water to the groundwater system. Fawcett's (1996) hydropedologic model concluded that seasonal saturation of the soil profile and the fracture volume was limited by the rate of groundwater discharge in the region, and that, once the subsoil and regolith crack volume was full, lateral flow through the A2 and Bt horizons dominated. Independently, Brouwer and Fitzpatrick (2000, 2002a, 2002b) described a toposequence in the EDT with three water tables: two perched freshwater systems – one in the deeper B horizon (0.8–1.0 m), one overlying the pallid saprolite (3.0–3.5 m) – and a permanent saline groundwater table on the top of the bedrock. Using the structural approach of Fritsch and Fitzpatrick (1994), they identified a 'hydromorphic domain' consisting of upper, middle, and lower 'hydromorphic soil systems associated with the 3 water tables, and a 'lateritic domain' which they associated with ancient hydrology. They stated that preferential flow in root channels and along planar structures in the subsoil were controlling water movement below the root zone of the existing vegetation, and contributing to recharge, mobilization of salt stores in the subsoil and regolith, and discharge of the deeper saline groundwater.

Fawcett (2004) investigated a field site in the EDT ('Merrifield', Fig. 3) by surveying three transects downslope through scalded discharge areas. Piezometers were installed at depths of 4–5 m along each transect and the valley floor. These were used, along with other regional piezometer and observation well data, to interpret groundwater flow paths, and to sample water for chemical analysis. Soil pits, push tubes, and hand augers were used to collect soil samples for analysis and descriptive data. A trench 14 m long and 6 m deep was excavated into the hillslope, through a discharge zone, in order to study the saprolite structure and geochemistry. A geoseismic survey by Fawcett (ibid., chapter 4) provided data to map the profile of the underlying saprolite to saprock boundary which was irregular and occurred at 2–15 m depth below ground surface. The irregular boundary was interpreted as the result of

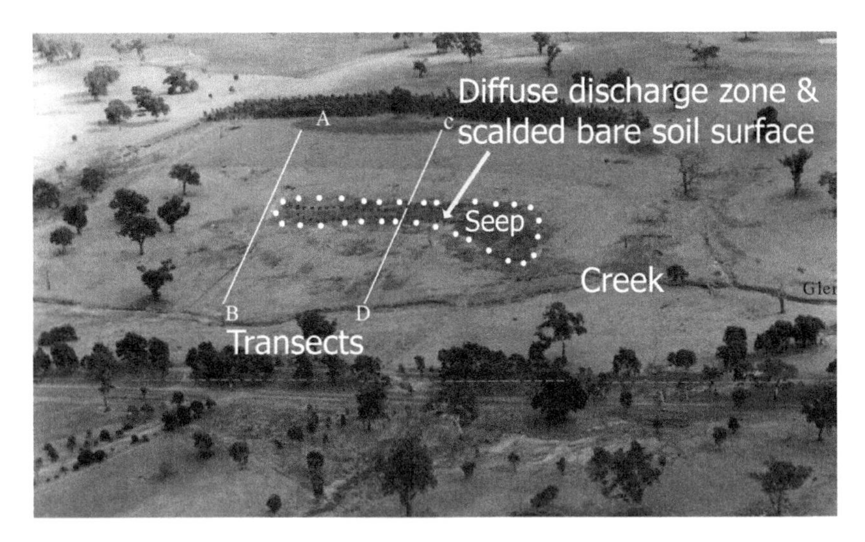

FIGURE 3 Photograph of part of the Merrifields research site in the Eastern Dundas Tablelands showing degraded areas with seeps, on the hillslopes, and the locations for two of the transects used to construct toposequences.

differential weathering resulting from the joint and fault structure of the rhyolite and ignimbrite.

Toposequences of soil horizons (morphology), soil electrical conductivity, pH and exchangeable sodium percentage, regolith, and groundwater were constructed from the transect data. In these toposequences, Fawcett (2004) described 5 'hydrological soil systems' for the A horizon (A1 and A2) and 9 systems for the subsoil. Simplifying this, Fawcett (2004) described a slope sequence of 4 'landscape units' in which different hydrologic processes dominated. These landscape units and their relationship to the nine land-surface units of Conacher and Dalrymple (1977) are summarized in Table 2 and illustrated in Fig. 4.

Fawcett (2004) measured the redox potential of discharge waters and groundwater in situ with a platinum electrode attached to a Micronta Auto-Range Digital Multimeter probe and using a calomel reference electrode. While the deep groundwater Eh values were positive (60–180 mV), there was a trend to lower Eh along the groundwater flow paths and negative Eh (−115 to −180 mV) in the discharge site.

Fawcett (2004) established that degradation within the Merrifield site discharge zones is increased by acidification associated with the development of acid sulfate soils; evaporative concentration of salt; and precipitation of iron minerals in soil pores and at the soil surface. The sodicity of soils in the discharge zones also contributes to their erodibility. The development of the acid sulfate soils is driven by the reduced nature of the groundwater system, providing a source of sulfur, as H_2S, to the discharge zone.

TABLE 2 Eastern Dundas Tablelands Landscape Units and Dominant Hydrologic Processes for Toposequences at the Merrifields Site (Summarized from Fawcett, 2004), and Equivalent Units in the Nine-Unit Land-surface Model of Conacher and Dalrymple (1977). See also Figs 2 and 4 and Table 1

Landscape unit (Fawcett, 2004)	Hydrologic processes described by Fawcett (2004) including the effect of land clearing	Land-surface unit equivalent after Conacher and Dalrymple (1977)
1 Tabletop	Soil hydrology is dominated by vertical water movement recharging the local groundwater flow system. Groundwater flow is horizontal (parallel to ground surface). Hydrology is assumed to have changed little since land clearing.	Dominantly a unit 1, but will behave as a unit 2 when throughflow in the A2 horizon is generated in saturated conditions. Small areas of a land-surface unit 3 (convex creep slope) occur on the crests linking the plateau and valley sides.
2 Mid-slope	Created by mass-wasting. Vertical recharge and throughflow from upslope into the local flow system, also contributing to the regional groundwater flow system. Some seasonal seepage of perched fresh (local) water. Throughflow and runoff are likely to have increased since clearing, thus affecting waterlogging and recharge to the local groundwater flow system.	Unit 2. Fawcett's (2004) description is of geomorphic development being due to mass wasting, but the contemporary processes are dominated by throughflow. In some positions on the landscape, this may be a unit 5 if mass movement is active.
3 Discharge zone: break of slope and mid-slope	This landscape unit is the interface between the hydrogeology of the regional, local, and seasonal groundwater flow systems. Discharge, seepage, freely flowing springs of saline groundwater. Land clearing has affected surface chemistry by exposure of the soil surface, therefore increasing iron precipitation, salt efflorescence, surface wash and deflation of the ground surface.	Without degradation, this would be a land-surface unit 2, dominated by throughflow and seepage. However, the erosion processes on degraded discharge areas transform this into a unit 5. (Anthropogeomorphic influences modify processes in land-surface units.)

(*Continued*)

TABLE 2 Eastern Dundas Tablelands Landscape Units and Dominant Hydrologic Processes for Toposequences at the Merrifields Site (Summarized from Fawcett, 2004), and Equivalent Units in the Nine-Unit Land-surface Model of Conacher and Dalrymple (1977). See also Figs 2 and 4 and Table 1—cont'd

Landscape unit (Fawcett, 2004)	Hydrologic processes described by Fawcett (2004) including the effect of land clearing	Land-surface unit equivalent after Conacher and Dalrymple (1977)
4 Creek flat or valley floor	Lateral groundwater flow dominates. Very little evidence of groundwater discharge except in the stream bed. A seasonally fluctuating shallow groundwater table underlies this unit. Increase in soil salinity due to redistribution of salt from degraded discharge zones upslope. Hydrogeology unchanged since clearing, surface hydrology affected by greater runoff and increased soil waterlogging. Thin layer of alluvial sediment attributed to post clearing erosion and redeposition from up valley sources.	Unit 6 or 7 (plus 8&9). The valley floor complex includes deposition from upslope (colluvial) and up valley (alluvial) as well as the incised stream bed.

The least severe acidification and lower salinity occurs within permanently wet and relatively undisturbed discharge zones; Fawcett (2004) and Gardner et al. (2004) hypothesized that free-flowing springs and ponding of discharge water, below the discharge area, allow the H_2S to dissipate to the atmosphere. They proposed that the severity of degradation in areas of diffuse discharge (rather than from flowing springs) is related to wetting and drying cycles – oxidation of H_2S to H_2SO_4 occurring within the soil profile during the drying phase.

The surface sealing of soil by iron is attributed to bacterial and redox processes associated with wetting and drying cycles. The red yellow gelatinous material consists of ferrihydrite within filaments of iron-oxidizing bacteria (*Gallionella* and *Leptothrix*). Ferrihydrite is held within the *Gallionella* and *Leptothrix* bacteria growing on discharge waters (Fitzpatrick et al., 1996) and is altered to less complex iron minerals as the gelatinous material dries. Subsequently, yellow-colored goethite and red-colored hematite deposits

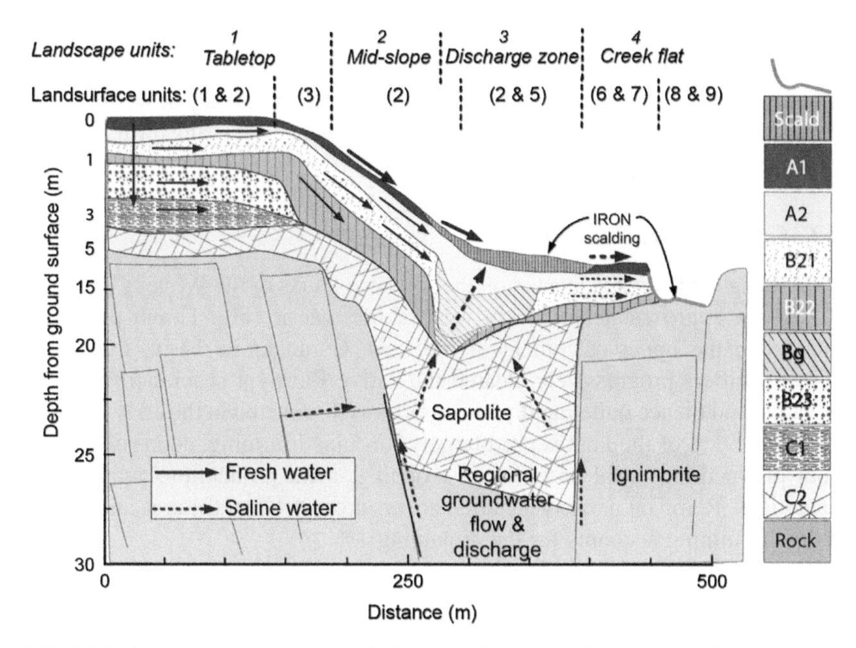

FIGURE 4 Schematic representation of principal soil and regolith horizons, and hydrology, for a hillslope in the Eastern Dundas Tablelands (Merrifields site). Landscape units (*italicized*) after Fawcett (2004). Land-surface units (number in parenthesis) after the system of Conacher and Dalrymple (1977). See also Tables 1 and 2. *(Color version online and in color plate)*

form on the soil surface (Fitzpatrick et al., 1996; Fitzpatrick and Self, 1997; Bigham et al., 2002).

4.1.5. Conceptual Model for Slope Hydrology and Degradation in the Eastern Dundas Tablelands

The conceptual model for the hydrology of the EDT is a relatively simple one of seasonal saturation of the soil profile, throughflow in the A2 and B horizons, and mixing of local, generally fresh, water with saline groundwater at discharge sites (Fig. 4). The scalding, which can be associated with extreme acidification, is a localized phenomenon and is largely due to changes in hydrology at and very close to the ground surface, particularly since land clearing.

4.1.6. Management of Degradation in the Eastern Dundas Tablelands

Both the historical evidence and contemporary analysis of the hydrology indicate that groundwater discharge in the EDT is largely unaffected by local changes in vegetation cover due to the overriding influence of the regional groundwater system. The degradation of discharge areas has been the result of vegetation removal and stock grazing in the discharge sites, which has allowed these sites to periodically dry out at the surface and to become excessively acidified. Maintaining plant cover, retaining water, or allowing free flow of groundwater from the seeps and springs will limit the frequency of wetting and drying at the surface, and thereby reduce the chance of scald formation or expansion of existing scalded areas.

4.2. Case Study: Pittong

The Pittong study area is located on the southern flanks of the Western Uplands of Victoria approximately 125 km west of Melbourne (Fig. 1) and occupies 6300 ha in the upper catchment of the Lake Corangamite basin. European pastoral settlers progressively cleared the native Eucalypt species from 1840 onwards, and thence quite rapidly after gold was discovered in the early 1850's. Currently, 95% of the land is used largely for mixed farming, comprising fine wool production, cereal and oilseed cropping, meat production, and timber plantations. Remnant native vegetation covers around 4% of the area and open-pit kaolin mining accounts for the remaining 1%.

4.2.1. Degradation Features at Pittong

Waterlogging is a general feature of the soils in the Pittong area. Salinity is largely confined to drainage lines and alluvial flats in this gently undulating landscape (Fig. 6). Saline groundwater discharge occurs on the upper slopes of broad convex drainage lines as discrete areas of saturated soil, rather than flowing springs. Salinity is evident as extensive areas of degraded pastures, with halophytic vegetation, mostly Spiny Rush, *Juncus acutus*, and Sea Barley Grass, *Hordeum leparinum*. The most severely affected areas may exhibit salt efflorescence in summer, are devoid of vegetation, and are prone to erosion. There are also environmental impacts of saline discharge from the Pittong area, via the Woady Yaloak River, into Lake Corangamite, a high-value Ramsar wetland, which has become increasingly saline and resulted in loss of aquatic biodiversity (Williams, 1995; Nicholson et al., 2006).

The extent of land salinity has been mapped by successive monitoring and research projects, using a combination of aerial photography, vegetation indicators (Matters and Bozon, 1995), and EM38 electromagnetic ground conductivity meter surveys (Geonics Ltd, Ontario, Canada) (Spies and Woodgate, 2005). Mapping in 2006 indicated that over 280 ha, or 4.5% of the Pittong area, is affected by salinity. In September 1996, a long-term land-salinity monitoring site (13 ha) was established in a small catchment in the area as part of the Victorian Dryland Salinity Monitoring Network (VDSMN site no. 9197) (Clark and Hekmeijer, 2004). The electromagnetic (EM38) surveys of the site show that the area of salinity (measured as $EC_a > 1$ dS/m in 0–60 cm soil depth) increased from 0.79 ha in September 1996 to 1.69 ha in September 2000, then reduced slightly to 1.48 ha in November 2007. This increase in salinity has occurred during a decade of lower than average rainfall and dropping water tables.

4.2.2. Hydrogeology Affecting the Pittong Study Area

The Pittong study area is a granite landscape situated within a complex geological setting in which the regional groundwater system provides the upward pressure for groundwater discharge in the region's catchments (Fig. 5).

Basement rocks of the Pittong area are folded quartz-rich Cambrian sandstones (arenites) and shales, intruded by granite bodies during the Late Devonian, resulting in an aureole of hornfels, approximately 1 km wide around the rim of the granites (Taylor et al., 1996). The hornfels are erosion resistant and generally form the watershed of the catchment, whereas the granites were deeply weathered during the late Palaeogene and early Neogene periods. The majority of the catchment comprises a thick (up to 30 m) regolith of sandy kaolin clay over the granite, but grades to thick (up to 10 m) grus (physically weathered granite residue) at the highest elevations, interspersed with outcropping granite tors. Some cappings of Tertiary alluvial gravels also occur in the upper catchment.

Salinity was attributed to increased recharge of local saline groundwater flow systems and rising water tables following clearing of the native vegetation (Nicholson et al., 1992). Subsequent research indicates that this is unlikely to be the sole explanation. Dahlhaus (2004) modeled regional groundwater flow directions using reduced water levels from groundwater bore data (Government of Victoria, 2010) for 5000 km^2 including an area to the north of Pittong and

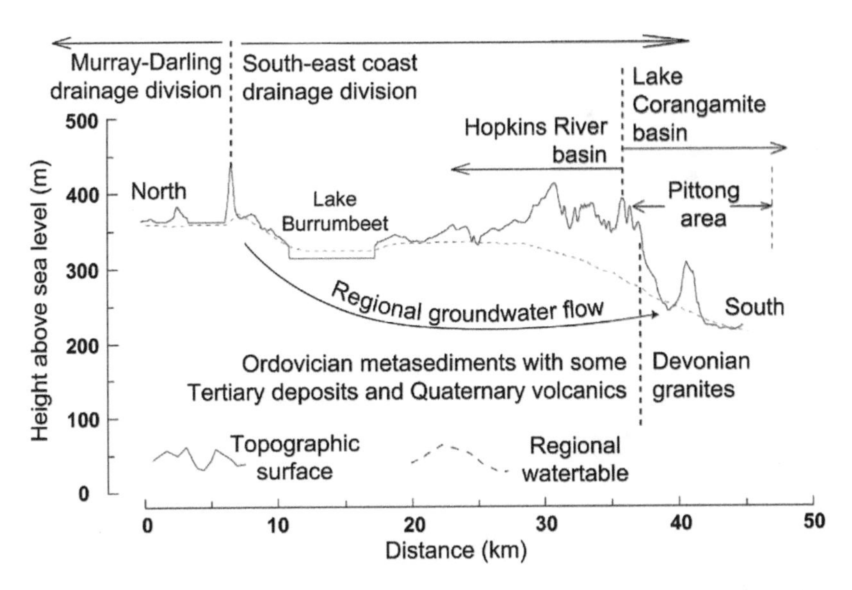

FIGURE 5 Schematic representation of the regional groundwater system driving saline groundwater discharge in the Pittong study area. Groundwater pressures in the Hopkins River basin to the north are driving groundwater discharge in the Pittong area (Lake Corangamite basin) to the south.

into the neighboring Hopkins River basin. Regional hydrogeological mapping (Bradley et al., 1994) assumes that groundwater in the unconfined surface aquifers of different lithologies is connected, so a coarse regional flow field was modeled using 715 bores and a grid cell size of 250 m. The model shows that groundwater flow from the Hopkins River basin contributes to discharge in the Pittong area of the Lake Corangamite basin (Fig. 5). If this is the case, groundwater discharge at Pittong cannot be controlled by management of recharge into the local flow systems (local surface water catchment) unless the groundwater levels in the neighboring river basin are also reduced.

4.2.3. Hydropedology and Conceptual Model for Pittong

A subcatchment of the Pittong area has been studied to investigate the relationship between groundwater discharge and soil salinization (Dahlhaus and MacEwan, 1997). Soil profiles were sampled in a 1900-m hillslope traverse from the crest below outcropping granite, to the subcatchment drainage line. The hillslope is undulating ranging from 3° to 15°. Soils were collected using push tubes (45 mm diameter) to a depth of 0.8 m, soil pits to 1.5 m depth, and hand augering (Edelmann 80 mm diameter) to 3.2 m (Conn, 1994). Additional soil samples were taken at 25-m intervals along a 400-m-long trench excavated for drainage installation on the lower section of the traverse (Fig. 6) in order to

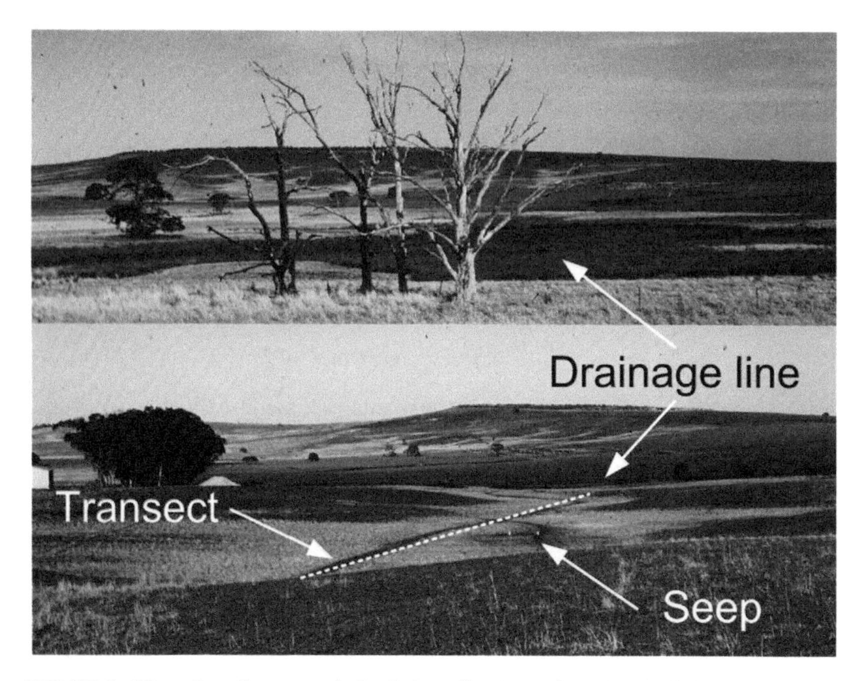

FIGURE 6 Views downslope towards the drainage line (upper image) and relative location of the survey transect and groundwater seep (lower image) at the Pittong research site. Transect is 400 m long. See also Fig. 7.

measure the electrical conductivity of the surface soil (0–0.1 m) and the subsoil at 1.0 m depth (Fig. 8). Electrical conductivity measurements were also made on standing water sampled from auger holes (Fig. 7).

Soils on the hillslope are all texture-contrast soils (Isbell, 2002), with sandy loam A1 horizons (75% sand; 10% silt; and 15% clay), sandy loam A2 (E) horizons (where present), and medium clay to heavy clay B horizons (40% sand; 12% silt; and 48% clay) (Conn, 1994). A2 horizons were absent from profiles sampled on the hill crest, a land-surface unit 3, convex creep slope (Conacher and Dalrymple, 1977), where soil profiles are shallow with coarse-grained physically weathered granite within 0.8 m of the ground surface. The major part of the hillslope traverse is interpreted as a land-surface unit 2, dominated by throughflow in the A2 horizons which thicken progressively downslope. The A2 horizons on these hillslopes are highly prone to winter waterlogging and are a hazard to farm traffic as they become extremely weak when saturated and result in bogging of farm vehicles. Subsoils are strongly structured (prismatic) and subsoil mottling and gley features are prominent, particularly in the lower gradient sections of the hillslope. Thick, dark gray (10YR3/1) organans and clay cutans cover the subsoil ped faces; ped interiors have light gray (10YR7/1) edges and yellowish-brown (10YR5/8) to red (10R5/8) cores. The redoximorphic features of the profile indicate periodic surface waterlogging and perched water on and in the B horizon, rather than a fluctuating groundwater (Lindbo et al., 2010; Vepraskas, 2004).

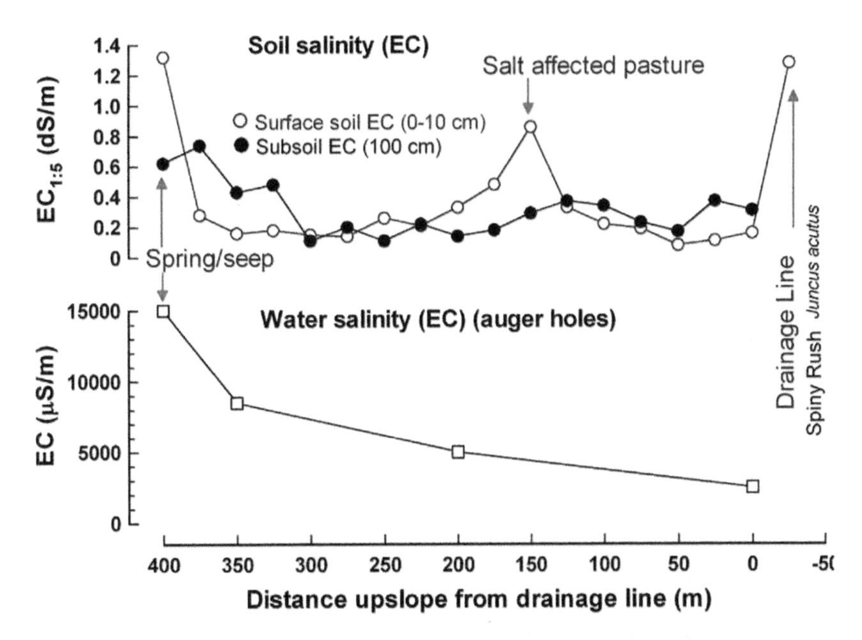

FIGURE 7 Electrical conductivity of soil and water for a 400-m transect extending from a point discharge site to a drainage line in the Pittong area (Upper Woady Yaloak catchment, see Fig. 6).

Electrical conductivity values for surface soil, deep subsoil, and water along a transect in a salt-affected section of the hillslope (Figs 6 and 7) show that the highest salinity values occur at, and close to, the seep, and decrease downslope. This pattern of salinity and the observed redoximorphic features support a hydrologic model of point discharge of saline groundwater being transmitted and diluted downslope via throughflow in the A2 of the land-surface unit 2 (Fig. 8).

4.2.4. Management of Degradation in the Pittong Area

Management of waterlogging and throughflow in the hillslopes can be achieved by shallow interceptor drains across slope or by subsurface pipe drainage. This will reduce the spread of salt from the seepage areas. However, without the flushing and dilution by freshwater throughflow, the soil salinity in the seep areas will increase potentially leading to scalding and erosion. One solution for Pittong has been to install a pipe drain at 1.0 m depth down the affected hillslope area and to connect a lateral pipe drain directly from the seepage area. Effectively, this diverts both the saline groundwater seepage and the throughflow into the pipe system, reducing surface discharge of groundwater and reducing waterlogging on the slope. There is a question of whether the net addition of salt into the drainage line and thence to the Woady Yaloak River by this management

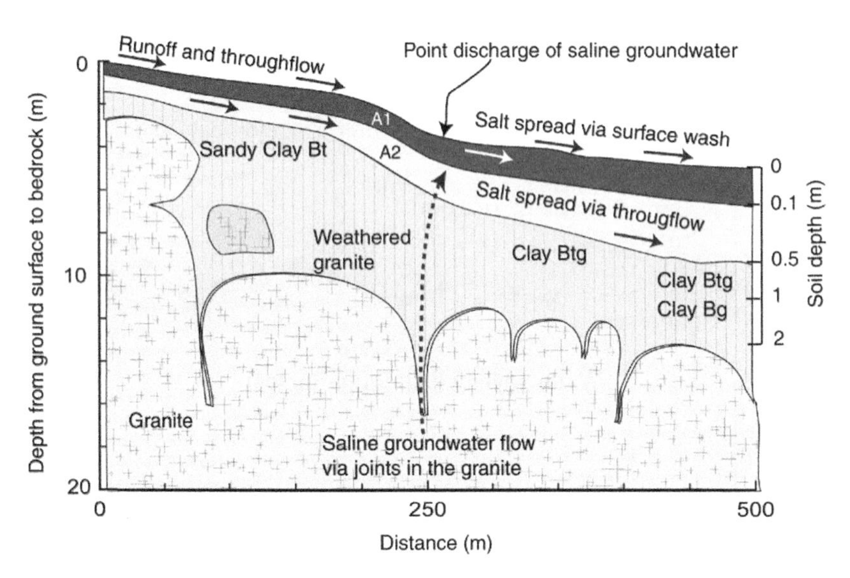

FIGURE 8 Schematic diagram illustrating the conceptual model to explain point source saline groundwater discharge into seeps and the spread of salt in mid- and lower slopes of the Pittong area. Topographic section is dominantly of a land-surface unit 2 (Conacher and Dalrymple, 1977). Soil-profile representation is grossly exaggerated for visual clarity of principal horizons; *Y*-axes indicate general depth to bedrock (left) and soil thickness (right). Saline groundwater flow is driven by the regional groundwater head beyond the local hillslope (shown in Fig. 5). *(Color version online)*

is greater or equal to the salt additions occurring without subsurface drainage; monitoring is required to answer this question in the long term.

4.3. Case Study: Heytesbury

The Heytesbury Settlement lies within a dissected coastal plain on the southern edge of the Western Plains of Victoria approximately 180 km southwest of Melbourne (Fig. 1) and consists of 40,000 ha cleared of trees between 1956 and 1971 to establish approximately 500 dairy farms. This was the last large-scale land clearing in Victoria and has been well documented by Fisher (1997). Tree clearing was achieved by two 240 hp bulldozers, each dragging 60 m of heavy chain attached to a 5-tonne steel land-clearing “High Ball”. Trees were windrowed and burnt. This operation was extremely disruptive and damaging to the soil profiles in the Heytesbury. The indigenous vegetation consisted of dense Stringybark forest (mainly *Eucalypt* spp.) on the valley slopes with 'grass tree plains' (*Xanthorrhoea* spp., *Melaleuca* spp., *Casuarina* spp., and *Banksia* spp.) on the sandy plateaus. During the early 1990s, it was estimated that the area of salinity was increasing by 9.1% annually in the Heytesbury Settlement, making it a priority target for management (Nicholson et al., 1992).

4.3.1. Soil and Land Degradation Features in the Heytesbury Settlement

Soil waterlogging, landslides, and small areas of salt-affected land compromise dairy pasture production in the Heytesbury Settlement. Waterlogging is extensive, occurring on all land-surface units throughout the winter, the main growing period for pasture. Treading by dairy cows, during the winter grazing season, results in pugging of the top 0.1 m of soil and compaction below the pugging depth. Salinity in the area presents as small areas of poorer pasture with more persistent waterlogging. The impact of salinity is masked to a degree by the general wetness of the area, which reduces the osmotic stress on plant species, and good ground cover is therefore maintained. Landslides are of various sizes and frequencies. Evidence from aerial photography taken from the 1940s (pre-settlement) to the 1990s (30 years post-settlement) is that the frequency of landslides has increased since clearing but the slips are generally smaller (Dahlhaus et al., 1996). Landslides occur at very low angles (6–14° slope) as complex slump-earthflows and translational failures (Dahlhaus and Miner, 1998) (Fig. 9).

4.3.2. Hydrogeology Affecting the Heytesbury Settlement Area

The Heytesbury is located in the Otway Basin, a sedimentary basin initiated in the Cretaceous and subsequently filled during the Palaeogene and Neogene with a series of paralic and marine sediments (Fig. 10). Three strata are significant for the hydrology of the area. The deepest of these, the Dilwyn Formation (Eocene), comprises sandy sediments (gravel, sand, silt, and clay) of deltaic and marine origins. The Dilwyn Formation is a major regional freshwater aquifer that is

FIGURE 9 Hillslope with hummocky landform due to landslips, on a dairy farm in the Heytesbury area. Trees have been planted on the upper slope as a management strategy to reduce waterlogging, landslips, and salinity.

recharged via outcrops on the flanks of the Otway Ranges. A thick (>200 m) (early Miocene) marine deposit, the Gellibrand Marl, overlies the Dilwyn Formation and functions as an aquitard. The Gellibrand Marl, a blue-gray marl varying to calcareous clay and silty, clayey limestone, is overlain in places by the Port Campbell Limestone (Late Miocene), a fine clayey limestone. The Dilwyn

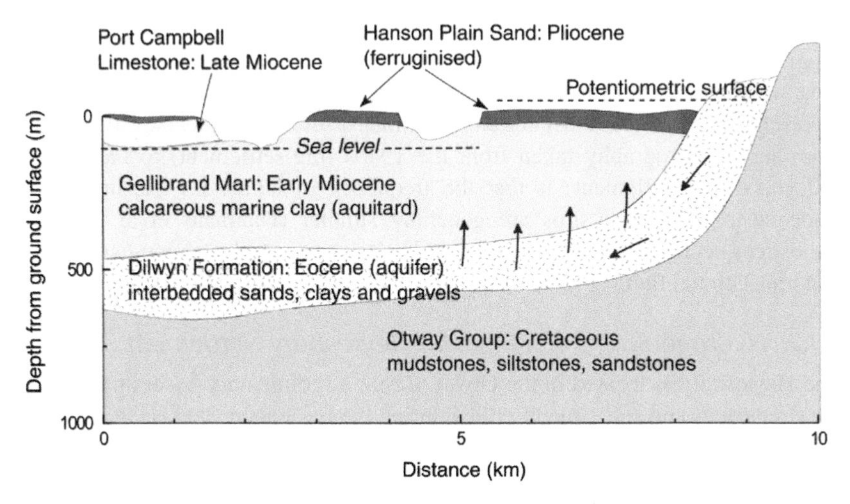

FIGURE 10 Schematic representation of the geology and regional groundwater system underlying the Heytesbury study area. *(Color version online)*

Formation dips steeply under the Gellibrand Marl and the potentiometric surface of the confined aquifer is approximately 20 m above the ground level throughout the Heytesbury Settlement resulting in the overlying confining bed being hydraulically pressurized from below (Bush, 2009) (Fig. 10).

The Hanson Plain Sand (Pliocene), a thin ($\leq$5 m) regressive-marine deposit of coarse to fine-grained sand with minor gravel, overlies the Gellibrand Marl and, to a lesser extent, the Port Campbell Limestone (Tickell et al., 1992). Across the northern section of the Heytesbury Settlement, the Hanson Plain Sand is found as a strong iron-cemented ferricrete; toward the coast the Hanson Plain Sand is found as a cemented, podsolized 'orstein' or 'coffee rock'. The contact boundary between the Hanson Plain Sand and the Gellibrand Marl is sharp, and functions as a throttle to downward movement of water (Fig. 11). Weathering of the Hanson Plain Sand ferricrete has resulted in distinct banding features or 'tiger mottles' (Fig. 12) similar to those observed in the Eastern Dundas Tablelands regolith. Residual products of this weathering are abundant ironstone gravels in the A2 horizons of soils northeast of the Heytesbury Settlement.

Local groundwater systems (salinity between 300 and 6000 mg/L) occur in the Gellibrand Marl, which is generally regarded as an aquitard (or at best, a low-yielding aquifer) with low hydraulic conductivities (Dahlhaus et al., 2002). The source of the salt in the groundwater is likely to be dominantly cyclic salt from rainfall (Downes, 1954), although there may still be some connate salt persisting from the marine depositional environment. Local groundwater flow systems also occur in isolated remnants of the Hanson Plain Sand on the plateaus.

FIGURE 11 Photograph showing the sharp boundary and strong texture contrast between the Hanson plain sand (overlying) and the Gellibrand Marl at 2.3 m below the ground surface. Both materials are strongly gleyed (dominantly light gray 5GY 7/1 with <20% yellow 10 YR 7/8 mottles).

FIGURE 12 'Tiger Mottles' in weathering ferruginized regolith (Hanson plain sand), 2 m depth. Similar features are seen in the Eastern Dundas Tablelands regolith. The banded features (red 10R 4/6 and gray 2.5Y 6/1) are due to the reduction and mobilization of iron on wetting fronts, or within (microscale) perched water tables. Illuviated clay is present within the gleyed (pale) bands. Pisoliths are forming as lithorelics from the weathered ferricrete in the oxidized bands. See also Figs 4 and 16. *(Color version online and in color plate)*

Seasonal groundwater discharge (generally fresh or of low salinity) is evident at spring lines along the unconformity between the Hanson Plain Sand and the underlying Gellibrand Marl.

4.3.3. Hydropedology and Conceptual Model for Hillslopes of the Heytesbury Area

Soils in the Heytesbury area have been studied at a range of scales; the most recent soil landform mapping for the region is at the 1:100,000 scale (Robinson et al., 2003). Hillslope soils were mapped at a scale of 1:10,000 for a dairy farm affected by salinity (Fig. 13), from which an initial conceptual model for soil distribution in the hillslopes and valleys was developed (MacEwan et al., 1996).

Hillslopes in the Heytesbury Settlement can be clearly classified according to the nine-unit land-surface model of Conacher and Dalrymple (1977) and related to soil types at farm scale and the occurrence of salinity. Soils on the slope crests, below the Hanson Plain Sand plateau, have developed in situ in the Gellibrand Marl and are characterized by a clay loam surface soil and heavy clay subsoil with soft calcium carbonate in the B horizon within 1.0 m of the surface ('Cooriemungle calcareous clay, CCC). All other soils have developed on materials transported by mass wasting (landslides) and alluvial processes. Depending on the origin of the transported materials (Hanson Plain Sand or

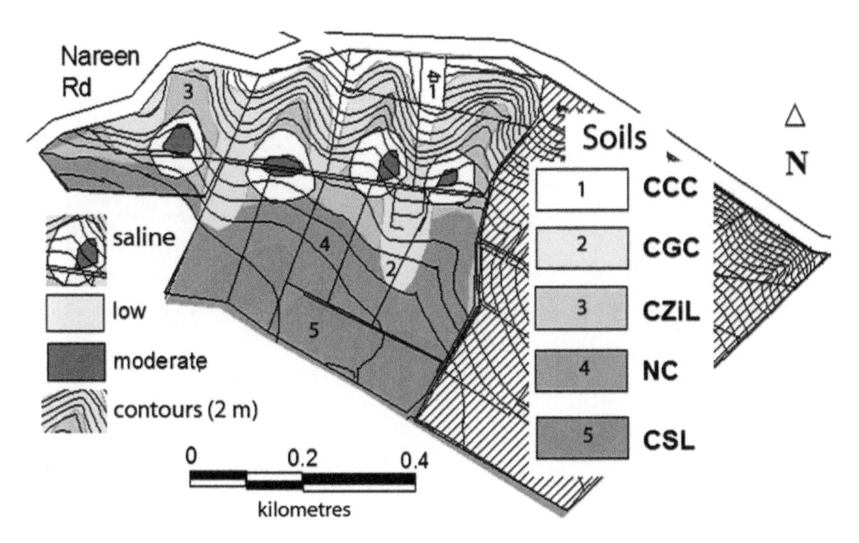

FIGURE 13 Soil and salinity map for a Heytesbury area dairy farm showing soil types, contours, and the locations of salt-affected pasture. See also Fig. 16 and text for explanation of soils legend. The soil symbols stand for local map units: CCC, Cooriemungle Calcareous Clay; CGC, Cooriemungle Gleyed Clay; CZiL, Cooriemungle Silt Loam; NC, Nareen Clay; and CSL, Cooriemungle Sandy Loam.

Gellibrand Marl), soils lower on the slopes and in the valleys contain variable surface textures (sandy loam to clay). Where sandier textures dominate, spodic horizons have developed below bleached A2 horizons, and subsoils are invariably gleyed, their textures ranging from silty clay loams to heavy clays. The clays have high shrink–swell properties and slickensides are prominent and large (up to a meter across at 1.5-meter depth).

Electromagnetic survey (EM38) was used to map apparent salinity. No salting was evident on the valley floor at this site and auger holes to 1.0 m depth remained dry. Salt-affected areas were apparent upslope at a relative elevation of 8.0 m above the valley floor (Fig. 13); nested piezometers at depths of 1.0, 10.0, and 20.0 m were installed to investigate groundwater pressure in one of these areas. The vertical groundwater gradients in this piezometer nest indicated declining groundwater head with depth. The 1.0-m piezometer showed a standing water level at 0.2 m above ground surface (artesian driven by local topographic head). The 5.0 and 20.0 m showed lower piezometric heads with depth, indicating that this was not a groundwater discharge area. MacEwan et al. (1996) proposed that the soil-salinization effects in this landscape position were most likely due to seasonal accumulation of runoff into concavities at the boundary between land-surface units 5 and 6 (landslides). In these positions, flow of surface water to the valley floor is impeded by surface topography and low soil hydraulic conductivity, leading to development of perched water tables, followed by evaporative concentration of salt in summer.

FIGURE 14 Pipeline route through the Heytesbury study area. Land-surface unit 3 (convex creep slope) in the close foreground, dropping away into land-surface units 5 (mass movement) and 6 (colluvial deposition). See also Fig. 16 and Table 1. *(Color version online)*

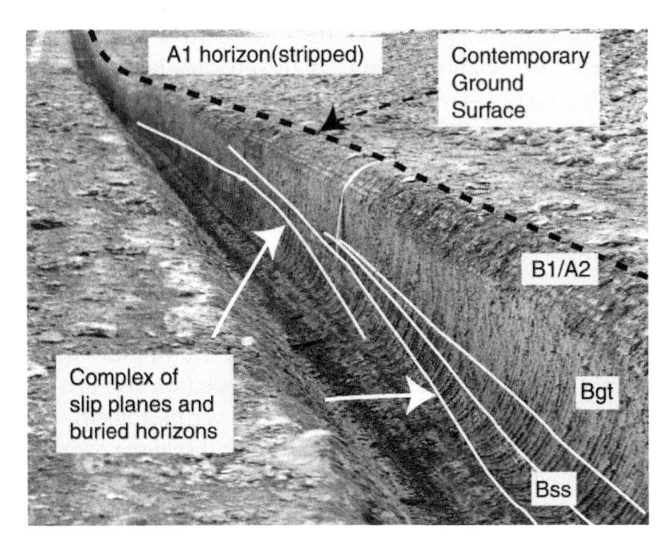

FIGURE 15 Trench section through a land-surface unit 5 (Table 1) on a hillslope in the Heytesbury study area (see also Fig. 14). The white lines show slip planes and buried soil horizons; the black dashed line follows the profile of the contemporary land surface. The soil is Cooriemungle gleyed clay (CGC, see also Figs 13 and 16). *(Color version online and in color plate)*

Soil investigations along a gas pipeline-installation trench 1.8 m deep by 0.8 m wide through the region (Figs 14 and 15) augmented the earlier conceptual model of the relationship between geomorphology, soils, and salinity for hillslopes and valley floors in the Heytesbury Settlement.

4.3.4. Management of Waterlogging and Salinity in the Heytesbury Area

Although the Dilwyn aquifer (fresh water) is artesian in the Heytesbury, the deeper groundwater pressures do not appear to be driving saline discharge in the area, which is presumed to be due to local geomorphic conditions on the hillslopes. It now appears that landslides and slip planes are strongly implicated in the occurrence of salinity on the Heytesbury hillslopes (Fig. 16). Discharge of saline water, exiting at toe slopes of landslides, occurs via the slip planes that provide preferential paths for movement of water through the Gellibrand Marl; degradation in the discharge areas is compounded by the waterlogged nature of

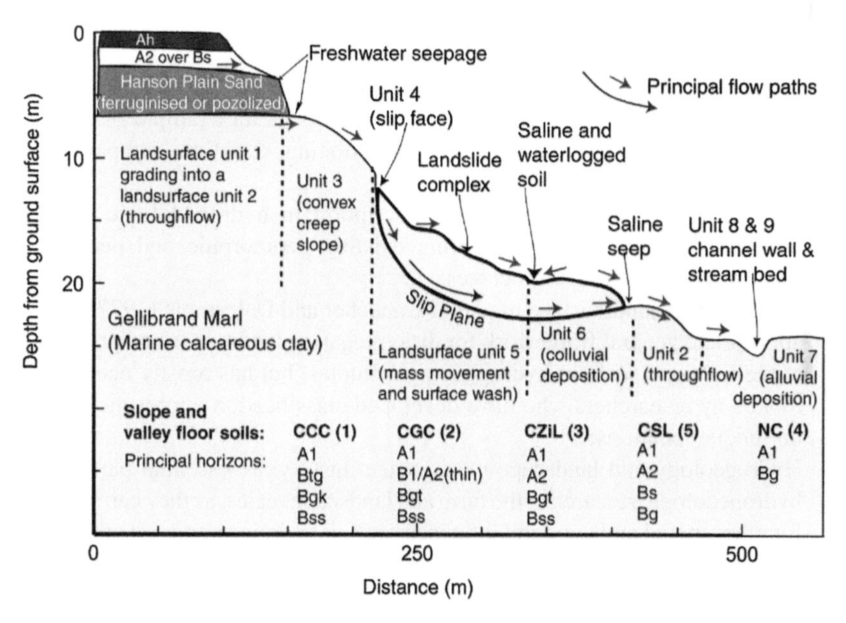

FIGURE 16 Simplified schematic of hillslope hydrology, landslips, soils, and salinity in the Heytesbury area. Discharge of groundwater occurs via weaknesses created by slip planes and results in seeps with low-level salinity at toe slopes; additionally, impeded drainage resulting from hummocky surface relief accentuates waterlogging and these locations develop low-level salinity due to evaporative concentration of salt and the lack of leaching. Land-surface units are based on Conacher and Dalrymple (1977) (Table 1). The land-surface unit sequence 5/6/2 may repeat on hillslopes (see Fig. 12). Principal soils on the slope and valley floor correspond to those shown in Fig. 13 (soil and salinity map). The soil symbols stand for local map units: CCC, Cooriemungle Calcareous Clay; CGC, Cooriemungle Gleyed Clay; CZiL, Cooriemungle Silt Loam; NC, Nareen Clay; and CSL, Cooriemungle Sandy Loam. *(Color version online)*

the soils. On-farm drainage systems using pipe and mole drainage (MacEwan et al., 1992) are appropriate for the relief of waterlogging and salinity on dairy farms in the Heytesbury as they both remove excess water from the profile and they reduce the hydraulic head acting on toe-slope discharge. Economic benefits are increased pasture production and utilization, which offset the cost of such installations within a few years, and maximization of land retained under pasture rather than set aside for trees planted for salinity control.

5. CONCLUSION

From the outset of each of our case studies, there have been questions concerning the role of land management in: (a) being the initial agent for land degradation and (b) the solution. Unraveling the local hydrologic processes from larger-scale regional processes has been seen as a key to determining whether local action could reduce groundwater-driven degradation or whether action would have to be at a much broader scale. Hydropedologic studies in Australia must be nested within broader scale contexts of hydrogeological systems. Main conclusions from the three case studies are:

- Each landscape has its own unique structure and controls but there are many common processes operating within them; each is an example of a soil–water–gravity system that determines partitioning, retention, and pathways for water, solutes, and sediments;
- A generic approach to landscape description in hydropedologic studies could be developed by augmenting existing geomorphic and pedologic models with hydrologic processes;
- The nine-unit land-surface model of Conacher and Dalrymple (1977) is one appropriate general framework for disaggregating landscapes and processes at the hillslope scale in hydropedologic studies, but has mostly been overlooked by researchers who have developed classification approaches based on unique locations;
- Hydrogeology and landscape-management history are essential partners to hydropedologic research at the farm and landscape scales, as they can account for other spatial and temporal influences on soil hydrology and soil condition.

REFERENCES

Bigham, J.M., Ciolkosz, E.J., Luxmoore, R.J. (Eds), 1993. Soil Color. SSSA Special Publication 31. Soil Science Society of America, Madison, WI.

Bigham, J.M., Fitzpatrick, R.W., Schulze, D.G., 2002. Iron oxides. In: Dixon, J.B., Schulze, D.G. (Eds.), Soil Mineralogy with Environmental Applications. Soil Science Society America Book Series No 7. SSSA, Madison, Wisconsin, pp. 323–366.

Bradley, J., Stanley, D.R., Mann, B.S., Chaplin, F., Foley, G., 1994. Ballarat Hydrogeological Map. Sheet SJ 54-8. 1:250,000 Murray Basin Hydrogeological Map Series. Australian Geological Survey Organisation, Canberra, ACT, Australia.

Brouwer, J., van de Graaff, R.H.M., 1989. Readjusting the water balance to combat dryland salting in southern Australia: changing the hydrology of a texture contrast soil by deep ripping. Agr. Water Manage. 14, 287–298.

Brouwer, J., Fitzpatrick, R.W., 2000. Characterisation of Nine Soils Down a Salt Affected Toposequence Near Gatum on the Dundas Tablelands in South-West Victoria. Technical Report 21/00. CSIRO Land and Water, Canberra, ACT, Australia.

Brouwer, J., Fitzpatrick, R.W., 2002a. Interpretation of morphological features in a salt-affected duplex soil toposequence with an altered soil water regime in western Victoria. Aust. J. Soil Res. 40, 903–926.

Brouwer, J., Fitzpatrick, R.W., 2002b. Restricting layers, flow paths and correlation between duration of soil saturation and soil morphological features along a hillslope with an altered soil water regime in western Victoria. Aust. J. Soil Res. 40, 927–946.

Bush, A.L., 2009. Physical and Chemical Hydrogeology of the Otway Basin, Southeast Australia. Ph.D. diss. School of Earth Sciences, Univ. of Melbourne, Melbourne, VIC, Australia.

Clark, R.M., Hekmeijer, P., 2004. Victorian Salinity Monitoring: Reassessment of Six Reference Discharge Monitoring Sites in the Corangamite Region. Centre for Land Protection research report 43. Primary Industries Research Victoria, Bendigo, VIC, Australia.

Commonwealth of Australia, 2004. Science Overcoming Salinity: Coordinating and Extending Science to Address the Nation's Salinity Problem. Report by the House of Representatives Standing Committee on Science and Innovation, Canberra, Australia.

Commonwealth of Australia, 2001. Australian Dryland Salinity Assessment 2000. Extent, Impacts, Processes, Monitoring and Management Options. Available at: http://nlwra.gov.au/files/products/national-land-and-water-resources-audit/pr010107/pr010107.pdf (verified 15 Mar. 2010). National Land and Water Resources Audit. Land and Water Australia, Canberra, ACT, Australia.

Conacher, A.J., Dalrymple, J.B., 1977. The nine unit landsurface model: an approach to pedogeomorphic research. Geoderma 18, 1–154.

Conn, A., 1994. Drainage Options for Waterlogging and Salinity Control in Granitic Soils at Pittong. Univ. of Ballarat, Ballarat, VIC, Australia.

Coram, J., Dyson, P., Evans, R., 2001. An Evaluation Framework for Dryland Salinity. Prepared for the National Land and Water Resources Audit. Available at: http://www.anra.gov.au/topics/salinity/pubs/national/evaluationframework.pdf (verified 15 Mar. 2010). Commonwealth of Australia, p. 123.

Cossens, B.L., 1999. Regional Groundwater Flow Systems of the Eastern Dundas Tableland, South-West Victoria. B. App. Sci. (Hons) diss. Univ. of Ballarat, Ballarat, VIC, Australia.

Dahlhaus, P.G., 2004. Salinity Risk in the Corangamite Region, Australia. Ph.D diss. School of the Environment, Flinders Univ, Adelaide, SA, Australia.

Dahlhaus, P.G., Miner, A.S., 1998. Stability of Clay Slopes in South West Victoria, Australia. In: Proceedings, Eighth International Congress, International Association for Engineering Geology and the Environment, 21–25 September 1998. A.A. Balkema, Vancouver, Canada, Rotterdam/Brookfield, pp. 1551–1556.

Dahlhaus, P.G., MacEwan, R.J., 1997. Dryland salinity in south west Victoria: challenging the myth. In: McNally, G.H. (Ed.), Collected Case Studies in Engineering Geology, Hydrogeology and Environmental Geology. 3rd Series. Environmental, Engineering and Hydrogeology Specialist Group, Geological Society of Australia, Sydney, pp. 165–180.

Dahlhaus, P.G., Buenen, B.J., MacEwan, R.J., 1996. Landslide studies in the Heytesbury region, south west Victoria. In: Soil Science – Raising the Profile. Proceedings of the Australian and New Zealand Soil Science Societies National Conference, The University of Melbourne, 1–4 July, 1996, vol. 3. Australian Society of Soil Science Inc., pp. 55–56.

Dahlhaus, P.G., Heislers, D.S., Dyson, P.R., 2002. Groundwater Flow Systems: Corangamite CMA. Consulting report CCMA 02/02. Corangamite Catchment Management Authority, Colac, VIC, Australia.

Dahlhaus, P.G., MacEwan, R.J., Nathan, E.L., Morand, V.J., 2000. Salinity on the southeastern Dundas Tableland, Victoria. Aust. J. Earth Sci. 47, 3–11.

Domenico, P.A., Schwartz, F.W., 1990. Physical and Chemical Hydrogeology. Wiley, New York.

Downes, R.G., 1954. Cyclic salt as a dominant factor in the genesis of soils in South-Eastern Australia. Aust. J. Agr. Res. 5, 448–464.

Eberbach, P.L., 2003. The eco-hydrology of partly cleared, native ecosystems in southern Australia: a review. Plant Soil 257, 357–369.

Fawcett, J., 1996. A Hydrological Investigation of Soil Waterlogging in the Bulart Region, South-eastern Dundas Tableland. B. App. Sci. (Hons.) diss. Univ. of Ballarat, Ballarat, VIC, Australia.

Fawcett, J., 2004. Processes and Implications of Scald Formation on the Eastern Dundas Tableland: A Case Study. Ph.D. diss. Univ. of Melbourne, Melbourne, VIC, Australia.

Fawcett, J., Fitzpatrick, R., Norton, R., 2008. Inland acid sulfate soils of the spring zones across the Eastern Dundas Tableland, south eastern Australia. In: Fitzpatrick, R., Shand, P. (Eds.), Inland Acid Sulfate Soil Systems across Australia. CRC LEME, Perth, Australia, pp. 268–280. CRC LEME Open File Report 249.

Fetter, C.W., 1980. Applied Hydrogeology. Merrill Publishing Company.

Fisher, H., 1997. Heytesbury: Once in a Lifetime. Spectrum Publications, Melbourne.

Fitzpatrick, R.W., Self, P.G., 1997. Iron oxyhydroxides, sulfides and oxyhydroxysulfates as indicators of acid sulfate weathering environments. In: Auerswald, K., Stanjek, H., Bigham, J.M. (Eds.), Soils and Environment. Soil Processes from Mineral to Landscape Scale, pp. 227–240. Adv. GeoEcol. 30.

Fitzpatrick, R.W., Fritsch, E., Self, P.G., 1996. Interpretation of soil features produced by ancient and modern processes in degraded landscapes: V development of saline sulfidic features in non-tidal seepage areas. Geoderma 69, 1–29.

Freeze, R.A., Cherry, J.A., 1979. Groundwater. Prentice-Hall.

Fritsch, E., Fitzpatrick, R.W., 1994. Interpretation of soil features produced by ancient and modern processes in degraded landscapes: I a new method for constructing conceptual soil-water-landscape models. Aust. J. Soil Res. 32, 889–907.

Gardner, W., Fawcett, J., Fitzpatrick, R., Norton, R., Trethowan, M., 2004. Chemical reduction causing land degradation. II detailed observations at a discharge site in the Eastern Dundas Tablelands, Victoria, Australia. Plant Soil 267, 85–95.

GHCMA, 2005. Glenelg Hopkins Salinity Plan 2005–2008. Glenelg Hopkins Catchment Management Authority, Hamilton. http://www.ghcma.vic.gov.au/imageandfileuploads/Glenelg%20Hopkins%20Salinity%20Plan%202005-2008.pdf

Gibbons, F.R., Downes, R.G., 1964. A Study of the Land in South-Western Victoria. Soil Conservation Authority, Kew, Victoria, Australia. Government of Victoria. 2010. Water Resources Data Warehouse. http://www.vicwaterdata.net/vicwaterdata/home.aspx

Government of Victoria. 2010. Water Resources Data Warehouse. http://www.vicwaterdata.net/vicwaterdata/home.aspx

Hatton, T.J., Nulsen, R.A., 1999. Towards achieving functional ecosystem mimicry with respect to water cycling in southern Australian agriculture. Agrofor. Sys. 45, 203–214.

Hills, E.S., 1939. The physiography of north-western Victoria. Proc. Royal Soc. Victoria 51, 297–323.

Isbell, R.F., 2002. The Australian Soil Classification System, revised ed. CSIRO Publishing, Melbourne, Australia.

Jerinic, F.L., 1993. A Hydrogeological Study of Dryland Salinity in the Bulart Region, Western Victoria. B. App. Sci. (Hons.) diss. Geology department, Univ. of Ballarat, Ballarat, VIC, Australia.

Joyce, E.B., Webb, J.A., Dahlhaus, P.G., Grimes, K.G., Hill, S.M., Kotsonis, A., J.MartinMitchell, M.M., Neilson, J.L., Orr, M.L., Peterson, J.A., Rosengren, N.J., Rowan, J.N., Rowe, R.K., Sargeant, I., Stone, T., Smith, B.L., White, S., Jenkin, J.J., 2003. Geomorphology. In: Birch, W.D. (Ed.), Geology of Victoria. 23 Geological Society of Australia Special Publication 23. Geol. Soc. of Australia, pp. 533–561 (Victoria Division).

Lin, H., 2003. Hydropedology: bridging disciplines, scales and data. Vadose Zone J. 2, 1–11.

Lin, H., 2010. Earth's critical zone and hydropedology: concepts, characteristics, and advances. Hydrol. Earth Syst. Sci. 14, 25–45.

Lindbo, D.L., Stolt, M.H., Vepraskas, M.J., 2010. Chapter 8-redoximorphic features. In: Stoops, G., Marcelino, V., Mees, F. (Eds.), Interpretation of Micromorphological Features of Soils and Regoliths. Elsevier, Amsterdam, pp. 129–147.

MacEwan, R.J., Gardner, W.K., Ellington, A., Hopkins, D.G., Bakker, A.C., 1992. Tile and mole drainage for control of waterlogging in duplex soils of south-eastern Australia. Aust. J. Exp. Ag. 32, 865–878.

MacEwan, R.J., 2005. "Tiger Mottles" Subsoil Weathering Features in a Ferruginised Regolith. Presentation to "Mapping Our New Horizons" Soil Science Society of America Sixty Nineth Annual Meeting. November 6–10, 2005. Available at: http://crops.confex.com/crops/2005am/techprogram/P7147.HTM. (verified 19 Mar. 2010). Soil Science Society of America, Salt Lake City.

MacEwan, R.J., Dahlhaus, P.G., Robertson, E.H., Eldridge, R.E., 1996. Waterlogging and dryland salinity as influenced by pedogeomorphic history in the Simpson area. In: Soil Science – Raising the Profile. Proceedings of the Australian and New Zealand Soil Science Societies National Conference, The University of Melbourne, 1–4 July, 1996, vol. 3. Australian Society of Soil Science Inc, pp. 145–146.

Matters, J., Bozon, J., 1995. Spotting Soil Salinity. A Victorian Field Guide to Salt Indicator Plants. Land Protection Division, Conservation and Natural Resources, Melbourne, Victoria.

McDonald, R.C., Isbell, R.F., Speight, J.G., Walker, J., Hopkins, M.S., 1990. Australian Soil and Land Survey. Field Handbook, second ed. Inkata, Melbourne.

Munroe, M., 1998. Salinity Discharge Mapping for the Dundas Tableland in the Glenelg Salinity Region. Department of Natural Resources and Environment, Victoria.

Narasimhan, T.N., 2005. Pedology: a hydrogeological perspective. Vadose Zone J. 4, 891–898.

Nathan, E., 2000. Giving salt some history and history some salt. Aust. Hist. Stud. 31 (115), 222–236.

Nathan, E.L., 1998. Dundas Landscapes and Dryland Salinity. M. App. Sci. diss. Univ. of Ballarat, Ballarat, VIC, Australia.

Nathan, E.L., 1999. Dryland salinity on the Dundas Tableland: an historical appraisal. Aust. Geogr. 30, 295–310.

National Research Council. (NRC), 2001. Basic Research Opportunities in Earth Science. National Academy Press, Washington DC, USA.

Nicholson, C., Dahlhaus, P.G., Anderson, G., Kelliher, C.K., Stephens, M., 2006. Corangamite Salinity Action Plan: 2005–2008. Corangamite Catchment Management Authority, Colac, Victoria.

Nicholson, C., Straw, W., Conroy, F., MacEwan, R.J., 1992. Restoring the Balance. Dryland Salinity Strategy for the Corangamite Salinity Region. Corangamite Salinity Forum, Colac, VIC, Australia.

Northcote, K.H., 1979. A Factual Key for the Recognition of Australian Soils. Rellim Technical Publications, Adelaide, South Australia.

Northcote, K.H., 1983. Soils, Soil Morphology and Soil Classification: Training Course Lectures. Rellim Technical Publications, Adelaide, South Australia.

Pachepsky, Y.A., Rawls, W.J., 2004. Development of Pedotransfer Functions in Soil Hydrology, Elsevier, Amsterdam.

Paine, M., 2003. Nature and Extent of Pliocene Strandlines on the Dundas Tableland, South-western Victoria. In: Roach, I.C. (Ed.), Advances in Regolith. CRC LEME, pp. 314–318.

Paine, M.D., Phang, C., 2005. Dundas Tableland, Victoria. In: Anand, R.R., De Broekert, P. (Eds.), Regolith Landscape Evolution across Australia, A Compilation of Regolith Landscape Case Studies with Regolith Landscape Evolution Models. Cooperative Research Centre for Landscape Environments and Mineral exploration, pp. 242–245.

Pannell, D.J., 2001. Dryland salinity: economic, scientific, social and policy dimensions. Aust. J. Agr. Resour. Ec. 45, 517–546.

Park, S.J., van de Giesen, N., 2004. Soil-landscape delineation to define spatial sampling domains for hillslope hydrology. J. Hydrol. 295, 28–46. DOI: 10.1016/j.jhydrol.2004.02.022.

Robinson, N., Rees, D., Reynard, K., MacEwan, R.J., Dahlhaus, P.G., Imhof, M., Boyle, G., Baxter, N., 2003. A Land Resource Assessment of the Corangamite Region. Department of Primary Industries, Bendigo, VIC, Australia.

Soil Survey Division Staff, 1993. Soil Survey Manual. Soil Conservation Service, U.S. Department of Agriculture Handbook 18. Available at: http://soils.usda.gov/technical/manual/ (verified 2 April 2010).

Soil Survey Staff, 2010. Keys to Soil Taxonomy, eleventh ed. USDA-Natural Resources Conservation Service, Washington, DC.

Spies, B., Woodgate, P., 2005. Salinity Mapping Methods in the Australian Context. Report Prepared for the Natural Resource Management Ministerial Council. Land and Water Australia. Department of the Environment and Heritage; and Agriculture, Fisheries and Forestry, Canberra, ACT, Australia.

Taylor, D.H., Whitehead, M.L., Olshina, A., Leonard, L.G., 1996. Ballarat 1:100,000 Map Geological Report. Geological Survey of Victoria Report 101. Energy and Minerals Victoria, Melbourne, VIC, Australia.

Ticehurst, J.L., Cresswell, H.P., McKenzie, N.J., Glover, M.R., 2007. Interpreting soil and topographic properties to conceptualise hillslope hydrology. Geoderma 137, 279–292.

Tickell, S.J., Edwards, J., Abele, C., 1992. Port Campbell Embayment 1:100,000 map and geological report. Geological Survey of Victoria, Report 95.

Tóth, J., 1962. A theoretical analysis of groundwater flow in small drainage basins. J. Geophys. Res. 68, 4795–4812.

Tóth, J., 1970. A conceptual model of groundwater regime and the hydrologic environment. J. Hydrol. 10, 164–176.

Tóth, J., 1999. Groundwater as a geological agent: an overview of the causes, processes, and manifestations. Hydrogeol. J. 7, 1–14.

van Bueren, M., Price, R.J., 2004. Breaking Ground. Key Findings from 10 years of Australia's National Dryland Salinity Program. Land and Water Australia, Canberra. Available at: http://ndsp.gov.au/library/scripts/14A86669-110B-4210-89B2-9FAD30715F8.pdf (verified 22 Mar 2010).

Vepraskas, M.J., 2004. Redoximorphic Features for Identifying Aquic Conditions. N. C. Agri. Res. Serv., Raleigh, NC. Tech. Bull. 301.

Walker, G., Gilfeder, M., Evans, R., Dyson, P., Stauffacher, M., 2003. Groundwater Flow Systems Framework. Essential Tools for Planning and Salinity Management. Murray-Darling Basin Commission, Canberra, ACT, Australia.

Williams, W.D., 1995. Lake Corangamite, Australia, a permanent saline lake: conservation and management issues. Lakes Reservoirs: Res. Manage. 1, 55–64.

Woof, C., 1994. Geological and Hydrogeological Mechanisms of Salinisation at Bulart. B. App. Sci. (Hons.) Diss. Geology department, Univ. of Ballarat, Ballarat, VIC, Australia.

Chapter 15

Hydropedology as a Powerful Tool for Environmental Policy and Regulations: Toward Sustainable Land Use, Management and Planning

J. Bouma[1]

ABSTRACT

Separate, relatively small disciplines have to define their particular "niche" in the societal debate on sustainable development because if they do not, they are bound to get lost in broad discussions covering interrelated and complex environmental, social, and economic aspects of sustainable development. Pedology and soil physics/hydrology have reached a development phase where an integrated approach to land and water management as proposed in hydropedology is more productive than continuing on two monodisciplinary tracks. Modern developments in the relation between science and society require transdisciplinarity including interaction with various stakeholders and policymakers. The recent Soil Protection Strategy of the European Union is used as a guide to define hydropedologic "niches" for research in terms of seven soil functions. In doing so, attention should be paid to: (i) the policy cycle, ranging from signaling a problem to evaluating research results; (ii) the "knowledge chain", ranging from K1 (tacit knowledge) to K5 (cutting-edge science) and back, and (iii) a systems analysis in space and time, defining various options for sustainable land use. Hydropedology research is needed on water regimes in undisturbed soils in the field and in watersheds, using an array of new monitoring and modeling technologies that do not implicitly assume soils to be isotropic and homogeneous.

1. INTRODUCTION

Science and society have a rather cumbersome and complicated relationship. There can be no doubt about the major impact of science over the years in improving living conditions of humankind. But many scientists feel that they

[1]Wageningen University, Spoorbaanweg 35, 3911CA Rhenen, The Netherlands
Emalil: johan.bouma@planet.nl

Hydropedology, Edited by H. Lin. DOI: 10.1016/B978-0-12-386941-8.00015-9

produce best results when well funded and when they can work in freedom and in pursuit of the truth, being guided by their curiosity. Society, on the other hand, is increasingly cost conscious and requires value for money leading to ever-stricter financial conditions and to complicated discussions on the "value" concept. All this creates tension: the way research is managed increasingly determines the amount of time that can be freely spent on policy-related research. The official policy line by the research community is, however, clear. The International Council for Science (ICSU) states, for example, in its strategic plan for 2006–2010: ..."*where science is used for the benefit of all...and where scientific knowledge is effectively linked to policy making*". Their goal is to "*strengthen the international science for the benefit to society*" (ICSU, 2005). The implicit suggestion here is that benefits of society are realized when scientific knowledge is effectively linked to policy making. All over the world, environmental policy has indeed resulted in rules and regulations and the question at hand is therefore whether research has indeed been successful in developing effective legislation. Regarding research, attention in this chapter will be restricted to hydropedology research, combining pedology and hydrology.

A recent example of major environmental legislation is the European Union (EU) Soil Protection Strategy (SPS), adopted in September 2006. The overall objective of this strategy is "*protection and sustainable use of soil by preventing further soil degradation by preserving its functions and by restoring degraded soil to a level of functionality consistent with current and intended use*". Nearly 60 soil scientists from countries within the EU initiated a four-year study to define research needs for this strategy (ECSSS, 2004). Their recommendations included 110 research priorities focused on eight threats to soils. This was not too convincing to policymakers as it appeared to reflect the scattered character of the soil research community and this example may therefore serve to illustrate the relevance of questions as to the effectiveness of research for policy making. In this chapter, the possible role and potential of hydropedology in defining effective environmental legislation will be investigated by: (i) exploring links between research and policy making when pursuing sustainable development, (ii) examining the role of hydropedology as a powerful tool in studies on sustainable development and illustrating the need for a systems analysis focused on soil functions, and (iii) considering the priority areas for hydropedologic research in relation to the seven soil functions.

2. LINKS BETWEEN RESEARCH AND POLICY MAKING WHEN PURSUING SUSTAINABLE DEVELOPMENT

Only some aspects of the complex relation between research and sustainability can be covered in the limited context of this chapter and attention will therefore be confined to (i) the policy cycle, (ii) different types of knowledge and expertise, (iii) dealing with uncertainty in a policy context, and (iv) the challenge to realize effective research on sustainable development.

2.1. The Policy Cycle

Policy formulation is not a single event but involves a complex trajectory in which the following phases can be distinguished (e.g. Bouma et al., 2007):

1) The *signaling* phase in which problems are identified, preferably based on specific characterization of actual conditions. Increasingly, signaling involves active engagement of various stakeholders often with strong and deviating opinions (e.g. Jacques et al., 2008; Bouma et al., 2011a).
2) The *design* phase in which solutions or alternative options for possible corrective action are defined based on research using existing information or results from new research initiated in the context of the particular problem being studied.
3) The *decision* phase in which a selection is made by policymakers or stakeholders of one particular solution presented in phase 2. The decision process often involves intense negotiations during the political decision process. The *decision* phase may result in specific rules and regulations that will guide later *implementation.* This may take considerable time and, in the meantime, a new dangerous contaminant may emerge (a case of *signaling*). Then, a decision has to be made as to whether a revised design and decision phase should be initiated based on this feedback or whether the original procedure should be continued to avoid delay. This, obviously, is a political decision but scientists can play an important role in pointing out consequences of either procedure.
4) The *implementation* phase in which the selected solution to the problem being considered is implemented in practice.
5) The *evaluation* phase in which the entire process is evaluated in terms of learning experiences. This phase may require observations and monitoring to document results obtained.

To be effective, the role of research in each of these phases should be different and researchers are well advised to be aware of this. *Signaling* can originate from different sources, for example, from various stakeholders, government agencies, or action groups. In addition, researchers can play an important role in signaling particular problems, preferably even before they are likely to occur. The interactive character of policy formulation these days requires a broad signaling process to the effect that a wide range of opinions is represented, even ones that appear rather weird at first sight. Researchers can play an important and active role here in gathering opinions and in providing supporting evidence. *Designing* is the classical role of research and many research projects are confined to this activity. In a democracy, *Deciding* is the prerogative of society but, in practice, policymakers or responsible stakeholders play a major role.

The role of researchers is very limited even though they sometimes play an important role in feeding the decision process at the right moment with crucial information. *Implementing* is usually not seen as part of the responsibility of researchers. They have done their research, a decision has been made to implement the results, and it is up to others to follow up. Still, researchers can play an important role in supporting the implementation process by making sure that deviations from the original plans do not occur. The time for new and interesting ideas that are likely to throw implementation off course is over. Researchers will have to realize that a measurable success of their work is only reached when at least some of their results are successfully implemented in the end. If all their research results too often wind up in a dusty drawer, their reputation is bound to suffer. Ensuring that their research findings are implemented and being involved in the implementation process in general, therefore, expresses well-understood self-interest. Finally, *evaluating* the entire process from signaling to implementation is very useful for all participants involved as a joint learning experience. Scientists have a particular responsibility here in applying objective reasoning to the evaluation of the outcomes that can provide positive feedback to efforts in future. The societal image of research suffers when too much research does not result in implementation. The impact of this harmful so-called: "knowledge paradox" can be lessened when systematically following the policy chain (Bouma, 2010). Of course, not every research project can lead to successful implementation but at least some projects should.

Most current research projects focus on *design* alone and a plea can be made to broaden the approach to the entire policy cycle, very much including *evaluation*. This will result in larger and more costly projects and also in more effective and successful research in the end. Of course, the policy cycle, as represented here, is a strong simplification of reality where different phases often occur simultaneously. Still, distinction of the phases may help to clarify and provide a systematic approach toward defining the role of research in society (see examples in Bouma et al., 2007). Recognizing the diverse interests and value judgments of the various partners involved in the policy cycle, attention has recently also been paid to the type of interaction processes, as such, and specifically to the question which type of knowledge is most relevant during various phases of the entire process (e.g. Bouma et al., 2011a).

2.2. Different Types of Knowledge and Expertise

Knowledge is the major tool (or better, the major weapon) of science. Different forms of knowledge are operational in the policy cycle. Current emphasis in research on publishing articles in international, refereed scientific journals implies a focus on one particular type of quantitative and mechanistic knowledge. But other types of knowledge are also quite important. The scheme of Fig. 1 defines five types of knowledge (K1–K5) in terms of two ranges of

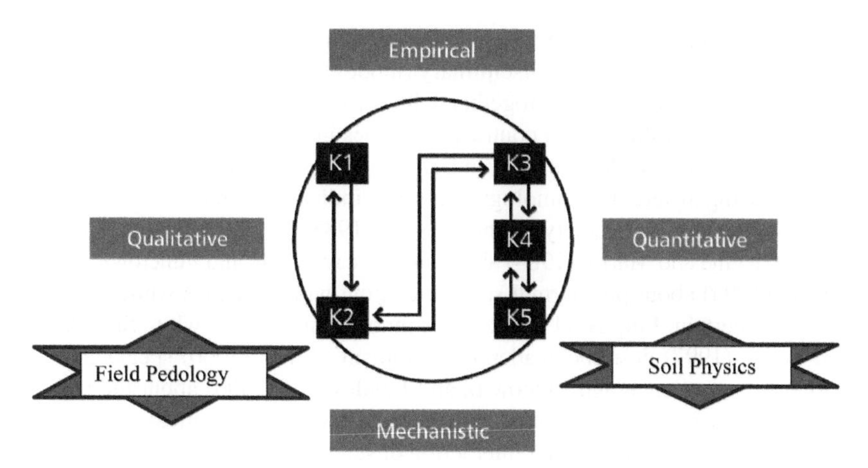

FIGURE 1 Knowledge diagram, defining five types of knowledge as a function of four criteria, as indicated. Soil survey activities or field pedology (K2) and soil physical research (K5) combine into hydropedology, as discussed in text. *(Modified from Bouma et al., 2008)*

characteristics: from qualitative to quantitative and from empirical to mechanistic. K1 represents user expertise, sometimes referred to as *tacit knowledge* . K2 is expert knowledge from practitioners or applied scientists. The still essentially empirical character of this knowledge is supported by a better understanding of the underlying processes and may still be considered *tacit* in nature. K3–K5 represents increasingly specific scientific knowledge. K3 represents e.g. empirical statistical relationships, while the underlying processes are increasingly expressed in quantitative terms in K4 by simple models and in K5, by cutting-edge science, a favorite entry point for the scientific journals mentioned above. Two types of arrows are shown in Fig. 1: from K1 to K5 indicating the use of tacit knowledge when planning research and from K5 to K1 indicating the flow of knowledge from research to practitioners and experts. The latter used to be considered in terms of extension services. Bouma and Droogers (1999) presented an example illustrating the K1–K5 scheme for determnining the soil moisture supply capacity.

The scheme of Fig. 1 focuses on field pedology, as indicated, as expressed by soil survey and its interpretations because hydropedology has particular links with soil survey interpretations. In pedology, there is also cutting-edge K5 research that is focused on soil genesis, for example, dealing with clay mineralogy, geochemical techniques, and modeling of pedogenesis. Such work may also be relevant for the soil functions but its focus is not primarily on applications.

Signaling, as discussed above, requires acknowledgment of K1, K2, and, evermore frequent, K3 knowledge of stakeholders who have access to Internet and are increasingly well informed.

During the last two decades, many papers and books have appeared covering changing relationships between science and society. The influential book by Gibbons et al. (1994) distinguished traditional monodisciplinary (mode-1) science versus transdisciplinary (mode-2) science in which scientists of different disciplines work together with various stakeholders and policymakers. Interdisciplinarity, in contrast, describes interaction between different scientific disciplines. Hessels and van Lente (2008) pointed out that many other authors, using different terminology, have also indicated increasing interaction between science and society. Bohme et al. (1983) speak about finalization science, Irvine and Martin (1984) about strategic research, and Functowicz and Ravetz (1993) about postnormal science dealing with decisions while knowledge is uncertain. Edquist (1997) defines innovation systems, while Slaughter and Leslie (1997) describe academic capitalism. Ziman (2000) speaks about postacademic science and Etzkowitz and Leydesdorff (2000) about the triple helix. Ten years later, we must conclude that many theoretical academic studies have appeared covering interaction of science with society, focusing on transdisciplinarity in various forms. But much less has happened in the real world where strategic plans and project requirements advocate the transdisciplinary approach, while many scientists stick to their old habit of doing monodisciplinary research, focusing on writing papers which is considerably less time consuming than getting bogged down in meandering discussions with a wide array of stakeholders and policymakers. This can be understood in the context of the current research climate where managers play an evermore important role but that does not change the facts. There is, unfortunately, a painful gap between words and deeds in the scientific community and this discrepancy is being noted by critical observers in the societal and political arena (e.g. Bouma et al., 2011a). Transdisciplinary science requires recognition, identification, and application of the entire knowledge chain K1–K5 and of K5–K1. Bouma (2001) made a plea for "research negotiation" where interaction of researchers with stakeholders or politicians starts at the K1 level and is gradually moved up the chain as the researcher points out the gains to be realized when investing in additional research as long, of course, as gains exceed costs. This procedure involves the stakeholders right from the start, takes them seriously, and is a more effective way to "sell" research as compared with the more traditional procedures where a project only involves research and reporting. Also, because of the need to publish results in scientific journals, research often applies K4 or K5 techniques, whereas K3 level work could have been adequate for some problems. Researchers, however, operate in an environment with ever-higher political and economic pressures and in a research culture that equates "success" with publishing in international refereed journals. This severely limits their flexibility and possibilities for time-consuming transdisciplinary research (e.g. Baveye and Philippe, 2008). Criticizing individual researchers is therefore too easy and it is more realistic to advocate new approaches to research management. Bouma et al. (2007) proposed *Communities of Scientific Practice* as a new way to deal with transdisciplinarity.

2.3. Dealing with Uncertainty in a Policy Context

There is, however, one remaining problem when dealing with knowledge to be obtained by research. The concept of "knowledge" can be defined as "fully internalized information as a result of a learning process". Information, in turn, is "data with a meaning and a context" while data itself is "a known quantity or entity". Just gathering data is not meaningful when it does not turn into information and knowledge. Moreover, the application of knowledge is bound to be more successful when applied with a certain degree of wisdom, which describes the indefinable ability to only take action when the time and place are right. A wise person knows not only how to listen and how to wait but also how to act decisively when needed. Uncertainty cannot always be solved by scientific research as was pointed out by van Asselt (2000), as cited by Bouma (2005), who distinguished seven forms of uncertainty, pointing out that only some of them can be resolved by science right now or, perhaps, even in future:

1) **Lack of measurements.** "We could know" if only we would be willing to spend the money to invest in existing or new measuring equipment. For example, new proximal and remote sensors offer exciting perspectives for soil science.
2) **Inaccurate measurements.** "We roughly know" and have no technical opportunity at this time to make better measurements. But when we invest in new techniques it is likely that such measurements can be made in future. This could mean either more measurements, reducing standard deviations of results, or the use of new, more reliable techniques.
3) **Cannot be measured yet but is felt to be real.** "We know what we don't know". For soils we can think of elusive issues such as "soil quality" or "resilience". One may expect that measurement in future may be possible, but not yet.
4. **Conflicting evidence.** "We don't know what we know". Different types of measurements of the same feature give different results. For soil science, we can think of chemical measurements based on different extraction techniques or different durations of shaking samples, yielding different numbers. Often, an arbitrary selection is made for one procedure to act as a standard. But this does not solve the basic problem.
5) **Lack of knowledge that may be gained in future.** "We don't know what we don't know yet". This positive attitude, which is supported by science history, represents blind trust in creative science and is an attractive and inspiring guide for any scientist.
6) **Lack of knowledge that is unlikely to be gained in future.** "We cannot know". One can think of measuring soil properties over large areas which will never result in a "complete" picture (e.g. Webster, 2000).
7) **Experiences of uncertainty beyond knowledge.** "We will never know". Here, religion may form the basis for "beliefs" beyond measurement that can be a major thrust for human happiness and fulfillment.

When interacting with stakeholders, uncertainties of types 5, 6, and 7 are common as people are motivated by certain experiences, ideologies, or religions. It is important to recognize this as forms of tacit K1 knowledge and be aware of the limits of the scientific endeavor to the first four (or perhaps five) sources of uncertainty (see also Bouma et al., 2011a).

2.4. Research on Sustainable Development

The integration of scientific research and policy is crucial for sustainable development where consideration not only of environmental but also of economic and social issues is required when studying a given problem. This creates complicated conditions that are difficult to handle not only because highly diverse groups of researchers are involved, each with their own research culture, but also because they have to communicate somehow with a quite diverse group of stakeholders and policymakers when following a desirable transdisciplinary approach. A full discussion of this important topic is beyond the scope of this chapter. However, one element can be stressed: the need for a clear identity of each of the disciplines being involved in terms of what contribution they can make to the overall discussion in the interdisciplinary groups studying sustainable development. The key question is: What are the core values that any given profession can contribute to the discussion on sustainable development? If such core values are not clear, the impact of that particular discipline in the discussions is likely to be minor and they may be sidelined. The question as to core values is also central to the debate on the relevance of hydropedology for studies dealing with sustainable development. We hypothesize that the core values for pedology and hydrology, when considered separately, are less relevant than when combined into hydropedology. This hypothesis will be evaluated in the next section.

3. HYDROPEDOLOGY AS A POWERFUL TOOL IN STUDIES ON SUSTAINABLE DEVELOPMENT

Pedology, as defined by Lin in the introduction of this book, is focused on the study of undisturbed soils in the field. Soil classification and mapping have been important activities for pedology, supported by basic chemical and mineralogical research on soil features important for classification. Soil survey interpretations for soil mapping units, shown at different spatial scales, have been used widely and successfully (e.g. Bouma et al., 2012). However, rather than define relative suitabilities for a wide variety of uses, relative limitations have been defined. This distinction is important because it implies that anything can be done anywhere but that serious problems may occur when "strong" limitations for any particular use are indicated. Even the qualification "slight limitation" suggests that successful use is not necessarily guaranteed. On the contrary, a statement that a given soil is suitable for a given use is absolute and may be subject to litigation when problems do occur (which is possible because so many factors other than soil factors determine whether or not a given use will be successful). Soil limitations clearly fit in the K2 category (Fig. 1). The same goes for the way in which hydrology is characterized in soil survey in terms of drainage classes, varying from well drained to very poorly drained, emphasizing mottling patterns in soils rather than physical measurements.

Unfortunately, the K2 approach is not adequate to answer questions raised in the sustainability debate where ecological and economic simulation models of planning agencies play an important role. Their data demands are quantitative. In reaction there has been research on the functional characterization of taxonomic soil types, using simulation models (e.g. Droogers and Bouma, 1997; Van Alphen and Stoorvogel, 2000; Sonneveld et al., 2002) and on digital mapping using innovative sampling including georeferencing, remote sensing, and spatial analysis applying geostatistics (e.g. Lagacherie and McBratney, 2007; Hartemink et al., 2008). These K5 developments are important and are now used to address key sustainability issues at the world level (Sanchez et al., 2009). Still, the major body of soil survey information is at the K2 level.

Hydrology, including the unsaturated zone in the soil studied by soil physicists, has developed into a quantitative, mechanistic science with a solid mathematical foundation in which simulation models are widely and successfully used, fitting in the K5 category of Fig. 1. The implicit, underlying assumption of the models, of soils being isotropic and homogeneous, is, however, increasingly problematic because many modeling results do not correspond well with evermore available field observations as monitoring and observation techniques improve. Proximal and remote-sensing techniques, in particular, offer new opportunities for monitoring (e.g. Bouma et al, 2011b). The common procedure of "model calibration" to forcefully match calculations with observations is too opportunistic, unscientific, and, therefore, unsatisfactory as the scale of observation used to calibrate the model does not allow unambiguous description of some of the processes. Recently, two prominent soil hydrologists, Don Nielsen (personal communication) and Keith Beven (2006), have made the following observations:

1) We take our equations (e.g. Darcy, Buckingham, and Richards) for granted. We are apathetic to explore the unknown. With the present knowledge, simulations of soil water within and across the natural landscape remain unverified.
2) There is a need to explore crucial, yet-to-be understood interacting physical, chemical, and biological soil processes and identify their dependency at different time and space scales.
3) Hypothetical studies of flow and transport in heterogeneous media have been widely reported but there have been very few detailed studies of field

sites. Nearly all transport models use the same small-scale laboratory homogeneous-domain theory to represent fluxes at larger scales. This explains major discrepancies between model predictions and reality.

In short, there is a major concern that hydrology may have moved too far from the field and this introduces a major reason to support the concept of hydropedology as it combines the best of the two worlds: K2 field expertise in field pedology, now increasingly supported by K5 digital mapping techniques, and the solid theoretical K5 basis of hydrology, also expressed by modeling expertise (e.g. Bouma et al., 2011b).

The above discussion suggests acceptance of the hypothesis that the combined expertise of pedology and hydrology (representing interdisciplinary cooperation) has more to offer to sustainability studies than the two separate disciplines. This, however, requires further analysis.

4. A FOCUS ON SYSTEMS ANALYSIS AND SOIL FUNCTIONS

4.1. Introduction

To avoid a rather sterile science-driven approach to the discussion of the sustainability issue, we return to the Soil Protection Strategy (SPS) of the European Union (EU) as an important policy document that, one may assume, reflects the desires of society and is therefore essentially demand driven (CEC, 2006). Earlier in 2000, the EU has proposed the legally binding Water Framework Directive (WFD) which is aimed at sustainable use of water resources in the EU (CEC, 2000; Moss, 2008). As stated in the introduction of this chapter, the focus of the SPS is also on sustainable use of the soil. The WFD and SPS provide two quite helpful policy signals for hydropedology. At the EU level, sustainable development of society is presented as a key issue covering a broad and overwhelming array of economic, environmental, and social issues. For a hydropedologist to be submerged into such broad discussions is not productive and most probably highly frustrating. By approving the WFD and the SPS, the EU, in fact, produces a welcome signal to the science community and more specifically to hydropedologists, that they can focus their work on sustainable use of water and soil under the implicit condition that their work will not be stand alone but will somehow fit into EU-wide, broader studies focused on sustainable development of society at large. In other words, whatever hydropedologists produce in future must fit into a broader sustainability context and must be of value there. This, then, presents an intriguing focus for hydropedology research that will be further explored in the next sections in terms of: (i) a primary focus on functions rather than on threats and (ii) exploring the potential of a systems analysis linking hydropedology to the overall sustainability debate.

4.2. Soil Functions versus Soil Threats

When defining research needs for the EU Soil Protection Strategy (ECSSS, 2004) a committee composed of soil scientists from all over the EU emphasized eight soil threats:

1) Erosion,
2) Local and diffuse contamination,
3) Loss of organic matter,
4) Loss of biodiversity,
5) Compaction and other physical soil degradation,
6) Salinization,
7) Floods and landslides, and
8) Sealing.

Research needs for each of these threats were defined for five clusters, applying the DPSIR system defining D(rivers) of land use change, the resulting P(ressures), the S(tate) of a given soil or landscape, I(mpact) of a given pressure, and R(esponse) of a system being subjected to pressures:

Cluster 1 Processes underlying soil functions and qualities.
Cluster 2 Spatial and temporal changes of soil processes and parameters (State: "S").
Cluster 3 Ecological, economical, and social drivers of soil threats (Drivers: "D" and Pressures: "P").
Cluster 4 Factors (threats) influencing soil eco-services (Impacts: "I").
Cluster 5 Strategies and operational procedures for soil protection (Responses: "R").

This resulted in some 110 "Priority Research Areas" for soil protection and the management of Europe's natural resources (ECSSS, 2004). Though conceptually solid and containing relevant and valid recommendations, the form of the analysis presented was not convincing to policymakers and stakeholders as it appeared to avoid making choices while catering to a general feeling that scientists are primarily interested to feather their own nests. The Royal Academy of Sciences of the Netherlands created a study group exploring the possibilities to derive a more operational system appealing to practitioners (KNAW, 2008; see also Bouma et al., 2008). One important difference with the ECSSS (2004) analysis was a primary emphasis on soil functions rather than on soil threats, both defined by CEC (2006). Seven functions are distinguished:

1) Production of food and other biomass;
2) Storing, filtering, and transformation of compounds;
3) Providing habitat and gene pool,
4) Providing the physical and cultural environment for humankind;
5) Source of raw materials;

6) Acting as a carbon pool; and
7) Archive of geological and archaeological heritage.

Soil functions are now widely accepted by the soil science community and are also used to define ecosystem services, a valuable concept when studying sustainable development. Aside from the fact that the term "soil threat" is unfortunate (soils do not threaten anyone but are threatened themselves), it also has a negative connotation which is not attractive. Rather than react to negative signals by trying to reverse developments and point them into a positive direction, it is preferable to formulate soil potentials in a positive way in terms of what might be possible, defining certain well-defined actions to be undertaken. In this context, soil functions are suitable to form the starting point for such an approach. Their character is neutral or positive rather than negative. But more important is the fact that "threats" are being studied by many disciplines while also many groups of stakeholders are directly involved. Soil scientists will somehow have to find their "niche" in all this when focusing on "threats". In contrast, soil scientists can exclusively focus on studying their soil functions which may help to define their "niche". As all functions are strongly determined by soil hydrology, studying soil functions should be a prime hydropedologic activity. Of course, the challenge remains to present results obtained in an effective manner to the larger interdisciplinary groups. Again, the assumption is that hydropedologic input is more relevant than the separate input of pedologists and hydrologists.

4.3. A Systems Approach Linking Soil Functions to the Sustainability Issue

The example to be discussed here originates from the Netherlands and deals with regional planning. A government publication in 2004 (VROM, 2004) introduced the "three-layer" model in Dutch law, requiring that regional land-use plans should systematically consider three elements before making new plans: (i) the dynamic soil/water regime of the region and associated ecological conditions, (ii) the transportation network on land and water, and (iii) the locations of settlements. This can be done at different spatial scales; an example is presented in Fig. 2 for the entire country (see also Bouma et al., 2008). One reason for this approach is the avoidance of unrealistic, futuristic plans that do not take into account actual conditions in terms of roads, canals, and settlements to the extent that such plans not only would be very costly but could also not be realized because of property conflicts. The addition of the dynamic water regimes of a region and its soil and ecological conditions is interesting. In the 1950's the feeling was that "anything can be done anywhere". Land reallotment plans were made in which meandering brooks were straightened, microtopography was flattened, and drainage systems were widely implemented. It turned out that changing natural water regimes came at a price because ecological equilibria

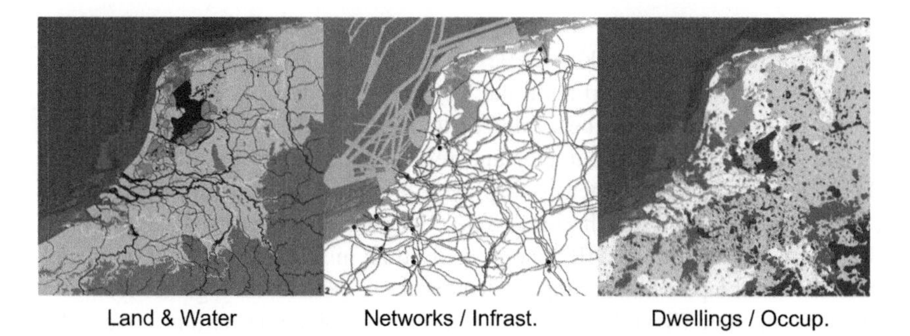

FIGURE 2 The three-layer model as used in Dutch land-use planning, requiring successive consideration of land-water dynamics and occurrence of infrastructure networks and occupation patterns before new plans are to be made. *(Color version online and in color plate)*

were disturbed and readjustment was quite costly. That is why the law now requires that the dynamic soil/water regime and the ecological conditions should be characterized first as the foundation for new plans to be made.

The scheme of the KNAW (2008), intended to structure soil research transparently and in a form that would effectively feed into the EU sustainability debate, consists of two diagrams. The first (Fig. 3) characterizes the State (S) of a given area of land. This can be at many spatial scales, a field, a farm, or a larger legal community, for instance, a province or a natural

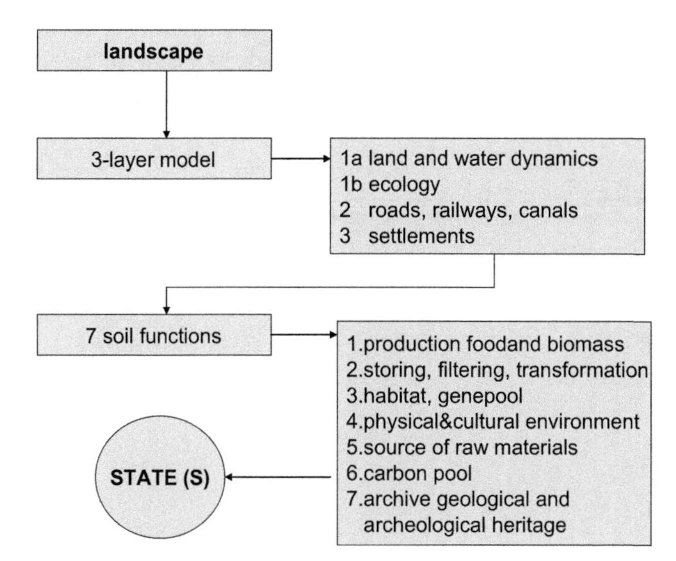

FIGURE 3 Diagram, illustrating definition of State (S) of any given piece of land, following a systematic characterization according to the 3-layer model and the seven soil functions of the European Soil Protection Strategy.

watershed. In line with the "three-layer" model, its three elements are described in general terms (at the K2–K3 level). Next, the seven soil functions are systematically defined, starting at the K2 level to be derived from existing soil surveys if available, and to be refined up to the K5 level as desired. Functions 4, 5, and 7 have a special character and require delineation of subareas where these functions are prominently displayed. Figure 4 starts with the present S (Spr), makes a link with the past (Spa), but particularly with possible future States (Sf). Here, the DPSIR scheme is used as in ECSSS, 2004. Drivers and pressures in the past had an impact, leading to responses, all together resulting in a current State (Spr). For the future, a series of planning options have to be compared, because there is, of course, no single land-use option that represents the one and only sustainable land-use plan. It is important to represent all ideas, even the improbable ones, to allow a comprehensive analysis in the signaling phase of the policy cycle. Different interests are at stake in each option and they have to be balanced, requiring a negotiating process. Each option is the result of certain tradeoffs between economic, environmental, and social demands. Production of maps, showing "what" would occur "where" in the area being considered when a given option is followed, is a powerful tool in the negotiation process which is an important part of the design phase in the policy cycle. As shown in Fig. 4, each option implies a certain response to the current situation, resulting in drivers, associated pressures, and an impact to be shown on a map. Each option is to be subjected to a cost–benefit analysis. Note that this economic analysis is suggested at the end of the process and not at the beginning, which is common now as economic analyses are often dominant when making land-use decisions, ignoring or de-emphasizing environmental and social aspects. Once decisions have been made, research can still be helpful

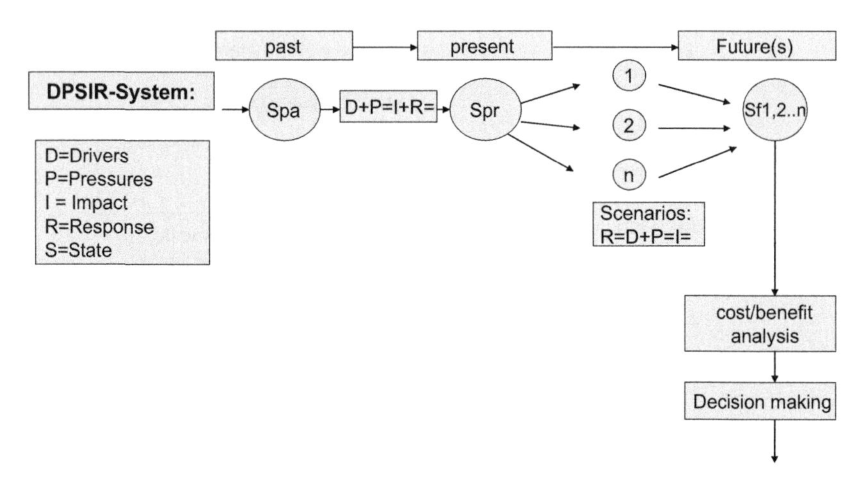

FIGURE 4 Schematic diagram representing development of future land-use scenario's, applying the DPSIR system. The State (S) expresses past, (Spa), present (Spr), and future conditions (Sf).

in the implementation phase of the policy cycle to check whether drivers, pressures, and impacts of a chosen option indeed materialize as foreseen.

A (partial) example of the proposed procedure for an area in the Netherlands was presented by Bouma et al. (2008). Issues mentioned in the five clusters and many of the 110 research priorities of ECSSS, 2004 are evident when defining the options for a given land area. However, in our approach they are fine-tuned for a given area rather than being generic, while they follow a logical operational sequence rather than an arbitrary listing by threat. Later, it will be important to collect and analyze experiences from different locations as part of the evaluation phase of the policy cycle. Scientists have a special role in the process defined above:

1) They assemble visions, interests, and values as articulated by various stakeholder groups and assist in merging those into options that can appeal to all and that may lead to reaching common goals. This is a time-consuming and often-frustrating activity, and assistance by communication experts or social scientists is often needed. Only some scientists should be involved with such activities, while others keep their focus on various forms of research, together constituting a "*Community of Scientific Practice*" (Bouma et al., 2008). However, the challenge to obtain a representative range of options for a given problem should not be completely delegated to third parties because meaningful results will only materialize in the end when the various interests are well represented and this requires direct involvement and commitment by the scientists. Recent examples of this complex approach have been presented by Bouma et al. (2011a).

2) They make simulations for each option at a selected K-level, depending on data availability, in terms of its environmental, economic, and social implications and present this to the decision makers, be it stakeholders or politicians.

So, rather than just answering specific and isolated questions by either stakeholders or policymakers, scientists (in this case hydropedologists) create a transparent picture of all possible options with internal tradeoffs. They are also in an excellent position to act as "knowledge brokers," making sure that the right knowledge is presented at the right time and place to allow continuous progress toward reaching the intended overall goals (e.g. Bouma et al., 2011a). This, ideally, will facilitate the negotiation process and avoid scientists being seen as spokesmen for either certain stakeholders or for policymakers, thus allowing scientific independence. The latter is crucial, because scientific curiosity and research often are primary drivers for innovation. An arbitrary example is the discussion about Dutch environmental rules and regulations where independent critical comments, based on scientific arguments, may offer new perspectives without being initiated by the policy or societal arena (Bouma, 2011a). Of course, this broad outline presents an idealized picture but it may serve to contribute to improving relations between science and society.

Clearly, the soil functions play a central role in the procedures outlined in Figs. 3 and 4, next to the DPSIR approach. The soil functions will therefore now be discussed in terms of important focal areas of hydropedologic research in future.

5. THE SEVEN SOIL FUNCTIONS: PRIORITY AREAS FOR HYDROPEDOLOGY RESEARCH

5.1. Introduction

Clearly, the seven soil functions of CEC (2006) cannot be characterized by soil scientists or hydrologists alone. Four of them represent key areas for hydropedology research: biomass production, filtering of compounds, habitat status, and the dynamics of the carbon pool. These are all intimately linked with water regimes. But water regimes are also key factors for the physical and cultural environment of humankind and for the archive of archeological heritage. The latter because the location of old settlements was strongly related to soil/water regimes in the natural landscape. Perhaps only the source of raw materials, such as gravel, is more determined by geological conditions but water regimes also play a crucial role when considering e.g. peat deposits.

The seven soil functions will now be discussed separately, indicating the particular potential for hydropedology research.

5.2. Soil Function 1: Production of Food and Other Biomass

Three key areas of hydropedology research can be distinguished that will enhance our understanding of biomass production: (i) better definition of soil input parameters for simulation modeling, (ii) importance of a pedological understanding in measuring input parameters, and (iii) characterizing regional flow regimes.

5.2.1. Soil Input into Simulation of Biomass Production

Many simulation models for plan growth are available. Two elements are particularly significant for hydropedology: (i) the contact between roots and soil and (ii) water in the rootzone in space and time. Some sophisticated crop growth simulation models such as the DSSAT family (Jones et al., 2002) have simple expressions for soil. They use the "available water" concept assuming water to be completely available only between two pressure heads: field capacity (often 0.3 bar) and wilting point (15 bar). (h2 and h4, respectively, in Fig. 5). Contact between soil and roots is assumed to be perfect, while the rootzone has a given thickness that may increase with time. Water in the rootzone can be extracted completely within the range of availability. Feddes et al. (1976) made the same assumptions except for introducing a flexible "sink-term" assuming maximum availability of water between two pressure heads (h2 and h3) that vary by crop and by evaporative demand, while decreasing

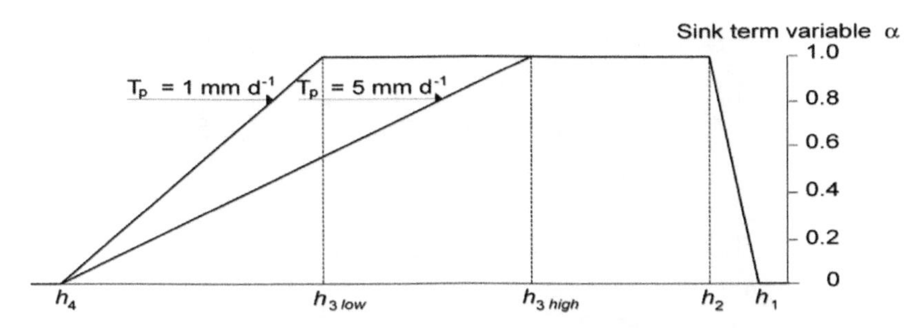

FIGURE 5 The sink term variable alpha (Feddes et al., 1976) which defines the reduction of potential transpiration as a function of the pressure head (h) in the soil and of the transpiration rate Tp.

gradually beyond these two points until h4 is reached. These authors also include effects of wetness by defining h1 as a "wet" pressure head beyond which no transpiration is possible (Fig. 5). Thus, the h2–h4 concept of "available water" is significantly refined.

Problems with some of the above-mentioned underlying assumptions have been observed: (i) The implicit assumption that contact between roots and soil is "perfect" in the rootzone is often not realistic. Soil surveyors have often observed concentration of roots on faces of structural elements ("peds") (Fig. 6). This implies that water at a given depth in the soil may be "available and extractable" according to the standard model but not "accessible" in reality. Droogers et al. (1997) showed this effect to be significant and presented a method to characterize "the accessibility" factor. Field observations confirm that sometimes plants wilt in structured soils, particularly at relatively high evaporation rates, while according to standard simulation models there should be plenty "available" water. (ii) A second problem arises as water infiltration and redistribution, as calculated in standard physical simulation models, is assumed to occur in a homogeneous and isotropic manner. In other words, the water content is the same at each depth at any given time, and water extraction occurs in the entire rootzone. This may more or less be correct in relatively homogeneous sands but not in structured soil. Even in sands, hydrophobicity results in heterogeneous infiltration patterns (e.g. Ritsema et al., 2005) with major effects on water uptake by roots. Bouma and Dekker (1978) have demonstrated infiltration patterns in dry clay soil, consisting of microrelief-governed infiltration at the soil surface into vertical, continuous cracks between peds, forming small bands of free water flowing vertically downwards along ped faces. Depths of 1 m or more were reached in just a few minutes and water may collect in dead-end pores at different depths in the soil (Fig. 6). The same effect was reported by van Stiphout et al. (1987) for silty soils with worm channels. The process of free water moving rapidly along macropores through an unsaturated soil matrix is called "bypass flow". A simple field measurement method has been developed to measure bypass flow (Booltink and Bouma, 2002b). A few deep roots may follow deep percolating water that accumulates

FIGURE 6 Vertical in situ vertical face of a prism at 80-cm depth in a clay soil shortly after a 10-mm shower, showing the effects of bypass flow resulting in free water being in contact with deep roots. *(Color version online)*

in pockets deep into the soil and this mechanism explains the common field observation that crops are not wilted on soils where water in the upper part of the rootzone is deemed to be not available anymore, according to standard theory which assumes that water infiltrates as a horizontal plain and wets the soil evenly. Clearly, not all current simulation models consider these important mechanisms and input from pedologists could help modelers to improve their representation of water uptake by roots. A hydropedologic example of simulation of water infiltration into structured soil, based on field observations described above, was presented by Hoogmoed and Bouma (1980) for infiltration in dry, cracked clay soil and by Bouma et al. (1982) for a soil with large vertical worm channels.

Taking heterogeneous and isotropic soil conditions into account, simulation of biomass productions is well possible with recent simulation models. This also offers an opportunity to define *soil quality* based on function 1. So far, a universally accepted measure for soil quality does not exist and this contrasts unfavorably with water and air that do have such squality measures. Bouma (2002) suggested the ratio between potential biomass production and water-limited production as a quality measure for a given soil at a given location.

There is a tendency in crop growth models to focus on plant aspects: evaporation at different growth stages, morphology of plants, etc. This is correct but the rest of the model should have a corresponding degree of detail for the entire model to be well balanced. If soil input only consists of estimates of available water and depth of rooting, data can easily be supplied by nonsoil scientists using available databases and pedotransfer functions relating soil texture, bulk density, and organic matter content to moisture retention (Pachepsky and Rawls, 2004). Soil scientists are not needed for such activities and in order to prove the relevance of their soil expertise, soil scientists should demonstrate that such simple models of water uptake provide incorrect and misleading results for many soils in the field and that more detailed research, as indicated above, is needed to allow better modeling. So far, soil scientists are far too relaxed about this and do not realize that easily accessible databases make their specialized expertise redundant if classical K2 approaches continue to be used by nonsoil scientists to generate input of soil data. This presents a basic problem. Not providing data for accessible databases is no option, not only because public funds are used to generate such data but also because this would be seen by society as elitist and self-serving. The only option is proactive action by showing that incorporation of hydropedologic data provides much better results than when using standard data. But how to prove that results are better? So far little monitoring data on crop yields and associated soil moisture regimes were available because field experiments are expensive. However, proximal- and remote-sensing techniques can be used effectively and at relatively low cost to show real crop transpiration rates as a function of time and place. Such measurements can be crucial in future to prove that investment in hydropedologic research has a low cost/benefit ratio (e.g. Bouma et al., 2011b).

5.2.2. *Using Soil Morphology to Estimate Sample Volumes and Sampling Strategies*

The measurement of the saturated conductivity (K-sat) of soils is crucial to determine the soil infiltration rate. Low rates may cause ponding of water and relatively long periods of soil wetness, strongly affecting plant growth. Standard methods for measuring K-sat do not define representative sample volumes and use standard cores. This may result in poor data. One way of using the large body of soil structure descriptions in pedological databases is to derive optimal sampling volumes of samples to be used for soil physical measurements. Cores with standard volumes yield unrepresentative results in structured soils. Anderson and Bouma (1973) showed that any K-sat of a B horizon of a silt loam soil could be measured by varying the height of the core. High cores yielding low values and low cores yielding high ones are explained by vertical continuity patterns of planar voids. Beyond a certain height, however, values became constant when a representative volume was reached. Empirically, a volume of at least 20 peds was considered to be representative.

An analysis of stained pore patterns in the silt loam soil indicated that macropore continuity determined K-sat and a simple statistical model could be used successfully to calculate K-sat values. The same was done later for a clay soil (Bouma et al., 1979). When peds become very large, samples should be large as well and for that purpose the column method was developed (see example in Dekker et al., 1984 and Booltink and Bouma, 2002a). Here, a large column was carefully excavated in situ and covered on the sides with plaster of Paris, allowing measurements of K-sat through an infiltrometer placed on top. When clear patterns occur in soils, such as bleached areas around vertical cracks, defining a single "representative" sample volume does not work and it will be necessary to subsample the soil, based on a morphological analysis. Bouma et al. (1989) sampled peds and bleached areas around peds separately in a boulder clay and obtained two rather distinct populations of K-sat that were used to calculate a representative K-sat for the horizon as a whole based on the relative percentages occupied by each population. Just placing relatively small cores at random in the horizon resulted in so much scatter that a meaningful interpretation was impossible. The conclusion that soil variability is large is not justified when using incorrect measurements! The much so-called variability is due to using the wrong methods at the wrong time and place.

5.2.3. Characterizing Regional Flow Regimes

The above discussion is not restricted to individual soil types but also applies at larger landscape scales. Soil maps show the spatial distribution of mapping units and their soil horizons and they can be used to provide data for regional hydrological models, which are currently often only calibrated on limited base-flow data (e.g. Bouma et al., 2011b). But many problems remain because soil surveys provide no data on the internal variability of the mapping units (e.g. Lahr and Kooistra, 2008). A detailed analysis is beyond the scope of this text but Vepraskas et al. (2006, 2009) have shown how pedological information on mottling patterns can be used to improve simulation of regional water regimes. Dekker et al. (1984) studied a heather area in the Netherlands with Spodosols with a hard and compact spodic horizon. K-sat of this horizon was never measured because cores could not be taken. As it was assumed that this horizon was impermeable, a land development plan was not approved because a planned breakup of the spodic horizon was expected to change local hydrology by allowing downward percolation of water and subsequent loss of assumed lateral water movement over the top of the spodic horizon toward local ponds. Measurements with the column method showed, however, that the spodic horizon was not impermeable and that water was moving along vertical crevices. Surface measurements indicated occurrence of hydrophobicity and lateral flow over the soil surface after showers. The original development plan could therefore be executed as lateral movement of water was assured, be it over the soil surface. When characterizing regional flow regimes, land-use change can be an important factor, significantly changing soil properties (see also examples in Section 5.7). For example, changes may occur in macropore density (Lindahl et al., 2009), soil distribution (Follain et al., 2009), soil physical properties (Frison et al., 2009), and physico-chemical properties (Montagne et al., 2008).

5.3. Soil Function 2: Storing, Filtering and Transformation of Compounds

As explained above, function 1 is often simply expressed in terms of K2 estimates of rooting depth and available water leading to unrepresentative results in many soils. In the same way, function 2 can simply be expressed by the cation-exchange capacity, organic matter content, and, perhaps, the air content at field capacity or some other simple measure for aeration. Again, as for function 1, such values can be derived from accessible databases and the role of the soil scientist in all of this remains therefore rather obscure. However, adequate characterization of function 2 needs a K4–5 approach to result in data that are truly representative for a wide range of soils, and hydropedologists can play a key role here. Two aspects will be discussed: (i) the importance of travel time in soil layers actively filtering and degrading compounds and (ii) the importance and implications of soil heterogeneity.

The importance of travel times is illustrated in Fig. 7. A solution with artificially high concentrations of poliovirus was percolated through a sand column at two flow rates. At 50 cm/day, a substantial quantity of virus left the column. At 5 cm/day, however, viruses were removed within 50 cm, as was observed when sampling the column after termination of the experiment. No virus particle ever left the column at this low flow rate. Slower travel times

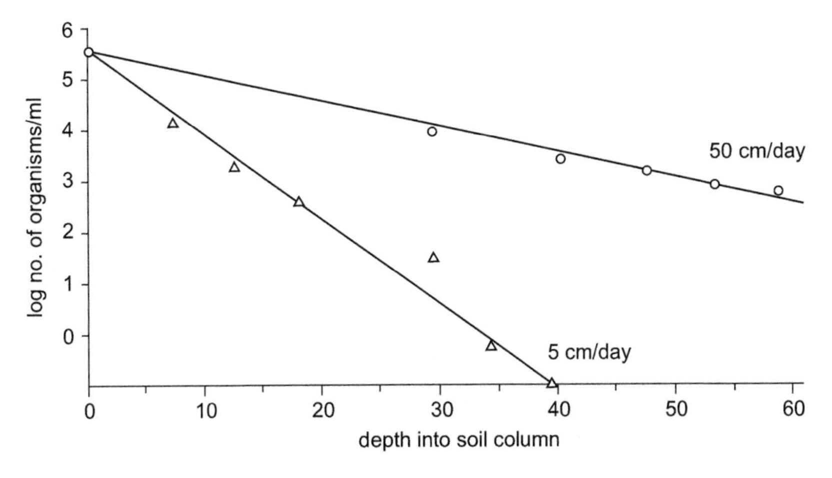

FIGURE 7 Purification of septic tank effluent spiked with high concentrations of poliovirus at two flow rates in a 60-cm-long sand column.

imply better and longer contact between the solution and the soil particles because the liquid moves through smaller pores as larger pores are filled with air in unsaturated soil. Each soil material has a characteristic relationship between travel times and purification and this information has been used to design innovative soil disposal systems of septic tank effluent (e.g. Bouma, 1979). Structured soils offer special problems. Water will generally pass slowly through and more rapidly around the structural elements (peds), while it only moves through the peds below a characteristic flow rate for each structure type. This has implications for purification as was demonstrated for a well-structured silt loam soil (Fig. 8). At flow rates of 5 mm/day, fecal coliforms passed through a 50-cm-long column, while they did not at 3 mm/day when the solution only moved through the peds and more rapid flow in relatively large pores around the peds could not occur, leading to longer contact times and better purification. Every structured ("pedal") soil material has a "critical" flow rate above which solute will flow not only through the soil matrix but also along macropores around peds resulting in relatively rapid movement of pollutants and little purification. Defining such "critical" flow rates for different structured soil materials would be an excellent future research theme for hydropedology.

Most simulation models for chemical transport in soil are based on the implicit assumption of physical flow models that the soil is isotropic and homogeneous. Some soils are more anisotropic than others but none of them are truly homogeneous. Young in NRC (2009) used modern scanning and

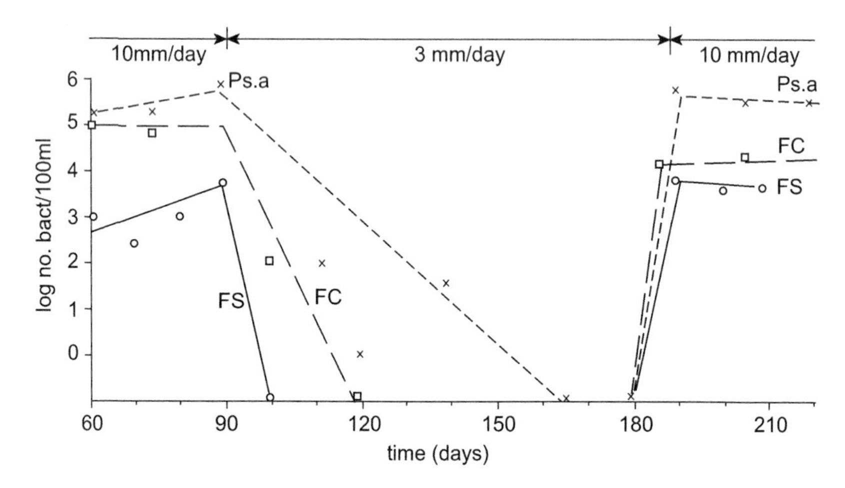

FIGURE 8 Purification of septic tank effluent spiked with high concentrations of FC (*Fecal Coliform*); FS (*Fecal Streptococcus*) and PSa (*Pseudomas Aeruginosa*) after infiltration into an undisturbed 60-cm-long sample of a well-structured B horizon in a silt loam soil. Two flow rates produce different purification as the high flux moves around and the low flux moves through the peds.

tracing techniques to show that often <0.01% of the soil volume is occupied by so-called biofilms that are active in chemical and microbiological transformations occurring in "hot spots". Bouma et al. (1979) showed, using dyes to trace vertical saturated flow patterns in a clay soil, that the soil volume directly in touch with the percolating solution was <0.1% (Fig. 9). Such phenomena are still more evident in unsaturated soils with bypass flow. Field observations of certain chemicals occurring in groundwater are common while models suggest this to be impossible. Hydropedologists can make a major contribution to the study of environmental problems by better defining flow patterns in soils and occurrence of hot spots, combining expertise of pedologists describing soils in detail at the K2–3 level and physicists knowing how to express such observations in quantitative terms allowing expressions in dynamic simulation models (K4–K5).

5.4. Function 3: Providing Habitat and Gene Pool

Habitat describes the living environment for plants and animals, which is strongly determined by climatic factors in terms of temperature, rainfall, and radiation. This function not only covers agricultural lands but also dunes, heathland, moorlands, bogs, and other specific ecological niches, all with characteristic soil profiles. Soil moisture contents and solute composition in space and time are important factors and remarks made for factors 1 and 2 apply here as well, offering again an ideal topic for hydropedology. The

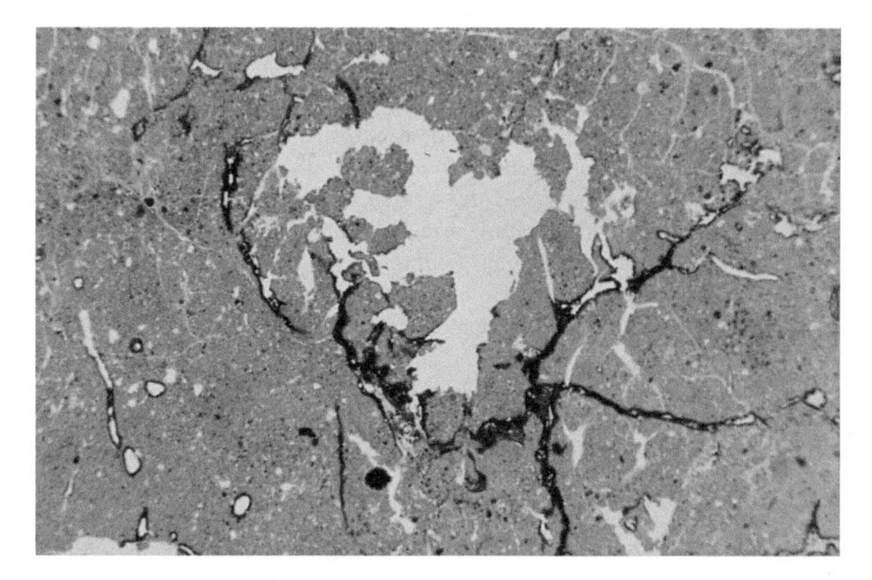

FIGURE 9 Horizontal section through a structured clay soil after saturated flow. Flow patterns are indicated with methylene blue dye and show that flow only occurs in part of the soil matrix as a function of macropore continuity. *(Color version online and in color plate)*

mentioning of the gene pool in the EU Soil Protection Strategy is interesting as it opens new as-yet hardly explored research opportunities. The DNA of soil biota and organic matter has been subject to evolution for millions of years and experts assume that this can be the source of new medicines in future.

5.5. Function 4: Providing the Physical and Cultural Environment for Humankind

This function is operational at the landscape level and acknowledges the importance of the physical environment for societal development and well-being over the centuries. Before the agricultural revolution, introducing chemical fertilizers, drainage, and irrigation, land use was strongly determined by dynamic, natural land, and water conditions. In relatively fertile river delta's of the world, settlements and arable land were established at the higher locations while meadows occupied the lower, wetter areas. Roads and waterways followed natural patterns in the landscape. In upland areas, settlements were often associated with sediments or rocks containing relatively high nutrient contents while the availability of water was always a concern and a governing factor for establishing settlements. The three-layer model for land-use planning in the Netherlands (Fig. 2) reflects such relationships. Modern agriculture has introduced many changes. Farm sizes increased, nutrient management was focused on the needs of agricultural crops or grassland, and water management was controlled. This has dramatically changed conditions in many areas. Currently, efforts are made to preserve some characteristic traditional landscapes as part of the general cultural heritage. These are called "National Landscapes" in the Netherlands of which 22 have been designated. Farmers receive payment for maintaining landscape elements, such as hedges. The general policy on National Landscapes does not advocate simple conservation but promotes "preservation by development". In other words, any possible development should be such that the unique character of the landscape is preserved. This often implies that the water and nutrient regime of an area can only be changed to a limited extent. Different regions have different regimes with corresponding vegetations. Here, hydropedology can play an important role in characterizing soil and water regimes in all phases of the policy cycle.

5.6. Function 5: Providing a Source of Raw Materials

This function is quite diverse, varying from excavating subsoil gravel to peat and sod for gardens and golf greens. Existing detailed soil surveys (K2) provide a good guide as to where certain deposits or vegetations occur. Hydropedologists can delineate certain areas in the signaling phase of the policy cycle, followed in the design phase by formulating alternative future land-use scenario's including different quantities and forms of excavation. In former soil surveys, areas were identified as being more or less suitable for excavation, considering technical and availability aspects only. Now excavations have to be viewed in a much broader societal context also considering environmental, ethical, and esthetical aspects. The procedure, deriving land-use options, as illustrated in Fig. 4 offers a suitable framework to address this new approach.

5.7. Function 6: Acting as a Carbon Pool

As the climate change discussion continues, there is increasing evidence that soils can be a major sink for carbon as certain forms of land management can result in a significant increase of the organic matter content. As the climate changes so does soil hydrology and ecology. When running scenario's about the effects of climate change on soil functions, such as the carbon pool, a hydropedologic approach is appropriate. Pedology alone would have no answers because their interpretations are based on past behavior. Hydrology would struggle to properly represent soil data in their models as discussed for function 1 above, as it is clear by now that different soils have quite different potentials to bind additional carbon. When working together they can present more realistic data to ecologists.

Estimates indicate that carbon storage in soil organic matter could contribute to 1.2 Gt carbon at world level, which is quite significant (Lal and Follet, 2009; Hillel and Rosenzweig, 2009). Dutch research has shown that organic matter contents in a given soil type can vary significantly as a function of soil management. Droogers et al. (1997) coined the term *phenoform* to express management-induced changes in organic matter content of a given soil series, the *genoform.* They showed for a prime agricultural soil in the Netherlands that organic matter contents varied from 5.0%, 3.3% to 1.7% when the land was permanent grassland, organic farming and high-tech arable. Bulk densities increased in the same sequence from 1.38 g/cm^3, to 1.47 g/cm^3 and 1.68 g/cm^3. Soil classifications were identical for the three phenoforms (mixed, mesic Typic Fluvaquent, according to Soil Survey Staff, 1998). A comparable study was made for the most common sandy soil in the Netherlands (a coarse loamy, siliceous, mesic Plagganthreptic Alorthod, according to Soil Survey Staff (1998)), again applying to all phenoforms (Sonneveld et al., 2002). Characteristic land uses were permanent grassland, reseeded grassland, and high-tech arable. Organic matter contents were 8.1%, 6.3%, and 4.8%, respectively, with corresponding bulk densities of 1.30 g/cm^3, 1.36 g/cm^3, and 1.48 g/cm^3. In both studies a simple regression equation was derived relating the organic matter content of topsoil to past land use with high correlation coefficients of 0.8. This offers the opportunity to go back to mapped areas of a given soil type and sample different fields with known management histories and use equations obtained to predict organic matter contents in future assuming certain types of management. Each soil type is bound to produce different relationships and these could be important for future soil characterization programs.

But the organic matter content is also important to improve the agricultural production potential as a higher organic matter content is associated with a higher water-holding capacity and stronger adsorption of nutrients. Increasing the organic matter content of agricultural soils can therefore also be a contribution toward global food security and soil function 1 (Lal and Follet, 2009).

5.8. Function 7: Archive of Geological and Archaeological Heritage

This last function has a similar origin as function 4 as the geological and archaeological heritage can be seen as part of the physical and cultural environment for humankind. Functions 4 and 7 are different in that function 4 can best be approached by the "preservation by development" concept, while "preservation" as such applies to function 7, be it that somehow typical geological and archeological features have to be integrated into the landscape in a satisfactory manner.

6. CONCLUSIONS

- Pedology and soil physics/hydrology have reached a point in their development where an integrated approach, as proposed in hydropedology, is bound to be more effective in making significant contributions to the sustainability debate in society as compared with continuing their monodisciplinary tracks. Hydropedology has therefore the potential to become a powerful tool for environmental policy and regulation.
- Relations between science and society are changing profoundly. Transdisciplinarity, including intensive interaction of scientists with stakeholders and policymakers, is required now but implies a risk that the disciplinary vigor of the profession will get lost in general discussions about sustainability including economic, environmental, and social visions. Hydropedology should therefore define characteristic research "niches" as a basis for both vital and relevant future research.
- The recently accepted Soil Protection Strategy for the European Union defines seven soil functions that are proposed to function as "niches" for hydropedology research, requiring a systems approach focusing on water and solute movement in undisturbed soils in the field. This includes innovative monitoring where proximal and remote-sensing techniques have high potential and modeling where implicit and misleading assumptions of soils being isotropic and homogeneous in space and time are abandoned. Research, thus defined, is likely to result in meaningful contributions to environmental rules and regulations when presented in a transparent way at the proper time and place.
- To become attractive for stakeholders and policymakers, contributions by soil research to environmental rules and regulations should be based on

a serious interaction of science with societal partners. In this context, attention is suggested for (i) systematic following of the policy cycle in research all the way from signaling to evaluating, (ii) applying the knowledge chain from tacit knowledge (K1) all the way to cutting edge-scientific knowledge (K5) and by feeding K3–K5 knowledge back into the chain, and (iii) defining land-use options, reflecting widely varying opinions as to the most desirable land use, from which selections can be made in the policy arena. Hydropedologists can thus put themselves in an attractive intermediate position by facilitating interaction between land users and policymakers without belonging to either camp.

REFERENCES

Anderson, J.L., Bouma, J., 1973. Relationships between hydraulic conductivity and morphometric data of an argillic horizon. Soil Sci. Soc. Am. Proc. 37, 408–413.

Asselt, M.B.A. van, 2000. Perspectives on Uncertainty and Risk: The Prima Approach to Decision Support. Kluwer Academic Publishers, Dordrecht, Netherlands.

Baveye, P., Philippe, C., 2008. Sticker Shock and Looming Tsunami: The High Cost of Academic Serials in Perspective. University of Toronto Press, North York, ON, Canada.

Beven, K., 2006. Searching for the holy grail of scientific hydrology. Hydrol. Earth Syst. Sci. 10, 609–618.

Bohme, G., van den Daele, W., Hohlfeld, R., Krohn, W., Schafer, W., 1983. Finalization in Science: The Social Orientation of Scientific Progress. Riedel Publ, Dordrecht.

Booltink, H.W.G., Bouma, J., 2002a. Steady flow soil column method. In: Dane, J.H., Topp, G.C. (Eds.), Methods of Soil Analysis. Part 4: Physical Methods. Soil Sci. Soc. of America Book Series no. 5, pp. 812–815.

Booltink, H.W.G., Bouma, J., 2002b. Suction crust infiltrometer and bypass flow. In: Dane, J.H., Topp, G.C. (Eds.), Methods of Soil Analysis. Part 4: Physical Methods. Soil Sci. Soc. of America Book Series No.5, pp. 926–937.

Bouma, J., 1979. Subsurface applications of sewage effluent. In: Beatty, M.T., Petersen, G.W., Swindale, L.D. (Eds.), Planning the Uses and Management of Land. Agronomy, 21. Am. Soc. of Agronomy, Madison, USA, pp. 665–703.

Bouma, J., 2001. The new role of soil science in a network society. Soil Sci. 166 (12), 874–879.

Bouma, J., 2002. Land quality indicators of sustainable land management across scales. Agric. Ecosyst. Environ. 88 (2), 129–136.

Bouma, J., 2005. Soil scientists in a changing world. Adv. Agron. 88, 67–97.

Bouma, J., 2010. Implications of the Knowledge Paradox for Soil Science. In: Advances in Agronomy, vol. 106. Academic Press, USA, pp. 143–171.

Bouma, J., 2011. Applying indicators, threshold values and proxies in environmental legislation: a case study for Dutch dairy farming. Environ. Sci. Policy 14, 231–238.

Bouma, J., Dekker, L.W., 1978. A case study on infiltration into dry clay soil. I morphological observations. Geoderma. 20, 27–40.

Bouma, J., Droogers, P., 1999. Comparing different methods for estimating the soil moisture supply capacity of a soil series subjected to different types of management. Geoderma. 92, 185–197.

Bouma, J., Jongerius, A., Schoonderbeek, D., 1979. Calculation of saturated hydraulic conductivity of some pedal clay soils with different macrostructure. Soil Sci. Soc. Am. J. 43 (2), 261–264.

Bouma, J., Belmans, C.F.M., Dekker, L.W., 1982. Water infiltration and redistribution in a silt loam subsoil with vertical worm channels. Soil Sci. Soc. Am. J. 46 (5), 917–921.

Bouma, J., Jongmans, A.G., Stein, A., Peek, G., 1989. Characterizing spatially variable hydraulic properties of a boulder clay deposit. Geoderma. 45, 19–31.

Bouma, J., Stoorvogel, J.J., Quiroz, R., Staal, S., Herrero, M., Immerzeel, W., Roetter, R.P., van den Bosch, H., Sterk, G., Rabbinge, R., Chater, S., 2007. Ecoregional research for development. Adv. Agron. 93, 258–313.

Bouma, J., de Vos, J.A., Sonneveld, M.P.W., Heuvelink, G.B.M., Stoorvogel, J.J., 2008. The role of scientists in multiscale land use analysis: lessons learned from Dutch communities of practice. Adv. Agron. 97, 175–238.

Bouma, J., van Altvorst, A.C., Eweg, R., Smeets, P.J.A.M., van Latesteijn, H.C., 2011a. The role of knowledge when studying innovation and the associated wicked sustainability problems in agriculture. Adv. Agron. 113, 285–314.

Bouma, J., Droogers, P., Sonneveld, M.P.W., Ritsema, C.J., Hunink, J.E., Immerzeel, W.W., Kauffman, S. 2011b. Hydropedological insights when considering catchment classification. Hydrol. Earth Syst. Sci., 15:1909–1919.

Bouma, J., Stoorvogel, J.J., Sonneveld, W.M.P., 2012. Land Evaluation for Landscape Units. In: Huang, P.M., Li, Y., Summer, M. (Eds.), Handbook of Soil Science, Second Edition. Chapter 34. CRC Press. Baco Raton, London, New York, pp. 1–34.

CEC (Commission of the European Community). 2000. Directive 2000/60/EC of the European Parliament and of the Council of 23-10-2000 establishing a Framework for Community Action in the field of Water Policy. Brussels.

CEC (Commission of the European Community). 2006. Directive of the European Parliament and of the Council, establishing a framework for the protection of soil and amending Directive 2004/35/EC. Brussels.

Dekker, L.W., Wosten, J.H.M., Bouma, J., 1984. Characterizing the soil moisture regime of a TYPIC Haplohumod. Geoderma. 34, 37–42.

Droogers, P., Bouma, J., 1997. Soil survey input in exploratory modelling of sustainable soil management practices. Soil Sci. Soc. Am. J. 61, 1704–1710.

Droogers, P., van der Meer, F.B.W., Bouma, J., 1997. Water accessibility to plant roots in different soil structures occurring in the same soil type. Plant Soil 188, 83–91.

ECSSS (European Confederation of Soil Science Societies), 2004. Scientific Basis for the Management of European Soil Resources. Research Agenda. Guthman-Peterson, Vienna.

Edquist, C., 1997. Systems of Innovation: Technologies, Institutions and Organisations. Pinter Publ, New York/London.

Etzkowitz, H., Leydesdorff, L., 2000. The dynamics of innovation: from national systems and "mode2" to a Triple helix of university–industry–government relations. Res. Policy 29 (2), 109–123.

Feddes, R.A., Kowalik, P., Kolynska-Malinka, K., Zaradny, H., 1976. Simulation of field water uptake by plants using a soil water dependent root extraction function. J. Hydrol. 31 (1/2), 13–26.

Follain, S., Walter, C., Bonté, P., Marguerie, D., Lefevre, I., 2009. A-horizon dynamics in a historical hedged landscape. Geoderma 150, 334–343.

Frison, A., Cousin, I., Montagne, D., Cornu, S., 2009. Soil hydraulic properties in relation to local rapid soil changes induced by field drainage: a case study. Eur. J. Soil Sci. 60, 662–670.

Functowicz, S., Ravetz, J., 1993. Science for the post-normal age. Futures 25, 735–755.

Gibbons, M., Limoges, C., Nowotny, H., Schwartzmann, S., Scott, P., Trow, M., 1994. The New Production of Knowledge: The Dynamics of Science and Research in Contemporary Societies. SAGE, London.

Hartemink, A.E., Bratney, A.Mc, Mendoca-Santos, M.L. (Eds.), 2008. Digital Soil Mapping with Limited Data. Springer Verlag, p. 445.

Hessels, L.K., van Lente, H., 2008. Re-thinking new knowledge production: a literature review and a research agenda. Res. Policy 37, 740–760.

Hillel, D., Rosenzweig, C., 2009. Carbon Exchange in the Terrestrial Domain and the Role of Agriculture, CSA News June V54NO6: pp. 5–11.

Hoogmoed, W.B., Bouma, J., 1980. A simulation model for predicting infiltration into cracked clay soil. Soil Sci. Soc. Am. J. 44 (3), 458–461.

ICSU (International Council for Science), 2005. Strategic plan for the International Council for Science 2006–2012. (www.icsu.org). Washington, D.C., USA.

Irvine, J., Martin, B.R., 1984. Foresight in Science: Picking the Winners. Frances Pinter, London.

Jacques, P.J., Dunlap, R.E., Freeman, M., 2008. The organisation of denial: conservative think tanks and environmental scepticism. Env. Pol. 17, 349–385.

Jones, J.W., Hoogenboom, G., Porter, C.H., Boote, K.J., Batchelor, W.D., Hunt, L.A., Wilkens, P.W., Singh, U., Gijsman, A.J., Ritchie, J.T., 2002. The DSSAT Cropping System Modelling. Elsevier Publ. Cie.

KNAW (Royal Dutch Academy of Arts, Sciences and Letters), 2008. Integrated Scientific Research for Spatial Planning. A Proposal for a Framework Programme. KNAW, Amsterdam (in Dutch).

Lagacherie, P., McBratney, A.B., 2007. Chapter 1. Spatial soil information systems and spatial soil inference systems: perspectives for digital soil mapping. In: Lagacherie, P., McBratney, A.B., Voltz, M. (Eds.), Digital Soil Mapping: An Introductory Perspective. Developments in Soil Science, vol 31. Elsevier, Amsterdam, pp. 3–24.

Lahr, J., Kooistra, L., 2008. Environmental risk mapping of pollutants: state of the art and communication aspects. Sci. Total Environ. 408, 3899–3907.

Lal, R., Follet, R.F. (Eds.), 2009. Soil Carbon Sequestration and the Greenhouse Effect. Soil Sci. Soc. Am. Special Publ. 57, Madison Wis USA.

Lindahl, A.M.L., Dubus, I.G., Jarvis, N.J., 2009. Site classification to predict the abundance of the deep-burrowing earthworm *Lumbricus terrestris* L. Vadose Zone J. 8, 911–915.

Montagne, D., Cornu, S., Le Forestier, L., Hardy, M., Josiere, O., Caner, L., Cousin, I., 2008. Impact of drainage on soil-forming mechanisms in a French Albeluvisol: input of mineralogical data in mass-balance modelling. Geoderma. 145, 426–438.

Moss, B., 2008. The water framework directive: total environment or political compromise? Sci. Total Environ. 400, 32–41.

National Research Council (NRC), 2009. Frontiers in Soil Science Research. Report of a Workshop. The National Academies Press, Washington, D.C., USA.

Pachepsky, Y.A., Rawls, W.J. (Eds.), 2004. Development of Pedotransfer Functions in Soil Hydrology. In: Developments in Soil Science 30. Elsevier, Amsterdam, the Netherlands.

Ritsema, C.J., van Dam, J.C., Dekker, L.W., Oostindie, K., 2005. A new modelling approach to simulate preferential flow and transport in water repellent porous media: model structure and validation. Aust. J. Soil Res. 43, 361–369.

Sanchez, P.A., et al., 2009. Digital soil map of the world. Science 325, 680–681 (August 7).

Slaughter, S., Leslie, L.L., 1997. Academic Capitalism: Politics, Policies and the Entrepreneurial University. The John Hopkins University Press, Baltimore.

Soil Survey Staff, 1998. Keys to Soil Taxonomy. US Gov. Printing Office, Wash. DC, USA, p. 328.

Sonneveld, M.P.W., Bouma, J., Veldkamp, A., 2002. Refining soil survey information for a Dutch soil series using land use history. Soil Use Manage. 18, 157–163.

van Stiphout, T.P.J., van Lanen, H.A.J., Boersma, O.H., Bouma, J., 1987. The effect of bypass flow and internal catchment of rain on the water regime in a clay loam grassland soil. J. Hydrol. 95 (1/2), 1–11.

Van Alphen, B.J., Stoorvogel, J.J., 2000. A functional approach to soil characterisation in support of precision agriculture. Soil Sci. Soc. Am. J. 64, 1706–1713.

Vepraskas, M.A., Huffman, R.L., Kreiser, G.S., 2006. Hydrologic models for altered landscapes. Geoderma 131, 287–298.

Vepraskas, M.A., Heitman, J.L., Austin, R.E., 2009. Future directions for hydropedology: quantifying imoacts of global change on land use. Hydrol. Earth Syst. Sci. 13, 1427–1438.

VROM, LNV, V&W and EZ (Ministeries of Environment, Agriculture, Traffic and Economics), 2004. Nota Ruimte: Ruimte voor ontwikkeling. (Space: opportunity for development). The Hague, Netherlands.

Webster, R., 2000. Is soil variation random? Geoderma 97, 149–163.

Ziman, J., 2000. Real Science: What it is and What it Means. Cambridge University Press, Cambridge.

Part III

Advances in Modeling, Mapping, and Coupling

Chapter 16

Soil Information in Hydrologic Models: Hard Data, Soft Data, and the Dialog between Experimentalists and Modelers

Michael Rinderer[1,*] and Jan Seibert[1,2]

ABSTRACT

For understanding and predicting rainfall–runoff processes in watersheds, soils and their hydraulic properties play a central role. Experimentalists observe and document hydric soil indicators in detail for more and more sites in various catchments. Modelers, on the other hand, try to break down natural process complexity into models that are based on simplified process descriptions. The challenge for both is to identify first-order controls of catchment hydrologic behavior, which helps to better understand the nonlinearity of natural systems. This chapter describes how both, experimentalists and modelers, can work together toward a better understanding and quantification of subsurface runoff processes. Specifically, this chapter addresses the following questions: (1) How are subsurface runoff processes represented in models of different complexity, ranging from simple conceptual ones to more complex physically based ones? (2) How can catchment-scale models be parametrized using point-scale measurements and existing model approaches originally developed for small scales (e.g. a soil column)? (3) Which information can be gained from soil surveying methods, including mapping approaches of hydric soil indicators? (4) Can decision schemes be useful to indicate dominant runoff processes in an objective way? Finally we describe the soft data concept as a possible way forward to enhance the dialog between experimentalists and modelers. Soft data refer to all kinds of qualitative or semi-quantitative information on pedologic and hydrologic processes and properties. These data can be made useful for modeling by applying fuzzy-logic-based functions to evaluate the degree of acceptance of model simulation outputs compared to experimentalists' field experience.

[1]Department of Geography, University of Zurich, Winterthurerstr 190, CH-8057 Zurich, Switzerland

[2]Department of Earth Sciences, Uppsala University, Villavägen 16, SE-752 36 Uppsala, Sweden

[*]Corresponding author: Email: michael.rinderer@geo.uzh.ch

Hydropedology, Edited by H. Lin. DOI: 10.1016/B978-0-12-386941-8.00016-0

1. DIFFERENT VIEWS ON SOIL HYDROLOGIC FUNCTION – AN INTRODUCTION

Experimentalists who have gained experience on rainfall–runoff processes during their fieldwork can confirm that the governing processes are complex and often show a nonlinear, threshold-type behavior, especially when considering the scale of an entire catchment (Tromp-van Meerveld and McDonnell, 2006). While experimentalists tend to document natural phenomena in detail, current models are often a gross simplification of this natural complexity. In other words, complex processes are expressed in the form of mathematical equations based on simplifying assumptions such as a steady-state groundwater response throughout the entire catchment or flow pathways determined by topography. Often these assumptions contradict with what is observed in the field. Improved discussions between experimentalists and modelers (Seibert and McDonnell, 2002; Weiler and McDonnell, 2004) have shown potential to improve process understanding and correct modeling for the right reasons (Kirchner, 2006; Klemes, 1986). Existing process conceptualizations being at odds with hydrometic, hydrochemical, and isotope measurements have forced hydrologists to reassess and extend existing process conceptualization and find new, more appropriate explanations of runoff generation (McGlynn et al., 2002). In this chapter, we focus on the potential of a dialog between experimentalists contributing their wealth of somehow intuitive field experience – called tacit knowledge – and the modelers, who are strong in conceptualizing and therefore structuring processes to distill natural complexity down to its first-order controls of runoff generation in catchments. We limit our discussion to subsurface runoff processes, which we define as saturated water flow phenomenon in the soil of a catchment either due to the rise of an existing water table into more transmissive soil layers (Fig. 1c) or due to the transient saturation above an impeding layer, soil–bedrock interface, or some zone of reduced permeability at depth (Fig. 1f,g), both causing a subsequent lateral flow component and often a more or less delayed response of stream flow discharge (Weiler et al., 2005). Included in our discussion are runoff processes due to exceeding infiltration capacity or saturation at the soil surface (Fig. 1a,b). The term soil water flow is used in circumstances when the process of redistribution of infiltrated rainfall in a soil profile through the soil matrix and/or through preferential flow pathways is meant. We discuss the potential of semi-quantitative or qualitative data (called soft data) and their value in model calibration and model structure optimization, especially in catchments with little or no data availability. In particular, we address the following questions:

1) How are soils represented in different types of hydrologic models? What kind of conceptual and physically based approaches do exist to simulate key processes of subsurface runoff?
2) How can soil routines used at the catchment scale be parametrized? What is the value and limitation of point measurements made at the plot scale and

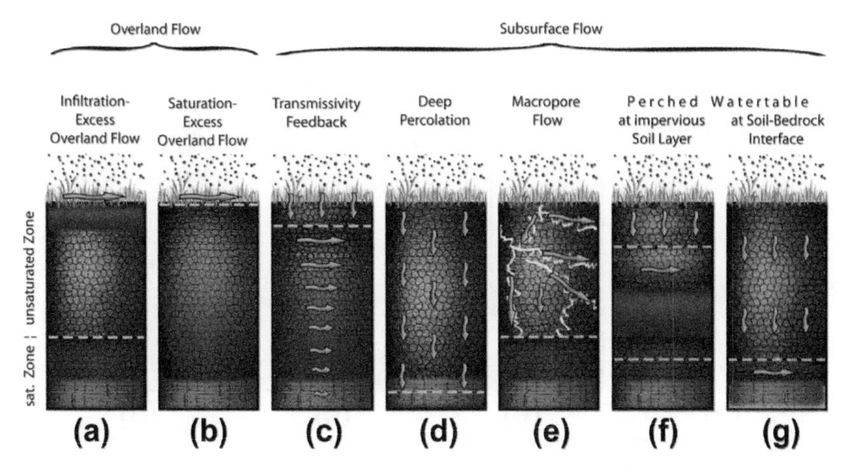

FIGURE 1 Different types of surface and subsurface runoff processes: If infiltration capacity (a) or storage capacity (b) of the soil layers at or near the surface is exceeded, overland flow is generated. As groundwater rises into more transmissive soil layers (c), lateral subsurface runoff increases. If the soil has a coarse texture and the bedrock is highly permeable, the groundwater table tends to be low and soil water can percolate deep down into the soil or bedrock (d). If distinct horizontal macropores exist, the flow rates toward the nearest stream channel can be high and subsurface runoff therefore responsive (e). If a water-restricting layer exists within the soil profile (f) or at the soil–bedrock interface (g), a perched water table might form from time to time and lateral subsurface flow can occur within the saturated soil layer. *(Color version online and in color plate)*

what kind of problems arise when upscaling this information to the catchment scale?

3) What kind of tacit knowledge as well as hard and soft indicators do already exist in soil survey data sets?
4) How can tacit knowledge and hard and soft indicators be linked in a systematic and objective way to indicate subsurface runoff processes?
5) How can the dialog between experimentalists and modelers be enhanced by utilizing soft data and tacit knowledge?

2. SOIL AND SOIL PARAMETERS IN HYDROLOGIC MODELS – THE MODELERS' POINT OF VIEW

2.1. Conceptualization of Subsurface Runoff Processes

This section briefly assembles and discusses the main modeling concepts currently used in conceptual and physically based hydrologic models for simulating subsurface runoff and soil water flow at the plot to catchment scale. This overview is not exhaustive but intends to raise the readers' awareness of the sometimes very simplistic view of modelers on subsurface runoff processes compared to the experimentalists' view, which focuses on the complexity of natural processes in all details (see Section 3).

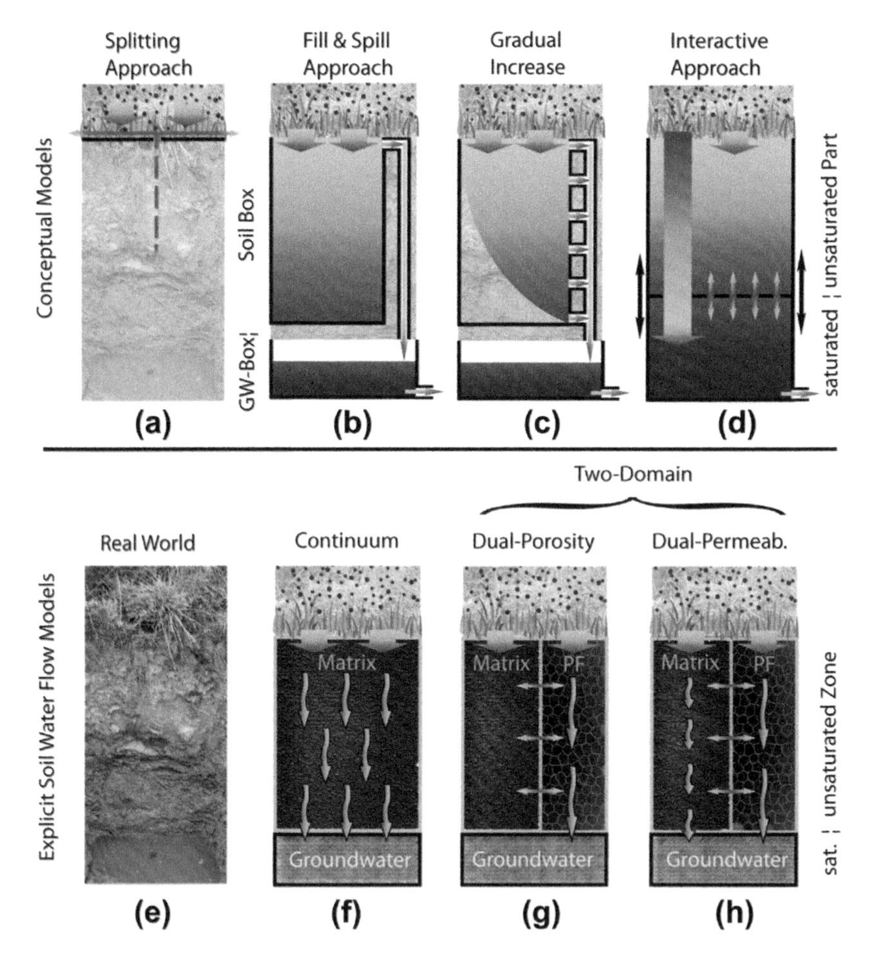

FIGURE 2 Conceptual and physically based modeling approaches of subsurface flow or soil storage response: Upper row shows conceptual modeling approaches of soil-water-storage outflow: (a) only accounting for surface runoff caused by infiltration excess; generating soil water outflow if bucket is (b) completely or (c) partly filled; and accounting for (d) interactions between saturated and unsaturated soil water storage. Lower row shows explicit soil water flow modeling approaches which can be categorized as (f) continuum approaches, which only account for matrix flow, and (g, h) two-domain approaches, which consider both matrix and preferential flow (PF). *(Color version online and in color plate)*

In very simple rainfall–runoff models, which only cover infiltration-excess overland flow (Figs. 1a and 2a), subsurface runoff is not simulated explicitly. The infiltration capacity of the soil surface determines the portion of rainfall, directly contributing to overland flow, which is of primary interest in many applied modeling cases to estimate flood discharge in small catchments. The simplest example of this model type is the rational method developed by Mulvaney (1851) or the unit hydrograph models based on Sherman (1932), which are still in use to estimate peak discharge (e.g. Hua et al., 2003; Rinderer et al., 2009).

Most hydrologic models do account explicitly for subsurface runoff processes. Conceptual models often mimic the property of the soil to store and release infiltrated water by a storage component with a defined storage–outflow relationship. The capacity for total water storage within a soil is dependent on its soil porositys and soil depth, but in most conceptual models only the dynamic storage is considered, which depends on drainable porosity. The soil storage must either first fill up to its maximum capacity before outflow commences (i.e. fill-and-spill) (Fig. 2b) or gradually increase its outflow (Fig. 2c). Stored water can be lost due to evaporation, to groundwater recharge, or to exfiltration. The latter is not represented in most conceptual models. Actual evaporation is usually estimated as a function of potential evaporation and the degree of soil saturation. Groundwater recharge can be simulated in a similar manner based on the amount of infiltrated water and the current soil-water-storage content. To account for interaction between saturated and unsaturated storage, the total soil water storage can be divided into two compartments with the boundary between them moving up or down according to the water budget in the storage (Fig. 2d) (Seibert et al., 2003; Seibert et al., 2011)).

In more complex models, not only the net soil water outflow from a bucket is computed, but also the vertical and/or lateral flux of water between soil units is simulated explicitly. Under the consideration of soil as a saturated, porous media with homogeneous properties, the velocity of flow through this media can be approximated by Darcy's law. Richards (1931) generalized Darcy's law to be applicable for unsaturated soil conditions by assuming the same linear relationship, but with the hydraulic conductivity varying nonlinearly according to the degree of soil saturation. Richards further combined the Darcy–Buckingham law with the continuity equation to result in a flow equation known as the Richards equation. To avoid computationally intense calculations – related to solving the Richards equation – the kinematic wave approach can be used as an approximation, which combines the continuity equation with a storage–flow relationship assuming a functional dependency between them.

Under natural conditions additional to matrix flow (Fig. 2f), often other types of soil water flow are observed. Therefore, more enhanced models – so-called two- or multidomain models (Köhne et al., 2009) – distinguish between rapid macroporous flow (e.g. flow through structured pores, cracks and fissures, and wormholes) and slow seepage through the soil matrix (Fig. 2g,h). Transfer terms account for the exchange between the flow domains. Domains are a conceptualization rather than a geometric separation of the soil volume. Two types of two- or multidomain models can be identified: Dual-porosity models (Fig. 2g), also known as mobile–immobile models (Armstrong et al., 2000), assume that the permeability of the soil matrix is very low due to poorly connected micropores and, thus, water in the matrix is immobile in the vertical direction. In contrast, dual-permeability models (Fig. 2h) allow for slow flux through the matrix in addition to the flux through macropores (Malone et al., 2004; Simunek and van Genuchten, 2008).

For more detailed mathematical descriptions of physically based soil water flow modeling, especially preferential flow, the reader is referred to the reviews of Gerke (2006) and Simunek et al. (2003). Köhne et al. (2009) provide a comprehensive review of applications of different modeling approaches undertaken during the last decade.

Detailed physically based models are primarily applied to studies at the scale of a soil column (length: 0.1 m–1 m) or a plot (soil profile: 1 m^2 to plot: 1000 m^2) (Köhne et al., 2009). However, if the same approaches are used at hillslope-scale or even catchment-scale studies (ha up to several tens of km^2), which often use large spatial units (grid cells) to represent the study area, some doubt arises as to whether or not these models can still be referred to as 'physically based', as expressed by several authors: "It seems to be an unresolved question if solving overland or groundwater flow equations for horizontal cell sizes of 100 m × 100 m, with average properties, can be considered a physically based approach" (Köhne et al., 2009:16). "The use of the Richards equation at the field and watershed scale is based more on pragmatism than on a sound physical basis" (Vereecken et al., 2007:1). As scale increases, other processes tend to dominate the overall flux or some local-scale effects become subordinate, causing a different behavior of the system. As a result, these systems exhibit complex nonlinear and, sometimes, threshold-type behavior that is difficult to describe with conventional methods (Tetzlaff et al., 2008; Tromp-van Meerveld and McDonnell, 2006).

2.2. Soil Property Estimation for Model Parametrization

The major challenges in soil parameter estimation for hydrologic modeling are related to scale issues and spatial variability. In this section, we briefly address (1) the challenge of upscaling soil hydraulic properties based on point- or plot-scale measurements and (2) interpolation or regionalization of point- and plot-scale measurements to locations with unknown soil parameters.

While parameters of soil properties are mainly measured at the scale of a soil core, for modeling estimates of parameter values are needed at larger spatial units such as grid cells or entire catchments. Parameters or models, which are used to simulate the response of large-scale systems without actually accounting for heterogeneity of properties and processes at the small scale are called effective parameters or models (Grayson and Blöschl, 2000). Environmental isotope studies revealed that catchment-scale soil hydraulic properties are different from those at soil core and plot scale (Bazemore et al., 1994). Catchment-average hydraulic conductivity is also found to increase with increasing catchment size (Blöschl and Sivapalan, 1995) and has been observed to be 2–15 times larger on the hillslope scale than values derived from soil core measurements (Brooks et al., 2004). A possible explanation is that with increasing spatial scale, the frequency and connectivity of preferential flowpaths are higher and, thus, average flow velocities are larger than in isolated soil cores. Therefore, a central question in hydrologic model parametrization is: How can point observations and small-scale process understanding be meaningfully linked to catchment-scale modeling? To obtain this, scaling is needed which is the procedure of determining effective parameters, properties, or mathematical descriptions needed on a target scale by incorporating information from a different scale. Two main categories of upscaling procedures exist (Vereecken et al., 2007): (1) Forward upscaling approaches derive target-scale parameters (in our case the catchment scale) from information on the spatial structure and variability of soil properties measured at a smaller scale (in our case, point or plot scale). (2) Inverse upscaling approaches derive parameters of the large-scale model and its large-scale model domain from (a) large-scale measurements (often remote-sensing data) or (b) from small-scale measurements, which are used as target variables in the calibration process of the model. In the latter case, it is assumed that the effective model structure or model equations describing the large-scale system behavior are known a priori.

Another challenge in model parametrization arises from the fact that we are unable to measure relevant soil parameters at all locations within a catchment over the entire depth of the soil profile. During the last decade, at least attempts have been made to capture the time-variant spatial structure of soil water flow and subsurface runoff processes using dense spatio-temporally distributed measurements: The Tarrawarra soil moisture data set (Australia) (Western and Grayson, 1998) is an example of an extensive measurement campaign to capture spatio-temporal soil moisture conditions of an entire catchment. It comprises more than 10,000 individual soil moisture observations collected during thirteen measuring campaigns within a two-year period. Soil moisture measurements were taken on a spatial resolution of 10×20 m within the 10.5-ha catchment using a mobile time-domain-reflectometry (TDR) instrument. Such surveys are expensive and time consuming; therefore, they are rare and only applicable in comparatively small catchments of a few 10th of hectares (Bogena et al., 2010).

Another central question in model parametrization is how to estimate soil parameter values of all locations within a catchment, based on a limited set of measurements. Interpolation and regionalization are tools to perform this task. Interpolation techniques, such as kriging or inverse distance weighting (IDW), use spatial distance as a weighting factor to estimate a parameter value based on neighboring measurement points. In addition, some other techniques incorporate auxiliary variables such as topographic indices or land use to estimate spatial patterns from point measurements (Lyon et al., 2010). Regionalization refers to methods which estimate variable values based on relations between the variable in question and known properties of the point or area in question. The relations have to be established based on observations of variables and properties of other sites prior to the regionalization procedure. In soil science pedotransfer functions are common to estimate soil hydraulic properties based on similarity of mapped soil types or other soil properties, which are easier or more common to be observed (Lin et al., 2006; Pachepsky et al., 2006). However, the spatially transferred values of soil hydraulic properties are not necessarily representative for catchment-scale effective values as they still might be derived from small-scale core samples (Lin et al., 1999). From a modeler's point of view, this inconsistency between point observations and effective parameters makes it difficult to evaluate whether or not the model is simulating internal runoff processes reasonably well. If careful evaluation and reasoning of the model structure are not undertaken, soil parameters of simple conceptual runoff models might also compensate for structural model errors and correlate with other characteristics in a catchment than those related to soils (Seibert, 1999).

3. TACIT KNOWLEDGE OF SUBSURFACE RUNOFF PROCESSES – THE EXPERIMENTALISTS' POINT OF VIEW

Tacit knowledge is expert knowledge that is internalized intuitively through repeated experience or observation (Hudson, 1992). It is inherent to its nature that tacit knowledge is difficult to express explicitly and to transfer from one expert to other persons (Polanyi, 1966). In fact, a major challenge of the dialog between the experimentalists and modelers in hydrologic science is the difficulty of (tacit) knowledge transfer and linking the experimentalists' observations to the governing mechanisms of the modelers' simulations. In the following part of this paper, we discuss a variety of hydric soil indicators, the experimentalists' tacit knowledge, and how to link it to subsurface runoff processes.

3.1. Indicating Dominant Subsurface Runoff Processes Based on Soil Survey Information

In this section, the potential of soil survey information to indicate dominant subsurface runoff processes is going to be discussed. The term "*dominant subsurface runoff processes*" is used for several reasons: (1) Soil survey information is not representative of seldom, temporary processes but is a fingerprint of *dominant*, prevailing, long-term, and persistent soil developing processes. (2) The dominant process concept (Grayson and Blöschl, 2000), which frames the following sections of this paper, was motivated by the recognition of being unable to observe and model all processes in detail, by the knowledge that only a few processes might be dominating the system behavior, and by the experience that simple models with only a few dominating parameters can be successful in modeling catchment runoff response (Sivakumar, 2004). The perceptual model, which comprises of the imagination of how the field evidences fit into dominant processes, is strongly influenced by the experimentalists' tacit knowledge. We begin this section with relevant information on subsurface runoff processes which can be derived from soil classes and their taxonomy used in soil mapping, continue with properties identifiable at a soil profile, and conclude with a discussion of small-scale hydromorphic features found in individual soil horizons, from which dominant subsurface runoff processes can be inferred. Having described the modelers' point of view on soil hydrologic function in Section 2, this section describes how experimentalists see soils and subsurface runoff processes.

Soil classes and their taxonomy shown in soil maps of various scales allow to indicate the degree and duration as well as the nature of soil saturation within a soil profile: Soils which are classified as gley soils are characterized by soil saturation in all layers from the upper boundary of saturation to a depth of 2 m or more by definition (Soil Survey Staff, 2006). They differ from soils classified as pseudo-gley insofar as they show saturation of one or more layers overlying a water-restricting, unsaturated horizon within 2 m of the surface (Soil Survey Staff, 2006). From this definition it is apparent that in pseudo-gley perched water tables must form and lateral subsurface flow can be expected in the soil horizons above the water-restricting layer (see Fig. 1f). In gley soils subsurface flow can be assumed to occur over the whole depth of the saturated soil profile. Both soil types are likely to show saturation-excess overland flow (Fig. 2b) as the groundwater table is generally high and therefore available storage capacity is rather small. This example shows that information on dominant subsurface runoff processes is inherent in soil classes and spatial distribution of dominant subsurface runoff processes can be roughly derived from soil maps, which might cover large areas of investigation (see also HOST classification in Section 3.2.2). Detailed soil surveying information often covers hydric properties. Natural drainage classes, for instance, describe the frequency and duration of wet periods within a year which have dominated soil formation at a particular site (Soil Survey Division Staff, 1993). Water-state annual patterns are describing soil water states of individual soil horizons or a standard depth zone over a year in tabular form and can show which soil horizons are saturated and therefore conductive at which time of the year. Inundation classes or inundation occurrences describe the degree, frequency, and duration of free water above the soil (Soil Survey Division Staff, 1993). Surface sealing, shrinking cracks, and indicators of hydrophobicity such as waxy coating of soil aggregates after a wild fire are recorded in a soil survey and can be used to anticipate infiltration capacity of soils and potential occurrence of infiltration-excess overland flow (Fig. 1a).

Soil horizons are the result of persistent subsurface flow and associated leaching, transport, and accumulation of material in different soil horizons. From soil hydraulic properties of individual soil horizons such as the saturated hydraulic conductivity – in both lateral and vertical directions – drainable porosity or bulk density, potential dominant subsurface runoff can be inferred. To estimate the storage capacity and the nature of onset of subsurface runoff, the depth to a water-restricting layer or a hydraulic discontinuity as well as depth to the soil–bedrock interface are important to consider. All this information is included in a description of a soil profile. If soil layers get saturated, highly conductive soil layers or preferential flow paths can be activated and, if connected efficiently, this can lead to a rapid contribution to stream flow discharge in a nearby channel (Seibert et al., 2009; Whipkey, 1965). Saturation occurs most likely at hydraulic discontinuities, either at water-restricting soil horizons (Fig. 1f) or at the soil–bedrock interface (Fig. 1g) (Freer et al., 2002; McDonnell, 1990). Water-restricting soil layers often form due to elluvial and illuvial processes which can be the result of persistent soil water flow driven by the hydraulic gradient and thus driven by topography. Confining flow trajectories due to topography, for instance, are the reason for constant soil water supply in depressions and thus influence soil formation there. For that reason, there is a feedback between topography, soil moisture state, and soil formation, which can be used to anticipate soil attributes at different topographic units within a landscape. A soil catena, for instance, maps this spatial dependency of soil properties and topography along a hillslope. While nowadays information on surface topography is available in detail for almost all catchments in form of digital elevation models (DEMs) there is hardly any information on the subsurface and bedrock topography, which is similarly important for anticipating dominant subsurface runoff processes (Freer et al., 2002).

When looking at individual soil horizons in more detail, a number of small-scale, visible hydric soil indicators can be observed which are characteristic to certain patterns of soil water movement. Soil redox features, also called soil mottling, are an example of soil morphological indicators which form by alternating reduction and oxidation due to saturation and drying which is associated with precipitation of Fe and Mn compounds in the soil (Soil Survey Staff, 1999). Redox concentration around macropores and redox depletion in the soil matrix could suggest that the macropores get drained and aired from time to time while the soil matrix stays more or less saturated constantly (Lin et al., 2008). The Soil Survey Staff at the US Department of Agriculture – National Resources Conservation Service (2010) published a set of such soil morphological characteristics ("field indicators of hydric soils") which allow a link to long-term persistent flow and transport processes.

3.2. Decision Schemes to Identify Dominant Runoff Processes

Linking hydric soil indicators, which can be identified in the course of a soil survey, with dominant runoff processes is often a matter of the experimentalists' tacit knowledge and thus subjective. Therefore, there is a need for rule-based decision schemes to link soil and land-use properties to dominant subsurface runoff processes in a structured and more objective way (Fig. 3). The spatial distribution of these dominant runoff processes can be expressed in the form of dominant runoff process maps. These maps structure the total area of a catchment into smaller, homogeneous units characterized by one dominant

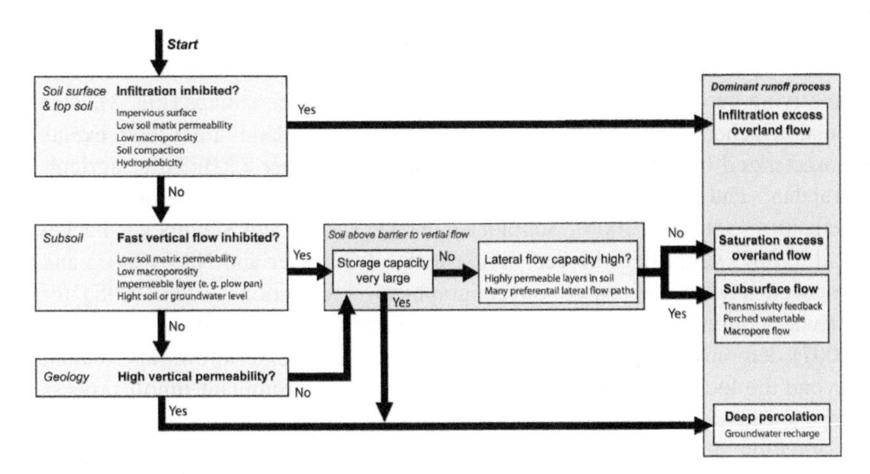

FIGURE 3 Schematic example of a decision scheme inferring dominant runoff processes from properties of top soil, subsoil, and geology. The decision tree is a means to capture the experimentalists' tacit knowledge so that it can be applied by others in an objective and thus comparable way. Modified based on Schmocker-Fackel, 2004.

runoff process. In the following, an overview of existing decision schemes structured into bottom–up and top–down approaches (as suggested by Schmocker-Fackel et al., 2007) will be given.

3.2.1. Bottom–up Approaches

Bottom–up approaches (e.g. Peschke et al., 1999; Scherrer and Naef, 2003; Schüler, 2005) are based on detailed soil surveying and/or artificial rainfall simulations performed on selected plots in the field. Processes, which have been identified to be dominant in the course of these detailed field investigations, are then assigned to other areas in the catchment having similar properties. To determine similarity, maps regarding soil type, land use, topography, and the drainage network are utilized (Peschke et al., 1999; Schmocker-Fackel et al., 2007). Markart et al. (2004b) present guidelines based on artificial rainfall simulations to derive infiltration capacity or surface runoff coefficients for various types of montane and alpine vegetation, including forest and pasture. The approach is different from the following as it is not considering subsurface processes but only infiltration capacity. However, it is an impressing example of experimentalists' aim to systematically build up process understanding in the form of an exceptional data set of 700 artificial rainfall simulations and to distill it down to a set of rules to indicate infiltration capacity (Markart et al., 2004a). Several decision schemes go further by inferring not only infiltration capacity, but also the dominant runoff processes (including subsurface runoff) from soil and land-cover or land-use properties.

Scherrer and Naef (2003) presented a decision scheme based on plot-scale artificial rainfall simulations and detailed observations of resulting surface and subsurface runoff processes. The decision scheme is structured like a soil profile considering the soil properties of the surface, topsoil, subsoil, and underlying geology and results in dominant runoff process maps (Fig. 3). This decision scheme was primarily developed for grassland areas and events characterized by high rainfall intensities but was further extended to agricultural land and forested areas incorporating both high and moderate rainfall intensities (Scherrer, 2006). Schmocker-Fackel (2004) and Schmocker-Fackel et al. (2007) simplified the decision scheme of Scherrer and Naef (2003) and proposed a workflow based on geographical information system (GIS) for deriving dominant runoff process maps utilizing existing soil data. Naef et al. (2007) demonstrated that this type of automated scheme allows application beyond the local scale and enables the generation of dominant runoff process maps for areas larger than 1000 km^2. As these authors admit, detailed geo-information such as soil maps – in their case on a scale of 1:5000 – are not widespread. Therefore, this approach is only applicable in certain data-rich regions. Schüler (2005) developed a similar decision scheme for forested areas using detailed forest management maps available in Rheinland-Pfalz, Germany (1:5000 to 1:10,000 scale). The decision scheme to assign a dominant runoff process is based on Scherrer and Naef (2003) and Scherrer (2006). Schüler (2005) further developed the approach by considering the delay of runoff process response as a function of the slope gradient. Steeper areas are expected to have fast responding surface and subsurface runoff processes, whereas flatter ones are expected to show a delayed response. In addition, line features such as drainage ditches, which can act as efficient connections between areas of runoff formation and the catchment outlet, are taken into account. The decision scheme by Waldenmeyer (2003) is based on detailed forest management maps and incorporates morphometric indices. While Scherrer and Naef (2003) and Schüler (2005) consider the runoff response of mapped dominant runoff processes to be invariant regardless of rainfall intensity, Waldenmeyer (2003) incorporates scenarios of various antecedent system conditions and precipitation intensities to account for varying runoff response.

3.2.2. Top–down Approaches

The bottom–up approaches discussed above require detailed soil- and land-use information to regionalize point observations to the catchment scale. Therefore, several authors (e.g. Tilch et al., 2006; Uhlenbrook, 2003) have followed a top–down approach. This type of approach starts with identifying hydrologically homogeneous units (HHU) based on existing geo-data (e.g. coarse DEMs and regional-scale soil and geological maps) or maps derived from aerial photos or other remote-sensing information (e.g. vegetation- and land-use maps). Compared to the maps used in bottom–up approaches, these data sets are generally coarser. HHUs derived by intersecting these data sets with a geographical information system (GIS) are then assumed to have hydrologically similar properties. Uhlenbrook (2003) and Tilch et al. (2006) used a top–down approach to delineate homogeneous landscape units in mesoscale catchments in Germany and Austria, respectively. Both authors performed field experiments in their catchments to assign hydrologically relevant properties to specific HHUs. Tilch et al. (2006) developed a regionalization approach to estimate additional necessary hydrologically relevant properties of the underlying sediments in the catchment based on DEM analysis. Transferability of such a top–down approach was tested successfully in a neighboring catchment (Uhlenbrook, 2003).

In Great Britain, a top–down approach was developed even applicable to indicate dominant runoff processes of very large areas up to hundred thousands of square kilometers. The Hydrology of Soil Types (HOST) classification (Boorman et al., 1995) is a soil-classification scheme based on regional soil maps of Great Britain (1:250.000 scale), to which different conceptual models of dominant subsurface runoff processes and pathways through the soil profile and/or parent material are assigned. The decision scheme first classifies soils into their physical setting by distinguishing between soils on a permeable parent material with mainly (1) deep groundwater tables, (2) shallow groundwater tables, and (3) soils and parent material with water-restricting layers within 1 m of the surface (Boorman et al., 1995). Within these three settings, 11 HOST response models are defined to account for differences in soil properties and wetness regimes. Each HOST response model can have further subdivisions regarding flow rate or flow mechanism, water-storage capacity, geology of parent material, saturated hydraulic conductivity, or artificial drainage. As not all combinations of HOST response models and subdivisions are realistic to occur and different combinations result in a similar hydrologic behavior, the set is limited to 29 combinations of so-called HOST classes. The HOST decision scheme was optimized by regressing base flow index (BFI) and the standard percentage runoff (SPR) – both characterizing catchment runoff response – to fractions of the various HOST classes occurring within the watershed of selected test catchments (Boorman et al., 1995). This HOST classification was then evaluated performing the same regression based on a set of 575 catchments in Great Britain, which resulted in a coefficient of determination of 0.79 and a standard error of 0.089 (Boorman et al., 1995). The results give reasonable confidence that the HOST classification can be a useful tool to estimate catchment-scale runoff response (BFI and SPR) for a very large area of investigation with lots of catchments incorporating the experts' process understanding. HOST maps with a spatial resolution of 1×1 km are available in digital form for the entire United Kingdom (data can be obtained from *Macaulay Land Use Research Institute*). In addition, lookup tables exist to convert more detailed soil maps to HOST classes (Boorman et al., 1995). Schneider et al. (2007) present a HOST-based hydrologic classification for all of Europe based on the Soil Geographical Database of Europe (SGDBE) scaled 1:1 million.

A criticism of the original work of Boorman et al. (1995) is that the HOST-based soil information cannot be directly linked to model parameters that are commonly used in rainfall–runoff modeling. For that reason, Dunn and Lilly (2001) present an approach linking soil parameters of a conceptual model to HOST classes. Monte Carlo simulations were performed to optimize parameter sets for two Scottish catchments of different character using the same meteorological input. Differences in optimal parameter sets of the two catchments were then assumed to reflect differences in fraction of HOST classes. Therefore, different parameter sets could be assigned to HOST classes, dominating in either the one or the other catchment.

3.2.3. Critical Thoughts and Remaining Challenges

Having mentioned bottom–up and top–down decision schemes to indicate dominant runoff processes for small-/mesoscale up to large-scale applications we now briefly discuss three critical challenges related to these approaches:

1) The results of field experiments that are representative of the governing processes at the point or plot scale are sometimes weak in representing the general hydrologic behavior of a larger scale (e.g. entire catchment). Scherrer and Naef (2003) report different hydrologic responses on soils with rather similar characteristics when performing their artificial rainfall simulations on selected plots. In fact, the influence of key processes emerged only in the context of the actual landscape position when considering the neighboring HHUs.
2) For many catchments of interest soil information does not exist and inferring dominant runoff processes from mapped soil indicators would be too cost- and labor-intensive. Therefore, various authors attempted to use indicator variables such as topography or vegetation in a preliminary working step to generate a first draft of a dominant runoff process map, which could then be evaluated in the field in a relatively short period of time (Meissl et al., 2009; Scherrer, 2006; Schmidt et al., 2000). Meissl et al. (2011) showed for an alpine study catchment that for 85% of a validation set of grid cells the same dominant runoff process could automatically be assigned as mapped in the field when using a classification and regression tree approach and the vegetation as the only explanatory variable. When incorporating both vegetation and topography, 98% of the validation set of grid cells could be classified correctly. The digital elevation model (DEM) can also be used to derive the degree of convergence and the slope gradient to estimate soil depths (Dahlke et al., 2009; Hjerdt et al., 2004) and thus storage capacity, which are good indications of potential dominant runoff processes (Scherrer, 2006).
3) A systematic comparison of the existing approaches to derive dominant runoff processes has not yet been published. This lack of comparative studies is partly due to the fact that the schemes require different sort and level of detail of geo-input data (soil- and land-use maps, DEM, etc.). Not all schemes differentiate between the same runoff processes and some do not consider

process intensities or runoff delay, respectively. Furthermore, the approaches are developed for different environments and are based on a limited number of field experiments representative for these environmental settings.

4. DISCUSSION

Having described the different points of view of modelers and experimentalists on subsurface runoff processes in the previous sections the question remains, how the dialog between experimentalists and modelers can actually be enhanced. In the following, we describe the potential of the Soft Data Concept (Fig. 4) and flexible box models to overcome the difficulty. We start with a general characterization of soft data.

4.1. The Soft Data Concept

Quantitative data, also called hard data, in soil surveying such as porosity or saturated hydraulic conductivity can directly be incorporated into quantitative models. In contrast, soft data are more difficult to consider in numerical modeling because of the following features: (1) Soft data are of qualitative, semi-quantitative, or categorical nature (e.g. natural drainage classes, inundation occurrence and hydric soil indicators). Soft data might (2) not meet

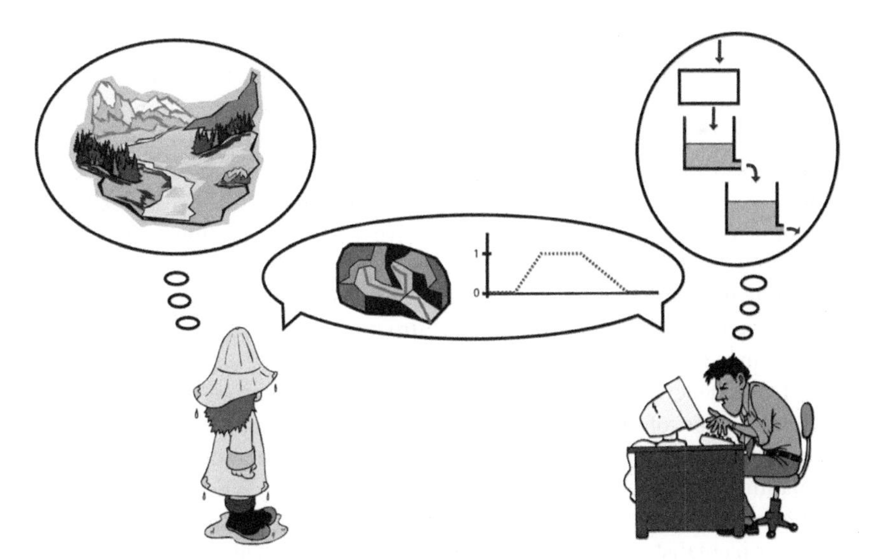

FIGURE 4 The Soft Data Concept as a tool to enhance the dialog between the experimentalists and modelers: Mapping is a means for synthesizing the experimentalists' tacit knowledge, gained in the course of field experience, which the modelers can distill down to an appropriate model structure. As qualitative or semi-quantitative data – called soft data – cannot directly be incorporated in the model calibration/validation procedure, fuzzy-member functions are used to assess the degree of acceptability of modeling results.

official quality standards and therefore cannot directly be utilized to calibrate a model without prior critical assessment: Information on water levels of historic flood events, for instance, which are based on old literature in archives (Schmocker-Fackel and Naef, 2010) or which are affected by unreliable rating curves, can still be very informative about the potential flood magnitude of extreme events. These rare events with a low probability of occurrence are normally not covered in existing time series of modern stream flow discharge measurements. Soft data might (3) first need the experts' interpretation to be applicable to modeling: Tracer observations, for instance, allow estimating catchment-scale mean transit times, which can be valuable for characterizing the hydrologic functioning of a catchment (Soulsby et al., 2010). Soft data can also (4) be affected by considerable spatial variability, often not resolved by the coarse spatial discretization of hydrologic models (subgrid variability): Data from point measurements such as groundwater levels or soil moisture data, used to calibrate and validate a model or its internal model state (Freer et al., 2004), are known to be affected by small-scale topographic and pedologic properties and therefore might not be directly comparable with modeling results on a coarser spatial resolution. The time series at hand might (5) not be continuous but only cover some events measured during a few field campaigns, thus the data are called "soft". Soft data might (6) also include indirect measurements or derived values of catchment-scale system states. Such soft information could be based on remotely-sensed data or upscaled from direct measurements using spatial interpolation, regionalization or other computational methods. Thermal images, for instance, might be useful to identify spatio-temporal patterns of soil moisture distribution or evaporation within a catchment (Ludwig et al., 2003). Another example is an estimate of catchment-wide storage based on field measurements, which can inform subsurface models (Seibert et al., 2011). Finally, it is important to recognize that (7) the experts' tacit knowledge is qualitative and subjective and therefore much harder to directly link to quantitative model results in an objective way.

Despite these limitations, soft data are particularly valuable for constraining model parameters and optimizing model structure in the calibration/validation process. In addition to quantitative goodness-of-fit criteria like the Nash–Sutcliffe efficiency (Nash and Sutcliffe, 1970), soft data can be useful to define a realistic range of possible parameter values in an automated calibration procedure (Winsemius et al., 2009). Soft data can be used to assess the acceptability of model simulations or parameter values. When the degree of acceptance is high the parameter values agree well with the field experience of the experimentalists. Fuzzy-logic based rules are applied to transform the qualitative information into a numeric value of acceptance (Seibert and McDonnell, 2002). For each type of soft data, which is going to be considered in the model calibration/validation procedure, a fuzzy-member function (Fig. 4) is defined to calculate the degree of acceptance (ranging from 0 to 1) of model results based on the experimentalists' tacit knowledge how the system should function. When using multiple soft data sources at once, the total degree of acceptance can be determined by calculating a mean (e.g. geometric mean) of individual acceptability scores. The goal of the model calibration procedure is then not only to result in a high goodness-of-fit score of modeled total catchment runoff but also to optimize the degree of acceptability of the internal model processes and, thus, to better agree with what the experimentalists observe in the field. Flexible box models, a type of conceptual model with tunable storage properties and components, allow optimizing the model structure to best capture all necessary processes identified by the experimentalists (Fenicia et al., 2008; Schmocker-Fackel, 2004; Uhlenbrook and Leibundgut, 2002). Soft data become particularly valuable for model exercises in ungauged catchments, where no hard data exist. In short, there may be a wealth of untapped data available for many catchments, which have not been utilized until now.

5. CONCLUDING REMARKS

For future progress in hydropedology a dialog between experimentalists and modelers offers great potential as both provide different views on investigating, understanding, and quantifying the first-order control of catchment runoff response. The experimentalists can offer a wealth of hard and soft facts, documented in soil surveys as well as their tacit knowledge gained during many years of field experience. A set of hydric soil indicators and decision schemes exist to systematically indicate dominant subsurface runoff processes. Mapping these dominant runoff processes has been identified as a potential way to capture the experimentalists' tacit knowledge and transfer it to the modelers, who are then able to give feedback based on their modeling results. The modelers can inform model calibration and parameter estimation in the course of the Soft Data Framework. This allows incorporating qualitative and semi-quantitative data, existing for not only gauged but also ungauged catchments. In this way, acceptability of model simulations is evaluated not only based on catchment runoff as a single integrated value but also based on additional information about internal system states and processes. The dialog between experimentalists and modelers should not necessarily aim at highly detailed observations and modeling of natural heterogeneity with evermore complex monitoring and modeling techniques. Instead, the goal should be to reveal first-order controls of natural hydrologic systems to better understand nonlinearity of runoff processes at the catchment scale (McDonnell, 2003).

ACKNOWLEDGMENTS

The authors thank Henry Lin, the editor of this book, Gertraud Meissl, Russell Smith, and two anonymous reviewers for their critical reviews and useful comments as well as Christine Rinderer for proofreading the manuscript.

REFERENCES

Armstrong, A.C., Matthews, A.M., Portwood, A.M., Leeds-Harrison, P.B., Jarvis, N.J., 2000. CRACK-NP: a pesticide leaching model for cracking clay soils. Agr. Water Manage. 44 (1–3), 183–199.

Bazemore, D.E., Eshleman, K.N., Hollenbeck, K.J., 1994. The role of soil water in stormflow generation in a forested headwater catchment: synthesis of natural tracer and hydrometric evidence. J. Hydrol. 162 (1–2), 47–75.

Blöschl, G., Sivapalan, M., 1995. Scale issues in hydrological modeling - a review. Hydrol. Process. 9 (3–4), 251–290.

Bogena, H.R., Herbst, M., Huisman, J.A., Rosenbaum, U., Weuthen, A., Vereecken, H., 2010. Potential of wireless sensor networks for measuring soil water content variability. Vadose Zone Journal 9 (4), 1002. Doi:10.2136/vzj2009.0173.

Boorman, D., Hollis, J., Lilly, A., 1995. Hydrology of Soil Types: a Hydrologically Based Classification of the Soils of the United Kingdom. Report 126. Institute of Hydrology, Wallingford.

Brooks, E.S., Boll, J., McDaniel, P.A., 2004. A hillslope-scale experiment to measure lateral saturated hydraulic conductivity. Water Resour. Res. 40 (4). Doi: 10.1029/2003wr002858.

Dahlke, H.E., Behrens, T., Seibert, J., Andersson, L., 2009. Test of statistical means for the extrapolation of soil depth point information using overlays of spatial environmental data and bootstrapping techniques. Hydrol. Process. 23 (21), 3017–3029. Doi: 10.1002/Hyp.7413.

Dunn, S.M., Lilly, A., 2001. Investigating the relationship between a soils classification and the spatial parameters of a conceptual catchment-scale hydrological model. J. Hydrol. 252 (1–4), 157–173.

Fenicia, F., McDonnell, J.J., Savenije, H.H.G., 2008. Learning from model improvement: on the contribution of complementary data to process understanding. Water Resour. Res. 44 (6). Doi: 10.1029/2007wr006386.

Freer, J., McMillan, H., McDonnell, J.J., Beven, K.J., 2004. Constraining dynamic TOPMODEL responses for imprecise water table information using fuzzy rule based performance measures. J. Hydrol. 291 (3–4), 254–277.

Freer, J., McDonnell, J.J., Beven, K.J., Peters, N.E., Burns, D.A., Hooper, R.P., Aulenbach, B., Kendall, C., 2002. The role of bedrock topography on subsurface storm flow. Water Resour. Res. 38 (12). Doi: 10.1029/2001wr000872.

Gerke, H.H., 2006. Preferential flow descriptions for structured soils. J. Plant Nutr. Soil Sci. 169 (3), 382–400. Doi: 10.1002/jpln.200521955.

Grayson, R., Blöschl, G., 2000. Spatial Patterns in Catchment Hydrology: Observations and Modelling. Cambridge University Press, Cambridge.

Hjerdt, K.N., McDonnell, J.J., Seibert, J., Rodhe, A., 2004. A new topographic index to quantify downslope controls on local drainage. Water Resour. Res. 40 (5). Doi: 10.1029/2004WR003130.

Hua, J.P., Liang, Z.M., Yu, Z.B., 2003. A modified rational formula for flood design in small basins. J. Am. Water Resour. Assoc. 39 (5), 1017–1025.

Hudson, B.D., 1992. The soil survey as paradigm-based science. Soil Sci. Soc. Am. J. 56 (3), 836–841.

Kirchner, J.W., 2006. Getting the right answers for the right reasons: linking measurements, analyses, and models to advance the science of hydrology. Water Resour. Res. 42 (3). Doi: 10.1029/2005wr004362.

Klemes, V., 1986. Dilettantism in hydrology - transition or destiny. Water Resour. Res. 22 (9), 177–188.

Köhne, J.M., Köhne, S., Simunek, J., 2009. A review of model applications for structured soils: (a) water flow and tracer transport. J. Contam. Hydrol. 104 (1–4), 4–35. Doi: 10.1016/j.jconhyd.2008.10.002.

Lin, H., Brooks, E., McDaniel, P., Boll, J., 2008. Hydropedology and surface/subsurface runoff processes. In: Anderson, M.G. (Ed.), Encyclopedia of Hydrological Sciences. Wiley, Chichester, Online Library, pp. 1–25.

Lin, H., Bouma, J., Pachepsky, Y., Western, A., Thompson, J., van Genuchten, R., Vogel, H.J., Lilly, A., 2006. Hydropedology: synergistic integration of pedology and hydrology. Water Resour. Res. 42 (5). Doi: 10.1029/2005wr004085.

Lin, H., McInnes, K.J., Wilding, L.P., Hallmark, C.T., 1999. Effects of soil morphology on hydraulic properties: II. Hydraulic pedotransfer functions. Soil Sci. Soc. Am. J. 63 (4), 955–961.

Ludwig, R., Probeck, M., Mauser, W., 2003. Mesoscale water balance modelling in the Upper Danube watershed using sub-scale land cover information derived from NOAA-AVHRR imagery and GIS-techniques. Physics and Chemistry of the Earth 28 (33–36), 1351–1364.

Lyon, S.W., Sorensen, R., Stendahl, J., Seibert, J., 2010. Using landscape characteristics to define an adjusted distance metric for improving kriging interpolations. Int. J. Geogr. Inf. Sci. 24 (5), 723–740. Doi: 10.1080/13658810903062487.

Malone, R.W., Ahuja, L.R., Ma, L.W., Wauchope, R.D., Ma, Q.L., Rojas, K.W., 2004. Application of the root zone water quality model (RZWQM) to pesticide fate and transport: an overview. Pest Manag. Sci. 60 (3), 205–221. Doi: 10.1002/Ps.789.

Markart, G., Kohl, B., Sotier, B., Schauer, T., Bunza, G., Stern, R., 2004b. Provisorische Geländeanleitung zur Abschätzung des Oberflächenabflussbeiwertes auf Alpinen Boden-/Vegetationseinheiten bei Konvektiven Starkregen (Version 1.0) – A Simple Code of Practice for Assessment of Surface Runoff Coefficients for Alpine Soil-/Vegetation Units in Torrential Rain (Version 1.0). BFW-Dokumentationen 3/2004. Bundesamt und Forschungszentrum für Wald, Vienna.

Markart, G., Kohl, B., Schauer, T., Sotier, B., Bunza, G., Stern, R., 2004a. Eine einfache Gelaendeanleitung zur Abschätzung des Oberflaechenabflussbeiwertes bei Starkregen. A simple code of practice for assessment of surface runoff coefficients in torrential rain. pp. 89–100. In: International Symposium INTERPRAEVENT (Ed.), International Symposium INTERPRAEVENT. 24.05. – 28.05. 2004. Riva del Garda/Trient/Italy.

McDonnell, J.J., 1990. A rationale for old water discharge through macropores in a steep, humid catchment. Water Resour. Res. 26 (11), 2821–2832.

McDonnell, J.J., 2003. Where does water go when it rains? Moving beyond the variable source area concept of rainfall-runoff response. Hydrol. Process. 17 (9), 1869–1875.

McGlynn, B.L., McDonnel, J.J., Brammer, D.D., 2002. A review of the evolving perceptual model of hillslope flowpaths at the Maimai catchments, New Zealand. J. Hydrol. 257 (1–4), 1–26.

Meissl, G., Geitner, C., Tusch, M., Stötter, J., 2009. Derivation of Dominant Runoff Processes in an Alpine Catchment using Vegetation and Relief Data. EGU General Assembly 2009. 19.04 – 24.04. 2009. Vienna.

Meissl, G., Geitner, C., Tusch, M., Schoeberl, F., Stoetter, J., 2011. Reliefparameter und abflusssteuernde flaecheneigenschaften: statistische analyse ihres zusammenhangs in einem kleinen alpinen einzugsgebiet. Zeitschrift für Geomorphologie 55 (3), 295–315.

Mulvaney, T.J., 1851. On the use of self-registering rain and flood gauges in making observations of the relations of rainfall and flood discharge in a given cathcment. P. I. Civil Eng. Ireland (4), 19–31.

Naef, F., Margreth, M., Schmocker-Fackel, P., Scherrer, S., 2007. Automatisch hergeleitete abflussprozesskarten - ein neues werkzeug zur abschätzung von hochwasserabflüssen. Wasser Energie Luft 99 (3), 267–272.

Nash, J.E., Sutcliffe, J.V., 1970. River flow forecasting through conceptual models part I – a discussion of principles. J. Hydrol. 10 (3), 282–290.

Pachepsky, Y.A., Rawls, W.J., Lin, H.S., 2006. Hydropedology and pedotransfer functions. Geoderma 131 (3–4), 308–316. Doi: 10.1016/j.geoderma.2005.03.012.

Peschke, G., Etzenberg, C., Müller, G., Töpfer, J., Zimmermann, S., 1999. Das Wissensbasierte System FLAB – Ein Instrument zur Rechnergestützten Bestimmung von Landschaftseinheiten mit Gleicher Abflussbilung. IHI-Schriften 10. Internationales Hochschulinstitut Zittau, Zittau.

Polanyi, M., 1966. The Tacit Dimension. Doubleday and Co., New York.

Richards, L.A., 1931. Capillary conduction of liquids through porous mediums. Journal of Applied Physics 1 (5), 318–333.

Rinderer, M., Jenewein, S., Senfter, S., Rickenmann, D., Schöberl, F., Stötter, J., Hegg, C., 2009. Runoff and bedload transport modelling for flood hazard assessment in small alpine catchments-the PROMABGIS model. In: Veulliet, E., Stötter, J., Weck-Hannemann, H. (Eds.), Sustainable Natural Hazard Management in Alpine Environments. Springer, Berlin, pp. 69–101.

Scherrer, S., 2006. Bestimmungsschlüssel zur Identifikation von Hochwasserrelevanten Flächen. Report 18/2006. Landesamt für Umwelt, Wasserwirtschaft und Gewerbeaufsicht, Rheinland-Pfalz.

Scherrer, S., Naef, F., 2003. A decision scheme to indicate dominant hydrological flow processes on temperate grassland. Hydrol. Process. 17 (2), 391–401. Doi: 10.1002/Hyp.1131.

Schmidt, J., Hennrich, K., Dikau, R., 2000. Scales and similarities in runoff processes with respect to geomorphometry. Hydrol. Process. 14 (11–12), 1963–1979.

Schmocker-Fackel, P., 2004. A Method to Delineate Runoff Processes in a Catchment and Its Implications for Runoff Simulations. PhD-thesis. Swiss Federal Institute of Technology ETH Zurich, Zurich.

Schmocker-Fackel, P., Naef, F., 2010. Changes in flood frequencies in Switzerland since 1500. Hydrol. Earth Syst. Sci. Discuss. 7 (1), 529–560.

Schmocker-Fackel, P., Naef, F., Scherrer, S., 2007. Identifying runoff processes on the plot and catchment scale. Hydrol. Earth Syst. Sci. 11 (2), 891–906.

Schneider, M., Brunner, F., Hollis, J., Stamm, C., 2007. Towards a hydrological classification of European soils: preliminary test of its predictive power for the base flow index using river discharge data. Hydrol. Earth Syst. Sci. 11 (4), 1501–1513.

Schüler, G., 2005. Herleitung von abflussrelevanten Fächen zur Steuerung von Wasserrückhaltemassnahmen im Wald. Freiburger Forstliche Forschung 62. Freiburg im Breisgau.

Seibert, J., 1999. Regionalisation of parameters for a conceptual rainfall-runoff model. Agr. Forest Meteorol. 98 (9), 279–293.

Seibert, J., Bishop, K., Nyberg, L., Rodhe, A., 2011. Water storage in a till catchment. I: Distributed modelling and relationship to runoff. Hydrol. Process. 25, 3937–3949.

Seibert, J., Rodhe, A., Bishop, K., 2003. Simulating interactions between saturated and unsaturated storage in a conceptual runoff model. Hydrol. Process. 17 (2), 379–390. Doi: 10.1002/Hyp.1130.

Seibert, J., McDonnell, J.J., 2002. On the dialog between experimentalist and modeler in catchment hydrology: use of soft data for multicriteria model calibration. Water Resour. Res. 38 (11). Doi: 10.1029/2001wr000978.

Seibert, J., Grabs, T., Kohler, S., Laudon, H., Winterdahl, M., Bishop, K., 2009. Linking soil- and stream-water chemistry based on a Riparian flow-concentration integration model. Hydrol. Earth Syst. Sci. 13 (12), 2287–2297.

Sherman, L., 1932. Streamflow from rainfall by the unit-graph method. Eng. News Rec. 108, 501–505.

Simunek, J., van Genuchten, M.T., 2008. Modeling nonequilibrium flow and transport processes using HYDRUS. Vadose Zone J. 7 (2), 782–797. Doi: 10.2136/Vzj2007.0074.

Simunek, J., Jarvis, N.J., van Genuchten, M.T., Gardenas, A., 2003. Review and comparison of models for describing non-equilibrium and preferential flow and transport in the vadose zone. J. Hydrol. 272 (1–4), 14–35. Doi: Pii S0022-1694(02)00252-4.

Sivakumar, B., 2004. Dominant processes concept in hydrology: moving forward. Hydrol. Process. 18 (12), 2349–2353. Doi: 10.1002/Hyp.5606.

Soil Survey Division Staff, 1993. Soil Survey Manual. United States Department of Agriculture – Soil Conservation Service, Washington D.C.

Soil Survey Staff, 1999. Soil Taxonomy: a Basic System of Soil Classification for Making and Interpreting Soil Surveys. United States Department of Agriculture - Natural Resources Conservation Service, Washonton D.C.

Soil Survey Staff, 2006. Keys to Soil Taxonomy. United States Department of Agriculture – Natural Resources Conservation Service, Washington, D.C.

Soil Survey Staff, 2010. Field Indicators of Hydric Soils in the United States. A Guide for Identifying and Delineating Hydric Soils. Version 7.0 ed. United States Department of Agriculture – Natural Resources Conservation Service, Washington D.C.

Soulsby, C., Tetzlaff, D., Hrachowitz, M., 2010. Are transit times useful process-based tools for flow prediction and classification in ungauged basins in montane regions? Hydrol. Process. 24 (12), 1685–1696. Doi: 10.1002/Hyp.7578.

Tetzlaff, D., McDonnell, J.J., Uhlenbrook, S., McGuire, K.J., Bogaart, P.W., Naef, F., Baird, A.J., Dunn, S.M., Soulsby, C., 2008. Conceptualizing catchment processes: simply too complex? Hydrol. Process. 22 (11), 1727–1730. Doi: 10.1002/Hyp.7069.

Tilch, N., Zillgens, B., Uhlenbrook, S., Leibundgut, C., Kirnbauer, R., Merz, B., 2006. GIS-gestützte Ausweisung von hydrologischen Umsatzräumen und Prozessen im Löhnersbach-Einzugsgebiet (Nördliche Grauwackenzone, Salzburger Land). Österreichische Wasser-und Abfallwirtschaft 58 (9), 141–151.

Tromp-van Meerveld, H.J., McDonnell, J.J., 2006. Threshold relations in subsurface stormflow: 2. The fill and spill hypothesis. Water Resour. Res. 42 (2). Doi: 10.1029/2004wr003800.

Uhlenbrook, S., 2003. An empirical approach for delineating spatial units with the same dominating runoff generation processes. Phys. Chem. Earth 28 (6–7), 297–303. Doi: 10.1016/S1474-7065(03)00041-X.

Uhlenbrook, S., Leibundgut, C., 2002. Process-oriented catchment modelling and multiple-response validation. Hydrol. Process. 16 (2), 423–440.

Vereecken, H., Kasteel, R., Vanderborght, J., Harter, T., 2007. Upscaling hydraulic properties and soil water flow processes in heterogeneous soils: a review. Vadose Zone J. 6 (1), 1–28. Doi: 10.2136/Vzj2006.0055.

Waldenmeyer, G., 2003. Abflussbildung und regionalisierung in einem forstlich genutzten einzugsgebiet (Dürreychtal, Nordschwarzwald). Karlsruher Schriften zur Geographie und Geoökologie 20. Karlsruhe.

Weiler, M., McDonnell, J., 2004. Virtual experiments: a new approach for improving process conceptualization in hillslope hydrology. J. Hydrol. 285 (1–4), 3–18. Doi: 10.1016/S0022-1694(03)00271-3.

Weiler, M., McDonnell, J.J., Tromp-Van Meerveld, I., Uchida, T., 2005. Subsurface stormflow. In: Anderson, G.M., McDonnell, J.J. (Eds.), Encyclopedia of Hydrological Sciences, pp. 1719–1732. Wiley, Chichester.

Western, A.W., Grayson, R.B., 1998. The tarrawarra data set: soil moisture patterns, soil characteristics, and hydrological flux measurements. Water Resour. Res. 34 (10), 2765–2768.

Whipkey, R.Z., 1965. Subsurface stormflow from forested slopes. Bulletin of the International Association of Scientific Hydrology 10 (2), 74–85.

Winsemius, H., Schaefli, B., Montanari, A., Savenije, H., 2009. On the calibration of hydrological models in ungauged basins: a framework for integrating hard and soft hydrological information. Water Resour. Res. 45 (12). Doi: 10.1029/2009WR007706.

Chapter 17

Hydrological Classifications of Soils and their Use in Hydrological Modeling

A. Lilly,[1,*] S.M. Dunn[1] and N.J. Baggaley[1]

ABSTRACT

Soils can have a profound effect on catchment hydrology by delaying runoff, storing and redistributing water; however, this varies considerably with soil type. The incorporation of soils information into catchment-scale hydrological models has many perceived obstacles. One is the specialist language used in soil science and another is the multitude of soil attributes that seem to influence catchment-scale hydrology. Soil hydrological classifications are one way that soil scientists can make their information more readily accessible to hydrologists; however, only a few have been developed thus far and these existing classifications could be further improved. Nevertheless, these classifications provide useful functional descriptions of the pathways of water movement through different soils and the dominant internal processes involved. While there still remains a challenge in translating these into quantitative parameters required by catchment-scale hydrological models, this problem is now more widely recognized and new methods are continually being explored. Continued efforts along this line of inquiry would undoubtedly help advance hydropedology.

1. INTRODUCTION

It is difficult to overestimate the role of soils in influencing the hydrological cycle. Soils delay runoff, store and redistribute water, and provide a supply of moisture for plant transpiration. However, these regulatory functions vary considerably with soil type and therefore the impact on the hydrological cycle and catchment runoff characteristics greatly depends on the nature and distribution of soils, the two principal components of a soil survey. Many hydrologists are wary of accessing or using soil survey information mainly due to the specialist language used by pedologists, the unwillingness of some soil surveyors

[1]The James Hutton Institute, Craigiebuckler, Aberdeen, AB15 8QH, Scotland, UK
[*]Corresponding author: Email: allan.lilly@hutton.ac.uk

Hydropedology, Edited by H. Lin. DOI: 10.1016/B978-0-12-386941-8.00017-2

to acknowledge uncertainty inherent in the map polygons, and the focus on traditional soil classification with limitless splits based on relatively minor morphological characteristics. The emergence of hydropedology as a discipline that links hydrology and pedology should help to overcome these barriers and move hydrological models away from treating soils as a black box to one where the nature and properties of soil can be explicitly embedded within the model.

In this chapter, we first summarize some research that shows how soil and soil properties influence the hydrological response of catchments. We then outline some attempts to classify soils specifically for use in predicting either catchment hydrology or flow pathways. Finally, we review some of the problems and possible solutions to the use of soils information in hydrological modeling.

2. IMPORTANCE OF SOILS FOR CATCHMENT HYDROLOGY

The hydrological response of a catchment to rainfall is a function of runoff rate, infiltration rate, and precipitation characteristics. These are in turn driven by topography, surface properties, and, importantly, the distribution of soil and vegetation within the catchment (Hawley et al., 1983; Aryal et al., 2003; Buttle et al., 2004). The redistribution of precipitation as runoff or infiltration is also determined by the antecedent soil moisture conditions (Menzaini et al., 2003; Anctil et al., 2008; Lin and Zhou et al., 2008).

It has been shown that the existence and spatial distribution of permanent and transient saturated areas can have a large impact on storm flow (Schmidt et al., 2000) and this spatial organization can influence both runoff and the export and delivery of chemicals to the water courses (Neumann et al., 2010).

The dynamics and spatial patterns of soil moisture have been the focus of research across a wide variety of landscapes and land uses. These studies span the globe, encompassing different climatic regions, and range from point-based measurements to remote sensing covering larger areas. They include small- and large-scale field campaigns, long-term temporal monitoring, remote sensing and geophysical techniques (Robinson et al., 2008).

As many of these studies tend to examine the hydrology associated with specific land uses, the following discussion outlines some of the conclusions drawn from an evaluation of the effect of soils on runoff, infiltration, and preferential flow pathways at a range of scales in forested, rangeland, arable, and montane catchments.

2.1. Forested Catchments

In forested landscapes, preferential flow and hydraulic connectivity between soils and surface waters have been shown to be important processes in a catchment's response to rainfall in many parts of the world including Shale Hills, USA (Lin and Zhou, 2008), Spain (Fitzjohn et al., 1998), Mont Saint Hilaire UNESCO Biosphere Reserve, Quebec (James and Roulet, 2007), and in Southern Italy (Buttafuoco et al., 2005).

At a landscape-scale, hydrological processes have also been shown to vary spatially within catchments as a function of soil texture, soil profile characteristics, depth to bedrock, and topographic position (Wang et al., 2002; Guntner et al., 2004; Buttafuocco et al., 2005; Lin et al., 2006, 2008).

The response of a soil varies temporally with antecedent moisture conditions and rainfall intensity (Lin et al., 2008). There is also hysteresis in the system due to cycles of drying and rewetting (Buttafuoco et al., 2005). Gunter et al. (2004) identified and mapped stable areas of wetness using soil characteristics such as gleying and the presence of hydrophilic plant species in the German Black Forest. Kim (2009) showed that the extent of saturated zones on a hillslope varied seasonally. These changes in soil wetness influenced hydraulic connectivity within the catchment and ultimately, stream response. This raises an important issue, that of the stability and persistence of preferential flow pathways and zones of wetness in relation to seasonal differences in rainfall.

James and Roulet (2007) used a geostatistical approach to examine connectivity in the forested humid Mont Saint Hilaire watershed. They measured the moisture contents of the upper 20 cm of the soil surface at over 300 points using TDR. Although they found that surface soil moisture contents had little influence on stream runoff, measurements of the water table fluctuations and moisture contents from vertically installed TDR probes suggested that there was greater connectivity within the subsoils. This was attributed to subsurface lateral flow associated with the development of perched water tables due to the presence of slowly permeable fragipans. This work highlights the need for a greater understanding of the subsurface soil properties in the interpretation of hillslope hydrology in addition to an understanding of surface topography and soil texture. It also points to the need to understand and take account of pedogenic processes. The presence of perched water tables can be readily identified by examining the distribution of the pattern of soil colors, in particular, mottling and gleying.

The pathway of water flow through different layers of the soil profile has also been shown to be important for the chemistry of receiving waters. Vestin et al. (2008a) found that the water chemistry reflected the chemistry of subsurface horizons of Podzols (Spodosols) and Arenosols (Psamments) in a boreal forest catchment in Sweden during low flows and the chemistry of surface horizons during high-flow events. The increased saturation in the riparian Arenosols (Psamments) was thought to be driving microbial activity and hence the cation and Dissolved Organic Carbon (DOC) concentrations (Vestin et al., 2008b).

2.2. Grassland and Mixed Rangeland Catchments

Several rangeland catchments, including the Australian Tarrawarra and associated catchments in Australia and New Zealand and the American Little Wattisha watersheds, have been the focus for the development of regression relationships between soil moisture, terrain, and soil attributes and how these change over time (Starks et al., 2006; Western et al., 2004; Brocca et al., 2009). These studies attempted to establish soil and landscape characteristics that explained the observed spatial distributions in soil moisture content. Both terrain and soils have been highlighted as covariates that explain the spatially distributed nature of surface soil moisture and appear to influence its temporal stability. It appears from many of these studies that although terrain is an easily measured covariate, it explains only part of the observed variation (Crave and Gascuel-Odoux, 1997; Wilson et al., 2004; Buttafuco et al., 2005). In five study catchments in Australia and New Zealand, Western et al. (2004) suggest that patterns of surface soil moisture, not explained by terrain, were due to vegetation and soil characteristics. The soil characteristics discussed include lateral flow in duplex soils, variations in soil texture and consequently water holding capacity, preferential flow due to pipe erosion, and percolation in deep, freely draining soils. In addition, they infer, from the results of geostatistical analysis, that these properties influence the distribution of soil moisture at different scales from those driven by terrain.

Recently, Empirical Orthogonal Functions (EOFs) have been used to analyze stable patterns of spatially distributed soil moisture and their relationship to spatio-temporal processes (Perry and Niemann, 2008; Korres et al., 2010). Korres et al. (2010) related these stable patterns to covariates such as soil attributes (for example, field capacity, stoniness, sand content, porosity, and soil organic carbon), land management, and terrain. This study showed that the relative importance of these covariates changed depending on the degree of overall catchment wetness.

Van Schaik (2009) examined the role of soil texture, position in the landscape, and vegetation in determining preferential flow in a semi-arid catchment in Spain. They found that the hillslope location and soil texture influenced the water regime at a landscape-scale within which the vegetation distribution introduced local-scale variation in soil porosity. The soils at this site varied in texture from loamy sand to silty loam and the inherent porosity was modified by the activity of both plants and animals. Lateral flow was observed along animal burrows and water stagnation noted at depth within the profile due to the presence of a clayey, more slowly permeable layer.

Van Tol et al. (2010) working in South Africa used differences in soil morphology to explain hillslope flow patterns, soil saturation, and streamflow generation processes. Their conceptual hillslope model was validated against tensiometric measurements. The authors postulated that all water contributing to streamflow at the foot of the slope must flow through a specific horizon, based on their observations. They then calculated the hydraulic conductivity of that horizon and the extent of this horizon and related this to stream base flow. Although the model underpredicted the streamflow, it did indicate that a large proportion of the streamflow was indeed passing through this soil layer and the presence of pale gley colors indicated that it was saturated for prolonged periods of time.

2.3. Arable Catchments

The focus in the predominantly arable catchments has been on the generation of runoff as a driver of streamflow and the associated diffuse pollutants it often carries (Freebairn et al., 2009). In northern Italy and Orgeval in France, antecedent moisture conditions have been shown to have a major influence on river flow and flooding in these areas (Menzaini et al., 2003; Anctil et al., 2008). Rainfall periodicity has also been shown to be an important factor in the hydrological response of soils and consequently in pattern development in Quebec (Parent et al., 2006).

Romshoo (2004) suggests that soils, microtopography, precipitation, and evapotranspiration drive the soil-moisture distribution in alluvial soils in a rain-fed paddy area in Thailand. In wet periods, the importance of increasing soil lateral hydraulic conductivity is highlighted in the development of connected zones of increased soil moisture. Lauzon et al. (2004) showed that layering in the soil has important consequences for the spatial patterns of soil moisture observed in the landscape and consequently on flow. Slowly permeable layers have been shown to influence runoff generation (De Lannoy et al., 2006). In the same field, these slowly permeable subsurface layers have also been shown to affect both hydrological dynamics and crop water availability and consequently crop yield (Gish et al., 2005; Gburek et al., 2006).

The dynamics in catchments that are highlighted in these temporal studies point to high resolution temporal data combined with remote-sensing data as being useful tools for applications such as flood forecasting (Brocca et al., 2010). These long-term temporal data sets need, as in grassland systems, to represent the soil moisture characteristics of the broader landscape. Brocca et al. (2010) suggest that one of their sites shows greater variability, due to a "coarse soil texture" but they do not consider the hydrological response of the soil profiles as a deterministic factor in soil moisture temporal stability (Vachaud et al., 1985). Giraldo et al. (2008) show that the variations between sites are not just due to land-use/land-cover characteristics and they suggest that "soil type", land use, and precipitation interact; however, they do not discuss specific soil properties which may be driving these interactions. In an arable test site in Germany, Korres et al. (2010) use EOFs to show the importance of soil texture and stone content as well as land management while Bachmair et al. (2009) identify the importance of soil structure, microtopography, land cover, and topsoil matrix characteristics in the development of preferential flow pathways. Underpinning land-management zones, and interacting with them, are the inherent soil and landscape properties of soil texture, soil profile, drainage characteristics and topographic position (Hawley et al., 1983; Bárdossy et al., 1998; Florinsky et al., 2002). Geophysical techniques have also been shown to be useful in defining within-field soil drainage and inherent subsurface variability (Liu et al., 2008; Robinson et al., 2009).

Thus, a conceptual understanding of landscape processes based on the interaction of soils, topography, and geomorphology is important in the understanding of soil-moisture distribution in the landscape. In a glaciated landscape in Bedfordshire, UK, where a high clay-content glacial till is found on ridges and glacio-fluvial sands and gravels dominate the valleys, soil-profile characteristics such as, permeability and texture, were shown to be the key factors explaining the spatial patterns of soil moisture. The soil moisture patterns were contrary to the expected based on topography alone, with wetter soils on the ridges and drier soils in the valley bottoms (Baggaley et al., 2009). The morphological characteristics of the soils, such as color, are a better guide to the spatial distribution of soil moisture than topography.

2.4. Montane Catchments

In unforested montane catchments, the relationships between soil hydrology and landscape parameters are often specific to particular landscapes. The influence of terrain on soil water content on steep hillslopes with shallow soils in the Italian Alps was debated by Penna et al. (2009). They concluded from measurements and modeling that the spatial distribution of soil moisture is largely controlled by soil drainage and root water uptake.

Holden (2009) shows that, in the UK, there is a relationship between macroporosity and runoff generation in Brown earths (also known as Cambisols or Inceptisols) and Stagnohumic gleys (or Gleysols or Aqualfs), with those areas with limited macroporosity having increased overland flow. However, in Peats (Histosols), overland flow appears to dominate, in spite of there being a distinctive pattern of macroporosity related to slope position. These peat soils are generally slowly permeable at shallow depths, which highlights the need to consider the whole soil profile in determining their influence on catchment hydrology rather than simply their component parts.

In the mountainous Feshie catchment in Scotland, glaciation has influenced soil patterns in the landscape that override the expected effect of terrain on soil hydrology where glacial drift dominates upper slopes and freely draining alluvial soils are found in valley bottoms (Soulsby et al., 2006). In these landscapes, a soil classification such as HOST (Boorman et al., 1995) provides an underpinning conceptual understanding of flow pathways that is crucial for application of an appropriate hydrological model (Tetzlaff et al., 2007, 2008).

2.5. Role of Soils in Catchment Hydrology

There are three common methods (time series analysis, geostatistics, and temporal stability analysis) used to analyze catchment response but often the soil component is not formalized when applying these methods but rather soil properties are linked to covariates, such as terrain and land use/land cover. These covariates are often available as high-resolution, continuous spatial fields that are easily and rapidly updated.

Conceptual and process-based modeling has been used to explore the interacting affects of pedology, climate, vegetation, and hillslope geometry on hydrology. Much of this work shows that there are complex nonlinear interactions which help to explain why simple indices can only account for a limited proportion of the spatial variation of a catchment response.

From the discussion above, it is clear that soil properties have a profound influence on catchment hydrology and that the horizontal and vertical distribution of soils and soil properties (as depicted on soil maps and in the soil profile) are also important.

Some of the key soil properties outlined in the studies considered above are:

- The influence of the nature and type of horizons within the soil profile, in particular, subsoil horizons and their depth/thickness, duplex (texture contrast) soils driving shallow lateral flow in the root zone, and lenses or layers of low permeability leading to perched water tables and subsurface flowpaths.
- Characteristics of the subsoil and the influence these have on the connectivity within the catchment. Specific examples highlighted in the studies outlined above include the occurrence of well-drained sandy soils where lateral flow is insignificant and the influence of the underlying geology.
- The importance of geomorphological processes in the development of a soil catena, specifically the effects of glaciation which may alter the hydrological expected response based on terrain alone.

Soil maps and pedological classification carry this type of information and should have a greater role in modeling catchment hydrology. The temporal variability in soil moisture content is also a key driver. However, it is equally clear that soil maps and soil classifications can present barriers to the use of soil data by hydrologists and that there is a perception that soil maps are simply two-dimensional, static assessments of a specific catchment characteristic. This view is often disputed by pedologists but there have only been a few attempts to translate this understanding of soils, soil mapping, and soil processes into soil hydrological classifications that are readily usable by hydrologists (for example, USDA, 1986; Quisenberry et al., 1993; Boorman et al., 1995; Schmocker-Fackel et al., 2007).

3. EXISTING SOIL HYDROLOGICAL CLASSIFICATIONS

As shown above, there are many factors that can affect wetness patterns, and hence, catchment hydrology, but the physical properties of soils and substrates, in particular, can exert a strong influence on the rate and distribution of infiltration and flow. Although many soil classification systems have soil wetness or drainage as a diagnostic criterion, few have purposely set out to characterize the rate and pathways of flow through the soil and substrate or characterize the storage capacity within the soil to aid in hydrological assessments. Despite considerable advances in computing power, scientific understanding, and improved models, it is still difficult to accurately predict river flows from classical soil physical data, such as moisture retention and hydraulic conductivity, for all but the smallest catchments. Indeed, Pachepsky et al. (2008), argue that, at the catchment or watershed scale, it is the structural (distribution of soil properties) and functional soil parameters that are important in determining river flows.

Although a relatively small component of a method to predict peak discharge, hydrographs, and catchment storage needs, the Hydrologic Soil Group (USDA, 1986) classifies all US soils into one of four groups depending on runoff potential and infiltration rate. Although described in terms of soil texture class, the classification of soils into Hydrologic Soil Groups does take account of the presence of slowly permeable layers, such as fragipans and depth to a water table. The presence of rock fragments, bulk density, and soil aggregation can also be used to classify the soils (NRCS, 2009). In essence, these morphological and physical features are used to assess the hydraulic conductivity of the least permeable layer. However, the low number of soil groups means that there will be a wide range of soil hydrological conditions represented by each of the 4 groups. Further, runoff rates and storage capacities of the soil, which have a strong influence on the catchment response, depend not only on the porosity of the least transmissive layer as described by NRCS (2009) but also on properties such as the total porosity of the soil and the substrate, the soil water regime, and presence of preferential flow pathways.

In order to predict a catchment response, these hydrologic soil groups are then used along with a 'curve number' to determine the hydrology of small watersheds. This curve number is derived from rainfall and runoff observations from many catchments throughout the United States of America. These observations are then standardized based on average or typical catchment wetness, median rainfall, and median runoff. The curve number can be varied according to land use, land management (both of which influence infiltration rates), and Hydrologic Soil Group. Having this ability to revise the curve number according to different land uses and land management systems is advantageous but the empirical nature of the methodology restricts its application to the prediction of hydrological indices only.

Regardless of this empiricism, authors in other countries have applied the curve numbers to their own catchments which are out with the geographical area, climatic and soil conditions for which it was intended; for example it has been applied in the very different environmental conditions of Australia, Botswana, China, Germany, Greece, India, Iran, Mexico and Turkey by various workers. In addition, despite the method being developed to predict hydrological flow indices in small, partly urbanized catchments, some authors have applied it to the prediction of P and sediment runoff (Elrashidi et al., 2003; Lindenschmidt et al., 2004). Although Garen and Moore (2005) advise against using this method for these purposes, its widespread use highlights the need for a method that allows the incorporation of soils data and information into hydrological models.

In 1993, Quisenberry et al. proposed a soil classification system for describing water and chemical transport through various soils. Although primarily restricted to the southeast coastal plain and foothills of the USA, they postulated that their system could be applied much more widely. The authors were critical of the assumption that water flow could always be successfully modeled using one-dimensional soil water simulation models that were solely based on piston-type replacement and uniform flow. When used to model chemical transport, these models gave the impression that soil was a highly effective buffer against potential pollutants; yet, observations in the field often contradicted this assumption. Instead, Quisenberry et al. (1993) observed that there were relatively stable flow pathways in some soils along which water and solutes could move and effectively bypass the bulk of the soil, reducing the opportunity for chemical buffering.

The classification itself was based primarily on the type of water movement through the soil (uniform or laminar microporous flow or macroporous, bypass, and preferential flow) and whether the flow pattern can be represented by the Richards equation (Richards, 1931). The underlying concept was to group soils that had similar flow pathways, regardless of the rate of water movement through the soil.

Quisenberry et al. (1993) argued that the degree of uniformity of flow in the topsoil was a function of soil aggregation and that this was, in turn, primarily a function of soil texture. However, they ignored the contribution of soil organic matter to the development and stability of aggregation, which may have limited the wider application of this classification. They identified four major groups based on the uniformity of flow (and the likelihood of piston-like displacement) and on the proportion of the topsoil intersected by water and solute flow and flow pathways.

It should be stressed that this classification system was not intended to be used to predict hydrological flows but to ascertain the ability of the soil to prevent potential contaminants from reaching groundwaters and to group soils according to their ability to be represented by one-dimensional soil-water simulation models. Perhaps this reliance on classifying according to uniformity of flow, the lack of information on some of the soils, and the limited geographical extent of the soils investigated limited the wider use of this classification. However, the underlying concepts of flow pathways, grouping soils by substrate, and a reliance on soil morphology (particularly soil structure) do have wider applicability.

Around the same time that Quisenberry et al. (1993) were developing their soil classification, a group of hydrologists and soil scientists in the UK were developing their own Hydrology of Soil Types (HOST) classification. Although the final report was published in 1995 (Boorman et al., 1995), this classification had begun its development in the mid-80s. What is important about this classification is that the underlying concepts are from both hydrology and soil science, reflecting the interdisciplinary team that was involved in its development. The classification was devised to predict river flows but based on the rate and pathways of water movement through the soil and, significantly, on the spatial distribution of soils within catchments. This latter aspect is crucial in determining the overall catchment response to rainfall. Pachepsky et al. (2008) likened the spatial distribution of soils to soil structure, albeit at the landscape or catchment-scale.

Initially, HOST developed as a twin-track approach to predicting base flow levels (that is, the proportion of river flow that is supplied by groundwater), with hydrologists developing multiple regression equations of soil properties against base flows from benchmark catchments that had relatively uniform soil types. The soil scientists on the other hand, were trying to preclassify the soils into homogenous groups based on morphology and relating the distribution of these to base flows in the same catchments (Boorman et al., 1995). Neither approach proved to be successful, often resulting in a classification with too many classes, little logic behind the groupings, and the inability to predict flows over the full range of those rivers dominated by base flow and those with more flashy catchments. Other environmental variables such as climate and slope indices were also tested but these did not substantially improve the regression equations. It was postulated that climate and slope were integral to soil development and, hence, their influence was already accounted for by the soil variables.

Although unable to predict base flows over the whole range experienced in the UK, multiple regression equations (where the Base Flow Index was the dependent variable and the proportions of soil groupings in the catchment were the independent variables) explained around 84% of the variance in Base flow Index (BFI, which is an annual index of the proportion of the river flow supplied by groundwater and calculated from hydrographs) indicating that the soil properties being tested (depth to gleying, depth to a slowly permeable layer, integrated air capacity, and the presence of an organic surface layer) were useful (Lilly et al., 1998; Lilly and Lin, 2004). However, this purely regression-based approach produced far too many unique combinations of soil properties for which BFIs could not be derived and too many improbable combinations of soil properties to be of any value.

In order to make sense of, and to systemize the results, a series of conceptual models were developed based on the results of multiple regression analyses (Boorman et al., 1995). Groups with similar predicted BFIs were amalgamated but only where it was deemed that the physical attributes of the soil and the flow pathways through the soil were similar. These models were then tested again and the process repeated. This critical review of the expected soils response in relation to the flow pathways and type of flow (intergranular or bypass) was key in developing the final classification. This resulted in 11 conceptual models that underpin the HOST classification (Fig. 1).

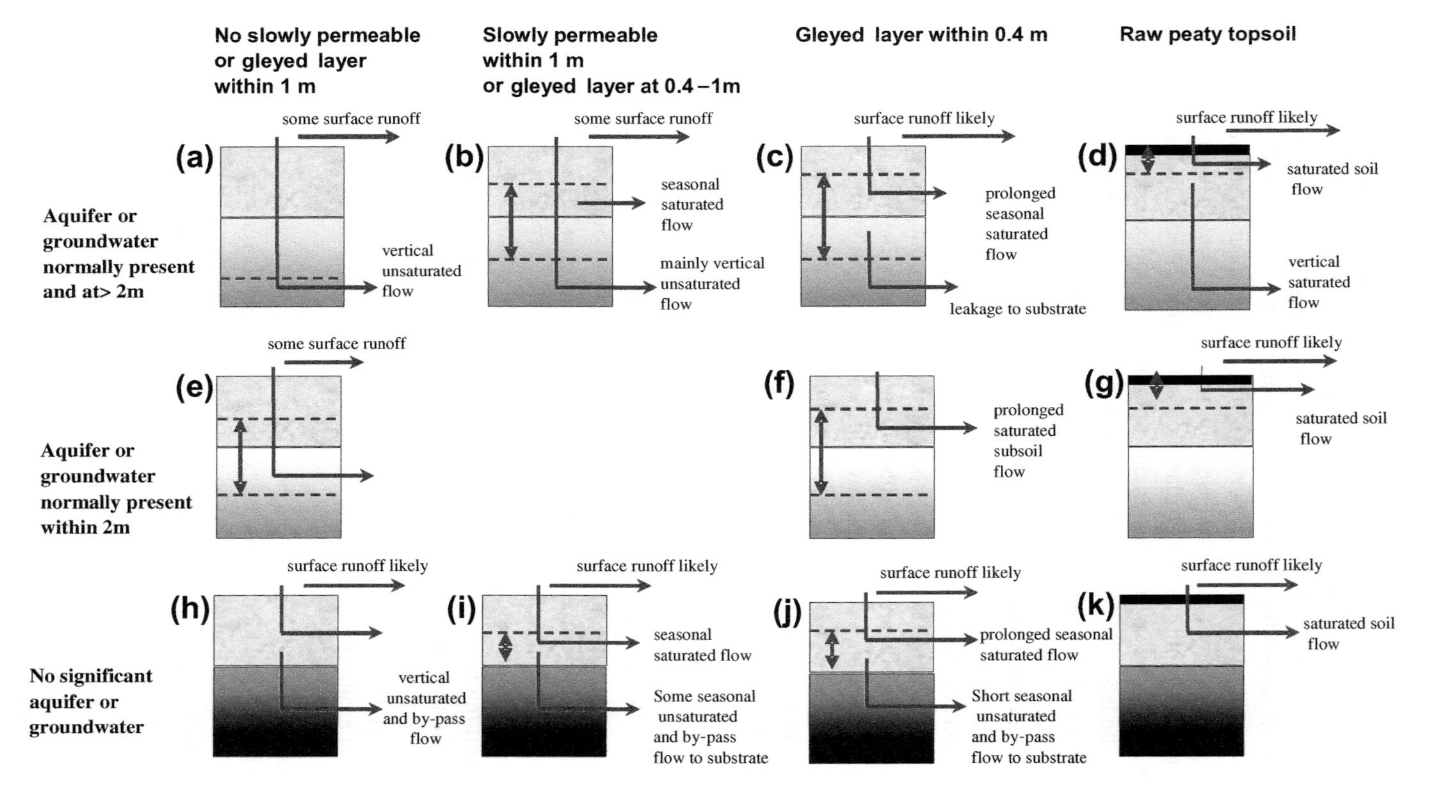

FIGURE 1 The Hydrology of Soil Types (HOST) classification of 11 conceptual models of flowpaths within the soil and substrate. *(From Boorman et al., 1995) (Color version online and in color plate)*

Another important step that made the HOST classification distinct and of particular value to hydrologists was the introduction of the three 'physical settings' that set the context for the interaction between the soil and groundwater. These grouped the substrates in which the soils developed into those that were slowly permeable without a true groundwater table, those that were permeable with deep groundwater (>2 m), and permeable substrates with shallow groundwater (<2 m). These were incorporated into the 11 conceptual models that underpin the HOST classification to give 29 HOST classes.

HOST describes water movement through soils and substrates with the simplest models either assuming vertical flow through permeable soils and substrates to deep groundwater that gives sustained low flows during dry periods or assuming lateral surface or near-surface flow over slowly permeable soils and substrates that give rise to rapid flood flow responses in the river network. Between these two extremes, the flow pathways are more complex and the effect of the presence and depth of slowly permeable layers on lateral flow and the extent of soil wetness become more significant. Within this broad classification, there are further subdivisions based on the flow mechanism based on the presence of bypass flow, the presence of laminar or uniform flow (similar to Quisenberry et al., 1993), hydraulic conductivity, substrate geology and soil water-storage capacity.

The 29 HOST classes can be spatially represented on current UK national-scale soil maps (1:250,000) to allow the calculation of catchment-scale flow indices but the classification itself is not scale-specific and can be applied from the profile to a reconnaissance-scale soil map. It does have certain limitations. Perhaps the most important in the context of climate change is that the classification is calibrated against historical low flows, which may not be relevant under current or changing climatic conditions. There is also no possibility of taking land use or land management into account. There is the potential for modern agricultural practices to induce subsoil compaction and hence limit the infiltration capacity of the soil (Jones et al., 2003). This will increase the propensity for runoff and decrease the base flow. More fundamentally, it was not always possible to obtain good estimates of BFI or Standard Percentage Runoff (SPR; defined as the total overland and near-surface runoff expressed as a fraction of the total annual precipitation) for each class.

Although HOST was developed for the UK, Schneider et al. (2007) reclassified the Soil Geographical Database of Europe (based on a 1:1,000,000 scale soil map and an associated soil attribute database) into HOST classes to test the ability of HOST to predict BFIs for European river catchments. The Soil Geographical Database of Europe holds soil attributes similar to those used in HOST but some reclassification was necessary to accommodate a wider range of soil types. This reclassified HOST was then used to predict base flows of rivers in England and Wales based on the European soil map. Despite the loss of information on the spatial distribution of soils due to the more coarse resolution of the European soil map, 68% of the variation in the BFIs of 21 UK catchments was explained compared with 79% of the original UK HOST classification (Boorman et al., 1995).

The authors then applied UK HOST BFI coefficients to European catchments and also calculated new BFI coefficients from European data using the reclassified HOST. They found that the UK HOST BFIs largely overpredicted flows of the European rivers, particularly in southern European (Mediterranean) catchments but, although the newly calculated BFIs were quantitatively different, their relative values remained largely similar. Unlike the UK HOST, Schneider et al. (2007) found that topography (in the form of a topographic index) improved the regression, particularly in southern European catchments, as did the introduction of climatic variables. This potentially indicates a geographical limit to the current HOST classification. Whereas the UK is largely a glaciated environment, southern European hill areas have been shaped by water erosion over millennia and have more well-developed landscape-scale flow pathways than the UK. Similarly, the UK has less climatic extremes than Europe, which may account for the increased influence of climate on BFIs in southern Europe.

Working in Switzerland, and based on some earlier work by Scherrer and Naef (2003), Schmocker-Fackel et al. (2007) developed a conceptual model of flow responses and related these to soil properties that were observable in the field such as topsoil compaction, the presence of plough pans, susceptibility to capping, the presence of slowly permeable layers, and the presence of preferential flow pathways. The flow responses were classified as Hortonian overland flow, saturation overland flow, subsurface flow, tile drain flow, and deep percolation, with all but the latter being subdivided to give 12 classes. An important benefit of this classification is that the effect of land management (such as the development of plough pans) on catchment hydrological response can be taken into account. The authors tested their conceptual model by installing a number of piezometers in a small Swiss catchment and observing their responses during a rainfall event. The authors noted differences between these flow response classes in the length of time a soil was at saturation, the height of the water table, as well as the time taken to reach saturation and to subsequently drain. In order to predict the flood response of different catchments, Schmocker-Fackel et al. (2007) developed a set of rules to classify the soils according to the flow responses based on information contained within soil, geology, land use, topographic maps and attribute databases. Soil classification, the presence of gleyic and stagnic features, soil texture, organic matter, and stone content were all used to classify the soils into these flow response classes. Despite some difficulties in finding suitable surrogates from the soil attribute data set for all the properties required in the model, the authors were able to derive reasonable estimates of runoff for two subcatchments. An important development in this work is that transient features such as surface capping and plough pans can be taken into account when predicting river flows. This introduces a much-needed temporal aspect to soil hydrological classifications.

A key feature of the above classification systems is that they rely heavily on the interpretation of soil morphological attributes to both explain and identify similarities in hydrological responses. While there is still a place for hydrological models that use soil water retention and hydraulic conductivity as soil parameters (whether measured directly or derived by pedotransfer functions), there is much to be gained by considering profile-scale soil morphology. There is a considerable amount of morphological data in various soil databases around the world and for many different climatic, topographic, and edaphic conditions and, if linked to suitable scale soil maps, can form an important component of hydropedologic advances. Although soil hydrological classifications provide useful functional descriptions of the principal runoff pathways and processes for different soil types, there remains a challenge in translating this representation into quantitative parameters typical of those required by catchment-scale hydrological models.

4. USING SOIL DATA AND SOIL HYDROLOGICAL CLASSIFICATIONS IN MODELING

Soil map and profile data are key components in the understanding and modeling of catchment-scale hydrological processes but many studies have identified scaling problems when attempting to use point measurements of soil physical properties, such as hydraulic conductivity, in catchment models (Wen et al., 1996; Christiaens and Feyen, 2001). The development of pedotransfer functions based on easily measurable soil properties, such as soil texture and particle size (Vereecken, 1995; Jarvis et al., 2002), is one approach that has been explored to address this issue. Pedotransfer functions have been found to successfully reproduce soil hydraulic behavior although problems have been encountered in systems where macropores play an important role in transport. Lin et al. (1999) examined how a range of soil structural information could be integrated in pedotransfer functions to account for macroporosity effects. Wagner et al. (2001) also advocate the integration of local structural information. However, the data required for such methods are less readily available and have not been commonly used in more recent approaches. Pedotransfer functions for preferential flow parameters are also discussed by Jarvis et al. (2011). Manus et al. (2009) applied a conceptual model to four small Mediterranean catchments using un-calibrated parameters derived from a regional evaluation with soil-parameter specifications based on an existing soil database. Locally developed pedotransfer functions based on soil texture and porosity were found to adequately characterize the catchments for simulation of peak discharges, although insufficient flow data were available for larger-scale evaluation. In general, more limited success has been found in using pedotransfer functions to parametrize models at larger catchment scales because of the difficulty in characterizing the complex spatial variability of processes. Consequently, most catchment-scale model applications resort to parameter calibration techniques, which makes it very difficult to retain realistic distinctions between functioning of different soil types.

Gan and Burges (2006) compared parameter values derived for 12 MOPEX (Model Parameter Estimation Experiment) river basins using calibration, with parameter values derived from soil-based algorithms by Koren et al. (2004). These soil-based algorithms were estimated by developing a set of physically based relationships between parameters of the Sacramento model and hydraulic properties predicted from soil texture to estimate *a priori* Sacramento model parameter values. Gan and Burges (2006) found major differences between the parameter values, including inconsistencies in the total storage capacity of the models, and found that the calibrated values generally gave a better hydrograph simulation. They then went on to explore whether the soil-based parameters could be used to help with transfer of model parameters between river basins by rescaling the calibrated parameters according to the soil-based parameters, but found this method to be ineffective. Failure of the soil-based parametrization in this instance was explained as the failure to account for elements of climatic, land use, and terrain characteristics which become embedded in the calibrated parameters.

Dunn and Lilly (2001) adopted a similar approach to explore the relationships between the UK HOST classification and the spatial parameters of a conceptual catchment-scale model. In this study, the HOST classification was used to qualitatively assign "high", "medium", or "low" values to model parameters (i.e. class pedotransfer functions) according to the functional behavior of various HOST classes. A procedure involving inverse modeling using Monte-Carlo simulations on two catchments was then used to quantify the relative differences between "high", "medium" and "low" for the different parameters. Three out of the four parameters tested were found to take consistent relative values for the two catchments tested, while problems in identifying consistent values for the fourth parameter were believed to be related to an underlying topographic control on its behavior. However, the approach was found to have only limited success when it was further tested by extrapolation to a third catchment without any parameter calibration (Dunn et al., 2003). Marechal and Holman (2005) also attempted to calibrate values for three parameters of the CRASH model for different HOST classes. Their study found that different values for the surface runoff, interflow, and base flow parameters could be identified for different HOST classes, but did not test the transferability of these parameter values to different catchments.

Greater success has been achieved in using soil classifications to derive parameters for large-scale model applications. Kling and Nachtnebel (2009) and Dunn et al. (2004) applied similar approaches to derive parameters for national-scale water balance models applied to Austria and Scotland respectively, through applying a multicatchment cross calibration approach. In both cases, indices of catchment flow were used to characterize catchment hydrographs and identify parameters for flowpath separation. Dunn et al. (2004) calculated the Base Flow Index (BFI) and Standard Percentage Runoff (SPR) at a 1 km^2 resolution using flow contributions simulated by the model over a one-year period. Storage capacity parameters for different soil types were derived using a pedotransfer methodology based on soil texture, bulk density, and organic matter content. Parameters of the model defining the flow-path separation were then adjusted until a good fit was obtained between the simulated values of BFI and SPR and values calculated independently for each soil map unit as part of the HOST classification of UK soils (Boorman et al., 1995). Kling and Nachtnebel (2009) also found that a measure of storage capacity combined with the BFI could be used to achieve a regional calibration of a monthly water balance model for Austria. As previously mentioned, Schneider et al. (2007) explored the utility of the BFI at larger scales through the reclassification of the Soil Geographical Database of Europe into UK HOST classes. UK BFI values were then used to predict the BFI for 103 catchments in Europe and cross-compared with measured values from discharge data. The poorer predictions for southern regions suggest decreasing influence of soil on the BFI in the warmer, drier regions.

5. SUMMARY AND FUTURE OUTLOOK

Although approaches such as those described above have had some successes in identifying consistent parameters for hydrological models, there are clear limitations in the scales and level of detail of the applications. Successes have generally been achieved either at small scales, where local data are available to support pedotransfer functions (Islam et al., 2006), or else at large-scales where coarser temporal resolutions have been utilized (Kling and Nachtnebel, 2009). In addition, validation of the internal functioning of soils within catchment models is difficult to achieve, because of the equifinality issue whereby different combinations of parameters may give equally good simulations of catchment-scale responses (Beven, 2006). The differences in behavior may be relatively subtle and difficult to disaggregate from other influences from terrain or land use, or from model error. While soil hydrological classifications can help reduce the number of spatial classes required for parameter assignment at the catchment-scale, models still have far more degrees of freedom than is compatible with the data available for parameter calibration and validation. This does not, however, imply that the heterogeneity should be neglected as the resulting differences can have important effects, as previously discussed.

The problem of soil parametrization in catchment models is starting to be more widely recognized not just for hydrological simulations but especially in the context of modeling water quality, and methods that build on the concepts outlined in this chapter are being more widely explored. For example, Barlund et al. (2009) recently approached the problem through attempting to parametrize a specific hydrological/water quality model (ICECREAM) for a range of typical mineral soils found in Finland, to enable a method for directly scaling outputs from their "classification" to larger scales. Essentially, this approach involves an extension of the concepts in developing pedotransfer functions and soil hydrological classifications but, importantly, makes a direct link to a dynamic water quality model rather than through an intermediate soil classification. In this way, it should be possible to achieve better integration of existing methods with dynamic models, such that model specific parameters (that are likely to vary depending on the modeling objective) can be estimated.

The methods developed by Koren et al. (2004) are also being enhanced by Zhang et al. (2011) to provide more automated approaches for processing soil survey and land cover data to estimate *a priori* distributed parameters for dynamic models. In particular, advances in remote sensing and measurement technologies have brought increasingly refined soil and land cover information that have the potential to improve the physical representation of *a priori* model parameters. While such developments will make the derivation of distributed catchment model inputs much simpler, there is a danger that they mask many of the uncertainties that remain in understanding how soils influence catchment hydrologic responses. At the other end of the scale, a recent study by Basu et al. (2010) found that intensively managed watersheds in the Midwest U.S. could be successfully modeled using a single value of soil water storage and gave equally good results as a much more complex distributed representation using the SWAT model.

It is clear that different approaches suit different situations, but recent studies generally indicate that a combination of enhanced spatial data together with more advanced data processing techniques ultimately should lead to better representation of soil functioning within catchment-scale hydrologic models. It has been shown that an understanding of soil morphology can also help identify and characterize these soil functions, reduce the amount of input data required and conceptualize flow pathways (Fig. 1).

REFERENCES

Anctil, F., Lauzon, N., Filion, M., 2008. Added gains of soil moisture content observations for streamflow predictions using neural networks. J. Hydrol. 359, 225–234.

Aryal, S.K., Mein, R.G., O'Loughlin, E.M., 2003. The concept of effective length in hillslopes: assessing the influence of climate and topography on the contributing areas of catchments. Hydrol. Process 17, 131–151.

Bachmair, S., Weiler, M., Nützmann, G., 2009. Controls of land use and soil structure on water movement: lessons for pollutant transfer through the unsaturated zone. J. Hydrol. 369, 241–252.

Baggaley, N., Mayr, T., Bellamy, P., 2009. Identification of key soil and terrain properties that influence the spatial variability of soil moisture throughout the growing season. Soil Use Manage. 25, 262–273.

Bárdossy, A., Lehmann, W., 1998. Spatial distribution of soil moisture in a small catchment. Part 1: geostatistical analysis. J. Hydrol. 206, 1–15.

Barlund, I., Tattari, S., Puustinen, M., Koskiaho, J., Yli-Halla, M., Posch, M., 2009. Soil parameter variability affecting simulated field-scale water balance, erosion and phosphorus losses. Agr. Food Sci. 18, 402–416.

Basu, N.B., Rao, P.S.C., Winzeler, H.E., Kumar, S., Owens, P., Merwade, V., 2010. Parsimonious modeling of hydrologic responses in engineered watersheds: structural heterogeneity versus functional homogeneity. Water Resour. Res. 46, 4. W04501.

Beven, K.J., 2006. A manifesto for the equifinality thesis. J. Hydrol. 320, 18–36.

Boorman, D.B., Hollis, J.M., Lilly, A., 1995. Hydrology of soil types: a hydrologically-based classification of the soils of the United Kingdom. Inst. of Hydrol. Rep. 126, Wallingford, UK.

Brocca, L., Melone, F., Moramarco, T., Morbidelli, R., 2009. Soil moisture temporal stability over experimental areas in Central Italy. Geoderma 148, 364–374.

Brocca, L., Melone, F., Moramarco, T., Morbidelli, R., 2010. Spatial-temporal variability of soil moisture and its estimation across scales. Water Resour. Res. 46, W02516.

Buttafuoco, G., Castrignano, A., Busoni, E., Dimase, A.C., 2005. Studying the spatial structure evolution of soil water content using multivariate geostatistics. J. Hydrol. 311, 202–218.

Buttle, J.M., Dillon, P.J., Eerkes, G.R., 2004. Hydrologic coupling of slopes, riparian zones and streams: an example from the Canadian Shield. J. Hydrol. 287, 161–177.

Christiaens, K., Feyen, J., 2001. Analysis of uncertainties associated with different methods to determine soil hydraulic properties and their propagation in the distributed hydrological MIKE SHE model. J. Hydrol. 246, 63–81.

Crave, A., Gascuel-odoux, C., 1997. The influence of topography on time and space distribution of soil surface water content. Hydrol. Process 11, 203–210.

De Lannoy, G.J.M., Verhoest, N.E.C., Houser, P.R., Gish, T.J., Van Meirvenne, M., 2006. Spatial and temporal characteristics of soil moisture in an intensively monitored agricultural field (OPE3). J. Hydrol. 331, 719–730.

Dunn, S.M., Lilly, A., DeGroote, J., Vinten, A.A., 2004. Nitrogen risk assessment model for Scotland: II. Hydrological transport and model testing. Hydrol. Earth Syst. Sci. 8, 205–219.

Dunn, S.M., Lilly, A., 2001. Investigating the relationship between a soils classification and the spatial parameters of a conceptual catchment-scale hydrological model. J. Hydrol. 252, 157–173.

Dunn, S.M., Soulsby, C., Lilly, A., 2003. Parameter identification for conceptual modelling using combined behavioural knowledge. Hydrol. Process 17, 329–343.

Elrashidi, M.A., Mays, M.D., Jones, P.E., 2003. A technique to estimate release characteristics and runoff phosphorus for agricultural land. Commun. Soil Sci. Plan. 34, 1759–1790.

Fitzjohn, C., Ternan, J.L., Williams, A.G., 1998. Soil moisture variability in a semi-arid gully catchment: implications for runoff and erosion control. Catena 32, 55–70.

Florinsky, I.V., Eilers, R.G., Manning, G.R., Fuller, L.G., 2002. Prediction of soil properties by digital terrain modelling. Environ. Modell. Softw. 17, 295–311.

Freebairn, D.M., Wockner, G.H., Hamilton, N.A., Rowland, P., 2009. Impact of soil conditions on hydrology and water quality for a brown clay in the north-eastern cereal zone of Australia. Aust. J. Soil Res. 47, 389–402.

Gan, T.Y., Burges, S.J., 2006. Assessment of soil-based and calibrated parameters of the Sacramento model and parameter transferability. J. Hydrol. 320, 117–131.

Garen, D.C., Moore, D.S., 2005. Curve number hydrology in water quality modeling: uses, abuses, and future directions. J. Am. Water Resour. As. 41, 377–388.

Gburek, W.J., Needelman, B.A., Srinivasan, M.S., 2006. Fragipan controls on runoff generation: hydropedological implications at landscape and watershed scales. Geoderma 131 (3–4), 330–344.

Giraldo, M.A., Bosch, D., Madden, M., Usery, L., Kvien, C., 2008. Landscape complexity and soil moisture variation in South Georgia, USA, for remote sensing applications. J. Hydrol. 357, 405–420.

Gish, T.J., Walthall, C.L., Daughtry, C.S.T., Kung, K.J.S., 2005. Using soil moisture and spatial yield patterns to identify subsurface flow pathways. J. Environ. Qual. 34, 274–286.

Guntner, A., Seibert, J., Uhlenbrook, S., 2004. Modeling spatial patterns of saturated areas: an evaluation of different terrain indices. Water Resour. Res. 40, W05114.

Hawley, M.E., Jackson, T.J., Mccuen, R.H., 1983. Surface soil-moisture variation on small agricultural watersheds. J. Hydrol. 62, 179–200.

Holden, J., 2009. Topographic controls upon soil macropore flow. Earth Surf. Process. 34, 345–351.

Islam, N., Wallender, W.W., Mitchell, J.P., S.Wicks, Howitt, R.E., 2006. Performance evaluation of methods for the estimation of soil hydraulic parameters and their suitability in a hydrologic model. Geoderma 134, 135–151.

James, A.L., Roulet, N.T., 2007. Investigating hydrologic connectivity and its association with threshold change in runoff response in a temperate forested watershed. Hydrol. Process. 21, 3391–3408.

Jarvis, N.J., Moeys, J., Hollis, J.M., 2011. Preferential flow in a hydropedological perspective. In: Lin, H.S. (Ed.), Hydropedology – Synergistic Integration of Pedology and Hydrology Dev. Soil. Sci.

Jarvis, N.J., Zavattaro, L., Rajkai, K., Reynolds, W.D., Olsen, P.A., McGechan, M., Mecke, M., Mohanty, B., Leeds-Harrison, P.B., Jacques, D., 2002. Indirect estimation of near-saturated hydraulic conductivity from readily available soil information. Geoderma 108, 1–17.

Jones, R.J.A., Spoor, G., Thomasson, A.J., 2003. Vulnerability of subsoils in Europe to compaction: a preliminary analysis. Soil Till. Res. 73, 131–143.

Kim, S., 2009. Characterization of soil moisture responses on a hillslope to sequential rainfall events during late autumn and spring. Water Resour. Res. 45, W09425.

Kling, H., Nachtnebel, H.P., 2009. A method for the regional estimation of runoff separation parameters for hydrological modeling. J. Hydrol. 364, 163–174.

Koren, V., Smith, M., Duan, Q., 2004. Use of a priori parameter estimates in the derivation of spatially consistent parameter sets of rainfall-runoff models. In: Duan, Q., et al. (Eds.), Calibration of Watershed Models. Water Science and Application, 6. AGU, Washington, DC, pp. 239–254.

Korres, W., Koyama, C.N., Fiener, P., Schneider, K., 2010. Analysis of surface soil moisture patterns in agricultural landscapes using empirical orthogonal functions. Hydrol. Earth Syst. Sci. 14, 751–764.

Lauzon, N., Anctil, F., Petrinovic, J., 2004. Characterization of soil moisture conditions at temporal scales from a few days to annual. Hydrol. Process. 18, 3235–3254.

Lilly, A., Lin, H.S., 2004. Using soil morphological attributes and soil structure in pedotransfer functions. In: Pachepsky, Y., Rawls, W. (Eds.), Development of Pedotransfer Functions in Soil Hydrology. Dev. Soil. Sci, vol. 30. Elsevier, Amsterdam, pp. 115–141.

Lilly, A., Boorman, D.B., Hollis, J.M., 1998. The development of a hydrological classification of UK soils and the inherent scale changes. Nutr. Cycl. Agroecosyst. 50, 299–302.

Lin, H., Zhou, X., 2008. Evidence of subsurface preferential flow using soil hydrologic monitoring in the Shale Hills catchment. Eur. J. Soil Sci. 59, 34–49.

Lin, H.S., McInnes, K.J., Wilding, L.P., Hallmark, C.T., 1999. Effects of soil morphology on hydraulic properties: II. Hydraulic pedotransfer functions. Soil Sci. Soc. Am. J. 63, 955–961.

Lin, H.S., Kogelmann, W., Walker, C., Bruns, M.A., 2006. Soil moisture patterns in a forested catchment: a hydropedological perspective. Geoderma 131, 345–368.

Lindenschmidt, K.E., Ollesch, G., Rode, M., 2004. Physically-based hydrological modelling for nonpoint dissolved phosphorus transport in small and medium-sized river basins. Hydrolog. Sci. J. 49, 495–510.

Liu, J.G., Pattey, E., Nolin, M.C., Miller, J.R., Ka, O., 2008. Mapping within-field soil drainage using remote sensing, DEM and apparent soil electrical conductivity. Geoderma. 143, 261–272.

Manus, C., Anquetin, S., Braud, I., Vandervaere, J.P., Creutin, J.D., Viallet, P., Gaume, E., 2009. A modeling approach to assess the hydrological response of small mediterranean catchments to the variability of soil characteristics in a context of extreme events. Hydrol. Earth Syst. Sci. 13, 79–97.

Maréchal, D., Holman, I.P., 2005. Development and application of a soil classification-based conceptual catchment-scale hydrological model. J. Hydrol. 312, 277–293.

Menziani, M., Pugnaghi, S., Vincenzi, S., Santangelo, R., 2003. Soil moisture monitoring in the Toce valley (Italy). Hydrol. Earth Syst. Sci. 7, 890–902.

Neumann, L., Western, A., Argent, R., 2010. The sensitivity of simulated flow and water quality response to spatial heterogeneity on a hillslope in the Tarrawarra catchment, Australia. Hydrol. Process. 24, 76–86.

NRCS, 2009. Hydrologic Soil Groups. Part 630 Hydrology, Chapter 7. National Engineering Handbook. United States Department of Agriculture - Natural Resources Conservation Service, Washington, DC. USA.

Pachepsky, Y., Gimenez, D., Lilly, A., Nemes, A., 2008. Promises of hydropedology. CAB Rev: Persp Agric, Vet. Sci. Nutr. Nat. Resour. 3 No. 040.

Parent, A.C., Anctil, F., Parent, L.E., 2006. Characterization of temporal variability in near-surface soil moisture at scales from 1 h to 2 weeks. J. Hydrol. 325, 56–66.

Penna, D., Borga, M., Norbiato, D., Fontana, G.D., 2009. Hillslope scale soil moisture variability in a steep alpine terrain. J. Hydrol. 364, 311–327.

Perry, M.A., Niemann, J.D., 2008. Generation of soil moisture patterns at the catchment scale by EOF interpolation. Hydrol. Earth Syst. Sci. 12, 39–53.

Quisenberry, V.L., Smith, B.R., Phillips, R.E., Scott, H.D., Nortcliff, S., 1993. A soil classification system for describing water and chemical transport. Soil Sci. 156, 306–315.

Richards, L.A., 1931. Capillary conduction of liquids through porous mediums. Phys. 1, 318–333.

Robinson, D.A., Campbell, C.S., Hopmans, J.W., Hornbuckle, B.K., Jones, S.B., Knight, R., Ogden, F., Selker, J., Wendroth, O., 2008. Soil moisture measurement for ecological and hydrological watershed-scale observatories: a review. Vadose Zone J. 7, 358–389.

Robinson, D.A., Lebron, I., Kocar, B., Phan, K., Sampson, M., Crook, N., Fendorf, S., 2009. Time-lapse geophysical imaging of soil moisture dynamics in tropical deltaic soils: an aid to interpreting hydrological and geochemical processes. Water Resour. Res. 45 W00D32.

Romshoo, S.A., 2004. Geostatistical analysis of soil moisture measurements and remotely sensed data at different spatial scales. Environ. Geol. 45 (3), 339–349.

Scherrer, S., Naef, F., 2003. A decision scheme to identify dominant flow processes at the plot-scale for the evaluation of contributing areas at the catchments-scale. Hydrol. Proc. 17, 391–401.

Schmidt, J., Hennrich, K., Dikau, R., 2000. Scales and similarities in runoff processes with respect to geomorphometry. Hydrol. Process. 14, 1963–1979.

Schmocker-Fackel, P., Naef, F., Scherrer, S., 2007. Identifying runoff processes on the plot and catchment scale. Hydrol. Earth Syst. Sci. 11, 891–906.

Schneider, M.K., Brunner, F., Hollis, J.M., Stamm, C., 2007. Towards a hydrological classification of European soils: preliminary test of its predictive power for the base flow index using river discharge data. Hydrol. Earth Syst. Sci. 11, 1501–1513.

Soulsby, C., Tetzlaff., D., Rodgers, P., Dunn, S., Waldron, S., 2006. Runoff processes, stream water residence times and controlling landscape characteristics in a mesoscale catchment: an initial evaluation. J. Hydrol. 325, 197–221.

Starks, P.J., Heathman, G.C., Jackson, T.J., Cosh, M.H., 2006. Temporal stability of soil moisture profile. J. Hydrol. 324, 400–411.

Tetzlaff, D., Soulsby, C., Waldron, S., Malcolm, I.A., Bacon, P.J., Dunn, S., Lilly, A., Youngson, A.F., 2007. Conceptualisation of runoff processes using tracers and GIS analysis in a nested mesoscale catchment. Hydrol. Process. 21, 1289–1307.

Tetzlaff, D., Uhlenbrook, S., Eppert, S., Soulsby, C., 2008. Does the incorporation of process conceptualization and tracer data improve the structure and performance of a simple rainfall-runoff model in a Scottish mesoscale catchment? Hydrol. Process. 22, 2461–2474.

USDA, 1986. Urban Hydrology for Small Watersheds. Technical Release 55 (TR55). Natural Resources Conservation Service and the Conservation Engineering Division. United States Department of Agriculture, Washington, DC.

Vachaud, G., Desilans, A.P., Balabanis, P., Vauclin, M., 1985. Temporal stability of spatially measured soil-water probability density-function. Soil Sci. Soc. Am. J. 49, 822–828.

van Schaik, N., 2009. Spatial variability of infiltration patterns related to site characteristics in a semi-arid watershed. Catena 78, 36–47.

Van Tol, J.J., le Roux, P.A.L., Hensley, M., Lorentz, S.A., 2010. Soil as indicator of hillslope hydrological behavior in the Weatherley catchment, Eastern Cape, South Africa. Water SA 36, 513–520.

Vereecken, H., 1995. Estimating the unsaturated hydraulic conductivity from theoretical-models using simple soil properties. Geoderma. 65, 81–92.

Vestin, J.L.K., Norström, S.H., Bylund, D., Lundstrom, U.S., 2008a. Soil solution and stream water chemistry in a forested catchment I: dynamics. Geoderma. 144, 256–270.

Vestin, J.L.K., Norström, S.H., Bylund, D., Mellander, P.E., Lundström, U.S., 2008b. Soil solution and stream water chemistry in a forested catchment II: influence of organic matter. Geoderma. 144, 271–278.

Wagner, B., Tarnawski, V.R., Hennings, V., Muller., U., Wessolek., G., Plagge, R., 2001. Evaluation of pedo-transfer functions for unsaturated soil hydraulic conductivity using an independent data set. Geoderma 102, 275–297.

Wang, H., Hall, C.A.S., Cornell, D., Hall, M.H.P., 2002. Spatial dependence and the relationship of soil organic carbon and soil moisture in the Luquillo experimental forest, pueto rico. Landscape Ecol. 17, 671–684.

Wen, X.H., Gómez-Hernández, J.J., 1996. Upscaling hydraulic conductivities in heterogeneous media: an overview. J. Hydrol. 183, 9–32.

Western, A.W., Zhou, S.L., Grayson, R.B., McMahon, T.A., Bloschl, G., Wilson, D.J., 2004. Spatial correlation of soil moisture in small catchments and its relationship to dominant spatial hydrological processes. J. Hydrol. 286, 113–134.

Wilson, D.J., Western, A.W., Grayson, R.B., 2004. Identifying and quantifying sources of variability in temporal and spatial soil moisture observations. Water Resour. Res. 40, WO2507.

Zhang, Z., Zhang, Y., Reed, S., Koren, V., 2011. An enhanced and automated approach for deriving a priori SAC-SMA parameters from the soil survey geographic database. Comput. Geosci. 37, 219–223.

Chapter 18

Subsurface Flow Networks at the Hillslope Scale: Detection and Modeling

Chris B. Graham[1,*] and Henry Lin[2]

ABSTRACT

Subsurface flow networks have been recognized in recent years as an important mechanism for water flow through various hillslopes and catchments. The formation, geometry, and dynamics of these networks, however, remain poorly understood and challenging to be characterized. This chapter provides a comprehensive review on recent advances in identifying, characterizing, and modeling subsurface preferential flow networks at the hillslope scale, including hydrologic connectivity and threshold behavior. While trenching, irrigation and tracer experiments, and geophysical imaging have been used to reveal some characteristics of these subsurface flow networks, much remains to be done in order to develop a network-based predictive hydrologic model. Two proposed conceptual models for subsurface lateral flow networks at the hillslope scale are 1) connected preferential flow pathways in the soil and 2) water flow along permeability contrasts such as the soil–bedrock interface and contrasting soil horizon interfaces. Current methods for modeling subsurface flow networks include topography-based flow routing, stochastic macropore modeling, and percolation-based approach. Better tools and techniques are needed for effective and efficient in situ determination of subsurface structures and flow dynamics with high enough spatial and temporal resolutions. This chapter also highlights unresolved questions, and suggests some possible ways forward, including the role that hydropedology plays.

1. INTRODUCTION

In recent years, the hydrologic community has come to a consensus that subsurface flow is generally dominated by preferential flow of various kinds (Jones, 2010; Lin, 2010; Uhlenbrook, 2006). The perception toward

[1]Boise State University

[2]Dept. of Ecosystem Science and Management, Pennsylvania State Univ., University Park, PA, USA

[*]Corresponding author: Email: chris.b.graham.hydro@gmail.com

Hydropedology, Edited by H. Lin. DOI: 10.1016/B978-0-12-386941-8.00018-6

preferential flow has changed from being the exception to the Darcy–Richards conceptualization of subsurface flow to being the rule in hydrologic systems. However, the network, initiation, duration, and controls of preferential flow and its dynamic interface with soil matrix remain elusive. Although significant research has gone into characterizing preferential flow over the past decades, one aspect that remains poorly understood is the nature and dynamics of the network of such preferential flows in the subsurface.

Subsurface preferential flow networks are the holistic connection of various preferential flow features (such as macropores, branching roots, bedrock fractures, and hydrophobicity) to route water through the subsurface, either vertically or laterally (Fig. 1). High water fluxes observed from preferential flow features suggest flow networks drain extensive upslope subsurface areas. The connectivity, extent, and capacity of these networks are critical controls on subsurface storm response in various landscapes, ranging from steep, forested hillslopes to flat, tile-drained agricultural fields.

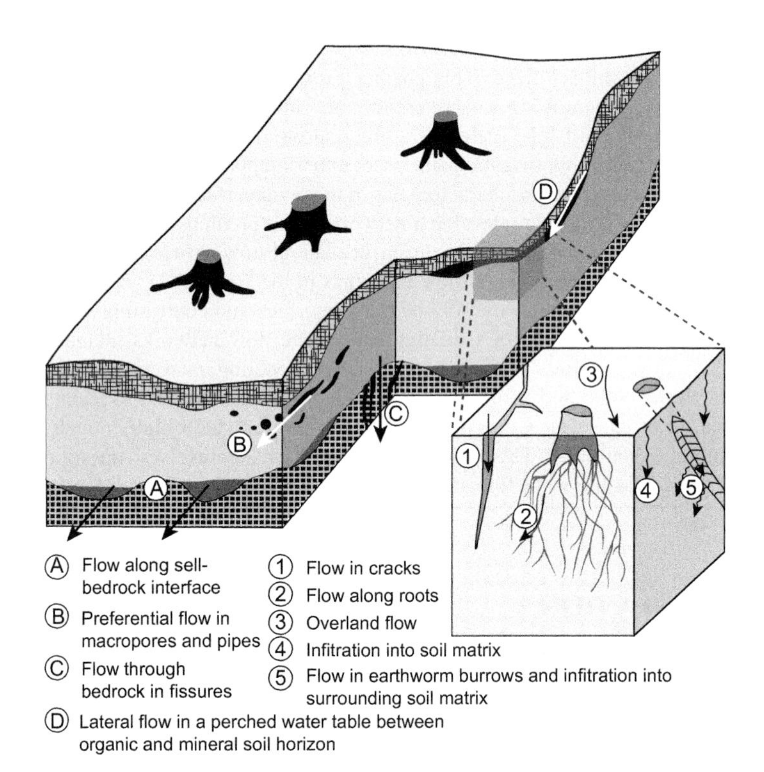

FIGURE 1 A conceptual model of subsurface preferential flow network (both vertical and lateral) in a forested hillslope, where lateral flow network consists of flow along the soil–bedrock interface, flow through macropores and along soil horizons, and flow through fractures in the underlying bedrock. *(From Sidle et al., 2001)*

Some of the difficulties in making progress in identifying and characterizing subsurface preferential flow networks are due to the common methodologies used in laboratory or field investigations. Measurements have generally been either point scale (i.e. soil moisture measured with TDR) or integrating entire hillslope segments (i.e. collection trenches at the base of hillslopes). There is a lack of mechanistic linkage between point-scale observations and integrated hillslope flow response. While these approaches have improved our understanding of subsurface flow processes, the natural complexity and heterogeneity of subsurface systems, combined with the inherent uncertainty in the contributing area and the flow network feeding the observed outflow, have hindered the progress in formulating a universal theory and robust predictive models.

In this chapter, we attempt to summarize recent advances in the field detection and modeling of subsurface flow networks with a special focus on the hillslope scale. It is hoped that such an effort can shed light on general characteristics and possible controls of the complex and dynamic subsurface preferential flow networks. We conclude with some suggestions for future research toward improving the understanding and prediction of subsurface flow networks.

2. PREFERENTIAL NETWORK DETECTION

Three primary experimental methods have been reported in the literature for characterizing and detecting subsurface preferential flow networks: 1) trenching and small-scale excavations, 2) hillslope-scale tracer experiments and excavations, and 3) nondestructive imaging or geophysical methods. In this section we will review these methods and the subsurface flow networks they have identified (Table 1).

2.1. Observations via Trenching and Small-Scale Excavations

Trenching is a method of investigating lateral subsurface flow processes where soil is excavated from the surface to a certain depth and lateral flow is monitored in a cross section of the hillslope which has a long history in hydrology (Whipkey, 1965). However, as reviewed by Beven and Germann (1982), observations of macropore flow in soils have a much longer history, dating back to 1850. Recently, detailed characterizations of soil macropore features have been increasingly conducted. For example, in two separate excavations, Noguchi and colleagues cataloged the density, diameters, gradient, and directions of all macropores observed at excavated pits in Hitachi Ohta (Noguchi et al., 1997b, 1999). Macropores were found to range from 2 to 60 mm in diameter, with 70% of macropores less than 8 cm long, and only 1% greater than 40 cm. The majority (55–70%) of the observed macropores were formed by live or decayed roots. While this kind of work identifies the characteristics of

TABLE 1 Selected Locations where Flow Networks have been Identified and Characterized, Including their Conceptual Network Model and Evidence

Site	Location	Network conceptual model	Evidence	Reference
Tatsunokuchi - yama	Japan	Flow along permeability contrast	Trench discharge and matric potential measurements	Tani, 1997
Hitachi Ohta	Japan	Connected macropore network	Discharge from monitored macropores	Noguchi et al., 2001; Sidle et al., 2001
Maimai	South Island, New Zealand	Flow along permeability contrast	Direct observation (excavation of subsurface flowpaths during irrigation)	Graham et al., 2010; Woods and Rowe, 1996
Panola	Georgia, USA	Flow along permeability contrast	Water-table patterns, soil-depth measurements, and observed spatial distribution of lateral flow	Freer et al., 1997, Tromp-van Meerveld and McDonnell, 2006
Los Alamos	New Mexico, USA	Connected macropore network	Sampling and observations at trench face	Newman et al., 2004; Newman et al., 1998

Snow Shoe Mine	Pennsylvania, USA	Connected macropore network	Excavation after dye applications, lateral subsurface flow monitoring	Guebert and Gardner, 2001
Russell Creek	British Columbia, Canada	Flow along permeability contrast	Direct observation (excavation of subsurface flowpaths during irrigation)	Anderson et al., 2010
Yellowknife	Northwest Territories, Canada	Flow along permeability contrast	Soil-depth measurements and observed spatial distribution of lateral subsurface flow	Spence and Woo, 2003
Plastic Lake	Ontario, Canada	Flow along permeability contrast	Observed spatial distribution of lateral flow at trench	Peters et al., 1995
Maesnant Peat	Wales, UK	Connected macropore network	Excavation after dye applications, lateral subsurface flow monitoring	Jones 1981; Jones et al., 1997; Jones 2010
Moore House Peat	North Pennines, UK	Connected macropore network	Ground-penetrating radar, tension infiltrometer experiments	Holden, 2004; Holden et al., 2002a,b

individual macropores and their small-scale connectivity, it does not reveal the actual flow connectivity through the macropores and the extent and functions of the subsurface flow networks from the upslope area in controlling the outflow downslope.

Only recently has subsurface lateral flow from individual and groups of active macropores at trench faces been isolated to give an understanding of the importance of preferential flow in subsurface flow networks. At Hitachi Ohta, Tsuboyama et al. (1994) excavated a trench in shallow soil at a hollow from the surface to the soil–bedrock interface. Lateral flows from the O–A horizon interface, the soil–bedrock interface, and from three groups of macropores were routed to five tipping buckets to gauge the relative contribution from each zone. The macropore network was shown to consist of short (2–20 cm) pipes with a variety of orientations, not all of which were parallel to the slope. Lateral subsurface flow was monitored at the Hitachi Ohta trench during small-scale irrigation experiments (Tsuboyama et al., 1994) and six precipitation events over the course of three months (Noguchi et al., 2001). While the majority of flow at the trench face occurred via the soil matrix above the soil–bedrock interface, a significant portion (up to 16%) of discharge was delivered via preferential flow from one set of macropores. This set of macropores were thought to be directly connected to a hydrologically active fracture in the bedrock, effectively extending the preferential flow network from the soil into the near-surface bedrock.

Extensive monitoring and mapping of soil piping have been performed by Jones and colleagues in the UK. Jones (1981, 2010) defined soil pipes as the largest macropores, where flows are generally turbulent and of sufficient size that water sculpts their form. The pipes studied by Jones and colleagues were generally confined to peat areas, especially in the UK. During monitoring of pipe discharge near a stream channel, Jones (1981, 1997) showed that soil pipes could contribute large volumes of water to the stream channel at velocities intermediate between those of overland flow and throughflow, and that the piping networks were extensive, reaching up to 150 m upslope of the stream channel (Fig. 2). The locations of the pipe networks were controlled by both surface topography and soil properties (Jones, 1988, 2004; Jones and Cottrell, 2007; Jones et al., 1997).

Since the mid-1990s, a series of larger trench systems have been built to obtain a better sense of the distribution of lateral subsurface flowpaths along the base of a hillslope. These include trenches at the Maimai Experimental Forest on the South Island, New Zealand, the Panola Mountain State Park in the outskirts of Atlanta, Georgia, the Los Alamos National Laboratory in New Mexico, and the Plastic Lake in Ontario, Canada, among others. This large-scale trenching refers to experiments where an excavation bisects the base of a hillslope, with lateral flow separated by horizons, morphologic feature, or in discrete segments. In general, these trenches extend 10s of meters, perpendicular to the presumed flowpaths down the hillslope.

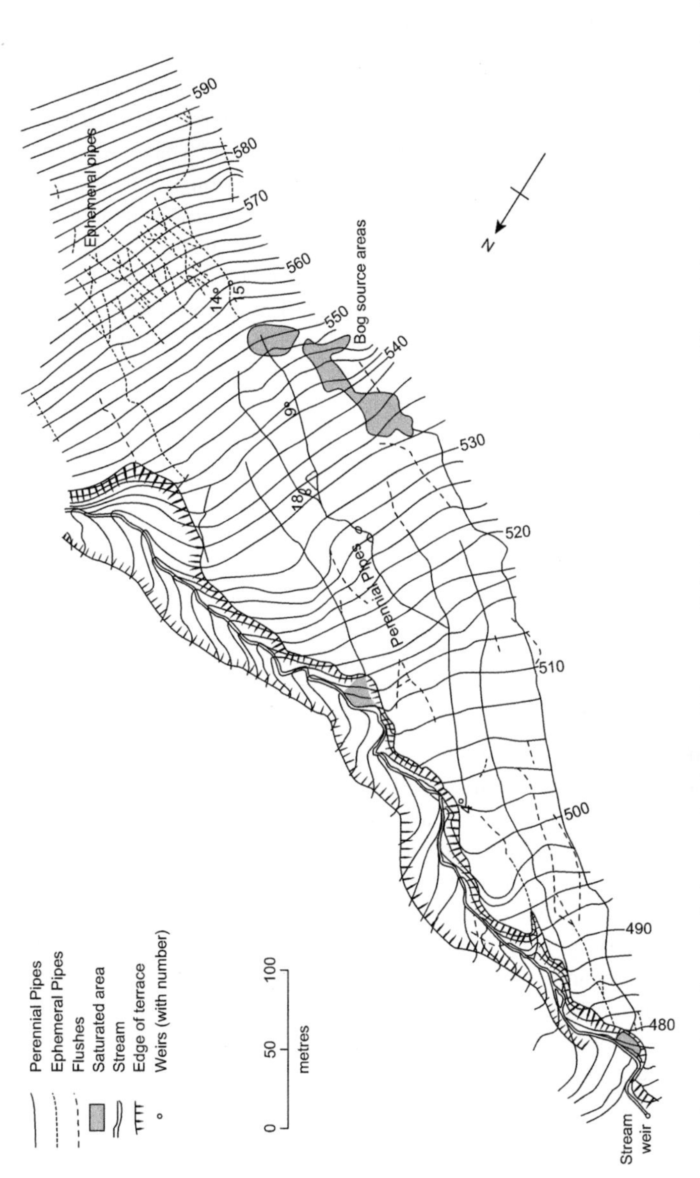

FIGURE 2 A map of perennial and seasonal pipes on the slopes of Masenant, UK. Soil pipes reached over 150 m upslope of the stream channel and provided significant contribution to stream flow. *(From Jones and Connelly, 2002)*

At the Maimai Experimental Forest, early irrigation experiments by Mosley (1979) demonstrated rapid flow through a macropore network, as evidenced by rapid breakthrough of a dye tracer from macropores in the soil and at the soil–bedrock interface of a series of small trenches 1-m downslope from application. Further work at the small trenches determined that discharge from the observed macropores was dominated by pre-event water (Sklash et al., 1986). McDonnell (1990) integrated the two observations with a "crack pipe" hypothesis, whereas water percolates via vertical cracks to the soil–bedrock interface, where it intersects the capillary fringe. With the addition of this event water, the soil moisture held in the capillary fringe is converted to phreatic water, which moves rapidly downslope via soil pipes generally confined to low in the soil profile. Extensive trenching at the base of a nearby hillslope by Woods and Rowe (1996) identified a wide variety of flow processes along a 30-m exposed trench face, including flow along the A–B horizon, flow from macropores embedded in the soil matrix, seepage from the soil matrix at the soil–bedrock interface, and flow through high permeability, high organic matter content soil in the A horizon. Analysis of flow volumes during storm response showed that subsurface lateral flow was concentrated in areas of high upslope contributing areas based on surface topography. Further analysis of the bedrock topography by Tromp van Meerveld and McDonnell (2005) suggested that bedrock topography might also route flow, though the relative lack of precision in either the surface or bedrock map made it difficult to reject either hypothesis. Later work by Graham et al. (2010; described in Section 2.2) confirmed that bedrock topography was the dominant control on the location of lateral subsurface flowpaths.

At the Panola Mountain State Park, an 20-m wide trench was excavated to the interface between sandy loam soils and the granite bedrock (Freer et al., 1997). The trench was separated into ten 2-m sections, and discharge from each section was measured with a tipping bucket flow meter. In addition to these sections, four individual macropores that were observed to contribute significant discharge were individually monitored. These macropores, which appeared to be decayed root channels, provided a significant amount of the measured discharge, up to 45% of storm flow (Freer et al., 2002). In addition, flow from the soil matrix generally occurred at the soil–bedrock interface, and the distribution of discharge was shown to coincide with the bedrock topography, rather than the surface topography. This indicates that, in addition to the apparent macropore flow network, an organized subsurface flow network at the soil–bedrock interface was active at the site.

At a high desert site in Northern New Mexico, Wilcox et al. (1997) built a trench system to capture flow from snowmelt and occasional precipitation events in a ponderosa pine hillslope. This trench system captured surface runoff, flow from A and B horizons, and lateral flow through the near-surface bedrock. Over the course of four years of monitoring, subsurface runoff was rare, occurring in six months out of 48, and accounted for 3–11% of the annual water balance. During snowmelt, however, this lateral flow accounted for up to 20% of the snowmelt. Additional analysis of isotopic, organic matter, and chloride concentrations of lateral flow indicated that not only was the majority of this water pre-event water, but also held in the larger soil pores and chemically distinct from water held in micropores (Newman et al., 1998). This indicates a two-domain flow system, where water moves laterally via a connected, macropore network, while a different pool consists of tightly held water in micropores in the soil profile. At an additional trench adjacent to the hillslope studied by Wilcox et al. (1997), Newman et al. (2004) observed patterns of soil wetting around a set of macropores during snowmelt. At the beginning of snowmelt, while soil moisture was low, macropore flow was constrained to decayed and live tree roots. As flow continued, the soil wetted both from the surface down and outward from the macropores. Late in the snowmelt, the halos around the macropores coalesced until the entire soil profile was saturated. At this point, the majority of flow remained confined to the observed macropores.

Peters et al. (1995) trenched two hillslopes at the Plastic Lake in south central Ontario, and separated lateral flow from the A–B horizon interface, matrix flow from the B horizon, and the soil–bedrock interface. They determined that nearly 100% of lateral subsurface flow occurred at the soil–bedrock interface through a thin layer of highly weathered and high permeable soil. Further work at the site by Buttle and McDonald (2002) confirmed that lateral flow through macropores in the soil profile was minor, and the majority of discharge was confined to a thin saturated area at the soil–bedrock interface. Buttle and McDonald (2002) then used stable isotope analysis of water collected at the trench to show that the lateral subsurface flow was dominated by pre-event water, indicating substantial mixing upslope.

Pit excavations and hillslope trenching have shown that high-volume lateral flow via preferential flow networks is the rule rather than the exception in natural systems. However, the extent, duration, and characteristics of the preferential flow network of the area feeding these macropores remain unclear.

2.2. Observations via Hillslope-Scale Tracer Experiments and Excavations

Despite the extensive observations of significant subsurface lateral flow both through macropores in the soil profile and along permeability contrasts such as the soil–bedrock interface, few studies had observed actual preferential flow network. To open this "black box" of hillslope hydrology, a more thorough investigation is needed.

Irrigation of water and dye tracers, followed by destructive sampling, has long been a technique used by hydrologists and soil scientists to identify vertical preferential flowpaths and flow networks at the pedon-scale. A number of studies have identified vertical preferential flow as the dominant pathway of water from the surface to depth at field sites from the tropics of Malaysia (Noguchi et al., 1997a), the mountains of Switzerland (Weiler and Flühler, 2004) and Japan

(Kitahara, 1989), to the agricultural land in the Netherlands (Bouma et al., 1982) and in the U.S. (Lin et al., 1996). While these experiments were useful in demonstrating the prevalence of vertical preferential flow, they did little to elucidate the characteristics of lateral subsurface lateral flow networks. Recently, a pair of studies has attempted to go beyond characterizing vertical preferential flow and macropore distribution at trench faces by performing hillslope-scale irrigation and excavations to trace lateral preferential flow networks.

Anderson et al. (2009, 2010) performed a combined irrigation and excavation experiment on a forested hillslope on the Vancouver Island, British Columbia, Canada. In this experiment, water was applied directly to the soil–bedrock interface via an excavated trench until discharge reached a steady state at a cut bank at the base of the hillslope (~30 m downslope of the application site). After steady-state conditions were met, brilliant blue dye was added to the application water, and irrigation continued until dye breakthrough was observed at the base of the hillslope. The dye tracer was observed very rapidly at the hillslope base, with initial breakthrough within 2 h, at an observed maximum velocity of over 100 m/h. After 14 h of drainage, the hillslope was excavated in 0.5-m increments upslope from the trench and the locations of the dye tracer were recorded. Analysis of the dye patterns in the excavated hillslope showed three lateral flow networks active in the hillslope corresponding to three dominant soil types: shallow clay and organic-rich soils (<100 cm, Ah and Bg horizons), poorly decomposed organic soils (thick O-horizon of 30–50 cm over mineral soil, Bf), and brown mineral soils (thin Ae horizon of 5 cm over 30–150 cm mineral soil, Bf) (Fig. 3). Flow through the clay and organic rich soils, found predominantly in the topographic hollows with high upslope contributing area, was confined to a connected network of preferential flow features such as soil pipes and areas of fine gravel. In the poorly decomposed organic soils, lateral flow was confined to the boundary between the organic horizon and the mineral soil surface, with water spread out along the permeability discontinuity between the soil horizons. In the brown mineral soils, where the upslope contributing area was low, lateral flow was observed both in the organic horizon and through coarse-textured near-surface soils. Some lateral flow was observed in pipes in the low-conductivity deep soils. From this irrigation and excavation experiment, it is clear that lateral flow at this site varies with local soil characteristics. When soil permeability was high in one horizon, water flowed through the matrix above a lower permeability horizon. When the permeability was low throughout the soil, water traveled through macropores and areas of isolated high-permeability soils.

At the Maimai hillslope where numerous earlier trenching studies were conducted, Graham et al. (2010) performed a similar, combined irrigation and excavation experiment to map the preferential flow network which was believed to facilitate subsurface lateral flow at the site. Previous work at the site indicated that lateral flow was dominated by macropore flow, both at the

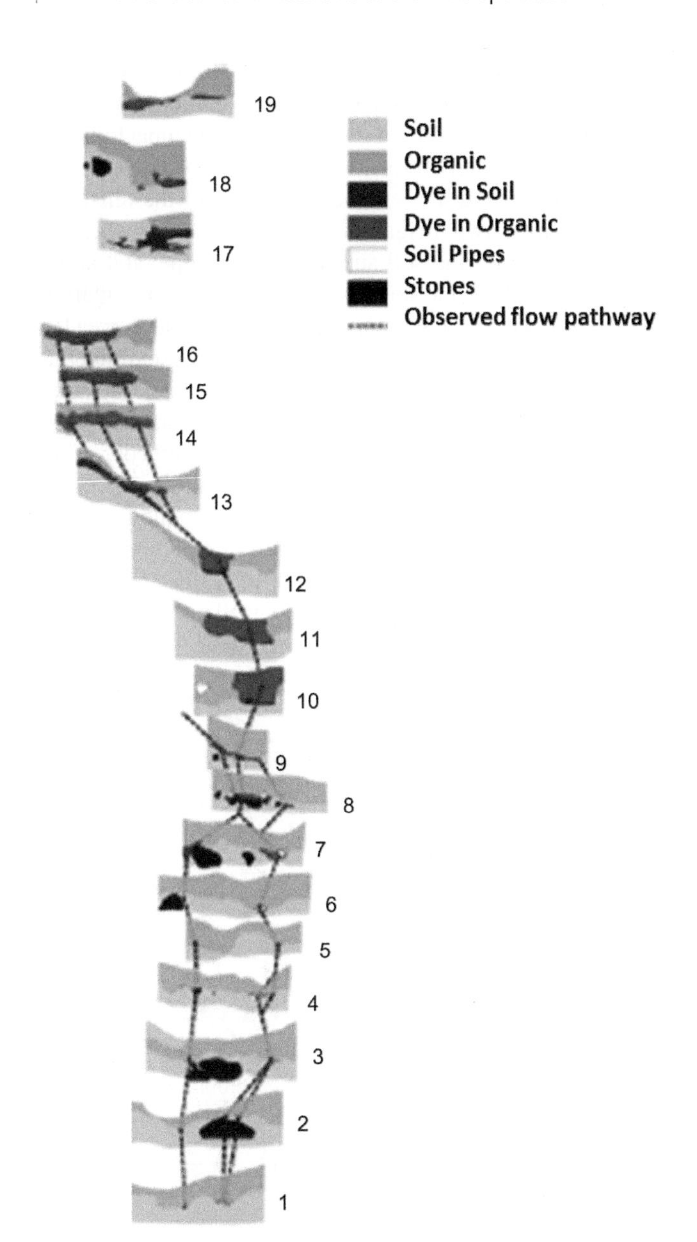

FIGURE 3 Subsurface flowpaths as identified in the Russell Creek irrigation and excavation experiment. 19 slices were excavated and lateral flowpaths were identified by blue dye tracer. The lateral flow network consists of flow through macropores, in high-permeability soil, and along the soil–bedrock interface. *(From Anderson et al., 2010) (Color version online and in color plate)*

soil–bedrock interface and throughout the soil profile (McDonnell, 1990; Woods and Rowe, 1996). In contrast to the hillslope explored by Anderson et al. (2009, 2010), the spatial variability of the soils on the Maimai hillslope was low in the area investigated, with more-or-less uniform, stony silt loams overlain by a 5-cm O-horizon throughout the excavated portion of the hillslope. Graham et al. (2010) performed two irrigation experiments, one where water was applied as a line source on the soil surface 4-m upslope, and another where water was applied directly to the soil–bedrock interface via a small trench 8-m upslope from the original trench system used by Woods and Rowe (1996). Once steady-state conditions were met, Graham et al. (2010) excavated the hillslope in 10–30-cm increments from the base of the hill, while water remained at a steady application rate, recording dyed flow locations as they progressed up toward the application site. For both application methods, within 50 cm downslope of the application site, subsurface lateral flow was concentrated at the soil–bedrock interface, where it was confined to a shallow gap between the bedrock surface and the lower soil boundary. Subsurface flowpaths were found to exhibit a dendritic pattern extending at least 8-m upslope from the base of the hillslope (Fig. 4), which were apparently controlled by small topographic

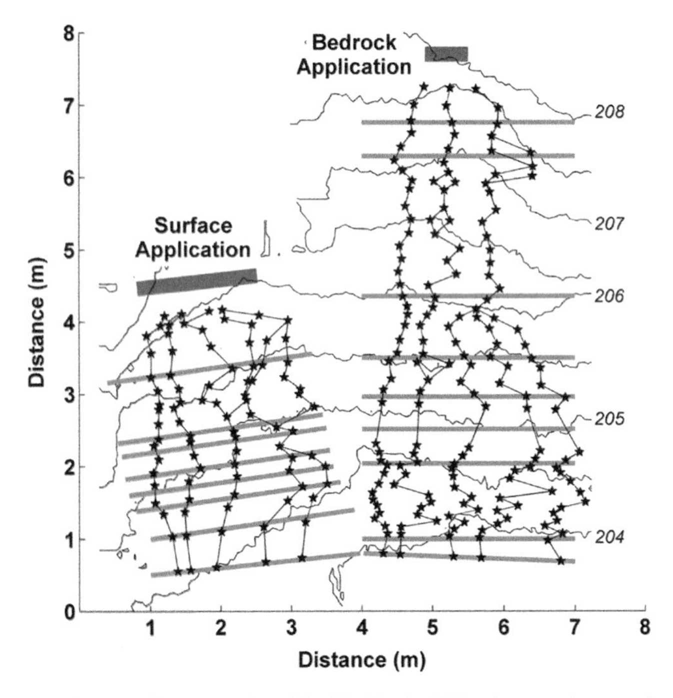

FIGURE 4 Preferential flow network as identified in the Maimai excavation experiment. Stars indicate observed concentrated lateral flow locations, while gray bars indicate locations of bromide tracer injection. Lateral flow was confined to a connected flow network at the soil–bedrock interface. *(From Graham et al., 2011)*

features (e.g. cobbles) in the bedrock surface. No significant flow through macropores above the soil–bedrock interface was observed greater than 50 cm below the application location. A bromide tracer was applied 17 times during the excavation, demonstrating that downslope velocities were high (30–40 m/h). These findings were similar to those of Anderson et al. (2009, 2010) in that the preferential flowpaths were continuous upslope, and flow velocities were higher than those predicted by soil properties.

The hillslope-scale experiments serve to expand the findings of many trenching and pedon-scale observations of lateral flow. While macropores have often been observed at trench faces and revealed to be primary conduits for rapid subsurface flow (e.g. Freer et al., 2002; Peters et al., 1995; Tani, 1997), the upslope extent of the contributing flow networks has generally been unknown. Measurements of individual macropores at trench faces have generally identified the connected length as extending much less than 5 m (e.g. Noguchi et al., 1999; Terajima et al., 1997). Through applying water with tracer directly to the soil–bedrock interface on a hillslope scale, both Anderson et al. (2009, 2010) and Graham et al. (2010) showed that these lateral preferential flow networks can be continuous through the extent of the hillslope.

While interesting, these experiments suffer from the same limitations of any true hydrologic experiments: boundary and edge effects. Many questions remain, such as 1) Can the artificial tracer applications truly reveal the nature of subsurface flow networks? 2) How do these trenching experiments, and the limitations of observations made at the point locations affect our understanding of the network nature of these subsurface flowpaths?

2.3. Observations via Nondestructive Imaging and Geophysical Methods

In recent years, with the increasing use of nondestructive microscopic imaging techniques in the laboratory (such as X-ray computed tomography (CT) and nuclear magnetic resonance (NMR)), widespread use of noninvasive geophysical and remote-sensing methods (such as ground penetrating radar (GPR), electromagnetic induction (EMI), and electrical resistivity tomography (ERT)), and growing arrays of sensor networks for in situ real-time monitoring (such as soil moisture and temperature monitoring networks), new opportunities arise for better documenting and understanding preferential flow occurrence and its networks. Only through such technological breakthroughs can we embody the detailed network-like flow and transport dynamics into new kinds of hydrologic models (Lin, 2010).

At the pore and core scales, X-ray computed tomography has been widely used to directly characterize the quantity, morphology, and continuity of macropores and related vertical preferential flow networks (Warner et al., 1989; Gantzer and Anderson, 2002; Luo et al., 2008). Three-dimensional visualization and quantification of soil macropores have been performed by synthesizing a series of two-dimensional CT images (Capowiez et al., 1998; Pierret et al., 2002). Sequential scans with and without a tracer allow the determination of active flow networks. Perret et al. (2000) applied medical CT to isolate the macropore domains from the soil matrix and monitored solute transport within an intact soil column (7.7 cm in diameter and 85 cm in length). Luo et al. (2008) showed that connected macropore networks could extend throughout soil columns 10 cm in diameter and 30 cm in length. The CT methods, however, are limited to laboratory studies using intact soil cores or columns.

At the pedon to hillslope scale, geophysical methods such as GPR (Holden, 2004; Yoder et al., 2001; Zhang et al., submitted for publication), EMI (Zhu et al., 2010), and ERT (Ward et al., 2010) have been used to observe subsurface flow networks in situ. van Overmeeren et al. (1997) tested GPR against capacitance probes and showed that the technology could accurately measure volumetric water content in the subsurface without excavation or other disturbances. Yoder et al. (2001) used both EMI and GPR to identify locations in an agricultural field where soil and bedrock features were conducive to promoting lateral preferential flow, though measurements were not made of actual water fluxes or stores. Similarly, Gish et al. (2005) mapped depth to restricting layers (clay lens) at an agricultural site using GPR and identified likely subsurface flow networks based on the topography of the restrictive layer.

At the Moor House Reserve, a peat-dominated catchment in the UK, Holden and colleagues used multiple noninvasive techniques to characterize soil piping in deep peat (Holden, 2004; Holden and Burt, 2002; Holden et al., 2002). Aided by the shallow surficial layer of the peat (~10 cm), Holden and Burt (2002) identified large pipe locations connecting bog pools to the stream channel via surface indicators such as depressions, vegetation changes, and occasional exfiltration of pipe water. Additionally, Holden et al. (2002) showed that GPR could identify large (15+ cm diameter) soil pipes in a peat wetland. Further work at the site using GPR and a salt tracer, Holden (2004) mapped subsurface flow networks at the hillslope scale (Fig. 5), showing that the subsurface flow network consisted of continuous pipes delivering water 100s of meters downslope, occasionally discharging water at the soil surface. These networks were shown to run roughly parallel with the slope, following topographic contours.

Zhang et al. (submitted for publication) used time-lapsed GPR, in combination with real-time soil moisture monitoring, at the hillslope scale to determine lateral flow pathways on a forested hillslope at the Shale Hills Catchment in central Pennsylvania. They showed that subsurface lateral flow was dominant in the shallow Weikert soil in a planar hillslope with 30% slope, whereas a combination of vertical macropore flow and lateral flow was dominant in the deep Rushtown soil located in a concave hillslope (swale) with 15% slope. The time-lapse GPR radargrams revealed the general infiltration wetting front and preferential flow pattern that were significantly different between the two types of hillslopes. Time-lapsed GPR was shown to be an attractive approach to help

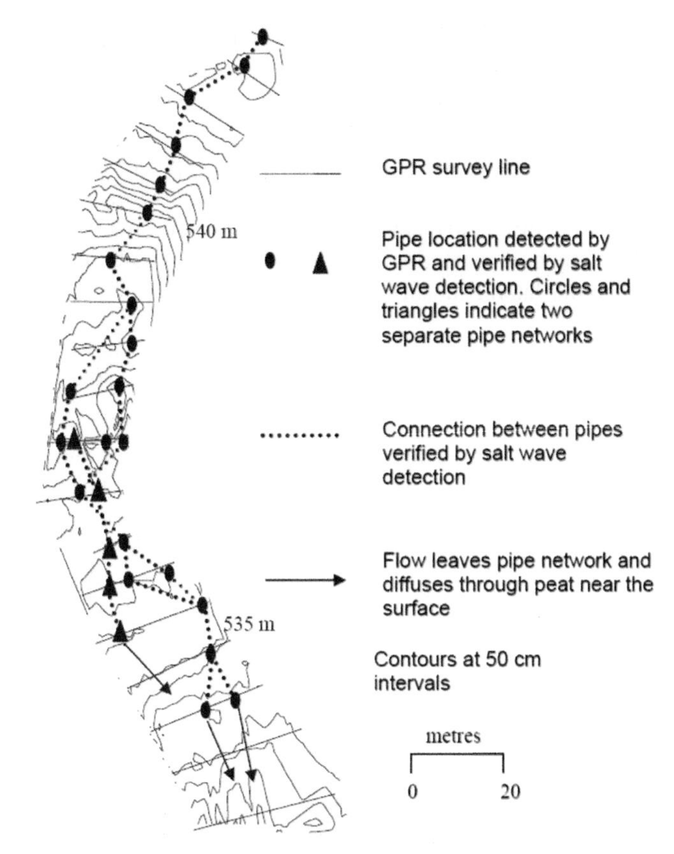

FIGURE 5 Subsurface flow network identified using ground-penetrating radar (GPR) and a salt tracer. *(From Holden, 2004))*

reveal hydrologic connectivity in the subsurface, which facilitates the formulation and testing of different conceptualizations of subsurface network modeling.

Depth-weighted soil electrical conductivity (ECa) measurements obtained from various EMI meters have been widely used to map different soil and hydrologic properties (James et al., 2003; Robinson et al., 2009; Sherlock and McDonnell, 2003). Repeated EMI surveys have been demonstrated by Zhu et al. (2010) to be useful in capturing the dynamics of soil moisture change and related subsurface flowpaths in the landscape. They showed that significantly higher ECa was detected in areas close to subsurface flowpaths in an agricultural hillslope, especially during the wetter periods.

Ward et al. (2010) used ERT through a combined tracer and monitoring approach to identify subsurface flowpaths in the hyporheic zone in a first-order stream meander in the Leading Ridge Watershed in central Pennsylvania. They applied a salt tracer to stream water, and used ERT to monitor fluxes through an abandoned stream bend. This allowed for a measurement of the areal extent of the hyporheic zone. While this technology has not yet been used to determine hillslope-scale subsurface flowpaths, Ward et al. (2010) have demonstrated the potential for ERT in hillslope hydrology.

The above-cited studies show the potential of geophysical techniques in probing the subsurface and parameterizing subsurface flow networks. Further work is needed to both identify new geophysical methods and determine how the spatial and temporal resolution limitations of EMI and GPR can be surmounted to pinpoint lateral subsurface flow networks.

3. HYDROLOGIC CONNECTIVITY AND THRESHOLD BEHAVIOR

While some aspects of subsurface flow networks have been identified and characterized through the experimental methods discussed above, a critical question remains: How do the observed networks transfer water from one area to another? This question is being addressed with a growing interest in hydrologic connectivity, defined by Stieglitz et al. (2003) as "*the condition by which disparate regions on the hillslope are linked via subsurface water flow.*" Most researchers have extended this definition to include connectivity from hillslopes to streams, and stream channels to each other. Pringle (2003) has emphasized the linkages of this concept to ecological variables, such as the fluxes of matter and energy through the ecosystem. Hydrologic connectivity has been identified as one of the key controls on subsurface lateral flow generation (Tromp-van Meerveld and McDonnell, 2006b), delivery of water from the hillslope to the stream (Jencso et al., 2009), and nutrient and sediment delivery (Akay and Fox, 2007). Hydrologic connectivity is a key indicator of the extent and stability of subsurface flow networks.

3.1. Connectivity in (Near-Surface) Soil Moisture

An early investigation of connectivity in the hydrologic community began with the work of a group of researchers working at the Tarrawarra agricultural watershed in southeastern Australia (Grayson and Western, 2001; Western et al., 1998, 1999a, 2004; Western and Blöschl, 1999; Western et al., 2001). Using a portable soil moisture measurement setup, this group measured surface soil moisture (upper 30 cm) on a 10 by 20 m grid in a 10.5-ha agricultural watershed. Over the course of two months, they made 13 soil moisture surveys of between 490 and 2056 points (Western et al., 1999b). They then defined connected areas as adjacent soil moisture measurement locations with high soil moisture (>0.43 m^3/m^3).

The work at Tarrawarra showed that as soil moisture increases, connectivity increases, until the majority of the catchment is connected with the outlet.

Further analysis of the Tarrawarra data set found a correlation between the observed soil moisture patterns with topographic indices such as upslope contributing area ($R^2 \sim 0.50$) and the topographic wetness index ($R^2 \sim 0.43$) during wet conditions, when connectivity measures were highest (Western et al., 1999a), but no correlation during dry conditions. The implicit assumption in the analysis of Western et al. (1999a) was that when areas in the catchment are connected with pathways of high soil moisture, lateral flow was enhanced, and a subsurface flow network was developed.

More recently, James and Roulet (2007) applied similar analysis to a steep, forested catchment in Quebec, Canada. There, they also measured near-surface soil moisture (20–30 cm), and linked the connectivity statistics developed by Western et al. (2001) to stream flow. James and Roulet (2007) found that aerial patterns of high soil moisture connectivity were similar throughout the year, though the extent of the connected area increased and decreased with precipitation and drought. The connection indices derived by Western et al. (2001) were poorly correlated to stream discharge, indicating no apparent connection between the hydrologic connectivity and system response. This work confirms the idea put forward by Tromp van Meerveld and McDonnell (2005), who argued that it is not surface soil moisture connectivity that is a control on lateral subsurface flow, but rather the connectivity of water tables at the soil–bedrock interface. These studies suggest that further research is needed to address how near-surface connectivity relates to lateral subsurface flow. If there is no near-surface lateral water flux, what is the causal mechanism for the observed patterns of connectivity?

3.2. Subsurface Hydrologic Connectivity

Spence and Woo (2003) developed the "fill and spill" hypothesis based on hydrometric and meteorological measurements made at the Pocket Lake catchment in the Northwest Territories, Canada. Spence and Woo (2003) determined that water from upslope areas moved laterally downslope and filled unrequited storage in the soils of the valley areas. Only when the storage in the valleys filled would lateral subsurface flow initiate, and the upslope areas would be connected hydrologically to the stream system.

Transferring the concept of "fill and spill" to the hillslope scale, Tromp van Meerveld and McDonnell (2006b) measured the transient water table above granitic bedrock at the trenched hillslope at the Panola Mountain State Park, Georgia, and compared patterns of water-table development with lateral subsurface flow. They found that, rather than connected patterns of surface soil moisture, it was the connection of the patches of saturated area at the soil–bedrock interface that controlled lateral subsurface flow. At this site, a minimum precipitation threshold of 55 mm was required to both fully connect the saturated areas from the hilltop to the hillslope base and create the subsequent order of magnitude increase in lateral subsurface storm flow (Fig. 6). This

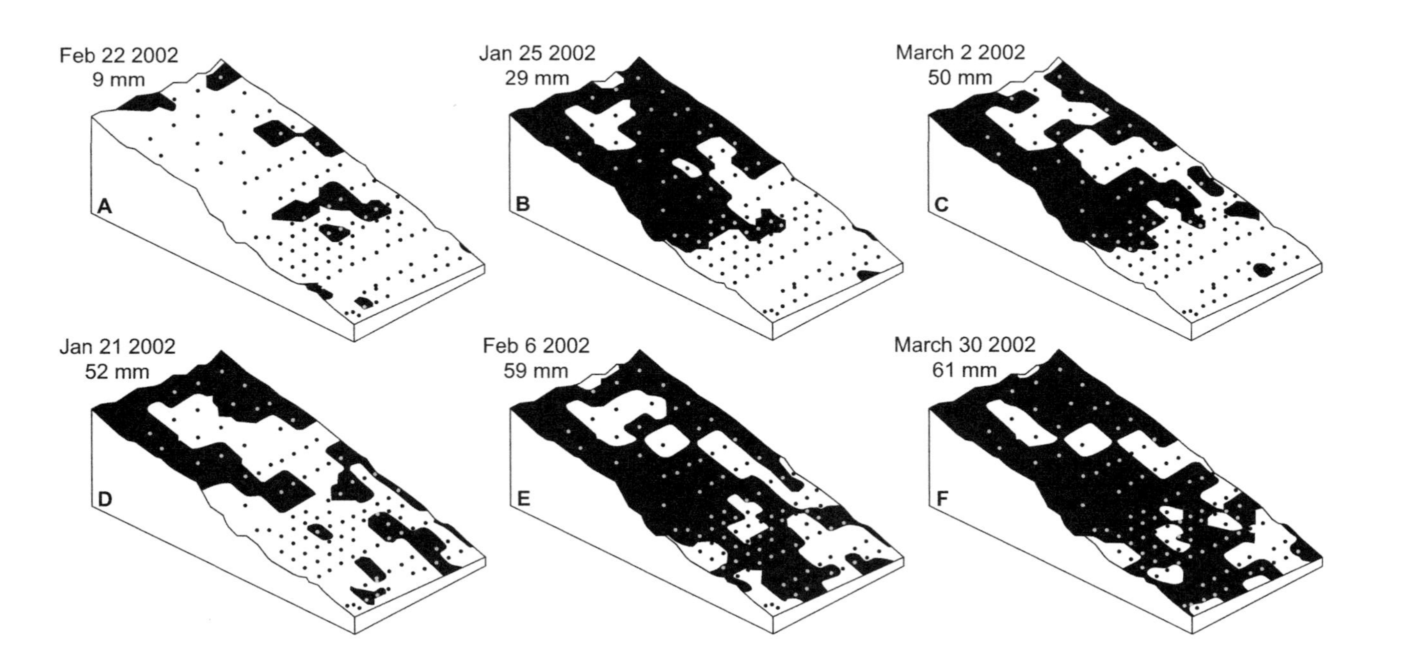

FIGURE 6 Connected subsurface saturated areas at the Panola hillslope with increased storm size, where a minimum of 55 mm precipitation is required to develop a connected subsurface network from the top to bottom and result in substantial lateral subsurface flow. Dots indicate locations of piezometers on hillslope, while black patches indicate measured saturated conditions for six storms of increasing magnitude. A connected subsurface flow network developed during the 59 mm precipitation event, resulting in an order of magnitude increase in lateral subsurface flow. *(From Tromp-van Meerveld and McDonnell, 2006b)*

work formed the basis of a percolation modeling of hillslope hydrology by Lehmann et al. (2007) (see Section 5.3).

Further analysis of subsurface hydrologic connectivity has been performed by Jencso et al. (2009), who measured transient water table 5 m and 10 m upslope of the stream channel at 24 transects in a steep forested catchment in the Tenderfoot Experimental Catchment in Montana. Wells were installed at the soil–bedrock interface, and water-table development was monitored for one year. Jencso et al. (2009) operationally defined hillslope hydrologic connectivity as temporally synced observed stream flow and measured water table. The duration of hydrologic connectivity was shown to be positively correlated with the size of the upslope contributing area ($R^2 \sim 0.91$). Using this relationship, the estimated hydrologic connectivity was plotted on the landscape, and the temporal probability distribution of stream flow was closely matched to the spatial probability distribution of connectivity.

Detty and McGuire (2009) investigated hydrologic connectivity with a well and soil moisture monitoring array in a forested catchment at the Hubbard Brook Experimental Forest in New Hampshire. They determined that water-table development was patchy and isolated during dry conditions in the growing season, when measured water table was confined to topographic hollows. During the dormant season, the majority of wells showed continuous water-table development extending throughout the instrumented area. Water-table persistence was positively correlated with topographic indices such as the upslope contributing area and the topographic wetness index proposed by Beven and Kirkby (1979).

Hydrologic connectivity is clearly a major control on lateral subsurface flow, encouraging additional questions to be answered, such as: How are areas of subsurface saturation connected? What are the characteristic travel and residence times of water in these connected areas? When measuring saturated areas, how does the choice of well depth affect the frequency and duration of hydrologic connectivity? Is there a connection between water-table distribution and soil moisture patterns?

3.3. Thresholds

An emergent behavior associated with hydrologic connectivity is that of thresholds in lateral subsurface flow. At many of the sites reviewed above, a threshold precipitation amount was needed before the initiation of substantial lateral subsurface flow. At the Panola instrumented hillslope, this threshold was 55 mm, though the threshold appeared to be larger for events with low initial soil moisture (Tromp-van Meerveld and McDonnell, 2006a). At two catchments in the Tatsunokuchi-yama Experimental Forest, Tani (1997) noted a threshold that varied with pre-event catchment discharge. After this threshold was reached, lateral discharge appeared to have a 1:1 relationship with additional precipitation. Detty and McGuire (2009) showed that the threshold observed at the Hubbard Brook Experimental Forest was a function of both precipitation and antecedent moisture status. Similar antecedent moisture-dependent thresholds have also been observed at the H.J. Andrews Experimental Forest in Oregon (Graham and McDonnell, 2010), the Maimai Experimental Catchments (Mosley, 1979), Pocket Lake in Canada (Spence and Woo, 2003), and elsewhere (Guebert and Gardner, 2001; Noguchi et al., 2001; Peters et al., 1995) (Fig. 7). These observed thresholds have been attributed to the amount of water needed to develop significant connectivity of upslope areas to the hillslope base or monitoring site.

The threshold behavior may be described with a simple equation (Graham and McDonnell, 2010):

$$Q = a(P - P_o), \tag{1}$$

where Q is storm total lateral discharge, a is the slope of the excess precipitation/discharge line, P is the storm total precipitation, and P_o is the precipitation threshold for lateral subsurface flow initiation (Fig. 7). The slope variable (a) describes how much additional precipitation above the threshold is expressed as lateral subsurface flow, and can reach as high as 1 (Tani, 1997), though at other sites the slope was significantly less (Graham and McDonnell, 2010; Tromp-van Meerveld and McDonnell, 2006a). The precipitation threshold prior to subsurface flow initiation has been shown to be similarly variable, ranging from near zero during wet conditions at the H.J. Andrews

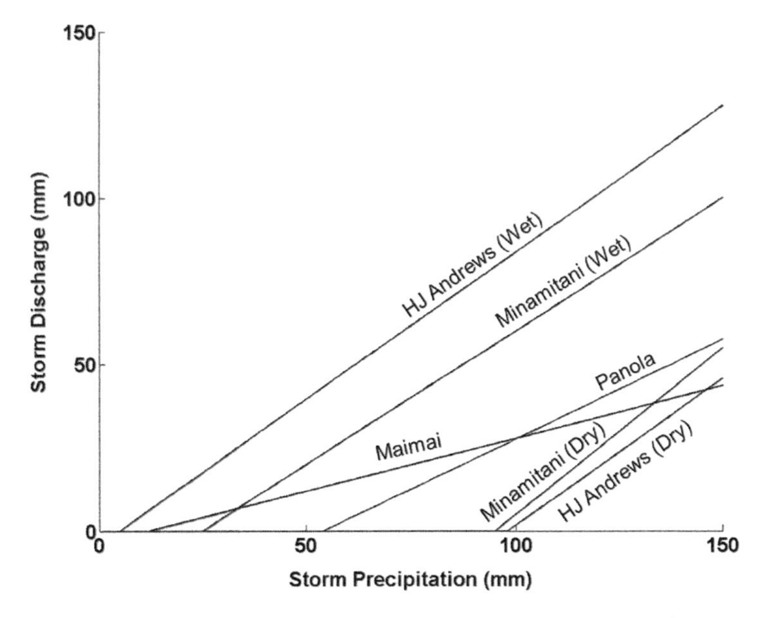

FIGURE 7 Schematic of threshold response at four catchments with lateral flow networks. Threshold precipitation volume is required to develop connected lateral flow network and initiate lateral subsurface flow. *(Data from Graham and McDonnell, 2010 for HJ Andrews and Maimai, Tromp-van Meerveld and McDonnell, 2006a, for Panola, and Tani, 1997, for Minamitani)*

(Graham and McDonnell, 2010) to nearly 100 mm when dry at the Minamitani (Tani, 1997) and the HJ Andrews (Graham and McDonnell, 2010).

Modeling by Graham and McDonnell (2010) demonstrated possible controls on the threshold parameters (a and P_o in Eqn (1)) using a virtual experiment approach based on findings at the Maimai instrumented hillslope of Woods and Rowe (1996). The slope parameter (a) was controlled by evapotranspiration and bedrock permeability, while the threshold was controlled by the time between events, evapotranspiration rate, bedrock permeability, and bedrock topography. The controls on thresholds remain unresolved at sites with different lateral flow networks, such as pipe-dominated systems or areas with impermeable bedrock.

These modeling and monitoring studies have shown the prevalence of thresholds, yet the controlling processes that determine the strength and magnitude of these thresholds remain unclear. Additional questions to be answered include: What do the observed thresholds reveal about underlying subsurface flow networks? How do soil properties (depth, layering, permeability, and hydraulic functions) affect observed thresholds? How do soil architecture and the distribution of various soils over the hillslope affect hydrologic connectivity and threshold behavior?

4. CONCEPTUAL MODELS OF SUBSURFACE FLOW NETWORKS

Based on the literature discussed above, two conceptual models of subsurface lateral flow networks have been suggested at the hillslope scale: 1) a model of connected preferential flow pathways in the soil (Lin, 2010; Sidle et al., 2001) and 2) a model of water flow along permeability contrasts such as the soil–bedrock interface and contrasting soil horizon interfaces (Lin and Zhou, 2008; Tromp-van Meerveld and McDonnell, 2006b).

4.1. Flow via a Series of Connected Preferential Flow Paths

In this conceptual model, subsurface flow is characterized by a series of preferential flow pathways (macropores, root channels, animal burrows, fractures in bedrock, and others) connected by nodes (such as areas of high-permeability soils) that are turned on and off based on hydrometric conditions. As soil moisture increases, these nodes wet up, and some will be turned on and thus be able to connect flow pathways and transmit water. After a critical number of nodes are activated, a flow network is formed that connects the entire hillslope, thus greatly increasing lateral discharge. The key here is that the nodes must be activated before lateral preferential flow down the hillslope can be initiated.

The above conceptual model was supported by the observations at the Hitachi Ohta that macropores tend to be short (<0.5 m), yet macropore flow was a significant component of lateral discharge (Noguchi et al., 1999, 2001).

This conceptual model was further tested via numeric experiments by Nieber and Sidle (2010), who used a numeric model of isolated macropores connected by soil matrix. They showed that subsurface lateral flow velocities and capacity were increased by the presence of these isolated macropores, while reducing positive pore pressures during transient saturated conditions.

Lin (2010) discussed the theory of evolving networks to shed light on a variety of preferential flow networks observed in field soils. Networks are abundant in soils (despite the difficulty of seeing them), such as root branching networks, mycorrhizal mycelial networks, animal borrowing networks, networks of cracks and fissures, and man-made drainage networks. Lin (2010) suggested that energy inputs cause flow networks to form in soils, and networks provide a means of minimizing energy dissipation (or equivalently, maximizing the efficiency of energy transfer). Like the energy of water flowing over the land surface that creates dendritic stream networks, water flowing through soils also creates network-like flowpaths in the subsurface. Some evidence has suggested that a similarity may exist between river dendritic structures and subsurface preferential flow networks (e.g. Deurer et al., 2003). As water moves through soils, changes in soil texture, structure, organic content, mineral species, biological activities, and other features will modify the resistance to the flow, causing change in flowpath to allow water to follow the least resistant path, thus resulting in a preferential flow network that has the least global flow resistance.

4.2. Flow at Permeability Contrasts

Another conceptual model is the fill-and-spill model of Spence and Woo (2003) and Tromp-van Meerveld and McDonnell (2006b). In this conceptualization, flow moves vertically via either the soil matrix or preferential flowpaths, then downslope through a connected preferential flow network located above the surface of a relatively impermeable bedrock or soil layer. Here the flowpath locations are dictated by the bedrock surface or low permeability water-restricting soil layer, as demonstrated by the apparent bedrock control of flow routing at the Panola (Freer et al., 2002) and Maimai hillslopes (Graham and McDonnell, 2010) and the strong impact of subsoil clay layer interface and the soil–bedrock interface in an agricultural hillslope (Zhu and Lin, 2009). While the nature of the subsurface flow network is unknown at the Panola and other nonexcavated hillslopes, at the Maimai Graham et al. (2010) showed that subsurface flow was via a connected network of pore space above the bedrock surface. These pore spaces seemed to be long-lived, as evidenced by organic staining on the bedrock surface and lower portion of the soil profile. Subsurface storage in bedrock topographic hollows must be filled before lateral flow could initiate (Spence and Woo, 2003; Tromp-van Meerveld and McDonnell, 2006b), leading to a connected network of pools (fill areas) connected by preferential flowpaths (spill areas).

This conceptual model was used to develop a simple numeric model of flow at the Maimai hillslope by Graham and McDonnell (2010), who were able to replicate threshold behavior and a 100-day hydrograph by assuming that the bedrock topography was parallel to the surface topography, and the lateral flow was confined to the bedrock surface. Additional modeling by Hopp and McDonnell (2009) replicated the hypothesized fill-and-spill behavior at the Panola with a Richards-equation-based hydrologic model (HYDRUS 3D) with no macropores but an impermeable bedrock.

While the works of Tromp-van Meerveld and McDonnell (2006b) and Graham and McDonnell (2010) occurred at areas where the underlying permeability contrast driving the fill-and-spill behavior was the bedrock surface, similar lateral flow routing has been observed in areas where the permeability contrast was caused by soil horizonation. Fragipans, duripans, clay layers, and other water-restricting soil layers have all been shown to contribute to subsurface lateral flow generation (Lin et al., 2008; McDaniel et al., 2008; Needelman et al., 2004; Whipkey, 1965; Zhu and Lin, 2009), and can all serve as controls on fill-and-spill-type lateral flow networks. At this point, the fill-and-spill hypothesis has not been explicitly demonstrated at a site where soil horizonation controls lateral flow initiation.

5. SUBSURFACE FLOW NETWORK MODELING

Numeric hydrologic models have been used to capture and develop the understanding of hillslope-scale hydrologic processes since Bergström and Forsman (1973). Only in recent years has the long-known network nature of subsurface lateral flow been incorporated into such models. In the last ten years, three methods for conceptualizing subsurface preferential flow networks have been developed, including 1) topography-based flow routing, 2) stochastic macropore modeling, and 3) percolation theory-based modeling. In the following, each modeling approach is summarized.

5.1. Topography-Based Lateral Flow

The vast majority of hydrologic modeling has used a topographic approach to simulate lateral flow routing. In these models, subsurface lateral flow occurs via topographic gradients, through the soil matrix without any special flow network. Models of this type include TOPMODEL (Beven and Freer, 2001; Beven and Kirkby, 1979), Soil Moisture Routing Model (Frankenberger et al., 1999), TOUGH2 (James and Roulet, 2007), HYDRUS 3D (Hopp and McDonnell, 2009; Simunek et al., 2008), and numerous other models. In these models, downslope flow is routed along a permeability contrast using one of a number of flow routing algorithms (see Seibert and McGlynn, 2007 for a review). Depending on the chosen flow routing algorithm (multiple or single direction), downslope flow networks can be diffuse or concentrated. Subsurface lateral flow is often induced by including anisotropy, either via soil layering (e.g. Hopp and McDonnell, 2009) or explicitly by assigning higher lateral than vertical hydraulic conductivity (e.g. Graham and McDonnell, 2010). Subsurface flow networks are also often accounted for by using an "effective" lateral hydraulic conductivity, which is generally higher than measured hydraulic conductivities (e.g. Saulnier et al., 1997).

While topography-based lateral flow models can generally capture stream and hillslope discharge, and occasionally soil moisture distribution, they are not designed to capture the subsurface flow network observed or inferred from field observations. Further research is needed to determine how the lack of an explicit subsurface flow network affects modeled lateral hydrologic processes, and how actual flow networks may vary from those inferred from topography gradients.

5.2. Stochastic Pipe Flow Modeling

Weiler and McDonnell (2007) developed a model (Hill-vi) to stochastically simulate subsurface lateral flow via a preferential flow network. They built a parsimonious model of subsurface lateral flow based on four assumptions: 1) soil pipe diameters are fixed within a narrow range (based on review by Uchida et al., 2001) but large enough that flow is not restricted by pipe diameter (based on the findings of Weiler, 2005), 2) pipes are short and discontinuous (based on findings of Anderson et al., 2009; Kitahara, 1993), 3) pipes are generally located near the soil–bedrock interface or soil permeability discontinuities (based on findings of Uchida et al., 2002), and 4) water flow is proportional to hydraulic head in pipe flow (based on laboratory findings of Sidle et al., 1995).

Pipe flow was then modeled based on these assumptions. Pipes were placed in random locations in the model domain based on a preset pipe density. The heights of the individual pipes above the soil–bedrock interface were determined based on a depth distribution determined by the user. Pipe flow is then determined by the following equation:

$$q_p(t) = k_p A^{0.5}(w(t) - z_p)^{\alpha}, \quad (2)$$

where $q_p(t)$ is the pipe flow at time t, k_p is a hydraulic conductivity parameter, A is the grid cell area, $w(t)$ is the water table height at time t, z_p is the vertical location of the pipe, and α is the slope of the log linear regression between pipe flow and hydraulic head. The range of pipe locations (z_p) and the hydraulic conductivity parameter k_p are used as calibration parameters in the model.

The Hill-vi model with pipe flow was tested on the Maimai experimental hillslope instrumented by Woods and Rowe (1996) (Fig. 8). The model was calibrated to one 55.8-mm precipitation storm event, and then validated against two other events. In addition to storm discharge, the model was tested against a hillslope tracer experiment at the site reported by Brammer (1996). Simulations were run with and without pipes. In general, a pipe network was needed to adequately represent the discharge time series at an event and

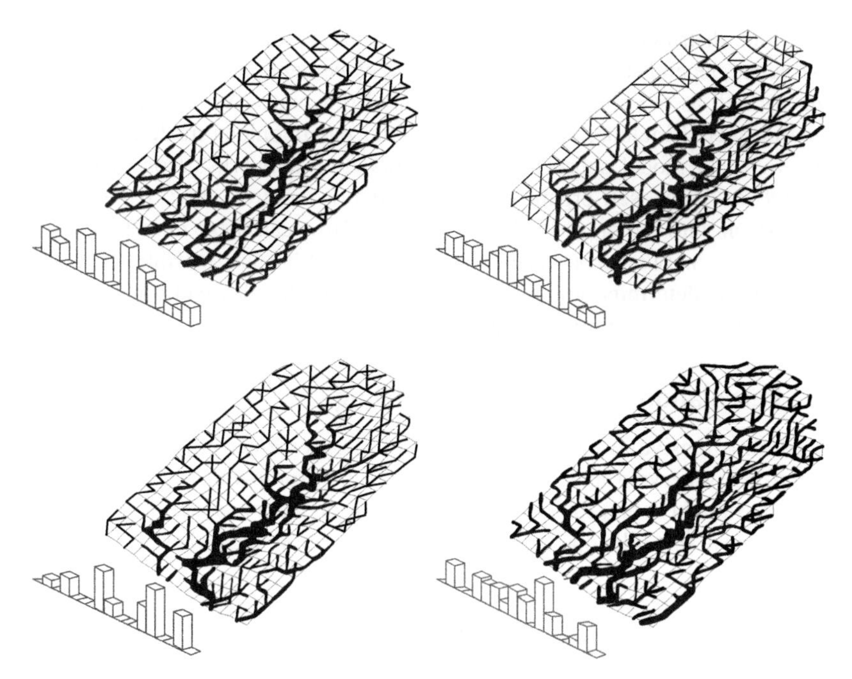

FIGURE 8 Four realizations of a stochastic model of preferential flow network in a hillslope using Hill-vi model. Thin black lines indicate soil pipes, while blue lines indicate connected lateral flowpaths. Red bars indicate flow volume from each grid cell along the hillslope base. *(From Weiler and McDonnell, 2007) (Color version online)*

annual time scale. Using different random arrangements of pipes, Weiler and McDonnell (2007) showed that, while the arrangement of pipes determined the exact location of the subsurface flowpaths, the geometry of the lateral flow network, and to some extent the locations of trench discharge, pipe arrangement did not impact the overall subsurface flow volumes or the general patterns of tracer breakthrough at the trench. In a later work, the Hill-vi model was applied to the Panola hillslope, where the influence of pipe locations and network geometry on discharge volumes was similarly low (Tromp van Meerveld and Weiler, 2008). To correctly model storm response at the Panola, both pipe flow and permeable bedrock components were needed.

The key findings from the Hill-vi modeling exercises include: 1) random placement of pipes leads to a network structure in the subsurface and 2) the location of pipes has little impact on water flux and tracer volume at the discharge outlet, though it does affect tracer breakthrough distribution spatially across the base of the hillslope.

A key limitation of the Hill-vi type of model is the random locations of the pipes, which have been shown experimentally to cluster at permeability contrasts (Graham et al., 2010) and near roots and other vegetation (Noguchi et al., 1997b). With the knowledge that soil pipes do not likely develop a random distribution, further work is needed to determine how network emergence depends on a more deterministic placement of soil pipes of various sizes as well as other nonrandom preferential flowpaths. Additional questions to be answered include: Are there additional parsimonious modeling formations that can model subsurface preferential flow networks? How would an ordered or even self-organized pipe network control flow patterns, in comparison with the prescribed random field? How can these models of preferential flow networks be verified in the field? What are feasible tools and techniques to determine soil pipes and other preferential flowpaths in hillslopes that can supply needed model parameters?

5.3. Percolation Theory Based Modeling

Based on extensive research at the Panola hillslope (Freer et al., 1997, 2002; Tromp-van Meerveld and McDonnell, 2006a,b; Tromp-van Meerveld et al., 2006), Lehmann et al. (2007) applied a percolation theory model of subsurface lateral flow at this hillslope (Fig. 9). While percolation theory has been used in

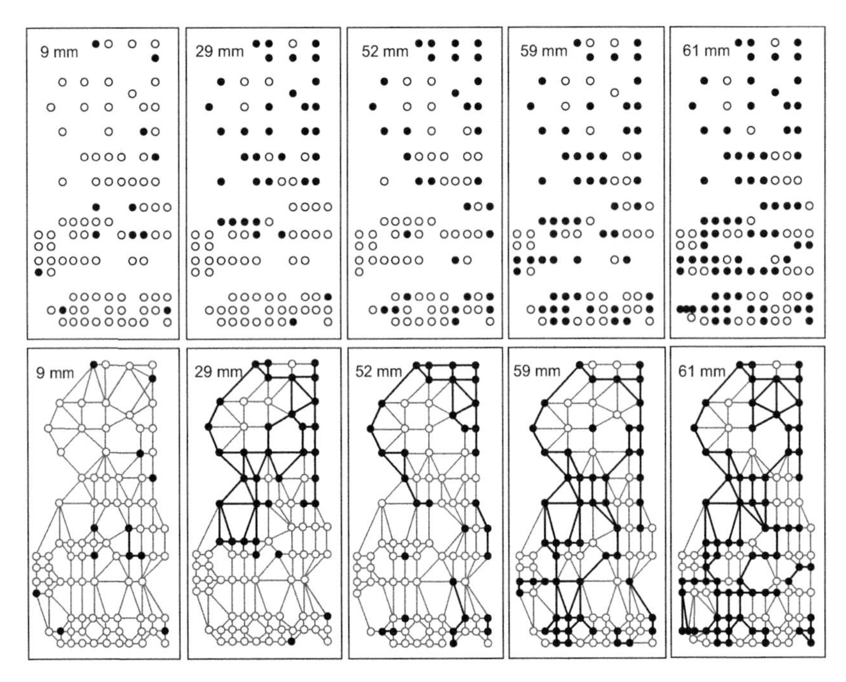

FIGURE 9 Panola percolation model, with increasing connectivity due to increased storm size. Dots indicate locations of piezometers on Panola hillslope, with open dots at dry locations, and filled dots at saturated locations. Increased storm size (from 9 to 61 mm) results in increased probability of well saturation. Between 52 and 59 mm a connected flow network develops from the ridge to the hillslope base. *(From Lehmann et al., 2007)*

soil science to describe vertical movement of water at the pore to core scales (Berkowitz and Ewing, 1998), the Lehmann et al. (2007) model was the first application of percolation theory to hillslope-scale lateral flow.

Percolation theory is based on the assumption that the system can be described as a set of "sites" (or nodes) connected by "bonds." In soil science, the sites are generally thought to be pore bodies, while the bonds are the pore throats (Berkowitz and Ewing, 1998). At the hillslope scale, Lehmann et al. (2007) defined the sites as areas with a transient water table, while the bonds were connections such as macropores or areas of connected high permeable soils (Fig. 9). Sites are considered occupied when there is a measured or modeled water table. Flow between sites occurs when the two sites are connected by hydrologic connection (a bond).

The key variables in percolation theory are the number and density of the sites, the number of bonds between sites (called the coordination number), and the occupation probability (i.e. the probability that a given site is saturated). The coordination number for a rectangular lattice can be up to 8, while a hexagonal lattice can have a maximum coordination number of 6. For coordination numbers less than the maximum, a model with a coordination number of eight is modified by having between 0 and 8 bonds randomly or systematically removed from each node until the average number of bonds per site equals the desired coordination number. In the case of Lehmann et al. (2007), the occupation probability is the probability that a water table is observed at a site. As the domain average occupation probability nears one, more sites are occupied by water table, and more sites become connected. Connected sites are termed clusters, and lateral flow is assumed to be dependent on the size of the cluster connected to the hillslope base. As the average occupational probability increases, cluster sizes (and lateral flow) increase until the percolation threshold is reached. The percolation threshold is the occupational probability at which it is statistically likely that a cluster will span the entire mesh, allowing for water to flow from one end of the domain to the other (e.g. from the ridge to the hill bottom). The percolation number is a function of the coordination number, and the mesh size, with smaller coordination numbers and larger meshes leading to higher percolation numbers.

In the application of percolation theory to the Panola hillslope, the model sites were associated with locations where the water table was monitored on the hillslope at the soil–bedrock interface (Lehmann et al., 2007). The probability that any two sites were connected by an active subsurface flowpath was a calibrated parameter ranging from 0 to 1. At each site, a random number was assigned to represent the storage capacity of the site. When precipitation was greater than the assigned storage capacity, water flow between connected sites initiated. This is analogous to the connectivity between nodes with states greater than the occupational probability. One hundred realizations were performed with different distributions of the random site storage capacities. Once the storage capacity of a site was reached, all water above the storage capacity would flow downslope via bonds to neighboring sites. For large events (high occupation probability), the majority of sites were filled, and widespread hillslope connectivity was achieved.

As the bedrock permeability at the Panola has been shown to be high and bedrock leakage is significant (Tromp-van Meerveld et al., 2006), 65% of precipitation was immediately assumed to be lost as deep seepage, and directly deducted from the soil moisture storage. By calibrating both the coordination number and the average storage capacity, Lehmann et al. (2007) were able to reproduce both the threshold response to precipitation observed at the site and the increased threshold during dry conditions, when the occupational probability decreased. In addition, the qualitative observations of connected areas upslope increasing with increased total storm precipitation were reproduced (Fig. 9).

While the model of Lehmann et al. (2007) used random distributions of site connections (preferential flowpaths) and site status (pre-event available storage), it is one of the few hillslope-scale models to use and define an explicit network of subsurface preferential flow (largely lateral flow). However, a number of questions remain: With the knowledge that preferential flow networks do not likely develop with a random distribution, how would such network emergence depend on the ordered placement of active sites? Can this model be parameterized a priori with field measurements? How can this model be verified in the field?

6. STATE OF THE SCIENCE AND A WAY FORWARD

It is clear that there is ample evidence for widespread existence of hillslope-scale preferential flow networks under a variety of soil, topographic, climatic, and geologic conditions. The rapid tracer fluxes over tens of meters observed by Brammer (1996), Nyberg et al. (1999), McGuire et al. (2007), and many others indicate that preferential flow networks act as high-velocity pathways of water, bypassing a large portion of lower permeable soils. Further observations of high flow volumes at trench faces by Peters et al. (1995), Woods and Rowe (1996), Freer et al. (1997), and many others indicate that these networks can drain large areas of hillslopes, often extending far upslope of measurement trenches. Long-term monitoring by Tani (1997), Tromp van Meerveld and McDonnell (2006a), Jencso et al. (2009), Graham and Lin (2011), and others have revealed some of the controls on the activation and dynamics of these subsurface flow networks.

Despite this considerable body of research, little remains known about the structure, extent, and evolution of the subsurface preferential flow networks that are assumed to be active under appropriate conditions. At a given site, the presence or absence of pipe or other macropore flow, flow along permeability contrasts, and flow through near-surface bedrock fractures often remain a guesswork based on limited observations (both temporally and spatially). While recent hillslope tracer applications and excavations performed by

Anderson et al., (2009, 2010) and Graham et al. (2010) and dense soil moisture monitoring networks performed by Western et al. (1999a,b), Lin (2006), and Graham and Lin (2011) have revealed the network structure at some hillslopes, more excavations and dense monitoring are needed to reveal realistic preferential flow networks and their dynamics across different climatic and soil conditions.

Many key questions remain about subsurface preferential flow networks, including:

1) What are the extent, connectivity, and structure of preferential flow networks in the subsurface?
2) How do our measurement techniques (trenching, excavating, and geophysical tools) affect our perception of the flow networks, and how can nondestructive methods be enhanced to provide higher spatial and temporal resolution of in situ subsurface structures and flow dynamics?
3) How do the current model conceptualizations of subsurface flow networks correspond with our evolving understanding of these networks?
4) Are there other methods for conceptualizing subsurface flow networks that might better capture dominant hydrologic processes?
5) How do these networks develop and evolve over time?
6) Which of the findings reviewed in this chapter are universal, and which are artifacts of the peculiarities of the chosen field sites and/or the investigation methods used?

We believe that to move forward, and to begin to answer the above questions, we need to expand the number of observations of subsurface preferential flow networks at multiple scales. Currently, there are simply not enough in situ observations and quantitative measurements of preferential flow networks to develop (or reject) a universal theory of subsurface flow networks. We found only very few instances where an attempt has been made to map the subsurface flow network, and even in these instances, artificial water applications raise questions on how representative the observations were of natural flow processes. In this regard, nondestructive mapping or imaging of opaque subsurface and flow networks in situ is highly desirable (Lin, 2010).

In addition, new modeling techniques need to be developed, to both represent established flow networks and predict the evolution of these networks. This would allow us to move away from the current assumptions regarding random placement of network components, and to develop theories on the deterministic placement of the nodes, connectivity, and dynamic changes of subsurface flow networks. In this regard, hydropedology principles and extensive knowledge and database developed in pedology could be better utilized (see other chapters in this book).

One possible source of new understanding along these lines may be the B2 hillslope evolution experiment, where a hillslope is being constructed from virgin soil, and soil structure will be allowed to evolve in a controlled environment over a relatively long time period (Hopp et al., 2009). This experiment will be helpful to the testing and development of hydrologic and geomorphologic models, as flow networks and corresponding evolution of hydrologic processes over the course of the experiment could be monitored in detail. However, possible artifacts (such as those induced by extensive insertions of large number of sensors into the constructed hillslope) might invoke some uncertainties.

Another desirable way forward is to enhance the use of nondestructive geophysical and remote-sensing tools to monitor in situ dynamic soil moisture and related hydrologic processes across space and time. The emergence of sensor networks and real-time data collection systems, coupled with improved noninvasive geophysical and remote-sensing investigations, could potentially provide increasingly spatially and temporally extensive data sets about the subsurface heterogeneity and related network-like flow dynamics. This could shed better light on the internal architecture of the subsurface and its controls on preferential flow networks. Better sensors and tools and improved techniques for effective and efficient in situ determination of subsurface structures and flow dynamics with high enough spatial and temporal resolutions are urgently needed. The current use of geophysical tools and remote sensing is limited by generally coarse spatial and temporal resolution and the fact that they do not measure properties of interest directly and thus uncertainties exist in inversion algorithms.

Hydropedology can make important contributions to the understanding and modeling of subsurface flow networks, among which are 1) the identification of water-restricting soil layers in different soil types, 2) various soil features potentially causing or reflecting preferential flow, 3) the diagnosis of soil features to answer "why-type" questions in hillslope/watershed hydrology, and 4) the possible links between preferential flow networks and soil genesis and distribution patterns in the landscape. For instance, Lin et al. (2008) have linked subsurface and surface runoff processes to pedologic understanding at the microscopic (macropores and aggregates), mesoscopic (horizons and pedons), and macroscopic (hillslopes and catchments) levels. Such links can and should be quantified and incorporated into hydrologic models to connect the forms and functions of hillslopes in a more realistic manner.

REFERENCES

Akay, O., Fox, G.A., 2007. Experimental investigations of direct connectivity between macropores and subsurface drains during infiltration. Soil Sci. Soc. Am. 71 (5), 1600–1606.

Anderson, A.E., Weiler, M., Alila, Y., Hudson, R.O., 2009. Subsurface flow velocities in a hillslope with lateral preferential flow. Water Resour. Res. 45 (11). 10.1029/2008WR007121.

Anderson, A.E., Weiler, M., Alila, Y., Hudson, R.O., 2010. Dye staining and excavation of a lateral preferential flow network. Hydrol. Earth Syst. Sci. 13 (6), 935–944.

Bergström, S., Forsman, A., 1973. Development of a conceptual deterministic rainfall-runoff model. Nordic Hydrol. 4 (3), 147–170.

Berkowitz, B., Ewing, R.P., 1998. Percolation theory and network modeling applications in soil physics. Surveys Geophys. 19 (1), 23–72.

Beven, K., Freer, J., 2001. A dynamic topmodel. Hydrol. Process. 15, 1993–2011.

Beven, K., Germann, P., 1982. Macropores and water flow in soils. Water Resour. Res. 18 (5), 1311–1325.

Beven, K., Kirkby, M.J., 1979. A physically-based variable contributing area model of basin hydrology. Hydrol. Sci. Bull. 24, 43–69.

Bouma, J., Belmans, C.F.M., Dekker, L.W., 1982. Water infiltration and redistribution in a silt loam subsoil with vertical worm channels. Soil Sci. Soc. Am. J. 46, 917–921.

Brammer, D., 1996. Hillslope Hydrology in a Small Forested Catchment, Maimai, New Zealand. M.S. Thesis, State University of New York College of Environmental Science and Forestry, Syracuse, p. 153.

Buttle, J.M., McDonald, D.J., 2002. Coupled vertical and lateral preferential flow on a forested slope. Water Resour. Res. 38 (5) 10.1029/2001WR000773.

Capowiez, Y., Pierret A., Daniel O., Monestiez, P., Kretzschmar A., 1998. 3D skeleton reconstructions of natural earthworm burrow systems using CAT scan images of soil cores. Biol. Fertil. Soils 27 (1), 51–59

Detty, J.M., McGuire, K.J., 2009. Topographic controls on shallow groundwater dynamics: implications of hydrologic connectivity between hillslopes and riparian zones in a till mantled catchment. Hydrol. Process. 24 (16), 2222–2236.

Deurer, M., Green, S.R., Clothier, B.E., Böttcher, J., Duijnisveld, W.H.M., 2003. Drainage networks in soils. A concept to describe bypass-flow pathways. J. Hydrol. 272 (1–4), 148–162.

Frankenberger, J.R., Brooks, E.S., Walter, M.T., Walter, M.F., Steenhuis, T.S., 1999. A GIS-based variable source area hydrology model. Hydrol. Process. 13 (6), 805–822.

Freer, J., et al., 1997. Topographic controls on subsurface storm flow at the hillslope scale for two hydrologically distinct small catchments. Hydrol. Process. 11 (9), 1347–1352.

Freer, J., et al., 2002. The role of bedrock topography on subsurface storm flow. Water Resour. Res. 38 (12).

Gantzer, C.J., Anderson, S.H., 2002. Computed tomographic measurement of macroporosity in chisel-disk and no-tillage seedbeds. Soil Till Res. 61 (1/2), 101–111.

Gish, T.J., Walthall, C.L., Daughtry, C.S.T., Kung, K.J.S., 2005. Using soil moisture and spatial yield patterns to identify subsurface flow pathways. J. Environ. Qual. 34, 274–286.

Graham, C., Lin, H., 2011. Controls and frequency of preferential flow occurrence: a 175-event analysis. Vadose Zone J. 10 (3), 816–831.

Graham, C.B., McDonnell, J.J., 2010. Hillslope threshold response to rainfall: (2) development and use of a macroscale model. J. Hydrol. doi:10.1016/j.jhydrol.2010.03.008.

Graham, C.B., Woods, R.A., McDonnell, J.J., 2010. Hillslope threshold response to rainfall: (1) a field based forensic approach. J. Hydrol. doi: 10.1016/j.jhydrol.2009.12.015.

Grayson, R., Western, A., 2001. Terrain and the distribution of soil moisture. Hydrol. Process. 15 (13), 2689–2690.

Guebert, M.D., Gardner, T.W., 2001. Macropore flow on a reclaimed surface mine: infiltration and hillslope hydrology. Geomorphology 39, 151–169.

Holden, J., 2004. Hydrological connectivity of soil pipes determined by ground-penetrating radar tracer detection. Earth Surf. Proc. Land. 29 (4), 437–442.

Holden, J., Burt, T.P., 2002. Piping and pipeflow in a deep peat catchment. Catena 48, 163–199.

Holden, J., Burt, T.P., Vilas, M., 2002. Application of ground-penetrating radar to the identification of subsurface piping in blanket peat. Earth Surf. Proc. Land. 27 (3), 235–249.

Hopp, L., et al., 2009. Hillslope hydrology under glass: confronting fundamental questions of soil-water-biota co-evolution at biosphere 2. Hydrol. Earth Syst. Sci. 13, 2105–2118.

Hopp, L., McDonnell, J.J., 2009. Connectivity at the hillslope scale: identifying interactions between storm size, bedrock permeability, slope angle and soil depth. J. Hydrol. 376 (3–4), 378–391.

James, A.L., Roulet, N.T., 2007. Investigating hydrologic connectivity and its association with threshold change in runoff response in a temperate forested watershed. Hydrol. Process. 21 (25), 3391–3408.

James, I., Waine, T., Bradley, R., Taylor, J., Godwin, R., 2003. Determination of soil type boundaries using electromagnetic induction scanning techniques. Biosyst. Eng. 86 (4), 421–430.

Jencso, K.G., et al., 2009. Hydrologic connectivity between landscapes and streams: transferring reach- and plot-scale understanding to the catchment scale. Water Resour. Res. 45 (4), W04428.

Jones, J.A.A., 1988. Modeling pipeflow contributions to stream runoff. Hydrol. Process. 2 (1), 1–17.

Jones, J.A.A., 2004. Implications of natural soil piping for basin management in upland Britain. Land Degrad. Dev. 15 (3), 325–349.

Jones, J.A.A., 2010. Soil piping and catchment response. Hydrol. Process. 24 (12), 1548–1566.

Jones, J.A.A., Cottrell, C., 2007. Long-term changes in stream bank soil pipes and the effects of afforestation. J. Geophys. Res. Earth Surf. 112 (F1).

Jones, J.A.A., 1981. The Nature of Soil Piping: A Review of Research. British Geomorphological Research Group Research Monograph 3. GeoBooks, Norwich, p. 301.

Jones, J.A.A., Connelly, L.J., 2002. A semi-distributed simulation model for natural pipeflow. J. Hydrol. 262 (1–4), 28–49.

Jones, J.A.A., Richardson, J.M., Jacob, H.J., 1997. Factors controlling the distribution of piping in Britain: a reconnaissance. Geomorphology 20 (3–4), 289–306.

Kitahara, H., 1989. Characteristics of pipe flow in a subsurface soil layer of a gentle hillside. (II) hydraulic properties of pipe. J. Jpn Forest Sci. (in Japanese) 71, 317–322.

Kitahara, H., 1993. Characteristics of Pipe Flow in Forested Slopes, vol. 212. IAHS Publication. 235–242.

Lehmann, P., Hinz, C., McGrath, G., Meerveld, H.J.T.-V., McDonnell, J.J., 2007. Rainfall threshold for hillslope outflow: an emergent property of flow pathway connectivity. Hydrol. Earth Syst. Sci. 11 (2), 1047–1063.

Lin, H., 2006. Temporal stability of soil moisture spatial pattern and subsurface preferential flow pathways in the shale hills catchment. Vadose Zone J. 5 (1), 317–340.

Lin, H., 2010. Linking principles of soil formation and flow regimes. J. Hydrol. 393 (1–2), 3–19.

Lin, H., McInnes, K., Wilding, L., Hallmark, C., 1996. Effective porosity and flow rate with infiltration at low tensions into a well-structured subsoil. Trans. ASAE 39 (1), 131–135.

Lin, H., Zhang, J., Andrews, D., Takagi, K., Doolittle, J., 2008. Hydropedologic investigations in the Shale hills catchment. Geochim. Cosmochim. Acta 72 (12) A552–A552.

Lin, H., Zhou, X., 2008. Evidence of subsurface preferential flow using soil hydrologic monitoring in the Shale hills catchment. Eur. J. Soil Sci. 59 (1), 34–49.

Luo, L., Lin, H., Halleck, P., 2007. Quantifying soil structure and preferential flow in intact soil using X-ray computed tomography. Soil Sci. Soc. Am. J. 72 (4), 1058–1069.

McDaniel, P.A., et al., 2008. Linking fragipans, perched water tables, and catchment-scale hydrological processes. Catena 73 (2), 166–173.

McDonnell, J.J., 1990. A rationale for old water discharge through macropores in a steep, humid catchment. Water Resour. Res. 26 (11), 2821–2832.

McGuire, K.J., McDonnell, J.J., Weiler, M.H., 2007. Integrating tracer experiments with modeling to infer water transit times. Adv. Water Resour. 30, 824–837.

Mosley, M.P., 1979. Streamflow generation in a forested watershed. Water Resour. Res. 15, 795–806.

Needelman, B.A., Gburek, W.J., Petersen, G.W., Sharpley, A.N., Kleinman, P.J.A., 2004. Surface runoff along two agricultural hillslopes with contrasting soils. Soil Sci. Soc. Am. J. 68, 914–923.

Newman, B.D., Campbell, A.R., Wilcox, B.P., 1998. Lateral subsurface flow pathways in a semiarid ponderosa pine hillslope. Water Resour. Res. 34 (12), 3485–3496.

Newman, B.D., Wilcox, B.P., Graham, R.C., 2004. Snowmelt-driven macropore flow and soil saturation in a semiarid forest. Hydrol. Process. 18 (5), 1035–1042.

Nieber, J.L., Sidle, R.C., 2010. How do disconnected macropores in sloping soils facilitate preferential flow? Hydrol. Process. 24 (12), 1582–1594.

Noguchi, S., Nik, A.R., Kasran, B., Tani, M., Sammori, T., 1997a. Soil physical properties and preferential flow pathways in tropical rain forest, Bukit Tarek, Pennisular Malaysia. J. Forest Res. 2, 115–120.

Noguchi, S., Tsuboyama, Y., Sidle, R.C., Hosoda, I., 1997b. Spatially distributed morphological characteristics of macropores in forest soils of Hitachi Ohta experimental Watershed, Japan. J. Forest Res. 2, 207–215.

Noguchi, S., Tsuboyama, Y., Sidle, R.C., Hosoda, I., 1999. Morphological characteristics of macropores and the distribution of preferential flow pathways in a forested slope segment. Soil Sci. Soc. Am. J. 63, 1413–1423.

Noguchi, S., Tsuboyama, Y., Sidle, R.C., Hosoda, I., 2001. Subsurface runoff characteristics from a forest hillslope soil profile including macropores, Hitachi Ohta, Japan. Hydrol. Process. 15, 2131–2149.

Nyberg, L., Rodhe, A., Bishop, K., 1999. Water transit times and flow paths from two line injections of super(3)H and super(36)Cl in a microcatchment at Gaardsjoen, Sweden. Hydrol. Process. 13 (11), 1557–1575.

Pierret, A., Capowiez, Y., Moran, C.J., Kretzschmar, A., 1999. X-ray computed tomography to quantify tree rooting spatial distributions. Geoderma 90 (3/4), 307–326.

Pierret, A., Capowiez, Y., Belzunces, L., Moran, CJ., 2002. 3D reconstruction and quantification of macropores using X-ray computed tomography and image analysis. Geoderma 106 (3/4), 247–271.

Peters, D.L., Buttle, J.M., Taylor, C.H., LaZerte, B.D., 1995. Runoff production in a forested, shallow soil, Canadian Shield basin. Water Resour. Res. 31 (5), 1291–1304.

Pringle, C., 2003. What is hydrologic connectivity and why is it ecologically important? Hydrol. Process. 17 (13), 2685–2689.

Robinson, D., et al., 2009. Time-lapse geophysical imaging of soil moisture dynamics in tropical deltaic soils: an aid to interpreting hydrological and geochemical processes. Water Resour. Res. 45.

Saulnier, G.-M., Beven, K., Obled, C., 1997. Digital elevation analysis for distributed hydrological modeling: reducing scale dependence in effective hydraulic conductivity values. Water Resour. Res. 33 (9), 2097–2101.

Seibert, J., McGlynn, B.L., 2007. A new triangular multiple flow direction algorithm for computing upslope areas from gridded digital elevation models. Water Resour. Res. 43.

Sherlock, M., McDonnell, J., 2003. A new tool for hillslope hydrologists: spatially distributed groundwater level and soilwater content measured using electromagnetic induction. Hydrol. Process. 17 (10), 1965–1977.

Sidle, R.C., Kitahara, H., Terajima, T., Nakai, Y., 1995. Experimental studies on the effects of pipeflow on throughflow partitioning. J. Hydrol. 165, 207–219.

Sidle, R.C., Noguchi, S., Tsuboyama, Y., Laursen, Karin, 2001. A conceptual model of preferential flow systems in forested hillslopes: evidence of self-organization. Hydrol. Process. 15 (10), 1675–1692.

Simunek, J., van Genuchten, M.T., Sejna, M., 2008. Development and applications of the hydrus and stanmod software packages and related codes. Vadose Zone J. 7 (2), 587–600.

Sklash, M.G., Stewart, M.K., Pearce, A.J., 1986. Storm runoff generation in humid headwater catchments: 2. A case study of hillslope and low-order stream response. Water Resour. Res. 22 (8), 1273–1282.

Spence, C., Woo, M.-K., 2003. Hydrology of subarctic Canadian shield: soil-filled valleys. J. Hydrol. 279 (1–4), 151–166.

Stieglitz, M., et al., 2003. An approach to understanding hydrologic connectivity on the hillslope and the implications for nutrient transport. Global Biogeochem. Cycles 17 (4), 1105. doi:10.1029/2003GB002041.

Tani, M., 1997. Runoff generation processes estimated from hydrological observations on a steep forested hillslope with a thin soil layer. J. Hydrol. 200, 84–109.

Terajima, T., Sakamoto, T., Nakai, Y., Kitamura, K., 1997. Suspended sediment discharge in subsurface flow from the head hollow of a small forested watershed, northern Japan. Earth Surf. Proc. Land. 22 (11), 987–1000.

Tromp van Meerveld, I., McDonnell, J.J., 2005. Comment to "spatial correlation of soil moisture in small catchments and its relationship to dominant spatial hydrological processes, Journal of Hydrology 286: 113–134". J. Hydrol. 303 (1–4), 307–312.

Tromp van Meerveld, I., Weiler, M., 2008. Hillslope dynamics modeled with increasing complexity. J. Hydrol. 361, 24–40.

Tromp-van Meerveld, H.J., McDonnell, J.J., 2006a. Threshold relations in subsurface stormflow: 1. A 147-storm analysis of the Panola hillslope. Water Resour. Res. 42.

Tromp-van Meerveld, H.J., McDonnell, J.J., 2006b. Threshold relations in subsurface stormflow: 2. The fill and spill hypothesis. Water Resour. Res. 42.

Tromp-van Meerveld, H.J., Peters, N.E., McDonnell, J.J., 2006. Effect of bedrock permeability on subsurface stormflow and the water balance of a trenched hillslope at the Panola mountain research watershed, Georgia, USA. Hydrol. Process. 21, 750–769.

Tsuboyama, Y., Sidle, R.C., Noguchi, S., Hosoda, I., 1994. Flow and solute transport through the soil matrix and macropores of a hillslope segment. Water Resour. Res. 30 (4), 879–890.

Uchida, T., Kosugi, K.I., Mizuyama, T., 2001. Effects of pipeflow on hydrological process and its relation to landslide: a review of pipeflow studies in forested headwater catchments. Hydrol. Process. 15, 2151–2174.

Uchida, T., Kosugi, K.I., Mizuyama, T., 2002. Effects of pipe flow and bedrock groundwater on runoff generation in a steep headwater catchment in Ashiu, central Japan. Water Resour. Res. 38 (7). doi:10.1029/2001WR000261.

Uhlenbrook, S., 2006. Catchment hydrology – a science in which all processes are preferential. Hydrol. Process. 20, 3581–3585.

van Overmeeren, R.A., Sariowan, S.V., Gehrels, J.C., 1997. Ground penetrating radar for determining volumetric soil water content; results of comparative measurements at two test sites. J. Hydrol. 197 (1–4), 316–338.

Ward, A.S., Gooseff, M.N., Singha, K., 2010. Imaging hyporheic zone solute transport using electrical resistivity. Hydrol. Process. 24 (7), 948–953.

Warner, G.S., Nieber, J.L., Moore, I.D., Geise, R.A., 1989. Characterizing macropores in soil by computed tomography. Soil Sci. Soc. Am. J. 53, 653–660.

Weiler, M., 2005. An infiltration model based on flow variability in macropores: development, sensitivity analysis and applications. J. Hydrol. 310 (1–4), 294–315.

Weiler, M., Flühler, H., 2004. Inferring flow types from dye patterns in macroporous soils. Geoderma 120 (1–2), 137–153.

Weiler, M., McDonnell, J.J., 2007. Conceptualizing lateral preferential flow and flow networks and simulating the effects on gauged and ungauged hillslopes. Water Resour. Res. 43.

Western, A.W., et al., 2004. Spatial correlation of soil moisture in small catchments and its relationship to dominant spatial hydrological processes. J. Hydrol. 286 (1–4), 113–134.

Western, A.W., Bloeschl, G., Grayson, R.B., 1998. Geostatistical characterisation of soil moisture patterns in the Tarrawarra catchment. J. Hydrol. 205 (1–2), 20–37.

Western, A.W., Blöschl, G., 1999. On the spatial scaling of soil moisture. J. Hydrol. 217, 203–224.

Western, A.W., Blöschl, G., Grayson, R.B., 2001. Toward capturing hydrologically significant connectivity in spatial patterns. Water Resour. Res. 37 (1), 83–97.

Western, A.W., Grayson, R.B., Blöschl, G., Willgoose, G.R., McMahon, T.A., 1999a. Observed spatial organization of soil moisture and its relation to terrain indices. Water Resour. Res. 35, 797–810.

Western, A.W., Grayson, R.B., Green, T.R., 1999b. The Tarrawarra project: high resolution spatial measurement, modelling and analysis of soil moisture and hydrological response. Hydrol. Process. 13, 633–652.

Whipkey, R.Z., 1965. Subsurface storm flow from forested slopes. Bull. Int. Assoc. Sci. Hydrol. 2, 74–85.

Wilcox, B.P., Newman, B.D., Bres, D., Davenport, D.W., Reid, K., 1997. Runoff from a semiarid ponderosa pine hillslope in New Mexico. Water Resour. Res. 33, 2301–2314.

Woods, R., Rowe, L., 1996. The changing spatial variability of subsurface flow across a hillside. J. Hydrol. (New Zealand) 35 (1), 51–86.

Yoder, R.E., Freeland, R.S., Ammons, J.T., Leonard, L.L., 2001. Mapping agricultural fields with GPR and EMI to identify offsite movement of agrochemicals. J. Appl. Geophys. 47 (3–4), 251–259.

Zhang, J., Lin, H.S., Doolittle, J. Subsurface lateral flow as revealed by combined ground penetrating radar and real-time soil moisture monitoring. Hydrol. Process. Submitted for publication.

Zhu, Q., Lin, H., 2009. Simulation and validation of concentrated subsurface lateral flow paths in an agricultural landscape. Hydrol. Earth Syst. Sci. 13 (8), 1503–1518.

Zhu, Q., Lin, H., Doolittle, J., 2010. Repeated electromagnetic induction surveys for improved soil mapping in an agricultural landscape. Soil Sci. Soc. Am. J. 74 (5), 1763–1774.

Chapter 19

Hydrologic Information in Pedologic Models

A. Samouëlian,[1,2,*] P. Finke,[3] Y. Goddéris[4] and S. Cornu[5]

ABSTRACT

In this chapter, we discuss hydrology as an input in modeling pedogenesis by comparing the advantages and limitations of different modeling approaches. Water is the main driving force of pedogenesis, as it is responsible for the transfer of solutes and particles and for most of the geochemical and biological reactions occurring in soils. It is thus indirectly responsible for soil texture and structure evolution that influences water transfer and residence time in soils. Two example models that combine geochemistry, hydrology, and plant growth modules are presented: SoilGen and WITCH. These two models simulate the evolution of some soil characteristics on the pedologic time scale reasonably well. However, in their present states of development, these models are not able to simulate any type of soil evolution; therefore, further improvements are needed to implement soil processes and the feedbacks among them. Parametrization and validation, on pedologic time scale, also remain a challenge.

1. INTRODUCTION

Modeling soil genesis is a long-term goal in soil sciences that has not been reached despite diverse attempts (Cornu et al., 2008). The first attempts at modeling pedogenesis were qualitative, as classified by Hoosbeek and Bryant (1992), and used two main types of modeling: the work by Dokuchaev and

[1]INRA, UR 0272 - Unité de Science du Sol, Centre de recherche d'Orléans, 2163 Avenue de la Pomme de Pin, CS 40001 Ardon, 45075 Orléans Cedex 2, France

[2]INRA, UMR Lisah – Unité 1221 Laboratoire d'études des Interactions Sol-Agrosystème-Hydrosystème (INRA-IRD-Montpellier Supagro), 2 place Pierre Viala 34060 Montpellier, France (present address)

[3]Dept. of Geology and Soil Science, Ghent University, Krijgslaan, 281, 9000 Ghent, Belgium

[4]LMTG, CNRS Observatoire Midi-Pyrénées, Toulouse, France

[5]INRA, UR119 – Unité de Géochimie des Sols et des Eaux, Europôle de l'Arbois, BP80, 13545 Aix en Provence cedex 4, France

*Corresponding author: Email: Anatja.Samouelian@supagro.inra.fr

Hydropedology, Edited by H. Lin. DOI: 10.1016/B978-0-12-386941-8.00019-6

Hilgard in the late 19th Century, which was cited by Jenny (1961) and Jenny (1941), was based on a functional–factorial approach of pedogenesis; and the energy model, which was proposed by Runge (1973), which was a more mechanistic model that considered, gravity to be as a source of energy. More recent works propose quantitative approaches, as classified by Hoosbeek and Byant (1992), of modeling pedogenesis (e.g. a mass balance model that is based on the solid phase of the soil, stochastic models, and models combining geochemistry, hydrology, and plant growth modules).

Hoosbeek and Byant (1992) provide a classification of pedological models that is based on their degree of computation (i.e. qualitative versus quantitative), their ability to consider time as a variable (i.e. capacity versus rate models), and the scale they are working on (i.e. infra pedon, pedon, and larger than the pedon). Samouëlian and Cornu (2008) proposed to separate these different attempts to model soil into two categories: the approaches that were mainly based on the solid phases of the soil that do not explicitly take into account water transfers and the approaches that were derived from other scientific fields that combine geochemistry, hydrology, and plant growth modules. These approaches differ in their purposes; some are used to quantify past processes and to interpret the soil spatial distribution, and the others are designed to simulate soil formation and predict soil evolution.

Currently, modeling pedogenesis is of increasing interest for soil-protection purposes (Hoosbeek and Bryant, 1992; Sommer et al., 2008; Lin, 2011). Because water governs most soil-forming processes via the transfer of solute and matter into the soil, transformation of soil constituents, and notably plant nutrition, we conclude, with Hoosbeek and Bryant (1992) that "*Water and solute transport and solution chemistry models are two key categories of sub-models for pedological models* ..." Therefore, in this chapter, we will emphasize pedogenesis modeling that is based on modular approaches and includes a hydrology module. Nevertheless, we briefly summarize the first approach that demonstrated the necessity of explicitly considering water transfer in soil into soil models to predict, through time, soil evolution that results from global change. To conclude, we will give two examples of models for pedogenesis from the current literature (Finke and Hutson, 2008; Godérris et al., 2006).

Hoosbeek and Bryant (1992) also stress the importance of the spatial scale at which the different models work. To predict soil evolution, a spatial scale ranging from the pedon to the landscape is most appropriate. Therefore, we will focus on models that are defined at this scale because Heuvelink and Pebesma (1999) stated that a "*model should be run at the support for which it was developed*" and added "*note that this is true for both the spatial and the temporal support of the model inputs.*" Therefore, much effort has been provided here to define the spatial and temporal scales for the described models.

In addition, while dealing with pedogenesis and hydrology, the time scale is of prime importance because a hydrologic module works on an event time scale while pedogenesis works on a tenth of years to millennia scale. The concept of soil functioning and pedogenesis that was extensively defined by Targulian and Krasilnikov (2007) indicated that while hydrology modules represent soil functioning at a date, *t*, pedogenesis represents the integration of all tiny changes that are induced by functioning along a period of several years or more. The sum of these tiny changes induces a significant change in the structure and other properties of soil, as was demonstrated by Montagne et al. (2008). This change, in turn, influences the hydrologic functioning of the soil. Therefore, using a modular approach for modeling pedogenesis and including a hydrologic module implies that interactions and feedbacks between soil phases and water transfer occur. This point remains a scientific challenge, because until now, only a few attempts, have been made to quantify these interactions. This idea will be developed later in this chapter.

2. BRIEF OVERVIEW OF MODELS THAT DO NOT EXPLICITLY CONSIDER WATER TRANSFER

Models that are based on the organization and distribution of solid phases have been developed for predicting and understanding the formation and spatial distribution of soils. These models can be subdivided into four main types (Samouëlian and Cornu, 2008): (i) the factorial approach that was derived from Jenny (1961), (ii) mass balance calculations, (iii) the energy model, and (iv) soil-formation functions. While the first two types of models are functional (as defined by Hoosbeek and Bryant, 1992), the last two are mechanical models.

Based on the factorial equation that was first proposed by Jenny (1941, 1961), statistical and geostatistical approaches were developed to link a soil profile (or soil properties) to climate (or climatic attributes), organisms (mainly land cover), topography, parent material, age or time, and space or geographic position (McBratney et al., 2003; Gray et al., 2009). The objectives of these works were to propose an efficient tool that predicted soil properties or soil types to aid in soil mapping by using the available data that were derived from digital-elevation models and spectral-reflectance bands from satellite imagery (McBratney et al., 2003). An additional objective was to reinterpret the available soil and climate databases (Gray et al., 2009).

In parallel, Phillips (1993, 1996) reinterpreted Jenny's (1941) approach to be a nonlinear dynamic system that explained the short-scale variability in soil, which could not be explained by the measured variations of the factors for soil formation but were related to slight variations during pedogenesis that were amplified with time. He demonstrated that local soil variations could lead to significant spatial variability in soil and showed that pedogenesis could be modeled as a nonlinear dynamic, or deterministic–chaotic system (Phillips,

1993, 2008; Ryzhova, 1996; Minasny and McBratney, 1999; D'Odorico, 2000; Furbish and Fagherazzi, 2001).

Mass-balance modeling allows for quantification of the transfer of elements and matter at the pedon scale during pedogenesis, and it has been widely used for quantification in geomorphology and weathering studies (e.g. Brimhall et al., 1985, 1988, 1991; Merritts et al., 1992; Chiquet et al., 2000; Montagne et al., 2008; Quénard et al., 2011). Gains or losses of elements from various soils are determined by comparing their stocks in the different soil horizons to those of the parent material. Interpretations of mass-balance modeling explain soil formation and quantify pedological processes that are deduced from elementary mass transfer.

The energy model is based on a concept that was first developed by Runge (1973), which indicates that soils are developed according to the amount of climatic energy that they receive and water flow that transits through the pedon. This approach was mainly conceptual, at first. However, Rasmussen et al. (2005, 2007) quantified the amounts of energy that were put into soil due to precipitation and net primary production, which gave a global quantity of energy that was received by the soil per year. Interestingly, this group was able to link the developmental stage of soil to this quantity of energy (Rasmussen et al., 2007); however, this model is static and only gives global information at the assumed pedon scale.

Soil-production functions are based on a more detailed mechanistic approach (e.g. Legros, 1984; Minasny and McBratney, 2001) and simulate the evolution of a soil system from parent material. These models are developed at the landscape scale (2D to 3D) or at the profile scale (1D) and are based on continuity equations. Soil-production functions described initially rock-fractioning mechanisms (i.e. physical weathering of parent material) (Dietrich et al., 1995; Heimsath et al., 1997; Minasny and McBratney, 1999) but more recently, chemical weathering (Humphreys and Wilkinson, 2007; Furbish and Fagherazzi, 2001; Minasny and McBratney, 2006; Brantley et al., 2008) and bioturbation (Salvador-Blanes et al., 2007; Yoo and Mudd, 2008) were also introduced into the soil-production functions.

Nevertheless, the impact of water on chemical, physical, and biological weathering is generally not taken into account in these modeling approaches. However, an exception is the model that was proposed by Saco et al. (2006), who modified the soil-production function by introducing a wetness index. By doing so, their model overcame an inconsistency that was found in previously published soil-production functions (Heimsath et al., 2001) and demonstrated that taking hydrology into account improved the approach. However, the impact of a change in hydrology due to climate change cannot be predicted from these types of models because only one compartment of the ecosystem is considered (i.e. the solid soil phases). Through climatic change and human activities, the water cycle is modified, which causes soil evolution (Montagne and Cornu, 2010) because water transfer is one of the key processes for soil evolution. Therefore, to predict the impact of climate change and/or human activities on soil evolution, modular-modeling approaches that contain a hydrologic module must be developed.

3. MODELS TO PREDICT SOIL EVOLUTION DUE TO CLIMATIC CHANGE, OR LAND USE

Because soil results from the interaction of biological (i.e. microbiological reactions, biogeochemical recycling, and bioturbation), physical (i.e. weathering, water- or wind-erosion, and the subsurface transfer of water, solutes, and particles), and chemical processes (i.e. weathering, complexation, precipitation/dissolution, redox reactions, ion exchange, and organic matter decay), we must consider these different types of processes when modeling soil evolution and simulating scenarios (Fig. 1). This may be achieved by coupling different modules that represent these different types of processes. Such modules exist for most of the listed processes in the literature, and different modules exist for a same soil process. However, coupling of modules was rarely used to model pedogenesis, but some exceptions exist: SoilGen by Finke and Hutson (2008); WITCH by Goddéris et al. (2006); and Time Split by Sommer et al. (2008). Some attempts in coupling transfer with geochemical processes (e.g. MIN3P, Mayer et al., 2002, and HP1, Jacques and Simunek, 2005) or plant module with hydrology (Ren et al., 2008) do exist in

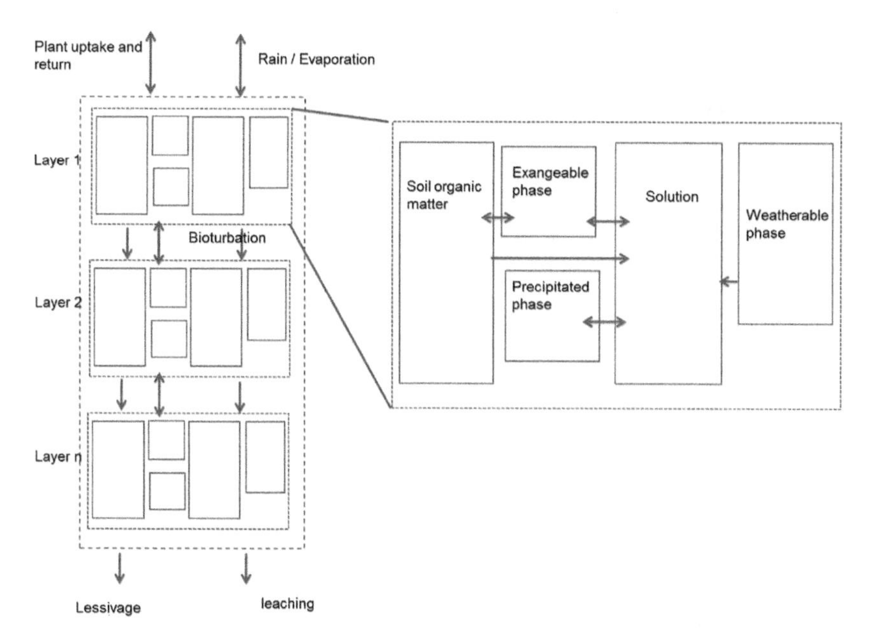

FIGURE 1 Pedogenesis modeling as a result of physical, chemical, and biological interactions.

the literature. However, these models were generally designed for environmental purposes (i.e. for soil and ecosystem functioning); therefore, feedbacks of soil functioning on soil solid phases (i.e. pedogenesis) are generally not completely described by these models. In the following section, we will discuss the advantages and limitations of the different types of modules for each type of process that exists in the literature.

3.1. Transfer Module

3.1.1. Water Transfer

Water transfer in soil occurs under variably saturated conditions because soils are part of the unsaturated zone. Models that predict water transfer in saturated media are beyond the scope of this chapter, despite the fact that soil can experience some degree of saturation in winter. Because other chapters in this book describe water transfer models in greater detail, we will only summarize the main features and discuss the advantages and limitations of the different approaches in terms of modeling pedogenesis.

The simplest way to model water transfer in soil is to apply a bucket model (e.g. ASPECTS, Rasse et al., 2001; CASA, Potter et al., 1993). In this model, water transfer is simulated using simple mathematical equations that consider the soil system to be a series of buckets that are filled sequentially. An advantage of this model is its relative ease of application, because it only requires a few parameters and simple initial conditions (Collon, 2003; Chahinian, 2004). This type of model also requires less calculation time than a mechanistic module, which is an advantage when modeling at pedogenic time scales.

Mechanistic approaches that were based on physical laws were also developed: HYDRUS (Simunek et al., 1992) or LEACHM (Hutson, 2003) were based on Richards' equation, which was mainly used for water transfer in the soil matrix, and mobile–immobile water models (e.g. MACRO, Jarvis, 1994), which focused on preferential transfers in soil macropores. Further implementations, which are also based on physical laws, permit the modeling of heat, CO_2, solute or, more recently, particle transfer. However, when considering pedogenesis, the solid phases of soil and their arrangements change significantly over time and therefore exhibit hydraulic properties evolution. Although it is well recognized that soil hydraulic properties evolve in space, their temporal changes are rarely taken into account and are, therefore, poorly known. In most models, soil hydraulic properties are considered to be constant parameters during the simulation and little experimental data exist on the temporal variability of soil properties, due to agricultural practices, in the surface horizon and at the annual scale.

When considered in models, the evolution of hydraulic properties is estimated via pedotransfer functions that were developed for example by Wösten et al., (2001) and are based in the best case on the properties of soil texture, density, and organic matter (Finke and Huston, 2008; Sommer et al., 2008).

3.1.2. Solute and Particle Transfer

Solute transfer is generally simulated by a convection–diffusion model (Feyen et al., 1998 and Simunek et al., 2003). When a part of the water is considered to be immobile, as in mobile–immobile models, solute transfer between these domains is simulated by a diffusion term, where, two supplementary parameters must be determined: the proportion of immobile water and the transfer coefficient from the immobile to the mobile water phase (Vanderborght et al., 1997).

To our knowledge, only particle transfers occurring through preferential transfers in macropores were simulated by an advection–dispersion equation (DeNovio et al., 2004; Simunek et al., 2006) or a convection–diffusion equation (for solutes), as in KDWP (Rousseau, 2003) or in MACRO (McGechan et al., 2002). More recently, Michel et al. (2010) developed a conceptual model for particle mobilization in macroporous soils that were based on the physical concept of pore drying, which induced the mobilization of a particle via deferential capillary stresses (Lehmann et al., 2008; Metzger and Tsotsas, 2005). These models consider attachment/detachment, film straining, and filtration processes but not the action of the physico-chemical processes that act on particle mobilization and transfer. In addition, these models do not simulate the change in soil structure that is induced by particle transfer because such a change has a feedback effect on the water transfer and, therefore, on particle transfer. This feedback is of particular importance when modeling long-term evolution as pedogenesis. This implies the development of soil structure module to predict hydraulic and transport parameters. This point remains a challenge for researchers in hydropedology.

3.2. Geochemical Modules

Several thermo-kinetic models exist in the literature, e.g. PHREEQC (Parkhurst and Appelo, 1999), Minteq (Gustafsson, 2001), CHESS (van der Lee, 1998), KINDIS (Madé et al., 1994), LEACHM (Hutson, 2003), WITCH (Goddéris et al., 2006), EQ3/6 (Wolery, 1992), and GEOCHEM (Parker et al., 1995). These models mainly simulate dissolution and precipitation of minerals, acido-basic reactions, oxydo-reduction, complexation, and ionic change. The various models simulate geochemical reactions with organic matter of soil differently. For example, a comparative study (Dudal and Gérard, 2004) demonstrated that the WHAM (Tipping and Hurley, 1992) and NICA-Donnan (Keiser and van Riemsdijk, 2002) models simulated these reactions better than the EQ3/6 (Wolery, 1992), GEOCHEM (Parker et al., 1995), MinteqA2 (Allison and Brown, 1995), and PHREEQC (Parkhurst and Appelo, 1999) models. These models also differ by their representation of the solid–solution interface and of the database of chemical parameters that they utilize.

Geochemical models were first used to simulate results that were obtained in lab experiments and to manage and protect the environment (e.g. to determine the chemistry of groundwater, acidification, pollutant transfer, management of radionucleides waste). For these particular purposes, a speciation simulation was developed. For pedogenesis, dissolution and precipitation of minerals are of greater importance and a few examples of such applications exist in the literature: weathering of granite as simulated by KINDISP (Probst et al., 2000); reverse modeling of surface water chemistry using PHREEQC (Lecomte et al., 2005); and weathering in a soil profile (Maher et al., 2009; Gérard et al., 2002). In these models, dissolution and precipitation are simulated by kinetic laws; however, in most cases, models do not allow for the formation of secondary minerals or solid solutions unless those minerals were introduced in trace amounts into the system. Therefore, these models do not allow for the appearance of a mineral that was not first introduced into the model. To overcome this limitation, modules of nucleation or crystalline growth useful in simulating nonequilibrium systems, were developed (e.g. the NANOKIN model (Noguera et al., 2006)). In these cases, the new mineral phase precipitates from a given level of oversaturation of the solution with respect to the solid phase and the feedback reactions between solution and mineral govern the evolution of the solution composition and, subsequently, the rate of mineral growth in the system.

To simulate geochemical processes in soils that are open systems, geochemical models must be coupled with a water-transfer module, as is the case for MIN3P (Mayer et al., 2002) or HP1 (PHREEQC + Hydrus1D; Jacques and Simunek, 2005). For example, MIN3P was used to simulate biogeochemical cycling of silicon in a temperate forest soil (Gérard et al., 2008), while HP1 was used to simulate interactions between major cations and heavy metals and the solid phase of a temperate climate Spodosol over a period of 30 years (Jacques et al., 2008). For pedogenesis, a link between the temporal evolution of the flow path and the residence time of the soil solution within the soil matrix is needed, which implies that the interactions and feedbacks should be integrated into a modeling framework. This was generally developed into a model that was dedicated to pedogenesis (see below) but for a restricted number of mineral–water interactions. In this case, the quantity of minerals is updated at each time step. One of the main questions that remain unanswered is the estimation of the mineral surface that interacts with water is a key parameter for geochemical modules and is highly dependent on the water pathway.

3.3. Module of Biology: Plants, Organic Carbon, and Bioturbation

3.3.1. Plant Module

In pedologic models, the plant module must simulate water and nutrient uptake; it also has to simulate the uptake of major elements from the soil (i.e. Si, Fe, and Al) and their restitution to the soil via litter and the root system. Because vegetation cover may vary over pedological time, plant models that simulate different types of cover (e.g. grass, forest, and agriculture), such as those used in the SoilGen (Finke and Hutson, 2008), Sheels (Ren et al., 2008), or Casa models (Potter et al., 1993), will be better adapted than plant modules that are specific for crop production, such as STICS (Brisson et al., 1998).

Both SoilGen and Sheels include a macroscopic description of the root water uptake because of the root shape factor, root distribution pattern, and moisture content that constrains transpiration (for SoilGen: or a soil-moisture constraint on transpiration and the evaporation rate for Sheels). SoilGen also simulates cation uptake by plants via a forcing function and simulates the return of organic matter, i.e. root and leaf litter, to the soil and its subsequent decomposition (Finke and Hutson, 2008).

Ren et al. (2008) compared two modeling approaches to estimate the net primary production of plant biomass. Modeling of the hydrologic cycle differs significantly between the two approaches. The Casa model (Potter et al., 1993) is a simple, bucket hydrology model, whereas the Sheels model (Ren et al., 2008) is a multilayer, 1D model that is based on the Richards equation. The result of using these models is that the estimation of net primary production is significantly improved when the hydrologic module is not oversimplified.

3.3.2. Organic Carbon Module

A large store of literature exists that compares soil carbon decomposition models (Smith et al., 1997; Zhang et al., 2008; McGill et al., 1996). Among the best known are ROTHC (Coleman and Jenkinson, 1995), CANDY (Franko et al., 1995), DAISY (Hansen et al., 1991), CENTURY (Parton, 1996), PATCIS (Fang and Moncrieff, 1999), NCSOIL (Molina, 1996), DNDC (Li et al., 1992), SOMM (Chertov and Komarov, 1996), ITE (Thornley and Verberne, 1989), and VERBERNE (Verberne, 1992). We will summarize the main features of these models.

All of the models are in the zero-dimension (i.e. without depth discretization), process-oriented, and multipool, and all were developed to model the biodegradability of the organic carbon of soil for a large range of ecosystems. RothC, CANDY, CENTURY, NCSOIL, and VERBERNE are applicable to all land uses; however, DNDC and DAISY were developed for grassland and arable land, and SOMM and ITE were developed for forest ecosystems. An important difference between these models is the number of pools they allow for and the interactions that are defined among the pools. Decomposition of the C pools is generally described by the law of first-order kinetics, meaning the different pools vary by their decomposition rate constants. Soil temperature and moisture are the most important variables for the decomposition rate (Singh and Gupta, 1977) and functions linking the decomposition rate to temperature and moisture have been defined in the literature. The studies by Bauer et al. (2008) have shown that the sensitivity of six soil carbon modules (CANDY, CENTURY, DAISY, PATCIS, ROTHC, and SOILCO2) to the soil temperature function was six to seven times higher than to the soil moisture function. This group concluded that the choice of the soil temperature- and moisture-reduction functions is a crucial factor for a reliable simulation of carbon turnover; however, no agreement currently exists on which of the different functions are best for simulation. Therefore, most models were developed to validate specific sites and data sets and could not be used for simulation of carbon turnover at any one site.

While soil moisture or water potential is taken into account in most C models, these variables are not sufficient to simulate the soil biology that governs the fate of C in soils. Other hydrologic characteristics, such as the soil structure, the proportion of the pore space filled with water, and the connectivity of the pore space, play a key role on the soil biology. Therefore, to be able to predict the impact of global change on the dynamics of soil carbon decomposition, carbon organic models must be coupled with other modules of the ecosystem such as soil water, nitrogen models, or plant growth (Smith et al., 1997; Bauer et al., 2008). However, a major obstacle in predicting future evolution is model calibration.

The prediction of SOM turnover at the global scale requires models that perform well under various environmental conditions (Bauer et al., 2008). Therefore, the interaction between multi-pool, SOM models and physical-based models that describe accurate soil transfer are indispensable. The coupled SOILCO2-ROTHC model (Herbst et al., 2008; Bauer et al., 2008) combined these two modeling approaches to improve modeling of carbon dynamics in soils. SoilGen also contains interactions between CO_2 production by SOM turnover, its transport in the soil, and its relation to inorganic carbon (i.e. calcite).

3.3.3. Module of Bioturbation

Until currently bioturbation was poorly modeled, because of the lack of quantification of this process. Using carbon isotopes as tracers, Elzein and Balesdent (1995) quantified bioturbation as a diffusive term and they determined the kinetics of this process in the studied soils. Salvador-Blanes et al. (2007) introduced bioturbation into the soil-production function and they used a sigmoid function that depended on the thickness of the soil. More recently, Jarvis et al. (2010) presented a general model that incorporated the effects of 'local' and 'nonlocal' biological mixing into the framework of the standard, physical (advective–dispersive) transport model. This model was successfully tested against measurements of the redistribution of caesium 137 (^{137}Cs) that were derived from the Chernobyl accident. Finke and Hutson (2008) used an incomplete mixing module to simulate bioturbation in SoilGen. In most cases, these modules must be calibrated on the site that is being considered.

3.3.4. Existing Attempts at Coupling Modules for Modeling Pedogenesis

Some attempts at coupling modules that could be used for modeling pedogenesis are described in the literature (SoilGen by Finke and Hutson, 2008; WITCH by Goddéris et al., 2006; Time Split by Sommer et al., 2008; MIN3P by Mayer et al., 2002; HP1 by Jacques and Simunek, 2005; and SHEELS by Ren et al., 2008), and in all of these studies a water-transfer module was coupled with a geochemical module with different degrees of detail. However, until now, simulations performed with MIN3P or HP1 focused on a short time evolution (a few years) and did not deal with feedback effects. The Time Split model was used to model soil-landscape development of Colluvic Regosol and included erosion, sedimentation, carbonate geochemistry, silicate weathering, and the clay-translocation process. Feedback effects due to dynamic changes in colluviums, the soil-landscape surface, and the subsurface structure were taken into account through the evolution of the distribution of hydraulic properties. This evolution occurred after a certain mass had been transported/translocated or a certain change in thickness of the soil horizons was reached (Sommer et al., 2008). SoilGen (Finke and Hutson, 2008), WITCH (Godérris et al., 2006), and SHEELS (Ren et al., 2008) were coupled with a plant module. SHEELS (Ren et al., 2008) is a multilayered, hydrologic model with a mechanistic description of the soil water fluxes that is coupled with a plant module. The plant module can simulate variation of the vegetation cover and land use at a seasonal time step. To our knowledge, simulations done with this model were only focused on soil carbon fluxes on an annual scale. SoilGen (Finke and Hutson, 2008) includes a physical transfer module (e.g. water, solutes, and particle) that is coupled with an organic matter module for different vegetation types, and WITCH (Goddéris et al., 2006) includes a detailed geochemical module that is coupled with a bucket hydrology model and a simplified plant module. In the following section, the SoilGen and WITCH models, which have been applied to pedogenesis, are presented in detail.

4. THE PEDOGENESIS MODEL SOILGEN

4.1. Model Framework

SoilGen (Finke and Hutson, 2008; Finke, 2011) was developed to simulate soil formation in parent materials of varying texture and mineralogy, possibly calcareous or gypsiferous, and in a range of climates that ranged from udic, ustic, xeric, and aridic soil moisture regimes and cryic, frigid, mesic, and thermic soil temperature regimes. The model cannot simulate podzolization and peat formation and is not intended for volcanic parent materials. Basically, the model is an enhanced, 1-D solute-flow model, where the horizontal area that is considered (i.e. spatial grain) is that of the soil pedon (near 1 m^2) and the vertical grain is a user input. Often, compartments that are near a 50-mm thickness are used with a user-input profile depth of between 1 and 3 m. Currently, the temporal extent has been a maximum of 15,000 years, and the time steps that are considered vary between 10^{-10} and 0.1 day. Because the model operates on large temporal extents (1000s of years), the soil properties of OC, texture, CEC, and calcite and gypsum content cannot be considered constant. This discrepancy has drawn attention toward the completeness of simulated processes and aptitude in models to deal with climate change, sedimentation, and erosion events. For an overview of the linkage between processes and time scales whereby processes interact, a reference is made to Finke and Hutson (2008: Fig. 1) and Finke (2011: Fig. 1).

4.2. Governing Laws or Equations

4.2.1. Transfer Module

The water, solute, and heat flow routines in SoilGen are based on the LEACHM model (Hutson, 2003), which solves the Richards' equation for transient vertical flow:

$$\frac{\partial h}{\partial t}C(\theta) = \frac{\partial}{\partial z}\left[K(\theta)\frac{\partial H}{\partial z}\right] - U(z,t), \tag{1}$$

where $C(\theta)$ is the differential water capacity $\partial\theta/\partial h$, θ is the volumetric water content (m^3/m^3), h is soil water pressure head (mm), $K(\theta)$ is hydraulic conductivity (mm/d), H is hydraulic head (mm), and $U(z,t)$ is a sink term that represents water that is lost at a depth, z and time, t by transpiration. The modified Campbell equations (Campbell et al., 1977; Hutson, 2003) provide the values for C and K from values of θ or h. Alternatively, the Van Genuchten equations (van Genuchten, 1980) can be used. Furthermore, calculated conductivities are strongly reduced in the case of frozen soil compartments.

Heat flow and temperature distribution are modeled by the following equation:

$$\frac{\partial T}{\partial t} = \frac{\partial}{\partial z}\left(\frac{K_t(\theta)}{\beta}\cdot\frac{\partial T}{\partial z}\right), \tag{2}$$

where T is the temperature (°C), $K_t(\theta)$ is the thermal conductivity ($J\ m^{-1}\ s^{-1}\ {}^{\circ}C^{-1}$), which is calculated at θ using the method presented by Wierenga et al. (1969), and β is the volumetric heat capacity.

The flow of soluble matter is simulated using a finite difference approximation to the convection–dispersion equation (CDE) for each soluble compound:

$$\frac{\partial(\theta C)}{\partial t} = \frac{\partial}{\partial z}[\theta D(\theta,q) - qC] \pm \Phi, \tag{3}$$

where C is the solute concentration (mg dm^{-3}), $D(\theta,q)$ is the apparent diffusion coefficient (mm^2 d^{-1}), q is the water flux (mm d^{-1}), and Φ is a sink term (mg dm^{-3} d^{-1}) that represents the plant uptake or release by mineralization of organic matter. Solute concentrations depend on chemical equilibria, weathering rates of primary minerals, decomposition rates of organic matter, and the partitioning between the exchange and solute phases.

The flow of CO_2 is simulated by an explicit numerical solution to the gas regime equation:

$$\varepsilon \cdot \frac{\partial c}{\partial t} = D(T)_{gs} \cdot \frac{\partial^2}{\partial z^2} + P(z,t), \tag{4}$$

where ε is the air-filled porosity, c is the CO_2 concentration (partial pressure) in the air of the soil, $P(z,t)$ is the CO_2 production, and $D(T)_{gs}$ is the gas diffusion coefficient in the soil (m^2 s^{-1}).

4.2.2. Biogeochemistry Module

SoilGen allocates cations and anions over five phases (i.e. the organic, exchange, precipitated, solution, and unweathered phases, Fig. 2). Ion transfer between the organic phase and the solution phase is simulated with a decomposition routine that is based on the RothC model (Jenkinson and Coleman, 1994), which describes the C cycle. Plant ions are obtained via the transpiration stream, which uses a forcing function that is dependent on the type of

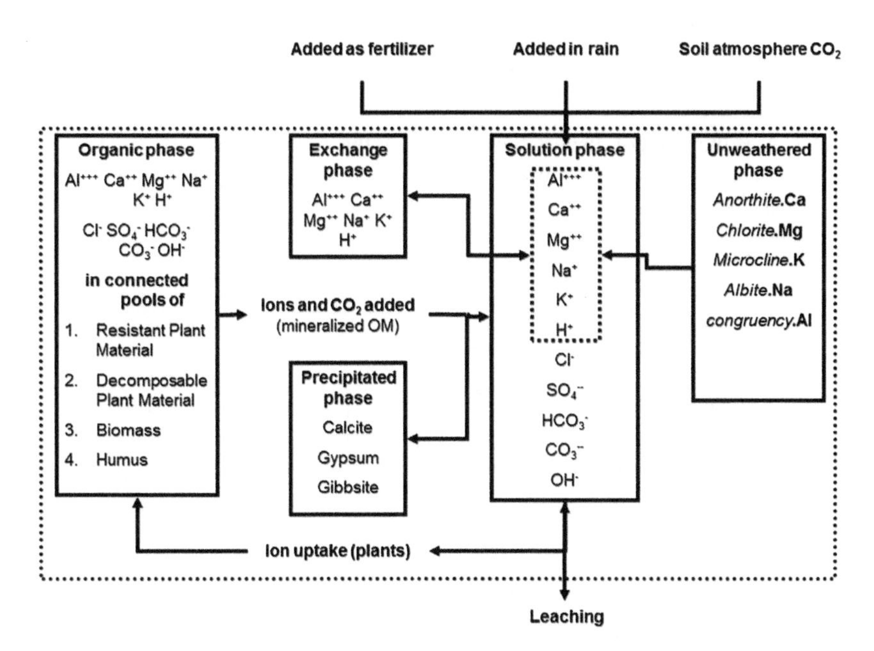

FIGURE 2 Ions and chemical phases in SoilGen.

vegetation (e.g. grass/scrubland, deciduous or coniferous forest, or agriculture). Chemical equilibria between the solution and precipitated phases are modified by the temperature-dependent Arrhenius correction and the Gapon relation is applied to describe partitioning between the exchange and solution phases. The transfer of cations between unweathered primary minerals and the solution phase is modeled as a first-order, pH-dependent process (van Grinsven, 1988), and soil CEC is variable in depth and time and is a function of OC and clay content (Foth and Ellis, 1996).

4.2.3. Other Processes

SoilGen simulates physical weathering by splitting particles where the splitting probability is a function of the hourly temperature gradient, d*T*/d*t* (Fig. 3):

$$P_s = \begin{cases} P_{s,\max} & \text{if } \dfrac{dT}{dt} > B \\ \dfrac{P_{s,\max} * \dfrac{dT}{dt}}{B} & \text{if } \dfrac{dT}{dt} \leq B \end{cases}, \quad (5)$$

where, $P_{s,\max}$ is the maximal split probability and B is a threshold temperature gradient over d*t*, where P_s becomes maximal. This physical weathering produces new material in the clay fraction.

The mobilization of clay particles is modeled to be a function of the impact of raindrops on the soil–air interface. The mass balance of dispersible particles at the surface layer is given by the following equation:

$$\frac{dA_s}{dt} = -D + P, \quad (6)$$

where, A_s is the mass of dispersible particles at the soil surface (g m^{-2}), D is the splash detachment rate (g m^{-2} h^{-1}), and P is the replenishment rate (g m^{-2} h^{-1}).

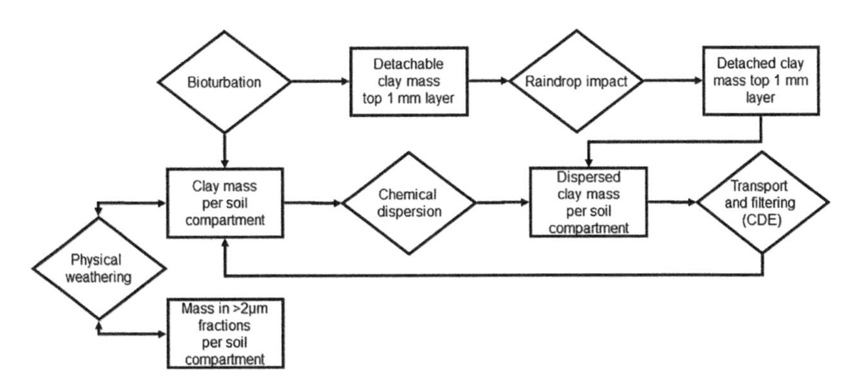

FIGURE 3 Interaction between clay migration, bioturbation, and physical weathering processes in SoilGen.

Additionally, clay mobilization is modeled as a dispersion/flocculation process, where the fraction of clay in a transportable, dispersed state, fDC, in every soil compartment is calculated by the following equation:

$$fDC = \{1 - (SC/CSC)\} * \theta_{macro} * fVC, \quad (7)$$

where, SC is the total electrolyte concentration (mmolc dm^{-3} water) that is calculated by the model per time step, CSC (mmolc dm^{-3} water) is the critical salt concentration at which soil clay mixtures remain flocculated, which is a function of clay mineralogy, pH, and SAR, θ_{macro} is the volume fraction of the soil that is occupied by macropores, and fVC is the fraction of the soil volume that is taken up by clay. Transport of dispersed clay is calculated by using the CDE and an additional sink term for flow straining.

Bioturbation is modeled as an incomplete mixing process that replenishes mobilized clay at the soil surface.

4.3. Input Data, Parameters

4.3.1. Initial Soil Conditions, Parameters and Pedotransfer Functions

The parent material for soil formation must be specified for each soil compartment by texture, bulk density, OC, calcite and gypsum content, and cations that are stored in the primary minerals. Such data may be derived from the C-horizons in the current profile if it is assumed that the current C-horizon is the initial state of the overlying soil horizons at the start of soil formation. Additionally, the initial chemical composition of solute and the exchange phase must be specified, and the initial soil temperature profile must be supplied. These data served as a starting point for the simulations that began during a previous year before the present and ended during the present year, when the current, observable soil was made.

The Gapon exchange coefficients are either inputs or calculated by the model using solute and exchange phase data. The parameters for clay migration and physical weathering that are obtained by calibration (Finke, 2011) are default but can be modified by the user. These values are assumed to be constant during the simulations, but physical characteristics of soil (h–θ–K) are updated annually using an internal pedotransfer function that is based on soil texture, OC, and bulk density as input (either Rawls and Brakensiek, 1985; or Wösten et al., 1999).

4.3.2. Boundary Conditions (Climate, Vegetation, and Bioturbation)

Climate and its associated vegetation development must be entered as a time series of annual precipitation, potential evapotranspiration, January and July temperatures, vegetation type, and annual litter input on or in the soil by leaves or roots. Optionally, a time series of atmospheric partial CO_2 pressures can be entered, but bioturbation (in kg 1000 ha^{-1} y^{-1} for the entire soil profile and as a depth-distribution function) must be specified. At the lower boundary, free drainage is assumed if a precipitation surplus occurs, or zero drainage is assumed if a drainage deficit occurs. By using the user interface, the depths of the perched water tables can also be set.

4.4. Model Limitations

SoilGen does not simulate water flow in structured media, but an estimation of flow in macropores is made for the calculation of clay migration. The major consequence of this choice is that stronger leaching in preferential pathways, such as ripening cracks or albeluvic tongues, is not simulated and, therefore leaching of the soil matrix is slightly overestimated (Sauer et al., 2011; Finke, 2011).

The current version of the model uses vegetation, litter C input, and bioturbation as boundary conditions, and thus, possible feedback mechanisms between the soil condition and vegetation or faunal activity are not represented adequately. Despite the paramount importance of the bioturbation process in soil formation, the amount and depth of bioturbation as a function of soil condition are still poorly documented.

A final limitation of this model is its runtime, which can reach several weeks of simulation time for soil development in deep soils over thousands of years. This is a serious drawback in model calibration.

4.5. Case Studies

4.5.1. Loess Parent Materials

SoilGen has been applied to loess soils in Belgium and Hungary using climate evolutions that began in the late glacial era (15,000 BP). A clear effect of differences in climate, when not including evolution, emerged and was confirmed by measurements (Finke and Hutson, 2008). Furthermore, a clear effect of the bioturbation process on the decalcification of topsoils was shown and bioturbation was shown to be an indicator of clay migration. Recently (Finke, 2011), the SoilGen model was applied to a toposequence in Belgian loess, which revealed a strong dependency on the slope gradient and slope aspect on (among others) decalcification, change of base saturation, and clay migration. This dependency is caused by differences in effective precipitation and evapotranspiration and is a function of local slope gradient and slope aspect. Bioturbation strongly affected the process of clay migration; and after calibration of the effective calcite dissolution constant, clay migration parameters, and physical weathering, the simulation results matched well with the measured data (i.e. Gowers' $d = 0.269$, where $0 =$ perfect similarity and $1 =$ total dissimilarity of measurements and simulations, as determined over 12 soil parameters; Finke, 2011). The model results were compared to the measured depth profiles of OC, Clay%, Silt%, Sand%, calcite%, CEC, Base Saturation, pH and exchangeable Ca, Mg, K, and Na. Recently, climosequences in the Chinese loess plateau were simulated with the purpose of calibrating the C module for semi-natural forests, and explorative studies were performed to simulate the genesis of paleosoil sequences in the Chinese loess.

The overall conclusion from these simulation studies was that it is most important to obtain the correct hydrology in terms of the boundary conditions of precipitation, potential evapotranspiration, and interception evaporation. Furthermore, after correction for interception, evaporation was performed and the calcite dissolution coefficient was calibrated accurately (Finke, 2011); the accuracy of simulated decalcification depths was still found to rely strongly on precipitation infiltrating the soil.

4.5.2. Marine Clay Parent Materials

Recently, SoilGen was applied to two chronosequences in Marine terraces in Norway that had evolved as a consequence of isostatic uplift after melting of the Scandinavian icecap. Dated sites ranged from 2100 BP to 11,050 BP. After the calibration of clay-migration parameters and physical weathering, the simulations compared well to the measurements (Gowers' $d=0.300$; Sauer et al., 2011), and the quality of the simulation varied between soil ages (i.e. the accuracy of the simulations for young (i.e. 2000–4000 years) soils was lower than that of old (i.e. 9000–11,000 years) soils). Given the textural variability in the parent materials, it was concluded that a correct initial profile and its associated h–θ–K relations is important to reach a good-quality simulation. Furthermore, an accurate description of soil hydrology, which includes an accurate estimate of interception evaporation, was determined to be important (Sauer et al., in review). Additionally, the occurrence and amount of preferential flow were expected to explain deviations between simulations and measurements in clay soils (Sauer et al., 2011).

4.6. SoilGen Summary

SoilGen is able to simulate the evolution of a large number of soil properties that are not constant at pedogenetic time scales. Nevertheless, case studies show that simulation quality can be improved by the following: (i) better process descriptions; (ii) taking into account heterogeneity within the pedon scale, which can lead to preferential flow; (iii) improving estimates of the initial situation; and (iv) improving the quality of boundary inputs – mainly hydrologic boundary conditions along the timeline and the role of (changing) vegetation. Importantly, a subset of these improvements is at the interface of hydropedology and biogeochemistry.

5. THE WITCH WEATHERING MODEL

5.1. Modeling Framework

The modeling method for the WITCH weathering model is based on two main hypotheses. First, the vegetal cover must be accounted for, not only from a geochemic point of view, but mostly because land plants control the hydrologic behavior of watersheds and weathering profiles. Second, experiment-derived kinetic rate laws can be used to describe mineral dissolution/precipitation in the natural environment. According to these two hypotheses, the model structure is a cascade of two models: (i) a model describing the productivity of the ground vegetation and the associated water fluxes and soil contents, which feed into (ii) the core of the WITCH model (Fig. 4), which

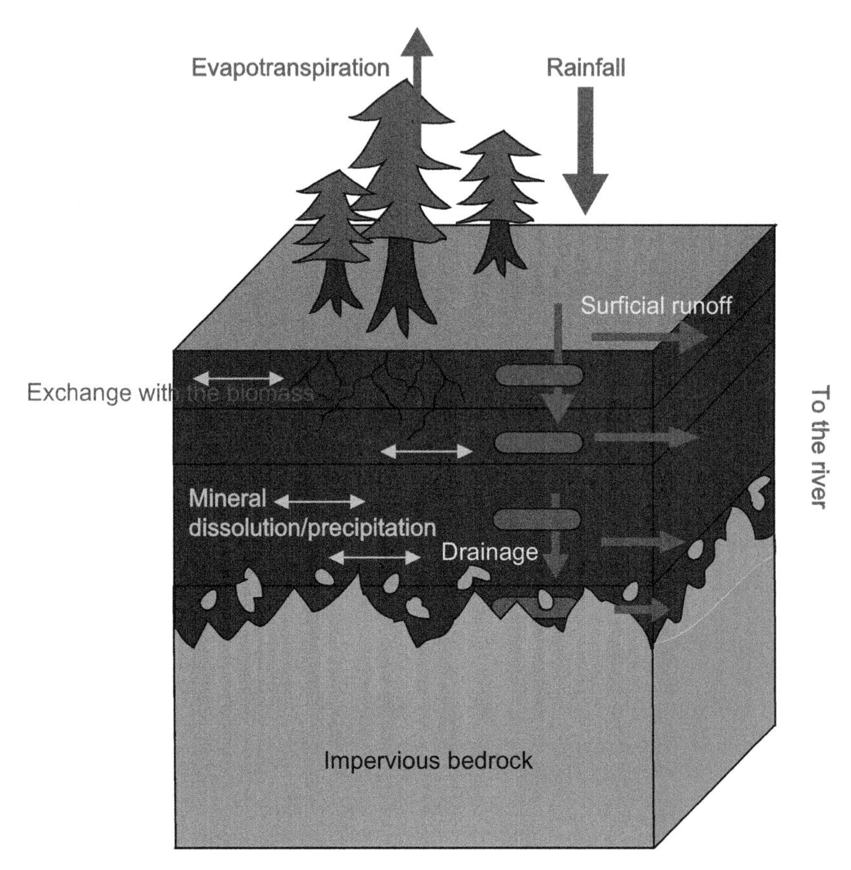

FIGURE 4 General description of the WITCH model. WITCH is a box model that calculates the chemical composition of the soil water and water fluxes, from the surface down to the impervious bedrock. The model equations describe the water–rock interactions and the exchange of major elements with the dead and living biomass. *(Color version online and in color plate)*

describes the dissolution/precipitation of minerals using experiment-derived kinetic rate laws. The output of the model cascade is the chemical composition of the water fluxes, including deep drainage and superficial runoff, which both feed into rivers, and the composition of the weathering profile solutions. Therefore, the WITCH model calculates the main cation production through weathering and the simultaneous atmospheric CO_2 consumption rate.

5.2. Driving Laws or Equations

5.2.1. Transfer Module

WITCH is a box model, whereby each box can represent the water content of a given soil horizon or any other uniform, well-mixed, unit of the watershed. For each of the water reservoirs, a set of mass balance equations are solved for the concentration of any given species, C_j, at each time step of the calculation:

$$\frac{\mathrm{d}(A \cdot z \cdot \theta \cdot C_j)}{\mathrm{d}t} = F_{win} - F_{wout} + \sum_{i=1}^{n_m} F^i_{\mathrm{weath},\,j} - \sum_{i=1}^{n_m} F^i_{\mathrm{prec},\,j} \pm R_j, \qquad (8)$$

where, A is the surface area of the considered reservoir, z is its thickness, and θ is its volumetric water content (i.e. the volume of water versus the total volume of the reservoir). F_{win} is the input flux inside the reservoir by water transport; and F_{wout} is the flux of elements being removed by water that is flowing out of the reservoir. $F^i_{\mathrm{weath},\,j}$ is the flux of elements being added by mineral dissolution, i, inside the box, the sum of which extends over the n_m minerals, while $F^i_{\mathrm{prec},\,j}$ is the removal of elements through mineral precipitation. Finally, R_j accounts for the exchange fluxes between the living and dead biomass and the aqueous solution or between the exchange complex and the aqueous solution.

A mass balance equation is also solved at each time step for the number of moles, Q_i, of each mineral, i, in each layer:

$$\frac{\mathrm{d}Q_i}{\mathrm{d}t} = F^i_{\mathrm{weath}} - F^i_{\mathrm{prec}}. \qquad (9)$$

This mass balance equation, which allows for the calculation of the time evolution of the mineralogic composition of the weathering profile, is only solved for long-term simulations (e.g. 10^3 years time scale). When the model is run at the decadal scale, Eqn (9) is set to zero in each layer so that the mineralogical composition remains constant. In that case, only the time evolution of the chemical composition of the soil water content and water fluxes are calculated.

WITCH does not include a hydrologic module, but $F_{w\ \mathrm{in}}$ and $F_{w\ \mathrm{out}}$ in Eqn (8) are linear functions of the water inflow and outflow of the reservoir. These fluxes, together with the water content of each reservoir, θ, are calculated offline by a hydrologic model or by the hydrologic module of a biospheric model. Consequently, the feedback between mineralogical evolution of a weathering profile, its porosity, and the water flow are not accounted for in the present version of the WITCH model.

In its published versions, and depending on study sites, the underground hydrology is calculated by using the following models:

1) The hydrologic sub-model of ASPECTS (Rasse et al., 2001), a numerical model that simulates the carbon productivity and water cycle in temperate forests (Goddéris et al., 2006);
2) A lumped, hydrologic box model that is designed for semi-arid tropical environments (Violette et al., 2010);
3) The hydrologic sub-model of the global biospheric model, LPJ (Sitch et al., 2003; Roelandt et al., 2010; Goddéris et al., 2009; Beaulieu et al., 2010);
4) The surface sub-model of an atmospheric, general circulation model that is coupled to a global vegetation model (Kaplan et al., 2003; Thomson and Pollard, 1997; Goddéris et al., 2010).

The modeled objects included the Strengbach small watershed (80 ha) in northeast France, the tropical watershed of Mule Hole located in South India (4.2 km^2), and large-scale continental watersheds such as the Orinoco or Mississippi rivers.

5.2.2. Biogeochemistry Module

5.2.2.1. Soil Geochemistry

Once the mass balance for the soil solutions has been solved, a chemical speciation is performed in each reservoir, which allows for the calculation of the concentrations of 16 chemical species (i.e. H_2CO_3, HCO_3^-, CO_3^{2-}, Al^{3+}, $AlOH^{2+}$, $Al(OH)_2^+$, $Al(OH)_4^-$, H_4SiO_4, $H_3SiO_4^-$, $H_2SiO_4^{2-}$, PO_4^{3-}, HPO_4^{2-}, $H_2PO_4^-$, H_3PO_4, and the organic ligands RCOOH and $RCOO^-$). Equilibrium constants at 25 °C and the enthalpies of reaction are given in Table 1 and pH is calculated by solving the charge balance equation in each reservoir.

The dissolution, $F^i_{\text{weath},\,j}$, and precipitation, $F^i_{\text{prec},\,j}$, terms for silicate minerals are calculated in each reservoir using kinetic laws and parameters that are derived from the transition state theory (TST; Eyring, 1935) and laboratory experiments. Within the framework of the TST, the rate of an elementary reaction is equal to the product of two terms: the concentration of the activated complex and the frequency with which these complexes cross the energy barrier that separates the reactants from the products. Assuming that the activated complexes can be formed by parallel reactions with H^+, OH^-, water and organic ligands, the overall dissolution rate of a mineral can be expressed by the following equation (Schott et al., 2009):

$$F_g = A'_g\left[\sum_l k_{l,g}\cdot\exp\left(\frac{-E^l_{a,g}}{R\cdot T}\right)\cdot a_l^{n_{l,g}}\cdot f_{inh}\right]\cdot\left(1-\Omega_g^{1/s}\right), \qquad (10)$$

TABLE 1 Equilibrium Constants and Enthalpies of Reactions Included in WITCH. The symbol ~ is Used Either When No Dependence on Temperatures is Calculated or When The Temperature Effect is Directly Included in the Equilibrium Constant. The Dependence of the K_{eq} on pH for Organic Ligands is a Parametric Equation from the SAFE Model of Sverdrup and Warfinge (1995)

Reaction	pK_{eq} (25 °C)	Enthalpy of reaction (kJ/mol)
$Al^{3+} + H_2O = AlOH^{2+} + H^+$	5	48.070
$Al^{3+} + 2H_2O = Al(OH)_2^+ + 2H^+$	10.1	125.1
$Al^{3+} + 4H_2O = Al(OH)_4^- + 4H^+$	22.7	177.0
$H_4SiO_4 = H^+ + H_3SiO_4^-$	9.83	~
$H_3SiO_4^- = H^+ + H_2SiO_4^{2-}$	13.17	~
$HPO_4^{2-} = H^+ + PO_4^{3-}$	11.77	~
$H_2PO_4^- = H^+ + HPO_4^{2-}$	7.2	~
$H_3PO_4 = H^+ + H_2PO_4^-$	2.12	~
$RCOOH + H_2O = RCOO^- + H_3O^+$	$0.96 + 0.9pH - 0.039pH^2$	~
Reaction	**pK_{eq} (as a function of temperature T in K)**	
$H_2CO_3 = HCO_3^- + H^+$	$3404/T + 0.0328T - 14.844$	~
$HCO_3^- = CO_3^{2-} + H^+$	$2902/T + 0.0238T - 6.498$	~
$H_2O = H^+ + OH^-$	$4471/T + 0.0171T - 6.075$	~
H_2CO_3 = kHenry PCO_2 (atm)	$-2386/T - 0.0153T + 14.844$	~

where, F_g is the overall dissolution rate of a mineral, g, inside a given layer. The summation accounts for the four, parallel, rate-controlling, elementary reactions that are assumed to describe the dissolution that is promoted by H^+, OH^-, H_2O and organic ligands for feldspars. In this relation, a_l and $n_{l,g}$ stand for the activity of the first species and the reaction order with respect to the first species, respectively. For organic ligands, a_l equals the activity of a generic organic species, $RCOO^-$, which is the conjugate of RCOOH. $k_{l,g}$ is the rate constant of the mineral, g, is the dissolution reaction that is promoted by species l, $E^l_{a,g}$ is the activation energy of this reaction, and f_{inh} stands for the inhibitory effects (i.e. by aqueous Al). The last factor in Eqn (4) describes the effect on the rate of departure from equilibrium where Ω_g is the solution saturation state with respect to mineral g ($\Omega_g = Q/K_g$, where Q is the activity quotient and K_g is the equilibrium constant for the mineral g dissociation reaction), and s, i.e. the Temkin's coefficient number, is the stoichiometric number of moles of activated complex that is formed from 1 mol of the mineral. Precipitation terms are calculated in the same manner, but in this case Ω_g is >1. All thermodynamic and kinetic parameters can be found in Tables 2 and 3. The total reactive surface is calculated for each reservoir from the grain-size fraction through the following phenomenological law (Sverdrup and Warfinge, 1995):

$$A' = (8.0x_{clay} + 2.2x_{silt} + 0.3x_{sand}) \cdot \rho, \quad (11)$$

where, x_{clay}, x_{silt}, and x_{sand} are the textural fractions of the soil clay, silt, and sand materials, respectively, with $x_{clay} + x_{silt} + x_{sand} = 1$ and ρ is the density of the considered layer in g m^{-3}. The total mineral surface is then distributed between the various minerals in each reservoir according to their respective volumetric abundances. When the time evolution of the mineralogical composition is calculated for long-term simulations, the reactive surface of each mineral, A'_g, is further made to be linearly dependent on its molar abundance.

Carbonate mineral dissolution/precipitation is included in WITCH and modeled within the framework of the transition state theory and surface coordination chemistry concepts (Pokrovsky et al., 1999; Pokrovsky and Schott, 1999, 2001). The calcite dissolution rate (in mol/m^2 of reactive surface/s) is described in WITCH by the following equation:

$$F_{cal} = \left(k_H^{cal} \cdot a_H + \frac{K_o}{1.10^{-5} + a_{CO_3}}\right) \cdot \left(1 - \Omega_{cal}^{1.0}\right), \quad (12)$$

where, k_H^{cal} equals $10^{-0.659}$ mol/m^2/s and k_0 equals 10^{-11} mol/m^2/s at 25 °C (Wollast, 1990) and the activation energies for the rate constants are set to 8.5 and 30 kJ/mol, respectively (Alkatan et al., 1998; Pokrovsky et al., 2009).

The dolomite dissolution rate equals:

$$F_{dol} = \left[k_H^{dol} \cdot a_H^{0.75} + k_{Mg}^{dol} \cdot \left(\frac{1.575 \cdot 10^{-9}}{1.575 \cdot 10^{-9} + 3.5 \cdot 10^{-5} + a_{CO_3} \cdot a_{ca}}\right)\right] \times \left(1 - \Omega_{dol}^{1.9}\right), \quad (13)$$

TABLE 2 Equilibrium Constants at 25 °C and Enthalpies of Reaction for Dissolution of Minerals Mentioned in Published Version of the WITCH Model. Unless Otherwise Specified, Data for Minerals and Aqueous Species are from SUPCRT (SPEQ03.DAT) Except for Al^{3+} (Castet et al., 1993; Wesolowski and Palmer, 1994) and H_4SiO_4 (Whalter and Helgeson, 1977)

Mineral	Dissolution reaction	pK_{eq}	ΔH_R^0(kJ/mol)
Anorthite[a]	$CaAl_2Si_2O_8 + 8H^+ \rightarrow Ca^{2+} + 2Al^{3+} + 2H_4SiO_4$	−24.95	−306.0
Albite	$NaAlSi_3O_8 + 4H_2O + H^+ \rightarrow Na^+ + Al^{3+} + 3H_4SiO_4$	−2.29	−73.82
Orthoclase	$KAlSi_3O_8 + 4H_2O + 4H^+ \rightarrow K^+ + Al^{3+} + 3H_4SiO_4$	0.022	−49.93
Quartz	$SiO_2+2H_2O \rightarrow H_4SiO_4$	3.98	25.06
Muscovite[b]	$KAl_3Si_3O_{10}(OH)_2 + 10H^+ \rightarrow K^+ + 3Al^{3+} + 3H_4SiO_4$	−12.70	−248.4
Biotite[c]	$KMg_{1.5}Fe_{1.5}AlSi3O_{10}(OH)_2 + 10H^+ \rightarrow K^+ + 1.5Mg^{2+} + 1.5Fe^{2+} + Al^{3+} + 3H_4SiO_4$	−32.87	−55.6
Kaolinite[b]	$Al_2Si_2O_5(OH)_4 + 6H^+ \rightarrow 2Al^{3+} + 2H_4SiO_4 + 2H_2O$	−7.43	−147.7
Halloysite[b]	$Al_2Si_2O_5(OH)_4 + 6H^+ \rightarrow 2Al^{3+} + 2H_4SiO_4 + 2H_2O$	−12.5	−166.6
Gibbsite[b]	$Al(OH)_3 + 3H^+ \rightarrow Al^{3+} + 3H_2O$	−7.74	−105.3
Gibbsite[b] (amorphous)	$Al(OH)_3 + 3H^+ \rightarrow Al^{3+} + 3H_2O$	−10.8	−110.9
Apatite[d]	$Ca_{10}(PO_4)_6F_2 + 12H^+ \rightarrow 10Ca^{2+} + 6H_2PO_4^- + 2F^-$	−49.99	−110
Ca-Smectite	$Si_4O_{10}(OH)_2Mg_{0.33}Al_{1.67}Ca_{0.165} + 6H^+ + 4H_2O \rightarrow 4H_4SiO_4 + 1.67Al^{3+} + 0.33Mg^{2+} + 0.165Ca^{2+}$		−81.65
Mg-Smectite	$Si_4O_{10}(OH)_2Mg_{0.33}Al_{1.67}Mg_{0.165} + 6H^+ + 4H_2O \rightarrow 4H_4SiO_4 + 1.67Al^{3+} + 0.50Mg^{2+}$		−86.44
Illite[e]	$Si_{3.4}O_{10}(OH)_2Mg_{0.02}Al_{2.38}Ca_{0.01}K_{0.44} + 7.64H^+ + 2H_2O \rightarrow 3.4H_4SiO_4 + 2.38Al^{3+} + 0.02Mg^{2+} + 0.01Ca^{2+} + 0.44K^+$	−6.82	−146.31

(Continued)

TABLE 2 Equilibrium Constants at 25 °C and Enthalpies of Reaction for Dissolution of Minerals Mentioned in Published Version of the WITCH Model. Unless Otherwise Specified, Data for Minerals and Aqueous Species are from SUPCRT (SPEQ03.DAT) Except for Al^{3+} (Castet et al., 1993; Wesolowski and Palmer, 1994) and H_4SiO_4 (Whalter and Helgeson, 1977)—cont'd

Mineral	Dissolution reaction	pK_{eq}	ΔH_R^0(kJ/mol)
Forsterite[b]	$Mg_2SiO_4 + 4H^+ \rightarrow 2Mg^{2+} + H_4SiO_4$	−28.31	−203.25
Andesine	$Na_{0.7}Ca_{0.3}Al_{1.3}Si_{2.708}O_8 + 2.8H_2O + 5.2H^+ \rightarrow 0.7Na^+ + 0.3Ca^{2+} + 1.3Al^{3+} + 2.7H_4SiO_4$	0.7 pK_{eq} (albite) + 0.3 pK_{eq} (anorthite)	
Labradorite	$Na_{0.4}Ca_{0.6}Al_{1.6}Si_{2.4}O_8 + 1.6H_2O + 6.4H^+ \rightarrow 0.4Na^+ + 0.6Ca^{2+} + 1.6Al^{3+} + 2.4H_4SiO_4$	0.4 pK_{eq} (albite) + 0.6 pK_{eq} (anorthite)	
Diopside	$CaMgSi_2O_6 + 4H^+ + 2H_2O \rightarrow Ca^{2+} + Mg^{2+} + 2H_4SiO_4$	~	~
Basaltic glass[b]	$SiAl_{0.33}Fe_{0.17}Ca_{0.23}Mg_{0.26}Na_{0.11}K_{0.01}O_{3.3}+2.6H^+ + 0.7H_2O \rightarrow 0.33Al^{3+} + 0.17Fe^{3+} + 0.23Ca^{2+} + 0.26Mg^{2+} + 0.11Na^+ + 0.01K^+ + H_4SiO_4$	2.71	14
Chlorite	$Mg_5Al_2Si_3O_{10}(OH)_8 + 16H^+ \rightarrow 5Mg^{2+} + 2Al^{3+} + 3H_4SiO_4 + 6H_2O$	−68.51	−644.5

[a] *Arnorsson and Stefansson (1999).*
[b] *Drever (1997).*
[c] *Biotite Gibbs free energy calculated in this study assuming a regular solid solution between annite and phlogopite (Fe atomic fraction in M1 sites of 0.9 and W=9 kJ, THERMOCALC) and end-member thermodynamic data from SUPCRT.*
[d] *Guidry and Mackenzie (2003).*
[e] *Illite Gibbs free energy was calculated based on the electronegativity scale method (Vieillard, 2000).*

TABLE 3 Kinetic Dissolution Constants at 25 °C ($mol/m^2/s$) and Activation Energy (kJ/mol) of the Dissolution Reactions Promoted by H^+, OH^-, Water (w), and Organic Ligands (L), as Used in Published Version of the WITCH Model. Also Shown is the Reaction Order with Respect to H^+ and OH^- Promoted Dissolution (n_H, n_{OH}). Symbol ~ Stands for Either "No Effect" or "No Value"

Mineral	pk_H E_aH (kJ/mol) n_H	pk_H E_aOH (kJ/mol) n_{OH}	pk_w E_aw (kJ/mol)	pk_L E_aL (kJ/mol)
Anorthite[a]	5 1 70	9.6 0.3 50	11.5 70	12.6 63.3
Albite[a]	9.50 60 0.5	9.95 50 0.3	12.60 67	12.96 59
Orthoclase[b]	9.65 60 0.5	10.70 50 0.3	12.85 67	~ ~
Quartz[c]	~ ~ ~	11.00 85 0.25	13.40 85	~ ~
Muscovite[b]	12.20 22 0.17	11.71 22 0.16	~ ~	~ ~
Biotite[b]	10.88 35 0.32	~ ~ ~	14.20 35	~ ~
Kaolinite[b]	12.45 50 0.38	10.74 40 0.73	14.43 55	~ ~
Gibbsite[d]	10.28 60 0.33	14.13 80 0.3	~ ~	~ ~
Apatite[e]	5.08 34.7 0.87	~ ~ ~	~ ~	~ ~
All smectite[f]	9.8 48 0.38	~ ~ ~	13.9 55	12.1 48.3

(Continued)

TABLE 3 Kinetic Dissolution Constants at 25 °C (mol/m^2/s) and Activation Energy (kJ/mol) of the Dissolution Reactions Promoted by H^+, OH^-, Water (w), and Organic Ligands (L), as Used in Published Version of the WITCH Model. Also Shown is the Reaction Order with Respect to H^+ and OH^- Promoted Dissolution (n_H, n_{OH}). Symbol ~ Stands for Either "No Effect" or "No Value"—cont'd

Mineral	pk_H E_aH (kJ/mol) n_H	pk_H E_aOH (kJ/mol) n_{OH}	pk_w E_aw (kJ/mol)	pk_L E_aL (kJ/mol)
Illite[g]	11.7	12.3	15	12.3
	46	67	14	48.3
	0.6	0.6		
Forsterite[h]	6.6	~	11	11.5
	75	~	75	75
	0.5	~		
Andesine[a]	12.40	9.8	12.4	12.96
	63	50	68	63.3
	0.5	0.3		
Labradorite[a]	8.28	9.7	11.5	12.6
	65	50	68	61
	0.7	0.3		
Diopside[i]	9.85	~	~	~
	42	~	~	~
	0.14	~		
Basaltic glass	6.70	9.45	10.35	~
	30	50	50	~
	0.5	0.175		
Chlorite	10.02	11.80	13.5	~
	40	40	40	~
	0.5	0.5		

[a] *fitting within the framework of Eqn (3) of data reported by Blum and Stillings (1995).*
[b] *fitting of data reported by Nagy (1995).*
[c] *Dove (1994).*
[d] *Mogollon et al. (1996).*
[e] *fitting of data reported by Guidry and Mackenzie (2003) and Chairat (2005).*
[f] *Holmqvist (2001).*
[g] *Köhler et al. (2003).*
[h] *Pokrovsky and Schott (2000).*
[i] *Schott et al. (1981).*

where, a_H, a_{CO_3}, and a_{ca} stand for the H^+, CO_3^{2-} and Ca^{2+} activities, k_H^{dol} equals 10^{-3} mol/m^2/s, and k_{Mg}^{dol} equals $10^{-8.2}$ mol/m^2/s at 25 °C. The activation energies are set to 40 and 60 kJ/mol, respectively (Pokrovsky et al., 1999; Pokrovsky and Schott, 1999, 2001).

5.2.2.2. Exchange with the Biomass

In some of the case studies that were performed with WITCH (Godérris et al., 2006, 2009), element exchange with the living (i.e. the uptake by roots) and dead (i.e. the release by decay) biomass is accounted for in a simple way. Uptake of elements (e.g. the main base cation, phosphorus) by land plants is set to be proportional to the gross primary productivity of carbon, which is calculated by the biospheric model and weighted by the measured base cation/C and P/C ratios in the living trees on the site, when data are available. Release of these elements is proportional to the release of carbon through the oxidation of organic matter, which is calculated using the biospheric model, within each reservoir (Goddéris et al., 2006; Roelandt et al., 2010).

5.2.2.3. Carbon

The WITCH model does not currently include a carbon budget equation. Therefore, the partial pressure of gaseous CO_2 is described in each reservoir as the production of underground CO_2, CO_2^{pr}, in the root zone (i.e. root respiration + heterotrophic respiration) is calculated by using the biospheric productivity model, and the gaseous soil CO_2 pressure is calculated by solving the diffusion equation for CO_2 (Van Bavel, 1951; Gwiazda and Broecker, 1994; Goddéris et al., 2010). The maximum CO_2 level at the basis of the root zone is given by the following equation:

$$PCO_2^{max} = PCO_2^{atm} + \frac{5.7 \cdot 10^{-7} \cdot CO_2^{pr} \cdot (root_{depth})^2}{2 \cdot \phi \cdot D_{CO_2}^{s}}, \quad (14)$$

where, $root_{depth}$ is the thickness of the root zone, ϕ is the porosity, $D_{CO_2}^{s}$ is the CO_2 diffusion coefficient in soil, and PCO_2^{atm} is the atmospheric CO_2 partial pressure. Once the water reservoir pH is calculated at each time step, a full carbonate speciation is solved, which allows for the estimation of the concentrations of H_2CO_3, HCO_3^-, and CO_3^{2-}.

The diffusion coefficient, $D_{CO_2}^{s}$, depends on the temperature of the study site, T, and the porosity, ϕ, and tortuosity, τ, are fixed to field values when measured, or to standard values if no data are available (Hillel, 1998):

$$\frac{D_{CO_2}^{s}}{D_{CO_2}^{a} \cdot \left[\frac{T}{273.15}\right]^2} = \phi \cdot \tau \cdot \quad (15)$$

$D_{CO_2}^{a}$ is the CO_2 diffusion coefficient in air and is set to 0.139 cm^2/s.

5.3. Input Data

5.3.1. Initial Soil Conditions, and Parameters

The key parameters of the WITCH model are the kinetics and thermodynamic constants that are required to evaluate the dissolution/precipitation rate of the minerals and the saturation state of the soil solutions with respect to the mineral phases. These parameters are summarized in Tables 2 and 3 and are fixed based on laboratory experiments. Other parameters that are required for performing the simulations are obtain from field studies and include the following: the grain size of the minerals that are in contact with the soil solutions, the porosity and tortuosity of the various model layers, the density of the materials, the cation exchange capacity, and the mineralogical composition of each layer. For long-term simulations (i.e. integration over several thousand years), this mineralogical composition is the initial composition of the material and is inferred from field studies. For short-term runs (i.e. several years and up to centuries), the mineralogical composition is assumed to be constant.

5.3.2. Boundary Condition

Boundary conditions include the following: the time evolution of volumetric water content, of the water fluxes between the reservoirs of the model, of the layer CO_2 levels, and of the exchange with biologic reservoirs. As explained above, all of these boundary conditions are calculated by using biospheric models.

5.4. Model Limitations

There are several major limitations to WITCH. First, there is no nucleation module for mineral phases, which means that if WITCH is calculating a chemical oversaturation for a given secondary mineral phases, that chemical will not precipitate unless the mineral is assumed to be present a priori. This limitation could cause the soil solutions to become unrealistically oversaturated with respect to a given mineral phase over the course of a simulation and could lead to errors in the dissolution rates of other mineral phases, which are assumed to be present, because the chemical composition and pH of the soil solution might be misevaluated. Generally, the solution to this problem is to assume that secondary mineral phases are not observed in the field because they are in very small amounts; therefore, a reactive surface can be calculated for these phases. In this way, precipitation can occur in cases where the soil solutions are oversaturated, which is particularly critical during dry episodes when the soil water content is drastically reduced.

Another limitation to this model is the absence of a description of the silica cycle in vegetation; only base cations and phosphorus are exchanged with the biomass. Therefore, future development of the model should include the role of phytoliths and amorphous phases, and efforts toward the integration of the role of the myccorhyzosphere in this model should be undertaken (Taylor et al., 2011; Bonneville et al., 2009).

5.5. Case Studies

5.5.1. Temperate Small Watershed

The first study that was performed with WITCH was the modeling of the weathering processes at play in a small, granitic watershed that was located in

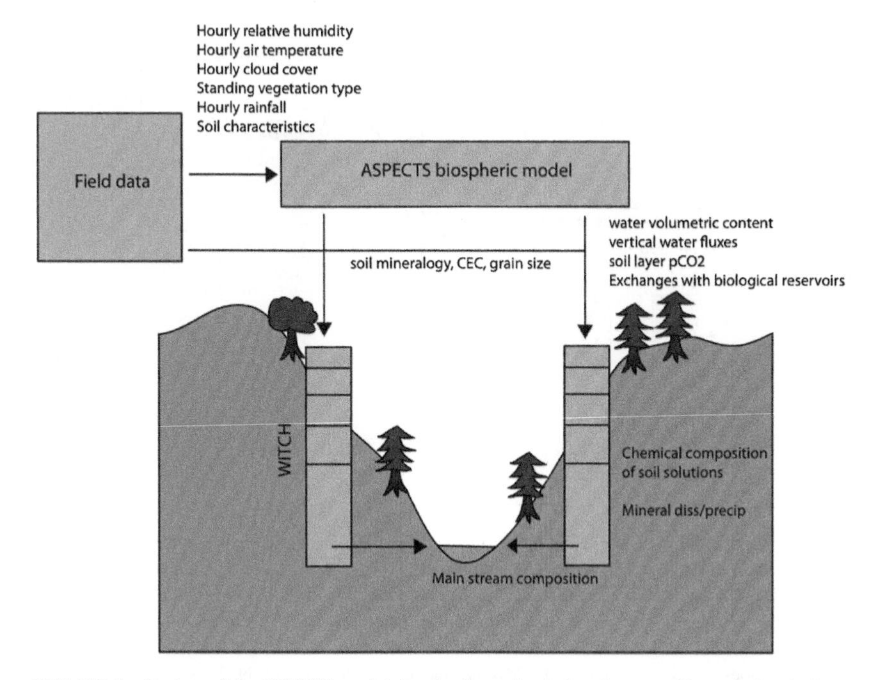

FIGURE 5 Design of the WITCH model for the Strengbach (northeastern France) simulations. WITCH is fed by the numerical model ASPECTS which describes the water and carbon cycle in a temperate forest. Only vertical fluxes are simulated on each slope of the watershed, and they feed the main stream. *(Color version online)*

a temperate environment (Strengbach catchment, Vosges mountains, France, 80 ha under forest) (Goddéris et al., 2006). In this area, the monthly average temperatures fluctuate between −2 °C and 14 °C, the annual rainfall is 1400 mm/yr, and the total runoff is around 850 mm/yr. The design of the modeling study is summarized in Fig. 5, and an example of the results is given in Fig. 6a. The ASPECTS model was used as the biospheric productivity model (Rasse et al., 2001). Two weathering profiles were modeled: one below the spruce trees, which covered 80% of the watershed surface, and one below the beech trees, which comprised the remaining 20% of the watershed surface. These two weathering profiles were assumed to be representative of the behavior of the weathering processes for all profiles below their respective spruce and beech trees. The up-scaling at the catchment scale was performed by determining the sum of the contribution from both calculated profiles, which was weighted by the area that was covered by the respective spruce and beech trees. The simulations were performed on a seasonal basis and as was extensively described by Goddéris et al. (2006), the model was able to reproduce the seasonal evolution of the chemical composition of the soil solutions and the mean annual export flux of the main base cations and silica. It was found that trace mineral dissolution (apatite) was responsible for up to 70% of the Ca^{2+} export at the watershed scale. Modeling results were highly sensitive to the solubility product of the mineral phases of clay mineral, in this

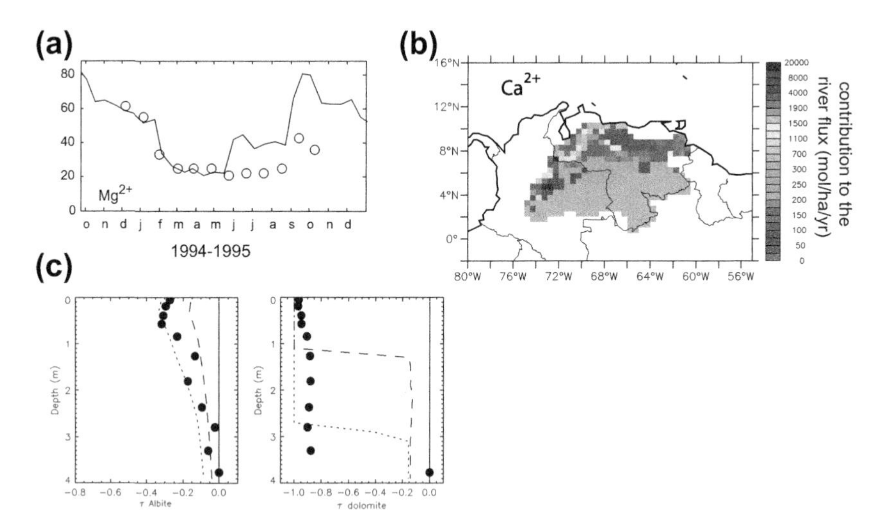

FIGURE 6 Examples of WITCH results. (a) Time evolution of the Mg^{2+} concentration from October 1994 to October 1995 in the soil solution at 60 cm depth on the south-facing slope of the watershed. Solid line: WITCH output. o: data. (From Goddéris et al., 2006); (b) Ca^{2+} production by each grid element of the Orinoco watershed as simulated by the WITCH model coupled to the LPJ global dynamic vegetation model. (From Roelandt et al., 2010); (c) Vertical mineralogical profile of a loess section located in the Mississippi river valley, 42°59' N. Dots stand for present day (0 kyr) data. All results are normalized to the quartz abundance. Solid line: model output at 10 kyr before present. Dashed line: model output at 5 kyr before present. Dotted line: model output at 0 kyr (end of the simulation). Left: albite. Right: dolomite. *(Color version online)*

case, smectites. Theoretically, well-crystallized phases would result in an underestimation of silica and Mg^{2+} export by a factor of three because smectite precipitation would be overestimated. To match the export of silica with that of Mg^{2+}, the solubility products of the smectitic phases must be increased by nearly an order of magnitude (i.e. a factor of eight) for this specific site. The model also predicts an overall dissolution of the smectite in the upper layers, down to about 20–30 cm, and an overall reprecipitation below those layers. This result has been confirmed by Li-isotope measurements (Lemarchand et al., 2010). A companion study has been performed on a tropical, small watershed (Mule Hole, South India, Violette et al., 2010). This study accounted for the horizontal transfer of water at the watershed scale and emphasized the key role that was played by a secondary mineral phase on the watershed-weathering budget.

5.5.2. A Climatic Transect: The Mississippi Valley Loess

These simulations were performed along a climatic transect from northeastern Iowa (42°59′ N) to south-central Louisiana (30°47′ N), and the modeled objects consisted of vertical profiles through the loess that was deposited about 23 to

10 kyr ago (i.e. the Peoria loess). Over the last 10 kyr, those pedons have been weathered. Starting from a uniform mineralogical composition, this composition evolves with depth, which depends on the evolution of the local climate. The WITCH model was used in its full mode, including the calculation for the time evolution of the mineralogical composition, to reproduce the differential evolution of two pedons over the last 10 kyr, which were located at the northern and southern ends of the valley. Vertical water fluxes, soil water content, and underground CO_2 production are calculated offline by the surface sub-model of the GENESIS 2.3 general circulation model (GCM), which is coupled to the BIOME4 vegetation model (Thomson and Pollard, 1997; Kaplan et al., 2003) (Figs. 6b and 7). Temperature evolution along the Mississippi valley is also provided by the GCM. Three simulations of the GCM have been performed at 10, 5 and 0 kyr before the present time, and the variables of interest were interpolated in time and space. The model was able to reproduce the time evolution of the mineralogical composition of the pedons and accounted for the differential climatic evolution from the north to the south of the valley over the last 10 kyr. These results also stress the key role that is played by the solubility product of clay minerals, which controls the saturation state of the draining waters with respect to albite. Second, the CO_2 level in soils, which originates

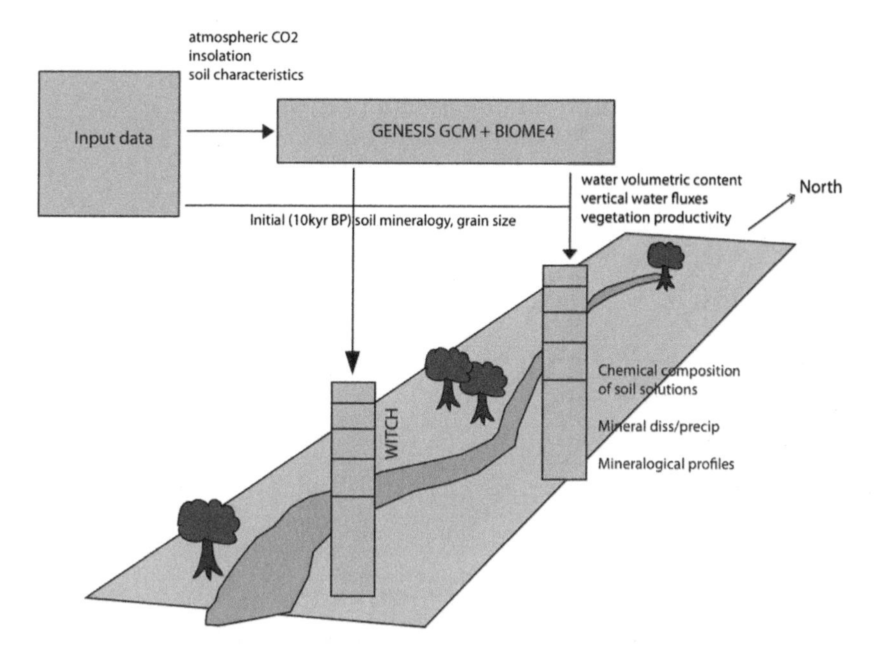

FIGURE 7 Design of the WITCH model for the Mississippi pedon simulations. Various vertical profiles are simulated along the Mississippi valley, and their mineralogical composition is calculated from 10 kyr before the present year to the present day. Water fluxes and contents are calculated by the coupled Genesis GCM and BIOME4 global vegetation model. See Goddéris et al. (2010) for more details. *(Color version online)*

from underground autotrophic and heterotrophic respirations, was found to be an important factor that controls weathering processes (Goddéris et al., 2010).

5.5.3. Continental Scale Watershed: The Orinoco

Fed by the hydrologic and carbon output of LPJ (Sitch et al., 2003), which is a dynamic global vegetation model (DGVM), the WITCH model was run over the entire Orinoco watershed (836,000 km^2) (Figs 6c and 8; Roelandt et al., 2010) at

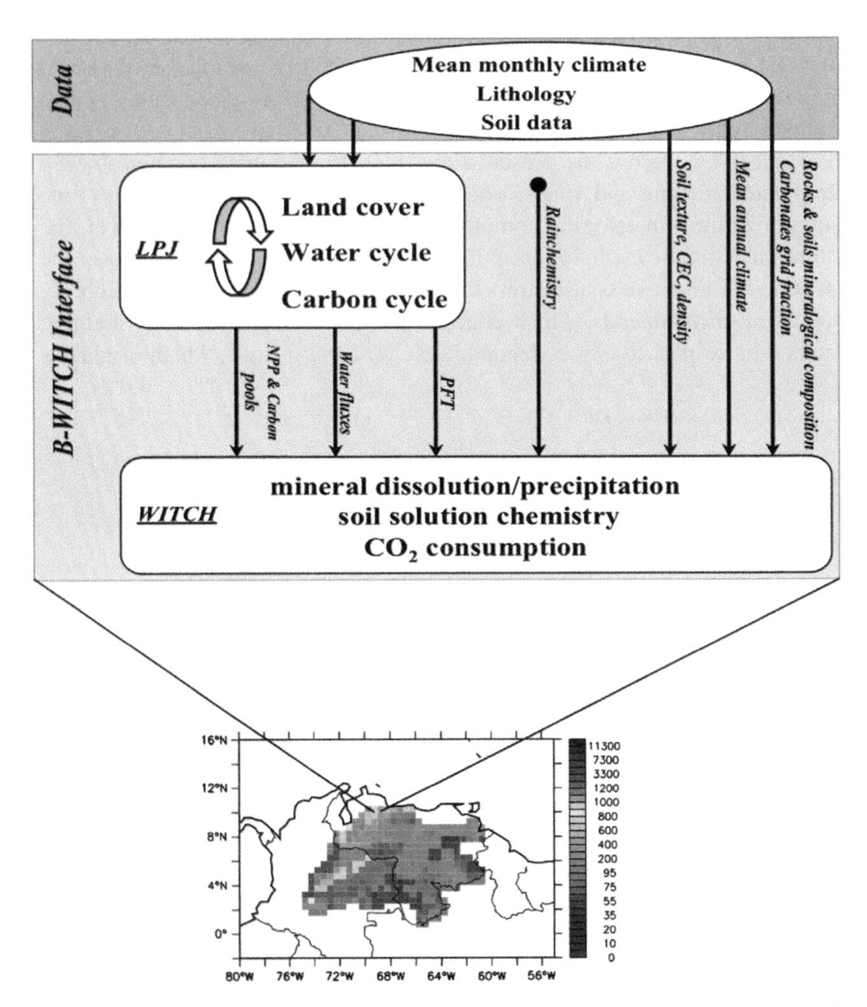

FIGURE 8 Design of the WITCH model for the Orinoco simulations. Each grid element is assumed to be disconnected from its neighbour, because of the low spatial resolution (0.5° long × 0.5° lat). Water fluxes and contents, soil respiration, and element uptake by plants are calculated for each grid cell by the LPJ global dynamic vegetation model. The WITCH model accounts for the distribution of lithology and soil mineralogy. See Roelandt et al. (2010) for more details. *(Color version online)*

a spatial resolution of 0.5° lat × 0.5° long. The vertical resolution was rather crude and was adapted to the geometry of the LPJ model: three vertical layers of 0–0.5 m, 0.5–1 m, 1–1.5 mm, and a saprolitic layer below 1.5 m; and the grid elements were not horizontally inter-connected, it is assumed that the spatial resolution is crude so all of the water that drains the grid element is directly transferred to the river through a spring. The input data included a lithological map (Amiotte-Suchet et al., 2003), which was used to define the mineralogical composition of the saprolitic layer, and a soil map for the mineralogical composition of the two upper layers (Batjes, 2005). Runs were performed on a mean, annual basis, which means that WITCH was forced to use annual mean values for the forcing functions and for the LPJ output. This model is able to reproduce the mean annual composition of the main stream, and a key finding of this study is that the first-order impact of land plants on weathering processes is the control that is exerted by the vegetation on the hydrology of the weathering profiles. In a sensitivity test where all of the vegetation was removed from the watershed, the main cation-weathering fluxes that were exported at the outlet (i.e. the CO_2 consumption) rose by 80% and this increase occurred despite a decrease in the DOC, the organic acid content of the soil layers, and the underground partial pressure of CO_2. The weathering fluxes were boosted by the dramatic decrease in evapotranspiration over the watershed, as vegetation vanished, thus, promoting the vertical drainage of the weathering profiles (Roelandt et al., 2010).

5.6. WITCH Summary

One of the main outcomes of the modeling exercises was that the hydrologic behavior of weathering profiles was a key controlling parameter of weathering reactions. Because this hydrologic behavior is largely dependent on the vegetation cover, our ability to accurately simulate the weathering budget of a watershed is dependent on our ability to correctly simulate the vegetation dynamics, especially in a context of global climatic change, and its impact on the superficial runoff and deep drainage. Our simulations also emphasize the critical interplay between water drainage (and climate) and the thermodynamic parameters of clay minerals. In all of our simulations, the dissolution/precipitation pattern of clay controls the silica budget of the weathering profiles and, therefore, the dissolution/precipitation pattern of silicate primary minerals. The solubility product of clay minerals (especially smectites) is unfortunately still poorly constrained. Future developments of the WITCH model include the implementation of an iron species speciation in soil solutions, and of a carbon budget equation (Serrano-Ortiz et al., 2010).

6. CONCLUSIONS AND PERSPECTIVES

The main limitations in pedogenic modeling are feedbacks, interactions, and cumulative effects between shorter and longer time scale changes, and all of these factors must be considered simultaneously (Lin, 2011). For example, the long-term effect of mineral precipitation or transfer of the solid phase by lessivage or bioturbation produces a change in soil structure and feedback to water and element fluxes. Additionally, evolution of the texture and structure of the solid phase influences transfer properties such as permeability. Increasing or decreasing permeability induces variations in contact time and contact surface between solid and liquid phases, and subsequently the evolution of soil solution. Therefore, it is necessary to improve the modeling of individual processes that take into account such feedbacks. Until now, only very few models have attempted to include feedback between soil structure and hydraulic properties. For example, SoilGen by Finke and Hutson (2008) adapts, at an annual time step, porosity and bulk density as a result of losses/gains of solid matter in the soil, and hydraulic properties through a pedotransfer function. Sommer et al. (2008) developed a "time split" concept that corresponded to a water and solute feedback after a critical threshold of transported mass. Although some feedbacks have been included in such models, their complete description is still limited by two difficulties: (i) the development of soil structure over time is still unaccounted for and (ii) isovolumetric assumptions are generally made and, in some cases, volume strain due to weathering or biological activity cannot be ignored. Schlutz et al. (2006) underlined that the incorporation of a soil structure dynamic module in a hydrologic model remains a challenge, particularly with respect to feedback effects of internal heterogeneity and preferential flow on the dynamics and development of soil structure.

Another difficulty is the validation of results from the models of pedogenesis. As long as the models are used for predicting future soil evolution, no validation reference can exist. Any existing validation is, therefore, based on past evolution trends (in particular, the analysis of chronosequences), and through the long-term monitoring of water quality and major soil elements (especially Si, Fe, and Al). However, as far as we know, such long-term data sets are rare and generally incomplete with respect to the parameters that are necessary for coupled geochemistry–transfer models.

REFERENCES

Alkatan, M., Oelkers, E.H., Dandurand, J.-L., Schott, J., 1998. An experimental study of calcite and limestone dissolution rates as a function of pH from −1 to 3 and temperature from 25 to 80 °C. Chem. Geol. 151, 199–214.

Allison, J.D., Brown, D.S., 1995. Minteqa2/Prodefa2. In: Loeppert, R.H., Schwab, A.P., Goldberg, S. (Eds.), A Geochemical Speciation Model and Interactive Preprocessor. Chemical Equilibrium and Reaction Models. Soil Sci. Soc. Am. and Am. Soc. Agronomy, Madison, WI, pp. 241–252.

Amiotte-Suchet, P., Probst, J.-L., Ludwig, W., 2003. World wide distribution of continental rock lithology: implications for atmospheric/soil CO_2 uptake by continental weathering and alkalinity river transport to the ocean. Global Biogeochem. Cycles 17. doi:10.1029/2002GB001891.

Arnorsson, S., Stefansson, A., 1999. Assessment of feldspar solubility constants in water in the range 0 ° to 350 °C at vapor saturation pressures. Am. J. Sci. 299, 173–209.

Batjes, N.H., 2005. ISRIC-WISE Global Data Set of Derived Soil Properties on a 0.5 by 0.5 degree Grid (version 3.0) ISRIC-World soil Inf. Rep. 2005/08. Available online.

Bauer, J., Herbst, M., Huisman, J.A., Weihermüller, L., Vereecken, H., 2008. Sensitivity of simulated soil heterotrophic respiration to temperature and moisture reduction functions. Geoderma 145, 17–27.

Beaulieu, E., Goddéris, Y., Labat, D., Roelandt, C., Oliva, P., Guerrero, B., 2010. Impact of atmospheric CO_2 levels on continental silicate weathering. Geochem. Geophys. Geosyst. 11. doi:10.1029/2010GC003078.

Blum, A.E., Stillings, L.L., 1995. Feldspar dissolution kinetics. Rev. Mineral. 31, 291–351.

Bonneville, S., Smits, M., Brown, A., Harrington, J., Leake, J., Brydson, R., Benning, L., 2009. Plant-driven fungal weathering: early stages of mineral alteration at the nanometer scale. Geology 37, 615–618.

Brantley, S.L., Bandstra, J., Moore, J., White, A.F., 2008. Modelling chemical depletion profiles in regolith. Geoderma 145, 495–504.

Brimhall, G.H., Alpers, C.N., Cunningham, A.B., 1985. Analysis of supergene ore-forming processes and ground water solute transport using mass balance principles. Econ. Geol. 80, 1227–1256.

Brimhall, G.H., Lewis, C.J., Ague, J.J., Dietrich, W.E., Hampel, J., Rix, P., 1988. Metal enrichment in bauxites by deposition of chemical mature eolian dust. Nature 333, 819–824.

Brimhall, G.H., Lewis, C.J., Ford, C., Bratt, J., Taylor, G., Warin, O., 1991. Quantitative geochemical approach to pedogenesis: importance of parent material reduction, volumetric expansion, and eolian influx in lateritization. Geoderma 51, 51–91.

Brisson, N., Mary, B., Ripoche, D., Jeuffroy, M.H., Ruget, F., Nicoullaud, B., Gate, P., Devienne-Barret, F., Antonioletti, R., Durr, C., Richard, G., Beaudoi, N., Recous, S., Tayot, X., Plenet, D., Cellier, P., Machet, J.-M., Meynard, J.M., Delécolle, R., 1998. STICS: a generic model for the simulation of crops and their water and nitrogen balances. 1. Theory and parameterization applied to wheat and corn. Agronomie 18, 311–346.

Campbell, C.A., Cameron, D.R., Nicholaichuk, W., Davidson, H.R., 1977. Effects of fertilizer N and soil moisture on growth, N content and moisture use by spring wheat. Can. J. Soil Sci. 57, 289–310.

Castet, S., Dandurand, J.-L., Schott, J., Gout, R., 1993. Boehmite solubility and aqueous aluminum speciation in hydrothermal solutions (90–350 °C): experimental study and modeling. Geochim. Cosmochim. Acta 57, 4869–4884.

Chahinian, N., 2004. Paramétrisation Multi-Critère et Multi-Échelle D'un Modèle Hydrologique Spatialisé de Crue en Milieu Agricole. (in French with English abstract) Ph. D. diss. Univ. of Montpellier, France.

Chairat, C., 2005. Etude Expérimentale de la Cinétique et des Mécanismes D'altération de Minéraux Apatitiques: Application au Comportement D'une Céramique de Confinement D'actinides Mineurs (in French with English abstract) Ph. D. diss. Univ. of Toulouse, France.

Chertov, O.G., Komarov, A.S., 1996. SOMM–a model of soil organic matter and nitrogen dynamics in terrestrial ecosystems. In: Powlson, D.S., et al. (Eds.), Evaluation of Soil Organic Matter Models Using Existing Long-Term Datasets. Springer-Verlag, Heidelberg, pp. 231–236.

Chiquet, A., Colin, F., Hamelin, B., Michard, A., Nahon, D., 2000. Chemical mass balance of calcrete genesis on the Toledo granite (Spain). Chemical. Geol. 170, 19–35.

Coleman, K., Jenkinson, D.S., 1995. RothC-26 3. A Model for the Turnover of Carbon in Soil: Model Description and Users Guide. ISBN 0951 4456 69.

Collon, P., 2003. Evolution de la Qualité de L'eau Dans les Mines Abandonnées du Bassin Ferrifère Lorrain. De L'expérimentation en Laboratoire à la Modélisation in Situ (in French with English abstract) Ph. D. diss. Univ. of Nancy, France.

Cornu, S., Samouelian, A., Phillips, J.D., 2008. Modelling pedogenesis: purpose and overview of the special issue. Geoderma 145, 399–400. doi:10.1016/j.geoderma.2008.02.003.

DeNovio, N.M., Saiers, J.E., Ryan, J.N., 2004. Colloid movement in unsaturated porous media: recent advances and future directions. 10.2113/3.2.338. Vadose Zone J. 3 (2), 338–351.

Dietrich, W.E., Reiss, R., Hsu, M., Montgomery, D.R., 1995. A process-based model for colluvial soil depth and shallow landsliding using digital elevation data. Hydrol. Process. 9, 383–400.

D'Odorico, P., 2000. A possible bistable evolution of soil thickness. J. Geophys. Res. Solid Earth B11, 25927–25935.

Dove, P.M., 1994. The dissolution kinetics of quartz in sodium chloride solutions at 25° and 300°. Am. J. Sci. 294, 665–712.

Drever, J.I., 1997. The Geochemistry of Natural Waters. Prentice Hall, Upper Saddle River, New Jersey. 07458.

Dudal, Y., Gérard, F., 2004. Accounting for natural organic matter in aqueous chemical equilibrium models: a review of the theories and applications. Earth Sci. Rev. 66, 199–216.

Elzein, A., Balesdent, J., 1995. Mechanistic simulation of vertical distribution of carbon concentrations and residence times in soils. Soil Sci. Soc. Am. J. 59 (5), 1328–1335.

Eyring, H., 1935. The activated complex in chemical reactions. Journal Chem. Phys. 3, 107–115.

Fang, C., Moncrieff, J.B., 1999. A model for soil CO_2 production and transport 1: model development. Agric. Forest Meteorol. 95, 225–236.

Feyen, J., Jacques, D., Timmerman, A., Vanderborght, J., 1998. Modelling water flow and solute transport in heterogeneous soils: a review of recent approaches. J. Agric. Eng. Res. 70, 231–256.

Finke, P.A., Hutson, J., 2008. Modelling soil genesis in calcareous löss. Geoderma 145, 462–479.

Finke, P.A., 2011. Modeling the genesis of Luvisols as a function of topographic position in loess parent material. Submitted to Quat. Int. http://dx.doi.org/10.1016/j.quaint.2011.10.016.

Foth, H.D., Ellis, B.G., 1996. Soil Fertility, second ed. CRC Press, Lewis, p. 304. ISBN 1-56670-243-7.

Franko, U., Oelschlagel, B., Schenk, S., 1995. Simulation of temperature, water and nitrogen dynamics using the model Candy 1. Ecol. Model. 81, 213–222.

Furbish, D.J., Fagherazzi, S., 2001. Stability of creeping soil and implications for hillslope evolution. Water Resour. Res. 37, 2607 (2001WR000239).

Gérard, F., François, M., Ranger, J., 2002. Processes controlling silica concentration in leaching and capillary soil polutions of an acidic brown forest soil (Rhône, France). Geoderma 107 (3–4), 197–226.

Gérard, F., Mayer, K.U., Hodson, M.J., Ranger, J., 2008. Modelling the biogeochemical cycle of silicon in soils: application to a temperate forest ecosystem. Geochim. Cosmochim. Acta 72, 741–758.

Goddéris, Y., François, L.M., Probst, A., Schott, J., Moncoulon, D., Labat, D., Viville, D., 2006. Modelling weathering processes at the catchment scale: the WITCH numerical model. Geochim. Cosmochim. Acta 70, 1128–1147.

Goddéris, Y., Roelandt, C., Schott, J., Pierret, M.-C., François, L.M., 2009. Towards an integrated model of weathering, climate, and biospheric processes. Rev. Mineral. Geochem. 70, 411–434.

Goddéris, Y., Williams, J.Z., Schott, J., Pollard, D., Brantley, S.L., 2010. Time evolution of the mineralogical composition of Mississippi valley loess over the last 10 kyr: climate and geochemical modeling. Geochim. Cosmochim. Acta 74, 6357–6374.

Gray, J.M., Humphreys, G.S., Deckers, J.A., 2009. Relationships in soil distribution revealed by a global soil database. Geoderma 150, 309–323.

Guidry, M.W., Mackenzie, F.T., 2003. Experimental study of igneous and sedimentary apatite dissolution: control of pH, distance from equilibrium, and temperature on dissolution rates. Geochim. Cosmochim. Acta 67, 2949–2963.

Gustafsson, J.P., 2001. Visual MINTEQ Version 4.0. http://www.lwr.kth.se/english/OurSoftware.

Gwiazda, R.H., Broecker, W.S., 1994. The separate and combined effects of temperature, soil pCO_2, and organic acidity on silicate weathering in the soil environment: formulation of a model and results. Global Biogeochem. Cycles 8, 141–155.

Hansen, S., Jensen, H.E., Nielsen, N.E., Svendsen, H., 1991. Simulation of nitrogen dynamics and biomass production in winter wheat using the Danish simulation model DAISY. Fert. Res. 27, 245–259.

Heimsath, A.M., Dietrich, W.E., Nishiizumi, K., Finkel, R.C., 1997. The soil production function and landscape equilibrium. Nature 388, 358–361.

Heimsath, A.M., Dietrich, W.E., Nishiizumi, K., Finkel, R.C., 2001. Stochastic processes of soil production and transport: erosion rates, topographic variation and cosmogenic nuclides in the Oregon coast range, earth surf. Process. Landforms 26, 531–552.

Herbst, M., Hellebrand, H.J., Bauer, J., Huisman, J.A., Vanderborght, J., Vereecken, H., 2008. Multiyear heterotrophic soil respiration: evaluation of a coupled CO_2 transport and carbon turnover model. Ecol. Model. 214 (2–4), 271–283.

Heuvelink, G.B.M., Pebesma, E.J., 1999. Spatial aggregation and soil process modelling. Geoderma 89, 45–67.

Hillel, D., 1998. Environmental Soil Physics. Academic Press.

Holmqvist, J., 2001. Modelling Chemical Weathering in Different Scales. Ph.D. thesis. Lund University, Lund.

Hoosbeek, M.R., Bryant, R.B., 1992. Towards the quantitative modeling of pedogenesis – a review. Geoderma 55, 183–210.

Humphreys, G.S., Wilkinson, M.T., 2007. The soil production function: a brief history and its rediscovery. Geoderma 137, 73–78. doi:10.1016/j.geoderma.2007.01.004.

Hutson, J.L., 2003. LEACHM – A Process-Based Model of Water and Solute Movement, Transformations, Plant Uptake and Chemical Reactions in the Unsaturated Zone. Version 4. Research Series No R03-1. Dept of Crop and Soil Sciences, Cornell University, Ithaca, NY.

Jacques, D., Simunek, J., 2005. User Manual of Multicomponent Variably-Saturated Flow and Transport Model HP1: Description, Verification, and Examples (Version 1.0). SCK-CEN, Mol, Belgium, p. 79.

Jacques, D., Simunek, J., Mallants, D., van Genuchten, M.Th., 2008. Modelling coupled water flow, solute transport and geochemical reactions affecting heavy metal migration in a podzol soil. Geoderma 145, 449–461.

Jarvis, N.J., 1994. The MACRO Model (Version 3.1). Technical Descrition and Sample Simulations. Reports and dissertations 19. Departement of Soil Science, Swedisch University of Agricultural Science, Uppsala, Sweden, p. 51.

Jarvis, N., Taylor, A., Larsboa, M., Etana, A., Rosen, K., 2010. Modelling the effects of bioturbation on the re-distribution of 137Cs in an undisturbed grassland soil. Eur. J. Soil Sci. 61, 24–34.

Jenkinson, D.S., Coleman, K., 1994. Calculating the annual input of organic matter to soil from measurements of total organic carbon and radiocarbon. Eur. J. Soil Sci. 45, 167–174.

Jenny, H., 1941. Factors of Soil Formation – A System of Quantitative Pedology. McGraw-Hill Book Company, p. 281.

Jenny, H., 1961. Derivation of state factor equations of soils and ecosystems. Soil Sci. Soc. Am. Proc. 25, 385–388.

Kaplan, J.O., Bigelow, N.H., Prentice, I.C., Harrison, S.P., Bartlein, P.J., Christensen, T.R., Cramer, W., Matveyeva, N.V., McGuire, A.D., Murray, D.F., Razzhivin, V.Y., Smith, B., Walker, D.A., Anderson, P.M., Andreev, A.A., Brubaker, L.B., Edwards, M.E., Lozhkin, A.V., 2003. Climate change and Artic ecosystems: 2. Modeling, paleodata-model comparisons and future projections. J. Geophys. Res. 108, 12.

Keiser, M.G., Van Riemsdijk, W.H., 2002. ECOSAT: A Computer Program for the Calculation of Speciation and Transport in Soil–Water Systems. Wageningen University, Wageningen, The Netherlands.

Köhler, S.J., Dufaud, F., Oelkers, E.H., 2003. An experimental study of illite dissolution kinetics as a function of pH from 1.4 to 12.4 and temperature from 5 to 50 °C. Geochim. Cosmochim. Acta 67, 3583–3594.

Lecomte, K.L., Pasquini, A.I., Depetris, P.J., 2005. Mineral weathering in a semiarid mountain river: its assessment through PHREEQC inverse modelling. Aqua. Geochem. 11, 173–194.

Legros, J.P., 1984. Introduction à l'étude de la simulation de l'évolution granulométrique du sol: présentation d'un modèle informatique. Bull. l'Assoc. Fr. Etude Sol. 2, 51–62.

Lehmann, P., Assouline, S., Or, D., 2008. Characteristic lengths affecting evaporativ drying of porous media. doi:10.1103/PhysRevE.77.056309. Phys. Rev. E 77 (056309).

Lemarchand, E., Chabaux, F., Vigier, N., Millot, R., Pierret, M.-C., 2010. Lithium isotope systematics in a forested granitic catchment (Strengbach, Vosges mountains, France). Geochim. Cosmochim. Acta 74, 4612–4628.

Li, C., Frolking, S., Frolking, T.A., 1992. A model of nitrous oxide evolution from soil driven by rainfall events, 1. Model structure and sensitivity. J. Geophys. Res. 97, 9759–9776.

Lin, H.S., 2011. Three principles of soil change and pedogenesis in time and space. Soil Sci. Soc. Am. J. 75 (6), 2049–2070.

Madé, B., Clément, A., Fritz, B., 1994. Modeling mineral/solution interactions: the thermodynamic and kinetic code KINDIS. Computers Geosci. 20 (9), 1347–1363.

Maher, K., Steefel, C.I., Whiite, A.F., Stonestrom, D.A., 2009. The role of reaction affinity and secondary minerals in regulating chemical weathering rates at the Santa Cruz soil chronosequence, California. Geochem. et Cosmochim. Acta 73, 2804–2831.

Mayer, K.U., Frind, E.O., Blowes, D.W., 2002. Multicomponent reactive transport modeling in variably saturated porous media using a generalized formulation for kinetically controlled reactions. Water Resources Research 38 (9), 1174. doi:10.1029/2001WR000862.

McBratney, A.B., Mendonça Santos, M.L., Minasny, B., 2003. On digital soil mapping. Geoderma 117, 3–52.

McGechan, M.B., Jarvis, N.J., Hooda, P.S., Vinten, A.J.A., 2002. Parameterization of the MACRO model to represent leaching of colloidally attached inorganic phosphorus following slurry spreading. Soil Use Manage. 18, 61–67.

McGill, W.B., 1996. Review and classification of ten soil organic matter (SOM) models. In: Powlson, D.S., Smith, P., Smith, J.U. (Eds.), Evaluation of Soil Organic Matter Models Using Existing Long-Term Datasets. NATO AS1 Series I, vol. 38. Springer-Verlag, Heidelberg, pp. 111–132.

Merritts, D.J., Chadwick, O.A., Hendricks, D.M., Brimhall, G.H., Lewis, C.J., 1992. The mass balance of soil evolution on late quaternary marine terraces, northern California. Geol. Soc. Am. Bull. 104, 1456–1470.

Metzger, T., Tsotsas, E., 2005. Influence of pore size distribution on drying kinetics: a simple capillary model. Dry. Technol. 23, 1797–1809.

Michel, E., Majdalani, S., Di-Pietro, L., 2010. How differential capillary stresses promote particle mobilization in macroporous soils: a novel conceptual model. Vadose Zone J. 9, 307–316.

Minasny, B., McBratney, A.B., 1999. A rudimentary mechanistic model for soil production and landscape development. Geoderma 90, 3–21.

Minasny, B., McBratney, A.B., 2001. A rudimentary mechanistic model for soil formation and landscape development: II. A two-dimensional model incorporating chemical weathering. Geoderma 103, 161–179.

Minasny, B., McBratney, A.B., 2006. Mechanistic soil-landscape modelling as an approach to developing pedogenetic classifaication. Geoderma 133, 138–149.

Mogollon, J.L., Ganor, J., Soler, J., Lasaga, A.C., 1996. Column experiments and the dissolution rate law of gibbsite. Am. J. Sci. 296, 729–765.

Molina, J.A.E., 1996. Description of the model NCSOIL. In: Powlson, D.S., Smith, P., Smith, J.U. (Eds.), Evaluation of Soil Organic Matter Models Using Existing Long-Term Datasets. NATO ASI Series I, vol. 38. Springer-Verlag, Heidelberg, pp. 269–274.

Montagne, M., Cornu, S., 2010. Do we need to include soil evolution module in models for prediction of future climate change? Clim. Change. 98, 75–86.

Montagne, D., Cornu, S., Le Forestier, L., Hardy, M., Josiere, O., Caner, L., Cousin, I., 2008. Mass balance modelling of the impact of drainage on elementary soil mechanisms in Albeluvisol: input of mineralogical data. Geoderma 145, 426–438.

Nagy, K.L., 1995. Dissolution and precipitation kinetics of sheet silicates. Rev. Mineral. 31, 173–233.

Noguera, C., Fritz, B., Clement, A., Baronnet, A., 2006. Nucleation, growth and ageing scenarios in closed systems II: dynamics of a new phase formation. J. Cryst. Growth 297 (1), 187–198.

Parker, D.R., Norvell, W.A., Chaney, R.L., 1995. GEOCHEM-PC. A chemical speciation program for IBM and compatible personal computers. In: Loeppert, R.H., Schwab, A.P., Goldberg, S. (Eds.), Chemical Equilibrium and Reaction Models. Soil Sci. Soc. Am. and Am. Soc. Agronomy, Madison, WI, pp. 253–269.

Parkhurst, D.L., Appelo, C.A.J., 1999. User's guide to PHREEQC (version 2) A computer program for speciation, batch-reaction, one dimensional transport and inverse geochemical calculation. USGS, Water-Resources Investigation Report 99–4249

Parton, W.J., 1996. The century model. In: Powlson, D.S., Smith, P., Smith, J.U. (Eds.), Evaluation of Soil Organic Matter Models Using Existing Long-Term Datasets. NATO ASI Series I, vol. 38. Springer-Verlag, Heidelberg, pp. 283–293.

Phillips, J.D., 1993. Stability implications of the state factor model of soils as a nonlinear dynamical system. Geoderma 58, 1–15.

Phillips, J.D., 2008. Soil system modelling and generation of field hypotheses. Geoderma 145, 419–425.

Phillips, J.D., Perry, D., Garbee, K., Carey, K., Stein, D., Morde, M.B., Sheehy, J.A., 1996. Deterministic uncertainty and complex pedogenesis in some Pleistocene dune soils. Geoderma 73, 147–164.

Pokrovsky, O.S., Schott, J., 1999. Processes at the magnesium-bearing carbonates/solution interface. II. Kinetics and mechanism of magnesite dissolution. Geochim. Cosmochim. Acta 63, 881–897.

Pokrovsky, O.S., Schott, J., 2000. Kinetics and mechanism of forsterite dissolution at 25 °C and pH from 1 to 12. Geochim. Cosmochim. Acta 64, 3313–3325.

Pokrovsky, O.S., Schott, J., 2001. Kinetics and mechanism of dolomite dissolution in neutral to alkaline solutions revisited. Am. J. Sci. 301, 597–626.

Pokrovsky, O.S., Schott, J., Thomas, F., 1999. Dolomite surface speciation and reactivity in aquatic systems. Geochim. Cosmochim. Acta 63, 3133–3143.

Pokrovsky, O.S., Golubev, S.V., Schott, J., Castillo, A., 2009. Calcite, dolomite, and magnesite dissolution kinetics in aqueous solutions at acid to circumneutral pH, 25 to 150 °C and 1 to 55 atm pCO_2: new constraints on CO_2 sequestration in sedimentary basins. Chem. Geol. 265, 20–32.

Potter, C.S., Randerson, J.T., Field, C.B., Matson, P.A., Vitousek, P.M., Mooney, H.A., Klooster, S.A., 1993. Terrestrial ecosystem production: a process model based on global satellite and surface data. Global Biogeochem. Cycles 7, 811–841.

Probst, A., Gh'mari, El, Aubert, A., Fritz, B., McNutt, R., 2000. Strontium as a tracer of weathering processes ina silicate catchment polluted by acid atmospheric inputs, Strengbach, France. Chem. Geol. 170, 203–219.

Quénard, L., Samouëlian, A., Laroche, B., Cornu, S., 2011. Lessivage as a major process of soil formation: a revisitation of existing dara. Geoderma 167–168, 135–147.

Rasmussen, C., Tabor, N.J., 2007. Applying a quantitative pedogenic energy model across a range of environmental gradients. Soil Sci. Soc. Am. J. 71 (6), 1719–1729.

Rasmussen, C., Southard, R.J., Horwath, W.R., 2005. Modelling energy inputs to predict pedogenic environments using regional environmental databases. Soil Sci. Soc. Am. J. 69, 1266–1274.

Rasse, D.P., François, L.M., Aubinet, M., Kowalski, A.S., Vande Walle, I., Laitat, E., Gérard, J.-C., 2001. Modelling short-term CO_2 fluxes and long term tree growth in temperate forests with ASPECTS. Ecol. Model. 141, 35–52.

Rawls, W.J., Brakensiek, D.L., 1985. Agricultural management effects on soil water retention. In: DeCoursey, D.G. (Ed.), Proceedings of the 1983 Natural Resources Modelling Symposium. U.S. Department of Agriculture, Agricultural Research Service, ARS-30, p. 532.

Ren, D., Leslie, L.M., Karoly, D.J., 2008. Sensitivity of an ecological model to soil moisture simulations from two different hydrological models. Meteorol. Atmos. Phys. 100, 87–99.

Roelandt, C., Goddéris, Y., Bonnet, M.-P., Sondag, F., 2010. Coupled modeing of biospheric and chemical weathering processes at the continental scale. Global Biogeochem. Cycles 24. doi:10.1029/2008GB003420.

Rousseau, M., 2003. Transport Préférentiel de Particules dans un sol Non Saturé: De L'expérimentation en Colonne Lysimétrique à L'élaboration d'un Modèle à Base Physique (in French with English abstract) Ph. D. diss. Univ. of Grenoble, France.

Runge, E.C.A., 1973. Soil development sequences and energy models. Soil Sci. 115, 183–193.

Ryzhova, I.M., 1996. Analysis of the feedback effects of ecosystems produced by changes in carbon-cycling parameters using mathematical models. Eurasian Soil Sci. 28, 44–52 (English translation).

Saco, P.M., Willgoose, G.R., Hancock, G.R., 2006. Spatial organization of soil depths using a landform evolution model. J. Geophys. Res. 111 (F02016). doi:10.1029/2005JF000351.

Salvador-Blanes, S., Minasny, B., McBratney, A.B., 2007. Modelling long-term in situ soil profile evolution: application to the genesis of soil profiles containing stone layers. EJSS 58 (6), 1535–1548.

Samouëlian, A., Cornu, S., 2008. Modelling the formation and evolution of soils, towards an initial synthesis. Geoderma 145, 401–409.

Sauer, D., Finke, P.A., Schülli-Maurer, I., Sperstad, R., Sørensen, R., Høeg, H.I., Stahr, K., 2011. Testing a soil development model against southern Norway soil chronosequences. Quat. Int., http://dx.doi.org/10.1016/j.quantint.2011.12.018.

Schott, J., Berner, R.A., Sjöberg, E.L., 1981. Mechanism of pyroxene and amphibole weathering: I. Experimental studies of iron-free minerals. Geochim. Cosmochim. Acta 45, 2123–2135.

Schott, J., Pokrovsky, O., Oelkers, E.H., 2009. The link between mineral dissolution/precipitation kinetics and solution chemistry. Rev. Mineral. Geochem. 70, 207–258.

Schultz, K., Seppelt, R., Zehe, E., H-Vogel, J., Attinger, S., 2006. Importance of spatial structure in advancing hydrological sciences. Water Res. Res. 42 (W03S03). doi:10.1029/2005WR004301.

Serrano-Ortiz, P., Roland, M., Sanchez-Moral, S., Janssens, I.A., Domingo, F., Goddéris, Y., Kowalski, A.S., 2010. Hidden, abiotic CO_2 flows and gaseous reservoirs in the terrestrial carbon cycle: review and perspectives. Agric. Forest Meteorol. 150, 321–329.

Simunek, J., Sejna, M., van Genuchten, M.Th., 1992. The hydrus-2D software package for simulating the two-dimensional movement of water, and multiple solutes in variably-saturated media version 1.2, U.S. Salinity Laboratory Agricultural Research Service, Riverside California Report n°126, p. 169.

Simunek, J., Jarvis, N.J., van Genuchten, M. Th., Gärdenäs, A., 2003. Review and comparison of models for describing non-equilibrium and preferential flow and transport in the vadose zone. J. Hydrol. 272, 14–35.

Simunek, J., He, C., Pang, L., Bradford, S.A., 2006. Colloid-facilitated solute transport in variably saturated porous media: numerical model and experimental verification. 10.2136/vzj2005.0151. Vadose Zone J. 5 (3), 1035–1047.

Singh, J.S., Gupta, S.R., 1977. Plant decomposition and soil respiration in terrestrial ecosystems. Bot. Rev. 43 (4), 499–528.

Sitch, S., Smith, B., Prentice, I.C., Arneth, A., Bondeau, A., Cramer, W., Kaplan, J.O., Levis, S., Lucht, W., Sykes, M.T., Thonicke, K., Venevsky, S., 2003. Evaluation of ecosystem dynamics, plant geography and terrestrial carbon cycling in the LPJ dynamic global vegetation model. Global Change Biol. 9, 161–185.

Smith, P., Smith, J.U., Powlson, D.S., McGill, W.B., Arah, J.R.M., Chertov, O.G., Coelman, K., Franko, U., Frolking, S., Jenkinson, D.S., Jensen, L.S., Kelly, R.H., Klein-Gunnewiek, H., Komarov, A.S., Li, C., Molina, J.A.E., Mueller, T., Parton, W.J., Thornley, J.H.M., Whitmore, A.P., 1997. A comparison of the performance of nine soil organic matter models using datasets from seven long-term experiments. Geoderma 81 (1–2), 153–225.

Sommer, M., Gerke, H., Deumlich, D., 2008. Modelling soil landscape genesis – a "time slip" approach for hummocky agricultural landscape. Geoderma 145, 480–493.

Sverdrup, H., Warfinge, P., 1995. Estmating field weathering rates using laboratory kinetics. Rev. Mineral. Geochem. 31, 485–541.

Targulian, V.O., Krasilnikov, P.V., 2007. Soil system and pedogenic processes: self-organization, time scales, and environmental significance. Catena 71, 373–381.

Taylor L., Banwart, S., Leake, J., Beerling, D., 2011. Modelling the evolutionary rise of the ectomycorrhyza on subsurface weathering environments and the geochemical carbon cycle. Am. J. Sci. 311, 369–403.

Thomson, S.L., Pollard, D., 1997. Greenland and Antarctic mass balances for present and doubled CO_2 from the GENESIS version 2 global climate model. J. Clim. 10, 871–900.

Thornley, J.H.M., Verberne, E.L.J., 1989. A model of nitrogen flows in grassland. Plant Cell Environ. 12, 863–886.

Tipping, E., Hurley, M.A., 1992. A unifying model of cation binding by humic substances. Geochim. Cosmochim. Acta 56, 3627–3641.

Van Bavel, C.H.M., 1951. A soil aeration theory based on diffusion. Soil Sci. 72, 33–46.

van der Lee, J., 1998. Thermodynamic and Mathematical Concepts for CHESS. Technical Report LHM/RD/98/39. CIG, Ecole des mines de Paris, Fontainebleau, France.

van Genuchten, M. Th, 1980. A closed-form equation for predicting the hydraulic conductivity of unsaturated soils. Soil Sci. Soc. Am. J. 44, 892–898.

Van Grinsven, J.J.M. 1988. Impact of Acid Atmospheric Deposition on Soils. Quantification of Chemical and Hydrological Processes. Ph.D. diss. Netherlands: Univ of Wageningen .

Vanderborght, J., Mallants, D., Vanclooster, M., Feyen, J., 1997. Parameter uncertainty in the mobile-immobile solute transport model. J. Hydrol. 190, 75–101.

Verberne, E.L.J., 1992. Simulation of Nitrogen and Water Balance in a System of Grassland and Soil. DLO-Instituut voor Bodemvruchtbaarheid, Oosterweg 92, Postbus 30003, 9750 RA Harem. p. 56.

Vieillard, P., 2000. A new method for the prediction of Gibbs free energies of formation of hydrated clay minerals based on the electronegativity scale. Clays Clay Miner. 48, 459–473.

Violette, A., Goddéris, Y., Maréchal, J.-C., Riotte, J., Oliva, P., Mohan Kumar, M.S., Sekhar, M., Braun, J.-J., 2010. Modelling the chemical weathering fluxes at the watershed scale in the Tropics (Mule Hole, South India): relative contribution of the smectite/kaolinite assemblage versus primary minerals. Chem. Geol. 277, 42–60.

Wesolowski, D.J., Palmer, D.A., 1994. Aluminium speciation and equilibria in aqueous solution: V. Gibbsite solubility at 50 °C and pH 3–9 in 0.1 molal NaCl solutions. Geochim. Cosmochim. Acta 58, 2947–2970.

Whalter, J.V., Helgeson, H.C., 1977. Calculation of the thermodynamic properties of aqueous silica and the solubility of quartz and its polymorphs at high pressures and temperatures. Am. J. Sci. 277, 1315–1351.

Wierenga, P.J., Nielsen, D.R., Hagan, R.M., 1969. Thermal properties of a soil based upon field and laboratory measurements. Soil Sci. Soc. Am. Proc. 33, 354–360.

Wolery, T.J., 1992. EQ3/6. Lawrence Livermore National Laboratory, Livermore, CA.

Wollast, R., 1990. Rate and Mechanism of Dissolution of Carbonates in the System $CaCO_3$–MgBauer et al., 2008.

Wösten, J.H.M., Lilly, A., Nemes, A., Le Bas, C., 1999. Development and use of a database of hydraulic properties of European soils. Geoderma 90, 169–185.

Wösten, J.H.M., Pachespsky, Y.A., Rawls, W.J., 2001. Pedotransfer functions: bridging the gap between available basic soil data and missing soil hydraulic characteristics. J. Hydrol. 251, 123–150.

Yoo, K., Mudd, S.M., 2008. Toward process-based modeling of geochemical soil formation across diverse landforms: a new mathematical framework. Geoderma 146, 248–260.

Zhang, C.F., Meng, F.-R., Bhatti, J.S., Trofymow, J.A., Arp, P.A., 2008. Modeling forest leaf-litter decomposition and N mineralization in litterbags, placed across Canada: A 5-model comparison.

Chapter 20

Modeling and Mapping Soil Spatial and Temporal Variability

D.J. Mulla[1]

ABSTRACT

This chapter reviews approaches for modeling and mapping spatial and temporal variability in soil properties, with special emphasis on soil water. Soil spatial variability can be modeled using a semivariogram, while temporal variability can be represented using time series analysis or mechanistic models. Soil properties differ in the magnitude of their spatial and temporal variability. A similar media theory can be used to model soil hydraulic characteristics at different scales. Soil databases are available at different scales of map resolution. Map resolution affects the spatial averaging of soil input factors used in hydrologic modeling, thereby affecting the accuracy of model predictions. Digital soil mapping is needed to provide high-quality, fine-resolution soil databases for modeling the spatial and temporal variability of soil functioning and ecosystem health. Digital soil mapping can be achieved using soil data, auxiliary data from landscape and terrain analysis, remote or proximal sensing, geostatistics, and pedotransfer function models.

1. MODELING SOIL–LANDSCAPE VARIABILITY

Soils vary with distance and time. This simple statement has profound implications for modeling hydrologic and biogeochemical cycles at spatial scales ranging from global all the way down to individual watersheds and hillslopes. The representation of this variability is of critical importance for improved understanding and management of food production and protection of soil, land, water, air, and wildlife resources. A variety of approaches are available for modeling soil and landscape variability. The first part of this chapter reviews some key approaches for modeling soil variability, including geostatistics and similar media scaling for spatial variability, and autocovariance and time series approaches for temporal variability. Approaches for improved mapping of

[1]Dept. of Soil, Water, and Climate, 1991 Upper Buford Circle, Univ. of Minnesota, St. Paul, MN, USA. Email: mulla003@umn.edu

Hydropedology, Edited by H. Lin. DOI: 10.1016/B978-0-12-386941-8.00020-4

variability are reviewed in the second part of this chapter. Soil databases at various scales provide useful information about soil hydraulic properties that are used for hydrologic modeling. Improved mapping of variability in soil hydraulic properties is possible using soil databases in conjunction with auxiliary data from terrain analysis, pedotransfer functions, and remote sensing.

1.1. Landscape Variability

Landscape variability plays an important role in soil formation and spatial variability of soil properties (Milne, 1936; Ruhe, 1969). Landscape geomorphology can be described in terms of slope position (e.g. summit, backslope, and footslope), or in terms of terrain attributes such as relief, slope, curvature, and flow accumulation (Moore et al., 1993). Landform classification has long been used to help describe and predict soil variability (Pennock et al., 1987).

The Soil Survey Staff (2006) used concepts from landscape geomorphology to delineate 278 Major Land Resource Areas (MLRAs) for the US. MLRAs represent coarse scale aggregations of soil–landscape combinations. They are useful for state, regional, and nationwide planning. Iwahashi and Pike (2007) have extended the approaches used to delineate MLRAs to the global scale, basing their approach on automated classification involving topographic slope, the arrangement of locations with low and high elevation (termed roughness or texture), and surface convexity. Thresholds for landscape classification in this approach vary with scale and location. Sixteen global topographic classes are identified, involving four broad combinations of roughness and convexity, coupled with four slope gradient classes.

1.2. Soil Spatial Variability

Representation of soil variability is achieved through a variety of approaches, depending on the scale associated with each approach. At fine scales, typical of hillslopes, pedologists and geomorphologists identify associations between landscape units and soil series. These soil associations (groupings of soil series within landscape elements) represent trends in soil properties due to changes in patterns of water runoff, infiltration, drainage, and storage as the landscape changes from summits to backslopes to footslopes. This concept is very useful for broader scale mapping of soils, because similar soil associations should occur wherever similar landscapes are found as long as variations in climate, vegetation, or geologic parent material are minimal. For example, in the Clarion–Nicollet–Webster soil association, the Clarion soil series typically occurs at the summit, the Nicollet soil occupies shoulder or backslope positions, and the Webster occurs at backslope or footslope positions.

Soil physicists, hydrologists, and pedologists are often less concerned with broad groupings or classifications of soils, and instead are interested in quantitative representations of the spatial variations in soil properties such as soil hydraulic conductivity and water retention characteristics (saturation, field capacity, permanent wilting point, water-holding capacity, and others). Typical representations of soil spatial variability at this scale include semivariogram parametrization, scaling or similitude relationships, and wavelets (Burrough et al., 1994; Logsdon et al., 2008). These representations are used to express various aspects of the spatial structure for the desired soil property, including the magnitude and directionality of spatial variability, the correlation scale (or range) for variability, and/or the spatial frequency of repeating patterns.

Soil mapping seeks to delineate soil units that are relatively homogeneous and are separated by sharp boundaries. In reality, soil properties are almost never completely uniform, even within soil map units (Heuvelink and Webster, 2001). Rather, they vary continuously in response to the impacts of broad variations in climate, landforms, and geologic parent material, variations in landscape elements, and variations caused by land-management practices. According to Burrough et al. (1994), spatial variation can take various functional forms, including relatively uniform values within soil map units followed by abrupt changes in mean value across soil boundaries. More commonly, soil properties vary smoothly and continuously in value with relatively low and uniform variance. A third functional form is relatively small nonrandom variations within soil units accompanied by either smooth or abrupt changes in mean across soil boundaries. Another form is for soil properties to exhibit abrupt shifts in variance across soil boundaries, with or without a change in mean. Finally, soil properties may exhibit large random variations that make it difficult to discern changes in mean value across soil mapping units.

It should be noted that for the same soil–landscape combination, different soil properties may obey different patterns in spatial variation (Mulla and McBratney, 2002). For example, soil saturated hydraulic conductivity may exhibit large random variations which make it difficult to identify changes in mean across soil mapping units. At the same location, spatial patterns in soil organic matter content may exhibit smooth, continuous variations that are closely linked with changes in landscape position.

1.2.1. Geostatistics

Geostatistics is a branch of applied statistics (Goovaerts, 1999) that is widely used in applications such as estimating spatial patterns in soil moisture, water retention, and hydraulic conductivity (Nielson et al., 1973; Mulla, 1988a; Bardossy and Lehmann, 1998). Geostatistics involves two broadly defined stages, namely, spatial modeling and spatial interpolation.

When there is a smooth continuous change in value of a soil property accompanied by relatively small variations that are not completely random (Cambardella et al., 1994), the soil property often exhibits spatial autocorrelation. In this case, soil properties sampled closer to one another are more similar in value than values separated by larger distances. Beyond a critical distance threshold, soil properties are no longer spatially correlated, and their relationship varies randomly. This type of spatial variation can be quantitatively modeled using a variety of techniques, the most common being semivariogram models (Mulla and McBratney, 2002).

Semivariograms are useful for modeling the change in variance of soil properties as separation distance between sampling locations increases (Goovaerts, 1999). The value of any soil property at location x_i can be represented as a random variable z_i. The semivariogram (γ) quantifies the spatial relationships between squared differences for z as a function of the distance (h) separating samples:

$$\gamma(h) = \left(\frac{1}{2n(h)}\right) \sum_{i=1}^{n(h)} [z_i - z_{i+h}]^2, \tag{1}$$

where $n(h)$ is the number of samples separated by distance h and z_{i+h} is the value of the soil property a distance h away from the location where sample z_i was sampled.

A semivariogram model is constructed by plotting individual values of semivariance as a function of separation distance, and fitting the points to one of several authorized functions. These functions typically include linear, spherical, or exponential models. Fitting parameters for these models can be used to identify the intercept on the y-axis (the nugget value), the distance at which spatial variation transitions from auto-correlated to random (the range), and the magnitude of maximum variance (the sill). An example semivariogram with a nugget of 0.2, a range of 75, and a sill of 1.0 is shown in Fig. 1. The ratio of the nugget to the sill indicates the proportion of variability that is attributed to randomness or uncertainty. Alternatively, subtracting this proportion from one gives the proportion of variability that is due to spatial autocorrelation, often used to infer the extent of spatial structure. For the example in Fig. 1, the proportion of variability that is structured is 0.8.

The general shape of semivariograms can be associated with the type of spatial variation exhibited by the soil property (Rossi et al., 1992). Theoretically, the sill equals the population variance. If a trend in mean value exists, the semivariogram model may not ever exhibit a sill; instead, the variance rises without bound as separation distance increases. This type of behavior is often associated with linear semivariogram models. On the other hand, there are situations where the semivariogram always exhibits the same value, regardless of separation distance. This type of behavior, associated with a horizontal linear semivariogram model that always equals the population variance, indicates that there is no spatial structure between samples. In other words, the variation is completely random.

When spatial structure exists, the range of the semivariogram is an important indicator of the correlation scale (Table 1). Soil properties vary tremendously in their ranges of spatial autocorrelation (Jury, 1986; Mulla

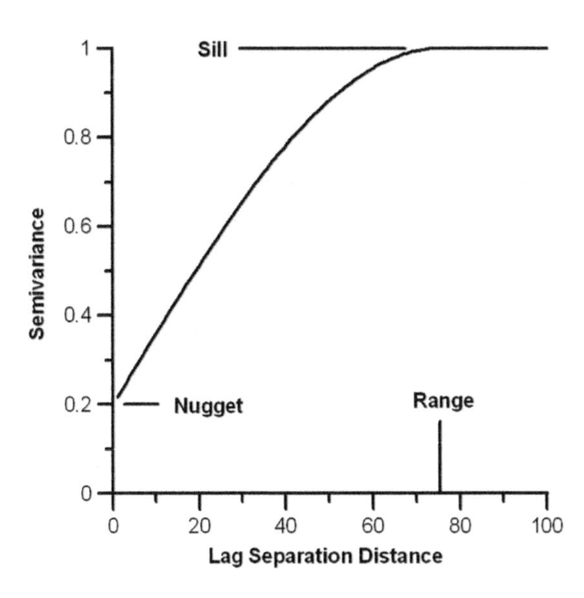

FIGURE 1 Typical spherical semivariogram model, showing the physical significance of model parameters nugget, sill, and range.

and McBratney, 2002). McBratney and Pringle (2001) reviewed 13 studies that estimated semivariograms for soil clay content. The range of the semivariogram for a majority of the studies varied between 25 and 60 m, indicating medium range variability. A review by Mulla and McBratney (2002) showed that soil-saturated hydraulic conductivity typically exhibits short-range

TABLE 1 Typical Values for the Range of the Semivariogram and Coefficient of Variation (CV) for Selected Soil Properties

Property	Range (m)	Spatial dependence	CV (%)	Magnitude of variability
Saturated hydraulic conductivity	1–34	Short range	48–352	High
%Sand	5–40	Short range	3–37	Low to moderate
%Clay	25–60	Short to moderate range	16–53	Moderate to high
Saturated water content	14–76	Short to moderate range	4–20	Low to moderate
Organic matter content	112–250	Long range	21–41	Moderate to high

variability (a range of from 1 to 34 m). Soil organic matter content, on the other hand, typically exhibits long-range variability (a range of from 112 to 250 m). The difference between the two latter magnitudes of spatial structure can be explained in two ways. First, the coefficient of variation (CV; Table 1) for saturated hydraulic conductivity (CV of from 48 to 352%) is much greater than for soil organic matter content (CV of from 21 to 41%). Second, saturated hydraulic conductivity is affected by local variations in soil structure, whereas soil organic matter content is typically affected by broad variations in soil moisture regime and vegetation patterns that are linked to hillslope geomorphology.

Once a spatial model such as a semivariogram has been developed for a single soil property, it is possible to use this model for spatial prediction at unsampled locations using geostatistical approaches (Goovaerts, 1999; Heuvelink and Webster, 2001). There are many options for spatial prediction using geostatistics; these include ordinary point kriging, block kriging, and universal kriging (Webster and Burgess, 1980). Interpolation of a soil property (Z_o^*) using ordinary punctual kriging at an unsampled location uses the following equation:

$$Z_o^* = \sum_{i=1}^{N} \lambda_i z_i, \tag{2}$$

where λ_i are weighting factors that are estimated based on distance from the unsampled location using the semivariogram. The weighting factors for the soil property are constrained to sum to unity, and techniques exist to calculate an estimation variance to gauge the accuracy of the interpolation process.

Factorial kriging (Goovaerts, 1999; Odeh and McBratney, 2000; Lopez-Granados et al., 2005) has gained popularity as a technique for spatial interpolation for a soil property that exhibits both local and regional scales of spatial correlation (nested scales of structure). Factorial kriging is useful for separately estimating and mapping the variability in a soil property at each of the nested scales of spatial structure. Bocchi et al. (2000) used factorial kriging to show that short-range soil property variations were dominated by soil texture, while long-range variations were dominated by soil organic matter content.

Regression kriging (Odeh and McBratney, 2000) is a simpler alternative to factorial kriging when both local spatial variations and regional trends exist. Regression kriging fits the regional trend with a standard regression model, and fits the residuals from the regional trend with a standard kriging model. Odeh and McBratney (2000) found that regional scale patterns in clay content in New South Wales were more accurately predicted using remote-sensing approaches combined with regression kriging than remote-sensing-based factorial kriging.

The main differences between kriging approaches arise from assumptions about stationarity of the mean and semivariogram function. Ordinary punctual kriging assumes that the interpolated property has a constant mean and that the semivariogram exists and depends only on separation distance (intrinsic stationarity). Block kriging is similar to ordinary punctual kriging, except that estimates are made in cells rather than points. Universal kriging assumes that the mean changes with location in a systematic fashion that can be determined using an iterative process to estimate the semivariogram. Factorial kriging and regression kriging are closely related to universal kriging, but use auxiliary variables and more sophisticated methods to determine the semivariogram and trends in the mean.

If spatial models for a primary soil property and a spatially correlated auxiliary variable (e.g. a landscape terrain attribute, a remotely sensed soil attribute, or a more densely sampled soil attribute) are available, along with the spatial model for cross-correlation between the primary and auxiliary variables, then it is possible to use these models along with spatial patterns in the auxiliary variable for spatial prediction of the primary variable at unsampled locations. The geostatistical techniques for achieving this process are known as cokriging (McBratney and Webster, 1983; Bhatti et al., 1991) or regression kriging (Odeh et al., 1995).

1.2.2. Similar Media Scaling

For the purposes of hydrologic modeling, it is often necessary to upscale soil hydraulic parameter measurements made at fine resolution (e.g. hillslopes) to the watershed scale. The fundamental theory on which this is based is known as similar media scaling theory (Warrick et al., 1977). Soil hydraulic properties are affected by soil particle-size distributions and pore-size distributions. Scaling assumes that soils with identical pore shapes, porosity, and pore-size or particle-size distributions will be geometrically similar at different scales if they differ only by a scaling factor (the characteristic length). The characteristic length scale for soils with a lognormal distribution of pore sizes is often estimated using mean pore radius (Tuli et al., 2001). Scaling factors are simply ratios of the characteristic length for a soil sample at a given location to the characteristic length of a reference soil sample.

To illustrate how similar media scaling works, consider values for soil water saturation (S_i), matric pressure (h_i), and hydraulic conductivity (K_i) measured at several locations in a field. These measured values are spatially variable. Scaled values S_m, h_m, and K_m corresponding to the measured values can then be estimated by using appropriate scaling factors b, α, and ω using the scaling relationships:

$$S_m = S_i^b, \tag{3}$$

$$h_m = \alpha h_i, \tag{4}$$

$$K_m = K_i/\omega^2. \tag{5}$$

The scaled soil hydraulic relationships will have significantly less spatial variability than the soil hydraulic relationships measured at multiple field locations.

Nasta et al. (2009) predicted scaled soil moisture retention characteristics for two small (~34 km^2) catchments in Italy based on multiple measurements of soil moisture retention, soil texture, and bulk density along transects within the watershed. They found that measuring soil particle-size distribution and water retention characteristics on 30 undisturbed samples was sufficient to estimate scaling coefficients to upscale the soil water retention curve to the entire watershed. Accuracy was improved when retention curves were estimated separately for coarser and finer textured soils. This approach could be used to identify watershed areas that have similar hydrologic responses.

One limitation of the similar media scaling approach is that it assumes that standard deviations for distribution functions of pore radius are identical for all soil textural classes (Kosugi and Hopmans, 1998). Das et al. (2005) scaled soil moisture characteristic curves from 247 soil cores collected across Indiana after grouping the data into seven distinct soil textural classes. This method performs better than scaling across soil textural classes, and shows that information about both pore-size radius and the standard deviation of pore radius frequency distributions is needed for accurate similar media scaling.

Si (2008) used scaling methods based on spectral and wavelet analyses to predict soil water retention characteristics. Spectral methods transform spatial patterns into the frequency domain, and are used to identify the dominant spatial frequencies at which properties show correlation (Mulla, 1988b). Random variations in soil properties will produce spectral signatures that have uniform variance at all frequencies. Wavelet methods are used to analyze variations in soil properties both by frequency and location. In wavelet analysis, windows of varying spatial length are used to filter the spatial data series and produce spectral estimates at different spatial scales. Wavelet analysis is more powerful than spectral analysis in examining spatial patterns in soil properties that change abruptly across soil boundaries. One limitation of spectral and wavelet analysis is that they both require regularly spaced sample data.

1.3. Soil Temporal Variability

Soil properties vary with time (Kachanoski and de Jong, 1988; Grayson and Western, 1998). This temporal variation is particularly important for soil moisture and antecedent moisture contents (Loague, 1992; Vinnickov et al., 1996), soil structure and permeability (Mulla et al., 1992; Zhou et al., 2008), and magnitudes of hydrologic discharge and biogeochemical flux. There is often an important interaction between management practices and the magnitude of temporal variation in soil properties, as shown by differences in temporal variability of soil aggregate stability in response to organic versus conventional farming practices (Mulla et al., 1992). Temporal variation in soil hydraulic properties can exceed the magnitude of spatial variation (Zhou et al., 2008). Temporal variation in soil properties can be modeled using mathematical, stochastic, or mechanistic process-based approaches. Temporal variation often has predictable annual, seasonal, or diurnal cycles. For example, a simple sinusoidal function is often used to represent the variation with time and depth of soil temperature in response to temporal variations in air temperature.

Stochastic representations of temporal variations in soil moisture could include first-order Markov processes, time series, or state-space approaches (Vinnickov et al., 1996; Heuvelink and Webster, 2001). At coarse scales, broad temporal variations in soil moisture can be modeled using an autocorrelation function, $r(h)$, that decays exponentially with time lag (Delworth and Manabe, 1988):

$$r(h) = \exp\left(-\frac{h}{T}\right), \tag{6}$$

where h is the time lag and T is a decay time scale constant that varies with latitude. At the equator, the decay time scale is roughly 1.2 months, while at upper latitudes it may be as high as 3 months. Alternatively, values for T can be estimated using the ratio between field capacity water content and potential evapotranspiration (Delworth and Manabe, 1988).

On finer spatial scales, temporal variability in soil moisture has been extensively studied (Heuvelink and Webster, 2001). Temporal variability in soil hydraulic properties is often greater in magnitude than spatial variability (van Es et al., 1999). Repeated measurements of soil water content at a fixed location can be modeled using time series analysis, the most common example being an autoregressive moving average (ARMA) approach. According to ARMA theory, the soil-water content at any time $Z(t)$ can be predicted from measurements at a previous time $Z(t-1)$ using

$$Z(t) = \mu + a[Z(t-1) - \mu] + \varepsilon(t), \tag{7}$$

where μ is the mean, a is a regression constant, and ε is uncorrelated random error. The autocovariance function for the time series of soil moisture measurements in this case will have an exponentially decreasing shape with time similar to the function described by Delworth and Manabe (1988) above.

An alternative to ARMA models is state-space models that seek to simultaneously describe both spatial and temporal variability. State-space models have been applied to the prediction of spatial and temporal variability in soil moisture at the field scale (Wendroth et al., 1999). In this approach, a Kalman filter is used to estimate the variance in soil moisture content with time, and this variance is used to update soil moisture estimates in time. State-space models, in contrast with ARMA models, can be used when the property to be estimated has missing values and has a mean that varies with distance or time (Morkoc et al., 1985).

Mechanistic approaches to estimating temporal variability typically attempt to describe the spatial variations in important soil hydrologic parameters (runoff curve number, hydraulic conductivity, moisture retention characteristics, available water-holding capacity, soil moisture content, etc.), and then predict temporal variations (Loague, 1992; Vepraskas and Caldwell, 2008) based on variations in climate (precipitation, radiation, temperature, humidity, wind speed, and others). These mechanistic models provide estimates for temporal variations in evapotranspiration, runoff, infiltration, drainage, and soil water storage, and their impact on greenhouse gas cycling, river discharge, groundwater recharge, and export of pollutants such as sediment, phosphorus, and nitrogen (Gowda et al., 2007; Nangia et al., 2010).

1.4. Soil Databases for Modeling

Soil databases are essential for providing input data to drive mechanistic models and predict critical processes such as crop growth and yield, aspects of the hydrologic cycle such as evapotranspiration, runoff and river discharge, infiltration, leaching, and groundwater recharge. They are also useful for simulating aspects of the biogeochemical cycle such as emissions and sequestration of greenhouse gases, nitrogen sources, sinks, and transformations, and impacts of sediment and phosphorus on aquatic ecosystem health and water quality. Soil databases are available at scales ranging from 1:100,000,000 (global) to 1:12,000 (county).

1.4.1. Global Databases

Global scale modeling of climate and biogeochemical cycling relies on spatially averaged soil properties. Global climate and hydrology models often require estimates of soil horizon thicknesses, soil texture, water-holding capacities, and organic carbon contents.

There have been several efforts to produce useable databases for modeling soil properties and related hydrologic or biogeochemical processes at the global scale (e.g. Webb et al., 2000; Yu et al., 2006). Often, these databases are derived using global or national databases for soil properties using low-resolution soil maps, supplemented by auxiliary information obtained using remote sensing, pedotransfer functions, and field sampling. The scale of resolution for these databases is typically 1°.

Batjes et al. (1994) and Dobos et al. (2005) outlined the efforts of the International Society of Soil Science (ISSS) and the International Soil Reference and Information Center (ISRIC) to develop the World Soil and Terrain Digital Database (SOTER) at a scale of 1:1,100,000. This effort relied on the FAO soil map of the world, the ISRIC Soil Information System (ISIS) database of 450 monoliths representing the FAO soil map, terrain mapping, and remote sensing. Each SOTER soil map unit is distinguished by a unique combination of landform (elevation, slope, relief, and dissection intensity), parent material, and soil component. The recent global availability of digital elevation models (DEMs) from the NASA Shuttle Radar Topography Mission (SRTM) provides topographic information for most of the world at a spatial resolution of 90 m (Farr and Kolbrick, 2000). SRTM DEMs have been used to speed up the analysis of landform units for development of a global SOTER database (Dobos et al., 2005).

Because the SOTER mapping is a slow process, an intermediate product known as the World Inventory of Soil Emission Potentials (WISE) has been developed at a scale of 1:5,000,000 or $0.5 \times 0.5°$. This database is particularly relevant to global modeling of soil erosion and greenhouse gas emissions (Gray et al., 2007). One other notable effort along these lines is the development of a Global Assessment of Soil Degradation (GLASOD). Loosely based on information provided by the FAO global soil map and ISIS databases, as well as on local information provided by 250 soil scientists working in 21 geographic regions around the world, GLASOD summarized the percent of mapped soils in each region that were degraded by water erosion, wind erosion, chemical deterioration, or physical deterioration (Oldeman et al., 1991).

1.4.1.1. Diagnostic Soil Horizons

Jenny (1941) identified five independent factors responsible for formation of different types of soils. These include parent material, climate, topography, biotic organisms, and time. Water movement plays a key role in the formation of soils and in their classification. For hydrologic modeling, water movement through and across the landscape differs significantly across soil orders. Aridisols typically form in regions that have relatively low amounts of precipitation. On the other hand, Oxisols are highly weathered as a result of heavy precipitation over long periods of time. The diagnostic argillic or spodic horizons that are used to classify Alfisols or Spodosols, respectively, are the direct result of water leaching clay or humic matter, respectively, downward from the surface.

In soil taxonomy (Soil Survey Staff, 1999) there are twelve widely recognized soil orders. These have been mapped globally at the scale of 1:130,000,000, and their spatial distribution is closely related to historical patterns in climate, vegetation, and geologic parent material.

The areas of ice-free land surface covered by soil orders vary considerably (USDA-NRCS, 2011). Poorly or weakly developed soil orders such as Entisols, Aridisols, and Inceptisols cover 38% of the ice-free land surface. Highly weathered soil orders such as Ultisols and Oxisols cover nearly 16%. Soil orders typically formed in forested regions such as Alfisols or Spodosols, having distinct argillic or spodic subsoil horizons, cover 12%. Organic matter-rich soils without permafrost such as Mollisols or Histosols cover 8%. Dark, organic-rich soils underlain by permafrost are classified as Gelisols; these cover nearly 9% of the ice-free land surface. Vertisols are the clay-enriched soil order that shrink and crack during dry periods, and swell during wet periods. Andisols formed in volcanic ash cover nearly 1% of the land surface.

As pointed out by Lin et al. (2008), there are many diagnostic subsoil horizons in Soil Taxonomy that result from the action of water. These include horizons cemented by silica (Duripans), humic matter and aluminum (Orsteins), calcium (Petrocalcic), gypsum (Petrogypsic), and iron (Placic). Soil horizons may also have a profound impact on the movement of water. For example, fragipans and argillic subsoil horizons can impede the infiltration of water, leading to ponding of water at the surface, or to lateral subsurface flow if these horizons are present on hillslopes (McDaniel et al., 2008). The qualitative information in diagnostic soil horizons is not widely used or incorporated into hydrologic modeling, leading to reduced accuracy in model parametrization of soil spatial variability.

1.4.2. US Soil Geographic Databases

The Soil Survey Staff (1995) developed three soil geographic databases for the US; each is differentiated by intensity and scale of mapping. These are the National Soil Geographic (NATSGO), State Soil Geographic (STATSGO), and Soil Survey Geographic (SSURGO) databases. The NATSGO database was derived to be consistent with Major Land Resource Area (MLRA) boundaries, and is mapped at a scale of 1:7,500,000. The STATSGO database represents map units consisting of groups (components) of soil associations at a scale of 1:250,000. The SSURGO database for the United States involves map units that correspond to over 18,000 individual soil series at scales ranging from 1:12,000 to 1:63,360.

1.4.2.1. Soil Properties for Model Input

Soil properties can be extracted from the NATSGO, STATSGO, and SSURGO soil databases. For each soil layer in a STATSGO component or a SSURGO soil series, 28 individual soil properties are provided, including texture, permeability, available water content, bulk density, and organic matter content. The lumping of spatial variability in soil properties represented by map units in these three soil databases increases from SSURGO to STATSGO to NATSGO. With SSURGO, a map unit may consist of a dominant soil series, while in STATSGO, a map unit may consist of a soil association made up of as many as 21 soil series. Hydrologic and biogeochemical models that derive soil input parameters from these databases will be better able to represent soil spatial variability with SSURGO than with STATSGO or NATSGO databases. Even models that rely on SSURGO soil databases underestimate the true extent of spatial variability in soil properties. This is because each SSURGO map unit can consist of up to three soil series, and the variability within and across these soil series, or for inclusions within them, is not represented.

1.5. Hydrologic Modeling

1.5.1. Representative Elementary Areas (REAs)

Spatial variability of soil factors has been difficult to account for in hydrologic and biogeochemical models. Wood et al. (1990) proposed that watersheds could be modeled by dividing them into small subbasins corresponding to the representative elementary area (REA). Soil properties used for input to the model are spatially averaged over the REA. Theoretically, the performance of hydrologic models should not improve if the area of subbasins is decreased below the REA. This approach was tested and confirmed using hydrologic modeling in several watersheds around the world. The area of an REA varied from 0.5 to 1 km^2 (50–100 ha). The REA was highly dependent on landscape characteristics, and did not seem to depend strongly on soil variability.

Franke et al. (2008) used digital elevation models (DEMs) to delineate landscape-based elementary hillslope areas (EHAs) that serve as hydrologic response units (HRUs) for watershed modeling. These EHAs are classified using terrain components so that they correspond to hillslope catenas. The average area of EHAs for a watershed in Spain was 1.8 km^2, which is consistent with the minimum area for an REA identified by Wood et al. (1990).

1.5.2. Impact of Soil Database Resolution on Modeling Accuracy

Hydrologic models require input data for a variety of factors, including climate, hydrologic channels, land use, elevation, and soils. Soil properties that are often needed for hydrologic modeling include horizon depths, and for each horizon, information on saturated hydraulic conductivity, available water content, bulk density, and soil moisture retention characteristics. When not directly available, soil texture is often used to indirectly estimate soil moisture retention characteristics (Campbell and Mulla, 1990). For hydrologic modeling in the US, the most widely used soil databases where information about soil properties is obtained include SSURGO and STATSGO (Sheshukov et al., 2009). SSURGO soil properties are averaged over areas no smaller than 2 ha. STATSGO soil properties are averaged over areas no smaller than 625 ha.

Lin et al. (2005) used hierarchical sampling to compare soil variability at the scales of STATSGO or SSURGO units to the hillslope and pedon scales. Their results showed that most of the variability in soil properties occurs at relatively fine scales (12 m or less). Juracek and Wolock (2002) studied the statistical differences between selected soil properties in the STATSGO and SSURGO databases for twelve Kansas watersheds averaging 1768 km^2 in area. They used spatial averaging techniques to compare soil permeability, percent clay content, and percent of area in soil hydrologic class B in the STATSGO and SSURGO databases at scales ranging from 0.01 to 400 km^2. The results showed that each soil property in the STATSGO database was strongly correlated (Pearson's r value ranging from 0.7 to 0.98) with the corresponding soil property in the SSURGO database over all scales, but the strength of correlation increased as the area where values were averaged increased. Differences between the two databases were minimal when the averaging area exceeded 25 km^2. They also found that the mean values for soil permeability and clay content were on average larger at each scale of averaging with the STATSGO than the SSURGO database, while mean values for percent soil hydrologic class B were smaller.

A number of authors have explored the accuracy of hydrologic models based on soil input data from SSURGO versus STATSGO databases (Gowda and Mulla, 2005; Geza and McCray, 2008). Sheshukov et al. (2009) studied the impact of using SSURGO versus STATSGO soil properties on runoff and soil erosion in a 7818 ha watershed located in Kansas. With the STATSGO database, the watershed was subdivided into three soil map units, whereas there were 18 soil map units in the watershed with the SSURGO database. They found that both runoff and soil erosion were over-predicted when the watershed was modeled using STATSGO soil properties rather than SSURGO soil properties. This was attributed to an overestimation of the area in the watershed occupied by poorly drained soils with the STATSGO relative to the SSURGO database.

Mednick (2010) found that hydrologic modeling with the long-term hydrologic impact assessment (L-THIA) model of 10-digit hydrologic unit code (HUC) watersheds in Wisconsin was more accurate when soil input data were obtained from SSURGO rather than STATSGO databases. These findings suggest that modeling accuracy suffers when lower-resolution soils data or greater spatial averaging is used in hydrologic models. A possible reason for this finding is that soil hydraulic properties tend to be overestimated when they are based on STATSGO-scale rather than SSURGO-scale databases.

2. MAPPING SOIL VARIABILITY

2.1. Digital Soil Mapping

Soil mapping at moderately high resolution (Order II) is common in the US and EU. Most existing paper soil maps in the developed world are now being converted to digital format, which allows individual rasters or grid cells to be queried for their attributes (or accessed for hydrologic and biogeochemical modeling) using GIS. McBratney et al. (2003) reviewed several approaches for producing digital soil maps using generalized linear models, classification and regression trees, neural networks, fuzzy systems, and geostatistics. Digital soil mapping (Fig. 2) relies on finding quantitative spatial relationships (through correlation or geostatistical analysis, and others) between a limited set of field data (horizon depths, soil properties, and others) and auxiliary variables derived from digital elevation models or remote sensing (McBratney et al., 2003; Dobos et al., 2006; Minasny et al., 2008). This first step results in spatial databases of soil properties as well as soil maps that contain useful information about diagnostic horizons, soil horizon depths, and others.

The second step in digital soil mapping is to conduct spatial modeling. Spatial modeling uses information from the soil databases and maps produced in the first step. This information is processed using techniques such as similar

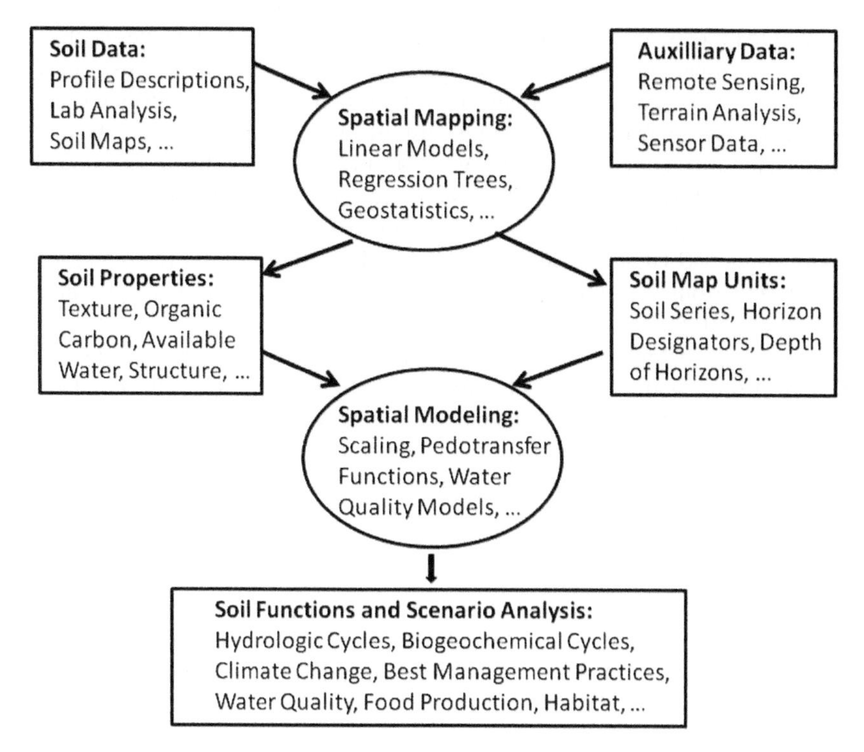

FIGURE 2 Procedural approaches and tools for digital soil mapping and modeling.

media scaling and/or pedotransfer functions to produce information that can be input into mechanistic models for hydrologic cycling, water quality, and biogeochemical cycling. Output from these models is used to infer the impact of soil functions (biomass production and attenuation of pollution) and their temporal variability on the environment, habitat, and economy. Scenario analysis can be used to optimize soil management practices that lead to improved ecosystem functioning.

Developing countries face significant challenges in deriving high-resolution soil maps, which are critically needed for better soil, land, and water resource planning and management. Several approaches have been proposed to overcome the barriers to high-resolution soil mapping in developing countries (Minasny et al., 2008; Stoorvogel et al., 2009). Many of these approaches start with the premise that soil variability is closely linked with patterns in landform geomorphology. Digital soil mapping offers the promise of helping developing countries produce soil maps for better natural resource management.

2.1.1. Terrain Attributes

There has been significant interest in using terrain attributes to predict spatial patterns in soil hydraulic properties, horizon depths, and organic matter content (Moore et al., 1993; Thompson et al., 2006). Terrain attributes such as elevation, slope, curvature, and upslope contributing area are often strongly correlated with spatial variations in soil properties. Terrain attributes are typically derived from DEMs using customized software such as "Terrain Analysis Using Digital Elevation Models" (TAUDEM) software version 5 (Tarboton, 2005) and ESRI's ArcGIS software version 9.3. DEMs are available in a variety of spatial grid cell data resolutions, ranging from 90 m for SRTM, 30 m for the US Geologic Survey's National Elevation Dataset (Gesch et al., 2009), and from 1 to 3 m for Light Detection and Ranging (LiDAR) data (Liu et al., 2005).

Terrain attributes have been used extensively to study topographic features of heterogeneous landscapes (Mulla, 1986, 1988b; Wilson and Gallant, 2000). Slope gradient is defined as the tangent of the slope angle (α). Specific catchment area (A_s) is the upslope contributing area that drains into any single cell per unit contour length. Its representation of drainage patterns makes it a valuable attribute for water-resource applications. The attribute A_s itself, as well as secondary attributes derived from it, have been used to predict overland runoff in a number of studies (Wilson and Gallant, 2000). Flow accumulation can be calculated using the $D\infty$ algorithm of flow routing in the 30-m analysis (Tarboton, 1997).

Profile curvature refers to the change in slope down a flow path; it represents the rate of change in gradient and is useful in identifying areas with potential flow velocity changes (Wilson and Gallant, 2000).

Stream Power Index (SPI) is a secondary terrain attribute that measures the erosive power of flowing water (Wilson and Gallant, 2000). Stream power itself is a misnomer; this index does not quantify the power of streams, but the power of overland flow. It is calculated based on

$$\text{SPI} = \ln[A_s \tan(\alpha)]. \tag{8}$$

The compound topographic index (CTI), also known as the topographic wetness index, is a secondary terrain attribute, which identifies areas on the landscape with a potential for ponding or saturation (Wilson and Gallant, 2000). It is calculated based on

$$\text{CTI} = \ln\left[\frac{A_s}{\tan(\alpha)}\right]. \tag{9}$$

2.1.2. *Functional Mapping of Hydrologically Critical Areas*

Hydrologic processes operating at the landscape scale include evapotranspiration, runoff, infiltration, preferential flow, interflow, and drainage, all of which are closely linked. Runoff occurs by several mechanisms (Lin et al., 2008). The first is infiltration excess overland flow (Hortonian runoff), which occurs mostly at summit positions when the intensity of precipitation exceeds the infiltration capacity of soil. The second is saturation excess overland flow (Dunne runoff), which occurs when rain falls where soil is saturated, mostly at lower slope positions. Soil saturation at lower slope positions can occur as a result of subsurface return flows (throughflow or interflow) from upper-slope positions (Freeze, 1972). An extension of saturation excess overland flow theory is runoff generated from partial contributing areas, also known as variable source areas (Dunne and Black, 1970), which are typically located in close proximity to stream channels. These small areas occur where the water table intersects the soil surface near the stream channel, giving rise to saturated soils, which produce runoff during rainstorms, regardless of the intensity of precipitation. Variable source areas (VSAs) are not static; they expand and contract in area over time, making it difficult to identify their location and model their impacts on hydrologic processes. Each of these three mechanisms is associated with specific combinations of soil and landscape features.

Significant research has been conducted to predict runoff mechanisms based on variability in soil and landscape factors (Liang and Xie, 2001; Zhu and Lin, 2009). Relatively less research has been conducted to identify impacts of pedologic attributes on runoff (Lin et al., 2008). There is a significant need to develop methods for identifying functionally different soil–landscape combinations that can be used to improve the accuracy of hydrologic modeling. With the advent of light detection and ranging (LiDAR)-derived DEMs, there has been significant progress in linking terrain analysis with hydrologic impacts of soil–landscape combinations. As an example, Galzki et al. (2011) used LiDAR-based terrain analysis to quickly identify critical areas within small watersheds that contributed large amounts of runoff and sediment pollution to surface waters in Minnesota. Figure 3 illustrates the combination of critical areas with high runoff potential (identified using SPI signatures) with surface soil-saturated hydraulic conductivity for a region in central Minnesota with many lakes. The most critical areas for runoff would have both high SPI values and low saturated hydraulic conductivity. Additional criteria, such as proximity to surface water, could be used to identify which of these high-risk portions of the landscape have the largest potential impacts on surface water-quality degradation.

Terrain modeling has been extensively used to improve hydrologic prediction (Quinn et al., 1991). Terrain attributes can be used to identify flow directions on the topography, which are needed for flow routing in hydrologic models. In addition, terrain attributes such as CTI and SPI indicate the potential for surface ponding and runoff of water on the landscape. Sloan and Moore (1984) showed that the hydraulic gradient for lateral subsurface water flow (q_{lat}) is dependent on slope (α), a parameter that is easily computed using terrain analysis:

$$q_{\mathrm{lat}} = 0.024\left[2\mathrm{SW} * K_s * \frac{\sin(\alpha)}{\varphi_d L}\right], \tag{10}$$

where SW is the drainable volume of soil, L the flow length, K_s the saturated hydraulic conductivity, and φ_d the drainable porosity. This expression shows that it is important to quantify the spatial variability in both terrain and soil

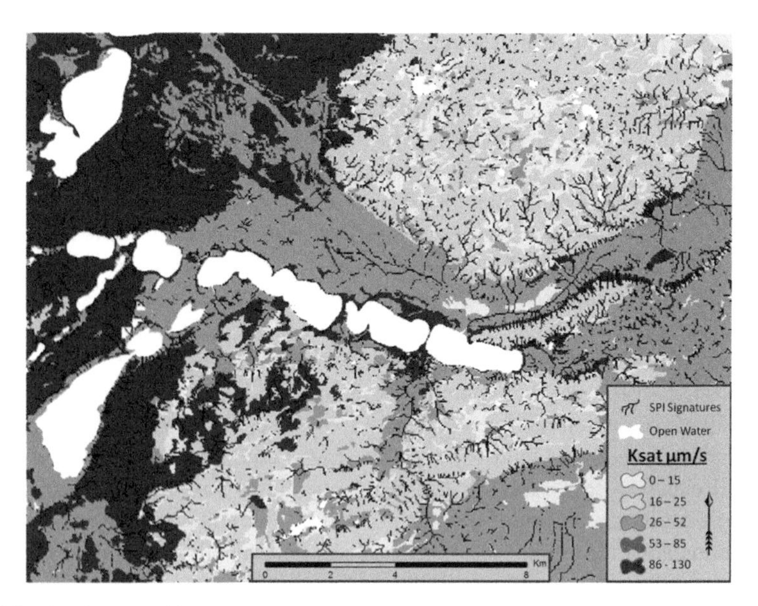

FIGURE 3 Overlay of Stream Power Index (SPI) signatures indicating areas of potentially high runoff with surface soil saturated hydraulic conductivity (Ksat) from SSURGO databases for the central lakes region of Minnesota. The landscape includes several large lakes (white).

properties such as saturated hydraulic conductivity in order to predict the spatial variability in soil water flow.

Soil–landscape combinations with a significant potential for saturation excess runoff and loss of sediments or phosphorus that are hydrologically connected with rivers, streams, ditches, or lakes are known as hydrologically sensitive areas (Walter et al., 2000) or critical source areas (Lyon et al., 2006). Lyon et al. (2004) developed a method based on terrain indices and soil properties to identify variable source areas. Briefly, this approach uses a traditional SCS-runoff curve method (based on land use and hydrologic soil class) to estimate the amount of runoff and the fraction of the watershed that generates this runoff. A modified CTI is then estimated for the entire watershed using

$$\mathrm{CTI}_m = \ln[A_s/(D * K_s * \tan(\alpha)], \qquad (11)$$

where D is soil depth and K_s is saturated hydraulic conductivity. Values for the fraction of the watershed producing runoff are then graphed as a function of values for CTI_m. A threshold CTI_m value is identified from the graph; above this threshold saturation excess runoff is generated, while below it infiltration occurs without saturation excess runoff. Comparison of locations for VSAs predicted using threshold CTI_m values with field measurements for two watersheds in New York and Australia showed good agreement (Lyon et al., 2004). Easton et al. (2008) built on this approach to develop a method for delineating spatially variable VSAs in the Soil and Water Assessment Tool (SWAT) and showed that it improved watershed runoff predictions relative to an approach that uses only the SCS-runoff curve method.

2.1.3. Pedotransfer Models

Pachepsky and Rawls (2003) reviewed approaches for estimating spatial patterns in soil hydraulic properties using pedotransfer functions. Pedotransfer functions for soil hydraulic properties are statistical predictions established using databases for soil texture, bulk density, and/or organic matter content (Saxton et al., 1986), although soil morphological features that reflect pedality and macroporosity can also be used (Lin et al., 1999). Pedotransfer functions for soil-moisture retention and hydraulic conductivity are typically much more accurate when they include both textural and structural attributes of soil.

In the US, pedotransfer functions for soil hydraulic properties are typically based on the extensive National Pedon Characterization (NPC) database compiled by the Soil Survey Staff (1997). This database is constructed from measurements on soil physical properties for nearly 15,000 soil pedons. Pedotransfer functions can either be used to estimate textural class-based soil hydraulic properties, or hydraulic properties that vary continuously in response to variations in soil texture or structure. While many pedotransfer functions for soil hydraulic properties are based on quantitative information contained in the NPC database, Pachepsky and Rawls (2003) introduced a regression tree methodology for estimating soil moisture retention characteristics from both quantitative information and qualitative information in the NPC database. Quantitative information is exemplified by soil textural class (based on laboratory determinations of particle-size distribution), while qualitative information is exemplified by sizes, shapes, and grades of structure in soil peds (e.g. medium size, blocky shape, and weak grade peds). Because soil texture and structural attributes of soil are often strongly correlated with variations in landscape position, Pachepsky and Rawls (2003) suggest that a regression tree-based pedotransfer function approach could be used along with topographic variables to estimate soil hydraulic properties across broad regions. This type of approach would contribute to the need to better define hydrologically sensitive areas for distributed hydrologic modeling.

Ferrer-Julia et al. (2004) constructed a database for saturated soil hydraulic conductivity for Spain using pedotransfer functions. Spain does not have a national soil survey, but Ferrer-Julia et al. (2004) based their pedotransfer functions on soil physical and chemical properties (texture, bulk density, and organic matter content) in an extensive database consisting of 2178 soil profiles and 7011 soil horizons. They divided this database into 19 representative soil taxonomic classes, and for each class derived a pedotransfer function for saturated hydraulic conductivity. Results for these soil classes were then amalgamated into a single pedotransfer function based only on percent sand content. The pedotransfer function for hydraulic conductivity data was fit to an exponential semivariogram (with a range of 50 km), and kriging was used to produce a map of saturated soil hydraulic conductivity for the entire country of Spain. Results from this study showed that the pedotransfer function concept is useful for developing broad country-scale databases for soil hydraulic properties using limited experimental databases of soil physical and chemical properties. These broad regional databases can be of great value for hydrologic modeling of runoff and infiltration.

Romano and Palladino (2002) showed that the prediction of soil water retention characteristics using conventional pedotransfer functions based on soil physical properties could be improved by accounting for the influences of terrain attributes. Hillslope attributes such as slope and aspect were particularly useful in improving the prediction of soil water retention. They interpreted this improvement in terms of the impact of slope and aspect on soil pedogenesis due to redistribution of water on the landscape. For example, soils with similar clay content on different slopes and aspects could differ in their soil water retention characteristics due to the influence of slope and aspect on soil pedogenesis, resulting in different clay mineralogies and pore-size distributions.

2.1.4. Remote Sensing

Remote-sensing approaches have been widely used for digital soil mapping (Dobos et al., 2006; Boettinger et al., 2008; Le Du-Blayo et al., 2008; Minasny et al., 2008) and to estimate soil properties including soil texture (Odeh and McBratney, 2000), soil drainage (Seelan et al., 2003), and soil organic carbon content (Rawlins et al., 2009). New aerial and satellite remote-sensing technologies such as Digital Globe's Quickbird provide detailed imagery at a spatial resolution of 3 m or finer to be recorded across a region. Imagery for a bare soil, combined with calibration using ground truth data, enables spatial variation in soil reflectance resulting from the variation in soil moisture, texture, or organic matter content to be identified accurately (Table 2). Proximal soil sensors for electrical conductivity, ground penetrating radar, or time domain reflectometry are also widely available for quantifying soil spatial variability (Viscarra Rossel and McBratney, 1998; Adamchuk et al., 2004). The combination of soil sample data, remotely or proximally sensed imagery of bare and vegetated fields, geostatistics, and pedotransfer functions is a potentially powerful approach for digital soil mapping (Lopez-Granados et al., 2005; Odeh and McBratney, 2000).

Remote and proximal sensors vary in their spatial resolution, spectral range, and use (Table 2). At the shortest wavelengths (or highest frequencies) are instruments such as gamma spectrometers. These have a moderate spatial resolution, and are useful for identifying spatial patterns in soil radioisotope concentrations. When these radioisotopes are spatially correlated with other soil properties, such as soil organic carbon content, they can be used for soil mapping (Rawlins et al., 2009). Satellite remote-sensing platforms such as Landsat and Quickbird or RapidEye are very widely used (Boettinger et al., 2008)

TABLE 2 Instruments for Remote and Proximal Sensing of Soils and Landscapes, Along with Their Spatial Resolution, Spectral Range, and Typical Applications

Instrument	Spatial resolution (m)	Spectral range	Typical applications
Gamma spectrometer	150–220	1.37–2.81 MeV	Soil radioisotope concentrations
Landsat	30	VIS, NIR, SWIR, TIR	Land use, vegetative cover, soil color, soil organic carbon, soil moisture
IKONOS Quickbird RapidEye GeoEye WorldView	4 0.6–2.4, 5, 1.6, 0.5	3 VIS, 1 NIR 3 VIS, 1 NIR 4 VIS, 1 NIR 3 VIS, 1 NIR 6 VIS, 2 NIR	Land use, vegetative cover, soil color, soil organic carbon, soil moisture
AVHRR	1090	VIS, NIR, SWIR, TIR	Cloud cover, land water boundaries, snow and ice cover, sea surface temperature
MODIS	250–1000	VIS, NIR, SWIR, TIR	Land use, vegetative cover, cloud cover, surface temperature
SRTM	90	3.1–5.6 cm	Digital elevation models
LiDAR	1–3	10 μm–250 nm	Digital elevation models, biomass
TDR	0.1	1 GHz	Soil moisture
GPR	0.2–0.5	100 MHz–1 GHz	Subsurface imaging, soil moisture
EM Resistance	0.5–1	10 kHz	Soil moisture, soil texture, salinity, soil organic carbon content

Abbreviations: AVHRR, Advanced Very High Resolution Radiometer; MODIS, Moderate Resolution Imaging Spectroradiometer; SRTM, Shuttle Radar Mission; LiDAR, Light Detection and Ranging; TDR, Time Domain Reflectometry; GPR, Ground Penetrating Radar; EM, Electromagnetic; VIS, Visible; NIR, Near Infrared; SWIR, Short Wave Infrared; TIR, Thermal Infrared

for mapping land use, land cover, soil color, soil organic carbon, and soil moisture. The spectral range of these platforms encompasses portions of the visible, near infrared, shortwave infrared, and thermal infrared regions. Landsat data are available at moderate spatial resolutions, while Quickbird, RapidEye,

GeoEye, and WorldView data are available at high resolution (Table 2). Satellite-, aerial-, or ground-based hyperspectral sensors provide continuous measurements of reflectance over the visible to near infrared spectrum at finer spectral resolution than the Landsat platform. These data can be used to calculate a wide variety of spectral indices (such as Normalized Difference Vegetative Index or NDVI), which are useful in estimating plant and soil characteristics (Miao et al., 2009).

Advanced Very High Resolution Radiometer (AVHRR) and Moderate Resolution Imaging Spectroradiometer (MODIS, Le Du-Blayo et al., 2008) satellite remote-sensing platforms cover the same spectral ranges as Landsat and Quickbird, but have a low spatial resolution, which may be suitable for mapping large areas, and providing information about cloud cover and surface temperature. SRTM and LiDAR are useful for developing digital elevation models, at moderate or high spatial resolutions, respectively.

Proximal sensors mounted on vehicles or carried by hand are useful for measuring a variety of soil properties at high spatial resolution. Time domain reflectometry (TDR) uses microwave frequencies to measure soil moisture content. Ground penetrating radar (GPR) uses radio waves for subsurface soil imaging. It is particularly useful for inferring depths and thicknesses of soil horizons, identifying the presence of diagnostic soil horizons, and assisting in soil classification (Doolittle and Collins, 1995). Electromagnetic (EM) resistance is a tool for rapid assessment of spatial patterns in soil moisture, clay content, and salinity (Adamchuk et al., 2004). Visible near infrared (VNIR) sensors has proven to be accurate for identifying spatial patterns in soil organic matter (Bilgili et al., 2011). The choice of a particular remote or proximal sensor technology depends on many factors, including cost, spatial coverage, temporal coverage, spatial resolution, spectral range, and intended application. They are invaluable at providing auxiliary information to assist with digital soil mapping.

3. CONCLUSIONS

Tools useful for modeling soil spatial or temporal variability include: geostatistics, similar media scaling, autoregressive moving averages, and state-space Kalman filtering. Geostatistics is capable of modeling spatial structure and using it to estimate a desired soil property at an unsampled location. Geostatistical approaches include ordinary punctual kriging, ordinary block kriging, cokriging, factorial kriging, and regression kriging. These approaches differ in their assumptions about stationarity of the mean value and semivariogram, and in their use of auxiliary data to improve spatial predictions.

Similar media scaling has been used in conjunction with digital soil databases to upscale soil hydraulic properties from the point location to the field or watershed scale. The method reduces the spatial variability in soil hydraulic properties through the use of characteristic length scale parameters estimated from pore-size distribution. Digital soil databases are available for hydrologic and biogeochemical modeling of soil function at different scales of map resolution; in the United States, typical scales of data availability include the soil map unit scale (SSURGO), state soil association scale (STATSGO), national (NATSGO), and global (SOTER) scales. As the scale of resolution becomes increasingly coarse (e.g. SSURGO versus STATSGO), the accuracy of using soil database parameters for predictions of spatial and temporal variability in soil moisture becomes subject to greater uncertainty. Most of the spatial variability in soil hydraulic properties occurs over fine scales, and this variability is difficult to predict using STATSGO database properties.

Digital soil mapping can be improved using auxiliary data from landscape and terrain analysis, remote or proximal sensing, geostatistics, and pedotransfer function models. Terrain analysis is becoming increasingly powerful with the advent of digital elevation models based on LiDAR data. With terrain analysis, it is possible to characterize primary (slope, aspect, curvature, or specific catchment area) or secondary (stream power or compound topographic wetness) features of the terrain that control both the flow and transport of water as well as soil development. Remote or proximal sensing has witnessed major advances in spatial, temporal, and spectral resolution over the last several decades. Spatial resolution has improved from 30 m to under 1 m with satellite or aerial imagery. Temporal resolution has improved from monthly to daily return frequencies. Spectral resolution has improved from a handful of visible wavelengths to hyperspectral imaging across the visible and near-infrared spectrum at increments as narrow as 1 nm. These advances allow for better unmixing and classification of pixels having diverse reflectance patterns from soil or vegetation, at higher spatial, temporal, and spectral resolution. The improved classification results in more accurate mapping of land use, land cover, soil color, soil organic carbon, soil moisture, and other soil–landscape features.

REFERENCES

Adamchuk, V.I., Hummel, J.W., Morgan, M.T., Upadhyaya, S.K., 2004. On-the-go soil sensors for precision agriculture. Comput. Electronics Agric. 44, 71–91.

Bardossy, A., Lehmann, W., 1998. Spatial distribution of soil moisture in a small catchment. Part I: geostatistical analysis. J. Hydrol. 206, 1–15.

Batjes, N.H., Van Engelen, V.W.P., Kauffman, J.H., Oldeman, L.R., 1994. Development of soil databases for global environmental modeling. Trans. 15th World Congress Soil Sci. (ISSS, Mexico).

Bhatti, A.U., Mulla, D.J., Frazier, B.E., 1991. Estimation of soil properties and wheat yields on complex eroded hills using geostatistics and thematic mapper images. Remote Sensing Environ. 37, 181–191.

Bilgili, A.V., Akbas, F., van Es, H.M., 2011. Combined use of hyperspectral VNIR spectroscopy and kriging methods to predict soil variables spatially. Precision Agric. 12, 395–420.

Bocchi, S., Castrignano, A, Fornaro, F., Maggiore, T., 2000. Application of factorial kriging for mapping soil variations at field scale. Eur. J. Agron. 13:295–308.

Boettinger, J.L., Ramsey, R.D., Bodily, J.M., Cole, N.J., Kienast-Brown, S., Nield, S.J., Saunders, A.M., Stum, A.K., 2008. Ch. 16. Landsat spectral data for digital soil mapping. In: Hartemink, A.E., McBratney, A., de Lourdes Mendonca-Santos, M. (Eds.), Digital Soil Mapping with Limited Data. Springer, p. 437.

Burrough, P.A., Bouma, J., Yates, S.R., 1994. The state of art in pedometrics. Geoderma 62, 311–362.

Cambardella, C.A., Moorman, T.B., Novak, J.M., Parkin, T.B., Karlen, D.L., Turco, R.F., Konopka, A.E., 1994. Field scale variability of soil properties in central Iowa soils. Soil Sci. Soc. Am. J. 58, 1501–1511.

Campbell, G.S., Mulla, D.J., 1990. Measurement of soil water content and potential. In: Stewart, B.A., Nielson, D.R. (Eds.), Irrigation of Agricultural Crops. A.S.A. Monograph #30. Am. Soc. Agronomy, Madison, Wisconsin, Ch. 6.

Das, B.S., Haws, N.W., Rao, P.S., 2005. Defining geometric similarity in soils. Vadose Zone J. 4, 264–270.

Delworth, T., Manabe, S., 1988. The influence of potential evapotranspiration on variabilities in simulated soil wetness and climate. J. Clim., 523–547.

Dobos, E., Daroussin, J., Montanarella, L., 2005. An SRTM-based Procedure to Delineate SOTER Terrain Units on 1:1 and 1:5 Million Scales. EUR 21571 EN. Office for Official Publications of the European Communities, Luxembourg, p. 55.

Dobos, E., Carré, F., Hengl, T., Reuter, H.I., Tóth, G., 2006. Digital Soil Mapping as a Support to Production of Functional Maps. EUR 22123 EN. Office for Official, Publications of the European Communities, Luxemburg, p. 68.

Doolittle, J.A., Collins, M.E., 1995. Use of soil information to determine application of ground penetrating radar. J. Appl. Geophys. 33, 101–108.

Dunne, T., Black, R.D., 1970. Partial area contributions to storm runoff in a small New England watershed. Water Resour. Res. 6, 1296–1311.

Easton, Z.M., Fuka, D.R., Walter, M.T., Cowan, D.M., Schneiderman, E.M., Steenhuis, T.S., 2008. Re-conceptualizing the soil and water assessment tool (SWAT) model to predict runoff from variable source areas. J. Hydrol. 348, 279–291.

Farr, T.G., Kolbrick, M., 2000. Shuttle radar topography mission produces a wealth of data. Am. Geophys. Union, EOS 81, 583–585.

Ferrer-Julia, M., Estrela Monreal, T., Sanchez del Corral Jimenez, A., Garcıa Melendez, E., 2004. Constructing a saturated hydraulic conductivity map of Spain using pedotransfer functions and spatial prediction. Geoderma 123, 257–277.

Franke, T., Guntner, A., Mamede, G., Muller, E.N., Bronstert, A., 2008. Automated catena-based discretization of landscapes for the derivation of hydrological modelling units. Intl. J. Geog. Info. Sci. 22, 111–132.

Freeze, R.A., 1972. Role of subsurface flow in generating surface runoff: 2. Upstream source areas. Water Resour. Res. 8, 1272–1283.

Galzki, J., Birr, A.S., Mulla, D.J., 2011. Identifying critical agricultural areas with 3-meter LiDAR elevation data for precision conservation. J. Soil Water Conserv. 66, 423–430.

Gesch, D., Evans, G., Mauck, J., Hutchinson, J., Carswell, W. J.,Jr. 2009. The National map—elevation: U.S. Geological Survey Fact Sheet 2009–3053. p. 4.

Geza, M., McCray, J.E., 2008. Effects of soil data resolution on SWAT model stream flow and water quality predictions. J. Environ. Manage. 88, 393–406.

Goovaerts, P., 1999. Geostatistics in soil science: state-of-the-art and perspectives. Geoderma 89, 1–45.

Gowda, P.H., Mulla, D.J., 2005. Scale effects of STATSGO vs. SSURGO soil databases on water quality predictions. In: Proc. Third Conf. Watershed Management to Meet Water Quality Standards and Emerging TMDL (Total Maximum Daily Load). Atlanta, Georgia, March 5–9, 2005. American Society of Agricultural Engineers, St. Joseph, MI.

Gowda, P.H., Dalzell, B.J., Mulla, D.J., 2007. Model based nitrate TMDLs for two agricultural watersheds in southern Minnesota. J. Am. Water Resour. Assoc. 43, 256–263.

Gray, J.M., Humphreys, G.S., Deckers, J.A., 2007. Use of a large world soil database for modeling the global soil distribution. Eurasian Soil Sci. 40, 928–933.

Grayson, R.B., Western, A.W., 1998. Towards a real estimation of soil water content from point measurements: time and space stability of mean response. J. Hydrol. 207, 68–82.

Heuvelink, G.B.M., Webster, R., 2001. Modeling soil variation: past, present, and future. Geoderma 100, 269–301.

Iwahashi, J., Pike, R.J., 2007. Automated classifications of topography from DEMs by an unsupervised nested-means algorithm and a three-part geometric signature. Geomorphology 86, 409–440.

Jenny, H., 1941. Factors of Soil Formation. A System of Quantitative Pedology. McGraw Hill Book Company, New York, NY, USA, p. 281.

Juracek, K.E., Wolock, D.M., 2002. Spatial and statistical differences between: 1:250,000- and 1:24,000 scale digital soil databases. J. Soil Water Conserv. 57, 89–94.

Jury, W.A., 1986. Spatial variability of soil properties. In: Hern, S.C., Melancon, S.M. (Eds.), Vadose Zone Modeling of Organic Pollutants. Lewis Publ., Chelsea, MI, pp. 245–269.

Kachanoski, R.G., de Jong, E., 1988. Scale dependence and the temporal persistence of spatial patterns of soil water storage. Water Resour. Res. 24, 85–91.

Kosugi, K., Hopmans, J.W., 1998. Scaling water retention curves for soils with log-normal pore-size distribution. Soil Sci. Soc. Am. J. 62, 1496–1505.

Le Du-Blayo, L., Gouery, P., Corpetti, T., Michel, K., . Lemercier, B., Walter, C., 2008. Ch. 30. Enhancing the use of remotely-sensed data and information for digital soilscape mapping. In: Hartemink, A.E., McBratney, A., de Lourdes Mendonca-Santos, M. (Eds.), Digital Soil Mapping with Limited Data. Springer, p. 437.

Liang, X., Xie, Z., 2001. A new surface runoff parameterization with subgrid scale soil heterogeneity for land surface models. Adv. Water Resour. 24, 2273–1193.

Lin, H.S., McInnes, K.J., Wilding, L.P., Hallmark, C.T., 1999. Effects of soil morphology on hydraulic properties:II. Hydraulic pedotransfer functions. Soil Sci. Soc. Am. J. 63, 955–961.

Lin, H., Wheeler, D., Bell, J., Wilding, L., 2005. Assessement of soil spatial variability at multiple scales. Ecol. Model. 182, 271–290.

Lin, H.S., Brooks, E., McDaniel, P., Boll, J., 2008. Hydropedology and surface/subsurface runoff processes. In: Anderson, M.G. (Ed.), Encyclopedia of Hydrologic Sciences. John Wiley & Sons, Ltd. doi: 10.1002/0470848944.hsa306.

Liu, X., Peterson, J., Zhang, Z., 2005. High-resolution DEM generated from LiDAR data for water resource management. In: MODSIM05 International Congress on Modelling and Simulation: Advances and Applications for Management and Decision Making, 12–15 Dec 2005, Melbourne, Australia.

Loague, K., 1992. Soil water content at R-5: part 2. Impact of antecedent conditions on rainfall-runoff simulations. J. Hydrol. 139, 253–261.

Logsdon, S., Perfect, E., Tarquis, A.M., 2008. Multiscale soil investigations: physical concepts and mathematical techniques. Vadose Zone J. 7, 453–455.

Lopez-Granados, F., Jurado-Exposito, M., Pena-Barragan, J.M., Garcia-Torres, L., 2005. Using geostatistical and remote sensing approaches for mapping soil properties. Eur. J. Agron. 23, 279–289.

Lyon, S.W., Walter, M.T., Gerard-Marchant, P., Steenhuis, T.S., 2004. Using a topographic index to distribute variable source area runoff predicted with the SCS curve-number equation. Hydrol. Process. 18, 2757–2771.

Lyon, S.W., McHale, M.R., Walter, M.T., Steenhuis, T.S., 2006. The impact of runoff generation mechanisms on the location of critical source areas. J. Am. Water Res. Assoc. 42, 793–804.

McBratney, A.B., Webster, R., 1983. Optimal interpolation and isarithmic mapping of soil properties: V. Co-regionalization and multiple sampling strategy. J. Soil Sci. 34, 137–162.

McBratney, A.B., Pringle, M.J., 2001. Estimating average and proportional variograms of soil properties and their potential use in precision agriculture. Prec. Ag. 1, 125–152.

McBratney, A.B., Mendonca Santos, M.L., Minasny, B., 2003. On digital soil mapping. Geoderma 117, 3–52.

McDaniel, P.A., Regan, M.P., Brooks, E., Boll, J., Barndt, S., Falen, A., Young, S.K., Hammel, J.E., 2008. Linking fragipans, perched water tables, and catchment-scale hydrological processes. Catena 73, 166–173.

Mednick, A.C., 2010. Does soil data resolution matter? State soil geographic database versus soil survey geographic database in rainfall-runoff modeling across Wisconsin. J. Soil Water Conserv. 65, 190–199.

Miao, Y., Mulla, D.J., Randall, G., Vetsch, J., Vintila, R., 2009. Combining chlorophyll meter readings and high spatial resolution remote sensing images for in-season site-specific nitrogen management of corn. Precision Agric. 10, 45–62.

Milne, G., 1936. A provisional soil map of east Africa. Eastern Africa agriculture research station. Amanin Memoirs, 1–34.

Minasny, B., McBratney, A.B., Lark, R.M., 2008. Digital soil mapping technologies for countries with sparse data infrastructures. Ch. 2. In: Hartemink, A.E., McBratney, A., de Lourdes Mendonca-Santos, M. (Eds.), Digital Soil Mapping with Limited Data. Springer, p. 437.

Moore, I.D., Gessler, P.E., Nielsen, G.A., Peterson, G.A., 1993. Soil attribute prediction using terrain analysis. Soil Sci. Soc. Amer. J. 57, 443–452.

Morkoc, F., Biggar, J.W., Nielsen, D.R., Rolston, D.E., 1985. Analysis of soil water content and temperature using state-space approach. Soil Sci. Soc. Am. J. 49, 798–803.

Mulla, D.J., 1986. Distribution of slope steepness in the Palouse region of Washington. Soil Sci. Soc. Amer. J. 50 (6), 1401–1405.

Mulla, D.J., 1988a. Estimating spatial patterns in water content, matric suction, and hydraulic conductivity. Soil Sci. Soc. Am. J. 52, 1547–1553.

Mulla, D.J., 1988b. Using geostatistics and spectral analysis to study spatial patterns in the topography of southeastern Washington State, U.S.A. Earth Surf. Process. Landforms 13, 389–405.

Mulla, D.J., Huyck, L.M., Reganold, J.P., 1992. Temporal variation in aggregate stability on conventional and alternative farms. Soil Sci. Soc. Am. J. 56, 1620–1624.

Mulla, D.J., McBratney, A.B., 2002. Soil spatial variability. Ch. 9. In: Warrick, A.W. (Ed.), Soil Physics Companion. CRC Press, Boca Raton, FL, p. 389.

Nangia, V., Mulla, D.J., Gowda, P.H., 2010. Precipitation changes impact stream discharge, nitrate-nitrogen load more than agricultural management changes. J. Environ. Qual. 39, 2063–2071.

Nasta, P., Kamai, T., Chirico, G.B., Hopmans, J.W., Romano, N., 2009. Scaling soil water retention functions using particle-size distribution. J. Hydrol. 374, 223–234.

Nielson, D.R., Biggar, J.W., Erh, K.T., 1973. Spatial variability of field-measured soil-water properties. Hilgardia 42, 215–259.

Odeh, I.O.A., McBratney, A.B., Chittleborough, D.J., 1995. Further results on prediction of soil properties from terrain attributes:heterotopic cokriging and regression-kriging. Geoderma 67, 215–225.
Odeh, I.O.A., McBratney, A.B., 2000. Using AVHRR images for spatial prediction of clay content in the lower Namoi Valley of eastern Australia. Geoderma 97, 237–254.
Oldeman, L.R., Hakkeling, R.T.A., Sombroek, W.G. 1991. World map of the status of human-induced soil degradation: an explanatory note. Wageningen: International Soil Reference and Information Centre; Nairobi: United Nations Environment Programme. -I11. Global Assessment of Soil Degradation GLASOD, October 1990, second revised edition October Publ. in cooperation with Winand Staring Centre, International Society of Soil Science, Food and Agricultural Organization of the United Nations, International Institute for Aerospace Survey and Earth Sciences.
Pachepsky, Y.A., Rawls, W.J., 2003. Soil structure and pedotransfer functions. Eur. J. Soil Sci. 53, 443–451.
Pennock, D.J., Zebarth, B.J., De Jong, E., 1987. Landform classification and soil distribution in hummocky terrain, Saskatchewan, Canada. Geoderma 40, 297–315.
Quinn, P., Beven, K., Chevallier, P., Planchon, O., 1991. The prediction of hillslope flow paths for distributed hydrological modeling using digital terrain models. Hydrol. Proc. 5, 59–79.
Rawlins, B.G., Marchant, B.P., Smyth, D., Scheib, C., Lark, R.M., Jordan, C., 2009. Airborne radiometric survey data and a DTM as covariates for regional scale mapping of soil organic carbon across Northern Ireland. Eur. J.Soil Sci. 60, 44–54.
Romano, N., Palladino, M., 2002. Prediction of soil water retention using soil physical data and terrain attributes. J. Hydrol. 265, 56–75.
Rossi, R.E., Mulla, D.J., Journel, A.G., Franz, E.H., 1992. Geostatistical interpretation of ecological phenomena: tools for modeling spatial dependence. Ecol. Monogr. 62, 277–314.
Ruhe, R.V., 1969. Quaternary Landscapes in Iowa. Iowa State University Press, Ames, IA.
Saxton, K.E., Rawls, W.J., Romberger, J.S., Papendick, R.I., 1986. Estimating generalized soil-water characteristics from texture. Sci. Soc. Am. J. 50, 1031–1036.
Seelan, S.K., Laguette, S., Casady, G.M., Seielstad, G.A., 2003. Remote sensing applications for precision agriculture: a learning community approach. Rem. Sens. Environ. 88, 157–169.
Sheshukov, A., Daggupati, P., Lee, M.C., Douglas-Mankin, K. 2009. ArcMap tool for pre-processing SSURGO soil database for ArcSWAT. Proc. 5th International SWAT Conference, Boulder, CO. Aug. 5–7.
Si, B.C., 2008. Spatial scaling analyses of soil physical properties: a review of spectral and wavelet methods. Vadose Zone J. 7, 547–562.
Sloan, P.G., Moore, I.D., 1984. Modeling subsurface stormflow on steeply sloping forested watersheds. Water Resour. Res. 20, 1815–1822.
Soil Survey Staff, 1995. Soil Survey Geographic (SSURGO) Data Base: Data Use Information. USDA-NRCS, Washington, DC, p. 31.
Soil Survey Staff, 1997. National Characterization Data. Soil Survey Laboratory. National Soil Survey Center, Natural Resources Conservation Service, Lincoln, NE.
Soil Survey Staff, 1999. Soil Taxonomy. Agricultural Handbook 436, second ed. USDA-NRCS, Washington, DC, p. 871.
Soil Survey Staff, 2006. Land Resource Regions and Major Land Resource Areas of the United States. In: Agriculture Handbook, revised ed., vol. 296. USDA, Soil Conservation Service, Washington, DC, p. 156.
Stoorvogel, J.J., Kempen, B., Heuvelink, G.B.M., de Bruin, S., 2009. Implementation and evaluation of existing knowledge for digital soil mapping in Senegal. Geoderma 149, 161–170.

Tarboton, D.G., 1997. A new method for the determination of flow directions and upslope areas in grid digital elevation models. Water Res. 33 (2), 309–319.

Tarboton, D.G. 2005. Terrain analysis using digital elevation models (TAUDEM) version 5. [Software and documentation online]. Available from: http://hydrology.usu.edu/taudem/taudem5.0/index.html/ (verified 28 Dec. 2010). Tarboton, D.G. Logan, UT.

Thompson, J.A., Pena-Yewtukhiw, E.M., Grove, J.H., 2006. Soil–landscape modeling across a physiographic region: topographic patterns and model transportability. Geoderma 133, 57–70.

Tuli, A., Kosugi, K., Hopmans, J.W., 2001. Simultaneous scaling of soil water retention and unsaturated soil hydraulic conductivity functions assuming lognormal pore-size distribution. Adv. Water Resour. 24, 677–688.

USDA-NRCS, 2011. The Twelve Orders of Soil Taxonomy. USDA-NRCS, Washington, DC. http://soils.usda.gov/technical/soil_orders/.

van Es, H.M., Ogden, C.B., Hill, R.L., Schindelbeck, R.R., Tsegaye, T., 1999. Integrated assessment of space, time, and management-related variability of soil hydraulic properties. Soil Sci. Soc. Am. J. 63, 1599–1607.

Vepraskas, M.J., Caldwell, P.V., 2008. Interpreting morphological features in wetland soils with a hydrologic model. Catena 73, 153–165.

Vinnickov, K.Y., Robock, A., Speranskaya, N.A., Schlosser, C.A., 1996. Scales of temporal and spatial variability of midlatitude soil moisture. J. Geophys. Res. 101, 7163–7174.

Viscarra Rossel, R.A., McBratney, A.B., 1998. Laboratory evaluation of a proximal sensing technique for simultaneous measurement of soil clay and water content. Geoderma 85, 19–39.

Walter, M.T., Walter, M.F., Brooks, E.S., Steenhuis, T.S., Boll, J., Weiler, K., 2000. Hydrologically sensitive areas: variable source area hydrology implications for water quality risk assessment. J. Soil Water Conserv. 55, 277–284.

Warrick, A.W., Mullen, G.J., Nielsen, D.R., 1977. Scaling field-measured soil hydraulic properties using a similar media concept. Water Resour. Res. 13, 355–362.

Webb, R.W., Rosenzweig, C.E., E. R. Levine. 2000. Global soil texture and derived water-holding capacities data set. Available from: http://www.daac.ornl.gov; from Oak Ridge National Laboratory Distributed Active Archive Center, Oak Ridge, Tennessee, U.S.A. doi:10.3334/ORNLDAAC/548.

Webster, R., Burgess, T.M., 1980. Optimal interpolation and isarithmic mapping of soil properties, III. Changing drift and universal kriging. J. Soil Sci. 31, 505–524.

Wendroth, O., Rogasik, H., Koszinski, S., Ritsema, C.J., Dekker, L.W., Nielsen, D.R., 1999. State-space prediction of field-scale soil water content time series in a sandy loam. Soil Till. Res. 50, 85–93.

Wilson, J.P., Gallant, J.C., 2000. Terrain Analysis: Principles and Applications. John Wiley & Sons, Inc., New York.

Wood, E.F., Sivapalan, M., Beven, K., 1990. Similarity and scale in catchment storm response. Rev. Geophys. 28, 1–18.

Yu, Z., Pollard, D., Chen, L., 2006. On continental scale hydrologic simulations with a coupled hydrologic model. J. Hydrol. 334, 110–124.

Zhou, X., Lin, H.S., White, E.A., 2008. Surface soil hydraulic properties in four soil series under different land uses and their temporal changes. Catena 73, 180–188.

Zhu, Q., Lin, H.S., 2009. Simulation and validation of subsurface lateral flow paths in an agricultural landscape. Hydrol. Earth Syst. Sci. Discuss. 6, 2893–2929.

Chapter 21

Digital Soil Mapping: Interactions with and Applications for Hydropedology

J.A. Thompson,[1,*] S. Roecker,[2] S. Grunwald[3] and P.R. Owens[4]

ABSTRACT

Spatial information on soils, particularly hydrologic and hydromorphic soil properties, is used to understand and assess soil water retention, flooding potential, erosion hazard, and depth to seasonal high water table. These properties influence soil use and management interpretations for construction, waste disposal, plant production, and water management. Observed soil characteristics (soil horizons and soil properties) serve as both evidence of past processes and an indicator of present processes. As such, hydrologic and hydropedologic factors are useful for understanding and predicting soil variability. Conversely, spatial information on soils can be a critical input to hydrologic and hydropedologic models. Digital soil mapping (DSM) has evolved from traditional soil survey to take advantage of advances in computing and geographic data handling, as well as increased availability of environmental covariate data from digital elevation models and remotely sensed imagery. Digital elevation models can be used to enhance the input to hydrologic and hydropedologic models and extrapolate outputs from these models.

[1]West Virginia University, Division of Plant and Soil Sciences, Morgantown, WV 26506-6108

[2]USDA-Natural Resources Conservation Service, Victorville MLRA Soil Survey Office, 14393 Park Avenue, Suite 200, Victorville, CA 92392

[3]University of Florida, Dept. of Soil and Water Science, 2169 McCarty Hall, Gainesville FL 32611

[4]Purdue University, Dept. of Agronomy, Lilly Hall of Life Sciences, 915 W. State Street, West Lafayette, IN 47907-2054

[*]Corresponding author: Email: james.thompson@mail.wvu.edu

Hydropedology, Edited by H. Lin. DOI: 10.1016/B978-0-12-386941-8.00021-6

1. SOIL MAPPING, SOIL SURVEY, AND THE VALUE OF SPATIAL SOIL INFORMATION

1.1. Pedology: Hydropedology and Pedometrics

Hydropedology is an integrative field of soil science, which incorporates the concepts of pedology, soil physics, and hydrology to understand soil–water interactions at various scales (Lin, 2003; Lin et al., 2005, 2006). Another integrative field that is rooted in pedology is pedometrics, which by contrast incorporates soil science, geographic information science, and statistics (Grunwald, 2006). Defined as "*the application of mathematical and statistical methods for the study of the distribution and genesis of soils*" (Heuvelink, 2003), pedometrics is concerned with quantifying soil variation in terms of its deterministic, stochastic, and semantic components. A subset of pedometrics is digital soil mapping (DSM), also referred to as predictive soil modeling (Scull et al., 2003) and quantitative soil survey (McKenzie and Ryan, 1999). Lagacherie (2008) has defined DSM as "*the creation and population of spatial soil information systems by numerical models inferring the spatial and temporal variations of soil types and soil properties from soil observations and knowledge and from related environmental variables.*" The explicit geographic nature of DSM aligns it with hydropedology because hydropedology has been advocated as a means to study the relationships between soils, landscapes, and hydrology (Lin et al., 2006). Therefore, DSM can provide effective linkages for the integration of pedology and hydrology. Lin (2011) referred to the development of such linkages between hydropedology and digital soil mapping as "*an exciting research area, which can improve the connection between spatial soil mapping and process-based modeling.*"

This chapter will examine the importance of hydropedology to advances in DSM through the expansion of our knowledge of pedogenesis and vertical and lateral soil variability at various scales; how DSM can support further advances in hydropedology through improved representation of spatial soil information; and how coevolution of these two fields may foster a greater understanding of – and sustainable use of – the Earth's Critical Zone.

1.2. Spatial Soil Information and Soil Survey

The perceived need for, as well as the demand for, spatial soil information is growing (McBratney et al., 2003, 2006; Lagacherie and McBratney, 2007; Hartemink and McBratney, 2008). Increasingly, detailed soil data from multiple counties and states are being viewed and analyzed together. To address resource issues ranging from local to global scales, environmental scientists and policymakers are seeking soil information that is more specific (soil properties) and more detailed (spatially explicit). These user's needs represent a challenge for soil scientists to provide new spatial soil information, particularly in a digital format that is readily incorporated into geographic information systems (GIS) and can be analyzed with other spatial data (Lagacherie and McBratney, 2007).

Soil surveys have been conducted to meet user needs by creating an inventory of local soil resources, and by educating the public about the role of soils in their environment (Brown and Miller, 1989). The primary objective of soil survey programs is relatively consistent: to delineate uniform management areas (i.e. map unit polygons) across the landscape and provide users with information on soil properties and soil interpretations for these map units to support generalized soil use and management decision making. Alternatively, as the Chief of the US Soil Survey Division put it, the role of the soil survey is "to get the facts about soils, to classify them, and to map them in ways that would furnish a sound basis for interpretation by other people" (Durana, 2002). However, more recently, users of soil survey information are asking questions of soil survey for which it was not intended to answer (Sanchez et al., 2009; Hartemink et al., 2010), such as calculation of C stocks, distributed hydrologic modeling, nutrient cycling and nutrient depletion, and examination of climate change.

Legacy soil maps (and associated tabular data) have been digitized to produce digital map products that represent the spatial extent of soil classes and soil properties (Grunwald et al., 2011). Most of these digitized maps use a vector format to represent polygons that consist of one or more soil types or classes (e.g. soil series or soil taxa), although some digitized soil maps utilize a raster format where each grid cell is assigned a soil class or a soil property value (Grunwald et al., 2011). Whether explicit (soil properties) or implicit (soil classes), these digital databases can be used to represent the spatial variability of hydrologic and hydromorphic properties. For example, the Soil Survey Geographic (SSURGO) database, which is the most detailed soil map data for the US, provides polygon soil class maps at scales of 1:12,000 to 1:125,000, and is 86% complete for the land area of the conterminous US (Soil Survey Staff, 2009). The map units represent one or more soil classes and are linked to tables of estimated property data for each component soil. The SSURGO data are most readily available through the USDA-NRCS Web Soil Survey (WSS). Summary statistics on WSS usage, particularly which properties and interpretations are most commonly requested (Table 1), indicate that soil–water properties are some of the most popular ones. Soil properties and soil ratings in the top 20 over a single year (January–December 2011) related to potential hydropedology applications or interpretations include the top three most requested ratings (hydrologic soil group, depth to water table, and drainage class) and eight of the top 20. If counting soil characteristics that are related to hydrologic properties or processes (e.g. percent sand, silt, and clay; soil organic matter content) this number of requested ratings is 12.

1.3. Use of Spatial Soil Information in Hydrology and Hydropedology

Legacy soil data are frequently used to visualize, analyze, or model the linkages between soil properties and hydrologic processes, or between soil variability

TABLE 1 Summary of Web Soil Survey (WSS) Usage for All of the Year 2011, Indicating the Rank of the Top 20 Queries Based on the Number of Individual Requests for Reporting of Each Property or Rating. Also Shown is the Number of Months Between January 2011 and December 2011 in which that Rating Appeared in the Top 50 of All Queries

Rank	Ratings	Count	Months in Top 50
1	Hydrologic Soil Group	12,2572	12
2	Depth to Water Table	93,514	12
3	Drainage Class	84,697	12
4	pH (1–1 Water)	80,581	12
5	K Factor, Whole Soil	77,317	12
6	Percent Clay	76,822	12
7	Flooding Frequency Class	76,453	12
8	Organic Matter	74,960	12
9	Corrosion of Concrete	73,209	12
10	Depth to Any Soil Restrictive Layer	73,067	12
11	Available Water Capacity	72,396	12
12	Saturated Hydraulic Conductivity (Ksat)	71,449	12
13	Percent Sand	71,429	12
14	Ponding Frequency Class	68,268	12
15	Plasticity Index	67,686	12
16	Percent Silt	67,641	12
17	T Factor	67,467	11
18	K Factor, Rock Free	66,898	10
19	Cation-Exchange Capacity (CEC-7)	66,693	9
20	Map Unit Name	66,383	11

and hydropedologic properties. For example, Turk and Graham (2011) used SSURGO and the smaller-scale State Soil Geographic (STATSGO) database to examine the extent and spatial distribution of vesicular horizons in the western US. Vesicular horizons are found at or near the soil surface, where the characteristic bubble-like vesicular pores influence surface hydrology by restricting infiltration, often encouraging surface runoff and ponding (Turk and Graham, 2011). As another example, Vepraskas et al. (2009) also used SSURGO data to examine hydromorphic properties and soil-use interpretations. Starting with measured soil water-table fluctuations for multiple soil series, which were combined with other historic climate data in a hydrologic model to compute long-term water-table fluctuations, Vepraskas et al. (2009) used the SSURGO data to extrapolate hydrologic modeling results and soil use and management interpretations (soil suitability for septic systems and the presence of wetland hydrology) across larger land areas. Vepraskas et al. (2009) went on to use these data and models to consider climate-change impacts on soil drainage class and, therefore, changes in soil use and management interpretations.

For hydrologic modeling, soil properties are one of the most fundamental input parameters of these models. For local-scale and watershed-scale modeling using models such as watershed modeling system (WMS) (Nelson et al., 1994), Soil Water Assessment Tool (SWAT) (Arnold and Allen, 1996), and TOPMODEL (Beven and Kirkby, 1979), SSURGO data are most commonly used as soil input parameters. Most of the hydrologic models are written so that input can be taken directly from the SSURGO data with minimal processing (Ogden et al., 2001). For continental and regional-scale climate, hydrology, and ecosystem modeling using models such as Variable Infiltration Capacity (VIC) (Liang et al., 1994, 1996), Noah Land Surface Model (Noah), and Common Land Model (CLM) (Bonan, 1998; Dai and Zeng, 1997; Dickinson et al., 1993), the 1-km data from the CONUS-Soil database for the Conterminous United States (Miller and White, 1998) are commonly used. This database is based on STATSGO, and it includes parameters such as soil texture class, depth to bedrock, bulk density, porosity, permeability, available water capacity, soil-hydrologic group, and curve numbers.

Land-surface climate models are used as large-scale hydrology models by parametrizing the land-surface schemes for water balance (Wood et al., 1998). The global land-surface models, such as CLM, MOSIAC model (Koster and Suarez, 1996), and VIC, use soil data from the Global Land Data Assimilation System (GLDAS). The soil data used in the GLDAS were derived from a global soil data set that includes fractions of sand, silt, and clay, and porosity, among other fields (Reynolds et al., 2000). These properties are spatially represented based on the FAO Soil Map of the World, which is correlated to a global database of over 1300 soil pedons. The spatial resolution of the GLDAS data is 1/4 or 1 degree. Soil properties drive much of the exchanges in the land-surface system and the global data sets are increasing the understanding of atmospheric feedbacks (Zaitchik et al., 2009).

1.4. The Case for DSM

While digitized soil maps are available for most of the world (Grunwald et al., 2011); for many areas those data are at a very small scale (1:1 million or coarser) and do not adequately represent soil variability in a format that is useful to non-pedologists (Sanchez et al., 2009). The majority of currently available digital soil maps are actually compilations of multiple legacy soil maps, which were initially produced as hard-copy maps and subsequently digitized (Grunwald et al., 2011). For example, although SSURGO and STATSGO are digital products, they are based on paper maps that were later converted to vector-based polygon maps. For SSURGO, the soil maps were originally produced as part of approximately 3,000 independent soil surveys, and these individual maps are of different vintages and different scales, they were created using different mapping concepts, and they often use different soil components and different estimated property data to represent the same soil–landscape features. Consequently, there are frequently artificial boundaries in the data associated with geopolitical boundaries caused by discontinuities in map unit composition and in estimated soil property data, which produce discontinuities in mapped soil properties and in soil use and management interpretations. All these emphasize the point that digitizing existing paper maps is not DSM.

Hartemink et al. (2010) listed the following limitations of most existing digitized soil survey maps: (i) they are static, (ii) they aggregate soil information into soil classes that are not readily compatible with quantitative applications, (iii) the information content has been overly generalized relative to the information on the regional soil resources that was collected to create the soil survey, (iv) they are improperly scaled, and (v) they represent the information as polygons that are not as readily combined with most other natural resource data, which are raster-based. Similarly, Zhu (2006) emphasized that the spatial and attribute generalization of soil spatial variation into discrete classes makes soil survey information incompatible with other forms of continuous spatial data for environmental modeling. All things considered, there is a tremendous potential for the DSM community to capitalize on the demand for better soil information by improving the quality of existing digital soil maps and directly creating raster-based soil data of functional soil properties for hydropedologic investigations and hydrologic modeling.

2. DIGITAL SOIL MAPPING—SPATIAL PREDICTION OF SOIL PROPERTIES AND TYPES

2.1. Review of Approaches to Soil Spatial Prediction

2.1.1. State Factor or clorpt Model

For the latter half of the 20th century, the scientific rationale for soil mapping has been the state factor, or *clorpt* model (Jenny, 1941, 1980), which was originally proposed by Dokucheav and Hiligard, and later fully articulated by

Hans Jenny (Hudson, 1992). The state factor model is expressed by the following equation:

$$S = f(\text{cl, o, r, p, } t, \ldots), \tag{1}$$

where soil (S) is considered to be a function of climate (cl), organisms (o), relief (r), and parent material (p) acting through time (*t*) (Jenny, 1941, 1980). The ellipsis (...) in the model is reserved for additional unique factors that may be locally significant, such as atmospheric deposition. The *clorpt* equation illustrates that by correlating soil attributes with observable differences in one or more of the state factors, a function (f) or model can be developed that explains the relationship between the two, which can be used to predict soil attributes at new locations when the state factors are known. An important distinction of the *clorpt* model is that "The factors are not formers, or creators, or forces; they are variables (state factors) that define the state of a soil system" (Jenny, 1961). This means that the factors do not constitute pedogenic processes, but are factors of the environmental system which condition processes. In order to bridge the gap between factors and processes, the *clorpt* model is supplemented by additional models that are useful at explaining various processes at different scales. A notable example is the catena concept (Milne, 1936), which attributes soil variation along a hillslope sequence to erosion and deposition, hydrology, and stratigraphy.

While Jenny developed quantitative relationships between soil attributes and the state factors one at a time (e.g. climofunctions), in conventional soil survey, soil scientists have to rationalize the local complexity of soil-landscapes in their entirety, and therefore develop conceptual models of the soil–landscape relationships. Soil scientists' derive their tacit knowledge of soil–landscape relationships by viewing the soil continuum at a number of opportune and purposive locations (McKenzie and Austin, 1993). To segment the soil continuum, soil scientists use natural boundaries, such as "topographic divides, contacts between different rocks or sediment, inflections in slope gradient or shape, and contacts between different landforms of different age, origin, and internal structure" (Wysocki et al., 2011). Other common boundaries used in soil survey include different vegetative communities, or predefined breaks in slope gradient that are deemed significant to land management. In order to articulate the landscape position of unique soil components within map units (e.g. soil delineations), soil scientists use stringently defined geomorphic descriptors (Schoenberger and Wysocki, 2008; USDA-NRCS, 2012). Due to the fact that soil components may only correspond with a portion of a given landform, there are numerous soil-geomorphic descriptors that are unique to soil science.

2.1.1.1. Hydrologic Processes of Soil Formation

While hydrology is not explicitly named as a state factor in the *clorpt* model, water movement and storage within pedons and across landscapes form the driving force of most pedogenesis (Buol et al., 2003). For this reason, Runge (1973) named water as one of the three factors in his alternative soil factorial model, in addition to organic matter production and time. In contrast to Jenny's (1941) state factor model, Runge incorporated elements of Simonson's (1959, 1978) process-systems model. Within the *clorpt* model, hydrologic factors – including atmospheric (precipitation amount and timing) and terrestrial (evapotranspiration, water-table fluctuations or lateral water flow, hydraulic gradient, or proximity to surface water) – have been considered to be represented by climate and relief, respectively (Jenny, 1941; Buol et al., 2003). This distinction is useful for separating regional and local influences on soil hydrology.

Therefore, hydrology is considered indirectly related to each of the state factors. Climatic variations in precipitation and evapotranspiration control the amount and timing of water additions to the land surface. Relief affects the redistribution of precipitation, either due to orographic effects or by concentrating runoff, and variation in evapotranspiration, due to differences in solar radiation. Organisms, particularly vegetation, not only respond to differences in available water (due to the effect of climate and relief), but also directly influence the amount of water through interception and consumptive water use. Overall, hydrology is not an independent state factor, but a collection of processes that are influenced by the state factors.

2.1.2. scorpan *Model – The Digital Soil Mapping Formula*

Recently, McBratney et al. (2003) have offered a revised formalization of the state factor model. This revised formula is expressed by the following equation:

$$S = f(s, c, o, r, p, a, n), \tag{2}$$

where S, a set of soil attributes (S_a) or classes (S_c), is considered a function of other known soil attributes or classes (s), climate (c), organisms (o), relief (r), parent materials (p), age or time (a), and spatial location or position (n). The *scorpan* equation also explicitly incorporates space (x,y coordinates) and time ($\sim t$). Thus, the *scorpan* equation can be expanded as follows:

$$\begin{aligned} S[x,y,\sim t] = f(&S[x,y,\sim t], c[x,y,\sim t], o[x,y,\sim t], r[x,y,\sim t], \\ &p[x,y,\sim t], a[x,y], [x,y]). \end{aligned} \tag{3}$$

This expansion indicates that *scorpan* is a geographic model, where the soil and factors are spatial layers that can be represented in a geographic information system.

The *scorpan* model deviates from *clorpt* in that it is intended for quantitative spatial prediction, rather than explanation (McBratney et al., 2003). This distinction justifies the inclusion of soil and space as factors, because soil attributes can be predicted from other soil attributes and spatial information. For example, many soil attributes are correlated and can thus be reasonably predicted from each other. Also, Tobler's first law of geography (Tobler, 1970) tells us that near things are spatially correlated, and can thus be predicted by their distance from their neighbors. To account for the soil factor, prior soil information can come from either published soil maps or the expert knowledge of soil surveyors. The space factors can come from indices of relative position, or by incorporating spatial autocorrelation (e.g. cokriging or regression kriging). While prior soil information is undoubtedly not an independent factor, the independence of spatial information is dubious. In most situations, spatial information likely accounts for relationships not captured by other factors (McBratney et al., 2003). Otherwise, from a metaphysical perspective, spatial information accounts for the random diffusion of particles trying to achieve a uniform state within a system (Hengl, 2009). The addition of space is not a new idea, as Jenny himself spent a great deal of time validating spatial soil relationships (Hudson, 1992). However, the revised formulation of *scorpan* recognizes the importance of prior soil information and spatial relationships to help explain soil spatial variation.

2.1.3. STEP-AWBH Model – A Space–Time Modeling Framework

Given the importance of anthropogenic forcings in determining observed soil properties, Grunwald et al. (2011) have proposed a new conceptual model for understanding soil properties for a pixel (p_x) of size x (width = length = x) at a specific location on Earth, at a given depth (z), and at the current time (t_c):

$$SA(z, p_x, t_c) = f\left\{\sum_{j}^{n}[S_j((z, p_x, t_c), T_j(p_x, t_c), E_j(p_x, t_c), P_j(p_x, t_c)]\right\};$$
$$\int_{i=0}^{m}\left\{\sum_{j}^{n}[A_j(p_x, t_i), W_j(p_x, t_i), B_j(p_x, t_i), H_j(p_x, t_i)]\right\}, \quad (4)$$

where the soil property of interest (*SA*) is a function of a number ($j = 0, 1, 2, \ldots, n$) of relatively static environmental factors (only at t_c): ancillary soil properties (*S*), topographic properties (*T*), ecological properties (*E*), and parent material properties (*P*), as well as a number ($j = 0, 1, 2, \ldots, n$) of dynamic environmental conditions (with values representing dynamics through time t_i with $i = 0, 1, 2, \ldots, m$): atmospheric properties (*A*), water properties (*W*), biotic properties (*B*), and human-induced forcings (*H*). The model is spatially explicit because it constrains all properties included in equation (4) to a specific pixel location. The model is temporally explicit as indicated by the inclusion of t (time) in equation (4). This recognizes the spatial variation and temporal evolution of STEP-AWBH properties that covary and coevolve with the target soil property SA. Similar to the *scorpan* model, the STEP-AWBH model reflects the new emphasis on existing soil information and spatial location as key attributes capable of providing predictive power in soil models.

The STEP-AWBH model separates hydrologic properties (*W*) from topographic (*T*) and climatic (*A*) factors, and formally includes anthropogenic properties (*H*). In previous factorial soil models, topography (relief) indirectly expressed the effects of hydrology on soil genesis. Yet, this is a simplification of reality that the STEP-AWBH model tries to overcome, where *T* represents topographic properties (e.g. elevation, slope gradient, slope curvature, and compound topographic index) that have been shown to be correlated with soil properties (see Grunwald, 2009). However, water flow and transport processes in soil depend on the interplay of *A* (atmospheric properties, such as precipitation arriving at the soil surface); soil surface conditions (e.g. salt crusts, residues, and density and composition of vegetation cover); internal soil characteristics that determine infiltration, percolation, and lateral flow processes (such as soil texture and soil organic matter); parent material (geologic formations) that control flow into the surface and deep aquifer; and topographic properties that enhance or subdue water flow at or close to the soil surface. Thus, in the STEP-AWBH model the hydrologic properties (*W*) are captured separately from *T*. Hydrologic properties can be measured using common physically based methods (e.g. to measure infiltration and soil moisture), assessed empirically (e.g. average water capacity or long-term upper and lower water table), or remotely (e.g. using soil moisture sensors such as those derived from the European Space Agency (ESA) Soil Moisture Ocean Salinity (SMOS) Earth Explorer mission or radar-derived soil moisture using RADARSAT). Specifically, satellite-derived hydrologic properties have emerged as a direct measurement set that provides pixel-specific values covering large regions. Since hydrologic properties vary in the space and time domain, they can be expressed as (i) a series of spatially and temporally explicit data inputs into equation (4), for example, soil moisture derived in monthly intervals over a 2- or 3-year time period for all pixels within a region or (ii) temporally aggregated sets (e.g. average, minimum, and peak soil moisture over a 5-year period) for all pixels within a region. Likewise, other AWBH variables can be entered into equation (4) using spatially and temporally explicit sets (e.g. temperatures at a discrete time, e.g. Jan. 1, 2011) or condensed/aggregated data over a period of time (e.g. mean annual temperatures 2000–2010). The *H* factor represents different anthropogenic forcings that can act across shorter or longer periods of time on SA(z,p_x,t_c) to shift SA into a different state, such as greenhouse gas emissions, contamination (e.g. an oil spill), disturbances, overgrazing, and others.

2.2. Digital Soil Mapping

The generic digital soil mapping approach takes the form of the *scorpan* model or STEP-AWBH. This approach is similar to that practiced in conventional soil mapping, except that the functional relationships between the soil attributes or classes and model factors are formulated using mathematical (e.g. expert rule-based or fuzzy logic models) or statistical models, rather than conceptual models (Ryan et al., 2000). These mathematical or statistical models are fitted or trained using geo-referenced soil data, expert knowledge, or preexisting soil maps.

The model factors (also known as environmental covariates and ancillary data) are represented by environmental layers contained in a GIS. Typically, preference is given to raster-based geographic data sets, such as derivatives of digital elevation models (DEM) and remotely sensed imagery (RSI), as they provide a dense grid of measured or interpolated values with which to correlate to soil attributes. Developing DSM models by correlating soil and environmental factors is an efficient quantitative spatial prediction approach if the factors are more easily attainable than the soil observations, and have a strong physical connection to soil attributes or classes (Gessler et al., 1995). McKenzie and Ryan (1999) cited the impetus for this approach due to the need to develop quantitative soil spatial prediction methods applicable at smaller scales, as opposed to methods based purely on spatial interpolation between soil observations.

2.2.1. Mathematical and Statistical Models

In order to predict the spatial distribution of soil properties or classes, various mathematical and statistical models can be used. A comprehensive review of their application in DSM is provided by McBratney et al. (2003). A similar review has taken place in ecological modeling (Guisan and Zimmermann, 2000). Regardless of the variety of models available, Austin et al. (2006) have stressed that in ecological modeling the most important consideration is not the statistical model employed, but the ecological knowledge and statistical skill of the analyst. Minasny and McBratney (2007) have likewise concluded that improved spatial prediction of soil characteristics will result from accumulating better soil data, rather than more sophisticated statistical models. Some of the most common mathematical and statistical methods applied in DSM are discussed here, such as kriging, generalized liner models, tree-based models, and fuzzy logic models.

2.2.2. Kriging

Kriging is a special case of the *scorpan* model, where only the *n* factor is considered. Like other standard forms of spatial interpolation, kriging develops predictions at new locations by modeling the spatial dependence between neighboring observations as a function of their distance. Therefore, by taking into account the distance between neighboring observations a locally weighted average is taken to estimate values at new points. However, unlike other forms of spatial interpolation, with kriging the spatial weights are estimated objectively with a statistical model, the variogram, rather than by an arbitrary mathematical function. In addition to interpolating predictions, kriging is able to estimate the variance at each point, which can be used to judge the spatial accuracy of the interpolation.

Kriging is a suitable method in the presence of spatial dependence. However, in many cases, soil properties are the result of environmental covariates. In such cases, it is beneficial to model the deterministic component of soil spatial variation as a function of the environmental covariates, and any residual stochastic component by kriging. Variants of kriging, which incorporates both deterministic and stochastic components, include cokriging and regression kriging. In comparison to other statistical and geostatistical models, Bishop and McBratney (2001) have demonstrated regression kriging to be superior. However, at the landscape scale, when the soil is not sampled at distances closer than the average range of spatial dependence, Scull et al. (2005) found multiple linear regression to be superior to regression kriging.

2.2.3. Generalized Linear Models

One of the most commonly used groups of regression and classification models are generalized linear models (GLM), which are a modified form of the classical linear model designed to handle situations in which the linear model's main assumptions are not met, with these assumptions being that the response is normally distributed with a constant variance and that the predictors combine additively on the response. Lane (2002) has advocated the use of GLM in soil science as opposed to transforming the linear model when these assumptions are not met, such as for binomial (presence/absence) and Poisson (counts) distributions. Transformations are typically used to modify the linear model to handle alternative distributions, but can affect the interpretation of additivity on the transformed scale, where statistics such as standard errors and variance ratios values should be used with caution (Webster, 2001). Like linear models, GLM have similar fitting procedures and diagnostics, so they can be likewise interpreted:

$$g(\mu) = \beta_0 + \beta x_1 + \ldots + \varepsilon. \tag{5}$$

To modify the classical linear model, GLM allow the response to belong to a wide range of exponential family distributions (e.g. Gaussian, binominal, Poisson, and gamma), and relate the response's mean to the model on scale where the effects combine additively through the link function, $g(\mu)$. The effect of the link function transforms the model to linearity, and maintains the response's range of values. More simply put, this transforms the model, rather than transforming the data to fit the model's assumptions. A consequence of modifying the linear model requires that the parameters be estimated iteratively by maximum likelihood, as opposed to being derived analytically as with least squares. Another consequence eliminates the ability to employ the analysis of variance. Instead, the analysis of deviance is used, which is a measure of the difference between the observations and the fitted model, which for Gaussian distributions equates to the residual sum of squares.

Aside from being able to handle multiple distributions, GLM have additional benefits, such as being able to use both categorical and continuous predictor variables. Also as with linear models, they allow interactions between the predictors and polynomial terms, so as to model more complex data structures. To identify such interactions, exploratory techniques such as tree-based models (Guisan et al., 2002) and coplots (McKenzie and Jacquier, 1997) may be used. The use of interactions however can increase colinearity within the model at the expense of identifying meaningful relationships between variables (Park and Vlek, 2002).

2.2.4. Tree-Based and Forest Models

Tree-based models or decision trees differ from GLM in that they do not make assumptions about the form of the data. Instead, they are often referred to as data driven, whereby the resulting model's structure is based on the data itself, rather than some assumed distribution. This can be seen as both an advantage and a disadvantage. For example, given a sizable data set trees can easily identify complex data structure. In the absence of a sizable data set, other parametric models (Maindonald and Braun, 2007) such as GLM are likely to provide better estimates, given that they make assumptions about the structure of the data.

To understand tree-based models it is best to discuss how they are grown. The standard method of tree construction develops a set of decision rules using binary partitioning, which repeatedly subdivides the response into two sets of increasingly more homogeneous groups until no further purity within the groups can be gained by splitting them. When plotted these decision rules resemble a tree. During each step of the tree's growth the partition of the response is based upon whatever split among the predictors creates the best fit. For continuous responses (e.g. regression trees) the splitting criteria used are the residual sum of squares, while for categorical responses (e.g. classification trees) there are a choice of three splitting criteria, all of which seek to optimize the proportion of correctly classified observations. After the tree is grown the final groups or leaves are labeled with the mean (e.g. regression trees) or majority (e.g. classification trees) response within the leaves. While growing a tree following this procedure can be simply automated, the decision of when to stop its growth requires the subjective intervention of the analyst. Ultimately, the process could continue until each observation is correctly classified. While this would accurately describe the given data set, it would overfit the existing data set, and therefore poorly predict new data. The idea is that as the tree grows, less and less reduction in deviance is gained with each split. So, the tree's overall accuracy would suffer little if it were pruned to a smaller number of leaves. To determine an optimum stopping point, cross-validation is used. This pruning method produces a plot of the number of leaves against the amount of deviance explained. The optimal stopping point is the location on the plot where the slope flattens out, or falls below one standard deviation of the minimum cross-validated error.

Because of the automated nature by which trees are grown, they are often useful for exploratory data analysis (Guisan et al., 2002), so as to indicate the relative importance and potential interaction between predictors. In addition, the results of a tree-based model are easily interpretable if the number of binary splits is small, and they allow a mixture of both continuous and categorical predictors. Despite the relative ease with which trees are constructed, Hastie et al. (2009) list three notable limitations. The first is that trees are inherently unstable due to their data-driven nature of construction. As such, any change in the data may produce a different tree. For this reason, research in DSM (Park and Vlek, 2001; Scull et al., 2005) has shown tree-based models to perform less well than parametric models when validated by an independent data set. A second limitation of trees is that for continuous responses they do not produce continuous predictions, but rather unrealistic stepped predictions. Still for noisy data sets, McKenzie and Ryan (1999) have suggested that this is not a problem. Lastly, Hastie et al. (2009) cite tree's inability to capture additive structure. Hastie et al. (2009) state that it is possible for trees to capture such structure with sufficient data, but that tree-based model construction process does not readily exploit such structure within data.

In an effort to overcome tree-based model's limitations, a number of alterations to the construction of trees have been proposed, such as boosting (Freund and Schapire, 1997), bagging (Breiman, 1996), and random forests (Breiman, 2001). Each distinct alteration creates a model comprised of multiple trees, generally termed an ensemble, or continuing with the use of tree metaphors a forest. By growing a forest rather than a single tree it is possible to take a majority or weighted vote among the trees, thereby increasing the accuracy and decreasing the sensitivity of the model. In boosting, a forest is grown by repeatedly reweighing the misclassified and correctly classified observations in the data set. In bagging, a forest is grown by taking repeated bootstrap samples of the observations in the data set. In random forest, a forest is grown by taking repeated bootstrap samples of the observations and predictors in the data set. While each of these ensemble methods typically generates better estimates than a single tree, they also come at the expense of their interpretability.

2.2.5. Fuzzy Logic

Fuzzy logic is an alternative to Boolean logic that determines the membership to a given class by either a 0 (no) or a 1 (yes). Fuzzy logic deals with the ambiguity of defining the soil–landscape continuum by allowing a soil to have partial membership to more than one class, on a scale between 0 and 1. Unlike kriging, regression, or classification, fuzzy logic is not truly a statistical model, because "it does not assess the accuracy of its predictions" (Heuvelink and Webster, 2001). The distinction between fuzzy logic and Boolean logic is that fuzzy logic is based on possibility theory, while Boolean logic is based on probability theory. In this way, fuzzy logic is a measure of a soil's similarity to a class, rather than its chance of belonging to it (Zhu, 2006). Zhu (2006) asserts that "soil classification is based on possibility, not probability", and, as such, fuzzy logic is a more appropriate approach for defining soil classes.

The advantage of fuzzy logic is that it allows for representing the continuous nature of the soil's both geographic distribution and attribute distinctness.

The most prominent application of fuzzy logic in DSM has been the SoLIM (Soil–Landscape Inference Model) model, developed by Zhu and Band (1994), Zhu (1997a,b), and Zhu et al. (1996, 1997). This approach uses the expert knowledge of an experienced soil scientist to formalize the relationship between soil characteristics and environmental covariates. The incorporation of soil scientists' expert knowledge though can be seen as both an advantage and a disadvantage (Scull et al., 2003). The advantage is that it can explicitly summarize a soil scientist's expert knowledge, which has been accumulated at great expense. The disadvantage is that a soil scientist's expert knowledge is subjective and lacks statistical grounds for inference.

2.3. State Factor Proxies

Historically, the *scorpan* or *clorpt* factors have been interpreted from aerial photography and topographic and geological maps. Recent advances in remote sensing and geomorphometry have made the quantification of such ancillary information more readily accessible in formats that can be manipulated with GIS. Perhaps the most easily quantified and directly correlated state factors are relief, as DEM derivatives have been the most popular predictor (McBratney et al., 2003). This is to be expected given that many soil-geomorphic and hydrologic landscape models are based on relief (Schaetzl and Anderson, 2005). However, in some cases, all the relevant state factor proxies may not exist for a given area, and therefore thematic (or vector-based) layers may need to be manually interpreted and digitized (MacMillan et al., 2010). In addition, not all state factors have predictors that are directly related to the factor they are trying to quantify; some are indirect. For example, direct estimates of age (or time) are typically absent from predictive models, unless manually incorporated (Noller, 2010). Instead, age has typically been indirectly inferred from parent material, relative landscape position (e.g. according to the principle of superposition), or by surface reflectance (e.g. dark surfaces indicate desert varnish). Below is a discussion of some of the most common environmental covariates derived from DEM and RSI.

2.3.1. Digital Elevation Models

Topography is seen to have a primary influence on local soil differentiation over time. Following the Simonson model (1959), topography is the surrogate for redistribution of water which provides energy for transfer, inputs, and outputs in the soil. In steeper terrain, topography also influences solar radiation and orographic effects. Along a hillslope, systematic soil variation is most commonly the result of the soil hydrology and erosion. This soil pattern was recognized by Milne (1936), and termed a catena. Others (Ruhe, 1975; Conacher and Dalrymple, 1977) have used this concept to segment or classify landforms into elements where different hillslope processes are dominant, which typically correspond with soil patterns. As such, hillslope models are meant to be universal, but in reality not all hillslope elements (e.g. shoulders, footslopes, or free faces) necessarily exist along any given hillslope. The measures used to differentiate hillslope elements are typically based on a land surface geometry (e.g. slope gradient and slope curvature) and relative position (e.g. slope length or contributing area). In other cases, neighborhood statistics (e.g. standard deviation or range within a moving window) have been used, particularly when discriminating macro-landforms (Dikau, 1989; Dobos et al., 2005). Thanks to advances in geomorphometry (or digital terrain analysis), such measures termed terrain attributes, land-surface parameters, or geomorphometric parameters can be readily calculated from a DEM. When included in predictive models, terrain attributes are used to infer hydrologic phenomena such as runoff potential, flow convergence or divergence, or the redistribution of soil material (Table 2). While some of these relationships are explicit, others are more implicit. For a detailed summary of soil-geomorphic processes associated with relief, see Schaetzl and Anderson (2005). What continues here is a brief summary of common DEM-derived parameters used within DSM that are relevant to hydropedology. For a detailed summary of geomorphometry, see Hengl and Reuter (2009).

2.3.1.1. Slope Gradient

The maximum rate of change in elevation (e.g. first derivative) is readily calculated from gridded DEM and is used to represent the hydraulic gradient acting upon overland and subsurface water flow through the influence of gravity (Gallant and Wilson, 2000) (Fig. 1a). A steeper slope gradient is associated with greater velocity of surface and subsurface flow, as well as greater potential of erosion and other translocation processes. Slope gradient is also a factor in several regional terrain attributes (Table 2), particularly those with hydrologic interpretations (e.g. topographic wetness index, stream power index, and sediment transport capacity index).

2.3.1.2. Slope Aspect, Solar Radiation, and Flow Direction

Slope aspect, or the compass direction that a point on the surface faces (up–down slope), is related to soil hydrology in two ways. Slope aspect is related to the amount of solar radiation received at a location, particularly when combined with slope gradient values (Wilson and Gallant, 2000; Böhner and Antionić, 2009). Insolation influences soil temperature and soil moisture content (Franzmeier et al., 1969; Chamran et al., 2002), which in turn can influence plant productivity (Hutchins et al., 1976), soil microbial activity (Abnee et al., 2004a and 2004b), and soil properties such as organic carbon content (Thompson and Kolka, 2005). When slope aspect is incorporated in predictive models, it is usually when the spatial extent of prediction encompasses large land areas with multiple hillslopes and a diversity of slope aspects. Despite this influence of slope aspect on soil processes and properties, it is not often used in predictive modeling because of the numerical

TABLE 2 Summary of Important DEM-Derived Terrain Attributes

Terrain attributes	Definition	Hydropedologic significance	References
Local – Geometric			
Slope gradient (degrees) (sg)	Maximum rate of change (degrees)	Flow velocity	Olaya, 2009
Slope aspect (sa)	Direction of slope gradient (degrees)	Flow direction	Olaya, 2009
Profile curvature (kp)	Curvature of slope gradient (radians)	Flow acceleration	Olaya, 2009
Tangential curvature (radians) (kt)	Curvature perpendicular to slope gradient (radians)	Flow convergence	Olaya, 2009
Slope shape	Classification of sg, kp, and kt into nine units	Segmentation of hillslope processes	Pennock et al., 1987
Multiresolution valley bottom flatness index	Measure of flatness and l owness	Depositional environments	Gallant and Dowling, 2003
Multiresolution ridge top flatness index	Measure of flatness and upness	Stable uplands settings	Gallant and Dowling, 2003
Local – statistical			
Standard deviation		Surface roughness or complexity	Wilson and Gallant, 2000
Range or relief		Potential energy	Wilson and Gallant, 2000
Percentile or rank		Landscape position	Wilson and Gallant, 2000
Regional			
Contributing area (ca)	Upslope area (meters^2)	Effective precipitation	Moore et al., 1991
Specific contributing area (sca)	ca/grid size (meters)	Effective precipitation	Moore et al., 1991
Topographic wetness index (twi)	ln (sca/sg)	Water accumulation or soil saturation	Moore et al., 1991

(*Continued*)

TABLE 2 Summary of Important DEM-Derived Terrain Attributes—cont'd

Terrain attributes	Definition	Hydropedologic significance	References
Stream power index (spi)	ln (sca*sg)	Soil erosion	Moore et al., 1991
Catchment height	Average upslope height	Potential energy	Moore et al., 1991
Catchment slope	Average upslope gradient	Flow velocity	Moore et al., 1991
Overland flow distance to stream	Relative topographic position (meters)	Potential energy	Tesfa et al., 2011
Vertical distance to channels	Relative topographic position (meters)	Cold air drainage	Bohner and Antonic, 2009
Normalized height	Relative topographic position (percent)	Cold air drainage	Bohner and Antonic, 2009
Solar radiation or insolation	Amount of incoming solar energy (kWh/mA^2)	Soil temperature and evapotranspiration	Bohner and Antonic, 2009

difficulties associated with slope aspect being a circular variable. As such, it is necessary to transform slope aspect to a linear scale, categorize slope aspect into classes, or instead use solar radiation. Estimation of solar radiation is more sophisticated, and therefore requires user inputs such as latitude, atmospheric properties, and temporal frequency (Fig. 1d). This extra effort is likely to be worthwhile, because solar radiation accounts for differences in topographic shadowing (from adjacent hills) and slope gradient (Hunckler and Schaetzl, 1997; Beaudette and O'Geen, 2009).

Slope aspect is also significant because it defines the flow direction of water and sediment movement through the landscape under the influence of gravity. Therefore, it is used to determine contributing area, which is a highly significant terrain attribute discussed below. While slope aspect is easily calculated, flow direction is more difficult as it must partition the proportion of flow between square grid cells. Various algorithms exist for determining flow direction, and they are classified simply as either a single flow direction (SFD) method (e.g. deterministic 8 or D8) (O'Callaghan and Mark, 1984) or a multiple flow direction (MFD) method (e.g. FD8 (Freeman, 1991), deterministic infinity or D∞ (Tarboton, 1997), and multiple deterministic infinity or MD∞ (Seibert and McGlynn, 2007)). For soil–landscape studies, MFD

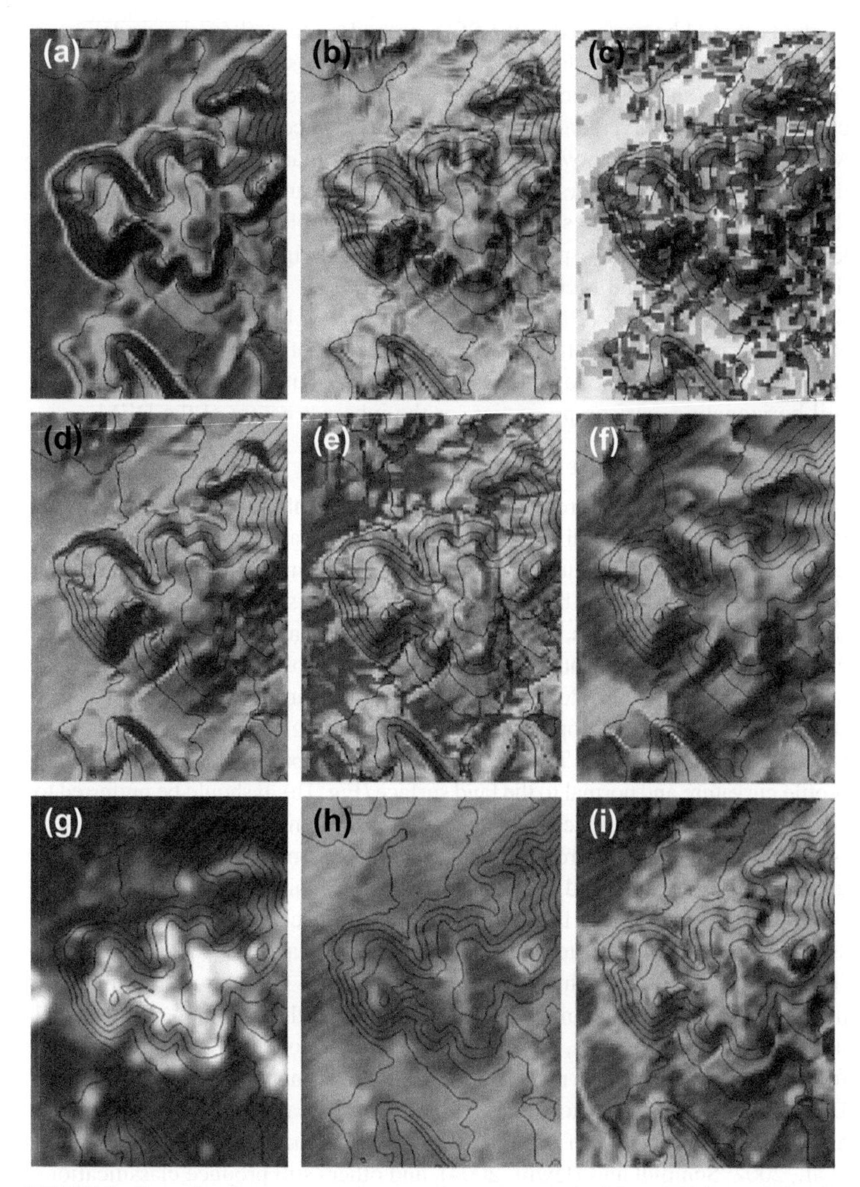

FIGURE 1 Example of common environmental covariates derived from DEM and RSI, with hillshade and 30 meter contour interval for affect. (a) Slope gradient, (b) Tangential curvature, (c) Slope shape, (d) Solar radiation, (e) Topographic wetness index, (f) Overland flow distance of channels, (g) False color composite (green, red, near infrared), (h) False color composite of tasseled cap components (brightness, greenness, wetness), (i) Normalized difference vegetation index. *(Color version online and in color plate)*

methods should be preferred to the SFD method, as they allow the modeling of flow dispersion along hillslopes. The SFD method is sufficient for determining river networks and watershed boundaries, but is inadequate on slopes because it cannot model flow divergence, and it therefore produces parallel flow lines. The MFD methods listed above produce largely similar results, but noticeable differences in flow dispersion due to each MFD-unique computation (and therefore assumptions of surface flow). For example, the FD8 algorithm is known to produce overdispersion, while the D∞ algorithm produces under-dispersion (Seibert and McGlynn, 2007). The MD∞ algorithm was shown by Seibert and McGlynn (2007) to be intermediary between FD8 and D∞. Regardless of the different theoretical approaches of each MFD to modeling surface flow, the hydropedogenic response to contributing area may vary due to soil horizonation, surface properties, or the inherent nature of the DEM.

2.3.1.3. Slope Curvature and Shape

Slope curvature measures are important because they influence local water flow in terms of convergence or divergence (tangential curvature) and acceleration or deceleration (profile curvature). There are various forms of slope curvature (e.g. second derivatives) that can be calculated from the land surface (Shary et al., 2002; Schmidt et al., 2003; Olaya, 2009), with only mean, unsphericity, and difference curvature being independent (Shary, 1995). Still the most common are profile and tangential curvature. Profile curvature is the curvature in the direction of the steepest slope (up–down slope) and normal (e.g. orthogonal or perpendicular, not vertical) to the land surface. Tangential curvature is perpendicular (across slope) to profile curvature and normal to the land surface (Fig. 1b). It should be preferred to contour curvature, with which is it often confused (Schmidt et al., 2003), because it is less sensitive to DEM errors. In many cases, the term plan curvature has been used for both tangential and contour curvatures (Schmidt et al., 2003).

Typically, profile and tangential curvature are classified into nine slope shapes, formed by the intersection of convex, linear, and concave surfaces (Ruhe, 1975) (Fig. 1c). This classification is useful for interpreting hillslope processes, such as soil and water redistribution, throughflow, and vertical infiltration. If recognizable, these forms can be used for communicating surface form, transferring research results, and serving as manageable land units. Many DEM-based approaches to soil and land classification have attempted to automate the production of these slope forms (Pennock et al., 1987; Florinsky et al., 2002; Schmidt and Hewitt, 2004), and others still produce classifications that largely resemble them (Burrough et al., 2000; Zhu et al., 2001). Still, any land classification is subjective and should correspond with recognizable ecological phenomena and environmental properties.

2.3.1.4. Contributing Area and Hydrologic Indices

Contributing area (CA), upslope area, or flow accumulation represents the area of land upslope of a specific contour length (e.g. grid size). As mentioned above, it is a product of flow direction. The value of this parameter has many process-based interpretations. Foremost, it can be used to estimate the amount of effective precipitation an area receives from upslope (assuming no infiltration or wind redistribution of snow). Second, it estimates the spatial connectivity of different positions within a watershed, and therefore can be used to infer landscape position (Gessler et al., 1995). Lastly, CA may be used to quantify flow convergence and divergence, irrespective of curvature. Given the multiple interpretations associated with CA, it is typically used for DSM applications, either alone or – more commonly – combined with other terrain attributes as an index related to soil wetness or erosion.

There are several hydrologic indices that are products of CA and other terrain attributes, including slope gradient and slope curvature (Table 2). The most widely utilized hydrologic index for both sampling design and model development is the topographic wetness index (TWI) or compound topographic index (Fig. 1e). The TWI is calculated as the natural logarithm of the ratio of specific contributing area (CA/grid size) (SCA) to slope gradient (Wilson and Gallant, 2000), and is an index of the likelihood of a cell to collect water. High values of TWI coincide with locations where larger amounts of water accumulate from upslope, but where this water is less likely to runoff because of flatter slope gradients, i.e. areas where water accumulates in the landscape. Consequently, TWI is considered a predictor of zones of soil saturation.

The stream power index (SPI), a measure of runoff erosivity, is the natural logarithm of the product of SCA and slope gradient (Wilson and Gallant, 2000). High values of SPI coincide with locations with high specific contributing area values and high slope gradient values, or areas where larger amounts of water flow in from upslope, and where this water is expected to continue to flow because of steeper slope gradients. Locations with higher SPI values are considered to be areas where water flows with higher energy and, consequently, where erosion and related landscape processes are more likely to occur.

2.3.1.5. Proximity to Drainage (Depression)

From a hydropedologic perspective, the flow path information used to calculate contributing area can also be used to calculate contextual variables. While most terrain attributes quantify the topographic conditions in the immediate vicinity (usually a 3×3 window of a gridded DEM), a number of terrain attributes have been derived that represent either a statistical summary of conditions upslope of a cell (e.g. mean upslope slope gradient or mean upslope plan curvature) (Gallant and Wilson, 2000) or an indicator of the spatial configuration between a cell and local peaks or pits (e.g. distance to the closest channel and elevation above the closest depression) (Olaya, 2009; Gruber and Peckham, 2009) (Table 2). The distance to the local drainage/depression can be calculated using either the Euclidean distance or the flow distance (Fig. 1f). Values of elevation above local drainage/depression and distance to local drainage/depression can be combined to calculate the slope gradient to the nearest drainage/depression. Such contextual terrain attributes are used in predictive modeling to quantify regional influence on soil processes and soil properties, particularly the regional gradient acting upon soil water and surface runoff. Bell et al. (1992, 1994) used these proximity measures to predict soil drainage class, while Arrouays et al. (1995) and Thompson and Kolka (2005) used such contextual terrain attributes to model soil organic C. One or more of these terrain attributes have also proven useful for predicting hydromorphic soil properties (Thompson et al., 1997; Chaplot et al., 2000) and soil classes (Lagacherie and Holmes, 1997; Moran and Bui, 2002).

2.3.2. Remotely Sensed Imagery

Remote sensing involves the use of sensors to measure the reflectance or emission of energy from a distance (Jensen, 2005). Historically within soil mapping, this has centered on the use of aerial photography, which captures images of reflected sunlight within the visible to near-infrared portion of the electromagnetic spectrum. Today, numerous sources of imagery (Fig. 1g) are available from various platforms, the most common being Earth-orbiting satellites. These satellites measure reflectance and emissions from a wide swath of the electromagnetic spectrum, with each having unique spectral and spatial resolutions tailored to their field and scale of interest, be it observing weather, oceans, or land. In principle, remote sensing is valuable because different earth-surface features have unique spectral signatures, as a result of their unique physical and chemical properties. The uniqueness of a given features spectral signature determines its ability to be identified correctly from afar.

Remote sensing can provide direct information on various *scorpan* or STEP-AWBH factors, including soil, organisms, and parent material. Also, indirect relationships can be established with other factors, such as climate and time. In humid environments, reflectance provides information primarily on vegetation, whereas in arid and agricultural environments where the soil surface is typically exposed, information on parent material and the soil can be inferred. In particular, in arid and agricultural environments some surface soil properties, such as mineralogy, texture, moisture, organic carbon, iron content, salinity, and carbonates, can be associated with specific adsorption features (Mulder et al., 2011). Due to the multitemporal frequency of satellites, they also offer further possibilities to observe changes in the land-surface over time, which is usually associated with the senescence of vegetation.

The wide array of potential information within RSI makes it an attractive, but complicated environmental covariate. For example, in many cases, it can be difficult to separate specific spectral signatures from the confounding effects of vegetation, topographic shadowing, or other factors. To compensate for these factors various image processing techniques have been developed. In the event that such techniques are insufficient, it is typically necessary to incorporate ancillary data, such as DEM derivatives, to distinguish seemly similar spectral signatures. One of the greatest difficulties is mapping human-modified landscapes, which contain a mosaic of patterns dominated by the effects of land use and cover rather than native vegetation. Therefore, in human-modified settings DEM derivatives are likely to provide better predictors of static soil properties (e.g. soil texture), unless predictions are constrained to similar land-use and cover types. Within human-modified landscapes, land use and cover may be used to develop conditional rules for inferring dynamic soil properties (e.g. organic matter).

The intent in the following section is to review derivatives of RSI that are commonly used as environmental covariates. For a thorough discussion of the application of RSI in DSM, see Boettinger et al. (2008) and Boettinger (2010). In contrast, Mulder et al. (2011) provides an exhaustive discussion of soil attributes that can be estimated using proximal and remote sensing for DSM. Full-length treatments on the subject of image processing can be found in Jensen (2005) and Schowengerdt (2007).

2.3.2.1. Spectral Transformations

Individual satellite bands can be used directly to infer soil characteristics, but commonly derivatives of RSI or spectral transformations are used. In contrast to spatial transforms, spectral transforms affect the multivariate or feature space, rather than the image space (e.g. x,y coordinates). Spectral transforms are applied to enhance specific spectral signatures and reduce data redundancy in the original bands. Common examples are band ratios, principal components, or tasseled-cap components.

Band ratios are image ratios where one band is divided by another. This is a nonlinear transformation that has the benefit of the reducing topographic shadowing inherent in the original bands, while enhancing absorption features contained in the denominator relative to the numerator, at the expense of increasing noise inherent in the original bands (Schowengerdt, 2007). Modulation or normalized difference ratios are a modification of simple ratios, which scale the ratio values between –1 and 1, and enhance the contrast of lower ratio values (Schowengerdt, 2007):

$$\text{Simple ratio} = \frac{\text{Band 1}}{\text{Band 2}}, \tag{6}$$

$$\text{Modulation or normalized difference ratio} = \frac{\text{Band 1} - \text{Band 2}}{\text{Band 1} + \text{Band 2}}. \tag{7}$$

One of the most common ratios is the normalized difference vegetation index (NDVI), which is a ratio of near infrared to red (Fig. 1i). This normalized ratio is used to assess the amount of healthy green vegetation on the land surface. Numerous other ratios have been utilized for enhancing various soil adsorption features in arid and semi-arid environments, such as carbonates, clay, iron content, organic matter, and grain size. An important caveat is that the physical interpretation of band ratios may not translate to different environments compared to those for which they were developed, due to the complex interaction of land-surface characteristics (e.g. vegetation and soils).

Principal components (PCs) are linear combinations of the original bands, designed to remove the collinearity (e.g. data redundancy) between them. PC analysis (PCA) involves rotating the original coordinate axes of the data to new orthogonal (e.g. perpendicular) axes that contain the maximum amount of information for each dimension. Because PCA is simply a mathematical rotation of the original bands, it is data dependent. Therefore, the physical interpretation of PC is unique to each particular satellite scene, and any correlation between the PC and soil characteristics is fortuitous. However, typically the first PC (e.g. PC 1) is typically dominated by the overall image brightness of the original bands, while the last PC is typically dominated by random noise.

Tasseled-cap components (TC) are similar to PC, a linear combination of the original bands (Fig. 1h). They differ from PC in that their rotation is fixed, rather than data dependent, which allows PC to maximize the variation in any given scene. Also, the TC rotation has been primarily developed for the Landsat satellites, and a handful of other satellites, whereas PC can be generated from any combination of DEM derivatives or RSI. The TC transform was originally developed by Kauth and Thomas (1976) in order to develop standardized components which have a common physical interpretation (Table 3).

3. DIGITAL SOIL MAPPING FOR HYDROPEDOLOGIC APPLICATIONS

Lin et al. (2006) describe how hydropedology provides a framework for "understanding, measurement, modeling, and prediction of water fluxes across the landscape." The key issues that require attention are structure, function, and scale. In the context of DSM, structure may be manifested in the prediction of the spatial arrangement of soil characteristics that influence hydrology (e.g. pixels or map units that represent soil properties, such as subsurface hydraulic conductivity; soil horizons, such as the depth to impermeable fragipan or bedrock; and soil classes, such as soil series with root-restricting layers or low available water-holding capacity). Observed structures are the result of the fluxes and processes that have occurred over time within the soil–landscape system, and therefore provide clues to current functions. DSM products provide a means to fully extend hydropedologic concepts and understanding in the spatial domain. In particular, such digital soil survey data may be used to derive hydromorphic properties and hydrologic processes suitable for (i) land-use and soil management decisions, (ii) input into hydrologic, ecological, or other earth-surface models, and (iii) estimates of soil characteristics at unsampled or unobserved locations.

3.1. Mapping Soil Properties that Influence Hydrology

Observed spatial and temporal variability in hydrologic response is attributable to landscape heterogeneity (Troch et al., 2009). McDonnell et al. (2007) have

TABLE 3 Summary of Common Spectral Transformations of Landsat 5 Thematic Mapper (TM) and Landsat 7 Enhanced Thematic Mapper (TM+)

Spectral tranformations	Definition	Interpretation	Scorpan factors	References
Image ratios				
Normalized difference vegetation index (NDVI)	(NIR − R)/ (NIR + R)	Vegetation abundance	Organisms, climate	
Soil adjusted vegetation index (SAVI)	(1 + L) ((NIR − R)/ (NIR + R + L))	Vegetation abundance in arid and semi-arid environments	Organisms, climate	Huete, 1988
Carbonate index	R/G	Carbonate radicals	Soil, parent material, age	Amen and Blaszczynski, 2001
Iron index	R/MIR_2	Ferrous iron	Soil, parent material, age	Amen and Blaszczynski, 2001
Clay index	MIR_1/MIR_2	Hydroxyl radical (e.g. clay)	Soil, parent material, age	Amen and Blaszczynski, 2001
Gypsic index	$(MIR_1 - MIR2)/(MIR1 + MIR2)$	Gypsic soils	Soil, parent material, age	Nield et al., 2007
Natric index	$(MIR_1 - NIR)/(MIR1 + NIR)$	Natric soils	Soil, parent material, age	Nield et al., 2007
Grain size index (GSI)	(R − B) (R + G + B)	Surface texture	Soil, parent material, age	Xiao et al., 2006
Principal components (PC)	**Mathematical rotation**			
PC_1		Image brightness		
$PC_{2\text{-}n}$		Variable		
PC_n		Random noise		
Tassel cap components (TC)	Fixed mathematical rotation			Crist et al., 1986; Huang et al., 2002

(*Continued*)

TABLE 3 Summary of Common Spectral Transformations of Landsat 5 Thematic Mapper (TM) and Landsat 7 Enhanced Thematic Mapper (TM+)—cont'd

Spectral tranformations	Definition	Interpretation	Scorpan factors	References
TC_1		Soil brightness	Soil, parent material	
TC_2		Greenness	Organisms, climate	
TC_3		Wetness (moisture)	Soil	
TC_4		Haze		

(B = Blue; G = Green; R = Red; NIR = Near infrared; MIR_1 = Mid-infrared 1; MIR_2 = Mid-infrared 2; L = Constant).

indicated that poor characterization of this landscape heterogeneity, including soil properties, has hindered the field of hydrology. Consequently, improving our understanding of the nature of soil variability across hillslopes and watersheds is expected to enhance our ability to model and predict hydrologic response (Troch et al., 2009). The role of spatial structures in hydrologic response and hydrologic modeling is discussed by Grayson and Bloshl (2000). In particular, Houser et al. (2000), Vertessy et al. (2000), and Western and Grayson (2000) illustrate the use of soil spatial data on modeling watershed-scale hydrologic response, and discuss the limitations of available soil data.

The spatial variability of soils and soil properties is recognized as having a significant influence on (i) the partitioning of precipitation between rainfall and runoff, (ii) lateral and vertical redistribution of subsurface water, and (iii) the spatial distribution of soil moisture at hillslope and catchment scales (Guntner and Bronstert, 2004). Both vertical and lateral variability in soil properties can influence hydrologic response, particularly at hillslope and catchment scales. Vertical changes in soil characteristics, particularly the presence of discontinuities and abrupt boundaries, will influence infiltration, percolation, and evapotranspiration. Furthermore, vertical heterogeneity (e.g. clay lenses; textural discontinuities, cemented horizons, or dense layers) can also promote lateral redistribution of soil water. Grayson and Bloshl (2000) provide a thorough examination of spatial patterns and spatial processes in hydrologic systems.

3.1.1. Soil Horizons

Certain soil horizons or horizon sequences are the result of hydrologic processes within the soil and serve as evidence of the magnitude and direction of water movement within the soil. Eluvial and illuvial processes translocate silicate clay minerals, iron oxides, humus, carbonates, and other soil constituents. It is because of these linkages between hydrology and pedogenesis that allow for the successful application of DSM approaches to predict the spatial distribution of soil horizons (Table 4), such as albic, argillic, spodic, and fragipan horizons. Conversely, the presence of these and other soil horizons (e.g. cemented layers) can influence soil–water–plant interactions and hydrologic processes by restricting root growth, storing more (or less) plant available water, promoting lateral water flow, or producing perched water tables.

There are a number of examples of the successful application of DSM to predict the occurrence of soil horizonation. Many have focused on the mapping of topsoil thickness (Pennock et al., 1987; Moore et al., 1993; Gessler et al., 1995; Zhu, 2000; Park et al., 2001; Hengl et al., 2004). Another common target property, and one that is important hydrologically relative to water-storage capacity and plant available water, is soil depth or depth to limiting layer (Odeh et al., 1994, 1995; McKenzie and Ryan, 1999; Sinowski and Aureswald, 1999; Ryan et al., 2000). Gessler et al. (1995) also developed models for the spatial prediction of the presence or absence of an albic horizon.

3.1.2. Soil Properties

Soil property maps developed using DSM efforts may have particular value in hydropedology as well as the environmental modeling community as input into models, or input into pedotransfer functions (PTFs) to develop the necessary property maps that can be used in such models. Texture and the related particle-size distribution greatly influence the rate of water flow through soils. This soil property is a key input parameter in all hydrologic models. Numerous DSM efforts have produced maps of soil textural properties (e.g. McKenzie and Austin, 1993; Moore et al., 1993; Arrouays et al., 1995; de Bruin and Stein, 1998; Oberthur et al., 1999; McBratney et al., 2000; Park and Vlek, 2002; Scull et al., 2005). For other important hydropedologic properties, such as soil structure, bulk density, and saturated hydraulic conductivity (Ksat), there are few examples of DSM approaches to mapping these properties.

Whole-profile soil properties integrate several soil properties to provide information on soil–water relationships and soil use and management interpretations. For example, soil water storage is a fundamental ecosystem service that is commonly determined from measured or estimated properties, including horizon thickness, soil texture, rock fragment content, and depth to root-limiting layer (if present). Quantifying soil water storage is crucial for understanding water transport. Commonly, runoff is predicted when the storage capacity for a given soil is exceeded. For climate modeling, information on soil water storage is needed to predict the amount of water available for evapotranspiration. The available water-holding capacity is influenced by depth to bedrock or depth to a hydraulically limiting layer. DSM products can improve the ability to represent these soil properties and deliver better information to

TABLE 4 Selected Examples of Studies Using DSM Approaches to Predict Soil Properties of Hydropedologic Significance

		Predictive factors			
Authors	Soil attribute or class	Terrain attributes	Remotely sensed imagery	Other	Spatial extent[a]
Soil Horizon and Profile Characteristics					
Moore et al. (1993)	Horizon thickness	X			Field
Zhu and Band (1994)	Horizon thickness	X	X		Local
Gessler et al. (1995)	Horizon thickness	X			Local
Boer et al. (1996)	Soil thickness	X			Local
McKenzie and Ryan (1999)	Soil thickness	X	X	Climate, geology	Local
Sinowski and Auerswald (1999)	Soil thickness	X			Local
Ryan et al. (2000)	Soil thickness	X	X	Climate	Local
Park et al. (2001)	Horizon thickness	X			Local
Hydromorphic Soil Properties					
Bell et al. (1992, 1994)	Soil drainage class	X			Local
Cialella et al. (1997)	Soil drainage class	X	X		Local
Thompson et al. (1997)	Hydromorphic index	X			Field
Chaplot et al. (2000)	Hydromorphic index	X			Field
Campling et al. (2002)	Soil drainage class	X	X		Local
Chaplot and Walter (2002)	Hydromorphic index	X			Regional
Peng et al. (2003)	Soil drainage class	X	X		Local

TABLE 4 Selected Examples of Studies Using DSM Approaches to Predict Soil Properties of Hydropedologic Significance—cont'd

		Predictive factors			
Authors	**Soil attribute or class**	**Terrain attributes**	**Remotely sensed imagery**	**Other**	**Spatial extent[a]**
Liu et al. (2008)	Soil drainage class	X	X	Proximal sensing	Field
Lemercier et al. (2012)	Soil drainage class	X	X	Geology	Local
Hydropedologic Properties					
McKenzie and Austin (1993)	Soil water content (−10 kPa, −1.5 Mpa	X	X		Local
Zheng et al. (1996)	Available water capacity	X			Local
Lark (1999)	Soil water content	X			Field
Ryan et al. (2000)	Available water capacity	X	X	Climate	Local
Pachepsky et al. (2001)	Soil water retention	X			Field
Romano and Palladino (2002)	Soil water retention	X			Local
Malone et al. (2009)	Available water capacity	X	X		Local
Other Soil Properties					
McKenzie and Austin (1993)	Bulk density; clay content	X	X		Local
Moore et al. (1993)	Organic matter; silt content; sand content	X			Field
Arrouays et al. (1995)	Organic matter content; clay content	X		Climate	Local
de Bruin and Stein (1998)	Clay content	X			Field

[a]*Spatial extents include: Field = <0.25 km^2, Local = 0.25–10^4 km^2, and Regional = 10^4–107 km^2.*

hydrologists. Examples of DSM efforts that have yielded predictions of soil water availability include McKenzie and Austin (1993), Zheng et al. (1996), Voltz et al. (1997), Lark (1999), Lagacherie and Voltz (2000), Ryan et al. (2000), Pachepsky et al. (2001), and Sommer et al. (2003).

3.2. Mapping Soil Properties Influenced by Hydrology

Hydropedologic properties have been a common product of pedometric modeling and DSM activities (Table 4) because (i) hydropedologic properties such as drainage class, available water content, and hydromorphic features are important for developing soil use and management interpretations and making land-use decisions and (ii) hydropedologic properties are strongly influenced by hillslope hydrologic processes, which are readily modeled with DEM-derived terrain attributes because of the relationship between topography and hillslope hydrologic processes. In fact, the earliest example of quantitative prediction of soil classes or properties noted by McBratney et al. (2003) in their comprehensive review of DSM is work by Troeh (1964) on the prediction of soil drainage classes using slope gradient and slope curvature. Bell et al. (1992, 1994), Campling et al. (2002), and Peng et al. (2003) also developed terrain-based statistical models to predict soil drainage classes. Similar soil–landscape modeling approaches have been used to predict soil hydromorphological properties (Thompson et al., 1997; Chaplot et al., 2000; Park et al., 2001), water content (McKenzie and Austin, 1993; Lark, 1999), available water capacity (Ryan et al., 2000), and water retention curve characteristics (Romano and Palladino, 2002).

The accumulation of SOC is often a function of hydrology, either local or regional, and is driven by climatic and topographic factors (Collins and Kuehl, 2001). Consequently, the spatial distribution of SOC is related to moisture availability for biomass production and/or soil saturation, which limits organic matter decomposition. Mapping SOC (or soil organic matter) using DSM approaches has been demonstrated on various scales by Arrouays et al. (1995, 1998), McKenzie and Ryan (1999), Bell et al. (2000), Gessler et al. (2000), Ryan et al. (2000), Chaplot et al. (2001), Hengl et al. (2004), Terra et al. (2004), Thompson and Kolka (2005), and Vasques et al. (2010).

There are many observed physical, chemical, and morphological properties that are the result of hydrologic conditions or processes within the soil. From the standpoint of soil use and management, soil hydromorphology is a primary factor in evaluating soil suitabilities and limitations. Redoximorphic features such as iron and manganese concentrations, iron depletions, and depleted matrix are directly related to the depth to and duration of saturated and reducing conditions in the soil, and provide evidence of where in the soil or landscape water accumulates and persists that had aggregated over long periods of time. The presence of redoximorphic features (kind, quantity, size, color, shape, and location) is used to establish soil drainage class and determine the occurrence of aquic conditions (Soil Survey Staff, 1999) and hydric soil conditions (USDA-NRCS, 2010). There are many examples of landscape-scale studies of soil wetness and hillslope hydrology that have scrutinized the relationship between hydrologic regimes (e.g. depth to and duration of observed water tables) and soil morphological features (e.g. Vepraskas and Wilding, 1983a,b; Thompson and Bell, 1996, 1998; Reuter and Bell, 2001, 2003; Jenkinson et al., 2002; D'Amore et al., 2004). As illustrated by McDaniel et al. (1992), Thompson et al. (1998), Noguchi et al. (1999), D'Amore et al. (2000), Lin (2006), and others, observed soil morphological properties – redoximorphic features, ped surface features, and pore architecture – at hillslope and catchment scales can be useful for interpreting the nature of water movement and areas of water storage. However, iron and manganese are not the only soil constituents that are influenced by hydrology. The distribution of carbon, nitrogen, calcium, phosphorus, and other elements can be related to hydrologic conditions and processes (Vasques et al., 2010). Thompson et al. (1997), Chaplot et al. (2000), and Chaplot and Walter (2002) have demonstrated that soil color indices that combine multiple soil color characteristics from a soil profile (e.g. color and thickness of the surface horizons and relative thickness of horizons with RMF) can be developed and, using DSM methods, used to map the extent of hydromorphic soils.

4. RESEARCH NEEDS AND FUTURE CONSIDERATIONS

As technology has become more accessible, allowing the common use of mathematical and statistical methods, a considerable degree of overlap between hydropedology and DSM has developed. Indeed, this trend is likely to continue as hydropedology seeks to analyze more sophisticated data sets. From this perspective, hydropedology, like other subdisciplines of soil science, can be seen as an application-oriented example of DSM that focuses on examining soil and water functions rather than methodological issues. Research needs and future considerations where hydropedology can benefit from DSM include capitalizing on existing soil databases, development of PTF, estimation of dynamic soil–water properties (including genoform and phenoform), estimation of model uncertainty, incorporation of depth (i.e. vertical anisotropy), examination of scaling of soil–water functions, and integration of computer software. As such, DSM products should reflect the input requirements for environmental models, such as those used in hydrology and hydropedology.

There are numerous opportunities for developments in DSM and hydropedology that will promote even greater interactions between these two fields. These needs are driven by the data and interpretation requirements both within the soil science community and among the external users of soil data, information, and knowledge. For example, DSM products must reflect the requirements for input into hydrologic models in terms of scale, spatial resolution, and representation of the soil–water continua. An immediate need is for

DSM approaches that capitalize on existing soil data (often referred to as legacy data), including maps, pedon data (characterization data and morphological descriptions), and point measurement data. A complement to the use of legacy data is the use of sensors for mapping soil properties at field scales as well as continental scales using proximal and remote-sensing systems. For both of these data sources, as well as traditional environmental covariate data, there is a need for development and refinement of mapping and modeling techniques.

In many developed countries, there is already a wealth of legacy soil information available. As such, there is increased effort to exploit existing point data (pedon characterization data and morphological descriptions), thematic maps, and expert knowledge to produce digital soil maps that represent soil properties and classes at scales (or resolutions) that meet the needs of the growing user community for soil geographic information (Mayr et al., 2010; Dobos et al., 2010). Although legacy data are readily available, their analyses are complicated because they were typically collected for different purposes, and therefore may be of inconsistent quality due to changing conventions, missing data, and limited geographic and environmental coverage. Therefore, significant effort is necessary to harmonize them for new purposes. It is difficult to estimate the bias in existing data, but preference is generally given to point data as they represent direct observations and measurements rather than aggregated data. Whatever format of legacy data is used, its source should be evaluated, and any subsequent results scrutinized according to current standards and assessed with independent data. Properly utilized legacy data can help answer new questions.

Due to the limited scope of legacy data, its primary utility may be serving as a covariate for *scorpan* or STEP-AWBH models. This approach is already commonplace. In a recent review, Grunwald (2009) found existing soil information used to develop PTF and soil spatial prediction functions (SSPF) in over 36 percent of DSM studies. PTF and taxotransfer functions (TTF) are one method that has been and is expected to continue to be a valuable tool for extending the utility of legacy data, particularly for hydropedology.

Scale and resolution are also critical issues, including the effects on predictions and the utility of these predictions. Most available soil map products are vector soil class maps, which depict polygons that represent one or more soil classes. The component soils of these map units, while not mapped spatially, often occur in specific landscape positions. These individual soil types within the delineated map units often have differing physical, chemical, biological, or morphological properties that produce different hydrologic responses. Knowledge of the location and extent of these contrasting soil classes or properties is important for accurately representing hydropedologic properties and soil-hydrologic responses. Hively et al. (2005) demonstrated that small undifferentiated areas within soil map unit delineations can be important areas of runoff generation and P loss processes. Furthermore, DSM products should also help bridge scales from pores to pedons, and from catenas to catchments.

Aggregation (generalization) of geographic space (e.g. to delineate soil map units) and/or attributes (e.g. to derive soil classes) have been used extensively in soil surveys, which is inherently ingrained in legacy data sets. Trends in DSM are evident to disaggregate larger map units into fine-grained pixels (coarse scale to fine scale) and to map continuous soil properties rather than soil classes or categorical properties. These trends are facilitated by geographic information technologies, soil and remote sensors, and advanced computational power which allow to develop soil prediction models with ancillary environmental data. Constraints imposed by data availability to characterize the spatial distribution and variability of soil properties has led to aggregation (lumping) of modeling hydrologic processes within subbasins, basins, or larger hydrologic units. Soil map units and hydrologic units often do not coincide and impose ambiguities on modeling the soil–water continuum. Considerations with regard to appropriate spatial and attribute scales are needed for successful future applications in hydropedology. This will require collaboration on characterizing spatial and attribute variability of parameters that characterize the soil–water continuum and joint investigation of scaling behavior. To match the map scale and resolution of space and attributes will be critical to interlink both disciplines – DSM and hydrology – to conjoin hydropedology.

DSM has put much emphasis on mapping the long-term responses (e.g. redoximorphic features, SOC) from natural and anthropogenic forcings through factorial models (e.g. *clorpt* and *scorpan*), often not explicitly accounting for the time component. In addition, assessment of spatial variability and distribution of stable soil properties has received much attention. In contrast, the main focus of hydrology has been on continuous measurements and/or simulations of hydrologic parameters to model water flux within the soil–landscape continuum. Thus, the time component has played a much more prominent role in hydrology than in soil mapping. Hydropedology is poised to bridge these space–time gaps and promote the harmonization of approaches and models accounting equally for both hydrologic and pedogenic factors and processes, e.g. spatial and temporal distribution and evolution of soil-hydrologic properties which drive processes, and vice versa. In particular, in the context of increasing concerns about how anthropogenic-induced forcings, such as global climate change and land-use shifts, will alter soil and water resources, more research on short- and long-term impacts are needed. The STEP-AWBH model proposed by Grunwald et al. (2011) is one such effort to explicitly address the importance of the temporal component in soil variability and soil mapping. Temporal databases, such as RSI from Earth-orbiting satellites, are expected to have useful applications for DSM and hydropedology. In particular, data from passive microwave sensors have been used to map the temporal variability in soil moisture (Hollenbeck et al., 1996; Jackson and Le Vine, 1996).

Whereas empirical and stochastic models to predict soil properties or classes are predominant in DSM, a mix of stochastic–deterministic (process-based) models is used to model hydrologic processes. Pedogenic simulation models are still sparse and difficult to validate due to the long time frames over which pedogenic processes operate, whereas hydrologic processes occur over shorter time intervals (e.g. rainfall–runoff events). This disparity in modeling approaches has led to separation rather than integration among soil science and hydrology disciplines. However, enhanced monitoring capabilities through sensors and long-term soil-hydrologic monitoring networks could facilitate to overcome these constraints. This will require concerted efforts and an investment in monitoring infrastructure. Establishing linkages between pedogenesis and hydrologic properties and processes (Lohse and Dietrich, 2005; Lin, 2010) and the development of quantitative models of pedogenesis (Rasmussen and Tabor, 2007; Minasny et al., 2008; Rasmussen et al., 2010) are important contributions that hydropedology can make to DSM.

An understanding of hydropedologic processes and relationships is expected to improve our ability to map and model soil–water spatial and temporal variability. Additionally, precise and accurate maps of soil properties and soil classes provide a means to both understand hydropedologic processes and functions at landscape and watershed scales (e.g. Lin, 2006), and extrapolate that knowledge across scales. The explicit consideration of hydropedology in DSM may encourage the development and adoption of quantitative maps and models of soil cover, soil structure, and soil horizonation such that modern soil map data can be more readily incorporated into water flow and transport models (Lin et al., 2006).

As soil scientists begin creating different soil data products, hydrologic models will be developed to use these products. In a study conducted by Basu et al. (2010), the researchers created the Threshold Exceedence LaGrangian Model (TELM) to predict stream discharge and peak flow in a tile-drained engineered landscape. This parsimonious model utilized basic soil input data to characterize a 700-km^2 watershed. This simple model based on soil filling and exceeding capacity performed as well as SWAT (Soil Water Assessment Tool) at characterizing stream flow. The setup time and calculation time for TELM were minimal compared to SWAT. Soil scientists working in DSM can play a significant role in the research area of stream flow prediction by working more closely with hydrologists.

Collaboration among DSM scientists and hydropedologists is expected to further our understanding of the complexities of the soil–water continuum within and across soil-landscapes. Digital soil maps and spatio-temporal databases are an important means by which DSM scientists and hydropedologists will communicate our knowledge with each other and with the greater scientific community. It will also be important that we have an understanding of modeler needs for input into hydrologic models. The linkage between hydropedology and pedometrics will be further strengthened as efforts advance from DSM to digital soil assessment (Carre et al., 2007).

5. SUMMARY

Creating synergies between hydropedology and DSM will promote new research endeavors and is expected to foster further advances in geospatial soil database development and process-based modeling (Lin, 2011). A primary linkage between hydropedology and DSM is the role of water in pedogenesis and soil–landscape processes that strongly influence spatial variability in soil properties. Hydrology influences pedology through soil-forming processes (additions, losses, transformations, and translocations), which led to the development of genetic horizons and features. However, soil properties, especially heterogeneity of soil properties, influence hydrology through soil–water interactions at the pore, pedon, and hillslope scales.

Many of the scientific questions regarding the Earth's Critical Zone require information on soils and soil variability. With the current technology, providing explicit spatial data has become possible. There is a tremendous research need to develop a greater understanding of the feedback between water redistribution on the landscape and the impact this has on soil properties. The deeper understanding of the feedback will provide scientists with tools to understand fundamental ecosystem relationships. DSM approaches seek to maximize information delivery from existing soils databases (including spatial and tabular, point and polygon) and develop new geospatial soil data products. These DSM products are expected to enhance spatial detail and to represent depth variation. Quantitative linkages between soil processes (hydropedology) and soil geography (DSM) are expected to yield more accurate and more precise soil maps, which will provide support for discovery-driven applications of hydropedology and contribute to resolving issues related to forcing, coupling, interfacing, and scaling (Lin, 2011).

REFERENCES

Abnee, A.C., Thompson, J.A., Kolka, R.K., D'Angelo, E.M., Coyne, M.S., 2004a. Landscape influences on potential soil respiration in a forested watershed of southeastern Kentucky. Environ. Manage. 33 (Suppl. 1), S160–S167.

Abnee, A.C., Thompson, J.A., Kolka, R.K., 2004b. Landscape modeling of in situ soil respiration in a forested watershed of southeastern Kentucky. Environ. Manage. 33 (Suppl. 1), S168–S175.

Amen, A., Blaszczynski, J., 2001. Integrated Landscape Analysis. U.S. Department of the Interior, Bureau of Land Management. National Science and Technology Center, Denver, CO, pp. 2–20.

Arnold, J.G., Allen, P.M., 1996. Estimating hydrologic budgets for three Illinois watersheds. J. Hydro. Amsterdam 176, 55–77.

Arrouays, D., Vion, I., Kicin, J.L., 1995. Spatial analysis and modeling of topsoil carbon storage in temperate forest humic loamy soils of France. Soil Sci. 159, 191–198.

Arrouays, D., Daroussin, J., Kicin, L., Hassika, P., 1998. Improving topsoil carbon storage prediction using a digital elevation model in temperate forest soils of France. Soil Sci. 163, 103–108.

Austin, M.P., Belbin, L., Meyers, J.A., Doherty, M.D., Luoto, M., 2006. Evaluation of statistical models used for predicting plant species distributions: role of artificial data and theory. Ecol. Model. 199, 197–216.

Basu, N.B., Rao, P.S.C., Winzeler, H.E., Kumar, S., Owens, P.R., Merwade, V., 2010. Parsimonious modeling of hydrologic responses in engineered watersheds: structural heterogeneity vs. functional homogeneity. Water Resour. Res. 46, W04501. http://www.agu.org/journals/wr/wr1004/2009WR007803/.

Beaudette, D.E., O'Geen, A.T., 2009. Quantifying the aspect effect: an application of solar radiation modeling for soil survey. Soil Sci. Soc. Am. J. 73, 1345–1352. doi:10.2136/sssaj2008.0229.

Bell, J.C., Cunningham, R.L., Havens, M.W., 1992. Calibration and validation of a soil-landscape model for predicting soil drainage class. Soil Sci. Soc. Am. J. 56, 1860–1866.

Bell, J.C., Cunningham, R.L., Havens, M.W., 1994. Soil drainage probability mapping using a soil-landscape model. Soil Sci. Soc. Am. J. 58, 464–470.

Bell, J.C., Grigal, D.F., Bates, P.C., et al., 2000. A soil–terrain model for estimating spatial patterns of soil organic carbon. In: Wilson, J.P. (Ed.), Terrain Analysis—Principles and Applications. John Wiley & Sons, New York, pp. 295–310.

Beven, K.J., Kirkby, M.J., 1979. A physically based variable contributing area model of catchment hydrology. Hydrol. Sci. Bull. 24, 43–69.

Bishop, T.F.A., McBratney, A.B., 2001. A comparison of prediction methods for the creation of field-extent soil property maps. Geoderma 103, 149–160.

Boer, M.M., Del Barrio, G., Puigdefabregas, J., 1996. Mapping soil depth classes in dry Mediterranean areas using terrain attributes derived from a digital elevation model. Geoderma 72, 99–118.

Boettinger, J.L., 2010. Environmental covariates for digital soil mapping in the Western USA. In: Boettinger, J.L., et al. (Eds.), Digital Soil Mapping: Bridging Research, Environmental Application, and Operation. Springer, Dordrecht, pp. 17–27.

Boettinger, J.L., Ramsey, R.D., Stum, A.K., Kienast-Brown, S., Nield, S.J., Saunders, A.M., Cole, N.J., Bodily, J.M., 2008. Landsat spectral data for digital soil mapping. In: Hatemink, A.E., McBratney, A.B., Mendonca-Santos, M.L. (Eds.), Digital Soil Mapping with Limited Data. Springer, Dordrecht, pp. 193–202.

Bohner, J., Antonic, O., 2009. Land-surface parameters specific to topo-climatology. In: Hengl, T., Reuter, H.I. (Eds.), Geomorphometry: Concepts, Software, Applications, Developments in Soil Science, vol. 33. Elsevier, Amsterdam, pp. 195–226.

Bonan, G.B., 1998. The land surface climatology of the NCAR land surface model coupled to the NCAR community climate model. J. Clim. 11, 1307–1326.

Breiman, L., 1996. Bagging predictors. Machine Learn. 26, 123–140.

Breiman, L., 2001. Random forest. Machine Learn. 45, 5–32.

Brown, R.B., Miller, G.A., 1989. Extending the use of soil survey information. J. Agron. Educ. 18, 32–36.

Buol, S.W., Southard, R.J., Graham, R.C., McDaniel, P.A., 2003. Soil Genesis and Classification, fifth ed. Blackwell/Iowa State Press, Ames, IA.

Burrough, P.A., van Gaans, P.F.M., MacMillan, R.A., 2000. High resolution landform classification using fuzzy-k means. Fuzzy Sets Syst. 113, 37–52.

Campling, P., Gobin, A., Feyen, J., 2002. Logistic modeling to spatially predict the probability of soil drainage classes. Soil Sci. Soc. Am. J. 66, 1390–1401.

Carré, F., McBratney, A.B., Mayr, T., Montanarella, L., 2007. Digital soil assessment: beyond DSM. Geoderma 142, 69–79.

Chamran, F., Gessler, P.E., Chadwick, O.A., 2002. Spatially explicit treatment of soil–water dynamics along a semiarid catena. Soil Sci. Soc. Am. J. 66, 1571–1583.

Chaplot, V., Walter, C. 2002. The suitability of quantitative soil – landscape models for predicting soil properties at a regional level. 7th World Congress of Soil Science, Bangkok, Thailand, August 14–21. Paper no. 2331.

Chaplot, V., Walter, C., Curmi, P., 2000. Improving soil hydromorphy prediction according to DEM resolution and available pedological data. Geoderma 97, 405–422.

Chaplot, V., Bernoux, M., Walter, C., Curmi, P., Herpin, U., 2001. Soil carbon storage prediction in temperate hydromorphic soils using a morphologic index and digital elevation model. Soil Sci. 166, 48–60.

Cialella, A.T., Dubayah, R., Lawrence, W., Levine, E., 1997. Predicting soil drainage class using remotely sensed and digital elevation data. Photogrammetric Engineering and Remote Sensing 63, 171–178.

Collins, M.E., Kuehl, R.J., 2001. Organic matter accumulation and organic soils. In: Richardson, J.L., Vepraskas, M.J. (Eds.), Wetland Soils: Their Genesis, Morphology, Hydrology, Landscapes, and Classification. CRC Press, Boca Raton, FL, pp. 137–162.

Conacher, A.J., Dalrymple, J.B., 1977. The nine unit landscape model: an approach to pedogeomorphic research. Geoderma 18, 1–154.

Crist, E.P., Laurin, R., Cicone, R.C., 1986. Vegetation and soils information contained in transformed Thematic Mapper data. In: IGARSS' 86, vol. ESA SP-254. ESA Publication Divison, Zurich, pp. 1465–1472.

Dai, Y., Zeng, Q., 1997. A land surface model (IAP94) for climate studies, part I: formulation and validation in off-line experiments. Adv. Atmos. Sci. 14, 443–460.

de Bruin, S., Stein, A., 1998. Soil–landscape modeling using fuzzy c-means clustering of attribute data derived from a Digital Elevation Model (DEM). Geoderma 83, 17–33.

Dickinson, R.E., Henderson-Sellers, A., Kennedy, P.J. 1993. Biosphere–Atmosphere Transfer Scheme (BATS) version 1e as coupled to the NCAR Community Climate Model. NCAR Tech. Note NCAR/TN-387+STR, p. 72.

Dikau, R., 1989. The application of a digital relief model to landform analysis. In: Raper, J.F. (Ed.), Three Dimensional Applications in Geographical Information Systems. Taylor and Francis, London, pp. 51–77.

Dobos, E., Daroussin, J., Montanarella, L., 2005. A SRTM-based Procedure to Delineate SOTER Terrain Units on 1:1 M and 1:5 M Scales. European Commission Report, EUR 21571. Office for Official Publications of the European Communities, Luxembourg.

Dobos, E., Bialko, T., Micheli, E., Kobza, J., 2010. Legacy soil data harmonization and database development. In: Boettinger, J.L., et al. (Eds.), Digital Soil Mapping: Bridging Research, Environmental Application, and Operation. Springer, Dordrecht, pp. 309–323.

Durana, P.J., 2002. Appendix A: chronology of the U.S. soil survey. In: Helms, D., Effland, A.B., Durana, P.J. (Eds.), Profiles in the History of the U.S. Soil Survey. Iowa State Press, Ames, Iowa, USA doi: 10.1002/9780470376959.app1.

D'Amore, D.V., Stewart, S.R., Huddleston, J.H., 2004. Saturation, reduction, and the formation of iron-manganese concentrations in the Jackson–Frazier wetland. Oregon. Soil Sci. Soc. Am. J. 68, 1012–1022.

D'Amore, D.V., Stewart, S.R., Huddleston, J.H., Glasmann, J.R., 2000. Stratigraphy and hydrology of the Jackson–Frazier wetland. Oregon. Soil Sci. Soc. Am. J. 64, 1535–1543.

Florinsky, I.V., Eilers, R.G., Manning, G.R., Fuller, L.G., 2002. Prediction of soil properties by digital terrain modeling. Environ. Model. Software 17, 295–311.

Franzmeier, D.P., Pedersen, E.J., Longwell, T.J., Byrne, J.G., Losche, C.K., 1969. Properties of some soils in the Cumberland Plateau as related to slope aspect and position. Soil Sci. Soc. Am. J. 33, 755–761.

Freeman, T.G., 1991. Calculating catchment area with divergent flow based on a regular grid. Comput. Geosciences 17, 413–422.

Freund, Y., Schapire, R.E., 1997. A decision-theoretic generalization of on-line learning and an application to boosting. J. Comput. Syst. Sci. 55 (1), 119–139.

Gallant, J.C., Dowling, T.I., 2003. A multiresolution index of valley bottom flatness for mapping depositional areas. Water Resour. Reseach 39 (12), 1347–1360.

Gallant, J.C., Wilson, J.P., 2000. Primary topographic attributes. In: Wilson, J.P., Gallant, J.C. (Eds.), Terrain Analysis: Principles and Applications. John Wiley & Sons, New York, pp. 51–85.

Gessler, P.E., Moore, I.D., McKenzie, N.J., Ryan, P.J., 1995. Soil–landscape modelling and spatial prediction of soil attributes. Int. J. Geographical Inf. Syst. 9, 421–432.

Gessler, P.E., Chadwick, O.A., Chamran, F., Althouse, L., Holmes, K., 2000. Modeling soil-landscape and ecosystem properties using terrain attributes. Soil Sci. Soc. Am. J. 64, 2046–2056.

Grayson, R., Bloschl, G. (Eds.), 2000. Spatial Pattern in Catchment Hydrology: Observations and Modeling. Cambridge Univ. Press, NewYork.

Gruber, S., Peckham, S., 2009. Land-surface parameters and objects in hydrology, pp. 171–194. In: Hengl, T., Reuter, H.I. (Eds.), Developments in Soil Science. Geomorphometry – Concepts, Software, Applications, vol. 33. Elsevier.

Grunwald, S. (Ed.), 2006. Environmental Soil-landscape Modeling – Geographic Information Technologies and Pedometrics. CRC Press, New York.

Grunwald, S., 2009. Multi-criteria characterization of recent digital soil mapping and modeling approaches. Geoderma 152, 195–207.

Grunwald, S., Thompson, J.A., Boettinger, J.L., 2011. Digital soil mapping and modeling at continental scales – finding solutions for global issues. Soil Sci. Soc. Am. J. (SSSA 75th Anniversary Special Paper) 75 (4), 1201–1213.

Guisan, A., Zimmermann, N.E., 2000. Predictive habitat distribution models in ecology. Ecol. Model. 135, 147–186.

Guisan, A., Edwards Jr., T.C., Hastie, T., 2002. Generalized linear and generalized additive models in studies of species distributions: setting the scene. Ecol. Model. 157, 89–100.

Guntner, A., Bronstert, A., 2004. Representation of landscape variability and lateral redistribution processes for large-scale hydrological modelling in semi-arid areas. J. Hydrol. 297, 136–161.

Hartemink, A.E., McBratney, A.B., 2008. A soil science renaissance. Geoderma 148, 123–129.

Hartemink, A.E., Hempel, J., Lagacherie, P., McBratney, A.B., McKenzie, N., MacMillan, R.A., Minasny, B., Montanarella, L., Mendonça Santos, M.L., Sanchez, P., Walsh, M., Zhang, G.L., et al., 2010. GlobalSoilMap.net: a new digital soil map of the world. In: Boettinger, J.L. (Ed.), Digital Soil Mapping: Bridging Research, Environmental Application, and Operation. Springer-Verlag, Dordrecht, the Netherlands, pp. 423–427.

Hastie, T.J., Tibshirani, R., Friedman, J., 2009. The Elements of Statistical Learning: Data Mining, Inference and Prediction. Springer Series in Statistics, second ed. Springer-Verlag, New York.

Hengl, T., 2009. A Practical Guide to Geostatistical Mapping, second ed. University of Amsterdam. www.lulc.com, p. 291.

Hengl, T., Reuter, H.I. (Eds.), 2009. Developments in Soil Science. Geomorphometry – Concepts, Software, Applications, vol. 33. Elsevier.

Hengl, T., Heuvelink, G.B.M., Stein, A., 2004. A generic framework for spatial prediction of soil variables based on regression kriging. Geoderma 120, 75–93.

Heuvelink, G., 2003. The definition of pedometrics. Pedometron 15, 11–12. Available from: http://www.pedometrics.org/pedometron/pedometron15.pdf (accessed 17.03.11.).

Heuvelink, G.B.M., Webster, R., 2001. Modeling soil variation: past, present, and future. Geoderma 100, 269–301.

Hively, W.D., Bryant, R.B., Fahey, T.J., 2005. Phosphorus concentrations in runoff from diverse locations on a New York dairy farm. J. Env. Qual. 34, 1224–1233.

Hollenbeck, K.J., Schmugge, T.J., Hornberger, G.M., Wang, J.R., 1996. Identifying soil hydraulic heterogeneity by detection of relative change in passive microwave remote sensing observations. Water Resour. Res. 32, 139–148.

Houser, P., Goodrich, D., Syed, K., 2000. Runoff, precipitation, and soil moisture at Walnut Gulch. In: Grayson, R., Bloschl, G. (Eds.), Spatial Pattern in Catchment Hydrology: Observations and Modeling. Cambridge Univ. Press, NewYork.

Huang, C., Wylie, B., Yang, L., Homer, C., Zylstra, G., 2002. Derivation of a tasseled cap transformation based on Landsat 7 at-satellite reflectance. Int. J. Remote Sensing 23 (8), 1741–1748.

Hudson, B.D., 1992. The soil survey as paradigm-based science. Soil Sci. Soc. Am. J. 56, 836–841.

Huete, A.R., 1988. A soil adjusted vegetation index (SAVI). Remote Sensing Environ. 25, 295–309.

Hunckler, R.V., Schaetzl, R.J., 1997. Spodosol development as affected by geomorphic aspect, Baraga County. Michigan. Soil Sci. Soc. Am. J. 61, 1105–1115.

Hutchins, R.B., Blevins, R.L., Hill, J.D., White, E.H., 1976. The influence of soils and microclimate on vegetation of forested slopes in eastern Kentucky. Soil Sci. 121, 234–241.

Jackson, T.J., Le Vine, D.E., 1996. Mapping surface soil moisture using an aircraft-based passive instrument: algorithm and example. J. Hydrol. 18, 85–99.

Jenkinson, B.J., Franzmeier, D.P., Lynn, W.C., 2002. Soil hydrology on an end moraine and a dissected till plain in west-central Indiana. Soil Sci. Soc. Am. J. 66, 1367–1376.

Jenny, H., 1941. Factors of Soil Formation: A System of Quantitative Pedology. McGraw-Hill, New York.

Jenny, H., 1961. Derivation of state factor equations of soils and ecosystems. Soil Sci. Soc. Am. Proc. 25, 385–388.

Jenny, H., 1980. The Soil Resources. Spring-Verlag, New York.

Jensen, J.R., 2005. Introductory Digital Image Processing. Prentice Hall, Upper Saddle River, NJ.

Kauth, R.J., Thomas, G.S., 1976. The tasselled cap – A graphic description of the spectral–temporal development of agricultural crops as seen by Landsat. In: Symposium on Machine Processing of Remotely Sensed Data, IEEE, vol. 76, pp. 41–51, (CH 1103-IMPRSD).

Koster, R.D., Suarez, M.J., 1996. Energy and Water Balance Calculations in the MOSAIC LSM. NASA Technical Memorandum 104606, pp. 9–76.

Lagacherie, P., 2008. Digital soil mapping: a State of the art, pp. 3–14. In: Hartemink, A.E., McBratney, A.B., Mendonca-Santos, M.L. (Eds.), Digital Soil Mapping with Limited Data. Springer, Dordrecht.

Lagacherie, P., Holmes, S., 1997. Addressing geographical data errors in a classification tree soil unit prediction. Int. J. Geographical Inf. Sci. 11, 183–198.

Lagacherie, P., McBratney, A.B., 2007. Spatial soil information systems and spatial soil inference systems: perspectives for digital soil mapping. In: Lagacherie, P., McBratney, A.B., Voltz, M. (Eds.), Digital Soil Mapping: An Introductory Perspective. Developments in Soil Science, vol. 31. Elsevier, Amsterdam, pp. 389–399.

Lagacherie, P., Voltz, M., 2000. Predicting soil properties over a region using sample information from a mapped reference area and digital elevation data: a conditional probability approach. Geoderma 97, 187–208.

Lane, P.W., 2002. Generalized linear models in soil science. Eur. J. Soil Sci. 53, 241–251.
Lark, R.M., 1999. Soil-landform relationships at within-field scales: an investigation using continuous classification. Geoderma 92, 141–165.
Lemercier, B., Lacoste, M., Loum, M., Walter, C., 2012. Extrapolation at regional scale of local soil knowledge using boosted classification trees: A two-step approach. Geoderma 171–172, 75–84.
Liang, X., Lettenmaier, D.P., Wood, E.F., Burges, S.J., 1994. A simple hydrologically based model of land surface water and energy fluxes for GSMs. J. Geophys. Res. 99 (D7), 14415–14428.
Liang, X., Lettenmaier, D.P., Wood, E.F., 1996. One-dimensional statistical dynamic representation of Subgrid spatial variability of precipitation in the two-layer variable infiltration capacity model. J. Geophys. Res. 101 (D16), 21403–21422.
Lin, H.S., 2003. Hydropedology: bridging disciplines, scales, and data. Vadose Zone J. 2, 1–11. doi:10.2113/2.1.1.
Lin, H.S., 2006. Temporal stability of soil moisture spatial pattern and subsurface preferential flow pathways in the Shale Hills Catchment. Vadose Zone J. 5, 317–340.
Lin, H.S., 2010. Linking proiniples of soil formation and flow regimes. J. Hydrology 393, 3–19.
Lin, H.S., 2011. Hydropedology: towards new insights into interactive pedologic and hydrologic processes across scales. J. Hydrology 406, 141–145. doi: 10.1016/j.jhydrol.2011.05.054.
Lin, H.S., Bouma, J., Wilding, L., Richardson, J., Kutilek, M., Nielsen, D., 2005. Advances in hydropedology. Adv. Agron. 85, 1–89.
Lin, H., Bouma, J., Pachepsky, Y., Western, A., Thompson, J., van Genuchten, R., Vogel, H.-J., Lilly, A., 2006. Hydropedology: synergistic integration of pedology and hydrology. Water Resour. Res. 42, W05301. doi:10.1029/2005WR004085.
Liu, J., Pattey, E., Nolin, M.C., Miller, J.R., Ka, O., 2008. Mapping with in-field soil drainage using remote sensing, DEM and apparent soil electrical conductivity. Geoderma 143, 261–272.
Lohse, K.A., Dietrich, W.E., 2005. Contrasting effects of soil development on hydrological properties and flow paths. Water Resour. Res. 41, W12419. doi: 10.1029/2004WR003403.
MacMillan, R.A., Moon, D.E., Coupé, R.A., Phillips, N., et al., 2010. Predictive Ecosystem Mapping (PEM) for 8.2 million ha of Forestland, British Columbia, Canada. In: Boettinger, J.L. (Ed.), Digital Soil Mapping: Bridging Research, Environmental Application, and Operation. Springer, Dordrecht, pp. 337–356.
Maindonald, J., Braun, J., 2007. Data Analysis and Graphics Using R: An Example-based Approach. Cambridge University Press, Cambridge.
Malone, B.P., McBratney, A.B., Minasny, B., Laslett, G.M., 2009. Mapping continuous depth functions of soil carbon storage and available water capacity. Geoderma 154 (1–2), 138–152.
Mayr, T., Rivas-Casado, M., Bellamy, P., Palmer, R., Zawadzka, J., Corstanje, R., 2010. Two methods for using legacy data in digital soil mapping. In: Boettinger, J.L., et al. (Eds.), Digital Soil Mapping: Bridging Research, Environmental Application, and Operation. Springer, Dordrecht, pp. 191–202.
McBratney, A.B., Odeh, I., Bishop, T., Dunbar, M.S., Shatar, T.M., 2000. An overview of pedometric techniques for use in soil survey. Geoderma 97, 293–327.
McBratney, A.B., Mendonça-Santos, M.L., Minasny, B., 2003. On digital soil mapping. Geoderma 117, 3–52.
McBratney, A.B., Minasny, B., Viscarra Rossel, R., 2006. Spectral soil analysis and inference systems: a powerful combination for solving the soil data crisis. Geoderma 136, 272–278.

McDaniel, P.A., Bathke, G.R., Buol, S.W., Cassel, D.K., Falen, A.L., 1992. Secondary manganese/iron ratios as pedochemical indicators of field-scale throughflow water movement. Soil Sci. Soc. Am. J. 56, 1211–1217.

McDonnell, J.J., Sivapalan, M., Vache, K., Dunn, S., Grant, G., Haggerty, R., Hinz, C., Hooper, R., Kirchner, J., Roderick, M.L., Selker, J., Weiler, M., 2007. Moving beyond heterogeneity and process complexity: a new vision for watershed hydrology. Water Resour. Res. 43, W07301.

McKenzie, N.J., Austin, M.P., 1993. A quantitative Australian approach to medium and small scale surveys based on soil stratigraphy and environmental correlation. Geoderma 57, 329–355.

McKenzie, N., Jacquier, D., 1997. Improving the field estimation of saturated hydraulic conductivity in soil survey. Aust. J. Soil Res. 35, 803–825.

McKenzie, N.J., Ryan, P.J., 1999. Spatial prediction of soil properties using environmental correlation. Geoderma 89, 67–94.

Miller, D.A., White, R.A., 1998. A conterminous United States multi-layer soil characteristics data set for regional climate and hydrology modeling. Earth Interact. 2.

Milne, G., 1936. Normal erosion as a factor in soil profile development. Nature 138, 548–549.

Minasny, B., McBratney, A.B., 2007. Spatial prediction of soil properties using EBLUP with Matérn covariance function. Geoderma 140, 324–336.

Minasny, B., McBratney, A.B., Salvador-Blanes, S., 2008. Quantitative models for pedoenesis—a review. Geoderma 144, 140–157.

Moore, I.D., Grayson, R.B., Ladson, A.R., 1991. Digital terrain modelling: a review of hydrological, geomorphological, and biological applications. Hydrol. Process. 5, 3–30.

Moore, I.D., Gessler, P.E., Nielsen, G.A., Peterson, G.A., 1993. Soil attribute prediction using terrain analysis. Soil Sci. Soc. America J. 57, 443–452.

Moran, C.J., Bui, E., 2002. Spatial data mining for enhanced soil map modelling. Int. J. Geogr. Inf. Sci. 16, 533–549.

Mulder, V.L., de Bruin, S., Schaepman, M.E., Mayr, T.R., 2011. The use of remote sensing in soil and terrain mapping – a review. Geoderma 162, 1–19.

Nelson, E.J., Jones, N.L., Miller, A.W., 1994. Algorithm for precise drainage-basin delineation. J. Hydraul. Eng. ASCE 120 (3), 298–312.

Nield, S.J., Boettinger, J.L., Ramsey, R.D., 2007. Digitally mapping gypsic and natric soil areas using Landsat ETM data. Soil Sci. Soc. America J. 71, 245–252.

Noguchi, S., Tsuboyama, Y., Sidle, R.C., Hosoda, I., 1999. Morphological characteristics of macropores and the distribution of preferential flow pathways in a forested slope segment. Soil Sci. Soc. Am. J. 63, 1413–1423.

Noller, J.S., 2010. Applying geochronology in predictive digital mapping of soils. In: Boettinger, J.L., et al. (Eds.), Digital Soil Mapping: Bridging Research, Environmental Application, and Operation. Springer, Dordrecht, pp. 43–53.

Oberthur, T., Goovaerts, P., Dobermann, A., 1999. Mapping soil texture classes using field texturing, particle size distribution and local knowledge by both conventional and geostatistical methods. Eur. J. Soil Sci. 50, 457–479.

Odeh, I.O.A., McBratney, A.B., Chittleborough, D.J., 1994. Spatial prediction of soil properties from landform attributes derived from a digital elevation model. Geoderma 63, 197–214.

Odeh, I.O.A., McBratney, A.B., Chittleborough, D.J., 1995. Further results on prediction of soil properties from terrain attributes: heterotopic cokriging and regression-kriging. Geoderma 67, 215–225.

Ogden, F.L., Garbrecht, J., DeBarry, P.A., Johnson, L.E., 2001. GIS and Distributed Watershed Models II. Publications from USDA-ARS / UNL Faculty. Paper 461. http://digitalcommons.unl.edu/usdaarsfacpub/461.

Olaya, V., 2009. Basic land-surface parameters. In: Hengl, T., Reuter, H.I. (Eds.), Developments in Soil Science. Geomorphometry – Concepts, Software, Applications, vol. 33. Elsevier.

O'Callaghan, J., Mark, D., 1984. The extraction of drainage networks from digital elevation data. Comput. Vis. Graphics Image Process. 28, 323–344.

Pachepsky, Y.A., Timlin, D.J., Rawls, W.J., 2001. Soil water retention as related to topographic variables. Soil Sci. Soc. Am. J. 65, 1787–1795.

Park, S.J., Vlek, L.G., 2002. Environmental correlation of three-dimensional soil spatial variability: a comparison of three environmental correlation techniques. Geoderma 109, 117–140.

Park, S.J., McSweeney, K., Lowery, B., 2001. Identification of the spatial distribution of soils using a process-based terrain characterization. Geoderma 103, 249–272.

Peng, W., Wheeler, D.B., Bell, J.C., Krusemark, M.G., 2003. Delineating patterns of soil drainage class on bare soils using remote sensing analyses. Geoderma 115, 261–279.

Pennock, D.J., Zebarth, B.J., De Jong, E., 1987. Landform classification and soil distribution in Hummocky terrain, Saskatchewan, Canada. Geoderma 40, 297–315.

Rasmussen, C., Tabor, N.J., 2007. Applying a quantitative pedogenic energy model across a range of environmental gradients. Soil Sci. Soc. Am. J. 71, 1719–1729.

Rasmussen, C., Troch, P.A., Chorover, J., Brooks, P., Pelletier, J., Huxman, T.E., 2010. An open system framework for integrating critical zone structure and function. Biogeochemistry doi:10.1007/s10533-010-9476-8.

Reuter, R.J., Bell, J.C., 2001. Soils and hydrology of a wet-sandy catena in east-central Minnesota. Soil Sci. Soc. Am. J. 65, 1559–1569.

Reuter, R.J., Bell, J.C., 2003. Hillslope hydrology and soil morphology for a wetland basin in south-central Minnesota. Soil Sci. Soc. Am. J. 67, 365–372.

Reynolds, C.A., Jackson, T.J., Rawls, W.J., 2000. Estimating soil water-holding capacities by linking the Food and Agriculture Organization soil map of the world with global pedon databases and continuous pedotransfer functions. Water Resour. Res. 36, 3653–3662.

Romano, N., Palladino, M., 2002. Prediction of soil water retention using soil physical data and terrain attributes. J. Hydrol. 265, 56–75.

Ruhe, R.V., 1975. Geomorphology. Houghton Mifflin, Boston, MA.

Runge, E.C.A., 1973. Soil development sequences and energy models. Soil Sci. 115, 183–193.

Ryan, P.J., McKenzie, N.J., O'Connell, D., Loughhead, A.N., Leppert, P.M., Jacquier, D., Ashton, L., 2000. Integrating forest soils information across scales: spatial prediction of soil properties under Australian forests. For. Ecol. Manage. 138, 139–157.

Sanchez, P.A., Ahamed, S., Carré, F., Hartemink, A.E., Hempel, J., Huising, J., et al., 2009. Digital soil map of the world. Science 325, 680–681. doi:10.1126/science.1175084.

Schaetzl, R.J., Anderson, S., 2005. Soils: Genesis and Geomorphology. Cambridge University Press, New York.

Schmidt, J., Hewitt, A.E., 2004. Fuzzy land element classification from DTMs based on geometry and terrain position. Geoderma 121, 234–256.

Schmidt, J., Evans, I.S., Brinkmann, J., 2003. Comparison of polynomial models for land surface curvature calculation. Int. J. Geographical Inf. Sci. 17, 797–814.

Schoenberger, P.J., Wysocki, D.A. (Eds.), 2008. Geomorphic Description System, Version 4.1. Natural Resources Conservation Service, National Soil Survey Center, Lincoln, NE.

Schowengerdt, R.A., 2007. Remote Sensing: Models and Methods for Image Processing, second ed. Elsevier, Burlington, MA.

Scull, P., Franklin, J., Chadwick, O.A., McArthur, D., 2003. Predictive soil mapping: a review. Prog. Phys. Geogr. 27, 171–197.

Scull, P., Okin, G., Chadwick, O.A., Franklin, J., 2005. A comparison of methods to predict soil surface texture in an alluvial basin. Prof. Geogr. 57 (3), 423–437.

Seibert, J., McGlynn, B., 2007. A new triangular multiple flow direction algorithm for computing upslope areas from gridded digital elevation models. Water Resour. Res. 43, W04501.

Shary, P.A., 1995. Land surface in gravity points classification by a complete system of curvatures. Math. Geol. 27, 373–390.

Shary, P.A., Sharayab, L.S., Mitusov, A.V., 2002. Fundamental quantitative methods of land surface analysis. Geoderma 107, 1–43.

Simonson, R.W., 1959. Outline of a geneialized theory of soil genesis. Soil Sci. Soc. Am. Proc. 23, 152–156.

Simonson, R.W., 1978. A multiple-process model of soil genesis. In: Mahanev, W.C. (Ed.), Quaternary Soils. Norwich. UK. Geo Abstracts, pp. 1–25.

Sinowski, W., Auerswald, K., 1999. Using relief parameters in a discriminant analysis to stratify geological areas with different spatial variability of soil properties. Geoderma 89, 113–128.

Soil Survey Staff, 1999. Soil Taxonomy: A Basic System of Soil Classification for Making and Interpreting Soil Surveys. USDA-SCS. Agriculture Handbook No. 436, second ed. US Government Printing Office, Washington, DC.

Soil Survey Staff, 2009. State Soil Survey (SSURGO) Database, Soil Data Mart Available from: http://soildatamart.nrcs.usda.gov/Default.aspx (accessed 12.30.09.).

Sommer, M., Wehrhan, M., Zipprich, M., Weller, U., Zu Castell, W., Ehrich, S., Tandler, B., Selige, T., 2003. Hierarchical data fusion for mapping soil units at field scale. Geoderma 112, 179–196.

Tarboton, D.G., 1997. A new method for the determination of flow directions and contributing areas in grid digital elevation models. Water Resour. Res. 33, 309–319.

Terra, J.A., Shaw, J.N., Reeves, D.W., Raper, R.L., van Santen, E., Mask, P.L., 2004. Soil carbon relationships with terrain attributes, electrical conductivity, and soil survey in a coastal plain landscape. Soil Sci. 169, 819–831.

Tesfa, T.K., Tarboton, D.G., Watson, D.W., Schreuders, K.A.T., Baker, M.E., Wallace, R.W., 2011. Extraction of hydrological proximity measures from DEMs using parallel processing. Environ. Model. Software 26, 1696–1709.

Thompson, J.A., Bell, J.C., 1996. Color index for identifying hydric soil conditions for seasonally saturated Mollisols in Minnesota. Soil Sci. Soc. Am. J. 60, 1979–1988.

Thompson, J.A., Bell, J.C., 1998. Hydric conditions and hydromorphic properties within a Mollisol catena in southeastern Minnesota. Soil Sci. Soc. Am. J. 62, 1116–1125.

Thompson, J.A., Kolka, R.K., 2005. Soil carbon storage estimation in a central hardwood forest watershed using quantitative soil-landscape modeling. Soil Sci. Soc. America J. 69, 1086–1093.

Thompson, J.A., Bell, J.C., Butler, C.A., 1997. Quantitative soil-landscape modeling for estimating the areal extent of hydromorphic soils. Soil Sci. Soc. Am. J. 61, 971–980.

Thompson, J.A., Bell, J.C., Zanner, C.W., 1998. Hydrology and hydric soil extent within a Mollisol catena in southeastern Minnesota. Soil Sci. Soc. Am. J. 62, 1126–1133.

Tobler, W., 1970. A computer movie simulating urban growth in the Detriot region. Econ. Geogr. 46 (2), 234–240.

Troch, P.A., Carrillo, G.A., Heidbuchel, I., Rajagopal, S., Switanek, M., Volkmann, T.H.M., Yaeger, M., 2009. Dealing with landscape heterogeneity in watershed hydrology: a review of recent progress toward new hydrological theory. Geogr. Comp. 3, 375–392.

Troeh, F.R., 1964. Landform parameters correlated to soil drainage. Soil Sci. Soc. Am. Proc. 28, 808–812.

Turk, J.K., Graham, R.C., 2011. Distribution and properties of vesicular horizons in the western United States. Soil Sci. Soc. Am. J. 75, 1449–1461. doi: 10.2136/sssaj2010.0445.

U.S. Department of Agriculture, Natural Resources Conservation Service, 2012. National Soil Survey Handbook, Title 430-VI Available from: http://soils.usda.gov/technical/handbook/ (accessed 02.05.12).

U.S. Department of Agriculture, Natural Resources Conservation Service, 2010. Field indicators of hydric soils in the United States, Version 7.0. In: Vasilas, L.M., Hurt, G.W., Noble, C.V. (Eds.), USDA, NRCS, in Cooperation with the National Technical Committee for Hydric Soils.

Vasques, G.M., Grunwald, S., Comerford, N.B., Sickman, J.O., 2010. Regional modeling of soil carbon at multiple depths within a subtropical watershed. Geoderma 156, 326–336.

Vepraskas, M.J., Wilding, L.P., 1983a. Aquic moisture regimes in soils with and without low chroma colors. Soil Sci. Soc. Am. J. 47, 280–285.

Vepraskas, M.J., Wilding, L.P., 1983b. Albic neoskeletons in argillic horizons as indices of seasonal saturation and reduction. Soil Sci. Soc. Am. J. 47, 1202–1208.

Vepraskas, M.J., Heitman, J.L., Austin, R.E., 2009. Future directions for hydropedology: quantifying impacts of global change on land use. Hydrol. Earth Syst. Sci. 13, 1427–1438.

Vertessy, R., Elsenbeer, H., Bessard, Y., Lack, A., 2000. Storm runoff generation at La Cuenca. In: Grayson, R., Bloschl, G. (Eds.), Spatial Pattern in Catchment Hydrology: Observations and Modeling. Cambridge Univ. Press, NewYork.

Voltz, M., Lagacherie, P., Louchart, X., 1997. Predicting soil properties over a region using sample information from a mapped reference area. Eur. J. Soil Sci. 48, 19–30.

Webster, R., 2001. Statistics to support soil research and their presentation. Eur. J. Soil Sci. 52, 331–340.

Western, A.W., Grayson, R.B., 2000. Soil moisture and runoff processes at Tarrawarra. In: Grayson, R.B., Bloschl, G. (Eds.), Spatial Patterns in Catchment Hydrology: Observations and Modelling. Cambridge Univ. Press, New York, pp. 209–246.

Wilson, J.P., Gallant, J.C., 2000. Secondary terrain attributes. In: Wilson, J.P., Gallant, J.C. (Eds.), Terrain Analysis. John Wiley and Sons, New York.

Wood, E.F., Liang, X., Lohmann, D., Lettenmaier, D.P., 1998. The project for intercomparison of land-surface parameterization schemes (PILPS) phase-2(c) Red-Arkansas River experiment, 1, Experiment description and summary intercomparisons. J. Glob. Planet. Change 19, 115–135.

Wysocki, D.A., Schoeneberger, P.J., Hirmas, D.R., LaGarry, H.E., 2011. Geomorphology of soil landscapes. In: Huang, et al. (Eds.), Handbook of Soil Sciences: Properties and Processes, second ed. CRC Press, Boca Raton, FL.

Xiao, J., Shen, Y., Tateishi, R., Bayaer, W., 2006. Development of topsoil grain size index for monitoring desertification in arid land using remote sensing. Int. J. Remote Sens. 27 (12), 2411–2422.

Zaitchik, B.F., Rodell, M., Olivera, F., 2009. Evaluation of the global land data assimilation system using global river discharge data and a source to sink routing scheme. Water Resour. Res. doi:10.1029/2009WR007811.

Zheng, D., Hunt, E.R., Running, S.W., 1996. Comparison of available soil water capacity estimated from topography and soil series information. Landsc. Ecol. 11, 3–14.

Zhu, A.X., 1997a. A similarity model for representing soil spatial information. Geoderma 77, 217–242.

Zhu, A.X., 1997b. Measuring uncertainty in class assignment for natural resource maps under fuzzy logic. Photogram. Eng. Remote Sens. 63, 1195–1202.

Zhu, A.X., 2000. Mapping soil landscape as spatial continua: the neural network approach. Water Resour. Res. 36, 663–677.

Zhu, A.X., 2006. Fuzzy logic models. In: Grunwald, S. (Ed.), Environmental Soil-Landscape Modeling – Geographic Information Technologies and Pedometrics. CRC, Boca Raton, FL, pp. 215–239.

Zhu, A.X., Band, L.E., 1994. A knowledge-based approach to data integration for soil mapping. Can. J. Remote Sens. 20, 408–418.

Zhu, A.X., Band, L.E., Dutton, B., Nimlos, T.J., 1996. Automated soil inference under fuzzy logic. Ecol. Model. 90, 123–145.

Zhu, A.X., Band, L.E., Vertessy, R., Dutton, dB., 1997. Derivation of soil properties using a soil land inference model (SoLIM). Soil Sci. Soc. Am. J. 61, 523–533.

Zhu, A.X., Hudson, B., Burt, J., Lubich, K., Simonson, D., 2001. Soil mapping using GIS, expert knowledge and fuzzy logic. Soil Sci. Soc. America J. 65, 1463–1472.

Chapter 22

Coupling Biogeochemistry and Hydropedology to Advance Carbon and Nitrogen Cycling Science

Michael J. Castellano,[1,*] David Bruce Lewis,[2]
Danielle M. Andrews[3] and Marshall D. McDaniel[3]

ABSTRACT

Biogeochemical processes are inextricably linked to hydropedology. Soil structure and water content interact to affect the transport and transformation of nutrients and carbon. These processes can affect ecosystem productivity, greenhouse gas emissions, as well as losses of dissolved nutrients and carbon. Advances in soil biogeochemistry research often rely on laboratory research and future approaches will continue to improve the integration with hydropedology by explicitly considering the interacting effects of soil structure and hydrology. Nevertheless, significant progress in coupling soil biogeochemistry and hydropedology has been made, particularly at the field scale, by incorporating advective processes into concepts of biogeochemical dynamics. This work has focused on "hot moments," which are brief periods of time that account for disproportionately large amounts of material transport and transformation. Hot moments often result from pulses of water availability and provide exceptional opportunities to couple soil biogeochemistry and hydropedology. Future soils research will benefit by drawing from stream ecology and biogeochemistry research principles that have a rich history of exploring interactions between material transport and transformation.

1. INTRODUCTION

Biogeochemical cycles – the transfer of materials and energy among biotic and abiotic pools – are dominated by microorganisms. Most simply, soil

[1]Dept. of Agronomy, Iowa State University, Ames, IA, USA
[2]Dept. of Integrative Biology, University of South Florida, Tampa, FL, USA
[3]Dept. of Crop and Soil Sciences, The Pennsylvania State University, University Park, PA, USA
[*]Corresponding author: Email: castellanomichaelj@gmail.com

Hydropedology, Edited by H. Lin. DOI: 10.1016/B978-0-12-386941-8.00022-8

microorganisms cannot function without the water, nutrients, and energy (C) that soil provides. However, more careful examination reveals that hydropedology interacts with soil biogeochemistry in many complex ways, often relating to the spatial and temporal variation in water, nutrients, and C.

Earth's biogeochemistry is largely controlled by cycles of oxidation–reduction (redox) reactions. Organisms use redox reactions to derive energy by transferring electrons from a reduced chemical species (electron donor) to an oxidized chemical species (electron acceptor) (Falkowski et al., 2008). In the soil, the rate and chemical pathways of these biogeochemical reactions are indirectly, but tightly linked to the interacting forces of water and soil structure.

The amount of energy that soil microorganisms can derive from oxidation–reduction reactions is often controlled by hydrology, and more specifically, its effect on material diffusion (Hedin et al., 1998). Aerobic metabolism dominates the biosphere because Oxygen reduction yields more available energy than the reduction of alternative electron acceptors. However, soils provide many locations and times when O_2 is not available due to high water contents and low air permeability. In these situations, less favorable electron acceptors are used. Interactions between soil water and structure control the availability of O_2 and other less favorable electron acceptors. Soil texture and structure affect air permeability. Moreover, O_2 is only slightly soluble in water, diffusing approximately 10^4 times more slowly through water than through air. Accordingly, in temporary or chronic situations of O_2 absence, facultative aerobic microorganisms and obligate anaerobic microorganisms begin to reduce alternative electron acceptors roughly following the order of free energy yield: nitrate (NO_3^-), manganic manganese (Mn(IV)), ferric iron (Fe(III)), sulfate (SO_4^{2-}), and carbon dioxide (CO_2). Monomers derived from plant- and microorganism-derived organic carbon are the largest source of electron donors (energy), but methane (CH_4), hydrogen (H_2), ammonium (NH_4^+), manganous manganese (Mn(II)), ferrous iron (Fe(II)), and hydrogen sulfide (H_2S) are also used.

In addition to affecting biogeochemical reactions through *gas* diffusion (e.g. O_2), soil water content further affects biogeochemical reactions through *solute* diffusion. As soils dry and water films thin, diffusion pathways become longer, more tortuous, and ultimately disconnected (Stark and Firestone, 1995). Pools of electron donors and acceptors in disconnected water films can be depleted at different rates. As a result, biogeochemical process rates often reach a maximum at soil water contents around field capacity because the interaction between gas and solute diffusion is optimal (Stark and Firestone, 1995; Davidson et al., 2000; Reichstein, 2003; Castellano et al., 2010, 2011).

Independent of soil water content, soil structure and aggregate dynamics can also impact biogeochemical cycling. Soil aggregate structures are well known to physically protect potentially mineralizable soil organic matter from microbial attack. Soil fauna, microorganisms, roots, mineralogy, and physical processes affect aggregate dynamics and the balance between soil organic matter decomposition and stabilization (Six et al., 2004). These properties and processes are affected by soil water content and subsequently affect soil water content.

Hydropedology offers an integrated framework to understand how hydrology and pedology interact to affect biogeochemistry. For example, soil structure and layering can lead to extremely heterogeneous hydrology that affects solute transport and transformation as well as the distribution of microorganisms (Fluhler et al., 1996; Bundt et al., 2001; Asano et al., 2006; Castellano and Kaye, 2009). Soil mineralogy and ion-exchange capacity can also affect the availability of solutes for transport and transformation (Lohse and Matson, 2005; Hobbie et al., 2007). Flow rates, as affected by soil structure and soil water content, can further affect biogeochemical processes (Asano et al., 2006). Such complex interactions between hydropedology and biogeochemistry are at least partially responsible for the large spatial variability in soil biogeochemical processes (Parkin, 1987, 1993; Castellano and Kaye, 2009).

This chapter reviews the potential and necessity to couple hydropedology and biogeochemistry. Opportunities abound in the laboratory and field across a range of spatial scales. The first section examines challenges in connecting laboratory-based biogeochemical research with patterns and processes observed in the field. This area is particularly challenging because evidence suggests that hydropedology affects biogeochemistry; however, soil structure is typically destroyed in laboratory procedures. Success in this area is critical to improve numerical, mechanistic models of ecosystem dynamics. Next, the progress of biogeochemical and hydrologic research themes, as they work toward integrating hydrology, biogeochemistry, and pedology will be reviewed. This area is growing rapidly from research rooted in hydrology (e.g. van Verseveld et al., 2008), soils (e.g. Sanderman and Amundson, 2008; Castellano et al., 2011), geology (e.g. Yoo et al., 2006), and ecosystem ecology (e.g. Lohse and Matson, 2005). However, we will focus on C and nitrogen cycling research that explicitly incorporates concepts of soil structure, hydrology, and biogeochemistry. Finally, future research directions will be discussed with an emphasis on an ecosystems perspective. Previous reviews have suggested that soil biogeochemistry can benefit from explicit inclusion of advective processes (Wagener et al., 1998; Fisher et al., 2004). Indeed, much can be gained by incorporating principles from stream ecology and aquatic biogeochemistry into soils research that aims to couple hydropedology and biogeochemistry. Stream ecologists have long included advective transport mechanisms into analyses of biogeochemistry, providing a framework to understand nutrient cycling across different flow paths and rates. Coupled analyses of soil biogeochemistry and hydropedology can build on this approach. We add to previous reviews by summarizing evidence for and approaches toward the successful transfer of stream ecology perspectives to the soil.

Coupled C and N cycles are the focus of this chapter. Due to high redox activity, broad relevance to global change, and research prominence, C and N provide a platform for introducing the existing research and future potential for coupling of hydropedology and biogeochemistry. However, hydropedology influences all aspects of biogeochemical cycling. Although the biophysical mechanisms affecting individual element cycles are generally similar, the relative importance of these mechanisms can differ for elements such as phosphorus that lack a significant gaseous phase and are dominated by physical rather than biological fluxes.

2. BRIDGING SCALES

Biogeochemistry and hydropedology share similar challenges in transferring research results among scales. Similar to classic soil physics research, laboratory-based soil biogeochemistry research has made important advances in understanding mechanistic, causal relationships between soil properties and biogeochemical processes. For example, solute diffusion limits microbial activity at high water potentials whereas cellular dehydration limits microbial activity at low water potentials (Stark and Firestone, 1995). However, a dynamic suite of variables interacts to control biogeochemical processes in the field. Under natural conditions, the relative importance of one independent variable typically changes as a function of many other independent variables. For example, large changes in soil water content (e.g. wilting point to field capacity) are expected to produce large changes in decomposition rates at high temperatures (>10 °C), but little-to-no change in decomposition rates at low temperatures (<10 °C; Reichstein et al., 2003). A numerical expression of this relationship is relatively simple; but as other independent variables such as organic matter quality, mineralogy, and soil structure are considered, complexity increases rapidly. In the laboratory, this complexity is typically sacrificed to identify interactions between a select few variables. On the other hand, field research cannot describe the full suite of ecosystem properties that controls soil biogeochemical processes.

2.1. Laboratory Insights

Laboratory-based biogeochemistry research has worked to identify relationships between soil water content and biogeochemical processes that are easily transferable to the field and consistent across diverse soil types. In general, this research has concluded that: 1) matric potential is difficult to measure, but it is a relatively consistent scalar of relative biogeochemical process rates across different soil types whereas, 2) volumetric water content is easy to measure, but it is not a consistent scalar of relative biogeochemical processes across soil types that differ in texture, structure, and bulk density (Sommers et al., 1981; Castellano et al., 2010, 2011). As a result, further work has searched for alternative measures of water content that are easy to measure and accurately scale biogeochemical processes across diverse soils (Table 1). This research has made significant progress despite mixed success.

Disruption of soil structure may be the most significant factor impeding the development of effective water scalars for biogeochemical processes. Similar to soil physics research, soil structure was destroyed in virtually all of the classic laboratory experiments that provide the basis for scaling biogeochemical processes from the laboratory to the field (Orchard and Cook, 1983; Linn and Doran, 1984; Fierer and Schimel, 2003). Disruption of soil structure can eliminate the potential to transfer laboratory research to the field by altering hydrologic properties, destroying aggregates, and homogenizing soil particles (Lin et al., 2005). Indeed, homogenization is typically the desired result of structure disruption and is accomplished with sieving (or other means). Homogenization is thought to reduce variation across replicates and allow greater focus on properties of interest. However, soil sieving increases reactant availability and process rates by increasing effective reaction area, maximizing diffusion potential, and liberating biologically available organic matter that is physically protected from microbes in aggregate structures (Franzluebbers,

TABLE 1 Common Measures of Soil Water Content that are Used to Scale Relative Rates of Biogeochemical Processes as a Function of Soil Water Content in Laboratory Experiments, Field Experiments, and Numerical Ecosystem Models (e.g. Linn and Doran, 1984; Franzluebbers, 1999; Reichstein et al., 2003; Del Grosso et al., 2005; Castellano et al., 2011). For Example, Numerical Ecosystem Models Often Assume Carbon Mineralization Rates are Maximized at Field Capacity and Decline with Decreases in Soil Water Content as defined by One of these Measures

Water scalar	Definition	Units
Volumetric water content	θ_v = cm^3 water cm^{-3} total soil volume	Proportion
Water filled pore space* (WFPS; Farquharson and Baldock, 2007)	$\mathrm{WFPS} = \theta_v/\phi = \theta^*_m \rho_d/(1-\rho_d/\rho_s)$	Proportion
Relative soil water content (RSWC; Reichstein et al., 2003)	$\mathrm{RSWC} = \theta_v/\theta_{\text{field capacity}}$	Proportion
Matric potential	ψ_m	Pressure (Pa)

Φ = total porosity (cm^3 pores cm^{-3} total soil).
θ_m = gravimetric water content (g water g^{-1} total soil).
ψ_m = matric potential.
ρ_d = bulk density (g soil particles cm^{-3} total soil volume).
ρ_s = particle density (g soil particles cm^{-3} soil particles).
$\theta_{\text{field capacity}} = \theta_v$ at field capacity.
** In soil physics literature, WFPS is often termed "Degree of Saturation".*

1999). Nevertheless, changes produced by the disruption of soil structure may be insignificant for some soil types, properties, and processes. Franzluebbers (1999) found that disruption of soil structure decreased N mineralization and increased C mineralization in short-term experiments although these effects were not apparent in long-term experiments.

Maintenance of soil structure from field sampling to laboratory experiments is gaining popularity in biogeochemistry research (e.g. Franzluebbers, 1999; Castellano et al., 2010, 2011). Moreover, soil disturbance has also been analyzed as an independent variable in an effort to understand the effect of soil structure and aggregate dynamics on organic matter mineralization (Plante et al., 2006). However, the selected volume of undisturbed soil cores used in biogeochemistry research is rarely discussed. In contrast, soil physicists often explicitly consider how undisturbed core volume affects soil processes. As a result, the concept of a representative elementary volume (REV) was developed to identify a volume of undisturbed soil that provides a representative characterization of in situ soil processes (Bear, 1972). From a hydrologic perspective, there is no doubt that such a concept is required: Bouma and Anderson (1973) demonstrated that saturated hydraulic conductivity can change as a function of core volume. Given the strong hydrologic controls on biogeochemical processes, biogeochemists should consider adoption of this concept. Does undisturbed core volume affect C and N mineralization rates as well as other biogeochemical processes? The answer to this question is unknown.

Lin et al. (2005) proposed that the REV should change as a function of soil structure, recommending that REV should increase with ped size. Empirical and theoretical analyses of soil structure, hydrology, and biogeochemistry support this proposal – soils dominated solely by sand or clay rather than a mixture of particle sizes typically have lower spatial variation in hydrologic and biogeochemical processes (Jarvis, 2007; Castellano and Kaye, 2009).

2.2. Scaling Biogeochemical Responses to Soil Water

Despite recent appreciation for the effect of soil structure on biogeochemical processes, basic soil physics principles are often simplified in biogeochemistry research. This becomes problematic as laboratory results are extrapolated to large scales and across soil types. One important example of this disconnect between hydropedology and biogeochemistry is the widespread use of water-filled pore space (WFPS, Table 1) as a scalar for the relative effect of soil water content on biogeochemical process rates across soil types (Fig. 1). The WFPS concept was developed to describe the relative response of biogeochemical processes to water availability and aeration across different soil types. However, soil structure significantly affects the relationship between WFPS and biogeochemical cycling, confounding the widespread applicability of these relationships (Fig. 2). Although many reports recognize the limitations of

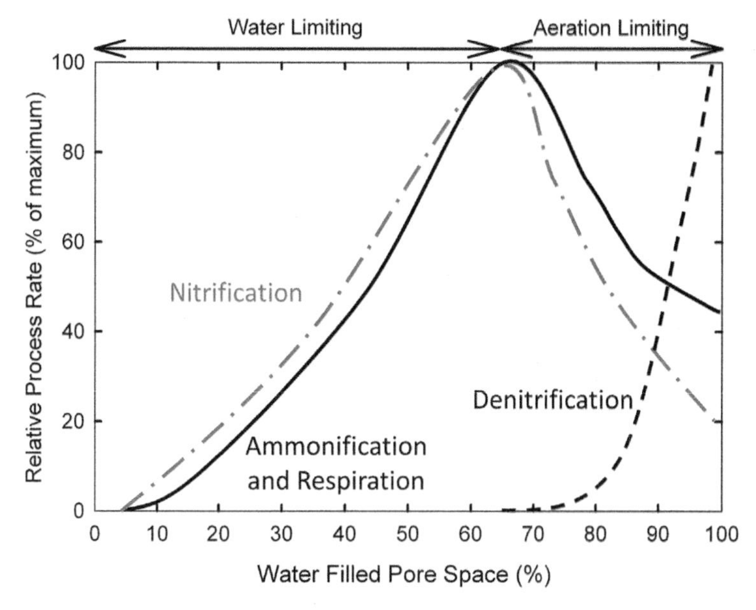

FIGURE 1 Conceptual relationship between percent water filled pore space (WFPS, Table 1) and relative biogeochemical process rates. *(Adapted from Linn and Doran, 1984, and Lohse et al., 2009)*

WFPS (Davidson et al., 2000; Farquharson and Baldock, 2007; Castellano et al., 2010), others suggest that WFPS is a reliable and extensively supported scalar of biogeochemical processes that can be substituted for matric potential (e.g. Linn and Doran, 1984; Lohse et al., 2009).

Soil physics principles demonstrate that WFPS cannot provide a consistent scalar of water or O_2 availability across different soil types and management scenarios. Percent WFPS is normalized for total porosity and therefore describes the proportion of water- and air-filled pores rather than the volume of water- and air-filled pores (on a total soil or individual pore basis; Table 1). Accordingly, WFPS is not directly proportional to gas or solute diffusivity, which is the key hydrologic control on biogeochemical process rates (Linn and Doran, 1984; Stark and Firestone, 1995; Farquharson and Baldock, 2007). Structurally distinct soils at an identical WFPS can have very different gas and solute diffusivities as well as matric potentials (Fig. 2). Soil structure and texture modulate the relationship between WFPS and biogeochemical process rates by affecting pore-size distribution (e.g. abundance of macropores) and bulk density. For example, if bulk density is increased while maintaining a constant WFPS, the volume of individual water-filled pores would decrease and matric potential would decrease (assuming that all pores decrease proportionally as the increase in bulk density reduces total porosity). Accordingly, as bulk density increases, maximum biogeochemical process rates (Fig. 1) typically occur at a higher percent WFPS because a greater percent of pores are water filled at field

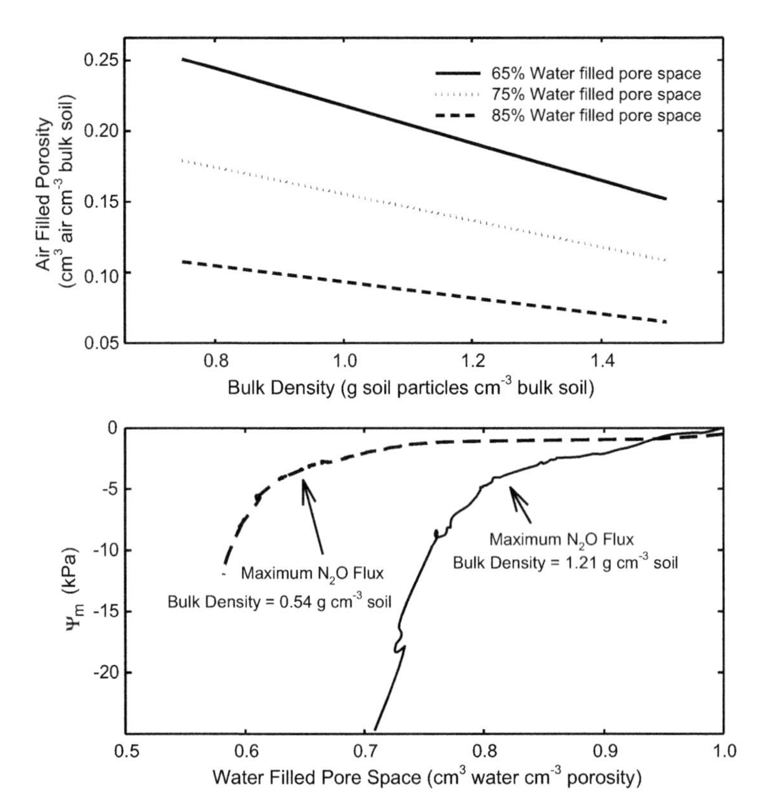

FIGURE 2 Top panel: interrelationships between water filled pore space (WFPS), volumetric air content, and bulk density. These data were simulated by modulating bulk density from 0.7 to 1.5 (g soil particles cm^{-3} total soil volume) at three different WFPS (65, 75, and 85%) and solving for air-filled porosity (see WFPS equation in Table 1). Across soils with different bulk density, aeration is not consistently proportional to WFPS of the soil (note different slopes of each line). Bottom panel: data from Castellano et al. (2010). Water filled pore space is not an effective substitute for matric potential ψ_m. Arrows indicate the location of relative maximum rate of nitrous oxide (N_2O) flux from two soils that have different bulk density (i.e. instantaneous flux rate divided by the maximum flux rate from the individual soil sample, see Castellano et al., 2010). Relative maximum rate of N_2O flux occurs at similar matric potential, but widely different WFPS. Data are empirically measured drainage curves for two soils with widely different bulk densities. *(Adapted from Castellano et al., 2010)*

capacity – the matric potential-based water content at which biogeochemical process rates are typically maximized (Fig. 2; Davidson et al., 2000; Reichstein et al., 2003; Castellano et al., 2010, 2011).

Published literature supports the fact that WFPS cannot consistently characterize biogeochemical processes in undisturbed soils. Using the search terms "nitrous oxide" and "water filled pore space", we searched the Thomson Reuters ISI Web of Knowledge for peer-reviewed publications that report nitrous oxide (N_2O) flux rate as a function of WFPS from undisturbed soil columns in a controlled laboratory setting. Although maximum N_2O flux rate is reported to occur between 60 and 70% WFPS (Fig. 1), this brief literature review demonstrates that maximum N_2O flux rate occurs across a wide range of percent WFPS and is positively correlated with soil bulk density (Fig. 3).

Compared to WFPS, matric potential and relative soil water content (RSWC; Reichstein, 2003; Castellano et al., 2011; Table 1) appear to be more consistent scalars of maximum biogeochemical process rates across soil types (e.g. Fig. 2). Matric potential appears to be a more consistent scalar because it describes the potential energy of soil water – in other words, the energy required for water uptake by plants and microbes (Sommers et al., 1981; Stark and Firestone, 1995; Castellano et al., 2010). This thermodynamic characterization of water availability would complement the widely used thermodynamic characterization of biogeochemical redox reactions based on free energy yield (Hedin et al., 1998). Relative soil water content is similar to matric potential because it normalizes water content against field capacity – the capacity for soil to hold water after gravitational drainage (Reichstein et al., 2003). Although field capacity is rarely well

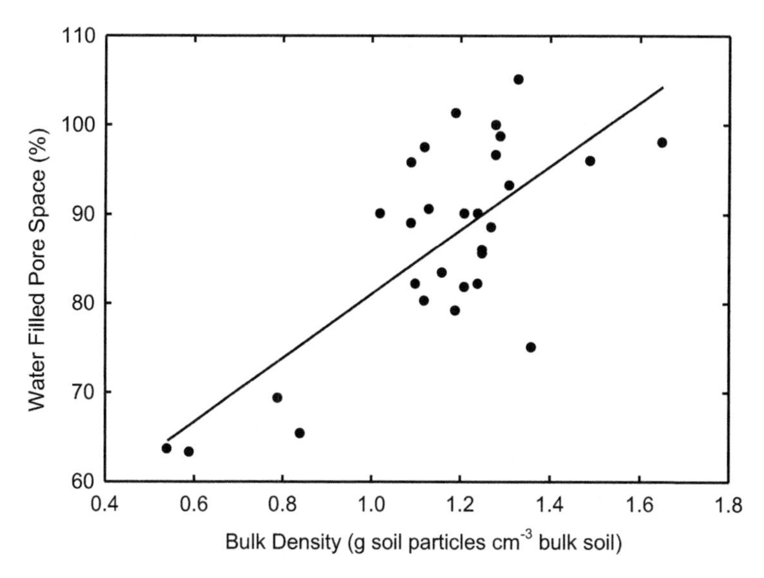

FIGURE 3 Relationship between bulk density and the water filled pore space (WFPS) at which instantaneous maximum nitrous oxide (N_2O) flux occurred. Data are summarized from controlled laboratory experiments that measured N_2O flux as a function of WFPS from undisturbed soil cores (del Prado et al., 2006; Ruser et al., 2006; Ball et al., 2008; Beare et al., 2009; Castellano et al., 2010). Samples represented the top 10–15 cm of soil. Solid line indicates linear relationship between bulk density and WFPS ($r^2 = 0.57$, $p < 0.01$, $n = 29$). Although WFPS measurements cannot exceed 100%, some data from these reports included values >100% WFPS due to measurement error.

defined, it is typically associated with a relatively narrow range of surface soil matric potential (generally −2.5 to −50 kPa, Cassel and Nielsen, 1986) and field capacity is widely used in ecosystem science. Several popular ecosystem models, including CENTURY, use RSWC (or slight variations that also explicitly incorporate the measure 'field capacity') to simulate C and N cycling (Reichstein et al., 2003; Del Grosso et al., 2005). A similar measure of water content known as "water holding capacity" is used in laboratory experiments as a proxy for matric potential and field capacity (Robertson et al., 1999). However, there is no definitive method to measure field capacity. Field capacity is a function of measurement depth and a number of soil properties including texture and structure (Cassel and Nielsen, 1986). Despite difficulties associated with the measurement of field capacity, ecosystem models have accurately simulated daily C and N mineralization as well as N_2O fluxes using RSWC (Reichstein, 2003; Jarecki et al., 2006). Nevertheless, ecosystem models are notoriously weak during "hot moments" of biogeochemical cycling (Jarecki et al., 2006; Groffman et al., 2009). "Hot moments" are short amounts of time that account for disproportionately large amounts of biogeochemical transport and transformation; they are typically the result of large increases in soil water content (Jarecki et al., 2006; Groffman et al., 2009). Further integration of biogeochemistry and hydropedology can help to improve modeling capabilities in this area.

Unfortunately, no measure of water content can scale relative rates of biogeochemical processes with total accuracy across soils that differ in structural properties including texture, bulk density, organic matter distribution, and pore-volume distribution. In particular, two properties inhibit the accurate characterization of relationships between water content and biogeochemical processes. First, pore-volume distribution can have a large effect on matric potential even at constant water contents and bulk density. For example, large macropores can drain rapidly while smaller intra-aggregate pore spaces maintain high water contents. Such a scenario would lead to highly heterogonous water contents and diffusion coefficients, supporting transfer of the REV concept to biogeochemical research (Lin et al., 2005). Second, the depth distribution of dissolved electron donors (largely organic C) and acceptors in combination with the depth at which water content is measured will affect the relationship between biogeochemical processes and any water scalar (Fig. 1). Because the depth distribution of electron donors and acceptors varies across soils, identifying a standard depth for water scalar measurement may not be an effective technique to standardize the measurement of biogeochemical process rates across soils.

Nevertheless, continued coupling of biogeochemistry and hydropedology is needed to understand ecosystem processes and improve our ability to predict the effects of new management practices and future climate on ecosystem structure and dynamics. Indeed, changing precipitation and temperature patterns are expected to be the most significant drivers of terrestrial ecosystem feedbacks to climate change (Field et al., 2007). Interactions between biogeochemistry and hydropedology could modulate these feedbacks.

3. NITROGEN BIOGEOCHEMISTRY AND THE ROLE OF HYDROPEDOLOGY

Greater integration of biological and hydrologic processes will be required to effectively couple biogeochemistry and hydropedology. Biogeochemistry and hydrology have a common goal to understand ecosystem processes and both disciplines have made major contributions to ecosystem science (Vitousek et al., 1982; Creed and Band, 1998). However, historical disciplinary distinctions between soil science, ecosystem science, and hydrology often result in research activities that independently focus on biological transformation processes or physical transport processes. As an example, Mitchell (2001) cites two independent studies from the Catskill Mountains (NY, USA) that examine processes affecting watershed NO_3^- export: one report highlights on the role of tree species in NO_3^- production (Lovett et al., 2002) while the second report highlights deep groundwater NO_3^- contributions to base flow (Burns et al., 1998). However, neither study can explain all variation in NO_3^- export. Hydropedology affects both of these N transformation and transport processes. Spatial and temporal variations in soil water content affect N transformation (Fig. 1) as well as N transport. In addition to distinct disciplinary perspectives, ecotype can also steer research foci. Because temperate forests typically export little dissolved N, most N biogeochemistry research in these systems has focused on microbial transformation processes that mediate N retention (Aber et al., 1998). In contrast, cereal cropping systems typically export large amounts of dissolved N; therefore, most research in these systems has focused on hydrologic processes that transport N (Dinnes et al., 2002). Nevertheless, all of these research perspectives are rooted in the conservation of mass where most simply, inputs − outputs = storage (Vitousek and Reiners, 1975). Greater integration of soil biogeochemistry and hydrology will improve solutions to this equation by advancing our ability to constrain the 'outputs' term. Interactions between soil biogeochemistry and hydrology affect dissolved and gaseous C and N outputs through transformation and transport processes. Carbon and N outputs must be transformed to gaseous or dissolved species to be transported out of the system. Hydropedology affects both transformation and transport processes.

3.1. Coupling Soil Biogeochemistry and Hydrology

Ecological explanations for N transport and transformation in forested watersheds have focused on biological processes (Mitchell, 2001; Lovett et al., 2002). Early work proposed vegetative uptake as the major ecosystem N sink (Aber et al., 1989). However, subsequent work showed that soil organic matter is also an important N sink (Nadelhoffer et al., 1999). Indeed, many unmanaged ecosystems retain a majority of N inputs in the soil, transforming mineral N (largely NO_3^- and NH_4^+) into relatively nonreactive stable organic N (Aber et al., 1989; Goodale et al., 2000). These results have led to the development of ecosystem N retention theory that focuses on soil-based mineral N consumption processes that mediate N retention. Major discoveries include: 1) negative correlations between soil C/N ratios and watershed NO_3^- exports, 2) positive correlations between soil C/N ratio and mineral N retention, and 3) predominance of dissolved organic N (DON) export rather than NO_3^- from pristine ecosystems (Perakis and Hedin, 2002). Explanations for these observations have identified the importance of biotic and abiotic processes. Biotic mechanisms focus on competition between plants and microbes for mineral N, resulting in tight cycling with little mineral N loss (Kaye and Hart, 1997). Abiotic mechanisms focus on the reaction of nitrite, NH_4^+, and labile organic N with aromatic ring structures of phenolic and lignitic soil organic matter resulting in the formation of decomposition-resistant compounds (Hättenschwiler and Vitousek, 2000), as well as the association of organic N with soil minerals (Hassink, 1997; Castellano et al., 2012). Both biotic and abiotic processes are C-dependent: the capacity of biotic mechanisms to sequester N depends on a surplus of microbe-available C and a deficit of microbe-available N. The capacity of abiotic mechanisms to sequester N depends on the availability of the C substrate as a reactant (particularly in the case of organic N stabilization through mineral association which occurs largely after microbial processing, Grandy and Neff, 2008). Nevertheless, these mechanisms cannot explain all variation in dissolved N export.

Hydropedology can explain additional variation in ecosystem N retention by linking ecological and hydrologic research perspectives. Current ecological research strongly suggests a role for hydropedology. Indeed, the ability of soil texture and mineralogy to affect solute transport and adsorption has been suggested to obscure or eliminate relationships between C/N ratios and nitrification or NO_3^- leaching (Lovett et al., 2004; Castellano et al., 2012). Moreover, ecosystem N retention theory is unable to explain "hot moments" of dissolved N losses that occur during intense precipitation events and account for disproportionately large N fluxes at pedon and watershed scales. In these situations, rapid flow likely exceeds the immobilization kinetics of dissolved N (Lovett and Goodale, 2011).

While ecological research has made significant progress describing soil-based N transformation mechanisms, hydrologic research has made significant progress toward understanding the effects of water source and flow path routing on ecosystem NO_3^-, DON, and dissolved organic C (DOC) fluxes (Hill, 1996). Variable source areas (VSAs) of water discharge from the landscape (e.g. vadose zone vs. groundwater) can contribute to differences in NO_3^-, DON, and DOC export based on the solute concentration and the proportional contribution of each source area to total watershed discharge (Inamdar et al., 2004). Flow paths can also contribute to differences in solute export by affecting flow rates and the interaction of electron donors and acceptors (Hill et al., 2000). A major discovery resulting from this work is the frequent observation of low DON, DOC, and NO_3^- export during base flow, contrasted by high export during intense precipitation events. As both total water discharge and concentrations are elevated during intense precipitation, these "hot moments" of export can account for significant portions of annual solute loads. This process, termed "flushing" (*sensu* Hornberger et al., 1994), is largely attributed to hydrologic mechanisms that rapidly transport NO_3^-, DON, and DOC from nutrient-rich surface soils to less biologically active subsoils and open water (Hagedorn et al., 2000; Dittman et al., 2007; van Verseveld et al., 2008). These mechanisms include: 1) the rise and fall of a transient water table that leaches NO_3^-, DON, and DOC from nutrient-rich surface soils to open waters and 2) the occurrence of rapid flow paths that transport NO_3^-, DON, and DOC vertically from nutrient-rich surface soils through less biologically active subsoils, then laterally downslope (Gaskin et al., 1989; Boyer et al., 1997). Rapid flow can occur in soils with coarse texture or an abundance of macropores. This mechanism assumes that rapid flow reduces contact time, allowing solutes to bypass biotic and abiotic sinks, traveling unabated from nutrient-rich surface soils to less biologically active subsoils and open waters (Hill, 1996; Mcguire and McDonnell, 2006; Dittman et al., 2007). However, NO_3^- flushing is not universally observed (Hill and Kemp, 1999). Moreover, a negative relationship between concentrations of DOC/N and NO_3^- in discharge at intra-storm and annual time scales is commonly observed (e.g. Inamdar et al., 2004; Dittman et al., 2007; van Verseveld et al., 2008). This negative relationship is attributed to the relative contribution of individual VSAs with different concentrations of DOC/N and NO_3^- to total watershed discharge. However, this relationship is not necessarily inconsistent with C-based N retention mechanisms and the transformation of NO_3^- to DOC/N compounds (Vitousek et al., 2002; Fig. 4).

Both ecology-based N retention and hydrology-based N transport mechanisms focus on processes that originate in nutrient-rich surface soils. Thus, the incorporation of ecosystem N retention theory and soil hydrology can help to determine *when, where, and why* ecosystems export dissolved N, potentially explaining the presence or absence of flushing patterns. Specifically, it is plausible that interactions between the magnitude and time scale of mineral N retention in soil organic matter could interact to affect dissolved N export: systems with high N retention efficiency over short time scales may not exhibit NO_3^- flushing. In contrast, systems with low N retention would be expected to lose significant amounts of NO_3^-, particularly during flushing events (Inamdar et al., 2004; van Verseveld et al., 2008). Moreover, C-based N retention mechanisms may also contribute to negative relationships between DOC/N and NO_3^- that are commonly reported during intense precipitation events. Aromatic

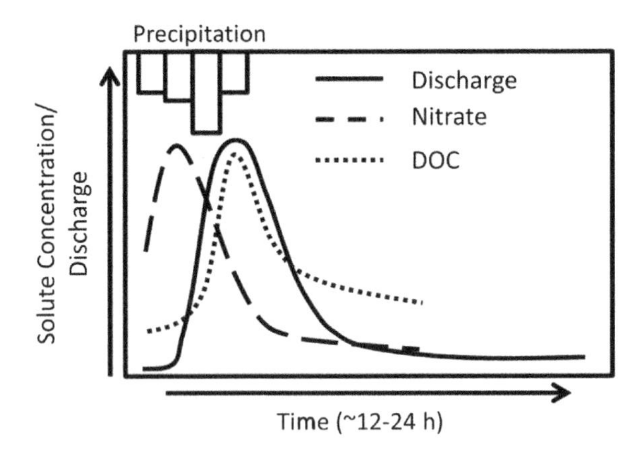

FIGURE 4 Two frequently observed biogeochemical patterns that occur during intense precipitation events: 1) large increases in solute concentrations and loads termed "flushing" and 2) peak nitrate (NO_3^-) flux preceding peak dissolved organic carbon (DOC) flux. Figure adapted from Inamdar et al. (2004). Hydrologic explanations for these patterns include flushing of DOC and NO_3^- from nutrient-rich surface soils as well as variable source area contributions to discharge. Biogeochemistry and ecosystem ecology explanations for these patterns include the carbon and nitrogen stoichiometry of NO_3^- immobilization.

DOC compounds may serve as an important sink for inorganic N. In fact, ecosystem export of NO_3^- may be limited by the transformation of inorganic N into DON compounds. This process may help to explain sustained N limitation in forests despite the cessation of net biomass accumulation and continued atmospheric deposition of N (Vitousek et al., 2002).

3.2. Integrating Hydropedology

Hydropedology can couple ecological and hydrologic perspectives on N biogeochemistry by incorporating relationships between soil structure, flow paths, and hydrology into the evaluation of N transport and transformation. At a broad scale, flow paths are well known to affect biogeochemical cycling by channeling water flow through areas with different reactant availabilities (Hill, 1996). For example, intersection of a hydrologic flow path with a C-rich riparian buffer can lead to large amounts of NO_3^- removal through denitrification and transformation into DON. However, at a smaller scale, soil structure can have a significant impact on biogeochemical cycling, potentially accounting for much of the unexplained variation in ecological and hydrologic processes discussed above. Soil structure can produce microsites where net N mineralization dominates and other nearby microsites where net N immobilization dominates. These soil microsites are well recognized as important controls on whole-soil biogeochemical cycling (Schimel and Bennett, 2004).

Soil water flow paths clearly represent soil microsites that affect biogeochemical cycling. Some flow paths can enhance mineral N retention, while others can facilitate mineral N transport to ground- and surface waters. Flow paths can affect biogeochemical processes by determining the availability of reactants (Bundt et al., 2001) and the flow rate (Shipitalo et al., 1990). Although soil water flow paths can be sorted into many categories that describe spatial heterogeneity and flow rate (e.g. fingering flow and funnel flow; Flury et al., 1994), we will focus on rapid, nonequilibrium preferential macropore flow paths and slower, relatively homogenous bulk soil matrix flow paths because they represent end points that are likely to differently affect biogeochemical processes. Macropore and matrix flow paths differ in several important physical and chemical properties that can affect N transport and transformation. However, due to the difficulty of separating preferential macropore and matrix flow in situ, little is known about how flow paths interact with environmental conditions and management scenarios to affect biogeochemistry.

Accordingly, there is no firm consensus regarding the effect of flow path on N transport (Jarvis, 2007). For example, relative to the bulk soil matrix, macropores have been shown to reduce and exacerbate mineral N transport. Macropores can reduce mineral N transport through at least two nonexclusive mechanisms. First, macropores typically have higher total organic C, C/N ratios, microbial biomass, and O_2 levels than the bulk soil matrix (Pankhurst et al., 2002). Moreover, DOC that is transported through macropores tends to be more humic than DOC transported through the soil matrix (Kaiser and Guggenberger, 2005; Sanderman and Amundson, 2008). These properties promote denitrification and N retention via C-based biotic and abiotic mechanisms, leading researchers to identify macropores as potential "hot spots" of biogeochemical cycling (Bundt et al., 2001). Second, macropores may be hydrologically disconnected from the bulk soil matrix where positive net N mineralization is more likely to occur due to lower C/N ratios (Asano et al., 2006). Accordingly, macropores may transport soil water that is relatively low in mineral N and perhaps more similar to precipitation than bulk soil matrix water (Larsson and Jarvis, 1999; Pankhurst et al., 2002).

In contrast, other work has determined that high O_2 levels in macropores can promote nitrification and increase NO_3^- levels relative to the bulk soil matrix (Hagedorn et al., 1999; Vinther et al., 1999). Additionally, macropores' low surface area to volume ratio can allow rapid flow and unabated transport of water and solutes from nutrient-rich biologically active surface soils to deep, relatively biologically inactive subsoils (Kung et al., 2000; Pampolino et al., 2000). Rapid flow and concomitant low contact time between solutes and soil particle surfaces may inhibit the retention of mineral N by exceeding mineral N immobilization kinetics (Lovett and Goodale, 2011) and thus allowing macropores to serve as conduits for mineral N transport (Muller, 2004). This process dominates in fluvial systems where rapid flow paths routinely reduce N cycling (Essington and Carpenter, 2000) and mineral N retention (Craig et al., 2008).

Ambiguous flow path controls on biogeochemical processes are further confounded by dynamic agricultural practices and environmental conditions. Agriculture can increase connections between macropores and the atmosphere, promoting relatively unaltered transport of precipitation to subsoils (Larsson and Jarvis, 1999). These connections can increase mineral N transport through macropores when fertilizer applications are broadcast over the soil surface rather than injected. In forests, however, mixing of precipitation and soil water in organic surface horizons can enrich macropore flow with dissolved N and C (Feyen et al., 1999). Flow-path effects on biogeochemical cycling and transport can also be affected by precipitation patterns and soil water content prior to flow-path activation (Gaskin et al., 1989; Shipitalo et al., 1990).

4. FUTURE DIRECTIONS

The potential for different soil water flow paths to account for unexplained variation in biogeochemical cycling and fluxes is gaining interest. Incorporation of soil water flow paths into N transport and transformation theories requires a strong foundation in hydropedology. Analyses of soil biogeochemistry processes have focused on laboratory and in situ field incubations of soil samples that reduce or eliminate material transport (Hart et al., 1994). This approach has worked well because soil transport processes are usually slow. In the absence of intense precipitation, advection is low and consequently the products of biogeochemical cycling do not travel far before re-use or adsorption (Wagener et al., 1998). Experimental evidence supports this perspective; exchange reactions, rather than advection, appears to control dissolved organic matter composition (Sanderman and Amundson, 2008). Thus, incubation methods have allowed soil ecologists to learn a great deal about biogeochemical cycling during times of low flow rates that dominate during steady state conditions. Watershed-scale hydrologic research confirms the generality of these findings; base flow periods, when soil water flow and advection are at a relatively steady state, account for a majority of annual N export (Pionke et al., 1996; Schilling and Zhang, 2004). Accordingly, soil-based N-retention mechanisms can explain significant portions of the variability in ecosystem NO_3^- export (e.g. Emmett et al., 1998; Lovett et al., 2002). However, hydrologic research also reveals that intense precipitation events rapidly transport soil nutrients and C, accounting for greater mass export per unit time (e.g. Pionke et al., 1996; Schilling and Zhang, 2004). In a Pennsylvania watershed, for example, 40% of annual NO_3^- export occurs during 10% of the time, a period defined as "storm flow" (Pionke et al., 1996). During these times, biogeochemical process rates are also elevated. These data implicate rapid nonequilibrium flow paths as ephemeral, yet important, controls on nutrient dynamics during intense precipitation events (e.g. van Verseveld et al., 2008).

4.1. Material Transport and Biogeochemistry

In streams, advection is too rapid to ignore. Flow velocity can have a large effect on biogeochemical cycling. Rapid flow "stretches" tight nutrient cycles across space. Accordingly, aquatic biogeochemists have incorporated advective transport into biogeochemical cycling with the development of a "nutrient spiraling" approach (Newbold et al., 1981). Building upon variables that commonly affect biogeochemical cycling in aquatic and terrestrial systems (e.g. temperature and substrate availability), nutrient spiraling explicitly couples physical transport and biological cycling (Fig. 5). The added transport dimension, spiral length, refers to the downstream distance a molecule travels before it is removed due to biological assimilation, transformation, or physical sorption. As flow velocity increases, spiral length increases nearly linearly (Newbold et al., 1981). However, spiral length can differ among chemical species. In North American headwater streams, the spiral length of NO_3^- is 5–10 times greater than NH_4 (Peterson et al., 2001). In other words, NO_3^- travels 5–10 times as long before uptake compared to NH_4^+.

The transfer of a nutrient spiraling approach to the soil has been proposed on at least two occasions (Wagener et al., 1998; Fisher et al., 2004). However, methods from aquatic ecosystems cannot be easily transferred to soils. Whereas advective material transport dominates in streams and diffusive material transport dominates in the bulk soil matrix, it is likely that convective transport

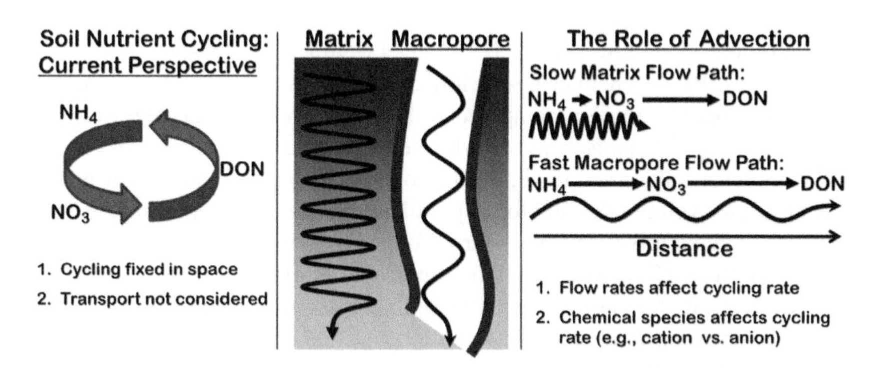

FIGURE 5 Left panel depicts the dominant modern paradigm of soil nutrient cycling. The middle and right panels depict how interactions between soil structure and advection affect nutrient transport and transformation. A simplified nitrogen cycle as affected by hydropedology is conceptualized. Few soil biogeochemical cycling concepts include advection (but see Sanderman and Amundson, 2008). However, advection processes differ among flow paths and can affect soil biogeochemistry. Nutrient spiraling approaches from aquatic ecology add a spatial component to the conventional soil biogeochemistry perspective that considers cycling rates in a fixed space. Spiral patterns indicate the relative distance a molecule travels before reuse. Flow paths and chemical species affect the mean distance molecules travel before reuse. For example, spiral patterns will be shorter and molecules will travel a shorter distance before reuse in relatively slow bulk soil matrix flow paths compared to faster flow paths such as macropores.

dominates material transfers in nonequilibrium soil water flow paths. Although significant exchange of soil water can occur between the bulk soil matrix and macropores during rapid, nonequilibrium flow (Shipitalo et al., 1990), recent evidence suggests that tightly bound soil water may not mix with more mobile water that is transported through the soil to groundwater and streams (Brooks et al., 2009). Although a lack of water mixing does not preclude material diffusion between these pools of water, this result does suggest that the transfer of nutrient spiraling to soils may be possible, particularly during intense precipitation events.

The role of advective transport in soil biogeochemical cycling is gaining appreciation. Significant research has demonstrated that intense precipitation can result in the rapid transport of solutes from nutrient-rich surface soils to the watershed outlet. However, few studies have examined the active cycling of solutes along flow paths. Among the early research to transfer the nutrient spiraling approach to the soil was a study that compared nutrient spiraling between an open stream channel and the saturated zone of a riparian zone soil (Lewis et al., 2006). The authors hypothesized that nutrient spirals would be longer in the open channel compared to the riparian zone. To test this hypothesis, they predicted that solute spatial variation (coefficient of variation) would be greater in the riparian zone, indicating more nutrient cycling per unit distance. Indeed, spatial variation of nutrient concentrations was 5–11 fold greater in the riparian zone. Predictably, actively cycled nutrients (e.g. NH_4^+) demonstrated greater spatial variation than more slowly cycled nutrients (e.g. Cl^-).

Based on the concept of conservative tracers, spatial variation in solute concentrations along flow paths is becoming a well-accepted proxy for nutrient cycling rates. Biogeochemicals with low cycling rates (e.g. Cl^-) have consistently low spatial variation compared to biogeochemicals with high cycling rates (e.g. NH_4^+; Manderscheid and Matzner, 1995; Asano et al., 2006; Lewis et al., 2006; Castellano and Kaye, 2009). Low spatial variation along flow paths indicates low cycling rates and long spiral lengths (e.g. the stream surface): in this case, solutes travel relatively far before transformation, resulting in low spatial variation. High spatial variation indicates short spiral lengths (e.g. the riparian zone): in this case, solutes are rapidly transformed over relatively short distances, resulting in high spatial variation. A similar approach takes advantage of natural variation in stable hydrogen isotope ratio of water (δD) to interpret variability in nutrient cycling along different soil water flow paths (Asano et al., 2006). The natural abundance of D can be used to index flow path length and contact time because temporal variation in throughfall, δD, is much greater than temporal variation in soil water, δD. Accordingly, as soil water mixes over time, variation in δD is reduced. Lysimeter samples that exhibit high temporal variation in δD are assumed to have more rapid flow (i.e. more advection and less diffusion) compared to lysimeter samples with lower temporal variation in δD (i.e. less advection and more diffusion). Asano et al. (2006) found that variation in δD explained significant variation in soil solution nutrient concentrations, confirming previous reports that flow paths can influence nutrient leaching. Moreover, DOC/N concentrations were positively correlated with δD variation, whereas NO_3^- concentrations were negatively correlated with δD variation suggesting that rapid flow paths are important conduits for DOC/N transport while slow flow paths are important sources of net nitrification. These results support DOC/N flushing hypotheses. However, the importance of DOC for NO_3^- immobilization remains unclear. It is possible that nitrification rates in rapid flow paths are high, but NO_3^- is rapidly transformed into DON. Further research making explicit links between flow-path transit times and biogeochemical processes are required. Although much of this work will index rather than measure flow rates, it is possible to measure the rate of DOC advection using the ^{14}C radioisotope (Sanderman and Amundson, 2008). Isotope methods, however, do not allow investigators to determine transformation rates except under strict assumptions that are only valid over short time periods (Hart et al., 1994); measureable transport is unlikely to occur over such short time scales.

4.2. Proximate Goals

Hydropedology clearly interacts with biogeochemistry to influence nutrient transport and transformation. Research is moving forward to understand interactions between flow rates, material transport processes, and biogeochemical transformation in the context of soil structure. These interactions are critical to predicting and managing the effects of intense precipitation on biogeochemical transport and transformation. Intense precipitation events represent times when coupled analyses of hydropedology and biogeochemistry are most tractable in the field and applicable to the advancement of scientific theories such as ecosystem N retention and solute flushing. Despite the short duration of intense precipitation events, they drive disproportionately large amounts of biogeochemical transport and transformation (McClain et al., 2003). These processes can have significant environmental impacts including water pollution and greenhouse gas emissions. Moreover, climate models indicate that the frequency of intense precipitation is increasing. Accordingly, there is a pressing need to advance the coupling of hydropedology and biogeochemistry.

5. CONCLUSIONS

- Biogeochemistry research must consider the impact of soil structure on spatial and temporal variations in soil water content.
- Transport rate of materials within the soil profile affects biogeochemical transformation rates. However, the abundance of biogeochemicals differs among flow paths with different transport rates. The location of biogeochemicals among flow paths with different transport rates will play an

important role in understanding variations in biogeochemical transformations and material transport.

- Intense precipitation events create hot moments of biogeochemical cycling and transport. These hot moments offer opportunities to link biogeochemistry and hydropedology with a high probability of success and direct relevance to global change.
- Nutrient spiraling principles from stream ecology and aquatic biogeochemistry literature can be transferred to soil flow paths to provide an initial conceptual framework that can be developed and expanded across soil types and management scenarios.

REFERENCES

Aber, J., Mcdowell, W., Nadelhoffer, K., Magill, A., 1998. Nitrogen saturation in temperate forest ecosystems. BioScience 48, 921–934.

Aber, J., Nadelhoffer, K., Steudler, P., 1989. Nitrogen saturation in northern forest ecosystems. BioScience 39, 378–386.

Asano, Y., Compton, J.E., Church, M.R., 2006. Hydrologic flowpaths influence inorganic and organic nutrient leaching in a forest soil. Biogeochemistry 81, 191–204. doi: 10.1007/s10533-006-9036-4.

Ball, B., Crichton, I., Horgan, G., 2008. Dynamics of upward and downward N_2O and CO_2 fluxes in ploughed or no-tilled soils in relation to water-filled pore space, compaction and crop presence. Soil and Tillage Research 101, 20–30. doi: 10.1016/j.still.2008.05.012.

Bear, J., 1972. Dynamics of Fluids in Porous Media. Elsevier, New York, NY.

Beare, M.H., Gregorich, E.G., St-Georges, P., 2009. Compaction effects on CO_2 and N_2O production during drying and rewetting of soil. Soil Biology and Biochemistry 41, 611–621. Elsevier Ltd. doi: 10.1016/j.soilbio.2008.12.024.

Bouma, J., Anderson, J.L., 1973. Relationships between soil structure characteristics and hydraulic conductivity. In: Bruce, R.R., Flach, K., Taylor, H.M. (Eds.), Field Soil Water Regime. SSSA special publication No. 5., Madison, WI, pp. 77–105.

Boyer, E.W., Hornberger, G.M., Bencala, K.E., Mcknight, D.M., 1997. Response characteristics of DOC flushing in an alpine catchment. Hydrol. Process. 11, 1635–1647.

Brooks, J.R., Barnard, H.R., Coulombe, R., McDonnell, J.J., 2009. Ecohydrologic separation of water between trees and streams in a Mediterranean climate. Nat. Geosci. 3, 100–104. Nature Publishing Group. doi: 10.1038/ngeo722.

Bundt, M., Widmer, F., Pesaro, M., Zeyer, J., Blaser, P., 2001. Preferential flow paths: biological "hot spots" in soils. Soil Biol. Biochem. 33, 729–738.

Burns, D.A., Murdoch, P.S., Lawrence, G.B., Michel, R.L., 1998. The effect of ground-water springs on NO_3^- concentrations during summer in Catskill Mountain streams. Water Resour. Res. 34, 1987–1996.

Cassel, D.K., Nielsen, D.R., 1986. Field capacity and available water capacity. In: Klute, A. (Ed.), Methods of Soil Analysis: Part 1. Physical and Mineralogical Methods, second ed. American Society of Agronomy, Madison, WI, pp. 901–926.

Castellano, M.J., Kaye, J.P., Lin, H., Schmidt, J.P., 2012. Linking carbon saturation concepts to nitrogen saturation and retention. Ecosystems. doi: 10.1007/s10021-011-9501-3.

Castellano, M.J., Schmidt, J.P., Kaye, J.P., Walker, C., Graham, C.B., Lin, H., Dell, C., 2011. Hydrological controls on heterotrophic soil respiration across an agricultural landscape. Geoderma 16, 273–280. doi: 10.1016/j.geoderma.2011.01.020.

Castellano, M.J., Schmidt, J.P., Kaye, J.P., Walker, C., Graham, C.B., Lin, H., Dell, C.J., 2010. Hydrological and biogeochemical controls on the timing and magnitude of nitrous oxide flux across an agricultural landscape. Global Change Biol. 16, 2711–2720. doi: 10.1111/j.1365-2486.2009.02116.x.

Castellano, M.J., Kaye, J.P., 2009. Global within-site variance in soil solution nitrogen and hydraulic conductivity are correlated with clay content. Ecosystems 12, 1343–1351. doi: 10.1007/s10021-009-9293-x.

Craig, L.S., Palmer, M.A., Richardson, D.C., Filoso, S., Bernhardt, E.S., Bledsoe, B.P., Doyle, M.W., Groffman, P.M., Hassett, B.A., Kaushal, S.S., Mayer, P.M., Smith, S.M., Wilcock, P.R., 2008. Stream restoration strategies for reducing river nitrogen loads. Front. Ecol. Environ. 6, 529–538. doi: 10.1890/070080.

Creed, I.F., Band, L.E., 1998. Export of nitrogen from catchments within a temperate forest: evidence for a unifying mechanism regulated by variable source area dynamics. Water Resour. 34, 3105–3120.

Davidson, E.A., Keller, M., Erickson, H.E., Verchot, L.V., Veldkamp, E., 2000. Testing a conceptual model of soil emissions of nitrous and nitric oxides. BioScience 50, 667–680.

Dinnes, D.L., Karlen, D.L., Jaynes, D.B., Kaspar, T.C., Hatfield, J.L., Colvin, T.S., Cambardella, C.A., 2002. Nitrogen management strategies to reduce nitrate leaching in tile-drained midwestern soils. Agronom. J. 94, 153–171.

Dittman, J.A., Driscoll, C.T., Groffman, P.M., Fahey, T.J., 2007. Dynamics of nitrogen and dissolved organic carbon at the Hubbard brook experimental forest. Ecology 88, 1153–1166.

Emmett, B.A., Boxman, D., Bredemeier, M., Gundersen, P., Kjønaas, O.J., Moldan, F., Schleppi, P., Tietema, A., Wright, R.F., 1998. Predicting the effects of atmospheric nitrogen deposition in conifer stands: evidence from the NITREX ecosystem-scale experiments. Ecosystems 1, 352–360.

Essington, T.E., Carpenter, S.R., 2000. Mini-review: nutrient cycling in lakes and streams: insights from a comparative analysis. Ecosystems 3, 131–143. doi: 10.1007/s100210000015.

Falkowski, P.G., Fenchel, T., Delong, E.F., 2008. The microbial engines that drive earth's biogeochemical cycles. Science 320, 1034–1039. doi: 10.1126/science.1153213.

Farquharson, R., Baldock, J., 2007. Concepts in modelling N_2O emissions from land use. Plant Soil 309, 147–167. doi: 10.1007/s11104-007-9485-0.

Feyen, H., Wunderli, H., Wydler, H., Papritz, A., 1999. A tracer experiment to study flow paths of water in a forest soil. J. Hydrol. 225, 155–167.

Field, C.B., Lobell, D.B., Peters, H.A., Chiariello, N.R., 2007. Feedbacks of terrestrial ecosystems to climate change. Ann. Rev. Environ. Resour. 32, 1–29. doi: 10.1146/annurev.energy.32.053006.141119.

Fierer, N., Schimel, J.P., 2003. A proposed mechanism for the pulse in carbon dioxide production commonly observed following the rapid rewetting of a dry soil. Soil Sci. Soc. Am. J. 67, 798. doi: 10.2136/sssaj2003.0798.

Fisher, S.G., Sponseller, R.A., Heffernan, J.B., 2004. Horizons in stream biogeochemistry: flowpaths to progress. Ecology 85, 2369–2379.

Fluhler, H., Durner, W., Flury, M., 1996. Lateral solute mixing processes a key for understanding field-scale transport of water and solutes. Geoderma 70, 165–183.

Flury, M., Fluhler, H., Jury, W.A., Leuenberger, J., 1994. Susceptibility of soils to preferential flow: A field study. Water Resources Research 30, 1945–1954.

Franzluebbers, A.J., 1999. Microbial activity in response to water- filled pore space of variably eroded southern Piedmont soils. Appl. Soil Ecol. 11, 91–109.

Gaskin, J.W., Dowd, J.F., Nutter, W.L., Swank, W.T., 1989. Vertical and lateral components of soil nutrient flux in a hillslope. J. Environ. Quality 18, 403–410.

Goodale, C.L., Aber, J.D., McDowell, W.H., 2000. The long-term effects of disturbance on organic and inorganic nitrogen export in the White Mountains, New Hampshire. Ecosystems 3, 433–450. doi: 10.1007/s100210000039.

Grandy, a S., Neff, J.C., 2008. Molecular C dynamics downstream: the biochemical decomposition sequence and its impact on soil organic matter structure and function. The Science of the Total Environment 404, 297–307. doi: 10.1016/j.scitotenv.2007.11.013.

Groffman, P.M., Butterbach-Bahl, K., Fulweiler, R.W., Gold, A.J., Morse, J.L., Stander, E.K., Tague, C., Tonitto, C., Vidon, P., 2009. Challenges to incorporating spatially and temporally explicit phenomena (hotspots and hot moments) in denitrification models. Biogeochemistry 93, 49–77. doi: 10.1007/s10533-008-9277-5.

Del Grosso, S.J., Parton, W.J., Mosier, A.R., Holland, E.A., Pendall, E., Schimel, D.S., Ojima, D.S., 2005. Modeling soil CO_2 emissions from ecosystems. Biogeochemistry 73, 71–91. doi: 10.1007/s10533-004-0898-z.

Hagedorn, F., Mohn, J., Schleppi, P., Flu, H., 1999. The role of rapid flow paths for nitrogen transformation in a forest soil: a field study with micro suction cups. Soil Sci. Soc. Am. J., 1915–1923.

Hagedorn, F., Schleppi, P., Waldner, P., Fluhler, H., 2000. Export of dissolved organic carbon and nitrogen from gleysol dominated catchments – the significance of water flow paths. Biogeochemistry 50, 137–161.

Hart, S.C., Stark, J.M., Davidson, E.A., Firestone, M.K., 1994. Nitrogen mineralization, immobilization, and nitrification. In: Methods of Soil Analysis, Part 2-Microbiological and Biochemical Properties. SSSA Book Series, vol. 5, pp. 985–1018. Madison, WI.

Hassink, J., 1997. The capacity of soils to preserve organic C and N by their association with clay and silt particles. Plant Soil 191, 77–87.

Hättenschwiler, S., Vitousek, P., 2000. The role of polyphenols in terrestrial ecosystem nutrient cycling. Trends Ecol. Evol. 15, 238–243.

Hedin, L.O., von Fischer, J.C., Ostrom, N.E., Kenneddy, B.P., Brown, M.G., Robertson, G.P., 1998. Thermodynamic constraints on nitrogen transformations and other biogeochemical processes at soil-stream interfaces. Ecology 79, 684–703.

Hill, A.R., 1996. Nitrate removal in stream riparian zones. J. Environ. Quality 25, 743. doi: 10.2134/jeq1996.00472425002500040014x.

Hill, A.R., Devito, K.J., Campagnolo, S., 2000. Subsurface denitrification in a forest riparian zone: interactions between hydrology and supplies of nitrate and organic carbon. Biogeochemistry, 193–223.

Hill, A.R., Kemp, W.A., 1999. Nitrogen chemistry of subsurface storm runoff on forested Canadian Shield hillslopes. Water Resour. 35, 811–821.

Hobbie, S.E., Ogdahl, M., Chorover, J., Chadwick, O.A., Oleksyn, J., Zytkowiak, R., Reich, P.B., 2007. Tree species effects on soil organic matter dynamics: the role of soil cation composition. Ecosystems 10, 999–1018. doi: 10.1007/s10021-007-9073-4.

Hornberger, G.M., Bencala, K.E., Mcknight, D.M., 1994. Hydrological controls on dissolved organic carbon during snowmelt in the snake river near Montezuma, CO. Biogeochemistry 25, 147–165.

Inamdar, S.P., Christopher, S.F., Mitchell, M.J., 2004. Export mechanisms for dissolved organic carbon and nitrate during summer storm events in a glaciated forested catchment in New York, USA. Hydrol. Process. 18, 2651–2661. doi: 10.1002/hyp.5572.

Jarecki, M.K., Parkin, T.B., Chan, A.S.K., Hatfield, J.L., Jones, R., 2006. Comparison of DAYCENT-simulated and measured nitrous oxide emissions from a corn field. J. Environ. Quality 37, 1685–1690. doi: 10.2134/jeq2007.0614.

Jarvis, N.J., 2007. A review of non-equilibrium water flow and solute transport in soil macropores: principles, controlling factors and consequences for water quality. Eur. J. Soil Sci. 58, 523–546. doi: 10.1111/j.1365-2389.2007.00915.x.

Kaiser, K., Guggenberger, G., 2005. Storm flow flushing in a structured soil changes the composition of dissolved organic matter leached into the subsoil. Geoderma 127, 177–187. doi: 10.1016/j.geoderma.2004.12.009.

Kaye, J.P., Hart, S.C., 1997. Competition for nitrogen between plants and soil microorganisms. Trends Ecol. Evol. 12, 139–143. doi: 10.1016/S0169-5347(97)01001-X.

Kung, K.-J.S, Steenhuis, T.S., Kladivko, E.J., Gish, T.J., Bubenzer, G., Bubenzer, G., Helling, C.S., 2000. Impact of preferential flow on the transport of adsorbing and non-adsorbing tracers. Soil Science Society of America Journal 64, 1290–1296.

Larsson, M.H., Jarvis, N.J., 1999. A dual-porosity model to quantify macropore flow effects on nitrate leaching. J. Environ. Quality 28, 1298–1307.

Lewis, D.B., Schade, J.D., Huth, A.K., Grimm, N.B., 2006. The spatial structure of variability in a semi-arid, fluvial ecosystem. Ecosystems 9, 386–397. doi: 10.1007/s10021-005-0161-z.

Lin, H., Bouma, J., Wilding, L.P., Richardson, J.L., Kutilik, M., Nielsen, D.R., 2005. Advances in hydropedology. Advan. Agronom. 85, 1–89.

Linn, D.W., Doran, J.W., 1984. Effect of water-filled pore space on carbon dioxide and nitrous oxide production in tilled and nontilled soils. Soil Sci. Soc. Am. J. 48, 1267–1272.

Lohse, K.A., Matson, P., 2005. Consequences of nitrogen additions for soil losses from wet tropical forests. Ecol. Appl. 15, 1629–1648.

Lohse, K.A., Brooks, P.D., McIntosh, J.C., Meixner, T., Huxman, T.E., 2009. Interactions between biogeochemistry and hydrologic systems. Annual Review of Environment and Resources 34, 65–96. doi: 10.1146/annurev.environ.33.031207.111141.

Lovett, G.M., Weathers, K.C., Arthur, M.A., Schultz, J.C., 2004. Nitrogen cycling in a northern hardwood forest: do species matter? Biogeochemistry 67, 289–308.

Lovett, G.M., Weathers, K.C., Arthur, M.A., 2002. Control of nitrogen loss from forested watersheds by soil carbon: nitrogen ratio and tree species composition. Ecosystems 5, 712–718. doi: 10.1007/s10021-002-0153-1.

Lovett, G.M., Goodale, C.L., 2011. A new conceptual model of nitrogen saturation based on experimental nitrogen addition to an Oak forest. Ecosystems 14, 615–631. doi: 10.1007/s10021-011-9432-z.

Manderscheid, B., Matzner, E., 1995. Spatial and temporal variation of soil solution chemistry and ion fluxes through the soil in a mature Norway Spruce (Picea abies (L.) Karst.) stand. Biogeochemistry 30, 99–114.

McClain, M.E., Boyer, E.W., Dent, C.L., Gergel, S.E., Grimm, N.B., Groffman, P.M., Hart, S.C., Harvey, J.W., Johnston, C.A., Mayorga, E., McDowell, W.H., Pinay, G., 2003. Biogeochemical hot spots and hot moments at the interface of terrestrial and aquatic ecosystems. Ecosystems 6, 301–312. doi: 10.1007/s10021-003-0161-9.

Mcguire, K., McDonnell, J., 2006. A review and evaluation of catchment transit time modeling. J. Hydrol. 330, 543–563. doi: 10.1016/j.jhydrol.2006.04.020.

Mitchell, M.J., 2001. Linkages of nitrate losses in watersheds to hydrological processes. Hydrol. Process. 15, 3305–3307. doi: 10.1002/hyp.503.

Muller, T., 2004. Soil organic matter turnover as a function of the soil clay content: consequences for model applications. Soil Biol. Biochem. 36, 877–888. doi:10.1016/j.soilbio.2003.12.015.

Nadelhoffer, K.J., Emmett, B.A., Gundersen, P., Kjùnaas, O.J., Koopmansk, C.J., Schleppi, P., Tietemak, A., Wright, R.F., 1999. Nitrogen deposition makes a minor contribution to carbon sequestration in temperate forests. Nature 398, 1997–2000.

Newbold, J.D., Elwood, J.W., O'Neill, R.V., Van Winkle, W., 1981. Measuring nutrient spiralling in streams. Canadian Journal of Fisheries and Aquatic Sciences 38, 860–863.

Orchard, V.A., Cook, F.J., 1983. Relationship between soil respiration and soil moisture. Soil Biol. Biochem. 15, 447–453.

Pampolino, M.F., Urushiyama, T., Hatano, R., 2000. Detection of nitrate leaching through bypass flow using pan lysimeter, suction cup and resin capsule. Soil Science and Plant Nutrition 46, 703–711.

Pankhurst, C.E., Pierret, A., Hawke, B.G., Kirby, J.M., 2002. Microbiological and chemical properties of soil associated with macropores at different depths in a red-duplex soil in NSW Australia. Plant Soil, 11–20.

Parkin, T.B., 1993. Spatial variability of microbial processes in soil - A Review. Journal of Environment Quality 22, 409–417.

Parkin, T.B., 1987. Soil microsites as a source of denitrification variability, 51, 1194–1199.

Perakis, S.S., Hedin, L.O., 2002. Nitrogen loss from unpolluted South American forests mainly via dissolved organic compounds. Nature 415, 416–419. doi:10.1038/415416a.

Peterson, B.J., Wollheim, W.M., Mulholland, P.J., Webster, J.R., Meyer, J.D., Tank, J.L., Martõ, E., Bowden, W.B., Valett, H.M., Hershey, A.E., McDowell, W.H., Dodds, W.K., Hamilton, S.K., Gregory, S., Morrall, D.D., 2001. Control of nitrogen export from watersheds by headwater streams. Science 292, 86–90.

Pionke, H.B., Gburek, W.J., Sharpley, A.N., Schnabel, R.R., 1996. Flow and nutrient export patterns for an agricultural. Water Resour. 32, 1795–1804.

Plante, A.F., Conant, R.T., Stewart, C.E., Paustian, K., Six, J., 2006. Impact of soil texture on the distribution of soil organic matter in physical and chemical fractions. Soil Sc. Soc. Am. J. 70, 287–296. doi:10.2136/sssaj2004.0363.

del Prado, A., Merino, P., Estavillo, J.M., Pinto, M., González-Murua, C., 2006. N_2O and NO emissions from different N sources and under a range of soil water contents. Nutrient Cycling in Agroecosystems 74, 229–243. doi: 10.1007/s10705-006-9001-6.

Reichstein, M., Rey, A., Freibauer, A., Tenhunen, J., Valentini, R., Banza, J., Casals, P., Cheng, Y., Grunzweig, J.M., Irvine, J., Joffre, R., Law, B.E., Loustau, D., Miglietta, F., Oechel, W., Ourcival, J., Pereira, J.S., Peressotti, A., Ponti, F., Qi, Y., Rambal, S., Rayment, M., Romanya, J., Rossi, F., Tedeschi, V., Tirone, G., Xu, M., Yakir, D., 2003. Modeling temporal and large-scale spatial variability of soil respiration from soil water availability, temperature and vegetation productivity indices. Global Biogeochemical Cycles 17. doi: 10.1029/2003GB002035.

Robertson, G.P., Wedin, D., Groffman, P.M., Blair, J.M., Holland, E.A., Nadelhoffer, K.J., Harris, D., 1999. Soil carbon and nitrogen availability: nitrogen mineralization, nitrification, and soil respiration potentials. In: Robertson, G.P., Coleman, D.C., Bledsoe, C.S., Sollins, P. (Eds.), Standard Soil Methods for Long-term Ecological Research. Oxford University Press, New York, pp. 258–271.

Ruser, R., Flessa, H., Russow, R., Schmidt, G., Buegger, F., Munch, J.C., 2006. Emission of N_2O, N_2 and CO_2 from soil fertilized with nitrate: effect of compaction, soil moisture and rewetting. Soil Biology 38, 263–274. doi: 10.1016/j.soilbio.2005.05.005.

Sanderman, J., Amundson, R., 2008. A comparative study of dissolved organic carbon transport and stabilization in California forest and grassland soils. Biogeochemistry 92, 41–59. doi:10.1007/s10533-008-9249-9.

Schilling, K., Zhang, Y.K., 2004. Baseflow contribution to nitrate-nitrogen export from a large agricultural watershed, USA. J. Hydrol. 295, 305–316.

Schimel, J.P., Bennett, J., 2004. Nitrogen mineralization: challenges of a changing paradigm. Ecology 85, 591–602.

Shipitalo, M.J., Edwards, W.M., Dick, W.A., Owens, L.B., 1990. Initial storm effects on macropore transport of surface-applied chemicals in no-till soil. Soil Sci. Soc. Am. J. 54, 1530–1536.

Six, J., Bossuyt, H., Degryze, S., Denef, K., 2004. A history of research on the link between (micro)aggregates, soil biota, and soil organic matter dynamics*1. Soil Till. Res. 79, 7–31. doi:10.1016/j.still.2004.03.008.

Sommers, L.E., Gilmour, C.M., Wildung, R.E., Beck, S.M., 1981. The effect of water potential on decomposition processes in soils. In: Parr, J.F., Gardner, W.R., Elliot, L.F. (Eds.), Water Potential Relations in Soil Microbiology. Soil Science Society of America Special Publication, Madison, WI Number 9.

Stark, J.M., Firestone, M.K., 1995. Mechanisms for soil moisture effects on activity of nitrifying bacteria. Appl. Environ. Microbiol. 61, 218–221.

van Verseveld, W.J., McDonnell, J.J., Lajtha, K., 2008. A mechanistic assessment of nutrient flushing at the catchment scale. J. Hydrol. 358, 268–287. doi:10.1016/j.jhydrol.2008.06.009.

Vinther, F.P., Eiland, F., Lind, A., Elsgaard, L., 1999. Microbial biomass and numbers of denitrifiers related to macropore channels in agricultural and forest soils. Soil Biol. Biochem. 31, 603–611.

Vitousek, P.M., Gosz, J.R., Grier, C.C., Melillo, J.M., Reiners, W.A., 1982. A comparative analysis of potential nitrification ans nitrate mobility in forest ecosystems. Ecol. Monographs 52, 155–177.

Vitousek, P.M., Hattenschwiler, S., Olander, L.P., Allison, S.D., 2002. Nitrogen and nature. Ambio 31, 97–101.

Vitousek, P.M., Reiners, W.A., 1975. Ecosystem succession and nutrient retention: a hypothesis. Bioscience 25, 376–381.

Wagener, S.M., Oswood, M.W. and Schimel, J.P. 1998. Rivers and soils: parallels in carbon and nutrient processing. Library.

Yoo, K., Amundson, R., Heimsath, A., Dietrich, W., 2006. Spatial patterns of soil organic carbon on hillslopes: integrating geomorphic processes and the biological C cycle. Geoderma 130, 47–65. doi: 10.1016/j.geoderma.2005.01.008.

Chapter 23

Coupling Ecohydrology and Hydropedology at Different Spatio-Temporal Scales in Water-Limited Ecosystems

Xiao-Yan Li,[1,2,*] Henry Lin[3] and Delphis F. Levia[4]

ABSTRACT

Ecohydrology and hydropedology are two interdisciplinary fields that can be coupled to improve the linkage between aboveground and belowground processes. Connecting multiple scales and identifying self-organizing structures between aboveground and underground processes remain a challenge for improved understanding of coupled soil hydrology and ecosystems. This chapter discusses some key connections between ecohydrology and hydropedology at different spatial and temporal scales in water-limited ecosystems. These connections include: 1) water flow from soil pore to plant stomata at the microscopic scale; 2) aboveground canopy precipitation partitioning and belowground preferential flow at the individual plant scale; 3) vegetation pattern and soil moisture distribution at the patch scale; and 4) surface and subsurface flow networks at the hillslope and catchment scales. Co-evolution of soil and vegetation during land degradation and rehabilitation is highlighted to show the connection between ecohydrology and hydropedology across temporal scales. We conclude this chapter by suggesting some future research needs.

[1]State Key Laboratory of Earth Surface Processes and Resource Ecology, Beijing Normal University, Beijing 100875, China

[2]College of Resources Science and Technology, Beijing Normal University, Xinjiekouwai Street 19, Beijing, China

[3]Dept. of Ecosystem Science and Management 116 ASI Building, The Pennsylvania State University, University Park, PA, USA

[4]Depts. of Geography and Plant and Soil Sciences, University of Delaware, Newark, DE, USA

[*]Corresponding author: Email: xyli@bnu.edu.cn

Hydropedology, Edited by H. Lin. DOI: 10.1016/B978-0-12-386941-8.00023-X

1. INTRODUCTION

Ecohydrology and hydropedology are two new scientific fields that are interconnected in terrestrial ecosystems. While there are some debates concerning the precise definition or the corpus of knowledge that constitutes ecohydrology (Bonell, 2002; Kundzewicz, 2002; Nuttle, 2002; Porporato and Rodriguez-Iturbe, 2002; Zalewski, 2002), most would agree that ecohydrology examines hydrologic mechanisms that underlie ecologic patterns and processes (Rodriguez-Iturbe, 2000). It seeks to elucidate: (1) how hydrologic processes influence the distribution, structure, function, and dynamics of ecosystems and (2) how feedbacks from biotic processes impact the water cycle (Newman et al., 2006). Hydropedology, on the other hand, investigates interactive pedologic and hydrologic processes and the landscape–soil–hydrology relationship across space and time (Lin, 2003; Lin et al., 2005, 2006). It aims to understand pedologic controls on hydrologic processes/properties and hydrologic impacts on soil formation and functions. Specifically, hydropedology seeks to elucidate: (1) how soil architecture and the distribution of soils over the landscape exert a first-order control on hydrologic processes (and associated biogeochemical and ecological dynamics) and (2) how landscape water (and the associated transport of energy, sediment, chemicals, and biomaterials by flowing water) influence soil genesis, evolution, variability, and functions (Lin et al., 2008). Ecohydrology addresses the interface between the biosphere and the hydrosphere, while hydropedology focuses on the interface between the pedosphere and the hydrosphere (Fig. 1). In the framework of the Earth's Critical Zone (NRC, 2001), water functions as the central link among all the interacting spheres in terrestrial ecosystems while the pedosphere acts as the central

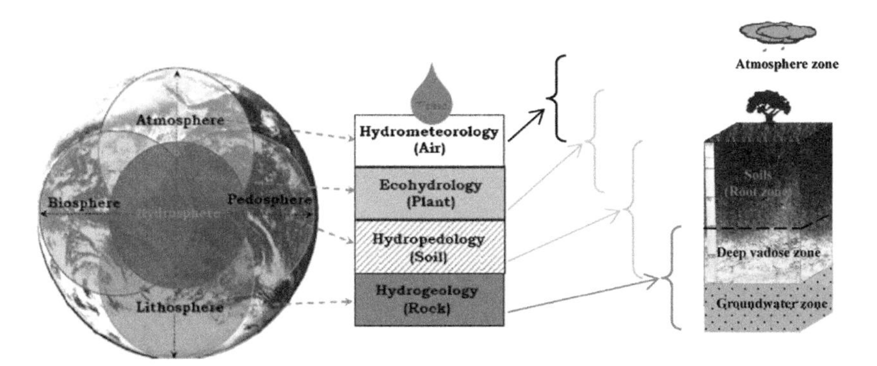

FIGURE 1 Interconnection between the pedosphere (soil), biosphere (plants), lithosphere (rock), hydrosphere (water), and atmosphere (air) in the Earth system. In the aboveground, hydropedology interfaces with ecohydrology and hydrometeorology to understand the feedback mechanisms between climate, vegetation, and soils. In the belowground, hydropedology interfaces with hydrogeology to study the interactions of water with solid earth (i.e. soils, rocks, and anything in between). *(Color version online and in color plate)*

juncture where all spheres interact (Fig. 1). The pedosphere is also the mediator of a variety of physical, chemical, and biological fluxes into and out of the Critical Zone. In the vertical dimension, hydropedology interfaces with ecohydrology and hydrometeorology to study the feedback mechanisms between climate, vegetation, and soils. Belowground, hydropedology, combined with hydrogeology, promotes an integrated systems approach to study the interactions of water with solid earth (i.e. soils, rocks, and anything in between) (Fig. 1).

Vegetation can be considered as a key link for connecting ecohydrology (mostly aboveground) and hydropedology (mostly belowground) in a soil–plant–water system at different scales. The aboveground and belowground biota interact with each other and can be powerful mutual drivers; however, the aboveground and belowground components have traditionally been considered in separation (Wardle et al., 2004). While the importance of coupling ecohydrology and hydropedology to enhance the understanding of aboveground–belowground relationships has been recognized (e.g. Li et al., 2009; Lin, 2010a), a cohesive framework of their linkages remains lacking (Breshears and Zou, 2006), in particular in arid and semi-arid environments where water is a critical limiting resource not only because of its scarcity but also because of its intermittency and unpredictable presence (Porporato and Rodriguez-Iturbe, 2002). In the dryland ecosystem of arid regions, rainfall is less than the potential evapotranspiration for the whole or part of the year, and conditions of either permanent or seasonal soil water deficit occur (D'Odorico and Porporato, 2006). The dryland ecosystem is in general water-controlled "with infrequent, discrete, and largely unpredictable water input" (Noy-Meir, 1973). Hence, this chapter discusses linkages between ecohydrology and hydropedology through examples at different spatial and temporal scales from water-limited ecosystems. The general principles discussed may be applicable to other terrestrial ecosystems. We are hopeful that such discussions can stimulate further studies and shed more light on future research directions.

2. WATER FLOW AT THE SOIL PORE AND PLANT STOMATA SCALE

Movement of water in the soil–plant–atmosphere system involves complex processes and interactions in and among the soil pore network, soil–root interface, xylem transport network, and leaf stomata (Fig. 2a). Spatial arrangement of soil pores and grains determines both energy and mass flow in the soil. Generally speaking, water moves from soil pores to plant roots following potential energy gradients. Once water has reached and entered the rooting system through a heterogeneous root membrane, water flows through a complex network within the xylem. It then experiences phase transition within the leaves, and exits to the atmosphere in the form of water vapor through leaf stomata (Katul et al., 2007) (Fig. 2a). The vapor molecules are

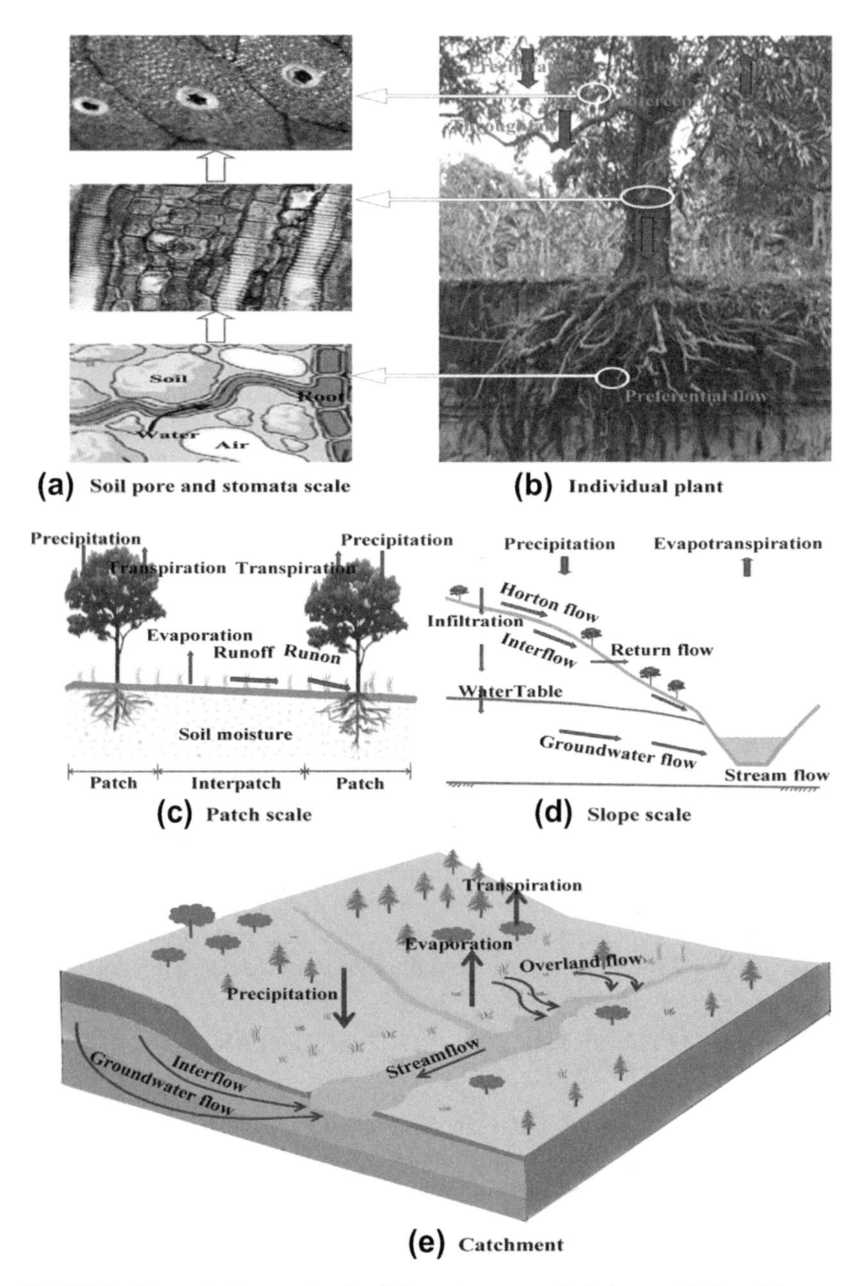

FIGURE 2 Schematic diagram showing linkages between ecohydrology and hydropedology at different spatial scales. (a) Upward water flow from the soil pore to the atmosphere through roots (a), xylem (b), and stomata (c). (b) Aboveground canopy partitioning of precipitation and belowground preferential flow in soils at the individual plant scale. (c) Vegetation pattern and its effects on soil moisture at the patch scale. (d) Surface and subsurface flow path at the hillslope scale. (e) Hydrologic processes and flow networks at the catchment scale. *(Color version online and in color plate)*

then transported by turbulent eddies from within the canopy into the free atmosphere. The transporting energy and sizes of these eddies are partially determined by complex interactions among canopy attributes (e.g. leaf area and height), forcing (e.g. geostrophic winds and weather patterns), and the heterogeneity of vegetation and landscape (Katul and Novick, 2009).

The flow of water is the unifying linkage between the plant, soil, and atmosphere. The small-scale processes of water and gas movement in a porous media are well studied from a physical or hydrodynamic point of view (Jarvis and McNaughton. 1986; Jury et al., 1991). However, their dynamic interfaces with biological systems remain poorly understood (Hopmans and Bristow, 2002). Fluxes at the interfaces (e.g. root–soil and plant–atmosphere) are mostly empirically derived. Much attention has been paid to water movement and chemical transport in the absence of biological impacts that include roots and microbes (Hopmans and Bristow, 2002). Therefore, the coupling of ecohydrology and hydropedology can shed light on the linkage between belowground soil–root interactions and aboveground plant–atmosphere interfaces.

It is generally accepted that plants regulate transpiration hydraulically through stomatal response to water pressure and/or biochemically through stomatal response to abscisic acid hormone (Siqueira et al., 2008). In water-limited ecosystems, water deficit can lead to an increase in stomatal density and a decrease in stomatal size (Spence et al., 1986; Martinez et al., 2007). In addition, hydraulic lift (also called hydraulic redistribution), referring to the transport of water from wetter into dryer soil areas through the actions of roots, is a mechanism that facilitates water movement through the soil–plant–atmosphere continuum and can delay the onset of plant-water stress (Caldwell et al., 1998; Brooks et al., 2002; Amenu and Kumar, 2008). Vertical water potential gradients in the soil and plants with shallow and deep roots are a prerequisite for hydraulic lift to occur (Scholz et al., 2010). Evidences of hydraulic lift have been reported for shrub, grasses, and trees, as well as for temperate, tropical, and desert ecosystems (Caldwell et al., 1998). One study by Scholz et al. (2010) in Brazilian savannas (Cerrado) demonstrated that the release of hydraulically lifted water by herbaceous plants contributed 98% to the partial daily recovery of soil water storage, whereas woody plants contributed the remaining 2%. This was attributed to the fact that herbaceous plants have abundance of fine roots in the upper soil. Hydraulic lift can replace 23% of evapotranspiration of the Brazilian savannas ecosystem during the peak of a dry season, suggesting that hydraulic lift has a great impact on water economy and other ecosystem processes (Scholz et al., 2010). Although the importance of hydraulic lift for arid ecosystem has been recognized, limited research efforts have been made to elucidate the basic coordinating mechanisms of stomata, root, and soil pores for enhancing hydraulic lift, and very few root water uptake models account for hydraulic lift in their formulations (Siqueira et al., 2008). This is a challenging task for the integrated study of ecohydrology and hydropedology.

While it is beyond the scope of this chapter to delve into the ecophysiological controls of water transport through the soil–plant–atmosphere system, it is important to note that the transport of water through plants is partly governed by xylem structure and stomatal regulation. Diffuse- and ring-porous xylem of angiosperms, for example, have higher water flux rates through the plant stem than the tracheid systems of gymnosperms, partly as a result of larger diameter of the conducting tissue and higher hydraulic conductivities (Milburn, 1979; Woodward, 1995; Tyree and Zimmermann, 2002). Stomatal control on foliar surfaces is a key control to water loss through transpiration. Schlerophyllous vegetation with small leaves and thick epicuticular waxes, for instance, are better able to limit water loss than those without wax (Juniper and Jeffree, 1983). It follows that water-limited ecosystems would be dominated by plants that achieve an optimal balance among rooting depth, xylem structure, and stomatal regulation.

3. CANOPY PARTITION OF PRECIPITATION AND PREFERENTIAL FLOW IN SOILS AT THE INDIVIDUAL PLANT SCALE

Vegetation canopies can affect the spatial distribution of water within the plant community in both vertical and horizontal dimensions. Partition of precipitation by vegetation canopy generally comprises three fractions (Fig. 2b): (1) interception, which is retained on the vegetation and is evaporated after or during rainfall; (2) throughfall, which reaches ground by passing directly through or dripping from tree canopies, and (3) stemflow, which flows to the ground via part of trunks or stems. Canopy interception generally prevents water from reaching the ground surface, but throughfall affects surface soil moisture distribution (both quantity and quality), while stemflow can alter the vertical distribution of water by quickly funneling water to the base of plants where it can infiltrate rapidly and preferentially into a deeper part of the soil (e.g. Devitt and Smith, 2002; Llorens and Domingo, 2007). A study by Li et al. (2009) reported that stemflow led to 1.5–3.2 times increase in infiltration depth, and as much as 10–140% increase in soil water content for *Sminthopsis psammophila* and *Helianthemum scoparium* as compared to that without stemflow. In arid and semi-arid regions, there is a general paucity of data on stemflow as well as their influential factors in various shrub species. Quantitative information on stemflow and throughfall processes and the final fate of this re-distributed water in the soil remain sparse (Martinez-Meza, 1994). Figure 3 summarizes the percentages of interception, throughfall, and stemflow of the incident gross precipitation by different trees, shrubs, and grasses based on the published literature. Under mean annual precipitation from 117 to 570 mm y^{-1}, the averages for interception, throughfall, and stemflow have been reported as $26.9 \pm 18.7\%$, $65.2 \pm 15.5\%$, and $11.5 \pm 11.9\%$ of total incident precipitation, respectively (Fig. 3). Grasses had a higher mean

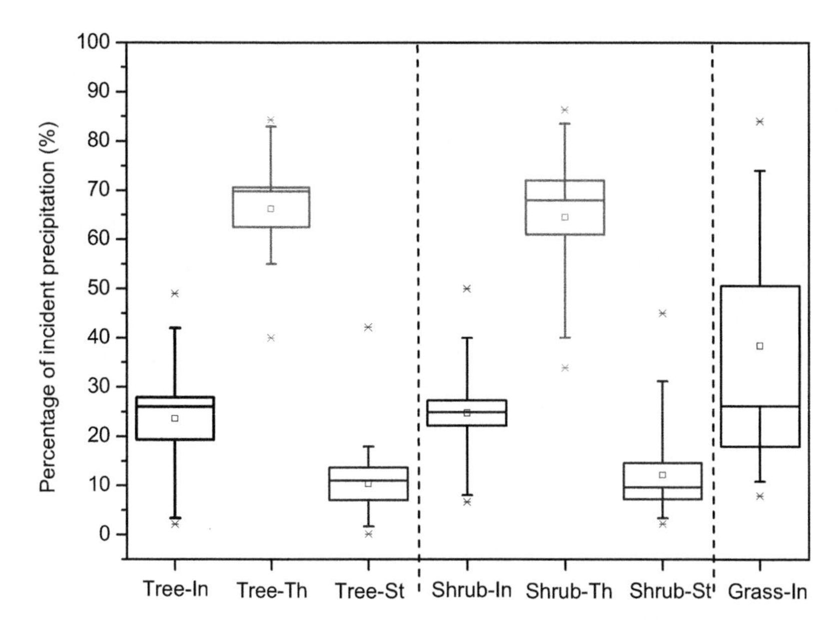

FIGURE 3 Box-and-whisker diagrams showing the median, 10th, 90th percentiles and standard deviation for interception (In), throughfall (Th), and stemflow (St) percentages for tree, shrub, and grass in the arid and semi-arid regions based on 41 published articles. (□) Represents mean value, (*) maximum and minimum value. There are 12, 22, and 7 data points for tree, shrub, and grass, respectively, under mean annual precipitation from 117 to 570 mm yr^{-1}. *(Color version online)*

value of interception (38.4 ± 32.3%) than trees (23.6 ± 14.9%) and shrubs (24.8 ± 12.9%), while no significant differences in interception, throughfall, and stemflow were found between trees and shrubs. Stemflow showed much greater variability (coefficient of variation >100%) than throughfall (coefficient of variation <30%) and interception (coefficient of variation <60%). Shrubs tend to have slightly higher stemflow (12.1 ± 12.4%) than those dominated by trees (10.3 ± 11.6%) (Fig. 3).

Li et al. (2009) demonstrated a clear linkage between aboveground ecohydrology (mainly stemflow) and belowground hydropedology (mainly preferential flow) in the desert shrubs at the individual plant scale. Johnson and Lehmann (2006) suggested that a plant canopy re-distributed hydrologic fluxes to the root zone through a double-funneling process: (1) the canopy first partitions rainfall into throughfall and stemflow, resulting in a spatial redistribution of water fluxes reaching the soil, and (2) stemflow that is delivered to the soil at the base of trees is further funneled by plant roots through belowground preferential flow of water. Deep infiltration of stemflow through preferential flow aids plants in withstanding droughts, while nutrient fluxes delivered in stemflow augment the quantities of available nutrients in the rhizosphere (Johnson and Lehmann, 2006). Levia and Frost (2003) reviewed the quantitative and qualitative information of stemflow in forested and agricultural ecosystems, and concluded that stemflow was a spatially localized point input of precipitation and solutes at the plant stem scale, which is of hydrologic and ecological significance. In arid and semi-arid regions, shrubs can concentrate more rainwater into the root area. For example, Navar (1993) reported that *Prosopis laevigata* and *Acacia farnesiana* shrubs had average stemflow funneling ratios (the ratio of rainfall amount delivered to the base of the tree to the rainfall that would have reached the ground were the tree not present) of 11, while ratios for *Diospyros texana* individuals averaged 58. Funneling ratios of scrubs varied with shrub species, with 25 for *Tamarix ramosissima* (Li et al., 2008b), 49 for *S. psammophila* and for *H. scoparium* (Li et al., 2009), 53 for *Reaumuria soongorica,* and 154 for *Caragana korshinskii* (Li et al., 2008b). Canopy partitioning of incident rainfall exerts a strong influence on nutrient fluxes delivered via stemflow. Comparing fluxes across a range of tree species and rainfall regimes shows larger stemflow fluxes of NO_3^- and K for species with greater stemflow partitioning, regardless of climate type (Johnson and Lehmann, 2006). Stemflow always results in spatial heterogeneity in soil-water fluxes and the flow of water through stem–root system often occurs as preferential flow in soils. Devitt and Smith (2002) reported that macropores formed by the root systems of woody shrubs may be an important conduit for downward water movement in desert soils. Li et al. (2009) confirmed that root channels were preferential pathways for most stemflow water moved into the soil, and stemflow was conducive to concentrate and store water in the deeper layers of the soil profiles, creating favorable soil-water conditions for plants to survive droughts under arid conditions.

Stemflow hydrology and preferential flow along roots are intimately linked, but direct integration in modeling is missing. Water utilization by plants is controlled by many factors including vegetation type, soil moisture storage and redistribution, depth of root and bedrock, and rainfall characteristics. As a result, the extent and importance of stemflow to plant-water utilization need further investigations. Integrated observations over a long period (such as successive years) and with a wide range of species are needed to quantify the spatio-temporal variability of rainfall partitioning by canopies, stems, roots, and soils, thus providing appropriate parameters needed for process modeling (Li, 2011).

Above and belowground biomass distribution pattern is another aspect relevant to coupling ecohydrology and hydropedology at the individual plant scale. Liang et al. (1989) estimated above- and below-ground biomass and water-use efficiency for five plant communities in north central Colorado, and found that the ratios of aboveground biomass (AGB) and belowground biomass (BGB) at the shrub sites were greater than that at grass sites (Table 1). This ratio ranged from 0.2 to 0.39 for shrub sites and only from 0.04 to 0.1 for grass and pasture sites. In addition, greater water-use efficiency values were found at

TABLE 1 **The Above- and Belowground Biomass (gm^{-2}) for Five Plant Communities in Water-Limited Ecosystems with a Mean Annual Precipitation of 311 mm at the Central Plains Experimental Range of Colorado. Water-use Efficiency is Determined as Aboveground Biomass Divided by Evapotranspiration**

Plant community	Sandy-loam shrub	Loam shrub	Clay-loam grass	Sandy-clay-loam half-shrub	Sandy-loam pasture
Aboveground Biomass (AGB) (g/m^2)					
Warm-season grass	140	66	49	101	13
Cool-season grass	10	78	23	5	118
Forbs	1	9	1	4	1
Shrubs	173	1209	0	28	9
Succulents	54	0	5	14	0
Total	**379**	**1362**	**77**	**150**	**142**
Belowground Biomass (BGB) (g/m^2)					
0–30 cm	1362	1948	1498	1469	1054
30–100 cm	492	1557	397	677	305
Total	**1854**	**3505**	**1895**	**2146**	**1359**
AGB:BGB ratio	***0.20***	***0.39***	***0.04***	***0.07***	***0.10***
Water-use Efficiency ($g/m^2/mm$)					
Growing season	**0.8**	**1.5**	**0.4**	**0.4**	**0.5**

Source: Liang et al., 1989

coarse-textured sites, and high biomass and high water-use efficiencies were related less to grass types than to the abundance of shrub. A study by Yang et al. (2010) reported that the relationship between BGB and AGB across China's grasslands could be characterized by a power function of $BGB = 17.13 \times AGB^{0.74}$ ($r^2 = 0.56$, $P < 0.001$). The slope (scaling exponent) of the allometric relationship across different grassland types was 1.02, with 95% confidence interval of 0.94–1.10. Ma et al. (2008) estimated that the meadow steppe had an AGB and BGB of 196.7 and 1385.2 gm^{-2}, respectively, significantly higher than the desert steppe with an AGB of 56.6 gm^{-2} and

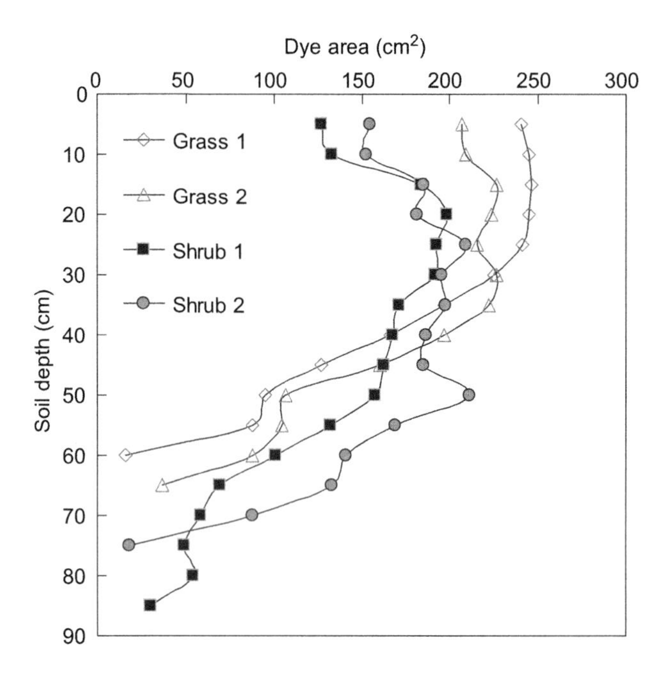

FIGURE 4 The dye-stained area in the vertical soil profile beneath the canopy of shrub *C. microphylla* and grass *Stipa* communities under double-ring infiltration of 30-liter water over 10 h in the Inner Mongolia grassland of north China. *(Color version online)*

a BGB of 301.0 gm^{-2}. These studies suggest that the distribution of AGB and BGB is a reflection of mixed interaction and feedback between vegetation types, soil texture and moisture, and precipitation. In the meantime, AGB and BGB in turn affect water flow in the soil profile and perhaps soil evolution. For example, the dye-stained area under shrub and grass reveals that water penetrates deeper in shrub than in grass due to the fact that roots usually distribute in deeper soil layers in shrubs, and thus more preferential flow is observed under shrubs as compared to grasses (Fig. 4).

4. VEGETATION PATTERN AND ITS EFFECTS ON WATER FLOW AND SOIL PROPERTIES AT THE PATCH SCALE

The vegetation of arid and semi-arid regions is usually patterned, consisting of patches with high plant cover interspersed in a low-cover or bare soil matrix (Aguiar and Sala, 1999) (Fig. 2c). This patchiness differs in scale and shape. Patterns reported in the literature include gaps, labyrinths, stripes ("tiger bush"), and spots ("leopard bush"), which occur in semi-arid and arid regions of Africa, Asia, Australia, and North America (Aguiar and Sala, 1999; Rietkerk et al., 2004). These vegetation patches can vary from small clumps of grasses (e.g. 0.5–2 m^2) to large groves of mulga (*Acacia aneura*) trees (e.g. 100–1000 m^2) (Ludwig et al., 2005). The general mechanism underlying this so-called self-organized patchiness is a positive feedback between plant growth and soil-water availability (Rietkerk et al., 2004). Vegetated patches and bare ground couple in a landscape mosaic of sources and sinks of water, sediments, and nutrients, with the bare ground serving as a runoff producing zone and vegetation as a water- and nutrient-concentrating zone (Reynolds et al., 1999; Wilcox et al., 2003). Ludwig et al. (2005) attributed ecohydrologic and hydropedologic interactions between vegetation patch and interpatch to the following aspects: (1) vegetation patches obstruct runoff and store runon (Fig. 2c); (2) runon enhances plant growth pulses; and (3) vegetation patches enhance soil infiltrability. Thus, the vegetated patches are characterized by greater water-storage capacity, increased soil organic carbon, larger nutrient inputs, greater soil biological activity, and higher net primary productivity than the adjacent inter-canopy area (Puigdefábregas, 2005), leading to the so-called "fertility islands" (Schlesinger and Pilmanis, 1998), "resource islands" (Reynolds et al., 1999), and "hydrologic islands" (Rango et al., 2006). Dunkerley (2000) reported that the water uptake rate within a shrub patch is at least 10% greater than the shrub interspace. Galle et al. (2001) estimated that the rainfall concentration factor (infiltration in the vegetated band divided by rainfall) was up to approximately 1.4 for *A. aneura* in Australia and 2–4 for tiger bush in Mexico and Niger. The enhanced infiltration rates under vegetated patches are due to improved soil aggregation and macroporosity related to biological activity (e.g. termites, ants, and earthworms are active in arid areas) and vegetation roots (Ludwig et al., 2005). Howes and Abrahams (2003) showed through modeling that, on average, runon infiltration accounts for 3–20% of the total infiltration under a shrub. Newman et al. (2010) stated that vegetation patches have substantially larger transpiration rates and lower evaporation/transpiration ratios than inter-canopy patches. Li et al. (2008a) reported that a spot-structured shrub patch in the Tengger Desert of China could trap 55% of the runoff from the crust patches, and therefore more than 75% of the sediments, 63% of soil carbon, 74% of nitrogen, and 45–73% of dissolved nutrients triggered by runoff from crust patches were delivered to shrub patches. Bisigato et al. (2009) reviewed that soil in a vegetation patch had 1.23–3.70 times more organic matter, 1.25–5.00 times more nitrogen, and 1.10–1.66 times more phosphorus than the soil of interpatch areas in Monte Desert. Different mechanisms have been proposed for higher concentrations of nutrients under plant cover: litterfall and death of roots of plants present in the patch, deposition of wind-blown soil and organic matter, washing of dust and nutrients deposited in the canopies by rain, and/or nutrient losses by erosion in the interpatch areas. However, the relative importance of these mechanisms has not been fully evaluated (Bisigato et al., 2009).

Preferential infiltration in the vegetated area and soil moisture redistribution have been recognized as two major processes influencing the establishment and persistence of patchy vegetation patterns in arid lands (Rietkerk et al., 2004; Ursino, 2009). These suggest that ecohydrology and hydropedology of arid environments are strongly coupled through available soil moisture, which acts as a vital moderator and control of processes affecting vegetation structure and ecosystem services. Soil moisture accumulates locally in the root zone and influences the root growth and the aboveground biomass organization (Ursino, 2009). On the one hand, soil moisture spatial distribution depends on soil properties, plant physiology, and seasonal and daily variability of rainfall (De Michele et al., 2008); on the other hand, soil moisture controls the movement of nutrients, the propagation of plant roots, and the activity of microbial populations that fuel the biogeochemical reactor in soils (Young et al., 2010). Few field studies have examined systematically how the spatial patterns of soil moisture influence transpiration patterns at the hillslope scale, and vice versa. Traditional approaches to the study of soils and their relation to plant growth have been somewhat compartmentalized, focusing on either the water aspects of fertility and nutrition or the plant response to resources and allocation. Future research thus needs to follow a more holistic approach that focuses on the linkages between resource pool dynamics, allocation processes, and feedback mechanisms (Young et al., 2010).

5. SURFACE AND SUBSURFACE FLOW AT THE HILLSLOPE AND CATCHMENT SCALES

Hillslopes are fundamental units of the hydrologic landscape, featuring heterogeneity in terrain characteristics, lithology, soils, land use, and vegetation. Runoff generation on hillslopes is thus highly variable and complex across space and time. The role of soils and plants has long been recognized as critical to rainfall-runoff processes in hillslopes and catchments (Fig. 2d). Thus, both ecohydrology and hydropedology possess significant potential in enhancing the understanding and prediction of rainfall-runoff processes.

The traditional view of surface runoff generation has been based on the "infiltration-excess" or "Hortonian" concept of runoff production (Horton, 1933), where surface runoff is the result of rainfall intensity exceeding infiltration capacity at the soil surface. This surface runoff occurs more commonly in arid and semi-arid regions, where rainfall intensities are frequently high and soil-infiltration capacity is reduced because of surface sealing or pavement (Yair and Lavee, 1985). An alternative runoff generation paradigm is a variable source area (Hewlett and Hibbert, 1967), which suggests "saturation-excess" runoff generation in localized (typically near-stream or topographic depression) areas of the landscape because of precipitation falling on soils that have little or no available storage, thus precluding infiltration. Consequently, soil properties and their spatial distributions, as well as their interactions with topography, land use, vegetation community, and other components of the hydrologic cycle (besides precipitation) become important (Lin et al., 2008).

At the catchment scale, various hydrologic storages (e.g. soil water, groundwater, and stream water) interact in complex nonlinear ways (Peters and Havstad, 2006) (Fig. 2e) and all processes are believed to be preferential (Uhlenbrook, 2006). The actual pathways from rainfall to streamflow usually involve a combination of surface and subsurface flow (Brooks et al., 1991), and many of them are related to biological factors including canopy, stem, and root impacts as described above. Water movement in upland watersheds from the soil surface to the stream is often described using the concept of translatory flow, which assumes that water entering the soil as precipitation would displace the "old" water that was present previously, pushing it deeper into the soil, and eventually into the stream (Horton and Hawkins, 1965). Soil is generally considered as a giant sponge that soaks up precipitation and slowly releases it to streams over time (Phillips, 2010). However, a recent study by Brooks et al. (2010) found that the water occupying the pore space of soils and used by trees does not participate in translatory flow. Soil water is assumed to be separated into two water worlds: mobile water, which eventually enters the stream, and tightly bound water used by plants (Brooks et al., 2010). This work would challenge the assumptions of translatory flow and the idea that plants and streams use the same water pools. To understand better the triggering mechanism for the often rapid soil water and groundwater contributions to flood runoff, Uhlenbrook (2006) suggested that the following needs to be better observed and quantified: (i) the role of connectivity of water bodies and flow pathways in the landscape; (ii) the role of the bedrock topography; (iii) the bedrock permeability; (iv) the storage versus permeability properties of the soil including mobile–immobile fractions and residence times of the water; and (v) the wave propagation (wave speed, celerity) in soil and groundwater versus real flow velocity.

Subsurface stormflow is another runoff producing mechanism operating in most upland terrains, especially in a humid environment and steep terrain with conductive soils or soils with a water-restricting layer. While some studies have documented subsurface stormflow as unsaturated flow in the vadose zone, most studies have shown that subsurface stormflow is saturated or near-saturated – due to the rise of the water table into more transmissive soil above or the transient saturation above an impeding layer, soil–bedrock interface, or zone of reduced permeability (e.g. fragipan, argillic horizon, and other dense layers in the soil profile) (Lin et al., 2008).

In recent years, growing evidence suggests that the subsurface flow network is a key to understanding rainfall-runoff processes including threshold behavior and that flow pathways often determine the hot spots and moments of biogeochemical processes (Lin et al., 2008). McClain et al. (2003) reported that hot spots change with scales for denitrification: the soil along root channels at the scale of soil column (1 m), topographic depressions at the scale of catena (10 m), the soil–stream interface at the scale of upland-stream toposequence (100 m), and the upland–wetland contact zone at the scale of sub-basin (10,000 m). The dynamic origin of network structures in soils and hydrologic systems and recurrent patterns of self-organization are the subjects of recent research and model development (e.g. Weiler and McDonnell, 2007; Lin et al., 2008). Networks are abundant in soils despite the difficulty of seeing them, such as root-branching networks, mycorrhizal mycelial networks, animal-borrowing networks, crack and fissure networks, and others (Lin, 2010b). Flow and transport networks in soils are formed by the forcing of soil formation, mainly climate and organisms. Cycles of wetting and drying, freezing and thawing, shrinking and swelling, coupled with organic matter accumulation and decomposition, biological activities, and chemical reactions lead to the formation of diverse soil aggregates and pore networks in the subsurface (Lin, 2010b). Some evidence has suggested that a similarity may exist between river dendritic structures and subsurface preferential flow networks (Deurer et al., 2003). In aboveground, tree branching, leaf architecture, and spatial vegetation distribution also exhibit a similar network structure. These suggest that aboveground vegetation and surface flow networks appear to have close connection to belowground root and subsurface flow networks. We believe that flow networks embedded in both surface and subsurface mosaics warrant further research, which will be another stimulating topic for coupling ecohydrology and hydropedology.

6. VEGETATION EVOLUTION AND SOIL DEVELOPMENT AT DIFFERENT TEMPORAL SCALES

Feedbacks between plants and soils play a major role in structuring the composition and evolution of plant communities. Plants can change biotic and abiotic characteristics of the soil in which they grow, which in turn affects their performance (Bever, 1994). Co-evolution of vegetation and soil therefore is essential to the coupling of ecohydrology and hydropedology. In the following, we use two examples of sand-dune stabilization and encroachment of woody shrubs into grasslands to illustrate vegetation–soil interactions and their feedbacks over time.

Vegetation evolution and soil development can be well demonstrated in a stabilized sand-dune area in the Shapotou region on the southeastern edge of the Tengger Desert of China. Planting of sand-binding vegetation (xerophytic trees and shrubs) successfully stabilized formerly mobile dunes since 1956, permitting the operation of the Baotou–Lanzhou railway (Fig. 5a). Community composition and structure of sand-binding vegetation and soil properties experienced remarkable changes since vegetation established. Most of the sand-binding vegetation began to degrade after 20 years, after which the initial woody species that were planted were gradually replaced by native herbaceous species (Li et al., 2010). The initial artificial community of *Hedysarum*

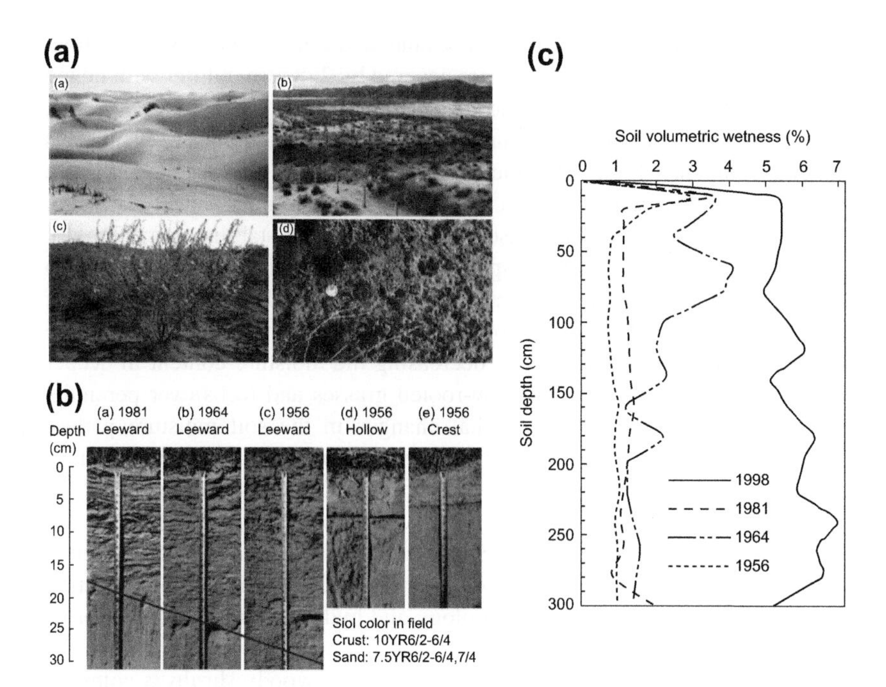

FIGURE 5 Co-evolution of vegetation and soil development during sand-dune stabilization by artificial vegetation planting. (a) changes in landscape ecosystem in the processes of shifting-sand fixation: a) shifting dunes; b) stabilized sand with straw checkerboards and vegetation; c) xerophytic shrub *C. korshinskii*; and d) biological soil crusts. (b) Changes in soil profile morphology on stabilized sand dunes with different ages of dune stabilization, where depth of the soil crust layer (denoted above the blue line) increased with age: a) leeward slope dunes stabilized in 1981; b) leeward slope dunes stabilized in 1965; c) leeward slope dunes stabilized in 1956; d) windward slope dunes stabilized in 1956; e) and hollow dunes stabilized in 1956. (c) Average soil moisture content from July to October 1998 at various dunes stabilized in different years, showing increased soil profile moisture storage with age of sand-dune stabilization. *(Modified from Xiao et al., 2009) (Color version online and in color plate)*

scoparium–C. korshinskii–Artemisia ordosica gradually evolved into a semi-natural community of *A. ordosica–C. korshinskii–Bassia dasyphylla–Eragrostis poaeoides* (Xiao et al., 2009). Correspondingly, soils there evolved from Entisols into Aridisols. The mean soil particle size changed from >0.2 to <0.08 mm in 0–20 cm soil depth with the succession from cultivated plants to natural vegetation. The profile of aeolian-sand soil in the cultivated vegetation area can be divided into a crust, a transitional layer, and the original shifting-sand layer (Fig. 5b). The soil crust is a little gray in color (10 YR 6/2–6/4) and mainly composed of silt and fine sand, and the transitional layer below is virtually the sand layer formed before being rehabilitated, with yellowish-brown color (10 YR 5/4–5/6), 15–20 cm thick, and has some roots (Fig. 5b). The soil crusts have a weak stratified structure formed by blowing sand, dustfall, and litter after the sandy land was rehabilitated by vegetation. The crusts around shrubs and herbs are covered by 2–8 mm of litter with lichen and mosses. Beneath it is an organic matter-stained layer, 5–10 cm in thickness. It has a compact and lumpy structure and contains large amounts of plant roots and undecomposed litter, and microorganism numbers in vegetated soil are a few tens of times higher than that in original shifting sand (Duan et al., 2004). The depth of the soil crust increases with ages of sand stabilization (Fig. 5b). The presence of the soil crust, especially moss and lichen crust, results in a remarkable reduction in infiltration as compared to mobile sand dune (Li et al., 2010), which would alter patterns of soil-water storage, increasing the moisture content near the soil surface while decreasing the moisture content in deeper layers (Fig. 5c). This favors shallow-rooted grasses and forbs over perennial shrubs. Li et al. (2010) reported that changes in the soil moisture pattern induced shifting of sand-binding vegetation from xerophytic shrub communities with higher coverage (35%) to complex communities dominated by shallow-rooted herbaceous species with low shrub coverage (9%). Thus, we speculate that the aboveground vegetation co-evolves with the belowground soil development during the processes of shifting-sand stabilization, indicating the importance for coupling ecohydrology and hydropedology at the different temporal scales.

In arid and semi-arid regions, encroachment of woody shrubs is common for grassland degradation (Asner et al., 2004). Invasion of grasslands by woody perennials may involve interactions between human-mediated factors, such as domestic livestock grazing and fire suppression, and natural variability in temperature and rainfall (McPherson, 1997). Potts et al. (2010) stated that the spatial and temporal dynamics of soil moisture in above- and belowground are altered by woody plants. Aboveground, woody plant canopies interact with characteristics of rainfall, such as rainfall intensity, to influence patterns of soil-water content through rainfall interception and shading the soil surface (Dunkerley, 2000). Woody plant litter that accumulates on the soil surface may increase soil-water infiltration (Ludwig et al., 2005) and reduce soil temperatures (Breshears et al., 1998), thereby prolonging periods of increased soil water. Belowground, the greater functional rooting depth of woody plants may decrease deep soil moisture drainage (Seyfried and Wilcox, 2006), while uptake and hydraulic redistribution by roots may further alter temporal variation of soil-water content beneath woody plants (Scott et al., 2008). Although individually these mechanisms are well described, the ways in which they interact to influence the overall patterns of soil moisture through time at different depths and between neighboring landscape patches remain poorly understood (Potts et al., 2010).

The above sand-dune stabilization and encroachment of woody shrubs into grasslands show that vegetation–soil feedbacks evolve with time. The feedback pathways toward positive or negative direction and the underlying mechanisms require further research. Coupling ecohydrology and hydropedology could offer a more systematic approach to study the co-evolution of vegetation change and soil development at different time scales. This has significant implications for ecological recovery under desertification.

7. SUMMARY AND FUTURE OUTLOOKS

This chapter has discussed key linkages between ecohydrology and hydropedology across multiple scales from the microscopic to the hillslope and catchment scales. Nonlinear dynamics across a range of scales show the changing dominant controls and the importance of interfaces for bridging scales. At the soil pore and stomata scale, the upward water flow from soils to roots and to leaves is controlled by soil properties, root morphology, plant species, and weather condition; at the individual plant scale, canopy rainfall partition and its contribution to water flow in the soil are controlled by rainfall characteristics, canopy architecture, plant variety, and soil types; at the patch scale, the distribution of soil water by patchy vegetation is controlled by soil properties, plant species, landscape features, and rainfall variability; at the hillslope and catchment scales, surface and subsurface flow distributions are affected by the variation in topography, soil conditions, vegetation patterns, hydrologic connectivity, disturbance, and management regimes. Cross-scale interactions among different linkages, however, remain uncertain and require further investigations. A challenge for bridging ecohydrology and hydropedology is to identify self-organizing principles and patterns across scales that can link aboveground and underground phenomena in the Critical Zone.

Re-vegetation for sand-dune stabilization and encroachment of woody shrubs into grasslands demonstrate the necessity for connecting ecohydrology and hydropedology at different temporal scales. Disappearance of woody plants in the process of sand-dune stabilization and encroachment of woody shrubs during grassland degradation suggest that there are likely self-organized and co-evolution patterns for vegetation succession and soil development. These can be used for restoration or remediation of degraded landscapes. The underlying mechanism and key interacting controls for vegetation degradation and recovery, however, require further research that would benefit from coupled ecohydrology and hydropedology.

There are other linkages between ecohydrology and hydropedology that are not discussed in this chapter, such as linking point measurements in soil profiles and individual plants to soil–landscape patterns and remote sensing footprints, characterizing soils associated with landform units to interpret and model hydrologic and ecosystem functions, connecting fire and drought stressors with ecohydrology–hydropedology interactions, and developing integrated models to explain vegetation response to soil heterogeneity under changing climate and land use. Overall, future research will need to permit a more comprehensive integration of ecohydrology and hydropedology, including:

- A fuller and more quantitative coupling of aboveground and belowground biomass distribution and its links to energy and water fluxes;
- Establishing linkages between observed vegetation and soil patterns across scales and their underlying controls;
- Elucidating co-evolution of soil development and vegetation succession over time; and
- Development, calibration, and validation of integrated models to couple aboveground and belowground phenomena across spatial and temporal scales.

ACKNOWLEDGMENTS

The study was supported by the National Science Foundation of China (NSFC 41025001 and 41130640), the Fundamental Research Funds for the Central Universities, and PCSIRT (No. IRT1108).

REFERENCES

Aguiar, M.R., Sala, O.E., 1999. Patch structure, dynamics and implications for the functioning of arid ecosystems. Tree 14 (7), 273–277.

Amenu, G.G., Kumar, P., 2008. A model for hydraulic redistribution incorporating coupled soil-root moisture transport. Hydrol. Earth Syst. Sci. 12, 55–74.

Asner, G.P., Elmore, A.J., Olander, L.P., Martin, R.E., Harris, A.T., 2004. Grazing systems, ecosystem responses, and global change. Annu. Rev. Environ. Resour. 29, 261–299.

Bever, J.D., 1994. Feedback between plants and their soil communities in an old field community. Ecology 75, 1965–1977.

Bisigato, A.J., Villagra, P.E., Ares, J.O., Rossi, B.E., 2009. Vegetation heterogeneity in Monte Desert ecosystems: a multi-scale approach linking patterns and processes. J. Arid Environ. 73, 182–191.

Bonell, M., 2002. Ecohydrology- a completely new idea? Hydrol. Sci. J. 47, 809–810.

Breshears, D.D., Zou, C.B., 2006. Bridging from ecology to hydrology and soil science through ecohydrology and hydropedology: the fundamental role of woody plant patterns on heterogeneity in soil water and carbon. American Geophysical Union 2006 Fall Meeting, 11–15 December, 2006 in San Francisco.

Breshears, D.D., Nyhan, W., Heil, C.E., Wilcox, B.P., 1998. Effects of woody plants on microclimate in a semiarid woodland: soil temperature and evaporation in canopy and open patches. Int. J. Plant Sci. 159, 1010–1017.

Brooks, K.N., Ffoiliott, P.F., Gregersen, H.M., Thames, J.L., 1991. Hydrology and the Management of Watersheds. Iowa State University Press, Ames, IA.

Brooks, J.R., Meinzer, F.C., Coulombe, R., Gregg, J., 2002. Hydraulic redistribution of soil water during summer drought in two contrasting Pacific Northwest coniferous forests. Tree Physiol. 22, 1107–1117.

Brooks, J.R., Barnard, H.R., Coulombe, R., McDonnell, J.J., 2010. Ecohydrologic separation of water between trees and streams in a Mediterranean climate. Nat. Geosci. 3, 100–104.

Caldwell, M.M., Dawson, T.E., Richards, J.H., 1998. Hydraulic lift: consequences of water efflux from the roots of plants. Oecologia 113, 151–161.

De Michele, C., Vezzoli, R., Pavlopoulos, H., Scholes, R.J., 2008. A minimal model of soil water-vegetation interactions forced by stochastic rainfall in water-limited ecosystem. Ecol. Model. 212, 397–407.

Deurer, M., Green, S.R., Clothier, B.E., Böttcher, J., Duijnisveld, W.H.M., 2003. Drainage networks in soils: a concept to describe bypass-flow pathways. J. Hydrol. 272, 148–162.

Devitt, D.A., Smith, S.D., 2002. Root channel macrospores enhance downward movement of water in a Mojave desert ecosystem. J. Arid Environ. 50, 99–108.

D'Odorico, P., Porporato, A., 2006. Ecohydrology of arid and semiarid ecosystems: an introduction. In: D'Odorico, P., Porporato, A. (Eds.), Dryland Ecohydrology. Springer, Dordrecht, The Netherlands, pp. 1–10.

Duan, Z.H., Xiao, H.L., Li, X.R., Dong, Z.B., Wang, G., 2004. Evolution of soil properties on stabilized sands in the Tengger Desert, China. Geomorphology 59, 237–246.

Dunkerley, D., 2000. Hydrologic effects of dryland shrubs: defining the spatial extent of modified soil water uptake rates at an Australian desert site. J. Arid Environ. 45, 159–172.

Galle, S., Brouwer, J., Delhoume, J.P., 2001. Soil water balance. In: Tongway, D.J., et al. (Eds.), Band Vegetation Patterning in Arid and Semiarid Environments. Ecological Studies, 149. Springer-Verlag, New York, pp. 77–104.

Hewlett, J.D., Hibbert, A.R., 1967. Factors affecting the response of small watersheds to precipitation in humid areas. In: Sopper, W.E., Lull, H.W. (Eds.), Forest Hydrology. Pergamon Press, N.Y., pp. 275–290.

Hopmans, J.W., Bristow, K.L., 2002. Current capabilities and future needs of root water and nutrient uptake modeling. Adv. Agron. 77, 103–183.

Horton, J.H., Hawkins, R.H., 1965. Flow path of rain from soil surface to water table. Soil Sci. 100, 377–383.

Horton, R.E., 1933. The role of infiltration in the hydrologic cycle. Trans. Amer. Geophys. Union 14, 446–460.

Howes, D.A., Abrahams, A.D., 2003. Modeling runoff and runon in a desert shrubland ecosystem, Jornada Basin, New Mexico. Geomorphology 53, 45–73.

Jarvis, P.G., McNaughton, K.G., 1986. Stomatal control of transpiration: scaling up from leaf to region. Adv. Ecol. Res. 15, 1–49.

Johnson, M.S., Lehmann, J., 2006. Double-funneling of trees: stemflow and root-induced preferential flow. Ecoscience 13 (3), 324–333.

Juniper, B.E., Jeffree, C.E., 1983. Plant Surfaces. Edward Arnold, London.

Jury, W.A., Gardner, W.R., Gardner, W.H., 1991. Soil Physics, fifth ed. John Wiley and Sons, Inc., New York.

Katul, G., Novick, K., 2009. Evapotranspiration. In: Likens, G.E. (Ed.), Encyclopedia of Inland Waters, vol. 1. Elsevier, Oxford, pp. 661–667.

Katul, G., Porporato, A., Oren, R., 2007. Stochastic dynamics of plant-water interactions. Annu. Rev. Ecol. Evol. Syst. 38, 767–791.

Kundzewicz, Z.W., 2002. Ecohydrology- seeking consensus on interpretation of the notion. Hydrol. Sci. J. 47, 799–804.

Levia, D.F., Frost, E.E., 2003. A review and evaluation of stemflow literature in the hydrologic and biogeochemical cycles of forested and agricultural ecosystems. J. Hydrol. 274, 1–29.

Li, X.J., Li, X.R., Song, W.M., Gao, Y.P., Zhang, J.G., Jia, R.L., 2008a. Effects of crust and shrub patches on runoff, sedimentation, and related nutrient (C, N) redistribution in the desertified steppe zone of the Tengger Desert, Northern China. Geomorphology 96, 221–232.

Li, X.Y., Liu, L.Y., Gao, S.Y., Ma, Y.J., Yang, Z.P., 2008b. Stemflow in three shrubs and its effect on soil water enhancement in semiarid loess region of China. Agric. For. Meteorol. 148 (10), 1501–1507.

Li, X.Y., Yang, Z.P., Li, Y.T., Lin, H., 2009. Connecting ecohydrology and hydropedology in desert shrubs: stemflow as a source of preferential flow in soils. Hydrol. Earth Syst. Sci. 13, 1133–1144.

Li, X.R., Tian, F., Jia, R.L., Zhang, Z.S., Liu, L.C., 2010. Do biological soil crusts determine vegetation changes in sandy deserts? Implications for managing artificial vegetation. Hydrol. Process. DOI: 10.1002/hyp.7791.

Li, X.Y., 2011. Hydrology and biogeochemistry of semiarid and arid regions. In: Levia, D.F., et al. (Eds.), Forest Hydrology and Biogeochemistry: Synthesis of Past Research and Future Directions. Ecological Studies Series, vol. 216, 285–299. Springer-Verlag, Heidelberg, Germany.

Liang, Y.M., Hazlett, D.L., Lauenroth, W.K., 1989. Biomass dynamics and water use efficiencies of five plant communities in the shortgrass steppe. Oecologia 80, 148–153.

Lin, H.S., Bouma, J., Wilding, L., Richardson, J., Kutilek, M., Nielsen, D., 2005. Advances in hydropedology. Adv. Agron. 85, 1–89.

Lin, H.S., Bouma, J., Pachepsky, Y., Western, A., Thompson, J., van Genuchten, M. Th., Vogel, H., Lilly, A., 2006. Hydropedology: synergistic integration of pedology and hydrology. Water Resource Res. 42, W05301. doi:10.1029/2005WR004085.

Lin, H.S., Brook, E., McDaniel, R., Boll, J., 2008. Hydropedology and surface/subsurface runoff processes. In: Anderson, M.G. (Ed.), Encyclopedia of Hydrologic Sciences. JohnWiley & Sons, Ltd. doi:10.1002/0470848944.hsa306.

Lin, H.S., 2003. Hydropedology: bridging disciplines, scales, and data. Vadose Zone. J. 2, 1–11.

Lin, H.S., 2010a. Earth's Critical Zone and hydropedology: concepts, characteristics, and advances. Hydrol. Earth Syst. Sci. 14, 25–45.

Lin, H.S., 2010b. Linking principles of soil formation and flow regimes. J. Hydrol. 393, 3–19. Journal of Hydrology. doi:10.1016/j.jhydrol.2010.02.013.

Llorens, P., Domingo, F., 2007. Rainfall partitioning by vegetation under Mediterranean conditions. A review of studies in Europe. J. Hydrol. 335, 37–54.

Ludwig, J.A., Wilcox, B.P., Breshears, D.D., Tongway, D.J., Imeson, A.C., 2005. Vegetation patches and runoff-erosion as interacting ecohydrological processes in semiarid landscapes. Ecology 86 (2), 288–297.

Ma, W.H., Yang, Y.H., He, J.S., Zeng, H., Fang, J.Y., 2008. Above- and belowground biomass in relation to environmental factors in temperate grasslands, Inner Mongolia. Sci. China Ser. C-Life Sci. 51 (3), 263–270.

Martinez, J.P., Silva, H., Ledent, J.F., Pinto, M., 2007. Effect of drought stress on the osmotic adjustment, cell wall elasticity and cell volume of six cultivars of common beans (Phaseolus vulgaris L.). Eur. J. Agron. 26, 30–38.

Martinez-Meza, 1994. Stemflow, Throughfall and Root Water Channelization by Three Arid Land Shrubs in Southern New Mexico. Ph.D. diss. New Mexico State University, Las Cruces, New Mexico, USA.

McClain, M.E., Boyer, E.W., Dent, C.L., Gergel, S.E., Grimm, N.B., Groffman, P.M., Hart, S.C., Harvey, J.W., Johnston, C.A., Mayorga, E., McDowell, W.H., Pinay, G., 2003. Biogeochemical hot spots and hot moments at the interface of terrestrial and aquatic ecosystems. Ecosystems 6, 301–312. DOI: 10.1007/s10021-003-0161-9.

McPherson, G.R., 1997. Ecology and Management of North American Savannas. University of Arizona Press, Tucson, AZ.

Milburn, J.A., 1979. Water Flow in Plants. Longman, New York.

National Research Council (NRC), 2001. Basic Research Opportunities in Earth Science. National Academy Press, Washington DC, USA.

Navar, J., 1993. The causes of stemflow variation in three semi-arid growing species of northeastern Mexico. J. Hydrol. 145, 165–190.

Newman, B.D., Wilcox, B.P., Archer, S.R., Breshears, D.D., Dahm, C.N., Duffy, C.J., McDowell, N.G., Phillips, F.M., Scanlon, B.R., Vivoni, E.R., 2006. Ecohydrology of water-limited environments: a scientific vision. Water Resour. Res. 42, W06302. doi:10.1029/2005WR004141.

Newman, B.D., Breshears, D.D., Gard, M.O., 2010. Evapotranspiration partitioning in a semiarid woodland: ecohydrologic heterogeneity and connectivity of vegetation patches. Vadose Zone J. 9, 561–572.

Noy-Meir, I., 1973. Desert ecosystems: environment and producers. Ann. Rev. Ecol. Syst. 4, 25–51.

Nuttle, W.K., 2002. Is ecohydrology one idea or many? Hydrol. Sci. J. 47, 805–807.

Peters, D.P.C., Havstad, K.M., 2006. Nonlinear dynamics in arid and semi-arid systems: interactions among drivers and processes across scales. J. Arid Environ. 65, 196–206.

Phillips, F.M., 2010. Soil-water bypass. Nat. Geosci. 3, 77–78.

Porporato, A., Rodriguez-Iturbe, I., 2002. Ecohydrology - a challenging multidisciplinary research perspective. Hydrol Sci. J. 47, 811–821.

Potts, D.L., Scott, R.L., Bayram, S., Carbonara, J., 2010. Woody plants modulate the temporal dynamics of soil moisture in a semi-arid mesquite savanna. Ecohydrol. 3, 20–27.

Puigdefábregas, J., 2005. The role of vegetation patterns in structuring runoff and sediment fluxes in drylands. Earth Surf. Process. Landforms 30, 133–147.

Rango, A., Tartowski, S.L., Laliberte, A., Wainwright, J., Parsons, A., 2006. Island of hydrologically enhanced biotic productivity in natural and managed arid ecosystems. J. Arid Environ. 65, 235–252.

Reynolds, J.F., Virginia, R.A., Kemp, P.A., de Soyza, A.G., Tremmel, D.C., 1999. Impact of drought on desert shrubs: effects of seasonality and degree of resource island development. Ecol. Monogr. 69, 69–106.

Rietkerk, M., Dekker, S.C., de Ruiter, P.C., van de Koppel, J., 2004. Self-organized patchiness and catastrophic shifts in ecosystems. Science 305, 1926–1929.

Rodriguez-Iturbe, I., 2000. Ecohydrology: a hydrologic perspective of climate-soil-vegetation dynamics. Water Resour. Res. 36, 3–9.

Schlesinger, W.H., Pilmanis, A.M., 1998. Plant–soil interactions in deserts. Biogeochemistry 42, 169–187.

Scholz, F.G., Bucci, S.J., Hoffmann, W.A., Meinzer, F.C., Goldstein, G., 2010. Hydraulic lift in a Neotropical savanna: experimental manipulation and model simulations. Agric. For. Meteorol. 150, 629–639.

Scott, R.L., Cable, W.L., Hultine, K.R., 2008. The ecohydrologic significance of hydraulic redistribution in a semiarid savanna. Water Resour. Res. 44, W02440. DOI: 10Ð1029/2007WR006149.

Seyfried, M.S., Wilcox, B.P., 2006. Soil water storage and rooting depth: key factors controlling recharge on rangelands. Hydrol. Process. 20, 3261–3275.

Siqueira, M., Katul, G., Porporato, A., 2008. Onset of water stress, hysteresis in plant conductance, and hydraulic lift: scaling soil water dynamics from millimeters to meters. Water Resour. Res. 44, W01432. doi:10.1029/2007WR006094.

Spence, R.D., Wu, H., Sharpe, P.J.H., Clark, K.G., 1986. Water stress effects on guard cell anatomy and the mechanical advantage of the epidermal cells. Plant Cell Environ. 9, 197–202.

Tromp-van Meerveld, H.J., McDonnell, J.J. On the interrelations between topography, soil depth, soil moisture, transpiration rates and species distribution at the hillslope scale. Advances in Water Resources 29, 293–310.

Tyree, M.T., Zimmermann, M.H., 2002. Xylem Structure and the Ascent of Sap, second ed. Springer, New York.

Uhlenbrook, S., 2006. Catchment hydrology-a science in which all processes are preferential. Hydrol. Process. 20, 3581–3585.

Ursino, N., 2009. Above and below ground biomass patterns in arid lands. Ecol. Model. 220, 1411–1418.

Wardle, D.A., Bardgett, R.D., Klironomos, J.N., Setälä, H., van der Putten, W.H., Wall, D.H., 2004. Ecological linkages between aboveground and belowground biota. Science 304, 1629–1633.

Weiler, M., McDonnell, J.J., 2007. Conceptualizing lateral preferential flow and flow networks and simulating the effects on gauged and ungauged hillslopes. Water Resour. Res. 43, W03403. DOI:10.1029/2006WR004867.

Wilcox, B.P., Breshears, D.D., Allen, C.D., 2003. Ecohydrology of a resource-conserving semiarid woodland: temporal and spatial scaling and disturbance. Ecol. Monogr. 73, 223–239.

Woodward, F.I., 1995. Ecophysiological controls of conifer distributions. In: Smith, W.K., Hinckley, T.M. (Eds.), Ecophysiology of Coniferous Forests. Academic Press, San Diego, pp. 79–94.

Xiao, H.L., Ren, J., Li, X.R., 2009. Effects of soil-plant system change on Eco-hydrology during Revegetation for mobile dune stabilization, Chinese arid desert. Sci. Cold Arid Regions 1 (3), 230–237.

Yair, A., Lavee, H., 1985. Runoff generation in arid and semi-arid zones. In: Anderson, M.G., Burt, T.P. (Eds.), Hydrological Forecasting. Wiley, Chichester, pp. 183–120.

Yang, Y.H., Fang, J.Y., Ma, W.H., Guo, D.L., Mohammat, A., 2010. Large-scale pattern of biomass partitioning across China's grasslands. Glob. Ecol. Biogeogr. 19, 268–277.

Young, M.H., Robinson, D.A., Ryel, R.J., 2010. Introduction to coupling soil science and hydrology with ecology: toward integrating landscape processes. Vadose Zone J. 9, 515–516.

Zalewski, M., 2002. Ecohydrology- the use of ecological and hydrological processes for sustainable management of water resources. Hydrol. Sci. J. 47, 823–832.

Chapter 24

Hydropedology: Summary and Outlook

Henry Lin[1]

ABSTRACT

This chapter summarizes the main messages of this book and highlights future opportunities and challenges for advancing hydropedology. The overview chapters in Part I of this book have discussed important foundations of hydropedology; the case studies in Part II have demonstrated the necessity and benefits of integrating pedologic and hydrologic expertise in landscape studies; and the perspectives in Part III have illustrated opportunities and challenges in hydropedology. Important issues for future progress in hydropedology include the need to 1) overcome conceptual and technological bottlenecks; 2) develop holistic, multiscale, and quantitative modeling frameworks; 3) build a global alliance of integrated databases and observatories; 4) apply hydropedology in societally important issues; and 5) improve interdisciplinary education of the next generation of soil scientists and hydrologists.

1. INTRODUCTION

"Our own civilization is now being tested in regard to its management of water as well as soil."

– Daniel Hillel (1991)

Hydropedology is an emerging interdisciplinary science with many unique aspects, as illustrated in this book, including:

- An emphasis on in situ soil architecture and the soil–landscape relationship to better understand interactive pedologic and hydrologic processes across scales;
- A focus on water storage and movement as the central thread for understanding complex physical, chemical, and biological processes in real-world soils;
- A holistic, systems approach to connect fast and slow soil processes (temporal domain) and to bridge soil structural and landscape units (spatial domain);

[1]Dept. of Ecosystem Science and Management, Pennsylvania State Univ., University Park, PA, USA. Email: henrylin@psu.edu

Hydropedology, Edited by H. Lin. DOI: 10.1016/B978-0-12-386941-8.00024-1

- An integration of soil formation and soil function, and a combined geographic and functional characterization of soils in diverse landscapes;
- An integrated, iterative system of mapping, monitoring, and modeling (3 M) for improved understanding and prediction of complex landscape–soil–water–vegetation relationships; and
- A foundation for ecosystem services and environmental sustainability.

While significant progress has been made in understanding flow and transport processes in soils from the pore to the watershed scales, fundamental governing principles in complex and heterogeneous real-world soils remain far from perfect. Consequently, our ability to realistically predict flow and transport in field soils remains limited. There is a pressing need to uncover new principles of flow and transport in real-world soils and to use this new foundation to develop the next generation of multiscale, distributed models that are capable of simulating fast and slow processes in heterogeneous but structured soils and landscapes.

This book has attempted to provide some new perspectives and illustrations of the emerging hydropedology paradigm, which seeks to integrate pedology (a unique subdiscipline of soil science and geoscience) with soil physics, hydrology, and other related bio- and geosciences to advance the systems understanding of complex landscape–soil–water–vegetation relationships. The main messages of this book are summarized in the following section, and possible paths to advance hydropedology are discussed in the subsequent section.

2. SUMMARY OF THE BOOK CHAPTERS

The core messages conveyed in Part I (overview chapters) and Part III (perspective chapters) of this book are summarized in Table 1. These chapters have discussed various aspects of the two fundamental questions of hydropedology: 1) the role of soil architecture and the soil–landscape relationship in hydrologic processes and 2) the role of hydrology in soil formation and soil function. The following themes have been particularly highlighted:

- The critical importance of in situ soil architecture across scales, and the intimate relationship between soil types and landscape settings (all Chapters).
- The significance and widespread occurrence of preferential flow in real-world soils (Chapters 1–4, 8–10, 12–18, and 22–23).
- The contribution of soil hydromorphology to infer in situ soil hydrology, and the importance of wet soils (including subaqueous soils) for carbon sequestration and other important applications (Chapters 5–6, 11, and 16–17).
- The fundamental control of topography and energy on hillslope processes and Critical Zone dynamics (Chapters 1–2, 7–14, and 16–23).
- The major influence of soil types and soil parameters in hydrologic modeling (Chapters 1–4, and 16–18).

- The essential role of hydrology in pedogenesis, soil modeling, and soil mapping (Chapters 1–2, 5–6, 9–14, and 19–21).
- The necessity of coupling hydropedology with other related interdisciplinary sciences (such as biogeochemistry and ecohydrology) as well as technological advances (such as geophysical tools and digital soil mapping) (Chapters 1–2, 7–8, 13–15, and 20–23).

Six case studies presented in Part II of this book are summarized in Table 2, which include 7 out of 12 soil orders in the U.S. *Soil Taxonomy* (Fig. 1). These case studies represent many years to decades of research in diverse soils and landscapes. The common theme of these case studies is the complex landscape–soil–water–vegetation relationship in the field setting. In each of these landscapes, geology, geomorphology, and soils are different and provide unique structural controls on surface and subsurface hydrology. Vegetation is the outcome of soils and hydrologic conditions, which has strong feedbacks that result in the co-evolution of the ecosystem. Development of landscape-based hydropedologic understanding of ecosystems has significant implications for effective land and water management, as demonstrated in these case studies.

In addition, Chapter 15 presents the application of hydropedology as a powerful tool for environmental policy and regulation (Table 1). Using several examples, this chapter has demonstrated that by integrating pedology and soil physics/hydrology, as proposed in the hydropedology paradigm, research can make a more effective contribution to environmental and sustainability issues than what has been provided in the past by traditional scientific fields.

Various chapters of this book may also be synthesized through the spectrum of soil wetness from subaqueous soils to dry soils, as shown in Fig. 2. Different soil moisture regimes, as defined in the U.S. *Soil Taxonomy* (Soil Survey Staff, 2010), can help understand some general trends of soil features with changing soil wetness conditions (Fig. 2). However, for most of the 12 soil orders, there are a range of soil moisture regimes possible within a soil order (Fig. 1), which depend on both regional climatic regime and local landscape setting. Thus, soil moisture regimes have been prominently used in various soil classification systems to distinguish diverse soil types, especially at the higher taxonomic levels such as orders, suborders, and great groups in the *Soil Taxonomy.*

3. OUTLOOK FOR ADVANCING HYDROPEDOLOGY

Like any emerging interdisciplinary science, there are both opportunities and challenges for its advancement. Some outlooks for advancing hydropedology are discussed below, which highlight selected key issues of bridging disciplines, scales, data, applications, and education. This discussion is intended to be illustrative rather than exhaustive, with the hope that this can stimulate more related discussions in the scientific community.

TABLE 1 Main Messages from Each Chapter of this Book, Except the Case Studies and the 1st and the Last Chapters

Chapter #	Focus	Main Points	Future Needs
Part I Overviews and Fundamentals			
2	Understanding soil architecture across scales	• Soils have complex architectural organizations as a result of natural soil-forming processes and anthropogenic impacts, which exhibit hierarchical organizations and encompass interlinked parts of solids, pores, and interfaces at each scale. • Two general levels can be used to facilitate the understanding and linkage of various scales of soil organization: 1) soil architecture within a soil profile (termed soil structural units, such as particles, pores, aggregates, horizons, and pedons) and 2) soil architecture in the landscape (termed soil-landscape units, such as catenae, soilscapes, soil sequences, and soil zones).	• A new era of soils research must be based on soil architecture that is the backbone of soil functions across scales in the landscape. • Continued technological and theoretical advances are needed to enhance in situ and noninvasive imaging/mapping and measurement of soil architecture and to quantitatively link soil architectural parameters to various soil functions.
3	Preferential flow in a pedological perspective	• Dominant flow and transport regimes are identified in the major soil types of the world, with spatial patterns of preferential flow at the landscape scale being far from random, but rather, having a clear deterministic component because of soil patterns. • An improved decision tree to classify susceptibility to macropore flow based on soils and site features is presented.	• Data for risk mapping and modeling are often too coarse for precise location of risk areas, thus combining traditional data sources with novel remote and proximal sensing and fuzzy classification can improve the spatial resolution and robustness of risk mapping. • Hydropedologic approaches offer considerable promise in supporting the parametrization of preferential flow and transport modeling.

4	Preferential flow dynamics and plant rooting systems	• A two-parameter Stokes flow is correlated with root density determined in situ in soils with stagnic properties. • Stokes approach to flow in permeable media is a candidate for dealing with preferential flow.	• Soil architecture can be inferred from the interpretation with Stokes flow theory of rapid infiltration and drainage at the soil profile scale. • In situ hydropedologic procedures are needed for parametrizing macropore flow at the soil profile scale.
5	Redoximorphic features as related to soil hydrology and hydric soil	• Hydromorphic features form under four specific conditions: organic matter must be present, microorganisms must be actively respiring and oxidizing organic matter, soil must be saturated, and dissolved oxygen must be removed. • There are seven rules for hydromorphic feature formation and interpretation that can be applied within a soil profile or toposequence to relate soil morphology to hydrology.	• Instrumentation must be in place in order to calibrate soil hydromorphic features to duration and frequency of saturation and reduction; • Relict hydromorphic features are difficult to identify and must be done cautiously.
6	Subaqueous soils: pedogenesis, mapping, and applications	• Subaqueous soils have been officially recognized and new taxa have been added to Entisols and Histosols in Soil Taxonomy to accommodate these soils. • Subaqueous soil surveys have become a resource management tool guiding use and management decisions for coastal, estuarine, and littoral ecosystems.	• Additional studies in different climates and settings should be undertaken to test subaqueous soils and further develop the link between these soils and particular interpretations. • Additional work is especially needed within the range of freshwater systems.
7	Quantifying processes governing soil-mantled hillslope evolution	• A diverse toolboxes are available for measuring physical and chemical weathering and soil erosion rates.	• Increase in overland flow downslope may mean that parent material beneath thicker soils is in contact with soil water that has traveled farther through soil and is not as able

(Continued)

TABLE 1 Main Messages from Each Chapter of this Book, Except the Case Studies and the 1st and the Last Chapters—Cont'd

Chapter #	Focus	Main Points	Future Needs
		• In the uplands studied, transition from dominant physical mixing to a dominance of overland flow with increasing distance downslope corresponds to the transition from thin to thick soils, and parent material strength increases with overlying soil thickness.	to weather as the water reacting with the parent material beneath thinner, upslope soils. • Hydrologic pathways drive the weathering processes of upland soils and saprolites and provide greater insight into how hillslope processes drive landscape evolution.
8	Thermodynamic limits in the Critical Zone and its relevance to hydropedology	• Processes within the Critical Zone are intimately coupled to each other by the flow and transformation of energy, and thermodynamic laws provide fundamental insights into directions, limits, evolutionary dynamics, and the fundamental importance of structures. • Critical Zone processes and fluxes can be formulated in terms of different forms of energy using sets of conjugate variables.	• Free energy should be quantified when measuring and modeling Critical Zone processes, including the types of energy represented, converted, exchanged, and dissipated. • Free energy quantification provides a more profound understanding of the Critical Zone and how it interacts with the Earth system, thus enabling us to build better predictive models of the Critical Zone.
Part II Case Studies and Application			
15	Hydropedology as a powerful tool for environmental policy and regulations	• Integrated pedology and soil physics/hydrology approach, as proposed in hydropedology, is bound to be more effective in making significant contributions to the sustainability debate in society.	• Hydropedology research is needed on water regimes in undisturbed soils in the field, using an array of new monitoring and modeling technologies that do not implicitly assume soils to be isotropic and homogeneous.

		• The Soil Protection Strategy of the European Union can be used as a guide to define hydropedologic "niches" for research in terms of seven soil functions.	• Hydropedologists can put themselves in an attractive intermediate position by facilitating interaction between land users and policymakers.
Part III Advances in Modeling, Mapping, and Coupling			
16	Soil information in hydrologic models: hard data, soft data, and the dialog between experimentalists and modelers	• Experimentalists and modelers need to work together towards a better understanding and quantification of the first order control of catchment subsurface runoff by utilizing hard data, soft facts, tacit knowledge, and mapping in soil survey databases. • Soft data framework and flexible box models may be used to enhance the dialog between experimentalists and modelers.	• A systematic comparison of existing approaches to derive dominant runoff processes has not yet been published. • Up-scaling soil hydraulic properties based on point- or plot-scale measurements and interpolation or regionalization of point- and plot-scale measurements to locations with unknown soil parameters remain unclear.
17	Hydrological classifications of soils and their use in hydrological modeling	• Soil hydrologic classifications can make soil data more accessible to hydrologists by providing functional descriptions of water flow pathways through different soils and the dominant internal processes involved. • Existing soil hydrologic classifications are rudimentary and need further improvements.	• Challenge remains in translating soil hydrologic classifications into quantitative parameters required by catchment-scale hydrologic models. • Quantitative use of soil morphology is desirable to help identify and characterize soil functions, reduce required input data, and conceptualize flow pathways.

(*Continued*)

TABLE 1 Main Messages from Each Chapter of this Book, Except the Case Studies and the 1st and the Last Chapters—Cont'd

Chapter #	Focus	Main Points	Future Needs
18	Subsurface flow networks at the hillslope scale: detection and modeling	• There is ample evidence for widespread existence of hillslope-scale preferential flow networks under a variety of soil, topographic, climatic, and geologic conditions. • Current methods for modeling subsurface flow networks include topography-based flow routing, stochastic macropore modeling, and percolation-based approach.	• While trenching, irrigation and tracer experiments, and geophysical imaging have revealed some characteristics of subsurface flow networks, much remains to be done in diverse soils and landscapes to develop a network-based hydrologic model. • Better tools and techniques are needed for effective and efficient in situ determination of subsurface structures and flow dynamics with high spatial and temporal resolutions.
19	Hydrologic information in pedologic models	• Hydrology is a critical input in modeling pedogenesis as water is the main driving force of soil formation and evolution. • Two example pedogenesis models that combine geochemistry, hydrology, and plant growth modules can simulate the evolution of some soil characteristics on pedological time scale but are not able to simulate any type of soil evolution.	• The main limitations in pedogenic modeling are feedbacks, interactions, and cumulative effects between shorter- and longer-time scale changes. • Validation of pedogenesis modeling results is difficult as a long time scale is involved and long-term monitoring datasets of water quality and major soil elements are rare and generally incomplete.

20	Modeling and mapping soil spatial and temporal variability	• Tools useful for modeling soil spatial or temporal variability include geostatistics, similar media scaling, autoregressive moving averages, and state-space Kalman filtering. • Digital soil mapping can be improved using auxiliary data from landscape and terrain analysis, remote or proximal sensing, geostatistics, and pedotransfer functions.	• There is a significant need to develop methods for identifying functionally different soil-landscape units that can be used to improve the accuracy of hydrologic modeling. • Digital soil mapping is needed to provide high-quality, fine-resolution soil databases for modeling the spatial and temporal variability of soil functioning and ecosystem health.
21	Digital soil mapping: interactions with and applications for hydropedology	• A primary linkage between hydropedology and digital soil mapping is the role of water in pedogenesis and soil—landscape processes that strongly influence spatial variability in soil properties. • Digital soil mapping products can enhance the input to hydrologic and hydropedologic models and extrapolate outputs from these models.	• Creating synergies between hydropedology and digital soil mapping promotes new research endeavors and is expected to foster further advances in geospatial soil database development and process-based modeling. • There is a need for a greater understanding of the feedback between water redistribution on the landscape and its impact on soil properties.
22	Coupling biogeochemistry and hydropedology to advance carbon and nitrogen cycling science	• Hot moments (brief time periods for disproportionately large amounts of nutrient transport and transformation) represent times when coupled analyses of hydropedology and biogeochemistry are most tractable in the field and applicable to advancing scientific theories such as ecosystem N retention and solute flushing. • Location of biogeochemicals among flow paths with different transport rates plays an	• Future soil biogeochemistry research needs to continue to improve the integration with hydropedology by explicitly considering the interacting effects of soil structure and hydrology. • Future soil research will benefit from stream ecology and biogeochemistry principles (e.g. nutrient spiraling) that have a rich history of exploring interactions between material transport and transformation.

(*Continued*)

TABLE 1 Main Messages from Each Chapter of this Book, Except the Case Studies and the 1st and the Last Chapters—Cont'd

Chapter #	Focus	Main Points	Future Needs
		important role in understanding variations in biogeochemical transport and transformations.	
23	Coupling ecohydrology and hydropedology at different spatio-temporal scales in water-limited ecosystems	• Key connections between ecohydrology and hydropedology include 1) water flow from soil pore to plant stomata at the microscopic scale; 2) aboveground canopy precipitation partitioning and belowground preferential flow at the individual plant scale; 3) vegetation pattern and soil moisture distribution at the patch scale; and 4) surface and subsurface flow networks at the hillslope and catchment scales. • Co-evolution of soil and vegetation during land degradation and rehabilitation shows linked ecohydrology and hydropedology across temporal scales.	• Bridging multiple scales and identifying self-organizing structures between aboveground and underground processes remain a challenge for improved understanding of coupled soil hydrology and ecosystems. • A fuller and more quantitative coupling of aboveground and belowground patterns and processes and their links to energy and water fluxes are desirable and beneficial.

3.1. Conceptual and Technical Bottlenecks

There are three important bottlenecks that need to be overcome for advancing hydropedology, which are briefly summarized in the following:

- *On the conceptual/theoretical front*: Going beyond small-scale physics and embracing a hierarchical framework is important to connect point-based observations to landscape processes. As Beven (2006) noted, "*Nearly all hydrologic, water quality, and sediment transport models use the same small scale laboratory homogeneous domain theory to represent integrated fluxes at the much larger scales of hillslope and catchment... This is the root of many discrepancies between model predictions and the reality.*" A possible paradigm shift from a continuum-based approach to a hierarchy-based approach, as discussed by Lin (2010a) and addressed in Chapter 2 of this book, could be one of the first steps to enhance the understanding, modeling, and scaling of subsurface hydrologic and pedologic processes. Soil physics and hydrology has traditionally applied findings from fluid mechanics, together with necessary constitutive relations, to develop sets of governing equations (much in the same way as atmospheric and ocean sciences have done). However, heterogeneities, structures, interfaces, roughness, and organisms in complex soil systems make the real-world soils deviate significantly from the continuum assumption. Complex networks of preferential flow, for example, pose a number of challenges to the *status quo* (Lin, 2010a).
- *On the time scale*: The reconciliation of fast and slow changes (e.g. biological vs. geological processes) is essential to understanding and modeling complex soil systems and their dynamics under changing climate and land use. As discussed in Chapter 1 of this book, fast and slow changes in multiphase soil systems are intertwined, with interactions and feedbacks between 1) fast and cyclic soil functioning processes involving mostly liquids, gases, and biota, and 2) slow and irreversible specific pedogenic processes involving predominantly solids. The principle of conservation plus evolution, as suggested by Lin (2011), can be used to reconcile these fast and slow changes over various time scales, where incomplete closure and partial irreversibility of many cyclic processes of soil functioning produce a range of residual solid products that are accumulated over time, giving rise to structured and informative soil profiles. Anthropogenic impacts have added another layer of complexity in understanding the coupled human–natural systems as the human time scale is quite different from the natural time scales.
- *On the technological front*: Landforms and vegetation can now be mapped with high resolution (e.g. using LiDAR and the WorldView satellite, respectively), but noninvasive and precision mapping of the subsurface across space and time remains a significant gap. Beven (2006) pointed out that the most important need to advance hydrology of the 21st century is to

TABLE 2 Summary of the Main Findings from Each Case Study Presented in Part II of this Book

Chapter #	Location	Focused feature	Soil orders	Landscape–soil–water–vegetation relationship	Implications/applications
9	Central Texas, USA	Caliche soils weathered from limestone and their catena in stepped riser–tread micro-terrains	Inceptisols, Mollisols	• Changes along the stepped riser–tread catena include decreases in vegetative cover, slope gradient, soil thickness, infiltration rate, water-storage capacity, biologic activity, and soil organic carbon content, but with increases in surface runoff and erosion. • Riser and tread microforms are independent hydropedologic units on stepped hillslopes, with long mean residence times of water and solutes, multiple cycles of water storage and bioremediation, on-land retention of eroded sediments, and high consumptive use by plants.	• Integrated landscape-based hydropedologic approach offers important insights into the real-world heterogeneous subsurface. • Interactive pedologic, geologic, biologic, and hydrologic processes across the landscape address the effectiveness of caliche soils for the treatment and disposal of on-site wastewaters in environmentally sensitive limestone landscapes.
10	Pacific Northwest, USA	Seasonally dry landscapes with water-restricting subsoil layers	Alfisols, Mollisols	• Argillic and fragipan horizons in north-facing wetter microclimates lead to seasonal perched water and rapid subsurface lateral flow, accelerating clay eluviation in E horizons and driving variable source area hydrology.	• Soil scientists, hydrologists, and engineers have integrated knowledge of soil formation and morphology with the understanding of hydrologic processes at the plot/field scale to the hillslope/catchment scale to better manage water resources.

				• At the hillslope scale, soil structural features largely control subsurface water movement, and long-term patterns in soil water movement in turn largely control soil formation.	• Development of conceptual hydropedology model has had profound implications for generating effective solutions to regional erosion, water quality, and groundwater recharge problems.
11	North American perhumid coastal temperate rainforest, USA to Canada	Hydropedologic units of soil, hydrology, and vegetation linked to dissolved organic matter biogeochemistry	Histosols, Inceptisols, Spodosols	• The concentration and quality of dissolved organic matter exported from watersheds vary according to soil types that have formed along a topographic gradient from flat organic soils to steep mineral soils. • Hydropedologic units can be delineated based on soil, hydrologic, and vegetation features to describe biogeochemical cycles at the watershed scale.	• Hydropedology is a flexible approach that infuses soil and wetland delineations with function by providing a link between soil hydrology and biogeochemistry, providing an approach to scale up biogeochemical fluxes from hydropedologic units to the watershed scale. • Hydropedologic units provide a response variable for predicting watershed functional alterations from changes in land use and climate.
12	Central Pennsylvania, USA	Soil moisture pattern and preferential flow dynamics	Inceptisols, Alfisols, Ultisols	• A steep-sloped forestland and a more gently rolling cropland showed some similar and different soil moisture spatial–temporal patterns because of the intertwined impacts from terrain features, soil types and properties, and vegetation dynamics. • Subsurface preferential flow was common in both landscapes, with the forestland having nearly doubled frequency of preferential flow occurrence (overall average	• Spatial complexity and temporal dynamics of soil moisture patterns in contrasting landscapes are controlled by the interactions of terrain, soil, and vegetation, and such interactions change with seasonal wetness, rainfall characteristics, and soil depth. • Hydropedologic features strongly influence landscape hydrology, and provide physical basis for distributed catchment modeling, biogeochemical hot spots and hot

(Continued)

TABLE 2 Summary of the Main Findings from Each Case Study Presented in Part II of this Book—Cont'd

Chapter #	Location	Focused feature	Soil orders	Landscape–soil–water–vegetation relationship	Implications/applications
				36% of all the precipitation events investigated) than that in the cropland (overall average 21%).	moments associated with preferential flow, and precision management of soil and water resources.
13	Central Pennsylvania, USA	Geophysical investigations of soil-landscape architecture and its impacts on subsurface flow	Inceptisols, Alfisols, Ultisols	• Relative difference in soil apparent electrical conductivity (ECa) across two contrasting landscapes remained relatively stable over time, indicating stable soil-landform units; while absolute changes in ECa over seasons indicated (to some degree) active zones of subsurface flow. Time-lapsed GPR identified subsurface preferential flow pathways and patterns, which showed significant difference between shallow and deep soils.	• Geophysical tools such GPR and EMI can be effectively used to aid in revealing complex soil-landscape architecture and its impacts on subsurface flow, especially at the intermediate scales of hillslopes and catchments and when used in time-lapse manner. • Geophysical tools can continue to help ease technological bottlenecks for subsurface investigation by improving their spatial and temporal resolutions and coupling with other advanced technologies.

				• South- vs. north-facing hillslopes and linear vs. concave slopes showed differences in EMI and GPR responses, which reflected the underlying differences in soil-landscape features. Depth to bedrock was highly variable across the two landscapes, but predictable patterns were revealed through extensive GPR surveys.	
14	South West Victoria, Australia	Managing hydrologically driven land degradation	Alfisols, Spodosols, Ultisols, Vertisols	• In each landscape, geological context, geomorphology, and soils are different and provide unique structural controls for the surface and subsurface hydrology. • The nine-unit landsurface model is an appropriate general framework for disaggregating landscapes and processes at the hillslope scale in hydropedologic studies.	• Hydropedologic studies in Australia have to be nested within broader scale contexts of hydrogeological systems. • Hydropedology, hydrogeology, and chemistry were successfully applied to the diagnosis of land degradation processes and the prescription of land management solutions.

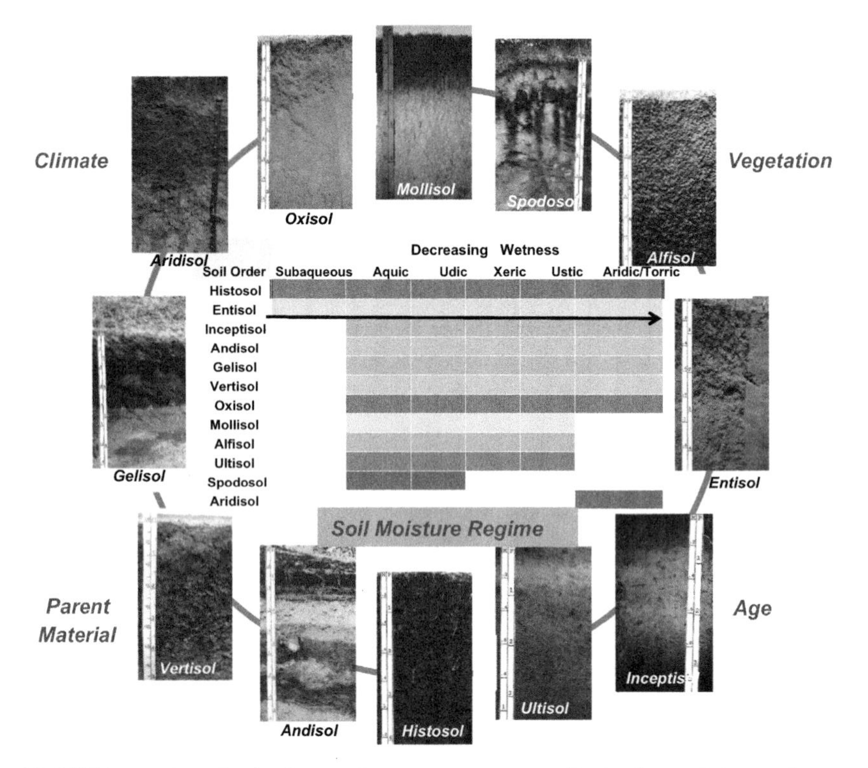

FIGURE 1 Twelve soil orders in the U.S. *Soil Taxonomy* grouped by dominant soil-forming factors (three soil orders in each group linked through a blue line) (soil profile photos courtesy of USDA-NRCS). The table in the center shows the range of soil moisture regimes possible in each of the 12 soil orders. Seven soil orders (Alfisols, Histosols, Inceptisols, Mollisols, Spodosols, Ultisols, and Vertisols) are included in the case studies reported in this book. *(Color version online and in color plate)*

provide techniques that can measure integrated fluxes and storages at useful scales. Lin (2010a) noted that characterizing complex subsurface heterogeneity is an important step in revealing the governing principles of subsurface processes. This is because predicting flow and transport in field soils using input–output relationships without adequate characterization of internal architecture is analogous to diagnosing a patient based on what this person takes in and excludes out of his/her body without a thorough understanding of human body's complex anatomy and circulatory systems. To close this gap, improved noninvasive and time-lapsed imaging/mapping techniques and various "smart" sensors/tracers need to be continuously developed to shed better light on the subsurface architecture and its impacts on flow paths, residence times, and reaction patterns. Continued development of transformative technologies for high-resolution subsurface imaging as well as spatially–temporally continuous monitoring of diverse soil properties and processes can significantly help open up the longstanding "black

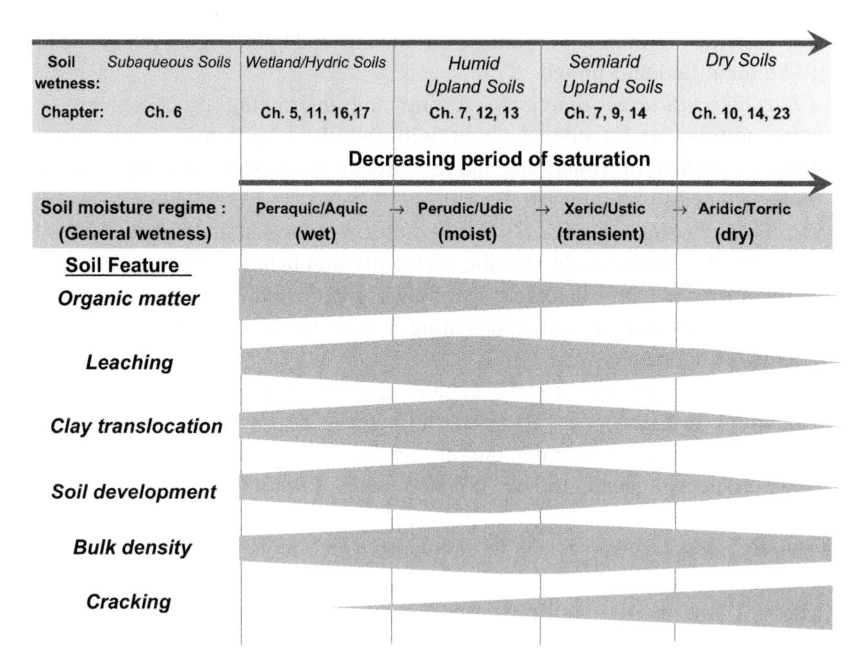

FIGURE 2 The spectrum of soil wetness from subaqueous soils to dry soils and the associated chapters in this book. Some general trends of selected soil features are shown with change in soil moisture regimes from wet to dry. These trends are only approximations (assuming other conditions being similar), but other soil-forming factors (such as soil temperature regimes, organisms, and parent materials) could complicate the general trends shown. In addition, the landscape setting (both regional and local) is also essential to appropriate interpretation of soil features. *(Color version online)*

box" of the subsurface. Enhanced techniques for soil dating is also important for better understanding of soil changes over time (see Chapter 7 of this book).

3.2. Holistic, Multiscale, and Quantitative Modeling Frameworks

A general framework for integrated hydropedologic studies was suggested by Lin et al. (2006). This conceptual framework has been applied by D'Amore et al. in Chapter 11 of this book to address regional dissolved organic matter biogeochemistry in relation to hydropedology. However, the quantification of this framework remains to be developed. Such a framework focuses on the following five key issues (Lin et al., 2006):

- *Identification of architectures*: System architecture creates constraints and conditions in which processes act and interact. Thus, the interrelationships between the hierarchical organizations of soils and hydrologic systems need to be identified and linked.

- *Characterization of functions*: Acting and interacting processes create preconditions and feedbacks that will modify a system's architecture. Hence, studies of functions and architectures of soils and hydrologic systems need to be integrated.
- *Bridging of scales*: A system's architecture and function interact at a variety of scales that define scale-specific variability and pattern. Optimal methods to quantify and communicate soil and hydrologic variability as a function of scale need to be improved, and scaling in soils needs to be correlated with scaling in hydrology.
- *Systems integration*: Models capable of integrating space and time (including fast and slow processes and heterogeneous soil-landscapes) are important for systems understanding and prediction. Pedologic modeling and prediction need to be coupled with hydrologic modeling and forecasting.
- *Human impacts*: Anthropogenic influences (such as land use and management) on soils and hydrologic systems are intimately linked. Thus, soil changes and hydrologic alterations under human impacts need to be addressed simultaneously.

Quantitative relationships between the architectures and functions of soils and hydrologic systems at different scales are the foundation for robust models. The challenge here is to build bridges that can connect different spatial and temporal scales in quantitative ways. This is a grand challenge across scientific disciplines and remains so in soil physics and hydrology (Jury et al., 2011). It is hoped that hydropedology can make a unique contribution to this longstanding endeavor.

3.3. A Global Alliance of Integrated Databases and Observatories

Integrated soils databases for multiscale modeling and a network of long-term observatories across geographic regions are important for the establishment of a global alliance for comprehensive field data collections and systematic coordination of databases. Towards such a possible alliance, many issues need to be addressed, such as:

- *Adapting an iterative approach of "monitoring, mapping, and modeling" (3 M)*: This 3 M system offers an integrated and evolutionary approach to quantitative assessment and forecasting, and provides an adaptive strategy for refining models and monitoring as knowledge and databases are accumulated. This iterative strategy blends together 1) real-time monitoring to capture temporal trends using sensor networks, in situ observations, remote sensing, field campaigns, and other approaches; 2) mapping landscape–soil–water–vegetation relationships to identify hierarchical spatial patterns via geospatial technologies, geophysical tools, ground-based surveys, and other means; and 3) multiscale modeling to integrate space and time through process-based and/or systems-oriented approaches. Both mapping and modeling can provide justifications and guidelines for field data gathering, including "what, where, and when" data should be collected, at what resolution, for how long, and for what purposes. In return, monitoring can provide feedbacks and ground-truthing to both mapping and modeling.

- *Precision mapping and functional characterization of soils*: The soil map is a common way to portray soil heterogeneity across the landscape and to provide soil input parameters to simulation models. Precision soil mapping is of increasing demand for site-specific applications; thus, mapping soils in greater detail and with higher precision are needed. Modern soil maps also need improvements in functional characterizations beyond taxonomic classifications. For example, derived and dynamic soil maps, tailored for a specific function or purpose, need to be produced for various practical applications. Hence, linking hydropedology and digital soil mapping is a necessary and promising direction that can improve coupled spatial mapping and process-based modeling of soils (see Chapter 21 of this book for further discussion on this topic).
- *Long-term monitoring*: Time scale of databases (e.g. duration, frequency, and extreme events) is critical to long-term understanding of the Earth system including the subsurface. In this time of accelerating global change, continuous observations are essential as they have the potential to reveal unexpected but relevant developments and processes. Long-term recording of soil health (e.g. through monitoring its "blood pressure" or soil water potential, temperature, water table, respiration, carbon, and other potentially key signs of global land change) is essential to ensure the sustainability of quality soils and ecosystem services. Long-term monitoring also provides necessary datasets for validating and refining the modeling of slow changes that occur in soil systems.

3.4. Societal Relevance and Environmental Sustainability

Numerous practical applications call for expertise in hydropedology, such as water quality and quantity, soil quality and quantity, land degradation, land-use planning, watershed management, wetland protection, nutrient cycling, contaminant prevention, waste disposal, precision agriculture, climate change, and ecosystem restoration. Some emerging applications where hydropedology can make a unique contribution are illustrated below:

- *Global food and water security*: Ecological and water footprints are becoming increasingly popular measures for determining the impact of goods and services on ecosystems and water resources, respectively, through accounting and markets. Ecological footprint (Rees, 1992; Wackernagel, 1994) represents the amount of biologically productive land necessary to supply resources a human population consumes and to assimilate associated wastes. Water footprint (Hoekstra, 2003; Deurer et al., 2011) focuses on water scarcity and water quality, with three components: the green-water footprint (including the soil water), the blue-water footprint (the net use of water resources), and the gray-water footprint (water quality impacts). Supermarket chains are developing ecological and/or water footprinting protocols for products, especially food, so that consumers can make informed purchase decisions based on the eco- and/or water-friendliness of products. Quantification of ecological and water footprints requires hydropedologic knowledge of different soil types in different parts of the world, how water flows through different soils and landscapes, and the impact of that on receiving waters, ecosystems, and products. Besides the land, the recognition and mapping of subaqueous soils in estuarine systems can also help the restoration of submerged aquatic vegetation and siting productive areas for aquaculture (see Chapter 6 of this book).

- *Environmental policy and regulations*: There are a number of environmental regulations pertaining to hydropedology, such as wetland protection, onlot sewage treatment, waste disposal, storm water management, nutrient management, and agrochemical application (Bouma, 2006; Lin, 2006). As articulated by Bouma in Chapter 15 of this book, hydropedology can become a powerful tool for environmental policy and regulation toward sustainable management of land. To become attractive to stakeholders and policymakers, contributions by hydropedology to environmental rules and regulations should be based on serious interaction of science with societal partners, and hydropedologists can put themselves in an attractive intermediate position by facilitating interaction between land users and policymakers (Bouma, 2006). The recently accepted Soil Protection Strategy for the European Union, for example, defines seven soil functions that require a systems approach with a focus on water and solute movement in field soils (see Chapter 15 of this book). Wilding et al. in Chapter 9 of this book also illustrate how misnomers regarding pedology, hydropedology, and landscape interactions have led to inappropriate and misleading government mandates concerning environmental policy, regulations, ordinances, and land management in the Central Texas Hill Country.
- *Soft engineering and hazard mitigation*: Hydropedology can contribute important knowledge to sustainable development that combines traditional "hard" engineering approaches with "soft," ecological service-based approaches that are aimed at utilizing natural ecosystems to balance man-made structures. This is evident in various hazard mitigations, such as erosion, hillslope stability, landslide, flooding, drought, waste disposal, and contamination remediation, where adequate understanding of

real-world landscape–soil–water–vegetation relationships can help the prevention and remediation of these hazards (Verstappen, 2011). Soft engineering, if well designed, is believed to be more sustainable as it requires less long-term maintenance costs due to natural ecosystems becoming more matured and stabilized over time, and soft engineering is also cheaper to implement as it makes use of what is already available in nature (Verstappen, 2011).

3.5. Next Generation of Soil Scientists and Hydrologists

Educating the next generation of hydropedologists who can solve interdisciplinary problems in the real world is essential to the future of hydropedology. Some important aspects are highlighted below in this regard:

- *Fundamental importance of fieldwork*: Fieldwork is a distinct aspect of hydropedology. In particular, field mapping and monitoring of the landscape–soil–water–vegetation relationship should be a fundamental skill for hydropedologists. Soil profile description and soil-landscape mapping, for instance, are essential to anyone who wishes to truly understand real-world soils, their architectures, and their functions in the landscape. Developing soil science field camps and enhancing field-based internships in soil science are essential to assure current students' and future faculty members' expertise with soils in the field context. Fieldwork ultimately serves as the initial drive of identifying real-world problems, formulating theories of natural processes, and validating models and their predictions.
- *Quantitative and advanced technical skills*: Quantitative skills (such as mathematics and simulation modeling) should an integral part of hydropedology education. It is also important that future hydropedologists possess skills in geospatial, 3-D and 4-D visualization, and other information technologies, as well as advanced instrumentation, to enable them to collect, visualize, analyze, and model spatial–temporal patterns of complex landscape–soil–water–vegetation dynamics across scales. Development and utilization of effective and compelling visualizations of complex soil systems can also help stimulate learning and enhance research.
- *Holistic and systems thinking*: Real-world soil systems consist of a wide variety of interacting parts and a web of processes that are constantly changing in the open and dissipative environment. This complexity requires a new way of thinking and a significantly improved scientific treatment of complex systems because classical analytical and statistical treatments often fail (von Bertalanffy, 1968; Weinberg, 1975; Lin, 2010b). Hydropedology education therefore should cover a broad spectrum of topics in pedology, soil physics, hydrology, as well as other related bio- and geosciences (such as geomorphology, geology, ecology, and other branches of soil science) and even philosophy and complex systems science to enable integrated holistic thinking. In addition, to solve practical problems,

integration of theory, field data, simulation models, expert judgments, along with policies and regulations, socio-economic factors, and ethics, is also important for interdisciplinary teams to achieve sustainable management of precious soil and water resources.

4. CONCLUDING REMARKS

As discussed throughout this book, hydropedology is an emerging interdisciplinary field that has significant potential to advance both soil science and hydrology, as well as other related bio- and geosciences. Grounded on an adequate understanding of in situ soil architecture and the soil–landscape relationship and the use of water as the central thread to link coupled processes, hydropedology can help guide more effective field data acquisition, multiscale knowledge integration, and model-based prediction of complex landscape–soil–water–vegetation relationships across space and time. A new era of soils research needs to be based on soil architecture that is the backbone of soil functions from the molecular level to the global pedosphere.

Hydropedology can make unique contributions to a wide variety of applications from the Earth's Critical Zone to extraterrestrial explorations, including the "7 + 1" roles of soils as described in Chapter 1 of this book. Exciting opportunities abound to advance the frontiers of hydropedology research, education, and applications toward sustainable ecosystems and human well-being.

ACKNOWLEDGMENTS

The author thanks Drs. Larry Wilding, Marcus Hardie, and Todd Caldwell for their review comments that have helped improve the quality of this paper.

REFERENCES

Beven, K., 2006. Searching for the holy grail of scientific hydrology: $Q_t = H(\underleftarrow{S}\ \underleftarrow{R}\ \Delta t)A$ as closure. Hydrol. Earth Syst. Sci. 10, 609–618.

Bouma, J., 2006. Hydropedology as a powerful tool for environmental policy research. Geoderma 131, 275–286.

Deurer, M., Green, S.R., Clothier, B.E., Mowat, A., 2011. Can product water footprints indicate the hydrological impact of primary production? – A case study of New Zealand kiwifruit. J. Hydrol. 408, 246–256.

Hillel, D., 1991. Out of the Earth – Civilization and the Life of the Soil. The Free Press, New York.

Hoekstra, A.Y. (Ed.), 2003. Virtual Water Trade: Proceedings of the International Expert Meeting on Virtual Water Trade. IHE Delft, the Netherlands.

Jury, W.A., Or, D., Pachepsky, Y., Vereecken, H., Hopmans, J.W., Ahuja, L.R., Clothier, B.E., Bristow, K.L., Kluitenberg, G.J., Moldrup, P., Simunek, J., vanGenuchten, M. Th., Horton, R.,

2011. Kirkham's legacy and contemporary challenges in soil physics research. Soil Sci. Soc. Am. J. 75, 1589–1601.

Lin, H.S., 2006. Hydropedology and modern soil survey applications. Soil Surv. Horiz. 47, 18–22.

Lin, H.S., 2010a. Linking principles of soil formation and flow regimes. J. Hydrol. 393, 3–19. doi:10.1016/j.jhydrol.2010.02.013.

Lin, H.S., 2010b. Earth's critical zone and hydropedology: concepts, characteristics, and advances. Hydrol. Earth Syst. Sci. 14, 25–45.

Lin, H.S., 2011. Three principles of soil change and pedogenesis in time and space. Soil Sci. Soc. Am. J. 75, 2049–2070.

Lin, H.S., Bouma, J., Pachepsky, Y., Western, A., Thompson, J., van Genuchten, M. Th., Vogel, H., Lilly, A., 2006. Hydropedology: synergistic integration of pedology and hydrology. Water Resour. Res. 42, W05301. doi:10.1029/2005WR004085.

Rees, William E, 1992. Ecological footprints and appropriated carrying capacity: what urban economics leaves out. Environ. Urban. 4, 121–130. doi:10.1177/095624789200400212.

Soil Survey Staff, 2010. Keys to Soil Taxonomy, eleventh ed. USDA NRCS, Washington, DC.

Verstappen, H Th, 2011. Natural disaster reduction and environmental management: a geomorphologist's view. Geogr. Fis. Din. Quat. 34, 55–64.

von Bertalanffy, L., 1968. General System Theory: Foundations, Development, Applications. George Braziller, New York.

Weinberg, G.M., 1975. An Introduction to General Systems Thinking. Silver Anniversary Edition. Dorset House Publishing, New York.

Wackernagel, M. 1994. Ecological Footprint and Appropriated Carrying Capacity: A Tool for Planning Toward Sustainability. PhD thesis, OCLC 41839429. School of Community and Regional Planning. The University of British Columbia, Vancouver, Canada.

Index

Note: Page numbers with "f" denote figures; "t" tables.

D

E

H

N

O

P

R

S

T

U

V

W

Color Plates

Chapter 1

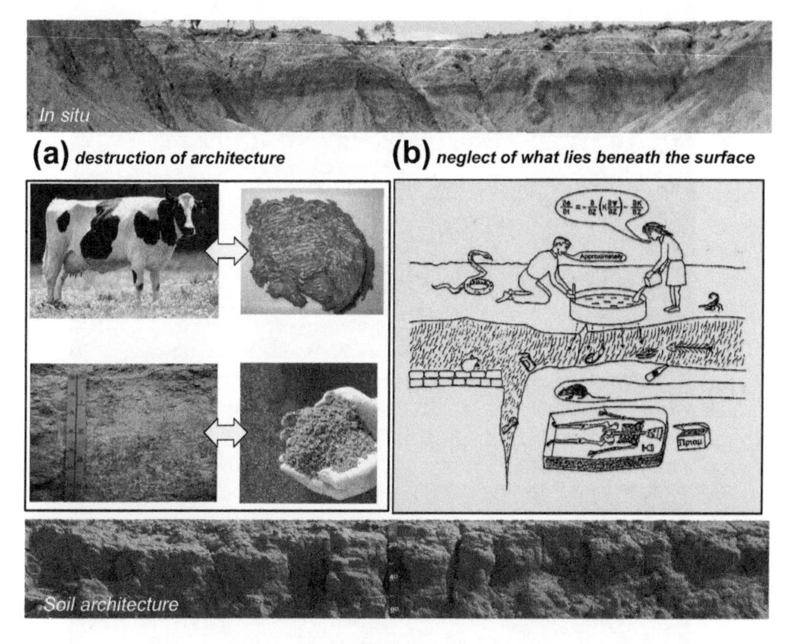

FIGURE 1

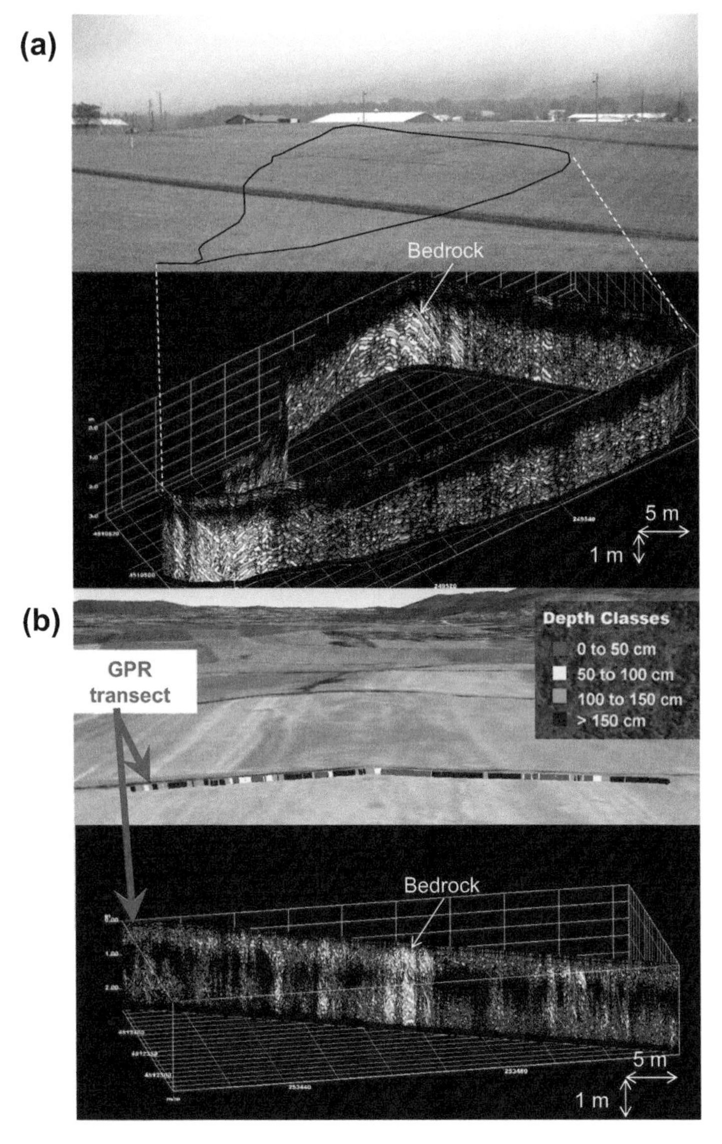
(a)
Bedrock
5 m
1 m
(b)
GPR transect
Depth Classes
0 to 50 cm
50 to 100 cm
100 to 150 cm
> 150 cm
Bedrock
5 m
1 m

FIGURE 2

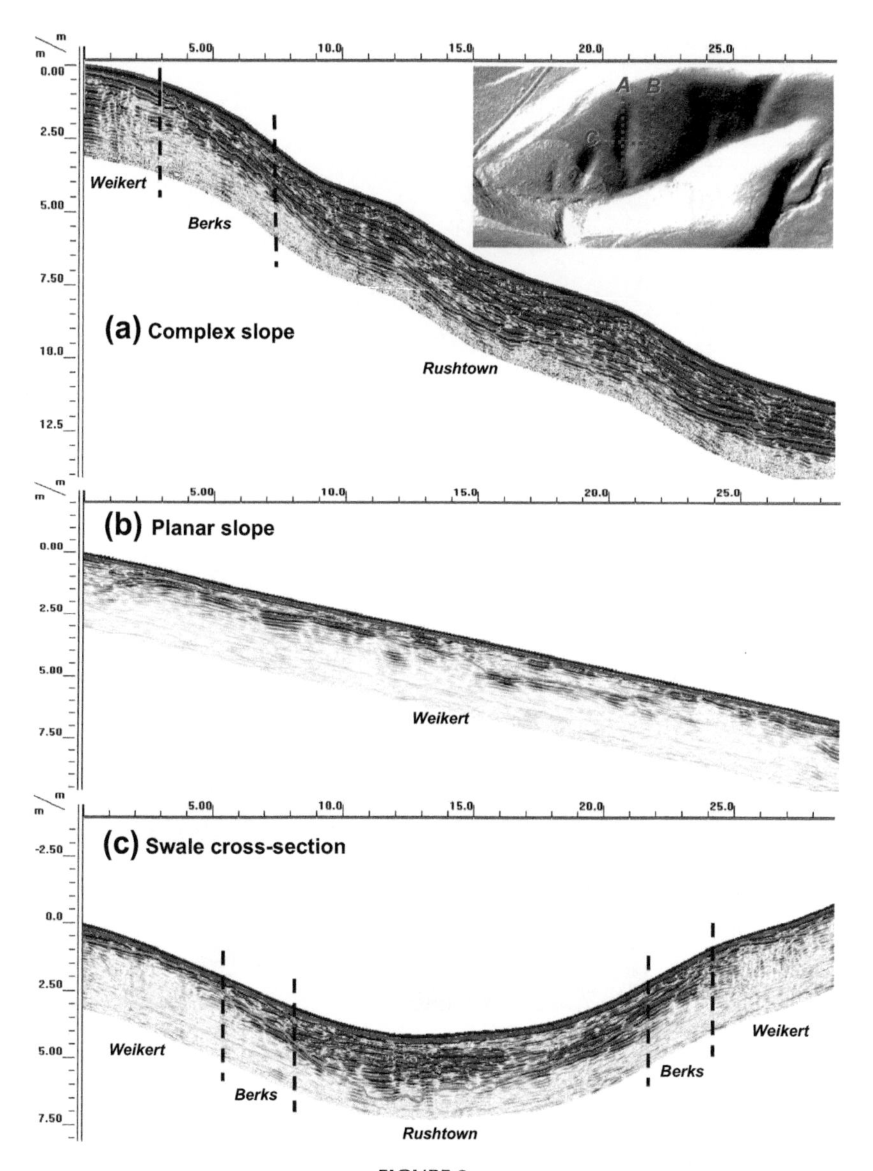

(a) Complex slope
Weikert
Berks
Rushtown
A
B
C
(b) Planar slope
Weikert
(c) Swale cross-section
Weikert
Berks
Rushtown
Berks
Weikert

FIGURE 3

(a)
Soilscape
(b)
Cross-
section
(c)
Macro-
morphology
(d)
Micro-
morphology
-0-
-1-
-2-
-3-
-4-
1 mm

FIGURE 4

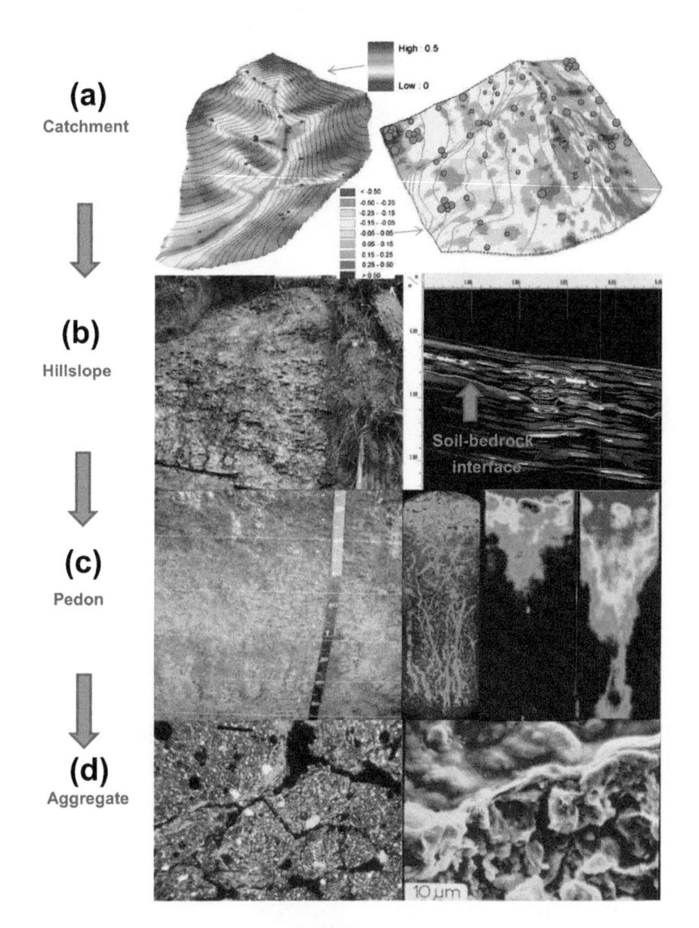

(a)
Catchment
High : 0.5
Low : 0
< -0.50
-0.50 - -0.25
-0.25 - -0.15
-0.15 - -0.05
-0.05 - 0.05
0.05 - 0.15
0.15 - 0.25
0.25 - 0.50
> 0.50
(b)
Hillslope
Soil-bedrock
interface
(c)
Pedon
(d)
Aggregate
10 µm

FIGURE 6

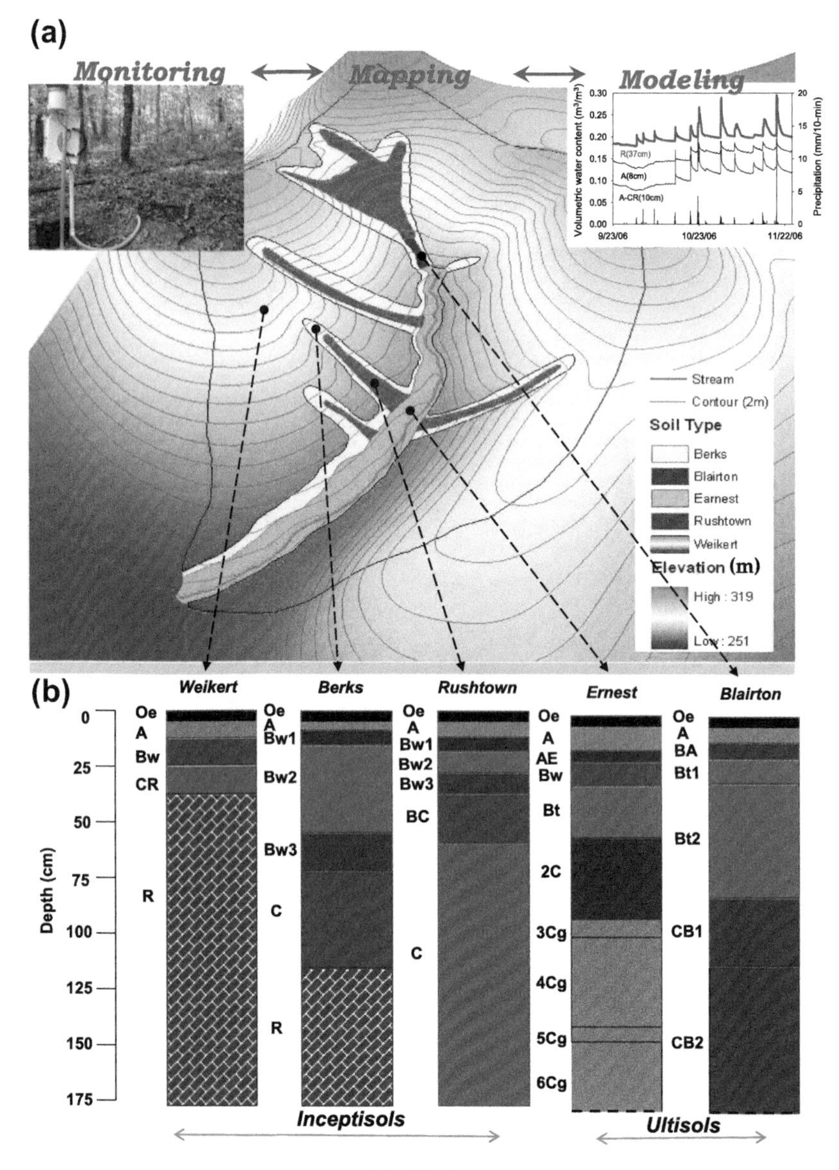
(a)
Monitoring
Mapping
Modeling
Volumetric water content (m3/m3)
Precipitation (mm/10-min)
R(37cm)
A(8cm)
A-CR(10cm)
9/23/06
10/23/06
11/22/06
Stream
Contour (2m)
Soil Type
Berks
Blairton
Earnest
Rushtown
Weikert
Elevation (m)
High : 319
Low : 251
(b)
Depth (cm)
Weikert
Berks
Rushtown
Ernest
Blairton
Oe
A
Bw
CR
R
Bw1
Bw2
Bw3
C
BC
AE
Bt
2C
3Cg
4Cg
5Cg
6Cg
BA
Bt1
Bt2
CB1
CB2
Inceptisols
Ultisols

FIGURE 11

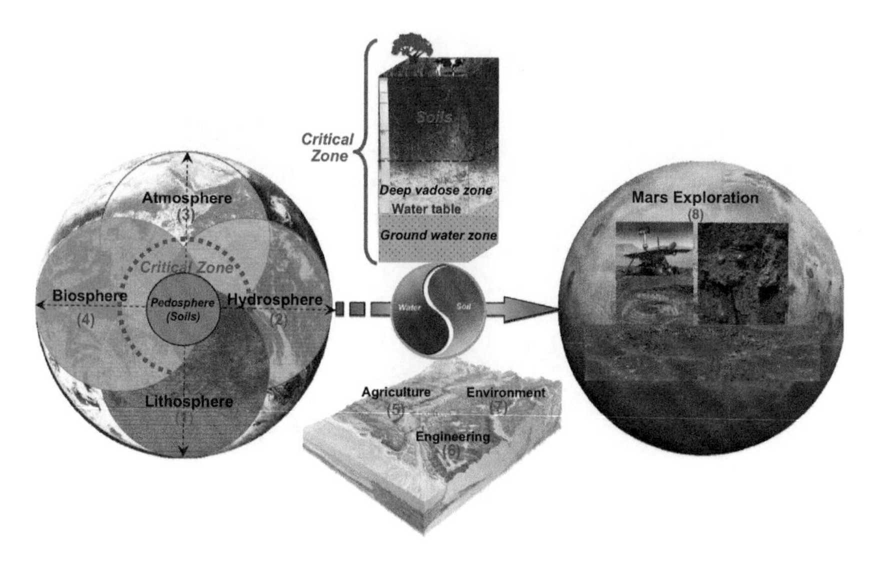

FIGURE 12

Chapter 2

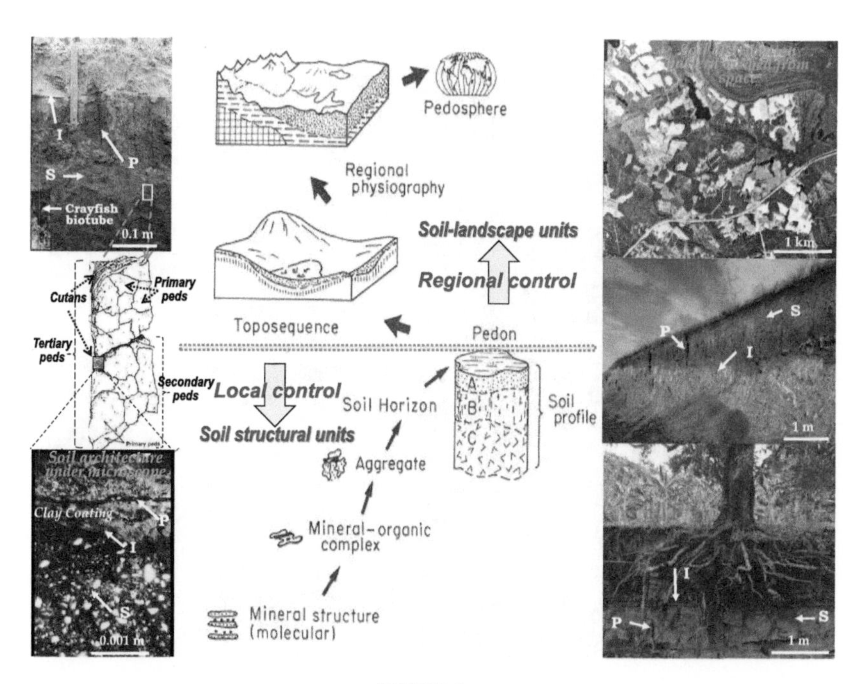

FIGURE 1

FIGURE 2

FIGURE 3

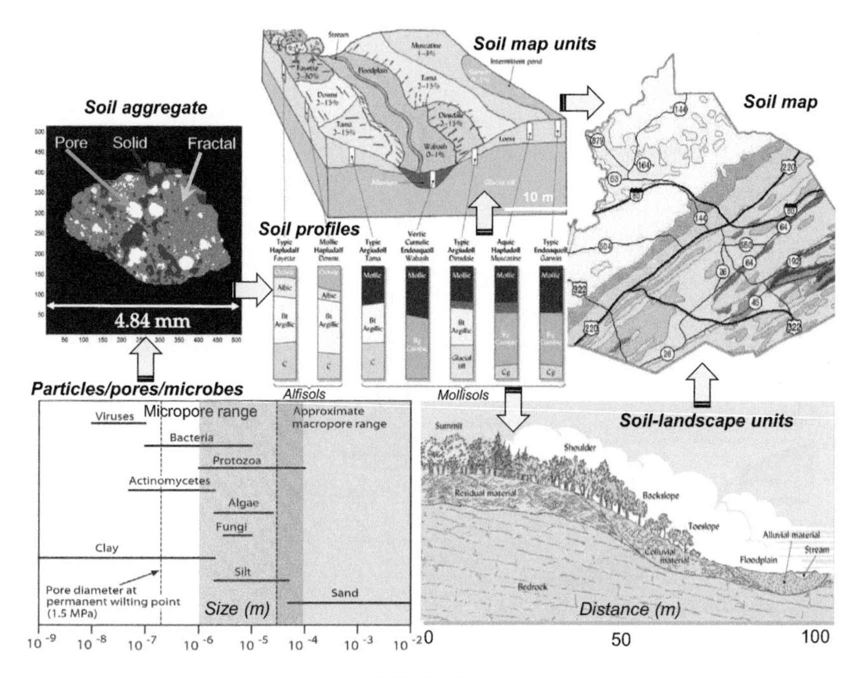

FIGURE 6

FIGURE 8

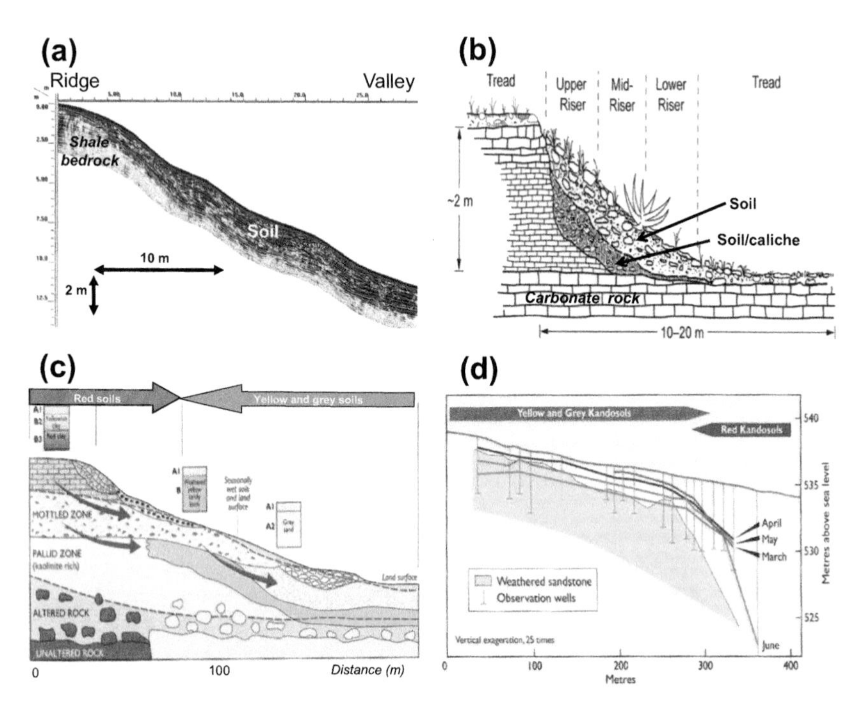
(a)
Ridge
Valley
Shale bedrock
Soil
10 m
2 m
(b)
Tread
Upper Riser
Mid-Riser
Lower Riser
Tread
~2 m
Soil
Soil/caliche
Carbonate rock
10–20 m
(c)
Red soils
Yellow and grey soils
A1
A2
MOTTLED ZONE
PALLID ZONE
(kaolinite rich)
ALTERED ROCK
UNALTERED ROCK
Land surface
0
100
Distance (m)
(d)
Yellow and Grey Kandosols
Red Kandosols
April
May
March
June
Weathered sandstone
Observation wells
Vertical exageration, 25 times
0
100
200
300
400
Metres
540
535
530
525
Metres above sea level

FIGURE 10

Chapter 5

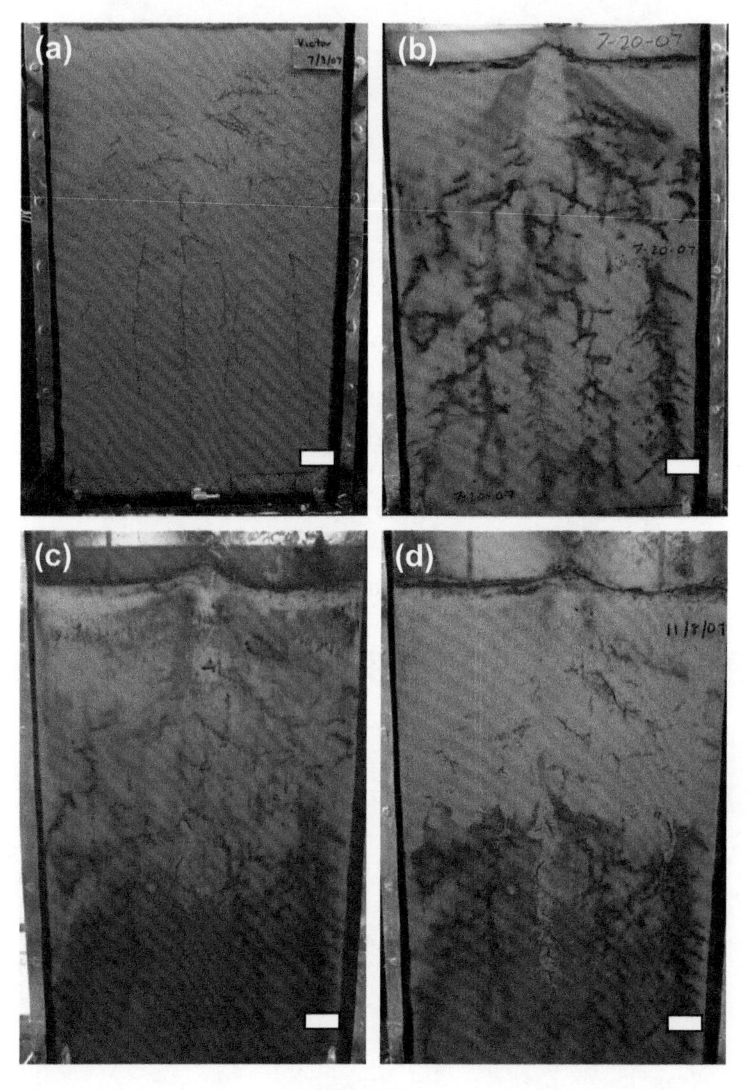

FIGURE 2

FIGURE 7

FIGURE 9

Chapter 6

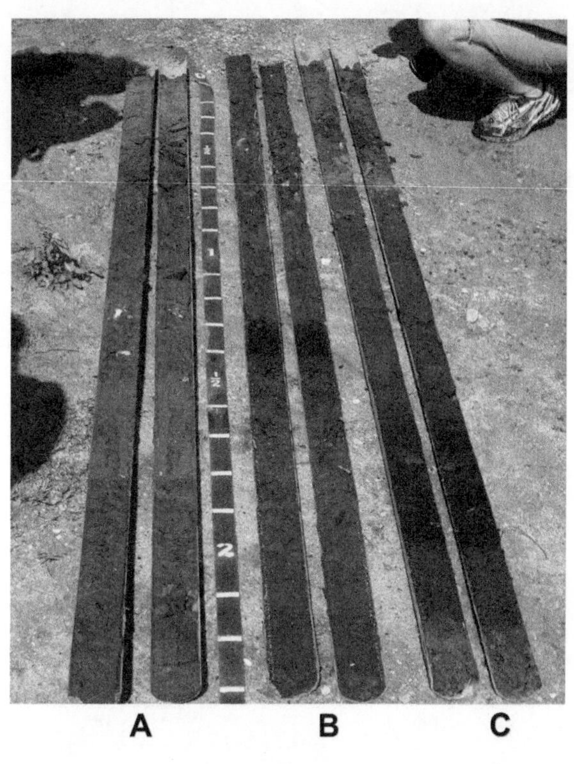

FIGURE 2

FIGURE 3

FIGURE 6

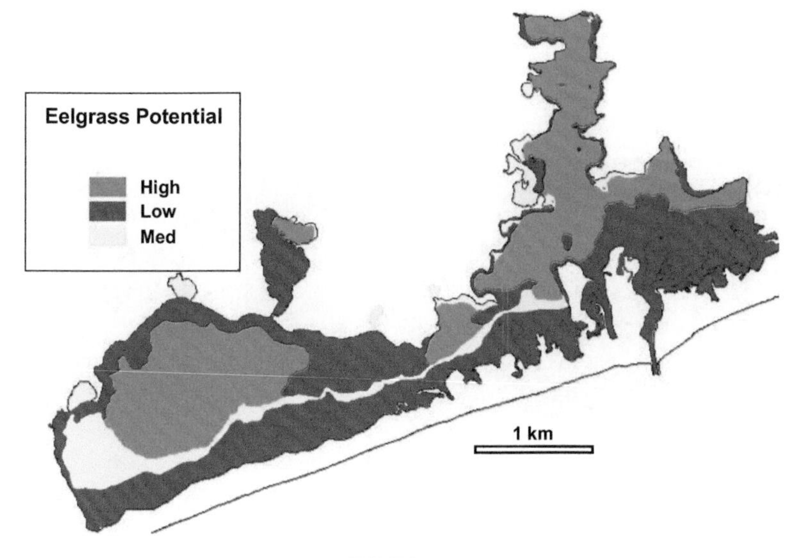

FIGURE 9

Chapter 9

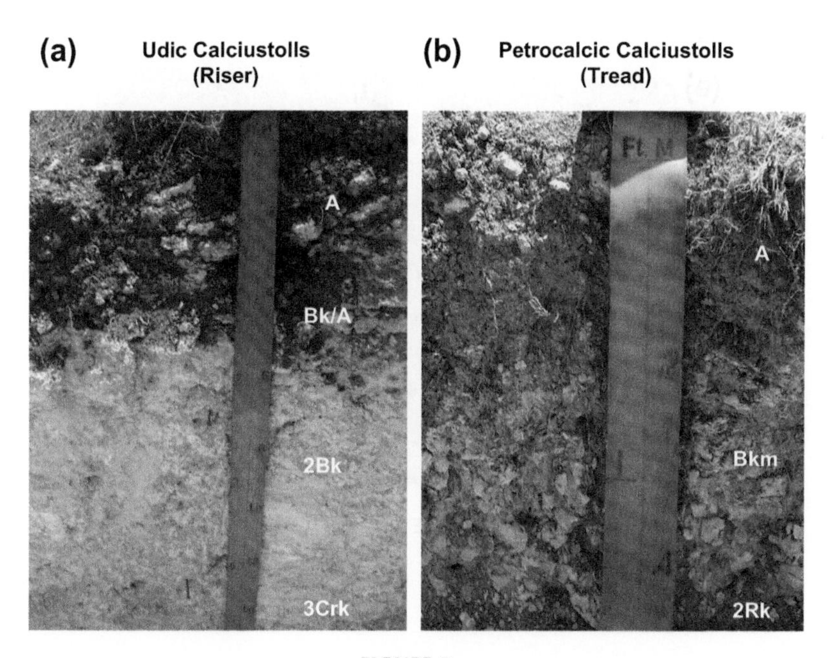

FIGURE 5

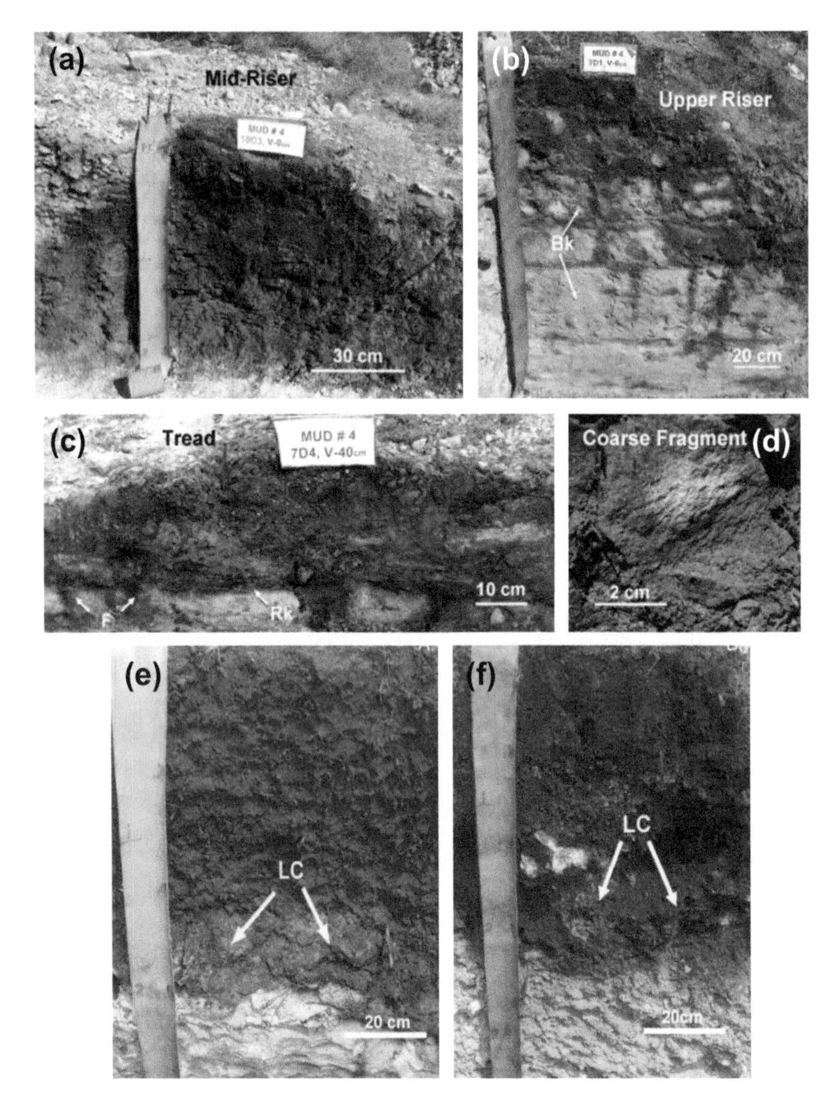
(a)
Mid-Riser
30 cm
(b)
Upper Riser
Bk
20 cm
(c)
Tread
MUD # 4
7D4, V-40cm
10 cm
Rk
(d)
Coarse Fragment
2 cm
(e)
LC
20 cm
(f)
LC
20cm

FIGURE 16

Chapter 10

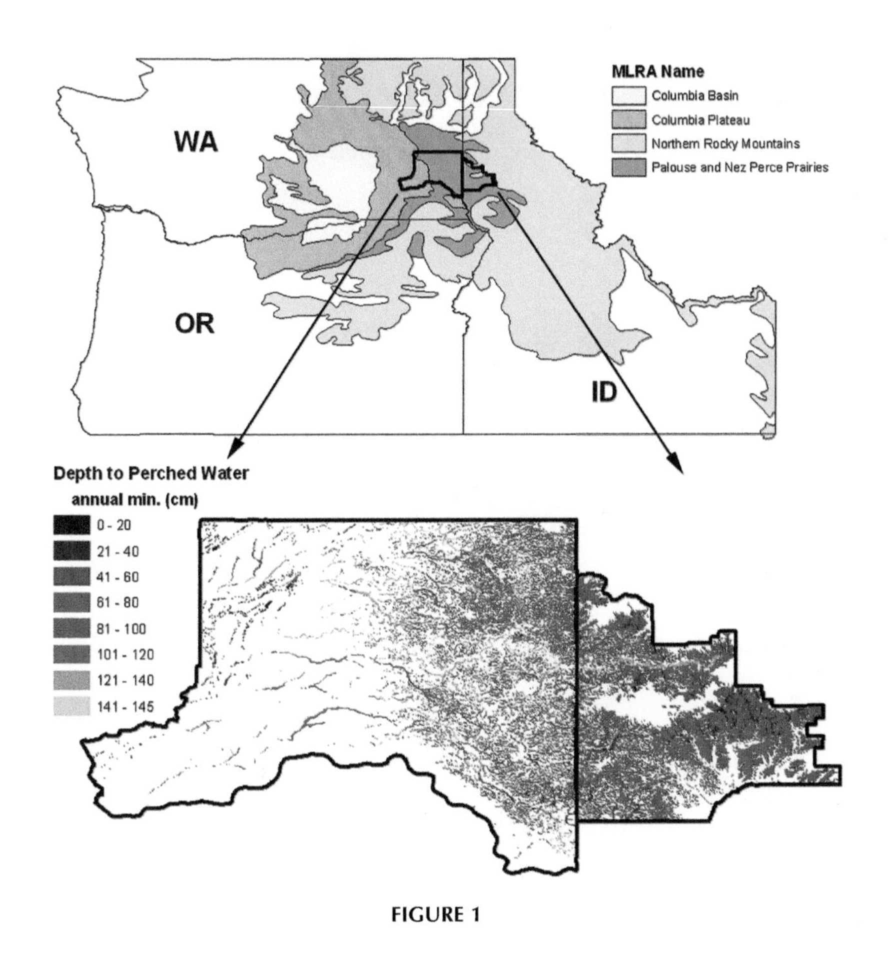

FIGURE 1

Mean Annual Precip.
mm
High: 1188
Low: 462
Soils
Haploxerolls
Argixerolls
Fragixeralfs
Vitricyands

FIGURE 2

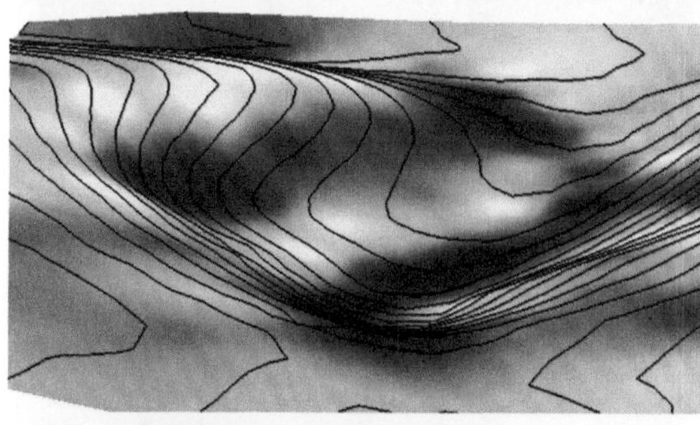

E Horizon Thickness

cm

High: 52

Low: 6

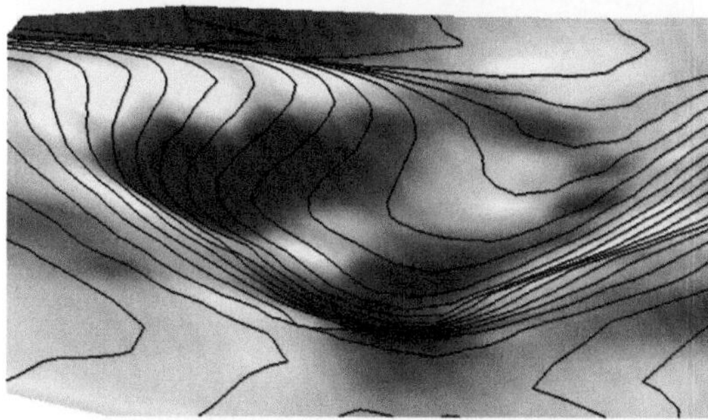

Depth to Fragipan Layer

cm

High: 118

Low: 29

FIGURE 5

Chapter 11

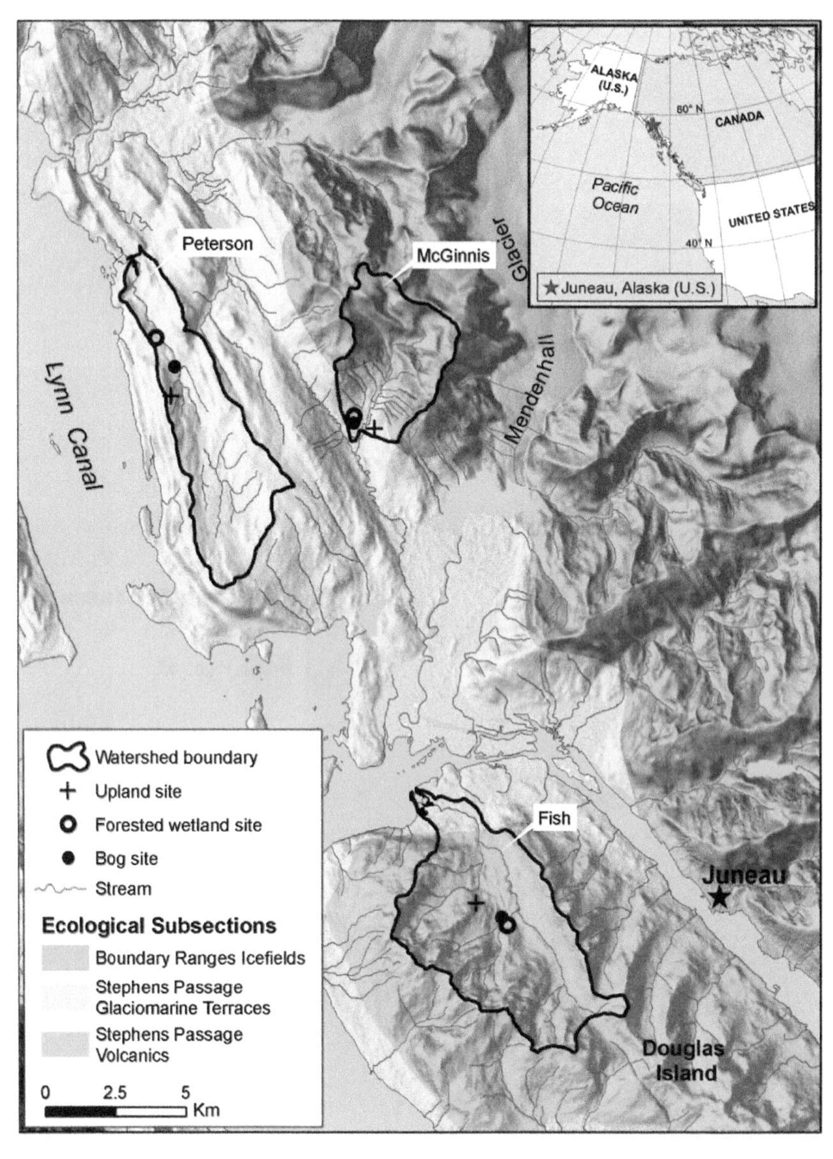

FIGURE 3

Chapter 12

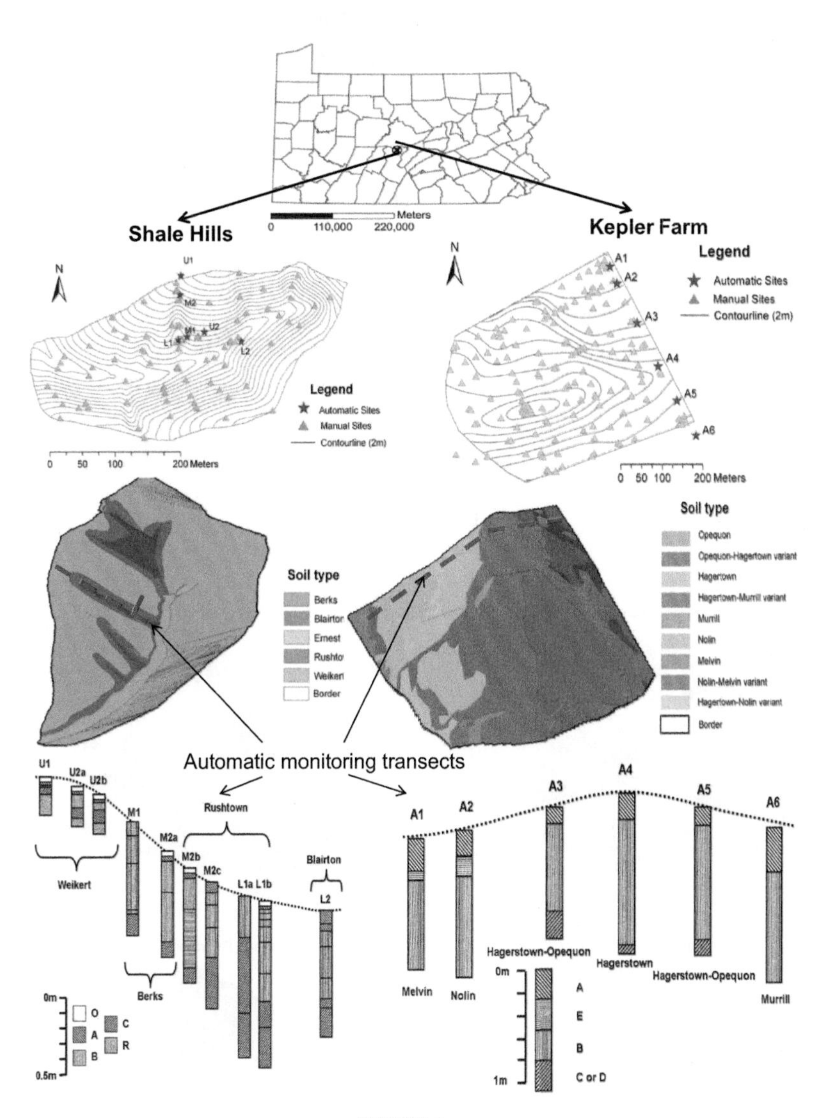

FIGURE 1

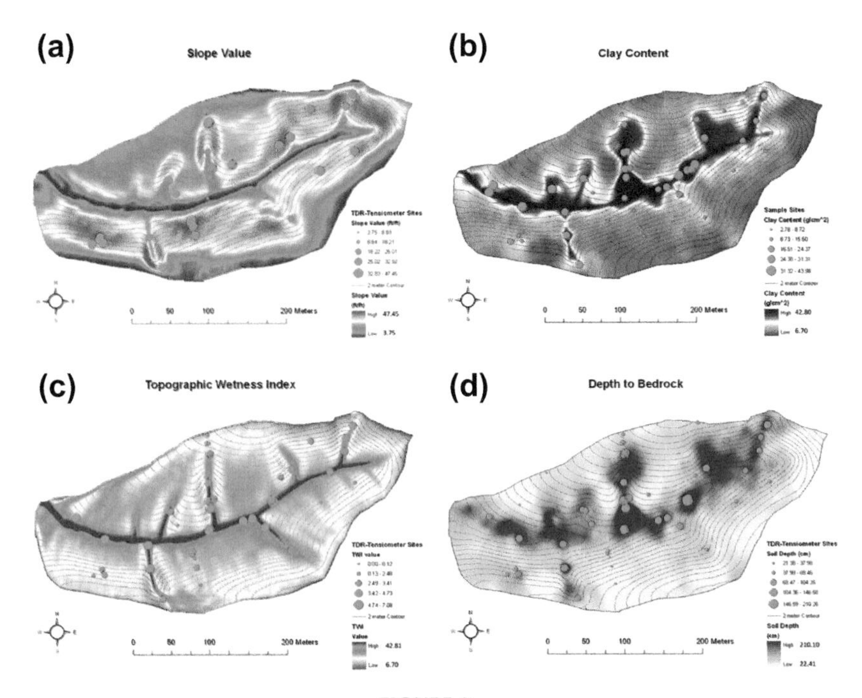

FIGURE 2

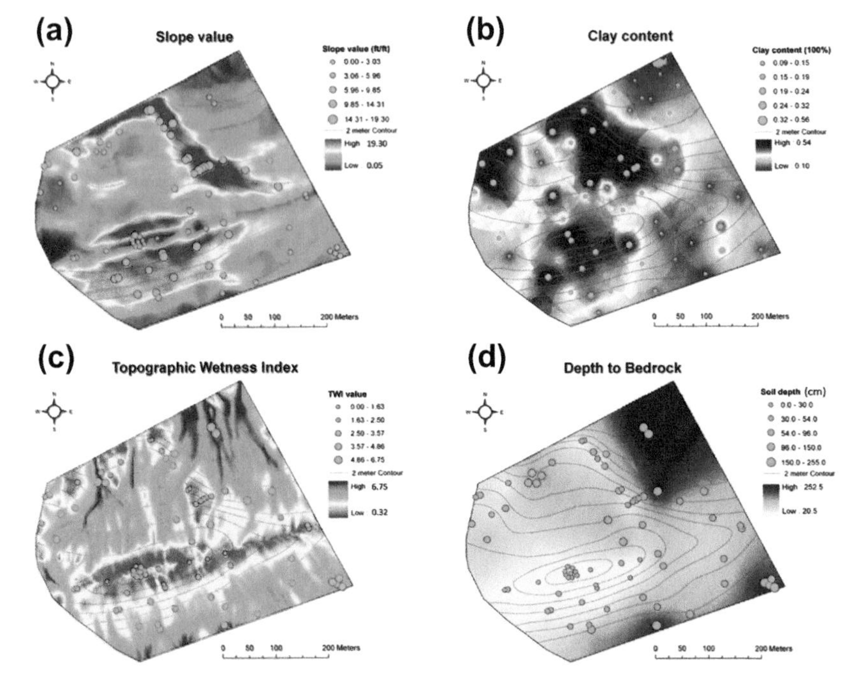

FIGURE 3

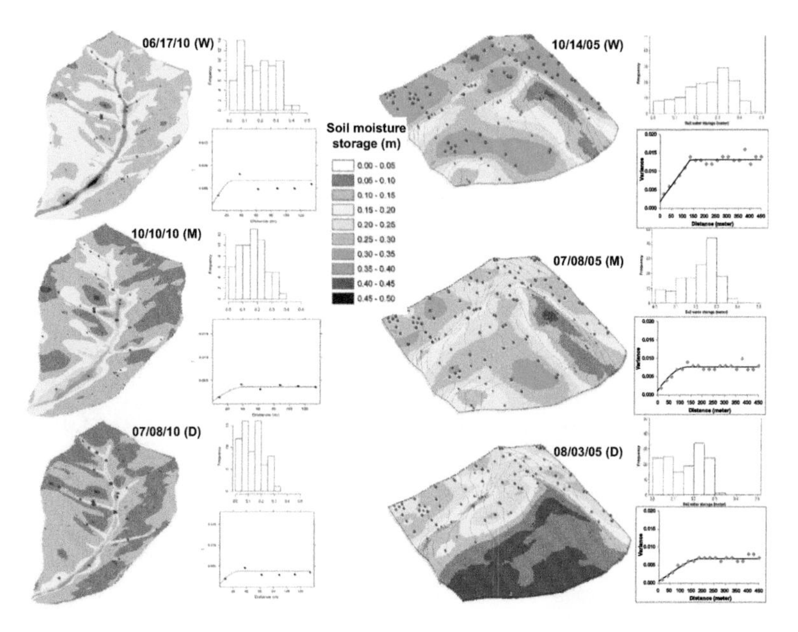
06/17/10 (W)
10/10/10 (M)
07/08/10 (D)
Soil moisture storage (m)
0.00 - 0.05
0.05 - 0.10
0.10 - 0.15
0.15 - 0.20
0.20 - 0.25
0.25 - 0.30
0.30 - 0.35
0.35 - 0.40
0.40 - 0.45
0.45 - 0.50
10/14/05 (W)
07/08/05 (M)
08/03/05 (D)
Variance
Distance (meter)

FIGURE 4

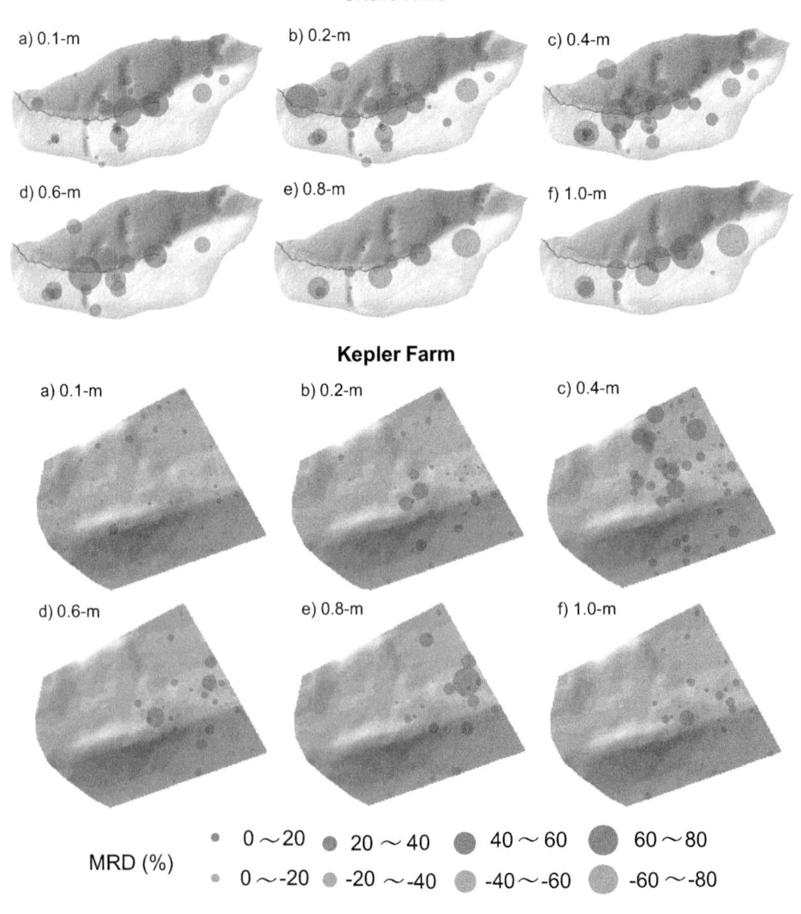
Shale Hills
a) 0.1-m
b) 0.2-m
c) 0.4-m
d) 0.6-m
e) 0.8-m
f) 1.0-m
Kepler Farm
a) 0.1-m
b) 0.2-m
c) 0.4-m
d) 0.6-m
e) 0.8-m
f) 1.0-m
MRD (%)
0 ~ 20
20 ~ 40
40 ~ 60
60 ~ 80
0 ~ -20
-20 ~ -40
-40 ~ -60
-60 ~ -80

FIGURE 6

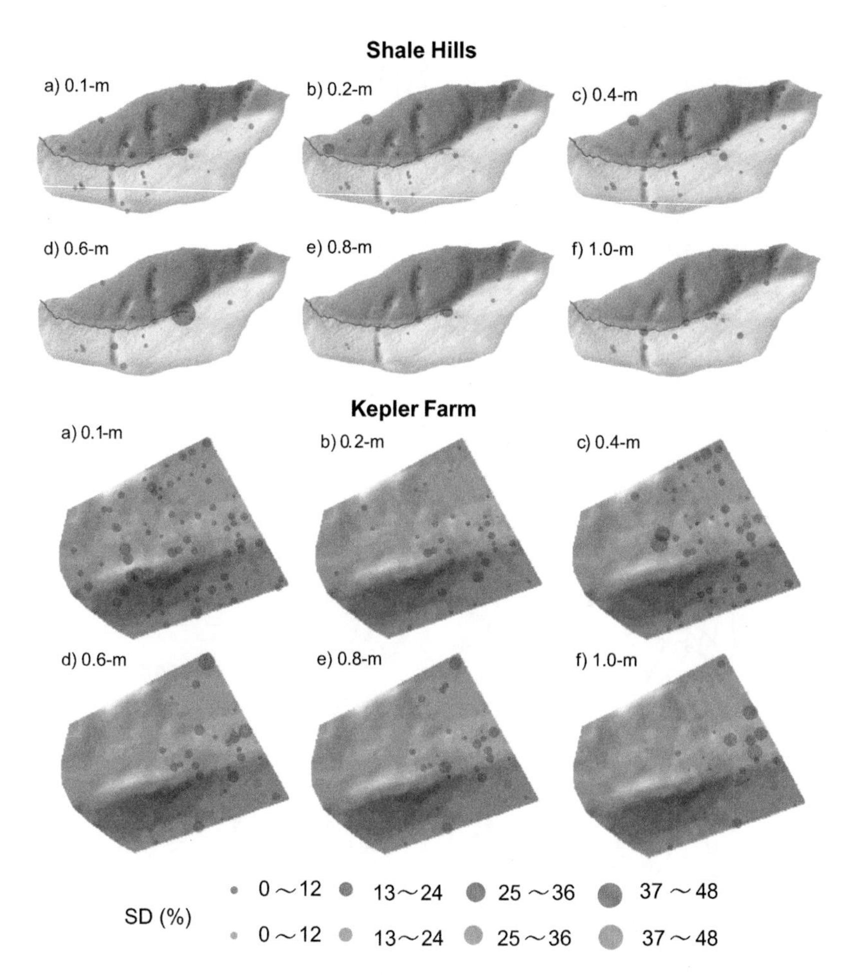
Shale Hills
a) 0.1-m
b) 0.2-m
c) 0.4-m
d) 0.6-m
e) 0.8-m
f) 1.0-m
Kepler Farm
a) 0.1-m
b) 0.2-m
c) 0.4-m
d) 0.6-m
e) 0.8-m
f) 1.0-m
SD (%)
0～12
13～24
25～36
37～48
0～12
13～24
25～36
37～48

FIGURE 7

Chapter 13

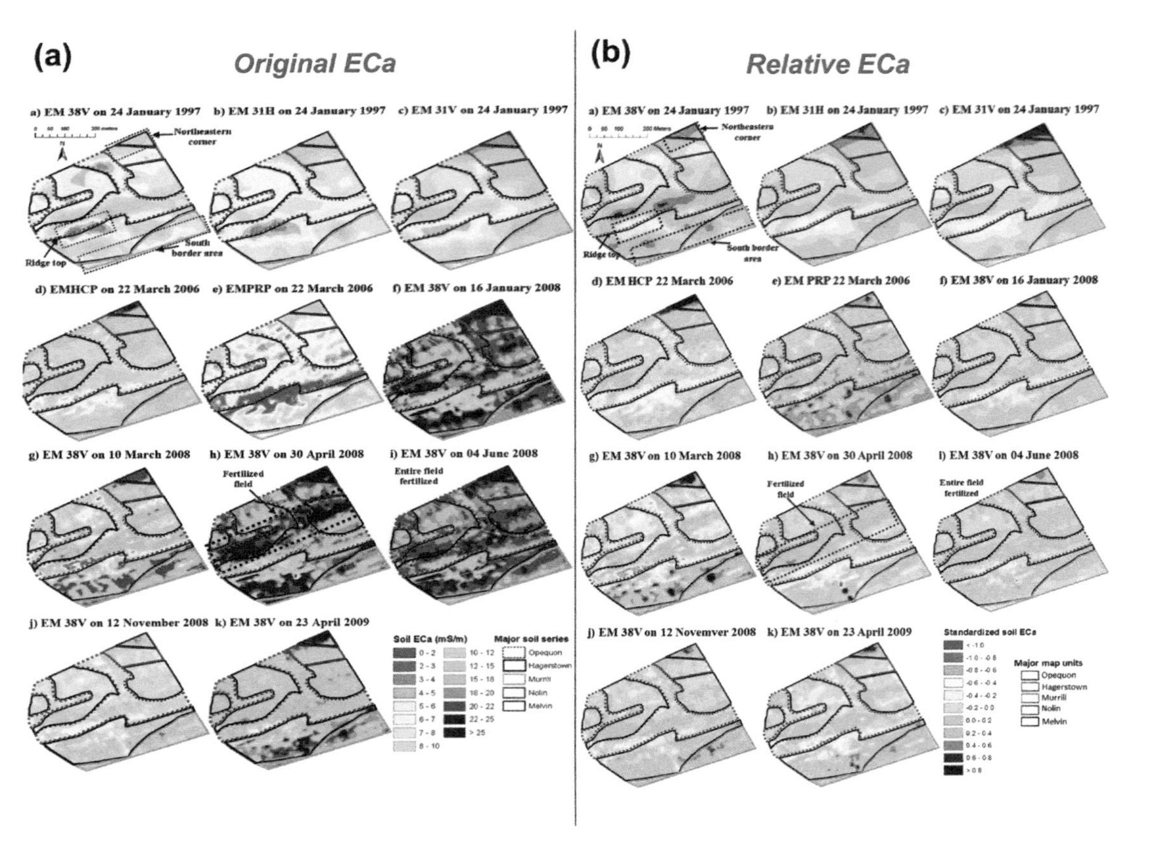

FIGURE 2

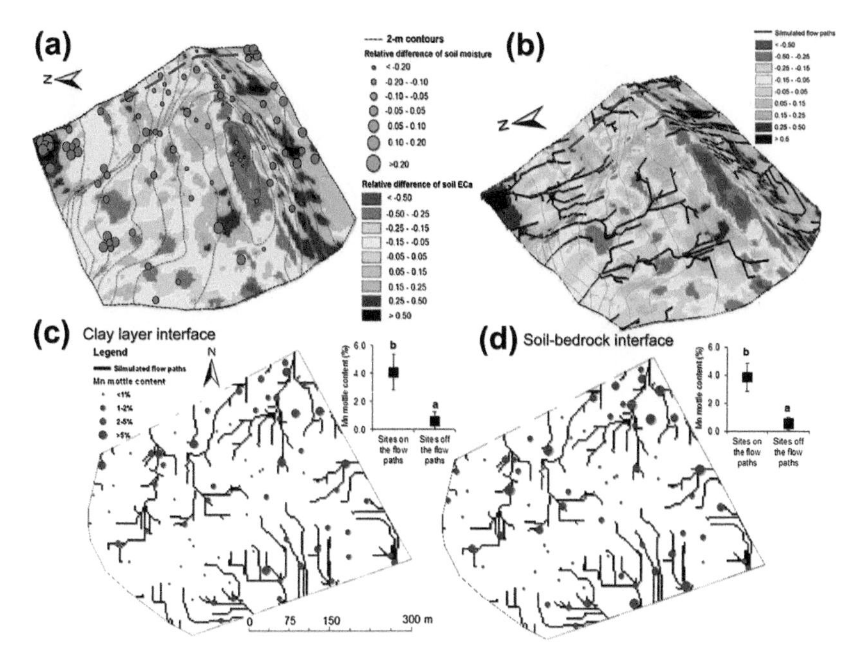

FIGURE 3

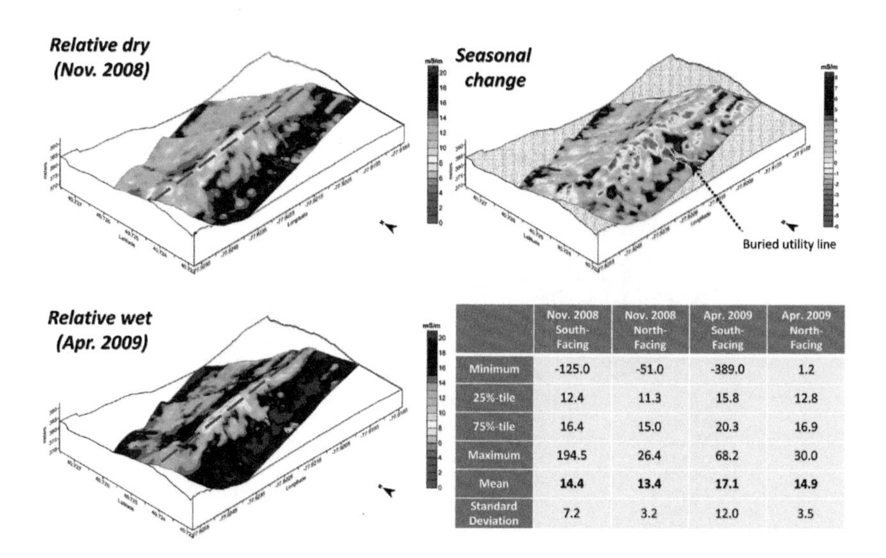

	Nov. 2008 South-Facing	Nov. 2008 North-Facing	Apr. 2009 South-Facing	Apr. 2009 North-Facing
Minimum	-125.0	-51.0	-389.0	1.2
25%-tile	12.4	11.3	15.8	12.8
75%-tile	16.4	15.0	20.3	16.9
Maximum	194.5	26.4	68.2	30.0
Mean	**14.4**	**13.4**	**17.1**	**14.9**
Standard Deviation	7.2	3.2	12.0	3.5

FIGURE 4

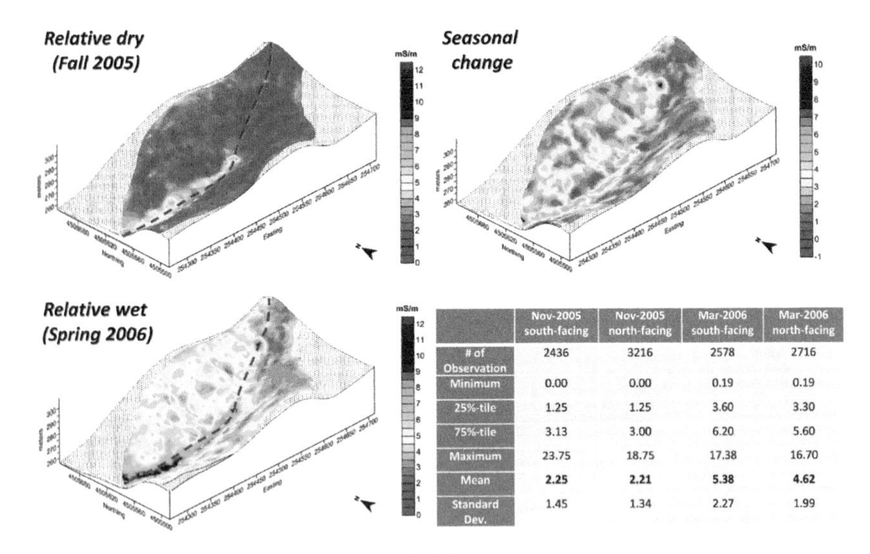

	Nov-2005 south-facing	Nov-2005 north-facing	Mar-2006 south-facing	Mar-2006 north-facing
# of Observation	2436	3216	2578	2716
Minimum	0.00	0.00	0.19	0.19
25%-tile	1.25	1.25	3.60	3.30
75%-tile	3.13	3.00	6.20	5.60
Maximum	23.75	18.75	17.38	16.70
Mean	**2.25**	**2.21**	**5.38**	**4.62**
Standard Dev.	1.45	1.34	2.27	1.99

FIGURE 5

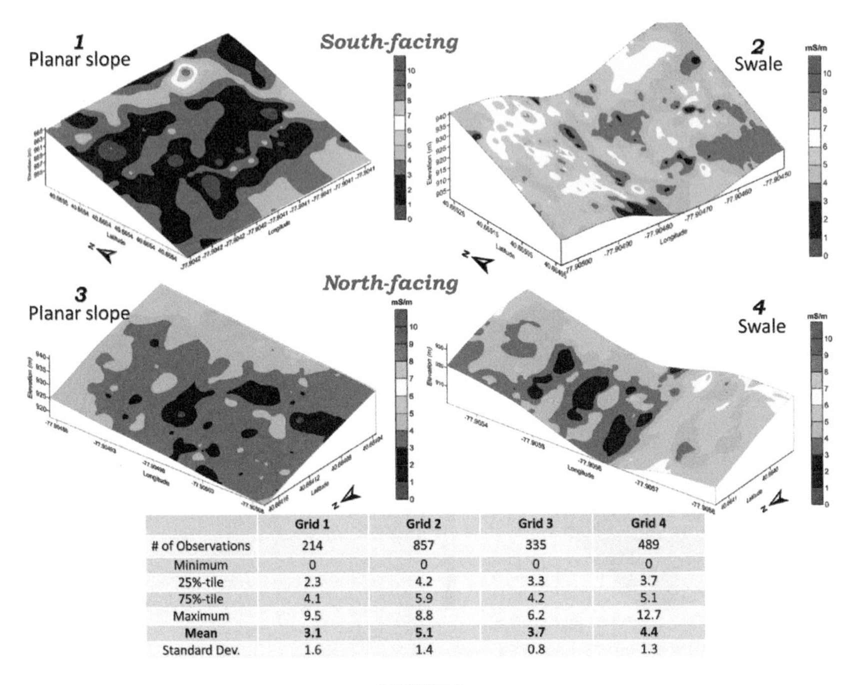

	Grid 1	Grid 2	Grid 3	Grid 4
# of Observations	214	857	335	489
Minimum	0	0	0	0
25%-tile	2.3	4.2	3.3	3.7
75%-tile	4.1	5.9	4.2	5.1
Maximum	9.5	8.8	6.2	12.7
Mean	**3.1**	**5.1**	**3.7**	**4.4**
Standard Dev.	1.6	1.4	0.8	1.3

FIGURE 6

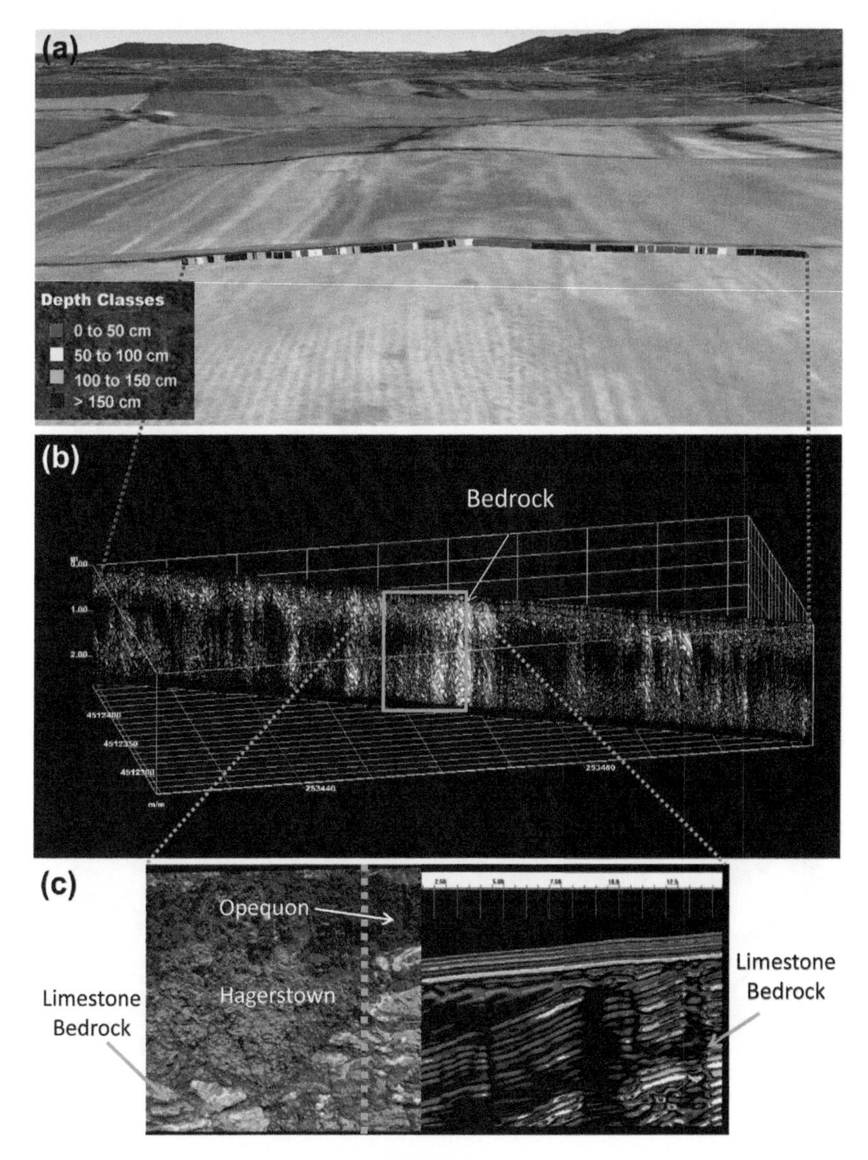

(a)
Depth Classes
0 to 50 cm
50 to 100 cm
100 to 150 cm
> 150 cm
(b)
Bedrock
(c)
Opequon
Hagerstown
Limestone Bedrock
Limestone Bedrock

FIGURE 7

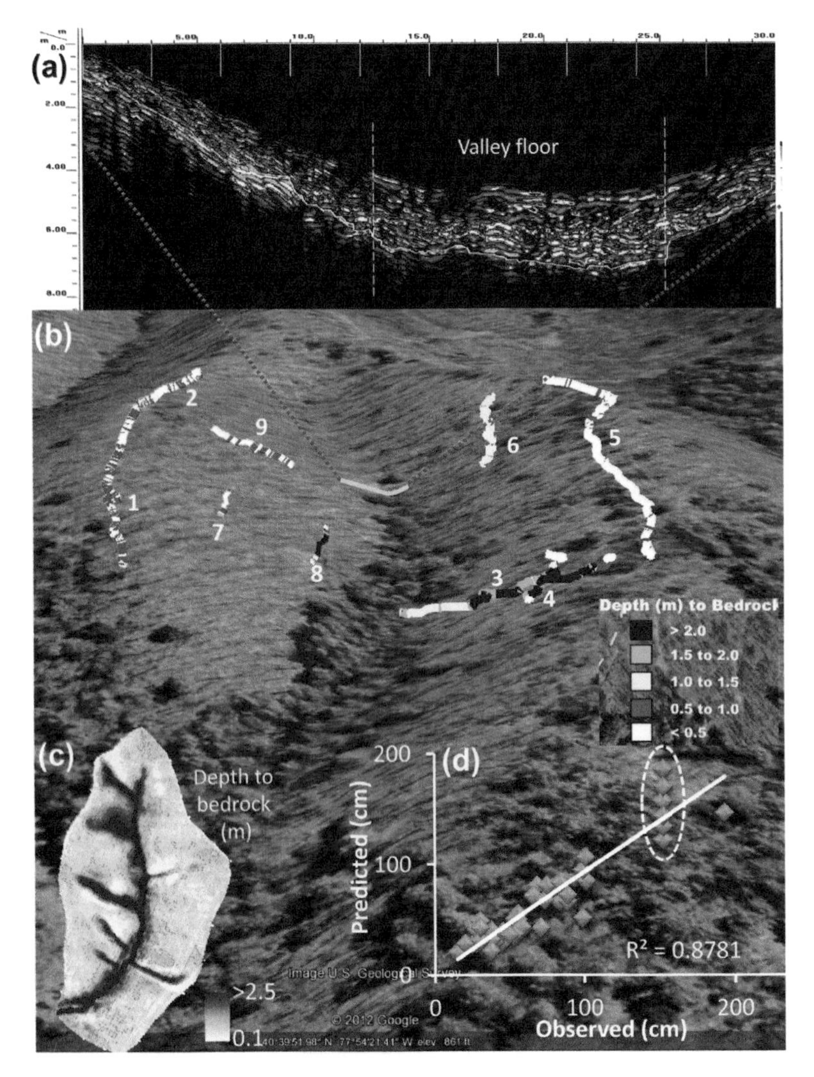

(a)
Valley floor
(b)
Depth (m) to Bedrock
> 2.0
1.5 to 2.0
1.0 to 1.5
0.5 to 1.0
< 0.5
(c)
Depth to bedrock (m)
>2.5
0.1
(d)
Predicted (cm)
Observed (cm)
R^2 = 0.8781
Image U.S. Geological Survey
© 2012 Google

FIGURE 8

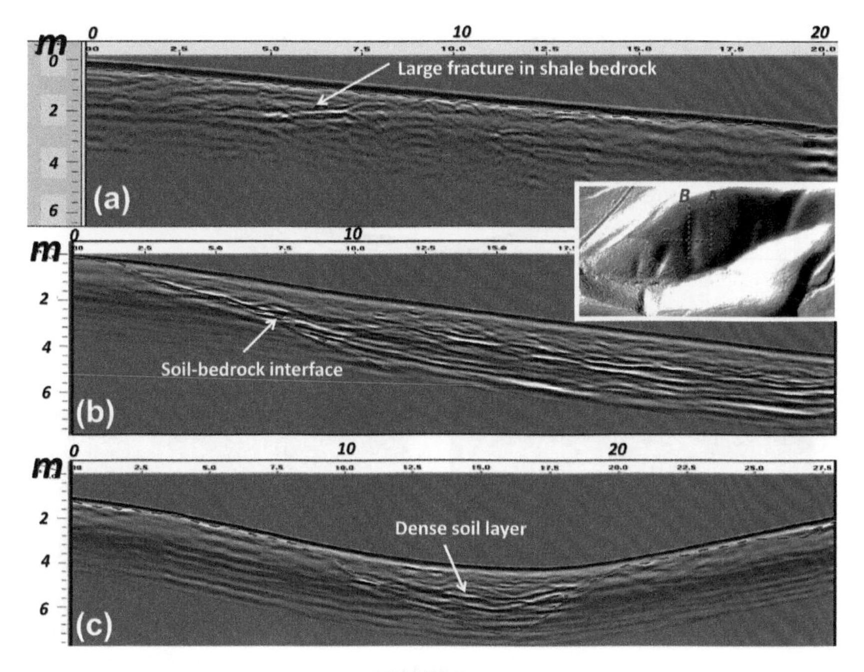

FIGURE 9

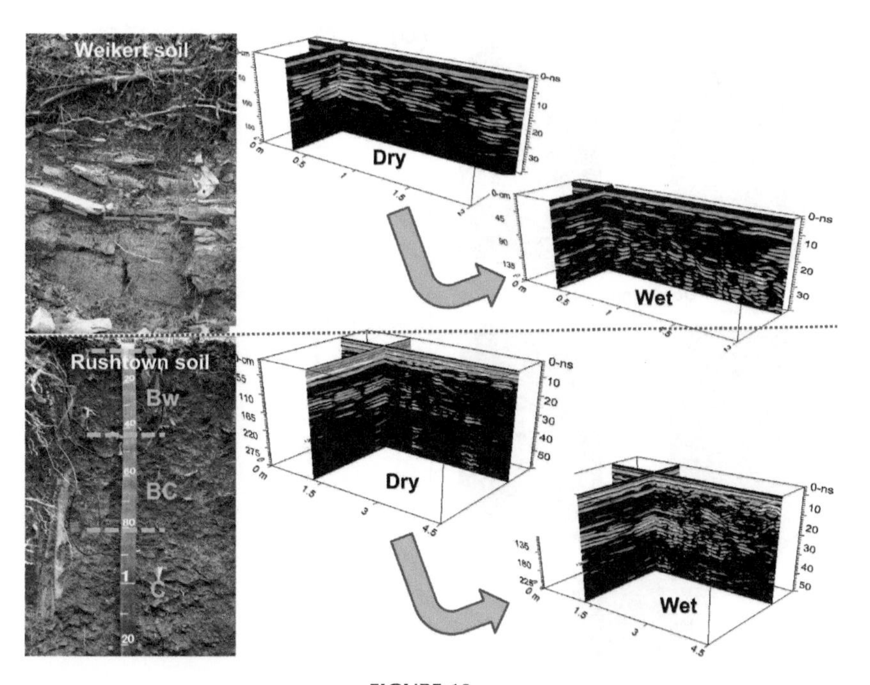

FIGURE 10

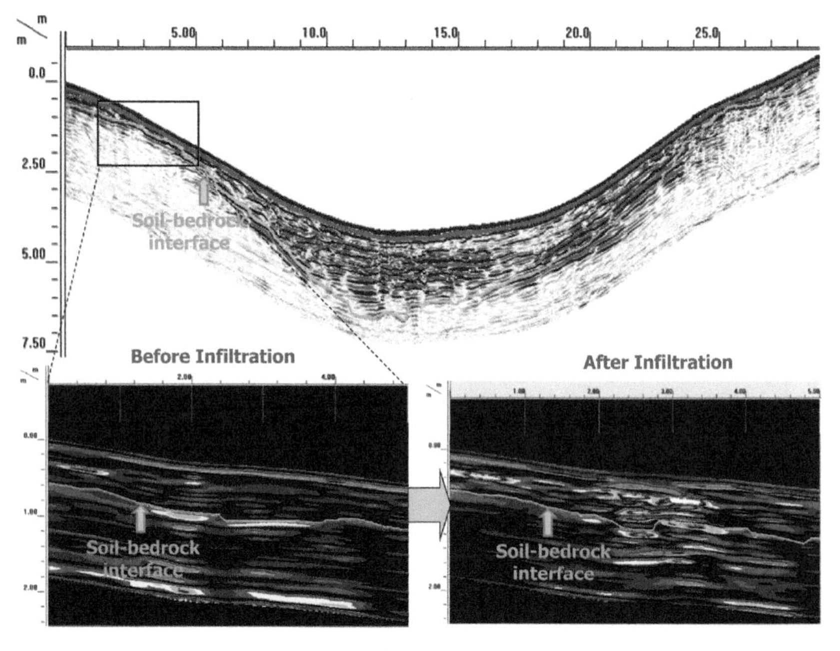

FIGURE 11

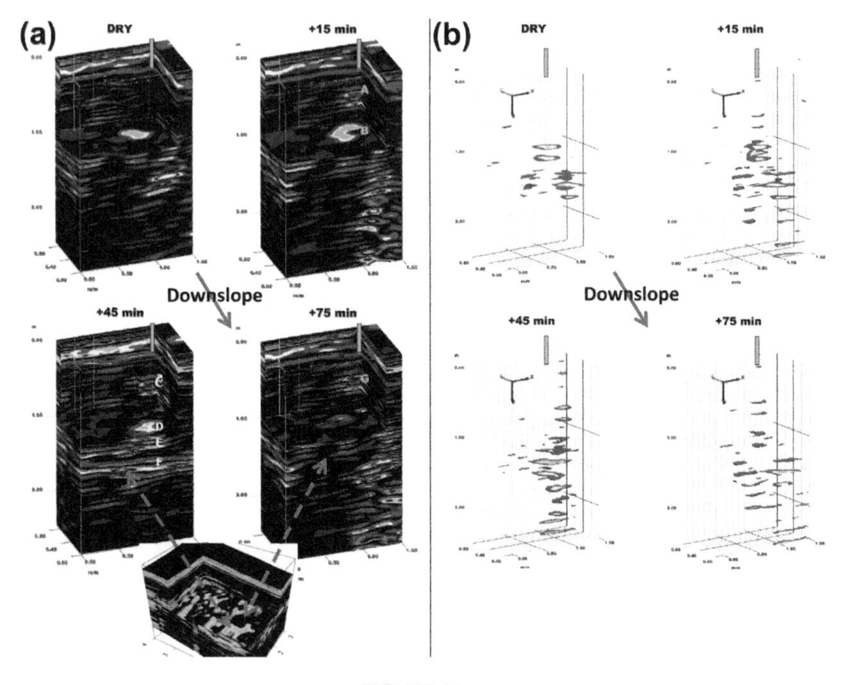

FIGURE 12

Chapter 14

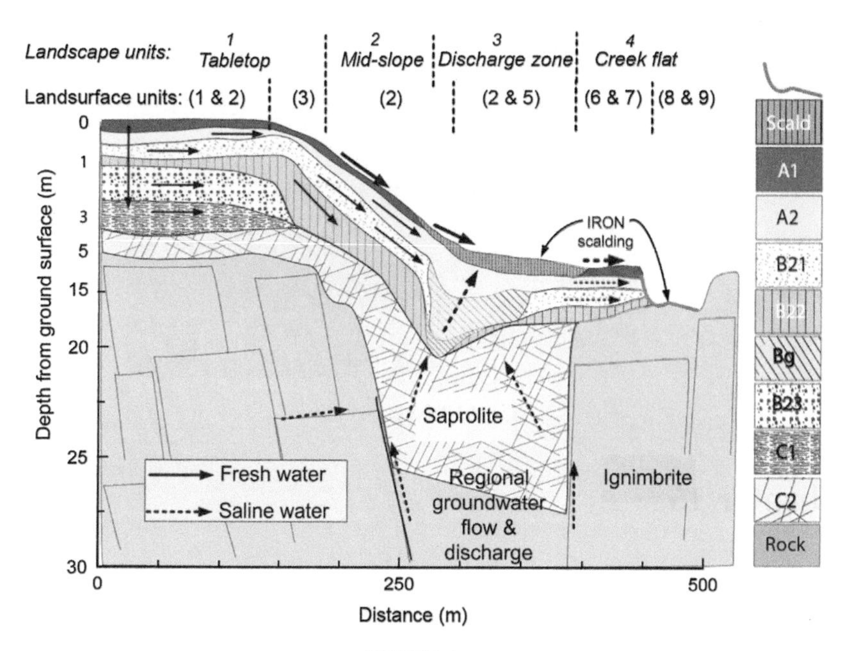

FIGURE 4

FIGURE 12

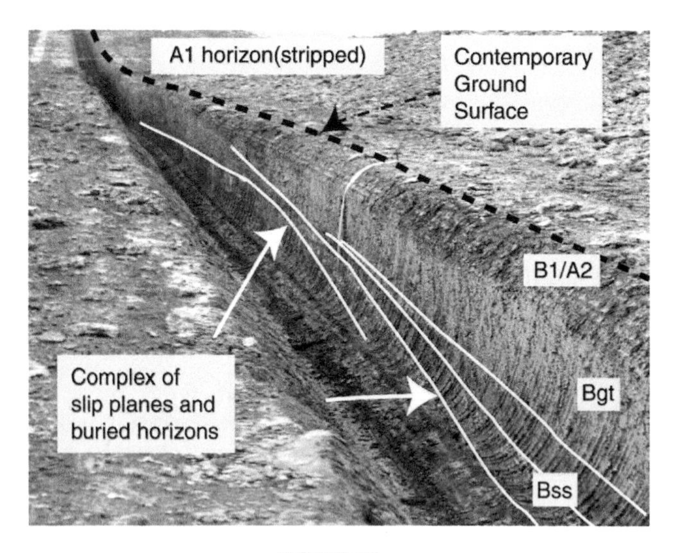

FIGURE 15

Chapter 15

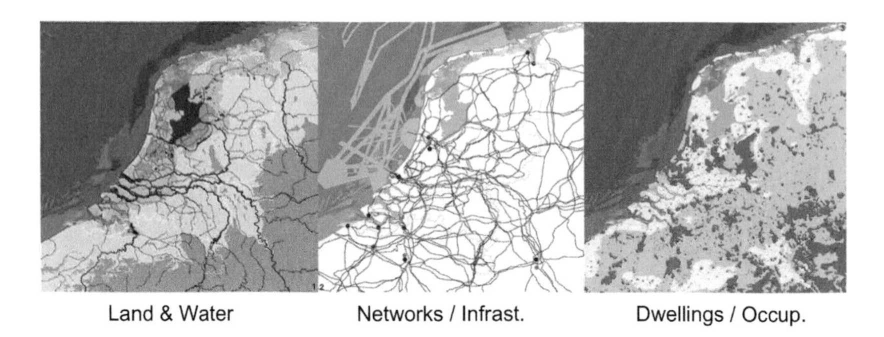

FIGURE 2

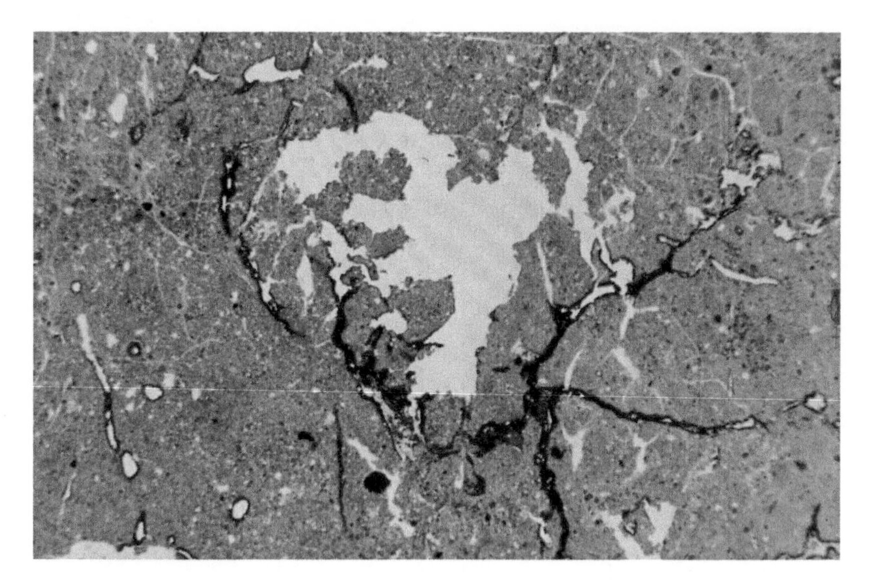

FIGURE 9

Chapter 16

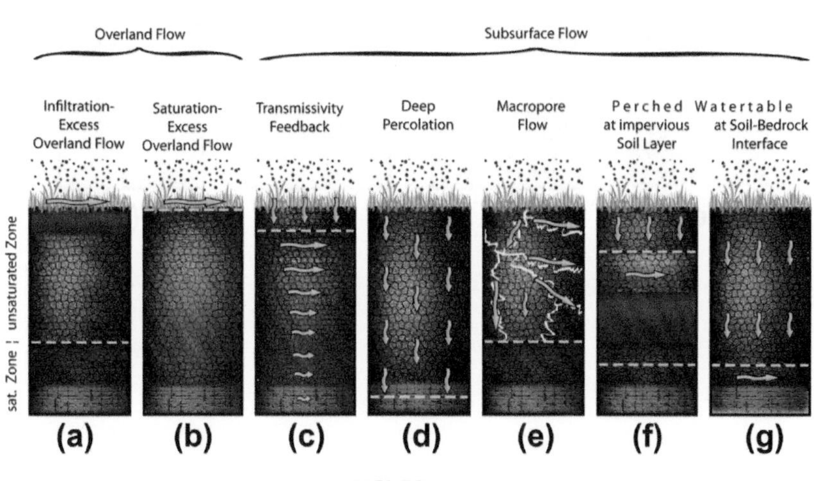

FIGURE 1

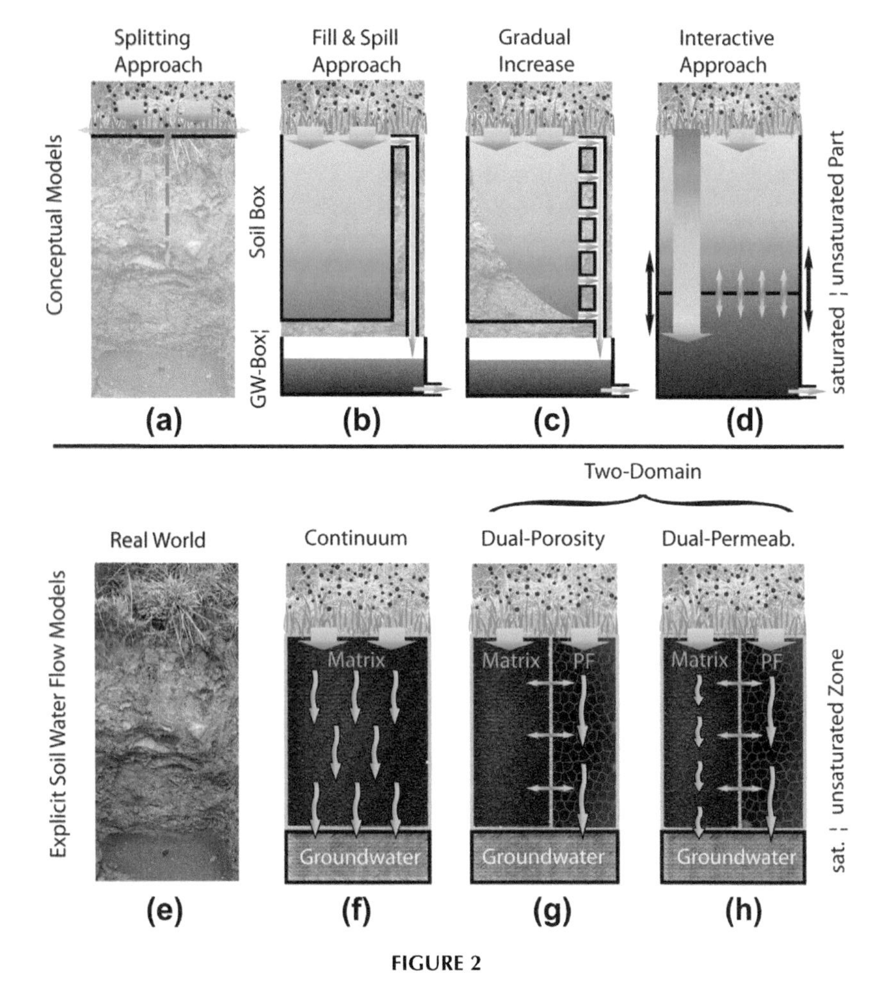
Splitting Approach
Fill & Spill Approach
Gradual Increase
Interactive Approach
Conceptual Models
Soil Box
GW-Box
saturated ¦ unsaturated Part
(a)
(b)
(c)
(d)
Two-Domain
Real World
Continuum
Dual-Porosity
Dual-Permeab.
Explicit Soil Water Flow Models
Matrix
PF
Groundwater
sat. ¦ unsaturated Zone
(e)
(f)
(g)
(h)

FIGURE 2

Chapter 17

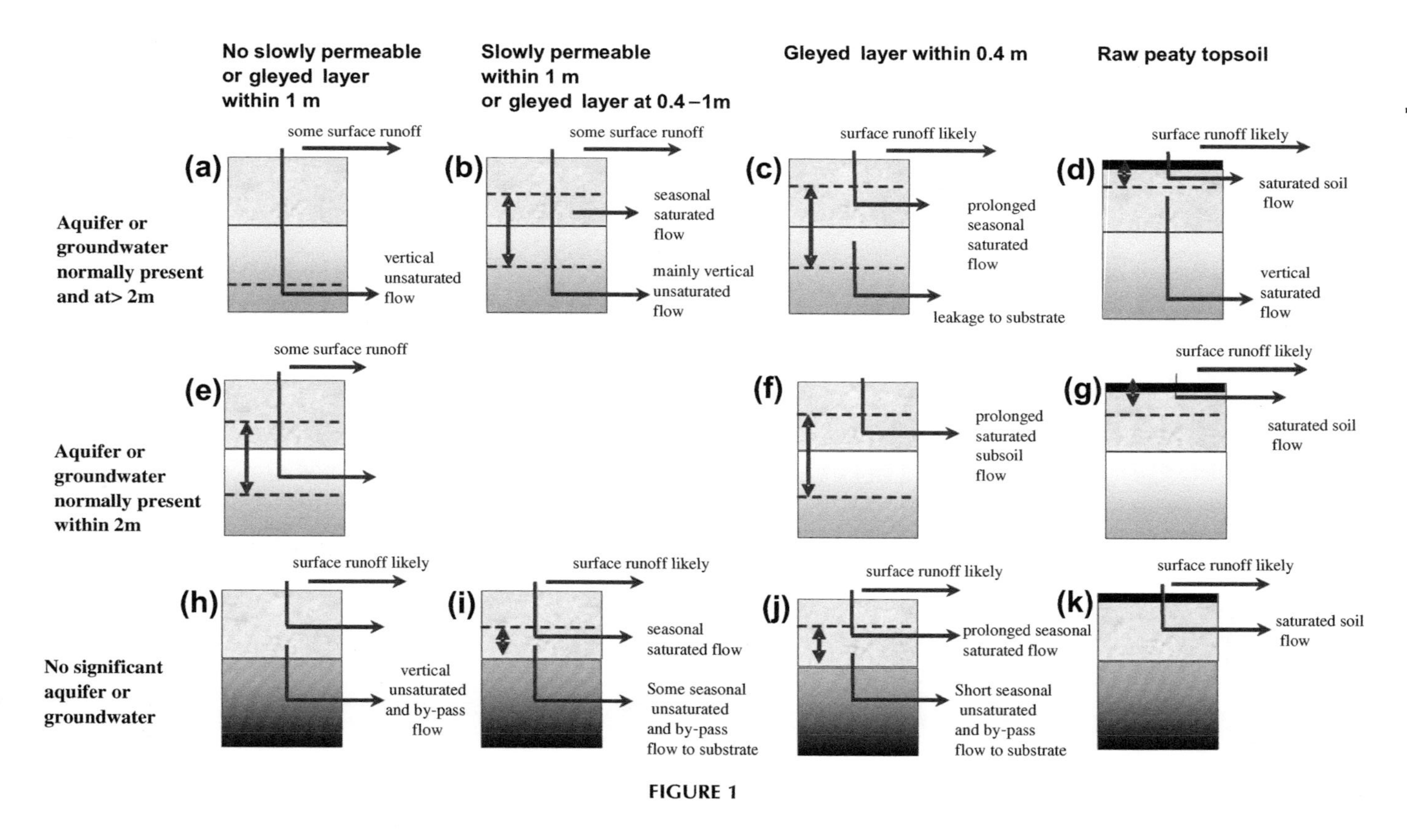

FIGURE 1

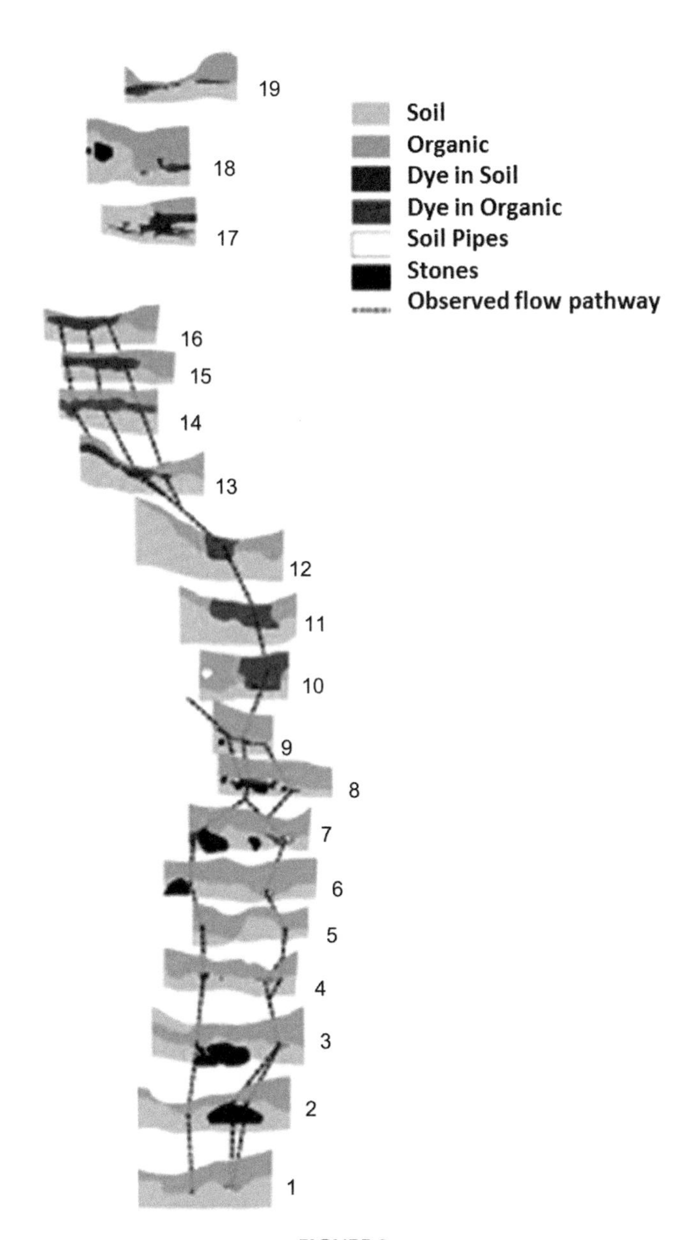
19
18
17
16
15
14
13
12
11
10
9
8
7
6
5
4
3
2
1
Soil
Organic
Dye in Soil
Dye in Organic
Soil Pipes
Stones
Observed flow pathway

FIGURE 3

Chapter 19

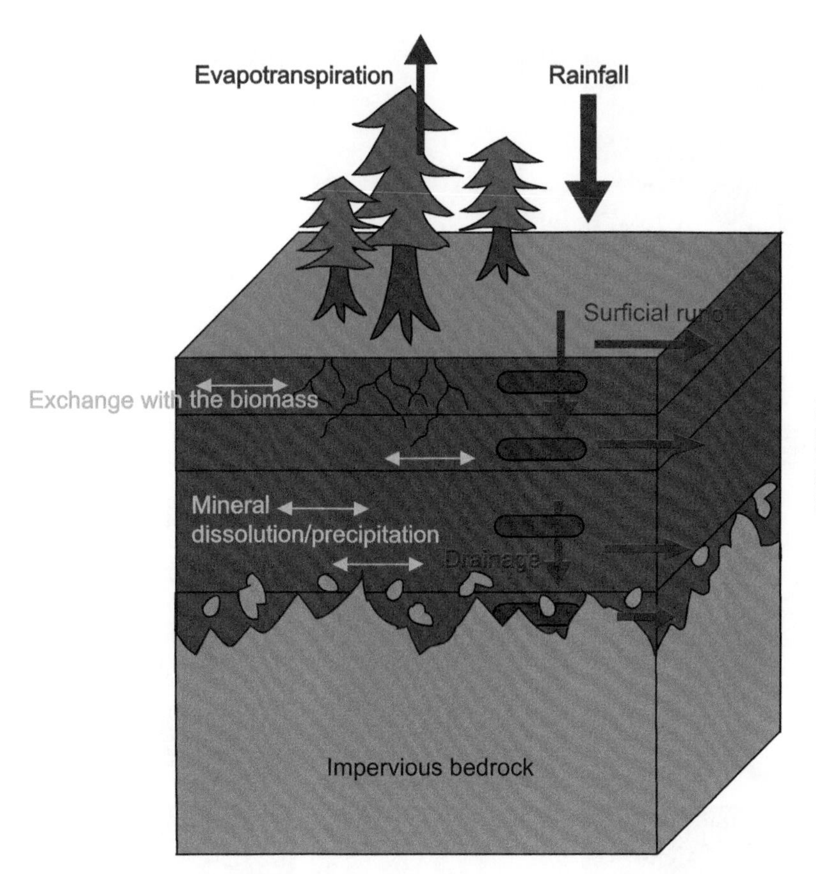

FIGURE 4

Chapter 21

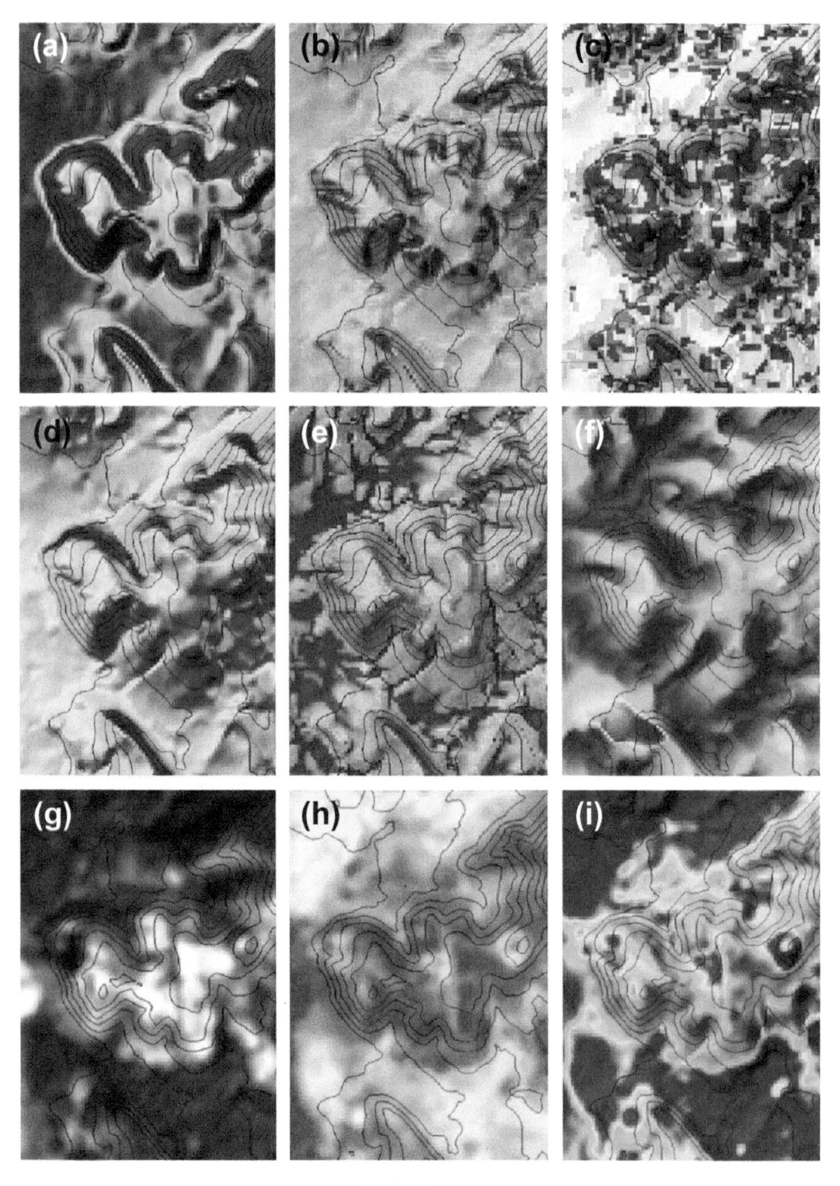

FIGURE 1

Chapter 23

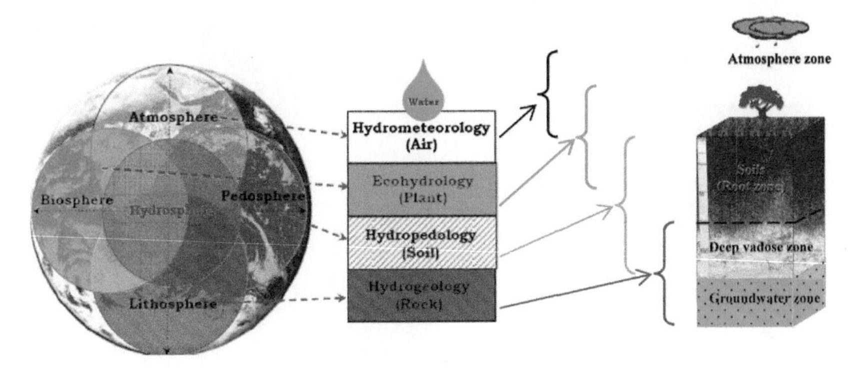

FIGURE 1

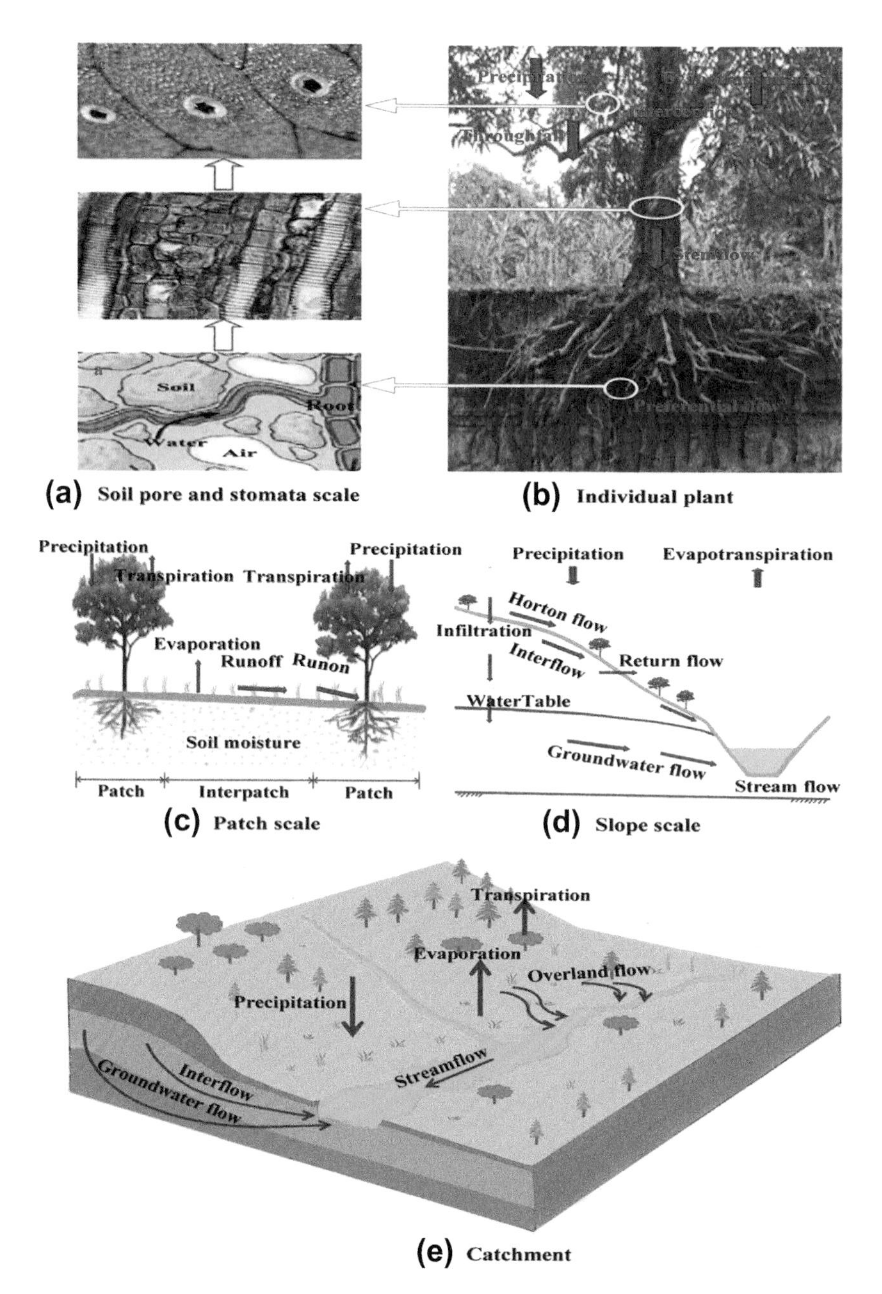

Soil
Root
Water
Air
(a) Soil pore and stomata scale
(b) Individual plant
Precipitation
Transpiration
Transpiration
Precipitation
Evaporation
Runoff
Runon
Soil moisture
Patch
Interpatch
Patch
(c) Patch scale
Precipitation
Evapotranspiration
Horton flow
Infiltration
Interflow
Return flow
WaterTable
Groundwater flow
Stream flow
(d) Slope scale
Transpiration
Evaporation
Overland flow
Precipitation
Streamflow
Interflow
Groundwater flow
(e) Catchment

FIGURE 2

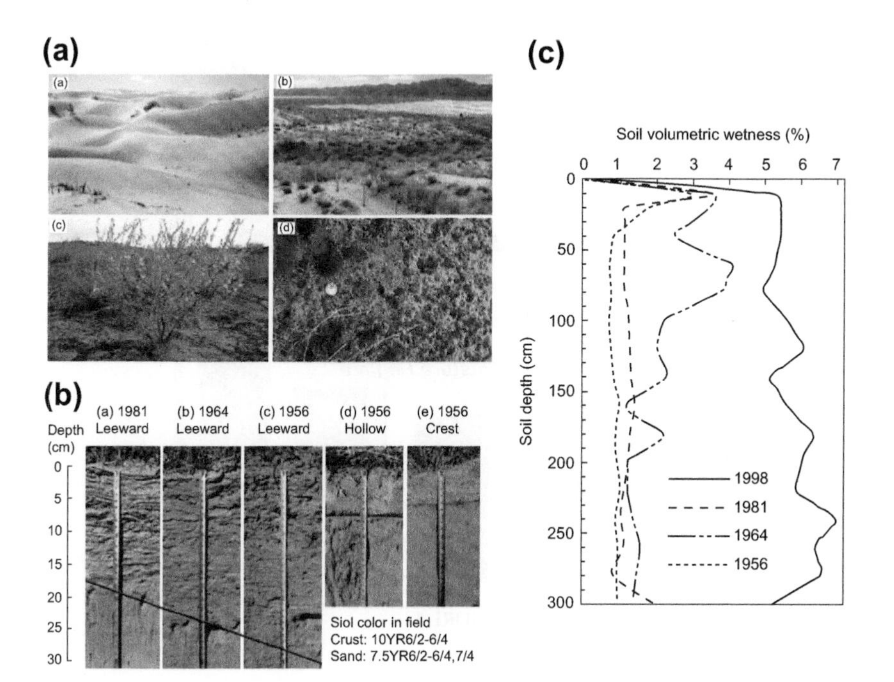

(a)
(a)
(b)
(c)
(d)
(b)
(a) 1981 Leeward
(b) 1964 Leeward
(c) 1956 Leeward
(d) 1956 Hollow
(e) 1956 Crest
Depth (cm)
0
5
10
15
20
25
30
Siol color in field
Crust: 10YR6/2-6/4
Sand: 7.5YR6/2-6/4,7/4
(c)
Soil volumetric wetness (%)
0 1 2 3 4 5 6 7
Soil depth (cm)
0
50
100
150
200
250
300
1998
1981
1964
1956

FIGURE 5

Chapter 24

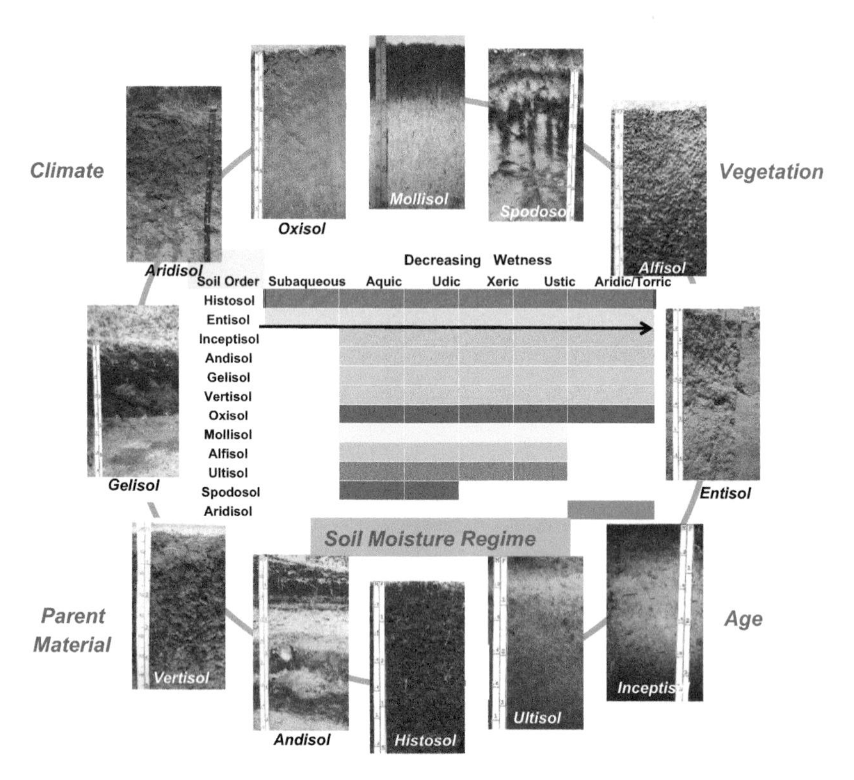

FIGURE 1

Printed and bound by CPI Group (UK) Ltd, Croydon, CR0 4YY
08/05/2025
01864840-0002